$\mathbf{F}[a, b]$	Space of all functions from $[a, b]$ to $\mathbb{R}$	271
$\mathbf{P}_n$	Space of polynomials of degree at most n; or 0	291
$\mathbf{D}[a, b]$	Space of all differentiable functions $[a, b]$ to $\mathbb{R}$	299
$U + W$ and $U \cap W$	Sum and intersection of subspaces U and W	317, 459
$E_a : \mathbf{P}_n \to \mathbb{R}$	Evaluation map at a	332
ker T and im T	Kernel and image of the transformation T	338
nullity T; rank T	Nullity and rank of the transformation T	339
$U^\perp$	Orthogonal complement of the subspace U	371, 419
$\text{proj}_U(X)$	Projection of X on the subspace U	372
$^{(r)}A$	Principal submatrices of A	386
r_k	Rayleigh quotients of A	394
$\langle Z, W \rangle$	Standard inner product in $\mathbb{C}^n$	397
Z^H	Conjugate transpose of the complex matrix Z	399
$\mathbb{Z}$	Set $\{0, \pm 1, \pm 2, \ldots\}$ of integers	408
$\mathbb{Z}_n$	Set of integers modulo n	408
$\text{wt}(\mathbf{w})$	Hamming weight of a word	412
$d(\mathbf{v}, \mathbf{w})$	Hamming distance between words	412
$C_B(\mathbf{v})$	Coordinate vector of the vector $\mathbf{v}$	437
$M_{DB}(T)$	Matrix of the transformation T	438
$\mathbf{L}(V, W)$	Space of linear transformations $V \to W$	444
$M_B(T)$	Matrix of the operator T	445
$P_{D \leftarrow B}$	Change matrix for bases B and D	446
$E_\lambda(T)$	Eigenspace of operator T with respect to λ	458
$V \oplus W$	Direct sum of V and W	459
$\langle \mathbf{v}, \mathbf{w} \rangle$	Inner product of vectors $\mathbf{v}$ and $\mathbf{w}$	468
$\|\mathbf{v}\|$ and $d(\mathbf{v}, \mathbf{w})$	Norm and distance of an inner product space	471
$\delta_k(x)$	Lagrange polynomials	479
$\text{proj}_U(\mathbf{v})$	Projection on an inner product space	482
$\mathcal{J}_n(\lambda)$	Jordan block corresponding to λ	519
i	Imaginary unit	523
$\bar{z}$ and $|z|$	Conjugate and absolute value of complex number z	525
$\arg z$	Argument of complex number z	527
$re^{i\theta}$	Polar form of complex number z	527
$p \Rightarrow q$	Statement p implies statement q	536
$p \Leftrightarrow q$	Statements p and q are logically equivalent	539

Linear Algebra
with Applications

SEVENTH EDITION

W. Keith Nicholson

Linear Algebra with Applications
Seventh Edition

Copyright © 2013, 2009, 2006, 2003 by McGraw-Hill Ryerson Limited, a Subsidiary of The McGraw-Hill Companies. Copyright © 1995 by PWS Publishing Company. Copyright © 1990 by PWS-KENT Publishing Company. Copyright © 1986 by PWS Publishers. All rights reserved. No part of this publication may be reproduced or transmitted in any form or by any means, or stored in a data base or retrieval system, without the prior written permission of McGraw-Hill Ryerson Limited, or in the case of photocopying or other reprographic copying, a licence from The Canadian Copyright Licensing Agency (Access Copyright). For an Access Copyright licence, visit www.accesscopyright.ca or call toll-free to 1-800-893-5777.

The Internet addresses listed in the text were accurate at the time of publication. The inclusion of a website does not indicate an endorsement by the authors or McGraw-Hill Ryerson, and McGraw-Hill Ryerson does not guarantee the accuracy of information presented at these sites.

ISBN-13: 978-0-07-040109-9
ISBN-10: 0-07-040109-8

4 5 6 7 8 9 WEB 1 9 8 7 6 5

Printed and bound in Canada

Care has been taken to trace ownership of copyright material contained in this text; however, the publisher will welcome any information that enables it to rectify any reference or credit for subsequent editions.

Director of Editorial: *Rhondda McNabb*
Senior Sponsoring Editor: *James Booty*
Marketing Manager: *Cathie LeFebvre*
Developmental Editor: *Erin Catto*
Supervising Editor: *Cathy Biribauer*
Senior Editorial Associate: *Stephanie Giles*
Copy Editor: *Armour Robert Templeton, First Folio Resource Group Inc.*
Production Coordinator: *Sheryl MacAdam*
Cover and Inside Design: *David Montle/Pixel Hive Studio*
Composition: *Tom Dart and Bradley T. Smith of First Folio Resource Group Inc.*
Cover Photo: *Image courtesy of Scott Norsworthy*
Printer: *Webcom*

Library and Archives Canada Cataloguing in Publication Data

Nicholson, W. Keith
Linear algebra with applications / W. Keith Nicholson. -- 7th ed.

Includes index.
ISBN 978-0-07-040109-9

1. Algebras, Linear--Textbooks. I. Title.

QA184.2.N53 2013   512'.5   C2012-904689-2

Contents

Chapter 1 Systems of Linear Equations 1
1.1 Solutions and Elementary Operations 1
1.2 Gaussian Elimination 9
1.3 Homogeneous Equations... 18
1.4 An Application to Network Flow 25
1.5 An Application to Electrical Networks 27
1.6 An Application to Chemical Reactions 29
Supplementary Exercises for Chapter 1 30

Chapter 2 Matrix Algebra 32
2.1 Matrix Addition, Scalar Multiplication, and Transposition 32
2.2 Equations, Matrices, and Transformations........... 42
2.3 Matrix Multiplication 56
2.4 Matrix Inverses 69
2.5 Elementary Matrices 83
2.6 Linear Transformations.. 91
2.7 LU-Factorization 103
2.8 An Application to Input-Output Economic Models.. 112
2.9 An Application to Markov Chains......................... 117
Supplementary Exercises for Chapter 2 124

Chapter 3 Determinants and Diagonalization 126
3.1 The Cofactor Expansion... 126
3.2 Determinants and Matrix Inverses 137
3.3 Diagonalization and Eigenvalues 150
3.4 An Application to Linear Recurrences................. 168
3.5 An Application to Systems of Differential Equations 173
3.6 Proof of the Cofactor Expansion Theorem......... 179
Supplementary Exercises for Chapter 3 183

Chapter 4 Vector Geometry 184
4.1 Vectors and Lines 184
4.2 Projections and Planes.. 198
4.3 More on the Cross Product................................... 213
4.4 Linear Operations on $\mathbb{R}^3$ 219
4.5 An Application to Computer Graphics................. 226
Supplementary Exercises for Chapter 4 228

Chapter 5 The Vector Space $\mathbb{R}^n$ 229
5.1 Subspaces and Spanning 229
5.2 Independence and Dimension............................... 236
5.3 Orthogonality.. 246
5.4 Rank of a Matrix 253
5.5 Similarity and Diagonalization 262
5.6 Best Approximation and Least Squares............... 273
5.7 An Application to Correlation and Variance 282
Supplementary Exercises for Chapter 5 287

Chapter 6 Vector Spaces 288
6.1 Examples and Basic Properties 288
6.2 Subspaces and Spanning Sets 296
6.3 Linear Independence and Dimension 303
6.4 Finite Dimensional Spaces.................................... 311
6.5 An Application to Polynomials 320
6.6 An Application to Differential Equations 325
Supplementary Exercises for Chapter 6 330

Chapter 7 Linear Transformations................331
7.1 Examples and Elementary Properties 331
7.2 Kernel and Image of a Linear Transformation ... 338
7.3 Isomorphisms and Composition............................ 347
7.4 A Theorem about Differential Equations............. 357
7.5 More on Linear Recurrences................................ 360

Chapter 8 Orthogonality 368
8.1 Orthogonal Complements and Projections 368
8.2 Orthogonal Diagonalization.................................. 376
8.3 Positive Definite Matrices 385
8.4 QR-Factorization 390
8.5 Computing Eigenvalues... 393
8.6 Complex Matrices 397
8.7 An Application to Linear Codes over Finite Fields 408
8.8 An Application to Quadratic Forms..................... 422
8.9 An Application to Constrained Optimization 431
8.10 An Application to Statistical Principal Component Analysis... 434

Chapter 9 Change of Basis 436
9.1 The Matrix of a Linear Transformation 436
9.2 Operators and Similarity 445
9.3 Invariant Subspaces and Direct Sums 454

Chapter 10 Inner Product Spaces 468
10.1 Inner Products and Norms 468
10.2 Orthogonal Sets of Vectors 477
10.3 Orthogonal Diagonalization 486
10.4 Isometries 493
10.5 An Application to Fourier Approximation 506

Chapter 11 Canonical Forms 510
11.1 Block Triangular Form 511
11.2 The Jordan Canonical Form 518

Appendix A Complex Numbers 523

Appendix B Proofs 536

Appendix C Mathematical Induction 541

Appendix D Polynomials 546
(Available in PDF format on Connect)

Selected Answers 550

Index 574

Preface

This textbook is an introduction to the ideas and techniques of linear algebra for first- or second-year students with a working knowledge of high school algebra. The contents have enough flexibility to present a traditional introduction to the subject, or to allow for a more applied course. Chapters 1–4 contain a one-semester course for beginners whereas Chapters 5–9 contain a second semester course (see the Suggested Course Outlines below). The text is primarily about real linear algebra with complex numbers being mentioned when appropriate (reviewed in Appendix A). Overall, the aim of the text is to achieve a balance among computational skills, theory, and applications of linear algebra. Calculus is not a prerequisite; places where it is mentioned may be omitted.

As a rule, students of linear algebra learn by studying examples and solving problems. Accordingly, the book contains a variety of exercises (over 1200, many with multiple parts), ordered as to their difficulty. In addition, more than 375 solved examples are included in the text, many of which are computational in nature.

The examples are also used to motivate (and illustrate) concepts and theorems, carrying the student from concrete to abstract. While the treatment is rigorous, proofs are presented at a level appropriate to the student and may be omitted with no loss of continuity. As a result, the book can be used to give a course that emphasizes computation and examples, or to give a more theoretical treatment (some longer proofs are deferred to the end of the Section).

Linear Algebra has application to the natural sciences, engineering, management, and the social sciences as well as mathematics. Consequently, 18 optional "applications" sections are included in the text introducing topics as diverse as electrical networks, economic models, Markov chains, linear recurrences, systems of differential equations, and linear codes over finite fields. Additionally some applications (for example linear dynamical systems, and directed graphs) are introduced in context. The applications sections appear at the end of the relevant chapters to encourage students to browse.

SUGGESTED COURSE OUTLINES

This text includes the basis for a two-semester course in linear algebra.

- Chapters 1–4 provide a standard one-semester course of 35 lectures, including linear equations, matrix algebra, determinants, diagonalization, and geometric vectors, with applications as time permits. At Calgary, we cover Sections 1.1–1.3, 2.1–2.6, 3.1–3.3 and 4.1–4.4, and the course is taken by all science and engineering students in their first semester. Prerequisites include a working knowledge of high school algebra (algebraic manipulations and some familiarity with polynomials); calculus is not required.
- Chapters 5–9 contain a second semester course including $\mathbb{R}^n$, abstract vector spaces, linear transformations (and their matrices), orthogonality, complex matrices (up to the spectral theorem) and applications. There is more material here than can be covered in one semester, and at Calgary we cover Sections 5.1–5.5, 6.1–6.4, 7.1–7.3, 8.1–8.6, and 9.1–9.3, with a couple of applications as time permits.
- Chapter 5 is a "bridging" chapter that introduces concepts like spanning, independence, and basis in the concrete setting of $\mathbb{R}^n$, before venturing into the abstract in Chapter 6. The duplication is balanced by the value of reviewing these notions, and it enables the student to focus in Chapter 6 on the new idea of an abstract system. Moreover, Chapter 5 completes the discussion of rank and diagonalization from earlier chapters, and includes a brief introduction to orthogonality in $\mathbb{R}^n$, which creates the possibility of a one-semester, matrix-oriented course covering Chapters 1–5 for students not wanting to study the abstract theory.

CHAPTER DEPENDENCIES

The following chart suggests how the material introduced in each chapter draws on concepts covered in certain earlier chapters. A solid arrow means that ready assimilation of ideas and techniques presented in the later chapter depends on familiarity with the earlier chapter. A broken arrow indicates that some reference to the earlier chapter is made but the chapter need not be covered.

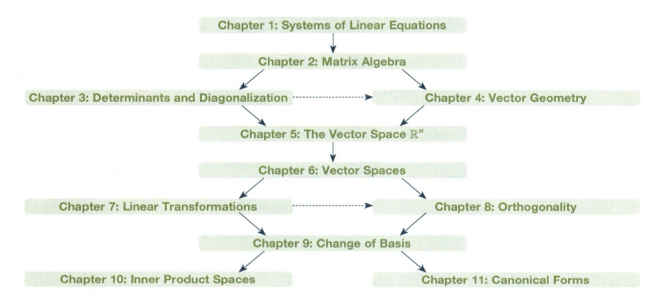

NEW IN THE SEVENTH EDITION

- **Vector notation.** Based on feedback from reviewers and current users, all vectors are denoted by boldface letters (used only in abstract spaces in earlier editions). Thus x becomes $\mathbf{x}$ in $\mathbb{R}^2$ and $\mathbb{R}^3$ (Chapter 4), and in $\mathbb{R}^n$ the column X becomes $\mathbf{x}$. Furthermore, the notation $[x_1 \ x_2 \ \ldots \ x_n]^T$ for vectors in $\mathbb{R}^n$ has been eliminated; instead we write vectors as n-tuples $(x_1, x_2, \ldots, x_n)$ or as columns $\begin{bmatrix} x_1 \\ x_2 \\ \vdots \\ x_n \end{bmatrix}$. The result is a uniform notation for vectors throughout the text.

- **Definitions.** Important ideas and concepts are identified in their given context for student's understanding. These are highlighted in the text when they are first discussed, identified in the left margin, and listed on the inside back cover for reference.

- **Exposition.** Several new margin diagrams have been included to clarify concepts, and the exposition has been improved to simplify and streamline discussion and proofs.

OTHER CHANGES

- Several new examples and exercises have been added.
- The motivation for the matrix inversion algorithm has been rewritten in Section 2.4.
- For geometric vectors in $\mathbb{R}^2$, addition (parallelogram law) and scalar multiplication now appear earlier (Section 2.2). The discussion of reflections in Section 2.6 has been simplified, and projections are now included.
- The example in Section 3.3, which illustrates that $\mathbf{x}$ in $\mathbb{R}^2$ is an eigenvector of A if, and only if, the line $\mathbb{R}_x$ is A-invariant, has been completely rewritten.
- The first part of Section 4.1 on vector geometry in $\mathbb{R}^2$ and $\mathbb{R}^3$ has also been rewritten and shortened.
- In Section 6.4 there are three improvements: Theorem 1 now shows that an independent set can be extended to a basis by adding vectors from *any* prescribed basis; the

proof that a spanning set can be cut down to a basis has been simplified (in Theorem 3); and in Theorem 4, the argument that independence is equivalent to spanning for a set $S \subseteq V$ with $|S| = \dim V$ has been streamlined and a new example added.

- In Section 8.1, the definition of projections has been clarified, as has the discussion of the nature of quadratic forms in $\mathbb{R}^2$.

HIGHLIGHTS OF THE TEXT

- **Two-stage definition of matrix multiplication.** First, in Section 2.2 matrix-vector products are introduced naturally by viewing the left side of a system of linear equations as a product. Second, matrix-matrix products are defined in Section 2.3 by taking the columns of a product AB to be A times the corresponding columns of B. This is motivated by viewing the matrix product as composition of maps (see next item). This works well pedagogically and the usual dot-product definition follows easily. As a bonus, the proof of associativity of matrix multiplication now takes four lines.
- **Matrices as transformations.** Matrix-column multiplications are viewed (in Section 2.2) as transformations $\mathbb{R}^n \to \mathbb{R}^m$. These maps are then used to describe simple geometric reflections and rotations in $\mathbb{R}^2$ as well as systems of linear equations.
- **Early linear transformations.** It has been said that vector spaces exist so that linear transformations can act on them—consequently these maps are a recurring theme in the text. Motivated by the matrix transformations introduced earlier, linear transformations $\mathbb{R}^n \to \mathbb{R}^m$ are defined in Section 2.6, their standard matrices are derived, and they are then used to describe rotations, reflections, projections, and other operators on $\mathbb{R}^2$.
- **Early diagonalization.** As requested by engineers and scientists, this important technique is presented in the first term using only determinants and matrix inverses (before defining independence and dimension). Applications to population growth and linear recurrences are given.
- **Early dynamical systems.** These are introduced in Chapter 3, and lead (via diagonalization) to applications like the possible extinction of species. Beginning students in science and engineering can relate to this because they can see (often for the first time) the relevance of the subject to the real world.
- **Bridging chapter.** Chapter 5 lets students deal with tough concepts (like independence, spanning, and basis) in the concrete setting of $\mathbb{R}^n$ before having to cope with abstract vector spaces in Chapter 6.
- **Examples.** The text contains over 375 worked examples, which present the main techniques of the subject, illustrate the central ideas, and are keyed to the exercises in each section.
- **Exercises.** The text contains a variety of exercises (nearly 1175, many with multiple parts), starting with computational problems and gradually progressing to more theoretical exercises. Exercises marked with a ♦ have an answer at the end of the book or in the Students Solution Manual (available online). There is a complete Solution Manual is available for instructors.
- **Applications.** There are optional applications at the end of most chapters (see the list below). While some are presented in the course of the text, most appear at the end of the relevant chapter to encourage students to browse.
- **Appendices.** Because complex numbers are needed in the text, they are described in Appendix A, which includes the polar form and roots of unity. Methods of proofs are discussed in Appendix B, followed by mathematical induction in Appendix C. A brief discussion of polynomials is included in Appendix D. All these topics are presented at the high-school level.
- **Self-Study.** This text is self-contained and therefore is suitable for self-study.
- **Rigour.** Proofs are presented as clearly as possible (some at the end of the section), but they are optional and the instructor can choose how much he or she wants to prove. However the proofs are there, so this text is more rigorous than most. Linear algebra provides one of the better venues where students begin to think logically and argue concisely. To this end, there are exercises that ask the student to "show" some simple implication, and others that ask her or him to either prove a given statement or give a counterexample. I personally present a few proofs in the first semester course and more in the second (see the Suggested Course Outlines).
- **Major Theorems.** Several major results are presented in the book. Examples: Uniqueness of the reduced row-echelon form; the cofactor expansion for determinants; the Cayley-Hamilton theorem; the Jordan canonical form; Schur's theorem on block triangular form; the principal axis and spectral theorems; and others. Proofs are included because the stronger students should at least be aware of what is involved.

ANCILLARY MATERIALS

CONNECT

McGraw-Hill Connect™ is a web-based assignment and assessment platform that gives students the means to better connect with their coursework, with their instructors, and with the important concepts that they will need to know for success now and in the future. With Connect, instructors can deliver assignments, quizzes and tests easily online. Students can practise important skills at their own pace and on their own schedule. With Connect, students also get 24/7 online access to an eBook—an online edition of the text—to aid them in successfully completing their work, wherever and whenever they choose.

INSTRUCTOR RESOURCES

Instructor Resources
- Instructor's Solutions Manual
- Partial Student's Solution Manual
- Computerized Test Bank

SUPERIOR LEARNING SOLUTIONS AND SUPPORT

The McGraw-Hill Ryerson team is ready to help you assess and integrate any of our products, technology, and services into your course for optimal teaching and learning performance. Whether it's helping your students improve their grades, or putting your entire course online, the McGraw-Hill Ryerson team is here to help you do it. Contact your *i*Learning Sales Specialist today to learn how to maximize all of McGraw-Hill Ryerson's resources!

For more information on the latest technology and Learning Solutions offered by McGraw-Hill Ryerson and its partners, please visit us online: **www.mcgrawhill.ca/he/solutions**.

CHAPTER SUMMARIES

Chapter 1: Systems of Linear Equations.

A standard treatment of gaussian elimination is given. The rank of a matrix is introduced via the row-echelon form, and solutions to a homogenous system are presented as linear combinations of basic solutions. Applications to network flows, electrical networks, and chemical reactions are provided.

Chapter 2: Matrix Algebra.

After a traditional look at matrix addition, scalar multiplication, and transposition in Section 2.1, matrix-vector multiplication is introduced in Section 2.2 by viewing the left side of a system of linear equations as the product $A\mathbf{x}$ of the coefficient matrix A with the column $\mathbf{x}$ of variables. The usual dot-product definition of a matrix-vector multiplication follows. Section 2.2 ends by viewing an $m \times n$ matrix A as a transformation $\mathbb{R}^n \to \mathbb{R}^m$. This is illustrated for $\mathbb{R}^2 \to \mathbb{R}^2$ by describing reflection in the x axis, rotation of $\mathbb{R}^2$ through $\frac{\pi}{2}$, shears, and so on.

In Section 2.3, the product of matrices A and B is defined by $AB = [A\mathbf{b}_1 \ A\mathbf{b}_2 \ \ldots \ A\mathbf{b}_n]$, where the $\mathbf{b}_i$ are the columns of B. A routine computation shows that this is the matrix of the transformation B followed by A. This observation is used frequently throughout the book, and leads to simple, conceptual proofs of the basic axioms of matrix algebra. Note that linearity is not required—all that is needed is some basic properties of matrix-vector multiplication developed in Section 2.2. Thus the usual arcane definition of matrix multiplication is split into two well motivated parts, each an important aspect of matrix algebra. Of course, this has the pedagogical advantage that the conceptual power of geometry can be invoked to illuminate and clarify algebraic techniques and definitions.

In Sections 2.4 and 2.5 matrix inverses are characterized, their geometrical meaning is explored, and block multiplication is introduced, emphasizing those cases needed later in the book. Elementary matrices are discussed, and the Smith normal form is derived. Then in Section 2.6, linear transformations $\mathbb{R}^n \to \mathbb{R}^m$ are defined and shown to be matrix transformations. The matrices of reflections, rotations, and projections in the plane are determined. Finally, matrix multiplication is related to directed graphs, matrix LU-factorization is introduced, and applications to economic models and Markov chains are presented.

Chapter 3: Determinants and Diagonalization.

The cofactor expansion is stated (proved by induction later) and used to define determinants inductively and to deduce the basic rules. The product and adjugate theorems are proved. Then the diagonalization algorithm is presented (motivated by an example about the possible extinction of a species of birds). As requested by our Engineering Faculty, this is done earlier than in most texts because it requires only determinants and matrix inverses, avoiding any need for subspaces, independence and dimension. Eigenvectors of a 2×2 matrix A are described geometrically (using the A-invariance of lines through the origin). Diagonalization is then used to study discrete linear dynamical systems and to discuss applications to linear recurrences and systems of differential equations. A brief discussion of Google PageRank is included.

Chapter 4: Vector Geometry.

Vectors are presented intrinsically in terms of length and direction, and are related to matrices via coordinates. Then vector operations are defined using matrices and shown to be the same as the corresponding intrinsic definitions. Next, dot products and projections are introduced to solve problems about lines and planes. This leads to the cross product. Then matrix transformations are introduced in $\mathbb{R}^3$, matrices of projections and reflections are derived, and areas and volumes are computed using determinants. The chapter closes with an application to computer graphics.

Chapter 5: The Vector Space $\mathbb{R}^n$.

Subspaces, spanning, independence, and dimensions are introduced in the context of $\mathbb{R}^n$ in the first two sections. Orthogonal bases are introduced and used to derive the expansion theorem. The basic properties of rank are presented and used to justify the definition given in Section 1.2. Then, after a rigorous study of diagonalization, best approximation and least squares are discussed. The chapter closes with an application to correlation and variance.

As in the sixth edition, this is a "bridging" chapter, easing the transition to abstract spaces. Concern about duplication with Chapter 6 is mitigated by the fact that this is the most difficult part of the course and many students welcome a repeat discussion of concepts like independence and spanning, albeit in the abstract setting. In a different direction, Chapters 1–5 could serve as a solid introduction to linear algebra for students not requiring abstract theory.

Chapter 6: Vector Spaces.

Building on the work on $\mathbb{R}^n$ in Chapter 5, the basic theory of abstract finite dimensional vector spaces is developed emphasizing new examples like matrices, polynomials and functions. This is the first acquaintance most students have had with an abstract system, so not having to deal with spanning, independence and dimension in the general context eases the transition to abstract thinking. Applications to polynomials and to differential equations are included.

Chapter 7: Linear Transformations.

General linear transformations are introduced, motivated by many examples from geometry, matrix theory, and calculus. Then kernels and images are defined, the dimension theorem is proved, and isomorphisms are discussed. The chapter ends with an application to linear recurrences. A proof is included that the order of a differential equation (with constant coefficients) equals the dimension of the space of solutions.

Chapter 8: Orthogonality.

The study of orthogonality in $\mathbb{R}^n$, begun in Chapter 5, is continued. Orthogonal complements and projections are defined and used to study orthogonal diagonalization. This leads to the principal axis theorem, the Cholesky factorization of a positive definite matrix, and QR-factorization. The theory is extended to $\mathbb{C}^n$ in Section 8.6 where hermitian and unitary matrices are discussed, culminating in Schur's theorem and the spectral theorem. A short proof of the Cayley-Hamilton theorem is also presented. In Section 8.7 the field Z_p of integers modulo p is constructed informally for any prime p, and codes are discussed over any finite field. The chapter concludes with applications to quadratic forms, constrained optimization, and statistical principal component analysis.

Chapter 9: Change of Basis.

The matrix of general linear transformation is defined and studied. In the case of an operator, the relationship between basis changes and similarity is revealed. This is illustrated by computing the matrix of a rotation about a line through the origin in $\mathbb{R}^3$. Finally, invariant subspaces and direct sums are introduced, related to similarity, and (as an example) used to show that every involution is similar to a diagonal matrix with diagonal entries ± 1.

Chapter 10: Inner Product Spaces.

General inner products are introduced and distance, norms, and the Cauchy-Schwarz inequality are discussed. The Gram-Schmidt algorithm is presented, projections are defined and the approximation theorem is proved (with an application to Fourier approximation). Finally, isometries are characterized, and distance preserving operators are shown to be composites of a translations and isometries.

Chapter 11: Canonical Forms.

The work in Chapter 9 is continued. Invariant subspaces and direct sums are used to derive the block triangular form. That, in turn, is used to give a compact proof of the Jordan canonical form. Of course the level is higher.

Appendices

In Appendix A, complex arithmetic is developed far enough to find nth roots. In Appendix B, methods of proof are discussed, while Appendix C presents mathematical induction. Finally, Appendix D describes the properties of polynomials in elementary terms.

LIST OF APPLICATIONS

- Network Flow (Section 1.4)
- Electrical Networks (Section 1.5)
- Chemical Reactions (Section 1.6)
- Directed Graphs (in Section 2.3)
- Input-Output Economic Models (Section 2.8)
- Markov Chains (Section 2.9)
- Polynomial Interpolation (in Section 3.2)
- Population Growth (Examples 1 and 10, Section 3.3)
- Google PageRank (in Section 3.3)
- Linear Recurrences (Section 3.4; see also Section 7.5)
- Systems of Differential Equations (Section 3.5)
- Computer Graphics (Section 4.5)
- Least Squares Approximation (in Section 5.6)
- Correlation and Variance (Section 5.7)
- Polynomials (Section 6.5)
- Differential Equations (Section 6.6)
- Linear Recurrences (Section 7.5)
- Error Correcting Codes (Section 8.7)
- Quadratic Forms (Section 8.8)
- Constrianed Optimization (Section 8.9)
- Statistical Principal Component Analysis (Section 8.10)
- Fourier Approximation (Section 10.5)

ACKNOWLEDGMENTS

Comments and suggestions that have been invaluable to the development of this edition were provided by a variety of reviewers, and I thank the following instructors:

Robert Andre
University of Waterloo

Dietrich Burbulla
University of Toronto

Dzung M. Ha
Ryerson University

Mark Solomonovich
Grant MacEwan

Fred Szabo
Concordia University

Edward Wang
Wilfred Laurier

Petr Zizler
Mount Royal University

It is a pleasure to recognize the contributions of several people to this book. First, I would like to thank a number of colleagues who have made suggestions that have improved this edition. Discussions with Thi Dinh and Jean Springer have been invaluable and many of their suggestions have been incorporated. Thanks are also due to Kristine Bauer and Clifton Cunningham for several conversations about the new way to look at matrix multiplication. I also wish to extend my thanks to Joanne Canape for being there when I have technical questions. Of course, thanks go to James Booty, Senior Sponsoring Editor, for his enthusiasm and effort in getting this project underway, to Sarah Fulton and Erin Catto, Developmental Editors, for their work on the editorial background of the book, and to Cathy Biribauer and Robert Templeton (First Folio Resource Group Inc.) and the rest of the production staff at McGraw-Hill Ryerson for their parts in the project. Thanks also go to Jason Nicholson for his help in various aspects of the book, particularly the Solutions Manual. Finally, I want to thank my wife Kathleen, without whose understanding and cooperation, this book would not exist.

W. Keith Nicholson
University of Calgary

1 Systems of Linear Equations

SECTION 1.1 Solutions and Elementary Operations

Practical problems in many fields of study—such as biology, business, chemistry, computer science, economics, electronics, engineering, physics and the social sciences—can often be reduced to solving a system of linear equations. Linear algebra arose from attempts to find systematic methods for solving these systems, so it is natural to begin this book by studying linear equations.

If a, b, and c are real numbers, the graph of an equation of the form

$$ax + by = c$$

is a straight line (if a and b are not both zero), so such an equation is called a *linear* equation in the variables x and y. However, it is often convenient to write the variables as $x_1, x_2, \ldots, x_n$, particularly when more than two variables are involved. An equation of the form

$$a_1x_1 + a_2x_2 + \cdots + a_nx_n = b$$

is called a **linear equation** in the n variables $x_1, x_2, \ldots, x_n$. Here $a_1, a_2, \ldots, a_n$ denote real numbers (called the **coefficients** of $x_1, x_2, \ldots, x_n$, respectively) and b is also a number (called the **constant term** of the equation). A finite collection of linear equations in the variables $x_1, x_2, \ldots, x_n$ is called a **system of linear equations** in these variables. Hence,

$$2x_1 - 3x_2 + 5x_3 = 7$$

is a linear equation; the coefficients of x_1, x_2, and x_3 are 2, -3, and 5, and the constant term is 7. Note that each variable in a linear equation occurs to the first power only.

Given a linear equation $a_1x_1 + a_2x_2 + \cdots + a_nx_n = b$, a sequence $s_1, s_2, \ldots, s_n$ of n numbers is called a **solution** to the equation if

$$a_1s_1 + a_2s_2 + \cdots + a_ns_n = b$$

that is, if the equation is satisfied when the substitutions $x_1 = s_1, x_2 = s_2, \ldots, x_n = s_n$ are made. A sequence of numbers is called **a solution to a system** of equations if it is a solution to every equation in the system.

For example, $x = -2, y = 5, z = 0$ and $x = 0, y = 4, z = -1$ are both solutions to the system

$$\begin{aligned} x + y + z &= 3 \\ 2x + y + 3z &= 1 \end{aligned}$$

Chapter 1 Systems of Linear Equations

A system may have no solution at all, or it may have a unique solution, or it may have an infinite family of solutions. For instance, the system $x + y = 2$, $x + y = 3$ has no solution because the sum of two numbers cannot be 2 and 3 simultaneously. A system that has no solution is called **inconsistent**; a system with at least one solution is called **consistent**. The system in the following example has infinitely many solutions.

EXAMPLE 1

Show that, for arbitrary values of s and t,

$$x_1 = t - s + 1$$
$$x_2 = t + s + 2$$
$$x_3 = s$$
$$x_4 = t$$

is a solution to the system

$$x_1 - 2x_2 + 3x_3 + x_4 = -3$$
$$2x_1 - x_2 + 3x_3 - x_4 = 0$$

Solution ▶ Simply substitute these values of x_1, x_2, x_3, and x_4 in each equation.

$$x_1 - 2x_2 + 3x_3 + x_4 = (t - s + 1) - 2(t + s + 2) + 3s + t = -3$$
$$2x_1 - x_2 + 3x_3 - x_4 = 2(t - s + 1) - (t + s + 2) + 3s - t = 0$$

Because both equations are satisfied, it is a solution for all choices of s and t.

The quantities s and t in Example 1 are called **parameters**, and the set of solutions, described in this way, is said to be given in **parametric form** and is called the **general solution** to the system. It turns out that the solutions to *every* system of equations (if there *are* solutions) can be given in parametric form (that is, the variables x_1, x_2, ... are given in terms of new independent variables s, t, etc.). The following example shows how this happens in the simplest systems where only one equation is present.

EXAMPLE 2

Describe all solutions to $3x - y + 2z = 6$ in parametric form.

Solution ▶ Solving the equation for y in terms of x and z, we get $y = 3x + 2z - 6$. If s and t are arbitrary then, setting $x = s$, $z = t$, we get solutions

$$x = s$$
$$y = 3s + 2t - 6 \quad s \text{ and } t \text{ arbitrary}$$
$$z = t$$

Of course we could have solved for x: $x = \frac{1}{3}(y - 2z + 6)$. Then, if we take $y = p$, $z = q$, the solutions are represented as follows:

$$x = \frac{1}{3}(p - 2q + 6)$$
$$y = p \quad\quad p \text{ and } q \text{ arbitrary}$$
$$z = q$$

The same family of solutions can "look" quite different!

SECTION 1.1 Solutions and Elementary Operations

When only two variables are involved, the solutions to systems of linear equations can be described geometrically because the graph of a linear equation $ax + by = c$ is a straight line if a and b are not both zero. Moreover, a point $P(s, t)$ with coordinates s and t lies on the line if and only if $as + bt = c$—that is when $x = s, y = t$ is a solution to the equation. Hence the solutions to a *system* of linear equations correspond to the points $P(s, t)$ that lie on *all* the lines in question.

In particular, if the system consists of just one equation, there must be infinitely many solutions because there are infinitely many points on a line. If the system has two equations, there are three possibilities for the corresponding straight lines:

1. *The lines intersect at a single point. Then the system has a* unique solution *corresponding to that point.*

2. *The lines are parallel (and distinct) and so do not intersect. Then the system has* no solution.

3. *The lines are identical. Then the system has* infinitely many solutions—one for each point on the (common) line.

These three situations are illustrated in Figure 1. In each case the graphs of two specific lines are plotted and the corresponding equations are indicated. In the last case, the equations are $3x - y = 4$ and $-6x + 2y = -8$, which have identical graphs.

When three variables are present, the graph of an equation $ax + by + cz = d$ can be shown to be a plane (see Section 4.2) and so again provides a "picture" of the set of solutions. However, this graphical method has its limitations: When more than three variables are involved, no physical image of the graphs (called hyperplanes) is possible. It is necessary to turn to a more "algebraic" method of solution.

Before describing the method, we introduce a concept that simplifies the computations involved. Consider the following system

$$\begin{aligned} 3x_1 + 2x_2 - x_3 + x_4 &= -1 \\ 2x_1 \quad\quad - x_3 + 2x_4 &= 0 \\ 3x_1 + x_2 + 2x_3 + 5x_4 &= 2 \end{aligned}$$

of three equations in four variables. The array of numbers[1]

$$\begin{bmatrix} 3 & 2 & -1 & 1 & | & -1 \\ 2 & 0 & -1 & 2 & | & 0 \\ 3 & 1 & 2 & 5 & | & 2 \end{bmatrix}$$

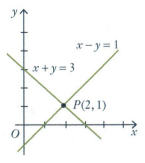

(a) Unique solution
($x = 2, y = 1$)

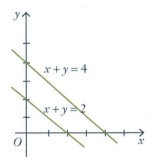

(b) No solution

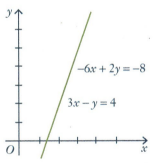

(c) Infinitely many solutions
($x = t, y = 3t - 4$)

■ FIGURE 1

occurring in the system is called the **augmented matrix** of the system. Each row of the matrix consists of the coefficients of the variables (in order) from the corresponding equation, together with the constant term. For clarity, the constants are separated by a vertical line. The augmented matrix is just a different way of describing the system of equations. The array of coefficients of the variables

$$\begin{bmatrix} 3 & 2 & -1 & 1 \\ 2 & 0 & -1 & 2 \\ 3 & 1 & 2 & 5 \end{bmatrix}$$

is called the **coefficient matrix** of the system and $\begin{bmatrix} -1 \\ 0 \\ 2 \end{bmatrix}$ is called the **constant matrix** of the system.

---

1 A rectangular array of numbers is called a **matrix**. Matrices will be discussed in more detail in Chapter 2.

Elementary Operations

The algebraic method for solving systems of linear equations is described as follows. Two such systems are said to be **equivalent** if they have the same set of solutions. A system is solved by writing a series of systems, one after the other, each equivalent to the previous system. Each of these systems has the same set of solutions as the original one; the aim is to end up with a system that is easy to solve. Each system in the series is obtained from the preceding system by a simple manipulation chosen so that it does not change the set of solutions.

As an illustration, we solve the system $x + 2y = -2$, $2x + y = 7$ in this manner. At each stage, the corresponding augmented matrix is displayed. The original system is

$$\begin{aligned} x + 2y &= -2 \\ 2x + y &= 7 \end{aligned} \qquad \begin{bmatrix} 1 & 2 & | & -2 \\ 2 & 1 & | & 7 \end{bmatrix}$$

First, subtract twice the first equation from the second. The resulting system is

$$\begin{aligned} x + 2y &= -2 \\ -3y &= 11 \end{aligned} \qquad \begin{bmatrix} 1 & 2 & | & -2 \\ 0 & -3 & | & 11 \end{bmatrix}$$

which is equivalent to the original (see Theorem 1). At this stage we obtain $y = -\frac{11}{3}$ by multiplying the second equation by $-\frac{1}{3}$. The result is the equivalent system

$$\begin{aligned} x + 2y &= -2 \\ y &= -\frac{11}{3} \end{aligned} \qquad \begin{bmatrix} 1 & 2 & | & -2 \\ 0 & 1 & | & -\frac{11}{3} \end{bmatrix}$$

Finally, we subtract twice the second equation from the first to get another equivalent system.

$$\begin{aligned} x &= \frac{16}{3} \\ y &= -\frac{11}{3} \end{aligned} \qquad \begin{bmatrix} 1 & 0 & | & \frac{16}{3} \\ 0 & 1 & | & -\frac{11}{3} \end{bmatrix}$$

Now *this* system is easy to solve! And because it is equivalent to the original system, it provides the solution to that system.

Observe that, at each stage, a certain operation is performed on the system (and thus on the augmented matrix) to produce an equivalent system.

Definition 1.1

*The following operations, called **elementary operations**, can routinely be performed on systems of linear equations to produce equivalent systems.*

(I) *Interchange two equations.*

(II) *Multiply one equation by a nonzero number.*

(III) *Add a multiple of one equation to a different equation.*

Theorem 1

Suppose that a sequence of elementary operations is performed on a system of linear equations. Then the resulting system has the same set of solutions as the original, so the two systems are equivalent.

The proof is given at the end of this section.

SECTION 1.1 Solutions and Elementary Operations

Elementary operations performed on a system of equations produce corresponding manipulations of the *rows* of the augmented matrix. Thus, multiplying a row of a matrix by a number k means multiplying *every entry* of the row by k. Adding one row to another row means adding *each entry* of that row to the corresponding entry of the other row. Subtracting two rows is done similarly. Note that we regard two rows as equal when corresponding entries are the same.

In hand calculations (and in computer programs) we manipulate the rows of the augmented matrix rather than the equations. For this reason we restate these elementary operations for matrices.

Definition 1.2

*The following are called **elementary row operations** on a matrix.*

(I) *Interchange two rows.*

(II) *Multiply one row by a nonzero number.*

(III) *Add a multiple of one row to a different row.*

In the illustration above, a series of such operations led to a matrix of the form

$$\begin{bmatrix} 1 & 0 & | & * \\ 0 & 1 & | & * \end{bmatrix}$$

where the asterisks represent arbitrary numbers. In the case of three equations in three variables, the goal is to produce a matrix of the form

$$\begin{bmatrix} 1 & 0 & 0 & | & * \\ 0 & 1 & 0 & | & * \\ 0 & 0 & 1 & | & * \end{bmatrix}$$

This does not always happen, as we will see in the next section. Here is an example in which it does happen.

EXAMPLE 3

Find all solutions to the following system of equations.

$$\begin{aligned} 3x + 4y + z &= 1 \\ 2x + 3y &= 0 \\ 4x + 3y - z &= -2 \end{aligned}$$

Solution ▶ The augmented matrix of the original system is

$$\begin{bmatrix} 3 & 4 & 1 & | & 1 \\ 2 & 3 & 0 & | & 0 \\ 4 & 3 & -1 & | & -2 \end{bmatrix}$$

To create a 1 in the upper left corner we could multiply row 1 through by $\frac{1}{3}$. However, the 1 can be obtained without introducing fractions by subtracting row 2 from row 1. The result is

$$\begin{bmatrix} 1 & 1 & 1 & | & 1 \\ 2 & 3 & 0 & | & 0 \\ 4 & 3 & -1 & | & -2 \end{bmatrix}$$

The upper left 1 is now used to "clean up" the first column, that is create zeros in the other positions in that column. First subtract 2 times row 1 from row 2 to obtain

$$\begin{bmatrix} 1 & 1 & 1 & | & 1 \\ 0 & 1 & -2 & | & -2 \\ 4 & 3 & -1 & | & -2 \end{bmatrix}$$

Next subtract 4 times row 1 from row 3. The result is

$$\begin{bmatrix} 1 & 1 & 1 & | & 1 \\ 0 & 1 & -2 & | & -2 \\ 0 & -1 & -5 & | & -6 \end{bmatrix}$$

This completes the work on column 1. We now use the 1 in the second position of the second row to clean up the second column by subtracting row 2 from row 1 and adding row 2 to row 3. For convenience, both row operations are done in one step. The result is

$$\begin{bmatrix} 1 & 0 & 3 & | & 3 \\ 0 & 1 & -2 & | & -2 \\ 0 & 0 & -7 & | & -8 \end{bmatrix}$$

Note that these manipulations *did not affect* the first column (the second row has a zero there), so our previous effort there has not been undermined. Finally we clean up the third column. Begin by multiplying row 3 by $-\frac{1}{7}$ to obtain

$$\begin{bmatrix} 1 & 0 & 3 & | & 3 \\ 0 & 1 & -2 & | & -2 \\ 0 & 0 & 1 & | & \frac{8}{7} \end{bmatrix}$$

Now subtract 3 times row 3 from row 1, and add 2 times row 3 to row 2 to get

$$\begin{bmatrix} 1 & 0 & 0 & | & -\frac{3}{7} \\ 0 & 1 & 0 & | & \frac{2}{7} \\ 0 & 0 & 1 & | & \frac{8}{7} \end{bmatrix}$$

The corresponding equations are $x = -\frac{3}{7}$, $y = \frac{2}{7}$, and $z = \frac{8}{7}$, which give the (unique) solution.

Every elementary row operation can be **reversed** by another elementary row operation of the same type (called its **inverse**). To see how, we look at types I, II, and III separately:

Type I   Interchanging two rows is reversed by interchanging them again.

Type II Multiplying a row by a nonzero number k is reversed by multiplying by $1/k$.

Type III Adding k times row p to a different row q is reversed by adding $-k$ times row p to row q (in the new matrix). Note that $p \neq q$ is essential here.

To illustrate the Type III situation, suppose there are four rows in the original matrix, denoted R_1, R_2, R_3, and R_4, and that k times R_2 is added to R_3. Then the reverse operation adds $-k$ times R_2, to R_3. The following diagram illustrates the effect of doing the operation first and then the reverse:

SECTION 1.1 Solutions and Elementary Operations

$$\begin{bmatrix} R_1 \\ R_2 \\ R_3 \\ R_4 \end{bmatrix} \to \begin{bmatrix} R_1 \\ R_2 \\ R_3 + kR_3 \\ R_4 \end{bmatrix} \to \begin{bmatrix} R_1 \\ R_2 \\ (R_3 + kR_2) - kR_2 \\ R_4 \end{bmatrix} = \begin{bmatrix} R_1 \\ R_2 \\ R_3 \\ R_4 \end{bmatrix}$$

The existence of inverses for elementary row operations and hence for elementary operations on a system of equations, gives:

PROOF OF THEOREM 1

Suppose that a system of linear equations is transformed into a new system by a sequence of elementary operations. Then every solution of the original system is automatically a solution of the new system because adding equations, or multiplying an equation by a nonzero number, always results in a valid equation. In the same way, each solution of the new system must be a solution to the original system because the original system can be obtained from the new one by another series of elementary operations (the inverses of the originals). It follows that the original and new systems have the same solutions. This proves Theorem 1.

EXERCISES 1.1

1. In each case verify that the following are solutions for all values of s and t.

   (a) $x = 19t - 35$
   $y = 25 - 13t$
   $z = t$
   is a solution of
   $2x + 3y + z = 5$
   $5x + 7y - 4z = 0$

   ◆(b) $x_1 = 2s + 12t + 13$
   $x_2 = s$
   $x_3 = -s - 3t - 3$
   $x_4 = t$
   is a solution of
   $2x_1 + 5x_2 + 9x_3 + 3x_4 = -1$
   $x_1 + 2x_2 + 4x_3 \qquad = 1$

2. Find all solutions to the following in parametric form in two ways.

   (a) $3x + y = 2$ ◆(b) $2x + 3y = 1$
   (c) $3x - y + 2z = 5$ ◆(b) $x - 2y + 5z = 1$

3. Regarding $2x = 5$ as the equation $2x + 0y = 5$ in two variables, find all solutions in parametric form.

◆4. Regarding $4x - 2y = 3$ as the equation $4x - 2y + 0z = 3$ in three variables, find all solutions in parametric form.

◆5. Find all solutions to the general system $ax = b$ of one equation in one variable (a) when $a = 0$ and (b) when $a \neq 0$.

6. Show that a system consisting of exactly one linear equation can have no solution, one solution, or infinitely many solutions. Give examples.

7. Write the augmented matrix for each of the following systems of linear equations.

   (a) $x - 3y = 5$ ◆(b) $x + 2y = 0$
   $2x + y = 1$ $y = 1$

   (c) $x - y + z = 2$ ◆(d) $x + y = 1$
   $x - z = 1$ $y + z = 0$
   $y + 2x = 0$ $z - x = 2$

---

2 A ◆ indicates that the exercise has an answer at the end of the book.

8. Write a system of linear equations that has each of the following augmented matrices.

    (a) $\begin{bmatrix} 1 & -1 & 6 & | & 0 \\ 0 & 1 & 0 & | & 3 \\ 2 & -1 & 0 & | & 1 \end{bmatrix}$ ◆(b) $\begin{bmatrix} 2 & -1 & 0 & | & -1 \\ -3 & 2 & 1 & | & 0 \\ 0 & 1 & 1 & | & 3 \end{bmatrix}$

9. Find the solution of each of the following systems of linear equations using augmented matrices.

    (a) $\begin{aligned} x - 3y &= 1 \\ 2x - 7y &= 3 \end{aligned}$ ◆(b) $\begin{aligned} x + 2y &= 1 \\ 3x + 4y &= -1 \end{aligned}$

    (c) $\begin{aligned} 2x + 3y &= -1 \\ 3x + 4y &= 2 \end{aligned}$ ◆(d) $\begin{aligned} 3x + 4y &= 1 \\ 4x + 5y &= -3 \end{aligned}$

10. Find the solution of each of the following systems of linear equations using augmented matrices.

    (a) $\begin{aligned} x + y + 2z &= -1 \\ 2x + y + 3z &= 0 \\ -2y + z &= 2 \end{aligned}$ ◆(b) $\begin{aligned} 2x + y + z &= -1 \\ x + 2y + z &= 0 \\ 3x - 2z &= 5 \end{aligned}$

11. Find all solutions (if any) of the following systems of linear equations.

    (a) $\begin{aligned} 3x - 2y &= 5 \\ -12x + 8y &= -20 \end{aligned}$ ◆(b) $\begin{aligned} 3x - 2y &= 5 \\ -12x + 8y &= 16 \end{aligned}$

12. Show that the system $\begin{cases} x + 2y - z = a \\ 2x + y + 3z = b \\ x - 4y + 9z = c \end{cases}$ is inconsistent unless $c = 2b - 3a$.

13. By examining the possible positions of lines in the plane, show that two equations in two variables can have zero, one, or infinitely many solutions.

14. In each case either show that the statement is true, or give an example[3] showing it is false.

    (a) If a linear system has n variables and m equations, then the augmented matrix has n rows.

    ◆(b) A consistent linear system must have infinitely many solutions.

    (c) If a row operation is done to a consistent linear system, the resulting system must be consistent.

    ◆(d) If a series of row operations on a linear system results in an inconsistent system, the original system is inconsistent.

15. Find a quadratic $a + bx + cx^2$ such that the graph of $y = a + bx + cx^2$ contains each of the points $(-1, 6)$, $(2, 0)$, and $(3, 2)$.

◆16. Solve the system $\begin{cases} 3x + 2y = 5 \\ 7x + 5y = 1 \end{cases}$ by changing variables $\begin{cases} x = 5x' - 2y' \\ y = -7x' + 3y' \end{cases}$ and solving the resulting equations for x' and y'.

◆17. Find a, b, and c such that
$$\frac{x^2 - x + 3}{(x^2 + 2)(2x - 1)} = \frac{ax + b}{x^2 + 2} + \frac{c}{2x - 1}$$
[*Hint*: Multiply through by $(x^2 + 2)(2x - 1)$ and equate coefficients of powers of x.]

18. A zookeeper wants to give an animal 42 mg of vitamin A and 65 mg of vitamin D per day. He has two supplements: the first contains 10% vitamin A and 25% vitamin D; the second contains 20% vitamin A and 25% vitamin D. How much of each supplement should he give the animal each day?

◆19. Workmen John and Joe earn a total of $24.60 when John works 2 hours and Joe works 3 hours. If John works 3 hours and Joe works 2 hours, they get $23.90. Find their hourly rates.

20. A biologist wants to create a diet from fish and meal containing 183 grams of protein and 93 grams of carbohydrate per day. If fish contains 70% protein and 10% carbohydrate, and meal contains 30% protein and 60% carbohydrate, how much of each food is required each day?

---

[3] Such an example is called a **counterexample**. For example, if the statement is that "all philosophers have beards", the existence of a non-bearded philosopher would be a counterexample proving that the statement is false. This is discussed again in Appendix B.

SECTION 1.2 Gaussian Elimination

The algebraic method introduced in the preceding section can be summarized as follows: Given a system of linear equations, use a sequence of elementary row operations to carry the augmented matrix to a "nice" matrix (meaning that the corresponding equations are easy to solve). In Example 3 Section 1.1, this nice matrix took the form

$$\begin{bmatrix} 1 & 0 & 0 & | & * \\ 0 & 1 & 0 & | & * \\ 0 & 0 & 1 & | & * \end{bmatrix}$$

The following definitions identify the nice matrices that arise in this process.

Definition 1.3   *A matrix is said to be in **row-echelon form** (and will be called a **row-echelon matrix**) if it satisfies the following three conditions:*

1. *All **zero rows** (consisting entirely of zeros) are at the bottom.*
2. *The first nonzero entry from the left in each nonzero row is a 1, called the **leading 1** for that row.*
3. *Each leading 1 is to the right of all leading 1s in the rows above it.*

*A row-echelon matrix is said to be in **reduced row-echelon form** (and will be called a **reduced row-echelon matrix**) if, in addition, it satisfies the following condition:*

4. *Each leading 1 is the only nonzero entry in its column.*

The row-echelon matrices have a "staircase" form, as indicated by the following example (the asterisks indicate arbitrary numbers).

$$\begin{bmatrix} 0 & 1 & * & * & * & * \\ 0 & 0 & 0 & 1 & * & * \\ 0 & 0 & 0 & 0 & 1 & * \\ 0 & 0 & 0 & 0 & 0 & 1 \\ 0 & 0 & 0 & 0 & 0 & 0 \end{bmatrix}$$

The leading 1s proceed "down and to the right" through the matrix. Entries above and to the right of the leading 1s are arbitrary, but all entries below and to the left of them are zero. Hence, a matrix in row-echelon form is in reduced form if, in addition, the entries directly above each leading 1 are all zero. Note that a matrix in row-echelon form can, with a few more row operations, be carried to reduced form (use row operations to create zeros above each leading one in succession, beginning from the right).

EXAMPLE 1

The following matrices are in row-echelon form (for any choice of numbers in *-positions).

$$\begin{bmatrix} 1 & * & * \\ 0 & 0 & 1 \end{bmatrix} \quad \begin{bmatrix} 0 & 1 & * & * \\ 0 & 0 & 1 & * \\ 0 & 0 & 0 & 0 \end{bmatrix} \quad \begin{bmatrix} 1 & * & * & * \\ 0 & 1 & * & * \\ 0 & 0 & 0 & 1 \end{bmatrix} \quad \begin{bmatrix} 1 & * & * \\ 0 & 1 & * \\ 0 & 0 & 1 \end{bmatrix}$$

The following, on the other hand, are in reduced row-echelon form.

$$\begin{bmatrix} 1 & * & 0 \\ 0 & 0 & 1 \end{bmatrix} \begin{bmatrix} 0 & 1 & 0 & * \\ 0 & 0 & 1 & * \\ 0 & 0 & 0 & 0 \end{bmatrix} \begin{bmatrix} 1 & 0 & * & 0 \\ 0 & 1 & * & 0 \\ 0 & 0 & 0 & 1 \end{bmatrix} \begin{bmatrix} 1 & 0 & 0 \\ 0 & 1 & 0 \\ 0 & 0 & 1 \end{bmatrix}$$

The choice of the positions for the leading 1s determines the (reduced) row-echelon form (apart from the numbers in *-positions).

The importance of row-echelon matrices comes from the following theorem.

Theorem 1

Every matrix can be brought to (reduced) row-echelon form by a sequence of elementary row operations.

In fact we can give a step-by-step procedure for actually finding a row-echelon matrix. Observe that while there are many sequences of row operations that will bring a matrix to row-echelon form, the one we use is systematic and is easy to program on a computer. Note that the algorithm deals with matrices in general, possibly with columns of zeros.

Gaussian[4] Algorithm[5]

Step 1. If the matrix consists entirely of zeros, stop—it is already in row-echelon form.

Step 2. Otherwise, find the first column from the left containing a nonzero entry (call it a), and move the row containing that entry to the top position.

Step 3. Now multiply the new top row by 1/a to create a leading 1.

Step 4. By subtracting multiples of that row from rows below it, make each entry below the leading 1 zero.

This completes the first row, and all further row operations are carried out on the remaining rows.

Step 5. Repeat steps 1–4 on the matrix consisting of the remaining rows.

The process stops when either no rows remain at step 5 or the remaining rows consist entirely of zeros.

Carl Friedrich Gauss. Photo © Corbis.

Observe that the gaussian algorithm is recursive: When the first leading 1 has been obtained, the procedure is repeated on the remaining rows of the matrix. This makes the algorithm easy to use on a computer. Notes that the solution to Example 3 Section 1.1 did not use the gaussian algorithm as written because the first leading 1 was not created by dividing row 1 by 3. The reason for this is that it avoids fractions. However, the general pattern is clear: Create the leading 1s from left to right, using each of them in turn to create zeros below it. Here are two more examples.

---

4   Carl Friedrich Gauss (1777–1855) ranks with Archimedes and Newton as one of the three greatest mathematicians of all time. He was a child prodigy and, at the age of 21, he gave the first proof that every polynomial has a complex root. In 1801 he published a timeless masterpiece, *Disquisitiones Arithmeticae*, in which he founded modern number theory. He went on to make groundbreaking contributions to nearly every branch of mathematics, often well before others rediscovered and published the results.

5   The algorithm was known to the ancient Chinese.

EXAMPLE 2

Solve the following system of equations.

$$\begin{aligned} 3x + y - 4z &= -1 \\ x \phantom{{}+y} + 10z &= 5 \\ 4x + y + 6z &= 1 \end{aligned}$$

Solution ▶ The corresponding augmented matrix is

$$\begin{bmatrix} 3 & 1 & -4 & | & -1 \\ 1 & 0 & 10 & | & 5 \\ 4 & 1 & 6 & | & 1 \end{bmatrix}$$

Create the first leading one by interchanging rows 1 and 2.

$$\begin{bmatrix} 1 & 0 & 10 & | & 5 \\ 3 & 1 & -4 & | & -1 \\ 4 & 1 & 6 & | & 1 \end{bmatrix}$$

Now subtract 3 times row 1 from row 2, and subtract 4 times row 1 from row 3. The result is

$$\begin{bmatrix} 1 & 0 & 10 & | & 5 \\ 0 & 1 & -34 & | & -16 \\ 0 & 1 & -34 & | & -19 \end{bmatrix}$$

Now subtract row 2 from row 3 to obtain

$$\begin{bmatrix} 1 & 0 & 10 & | & 5 \\ 0 & 1 & -34 & | & -16 \\ 0 & 0 & 0 & | & -3 \end{bmatrix}$$

This means that the following system of equations

$$\begin{aligned} x \phantom{{}+y} + 10z &= 5 \\ y - 34z &= -16 \\ 0 &= -3 \end{aligned}$$

is equivalent to the original system. In other words, the two have the *same* solutions. But this last system clearly has *no* solution (the last equation requires that x, y and z satisfy $0x + 0y + 0z = -3$, and no such numbers exist). Hence the original system has *no* solution.

EXAMPLE 3

Solve the following system of equations.

$$\begin{aligned} x_1 - 2x_2 - x_3 + 3x_4 &= 1 \\ 2x_1 - 4x_2 + x_3 \phantom{{}+3x_4} &= 5 \\ x_1 - 2x_2 + 2x_3 - 3x_4 &= 4 \end{aligned}$$

Solution ▶ The augmented matrix is

$$\begin{bmatrix} 1 & -2 & -1 & 3 & | & 1 \\ 2 & -4 & 1 & 0 & | & 5 \\ 1 & -2 & 2 & -3 & | & 4 \end{bmatrix}$$

Subtracting twice row 1 from row 2 and subtracting row 1 from row 3 gives

$$\begin{bmatrix} 1 & -2 & -1 & 3 & | & 1 \\ 0 & 0 & 3 & -6 & | & 3 \\ 0 & 0 & 3 & -6 & | & 3 \end{bmatrix}$$

Now subtract row 2 from row 3 and multiply row 2 by $\frac{1}{3}$ to get

$$\begin{bmatrix} 1 & -2 & -1 & 3 & | & 1 \\ 0 & 0 & 1 & -2 & | & 1 \\ 0 & 0 & 0 & 0 & | & 0 \end{bmatrix}$$

This is in row-echelon form, and we take it to reduced form by adding row 2 to row 1:

$$\begin{bmatrix} 1 & -2 & 0 & 1 & | & 2 \\ 0 & 0 & 1 & -2 & | & 1 \\ 0 & 0 & 0 & 0 & | & 0 \end{bmatrix}$$

The corresponding system of equations is

$$\begin{aligned} x_1 - 2x_2 \quad\quad + x_4 &= 2 \\ x_3 - 2x_4 &= 1 \\ 0 &= 0 \end{aligned}$$

The leading ones are in columns 1 and 3 here, so the corresponding variables x_1 and x_3 are called leading variables. Because the matrix is in reduced row-echelon form, these equations can be used to solve for the leading variables in terms of the nonleading variables x_2 and x_4. More precisely, in the present example we set $x_2 = s$ and $x_4 = t$ where s and t are arbitrary, so these equations become

$$x_1 - 2s + t = 2 \quad \text{and} \quad x_3 - 2t = 1.$$

Finally the solutions are given by

$$\begin{aligned} x_1 &= 2 + 2s - t \\ x_2 &= s \\ x_3 &= 1 + 2t \\ x_4 &= t \end{aligned}$$

where s and t are arbitrary.

The solution of Example 3 is typical of the general case. To solve a linear system, the augmented matrix is carried to reduced row-echelon form, and the variables corresponding to the leading ones are called **leading variables**. Because the matrix is in reduced form, each leading variable occurs in exactly one equation, so that equation can be solved to give a formula for the leading variable in terms of the nonleading variables. It is customary to call the nonleading variables "free" variables, and to label them by new variables $s, t, \ldots$, called **parameters**. Hence, as in Example 3, every variable x_i is given by a formula in terms of the parameters s

SECTION 1.2 Gaussian Elimination

and t. Moreover, every choice of these parameters leads to a solution to the system, and every solution arises in this way. This procedure works in general, and has come to be called

Gaussian Elimination

To solve a system of linear equations proceed as follows:
1. *Carry the augmented matrix to a reduced row-echelon matrix using elementary row operations.*
2. *If a row $[0\ 0\ 0\ \cdots\ 0\ 1]$ occurs, the system is inconsistent.*
3. *Otherwise, assign the nonleading variables (if any) as parameters, and use the equations corresponding to the reduced row-echelon matrix to solve for the leading variables in terms of the parameters.*

There is a variant of this procedure, wherein the augmented matrix is carried only to row-echelon form. The nonleading variables are assigned as parameters as before. Then the last equation (corresponding to the row-echelon form) is used to solve for the last leading variable in terms of the parameters. This last leading variable is then substituted into all the preceding equations. Then, the second last equation yields the second last leading variable, which is also substituted back. The process continues to give the general solution. This procedure is called **back-substitution**. This procedure can be shown to be numerically more efficient and so is important when solving very large systems.[6]

EXAMPLE 4

Find a condition on the numbers a, b, and c such that the following system of equations is consistent. When that condition is satisfied, find all solutions (in terms of a, b, and c).

$$x_1 + 3x_2 + x_3 = a$$
$$-x_1 - 2x_2 + x_3 = b$$
$$3x_1 + 7x_2 - x_3 = c$$

Solution ▶ We use gaussian elimination except that now the augmented matrix

$$\begin{bmatrix} 1 & 3 & 1 & | & a \\ -1 & -2 & 1 & | & b \\ 3 & 7 & -1 & | & c \end{bmatrix}$$

has entries a, b, and c as well as known numbers. The first leading one is in place, so we create zeros below it in column 1:

$$\begin{bmatrix} 1 & 3 & 1 & | & a \\ 0 & 1 & 2 & | & a+b \\ 0 & -2 & -4 & | & c-3a \end{bmatrix}$$

---

[6] With n equations where n is large, gaussian elimination requires roughly $n^3/2$ multiplications and divisions, whereas this number is roughly $n^3/3$ if back substitution is used.

The second leading 1 has appeared, so use it to create zeros in the rest of column 2:

$$\begin{bmatrix} 1 & 0 & -5 & | & -2a-3b \\ 0 & 1 & 2 & | & a+b \\ 0 & 0 & 0 & | & c-a+2b \end{bmatrix}$$

Now the whole solution depends on the number $c - a + 2b = c - (a - 2b)$. The last row corresponds to an equation $0 = c - (a - 2b)$. If $c \neq a - 2b$, there is *no* solution (just as in Example 2). Hence:

The system is consistent if and only if $c = a - 2b$.

In this case the last matrix becomes

$$\begin{bmatrix} 1 & 0 & -5 & | & -2a-3b \\ 0 & 1 & 2 & | & a+b \\ 0 & 0 & 0 & | & 0 \end{bmatrix}$$

Thus, if $c = a - 2b$, taking $x_3 = t$ where t is a parameter gives the solutions

$$x_1 = 5t - (2a + 3b) \quad x_2 = (a+b) - 2t \quad x_3 = t.$$

Rank

It can be proven that the *reduced* row-echelon form of a matrix A is uniquely determined by A. That is, no matter which series of row operations is used to carry A to a reduced row-echelon matrix, the result will always be the same matrix. (A proof is given at the end of Section 2.5.) By contrast, this is not true for row-echelon matrices: Different series of row operations can carry the same matrix A to *different* row-echelon matrices. Indeed, the matrix $A = \begin{bmatrix} 1 & -1 & 4 \\ 2 & -1 & 2 \end{bmatrix}$ can be carried (by one row operation) to the row-echelon matrix $\begin{bmatrix} 1 & -1 & 4 \\ 0 & 1 & -6 \end{bmatrix}$, and then by another row operation to the (reduced) row-echelon matrix $\begin{bmatrix} 1 & 0 & -2 \\ 0 & 1 & -6 \end{bmatrix}$. However, it *is* true that the number r of leading 1s must be the same in each of these row-echelon matrices (this will be proved in Chapter 5). Hence, the number r depends only on A and not on the way in which A is carried to row-echelon form.

Definition 1.4 *The **rank** of matrix A is the number of leading 1s in any row-echelon matrix to which A can be carried by row operations.*

EXAMPLE 5

Compute the rank of $A = \begin{bmatrix} 1 & 1 & -1 & 4 \\ 2 & 1 & 3 & 0 \\ 0 & 1 & -5 & 8 \end{bmatrix}$.

SECTION 1.2 Gaussian Elimination

Solution ▶ The reduction of A to row-echelon form is

$$A = \begin{bmatrix} 1 & 1 & -1 & 4 \\ 2 & 1 & 3 & 0 \\ 0 & 1 & -5 & 8 \end{bmatrix} \to \begin{bmatrix} 1 & 1 & -1 & 4 \\ 0 & -1 & 5 & -8 \\ 0 & 1 & -5 & 8 \end{bmatrix} \to \begin{bmatrix} 1 & 1 & -1 & 4 \\ 0 & 1 & -5 & 8 \\ 0 & 0 & 0 & 0 \end{bmatrix}$$

Because this row-echelon matrix has two leading 1s, rank $A = 2$.

Suppose that rank $A = r$, where A is a matrix with m rows and n columns. Then $r < m$ because the leading 1s lie in different rows, and $r < n$ because the leading 1s lie in different columns. Moreover, the rank has a useful application to equations. Recall that a system of linear equations is called consistent if it has at least one solution.

Theorem 2

Suppose a system of m equations in n variables is consistent, and that the rank of the augmented matrix is r.

(1) The set of solutions involves exactly $n - r$ parameters.

(2) If $r < n$, the system has infinitely many solutions.

(3) If $r = n$, the system has a unique solution.

PROOF

The fact that the rank of the augmented matrix is r means there are exactly r leading variables, and hence exactly $n - r$ nonleading variables. These nonleading variables are all assigned as parameters in the gaussian algorithm, so the set of solutions involves exactly $n - r$ parameters. Hence if $r < n$, there is at least one parameter, and so infinitely many solutions. If $r = n$, there are no parameters and so a unique solution.

Theorem 2 shows that, for any system of linear equations, exactly three possibilities exist:

1. **No solution.** This occurs when a row $[0 \ 0 \ \cdots \ 0 \ 1]$ occurs in the row-echelon form. This is the case where the system is inconsistent.

2. **Unique solution.** This occurs when every variable is a leading variable.

3. **Infinitely many solutions.** This occurs when the system is consistent and there is at least one nonleading variable, so at least one parameter is involved.

EXAMPLE 6

Suppose the matrix A in Example 5 is the augmented matrix of a system of $m = 3$ linear equations in $n = 3$ variables. As rank $A = r = 2$, the set of solutions will have $n - r = 1$ parameter. The reader can verify this fact directly.

Many important problems involve **linear inequalities** rather than **linear equations**. For example, a condition on the variables x and y might take the form of an inequality $2x - 5y \leq 4$ rather than an equality $2x - 5y = 4$. There is a technique (called the **simplex algorithm**) for finding solutions to a system of such inequalities that maximizes a function of the form $p = ax + by$ where a and b are fixed constants. This procedure involves gaussian elimination techniques, and the interested reader can find an introduction on Connect by visiting www.mcgrawhill.ca/college/nicholson and then selecting this text.

EXERCISES 1.2

1. Which of the following matrices are in reduced row-echelon form? Which are in row-echelon form?

   (a) $\begin{bmatrix} 1 & -1 & 2 \\ 0 & 0 & 0 \\ 0 & 0 & 1 \end{bmatrix}$ ◆(b) $\begin{bmatrix} 2 & 1 & -1 & 3 \\ 0 & 0 & 0 & 0 \end{bmatrix}$

   (c) $\begin{bmatrix} 1 & -2 & 3 & 5 \\ 0 & 0 & 0 & 1 \end{bmatrix}$ ◆(d) $\begin{bmatrix} 1 & 0 & 0 & 3 & 1 \\ 0 & 0 & 0 & 1 & 1 \\ 0 & 0 & 0 & 0 & 1 \end{bmatrix}$

   (e) $\begin{bmatrix} 1 & 1 \\ 0 & 1 \end{bmatrix}$ ◆(f) $\begin{bmatrix} 0 & 0 & 1 \\ 0 & 0 & 1 \\ 0 & 0 & 1 \end{bmatrix}$

2. Carry each of the following matrices to reduced row-echelon form.

   (a) $\begin{bmatrix} 0 & -1 & 2 & 1 & 2 & 1 & -1 \\ 0 & 1 & -2 & 2 & 7 & 2 & 4 \\ 0 & -2 & 4 & 3 & 7 & 1 & 0 \\ 0 & 3 & -6 & 1 & 6 & 4 & 1 \end{bmatrix}$

   ◆(b) $\begin{bmatrix} 0 & -1 & 3 & 1 & 3 & 2 & 1 \\ 0 & -2 & 6 & 1 & -5 & 0 & -1 \\ 0 & 3 & -9 & 2 & 4 & 1 & -1 \\ 0 & 1 & -3 & -1 & 3 & 0 & 1 \end{bmatrix}$

3. The augmented matrix of a system of linear equations has been carried to the following by row operations. In each case solve the system.

   (a) $\left[\begin{array}{cccccc|c} 1 & 2 & 0 & 3 & 1 & 0 & -1 \\ 0 & 0 & 1 & -1 & 1 & 0 & 2 \\ 0 & 0 & 0 & 0 & 0 & 1 & 3 \\ 0 & 0 & 0 & 0 & 0 & 0 & 0 \end{array}\right]$

   ◆(b) $\left[\begin{array}{ccccc|c} 1 & -2 & 0 & 2 & 0 & 1 & 1 \\ 0 & 0 & 1 & 5 & 0 & -3 & -1 \\ 0 & 0 & 0 & 0 & 1 & 6 & 1 \\ 0 & 0 & 0 & 0 & 0 & 0 & 0 \end{array}\right]$

   (c) $\left[\begin{array}{cccc|c} 1 & 2 & 1 & 3 & 1 & 1 \\ 0 & 1 & -1 & 0 & 1 & 1 \\ 0 & 0 & 0 & 1 & -1 & 0 \\ 0 & 0 & 0 & 0 & 0 & 0 \end{array}\right]$

   ◆(d) $\left[\begin{array}{ccccc|c} 1 & -1 & 2 & 4 & 6 & 2 \\ 0 & 1 & 2 & 1 & -1 & -1 \\ 0 & 0 & 0 & 1 & 0 & 1 \\ 0 & 0 & 0 & 0 & 0 & 0 \end{array}\right]$

4. Find all solutions (if any) to each of the following systems of linear equations.

   (a) $\begin{aligned} x - 2y &= 1 \\ 4y - x &= -2 \end{aligned}$ ◆(b) $\begin{aligned} 3x - y &= 0 \\ 2x - 3y &= 1 \end{aligned}$

   (c) $\begin{aligned} 2x + y &= 5 \\ 3x + 2y &= 6 \end{aligned}$ ◆(d) $\begin{aligned} 3x - y &= 2 \\ 2y - 6x &= -4 \end{aligned}$

   (e) $\begin{aligned} 3x - y &= 4 \\ 2y - 6x &= 1 \end{aligned}$ ◆(f) $\begin{aligned} 2x - 3y &= 5 \\ 3y - 2x &= 2 \end{aligned}$

5. Find all solutions (if any) to each of the following systems of linear equations.

   (a) $\begin{aligned} x + y + 2z &= 8 \\ 3x - y + z &= 0 \\ -x + 3y + 4z &= -4 \end{aligned}$ ◆(b) $\begin{aligned} -2x + 3y + 3z &= -9 \\ 3x - 4y + z &= 5 \\ -5x + 7y + 2z &= -14 \end{aligned}$

   (c) $\begin{aligned} x + y - z &= 10 \\ -x + 4y + 5z &= -5 \\ x + 6y + 3z &= 15 \end{aligned}$ ◆(d) $\begin{aligned} x + 2y - z &= 2 \\ 2x + 5y - 3z &= 1 \\ x + 4y - 3z &= 3 \end{aligned}$

   (e) $\begin{aligned} 5x + y &= 2 \\ 3x - y + 2z &= 1 \\ x + y - z &= 5 \end{aligned}$ ◆(f) $\begin{aligned} 3x - 2y + z &= -2 \\ x - y + 3z &= 5 \\ -x + y + z &= -1 \end{aligned}$

   (g) $\begin{aligned} x + y + z &= 2 \\ x + z &= 1 \\ 2x + 5y + 2z &= 7 \end{aligned}$ ◆(h) $\begin{aligned} x + 2y - 4z &= 10 \\ 2x - y + 2z &= 5 \\ x + y - 2z &= 7 \end{aligned}$

6. Express the last equation of each system as a sum of multiples of the first two equations. [*Hint*: Label the equations, use the gaussian algorithm.]

(a) $x_1 + x_2 + x_3 = 1$
$2x_1 - x_2 + 3x_3 = 3$
$x_1 - 2x_2 + 2x_3 = 2$

◆(b) $x_1 + 2x_2 - 3x_3 = -3$
$x_1 + 3x_2 - 5x_3 = 5$
$x_1 - 2x_2 + 5x_3 = -35$

7. Find all solutions to the following systems.

(a) $3x_1 + 8x_2 - 3x_3 - 14x_4 = 2$
$2x_1 + 3x_2 - x_3 - 2x_4 = 1$
$x_1 - 2x_2 + x_3 + 10x_4 = 0$
$x_1 + 5x_2 - 2x_3 - 12x_4 = 1$

◆(b) $x_1 - x_2 + x_3 - x_4 = 0$
$-x_1 + x_2 + x_3 + x_4 = 0$
$x_1 + x_2 - x_3 + x_4 = 0$
$x_1 + x_2 + x_3 + x_4 = 0$

(c) $x_1 - x_2 + x_3 - 2x_4 = 1$
$-x_1 + x_2 + x_3 + x_4 = -1$
$-x_1 + 2x_2 + 3x_3 - x_4 = 2$
$x_1 - x_2 + 2x_3 + x_4 = 1$

◆(d) $x_1 + x_2 + 2x_3 - x_4 = 4$
$3x_2 - x_3 + 4x_4 = 2$
$x_1 + 2x_2 - 3x_3 + 5x_4 = 0$
$x_1 + x_2 - 5x_3 + 6x_4 = -3$

8. In each of the following, find (if possible) conditions on a and b such that the system has no solution, one solution, and infinitely many solutions.

(a) $x - 2y = 1$
$ax + by = 5$

◆(b) $x + by = -1$
$ax + 2y = 5$

(c) $x - by = -1$
$x + ay = 3$

◆(d) $ax + y = 1$
$2x + y = b$

9. In each of the following, find (if possible) conditions on a, b, and c such that the system has no solution, one solution, or infinitely many solutions.

(a) $3x + y - z = a$
$x - y + 2z = b$
$5x + 3y - 4z = c$

◆(b) $2x + y - z = a$
$2y + 3z = b$
$x - z = c$

(c) $-x + 3y + 2z = -8$
$x + z = 2$
$3x + 3y + az = b$

◆(d) $x + ay = 0$
$y + bz = 0$
$z + cx = 0$

(e) $3x - y + 2z = 3$
$x + y - z = 2$
$2x - 2y + 3z = b$

◆(f) $x + ay - z = 1$
$-x + (a-2)y + z = -1$
$2x + 2y + (a-2)z = 1$

◆10. Find the rank of each of the matrices in Exercise 1.

11. Find the rank of each of the following matrices.

(a) $\begin{bmatrix} 1 & 1 & 2 \\ 3 & -1 & 1 \\ -1 & 3 & 4 \end{bmatrix}$

◆(b) $\begin{bmatrix} -2 & 3 & 3 \\ 3 & -4 & 1 \\ -5 & 7 & 2 \end{bmatrix}$

(c) $\begin{bmatrix} 1 & 1 & -1 & 3 \\ -1 & 4 & 5 & -2 \\ 1 & 6 & 3 & 4 \end{bmatrix}$

◆(d) $\begin{bmatrix} 3 & -2 & 1 & -2 \\ 1 & -1 & 3 & 5 \\ -1 & 1 & 1 & -1 \end{bmatrix}$

(e) $\begin{bmatrix} 1 & 2 & -1 & 0 \\ 0 & a & 1-a & a^2+1 \\ 1 & 2-a & -1 & -2a^2 \end{bmatrix}$

◆(f) $\begin{bmatrix} 1 & 1 & 2 & a^2 \\ 1 & 1-a & 2 & 0 \\ 2 & 2-a & 6-a & 4 \end{bmatrix}$

12. Consider a system of linear equations with augmented matrix A and coefficient matrix C. In each case either prove the statement or give an example showing that it is false.

(a) If there is more than one solution, A has a row of zeros.

◆(b) If A has a row of zeros, there is more than one solution.

(c) If there is no solution, the row-echelon form of C has a row of zeros.

◆(d) If the row-echelon form of C has a row of zeros, there is no solution.

(e) There is no system that is inconsistent for every choice of constants.

◆(f) If the system is consistent for some choice of constants, it is consistent for every choice of constants.

Now assume that the augmented matrix A has 3 rows and 5 columns.

(g) If the system is consistent, there is more than one solution.

♦(h) The rank of A is at most 3.

(i) If rank $A = 3$, the system is consistent.

(j) If rank $C = 3$, the system is consistent.

13. Find a sequence of row operations carrying
$$\begin{bmatrix} b_1+c_1 & b_2+c_2 & b_3+c_3 \\ c_1+a_1 & c_2+a_2 & c_3+a_3 \\ a_1+b_1 & a_2+b_2 & a_3+b_3 \end{bmatrix} \text{ to } \begin{bmatrix} a_1 & a_2 & a_3 \\ b_1 & b_2 & b_3 \\ c_1 & c_2 & c_3 \end{bmatrix}$$

14. In each case, show that the reduced row-echelon form is as given.

    (a) $\begin{bmatrix} p & 0 & a \\ b & 0 & 0 \\ q & c & r \end{bmatrix}$ with $abc \neq 0$; $\begin{bmatrix} 1 & 0 & 0 \\ 0 & 1 & 0 \\ 0 & 0 & 1 \end{bmatrix}$

    ♦(b) $\begin{bmatrix} 1 & a & b+c \\ 1 & b & c+a \\ 1 & c & a+b \end{bmatrix}$ where $c \neq a$ or $b \neq a$; $\begin{bmatrix} 1 & 0 & * \\ 0 & 1 & * \\ 0 & 0 & 0 \end{bmatrix}$

15. Show that $\begin{cases} ax + by + cz = 0 \\ a_1x + b_1y + c_1z = 0 \end{cases}$ always has a solution other than $x = 0, y = 0, z = 0$.

16. Find the circle $x^2 + y^2 + ax + by + c = 0$ passing through the following points.

    (a) $(-2, 1)$, $(5, 0)$, and $(4, 1)$

    ♦(b) $(1, 1)$, $(5, -3)$, and $(-3, -3)$

17. Three Nissans, two Fords, and four Chevrolets can be rented for \$106 per day. At the same rates two Nissans, four Fords, and three Chevrolets cost \$107 per day, whereas four Nissans, three Fords, and two Chevrolets cost \$102 per day. Find the rental rates for all three kinds of cars.

♦18. A school has three clubs and each student is required to belong to exactly one club. One year the students switched club membership as follows:

Club A. $\frac{4}{10}$ remain in A, $\frac{1}{10}$ switch to B, $\frac{5}{10}$ switch to C.

Club B. $\frac{7}{10}$ remain in B, $\frac{2}{10}$ switch to A, $\frac{1}{10}$ switch to C.

Club C. $\frac{6}{10}$ remain in C, $\frac{2}{10}$ switch to A, $\frac{2}{10}$ switch to B.

If the fraction of the student population in each club is unchanged, find each of these fractions.

19. Given points (p_1, q_1), (p_2, q_2), and (p_3, q_3) in the plane with $p_1, p_2,$ and p_3 distinct, show that they lie on some curve with equation $y = a + bx + cx^2$. [*Hint*: Solve for $a, b,$ and c.]

20. The scores of three players in a tournament have been lost. The only information available is the total of the scores for players 1 and 2, the total for players 2 and 3, and the total for players 3 and 1.

    (a) Show that the individual scores can be rediscovered.

    (b) Is this possible with four players (knowing the totals for players 1 and 2, 2 and 3, 3 and 4, and 4 and 1)?

21. A boy finds \$1.05 in dimes, nickels, and pennies. If there are 17 coins in all, how many coins of each type can he have?

22. If a consistent system has more variables than equations, show that it has infinitely many solutions. [*Hint*: Use Theorem 2.]

SECTION 1.3 Homogeneous Equations

A system of equations in the variables $x_1, x_2, \ldots, x_n$ is called **homogeneous** if all the constant terms are zero—that is, if each equation of the system has the form

$$a_1x_1 + a_2x_2 + \cdots + a_nx_n = 0$$

Clearly $x_1 = 0, x_2 = 0, \ldots, x_n = 0$ is a solution to such a system; it is called the **trivial solution**. Any solution in which at least one variable has a nonzero value is called a **nontrivial solution**. Our chief goal in this section is to give a useful condition for a homogeneous system to have nontrivial solutions. The following example is instructive.

SECTION 1.3 Homogeneous Equations

EXAMPLE 1

Show that the following homogeneous system has nontrivial solutions.

$$\begin{aligned} x_1 - x_2 + 2x_3 - x_4 &= 0 \\ 2x_1 + 2x_2 + x_4 &= 0 \\ 3x_1 + x_2 + 2x_3 - x_4 &= 0 \end{aligned}$$

Solution ▶ The reduction of the augmented matrix to reduced row-echelon form is outlined below.

$$\begin{bmatrix} 1 & -1 & 2 & 1 & | & 0 \\ 2 & 2 & 0 & -1 & | & 0 \\ 3 & 1 & 2 & 1 & | & 0 \end{bmatrix} \to \begin{bmatrix} 1 & -1 & 2 & 1 & | & 0 \\ 0 & 4 & -4 & -3 & | & 0 \\ 0 & 4 & -4 & -2 & | & 0 \end{bmatrix} \to \begin{bmatrix} 1 & 0 & 1 & 0 & | & 0 \\ 0 & 1 & -1 & 0 & | & 0 \\ 0 & 0 & 0 & 1 & | & 0 \end{bmatrix}$$

The leading variables are x_1, x_2, and x_4, so x_3 is assigned as a parameter—say $x_3 = t$. Then the general solution is $x_1 = -t$, $x_2 = t$, $x_3 = t$, $x_4 = 0$. Hence, taking $t = 1$ (say), we get a nontrivial solution: $x_1 = -1$, $x_2 = 1$, $x_3 = 1$, $x_4 = 0$.

The existence of a nontrivial solution in Example 1 is ensured by the presence of a parameter in the solution. This is due to the fact that there is a *nonleading* variable (x_3 in this case). But there *must* be a nonleading variable here because there are four variables and only three equations (and hence at *most* three leading variables). This discussion generalizes to a proof of the following fundamental theorem.

Theorem 1

If a homogeneous system of linear equations has more variables than equations, then it has a nontrivial solution (in fact, infinitely many).

PROOF

Suppose there are m equations in n variables where $n > m$, and let R denote the reduced row-echelon form of the augmented matrix. If there are r leading variables, there are $n - r$ nonleading variables, and so $n - r$ parameters. Hence, it suffices to show that $r < n$. But $r \leq m$ because R has r leading 1s and m rows, and $m < n$ by hypothesis. So $r \leq m < n$, which gives $r < n$.

Note that the converse of Theorem 1 is not true: if a homogeneous system has nontrivial solutions, it need not have more variables than equations (the system $x_1 + x_2 = 0$, $2x_1 + 2x_2 = 0$ has nontrivial solutions but $m = 2 = n$.)

Theorem 1 is very useful in applications. The next example provides an illustration from geometry.

EXAMPLE 2

We call the graph of an equation $ax^2 + bxy + cy^2 + dx + ey + f = 0$ a **conic** if the numbers a, b, and c are not all zero. Show that there is at least one conic through any five points in the plane that are not all on a line.

Solution ▶ Let the coordinates of the five points be (p_1, q_1), (p_2, q_2), (p_3, q_3), (p_4, q_4), and (p_5, q_5). The graph of $ax^2 + bxy + cy^2 + dx + ey + f = 0$ passes through (p_i, q_i) if

$$ap_i^2 + bp_iq_i + cq_i^2 + dp_i + eq_i + f = 0$$

This gives five equations, one for each i, linear in the six variables a, b, c, d, e, and f. Hence, there is a nontrivial solution by Theorem 1. If $a = b = c = 0$, the five points all lie on the line $dx + ey + f = 0$, contrary to assumption. Hence, one of a, b, c is nonzero.

Linear Combinations and Basic Solutions

Naturally enough, two columns are regarded as **equal** if they have the same number of entries and corresponding entries are the same. Let **x** and **y** be columns with the same number of entries. As for elementary row operations, their **sum x + y** is obtained by adding corresponding entries and, if k is a number, the **scalar product** $k\mathbf{x}$ is defined by multiplying each entry of **x** by k. More precisely:

$$\text{If } \mathbf{x} = \begin{bmatrix} x_1 \\ x_2 \\ \vdots \\ x_n \end{bmatrix} \text{ and } \mathbf{y} = \begin{bmatrix} y_1 \\ y_2 \\ \vdots \\ y_n \end{bmatrix} \text{ then } \mathbf{x} + \mathbf{y} = \begin{bmatrix} x_1 + y_1 \\ x_2 + y_2 \\ \vdots \\ x_n + y_n \end{bmatrix} \text{ and } k\mathbf{x} = \begin{bmatrix} kx_1 \\ kx_2 \\ \vdots \\ kx_n \end{bmatrix}.$$

A sum of scalar multiples of several columns is called a **linear combination** of these columns. For example, $s\mathbf{x} + t\mathbf{y}$ is a linear combination of **x** and **y** for any choice of numbers s and t.

EXAMPLE 3

If $\mathbf{x} = \begin{bmatrix} 3 \\ -2 \end{bmatrix}$ and $\mathbf{y} = \begin{bmatrix} -1 \\ 1 \end{bmatrix}$ then $2\mathbf{x} + 5\mathbf{y} = \begin{bmatrix} 6 \\ -4 \end{bmatrix} + \begin{bmatrix} -5 \\ 5 \end{bmatrix} = \begin{bmatrix} 1 \\ 1 \end{bmatrix}$.

EXAMPLE 4

Let $\mathbf{x} = \begin{bmatrix} 1 \\ 0 \\ 1 \end{bmatrix}$, $\mathbf{y} = \begin{bmatrix} 2 \\ 1 \\ 0 \end{bmatrix}$ and $\mathbf{z} = \begin{bmatrix} 3 \\ 1 \\ 1 \end{bmatrix}$. If $\mathbf{v} = \begin{bmatrix} 0 \\ -1 \\ 2 \end{bmatrix}$ and $\mathbf{w} = \begin{bmatrix} 1 \\ 1 \\ 1 \end{bmatrix}$, determine whether **v** and **w** are linear combinations of **x** and **y**.

Solution ▶ For **v**, we must determine whether numbers r, s, and t exist such that $\mathbf{v} = r\mathbf{x} + s\mathbf{y} + t\mathbf{z}$, that is, whether

SECTION 1.3 Homogeneous Equations

$$\begin{bmatrix} 0 \\ -1 \\ 2 \end{bmatrix} = r \begin{bmatrix} 1 \\ 0 \\ 1 \end{bmatrix} + s \begin{bmatrix} 2 \\ 1 \\ 0 \end{bmatrix} + t \begin{bmatrix} 3 \\ 1 \\ 1 \end{bmatrix} = \begin{bmatrix} r + 2s + 3t \\ s + t \\ r + t \end{bmatrix}.$$

Equating corresponding entries gives a system of linear equations $r + 2s + 3t = 0$, $s + t = -1$, and $r + t = 2$ for r, s, and t. By gaussian elimination, the solution is $r = 2 - k$, $s = -1 - k$, and $t = k$ where k is a parameter. Taking $k = 0$, we see that $\mathbf{v} = 2\mathbf{x} - \mathbf{y}$ is indeed a linear combination of $\mathbf{x}$, $\mathbf{y}$, and $\mathbf{z}$.

Turning to $\mathbf{w}$, we again look for r, s, and t such that $\mathbf{w} = r\mathbf{x} + s\mathbf{y} + t\mathbf{z}$; that is,

$$\begin{bmatrix} 1 \\ 1 \\ 1 \end{bmatrix} = r \begin{bmatrix} 1 \\ 0 \\ 1 \end{bmatrix} + s \begin{bmatrix} 2 \\ 1 \\ 0 \end{bmatrix} + t \begin{bmatrix} 3 \\ 1 \\ 1 \end{bmatrix} = \begin{bmatrix} r + 2s + 3t \\ s + t \\ r + t \end{bmatrix},$$

leading to equations $r + 2s + 3t = 1$, $s + t = 1$, and $r + t = 1$ for real numbers r, s, and t. But this time there is *no* solution as the reader can verify, so $\mathbf{w}$ is *not* a linear combination of $\mathbf{x}$, $\mathbf{y}$, and $\mathbf{z}$.

Our interest in linear combinations comes from the fact that they provide one of the best ways to describe the general solution of a homogeneous system of linear equations. When solving such a system with n variables $x_1, x_2, \ldots, x_n$, write the variables as a column[7] matrix: $\mathbf{x} = \begin{bmatrix} x_1 \\ x_2 \\ \vdots \\ x_n \end{bmatrix}$. The trivial solution is denoted $\mathbf{0} = \begin{bmatrix} 0 \\ 0 \\ \vdots \\ 0 \end{bmatrix}$.

As an illustration, the general solution in Example 1 is $x_1 = -t$, $x_2 = t$, $x_3 = t$, and $x_4 = 0$, where t is a parameter, and we would now express this by saying that the general solution is $\mathbf{x} = \begin{bmatrix} -t \\ t \\ t \\ 0 \end{bmatrix}$, where t is arbitrary.

Now let $\mathbf{x}$ and $\mathbf{y}$ be two solutions to a homogeneous system with n variables. Then any linear combination $s\mathbf{x} + t\mathbf{y}$ of these solutions turns out to be again a solution to the system. More generally:

Any linear combination of solutions to a homogeneous system is again a solution. (∗)

In fact, suppose that a typical equation in the system is

$a_1 x_1 + a_2 x_2 + \cdots + a_n x_n = 0$, and suppose that $\mathbf{x} = \begin{bmatrix} x_1 \\ x_2 \\ \vdots \\ x_n \end{bmatrix}$ and $\mathbf{y} = \begin{bmatrix} y_1 \\ y_2 \\ \vdots \\ y_n \end{bmatrix}$ are

solutions. Then $a_1 x_1 + a_2 x_2 + \cdots + a_n x_n = 0$ and $a_1 y_1 + a_2 y_2 + \cdots + a_n y_n = 0$.

Hence $s\mathbf{x} + t\mathbf{y} = \begin{bmatrix} sx_1 + ty_1 \\ sx_2 + ty_2 \\ \vdots \\ sx_n + ty_n \end{bmatrix}$ is also a solution because

---

[7] The reason for using columns will be apparent later.

$$a_1(sx_1 + ty_1) + a_2(sx_2 + ty_2) + \cdots + a_n(sx_n + ty_n)$$
$$= [a_1(sx_1) + a_2(sx_2) + \cdots + a_n(sx_n)] + [a_1(ty_1) + a_2(ty_2) + \cdots + a_n(ty_n)]$$
$$= s(a_1x_1 + a_2x_2 + \cdots + a_nx_n) + t(a_1y_1 + a_2y_2 + \cdots + a_ny_n)$$
$$= s(0) + t(0)$$
$$= 0.$$

A similar argument shows that (∗) is true for linear combinations of more than two solutions.

The remarkable thing is that *every* solution to a homogeneous system is a linear combination of certain particular solutions and, in fact, these solutions are easily computed using the gaussian algorithm. Here is an example.

EXAMPLE 5

Solve the homogeneous system with coefficient matrix

$$A = \begin{bmatrix} 1 & -2 & 3 & -2 \\ -3 & 6 & 1 & 0 \\ -2 & 4 & 4 & -2 \end{bmatrix}$$

Solution ▶ The reduction of the augmented matrix to reduced form is

$$\begin{bmatrix} 1 & -2 & 3 & -2 & | & 0 \\ -3 & 6 & 1 & 0 & | & 0 \\ -2 & 4 & 4 & -2 & | & 0 \end{bmatrix} \rightarrow \begin{bmatrix} 1 & -2 & 0 & -\tfrac{1}{5} & | & 0 \\ 0 & 0 & 1 & -\tfrac{3}{5} & | & 0 \\ 0 & 0 & 0 & 0 & | & 0 \end{bmatrix}$$

so the solutions are $x_1 = 2s + \tfrac{1}{5}t$, $x_2 = s$, $x_3 = \tfrac{3}{5}t$, and $x_4 = t$ by gaussian elimination. Hence we can write the general solution $\mathbf{x}$ in the matrix form

$$\mathbf{x} = \begin{bmatrix} x_1 \\ x_2 \\ x_3 \\ x_4 \end{bmatrix} = \begin{bmatrix} 2s + \tfrac{1}{5}t \\ s \\ \tfrac{3}{5}t \\ t \end{bmatrix} = s \begin{bmatrix} 2 \\ 1 \\ 0 \\ 0 \end{bmatrix} + t \begin{bmatrix} \tfrac{1}{5} \\ 0 \\ \tfrac{3}{5} \\ 1 \end{bmatrix} = s\mathbf{x}_1 + t\mathbf{x}_2$$

where $\mathbf{x}_1 = \begin{bmatrix} 2 \\ 1 \\ 0 \\ 0 \end{bmatrix}$ and $\mathbf{x}_1 = \begin{bmatrix} \tfrac{1}{5} \\ 0 \\ \tfrac{3}{5} \\ 1 \end{bmatrix}$ are particular solutions determined by the gaussian algorithm.

The solutions $\mathbf{x}_1$ and $\mathbf{x}_2$ in Example 5 are denoted as follows:

Definition 1.5 *The gaussian algorithm systematically produces solutions to any homogeneous linear system, called* **basic solutions***, one for every parameter.*

Moreover, the algorithm gives a routine way to express *every* solution as a linear combination of basic solutions as in Example 5, where the general solution $\mathbf{x}$ becomes

SECTION 1.3 Homogeneous Equations

$$\mathbf{x} = s\begin{bmatrix} 2 \\ 1 \\ 0 \\ 0 \end{bmatrix} + t\begin{bmatrix} \frac{1}{5} \\ 0 \\ \frac{3}{5} \\ 1 \end{bmatrix} = s\begin{bmatrix} 2 \\ 1 \\ 0 \\ 0 \end{bmatrix} + \frac{1}{5}t\begin{bmatrix} 1 \\ 0 \\ 3 \\ 5 \end{bmatrix}$$

Hence by introducing a new parameter $r = t/5$ we can multiply the original basic solution $\mathbf{x}_2$ by 5 and so eliminate fractions. For this reason:

Any nonzero scalar multiple of a basic solution will still be called a basic solution.

In the same way, the gaussian algorithm produces basic solutions to *every* homogeneous system, one for each parameter (there are *no* basic solutions if the system has only the trivial solution). Moreover every solution is given by the algorithm as a linear combination of these basic solutions (as in Example 5). If A has rank r, Theorem 2 Section 1.2 shows that there are exactly $n - r$ parameters, and so $n - r$ basic solutions. This proves:

Theorem 2

Let A be an $m \times n$ matrix of rank r, and consider the homogeneous system in n variables with A as coefficient matrix. Then:

1. The system has exactly $n - r$ basic solutions, one for each parameter.
2. Every solution is a linear combination of these basic solutions.

EXAMPLE 6

Find basic solutions of the homogeneous system with coefficient matrix A, and express every solution as a linear combination of the basic solutions, where

$$A = \begin{bmatrix} 1 & -3 & 0 & 2 & 2 \\ -2 & 6 & 1 & 2 & -5 \\ 3 & -9 & -1 & 0 & 7 \\ -3 & 9 & 2 & 6 & -8 \end{bmatrix}$$

Solution ▶ The reduction of the augmented matrix to reduced row-echelon form is

$$\begin{bmatrix} 1 & -3 & 0 & 2 & 2 & | & 0 \\ -2 & 6 & 1 & 2 & -5 & | & 0 \\ 3 & -9 & -1 & 0 & 7 & | & 0 \\ -3 & 9 & 2 & 6 & -8 & | & 0 \end{bmatrix} \rightarrow \begin{bmatrix} 1 & -3 & 0 & 2 & 2 & | & 0 \\ 0 & 0 & 1 & 6 & -1 & | & 0 \\ 0 & 0 & 0 & 0 & 0 & | & 0 \\ 0 & 0 & 0 & 0 & 0 & | & 0 \end{bmatrix}$$

so the general solution is $x_1 = 3r - 2s - 2t$, $x_2 = r$, $x_3 = -6s + t$, $x_4 = s$, and $x_5 = t$ where r, s, and t are parameters. In matrix form this is

$$\mathbf{x} = \begin{bmatrix} x_1 \\ x_2 \\ x_3 \\ x_4 \\ x_5 \end{bmatrix} = \begin{bmatrix} 3r - 2s - 2t \\ r \\ -6s + t \\ s \\ t \end{bmatrix} = r\begin{bmatrix} 3 \\ 1 \\ 0 \\ 0 \\ 0 \end{bmatrix} + s\begin{bmatrix} -2 \\ 0 \\ -6 \\ 1 \\ 0 \end{bmatrix} + t\begin{bmatrix} -2 \\ 0 \\ 1 \\ 0 \\ 1 \end{bmatrix}$$

Hence basic solutions are $\mathbf{x}_1 = \begin{bmatrix} 3 \\ 1 \\ 0 \\ 0 \\ 0 \end{bmatrix}$, $\mathbf{x}_2 = \begin{bmatrix} -2 \\ 0 \\ -6 \\ 1 \\ 0 \end{bmatrix}$, and $\mathbf{x}_3 = \begin{bmatrix} -2 \\ 0 \\ 1 \\ 0 \\ 1 \end{bmatrix}$.

EXERCISES 1.3

1. Consider the following statements about a system of linear equations with augmented matrix A. In each case either prove the statement or give an example for which it is false.

   (a) If the system is homogeneous, every solution is trivial.

   ◆(b) If the system has a nontrivial solution, it cannot be homogeneous.

   (c) If there exists a trivial solution, the system is homogeneous.

   ◆(d) If the system is consistent, it must be homogeneous.

   Now assume that the system is homogeneous.

   (e) If there exists a nontrivial solution, there is no trivial solution.

   ◆(f) If there exists a solution, there are infinitely many solutions.

   (g) If there exist nontrivial solutions, the row-echelon form of A has a row of zeros.

   ◆(h) If the row-echelon form of A has a row of zeros, there exist nontrivial solutions.

   (i) If a row operation is applied to the system, the new system is also homogeneous.

2. In each of the following, find all values of a for which the system has nontrivial solutions, and determine all solutions in each case.

   (a) $\begin{aligned} x - 2y + z &= 0 \\ x + ay - 3z &= 0 \\ -x + 6y - 5z &= 0 \end{aligned}$ ◆(b) $\begin{aligned} x + 2y + z &= 0 \\ x + 3y + 6z &= 0 \\ 2x + 3y + az &= 0 \end{aligned}$

   (c) $\begin{aligned} x + y - z &= 0 \\ ay - z &= 0 \\ x + y + az &= 0 \end{aligned}$ ◆(d) $\begin{aligned} ax + y + z &= 0 \\ x + y - z &= 0 \\ x + y + az &= 0 \end{aligned}$

3. Let $\mathbf{x} = \begin{bmatrix} 2 \\ 1 \\ -1 \end{bmatrix}$, $\mathbf{y} = \begin{bmatrix} 1 \\ 0 \\ 1 \end{bmatrix}$, and $\mathbf{z} = \begin{bmatrix} 1 \\ 1 \\ -2 \end{bmatrix}$. In each case, either write $\mathbf{v}$ as a linear combination of $\mathbf{x}$, $\mathbf{y}$, and $\mathbf{z}$, or show that it is not such a linear combination.

   (a) $\mathbf{v} = \begin{bmatrix} 0 \\ 1 \\ -3 \end{bmatrix}$ ◆(b) $\mathbf{v} = \begin{bmatrix} 4 \\ 3 \\ -4 \end{bmatrix}$

   (c) $\mathbf{v} = \begin{bmatrix} 3 \\ 1 \\ 0 \end{bmatrix}$ ◆(d) $\mathbf{v} = \begin{bmatrix} 3 \\ 0 \\ 3 \end{bmatrix}$

4. In each case, either express $\mathbf{y}$ as a linear combination of $\mathbf{a}_1$, $\mathbf{a}_2$, and $\mathbf{a}_3$, or show that it is not such a linear combination. Here:

   $\mathbf{a}_1 = \begin{bmatrix} -1 \\ 3 \\ 0 \\ 1 \end{bmatrix}$, $\mathbf{a}_2 = \begin{bmatrix} 3 \\ 1 \\ 2 \\ 0 \end{bmatrix}$, and $\mathbf{a}_3 = \begin{bmatrix} 1 \\ 1 \\ 1 \\ 1 \end{bmatrix}$

   (a) $\mathbf{y} = \begin{bmatrix} 1 \\ 2 \\ 4 \\ 0 \end{bmatrix}$ (b) $\mathbf{x} = \begin{bmatrix} -1 \\ 9 \\ 2 \\ 6 \end{bmatrix}$

5. For each of the following homogeneous systems, find a set of basic solutions and express the general solution as a linear combination of these basic solutions.

   (a) $\begin{aligned} x_1 + 2x_2 - x_3 + 2x_4 + x_5 &= 0 \\ x_1 + 2x_2 + 2x_3 + x_5 &= 0 \\ 2x_1 + 4x_2 - 2x_3 + 3x_4 + x_5 &= 0 \end{aligned}$

   ◆(b) $\begin{aligned} x_1 + 2x_2 - x_3 + x_4 + x_5 &= 0 \\ -x_1 - 2x_2 + 2x_3 + x_5 &= 0 \\ -x_1 - 2x_2 + 3x_3 + x_4 + 3x_5 &= 0 \end{aligned}$

(c) $\begin{aligned} x_1 + x_2 - x_3 + 2x_4 + x_5 &= 0 \\ x_1 + 2x_2 - x_3 + x_4 + x_5 &= 0 \\ 2x_1 + 3x_2 - x_3 + 2x_4 + x_5 &= 0 \\ 4x_1 + 5x_2 - 2x_3 + 5x_4 + 2x_5 &= 0 \end{aligned}$

◆(d) $\begin{aligned} x_1 + x_2 - 2x_3 - 2x_4 + 2x_5 &= 0 \\ 2x_1 + 2x_2 - 4x_3 - 4x_4 + x_5 &= 0 \\ x_1 - x_2 + 2x_3 + 4x_4 + x_5 &= 0 \\ -2x_1 - 4x_2 + 8x_3 + 10x_4 + x_5 &= 0 \end{aligned}$

6. (a) Does Theorem 1 imply that the system $\begin{cases} -z + 3y = 0 \\ 2x - 6y = 0 \end{cases}$ has nontrivial solutions? Explain.

◆(b) Show that the converse to Theorem 1 is not true. That is, show that the existence of nontrivial solutions does *not* imply that there are more variables than equations.

7. In each case determine how many solutions (and how many parameters) are possible for a homogeneous system of four linear equations in six variables with augmented matrix A. Assume that A has nonzero entries. Give all possibilities.

   (a) Rank $A = 2$.

   ◆(b) Rank $A = 1$.

   (c) A has a row of zeros.

   ◆(d) The row-echelon form of A has a row of zeros.

8. The graph of an equation $ax + by + cz = 0$ is a plane through the origin (provided that not all of a, b, and c are zero). Use Theorem 1 to show that two planes through the origin have a point in common other than the origin $(0, 0, 0)$.

9. (a) Show that there is a line through any pair of points in the plane. [*Hint*: Every line has equation $ax + by + c = 0$, where a, b, and c are not all zero.]

   ◆(b) Generalize and show that there is a plane $ax + by + cz + d = 0$ through any three points in space.

10. The graph of $a(x^2 + y^2) + bx + cy + d = 0$ is a circle if $a \neq 0$. Show that there is a circle through any three points in the plane that are not all on a line.

◆11. Consider a homogeneous system of linear equations in n variables, and suppose that the augmented matrix has rank r. Show that the system has nontrivial solutions if and only if $n > r$.

12. If a consistent (possibly nonhomogeneous) system of linear equations has more variables than equations, prove that it has more than one solution.

SECTION 1.4 An Application to Network Flow

There are many types of problems that concern a network of conductors along which some sort of flow is observed. Examples of these include an irrigation network and a network of streets or freeways. There are often points in the system at which a net flow either enters or leaves the system. The basic principle behind the analysis of such systems is that the total flow into the system must equal the total flow out. In fact, we apply this principle at every junction in the system.

Junction Rule

At each of the junctions in the network, the total flow into that junction must equal the total flow out.

This requirement gives a linear equation relating the flows in conductors emanating from the junction.

EXAMPLE 1

A network of one-way streets is shown in the accompanying diagram. The rate of flow of cars into intersection A is 500 cars per hour, and 400 and 100 cars per hour emerge from B and C, respectively. Find the possible flows along each street.

Solution ▶ Suppose the flows along the streets are f_1, f_2, f_3, f_4, f_5, and f_6 cars per hour in the directions shown. Then, equating the flow in with the flow out at each intersection, we get

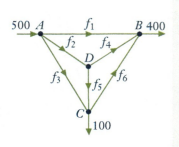

$$\begin{aligned}
\text{Intersection } A \quad & 500 = f_1 + f_2 + f_3 \\
\text{Intersection } B \quad & f_1 + f_4 + f_6 = 400 \\
\text{Intersection } C \quad & f_3 + f_5 = f_6 + 100 \\
\text{Intersection } D \quad & f_2 = f_4 + f_5
\end{aligned}$$

These give four equations in the six variables $f_1, f_2, \ldots, f_6$.

$$\begin{aligned}
f_1 + f_2 + f_3 \quad\quad\quad\quad\quad &= 500 \\
f_1 \quad\quad + f_4 \quad\quad + f_6 &= 400 \\
f_3 \quad\quad + f_5 - f_6 &= 100 \\
f_2 \quad\quad - f_4 - f_5 \quad\quad &= 0
\end{aligned}$$

The reduction of the augmented matrix is

$$\begin{bmatrix} 1 & 1 & 1 & 0 & 0 & 0 & | & 500 \\ 1 & 0 & 0 & 1 & 0 & 1 & | & 400 \\ 0 & 0 & 1 & 0 & 1 & -1 & | & 100 \\ 0 & 1 & 0 & -1 & -1 & 0 & | & 0 \end{bmatrix} \rightarrow \begin{bmatrix} 1 & 0 & 0 & 1 & 0 & 1 & | & 400 \\ 0 & 1 & 0 & -1 & -1 & 0 & | & 0 \\ 0 & 0 & 1 & 0 & 1 & -1 & | & 100 \\ 0 & 0 & 0 & 0 & 0 & 0 & | & 0 \end{bmatrix}$$

Hence, when we use f_4, f_5, and f_6 as parameters, the general solution is

$$f_1 = 400 - f_4 - f_6 \quad f_2 = f_4 + f_5 \quad f_3 = 100 - f_5 + f_6$$

This gives all solutions to the system of equations and hence all the possible flows.

Of course, not all these solutions may be acceptable in the real situation. For example, the flows $f_1, f_2, \ldots, f_6$ are all *positive* in the present context (if one came out negative, it would mean traffic flowed in the opposite direction). This imposes constraints on the flows: $f_1 \geq 0$ and $f_3 \geq 0$ become

$$f_4 + f_6 \leq 400 \quad f_5 - f_6 \leq 100$$

Further constraints might be imposed by insisting on maximum values on the flow in each street.

EXERCISES 1.4

1. Find the possible flows in each of the following networks of pipes.

   (a)

   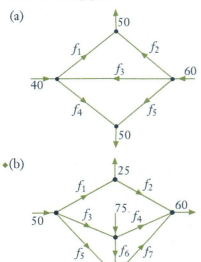

   ◆(b)

2. A proposed network of irrigation canals is described in the accompanying diagram. At peak demand, the flows at interchanges A, B, C, and D are as shown.

   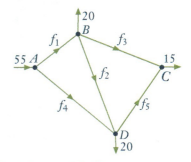

   (a) Find the possible flows.

   ◆(b) If canal BC is closed, what range of flow on AD must be maintained so that no canal carries a flow of more than 30?

3. A traffic circle has five one-way streets, and vehicles enter and leave as shown in the accompanying diagram.

   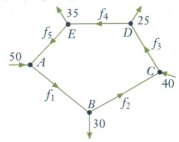

   (a) Compute the possible flows.

   ◆(b) Which road has the heaviest flow?

SECTION 1.5   An Application to Electrical Networks[8]

In an electrical network it is often necessary to find the current in amperes (A) flowing in various parts of the network. These networks usually contain resistors that retard the current. The resistors are indicated by a symbol ─⩔⩔─, and the resistance is measured in ohms (Ω). Also, the current is increased at various points by voltage sources (for example, a battery). The voltage of these sources is measured in volts (V), and they are represented by the symbol ⊣⊦. We assume these voltage sources have no resistance. The flow of current is governed by the following principles.

Ohm's Law

The current I and the voltage drop V across a resistance R are related by the equation $V = RI$.

---

[8] This section is independent of Section 1.4

Kirchhoff's Laws

1. *(Junction Rule)* The current flow into a junction equals the current flow out of that junction.
2. *(Circuit Rule)* The algebraic sum of the voltage drops (due to resistances) around any closed circuit of the network must equal the sum of the voltage increases around the circuit.

When applying rule 2, select a direction (clockwise or counterclockwise) around the closed circuit and then consider all voltages and currents positive when in this direction and negative when in the opposite direction. This is why the term *algebraic sum* is used in rule 2. Here is an example.

EXAMPLE 1

Find the various currents in the circuit shown.

Solution ▶ First apply the junction rule at junctions A, B, C, and D to obtain

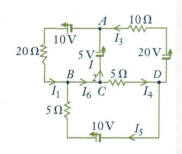

Junction A	$I_1 = I_2 + I_3$
Junction B	$I_6 = I_1 + I_5$
Junction C	$I_2 + I_4 = I_6$
Junction D	$I_3 + I_5 = I_4$

Note that these equations are not independent (in fact, the third is an easy consequence of the other three).

Next, the circuit rule insists that the sum of the voltage increases (due to the sources) around a closed circuit must equal the sum of the voltage drops (due to resistances). By Ohm's law, the voltage loss across a resistance R (in the direction of the current I) is RI. Going counterclockwise around three closed circuits yields

Upper left	$10 + 5 = 20 I_1$
Upper right	$-5 + 20 = 10 I_3 + 5 I_4$
Lower	$-10 = -20 I_5 - 5 I_4$

Hence, disregarding the redundant equation obtained at junction C, we have six equations in the six unknowns $I_1, \dots, I_6$. The solution is

$$I_1 = \tfrac{15}{20} \quad I_4 = \tfrac{28}{20}$$
$$I_2 = \tfrac{-1}{20} \quad I_5 = \tfrac{12}{20}$$
$$I_3 = \tfrac{16}{20} \quad I_6 = \tfrac{27}{20}$$

The fact that I_2 is negative means, of course, that this current is in the opposite direction, with a magnitude of $\tfrac{1}{20}$ amperes.

EXERCISES 1.5

In Exercises 1–4, find the currents in the circuits.

1.

2.

3.

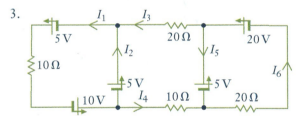

4. All resistances are 10 Ω.

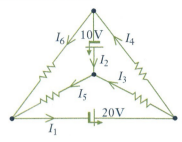

5. Find the voltage x such that the current $I_1 = 0$.

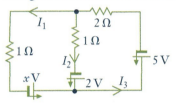

SECTION 1.6 An Application to Chemical Reactions

When a chemical reaction takes place a number of molecules combine to produce new molecules. Hence, when hydrogen H_2 and oxygen O_2 molecules combine, the result is water H_2O. We express this as

$$H_2 + O_2 \to H_2O$$

Individual atoms are neither created nor destroyed, so the number of hydrogen and oxygen atoms going into the reaction must equal the number coming out (in the form of water). In this case the reaction is said to be *balanced*. Note that each hydrogen molecule H_2 consists of two atoms as does each oxygen molecule O_2, while a water molecule H_2O consists of two hydrogen atoms and one oxygen atom. In the above reaction, this requires that twice as many hydrogen molecules enter the reaction; we express this as follows:

$$2H_2 + O_2 \to 2H_2O$$

This is now balanced because there are 4 hydrogen atoms and 2 oxygen atoms on each side of the reaction.

EXAMPLE 1

Balance the following reaction for burning octane C_8H_{18} in oxygen O_2:

$$C_8H_{18} + O_2 \rightarrow CO_2 + H_2O$$

where CO_2 represents carbon dioxide. We must find positive integers x, y, z, and w such that

$$xC_8H_{18} + yO_2 \rightarrow zCO_2 + wH_2O$$

Equating the number of carbon, hydrogen, and oxygen atoms on each side gives $8x = z$, $18x = 2w$ and $2y = 2z + w$, respectively. These can be written as a homogeneous linear system

$$\begin{aligned} 8x \quad\quad - z \quad\quad &= 0 \\ 18x \quad\quad\quad\quad - 2w &= 0 \\ 2y - 2z - w &= 0 \end{aligned}$$

which can be solved by gaussian elimination. In larger systems this is necessary but, in such a simple situation, it is easier to solve directly. Set $w = t$, so that $x = \frac{1}{9}t$, $z = \frac{8}{9}t$, $2y = \frac{16}{9}t + t = \frac{25}{9}t$. But $x, y, z,$ and w must be positive integers, so the smallest value of t that eliminates fractions is 18. Hence, $x = 2$, $y = 25$, $z = 16$, and $w = 18$, and the balanced reaction is

$$2C_8H_{18} + 25O_2 \rightarrow 16CO_2 + 18H_2O$$

The reader can verify that this is indeed balanced.

It is worth noting that this problem introduces a new element into the theory of linear equations: the insistence that the solution must consist of positive integers.

EXERCISES 1.6

In each case balance the chemical reaction.

1. $CH_4 + O_2 \rightarrow CO_2 + H_2O$. This is the burning of methane CH_4.

♦2. $NH_3 + CuO \rightarrow N_2 + Cu + H_2O$. Here NH_3 is ammonia, CuO is copper oxide, Cu is copper, and N_2 is nitrogen.

3. $CO_2 + H_2O \rightarrow C_6H_{12}O_6 + O_2$. This is called the photosynthesis reaction—$C_6H_{12}O_6$ is glucose.

♦4. $Pb(N_3)_2 + Cr(MnO_4)_2 \rightarrow Cr_2O_3 + MnO_2 + Pb_3O_4 + NO$.

SUPPLEMENTARY EXERCISES FOR CHAPTER 1

1. We show in Chapter 4 that the graph of an equation $ax + by + cz = d$ is a plane in space when not all of a, b, and c are zero.

    (a) By examining the possible positions of planes in space, show that three equations in three variables can have zero, one, or infinitely many solutions.

    ♦(b) Can two equations in three variables have a unique solution? Give reasons for your answer.

2. Find all solutions to the following systems of linear equations.

    (a) $\begin{aligned} x_1 + x_2 + x_3 - x_4 &= 3 \\ 3x_1 + 5x_2 - 2x_3 + x_4 &= 1 \\ -3x_1 - 7x_2 + 7x_3 - 5x_4 &= 7 \\ x_1 + 3x_2 - 4x_3 + 3x_4 &= -5 \end{aligned}$

◆(b) $x_1 + 4x_2 - x_3 + x_4 = 2$
$3x_1 + 2x_2 + x_3 + 2x_4 = 5$
$x_1 - 6x_2 + 3x_3 = 1$
$x_1 + 14x_2 - 5x_3 + 2x_4 = 3$

3. In each case find (if possible) conditions on a, b, and c such that the system has zero, one, or infinitely many solutions.

   (a) $x + 2y - 4z = 4$
   $3x - y + 13z = 2$
   $4x + y + a^2 z = a + 3$

   ◆(b) $x + y + 3z = a$
   $ax + y + 5z = 4$
   $x + ay + 4z = a$

◆4. Show that any two rows of a matrix can be interchanged by elementary row transformations of the other two types.

5. If $ad \neq bc$, show that $\begin{bmatrix} a & b \\ c & d \end{bmatrix}$ has reduced row-echelon form $\begin{bmatrix} 1 & 0 \\ 0 & 1 \end{bmatrix}$.

◆6. Find a, b, and c so that the system

   $x + ay + cz = 0$
   $bx + cy - 3z = 1$
   $ax + 2y + bz = 5$

   has the solution $x = 3, y = -1, z = 2$.

7. Solve the system

   $x + 2y + 2z = -3$
   $2x + y + z = -4$
   $x - y + iz = i$

   where $i^2 = -1$. [See Appendix A.]

◆8. Show that the *real* system

   $\begin{cases} x + y + z = 5 \\ 2x - y - z = 1 \\ -3x + 2y + 2z = 0 \end{cases}$

   has a *complex* solution: $x = 2, y = i, z = 3 - i$ where $i^2 = -1$. Explain. What happens when such a real system has a unique solution?

9. A man is ordered by his doctor to take 5 units of vitamin A, 13 units of vitamin B, and 23 units of vitamin C each day. Three brands of vitamin pills are available, and the number of units of each vitamin per pill are shown in the accompanying table.

	Vitamin		
Brand	A	B	C
1	1	2	4
2	1	1	3
3	0	1	1

   (a) Find all combinations of pills that provide exactly the required amount of vitamins (no partial pills allowed).

   ◆(b) If brands 1, 2, and 3 cost 3¢, 2¢, and 5¢ per pill, respectively, find the least expensive treatment.

10. A restaurant owner plans to use x tables seating 4, y tables seating 6, and z tables seating 8, for a total of 20 tables. When fully occupied, the tables seat 108 customers. If only half of the x tables, half of the y tables, and one-fourth of the z tables are used, each fully occupied, then 46 customers will be seated. Find x, y, and z.

11. (a) Show that a matrix with two rows and two columns that is in reduced row-echelon form must have one of the following forms:

    $\begin{bmatrix} 1 & 0 \\ 0 & 1 \end{bmatrix} \begin{bmatrix} 0 & 1 \\ 0 & 0 \end{bmatrix} \begin{bmatrix} 0 & 0 \\ 0 & 0 \end{bmatrix} \begin{bmatrix} 1 & * \\ 0 & 0 \end{bmatrix}$

    [*Hint*: The leading 1 in the first row must be in column 1 or 2 or not exist.]

    (b) List the seven reduced row-echelon forms for matrices with two rows and three columns.

    (c) List the four reduced row-echelon forms for matrices with three rows and two columns.

12. An amusement park charges $7 for adults, $2 for youths, and $0.50 for children. If 150 people enter and pay a total of $100, find the numbers of adults, youths, and children. [*Hint*: These numbers are nonnegative *integers*.]

13. Solve the following system of equations for x and y.

    $x^2 + xy - y^2 = 1$
    $2x^2 - xy + 3y^2 = 13$
    $x^2 + 3xy + 2y^2 = 0$

    [*Hint*: These equations are linear in the new variables $x_1 = x^2$, $x_2 = xy$, and $x_3 = y^2$.]

Matrix Algebra

2

In the study of systems of linear equations in Chapter 1, we found it convenient to manipulate the augmented matrix of the system. Our aim was to reduce it to row-echelon form (using elementary row operations) and hence to write down all solutions to the system. In the present chapter we consider matrices for their own sake. While some of the motivation comes from linear equations, it turns out that matrices can be multiplied and added and so form an algebraic system somewhat analogous to the real numbers. This "matrix algebra" is useful in ways that are quite different from the study of linear equations. For example, the geometrical transformations obtained by rotating the euclidean plane about the origin can be viewed as multiplications by certain 2×2 matrices. These "matrix transformations" are an important tool in geometry and, in turn, the geometry provides a "picture" of the matrices. Furthermore, matrix algebra has many other applications, some of which will be explored in this chapter. This subject is quite old and was first studied systematically in 1858 by Arthur Cayley.[1]

Arthur Cayley. Photo © Corbis.

SECTION 2.1   Matrix Addition, Scalar Multiplication, and Transposition

A rectangular array of numbers is called a **matrix** (the plural is **matrices**), and the numbers are called the **entries** of the matrix. Matrices are usually denoted by uppercase letters: A, B, C, and so on. Hence,

$$A = \begin{bmatrix} 1 & 2 & -1 \\ 0 & 5 & 6 \end{bmatrix} \quad B = \begin{bmatrix} 1 & -1 \\ 0 & 2 \end{bmatrix} \quad C = \begin{bmatrix} 1 \\ 3 \\ 2 \end{bmatrix}$$

are matrices. Clearly matrices come in various shapes depending on the number of **rows** and **columns**. For example, the matrix A shown has 2 rows and 3 columns. In general, a matrix with m rows and n columns is referred to as an $m \times n$ **matrix** or as having **size** $m \times n$. Thus matrices A, B, and C above have sizes 2×3, 2×2, and 3×1, respectively. A matrix of size $1 \times n$ is called a **row matrix**, whereas one of

---

[1] Arthur Cayley (1821–1895) showed his mathematical talent early and graduated from Cambridge in 1842 as senior wrangler. With no employment in mathematics in view, he took legal training and worked as a lawyer while continuing to do mathematics, publishing nearly 300 papers in fourteen years. Finally, in 1863, he accepted the Sadlerian professorship at Cambridge and remained there for the rest of his life, valued for his administrative and teaching skills as well as for his scholarship. His mathematical achievements were of the first rank. In addition to originating matrix theory and the theory of determinants, he did fundamental work in group theory, in higher-dimensional geometry, and in the theory of invariants. He was one of the most prolific mathematicians of all time and produced 966 papers.

SECTION 2.1 Matrix Addition, Scalar Multiplication, and Transposition

size $m \times 1$ is called a **column matrix**. Matrices of size $n \times n$ for some n are called **square** matrices.

Each entry of a matrix is identified by the row and column in which it lies. The rows are numbered from the top down, and the columns are numbered from left to right. Then the **(i, j)-entry** of a matrix is the number lying simultaneously in row i and column j. For example,

$$\text{The } (1, 2)\text{-entry of } \begin{bmatrix} 1 & -1 \\ 0 & 1 \end{bmatrix} \text{ is } -1.$$

$$\text{The } (2, 3)\text{-entry of } \begin{bmatrix} 1 & 2 & -1 \\ 0 & 5 & 6 \end{bmatrix} \text{ is } 6.$$

A special notation is commonly used for the entries of a matrix. If A is an $m \times n$ matrix, and if the (i, j)-entry of A is denoted as a_{ij}, then A is displayed as follows:

$$A = \begin{bmatrix} a_{11} & a_{12} & a_{13} & \cdots & a_{1n} \\ a_{21} & a_{22} & a_{23} & \cdots & a_{2n} \\ \vdots & \vdots & \vdots & & \vdots \\ a_{m1} & a_{m2} & a_{m3} & \cdots & a_{mn} \end{bmatrix}$$

This is usually denoted simply as $A = [a_{ij}]$. Thus a_{ij} is the entry in row i and column j of A. For example, a 3×4 matrix in this notation is written

$$A = \begin{bmatrix} a_{11} & a_{12} & a_{13} & a_{14} \\ a_{21} & a_{22} & a_{23} & a_{24} \\ a_{31} & a_{32} & a_{33} & a_{34} \end{bmatrix}$$

It is worth pointing out a convention regarding rows and columns: *Rows are mentioned before columns.* For example:

- *If a matrix has size $m \times n$, it has m rows and n columns.*
- *If we speak of the (i, j)-entry of a matrix, it lies in row i and column j.*
- *If an entry is denoted a_{ij}, the first subscript i refers to the row and the second subscript j to the column in which a_{ij} lies.*

Two points (x_1, y_1) and (x_2, y_2) in the plane are equal if and only if[2] they have the same coordinates, that is $x_1 = x_2$ and $y_1 = y_2$. Similarly, two matrices A and B are called **equal** (written $A = B$) if and only if:

1. They have the same size.
2. Corresponding entries are equal.

If the entries of A and B are written in the form $A = [a_{ij}]$, $B = [b_{ij}]$, described earlier, then the second condition takes the following form:

$$[a_{ij}] = [b_{ij}] \quad \text{means } a_{ij} = b_{ij} \text{ for all } i \text{ and } j$$

EXAMPLE 1

Given $A = \begin{bmatrix} a & b \\ c & d \end{bmatrix}$, $B = \begin{bmatrix} 1 & 2 & -1 \\ 3 & 0 & 1 \end{bmatrix}$, and $C = \begin{bmatrix} 1 & 0 \\ -1 & 2 \end{bmatrix}$, discuss the possibility that $A = B$, $B = C$, $A = C$.

---

[2] If p and q are statements, we say that p implies q if q is true whenever p is true. Then "p if and only if q" means that both p implies q and q implies p. See Appendix B for more on this.

Solution ▶ $A = B$ is impossible because A and B are of different sizes: A is 2×2 whereas B is 2×3. Similarly, $B = C$ is impossible. But $A = C$ is possible provided that corresponding entries are equal: $\begin{bmatrix} a & b \\ c & d \end{bmatrix} = \begin{bmatrix} 1 & 0 \\ -1 & 2 \end{bmatrix}$ means $a = 1, b = 0, c = -1,$ and $d = 2$.

Matrix Addition

Definition 2.1 *If A and B are matrices of the same size, their **sum** $A + B$ is the matrix formed by adding corresponding entries.*

If $A = [a_{ij}]$ and $B = [b_{ij}]$, this takes the form

$$A + B = [a_{ij} + b_{ij}]$$

Note that addition is *not* defined for matrices of different sizes.

EXAMPLE 2

If $A = \begin{bmatrix} 2 & 1 & 3 \\ -1 & 2 & 0 \end{bmatrix}$ and $B = \begin{bmatrix} 1 & 1 & -1 \\ 2 & 0 & 6 \end{bmatrix}$, compute $A + B$.

Solution ▶ $A + B = \begin{bmatrix} 2+1 & 1+1 & 3-1 \\ -1+2 & 2+0 & 0+6 \end{bmatrix} = \begin{bmatrix} 3 & 2 & 2 \\ 1 & 2 & 6 \end{bmatrix}$.

EXAMPLE 3

Find a, b, and c if $[a \ b \ c] + [c \ a \ b] = [3 \ 2 \ -1]$.

Solution ▶ Add the matrices on the left side to obtain

$$[a + c \quad b + a \quad c + b] = [3 \ 2 \ -1]$$

Because corresponding entries must be equal, this gives three equations: $a + c = 3$, $b + a = 2$, and $c + b = -1$. Solving these yields $a = 3$, $b = -1$, $c = 0$.

If A, B, and C are any matrices *of the same size*, then

$$A + B = B + A \qquad \text{(commutative law)}$$
$$A + (B + C) = (A + B) + C) \qquad \text{(associative law)}$$

In fact, if $A = [a_{ij}]$ and $B = [b_{ij}]$, then the (i, j)-entries of $A + B$ and $B + A$ are, respectively, $a_{ij} + b_{ij}$ and $b_{ij} + a_{ij}$. Since these are equal for all i and j, we get

$$A + B = [a_{ij} + b_{ij}] = [b_{ij} + a_{ij}] = B + A$$

The associative law is verified similarly.

The $m \times n$ matrix in which every entry is zero is called the $m \times n$ **zero matrix** and is denoted as 0 (or 0_{mn} if it is important to emphasize the size). Hence,

$$0 + X = X$$

SECTION 2.1 *Matrix Addition, Scalar Multiplication, and Transposition*

holds for all $m \times n$ matrices X. The **negative** of an $m \times n$ matrix A (written $-A$) is defined to be the $m \times n$ matrix obtained by multiplying each entry of A by -1. If $A = [a_{ij}]$, this becomes $-A = [-a_{ij}]$. Hence,

$$A + (-A) = 0$$

holds for all matrices A where, of course, 0 is the zero matrix of the same size as A.

A closely related notion is that of subtracting matrices. If A and B are two $m \times n$ matrices, their **difference** $A - B$ is defined by

$$A - B = A + (-B)$$

Note that if $A = [a_{ij}]$ and $B = [b_{ij}]$, then

$$A - B = [a_{ij}] + [-b_{ij}] = [a_{ij} - b_{ij}]$$

is the $m \times n$ matrix formed by *subtracting* corresponding entries.

EXAMPLE 4

Let $A = \begin{bmatrix} 3 & -1 & 0 \\ 1 & 2 & -4 \end{bmatrix}$, $B = \begin{bmatrix} 1 & -1 & 1 \\ -2 & 0 & 6 \end{bmatrix}$, and $C = \begin{bmatrix} 1 & 0 & -2 \\ 3 & 1 & 1 \end{bmatrix}$. Compute $-A$, $A - B$, and $A + B - C$.

Solution ▶ $-A = \begin{bmatrix} -3 & 1 & 0 \\ -1 & -2 & 4 \end{bmatrix}$

$A - B = \begin{bmatrix} 3 - 1 & -1 - (-1) & 0 - 1 \\ 1 - (-2) & 2 - 0 & -4 - 6 \end{bmatrix} = \begin{bmatrix} 2 & 0 & -1 \\ 3 & 2 & -10 \end{bmatrix}$

$A + B - C = \begin{bmatrix} 3 + 1 - 1 & -1 - 1 - 0 & 0 + 1 - (-2) \\ 1 - 2 - 3 & 2 + 0 - 1 & -4 + 6 - 1 \end{bmatrix} = \begin{bmatrix} 3 & -2 & 3 \\ -4 & 1 & 1 \end{bmatrix}$

EXAMPLE 5

Solve $\begin{bmatrix} 3 & 2 \\ -1 & 1 \end{bmatrix} + X = \begin{bmatrix} 1 & 0 \\ -1 & 2 \end{bmatrix}$, where X is a matrix.

Solution ▶ We solve a numerical equation $a + x = b$ by subtracting the number a from both sides to obtain $x = b - a$. This also works for matrices. To solve $\begin{bmatrix} 3 & 2 \\ -1 & 1 \end{bmatrix} + X = \begin{bmatrix} 1 & 0 \\ -1 & 2 \end{bmatrix}$, simply subtract the matrix $\begin{bmatrix} 3 & 2 \\ -1 & 1 \end{bmatrix}$ from both sides to get

$$X = \begin{bmatrix} 1 & 0 \\ -1 & 2 \end{bmatrix} - \begin{bmatrix} 3 & 2 \\ -1 & 1 \end{bmatrix} = \begin{bmatrix} 1 - 3 & 0 - 2 \\ -1 - (-1) & 2 - 1 \end{bmatrix} = \begin{bmatrix} -2 & -2 \\ 0 & 1 \end{bmatrix}$$

The reader should verify that this matrix X does indeed satisfy the original equation.

The solution in Example 5 solves the single matrix equation $A + X = B$ directly via matrix subtraction: $X = B - A$. This ability to work with matrices as entities lies at the heart of matrix algebra.

It is important to note that the sizes of matrices involved in some calculations are often determined by the context. For example, if

$$A + C = \begin{bmatrix} 1 & 3 & -1 \\ 2 & 0 & 1 \end{bmatrix}$$

then A and C must be the same size (so that $A + C$ makes sense), and that size must be 2×3 (so that the sum is 2×3). For simplicity we shall often omit reference to such facts when they are clear from the context.

Scalar Multiplication

In gaussian elimination, multiplying a row of a matrix by a number k means multiplying *every* entry of that row by k.

Definition 2.2 *More generally, if A is any matrix and k is any number, the **scalar multiple** kA is the matrix obtained from A by multiplying each entry of A by k.*

If $A = [a_{ij}]$, this is

$$kA = [ka_{ij}]$$

Thus $1A = A$ and $(-1)A = -A$ for any matrix A.

The term *scalar* arises here because the set of numbers from which the entries are drawn is usually referred to as the set of scalars. We have been using real numbers as scalars, but we could equally well have been using complex numbers.

EXAMPLE 6

If $A = \begin{bmatrix} 3 & -1 & 4 \\ 2 & 0 & 1 \end{bmatrix}$ and $B = \begin{bmatrix} 1 & 2 & -1 \\ 0 & 3 & 2 \end{bmatrix}$, compute $5A$, $\frac{1}{2}B$, and $3A - 2B$.

Solution ▶ $5A = \begin{bmatrix} 15 & -5 & 20 \\ 10 & 0 & 30 \end{bmatrix}$, $\frac{1}{2}B = \begin{bmatrix} \frac{1}{2} & 1 & -\frac{1}{2} \\ 0 & \frac{3}{2} & 1 \end{bmatrix}$

$3A - 2B = \begin{bmatrix} 9 & -3 & 12 \\ 6 & 0 & 18 \end{bmatrix} - \begin{bmatrix} 2 & 4 & -2 \\ 0 & 6 & 4 \end{bmatrix} = \begin{bmatrix} 7 & -7 & 14 \\ 6 & -6 & 14 \end{bmatrix}$

If A is any matrix, note that kA is the same size as A for all scalars k. We also have

$$0A = 0 \quad \text{and} \quad k0 = 0$$

because the zero matrix has every entry zero. In other words, $kA = 0$ if either $k = 0$ or $A = 0$. The converse of this statement is also true, as Example 7 shows.

EXAMPLE 7

If $kA = 0$, show that either $k = 0$ or $A = 0$.

Solution ▶ Write $A = [a_{ij}]$ so that $kA = 0$ means $ka_{ij} = 0$ for all i and j. If $k = 0$, there is nothing to do. If $k \neq 0$, then $ka_{ij} = 0$ implies that $a_{ij} = 0$ for all i and j; that is, $A = 0$.

SECTION 2.1 Matrix Addition, Scalar Multiplication, and Transposition

For future reference, the basic properties of matrix addition and scalar multiplication are listed in Theorem 1.

Theorem 1

Let A, B, and C denote arbitrary $m \times n$ matrices where m and n are fixed. Let k and p denote arbitrary real numbers. Then

1. $A + B = B + A$.
2. $A + (B + C) = (A + B) + C$.
3. There is an $m \times n$ matrix 0, such that $0 + A = A$ for each A.
4. For each A there is an $m \times n$ matrix, $-A$, such that $A + (-A) = 0$.
5. $k(A + B) = kA + kB$.
6. $(k + p)A = kA + pA$.
7. $(kp)A = k(pA)$.
8. $1A = A$.

PROOF

Properties 1–4 were given previously. To check property 5, let $A = [a_{ij}]$ and $B = [b_{ij}]$ denote matrices of the same size. Then $A + B = [a_{ij} + b_{ij}]$, as before, so the (i, j)-entry of $k(A + B)$ is

$$k(a_{ij} + b_{ij}) = ka_{ij} + kb_{ij}$$

But this is just the (i, j)-entry of $kA + kB$, and it follows that $k(A + B) = kA + kB$. The other properties can be similarly verified; the details are left to the reader.

The properties in Theorem 1 enable us to do calculations with matrices in much the same way that numerical calculations are carried out. To begin, property 2 implies that the sum $(A + B) + C = A + (B + C)$ is the same no matter how it is formed and so is written as $A + B + C$. Similarly, the sum $A + B + C + D$ is independent of how it is formed; for example, it equals both $(A + B) + (C + D)$ and $A + [B + (C + D)]$. Furthermore, property 1 ensures that, for example, $B + D + A + C = A + B + C + D$. In other words, the *order* in which the matrices are added does not matter. A similar remark applies to sums of five (or more) matrices.

Properties 5 and 6 in Theorem 1 are called distributive laws for scalar multiplication, and they extend to sums of more than two terms. For example,

$$k(A + B - C) = kA + kC - kC$$
$$(k + p - m)A = kA + pA - mA$$

Similar observations hold for more than three summands. These facts, together with properties 7 and 8, enable us to simplify expressions by collecting like terms, expanding, and taking common factors in exactly the same way that algebraic expressions involving variables and real numbers are manipulated. The following example illustrates these techniques.

EXAMPLE 8

Simplify $2(A + 3C) - 3(2C - B) - 3[2(2A + B - 4C) - 4(A - 2C)]$ where A, B, and C are all matrices of the same size.

Solution ▶ The reduction proceeds as though A, B, and C were variables.

$$2(A + 3C) - 3(2C - B) - 3[2(2A + B - 4C) - 4(A - 2C)]$$
$$= 2A + 6C - 6C + 3B - 3[4A + 2B - 8C - 4A + 8C]$$
$$= 2A + 3B - 3[2B]$$
$$= 2A - 3B$$

Transpose of a Matrix

Many results about a matrix A involve the *rows* of A, and the corresponding result for columns is derived in an analogous way, essentially by replacing the word *row* by the word *column* throughout. The following definition is made with such applications in mind.

Definition 2.3 *If A is an $m \times n$ matrix, the **transpose** of A, written A^T, is the $n \times m$ matrix whose rows are just the columns of A in the same order.*

In other words, the first row of A^T is the first column of A (that is it consists of the entries of column 1 in order). Similarly the second row of A^T is the second column of A, and so on.

EXAMPLE 9

Write down the transpose of each of the following matrices.

$$A = \begin{bmatrix} 1 \\ 3 \\ 2 \end{bmatrix} \quad B = [5 \ 2 \ 6] \quad C = \begin{bmatrix} 1 & 2 \\ 3 & 4 \\ 5 & 6 \end{bmatrix} \quad D = \begin{bmatrix} 3 & 1 & -1 \\ 1 & 3 & 2 \\ -1 & 2 & 1 \end{bmatrix}$$

Solution ▶ $A^T = [1 \ 3 \ 2]$, $B^T = \begin{bmatrix} 5 \\ 2 \\ 6 \end{bmatrix}$, $C^T = \begin{bmatrix} 1 & 3 & 5 \\ 2 & 4 & 6 \end{bmatrix}$, and $D^T = D$.

If $A = [a_{ij}]$ is a matrix, write $A^T = [b_{ij}]$. Then b_{ij} is the jth element of the ith row of A^T and so is the jth element of the ith *column* of A. This means $b_{ij} = a_{ji}$, so the definition of A^T can be stated as follows:

$$\text{If } A = [a_{ij}], \text{ then } A^T = [a_{ji}] \qquad (*)$$

This is useful in verifying the following properties of transposition.

SECTION 2.1 *Matrix Addition, Scalar Multiplication, and Transposition*

Theorem 2

Let A and B denote matrices of the same size, and let k denote a scalar.
1. If A is an $m \times n$ matrix, then A^T is an $n \times m$ matrix.
2. $(A^T)^T = A$.
3. $(kA)^T = kA^T$.
4. $(A + B)^T = A^T + B^T$.

PROOF

Property 1 is part of the definition of A^T, and property 2 follows from (*). As to property 3: If $A = [a_{ij}]$, then $kA = [ka_{ij}]$, so (*) gives
$$(kA)^T = [ka_{ji}] = k[a_{ji}] = kA^T$$
Finally, if $B = [b_{ij}]$, then $A + B = [c_{ij}]$ where $c_{ij} = a_{ij} + b_{ij}$ Then (*) gives property 4:
$$(A + B)^T = [c_{ij}]^T = [c_{ji}] = [a_{ji} + b_{ji}] = [a_{ji}] + [b_{ji}] = A^T + B^T$$

There is another useful way to think of transposition. If $A = [a_{ij}]$ is an $m \times n$ matrix, the elements a_{11}, a_{22}, a_{33}, ... are called the **main diagonal** of A. Hence the main diagonal extends down and to the right from the upper left corner of the matrix A; it is shaded in the following examples:

$$\begin{bmatrix} a_{11} & a_{12} \\ a_{21} & a_{22} \\ a_{31} & a_{32} \end{bmatrix} \quad \begin{bmatrix} a_{11} & a_{12} & a_{13} \\ a_{21} & a_{22} & a_{23} \end{bmatrix} \quad \begin{bmatrix} a_{11} & a_{12} & a_{13} \\ a_{21} & a_{22} & a_{23} \\ a_{31} & a_{32} & a_{33} \end{bmatrix} \quad \begin{bmatrix} a_{11} \\ a_{21} \end{bmatrix}$$

Thus forming the transpose of a matrix A can be viewed as "flipping" A about its main diagonal, or as "rotating" A through $180°$ about the line containing the main diagonal. This makes property 2 in Theorem 2 transparent.

EXAMPLE 10

Solve for A if $\left(2A^T - 3\begin{bmatrix} 1 & 2 \\ -1 & 1 \end{bmatrix}\right)^T = \begin{bmatrix} 2 & 3 \\ -1 & 2 \end{bmatrix}$.

Solution ▶ Using Theorem 2, the left side of the equation is
$$\left(2A^T - 3\begin{bmatrix} 1 & 2 \\ -1 & 1 \end{bmatrix}\right)^T = 2(A^T)^T - 3\begin{bmatrix} 1 & 2 \\ -1 & 1 \end{bmatrix}^T = 2A - 3\begin{bmatrix} 1 & -1 \\ 2 & 1 \end{bmatrix}$$
Hence the equation becomes
$$2A - 3\begin{bmatrix} 1 & -1 \\ 2 & 1 \end{bmatrix} = \begin{bmatrix} 2 & 3 \\ -1 & 2 \end{bmatrix}$$
Thus $2A = \begin{bmatrix} 2 & 3 \\ -1 & 2 \end{bmatrix} + 3\begin{bmatrix} 1 & -1 \\ 2 & 1 \end{bmatrix} = \begin{bmatrix} 5 & 0 \\ 5 & 5 \end{bmatrix}$, so finally $A = \frac{1}{2}\begin{bmatrix} 5 & 0 \\ 5 & 5 \end{bmatrix} = \frac{5}{2}\begin{bmatrix} 1 & 0 \\ 1 & 1 \end{bmatrix}$.

Note that Example 10 can also be solved by first transposing both sides, then solving for A^T, and so obtaining $A = (A^T)^T$. The reader should do this.

The matrix D in Example 9 has the property that $D = D^T$. Such matrices are important; a matrix A is called **symmetric** if $A = A^T$. A symmetric matrix A is necessarily square (if A is $m \times n$, then A^T is $n \times m$, so $A = A^T$ forces $n = m$). The name comes from the fact that these matrices exhibit a symmetry about the main diagonal. That is, entries that are directly across the main diagonal from each other are equal.

For example, $\begin{bmatrix} a & b & c \\ b' & d & e \\ c' & e' & f \end{bmatrix}$ is symmetric when $b = b'$, $c = c'$, and $e = e'$.

EXAMPLE 11

If A and B are symmetric $n \times n$ matrices, show that $A + B$ is symmetric.

Solution ▶ We have $A^T = A$ and $B^T = B$, so, by Theorem 2, we have $(A + B)^T = A^T + B^T = A + B$. Hence $A + B$ is symmetric.

EXAMPLE 12

Suppose a square matrix A satisfies $A = 2A^T$. Show that necessarily $A = 0$.

Solution ▶ If we iterate the given equation, Theorem 2 gives
$$A = 2A^T = 2[2A^T]^T = 2[2(A^T)^T] = 4A$$
Subtracting A from both sides gives $3A = 0$, so $A = \tfrac{1}{3}(3A) = \tfrac{1}{3}(0) = 0$.

EXERCISES 2.1

1. Find a, b, c, and d if

   (a) $\begin{bmatrix} a & b \\ c & d \end{bmatrix} = \begin{bmatrix} c - 3d & -d \\ 2a + d & a + b \end{bmatrix}$

   ◆(b) $\begin{bmatrix} a - b & b - c \\ c - d & d - a \end{bmatrix} = 2\begin{bmatrix} 1 & 1 \\ -3 & 1 \end{bmatrix}$

   (c) $3\begin{bmatrix} a \\ b \end{bmatrix} + 2\begin{bmatrix} b \\ a \end{bmatrix} = \begin{bmatrix} 1 \\ 2 \end{bmatrix}$

   ◆(d) $\begin{bmatrix} a & b \\ c & d \end{bmatrix} = \begin{bmatrix} b & c \\ d & a \end{bmatrix}$

2. Compute the following:

   (a) $\begin{bmatrix} 3 & 2 & 1 \\ 5 & 1 & 0 \end{bmatrix} - 5\begin{bmatrix} 3 & 0 & -2 \\ 1 & -1 & 2 \end{bmatrix}$

   ◆(b) $3\begin{bmatrix} 3 \\ -1 \end{bmatrix} - 5\begin{bmatrix} 6 \\ 2 \end{bmatrix} + 7\begin{bmatrix} 1 \\ -1 \end{bmatrix}$

   (c) $\begin{bmatrix} -2 & 1 \\ 3 & 2 \end{bmatrix} - 4\begin{bmatrix} 1 & -2 \\ 0 & -1 \end{bmatrix} + 3\begin{bmatrix} 2 & -3 \\ -1 & -2 \end{bmatrix}$

   ◆(d) $[3 \; -1 \; 2] - 2[9 \; 3 \; 4] + [3 \; 11 \; -6]$

   (e) $\begin{bmatrix} 1 & -5 & 4 & 0 \\ 2 & 1 & 0 & 6 \end{bmatrix}^T$

   ◆(f) $\begin{bmatrix} 0 & -1 & 2 \\ 1 & 0 & -4 \\ -2 & 4 & 0 \end{bmatrix}^T$

   (g) $\begin{bmatrix} 3 & -1 \\ 2 & 1 \end{bmatrix} - 2\begin{bmatrix} 1 & -2 \\ 1 & 1 \end{bmatrix}^T$

   ◆(h) $3\begin{bmatrix} 2 & 1 \\ -1 & 0 \end{bmatrix}^T - 2\begin{bmatrix} 1 & -1 \\ 2 & 3 \end{bmatrix}$

3. Let $A = \begin{bmatrix} 2 & 1 \\ 0 & -1 \end{bmatrix}$, $B = \begin{bmatrix} 3 & -1 & 2 \\ 0 & 1 & 4 \end{bmatrix}$, $C = \begin{bmatrix} 3 & -1 \\ 2 & 0 \end{bmatrix}$,

   $D = \begin{bmatrix} 1 & 3 \\ -1 & 0 \\ 1 & 4 \end{bmatrix}$, and $E = \begin{bmatrix} 1 & 0 & 1 \\ 0 & 1 & 0 \end{bmatrix}$.

   Compute the following (where possible).

   (a) $3A - 2B$ ◆(b) $5C$

   (c) $3E^T$ ◆(d) $B + D$

   (e) $4A^T - 3C$ ◆(f) $(A + C)^T$

   (g) $2B - 3E$ ◆(h) $A - D$

   (i) $(B - 2E)^T$

4. Find A if:

   (a) $5A - \begin{bmatrix} 1 & 0 \\ 2 & 3 \end{bmatrix} = 3A - \begin{bmatrix} 5 & 2 \\ 6 & 1 \end{bmatrix}$

   ◆(b) $3A + \begin{bmatrix} 2 \\ 1 \end{bmatrix} = 5A - 2\begin{bmatrix} 3 \\ 0 \end{bmatrix}$

5. Find A in terms of B if:

   (a) $A + B = 3A + 2B$

   ◆(b) $2A - B = 5(A + 2B)$

6. If $X, Y, A,$ and B are matrices of the same size, solve the following equations to obtain X and Y in terms of A and B.

   (a) $5X + 3Y = A$
       $2X + Y = B$

   ◆(b) $4X + 3Y = A$
        $5X + 4Y = B$

7. Find all matrices X and Y such that:

   (a) $3X - 2Y = [3 \ -1]$

   ◆(b) $2X - 5Y = [1 \ 2]$

8. Simplify the following expressions where A, B, and C are matrices.

   (a) $2[9(A - B) + 7(2B - A)]$
       $- 2[3(2B + A) - 2(A + 3B) - 5(A + B)]$

   ◆(b) $5[3(A - B + 2C) - 2(3C - B) - A]$
        $+ 2[3(3A - B + C) + 2(B - 2A) - 2C]$

9. If A is any 2×2 matrix, show that:

   (a) $A = a\begin{bmatrix} 1 & 0 \\ 0 & 0 \end{bmatrix} + b\begin{bmatrix} 0 & 1 \\ 0 & 0 \end{bmatrix} + c\begin{bmatrix} 0 & 0 \\ 1 & 0 \end{bmatrix} + d\begin{bmatrix} 0 & 0 \\ 0 & 1 \end{bmatrix}$
   for some numbers $a, b, c,$ and d.

   ◆(b) $A = p\begin{bmatrix} 1 & 0 \\ 0 & 1 \end{bmatrix} + q\begin{bmatrix} 1 & 1 \\ 0 & 0 \end{bmatrix} + r\begin{bmatrix} 1 & 0 \\ 1 & 0 \end{bmatrix} + s\begin{bmatrix} 0 & 1 \\ 1 & 0 \end{bmatrix}$
   for some numbers $p, q, r,$ and s.

10. Let $A = [1 \ 1 \ -1]$, $B = [0 \ 1 \ 2]$, and $C = [3 \ 0 \ 1]$. If $rA + sB + tC = 0$ for some scalars $r, s,$ and t, show that necessarily $r = s = t = 0$.

11. (a) If $Q + A = A$ holds for every $m \times n$ matrix A, show that $Q = 0_{mn}$.

    ◆(b) If A is an $m \times n$ matrix and $A + A' = 0_{mn}$, show that $A' = -A$.

12. If A denotes an $m \times n$ matrix, show that $A = -A$ if and only if $A = 0$.

13. A square matrix is called a **diagonal** matrix if all the entries off the main diagonal are zero. If A and B are diagonal matrices, show that the following matrices are also diagonal.

    (a) $A + B$ ◆(b) $A - B$

    (c) kA for any number k

14. In each case determine all s and t such that the given matrix is symmetric:

    (a) $\begin{bmatrix} 1 & s \\ -2 & t \end{bmatrix}$ ◆(b) $\begin{bmatrix} s & t \\ st & 1 \end{bmatrix}$

    (c) $\begin{bmatrix} s & 2s & st \\ t & -1 & s \\ t & s^2 & s \end{bmatrix}$ ◆(d) $\begin{bmatrix} 2 & s & t \\ 2s & 0 & s+t \\ 3 & 3 & t \end{bmatrix}$

15. In each case find the matrix A.

    (a) $\left(A + 3\begin{bmatrix} 1 & -1 & 0 \\ 1 & 2 & 4 \end{bmatrix}\right)^T = \begin{bmatrix} 2 & 1 \\ 0 & 5 \\ 3 & 8 \end{bmatrix}$

    ◆(b) $\left(3A^T + 2\begin{bmatrix} 1 & 0 \\ 0 & 2 \end{bmatrix}\right)^T = \begin{bmatrix} 8 & 0 \\ 3 & 1 \end{bmatrix}$

    (c) $(2A - 3[1 \ 2 \ 0])^T = 3A^T + [2 \ 1 \ -1]^T$

    ◆(d) $\left(2A^T - 5\begin{bmatrix} 1 & 0 \\ -1 & 2 \end{bmatrix}\right)^T = 4A - 9\begin{bmatrix} 1 & 1 \\ -1 & 0 \end{bmatrix}$

16. Let A and B be symmetric (of the same size). Show that each of the following is symmetric.

    (a) $(A - B)$

    ◆(b) kA for any scalar k

17. Show that $A + A^T$ is symmetric for *any* square matrix A.

18. If A is a square matrix and $A = kA^T$ where $k \neq \pm 1$, show that $A = 0$.

19. In each case either show that the statement is true or give an example showing it is false.

    (a) If $A + B = A + C$, then B and C have the same size.

    ◆(b) If $A + B = 0$, then $B = 0$.

    (c) If the $(3, 1)$-entry of A is 5, then the $(1, 3)$-entry of A^T is -5.

    ◆(d) A and A^T have the same main diagonal for every matrix A.

    (e) If B is symmetric and $A^T = 3B$, then $A = 3B$.

    ◆(f) If A and B are symmetric, then $kA + mB$ is symmetric for any scalars k and m.

20. A square matrix W is called **skew-symmetric** if $W^T = -W$. Let A be any square matrix.

    (a) Show that $A - A^T$ is skew-symmetric.

    (b) Find a symmetric matrix S and a skew-symmetric matrix W such that $A = S + W$.

    ♦(c) Show that S and W in part (b) are uniquely determined by A.

21. If W is skew-symmetric (Exercise 20), show that the entries on the main diagonal are zero.

22. Prove the following parts of Theorem 1.

    (a) $(k + p)A = kA + pA$

    ♦(b) $(kp)A = k(pA)$

23. Let $A, A_1, A_2, \ldots, A_n$ denote matrices of the same size. Use induction on n to verify the following extensions of properties 5 and 6 of Theorem 1.

    (a) $k(A_1 + A_2 + \cdots + A_n)$
    $= kA_1 + kA_2 + \cdots + kA_n$ for any number k

    (b) $(k_1 + k_2 + \cdots + k_n)A$
    $= k_1A + k_2A + \cdots + k_nA$ for any numbers $k_1, k_2, \ldots, k_n$

24. Let A be a square matrix. If $A = pB^T$ and $B = qA^T$ for some matrix B and numbers p and q, show that either $A = 0 = B$ or $pq = 1$. [*Hint*: Example 7.]

SECTION 2.2 Equations, Matrices, and Transformations

Up to now we have used matrices to solve systems of linear equations by manipulating the rows of the augmented matrix. In this section we introduce a different way of describing linear systems that makes more use of the coefficient matrix of the system and leads to a useful way of "multiplying" matrices.

Vectors

It is a well-known fact in analytic geometry that two points in the plane with coordinates (a_1, a_2) and (b_1, b_2) are equal if and only if $a_1 = b_1$ and $a_2 = b_2$. Moreover, a similar condition applies to points (a_1, a_2, a_3) in space. We extend this idea as follows.

An ordered sequence $(a_1, a_2, \ldots, a_n)$ of real numbers is called an **ordered n-tuple**. The word "ordered" here reflects our insistence that two ordered n-tuples are equal if and only if corresponding entries are the same. In other words,

$$(a_1, a_2, \ldots, a_n) = (b_1, b_2, \ldots, b_n) \quad \text{if and only if} \quad a_1 = b_1, a_2 = b_2, \ldots, \text{and } a_n = b_n.$$

Thus the ordered 2-tuples and 3-tuples are just the ordered pairs and triples familiar from geometry.

Definition 2.4 *Let $\mathbb{R}$ denote the set of all real numbers. The set of all ordered n-tuples from $\mathbb{R}$ has a special notation:*

$\mathbb{R}^n$ denotes the set of all ordered n-tuples of real numbers.

There are two commonly used ways to denote the n-tuples in $\mathbb{R}^n$: As rows $(r_1, r_2, \ldots, r_n)$ or columns $\begin{bmatrix} r_1 \\ r_2 \\ \vdots \\ r_n \end{bmatrix}$; the notation we use depends on the context. In any event they are called **vectors** or **n-vectors** and will be denoted using bold type such as **x** or **v**. For example, an $m \times n$ matrix A will be written as a row of columns:

$$A = [\mathbf{a}_1 \; \mathbf{a}_2 \; \cdots \; \mathbf{a}_n] \text{ where } \mathbf{a}_j \text{ denotes column } j \text{ of } A \text{ for each } j.$$

If $\mathbf{x}$ and $\mathbf{y}$ are two n-vectors in $\mathbb{R}^n$, it is clear that their matrix sum $\mathbf{x} + \mathbf{y}$ is also in $\mathbb{R}^n$ as is the scalar multiple $k\mathbf{x}$ for any real number k. We express this observation by saying that $\mathbb{R}^n$ is **closed** under addition and scalar multiplication. In particular, all the basic properties in Theorem 1 Section 2.1 are true of these n-vectors. These properties are fundamental and will be used frequently below without comment. As for matrices in general, the $n \times 1$ zero matrix is called the **zero n-vector** in $\mathbb{R}^n$ and, if $\mathbf{x}$ is an n-vector, the n-vector $-\mathbf{x}$ is called the **negative x**.

Of course, we have already encountered these n-vectors in Section 1.3 as the solutions to systems of linear equations with n variables. In particular we defined the notion of a linear combination of vectors and showed that a linear combination of solutions to a homogeneous system is again a solution. Clearly, a linear combination of n-vectors in $\mathbb{R}^n$ is again in $\mathbb{R}^n$, a fact that we will be using.

Matrix-Vector Multiplication

Given a system of linear equations, the left sides of the equations depend only on the coefficient matrix A and the column $\mathbf{x}$ of variables, and not on the constants. This observation leads to a fundamental idea in linear algebra: We view the left sides of the equations as the "product" $A\mathbf{x}$ of the matrix A and the vector $\mathbf{x}$. This simple change of perspective leads to a completely new way of viewing linear systems—one that is very useful and will occupy our attention throughout this book.

To motivate the definition of the "product" $A\mathbf{x}$, consider first the following system of two equations in three variables:

$$ax_1 + bx_2 + cx_3 = b_1$$
$$a'x_1 + b'x_2 + c'x_3 = b_1 \quad (*)$$

and let $A = \begin{bmatrix} a & b & c \\ a' & b' & c' \end{bmatrix}$, $\mathbf{x} = \begin{bmatrix} x_1 \\ x_2 \\ x_3 \end{bmatrix}$, and $\mathbf{b} = \begin{bmatrix} b_1 \\ b_2 \end{bmatrix}$ denote the coefficient matrix, the variable matrix, and the constant matrix, respectively. The system $(*)$ can be expressed as a single vector equation

$$\begin{bmatrix} ax_1 + bx_2 + cx_3 \\ a'x_1 + b'x_2 + c'x_3 \end{bmatrix} = \begin{bmatrix} b_1 \\ b_2 \end{bmatrix},$$

which in turn can be written as follows:

$$x_1 \begin{bmatrix} a \\ a' \end{bmatrix} + x_2 \begin{bmatrix} b \\ b' \end{bmatrix} + x_3 \begin{bmatrix} c \\ c' \end{bmatrix} = \begin{bmatrix} b_1 \\ b_2 \end{bmatrix}.$$

Now observe that the vectors appearing on the left side are just the columns

$$\mathbf{a}_1 = \begin{bmatrix} a \\ a' \end{bmatrix}, \mathbf{a}_2 = \begin{bmatrix} b \\ b' \end{bmatrix}, \text{ and } \mathbf{a}_3 = \begin{bmatrix} c \\ c' \end{bmatrix}$$

of the coefficient matrix A. Hence the system $(*)$ takes the form

$$x_1\mathbf{a}_1 + x_2\mathbf{a}_2 + x_3\mathbf{a}_3 = B. \quad (**)$$

This shows that the system $(*)$ has a solution if and only if the constant matrix $\mathbf{b}$ is a linear combination[3] of the columns of A, and that in this case the entries of the solution are the coefficients x_1, x_2, and x_3 in this linear combination.

Moreover, this holds in general. If A is any $m \times n$ matrix, it is often convenient to view A as a row of columns. That is, if $\mathbf{a}_1, \mathbf{a}_2, \ldots, \mathbf{a}_n$ are the columns of A, we write

$$A = [\mathbf{a}_1 \; \mathbf{a}_2 \; \cdots \; \mathbf{a}_n]$$

and say that $A = [\mathbf{a}_1 \; \mathbf{a}_2 \; \cdots \; \mathbf{a}_n]$ is *given in terms of its columns*.

---

[3] Linear combinations were introduced in Section 1.3 to describe the solutions of homogeneous systems of linear equations. They will be used extensively in what follows.

Now consider any system of linear equations with $m \times n$ coefficient matrix A. If **b** is the constant matrix of the system, and if $\mathbf{x} = \begin{bmatrix} x_1 \\ x_2 \\ \vdots \\ x_n \end{bmatrix}$ is the matrix of variables then, exactly as above, the system can be written as a single vector equation

$$x_1\mathbf{a}_1 + x_2\mathbf{a}_2 + \cdots + x_n\mathbf{a}_n = \mathbf{b}. \qquad (***)$$

EXAMPLE 1

Write the system $\begin{cases} 3x_1 + 2x_2 - 4x_3 = 0 \\ x_1 - 3x_2 + x_3 = 3 \\ x_2 - 5x_3 = -1 \end{cases}$ in the form given in $(***)$.

Solution ▶ $x_1 \begin{bmatrix} 3 \\ 1 \\ 0 \end{bmatrix} + x_2 \begin{bmatrix} 2 \\ -3 \\ 1 \end{bmatrix} + x_3 \begin{bmatrix} -4 \\ 1 \\ -5 \end{bmatrix} = \begin{bmatrix} 0 \\ 3 \\ -1 \end{bmatrix}$.

As mentioned above, we view the left side of $(***)$ as the *product* of the matrix A and the vector **x**. This basic idea is formalized in the following definition:

Definition 2.5

Matrix-Vector Products Let $A = [\mathbf{a}_1 \; \mathbf{a}_2 \; \cdots \; \mathbf{a}_n]$ be an $m \times n$ matrix, written in terms of its columns $\mathbf{a}_1, \mathbf{a}_2, \ldots, \mathbf{a}_n$. If $\mathbf{x} = \begin{bmatrix} x_1 \\ x_2 \\ \vdots \\ x_n \end{bmatrix}$ is any n-vector, the **product** $A\mathbf{x}$ is defined to be the m-vector given by:

$$A\mathbf{x} = x_1\mathbf{a}_1 + x_1\mathbf{a}_2 + x_1\mathbf{a}_n.$$

In other words, if A is $m \times n$ and **x** is an n-vector, the product $A\mathbf{x}$ is the linear combination of the columns of A where the coefficients are the entries of **x** (in order).

Note that if A is an $m \times n$ matrix, the product $A\mathbf{x}$ is only defined if **x** is an n-vector and then the vector $A\mathbf{x}$ is an m-vector because this is true of each column $\mathbf{a}_j$ of A. But in this case the *system* of linear equations with coefficient matrix A and constant vector **b** takes the form of a *single* matrix equation

$$A\mathbf{x} = \mathbf{b}.$$

The following theorem combines Definition 2.5 and equation $(***)$ and summarizes the above discussion. Recall that a system of linear equations is said to be *consistent* if it has at least one solution.

Theorem 1

(1) Every system of linear equations has the form $A\mathbf{x} = \mathbf{b}$ where A is the coefficient matrix, **b** is the constant matrix, and X is the matrix of variables.

(2) The system $A\mathbf{x} = \mathbf{b}$ is consistent if and only if **b** is a linear combination of the columns of A.

(3) If $\mathbf{a}_1, \mathbf{a}_2, \ldots, \mathbf{a}_n$ are the columns of A and if $\mathbf{x} = \begin{bmatrix} x_1 \\ x_2 \\ \vdots \\ x_n \end{bmatrix}$, then **x** is a solution to the linear system $A\mathbf{x} = \mathbf{b}$ if and only if $x_1, x_2, \ldots, x_n$ are a solution of the vector equation $x_1\mathbf{a}_1 + x_2\mathbf{a}_2 + \cdots + x_n\mathbf{a}_n = \mathbf{b}$.

SECTION 2.2 *Equations, Matrices, and Transformations*

A system of linear equations in the form $A\mathbf{x} = \mathbf{b}$ as in (1) of Theorem 1 is said to be written in **matrix form**. This is a useful way to view linear systems as we shall see.

Theorem 1 transforms the problem of solving the linear system $A\mathbf{x} = \mathbf{b}$ into the problem of expressing the constant matrix B as a linear combination of the columns of the coefficient matrix A. Such a change in perspective is very useful because one approach or the other may be better in a particular situation; the importance of the theorem is that there is a choice.

EXAMPLE 2

If $A = \begin{bmatrix} 2 & -1 & 3 & 5 \\ 0 & 2 & -3 & 1 \\ -3 & 4 & 1 & 2 \end{bmatrix}$ and $\mathbf{x} = \begin{bmatrix} 2 \\ 1 \\ 0 \\ -2 \end{bmatrix}$, compute $A\mathbf{x}$.

Solution ▶ By Definition 2.5: $A\mathbf{x} = 2\begin{bmatrix} 2 \\ 0 \\ -3 \end{bmatrix} + 1\begin{bmatrix} -1 \\ 2 \\ 4 \end{bmatrix} + 0\begin{bmatrix} 3 \\ -3 \\ 1 \end{bmatrix} - 2\begin{bmatrix} 5 \\ 1 \\ 2 \end{bmatrix} = \begin{bmatrix} -7 \\ 0 \\ -6 \end{bmatrix}$.

EXAMPLE 3

Given columns $\mathbf{a}_1, \mathbf{a}_2, \mathbf{a}_3$, and $\mathbf{a}_4$ in $\mathbb{R}^3$, write $2\mathbf{a}_1 - 3\mathbf{a}_2 + 5\mathbf{a}_3 + \mathbf{a}_4$ in the form $A\mathbf{x}$ where A is a matrix and $\mathbf{x}$ is a vector.

Solution ▶ Here the column of coefficients is $\mathbf{x} = \begin{bmatrix} 2 \\ -3 \\ 5 \\ 1 \end{bmatrix}$. Hence Definition 2.5 gives

$$A\mathbf{x} = 2\mathbf{a}_1 - 3\mathbf{a}_2 + 5\mathbf{a}_3 + \mathbf{a}_4$$

where $A = [\mathbf{a}_1 \ \mathbf{a}_2 \ \mathbf{a}_3 \ \mathbf{a}_4]$ is the matrix with $\mathbf{a}_1, \mathbf{a}_2, \mathbf{a}_3$, and $\mathbf{a}_4$ as its columns.

EXAMPLE 4

Let $A = [\mathbf{a}_1 \ \mathbf{a}_2 \ \mathbf{a}_3 \ \mathbf{a}_4]$ be the 3×4 matrix given in terms of its columns $\mathbf{a}_1 = \begin{bmatrix} 2 \\ 0 \\ -1 \end{bmatrix}$, $\mathbf{a}_2 = \begin{bmatrix} 1 \\ 1 \\ 1 \end{bmatrix}$, $\mathbf{a}_4 = \begin{bmatrix} 3 \\ -1 \\ -3 \end{bmatrix}$, and $\mathbf{a}_4 = \begin{bmatrix} 3 \\ 1 \\ 0 \end{bmatrix}$. In each case below, either express $\mathbf{b}$ as a linear combination of $\mathbf{a}_1, \mathbf{a}_2, \mathbf{a}_3$, and $\mathbf{a}_4$, or show that it is not such a linear combination. Explain what your answer means for the corresponding system $A\mathbf{x} = \mathbf{b}$ of linear equations.

(a) $\mathbf{b} = \begin{bmatrix} 1 \\ 2 \\ 3 \end{bmatrix}$ (b) $\mathbf{b} = \begin{bmatrix} 4 \\ 2 \\ 1 \end{bmatrix}$.

Solution ▶ By Theorem 1, $\mathbf{b}$ is a linear combination of $\mathbf{a}_1, \mathbf{a}_2, \mathbf{a}_3$, and $\mathbf{a}_4$ if and only if the system $A\mathbf{x} = \mathbf{b}$ is consistent (that is, it has a solution). So in each case we carry the augmented matrix $[A|\mathbf{b}]$ of the system $A\mathbf{x} = \mathbf{b}$ to reduced form.

(a) Here $\begin{bmatrix} 2 & 1 & 3 & 3 & | & 1 \\ 0 & 1 & -1 & 1 & | & 2 \\ -1 & 1 & -3 & 0 & | & 3 \end{bmatrix} \to \begin{bmatrix} 1 & 0 & 2 & 1 & | & 0 \\ 0 & 1 & -1 & 1 & | & 0 \\ 0 & 0 & 0 & 0 & | & 1 \end{bmatrix}$, so the system $A\mathbf{x} = \mathbf{b}$ has no solution in this case. Hence $\mathbf{b}$ is *not* a linear combination of $\mathbf{a}_1, \mathbf{a}_2, \mathbf{a}_3,$ and $\mathbf{a}_4$.

(b) Now $\begin{bmatrix} 2 & 1 & 3 & 3 & | & 4 \\ 0 & 1 & -1 & 1 & | & 2 \\ -1 & 1 & -3 & 0 & | & 1 \end{bmatrix} \to \begin{bmatrix} 1 & 0 & 2 & 1 & | & 1 \\ 0 & 1 & -1 & 1 & | & 2 \\ 0 & 0 & 0 & 0 & | & 0 \end{bmatrix}$, so the system $A\mathbf{x} = \mathbf{b}$ is consistent.

Thus $\mathbf{b}$ is a linear combination of $\mathbf{a}_1, \mathbf{a}_2, \mathbf{a}_3,$ and $\mathbf{a}_4$ in this case. In fact the general solution is $x_1 = 1 - 2s - t$, $x_2 = 2 + s - t$, $x_3 = s$, and $x_4 = t$ where s and t are arbitrary parameters. Hence $x_1\mathbf{a}_1 + x_2\mathbf{a}_2 + x_3\mathbf{a}_3 + x_4\mathbf{a}_4 = \mathbf{b} = \begin{bmatrix} 4 \\ 2 \\ 1 \end{bmatrix}$ for *any* choice of s and t. If we take $s = 0$ and $t = 0$, this becomes $\mathbf{a}_1 + 2\mathbf{a}_2 = \mathbf{b}$, whereas taking $s = 1 = t$ gives $-2\mathbf{a}_1 + 2\mathbf{a}_2 + \mathbf{a}_3 + \mathbf{a}_4 = \mathbf{b}$.

EXAMPLE 5

Taking A to be the zero matrix, we have $0\mathbf{x} = \mathbf{0}$ for all vectors $\mathbf{x}$ by Definition 2.5 because every column of the zero matrix is zero. Similarly, $A\mathbf{0} = \mathbf{0}$ for all matrices A because every entry of the zero vector is zero.

EXAMPLE 6

If $I = \begin{bmatrix} 1 & 0 & 0 \\ 0 & 1 & 0 \\ 0 & 0 & 1 \end{bmatrix}$, show that $I\mathbf{x} = \mathbf{x}$ for any vector $\mathbf{x}$ in $\mathbb{R}^3$.

Solution ▶ If $\mathbf{x} = \begin{bmatrix} x_1 \\ x_2 \\ x_3 \end{bmatrix}$ then Definition 2.5 gives

$$I\mathbf{x} = x_1\begin{bmatrix} 1 \\ 0 \\ 0 \end{bmatrix} + x_2\begin{bmatrix} 0 \\ 1 \\ 0 \end{bmatrix} + x_3\begin{bmatrix} 0 \\ 0 \\ 1 \end{bmatrix} = \begin{bmatrix} x_1 \\ 0 \\ 0 \end{bmatrix} + \begin{bmatrix} 0 \\ x_2 \\ 0 \end{bmatrix} + \begin{bmatrix} 0 \\ 0 \\ x_3 \end{bmatrix} = \begin{bmatrix} x_1 \\ x_2 \\ x_3 \end{bmatrix} = \mathbf{x}.$$

The matrix I in Example 6 is called the 3×3 **identity matrix**, and we will encounter such matrices again in Example 11 below.

Before proceeding, we develop some algebraic properties of matrix-vector multiplication that are used extensively throughout linear algebra.

Theorem 2

Let A and B be $m \times n$ matrices, and let $\mathbf{x}$ and $\mathbf{y}$ be n-vectors in $\mathbb{R}^n$. Then:

(1) $A(\mathbf{x} + \mathbf{y}) = A\mathbf{x} + A\mathbf{y}$.

(2) $A(a\mathbf{x}) = a(A\mathbf{x}) = (aA)\mathbf{x}$ for all scalars a.

(3) $(A + B)\mathbf{x} = A\mathbf{x} + B\mathbf{x}$.

SECTION 2.2 *Equations, Matrices, and Transformations*

> **PROOF**
>
> We prove (3); the other verifications are similar and are left as exercises. Let $A = [\mathbf{a}_1 \ \mathbf{a}_2 \ \cdots \ \mathbf{a}_n]$ and $B = [\mathbf{b}_1 \ \mathbf{b}_2 \ \cdots \ \mathbf{b}_n]$ be given in terms of their columns. Since adding two matrices is the same as adding their columns, we have
>
> $$A + B = [\mathbf{a}_1 + \mathbf{b}_1 \ \ \mathbf{a}_2 + \mathbf{b}_1 \ \ \cdots \ \ \mathbf{a}_n + \mathbf{b}_n]$$
>
> If we write $\mathbf{x} = \begin{bmatrix} x_1 \\ x_2 \\ \vdots \\ x_n \end{bmatrix}$ Definition 2.5 gives
>
> $$\begin{aligned}(A + B)\mathbf{x} &= x_1(\mathbf{a}_1 + \mathbf{b}_1) + x_2(\mathbf{a}_2 + \mathbf{b}_2) + \cdots + x_n(\mathbf{a}_n + \mathbf{b}_n) \\ &= (x_1\mathbf{a}_1 + x_2\mathbf{a}_2 + \cdots + x_n\mathbf{a}_n) + (x_1\mathbf{b}_1 + x_2\mathbf{b}_2 + \cdots + x_n\mathbf{b}_n) \\ &= A\mathbf{x} + B\mathbf{x}.\end{aligned}$$

Theorem 2 allows matrix-vector computations to be carried out much as in ordinary arithmetic. For example, for any $m \times n$ matrices A and B and any n-vectors $\mathbf{x}$ and $\mathbf{y}$, we have:

$$A(2\mathbf{x} - 5\mathbf{y}) = 2A\mathbf{x} - 5A\mathbf{y} \quad \text{and} \quad (3A - 7B)\mathbf{x} = 3A\mathbf{x} - 7B\mathbf{x}$$

We will use such manipulations throughout the book, often without mention.

Theorem 2 also gives a useful way to describe the solutions to a system

$$A\mathbf{x} = \mathbf{b}$$

of linear equations. There is a related system

$$A\mathbf{x} = \mathbf{0}$$

called the **associated homogeneous system**, obtained from the original system $A\mathbf{x} = \mathbf{b}$ by replacing all the constants by zeros. Suppose $\mathbf{x}_1$ is a solution to $A\mathbf{x} = \mathbf{b}$ and $\mathbf{x}_0$ is a solution to $A\mathbf{x} = \mathbf{0}$ (that is $A\mathbf{x}_1 = \mathbf{b}$ and $A\mathbf{x}_0 = \mathbf{0}$). Then $\mathbf{x}_1 + \mathbf{x}_0$ is another solution to $A\mathbf{x} = \mathbf{b}$. Indeed, Theorem 2 gives

$$A(\mathbf{x}_1 + \mathbf{x}_0) = A\mathbf{x}_1 + A\mathbf{x}_0 = \mathbf{b} + \mathbf{0} = \mathbf{b}$$

This observation has a useful converse.

Theorem 3

Suppose $\mathbf{x}_1$ is any particular solution to the system $A\mathbf{x} = \mathbf{b}$ of linear equations. Then every solution $\mathbf{x}_2$ to $A\mathbf{x} = \mathbf{b}$ has the form

$$\mathbf{x}_2 = \mathbf{x}_0 + \mathbf{x}_1$$

for some solution $\mathbf{x}_0$ of the associated homogeneous system $A\mathbf{x} = \mathbf{0}$.

> **PROOF**
>
> Suppose $\mathbf{x}_2$ is also a solution to $A\mathbf{x} = \mathbf{b}$, so that $A\mathbf{x}_2 = \mathbf{b}$. Write $\mathbf{x}_0 = \mathbf{x}_2 - \mathbf{x}_1$. Then $\mathbf{x}_2 = \mathbf{x}_0 + \mathbf{x}_1$ and, using Theorem 2, we compute
>
> $$A\mathbf{x}_0 = A(\mathbf{x}_2 - \mathbf{x}_1) = A\mathbf{x}_2 - A\mathbf{x}_1 = \mathbf{b} - \mathbf{b} = \mathbf{0}.$$
>
> Hence $\mathbf{x}_0$ is a solution to the associated homogeneous system $A\mathbf{x} = \mathbf{0}$.

Note that gaussian elimination provides one such representation.

EXAMPLE 7

Express every solution to the following system as the sum of a specific solution plus a solution to the associated homogeneous system.

$$\begin{aligned} x_1 - x_2 - x_3 + 3x_4 &= 2 \\ 2x_1 - x_2 - 3x_3 + 4x_4 &= 6 \\ x_1 - 2x_3 + x_4 &= 4 \end{aligned}$$

Solution ▶ Gaussian elimination gives $x_1 = 4 + 2s - t$, $x_2 = 2 + s + 2t$, $x_3 = s$, and $x_4 = t$ where s and t are arbitrary parameters. Hence the general solution can be written

$$\mathbf{x} = \begin{bmatrix} x_1 \\ x_2 \\ x_3 \\ x_4 \end{bmatrix} = \begin{bmatrix} 4 + 2s - t \\ 2 + s + 2t \\ s \\ t \end{bmatrix} = \begin{bmatrix} 4 \\ 2 \\ 0 \\ 0 \end{bmatrix} + \left(s \begin{bmatrix} 2 \\ 1 \\ 1 \\ 0 \end{bmatrix} + t \begin{bmatrix} -1 \\ 2 \\ 0 \\ 1 \end{bmatrix} \right).$$

Thus $\mathbf{x} = \begin{bmatrix} 4 \\ 2 \\ 0 \\ 0 \end{bmatrix}$ is a particular solution (where $s = 0 = t$), and $\mathbf{x}_0 = s \begin{bmatrix} 2 \\ 1 \\ 1 \\ 0 \end{bmatrix} + t \begin{bmatrix} -1 \\ 2 \\ 0 \\ 1 \end{bmatrix}$

gives *all* solutions to the associated homogeneous system. (To see why this is so, carry out the gaussian elimination again but with all the constants set equal to zero.)

The Dot Product

Definition 2.5 is not always the easiest way to compute a matrix-vector product $A\mathbf{x}$ because it requires that the columns of A be explicitly identified. There is another way to find such a product which uses the matrix A as a whole with no reference to its columns, and hence is useful in practice. The method depends on the following notion.

Definition 2.6 If $(a_1, a_2, \ldots, a_n)$ and $(b_1, b_2, \ldots, b_n)$ are two ordered n-tuples, their **dot product** is defined to be the number

$$a_1 b_1 + a_2 b_2 + \cdots + a_n b_n$$

obtained by multiplying corresponding entries and adding the results.

To see how this relates to matrix products, let A denote a 3×4 matrix and let $\mathbf{x}$ be a 4-vector. Writing

$$\mathbf{x} = \begin{bmatrix} x_1 \\ x_2 \\ x_3 \\ x_4 \end{bmatrix} \quad \text{and} \quad A = \begin{bmatrix} a_{11} & a_{12} & a_{13} & a_{14} \\ a_{21} & a_{22} & a_{23} & a_{24} \\ a_{31} & a_{32} & a_{33} & a_{34} \end{bmatrix}$$

in the notation of Section 2.1, we compute

$$A\mathbf{x} = \begin{bmatrix} a_{11} & a_{12} & a_{13} & a_{14} \\ a_{21} & a_{22} & a_{23} & a_{24} \\ a_{31} & a_{32} & a_{33} & a_{34} \end{bmatrix} \begin{bmatrix} x_1 \\ x_2 \\ x_3 \\ x_4 \end{bmatrix} = x_1 \begin{bmatrix} a_{11} \\ a_{21} \\ a_{31} \end{bmatrix} + x_2 \begin{bmatrix} a_{12} \\ a_{22} \\ a_{32} \end{bmatrix} + x_3 \begin{bmatrix} a_{13} \\ a_{23} \\ a_{33} \end{bmatrix} + x_4 \begin{bmatrix} a_{14} \\ a_{24} \\ a_{34} \end{bmatrix}$$

$$= \begin{bmatrix} a_{11}x_1 & a_{12}x_2 & a_{13}x_3 & a_{14}x_4 \\ a_{21}x_1 & a_{22}x_2 & a_{23}x_3 & a_{24}x_4 \\ a_{31}x_1 & a_{32}x_2 & a_{33}x_3 & a_{34}x_4 \end{bmatrix}.$$

SECTION 2.2 Equations, Matrices, and Transformations

From this we see that each entry of $A\mathbf{x}$ is the dot product of the corresponding row of A with $\mathbf{x}$. This computation goes through in general, and we record the result in Theorem 4.

Theorem 4

Dot Product Rule
Let A be an $m \times n$ matrix and let $\mathbf{x}$ be an n-vector. Then each entry of the vector $A\mathbf{x}$ is the dot product of the corresponding row of A with $\mathbf{x}$.

This result is used extensively throughout linear algebra.

If A is $m \times n$ and $\mathbf{x}$ is an n-vector, the computation of $A\mathbf{x}$ by the dot product rule is simpler than using Definition 2.5 because the computation can be carried out directly with no explicit reference to the columns of A (as in Definition 2.5). The first entry of $A\mathbf{x}$ is the dot product of row 1 of A with $\mathbf{x}$. In hand calculations this is computed by going *across* row one of A, going *down* the column $\mathbf{x}$, multiplying corresponding entries, and adding the results. The other entries of $A\mathbf{x}$ are computed in the same way using the other rows of A with the column $\mathbf{x}$.

In general, compute entry k of $A\mathbf{x}$ as follows (see the diagram):

> Go *across* row k of A and *down* column $\mathbf{x}$, multiply corresponding entries, and add the results.

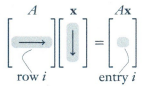

As an illustration, we rework Example 2 using the dot product rule instead of Definition 2.5.

EXAMPLE 8

If $A = \begin{bmatrix} 2 & -1 & 3 & 5 \\ 0 & 2 & -3 & 1 \\ -3 & 4 & 1 & 2 \end{bmatrix}$ and $\mathbf{x} = \begin{bmatrix} 2 \\ 1 \\ 0 \\ -2 \end{bmatrix}$, compute $A\mathbf{x}$.

Solution ▶ The entries of $A\mathbf{x}$ are the dot products of the rows of A with $\mathbf{x}$:

$$A\mathbf{x} = \begin{bmatrix} 2 & -1 & 3 & 5 \\ 0 & 2 & -3 & 1 \\ -3 & 4 & 1 & 2 \end{bmatrix} \begin{bmatrix} 2 \\ 1 \\ 0 \\ -2 \end{bmatrix} = \begin{bmatrix} 2\cdot 2 + (-1)1 + 3\cdot 0 + 5(-2) \\ 0\cdot 2 + 2\cdot 1 + (-3)0 + 1(-2) \\ (-3)2 + 4\cdot 1 + 1\cdot 0 + 2(-2) \end{bmatrix} = \begin{bmatrix} -7 \\ 0 \\ -6 \end{bmatrix}.$$

Of course, this agrees with the outcome in Example 2.

EXAMPLE 9

Write the following system of linear equations in the form $A\mathbf{x} = \mathbf{b}$.

$$\begin{aligned} 5x_1 - x_2 + 2x_3 + x_4 - 3x_5 &= 8 \\ x_1 + x_2 + 3x_3 - 5x_4 + 2x_5 &= -2 \\ -x_1 + x_2 - 2x_3 \quad\quad - 3x_5 &= 0 \end{aligned}$$

Solution ▶ Write $A = \begin{bmatrix} 5 & -1 & 2 & 1 & -3 \\ 1 & 1 & 3 & -5 & 2 \\ -1 & 1 & -2 & 0 & -3 \end{bmatrix}$, $\mathbf{b} = \begin{bmatrix} 8 \\ -2 \\ 0 \end{bmatrix}$, and $\mathbf{x} = \begin{bmatrix} x_1 \\ x_2 \\ x_3 \\ x_4 \\ x_5 \end{bmatrix}$. Then the dot product rule gives $A\mathbf{x} = \begin{bmatrix} 5x_1 - x_2 + 2x_3 + x_4 - 3x_5 \\ x_1 + x_2 + 3x_3 - 5x_4 + 2x_5 \\ -x_1 + x_2 - 2x_3 - 3x_5 \end{bmatrix}$, so the entries of $A\mathbf{x}$ are the left sides of the equations in the linear system. Hence the system becomes $A\mathbf{x} = \mathbf{b}$ because matrices are equal if and only corresponding entries are equal.

EXAMPLE 10

If A is the zero $m \times n$ matrix, then $A\mathbf{x} = \mathbf{0}$ for each n-vector $\mathbf{x}$.

Solution ▶ For each k, entry k of $A\mathbf{x}$ is the dot product of row k of A with $\mathbf{x}$, and this is zero because row k of A consists of zeros.

Definition 2.7 For each $n > 2$, the **identity matrix** I_n is the $n \times n$ matrix with 1s on the main diagonal (upper left to lower right), and zeros elsewhere.

The first few identity matrices are

$$I_2 = \begin{bmatrix} 1 & 0 \\ 0 & 1 \end{bmatrix}, \quad I_3 = \begin{bmatrix} 1 & 0 & 0 \\ 0 & 1 & 0 \\ 0 & 0 & 1 \end{bmatrix}, \quad I_4 = \begin{bmatrix} 1 & 0 & 0 & 0 \\ 0 & 1 & 0 & 0 \\ 0 & 0 & 1 & 0 \\ 0 & 0 & 0 & 1 \end{bmatrix}, \ldots$$

In Example 6 we showed that $I_3\mathbf{x} = \mathbf{x}$ for each 3-vector $\mathbf{x}$ using Definition 2.5. The following result shows that this holds in general, and is the reason for the name.

EXAMPLE 11

For each $n \geq 2$ we have $I_n\mathbf{x} = \mathbf{x}$ for each n-vector $\mathbf{x}$ in $\mathbb{R}^n$.

Solution ▶ We verify the case $n = 4$. Given the 4-vector $\mathbf{x} = \begin{bmatrix} x_1 \\ x_2 \\ x_3 \\ x_4 \end{bmatrix}$ the dot product rule gives

$$I_4\mathbf{x} = \begin{bmatrix} 1 & 0 & 0 & 0 \\ 0 & 1 & 0 & 0 \\ 0 & 0 & 1 & 0 \\ 0 & 0 & 0 & 1 \end{bmatrix} \begin{bmatrix} x_1 \\ x_2 \\ x_3 \\ x_4 \end{bmatrix} = \begin{bmatrix} x_1 + 0 + 0 + 0 \\ 0 + x_2 + 0 + 0 \\ 0 + 0 + x_3 + 0 \\ 0 + 0 + 0 + x_4 \end{bmatrix} = \begin{bmatrix} x_1 \\ x_2 \\ x_3 \\ x_4 \end{bmatrix} = \mathbf{x}.$$

In general, $I_n\mathbf{x} = \mathbf{x}$ because entry k of $I_n\mathbf{x}$ is the dot product of row k of I_n with $\mathbf{x}$, and row k of I_n has 1 in position k and zeros elsewhere.

EXAMPLE 12

Let $A = [\mathbf{a}_1 \ \mathbf{a}_2 \ \cdots \ \mathbf{a}_n]$ be any $m \times n$ matrix with columns $\mathbf{a}_1, \mathbf{a}_2, \ldots, \mathbf{a}_n$. If $\mathbf{e}_j$ denotes column j of the $n \times n$ identity matrix I_n, then $A\mathbf{e}_j = \mathbf{a}_j$ for each $j = 1, 2, \ldots, n$.

Solution ▶ Write $\mathbf{e}_j = \begin{bmatrix} t_1 \\ t_2 \\ \vdots \\ t_n \end{bmatrix}$ where $t_j = 1$, but $t_i = 0$ for all $i \neq j$. Then Theorem 4 gives

$$A\mathbf{e}_j = t_1\mathbf{a}_1 + \cdots + t_j\mathbf{a}_j + \cdots + t_n\mathbf{a}_n = 0 + \cdots + \mathbf{a}_j + \cdots + 0 = \mathbf{a}_j.$$

Example 12 will be referred to later; for now we use it to prove

Theorem 5

Let A and B be $m \times n$ matrices. If $A\mathbf{x} = B\mathbf{x}$ for all $\mathbf{x}$ in $\mathbb{R}^n$, then $A = B$.

PROOF

Write $A = [\mathbf{a}_1 \ \mathbf{a}_2 \ \cdots \ \mathbf{a}_n]$ and $B = [\mathbf{b}_1 \ \mathbf{b}_2 \ \cdots \ \mathbf{b}_n]$ and in terms of their columns. It is enough to show that $\mathbf{a}_k = \mathbf{b}_k$ holds for all k. But we are assuming that $A\mathbf{e}_k = B\mathbf{e}_k$, which gives $\mathbf{a}_k = \mathbf{b}_k$ by Example 12.

We have introduced matrix-vector multiplication as a new way to think about systems of linear equations. But it has several other uses as well. It turns out that many geometric operations can be described using matrix multiplication, and we now investigate how this happens. As a bonus, this description provides a geometric "picture" of a matrix by revealing the effect on a vector when it is multiplied by A. This "geometric view" of matrices is a fundamental tool in understanding them.

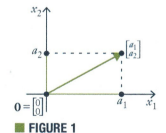
■ FIGURE 1

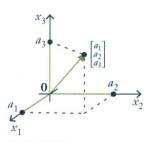
■ FIGURE 2

Transformations

The set $\mathbb{R}^2$ has a geometrical interpretation as the euclidean plane where a vector $\begin{bmatrix} a_1 \\ a_2 \end{bmatrix}$ in $\mathbb{R}^2$ represents the point (a_1, a_2) in the plane (see Figure 1). In this way we regard $\mathbb{R}^2$ as the set of all points in the plane. Accordingly, we will refer to vectors in $\mathbb{R}^2$ as points, and denote their coordinates as a column rather than a row. To enhance this geometrical interpretation of the vector $\begin{bmatrix} a_1 \\ a_2 \end{bmatrix}$, it is denoted graphically by an arrow from the origin $\begin{bmatrix} 0 \\ 0 \end{bmatrix}$ to the vector as in Figure 1.

Similarly we identify $\mathbb{R}^3$ with 3-dimensional space by writing a point (a_1, a_2, a_3) as the vector $\begin{bmatrix} a_1 \\ a_2 \\ a_3 \end{bmatrix}$ in $\mathbb{R}^3$, again represented by an arrow[4] from the origin to the point as in Figure 2. In this way the terms "point" and "vector" mean the same thing in the plane or in space.

We begin by describing a particular geometrical transformation of the plane $\mathbb{R}^2$.

---

4 This "arrow" representation of vectors in $\mathbb{R}^2$ and $\mathbb{R}^3$ will be used extensively in Chapter 4.

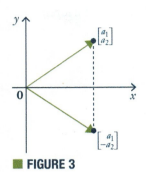

FIGURE 3

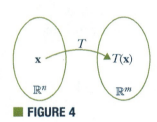

FIGURE 4

EXAMPLE 13

Consider the transformation of $\mathbb{R}^2$ given by *reflection* in the x axis. This operation carries the vector $\begin{bmatrix} a_1 \\ a_2 \end{bmatrix}$ to its reflection $\begin{bmatrix} a_1 \\ -a_2 \end{bmatrix}$ as in Figure 3. Now observe that

$$\begin{bmatrix} a_1 \\ -a_2 \end{bmatrix} = \begin{bmatrix} 1 & 0 \\ 0 & -1 \end{bmatrix} \begin{bmatrix} a_1 \\ a_2 \end{bmatrix}$$

so reflecting $\begin{bmatrix} a_1 \\ a_2 \end{bmatrix}$ in the x axis can be achieved by multiplying by the matrix $\begin{bmatrix} 1 & 0 \\ 0 & -1 \end{bmatrix}$.

If we write $A = \begin{bmatrix} 1 & 0 \\ 0 & -1 \end{bmatrix}$, Example 13 shows that reflection in the x axis carries each vector $\mathbf{x}$ in $\mathbb{R}^2$ to the vector $A\mathbf{x}$ in $\mathbb{R}^2$. It is thus an example of a function

$$T : \mathbb{R}^2 \to \mathbb{R}^2 \quad \text{where} \quad T(\mathbf{x}) = A\mathbf{x} \text{ for all } \mathbf{x} \text{ in } \mathbb{R}^2.$$

As such it is a generalization of the familiar functions $f : \mathbb{R} \to \mathbb{R}$ that carry a *number* x to another real *number* $f(x)$.

More generally, functions $T : \mathbb{R}^n \to \mathbb{R}^m$ are called **transformations** from $\mathbb{R}^n$ to $\mathbb{R}^m$. Such a transformation T is a rule that assigns to every vector $\mathbf{x}$ in $\mathbb{R}^n$ a uniquely determined vector $T(\mathbf{x})$ in $\mathbb{R}^m$ called the **image** of $\mathbf{x}$ under T. We denote this state of affairs by writing

$$T : \mathbb{R}^n \to \mathbb{R}^m \quad \text{or} \quad \mathbb{R}^n \xrightarrow{T} \mathbb{R}^m$$

The transformation T can be visualized as in Figure 4.

To describe a transformation $T : \mathbb{R}^n \to \mathbb{R}^m$ we must specify the vector $T(\mathbf{x})$ in $\mathbb{R}^m$ for every $\mathbf{x}$ in $\mathbb{R}^n$. This is referred to as **defining** T, or as specifying the **action** of T. Saying that the action *defines* the transformation means that we regard two transformations $S : \mathbb{R}^n \to \mathbb{R}^m$ and $T : \mathbb{R}^n \to \mathbb{R}^m$ as **equal** if they have the **same action**; more formally

$$S = T \quad \text{if and only if} \quad S(\mathbf{x}) = T(\mathbf{x}) \text{ for all } \mathbf{x} \text{ in } \mathbb{R}^n.$$

Again, this what we mean by $f = g$ where $f, g : \mathbb{R} \to \mathbb{R}$ are ordinary functions.

Functions $f : \mathbb{R} \to \mathbb{R}$ are often described by a formula, examples being $f(x) = x^2 + 1$ and $f(x) = \sin x$. The same is true of transformations; here is an example.

EXAMPLE 14

The formula $T \begin{bmatrix} x_1 \\ x_2 \\ x_3 \\ x_4 \end{bmatrix} = \begin{bmatrix} x_1 + x_2 \\ x_2 + x_3 \\ x_3 + x_4 \end{bmatrix}$ defines a transformation $\mathbb{R}^4 \to \mathbb{R}^3$.

Example 13 suggests that matrix multiplication is an important way of defining transformations $\mathbb{R}^n \to \mathbb{R}^m$. If A is any $m \times n$ matrix, multiplication by A gives a transformation

$$T_A : \mathbb{R}^n \to \mathbb{R}^m \quad \text{defined by} \quad T_A(\mathbf{x}) = A\mathbf{x} \text{ for every } \mathbf{x} \text{ in } \mathbb{R}^n.$$

SECTION 2.2 Equations, Matrices, and Transformations

Definition 2.8 T_A is called the **matrix transformation induced** by A.

Thus Example 13 shows that reflection in the x axis is the matrix transformation $\mathbb{R}^2 \to \mathbb{R}^2$ induced by the matrix $\begin{bmatrix} 1 & 0 \\ 0 & -1 \end{bmatrix}$. Also, the transformation $R : \mathbb{R}^4 \to \mathbb{R}^3$ in Example 13 is the matrix transformation induced by the matrix

$$A = \begin{bmatrix} 1 & 1 & 0 & 0 \\ 0 & 1 & 1 & 0 \\ 0 & 0 & 1 & 1 \end{bmatrix} \text{ because } \begin{bmatrix} 1 & 1 & 0 & 0 \\ 0 & 1 & 1 & 0 \\ 0 & 0 & 1 & 1 \end{bmatrix} \begin{bmatrix} x_1 \\ x_2 \\ x_3 \\ x_4 \end{bmatrix} = \begin{bmatrix} x_1 + x_2 \\ x_2 + x_3 \\ x_3 + x_4 \end{bmatrix}.$$

EXAMPLE 15

Let $R_{\frac{\pi}{2}} : \mathbb{R}^4 \to \mathbb{R}^3$ denote counterclockwise rotation about the origin through $\frac{\pi}{2}$ radians (that is, 90°).[5] Show that $R_{\frac{\pi}{2}}$ is induced by the matrix $\begin{bmatrix} 0 & -1 \\ 1 & 0 \end{bmatrix}$.

Solution ▶ The effect of $R_{\frac{\pi}{2}}$ is to rotate the vector $\mathbf{x} = \begin{bmatrix} a \\ b \end{bmatrix}$ counterclockwise through $\frac{\pi}{2}$ to produce the vector $R_{\frac{\pi}{2}}(\mathbf{x})$ shown in Figure 5. Since triangles **0px** and **0q**$R_{\frac{\pi}{2}}(\mathbf{x})$ are identical, we obtain $R_{\frac{\pi}{2}}(\mathbf{x}) = \begin{bmatrix} -b \\ a \end{bmatrix}$. But

$$\begin{bmatrix} -b \\ a \end{bmatrix} = \begin{bmatrix} 0 & -1 \\ 1 & 0 \end{bmatrix} \begin{bmatrix} a \\ b \end{bmatrix},$$ so we obtain $R_{\frac{\pi}{2}}(\mathbf{x}) = A\mathbf{x}$ for all $\mathbf{x}$ in $\mathbb{R}^2$ where $A = \begin{bmatrix} 0 & -1 \\ 1 & 0 \end{bmatrix}$.

In other words, $R_{\frac{\pi}{2}}$ is the matrix transformation induced by A.

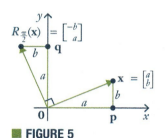

FIGURE 5

If A is the $m \times n$ zero matrix, then A induces the transformation

$$T : \mathbb{R}^n \to \mathbb{R}^m \quad \text{given by} \quad T(\mathbf{x}) = A\mathbf{x} = \mathbf{0} \text{ for all } \mathbf{x} \text{ in } \mathbb{R}^n.$$

This is called the **zero transformation**, and is denoted $T = 0$.

Another important example is the **identity transformation**

$$1_{\mathbb{R}^n} : \mathbb{R}^n \to \mathbb{R}^n \quad \text{given by} \quad 1_{\mathbb{R}^n}(\mathbf{x}) = \mathbf{x} \text{ for all } \mathbf{x} \text{ in } \mathbb{R}^n.$$

That is, the action of $1_{\mathbb{R}^n}$ on $\mathbf{x}$ is to do nothing to it. If I_n denotes the $n \times n$ identity matrix, we showed in Example 11 that $I_n \mathbf{x} = \mathbf{x}$ for all $\mathbf{x}$ in $\mathbb{R}^n$. Hence $1_{\mathbb{R}^n}(\mathbf{x}) = I_n \mathbf{x}$ for all $\mathbf{x}$ in $\mathbb{R}^n$; that is, the identity matrix I_n induces the identity transformation.

Here are two more examples of matrix transformations with a clear geometric description.

EXAMPLE 16

If $a > 0$, the matrix transformation $T \begin{bmatrix} x \\ y \end{bmatrix} = \begin{bmatrix} ax \\ y \end{bmatrix}$ induced by the matrix $A = \begin{bmatrix} a & 0 \\ 0 & 1 \end{bmatrix}$ is called an ***x*-expansion** of $\mathbb{R}^2$ if $a > 1$, and an ***x*-compression** if $0 < a < 1$. The reason for the names is clear in the diagram below. Similarly, if $b > 0$ the matrix $\begin{bmatrix} 1 & 0 \\ 0 & b \end{bmatrix}$ gives rise to ***y*-expansions** and ***y*-compressions**.

---

[5] *Radian measure* for angles is based on the fact that 360° equals 2π radians. Hence $\pi = 180°$ and $\frac{\pi}{2} = 90°$.

54 Chapter 2 Matrix Algebra

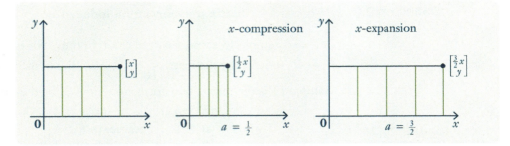

EXAMPLE 17

If a is a number, the matrix transformation $T\begin{bmatrix}x\\y\end{bmatrix} = \begin{bmatrix}x+ay\\y\end{bmatrix}$ induced by the matrix $A = \begin{bmatrix}1 & a\\0 & 1\end{bmatrix}$ is called an *x*-**shear** of $\mathbb{R}^2$ (**positive** if $a > 0$ and **negative** if $a < 0$). Its effect is illustrated below when $a = \frac{1}{4}$ and $a = -\frac{1}{4}$.

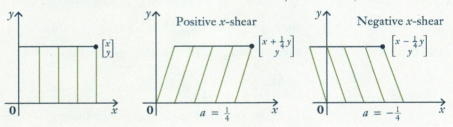

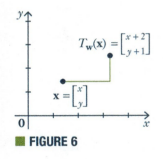

■ **FIGURE 6**

We hasten to note that there are important geometric transformations that are *not* matrix transformations. For example, if **w** is a fixed column in $\mathbb{R}^n$, define the transformation $T_\mathbf{w} : \mathbb{R}^n \to \mathbb{R}^n$ by

$$T_\mathbf{w}(\mathbf{x}) = \mathbf{x} + \mathbf{w} \quad \text{for all } \mathbf{x} \text{ in } \mathbb{R}^n.$$

Then $T_\mathbf{w}$ is called **translation** by **w**. In particular, if $\mathbf{w} = \begin{bmatrix}2\\1\end{bmatrix}$ in $\mathbb{R}^2$, the effect of $T_\mathbf{w}$ on $\begin{bmatrix}x\\y\end{bmatrix}$ is to translate it two units to the right and one unit up (see Figure 6).

The translation $T_\mathbf{w}$ is not a matrix transformation unless $\mathbf{w} = \mathbf{0}$. Indeed, if $T_\mathbf{w}$ were induced by a matrix A, then $A\mathbf{x} = T_\mathbf{w}(\mathbf{x}) = \mathbf{x} + \mathbf{w}$ would hold for every **x** in $\mathbb{R}^n$. In particular, taking $\mathbf{x} = \mathbf{0}$ gives $\mathbf{w} = A\mathbf{0} = \mathbf{0}$.

EXERCISES 2.2

1. In each case find a system of equations that is equivalent to the given vector equation. (Do not solve the system.)

   (a) $x_1\begin{bmatrix}2\\-3\\0\end{bmatrix} + x_2\begin{bmatrix}1\\1\\4\end{bmatrix} + x_3\begin{bmatrix}2\\0\\-1\end{bmatrix} = \begin{bmatrix}5\\6\\-3\end{bmatrix}$

   ◆(b) $x_1\begin{bmatrix}1\\0\\1\\0\end{bmatrix} + x_2\begin{bmatrix}-3\\8\\2\\1\end{bmatrix} + x_3\begin{bmatrix}-3\\0\\2\\2\end{bmatrix} + x_4\begin{bmatrix}3\\2\\0\\-2\end{bmatrix} = \begin{bmatrix}5\\1\\2\\0\end{bmatrix}$

2. In each case find a vector equation that is equivalent to the given system of equations. (Do not solve the equation.)

   (a) $\begin{aligned}x_1 - x_2 + 3x_3 &= 5\\-3x_1 + x_2 + x_3 &= -6\\5x_1 - 8x_2 &= 9\end{aligned}$

   ◆(b) $\begin{aligned}x_1 - 2x_2 - x_3 + x_4 &= 5\\-x_1 + x_3 - 2x_4 &= -3\\2x_1 - 2x_2 + 7x_3 &= 8\\3x_1 - 4x_2 + 9x_3 - 2x_4 &= 12\end{aligned}$

3. In each case compute $A\mathbf{x}$ using: (i) Definition 2.5. (ii) Theorem 4.

  (a) $A = \begin{bmatrix} 3 & -2 & 0 \\ 5 & -4 & 1 \end{bmatrix}$ and $\mathbf{x} = \begin{bmatrix} x_1 \\ x_2 \\ x_3 \end{bmatrix}$.

  ◆(b) $A = \begin{bmatrix} 1 & 2 & 3 \\ 0 & -4 & 5 \end{bmatrix}$ and $\mathbf{x} = \begin{bmatrix} x_1 \\ x_2 \\ x_3 \end{bmatrix}$.

  (c) $A = \begin{bmatrix} -2 & 0 & 5 & 4 \\ 1 & 2 & 0 & 3 \\ -5 & 6 & -7 & 8 \end{bmatrix}$ and $\mathbf{x} = \begin{bmatrix} x_1 \\ x_2 \\ x_3 \\ x_4 \end{bmatrix}$.

  ◆(d) $A = \begin{bmatrix} 3 & -4 & 1 & 6 \\ 0 & 2 & 1 & 5 \\ -8 & 7 & -3 & 0 \end{bmatrix}$ and $\mathbf{x} = \begin{bmatrix} x_1 \\ x_2 \\ x_3 \\ x_4 \end{bmatrix}$.

4. Let $A = [\mathbf{a}_1 \ \mathbf{a}_2 \ \mathbf{a}_3 \ \mathbf{a}_4]$ be the 3×4 matrix given in terms of its columns

   $\mathbf{a}_1 = \begin{bmatrix} 1 \\ 1 \\ -1 \end{bmatrix}$, $\mathbf{a}_2 = \begin{bmatrix} 3 \\ 0 \\ 2 \end{bmatrix}$, $\mathbf{a}_3 = \begin{bmatrix} 2 \\ -1 \\ 3 \end{bmatrix}$, and $\mathbf{a}_4 = \begin{bmatrix} 0 \\ -3 \\ 5 \end{bmatrix}$.

   In each case either express $\mathbf{b}$ as a linear combination of $\mathbf{a}_1$, $\mathbf{a}_2$, $\mathbf{a}_3$, and $\mathbf{a}_4$, or show that it is not such a linear combination. Explain what your answer means for the corresponding system $A\mathbf{x} = \mathbf{b}$ of linear equations.

   (a) $\mathbf{b} = \begin{bmatrix} 0 \\ 3 \\ 5 \end{bmatrix}$ (b) $\mathbf{b} = \begin{bmatrix} 4 \\ 1 \\ 1 \end{bmatrix}$

5. In each case, express every solution of the system as a sum of a specific solution plus a solution of the associated homogeneous system.

  (a) $\begin{array}{l} x + y + z = 2 \\ 2x + y = 3 \\ x - y - 3z = 0 \end{array}$ ◆(b) $\begin{array}{l} x - y - 4z = -4 \\ x + 2y + 5z = 2 \\ x + y + 2z = 0 \end{array}$

  (c) $\begin{array}{l} x_1 + x_2 - x_3 - 5x_5 = 2 \\ x_2 + x_3 - 4x_5 = -1 \\ x_2 + x_3 + x_4 - x_5 = -1 \\ 2x_1 - 4x_3 + x_4 + x_5 = 6 \end{array}$

  ◆(d) $\begin{array}{l} 2x_1 + x_2 - x_3 - x_4 = -1 \\ 3x_1 + x_2 + x_3 - 2x_4 = -2 \\ -x_1 - x_2 + 2x_3 + x_4 = 2 \\ -2x_1 - x_2 + 2x_4 = 3 \end{array}$

◆6. If $\mathbf{x}_0$ and $\mathbf{x}_1$ are solutions to the homogeneous system of equations $A\mathbf{x} = \mathbf{0}$, use Theorem 2 to show that $s\mathbf{x}_0 + t\mathbf{x}_1$ is also a solution for any scalars s and t (called a **linear combination** of $\mathbf{x}_0$ and $\mathbf{x}_1$).

7. Assume that $A \begin{bmatrix} 1 \\ -1 \\ 2 \end{bmatrix} = \mathbf{0} = A \begin{bmatrix} 2 \\ 0 \\ 3 \end{bmatrix}$. Show that $\mathbf{x}_0 = \begin{bmatrix} 2 \\ -1 \\ 3 \end{bmatrix}$ is a solution to $A\mathbf{x} = \mathbf{b}$. Find a two-parameter family of solutions to $A\mathbf{x} = \mathbf{b}$.

8. In each case write the system in the form $A\mathbf{x} = \mathbf{b}$, use the gaussian algorithm to solve the system, and express the solution as a particular solution plus a linear combination of basic solutions to the associated homogeneous system $A\mathbf{x} = \mathbf{0}$.

  (a) $\begin{array}{l} x_1 - 2x_2 + x_3 + 4x_4 - x_5 = 8 \\ -2x_1 + 4x_2 + x_3 - 2x_4 - 4x_5 = -1 \\ 3x_1 - 6x_2 + 8x_3 + 4x_4 - 13x_5 = 1 \\ 8x_1 - 16x_2 + 7x_3 + 12x_4 - 6x_5 = 11 \end{array}$

  ◆(b) $\begin{array}{l} x_1 - 2x_2 + x_3 + 2x_4 + 3x_5 = -4 \\ -3x_1 + 6x_2 - 2x_3 - 3x_4 - 11x_5 = 11 \\ -2x_1 + 4x_2 - x_3 + x_4 - 8x_5 = 7 \\ -x_1 + 2x_2 + 3x_4 - 5x_5 = 3 \end{array}$

9. Given vectors

   $\mathbf{a}_1 = \begin{bmatrix} 1 \\ 0 \\ 1 \end{bmatrix}$, $\mathbf{a}_2 = \begin{bmatrix} 1 \\ 1 \\ 0 \end{bmatrix}$, and $\mathbf{a}_3 = \begin{bmatrix} 0 \\ -1 \\ 1 \end{bmatrix}$, find a vector

   $\mathbf{b}$ that is *not* a linear combination of $\mathbf{a}_1$, $\mathbf{a}_2$, and $\mathbf{a}_3$. Justify your answer. [*Hint*: Part (2) of Theorem 1.]

10. In each case either show that the statement is true, or give an example showing that it is false.

  (a) $\begin{bmatrix} 3 \\ 2 \end{bmatrix}$ is a linear combination of $\begin{bmatrix} 1 \\ 0 \end{bmatrix}$ and $\begin{bmatrix} 0 \\ 1 \end{bmatrix}$.

  ◆(b) If $A\mathbf{x}$ has a zero entry, then A has a row of zeros.

  (c) If $A\mathbf{x} = \mathbf{0}$ where $\mathbf{x} \neq \mathbf{0}$, then $A = 0$.

  ◆(d) Every linear combination of vectors in $\mathbb{R}^n$ can be written in the form $A\mathbf{x}$.

  (e) If $A = [\mathbf{a}_1 \ \mathbf{a}_2 \ \mathbf{a}_3]$ in terms of its columns, and if $\mathbf{b} = 3\mathbf{a}_1 - 2\mathbf{a}_2$, then the system $A\mathbf{x} = \mathbf{b}$ has a solution.

  ◆(f) If $A = [\mathbf{a}_1 \ \mathbf{a}_2 \ \mathbf{a}_3]$ in terms of its columns, and if the system $A\mathbf{x} = \mathbf{b}$ has a solution, then $\mathbf{b} = s\mathbf{a}_1 + t\mathbf{a}_2$ for some s, t.

  (g) If A is $m \times n$ and $m < n$, then $A\mathbf{x} = \mathbf{b}$ has a solution for every column $\mathbf{b}$.

  ◆(h) If $A\mathbf{x} = \mathbf{b}$ has a solution for some column $\mathbf{b}$, then it has a solution for every column $\mathbf{b}$.

  (i) If $\mathbf{x}_1$ and $\mathbf{x}_2$ are solutions to $A\mathbf{x} = \mathbf{b}$, then $\mathbf{x}_1 - \mathbf{x}_2$ is a solution to $A\mathbf{x} = \mathbf{0}$.

(j) Let $A = [\mathbf{a}_1 \; \mathbf{a}_2 \; \mathbf{a}_3]$ in terms of its columns. If $\mathbf{a}_3 = s\mathbf{a}_1 + t\mathbf{a}_2$, then $A\mathbf{x} = \mathbf{0}$, where $\mathbf{x} = \begin{bmatrix} s \\ t \\ -1 \end{bmatrix}$.

11. Let $T: \mathbb{R}^2 \to \mathbb{R}^2$ be a transformation. In each case show that T is induced by a matrix and find the matrix.

    (a) T is a reflection in the y axis.

    ◆(b) T is a reflection in the line $y = x$.

    (c) T is a reflection in the line $y = -x$.

    ◆(d) T is a clockwise rotation through $\frac{\pi}{2}$.

12. The **projection** $P: \mathbb{R}^3 \to \mathbb{R}^2$ is defined by $P\begin{bmatrix} x \\ y \\ z \end{bmatrix} = \begin{bmatrix} x \\ y \end{bmatrix}$ for all $\begin{bmatrix} x \\ y \\ z \end{bmatrix}$ in $\mathbb{R}^3$. Show that P is induced by a matrix and find the matrix.

13. Let $T: \mathbb{R}^3 \to \mathbb{R}^3$ be a transformation. In each case show that T is induced by a matrix and find the matrix.

    (a) T is a reflection in the x-y plane.

    ◆(b) T is a reflection in the y-z plane.

14. Fix $a > 0$ in $\mathbb{R}$, and define $T_a: \mathbb{R}^4 \to \mathbb{R}^4$ by $T_a(\mathbf{x}) = a\mathbf{x}$ for all $\mathbf{x}$ in $\mathbb{R}^4$. Show that T is induced by a matrix and find the matrix. [T is called a **dilation** if $a > 1$ and a **contraction** if $a < 1$.]

15. Let A be $m \times n$ and let $\mathbf{x}$ be in $\mathbb{R}^n$. If A has a row of zeros, show that $A\mathbf{x}$ has a zero entry.

◆16. If a vector B is a linear combination of the columns of A, show that the system $A\mathbf{x} = \mathbf{b}$ is consistent (that is, it has at least one solution.)

17. If a system $A\mathbf{x} = \mathbf{b}$ is inconsistent (no solution), show that $\mathbf{b}$ is not a linear combination of the columns of A.

18. Let $\mathbf{x}_1$ and $\mathbf{x}_2$ be solutions to the homogeneous system $A\mathbf{x} = \mathbf{0}$.

    (a) Show that $\mathbf{x}_1 + \mathbf{x}_2$ is a solution to $A\mathbf{x} = \mathbf{0}$.

    ◆(b) Show that $t\mathbf{x}_1$ is a solution to $A\mathbf{x} = \mathbf{0}$ for any scalar t.

19. Suppose $\mathbf{x}_1$ is a solution to the system $A\mathbf{x} = \mathbf{b}$. If $\mathbf{x}_0$ is any nontrivial solution to the associated homogeneous system $A\mathbf{x} = \mathbf{0}$, show that $\mathbf{x}_1 + t\mathbf{x}_0$, t a scalar, is an infinite one parameter family of solutions to $A\mathbf{x} = \mathbf{b}$. [*Hint*: Example 7 Section 2.1.]

20. Let A and B be matrices of the same size. If $\mathbf{x}$ is a solution to both the system $A\mathbf{x} = \mathbf{0}$ and the system $B\mathbf{x} = \mathbf{0}$, show that $\mathbf{x}$ is a solution to the system $(A + B)\mathbf{x} = \mathbf{0}$.

21. If A is $m \times n$ and $A\mathbf{x} = \mathbf{0}$ for every $\mathbf{x}$ in $\mathbb{R}^n$, show that $A = 0$ is the zero matrix. [*Hint*: Consider $A\mathbf{e}_j$ where $\mathbf{e}_j$ is the jth column of I_n; that is, $\mathbf{e}_j$ is the vector in $\mathbb{R}^n$ with 1 as entry j and every other entry 0.]

◆22. Prove part (1) of Theorem 2.

23. Prove part (2) of Theorem 2.

SECTION 2.3 Matrix Multiplication

In Section 2.2 matrix-vector products were introduced. If A is an $m \times n$ matrix, the product $A\mathbf{x}$ was defined for any n-column $\mathbf{x}$ in $\mathbb{R}^n$ as follows: If $A = [\mathbf{a}_1 \; \mathbf{a}_2 \; \cdots \; \mathbf{a}_n]$ where the A_j are the columns of A, and if $\mathbf{x} = \begin{bmatrix} x_1 \\ x_2 \\ \vdots \\ x_n \end{bmatrix}$, Definition 2.5 reads

$$A\mathbf{x} = x_1\mathbf{a}_1 + x_2\mathbf{a}_2 + \cdots + x_n\mathbf{a}_n \qquad (*)$$

This was motivated as a way of describing systems of linear equations with coefficient matrix A. Indeed every such system has the form $A\mathbf{x} = \mathbf{b}$ where $\mathbf{b}$ is the column of constants.

In this section we extend this matrix-vector multiplication to a way of multiplying matrices in general, and then investigate matrix algebra for its own sake. While it shares several properties of ordinary arithmetic, it will soon become clear that matrix arithmetic is different in a number of ways.

Matrix multiplication is closely related to composition of transformations.

Composition and Matrix Multiplication

Sometimes two transformations "link" together as follows:

$$\mathbb{R}^k \xrightarrow{T} \mathbb{R}^n \xrightarrow{S} \mathbb{R}^m.$$

In this case we can apply T first and then apply S, and the result is a new transformation

$$S \circ T : \mathbb{R}^k \to \mathbb{R}^m,$$

called the **composite** of S and T, defined by

$$(S \circ T)(\mathbf{x}) = S[T(\mathbf{x})] \quad \text{for all } \mathbf{x} \text{ in } \mathbb{R}^k.$$

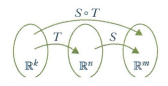

The action of $S \circ T$ can be described as "first T then S" (note the order!)[6]. This new transformation is described in the diagram. The reader will have encountered composition of ordinary functions: For example, consider $\mathbb{R} \xrightarrow{g} \mathbb{R} \xrightarrow{f} \mathbb{R}$ where $f(x) = x^2$ and $g(x) = x + 1$ for all x in $\mathbb{R}$. Then

$$(f \circ g)(x) = f[g(x)] = f(x + 1) = (x + 1)^2$$
$$(g \circ f)(x) = g[f(x)] = g(x^2) = x^2 + 1.$$

for all x in $\mathbb{R}$.

Our concern here is with matrix transformations. Suppose that A is an $m \times n$ matrix and B is an $n \times k$ matrix, and let $\mathbb{R}^k \xrightarrow{T_B} \mathbb{R}^n \xrightarrow{T_A} \mathbb{R}^m$ be the matrix transformations induced by B and A respectively, that is:

$$T_A(\mathbf{x}) = B\mathbf{x} \text{ for all } \mathbf{x} \text{ in } \mathbb{R}^k \quad \text{and} \quad T_A(\mathbf{y}) = B\mathbf{y} \text{ for all } \mathbf{y} \text{ in } \mathbb{R}^n.$$

Write $B = [\mathbf{b}_1 \ \mathbf{b}_2 \ \cdots \ \mathbf{b}_k]$ where $\mathbf{b}_j$ denotes column j of B for each j. Hence each $\mathbf{b}_j$ is an n-vector (B is $n \times k$) so we can form the matrix-vector product $A\mathbf{b}_j$. In particular, we obtain an $m \times k$ matrix

$$[A\mathbf{b}_1 \ A\mathbf{b}_1 \ \cdots \ A\mathbf{b}_k]$$

with columns $A\mathbf{b}_1, A\mathbf{b}_2, \cdots, A\mathbf{b}_k$. Now compute $(T_A \circ T_B)(\mathbf{x})$ for any $\mathbf{x} = \begin{bmatrix} x_1 \\ x_2 \\ \vdots \\ x_k \end{bmatrix}$ in $\mathbb{R}^k$:

$$
\begin{aligned}
(T_A \circ T_B)(\mathbf{x}) &= T_A[T_B(\mathbf{x})] && \text{Definition of } T_A \circ T_B \\
&= A(B\mathbf{x}) && A \text{ and } B \text{ induce } T_A \text{ and } T_B \\
&= A(x_1\mathbf{b}_1 + x_2\mathbf{b}_2 + \cdots + x_k\mathbf{b}_k) && \text{Equation } (*) \text{ above} \\
&= A(x_1\mathbf{b}_1) + A(x_2\mathbf{b}_2) + \cdots + A(x_k\mathbf{b}_k) && \text{Theorem 2, Section 2.2} \\
&= x_1(A\mathbf{b}_1) + x_2(A\mathbf{b}_2) + \cdots + x_k(A\mathbf{b}_k) && \text{Theorem 2, Section 2.2} \\
&= [A\mathbf{b}_1 \ A\mathbf{b}_2 \ \cdots \ A\mathbf{b}_k]\mathbf{x}. && \text{Equation } (*) \text{ above}
\end{aligned}
$$

Because $\mathbf{x}$ was an arbitrary vector in $\mathbb{R}^n$, this shows that $T_A \circ T_B$ is the matrix transformation induced by the matrix $[A\mathbf{b}_1, A\mathbf{b}_2, \cdots, A\mathbf{b}_n]$. This motivates the following definition.

Definition 2.9

Matrix Multiplication

Let A be an $m \times n$ matrix, let B be an $n \times k$ matrix, and write $B = [\mathbf{b}_1 \ \mathbf{b}_2 \ \cdots \ \mathbf{b}_k]$ where $\mathbf{b}_j$ is column j of B for each j. The product matrix AB is the $m \times k$ matrix defined as follows:

$$AB = A[\mathbf{b}_1, \mathbf{b}_2, \cdots, \mathbf{b}_k] = [A\mathbf{b}_1, A\mathbf{b}_2, \cdots, A\mathbf{b}_k]$$

---

[6] When reading the notation $S \circ T$, we read S first and then T even though the action is "first T then S". This annoying state of affairs results because we write $T(\mathbf{x})$ for the effect of the transformation T on $\mathbf{x}$, with T on the left. If we wrote this instead as $(\mathbf{x})T$, the confusion would not occur. However the notation $T(\mathbf{x})$ is well established.

Thus the product matrix AB is given in terms of its columns $A\mathbf{b}_1, A\mathbf{b}_2, \ldots, A\mathbf{b}_n$: Column j of AB is the matrix-vector product $A\mathbf{b}_j$ of A and the corresponding column $\mathbf{b}_j$ of B. Note that each such product $A\mathbf{b}_j$ makes sense by Definition 2.5 because A is $m \times n$ and each $\mathbf{b}_j$ is in $\mathbb{R}^n$ (since B has n rows). Note also that if B is a column matrix, this definition reduces to Definition 2.5 for matrix-vector multiplication.

Given matrices A and B, Definition 2.9 and the above computation give

$$A(B\mathbf{x}) = [A\mathbf{b}_1 \ A\mathbf{b}_2 \ \cdots \ A\mathbf{b}_n]\mathbf{x} = (AB)\mathbf{x}$$

for all $\mathbf{x}$ in $\mathbb{R}^k$. We record this for reference.

Theorem 1

Let A be an $m \times n$ matrix and let B be an $n \times k$ matrix. Then the product matrix AB is $m \times k$ and satisfies

$$A(B\mathbf{x}) = (AB)\mathbf{x} \quad \text{for all } \mathbf{x} \text{ in } \mathbb{R}^k.$$

Here is an example of how to compute the product AB of two matrices using Definition 2.9.

EXAMPLE 1

Compute AB if $A = \begin{bmatrix} 2 & 3 & 5 \\ 1 & 4 & 7 \\ 0 & 1 & 8 \end{bmatrix}$ and $B = \begin{bmatrix} 8 & 9 \\ 7 & 2 \\ 6 & 1 \end{bmatrix}$.

Solution ▶ The columns of B are $\mathbf{b}_1 = \begin{bmatrix} 8 \\ 7 \\ 6 \end{bmatrix}$ and $\mathbf{b}_2 = \begin{bmatrix} 9 \\ 2 \\ 1 \end{bmatrix}$, so Definition 2.5 gives

$$A\mathbf{b}_1 = \begin{bmatrix} 2 & 3 & 5 \\ 1 & 4 & 7 \\ 0 & 1 & 8 \end{bmatrix}\begin{bmatrix} 8 \\ 7 \\ 6 \end{bmatrix} = \begin{bmatrix} 67 \\ 78 \\ 55 \end{bmatrix} \text{ and } A\mathbf{b}_2 = \begin{bmatrix} 2 & 3 & 5 \\ 1 & 4 & 7 \\ 0 & 1 & 8 \end{bmatrix}\begin{bmatrix} 9 \\ 2 \\ 1 \end{bmatrix} = \begin{bmatrix} 29 \\ 24 \\ 10 \end{bmatrix}.$$

Hence Definition 2.9 above gives $AB = [A\mathbf{b}_1 \ A\mathbf{b}_2] = \begin{bmatrix} 67 & 29 \\ 78 & 24 \\ 55 & 10 \end{bmatrix}$.

While Definition 2.9 is important, there is another way to compute the matrix product AB that gives a way to calculate each individual entry. In Section 2.2 we defined the dot product of two n-tuples to be the sum of the products of corresponding entries. We went on to show (Theorem 4 Section 2.2) that if A is an $m \times n$ matrix and $\mathbf{x}$ is an n-vector, then entry j of the product $A\mathbf{x}$ is the dot product of row j of A with $\mathbf{x}$. This observation was called the "dot product rule" for matrix-vector multiplication, and the next theorem shows that it extends to matrix multiplication in general.

SECTION 2.3 Matrix Multiplication

Theorem 2

Dot Product Rule
Let A and B be matrices of sizes $m \times n$ and $n \times k$, respectively. Then the (i, j)-entry of AB is the dot product of row i of A with column j of B.

PROOF

Write $B = [\mathbf{b}_1 \ \mathbf{b}_2 \ \cdots \ \mathbf{b}_n]$ in terms of its columns. Then $A\mathbf{b}_j$ is column j of AB for each j. Hence the (i, j)-entry of AB is entry i of $A\mathbf{b}_j$, which is the dot product of row i of A with $\mathbf{b}_j$. This proves the theorem.

Thus to compute the (i, j)-entry of AB, proceed as follows (see the diagram):

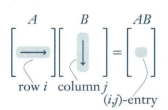

Go *across* row i of A, and *down* column j of B,
multiply corresponding entries, and add the results.

Note that this requires that the rows of A must be the same length as the columns of B. The following rule is useful for remembering this and for deciding the size of the product matrix AB.

Compatibility Rule

Let A and B denote matrices. If A is $m \times n$ and B is $n' \times k$, the product AB can be formed if and only if $n = n'$. In this case the size of the product matrix AB is $m \times k$, and we say that AB is **defined**, or that A and B are **compatible** for multiplication.

The diagram provides a useful mnemonic for remembering this. We adopt the following convention:

Convention

Whenever a product of matrices is written, it is tacitly assumed that the sizes of the factors are such that the product is defined.

To illustrate the dot product rule, we recompute the matrix product in Example 1.

EXAMPLE 2

Compute AB if $A = \begin{bmatrix} 2 & 3 & 5 \\ 1 & 4 & 7 \\ 0 & 1 & 8 \end{bmatrix}$ and $B = \begin{bmatrix} 8 & 9 \\ 7 & 2 \\ 6 & 1 \end{bmatrix}$.

Solution ▶ Here A is 3×3 and B is 3×2, so the product matrix AB is defined and will be of size 3×2. Theorem 2 gives each entry of AB as the dot product of the corresponding row of A with the corresponding column of B; that is,

$$AB = \begin{bmatrix} 2 & 3 & 5 \\ 1 & 4 & 7 \\ 0 & 1 & 8 \end{bmatrix} \begin{bmatrix} 8 & 9 \\ 7 & 2 \\ 6 & 1 \end{bmatrix} = \begin{bmatrix} 2\cdot 8 + 3\cdot 7 + 5\cdot 6 & 2\cdot 9 + 3\cdot 2 + 5\cdot 1 \\ 1\cdot 8 + 4\cdot 7 + 7\cdot 6 & 1\cdot 9 + 4\cdot 2 + 7\cdot 1 \\ 0\cdot 8 + 1\cdot 7 + 8\cdot 6 & 0\cdot 9 + 1\cdot 2 + 8\cdot 1 \end{bmatrix} = \begin{bmatrix} 67 & 29 \\ 78 & 24 \\ 55 & 10 \end{bmatrix}.$$

Of course, this agrees with Example 1.

EXAMPLE 3

Compute the (1, 3)- and (2, 4)-entries of AB where

$$A = \begin{bmatrix} 3 & -1 & 2 \\ 0 & 1 & 4 \end{bmatrix} \text{ and } B = \begin{bmatrix} 2 & 1 & 6 & 0 \\ 0 & 2 & 3 & 4 \\ -1 & 0 & 5 & 8 \end{bmatrix}.$$

Then compute AB.

Solution ▶ The (1, 3)-entry of AB is the dot product of row 1 of A and column 3 of B (highlighted in the following display), computed by multiplying corresponding entries and adding the results.

$$\begin{bmatrix} 3 & -1 & 2 \\ 0 & 1 & 4 \end{bmatrix} \begin{bmatrix} 2 & 1 & 6 & 0 \\ 0 & 2 & 3 & 4 \\ -1 & 0 & 5 & 8 \end{bmatrix} \quad (1, 3)\text{-entry} = 3 \cdot 6 + (-1) \cdot 3 + 2 \cdot 5 = 25$$

Similarly, the (2, 4) entry of AB involves row 2 of A and column 4 of B.

$$\begin{bmatrix} 3 & -1 & 2 \\ 0 & 1 & 4 \end{bmatrix} \begin{bmatrix} 2 & 1 & 6 & 0 \\ 0 & 2 & 3 & 4 \\ -1 & 0 & 5 & 8 \end{bmatrix} \quad (2, 4)\text{-entry} = 0 \cdot 0 + 1 \cdot 4 + 4 \cdot 8 = 36$$

Since A is 2×3 and B is 3×4, the product is 2×4.

$$AB = \begin{bmatrix} 3 & -1 & 2 \\ 0 & 1 & 4 \end{bmatrix} \begin{bmatrix} 2 & 1 & 6 & 0 \\ 0 & 2 & 3 & 4 \\ -1 & 0 & 5 & 8 \end{bmatrix} = \begin{bmatrix} 4 & 1 & 25 & 12 \\ -4 & 2 & 23 & 36 \end{bmatrix}$$

EXAMPLE 4

If $A = [1 \ 3 \ 2]$ and $B = \begin{bmatrix} 5 \\ 6 \\ 4 \end{bmatrix}$, compute A^2, AB, BA, and B^2 when they are defined.[7]

Solution ▶ Here, A is a 1×3 matrix and B is a 3×1 matrix, so A^2 and B^2 are not defined. However, the rule reads

$$\begin{matrix} A & B \\ 1 \times 3 & 3 \times 1 \end{matrix} \quad \text{and} \quad \begin{matrix} B & A \\ 3 \times 1 & 1 \times 3 \end{matrix}$$

so both AB and BA can be formed and these are 1×1 and 3×3 matrices, respectively.

$$AB = [1 \ 3 \ 2] \begin{bmatrix} 5 \\ 6 \\ 4 \end{bmatrix} = [1 \cdot 5 + 3 \cdot 6 + 2 \cdot 4] = [31]$$

$$BA = \begin{bmatrix} 5 \\ 6 \\ 4 \end{bmatrix} [1 \ 3 \ 2] = \begin{bmatrix} 5 \cdot 1 & 5 \cdot 3 & 5 \cdot 2 \\ 6 \cdot 1 & 6 \cdot 3 & 6 \cdot 2 \\ 4 \cdot 1 & 4 \cdot 3 & 4 \cdot 2 \end{bmatrix} = \begin{bmatrix} 5 & 15 & 10 \\ 6 & 18 & 12 \\ 4 & 12 & 8 \end{bmatrix}$$

Unlike numerical multiplication, matrix products AB and BA *need not be equal*. In fact they need not even be the same size, as Example 4 shows. It turns out to be rare that $AB = BA$ (although it is by no means impossible), and A and B are said to **commute** when this happens.

---

[7] As for numbers, we write $A^2 = A \cdot A$, $A^3 = A \cdot A \cdot A$, etc. Note that A^2 is defined if and only if A is of size $n \times n$ for some n.

SECTION 2.3 Matrix Multiplication

EXAMPLE 5

Let $A = \begin{bmatrix} 6 & 9 \\ -4 & -6 \end{bmatrix}$ and $B = \begin{bmatrix} 1 & 2 \\ -1 & 0 \end{bmatrix}$. Compute A^2, AB, BA.

Solution ▶ $A^2 = \begin{bmatrix} 6 & 9 \\ -4 & -6 \end{bmatrix}\begin{bmatrix} 6 & 9 \\ -4 & -6 \end{bmatrix} = \begin{bmatrix} 0 & 0 \\ 0 & 0 \end{bmatrix}$, so $A^2 = 0$ can occur even if $A \neq 0$. Next,

$$AB = \begin{bmatrix} 6 & 9 \\ -4 & -6 \end{bmatrix}\begin{bmatrix} 1 & 2 \\ -1 & 0 \end{bmatrix} = \begin{bmatrix} -3 & 12 \\ 2 & -8 \end{bmatrix}$$

$$BA = \begin{bmatrix} 1 & 2 \\ -1 & 0 \end{bmatrix}\begin{bmatrix} 6 & 9 \\ -4 & -6 \end{bmatrix} = \begin{bmatrix} -2 & -3 \\ -6 & -9 \end{bmatrix}$$

Hence $AB \neq BA$, even though AB and BA are the same size.

EXAMPLE 6

If A is any matrix, then $IA = A$ and $AI = A$, and where I denotes an identity matrix of a size so that the multiplications are defined.

Solution ▶ These both follow from the dot product rule as the reader should verify. For a more formal proof, write $A = [\mathbf{a}_1 \ \mathbf{a}_2 \ \cdots \ \mathbf{a}_n]$ where $\mathbf{a}_j$ is column j of A. Then Definition 2.9 and Example 11 Section 2.2 give

$$IA = [I\mathbf{a}_1 \ I\mathbf{a}_2 \ \cdots \ I\mathbf{a}_n] = [\mathbf{a}_1 \ \mathbf{a}_2 \ \cdots \ \mathbf{a}_n] = A$$

If $\mathbf{e}_j$ denotes column j of i, then $A\mathbf{e}_j = \mathbf{a}_j$ for each j by Example 12 Section 2.2. Hence Definition 2.9 gives:

$$AI = A[\mathbf{e}_1 \ \mathbf{e}_2 \ \cdots \ \mathbf{e}_n] = [A\mathbf{e}_1 \ A\mathbf{e}_2 \ \cdots \ A\mathbf{e}_n] = [\mathbf{a}_1 \ \mathbf{a}_2 \ \cdots \ \mathbf{a}_n] = A$$

The following theorem collects several results about matrix multiplication that are used everywhere in linear algebra.

Theorem 3

Assume that a is any scalar, and that A, B, and C are matrices of sizes such that the indicated matrix products are defined. Then:
1. $IA = A$ and $AI = A$ where I denotes an identity matrix.
2. $A(BC) = (AB)C$.
3. $A(B + C) = AB + AC$.
4. $(B + C)A = BA + CA$.
5. $a(AB) = (aA)B = A(aB)$.
6. $(AB)^T = B^T A^T$.

> **PROOF**
>
> (1) is Example 6; we prove (2), (4), and (6) and leave (3) and (5) as exercises.
>
> (2) If $C = [\mathbf{c}_1 \ \mathbf{c}_2 \ \cdots \ \mathbf{c}_k]$ in terms of its columns, then $BC = [B\mathbf{c}_1 \ B\mathbf{c}_2 \ \cdots \ B\mathbf{c}_k]$ by Definition 2.9, so
>
> $$\begin{aligned} A(BC) &= [A(B\mathbf{c}_1) \ A(B\mathbf{c}_2) \ \cdots \ A(B\mathbf{c}_k)] & \text{Definition 2.9} \\ &= [(AB)\mathbf{c}_1 \ (AB)\mathbf{c}_2 \ \cdots \ (AB)\mathbf{c}_k] & \text{Theorem 1} \\ &= (AB)C & \text{Definition 2.9} \end{aligned}$$
>
> (4) We know (Theorem 2 Section 2.2) that $(B + C)\mathbf{x} = B\mathbf{x} + C\mathbf{x}$ holds for every column $\mathbf{x}$. If we write $A = [\mathbf{a}_1 \ \mathbf{a}_2 \ \cdots \ \mathbf{a}_n]$ in terms of its columns, we get
>
> $$\begin{aligned} (B+C)A &= [(B+C)\mathbf{a}_1 \ (B+C)\mathbf{a}_2 \ \cdots \ (B+C)\mathbf{a}_n] & \text{Definition 2.9} \\ &= [B\mathbf{a}_1 + C\mathbf{a}_1 \ B\mathbf{a}_2 + C\mathbf{a}_2 \ \cdots \ B\mathbf{a}_n + C\mathbf{a}_n] & \text{Theorem 2 Section 2.2} \\ &= [B\mathbf{a}_1 \ B\mathbf{a}_2 \ \cdots \ B\mathbf{a}_n] + [C\mathbf{a}_1 \ C\mathbf{a}_2 \ \cdots \ C\mathbf{a}_n] & \text{Adding Columns} \\ &= BA + CA & \text{Definition 2.9} \end{aligned}$$
>
> (6) As in Section 2.1, write $A = [a_{ij}]$ and $B = [b_{ij}]$, so that $A^T = [a'_{ij}]$ and $B^T = [b'_{ij}]$ where $a'_{ij} = a_{ji}$ and $b'_{ji} = b_{ij}$ for all i and j. If c_{ij} denotes the (i, j)-entry of $B^T A^T$, then c_{ij} is the dot product of row i of B^T with column j of A^T. Since row i of B^T is $[b'_{i1} \ b'_{i2} \ b'_{im}]$ and column j of A^T is $[a'_{i1} \ a'_{i2} \ a'_{mj}]$, we obtain
>
> $$\begin{aligned} c_{ij} &= b'_{i1}a'_{1j} + b'_{i2}a'_{2j} + \cdots + b'_{im}a'_{mj} \\ &= b_{1i}a_{j1} + b_{2i}a_{j2} + \cdots + b_{mi}a_{jm} \\ &= a_{j1}b_{1i} + a_{j2}b_{2i} + \cdots + a_{jm}b_{mi}. \end{aligned}$$
>
> But this is the dot product of row j of A with column i of B; that is, the (j, i)-entry of AB; that is, the (i, j)-entry of $(AB)^T$. This proves (6).

Property 2 in Theorem 3 is called the **associative law** of matrix multiplication. It asserts that $A(BC) = (AB)C$ holds for all matrices (if the products are defined). Hence this product is the same no matter how it is formed, and so is written simply as ABC. This extends: The product $ABCD$ of four matrices can be formed several ways—for example, $(AB)(CD)$, $[A(BC)]D$, and $A[B(CD)]$—but the associative law implies that they are all equal and so are written as $ABCD$. A similar remark applies in general: Matrix products can be written unambiguously with no parentheses.

However, a note of caution about matrix multiplication must be taken: The fact that AB and BA need *not* be equal means that the *order* of the factors is important in a product of matrices. For example $ABCD$ and $ADCB$ may *not* be equal.

Warning

If the order of the factors in a product of matrices is changed, the product matrix may change (or may not be defined). Ignoring this warning is a source of many errors by students of linear algebra!

Properties 3 and 4 in Theorem 3 are called **distributive laws**. They assert that $A(B + C) = AB + AC$ and $(B + C)A = BA + CA$ hold whenever the sums and products are defined. These rules extend to more than two terms and, together with Property 5, ensure that many manipulations familiar from ordinary algebra extend to matrices. For example

$$A(2B - 3C + D - 5E) = 2AB - 3AC + AD - 5AE$$
$$(A + 3C - 2D)B = AB + 3CB - 2DB$$

SECTION 2.3 Matrix Multiplication

Note again that the warning is in effect: For example $A(B - C)$ need *not* equal $AB - CA$. These rules make possible a lot of simplification of matrix expressions.

EXAMPLE 7

Simplify the expression $A(BC - CD) + A(C - B)D - AB(C - D)$.

Solution ▶ $A(BC - CD) + A(C - B)D - AB(C - D)$
$= A(BC) - A(CD) + (AC - AB)D - (AB)C + (AB)D$
$= ABC - ACD + ACD - ABD - ABC + ABD$
$= 0$.

Examples 8 and 9 below show how we can use the properties in Theorem 2 to deduce other facts about matrix multiplication. Matrices A and B are said to **commute** if $AB = BA$.

EXAMPLE 8

Suppose that A, B, and C are $n \times n$ matrices and that both A and B commute with C; that is, $AC = CA$ and $BC = CB$. Show that AB commutes with C.

Solution ▶ Showing that AB commutes with C means verifying that $(AB)C = C(AB)$. The computation uses the associative law several times, as well as the given facts that $AC = CA$ and $BC = CB$.

$$(AB)C = A(BC) = A(CB) = (AC)B = (CA)B = C(AB)$$

EXAMPLE 9

Show that $AB = BA$ if and only if $(A - B)(A + B) = A^2 - B^2$.

Solution ▶ The following always holds:

$$(A - B)(A + B) = A(A + B) - B(A + B) = A^2 + AB - BA - B^2 \quad (*)$$

Hence if $AB = BA$, then $(A - B)(A + B) = A^2 - B^2$ follows. Conversely, if this last equation holds, then equation $(*)$ becomes

$$A^2 - B^2 = A^2 + AB - BA - B^2$$

This gives $0 = AB - BA$, and $AB = BA$ follows.

In Section 2.2 we saw (in Theorem 1) that every system of linear equations has the form

$$A\mathbf{x} = \mathbf{b}$$

where A is the coefficient matrix, $\mathbf{x}$ is the column of variables, and $\mathbf{b}$ is the constant matrix. Thus the *system* of linear equations becomes a single matrix equation. Matrix multiplication can yield information about such a system.

EXAMPLE 10

Consider a system $A\mathbf{x} = \mathbf{b}$ of linear equations where A is an $m \times n$ matrix. Assume that a matrix C exists such that $CA = I_n$. If the system $A\mathbf{x} = \mathbf{b}$ *has* a solution, show that this solution must be $C\mathbf{b}$. Give a condition guaranteeing that $C\mathbf{b}$ *is in fact* a solution.

Solution ▶ Suppose that $\mathbf{x}$ is any solution to the system, so that $A\mathbf{x} = \mathbf{b}$. Multiply both sides of this matrix equation by C to obtain, successively,

$$C(A\mathbf{x}) = C\mathbf{b}, \quad (CA)\mathbf{x} = C\mathbf{b}, \quad I_n\mathbf{x} = C\mathbf{b}, \quad \mathbf{x} = C\mathbf{b}$$

This shows that *if* the system has a solution $\mathbf{x}$, then that solution must be $\mathbf{x} = C\mathbf{b}$, as required. But it does *not* guarantee that the system *has* a solution. However, if we write $\mathbf{x}_1 = C\mathbf{b}$, then

$$A\mathbf{x}_1 = A(C\mathbf{b}) = (AC)\mathbf{b}.$$

Thus $\mathbf{x}_1 = C\mathbf{b}$ will be a solution if the condition $AC = I_m$ is satisfied.

The ideas in Example 10 lead to important information about matrices; this will be pursued in the next section.

Block Multiplication

Definition 2.10 *It is often useful to consider matrices whose entries are themselves matrices (called* **blocks**). *A matrix viewed in this way is said to be* **partitioned into blocks**.

For example, writing a matrix B in the form

$$B = [\mathbf{b}_1 \; \mathbf{b}_2 \; \cdots \; \mathbf{b}_k] \text{ where the } \mathbf{b}_j \text{ are the columns of } B$$

is such a block partition of B. Here is another example.

Consider the matrices

$$A = \begin{bmatrix} 1 & 0 & 0 & 0 & 0 \\ 0 & 1 & 0 & 0 & 0 \\ 2 & -1 & 4 & 2 & 1 \\ 3 & 1 & -1 & 7 & 5 \end{bmatrix} = \begin{bmatrix} I_2 & 0_{23} \\ P & Q \end{bmatrix} \quad \text{and} \quad B = \begin{bmatrix} 4 & -2 \\ 5 & 6 \\ 7 & 3 \\ -1 & 0 \\ 1 & 6 \end{bmatrix} = \begin{bmatrix} X \\ Y \end{bmatrix}$$

where the blocks have been labelled as indicated. This is a natural way to partition A into blocks in view of the blocks I_2 and 0_{23} that occur. This notation is particularly useful when we are multiplying the matrices A and B because the product AB can be computed in block form as follows:

$$AB = \begin{bmatrix} I & 0 \\ P & Q \end{bmatrix} \begin{bmatrix} X \\ Y \end{bmatrix} = \begin{bmatrix} IX + 0Y \\ PX + QY \end{bmatrix} = \begin{bmatrix} X \\ PX + QY \end{bmatrix} = \begin{bmatrix} 4 & -2 \\ 5 & 6 \\ 30 & 8 \\ 8 & 27 \end{bmatrix}$$

This is easily checked to be the product AB, computed in the conventional manner.

In other words, *we can compute the product AB by ordinary matrix multiplication, using blocks as entries*. The only requirement is that the blocks be **compatible**. That is, *the sizes of the blocks must be such that all (matrix) products of blocks that occur make sense*. This means that the number of columns in each block of A must equal the number of rows in the corresponding block of B.

SECTION 2.3 Matrix Multiplication

> **Theorem 4**
>
> **Block Multiplication**
> If matrices A and B are partitioned compatibly into blocks, the product AB can be computed by matrix multiplication using blocks as entries.

We omit the proof.

We have been using two cases of block multiplication. If $B = [\mathbf{b}_1 \ \mathbf{b}_2 \ \cdots \ \mathbf{b}_k]$ is a matrix where the $\mathbf{b}_j$ are the columns of B, and if the matrix product AB is defined, then we have

$$AB = A[\mathbf{b}_1 \ \mathbf{b}_2 \ \cdots \ \mathbf{b}_k] = [A\mathbf{b}_1 \ A\mathbf{b}_2 \ \cdots \ A\mathbf{b}_k].$$

This is Definition 2.9 and is a block multiplication where $A = [A]$ has only one block. As another illustration,

$$B\mathbf{x} = [\mathbf{b}_1 \ \mathbf{b}_2 \ \cdots \ \mathbf{b}_k]\begin{bmatrix} x_1 \\ x_2 \\ \vdots \\ x_k \end{bmatrix} = x_1\mathbf{b}_1 + x_2\mathbf{b}_2 + \cdots + x_k\mathbf{b}_k].$$

where $\mathbf{x}$ is any $k \times 1$ column matrix (this is Definition 2.5).

It is not our intention to pursue block multiplication in detail here. However, we give one more example because it will be used below.

> **Theorem 5**
>
> Suppose matrices $A = \begin{bmatrix} B & X \\ 0 & C \end{bmatrix}$ and $A_1 = \begin{bmatrix} B_1 & X_1 \\ 0 & C_1 \end{bmatrix}$ are partitioned as shown where B and B_1 are square matrices of the same size, and C and C_1 are also square of the same size. These are compatible partitionings and block multiplication gives
>
> $$AA_1 = \begin{bmatrix} B & X \\ 0 & C \end{bmatrix}\begin{bmatrix} B_1 & X_1 \\ 0 & C_1 \end{bmatrix} = \begin{bmatrix} BB_1 & BX_1 + XC_1 \\ 0 & CC_1 \end{bmatrix}.$$

> **EXAMPLE 11**
>
> Obtain a formula for A^k where $A = \begin{bmatrix} I & X \\ 0 & 0 \end{bmatrix}$ is square and I is an identity matrix.
>
> **Solution ▶** We have $A^2 = \begin{bmatrix} I & X \\ 0 & 0 \end{bmatrix}\begin{bmatrix} I & X \\ 0 & 0 \end{bmatrix} = \begin{bmatrix} I^2 & IX + X0 \\ 0 & 0^2 \end{bmatrix} = \begin{bmatrix} I & X \\ 0 & 0 \end{bmatrix} = A$. Hence $A^3 = AA^2 = AA = A^2 = A$. Continuing in this way, we see that $A^k = A$ for every $k \geq 1$.

Block multiplication has theoretical uses as we shall see. However, it is also useful in computing products of matrices in a computer with limited memory capacity. The matrices are partitioned into blocks in such a way that each product of blocks can be handled. Then the blocks are stored in auxiliary memory and their products are computed one by one.

Directed Graphs

The study of directed graphs illustrates how matrix multiplication arises in ways other than the study of linear equations or matrix transformations.

A **directed graph** consists of a set of points (called **vertices**) connected by arrows (called **edges**). For example, the vertices could represent cities and the edges available flights. If the graph has n vertices $v_1, v_2, \ldots, v_n$, the **adjacency matrix** $A = [a_{ij}]$ is the $n \times n$ matrix whose (i, j)-entry a_{ij} is 1 if there is an edge from v_j to v_i (note the order), and zero otherwise. For example, the adjacency matrix of the directed graph shown is $A = \begin{bmatrix} 1 & 1 & 0 \\ 1 & 0 & 1 \\ 1 & 0 & 0 \end{bmatrix}$. A **path of length** r (or an r-path) from vertex j to vertex i is a sequence of r edges leading from v_j to v_i. Thus $v_1 \to v_2 \to v_1 \to v_1 \to v_3$ is a 4-path from v_1 to v_3 in the given graph. The edges are just the paths of length 1, so the (i, j)-entry a_{ij} of the adjacency matrix A is the number of 1-paths from v_j to v_i. This observation has an important extension:

Theorem 6

If A is the adjacency matrix of a directed graph with n vertices, then the (i, j)-entry of A^r is the number of r-paths $v_j \to v_i$.

As an illustration, consider the adjacency matrix A in the graph shown. Then

$$A = \begin{bmatrix} 1 & 1 & 0 \\ 1 & 0 & 1 \\ 1 & 0 & 0 \end{bmatrix}, \quad A^2 = \begin{bmatrix} 2 & 1 & 1 \\ 2 & 1 & 0 \\ 1 & 1 & 0 \end{bmatrix}, \quad \text{and} \quad A^3 = \begin{bmatrix} 4 & 2 & 1 \\ 3 & 2 & 1 \\ 2 & 1 & 1 \end{bmatrix}.$$

Hence, since the $(2, 1)$-entry of A^2 is 2, there are two 2-paths $v_1 \to v_2$ (in fact $v_1 \to v_1 \to v_2$ and $v_1 \to v_3 \to v_2$). Similarly, the $(2, 3)$-entry of A^2 is zero, so there are *no* 2-paths $v_3 \to v_2$, as the reader can verify. The fact that no entry of A^3 is zero shows that it is possible to go from any vertex to any other vertex in exactly three steps.

To see why Theorem 6 is true, observe that it asserts that

$$\text{the } (i, j)\text{-entry of } A^r \text{ equals the number of } r\text{-paths } v_j \to v_i \tag{$*$}$$

holds for each $r \geq 1$. We proceed by induction on r (see Appendix C). The case $r = 1$ is the definition of the adjacency matrix. So assume inductively that ($*$) is true for some $r \geq 1$; we must prove that ($*$) also holds for $r + 1$. But every $(r + 1)$-path $v_j \to v_i$ is the result of an r-path $v_j \to v_k$ for some k, followed by a 1-path $v_k \to v_i$. Writing $A = [a_{ij}]$ and $A^r = [b_{ij}]$, there are b_{kj} paths of the former type (by induction) and a_{ik} of the latter type, and so there are $a_{ik}b_{kj}$ such paths in all. Summing over k, this shows that there are

$$a_{i1}b_{1j} + a_{i2}b_{2j} + \cdots + a_{in}b_{nj} \quad (r + 1)\text{-paths } v_j \to v_i.$$

But this sum is the dot product of the ith row $[a_{i1} \ a_{i2} \ \cdots \ a_{in}]$ of A with the jth column $[b_{1j} \ b_{2j} \ \cdots \ b_{nj}]^T$ of A^r. As such, it is the (i, j)-entry of the matrix product $A^r A = A^{r+1}$. This shows that ($*$) holds for $r + 1$, as required.

EXERCISES 2.3

1. Compute the following matrix products.

   (a) $\begin{bmatrix} 1 & 3 \\ 0 & -2 \end{bmatrix}\begin{bmatrix} 2 & -1 \\ 0 & 1 \end{bmatrix}$

   (b) $\begin{bmatrix} 1 & -1 & 2 \\ 2 & 0 & 4 \end{bmatrix}\begin{bmatrix} 2 & 3 & 1 \\ 1 & 9 & 7 \\ -1 & 0 & 2 \end{bmatrix}$

   (c) $\begin{bmatrix} 5 & 0 & -7 \\ 1 & 5 & 9 \end{bmatrix}\begin{bmatrix} 3 \\ 1 \\ -1 \end{bmatrix}$

   (d) $[1 \;\; 3 \;\; -3]\begin{bmatrix} 3 & 0 \\ -2 & 1 \\ 0 & 6 \end{bmatrix}$

   (e) $\begin{bmatrix} 1 & 0 & 0 \\ 0 & 1 & 0 \\ 0 & 0 & 1 \end{bmatrix}\begin{bmatrix} 3 & -2 \\ 5 & -7 \\ 9 & 7 \end{bmatrix}$

   (f) $[1 \;\; -1 \;\; 3]\begin{bmatrix} 2 \\ 1 \\ -8 \end{bmatrix}$

   (g) $\begin{bmatrix} 2 \\ 1 \\ -7 \end{bmatrix}[1 \;\; -1 \;\; 3]$

   (h) $\begin{bmatrix} 3 & 1 \\ 5 & 2 \end{bmatrix}\begin{bmatrix} 2 & -1 \\ -5 & 3 \end{bmatrix}$

   (i) $\begin{bmatrix} 2 & 3 & 1 \\ 5 & 7 & 4 \end{bmatrix}\begin{bmatrix} a & 0 & 0 \\ 0 & b & 0 \\ 0 & 0 & c \end{bmatrix}$

   (j) $\begin{bmatrix} a & 0 & 0 \\ 0 & b & 0 \\ 0 & 0 & c \end{bmatrix}\begin{bmatrix} a' & 0 & 0 \\ 0 & b' & 0 \\ 0 & 0 & c' \end{bmatrix}$

2. In each of the following cases, find all possible products A^2, AB, AC, and so on.

   (a) $A = \begin{bmatrix} 1 & 2 & 3 \\ -1 & 0 & 0 \end{bmatrix}$, $B = \begin{bmatrix} 1 & -2 \\ \frac{1}{2} & 3 \end{bmatrix}$, $C = \begin{bmatrix} -1 & 0 \\ 2 & 5 \\ 0 & 5 \end{bmatrix}$

   (b) $A = \begin{bmatrix} 1 & 2 & 4 \\ 0 & 1 & -1 \end{bmatrix}$, $B = \begin{bmatrix} -1 & 6 \\ 1 & 0 \end{bmatrix}$, $C = \begin{bmatrix} 2 & 0 \\ -1 & 1 \\ 1 & 2 \end{bmatrix}$

3. Find a, b, a_1, and b_1 if:

   (a) $\begin{bmatrix} a & b \\ a_1 & b_1 \end{bmatrix}\begin{bmatrix} 3 & -5 \\ -1 & 2 \end{bmatrix} = \begin{bmatrix} 1 & -1 \\ 2 & 0 \end{bmatrix}$

   (b) $\begin{bmatrix} 2 & 1 \\ -1 & 2 \end{bmatrix}\begin{bmatrix} a & b \\ a_1 & b_1 \end{bmatrix} = \begin{bmatrix} 7 & 2 \\ -1 & 4 \end{bmatrix}$

4. Verify that $A^2 - A - 6I = 0$ if:

   (a) $\begin{bmatrix} 3 & -1 \\ 0 & -2 \end{bmatrix}$

   (b) $\begin{bmatrix} 2 & 2 \\ 2 & -1 \end{bmatrix}$

5. Given $A = \begin{bmatrix} 1 & -1 \\ 0 & 1 \end{bmatrix}$, $B = \begin{bmatrix} 1 & 0 & -2 \\ 3 & 1 & 0 \end{bmatrix}$, $C = \begin{bmatrix} 1 & 0 \\ 2 & 1 \\ 5 & 8 \end{bmatrix}$,

   and $D = \begin{bmatrix} 3 & -1 & 2 \\ 1 & 0 & 5 \end{bmatrix}$, verify the following facts from Theorem 1.

   (a) $A(B - D) = AB - AD$

   (b) $A(BC) = (AB)C$

   (c) $(CD)^T = D^T C^T$

6. Let A be a 2×2 matrix.

   (a) If A commutes with $\begin{bmatrix} 0 & 1 \\ 0 & 0 \end{bmatrix}$, show that $A = \begin{bmatrix} a & b \\ 0 & a \end{bmatrix}$ for some a and b.

   (b) If A commutes with $\begin{bmatrix} 0 & 0 \\ 1 & 0 \end{bmatrix}$, show, that $A = \begin{bmatrix} a & 0 \\ c & a \end{bmatrix}$ for some a and c.

   (c) Show that A commutes with every 2×2 matrix if and only if $A = \begin{bmatrix} a & 0 \\ 0 & a \end{bmatrix}$ for some a.

7. (a) If A^2 can be formed, what can be said about the size of A?

   (b) If AB and BA can both be formed, describe the sizes of A and B.

   (c) If ABC can be formed, A is 3×3, and C is 5×5, what size is B?

8. (a) Find two 2×2 matrices A such that $A^2 = 0$.

   (b) Find three 2×2 matrices A such that (i) $A^2 = I$; (ii) $A^2 = A$.

   (c) Find 2×2 matrices A and B such that $AB = 0$ but $BA \neq 0$.

9. Write $P = \begin{bmatrix} 1 & 0 & 0 \\ 0 & 0 & 1 \\ 0 & 1 & 0 \end{bmatrix}$, and let A be $3 \times n$ and B be $m \times 3$.

   (a) Describe PA in terms of the rows of A.

   (b) Describe BP in terms of the columns of B.

10. Let A, B, and C be as in Exercise 5. Find the $(3, 1)$-entry of CAB using exactly six numerical multiplications.

11. Compute AB, using the indicated block partitioning.

    $A = \left[\begin{array}{cc|cc} 2 & -1 & 3 & 1 \\ 1 & 0 & 1 & 2 \\ \hline 0 & 0 & 1 & 0 \\ 0 & 0 & 0 & 1 \end{array}\right]$ $B = \left[\begin{array}{cc|c} 1 & 2 & 0 \\ -1 & 0 & 0 \\ \hline 0 & 5 & 1 \\ 1 & -1 & 0 \end{array}\right]$

12. In each case give formulas for all powers A, A^2, A^3, ... of A using the block decomposition indicated.

(a) $A = \begin{bmatrix} 1 & 0 & 0 \\ 1 & 1 & -1 \\ 1 & -1 & 1 \end{bmatrix}$
◆(b) $A = \begin{bmatrix} 1 & -1 & 2 & -1 \\ 0 & 1 & 0 & 0 \\ 0 & 0 & -1 & 1 \\ 0 & 0 & 0 & 1 \end{bmatrix}$

13. Compute the following using block multiplication (all blocks are $k \times k$).
    (a) $\begin{bmatrix} I & X \\ -Y & I \end{bmatrix}\begin{bmatrix} I & 0 \\ Y & I \end{bmatrix}$
    ◆(b) $\begin{bmatrix} I & X \\ 0 & I \end{bmatrix}\begin{bmatrix} I & -X \\ 0 & I \end{bmatrix}$
    (c) $[I\ X][I\ X]^T$
    ◆(d) $[I\ X^T][-X\ I]^T$
    (e) $\begin{bmatrix} I & X \\ 0 & -I \end{bmatrix}^n$, any $n \geq 1$
    ◆(f) $\begin{bmatrix} 0 & X \\ I & 0 \end{bmatrix}^n$, any $n \geq 1$

14. Let A denote an $m \times n$ matrix.
    (a) If $AX = 0$ for every $n \times 1$ matrix X, show that $A = 0$.
    ◆(b) If $YA = 0$ for every $1 \times m$ matrix Y, show that $A = 0$.

15. (a) If $U = \begin{bmatrix} 1 & 2 \\ 0 & -1 \end{bmatrix}$, and $AU = 0$, show that $A = 0$.
    (b) Let U be such that $AU = 0$ implies that $A = 0$. If $PU = QU$, show that $P = Q$.

16. Simplify the following expressions where A, B, and C represent matrices.
    (a) $A(3B - C) + (A - 2B)C + 2B(C + 2A)$
    ◆(b) $A(B + C - D) + B(C - A + D) - (A + B)C + (A - B)D$
    (c) $AB(BC - CB) + (CA - AB)BC + CA(A - B)C$
    ◆(d) $(A - B)(C - A) + (C - B)(A - C) + (C - A)^2$

17. If $A = \begin{bmatrix} a & b \\ c & d \end{bmatrix}$ where $a \neq 0$, show that A factors in the form $A = \begin{bmatrix} 1 & 0 \\ x & 1 \end{bmatrix}\begin{bmatrix} y & z \\ 0 & w \end{bmatrix}$.

18. If A and B commute with C, show that the same is true of:
    (a) $A + B$
    ◆(b) kA, k any scalar

19. If A is any matrix, show that both AA^T and A^TA are symmetric.

◆20. If A and B are symmetric, show that AB is symmetric if and only if $AB = BA$.

21. If A is a 2×2 matrix, show that $A^TA = AA^T$ if and only if A is symmetric or $A = \begin{bmatrix} a & b \\ -b & a \end{bmatrix}$ for some a and b.

22. (a) Find all symmetric 2×2 matrices A such that $A^2 = 0$.
    ◆(b) Repeat (a) if A is 3×3.
    (c) Repeat (a) if A is $n \times n$.

23. Show that there exist no 2×2 matrices A and B such that $AB - BA = I$. [*Hint*: Examine the (1, 1)- and (2, 2)-entries.]

◆24. Let B be an $n \times n$ matrix. Suppose $AB = 0$ for some nonzero $m \times n$ matrix A. Show that no $n \times n$ matrix C exists such that $BC = I$.

25. An autoparts manufacturer makes fenders, doors, and hoods. Each requires assembly and packaging carried out at factories: Plant 1, Plant 2, and Plant 3. Matrix A below gives the number of hours for assembly and packaging, and matrix B gives the hourly rates at the three plants. Explain the meaning of the (3, 2)-entry in the matrix AB. Which plant is the most economical to operate? Give reasons.

	Assembly	Packaging
Fenders	12	2
Doors	21	3
Hoods	10	2

$= A$

	Plant 1	Plant 2	Plant 3
Assembly	21	18	20
Packaging	14	10	13

$= B$

◆26. For the directed graph at the right, find the adjacency matrix A, compute A^3, and determine the number of paths of length 3 from v_1 to v_4 and from v_2 to v_3.

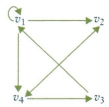

27. In each case either show the statement is true, or give an example showing that it is false.
    (a) If $A^2 = I$, then $A = I$.
    ◆(b) If $AJ = A$, then $J = I$.
    (c) If A is square, then $(A^T)^3 = (A^3)^T$.
    ◆(d) If A is symmetric, then $I + A$ is symmetric.
    (e) If $AB = AC$ and $A \neq 0$, then $B = C$.
    ◆(f) If $A \neq 0$, then $A^2 \neq 0$.
    (g) If A has a row of zeros, so also does BA for all B.

◆(h) If A commutes with $A + B$, then A commutes with B.

(i) If B has a column of zeros, so also does AB.

◆(j) If AB has a column of zeros, so also does B.

(k) If A has a row of zeros, so also does AB.

◆(l) If AB has a row of zeros, so also does A.

28. (a) If A and B are 2×2 matrices whose rows sum to 1, show that the rows of AB also sum to 1.

    ◆(b) Repeat part (a) for the case where A and B are $n \times n$.

29. Let A and B be $n \times n$ matrices for which the systems of equations $A\mathbf{x} = \mathbf{0}$ and $B\mathbf{x} = \mathbf{0}$ each have only the trivial solution $\mathbf{x} = \mathbf{0}$. Show that the system $(AB)\mathbf{x} = \mathbf{0}$ has only the trivial solution.

30. The **trace** of a square matrix A, denoted tr A, is the sum of the elements on the main diagonal of A. Show that, if A and B are $n \times n$ matrices:

    (a) $\text{tr}(A + B) = \text{tr } A + \text{tr } B$.

    ◆(b) $\text{tr}(kA) = k \, \text{tr}(A)$ for any number k.

    (c) $\text{tr}(A^T) = \text{tr}(A)$.

    (d) $\text{tr}(AB) = \text{tr}(BA)$.

    ◆(e) $\text{tr}(AA^T)$ is the sum of the squares of all entries of A.

31. Show that $AB - BA = I$ is impossible. [*Hint*: See the preceding exercise.]

32. A square matrix P is called an **idempotent** if $P^2 = P$. Show that:

    (a) 0 and I are idempotents.

    (b) $\begin{bmatrix} 1 & 1 \\ 0 & 0 \end{bmatrix}$, $\begin{bmatrix} 1 & 0 \\ 1 & 0 \end{bmatrix}$, and $\frac{1}{2}\begin{bmatrix} 1 & 1 \\ 1 & 1 \end{bmatrix}$, are idempotents.

    (c) If P is an idempotent, so is $I - P$. Show further that $P(I - P) = 0$.

    (d) If P is an idempotent, so is P^T.

    ◆(e) If P is an idempotent, so is $Q = P + AP - PAP$ for any square matrix A (of the same size as P).

    (f) If A is $n \times m$ and B is $m \times n$, and if $AB = I_n$, then BA is an idempotent.

33. Let A and B be $n \times n$ **diagonal matrices** (all entries off the main diagonal are zero).

    (a) Show that AB is diagonal and $AB = BA$.

    (b) Formulate a rule for calculating XA if X is $m \times n$.

    (c) Formulate a rule for calculating AY if Y is $n \times k$.

34. If A and B are $n \times n$ matrices, show that:

    (a) $AB = BA$ if and only if
    $$(A + B)^2 = A^2 + 2AB + B^2.$$

    ◆(b) $AB = BA$ if and only if
    $$(A + B)(A - B) = (A - B)(A + B).$$

35. In Theorem 3, prove

    (a) part 3;   ◆(b) part 5.

◆36. (V. Camillo) Show that the product of two reduced row-echelon matrices is also reduced row-echelon.

SECTION 2.4 Matrix Inverses

Three basic operations on matrices, addition, multiplication, and subtraction, are analogs for matrices of the same operations for numbers. In this section we introduce the matrix analog of numerical division.

To begin, consider how a numerical equation

$$ax = b$$

is solved when a and b are known numbers. If $a = 0$, there is no solution (unless $b = 0$). But if $a \neq 0$, we can multiply both sides by the inverse $a^{-1} = \frac{1}{a}$ to obtain the solution $x = a^{-1}b$. Of course multiplying by a^{-1} is just dividing by a, and the property of a^{-1} that makes this work is that $a^{-1}a = 1$. Moreover, we saw in Section 2.2 that the role that 1 plays in arithmetic is played in matrix algebra by the identity matrix I.

Definition 2.11

This suggests the following definition. If A is a square matrix, a matrix B is called an **inverse** of A if and only if

$$AB = I \quad \text{and} \quad BA = I$$

A matrix A that has an inverse is called an **invertible matrix**.[8]

EXAMPLE 1

Show that $B = \begin{bmatrix} -1 & 1 \\ 1 & 0 \end{bmatrix}$ is an inverse of $A = \begin{bmatrix} 0 & 1 \\ 1 & 1 \end{bmatrix}$.

Solution ▶ Compute AB and BA.

$$AB = \begin{bmatrix} 0 & 1 \\ 1 & 1 \end{bmatrix}\begin{bmatrix} -1 & 1 \\ 1 & 0 \end{bmatrix} = \begin{bmatrix} 1 & 0 \\ 0 & 1 \end{bmatrix} \quad BA = \begin{bmatrix} -1 & 1 \\ 1 & 0 \end{bmatrix}\begin{bmatrix} 0 & 1 \\ 1 & 1 \end{bmatrix} = \begin{bmatrix} 1 & 0 \\ 0 & 1 \end{bmatrix}$$

Hence $AB = I = BA$, so B is indeed an inverse of A.

EXAMPLE 2

Show that $A = \begin{bmatrix} 0 & 0 \\ 1 & 3 \end{bmatrix}$ has no inverse.

Solution ▶ Let $B = \begin{bmatrix} a & b \\ c & d \end{bmatrix}$ denote an arbitrary 2×2 matrix. Then

$$AB = \begin{bmatrix} 0 & 0 \\ 1 & 3 \end{bmatrix}\begin{bmatrix} a & b \\ c & d \end{bmatrix} = \begin{bmatrix} 0 & 0 \\ a + 3c & b + 3d \end{bmatrix}$$

so AB has a row of zeros. Hence AB cannot equal I for any B.

The argument in Example 2 shows that no zero matrix has an inverse. But Example 2 also shows that, unlike arithmetic, *it is possible for a nonzero matrix to have no inverse*. However, if a matrix *does* have an inverse, it has only one.

Theorem 1

If B and C are both inverses of A, then $B = C$.

PROOF

Since B and C are both inverses of A, we have $CA = I = AB$.
Hence $B = IB = (CA)B = C(AB) = CI = C$.

If A is an invertible matrix, the (unique) inverse of A is denoted A^{-1}. Hence A^{-1} (when it exists) is a square matrix of the same size as A with the property that

$$AA^{-1} = I \quad \text{and} \quad A^{-1}A = I$$

---

[8] Only square matrices have inverses. Even though it is plausible that nonsquare matrices A and B could exist such that $AB = I_m$ and $BA = I_n$, where A is $m \times n$ and B is $n \times m$, we claim that this forces $n = m$. Indeed, if $m < n$ there exists a nonzero column **x** such that $A\mathbf{x} = \mathbf{0}$ (by Theorem 1 Section 1.3), so $\mathbf{x} = I_n\mathbf{x} = (BA)\mathbf{x} = B(A\mathbf{x}) = B(\mathbf{0}) = \mathbf{0}$, a contradiction. Hence $m \geq n$. Similarly, the condition $AB = I_m$ implies that $n \geq m$. Hence $m = n$ so A is square.

SECTION 2.4 *Matrix Inverses*

These equations characterize A^{-1} in the following sense: If somehow a matrix B can be found such that $AB = I = BA$, then A is invertible and B is the inverse of A; in symbols, $B = A^{-1}$. This gives us a way of verifying that the inverse of a matrix exists. Examples 3 and 4 offer illustrations.

EXAMPLE 3

If $A = \begin{bmatrix} 0 & -1 \\ 1 & -1 \end{bmatrix}$, show that $A^3 = I$ and so find A^{-1}.

Solution ▶ We have $A^2 = \begin{bmatrix} 0 & -1 \\ 1 & -1 \end{bmatrix}\begin{bmatrix} 0 & -1 \\ 1 & -1 \end{bmatrix} = \begin{bmatrix} -1 & 1 \\ -1 & 0 \end{bmatrix}$, and so

$$A^3 = A^2 A = \begin{bmatrix} -1 & 1 \\ -1 & 0 \end{bmatrix}\begin{bmatrix} 0 & -1 \\ 1 & -1 \end{bmatrix} = \begin{bmatrix} 1 & 0 \\ 0 & 1 \end{bmatrix} = I$$

Hence $A^3 = I$, as asserted. This can be written as $A^2 A = I = AA^2$, so it shows that A^2 is the inverse of A. That is, $A^{-1} = A^2 = \begin{bmatrix} -1 & 1 \\ -1 & 0 \end{bmatrix}$.

The next example presents a useful formula for the inverse of a 2×2 matrix $A = \begin{bmatrix} a & b \\ c & d \end{bmatrix}$. To state it, we define the **determinant** $\det A$ and the **adjugate** $\operatorname{adj} A$ of the matrix A as follows:

$$\det\begin{bmatrix} a & b \\ c & d \end{bmatrix} = ad - bc, \quad \text{and} \quad \operatorname{adj}\begin{bmatrix} a & b \\ c & d \end{bmatrix} = \begin{bmatrix} d & -b \\ -c & a \end{bmatrix}$$

EXAMPLE 4

If $A = \begin{bmatrix} a & b \\ c & d \end{bmatrix}$, show that A has an inverse if and only if $\det A \neq 0$, and in this case

$$A^{-1} = \frac{1}{\det A} \operatorname{adj} A$$

Solution ▶ For convenience, write $e = \det A = ad - bc$ and $B = \operatorname{adj} A = \begin{bmatrix} d & -b \\ -c & a \end{bmatrix}$.
Then $AB = eI = BA$ as the reader can verify. So if $e \neq 0$, scalar multiplication by $1/e$ gives $A(\frac{1}{e}B) = I = (\frac{1}{e}B)A$. Hence A is invertible and $A^{-1} = \frac{1}{e}B$. Thus it remains only to show that if A^{-1} exists, then $e \neq 0$.

We prove this by showing that assuming $e = 0$ leads to a contradiction. In fact, if $e = 0$, then $AB = eI = 0$, so left multiplication by A^{-1} gives $A^{-1}AB = A^{-1}0$; that is, $IB = 0$, so $B = 0$. But this implies that a, b, c, and d are all zero, so $A = 0$, contrary to the assumption that A^{-1} exists.

As an illustration, if $A = \begin{bmatrix} 2 & 4 \\ -3 & 8 \end{bmatrix}$ then $\det A = 2 \cdot 8 - 4 \cdot (-3) = 28 \neq 0$. Hence A is invertible and $A^{-1} = \frac{1}{\det A} \operatorname{adj} A = \frac{1}{28}\begin{bmatrix} 8 & -4 \\ 3 & 2 \end{bmatrix}$, as the reader is invited to verify.

The determinant and adjugate will be defined in Chapter 3 for any square matrix, and the conclusions in Example 4 will be proved in full generality.

Inverses and Linear Systems

Matrix inverses can be used to solve certain systems of linear equations. Recall that a *system* of linear equations can be written as a *single* matrix equation

$$A\mathbf{x} = \mathbf{b}$$

where A and $\mathbf{b}$ are known matrices and $\mathbf{x}$ is to be determined. If A is invertible, we multiply each side of the equation on the left by A^{-1} to get

$$A^{-1}A\mathbf{x} = A^{-1}\mathbf{b}$$
$$I\mathbf{x} = A^{-1}\mathbf{b}$$
$$\mathbf{x} = A^{-1}\mathbf{b}$$

This gives the solution to the system of equations (the reader should verify that $\mathbf{x} = A^{-1}\mathbf{b}$ really does satisfy $A\mathbf{x} = \mathbf{b}$). Furthermore, the argument shows that if $\mathbf{x}$ is *any* solution, then necessarily $\mathbf{x} = A^{-1}\mathbf{b}$, so the solution is unique. Of course the technique works only when the coefficient matrix A has an inverse. This proves Theorem 2.

Theorem 2

Suppose a system of n equations in n variables is written in matrix form as

$$A\mathbf{x} = \mathbf{b}$$

If the $n \times n$ coefficient matrix A is invertible, the system has the unique solution

$$\mathbf{x} = A^{-1}\mathbf{b}$$

EXAMPLE 5

Use Example 4 to solve the system $\begin{cases} 5x_1 - 3x_2 = -4 \\ 7x_1 + 4x_2 = 8 \end{cases}$.

Solution ▶ In matrix form this is $A\mathbf{x} = \mathbf{b}$ where $A = \begin{bmatrix} 5 & -3 \\ 7 & 4 \end{bmatrix}$, $\mathbf{x} = \begin{bmatrix} x_1 \\ x_2 \end{bmatrix}$, and $\mathbf{b} = \begin{bmatrix} -4 \\ 8 \end{bmatrix}$. Then $\det A = 5 \cdot 4 - (-3) \cdot 7 = 41$, so A is invertible and $A^{-1} = \frac{1}{41}\begin{bmatrix} 4 & 3 \\ -7 & 5 \end{bmatrix}$ by Example 4. Thus Theorem 2 gives

$$\mathbf{x} = A^{-1}\mathbf{b} = \frac{1}{41}\begin{bmatrix} 4 & 3 \\ -7 & 5 \end{bmatrix}\begin{bmatrix} -4 \\ 8 \end{bmatrix} = \frac{1}{41}\begin{bmatrix} 8 \\ 68 \end{bmatrix},$$

so the solution is $x_1 = \frac{8}{41}$ and $x_2 = \frac{68}{41}$.

An Inversion Method

If a matrix A is $n \times n$ and invertible, it is desirable to have an efficient technique for finding the inverse matrix A^{-1}. In fact, we can determine A^{-1} from the equation

$$AA^{-1} = I_n$$

Write A^{-1} in terms of its columns as $A^{-1} = [\mathbf{x}_1 \ \mathbf{x}_2 \ \cdots \ \mathbf{x}_n]$, where the columns $\mathbf{x}_j$ are to be determined. Similarly, write $I_n = [\mathbf{e}_1, \mathbf{e}_2, ..., \mathbf{e}_n]$ in terms of its columns. Then (using Definition 2.9) the condition $AA^{-1} = I$ becomes

$$[A\mathbf{x}_1 \ A\mathbf{x}_2 \ \cdots \ A\mathbf{x}_n] = [\mathbf{e}_1 \ \mathbf{e}_2 \ \cdots \ \mathbf{e}_n]$$

SECTION 2.4 Matrix Inverses

Equating columns gives

$$A\mathbf{x}_j = \mathbf{e}_j \quad \text{for each } j = 1, 2, \ldots, n.$$

These are systems of linear equations for the $\mathbf{x}_j$, each with A as a coefficient matrix. Since A is invertible, each system has a unique solution by Theorem 2. But this means that the reduced row-echelon form R of A cannot have a row of zeros, so $R = I_n$ (R is square). Hence there is a sequence of elementary row operations carrying $A \to I_n$. This sequence carries the augmented matrix of each system $A\mathbf{x}_j = \mathbf{e}_j$ to reduced row-echelon form:

$$[A \mid \mathbf{e}_j] \to [I_n \mid \mathbf{x}_j] \quad \text{for each } j = 1, 2, \ldots, n.$$

This determines the solutions $\mathbf{x}_j$, and hence determines $A^{-1} = [\mathbf{x}_1 \; \mathbf{x}_2 \; \cdots \; \mathbf{x}_n]$. But the fact that the *same* sequence $A \to I_n$ works for each j means that we can do all these calculations simultaneously by applying the elementary row operations to the double matrix $[A \; I]$:

$$[A \; I] \to [I \; A^{-1}].$$

This is the desired algorithm.

Matrix Inversion Algorithm

If A is an invertible (square) matrix, there exists a sequence of elementary row operations that carry A to the identity matrix I of the same size, written $A \to I$. This same series of row operations carries I to A^{-1}; that is, $I \to A^{-1}$. The algorithm can be summarized as follows:

$$[A \; I] \to [I \; A^{-1}]$$

where the row operations on A and I are carried out simultaneously.

EXAMPLE 6

Use the inversion algorithm to find the inverse of the matrix

$$A = \begin{bmatrix} 2 & 7 & 1 \\ 1 & 4 & -1 \\ 1 & 3 & 0 \end{bmatrix}$$

Solution ▶ Apply elementary row operations to the double matrix

$$[A \; I] = \begin{bmatrix} 2 & 7 & 1 & | & 1 & 0 & 0 \\ 1 & 4 & -1 & | & 0 & 1 & 0 \\ 1 & 3 & 0 & | & 0 & 0 & 1 \end{bmatrix}$$

so as to carry A to I. First interchange rows 1 and 2.

$$\begin{bmatrix} 1 & 4 & -1 & | & 0 & 1 & 0 \\ 2 & 7 & 1 & | & 1 & 0 & 0 \\ 1 & 3 & 0 & | & 0 & 0 & 1 \end{bmatrix}$$

Next subtract 2 times row 1 from row 2, and subtract row 1 from row 3.

$$\begin{bmatrix} 1 & 4 & -1 & | & 0 & 1 & 0 \\ 0 & -1 & 3 & | & 1 & -2 & 0 \\ 0 & -1 & 1 & | & 0 & -1 & 1 \end{bmatrix}$$

Continue to reduced row-echelon form.

$$\begin{bmatrix} 1 & 0 & 11 & | & 4 & -7 & 0 \\ 0 & 1 & -3 & | & -1 & 2 & 0 \\ 0 & 0 & -2 & | & -1 & 1 & 1 \end{bmatrix}$$

$$\begin{bmatrix} 1 & 0 & 0 & | & \frac{-3}{2} & \frac{-3}{2} & \frac{11}{2} \\ 0 & 1 & 0 & | & \frac{1}{2} & \frac{1}{2} & \frac{-3}{2} \\ 0 & 0 & 1 & | & \frac{-1}{2} & \frac{1}{2} & \frac{-1}{2} \end{bmatrix}$$

Hence $A^{-1} = \frac{1}{2}\begin{bmatrix} -3 & -3 & 11 \\ 1 & 1 & -3 \\ 1 & -1 & -1 \end{bmatrix}$, as is readily verified.

Given any $n \times n$ matrix A, Theorem 1 Section 1.2 shows that A can be carried by elementary row operations to a matrix R in reduced row-echelon form. If $R = I$, the matrix A is invertible (this will be proved in the next section), so the algorithm produces A^{-1}. If $R \neq I$, then R has a row of zeros (it is square), so no system of linear equations $A\mathbf{x} = \mathbf{b}$ can have a unique solution. But then A is not invertible by Theorem 2. Hence, the algorithm is effective in the sense conveyed in Theorem 3.

Theorem 3

If A is an $n \times n$ matrix, either A can be reduced to I by elementary row operations or it cannot. In the first case, the algorithm produces A^{-1}; in the second case, A^{-1} does not exist.

Properties of Inverses

The following properties of an invertible matrix are used everywhere.

EXAMPLE 7

Cancellation Laws Let A be an invertible matrix. Show that:

(1) If $AB = AC$, then $B = C$.
(2) If $BA = CA$, then $B = C$.

Solution ▶ Given the equation $AB = AC$, left multiply both sides by A^{-1} to obtain $A^{-1}AB = A^{-1}AC$. This is $IB = IC$, that is $B = C$. This proves (1) and the proof of (2) is left to the reader.

Properties (1) and (2) in Example 7 are described by saying that an invertible matrix can be "left cancelled" and "right cancelled", respectively. Note however that "mixed" cancellation does not hold in general: If A is invertible and $AB = CA$, then B and C may *not* be equal, even if both are 2×2. Here is a specific example:

$A = \begin{bmatrix} 1 & 1 \\ 0 & 1 \end{bmatrix}, B = \begin{bmatrix} 0 & 0 \\ 1 & 2 \end{bmatrix}$, and $C = \begin{bmatrix} 1 & 1 \\ 1 & 1 \end{bmatrix}$.

SECTION 2.4 Matrix Inverses

Sometimes the inverse of a matrix is given by a formula. Example 4 is one illustration; Examples 8 and 9 provide two more. The idea in both cases is that, given a square matrix A, if a matrix B can be found such that $AB = I = BA$, then A is invertible and $A^{-1} = B$.

EXAMPLE 8

If A is an invertible matrix, show that the transpose A^T is also invertible. Show further that the inverse of A^T is just the transpose of A^{-1}; in symbols, $(A^T)^{-1} = (A^{-1})^T$.

Solution ▶ A^{-1} exists (by assumption). Its transpose $(A^{-1})^T$ is the candidate proposed for the inverse of A^T. Using Theorem 3 Section 2.3, we test it as follows:

$$A^T(A^{-1})^T = (A^{-1}A)^T = I^T = I$$
$$(A^{-1})^T A^T = (AA^{-1})^T = I^T = I$$

Hence $(A^{-1})^T$ is indeed the inverse of A^T; that is, $(A^T)^{-1} = (A^{-1})^T$.

EXAMPLE 9

If A and B are invertible $n \times n$ matrices, show that their product AB is also invertible and $(AB)^{-1} = B^{-1}A^{-1}$.

Solution ▶ We are given a candidate for the inverse of AB, namely $B^{-1}A^{-1}$. We test it as follows:

$$(B^{-1}A^{-1})(AB) = B^{-1}(A^{-1}A)B = B^{-1}IB = B^{-1}B = I$$
$$(AB)(B^{-1}A^{-1}) = A(BB^{-1})A^{-1} = AIA^{-1} = AA^{-1} = I$$

Hence $B^{-1}A^{-1}$ is the inverse of AB; in symbols, $(AB)^{-1} = B^{-1}A^{-1}$.

We now collect several basic properties of matrix inverses for reference.

Theorem 4

All the following matrices are square matrices of the same size.
1. I is invertible and $I^{-1} = I$.
2. If A is invertible, so is A^{-1}, and $(A^{-1})^{-1} = A$.
3. If A and B are invertible, so is AB, and $(AB)^{-1} = B^{-1}A^{-1}$.
4. If $A_1, A_2, \ldots, A_k$ are all invertible, so is their product $A_1 A_2 \cdots A_k$, and $(A_1 A_2 \cdots A_k)^{-1} = A_k^{-1} \cdots A_2^{-1} A_1^{-1}$.
5. If A is invertible, so is A^k for any $k \geq 1$, and $(A^k)^{-1} = (A^{-1})^k$.
6. If A is invertible and $a \neq 0$ is a number, then aA is invertible and $(aA)^{-1} = \frac{1}{a}A^{-1}$.
7. If A is invertible, so is its transpose A^T, and $(A^T)^{-1} = (A^{-1})^T$.

PROOF

1. This is an immediate consequence of the fact that $I^2 = I$.
2. The equations $AA^{-1} = I = A^{-1}A$ show that A is the inverse of A^{-1}; in symbols, $(A^{-1})^{-1} = A$.
3. This is Example 9.
4. Use induction on k. If $k = 1$, there is nothing to prove, and if $k = 2$, the result is property 3. If $k > 2$, assume inductively that $(A_1 A_2 \cdots A_{k-1})^{-1} = A^{-1}_{k-1} \cdots A_2^{-1} A_1^{-1}$. We apply this fact together with property 3 as follows:
$$[A_1 A_2 \cdots A_{k-1} A_k]^{-1} = [(A_1 A_2 \cdots A_{k-1}) A_k]^{-1}$$
$$= A_k^{-1} (A_1 A_2 \cdots A_{k-1})^{-1}$$
$$= A_k^{-1} (A_{k-1}^{-1} \cdots A_2^{-1} A_1^{-1})$$
So the proof by induction is complete.
5. This is property 4 with $A_1 = A_2 = \cdots = A_k = A$.
6. This is left as Exercise 29.
7. This is Example 8.

The reversal of the order of the inverses in properties 3 and 4 of Theorem 4 is a consequence of the fact that matrix multiplication is not commutative. Another manifestation of this comes when matrix equations are dealt with. If a matrix equation $B = C$ is given, it can be *left-multiplied* by a matrix A to yield $AB = AC$. Similarly, *right-multiplication* gives $BA = CA$. However, we cannot mix the two: If $B = C$, it need *not* be the case that $AB = CA$ even if A is invertible, for example, $A = \begin{bmatrix} 1 & 1 \\ 0 & 1 \end{bmatrix}$, $B = \begin{bmatrix} 0 & 0 \\ 1 & 0 \end{bmatrix} = C$.

Part 7 of Theorem 4 together with the fact that $(A^T)^T = A$ gives

Corollary 1

A square matrix A is invertible if and only if A^T is invertible.

EXAMPLE 10

Find A if $(A^T - 2I)^{-1} = \begin{bmatrix} 2 & 1 \\ -1 & 0 \end{bmatrix}$.

Solution ▶ By Theorem 4(2) and Example 4, we have
$$(A^T - 2I) = [(A^T - 2I)^{-1}]^{-1} = \begin{bmatrix} 2 & 1 \\ -1 & 0 \end{bmatrix}^{-1} = \begin{bmatrix} 0 & -1 \\ 1 & 2 \end{bmatrix}$$

Hence $A^T = 2I + \begin{bmatrix} 0 & -1 \\ 1 & 2 \end{bmatrix} = \begin{bmatrix} 2 & -1 \\ 1 & 4 \end{bmatrix}$, so $A = \begin{bmatrix} 2 & 1 \\ -1 & 4 \end{bmatrix}$ by Theorem 4(2).

SECTION 2.4 Matrix Inverses

The following important theorem collects a number of conditions all equivalent[9] to invertibility. It will be referred to frequently below.

Theorem 5

Inverse Theorem
The following conditions are equivalent for an $n \times n$ matrix A:

1. A is invertible.
2. The homogeneous system $A\mathbf{x} = \mathbf{0}$ has only the trivial solution $\mathbf{x} = \mathbf{0}$.
3. A can be carried to the identity matrix I_n by elementary row operations.
4. The system $A\mathbf{x} = \mathbf{b}$ has at least one solution $\mathbf{x}$ for every choice of column $\mathbf{b}$.
5. There exists an $n \times n$ matrix C such that $AC = I_n$.

PROOF

We show that each of these conditions implies the next, and that (5) implies (1).

(1) $\Rightarrow$ (2). If A^{-1} exists, then $A\mathbf{x} = \mathbf{0}$ gives $\mathbf{x} = I_n\mathbf{x} = A^{-1}A\mathbf{x} = A^{-1}\mathbf{0} = \mathbf{0}$.

(2) $\Rightarrow$ (3). Assume that (2) is true. Certainly $A \to R$ by row operations where R is a reduced, row-echelon matrix. It suffices to show that $R = I_n$. Suppose that this is not the case. Then R has a row of zeros (being square). Now consider the augmented matrix $[A \mid \mathbf{0}]$ of the system $A\mathbf{x} = \mathbf{0}$. Then $[A \mid \mathbf{0}] \to [R \mid \mathbf{0}]$ is the reduced form, and $[R \mid \mathbf{0}]$ also has a row of zeros. Since R is square there must be at least one nonleading variable, and hence at least one parameter. Hence the system $A\mathbf{x} = \mathbf{0}$ has infinitely many solutions, contrary to (2). So $R = I_n$ after all.

(3) $\Rightarrow$ (4). Consider the augmented matrix $[A \mid \mathbf{b}]$ of the system $A\mathbf{x} = \mathbf{b}$. Using (3), let $A \to I_n$ by a sequence of row operations. Then these same operations carry $[A \mid \mathbf{b}] \to [I_n \mid \mathbf{c}]$ for some column $\mathbf{c}$. Hence the system $A\mathbf{x} = \mathbf{b}$ has a solution (in fact unique) by gaussian elimination. This proves (4).

(4) $\Rightarrow$ (5). Write $I_n = [\mathbf{e}_1 \; \mathbf{e}_2 \; \cdots \; \mathbf{e}_n]$ where $\mathbf{e}_1, \mathbf{e}_2, \ldots, \mathbf{e}_n$ are the columns of I_n. For each $j = 1, 2, \ldots, n$, the system $A\mathbf{x} = \mathbf{e}_j$ has a solution $\mathbf{c}_j$ by (4), so $A\mathbf{c}_j = \mathbf{e}_j$. Now let $C = [\mathbf{c}_1 \; \mathbf{c}_2 \; \cdots \; \mathbf{c}_n]$ be the $n \times n$ matrix with these matrices $\mathbf{c}_j$ as its columns. Then Definition 2.9 gives (5):

$$AC = A[\mathbf{c}_1 \; \mathbf{c}_2 \; \cdots \; \mathbf{c}_n] = [A\mathbf{c}_1 \; A\mathbf{c}_2 \; \cdots \; A\mathbf{c}_n] = [\mathbf{e}_1 \; \mathbf{e}_2 \; \cdots \; \mathbf{e}_n] = I_n$$

(5) $\Rightarrow$ (1). Assume that (5) is true so that $AC = I_n$ for some matrix C. Then $C\mathbf{x} = \mathbf{0}$ implies $\mathbf{x} = \mathbf{0}$ (because $\mathbf{x} = I_n\mathbf{x} = AC\mathbf{x} = A\mathbf{0} = \mathbf{0}$). Thus condition (2) holds for the matrix C rather than A. Hence the argument above that (2) $\Rightarrow$ (3) $\Rightarrow$ (4) $\Rightarrow$ (5) (with A replaced by C) shows that a matrix C' exists such that $CC' = I_n$. But then

$$A = AI_n = A(CC') = (AC)C' = I_nC' = C'$$

Thus $CA = CC' = I_n$ which, together with $AC = I_n$, shows that C is the inverse of A. This proves (1).

---

[9] If p and q are statements, we say that p **implies** q (written $p \Rightarrow q$) if q is true whenever p is true. The statements are called **equivalent** if both $p \Rightarrow q$ and $q \Rightarrow p$ (written $p \Leftrightarrow q$, spoken "p if and only if q"). See Appendix B.

The proof of (5) ⇒ (1) in Theorem 5 shows that if $AC = I$ for square matrices, then necessarily $CA = I$, and hence that C and A are inverses of each other. We record this important fact for reference.

Corollary 1

If A and C are square matrices such that $AC = I$, then also $CA = I$. In particular, both A and C are invertible, $C = A^{-1}$, and $A = C^{-1}$.

Observe that Corollary 2 is false if A and C are not square matrices. For example, we have

$$\begin{bmatrix} 1 & 2 & 1 \\ 1 & 1 & 1 \end{bmatrix} \begin{bmatrix} -1 & 1 \\ 1 & -1 \\ 0 & 1 \end{bmatrix} = I_2 \quad \text{but} \quad \begin{bmatrix} -1 & 1 \\ 1 & -1 \\ 0 & 1 \end{bmatrix} \begin{bmatrix} 1 & 2 & 1 \\ 1 & 1 & 1 \end{bmatrix} \neq I_3$$

In fact, it is verified in the footnote on page 70 that if $AB = I_m$ and $BA = I_n$, where A is $m \times n$ and B is $n \times m$, then $m = n$ and A and B are (square) inverses of each other.

An $n \times n$ matrix A has rank n if and only if (3) of Theorem 5 holds. Hence

Corollary 2

An $n \times n$ matrix A is invertible if and only if $\operatorname{rank} A = n$.

Here is a useful fact about inverses of block matrices.

EXAMPLE 11

Let $P = \begin{bmatrix} A & X \\ 0 & B \end{bmatrix}$ and $Q = \begin{bmatrix} A & 0 \\ Y & B \end{bmatrix}$ be block matrices where A is $m \times m$ and B is $n \times n$ (possibly $m \neq n$).

(a) Show that P is invertible if and only if A and B are both invertible. In this case, show that $P^{-1} = \begin{bmatrix} A^{-1} & -A^{-1}XB^{-1} \\ 0 & B^{-1} \end{bmatrix}$.

(b) Show that Q is invertible if and only if A and B are both invertible. In this case, show that $Q^{-1} = \begin{bmatrix} A^{-1} & 0 \\ -B^{-1}YA^{-1} & B^{-1} \end{bmatrix}$.

Solution ▶ We do (a) and leave (b) for the reader.

(a) If A^{-1} and B^{-1} both exist, write $R = \begin{bmatrix} A^{-1} & -A^{-1}XB^{-1} \\ 0 & B^{-1} \end{bmatrix}$. Using block multiplication, one verifies that $PR = I_{m+n} = RP$, so P is invertible, and $P^{-1} = R$. Conversely, suppose that P is invertible, and write $P^{-1} = \begin{bmatrix} C & V \\ W & D \end{bmatrix}$ in block form, where C is $m \times m$ and D is $n \times n$. Then the equation $PP^{-1} = I_{n+m}$ becomes

$$\begin{bmatrix} A & X \\ 0 & B \end{bmatrix} \begin{bmatrix} C & V \\ W & D \end{bmatrix} = \begin{bmatrix} AC + XW & AV + XD \\ BW & BD \end{bmatrix} = I_{m+n} = \begin{bmatrix} I_m & 0 \\ 0 & I_n \end{bmatrix}$$

using block notation. Equating corresponding blocks, we find

$$AC + XW = I_m, \quad BW = 0, \quad \text{and} \quad BD = I_n$$

Hence B is invertible because $BD = I_n$ (by Corollary 1), then $W = 0$ because $BW = 0$, and finally, $AC = I_m$ (so A is invertible, again by Corollary 1).

Inverses of Matrix Transformations

Let $T = T_A : \mathbb{R}^n \to \mathbb{R}^n$ denote the matrix transformation induced by the $n \times n$ matrix A. Since A is square, it may very well be invertible, and this leads to the question:

What does it mean geometrically for T that A is invertible?

To answer this, let $T' = T_{A^{-1}} : \mathbb{R}^n \to \mathbb{R}^n$ denote the transformation induced by A^{-1}. Then

$$T'[T(\mathbf{x})] = A^{-1}[A\mathbf{x}] = I\mathbf{x} = \mathbf{x}$$
$$T[T'(\mathbf{x})] = A[A^{-1}\mathbf{x}] = I\mathbf{x} = \mathbf{x}$$
for all $\mathbf{x}$ in $\mathbb{R}^n$ \hfill (∗)

The first of these equations asserts that, if T carries $\mathbf{x}$ to a vector $T(\mathbf{x})$, then T' carries $T(\mathbf{x})$ right back to $\mathbf{x}$; that is T' "reverses" the action of T. Similarly T "reverses" the action of T'. Conditions (∗) can be stated compactly in terms of composition:

$$T' \circ T = 1_{\mathbb{R}^n} \quad \text{and} \quad T \circ T' = 1_{\mathbb{R}^n} \quad (**)$$

When these conditions hold, we say that the matrix transformation T' is an **inverse** of T, and we have shown that if the matrix A of T is invertible, then T has an inverse (induced by A^{-1}).

The converse is also true: If T has an inverse, then its matrix A must be invertible. Indeed, suppose $S : \mathbb{R}^n \to \mathbb{R}^n$ is any inverse of T, so that $S \circ T = 1_{\mathbb{R}^n}$ and $T \circ S = 1_{\mathbb{R}^n}$. If B is the matrix of S, we have

$$BA\mathbf{x} = S[T(\mathbf{x})] = (S \circ T)(\mathbf{x}) = 1_{\mathbb{R}^n}(\mathbf{x}) = \mathbf{x} = I_n\mathbf{x} \quad \text{for all } \mathbf{x} \text{ in } \mathbb{R}^n$$

It follows by Theorem 5 Section 2.2 that $BA = I_n$, and a similar argument shows that $AB = I_n$. Hence A is invertible with $A^{-1} = B$. Furthermore, the inverse transformation S has matrix A^{-1}, so $S = T'$ using the earlier notation. This proves the following important theorem.

Theorem 6

Let $T : \mathbb{R}^n \to \mathbb{R}^n$ denote the matrix transformation induced by an $n \times n$ matrix A. Then

A is invertible if and only if T has an inverse.

In this case, T has exactly one inverse (which we denote as T^{-1}), and $T^{-1} : \mathbb{R}^n \to \mathbb{R}^n$ is the transformation induced by the matrix A^{-1}. In other words

$$(T_A)^{-1} = T_{A^{-1}}$$

The geometrical relationship between T and T^{-1} is embodied in equations (∗) above:

$$T^{-1}[T(\mathbf{x})] = \mathbf{x} \quad \text{and} \quad T[T^{-1}(\mathbf{x})] = \mathbf{x} \quad \text{for all } \mathbf{x} \text{ in } \mathbb{R}^n$$

These equations are called the **fundamental identities** relating T and T^{-1}. Loosely speaking, they assert that each of T and T^{-1} "reverses" or "undoes" the action of the other.

This geometric view of the inverse of a linear transformation provides a new way to find the inverse of a matrix A. More precisely, if A is an invertible matrix, we proceed as follows:

1. Let T be the linear transformation induced by A.
2. Obtain the linear transformation T^{-1} which "reverses" the action of T.
3. Then A^{-1} is the matrix of T^{-1}.

Here is an example.

EXAMPLE 12

Find the inverse of $A = \begin{bmatrix} 0 & 1 \\ 1 & 0 \end{bmatrix}$ by viewing it as a linear transformation $\mathbb{R}^2 \to \mathbb{R}^2$.

Solution ▸ If $\mathbf{x} = \begin{bmatrix} x \\ y \end{bmatrix}$ the vector $A\mathbf{x} = \begin{bmatrix} 0 & 1 \\ 1 & 0 \end{bmatrix}\begin{bmatrix} x \\ y \end{bmatrix} = \begin{bmatrix} y \\ x \end{bmatrix}$ is the result of reflecting $\mathbf{x}$ in the line $y = x$ (see the diagram). Hence, if $Q_1 : \mathbb{R}^2 \to \mathbb{R}^2$ denotes reflection in the line $y = x$, then A is the matrix of Q_1. Now observe that Q_1 reverses itself because reflecting a vector $\mathbf{x}$ twice results in $\mathbf{x}$. Consequently $Q_1^{-1} = Q_1$. Since A^{-1} is the matrix of Q_1^{-1} and A is the matrix of Q, it follows that $A^{-1} = A$. Of course this conclusion is clear by simply observing directly that $A^2 = I$, but the geometric method can often work where these other methods may be less straightforward.

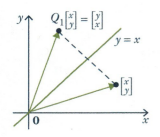

EXERCISES 2.4

1. In each case, show that the matrices are inverses of each other.

   (a) $\begin{bmatrix} 3 & 5 \\ 1 & 2 \end{bmatrix}, \begin{bmatrix} 2 & -5 \\ -1 & 3 \end{bmatrix}$

   (b) $\begin{bmatrix} 3 & 0 \\ 1 & -4 \end{bmatrix}, \frac{1}{2}\begin{bmatrix} 4 & 0 \\ 1 & -3 \end{bmatrix}$

   (c) $\begin{bmatrix} 1 & 2 & 0 \\ 0 & 2 & 3 \\ 1 & 3 & 1 \end{bmatrix}, \begin{bmatrix} 7 & 2 & -6 \\ -3 & -1 & 3 \\ 2 & 1 & -2 \end{bmatrix}$

   (d) $\begin{bmatrix} 3 & 0 \\ 0 & 5 \end{bmatrix}, \begin{bmatrix} \frac{1}{3} & 0 \\ 0 & \frac{1}{5} \end{bmatrix}$

2. Find the inverse of each of the following matrices.

   (a) $\begin{bmatrix} 1 & -1 \\ -1 & 3 \end{bmatrix}$

   ◆(b) $\begin{bmatrix} 4 & 1 \\ 3 & 2 \end{bmatrix}$

   (c) $\begin{bmatrix} 1 & 0 & -1 \\ 3 & 2 & 0 \\ -1 & -1 & 0 \end{bmatrix}$

   ◆(d) $\begin{bmatrix} 1 & -1 & 2 \\ -5 & 7 & -11 \\ -2 & 3 & -5 \end{bmatrix}$

   (e) $\begin{bmatrix} 3 & 5 & 0 \\ 3 & 7 & 1 \\ 1 & 2 & 1 \end{bmatrix}$

   ◆(f) $\begin{bmatrix} 3 & 1 & -1 \\ 2 & 1 & 0 \\ 1 & 5 & -1 \end{bmatrix}$

   (g) $\begin{bmatrix} 2 & 4 & 1 \\ 3 & 3 & 2 \\ 4 & 1 & 4 \end{bmatrix}$

   ◆(h) $\begin{bmatrix} 3 & 1 & -1 \\ 5 & 2 & 0 \\ 1 & 1 & -1 \end{bmatrix}$

   (i) $\begin{bmatrix} 3 & 1 & 2 \\ 1 & -1 & 3 \\ 1 & 2 & 4 \end{bmatrix}$

   ◆(j) $\begin{bmatrix} -1 & 4 & 5 & 2 \\ 0 & 0 & 0 & -1 \\ 1 & -2 & -2 & 0 \\ 0 & -1 & -1 & 0 \end{bmatrix}$

   (k) $\begin{bmatrix} 1 & 0 & 7 & 5 \\ 0 & 1 & 3 & 6 \\ 1 & -1 & 5 & 2 \\ 1 & -1 & 5 & 1 \end{bmatrix}$

   ◆(l) $\begin{bmatrix} 1 & 2 & 0 & 0 & 0 \\ 0 & 1 & 3 & 0 & 0 \\ 0 & 0 & 1 & 5 & 0 \\ 0 & 0 & 0 & 1 & 7 \\ 0 & 0 & 0 & 0 & 1 \end{bmatrix}$

3. In each case, solve the systems of equations by finding the inverse of the coefficient matrix.

   (a) $3x - y = 5$
       $2x + 2y = 1$

   ◆(b) $2x - 3y = 0$
        $x - 4y = 1$

(c) $x + y + 2z = 5$
$ x + y + z = 0$
$ x + 2y + 4z = -2$

◆(d) $x + 4y + 2z = 1$
$ 2x + 3y + 3z = -1$
$ 4x + y + 4z = 0$

4. Given $A^{-1} = \begin{bmatrix} 1 & -1 & 3 \\ 2 & 0 & 5 \\ -1 & 1 & 0 \end{bmatrix}$:

   (a) Solve the system of equations $A\mathbf{x} = \begin{bmatrix} 1 \\ -1 \\ 3 \end{bmatrix}$.

   ◆(b) Find a matrix B such that $AB = \begin{bmatrix} 1 & -1 & 2 \\ 0 & 1 & 1 \\ 1 & 0 & 0 \end{bmatrix}$.

   (c) Find a matrix C such that $CA = \begin{bmatrix} 1 & 2 & -1 \\ 3 & 1 & 1 \end{bmatrix}$.

5. Find A when

   (a) $(3A)^{-1} = \begin{bmatrix} 1 & -1 \\ 0 & 1 \end{bmatrix}$

   ◆(b) $(2A)^T = \begin{bmatrix} 1 & -1 \\ 2 & 3 \end{bmatrix}^{-1}$

   (c) $(I + 3A)^{-1} = \begin{bmatrix} 2 & 0 \\ 1 & -1 \end{bmatrix}$

   ◆(d) $(I - 2A^T)^{-1} = \begin{bmatrix} 2 & 1 \\ 1 & 1 \end{bmatrix}$

   (e) $\left(A\begin{bmatrix} 1 & -1 \\ 0 & 1 \end{bmatrix}\right)^{-1} = \begin{bmatrix} 2 & 3 \\ 1 & 1 \end{bmatrix}$

   ◆(f) $\left(\begin{bmatrix} 1 & 0 \\ 2 & 1 \end{bmatrix} A \right)^{-1} = \begin{bmatrix} 1 & 0 \\ 2 & 2 \end{bmatrix}$

   (g) $(A^T - 2I)^{-1} = 2\begin{bmatrix} 1 & 1 \\ 2 & 3 \end{bmatrix}$

   ◆(h) $(A^{-1} - 2I)^T = -2\begin{bmatrix} 1 & 1 \\ 1 & 0 \end{bmatrix}$

6. Find A when:

   (a) $A^{-1} = \begin{bmatrix} 1 & -1 & 3 \\ 2 & 1 & 1 \\ 0 & 2 & -2 \end{bmatrix}$

   ◆(b) $A^{-1} = \begin{bmatrix} 0 & 1 & -1 \\ 1 & 2 & 1 \\ 1 & 0 & 1 \end{bmatrix}$

7. Given $\begin{bmatrix} x_1 \\ x_2 \\ x_3 \end{bmatrix} = \begin{bmatrix} 3 & -1 & 2 \\ 1 & 0 & 4 \\ 2 & 1 & 0 \end{bmatrix} \begin{bmatrix} y_1 \\ y_2 \\ y_3 \end{bmatrix}$ and

   $\begin{bmatrix} z_1 \\ z_2 \\ z_3 \end{bmatrix} = \begin{bmatrix} 1 & -1 & 1 \\ 2 & -3 & 0 \\ -1 & 1 & -2 \end{bmatrix} \begin{bmatrix} y_1 \\ y_2 \\ y_3 \end{bmatrix}$, express the variables $x_1, x_2,$ and x_3 in terms of $z_1, z_2,$ and z_3.

8. (a) In the system $\begin{array}{l} 3x + 4y = 7 \\ 4x + 5y = 1 \end{array}$, substitute the new variables x' and y' given by $\begin{array}{l} x = -5x' + 4y' \\ y = 4x' - 3y' \end{array}$. Then find x and y.

◆(b) Explain part (a) by writing the equations as $A\begin{bmatrix} x \\ y \end{bmatrix} = \begin{bmatrix} 7 \\ 1 \end{bmatrix}$ and $\begin{bmatrix} x \\ y \end{bmatrix} = B\begin{bmatrix} x' \\ y' \end{bmatrix}$. What is the relationship between A and B?

9. In each case either prove the assertion or give an example showing that it is false.

   (a) If $A \neq 0$ is a square matrix, then A is invertible.

   ◆(b) If A and B are both invertible, then $A + B$ is invertible.

   (c) If A and B are both invertible, then $(A^{-1}B)^T$ is invertible.

   ◆(d) If $A^4 = 3I$, then A is invertible.

   (e) If $A^2 = A$ and $A \neq 0$, then A is invertible.

   ◆(f) If $AB = B$ for some $B \neq 0$, then A is invertible.

   (g) If A is invertible and skew symmetric $(A^T = -A)$, the same is true of A^{-1}.

   ◆(h) If A^2 is invertible, then A is invertible.

   (i) If $AB = I$, then A and B commute.

10. (a) If A, B, and C are square matrices and $AB = I = CA$, show that A is invertible and $B = C = A^{-1}$.

    ◆(b) If $C^{-1} = A$, find the inverse of C^T in terms of A.

11. Suppose $CA = I_m$, where C is $m \times n$ and A is $n \times m$. Consider the system $A\mathbf{x} = \mathbf{b}$ of n equations in m variables.

    (a) Show that this system has a unique solution CB if it is consistent.

    ◆(b) If $C = \begin{bmatrix} 0 & -5 & 1 \\ 3 & 0 & -1 \end{bmatrix}$ and $A = \begin{bmatrix} 2 & -3 \\ 1 & -2 \\ 6 & -10 \end{bmatrix}$, find $\mathbf{x}$ (if it exists) when

    (i) $\mathbf{b} = \begin{bmatrix} 1 \\ 0 \\ 3 \end{bmatrix}$; and (ii) $\mathbf{b} = \begin{bmatrix} 7 \\ 4 \\ 22 \end{bmatrix}$.

12. Verify that $A = \begin{bmatrix} 1 & -1 \\ 0 & 2 \end{bmatrix}$ satisfies $A^2 - 3A + 2I = 0$, and use this fact to show that $A^{-1} = \frac{1}{2}(3I - A)$.

13. Let $Q = \begin{bmatrix} a & -b & -c & -d \\ b & a & -d & c \\ c & d & a & -b \\ d & -c & b & a \end{bmatrix}$. Compute QQ^T and so find Q^{-1} if $Q \neq 0$.

14. Let $U = \begin{bmatrix} 0 & 1 \\ 1 & 0 \end{bmatrix}$. Show that each of U, $-U$, and $-I_2$ is its own inverse and that the product of any two of these is the third.

15. Consider $A = \begin{bmatrix} 1 & 1 \\ -1 & 0 \end{bmatrix}$, $B = \begin{bmatrix} 0 & -1 \\ 1 & 0 \end{bmatrix}$, $C = \begin{bmatrix} 0 & 1 & 0 \\ 0 & 0 & 1 \\ 5 & 0 & 0 \end{bmatrix}$.

    Find the inverses by computing (a) A^6; ◆(b) B^4; and (c) C^3.

◆16. Find the inverse of $\begin{bmatrix} 1 & 0 & 1 \\ c & 1 & c \\ 3 & c & 2 \end{bmatrix}$ in terms of c.

17. If $c \neq 0$, find the inverse of $\begin{bmatrix} 1 & -1 & 1 \\ 2 & -1 & 2 \\ 0 & 2 & c \end{bmatrix}$ in terms of c.

18. Show that A has no inverse when:

    (a) A has a row of zeros.

    ◆(b) A has a column of zeros.

    (c) each row of A sums to 0. [*Hint*: Theorem 5(2).]

    ◆(d) each column of A sums to 0.
    [*Hint*: Corollary 2, Theorem 4.]

19. Let A denote a square matrix.

    (a) Let $YA = 0$ for some matrix $Y \neq 0$. Show that A has no inverse. [*Hint*: Corollary 2, Theorem 4.]

    (b) Use part (a) to show that
    (i) $\begin{bmatrix} 1 & -1 & 1 \\ 0 & 1 & 1 \\ 1 & 0 & 2 \end{bmatrix}$; and ◆(ii) $\begin{bmatrix} 2 & 1 & -1 \\ 1 & 1 & 0 \\ 1 & 0 & -1 \end{bmatrix}$
    have no inverse.

    [*Hint*: For part (ii) compare row 3 with the difference between row 1 and row 2.]

20. If A is invertible, show that

    (a) $A^2 \neq 0$.   ◆(b) $A^k \neq 0$ for all $k = 1, 2, \ldots$.

21. Suppose $AB = 0$, where A and B are square matrices. Show that:

    (a) If one of A and B has an inverse, the other is zero.

    ◆(b) It is impossible for both A and B to have inverses.

    (c) $(BA)^2 = 0$.

◆22. Find the inverse of the X-expansion in Example 16 Section 2.2 and describe it geometrically.

23. Find the inverse of the shear transformation in Example 17 Section 2.2 and describe it geometrically.

24. In each case assume that A is a square matrix that satisfies the given condition. Show that A is invertible and find a formula for A^{-1} in terms of A.

    (a) $A^3 - 3A + 2I = 0$.

    ◆(b) $A^4 + 2A^3 - A - 4I = 0$.

25. Let A and B denote $n \times n$ matrices.

    (a) If A and AB are invertible, show that B is invertible using only (2) and (3) of Theorem 4.

    ◆(b) If AB is invertible, show that both A and B are invertible using Theorem 5.

26. In each case find the inverse of the matrix A using Example 11.

    (a) $A = \begin{bmatrix} -1 & 1 & 2 \\ 0 & 2 & -1 \\ 0 & 1 & -1 \end{bmatrix}$ ◆(b) $A = \begin{bmatrix} 3 & 1 & 0 \\ 5 & 2 & 0 \\ 1 & 3 & -1 \end{bmatrix}$

    (c) $A = \begin{bmatrix} 3 & 4 & 0 & 0 \\ 2 & 3 & 0 & 0 \\ 1 & -1 & 1 & 3 \\ 3 & 1 & 1 & 4 \end{bmatrix}$

    ◆(d) $A = \begin{bmatrix} 2 & 1 & 5 & 2 \\ 1 & 1 & -1 & 0 \\ 0 & 0 & 1 & -1 \\ 0 & 0 & 1 & -2 \end{bmatrix}$

27. If A and B are invertible symmetric matrices such that $AB = BA$, show that A^{-1}, AB, AB^{-1}, and $A^{-1}B^{-1}$ are also invertible and symmetric.

28. Let A be an $n \times n$ matrix and let I be the $n \times n$ identity matrix.

    (a) If $A^2 = 0$, verify that $(I - A)^{-1} = I + A$.

    (b) If $A^3 = 0$, verify that $(I - A)^{-1} = I + A + A^2$.

    (c) Find the inverse of $\begin{bmatrix} 1 & 2 & -1 \\ 0 & 1 & 3 \\ 0 & 0 & 1 \end{bmatrix}$.

    ◆(d) If $A^n = 0$, find the formula for $(I - A)^{-1}$.

29. Prove property 6 of Theorem 4: If A is invertible and $a \neq 0$, then aA is invertible and $(aA)^{-1} = \frac{1}{a}A^{-1}$.

30. Let A, B, and C denote $n \times n$ matrices. Using only Theorem 4, show that:
    (a) If A, C, and ABC are all invertible, B is invertible.
    ◆(b) If AB and BA are both invertible, A and B are both invertible.

31. Let A and B denote invertible $n \times n$ matrices.
    (a) If $A^{-1} = B^{-1}$, does it mean that $A = B$? Explain.
    (b) Show that $A = B$ if and only if $A^{-1}B = I$.

32. Let A, B, and C be $n \times n$ matrices, with A and B invertible. Show that
    ◆(a) If A commutes with C, then A^{-1} commutes with C.
    (b) If A commutes with B, then A^{-1} commutes with B^{-1}.

33. Let A and B be square matrices of the same size.
    (a) Show that $(AB)^2 = A^2B^2$ if $AB = BA$.
    ◆(b) If A and B are invertible and $(AB)^2 = A^2B^2$, show that $AB = BA$.
    (c) If $A = \begin{bmatrix} 1 & 0 \\ 0 & 0 \end{bmatrix}$ and $B = \begin{bmatrix} 1 & 1 \\ 0 & 0 \end{bmatrix}$, show that $(AB)^2 = A^2B^2$ but $AB \neq BA$.

◆34. Let A and B be $n \times n$ matrices for which AB is invertible. Show that A and B are both invertible.

35. Consider $A = \begin{bmatrix} 1 & 3 & -1 \\ 2 & 1 & 5 \\ 1 & -7 & 13 \end{bmatrix}$, $B = \begin{bmatrix} 1 & 1 & 2 \\ 3 & 0 & -3 \\ -2 & 5 & 17 \end{bmatrix}$.
    (a) Show that A is not invertible by finding a nonzero 1×3 matrix Y such that $YA = 0$.
      [Hint: Row 3 of A equals 2(row 2) − 3 (row 1).]
    ◆(b) Show that B is not invertible.
      [Hint: Column 3 = 3(column 2) − column 1.]

36. Show that a square matrix A is invertible if and only if it can be left-cancelled: $AB = AC$ implies $B = C$.

37. If $U^2 = I$, show that $I + U$ is not invertible unless $U = I$.

38. (a) If J is the 4×4 matrix with every entry 1, show that $I - \frac{1}{2}J$ is self-inverse and symmetric.
    ◆(b) If X is $n \times m$ and satisfies $X^TX = I_m$, show that $I_n - 2XX^T$ is self-inverse and symmetric.

39. An $n \times n$ matrix P is called an idempotent if $P^2 = P$. Show that:
    (a) I is the only invertible idempotent.
    ◆(b) P is an idempotent if and only if $I - 2P$ is self-inverse.
    (c) U is self-inverse if and only if $U = I - 2P$ for some idempotent P.
    (d) $I - aP$ is invertible for any $a \neq 1$, and $(I - aP)^{-1} = I + \left(\frac{a}{1-a}\right)P$.

40. If $A^2 = kA$, where $k \neq 0$, show that A is invertible if and only if $A = kI$.

41. Let A and B denote $n \times n$ invertible matrices.
    (a) Show that $A^{-1} + B^{-1} = A^{-1}(A + B)B^{-1}$.
    ◆(b) If $A + B$ is also invertible, show that $A^{-1} + B^{-1}$ is invertible and find a formula for $(A^{-1} + B^{-1})^{-1}$.

42. Let A and B be $n \times n$ matrices, and let I be the $n \times n$ identity matrix.
    (a) Verify that $A(I + BA) = (I + AB)A$ and that $(I + BA)B = B(I + AB)$.
    (b) If $I + AB$ is invertible, verify that $I + BA$ is also invertible and that $(I + BA)^{-1} = I - B(I + AB)^{-1}A$.

SECTION 2.5 Elementary Matrices

It is now clear that elementary row operations are important in linear algebra: They are essential in solving linear systems (using the gaussian algorithm) and in inverting a matrix (using the matrix inversion algorithm). It turns out that they can be performed by left multiplying by certain invertible matrices. These matrices are the subject of this section.

Chapter 2   Matrix Algebra

Definition 2.12 An $n \times n$ matrix E is called an **elementary matrix** if it can be obtained from the identity matrix I_n by a single elementary row operation (called the operation **corresponding** to E). We say that E is of type I, II, or III if the operation is of that type (see page 6).

Hence
$$E_1 = \begin{bmatrix} 0 & 1 \\ 1 & 0 \end{bmatrix}, \quad E_2 = \begin{bmatrix} 1 & 0 \\ 0 & 9 \end{bmatrix}, \quad \text{and} \quad E_3 = \begin{bmatrix} 1 & 5 \\ 0 & 1 \end{bmatrix},$$
are elementary of types I, II, and III, respectively, obtained from the 2×2 identity matrix by interchanging rows 1 and 2, multiplying row 2 by 9, and adding 5 times row 2 to row 1.

Suppose now that a matrix $A = \begin{bmatrix} a & b & c \\ p & q & r \end{bmatrix}$ is left multiplied by the above elementary matrices E_1, E_2, and E_3. The results are:

$$E_1 A = \begin{bmatrix} 0 & 1 \\ 1 & 0 \end{bmatrix}\begin{bmatrix} a & b & c \\ p & q & r \end{bmatrix} = \begin{bmatrix} p & q & r \\ a & b & c \end{bmatrix}$$

$$E_2 A = \begin{bmatrix} 1 & 0 \\ 0 & 9 \end{bmatrix}\begin{bmatrix} a & b & c \\ p & q & r \end{bmatrix} = \begin{bmatrix} a & b & c \\ 9p & 9q & 9r \end{bmatrix}$$

$$E_3 A = \begin{bmatrix} 1 & 5 \\ 0 & 1 \end{bmatrix}\begin{bmatrix} a & b & c \\ p & q & r \end{bmatrix} = \begin{bmatrix} a+5p & b+5q & c+5r \\ p & q & r \end{bmatrix}$$

In each case, left multiplying A by the elementary matrix has the *same* effect as doing the corresponding row operation to A. This works in general.

Lemma 1[10]

If an elementary row operation is performed on an $m \times n$ matrix A, the result is EA where E is the elementary matrix obtained by performing the same operation on the $m \times m$ identity matrix.

PROOF

We prove it for operations of type III; the proofs for types I and II are left as exercises. Let E be the elementary matrix corresponding to the operation that adds k times row p to row $q \neq p$. The proof depends on the fact that each row of EA is equal to the corresponding row of E times A. Let $K_1, K_2, \ldots, K_m$ denote the rows of I_m. Then row i of E is K_i if $i \neq q$, while row q of E is $K_q + kK_p$. Hence:

If $i \neq q$ then row i of $EA = K_i A = $ (row i of A).

Row q of $EA = (K_q + kK_p)A = K_q A + k(K_p A)$
$\phantom{\text{Row } q \text{ of } EA} = $ (row q of A) plus k (row p of A).

Thus EA is the result of adding k times row p of A to row q, as required.

The effect of an elementary row operation can be reversed by another such operation (called its inverse) which is also elementary of the same type (see the discussion following Example 3 Section 1.1). It follows that each elementary matrix E is invertible. In fact, if a row operation on I produces E, then the inverse operation carries E back to I. If F is the elementary matrix corresponding to the inverse operation, this means $FE = I$ (by Lemma 1). Thus $F = E^{-1}$ and we have proved

---
10   A *lemma* is an auxiliary theorem used in the proof of other theorems.

Lemma 2

Every elementary matrix E is invertible, and E^{-1} is also a elementary matrix (of the same type). Moreover, E^{-1} corresponds to the inverse of the row operation that produces E.

The following table gives the inverse of each type of elementary row operation:

Type	Operation	Inverse Operation
I	Interchange rows p and q	Interchange rows p and q
II	Multiply row p by $k \neq 0$	Multiply row p by $1/k$
III	Add k times row p to row $q \neq p$	Subtract k times row p from row q

Note that elementary matrices of type I are self-inverse.

EXAMPLE 1

Find the inverse of each of the elementary matrices

$$E_1 = \begin{bmatrix} 0 & 1 & 0 \\ 1 & 0 & 0 \\ 0 & 0 & 1 \end{bmatrix}, \quad E_2 = \begin{bmatrix} 1 & 0 & 0 \\ 0 & 1 & 0 \\ 0 & 0 & 9 \end{bmatrix}, \quad \text{and} \quad E_3 = \begin{bmatrix} 1 & 0 & 5 \\ 0 & 1 & 0 \\ 0 & 0 & 1 \end{bmatrix}.$$

Solution ▶ E_1, E_2, and E_3 are of Type I, II, and III respectively, so the table gives

$$E_1^{-1} = \begin{bmatrix} 0 & 1 & 0 \\ 1 & 0 & 0 \\ 0 & 0 & 1 \end{bmatrix} = E_1, \quad E_2^{-1} = \begin{bmatrix} 1 & 0 & 0 \\ 0 & 1 & 0 \\ 0 & 0 & \frac{1}{9} \end{bmatrix}, \quad \text{and} \quad E_3^{-1} = \begin{bmatrix} 1 & 0 & -5 \\ 0 & 1 & 0 \\ 0 & 0 & 1 \end{bmatrix}.$$

Inverses and Elementary Matrices

Suppose that an $m \times n$ matrix A is carried to a matrix B (written $A \to B$) by a series of k elementary row operations. Let $E_1, E_2, \ldots, E_k$ denote the corresponding elementary matrices. By Lemma 1, the reduction becomes

$$A \to E_1 A \to E_2 E_1 A \to E_3 E_2 E_1 A \to \cdots \to E_k E_{k-1} \cdots E_2 E_1 A = B$$

In other words,

$$A \to UA = B \quad \text{where} \quad U = E_k E_{k-1} \cdots E_2 E_1$$

The matrix $U = E_k E_{k-1} \cdots E_2 E_1$ is invertible, being a product of invertible matrices by Lemma 2. Moreover, U can be computed without finding the E_i as follows: If the above series of operations carrying $A \to B$ is performed on I_m in place of A, the result is $I_m \to UI_m = U$. Hence this series of operations carries the block matrix $[A \ I_m] \to [B \ U]$. This, together with the above discussion, proves

Theorem 1

Suppose A is $m \times n$ and $A \to B$ by elementary row operations.

1. *$B = UA$ where U is an $m \times m$ invertible matrix.*
2. *U can be computed by $[A \ I_m] \to [B \ U]$ using the operations carrying $A \to B$.*
3. *$U = E_k E_{k-1} \cdots E_2 E_1$ where $E_1, E_1, \ldots, E_k$ are the elementary matrices corresponding (in order) to the elementary row operations carrying A to B.*

EXAMPLE 2

If $A = \begin{bmatrix} 2 & 3 & 1 \\ 1 & 2 & 1 \end{bmatrix}$, express the reduced row-echelon form R of A as $R = UA$ where U is invertible.

Solution ▶ Reduce the double matrix $[A\ I] \to [R\ U]$ as follows:

$$[A\ I] = \begin{bmatrix} 2 & 3 & 1 & | & 1 & 0 \\ 1 & 2 & 1 & | & 0 & 1 \end{bmatrix} \to \begin{bmatrix} 1 & 2 & 1 & | & 0 & 1 \\ 2 & 3 & 1 & | & 1 & 0 \end{bmatrix} \to \begin{bmatrix} 1 & 2 & 1 & | & 0 & 1 \\ 0 & -1 & -1 & | & 1 & -2 \end{bmatrix}$$

$$\to \begin{bmatrix} 1 & 0 & -1 & | & 2 & -3 \\ 0 & 1 & 1 & | & -1 & 2 \end{bmatrix}$$

Hence $R = \begin{bmatrix} 1 & 0 & -1 \\ 0 & 1 & 1 \end{bmatrix}$ and $U = \begin{bmatrix} 2 & -3 \\ -1 & 2 \end{bmatrix}$.

Now suppose that A is invertible. We know that $A \to I$ by Theorem 5 Section 2.4, so taking $B = I$ in Theorem 1 gives $[A\ I] \to [I\ U]$ where $I = UA$. Thus $U = A^{-1}$, so we have $[A\ I] \to [I\ A^{-1}]$. This is the matrix inversion algorithm, derived (in another way) in Section 2.4. However, more is true: Theorem 1 gives $A^{-1} = U = E_k E_{k-1} \cdots E_2 E_1$ where $E_1, E_2, \ldots, E_k$ are the elementary matrices corresponding (in order) to the row operations carrying $A \to I$. Hence

$$A = (A^{-1})^{-1} = (E_k E_{k-1} \cdots E_2 E_1)^{-1} = E_1^{-1} E_2^{-1} \cdots E_{k-1}^{-1} E_k^{-1}. \tag{*}$$

By Lemma 2, this shows that every invertible matrix A is a product of elementary matrices. Since elementary matrices are invertible (again by Lemma 2), this proves the following important characterization of invertible matrices.

Theorem 2

A square matrix is invertible if and only if it is a product of elementary matrices.

It follows that $A \to B$ by row operations if and only if $B = UA$ for some invertible matrix B. In this case we say that A and B are **row-equivalent**. (See Exercise 17.)

EXAMPLE 3

Express $A = \begin{bmatrix} -2 & 3 \\ 1 & 0 \end{bmatrix}$ as a product of elementary matrices.

Solution ▶ Using Lemma 1, the reduction of $A \to I$ is as follows:

$$A = \begin{bmatrix} -2 & 3 \\ 1 & 0 \end{bmatrix} \to E_1 A = \begin{bmatrix} 1 & 0 \\ -2 & 3 \end{bmatrix} \to E_2 E_1 A = \begin{bmatrix} 1 & 0 \\ 0 & 3 \end{bmatrix} \to E_3 E_2 E_1 A = \begin{bmatrix} 1 & 0 \\ 0 & 1 \end{bmatrix}$$

where the corresponding elementary matrices are

$$E_1 = \begin{bmatrix} 0 & 1 \\ 1 & 0 \end{bmatrix}, \quad E_2 = \begin{bmatrix} 1 & 0 \\ 2 & 1 \end{bmatrix}, \quad E_3 = \begin{bmatrix} 1 & 0 \\ 0 & \frac{1}{3} \end{bmatrix}$$

Hence $(E_3 E_2 E_1)A = I$, so:

$$A = (E_3 E_2 E_1)^{-1} = E_1^{-1} E_2^{-1} E_3^{-1} = \begin{bmatrix} 0 & 1 \\ 1 & 0 \end{bmatrix} \begin{bmatrix} 1 & 0 \\ -2 & 1 \end{bmatrix} \begin{bmatrix} 1 & 0 \\ 0 & 3 \end{bmatrix}.$$

Smith Normal Form

Let A be an $m \times n$ matrix of rank r, and let R be the reduced row-echelon form of A. Theorem 1 shows that $R = UA$ where U is invertible, and that U can be found from $[A \ I_m] \to [R \ U]$.

The matrix R has r leading ones (since $rank\ A = r$) so, as R is reduced, the $n \times m$ matrix R^T contains each row of I_r in the first r columns. Thus row operations will carry $R^T \to \begin{bmatrix} I_r & 0 \\ 0 & 0 \end{bmatrix}_{n \times m}$. Hence Theorem 1 (again) shows that $\begin{bmatrix} I_r & 0 \\ 0 & 0 \end{bmatrix}_{n \times m} = U_1 R^T$ where U_1 is an $n \times n$ invertible matrix. Writing $V = U_1^T$, we obtain

$$UAV = RV = RU_1^T = (U_1 R^T)^T = \left(\begin{bmatrix} I_r & 0 \\ 0 & 0 \end{bmatrix}_{n \times m}\right)^T = \begin{bmatrix} I_r & 0 \\ 0 & 0 \end{bmatrix}_{m \times n}.$$

Moreover, the matrix $U_1 = V^T$ can be computed by $[R^T \ I_n] \to \left[\begin{bmatrix} I_r & 0 \\ 0 & 0 \end{bmatrix}_{n \times m} V^T\right]$. This proves

Theorem 3

Let A be an $m \times n$ matrix of rank r. There exist invertible matrices U and V of size $m \times m$ and $n \times n$, respectively, such that

$$UAV = \begin{bmatrix} I_r & 0 \\ 0 & 0 \end{bmatrix}_{m \times n}.$$

Moreover, if R is the reduced row-echelon form of A, then:

1. U can be computed by $[A \ I_m] \to [R \ U]$;
2. V can be computed by $[R^T \ I_n] \to \left[\begin{bmatrix} I_r & 0 \\ 0 & 0 \end{bmatrix}_{n \times m} V^T\right]$.

If A is an $m \times n$ matrix of rank r, the matrix $\begin{bmatrix} I_r & 0 \\ 0 & 0 \end{bmatrix}$ is called the **Smith normal form**[11] of A. Whereas the reduced row-echelon form of A is the "nicest" matrix to which A can be carried by row operations, the Smith canonical form is the "nicest" matrix to which A can be carried by *row and column* operations. This is because doing row operations to R^T amounts to doing *column* operations to R and then transposing.

EXAMPLE 4

Given $A = \begin{bmatrix} 1 & -1 & 1 & 2 \\ 2 & -2 & 1 & -1 \\ -1 & 1 & 0 & 3 \end{bmatrix}$, find invertible matrices U and V such that $UAV = \begin{bmatrix} I_r & 0 \\ 0 & 0 \end{bmatrix}$, where $r = rank\ A$.

Solution ▶ The matrix U and the reduced row-echelon form R of A are computed by the row reduction $[A \ I_3] \to [R \ U]$:

---

11 Named after Henry John Stephen Smith (1826–83).

$$\begin{bmatrix} 1 & -1 & 1 & 2 & | & 1 & 0 & 0 \\ 2 & -2 & 1 & -1 & | & 0 & 1 & 0 \\ -1 & 1 & 0 & 3 & | & 0 & 0 & 1 \end{bmatrix} \to \begin{bmatrix} 1 & -1 & 0 & -3 & | & -1 & 1 & 0 \\ 0 & 0 & 1 & 5 & | & 2 & -1 & 0 \\ 0 & 0 & 0 & 0 & | & -1 & 1 & 1 \end{bmatrix}$$

Hence

$$R = \begin{bmatrix} 1 & -1 & 0 & -3 \\ 0 & 0 & 1 & 5 \\ 0 & 0 & 0 & 0 \end{bmatrix} \quad \text{and} \quad U = \begin{bmatrix} -1 & 1 & 0 \\ 2 & -1 & 0 \\ -1 & 1 & 1 \end{bmatrix}$$

In particular, $r = \text{rank } R = 2$. Now row-reduce $[R^T \ I_4] \to \left[\begin{bmatrix} I_r & 0 \\ 0 & 0 \end{bmatrix} \ V^T\right]$:

$$\begin{bmatrix} 1 & 0 & 0 & | & 1 & 0 & 0 & 0 \\ -1 & 0 & 0 & | & 0 & 1 & 0 & 0 \\ 0 & 1 & 0 & | & 0 & 0 & 1 & 0 \\ -3 & 5 & 0 & | & 0 & 0 & 0 & 1 \end{bmatrix} \to \begin{bmatrix} 1 & 0 & 0 & | & 1 & 0 & 0 & 0 \\ 0 & 1 & 0 & | & 0 & 0 & 1 & 0 \\ 0 & 0 & 0 & | & 1 & 1 & 0 & 0 \\ 0 & 0 & 0 & | & 3 & 0 & -5 & 1 \end{bmatrix}.$$

whence

$$V^T = \begin{bmatrix} 1 & 0 & 0 & 0 \\ 0 & 0 & 1 & 0 \\ 1 & 1 & 0 & 0 \\ 3 & 0 & -5 & 1 \end{bmatrix} \quad \text{so} \quad V = \begin{bmatrix} 1 & 0 & 1 & 3 \\ 0 & 0 & 1 & 0 \\ 0 & 1 & 0 & -5 \\ 0 & 0 & 0 & 1 \end{bmatrix}$$

Then $UAV = \begin{bmatrix} I_2 & 0 \\ 0 & 0 \end{bmatrix}$ as is easily verified.

Uniqueness of the Reduced Row-echelon Form

In this short subsection, Theorem 1 is used to prove the following important theorem.

Theorem 4

If a matrix A is carried to reduced row-echelon matrices R and S by row operations, then $R = S$.

PROOF

Observe first that $UR = S$ for some invertible matrix U (by Theorem 1 there exist invertible matrices P and Q such that $R = PA$ and $S = QA$; take $U = QP^{-1}$). We show that $R = S$ by induction on the number m of rows of R and S. The case $m = 1$ is left to the reader. If R_j and S_j denote column j in R and S respectively, the fact that $UR = S$ gives

$$UR_j = S_j \quad \text{for each } j. \tag{$*$}$$

Since U is invertible, this shows that R and S have the same zero columns. Hence, by passing to the matrices obtained by deleting the zero columns from R and S, we may assume that R and S have no zero columns.

But then the first column of R and S is the first column of I_m because R and S are row-echelon so $(*)$ shows that the first column of U is column 1 of I_m. Now write U, R, and S in block form as follows.

SECTION 2.5 Elementary Matrices

$$U = \begin{bmatrix} 1 & X \\ 0 & V \end{bmatrix}, \quad R = \begin{bmatrix} 1 & X \\ 0 & R' \end{bmatrix}, \quad \text{and} \quad S = \begin{bmatrix} 1 & Z \\ 0 & S' \end{bmatrix}.$$

Since $UR = S$, block multiplication gives $VR' = S'$ so, since V is invertible (U is invertible) and both R' and S' are reduced row-echelon, we obtain $R' = S'$ by induction. Hence R and S have the same number (say r) of leading 1s, and so both have $m-r$ zero rows.

In fact, R and S have leading ones in the same columns, say r of them. Applying (∗) to these columns shows that the first r columns of U are the first r columns of I_m. Hence we can write U, R, and S in block form as follows:

$$U = \begin{bmatrix} I_r & M \\ 0 & W \end{bmatrix}, \quad R = \begin{bmatrix} R_1 & R_2 \\ 0 & 0 \end{bmatrix}, \quad \text{and} \quad S = \begin{bmatrix} S_1 & S_2 \\ 0 & 0 \end{bmatrix}$$

where R_1 and S_1 are $r \times r$. Then block multiplication gives $UR = R$; that is, $S = R$. This completes the proof.

EXERCISES 2.5

1. For each of the following elementary matrices, describe the corresponding elementary row operation and write the inverse.

   (a) $E = \begin{bmatrix} 1 & 0 & 3 \\ 0 & 1 & 0 \\ 0 & 0 & 1 \end{bmatrix}$ ♦(b) $E = \begin{bmatrix} 0 & 0 & 1 \\ 0 & 1 & 0 \\ 1 & 0 & 0 \end{bmatrix}$

   (c) $E = \begin{bmatrix} 1 & 0 & 0 \\ 0 & \frac{1}{2} & 0 \\ 0 & 0 & 1 \end{bmatrix}$ ♦(d) $E = \begin{bmatrix} 1 & 0 & 0 \\ -2 & 1 & 0 \\ 0 & 0 & 1 \end{bmatrix}$

   (e) $E = \begin{bmatrix} 0 & 1 & 0 \\ 1 & 0 & 0 \\ 0 & 0 & 1 \end{bmatrix}$ ♦(f) $E = \begin{bmatrix} 1 & 0 & 0 \\ 0 & 1 & 0 \\ 0 & 0 & 5 \end{bmatrix}$

2. In each case find an elementary matrix E such that $B = EA$.

   (a) $A = \begin{bmatrix} 2 & 1 \\ 3 & -1 \end{bmatrix}, B = \begin{bmatrix} 2 & 1 \\ 1 & -2 \end{bmatrix}$

   ♦(b) $A = \begin{bmatrix} -1 & 2 \\ 0 & 1 \end{bmatrix}, B = \begin{bmatrix} 1 & -2 \\ 0 & 1 \end{bmatrix}$

   (c) $A = \begin{bmatrix} 1 & 1 \\ -1 & 2 \end{bmatrix}, B = \begin{bmatrix} -1 & 2 \\ 1 & 1 \end{bmatrix}$

   ♦(d) $A = \begin{bmatrix} 4 & 1 \\ 3 & 2 \end{bmatrix}, B = \begin{bmatrix} 1 & -1 \\ 3 & 2 \end{bmatrix}$

   (e) $A = \begin{bmatrix} -1 & 1 \\ 1 & -1 \end{bmatrix}, B = \begin{bmatrix} -1 & 1 \\ -1 & 1 \end{bmatrix}$

   ♦(f) $A = \begin{bmatrix} 2 & 1 \\ -1 & 3 \end{bmatrix}, B = \begin{bmatrix} -1 & 3 \\ 2 & 1 \end{bmatrix}$

3. Let $A = \begin{bmatrix} 1 & 2 \\ -1 & 1 \end{bmatrix}$ and $C = \begin{bmatrix} -1 & 1 \\ 2 & 1 \end{bmatrix}$.

   (a) Find elementary matrices E_1 and E_2 such that $C = E_2 E_1 A$.

   ♦(b) Show that there is *no* elementary matrix E such that $C = EA$.

4. If E is elementary, show that A and EA differ in at most two rows.

5. (a) Is I an elementary matrix? Explain.

   ♦(b) Is 0 an elementary matrix? Explain.

6. In each case find an invertible matrix U such that $UA = R$ is in reduced row-echelon form, and express U as a product of elementary matrices.

   (a) $A = \begin{bmatrix} 1 & -1 & 2 \\ -2 & 1 & 0 \end{bmatrix}$ ♦(b) $A = \begin{bmatrix} 1 & 2 & 1 \\ 5 & 12 & -1 \end{bmatrix}$

   (c) $A = \begin{bmatrix} 1 & 2 & -1 & 0 \\ 3 & 1 & 1 & 2 \\ 1 & -3 & 3 & 2 \end{bmatrix}$ ♦(d) $A = \begin{bmatrix} 2 & 1 & -1 & 0 \\ 3 & -1 & 2 & 1 \\ 1 & -2 & 3 & 1 \end{bmatrix}$

7. In each case find an invertible matrix U such that $UA = B$, and express U as a product of elementary matrices.

   (a) $A = \begin{bmatrix} 2 & 1 & 3 \\ -1 & 1 & 2 \end{bmatrix}, B = \begin{bmatrix} 1 & -1 & -2 \\ 3 & 0 & 1 \end{bmatrix}$

   ♦(b) $A = \begin{bmatrix} 2 & -1 & 0 \\ 1 & 1 & 1 \end{bmatrix}, B = \begin{bmatrix} 3 & 0 & 1 \\ 2 & -1 & 0 \end{bmatrix}$

8. In each case factor A as a product of elementary matrices.

   (a) $A = \begin{bmatrix} 1 & 1 \\ 2 & 1 \end{bmatrix}$ ◆(b) $A = \begin{bmatrix} 2 & 3 \\ 1 & 2 \end{bmatrix}$

   (c) $A = \begin{bmatrix} 1 & 0 & 2 \\ 0 & 1 & 1 \\ 2 & 1 & 6 \end{bmatrix}$ ◆(d) $A = \begin{bmatrix} 1 & 0 & -3 \\ 0 & 1 & 4 \\ -2 & 2 & 15 \end{bmatrix}$

9. Let E be an elementary matrix.

   (a) Show that E^T is also elementary of the same type.

   (b) Show that $E^T = E$ if E is of type I or II.

◆10. Show that every matrix A can be factored as $A = UR$ where U is invertible and R is in reduced row-echelon form.

11. If $A = \begin{bmatrix} 1 & 2 \\ 1 & -3 \end{bmatrix}$ and $B = \begin{bmatrix} 5 & 2 \\ -5 & -3 \end{bmatrix}$, find an elementary matrix F such that $AF = B$. [*Hint*: See Exercise 9.]

12. In each case find invertible U and V such that $UAV = \begin{bmatrix} I_r & 0 \\ 0 & 0 \end{bmatrix}$, where $r = \operatorname{rank} A$.

   (a) $A = \begin{bmatrix} 1 & 1 & -1 \\ -2 & -2 & 4 \end{bmatrix}$ ◆(b) $A = \begin{bmatrix} 3 & 2 \\ 2 & 1 \end{bmatrix}$

   (c) $A = \begin{bmatrix} 1 & -1 & 2 & 1 \\ 2 & -1 & 0 & 3 \\ 0 & 1 & -4 & 1 \end{bmatrix}$ ◆(d) $A = \begin{bmatrix} 1 & 1 & 0 & -1 \\ 3 & 2 & 1 & 1 \\ 1 & 0 & 1 & 3 \end{bmatrix}$

13. Prove Lemma 1 for elementary matrices of:

   (a) type I;   (b) type II.

14. While trying to invert A, $[A\ I]$ is carried to $[P\ Q]$ by row operations. Show that $P = QA$.

15. If A and B are $n \times n$ matrices and AB is a product of elementary matrices, show that the same is true of A.

◆16. If U is invertible, show that the reduced row-echelon form of a matrix $[U\ A]$ is $[I\ U^{-1}A]$.

17. Two matrices A and B are called **row-equivalent** (written $A \overset{r}{\sim} B$) if there is a sequence of elementary row operations carrying A to B.

   (a) Show that $A \overset{r}{\sim} B$ if and only if $A = UB$ for some invertible matrix U.

   ◆(b) Show that:

   (i) $A \overset{r}{\sim} A$ for all matrices A.

   (ii) If $A \overset{r}{\sim} B$, then $B \overset{r}{\sim} A$.

   (iii) If $A \overset{r}{\sim} B$ and $B \overset{r}{\sim} C$, then $A \overset{r}{\sim} C$.

   (c) Show that, if A and B are both row-equivalent to some third matrix, then $A \overset{r}{\sim} B$.

   (d) Show that $\begin{bmatrix} 1 & -1 & 3 & 2 \\ 0 & 1 & 4 & 1 \\ 1 & 0 & 8 & 6 \end{bmatrix}$ and $\begin{bmatrix} 1 & -1 & 4 & 5 \\ -2 & 1 & -11 & -8 \\ -1 & 2 & 2 & 2 \end{bmatrix}$
   are row-equivalent. [*Hint*: Consider (c) and Theorem 1 Section 1.2.]

18. If U and V are invertible $n \times n$ matrices, show that $U \overset{r}{\sim} V$. (See Exercise 17.)

19. (See Exercise 17.) Find all matrices that are row-equivalent to:

   (a) $\begin{bmatrix} 0 & 0 & 0 \\ 0 & 0 & 0 \end{bmatrix}$ ◆(b) $\begin{bmatrix} 0 & 0 & 0 \\ 0 & 0 & 1 \end{bmatrix}$

   (c) $\begin{bmatrix} 1 & 0 & 0 \\ 0 & 1 & 0 \end{bmatrix}$ ◆(d) $\begin{bmatrix} 1 & 2 & 0 \\ 0 & 0 & 1 \end{bmatrix}$

20. Let A and B be $m \times n$ and $n \times m$ matrices, respectively. If $m > n$, show that AB is not invertible. [*Hint*: Use Theorem 1 Section 1.3 to find $\mathbf{x} \neq \mathbf{0}$ with $B\mathbf{x} = \mathbf{0}$.]

21. Define an *elementary column operation* on a matrix to be one of the following: (I) Interchange two columns. (II) Multiply a column by a nonzero scalar. (III) Add a multiple of a column to another column. Show that:

   (a) If an elementary column operation is done to an $m \times n$ matrix A, the result is AF, where F is an $n \times n$ elementary matrix.

   (b) Given any $m \times n$ matrix A, there exist $m \times m$ elementary matrices $E_1, \ldots, E_k$ and $n \times n$ elementary matrices $F_1, \ldots, F_p$ such that, in block form,

   $$E_k \cdots E_1 A F_1 \cdots F_p = \begin{bmatrix} I_r & 0 \\ 0 & 0 \end{bmatrix}.$$

22. Suppose B is obtained from A by:

   (a) interchanging rows i and j;

   ◆(b) multiplying row i by $k \neq 0$;

   (c) adding k times row i to row j $(i \neq j)$.

   In each case describe how to obtain B^{-1} from A^{-1}. [*Hint*: See part (a) of the preceding exercise.]

23. Two $m \times n$ matrices A and B are called **equivalent** (written $A \overset{e}{\sim} B$ if there exist

invertible matrices U and V (sizes $m \times m$ and $n \times n$) such that $A = UBV$.

(a) Prove the following the properties of equivalence.

   (i) $A \stackrel{e}{\sim} A$ for all $m \times n$ matrices A.

   (ii) If $A \stackrel{e}{\sim} B$, then $B \stackrel{e}{\sim} A$.

   (iii) If $A \stackrel{e}{\sim} B$ and $B \stackrel{e}{\sim} C$, then $A \stackrel{e}{\sim} C$.

(b) Prove that two $m \times n$ matrices are equivalent if they have the same rank. [*Hint*: Use part (a) and Theorem 3.]

SECTION 2.6 Linear Transformations

If A is an $m \times n$ matrix, recall that the transformation $T_A : \mathbb{R}^n \to \mathbb{R}^m$ defined by

$$T_A(\mathbf{x}) = A\mathbf{x} \text{ for all } \mathbf{x} \text{ in } \mathbb{R}^n$$

is called the *matrix transformation induced* by A. In Section 2.2, we saw that many important geometric transformations were in fact matrix transformations. These transformations can be characterized in a different way. The new idea is that of a linear transformation, one of the basic notions in linear algebra. We define these transformations in this section, and show that they are really just the matrix transformations looked at in another way. Having these two ways to view them turns out to be useful because, in a given situation, one perspective or the other may be preferable.

Linear Transformations

Definition 2.13 *A transformation* $T : \mathbb{R}^n \to \mathbb{R}^m$ *is called a* **linear transformation** *if it satisfies the following two conditions for all vectors* $\mathbf{x}$ *and* $\mathbf{y}$ *in* $\mathbb{R}^n$ *and all scalars* a:

T1 $T(\mathbf{x} + \mathbf{y}) = T(\mathbf{x}) + T(\mathbf{y})$

T2 $T(a\mathbf{x}) = aT(\mathbf{x})$

Of course, $\mathbf{x} + \mathbf{y}$ and $a\mathbf{x}$ here are computed in $\mathbb{R}^n$, while $T(\mathbf{x}) + T(\mathbf{y})$ and $aT(\mathbf{x})$ are in $\mathbb{R}^m$. We say that T *preserves addition* if T1 holds, and that T *preserves scalar multiplication* if T2 holds. Moreover, taking $a = 0$ and $a = -1$ in T2 gives

$$T(\mathbf{0}) = \mathbf{0} \quad \text{and} \quad T(-\mathbf{x}) = -T(\mathbf{x})$$

Hence T preserves the zero vector and the negative of a vector. Even more is true.

Recall that a vector $\mathbf{y}$ in $\mathbb{R}^n$ is called a **linear combination** of vectors $\mathbf{x}_1, \mathbf{x}_2, \ldots, \mathbf{x}_k$ if $\mathbf{y}$ has the form

$$\mathbf{y} = a_1\mathbf{x}_1 + a_2\mathbf{x}_2 + \cdots + a_k\mathbf{x}_k$$

for some scalars $a_1, a_2, \ldots, a_k$. Conditions T1 and T2 combine to show that every linear transformation T *preserves linear combinations* in the sense of the following theorem.

Theorem 1

If $T : \mathbb{R}^n \to \mathbb{R}^m$ is a linear transformation, then for each $k = 1, 2, \ldots$

$$T(a_1\mathbf{x}_1 + a_2\mathbf{x}_2 + \cdots + a_k\mathbf{x}_k) = a_1T(\mathbf{x}_1) + a_2T(\mathbf{x}_2) + \cdots + a_kT(\mathbf{x}_k)$$

for all scalars a_i and all vectors $\mathbf{x}_i$ in $\mathbb{R}^n$.

PROOF

If $k = 1$, it reads $T(a_1\mathbf{x}_1) = a_1 T(\mathbf{x}_1)$ which is Condition T1. If $k = 2$, we have

$$\begin{aligned} T(a_1\mathbf{x}_1 + a_2\mathbf{x}_2) &= T(a_1\mathbf{x}_1) + T(a_2\mathbf{x}_2) & \text{by Condition T1} \\ &= a_1 T(\mathbf{x}_1) + a_2 T(\mathbf{x}_2) & \text{by Condition T2} \end{aligned}$$

If $k = 3$, we use the case $k = 2$ to obtain

$$\begin{aligned} T(a_1\mathbf{x}_1 + a_2\mathbf{x}_2 + a_3\mathbf{x}_3) &= T[(a_1\mathbf{x}_1 + a_2\mathbf{x}_2) + a_3\mathbf{x}_3] & \text{collect terms} \\ &= T(a_1\mathbf{x}_1 + a_2\mathbf{x}_2) + T(a_3\mathbf{x}_3) & \text{by Condition T1} \\ &= [a_1 T(\mathbf{x}_1) + a_2 T(\mathbf{x}_2)] + T(a_3\mathbf{x}_3) & \text{by the case } k = 2 \\ &= [a_1 T(\mathbf{x}_1) + a_2 T(\mathbf{x}_2)] + a_3 T(\mathbf{x}_3) & \text{by Condition T2} \end{aligned}$$

The proof for any k is similar, using the previous case $k - 1$ and Conditions T1 and T2.

The method of proof in Theorem 1 is called *mathematical induction* (Appendix C).

Theorem 1 shows that if T is a linear transformation and $T(\mathbf{x}_1), T(\mathbf{x}_2), \ldots, T(\mathbf{x}_k)$ are all known, then $T(\mathbf{y})$ can be easily computed for any linear combination $\mathbf{y}$ of $\mathbf{x}_1, \mathbf{x}_2, \ldots, \mathbf{x}_k$. This is a very useful property of linear transformations, and is illustrated in the next example.

EXAMPLE 1

If $T: \mathbb{R}^2 \to \mathbb{R}^2$ is a linear transformation, $T\begin{bmatrix}1\\1\end{bmatrix} = \begin{bmatrix}2\\-3\end{bmatrix}$ and $T\begin{bmatrix}1\\-2\end{bmatrix} = \begin{bmatrix}5\\1\end{bmatrix}$, find $T\begin{bmatrix}4\\3\end{bmatrix}$.

Solution ▶ Write $\mathbf{z} = \begin{bmatrix}4\\3\end{bmatrix}$, $\mathbf{x} = \begin{bmatrix}1\\1\end{bmatrix}$, and $\mathbf{y} = \begin{bmatrix}1\\-2\end{bmatrix}$ for convenience. Then we know $T(\mathbf{x})$ and $T(\mathbf{y})$ and we want $T(\mathbf{z})$, so it is enough by Theorem 1 to express $\mathbf{z}$ as a linear combination of $\mathbf{x}$ and $\mathbf{y}$. That is, we want to find numbers a and b such that $\mathbf{z} = a\mathbf{x} + b\mathbf{y}$. Equating entries gives two equations $4 = a + b$ and $3 = a - 2b$. The solution is, $a = \frac{11}{3}$ and $b = \frac{1}{3}$, so $\mathbf{z} = \frac{11}{3}\mathbf{x} + \frac{1}{3}\mathbf{y}$. Thus Theorem 1 gives

$$T(\mathbf{z}) = \tfrac{11}{3}T(\mathbf{x}) + \tfrac{1}{3}T(\mathbf{y}) = \tfrac{11}{3}\begin{bmatrix}2\\-3\end{bmatrix} + \tfrac{1}{3}\begin{bmatrix}5\\1\end{bmatrix} = \tfrac{1}{3}\begin{bmatrix}27\\-32\end{bmatrix}$$

This is what we wanted.

EXAMPLE 2

If A is $m \times n$, the matrix transformation $T_A: \mathbb{R}^n \to \mathbb{R}^m$, A, is a linear transformation.

Solution ▶ We have $T_A(\mathbf{x}) = A\mathbf{x}$ for all $\mathbf{x}$ in $\mathbb{R}^n$, so Theorem 2 Section 2.2 gives

$$T_A(\mathbf{x} + \mathbf{y}) = A(\mathbf{x} + \mathbf{y}) = A\mathbf{x} + A\mathbf{y} = T_A(\mathbf{x}) + T_A(\mathbf{y})$$

and

$$T_A(a\mathbf{x}) = A(a\mathbf{x}) = a(A\mathbf{x}) = aT_A(\mathbf{x})$$

hold for all $\mathbf{x}$ and $\mathbf{y}$ in $\mathbb{R}^n$ and all scalars a. Hence T_A satisfies T1 and T2, and so is linear.

SECTION 2.6 Linear Transformations

The remarkable thing is that the *converse* of Example 2 is true: Every linear transformation $T: \mathbb{R}^n \to \mathbb{R}^m$ is actually a matrix transformation. To see why, we define the **standard basis** of $\mathbb{R}^n$ to be the set of columns

$$\{\mathbf{e}_1, \mathbf{e}_2, \ldots, \mathbf{e}_n\}$$

of the identity matrix I_n. Then each $\mathbf{e}_i$ is in $\mathbb{R}^n$ and every vector $\mathbf{x} = \begin{bmatrix} x_1 \\ x_2 \\ \vdots \\ x_n \end{bmatrix}$ in $\mathbb{R}^n$ is a linear combination of the $\mathbf{e}_i$. In fact:

$$\mathbf{x} = x_1 \mathbf{e}_1 + x_2 \mathbf{e}_2 + \cdots + x_n \mathbf{e}_n$$

as the reader can verify. Hence Theorem 1 shows that

$$T(\mathbf{x}) = T(x_1 \mathbf{e}_1 + x_2 \mathbf{e}_2 + \cdots + x_n \mathbf{e}_n) = x_1 T(\mathbf{e}_1) + x_2 T(\mathbf{e}_2) + \cdots + x_n T(\mathbf{e}_n)$$

Now observe that each $T(\mathbf{e}_i)$ is a column in $\mathbb{R}^m$, so

$$A = [T(\mathbf{e}_1) \; T(\mathbf{e}_2) \; \cdots \; T(\mathbf{e}_n)]$$

is an $m \times n$ matrix. Hence we can apply Definition 2.5 to get

$$T(\mathbf{x}) = x_1 T(\mathbf{e}_1) + x_2 T(\mathbf{e}_2) + \cdots + x_n T(\mathbf{e}_n) = [T(\mathbf{e}_1) \; T(\mathbf{e}_2) \; \cdots \; T(\mathbf{e}_n)] \begin{bmatrix} x_1 \\ x_2 \\ \vdots \\ x_n \end{bmatrix} = A\mathbf{x}.$$

Since this holds for every $\mathbf{x}$ in $\mathbb{R}^n$, it shows that T is the matrix transformation induced by A, and so proves most of the following theorem.

Theorem 2

Let $T: \mathbb{R}^n \to \mathbb{R}^m$ be a transformation.
1. T is linear if and only if it is a matrix transformation.
2. In this case $T = T_A$ is the matrix transformation induced by a unique $m \times n$ matrix A, given in terms of its columns by

$$A = [T(\mathbf{e}_1) \; T(\mathbf{e}_2) \; \cdots \; T(\mathbf{e}_n)]$$

where $\{\mathbf{e}_1, \mathbf{e}_2, \ldots, \mathbf{e}_n\}$ is the standard basis of $\mathbb{R}^n$.

PROOF

It remains to verify that the matrix A is unique. Suppose that T is induced by another matrix B. Then $T(\mathbf{x}) = B\mathbf{x}$ for all $\mathbf{x}$ in $\mathbb{R}^n$. But $T(\mathbf{x}) = A\mathbf{x}$ for each $\mathbf{x}$, so $B\mathbf{x} = A\mathbf{x}$ for every $\mathbf{x}$. Hence $A = B$ by Theorem 5 Section 2.2.

Hence we can speak of *the* matrix of a linear transformation. Because of Theorem 2 we may (and shall) use the phrases "linear transformation" and "matrix transformation" interchangeably.

EXAMPLE 3

Define $T_1: \mathbb{R}^3 \to \mathbb{R}^2$ by $T\begin{bmatrix} x_1 \\ x_2 \\ x_3 \end{bmatrix} = \begin{bmatrix} x_1 \\ x_2 \end{bmatrix}$ for all $\begin{bmatrix} x_1 \\ x_2 \\ x_3 \end{bmatrix}$ in $\mathbb{R}^3$. Show that T is a linear transformation and use Theorem 2 to find its matrix.

Solution ▶ Write $\mathbf{x} = \begin{bmatrix} x_1 \\ x_2 \\ x_3 \end{bmatrix}$ and $\mathbf{y} = \begin{bmatrix} y_1 \\ y_2 \\ y_3 \end{bmatrix}$, so that $\mathbf{x} + \mathbf{y} = \begin{bmatrix} x_1 + y_1 \\ x_2 + y_2 \\ x_3 + y_3 \end{bmatrix}$. Hence

$$T(\mathbf{x} + \mathbf{y}) = \begin{bmatrix} x_1 + y_1 \\ x_2 + y_2 \end{bmatrix} = \begin{bmatrix} x_1 \\ x_2 \end{bmatrix} + \begin{bmatrix} y_1 \\ y_2 \end{bmatrix} = T(\mathbf{x}) + T(\mathbf{y})$$

Similarly, the reader can verify that $T(a\mathbf{x}) = aT(\mathbf{x})$ for all a in $\mathbb{R}$, so T is a linear transformation. Now the standard basis of $\mathbb{R}^3$ is

$$\mathbf{e}_1 = \begin{bmatrix} 1 \\ 0 \\ 0 \end{bmatrix}, \quad \mathbf{e}_2 = \begin{bmatrix} 0 \\ 1 \\ 0 \end{bmatrix}, \quad \text{and} \quad \mathbf{e}_3 = \begin{bmatrix} 0 \\ 0 \\ 1 \end{bmatrix}$$

so, by Theorem 2, the matrix of T is

$$A = [T(\mathbf{e}_1) \; T(\mathbf{e}_2) \; T(\mathbf{e}_3)] = \begin{bmatrix} 1 & 0 & 0 \\ 0 & 1 & 0 \end{bmatrix}.$$

Of course, the fact that $T\begin{bmatrix} x_1 \\ x_2 \\ x_3 \end{bmatrix} = \begin{bmatrix} x_1 \\ x_2 \end{bmatrix} = \begin{bmatrix} 1 & 0 & 0 \\ 0 & 1 & 0 \end{bmatrix} \begin{bmatrix} x_1 \\ x_2 \\ x_3 \end{bmatrix}$ shows directly that T is a matrix transformation (hence linear) and reveals the matrix.

To illustrate how Theorem 2 is used, we rederive the matrices of the transformations in Examples 13 and 15 in Section 2.2.

EXAMPLE 4

Let $Q_0: \mathbb{R}^2 \to \mathbb{R}^2$ denote reflection in the x axis (as in Example 13 Section 2.2) and let $R_{\frac{\pi}{2}}: \mathbb{R}^2 \to \mathbb{R}^2$ denote counterclockwise rotation through $\frac{\pi}{2}$ about the origin (as in Example 15 Section 2.2). Use Theorem 2 to find the matrices of Q_0 and $R_{\frac{\pi}{2}}$.

Solution ▶ Observe that Q_0 and $R_{\frac{\pi}{2}}$ are linear by Example 2 (they are matrix transformations), so Theorem 2 applies to them. The standard basis of $\mathbb{R}^2$ is $\{\mathbf{e}_1, \mathbf{e}_2\}$ where $\mathbf{e}_1 = \begin{bmatrix} 1 \\ 0 \end{bmatrix}$ points along the positive x axis, and $\mathbf{e}_2 = \begin{bmatrix} 0 \\ 1 \end{bmatrix}$ points along the positive y axis (see Figure 1).

The reflection of $\mathbf{e}_1$ in the x axis is $\mathbf{e}_1$ itself because $\mathbf{e}_1$ points along the x axis, and the reflection of $\mathbf{e}_2$ in the x axis is $-\mathbf{e}_2$ because $\mathbf{e}_2$ is perpendicular to the x axis. In other words, $Q_0(\mathbf{e}_1) = \mathbf{e}_1$ and $Q_0(\mathbf{e}_2) = -\mathbf{e}_2$. Hence Theorem 2 shows that the matrix of Q_0 is

$$[Q_0(\mathbf{e}_1) \; Q_0(\mathbf{e}_2)] = [\mathbf{e}_1 \; -\mathbf{e}_2] = \begin{bmatrix} 1 & 0 \\ 0 & -1 \end{bmatrix}$$

which agrees with Example 13 Section 2.2.

Similarly, rotating $\mathbf{e}_1$ through $\frac{\pi}{2}$ counterclockwise about the origin produces $\mathbf{e}_2$, and rotating $\mathbf{e}_2$ through $\frac{\pi}{2}$ counterclockwise about the origin gives $-\mathbf{e}_1$. That is, $R_{\frac{\pi}{2}}(\mathbf{e}_1) = \mathbf{e}_2$ and $R_{\frac{\pi}{2}}(\mathbf{e}_2) = -\mathbf{e}_1$. Hence, again by Theorem 2, the matrix of $R_{\frac{\pi}{2}}$ is

$$[R_{\frac{\pi}{2}}(\mathbf{e}_1) \; R_{\frac{\pi}{2}}(\mathbf{e}_2)] = [\mathbf{e}_2 \; -\mathbf{e}_1] = \begin{bmatrix} 0 & -1 \\ 1 & 0 \end{bmatrix}$$

agreeing with Example 15 Section 2.2.

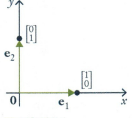

FIGURE 1

SECTION 2.6 Linear Transformations

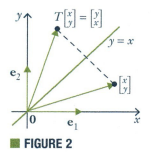

FIGURE 2

EXAMPLE 5

Let $Q_1: \mathbb{R}^2 \to \mathbb{R}^2$ denote reflection in the line $y = x$. Show that Q_1 is a matrix transformation, find its matrix, and use it to illustrate Theorem 2.

Solution ▸ Figure 2 shows that $Q_1 \begin{bmatrix} x \\ y \end{bmatrix} = \begin{bmatrix} y \\ x \end{bmatrix}$. Hence $Q_1 \begin{bmatrix} x \\ y \end{bmatrix} = \begin{bmatrix} 0 & 1 \\ 1 & 0 \end{bmatrix} \begin{bmatrix} y \\ x \end{bmatrix}$,

so Q_1 is the matrix transformation induced by the matrix $A = \begin{bmatrix} 0 & 1 \\ 1 & 0 \end{bmatrix}$.

Hence Q_1 is linear (by Example 2) and so Theorem 2 applies. If $\mathbf{e}_1 = \begin{bmatrix} 1 \\ 0 \end{bmatrix}$ and $\mathbf{e}_1 = \begin{bmatrix} 0 \\ 1 \end{bmatrix}$ are the standard basis of $\mathbb{R}^2$, then it is clear geometrically that $Q_1(\mathbf{e}_1) = \mathbf{e}_2$ and $Q_1(\mathbf{e}_2) = \mathbf{e}_1$. Thus (by Theorem 2) the matrix of Q_1 is $[Q_1(\mathbf{e}_1) \; Q_1(\mathbf{e}_2)] = [\mathbf{e}_2 \; \mathbf{e}_1] = A$ as before.

Recall that, given two "linked" transformations
$$\mathbb{R}^k \xrightarrow{T} \mathbb{R}^n \xrightarrow{S} \mathbb{R}^m,$$
we can apply T first and then apply S, and so obtain a new transformation
$$S \circ T : \mathbb{R}^k \to \mathbb{R}^m,$$
called the **composite** of S and T, defined by
$$(S \circ T)(\mathbf{x}) = S[T(\mathbf{x})] \text{ for all } \mathbf{x} \text{ in } \mathbb{R}^k.$$
If S and T are linear, the action of $S \circ T$ can be computed by multiplying their matrices.

Theorem 3

Let $\mathbb{R}^k \xrightarrow{T} \mathbb{R}^n \xrightarrow{S} \mathbb{R}^m$, be linear transformations, and let A and B be the matrices of S and T respectively. Then $S \circ T$ is linear with matrix AB.

PROOF

$(S \circ T)(\mathbf{x}) = S[T(\mathbf{x})] = A[B\mathbf{x}] = (AB)\mathbf{x}$ for all $\mathbf{x}$ in $\mathbb{R}^k$.

Theorem 3 shows that the action of the composite $S \circ T$ is determined by the matrices of S and T. But it also provides a very useful interpretation of matrix multiplication. If A and B are matrices, the product matrix AB induces the transformation resulting from first applying B and then applying A. Thus the study of matrices can cast light on geometrical transformations and vice-versa. Here is an example.

EXAMPLE 6

Show that reflection in the x axis followed by rotation through $\frac{\pi}{2}$ is reflection in the line $y = x$.

Solution ▶ The composite in question is $R_{\frac{\pi}{2}} \circ Q_0$ where Q_0 is reflection in the x axis and $R_{\frac{\pi}{2}}$ is rotation through $\frac{\pi}{2}$. By Example 4, $R_{\frac{\pi}{2}}$ has matrix $A = \begin{bmatrix} 0 & -1 \\ 1 & 0 \end{bmatrix}$ and Q_0 has matrix $B = \begin{bmatrix} 1 & 0 \\ 0 & -1 \end{bmatrix}$. Hence Theorem 3 shows that the matrix of $R_{\frac{\pi}{2}} \circ Q_0$ is $AB = \begin{bmatrix} 0 & -1 \\ 1 & 0 \end{bmatrix} \begin{bmatrix} 1 & 0 \\ 0 & -1 \end{bmatrix} = \begin{bmatrix} 0 & 1 \\ 1 & 0 \end{bmatrix}$, which is the matrix of reflection in the line $y = x$ by Example 3.

This conclusion can also be seen geometrically. Let **x** be a typical point in $\mathbb{R}^2$, and assume that **x** makes an angle α with the positive x axis. The effect of first applying Q_0 and then applying $R_{\frac{\pi}{2}}$ is shown in Figure 3. The fact that $R_{\frac{\pi}{2}}[Q_0(\mathbf{x})]$ makes the angle α with the positive y axis shows that $R_{\frac{\pi}{2}}[Q_0(\mathbf{x})]$ is the reflection of **x** in the line $y = x$.

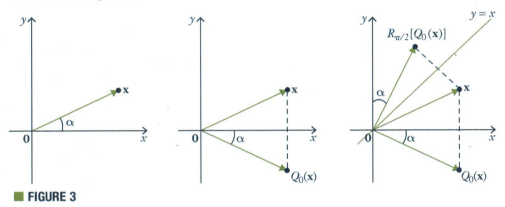

■ **FIGURE 3**

In Theorem 3, we saw that the matrix of the composite of two linear transformations is the product of their matrices (in fact, matrix products were defined so that this is the case). We are going to apply this fact to rotations, reflections, and projections in the plane. Before proceeding, we pause to present useful geometrical descriptions of vector addition and scalar multiplication in the plane, and to give a short review of angles and the trigonometric functions.

Some Geometry

As we have seen, it is convenient to view a vector **x** in $\mathbb{R}^2$ as an arrow from the origin to the point **x** (see Section 2.2). This enables us to visualize what sums and scalar multiples mean geometrically. For example consider $\mathbf{x} = \begin{bmatrix} 1 \\ 2 \end{bmatrix}$ in $\mathbb{R}^2$. Then $2\mathbf{x} = \begin{bmatrix} 2 \\ 4 \end{bmatrix}$, $\frac{1}{2}\mathbf{x} = \begin{bmatrix} \frac{1}{2} \\ 1 \end{bmatrix}$ and $-\frac{1}{2}\mathbf{x} = \begin{bmatrix} -\frac{1}{2} \\ -1 \end{bmatrix}$, and these are shown as arrows in Figure 4.

Observe that the arrow for $2\mathbf{x}$ is twice as long as the arrow for **x** and in the same direction, and that the arrows for $\frac{1}{2}\mathbf{x}$ is also in the same direction as the arrow for **x**, but only half as long. On the other hand, the arrow for $-\frac{1}{2}\mathbf{x}$ is half as long as the arrow for **x**, but in the *opposite* direction. More generally, we have the following geometrical description of scalar multiplication in $\mathbb{R}^2$:

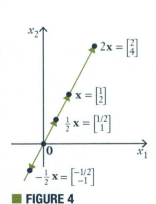

■ **FIGURE 4**

SECTION 2.6 Linear Transformations

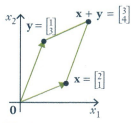

FIGURE 5

Scalar Multiple Law

Let **x** *be a vector in* $\mathbb{R}^2$. *The arrow for* $k\mathbf{x}$ *is* $|k|$ *times*[12] *as long as the arrow for* **x**, *and is in the same direction as the arrow for* **x** *if* $k > 0$, *and in the opposite direction if* $k < 0$.

Now consider two vectors $\mathbf{x} = \begin{bmatrix} 2 \\ 1 \end{bmatrix}$ and $\mathbf{y} = \begin{bmatrix} 1 \\ 3 \end{bmatrix}$ in $\mathbb{R}^2$. They are plotted in Figure 5 along with their sum $\mathbf{x} + \mathbf{y} = \begin{bmatrix} 3 \\ 4 \end{bmatrix}$. It is a routine matter to verify that the four points **0**, **x**, **y**, and $\mathbf{x} + \mathbf{y}$ form the vertices of a **parallelogram**–that is opposite sides are parallel and of the same length. (The reader should verify that the side from **0** to **x** has slope of $\frac{1}{2}$, as does the side from **y** to $\mathbf{x} + \mathbf{y}$, so these sides are parallel.) We state this as follows:

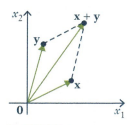

FIGURE 6

Parallelogram Law

Consider vectors **x** *and* **y** *in* $\mathbb{R}^2$. *If the arrows for* **x** *and* **y** *are drawn (see Figure 6), the arrow for* $\mathbf{x} + \mathbf{y}$ *corresponds to the fourth vertex of the parallelogram determined by the points* **x**, **y**, *and* **0**.

We will have more to say about this in Chapter 4.

We now turn to a brief review of angles and the trigonometric functions. Recall that an angle θ is said to be in **standard position** if it is measured counterclockwise from the positive x axis (as in Figure 7). Then θ uniquely determines a point **p** on the **unit circle** (radius 1, centre at the origin). The **radian** measure of θ is the length of the arc on the unit circle from the positive x axis to **p**. Thus $360° = 2\pi$ radians, $180° = \pi$, $90° = \frac{\pi}{2}$, and so on.

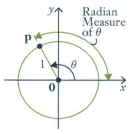

FIGURE 7

The point **p** in Figure 7 is also closely linked to the trigonometric functions **cosine** and **sine**, written $\cos \theta$ and $\sin \theta$ respectively. In fact these functions are *defined* to be the x and y coordinates of **p**; that is $\mathbf{p} = \begin{bmatrix} \cos \theta \\ \sin \theta \end{bmatrix}$. This defines $\cos \theta$ and $\sin \theta$ for the arbitrary angle θ (possibly negative), and agrees with the usual values when θ is an acute angle $\left(0 \leq \theta \leq \frac{\pi}{2}\right)$ as the reader should verify. For more discussion of this, see Appendix A.

Rotations

We can now describe rotations in the plane. Given an angle θ, let

$$R_\theta : \mathbb{R}^2 \to \mathbb{R}^2$$

denote counterclockwise rotation of $\mathbb{R}^2$ about the origin through the angle θ. The action of R_θ is depicted in Figure 8. We have already looked at $R_{\frac{\pi}{2}}$ (in Example 15 Section 2.2) and found it to be a matrix transformation. It turns out that R_θ is a matrix transformation for *every* angle θ (with a simple formula for the matrix), but it is not clear how to find the matrix. Our approach is to first establish the (somewhat surprising) fact that R_θ is linear, and then obtain the matrix from Theorem 2.

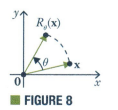

FIGURE 8

Let **x** and **y** be two vectors in $\mathbb{R}^2$. Then $\mathbf{x} + \mathbf{y}$ is the diagonal of the parallelogram determined by **x** and **y** as in Figure 9. The effect of R_θ is to rotate the *entire* parallelogram to obtain the new parallelogram determined by $R_\theta(\mathbf{x})$ and $R_\theta(\mathbf{y})$, with diagonal $R_\theta(\mathbf{x} + \mathbf{y})$. But this diagonal is $R_\theta(\mathbf{x}) + R_\theta(\mathbf{y})$ by the parallelogram law (applied to the new parallelogram). It follows that

$$R_\theta(\mathbf{x} + \mathbf{y}) = R_\theta(\mathbf{x}) + R_\theta(\mathbf{y}).$$

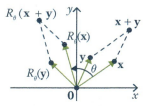

FIGURE 9

---

[12] If k is a real number, $|k|$ denotes the **absolute value** of k; that is, $|k| = k$ if $k \geq 0$ and $|k| = -k$ if $k < 0$.

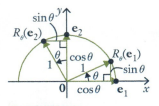

FIGURE 10

A similar argument shows that $R_\theta(a\mathbf{x}) = aR_\theta(\mathbf{x})$ for any scalar a, so $R_\theta: \mathbb{R}^2 \to \mathbb{R}^2$ is indeed a linear transformation.

With linearity established we can find the matrix of R_θ. Let $\mathbf{e}_1 = \begin{bmatrix} 1 \\ 0 \end{bmatrix}$ and $\mathbf{e}_2 = \begin{bmatrix} 0 \\ 1 \end{bmatrix}$ denote the standard basis of $\mathbb{R}^2$. By Figure 10 we see that

$$R_\theta(\mathbf{e}_1) = \begin{bmatrix} \cos\theta \\ \sin\theta \end{bmatrix} \quad \text{and} \quad R_\theta(\mathbf{e}_2) = \begin{bmatrix} -\sin\theta \\ \cos\theta \end{bmatrix}.$$

Hence Theorem 2 shows that R_θ is induced by the matrix

$$[R_\theta(\mathbf{e}_1) \; R_\theta(\mathbf{e}_2)] = \begin{bmatrix} \cos\theta & -\sin\theta \\ \sin\theta & \cos\theta \end{bmatrix}.$$

We record this as

Theorem 4

The rotation $R_\theta : \mathbb{R}^2 \to \mathbb{R}^2$ is the linear transformation with matrix $\begin{bmatrix} \cos\theta & -\sin\theta \\ \sin\theta & \cos\theta \end{bmatrix}$.

For example, $R_{\frac{\pi}{2}}$ and R_π have matrices $\begin{bmatrix} 0 & -1 \\ 1 & 0 \end{bmatrix}$ and $\begin{bmatrix} -1 & 0 \\ 0 & -1 \end{bmatrix}$, respectively, by Theorem 4. The first of these confirms the result in Example 15 Section 2.2. The second shows that rotating a vector $\mathbf{x} = \begin{bmatrix} x \\ y \end{bmatrix}$ through the angle π results in

$$R_\pi(\mathbf{x}) = \begin{bmatrix} -1 & 0 \\ 0 & -1 \end{bmatrix}\begin{bmatrix} x \\ y \end{bmatrix} = \begin{bmatrix} -x \\ -y \end{bmatrix} = -\mathbf{x}.$$ Thus applying R_π is the same as negating $\mathbf{x}$, a fact that is evident without Theorem 4.

EXAMPLE 7

Let θ and ϕ be angles. By finding the matrix of the composite $R_\theta \circ R_\phi$, obtain expressions for $\cos(\theta + \phi)$ and $\sin(\theta + \phi)$.

Solution ▶ Consider the transformations $\mathbb{R}^2 \xrightarrow{R_\phi} \mathbb{R}^2 \xrightarrow{R_\theta} \mathbb{R}^2$. Their composite $R_\theta \circ R_\phi$ is the transformation that first rotates the plane through ϕ and then rotates it through θ, and so is the rotation through the angle $\theta + \phi$ (see Figure 11). In other words

$$R_{\theta+\phi} = R_\theta \circ R_\phi.$$

Theorem 3 shows that the corresponding equation holds for the matrices of these transformations, so Theorem 4 gives:

$$\begin{bmatrix} \cos(\theta+\phi) & -\sin(\theta+\phi) \\ \sin(\theta+\phi) & \cos(\theta+\phi) \end{bmatrix} = \begin{bmatrix} \cos\theta & -\sin\theta \\ \sin\theta & \cos\theta \end{bmatrix}\begin{bmatrix} \cos\phi & -\sin\phi \\ \sin\phi & \cos\phi \end{bmatrix}$$

If we perform the matrix multiplication on the right, and then compare first column entries, we obtain

$$\cos(\theta + \phi) = \cos\theta\cos\phi - \sin\theta\sin\phi$$
$$\sin(\theta + \phi) = \sin\theta\cos\phi + \cos\theta\sin\phi$$

These are the two basic identities from which most of trigonometry can be derived.

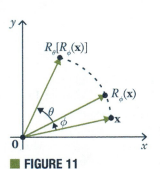

FIGURE 11

Reflections

The line through the origin with slope m has equation $y = mx$, and we let $Q_m: \mathbb{R}^2 \to \mathbb{R}^2$ denote reflection in the line $y = mx$.

This transformation is described geometrically in Figure 12. In words, $Q_m(\mathbf{x})$ is the "mirror image" of $\mathbf{x}$ in the line $y = mx$. If $m = 0$ then Q_0 is reflection in the x axis, so we already know Q_0 is linear. While we could show directly that Q_m is linear (with an argument like that for R_θ), we prefer to do it another way that is instructive and derives the matrix of Q_m directly without using Theorem 2.

Let θ denote the angle between the positive x axis and the line $y = mx$. The key observation is that the transformation Q_m can be accomplished in three steps: First rotate through $-\theta$ (so our line coincides with the x axis), then reflect in the x axis, and finally rotate back through θ. In other words:

$$Q_m = R_\theta \circ Q_0 \circ R_{-\theta}$$

Since $R_{-\theta}$, Q_0, and R_θ are all linear, this (with Theorem 3) shows that Q_m is linear and that is matrix is the product of the matrices of R_θ, Q_0, and $R_{-\theta}$. If we write $c = \cos \theta$ and $s = \sin \theta$ for simplicity, then the matrices of R_θ, $R_{-\theta}$, and Q_0 are

$$\begin{bmatrix} c & -s \\ s & c \end{bmatrix}, \quad \begin{bmatrix} c & s \\ -s & c \end{bmatrix}, \quad \text{and} \quad \begin{bmatrix} 1 & 0 \\ 0 & -1 \end{bmatrix} \text{ respectively.}^{13}$$

Hence, by Theorem 3, the matrix of $Q_m = R_\theta \circ Q_0 \circ R_{-\theta}$ is

$$\begin{bmatrix} c & -s \\ s & c \end{bmatrix} \begin{bmatrix} 1 & 0 \\ 0 & -1 \end{bmatrix} \begin{bmatrix} c & s \\ -s & c \end{bmatrix} = \begin{bmatrix} c^2 - s^2 & 2sc \\ 2sc & s^2 - c^2 \end{bmatrix}.$$

We can obtain this matrix in terms of m alone. Figure 13 shows that

$$\cos \theta = \frac{1}{\sqrt{1 + m^2}} \quad \text{and} \quad \sin \theta = \frac{m}{\sqrt{1 + m^2}},$$

so the matrix $\begin{bmatrix} c^2 - s^2 & 2sc \\ 2sc & s^2 - c^2 \end{bmatrix}$ of Q_m becomes $\dfrac{1}{1 + m^2} \begin{bmatrix} 1 - m^2 & 2m \\ 2m & m^2 - 1 \end{bmatrix}.$

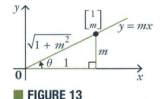
FIGURE 13

Theorem 5

Let Q_m denote reflection in the line $y = mx$. Then Q_m is a linear transformation with matrix $\dfrac{1}{1 + m^2} \begin{bmatrix} 1 - m^2 & 2m \\ 2m & m^2 - 1 \end{bmatrix}.$

Note that if $m = 0$, the matrix in Theorem 5 becomes $\begin{bmatrix} 1 & 0 \\ 0 & -1 \end{bmatrix}$, as expected. Of course this analysis fails for reflection in the y axis because vertical lines have no slope. However it is an easy exercise to verify that reflection in the y axis is indeed linear with matrix $\begin{bmatrix} -1 & 0 \\ 0 & 1 \end{bmatrix}.^{14}$

EXAMPLE 8

Let $T: \mathbb{R}^2 \to \mathbb{R}^2$ be rotation through $-\frac{\pi}{2}$ followed by reflection in the y axis. Show that T is a reflection in a line through the origin and find the line.

---

13. The matrix of $R_{-\theta}$ comes from the matrix of R_θ using the fact that, for all angles θ, $\cos(-\theta) = \cos \theta$ and $\sin(-\theta) = -\sin(\theta)$.

14. Note that $\begin{bmatrix} -1 & 0 \\ 0 & 1 \end{bmatrix} = \lim_{m \to \infty} \dfrac{1}{1 + m^2} \begin{bmatrix} 1 - m^2 & 2m \\ 2m & m^2 - 1 \end{bmatrix}.$

Solution ▶ The matrix of $R_{-\frac{\pi}{2}}$ is $\begin{bmatrix} \cos(-\frac{\pi}{2}) & -\sin(-\frac{\pi}{2}) \\ \sin(-\frac{\pi}{2}) & \cos(-\frac{\pi}{2}) \end{bmatrix} = \begin{bmatrix} 0 & 1 \\ -1 & 0 \end{bmatrix}$ and the matrix of reflection in the y axis is $\begin{bmatrix} -1 & 0 \\ 0 & 1 \end{bmatrix}$. Hence the matrix of T is

$$\begin{bmatrix} -1 & 0 \\ 0 & 1 \end{bmatrix}\begin{bmatrix} 0 & 1 \\ -1 & 0 \end{bmatrix} = \begin{bmatrix} 0 & -1 \\ -1 & 0 \end{bmatrix}$$ and this is reflection in the line $y = -x$ (take $m = -1$ in Theorem 5).

Projections

The method in the proof of Theorem 5 works more generally. Let $P_m : \mathbb{R}^2 \to \mathbb{R}^2$ denote projection on the line $y = mx$. This transformation is described geometrically in Figure 14. If $m = 0$, then $P_0\begin{bmatrix} x \\ y \end{bmatrix} = \begin{bmatrix} x \\ 0 \end{bmatrix}$ for all $\begin{bmatrix} x \\ y \end{bmatrix}$ in $\mathbb{R}^2$, so P_0 is linear with matrix $\begin{bmatrix} 1 & 0 \\ 0 & 0 \end{bmatrix}$. Hence the argument above for Q_m goes through for P_m. First observe that

$$P_m = R_\theta \circ P_0 \circ R_{-\theta}$$

as before. So, P_m is linear with matrix

$$\begin{bmatrix} c & -s \\ s & c \end{bmatrix}\begin{bmatrix} 1 & 0 \\ 0 & 0 \end{bmatrix}\begin{bmatrix} c & s \\ -s & c \end{bmatrix} = \begin{bmatrix} c^2 & sc \\ sc & s^2 \end{bmatrix}$$

where $c = \cos\theta = \dfrac{1}{\sqrt{1+m^2}}$ and $s = \sin\theta = \dfrac{m}{\sqrt{1+m^2}}$. This gives:

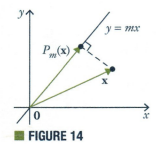

FIGURE 14

Theorem 6

Let $P_m : \mathbb{R}^2 \to \mathbb{R}^2$ be projection on the line $y = mx$. Then P_m is a linear transformation with matrix $\dfrac{1}{1+m^2}\begin{bmatrix} 1 & m \\ m & m^2 \end{bmatrix}$.

Again, if $m = 0$, then the matrix in Theorem 6 reduces to $\begin{bmatrix} 1 & 0 \\ 0 & 0 \end{bmatrix}$ as expected. As the y axis has no slope, the analysis fails for projection on the y axis, but this transformation is indeed linear with matrix $\begin{bmatrix} 0 & 0 \\ 0 & 1 \end{bmatrix}$ as is easily verified.

EXAMPLE 9

Given $\mathbf{x}$ in $\mathbb{R}^2$, write $\mathbf{y} = P_m(\mathbf{x})$. The fact that $\mathbf{y}$ lies on the line $y = mx$ means that $P_m(\mathbf{y}) = \mathbf{y}$. But then

$$(P_m \circ P_m)(\mathbf{x}) = P_m(\mathbf{y}) = \mathbf{y} = P_m(\mathbf{x}) \text{ for all } \mathbf{x} \text{ in } \mathbb{R}^2\text{, that is, } P_m \circ P_m = P_m.$$

In particular, if we write the matrix of P_m as $A = \dfrac{1}{1+m^2}\begin{bmatrix} 1 & m \\ m & m^2 \end{bmatrix}$, then $A^2 = A$. The reader should verify this directly.

SECTION 2.6 Linear Transformations

EXERCISES 2.6

1. Let $T_\theta : \mathbb{R}^3 \to \mathbb{R}^2$ be a linear transformation.

   (a) Find $T\begin{bmatrix} 8 \\ 3 \\ 7 \end{bmatrix}$ if $T\begin{bmatrix} 1 \\ 0 \\ -1 \end{bmatrix} = \begin{bmatrix} 2 \\ 3 \end{bmatrix}$ and $T\begin{bmatrix} 2 \\ 1 \\ 3 \end{bmatrix} = \begin{bmatrix} -1 \\ 0 \end{bmatrix}$.

   ◆(b) Find $T\begin{bmatrix} 5 \\ 6 \\ -13 \end{bmatrix}$ if $T\begin{bmatrix} 3 \\ 2 \\ -1 \end{bmatrix} = \begin{bmatrix} 3 \\ 5 \end{bmatrix}$ and $T\begin{bmatrix} 2 \\ 0 \\ 5 \end{bmatrix} = \begin{bmatrix} -1 \\ 2 \end{bmatrix}$.

2. Let $T_\theta : \mathbb{R}^4 \to \mathbb{R}^3$ be a linear transformation.

   (a) Find $T\begin{bmatrix} 1 \\ 3 \\ -2 \\ -3 \end{bmatrix}$ if $T\begin{bmatrix} 1 \\ 1 \\ 0 \\ -1 \end{bmatrix} = \begin{bmatrix} 2 \\ 3 \\ -1 \end{bmatrix}$ and
   $T\begin{bmatrix} 0 \\ -1 \\ 1 \\ 1 \end{bmatrix} = \begin{bmatrix} 5 \\ 0 \\ 1 \end{bmatrix}$.

   ◆(b) Find $T\begin{bmatrix} 5 \\ -1 \\ 2 \\ -4 \end{bmatrix}$ if $T\begin{bmatrix} 1 \\ 1 \\ 1 \\ 1 \end{bmatrix} = \begin{bmatrix} 5 \\ 1 \\ -3 \end{bmatrix}$ and
   $T\begin{bmatrix} -1 \\ 1 \\ 0 \\ 2 \end{bmatrix} = \begin{bmatrix} 2 \\ 0 \\ 1 \end{bmatrix}$.

3. In each case assume that the transformation T is linear, and use Theorem 2 to obtain the matrix A of T.

   (a) $T : \mathbb{R}^2 \to \mathbb{R}^2$ is reflection in the line $y = -x$.

   ◆(b) $T : \mathbb{R}^2 \to \mathbb{R}^2$ is given by $T(\mathbf{x}) = -\mathbf{x}$ for each $\mathbf{x}$ in $\mathbb{R}^2$.

   (c) $T : \mathbb{R}^2 \to \mathbb{R}^2$ is clockwise rotation through $\frac{\pi}{4}$.

   ◆(d) $T : \mathbb{R}^2 \to \mathbb{R}^2$ is counterclockwise rotation through $\frac{\pi}{4}$.

4. In each case use Theorem 2 to obtain the matrix A of the transformation T. You may assume that T is linear in each case.

   (a) $T : \mathbb{R}^3 \to \mathbb{R}^3$ is reflection in the x-z plane.

   ◆(b) $T : \mathbb{R}^3 \to \mathbb{R}^3$ is reflection in the y-z plane.

5. Let $T : \mathbb{R}^n \to \mathbb{R}^m$ be a linear transformation.

   (a) If $\mathbf{x}$ is in $\mathbb{R}^n$, we say that $\mathbf{x}$ is in the *kernel* of T if $T(\mathbf{x}) = \mathbf{0}$. If $\mathbf{x}_1$ and $\mathbf{x}_2$ are both in the kernel of T, show that $a\mathbf{x}_1 + b\mathbf{x}_2$ is also in the kernel of T for all scalars a and b.

   ◆(b) If $\mathbf{y}$ is in $\mathbb{R}^n$, we say that $\mathbf{y}$ is in the *image* of T if $\mathbf{y} = T(\mathbf{x})$ for some $\mathbf{x}$ in $\mathbb{R}^n$. If $\mathbf{y}_1$ and $\mathbf{y}_2$ are both in the image of T, show that $a\mathbf{y}_1 + b\mathbf{y}_2$ is also in the image of T for all scalars a and b.

6. Use Theorem 2 to find the matrix of the **identity transformation** $1_{\mathbb{R}^n} : \mathbb{R}^n \to \mathbb{R}^n$ defined by $1_{\mathbb{R}^n} : (\mathbf{x}) = \mathbf{x}$ for each $\mathbf{x}$ in $\mathbb{R}^n$.

7. In each case show that $T : \mathbb{R}^2 \to \mathbb{R}^2$ is not a linear transformation.

   (a) $T\begin{bmatrix} x \\ y \end{bmatrix} = \begin{bmatrix} xy \\ 0 \end{bmatrix}$.   ◆(b) $T\begin{bmatrix} x \\ y \end{bmatrix} = \begin{bmatrix} 0 \\ y^2 \end{bmatrix}$

8. In each case show that T is either reflection in a line or rotation through an angle, and find the line or angle.

   (a) $T\begin{bmatrix} x \\ y \end{bmatrix} = \frac{1}{5}\begin{bmatrix} -3x + 4y \\ 4x + 3y \end{bmatrix}$.

   ◆(b) $T\begin{bmatrix} x \\ y \end{bmatrix} = \frac{1}{\sqrt{2}}\begin{bmatrix} x + y \\ -x + y \end{bmatrix}$.

   (c) $T\begin{bmatrix} x \\ y \end{bmatrix} = \frac{1}{\sqrt{3}}\begin{bmatrix} x - \sqrt{3}y \\ \sqrt{3}x + y \end{bmatrix}$.

   ◆(d) $T\begin{bmatrix} x \\ y \end{bmatrix} = -\frac{1}{10}\begin{bmatrix} 8x + 6y \\ 6x - 8y \end{bmatrix}$.

9. Express reflection in the line $y = -x$ as the composition of a rotation followed by reflection in the line $y = x$.

10. In each case find the matrix of $T : \mathbb{R}^3 \to \mathbb{R}^3$:

    (a) T is rotation through θ about the x axis (from the y axis to the z axis).

    ◆(b) T is rotation through θ about the y axis (from the x axis to the z axis).

11. Let $T_\theta : \mathbb{R}^2 \to \mathbb{R}^2$ denote reflection in the line making an angle θ with the positive x axis.

    (a) Show that the matrix of T_θ is $\begin{bmatrix} \cos 2\theta & \sin 2\theta \\ \sin 2\theta & -\cos 2\theta \end{bmatrix}$ for all θ.

    (b) Show that $T_\theta \circ R_{2\phi} = T_{\theta - \phi}$ for all θ and ϕ.

12. In each case find a rotation or reflection that equals the given transformation.

    (a) Reflection in the y axis followed by rotation through $\frac{\pi}{2}$.

    ◆(b) Rotation through π followed by reflection in the x axis.

(c) Rotation through $\frac{\pi}{2}$ followed by reflection in the line $y = x$.

◆(d) Reflection in the x axis followed by rotation through $\frac{\pi}{2}$.

(e) Reflection in the line $y = x$ followed by reflection in the x axis.

◆(f) Reflection in the x axis followed by reflection in the line $y = x$.

13. Let R and S be matrix transformations $\mathbb{R}^n \to \mathbb{R}^m$ induced by matrices A and B respectively. In each case, show that T is a matrix transformation and describe its matrix in terms of A and B.

(a) $T(\mathbf{x}) = R(\mathbf{x}) + S(\mathbf{x})$ for all $\mathbf{x}$ in $\mathbb{R}^n$.

◆(b) $T(\mathbf{x}) = aR(\mathbf{x})$ for all $\mathbf{x}$ in $\mathbb{R}^n$ (where a is a fixed real number).

14. Show that the following hold for all linear transformations $T : \mathbb{R}^n \to \mathbb{R}^m$:

(a) $T(\mathbf{0}) = \mathbf{0}$. ◆(b) $T(-\mathbf{x}) = -T(\mathbf{x})$ for all $\mathbf{x}$ in $\mathbb{R}^n$.

15. The transformation $T : \mathbb{R}^n \to \mathbb{R}^m$ defined by $T(\mathbf{x}) = \mathbf{0}$ for all $\mathbf{x}$ in $\mathbb{R}^n$ is called the **zero transformation**.

(a) Show that the zero transformation is linear and find its matrix.

(b) Let $\mathbf{e}_1, \mathbf{e}_2, \ldots, \mathbf{e}_n$ denote the columns of the $n \times n$ identity matrix. If $T : \mathbb{R}^n \to \mathbb{R}^m$ is linear and $T(\mathbf{e}_i) = \mathbf{0}$ for each i, show that T is the zero transformation. [*Hint*: Theorem 1.]

16. Write the elements of $\mathbb{R}^n$ and $\mathbb{R}^m$ as rows. If A is an $m \times n$ matrix, define $T : \mathbb{R}^m \to \mathbb{R}^n$ by $T(\mathbf{y}) = \mathbf{y}A$ for all rows $\mathbf{y}$ in $\mathbb{R}^m$. Show that:

(a) T is a linear transformation.

(b) the rows of A are $T(\mathbf{f}_1), T(\mathbf{f}_2), \ldots, T(\mathbf{f}_m)$ where $\mathbf{f}_i$ denotes row i of I_m. [*Hint*: Show that $\mathbf{f}_i A$ is row i of A.]

17. Let $S : \mathbb{R}^n \to \mathbb{R}^n$ and $T : \mathbb{R}^n \to \mathbb{R}^n$ be linear transformations with matrices A and B respectively.

(a) Show that $B^2 = B$ if and only if $T^2 = T$ (where T^2 means $T \circ T$).

◆(b) Show that $B^2 = I$ if and only if $T^2 = 1_{\mathbb{R}^n}$.

(c) Show that $AB = BA$ if and only if $S \circ T = T \circ S$.

[*Hint*: Theorem 3.]

18. Let $Q_0 : \mathbb{R}^2 \to \mathbb{R}^2$ be reflection in the x axis, let $Q_1 : \mathbb{R}^2 \to \mathbb{R}^2$ be reflection in the line $y = x$, let $Q_{-1} : \mathbb{R}^2 \to \mathbb{R}^2$ be reflection in the line $y = -x$, and let $R_{\frac{\pi}{2}} : \mathbb{R}^2 \to \mathbb{R}^2$ be counterclockwise rotation through $\frac{\pi}{2}$.

(a) Show that $Q_1 \circ R_{\frac{\pi}{2}} = Q_0$.

◆(b) Show that $Q_1 \circ Q_2 = R_{\frac{\pi}{2}}$.

(c) Show that $R_{\frac{\pi}{2}} \circ Q_0 = Q_1$.

◆(d) Show that $Q_0 \circ R_{\frac{\pi}{2}} = Q_{-1}$.

19. For any slope m, show that:

(a) $Q_m \circ P_m = P_m$ (b) $P_m \circ Q_m = P_m$

◆20. Define $T : \mathbb{R}^n \to \mathbb{R}$ by $T(x_1, x_2, \ldots, x_n) = x_1 + x_2 + \cdots + x_n$. Show that T is a linear transformation and find its matrix.

21. Given c in $\mathbb{R}$, define $T_c : \mathbb{R}^n \to \mathbb{R}$ by $T_c(\mathbf{x}) = c\mathbf{x}$ for all $\mathbf{x}$ in $\mathbb{R}^n$. Show that T_c is a linear transformation and find its matrix.

22. Given vectors $\mathbf{w}$ and $\mathbf{x}$ in $\mathbb{R}^n$, denote their dot product by $\mathbf{w} \cdot \mathbf{x}$.

(a) Given $\mathbf{w}$ in $\mathbb{R}^n$, define $T_\mathbf{w} : \mathbb{R}^n \to \mathbb{R}$ by $T_\mathbf{w}(\mathbf{x}) = \mathbf{w} \cdot \mathbf{x}$ for all $\mathbf{x}$ in $\mathbb{R}^n$. Show that $T_\mathbf{w}$ is a linear transformation.

◆(b) Show that *every* linear transformation $T : \mathbb{R}^n \to \mathbb{R}$ is given as in (a); that is $T = T_\mathbf{w}$ for some $\mathbf{w}$ in $\mathbb{R}^n$.

23. If $\mathbf{x} \neq \mathbf{0}$ and $\mathbf{y}$ are vectors in $\mathbb{R}^n$, show that there is a linear transformation $T : \mathbb{R}^n \to \mathbb{R}^n$ such that $T(\mathbf{x}) = \mathbf{y}$. [*Hint*: By Definition 2.5, find a matrix A such that $A\mathbf{x} = \mathbf{y}$.]

24. Let $\mathbb{R}^n \xrightarrow{T} \mathbb{R}^m \xrightarrow{S} \mathbb{R}^k$ be two linear transformations. Show directly that $S \circ T$ is linear. That is:

(a) Show that $(S \circ T)(\mathbf{x} + \mathbf{y}) = (S \circ T)\mathbf{x} + (S \circ T)\mathbf{y}$ for all $\mathbf{x}, \mathbf{y}$ in $\mathbb{R}^n$.

◆(b) Show that $(S \circ T)(a\mathbf{x}) = a[(S \circ T)\mathbf{x}]$ for all $\mathbf{x}$ in $\mathbb{R}^n$ and all a in $\mathbb{R}$.

25. Let $\mathbb{R}^n \xrightarrow{T} \mathbb{R}^m \xrightarrow{S} \mathbb{R}^k \xrightarrow{R} \mathbb{R}^k$ be linear transformations. Show that $R \circ (S \circ T) = (R \circ S) \circ T$ by showing directly that $[R \circ (S \circ T)](\mathbf{x}) = [(R \circ S) \circ T](\mathbf{x})$ holds for each vector $\mathbf{x}$ in $\mathbb{R}^n$.

SECTION 2.7 LU-Factorization[15]

The solution to a system $A\mathbf{x} = \mathbf{b}$ of linear equations can be solved quickly if A can be factored as $A = LU$ where L and U are of a particularly nice form. In this section we show that gaussian elimination can be used to find such factorizations.

Triangular Matrices

As for square matrices, if $A = [a_{ij}]$ is an $m \times n$ matrix, the elements $a_{11}, a_{22}, a_{33}, \ldots$ form the **main diagonal** of A. Then A is called **upper triangular** if every entry below and to the left of the main diagonal is zero. Every row-echelon matrix is upper triangular, as are the matrices

$$\begin{bmatrix} 1 & -1 & 0 & 3 \\ 0 & 2 & 1 & 1 \\ 0 & 0 & -3 & 0 \end{bmatrix} \begin{bmatrix} 0 & 2 & 1 & 0 & 5 \\ 0 & 0 & 0 & 3 & 1 \\ 0 & 0 & 1 & 0 & 1 \end{bmatrix} \begin{bmatrix} 1 & 1 & 1 \\ 0 & -1 & 1 \\ 0 & 0 & 0 \\ 0 & 0 & 0 \end{bmatrix}$$

By analogy, a matrix A is called **lower triangular** if its transpose is upper triangular, that is if each entry above and to the right of the main diagonal is zero. A matrix is called **triangular** if it is upper or lower triangular.

EXAMPLE 1

Solve the system

$$\begin{aligned} x_1 + 2x_2 - 3x_3 - x_4 + 5x_5 &= 3 \\ 5x_3 + x_4 + x_5 &= 8 \\ 2x_5 &= 6 \end{aligned}$$

where the coefficient matrix is upper triangular.

Solution ▶ As in gaussian elimination, let the "non-leading" variables be parameters: $x_2 = s$ and $x_4 = t$. Then solve for x_5, x_3, and x_1 in that order as follows. The last equation gives

$$x_5 = \tfrac{6}{2} = 3$$

Substitution into the second last equation gives

$$x_3 = 1 - \tfrac{1}{5}t$$

Finally, substitution of both x_5 and x_3 into the first equation gives

$$x_1 = -9 - 2s + \tfrac{2}{5}t.$$

The method used in Example 1 is called **back substitution** because later variables are substituted into earlier equations. It works because the coefficient matrix is upper triangular. Similarly, if the coefficient matrix is lower triangular the system can be solved by **forward substitution** where earlier variables are substituted into later equations. As observed in Section 1.2, these procedures are more efficient than gaussian elimination.

---

[15] This section is not used later and so may be omitted with no loss of continuity.

Now consider a system $A\mathbf{x} = \mathbf{b}$ where A can be factored as $A = LU$ where L is lower triangular and U is upper triangular. Then the system $A\mathbf{x} = \mathbf{b}$ can be solved in two stages as follows:

1. First solve $L\mathbf{y} = \mathbf{b}$ for $\mathbf{y}$ by forward substitution.
2. Then solve $U\mathbf{x} = \mathbf{y}$ for $\mathbf{x}$ by back substitution.

Then $\mathbf{x}$ is a solution to $A\mathbf{x} = \mathbf{b}$ because $A\mathbf{x} = LU\mathbf{x} = L\mathbf{y} = \mathbf{b}$. Moreover, every solution $\mathbf{x}$ arises this way (take $\mathbf{y} = U\mathbf{x}$). Furthermore the method adapts easily for use in a computer.

This focuses attention on efficiently obtaining such factorizations $A = LU$. The following result will be needed; the proof is straightforward and is left as Exercises 7 and 8.

Lemma 1

Let A and B denote matrices.
1. If A and B are both lower (upper) triangular, the same is true of AB.
2. If A is $n \times n$ and lower (upper) triangular, then A is invertible if and only if every main diagonal entry is nonzero. In this case A^{-1} is also lower (upper) triangular.

LU-Factorization

Let A be an $m \times n$ matrix. Then A can be carried to a row-echelon matrix U (that is, upper triangular). As in Section 2.5, the reduction is

$$A \to E_1 A \to E_2 E_1 A \to E_3 E_2 E_1 A \to \cdots \to E_k E_{k-1} \cdots E_2 E_1 A = U$$

where $E_1, E_2, \ldots, E_k$ are elementary matrices corresponding to the row operations used. Hence

$$A = LU$$

where $L = (E_k E_{k-1} \cdots E_2 E_1)^{-1} = E_1^{-1} E_2^{-1} \cdots E_{k-1}^{-1} E_k^{-1}$. If we do not insist that U is reduced then, except for row interchanges, none of these row operations involve adding a row to a row *above* it. Thus, if no row interchanges are used, all the E_i are *lower* triangular, and so L is lower triangular (and invertible) by Lemma 1. This proves the following theorem. For convenience, let us say that A can be **lower reduced** if it can be carried to row-echelon form using no row interchanges.

Theorem 1

If A can be lower reduced to a row-echelon matrix U, then

$$A = LU$$

where L is lower triangular and invertible and U is upper triangular and row-echelon.

Definition 2.14   A factorization $A = LU$ as in Theorem 1 is called an **LU-factorization** of A.

Such a factorization may not exist (Exercise 4) because A cannot be carried to row-echelon form using no row interchange. A procedure for dealing with this situation will be outlined later. However, if an LU-factorization $A = LU$ does exist, then the gaussian algorithm gives U and also leads to a procedure for finding L.

SECTION 2.7 LU-Factorization

Example 2 provides an illustration. For convenience, the first nonzero column from the left in a matrix A is called the **leading column** of A.

EXAMPLE 2

Find an LU-factorization of $A = \begin{bmatrix} 0 & 2 & -6 & -2 & 4 \\ 0 & -1 & 3 & 3 & 2 \\ 0 & -1 & 3 & 7 & 10 \end{bmatrix}$.

Solution ▶ We lower reduce A to row-echelon form as follows:

$$A = \begin{bmatrix} 0 & ② & -6 & -2 & 4 \\ 0 & -1 & 3 & 3 & 2 \\ 0 & -1 & 3 & 7 & 10 \end{bmatrix} \to \begin{bmatrix} 0 & 1 & -3 & -1 & 2 \\ 0 & 0 & 0 & ② & 4 \\ 0 & 0 & 0 & ⑥ & 12 \end{bmatrix} \to \begin{bmatrix} 0 & 1 & -3 & -1 & 2 \\ 0 & 0 & 0 & 1 & 2 \\ 0 & 0 & 0 & 0 & 0 \end{bmatrix} = U$$

The circled columns are determined as follows: The first is the leading column of A, and is used (by lower reduction) to create the first leading 1 and create zeros below it. This completes the work on row 1, and we repeat the procedure on the matrix consisting of the remaining rows. Thus the second circled column is the leading column of this smaller matrix, which we use to create the second leading 1 and the zeros below it. As the remaining row is zero here, we are finished. Then $A = LU$ where

$$L = \begin{bmatrix} 2 & 0 & 0 \\ -1 & 2 & 0 \\ -1 & 6 & 1 \end{bmatrix}.$$

This matrix L is obtained from I_3 by replacing the bottom of the first two columns by the circled columns in the reduction. Note that the rank of A is 2 here, and this is the number of circled columns.

The calculation in Example 2 works in general. There is no need to calculate the elementary matrices E_i, and the method is suitable for use in a computer because the circled columns can be stored in memory as they are created. The procedure can be formally stated as follows:

LU-Algorithm

Let A be an $m \times n$ matrix of rank r, and suppose that A can be lower reduced to a row-echelon matrix U. Then $A = LU$ where the lower triangular, invertible matrix L is constructed as follows:

1. If $A = 0$, take $L = I_m$ and $U = 0$.
2. If $A \neq 0$, write $A_1 = A$ and let $\mathbf{c}_1$ be the leading column of A_1. Use $\mathbf{c}_1$ to create the first leading 1 and create zeros below it (using lower reduction). When this is completed, let A_2 denote the matrix consisting of rows 2 to m of the matrix just created.
3. If $A_2 \neq 0$, let $\mathbf{c}_2$ be the leading column of A_2 and repeat Step 2 on A_2 to create A_3.
4. Continue in this way until U is reached, where all rows below the last leading 1 consist of zeros. This will happen after r steps.
5. Create L by placing $\mathbf{c}_1, \mathbf{c}_2, \ldots, \mathbf{c}_r$ at the bottom of the first r columns of I_m.

A proof of the LU-algorithm is given at the end of this section.

LU-factorization is particularly important if, as often happens in business and industry, a series of equations $A\mathbf{x} = B_1, A\mathbf{x} = B_2, \ldots, A\mathbf{x} = B_k$, must be solved, each with the same coefficient matrix A. It is very efficient to solve the first system by gaussian elimination, simultaneously creating an LU-factorization of A, and then using the factorization to solve the remaining systems by forward and back substitution.

EXAMPLE 3

Find an LU-factorization for $A = \begin{bmatrix} 5 & -5 & 10 & 0 & 5 \\ -3 & 3 & 2 & 2 & 1 \\ -2 & 2 & 0 & -1 & 0 \\ 1 & -1 & 10 & 2 & 5 \end{bmatrix}$.

Solution ▶ The reduction to row-echelon form is

$$\begin{bmatrix} 5 & -5 & 10 & 0 & 5 \\ -3 & 3 & 2 & 2 & 1 \\ -2 & 2 & 0 & -1 & 0 \\ 1 & -1 & 10 & 2 & 5 \end{bmatrix} \to \begin{bmatrix} 1 & -1 & 2 & 0 & 1 \\ 0 & 0 & 8 & 2 & 4 \\ 0 & 0 & 4 & -1 & 2 \\ 0 & 0 & 8 & 2 & 4 \end{bmatrix}$$

$$\to \begin{bmatrix} 1 & -1 & 2 & 0 & 1 \\ 0 & 0 & 1 & \frac{1}{4} & \frac{1}{2} \\ 0 & 0 & 0 & -2 & 0 \\ 0 & 0 & 0 & 0 & 0 \end{bmatrix}$$

$$\to \begin{bmatrix} 1 & -1 & 2 & 0 & 1 \\ 0 & 0 & 1 & \frac{1}{4} & \frac{1}{2} \\ 0 & 0 & 0 & 1 & 0 \\ 0 & 0 & 0 & 0 & 0 \end{bmatrix} = U$$

If U denotes this row-echelon matrix, then $A = LU$, where

$$L = \begin{bmatrix} 5 & 0 & 0 & 0 \\ -3 & 8 & 0 & 0 \\ -2 & 4 & -2 & 0 \\ 1 & 8 & 0 & 1 \end{bmatrix}$$

The next example deals with a case where no row of zeros is present in U (in fact, A is invertible).

EXAMPLE 4

Find an LU-factorization for $A = \begin{bmatrix} 2 & 4 & 2 \\ 1 & 1 & 2 \\ -1 & 0 & 2 \end{bmatrix}$.

Solution ▶ The reduction to row-echelon form is

$$\begin{bmatrix} 2 & 4 & 2 \\ 1 & 1 & 2 \\ -1 & 0 & 2 \end{bmatrix} \to \begin{bmatrix} 1 & 2 & 1 \\ 0 & -1 & 1 \\ 0 & 2 & 3 \end{bmatrix} \to \begin{bmatrix} 1 & 2 & 1 \\ 0 & 1 & -1 \\ 0 & 0 & 5 \end{bmatrix} \to \begin{bmatrix} 1 & 2 & 1 \\ 0 & 1 & -1 \\ 0 & 0 & 1 \end{bmatrix} = U$$

SECTION 2.7 LU-Factorization

Hence $A = LU$ where $L = \begin{bmatrix} 2 & 0 & 0 \\ 1 & -1 & 0 \\ -1 & 2 & 5 \end{bmatrix}$.

There are matrices (for example $\begin{bmatrix} 0 & 1 \\ 1 & 0 \end{bmatrix}$) that have no LU-factorization and so require at least one row interchange when being carried to row-echelon form via the gaussian algorithm. However, it turns out that, if all the row interchanges encountered in the algorithm are carried out first, the resulting matrix requires no interchanges and so has an LU-factorization. Here is the precise result.

Theorem 2

Suppose an $m \times n$ matrix A is carried to a row-echelon matrix U via the gaussian algorithm. Let $P_1, P_2, \ldots, P_s$ be the elementary matrices corresponding (in order) to the row interchanges used, and write $P = P_s \cdots P_2 P_1$. (If no interchanges are used take $P = I_m$.) Then:

1. *PA is the matrix obtained from A by doing these interchanges (in order) to A.*
2. *PA has an LU-factorization.*

The proof is given at the end of this section.

A matrix P that is the product of elementary matrices corresponding to row interchanges is called a **permutation matrix**. Such a matrix is obtained from the identity matrix by arranging the rows in a different order, so it has exactly one 1 in each row and each column, and has zeros elsewhere. We regard the identity matrix as a permutation matrix. The elementary permutation matrices are those obtained from I by a single row interchange, and every permutation matrix is a product of elementary ones.

EXAMPLE 5

If $A = \begin{bmatrix} 0 & 0 & -1 & 2 \\ -1 & -1 & 1 & 2 \\ 2 & 1 & -3 & 6 \\ 0 & 1 & -1 & 4 \end{bmatrix}$, find a permutation matrix P such that PA has an LU-factorization, and then find the factorization.

Solution ▶ Apply the gaussian algorithm to A:

$$A \xrightarrow{*} \begin{bmatrix} -1 & -1 & 1 & 2 \\ 0 & 0 & -1 & 2 \\ 2 & 1 & -3 & 6 \\ 0 & 1 & -1 & 4 \end{bmatrix} \rightarrow \begin{bmatrix} 1 & 1 & -1 & -2 \\ 0 & 0 & -1 & 2 \\ 0 & -1 & -1 & 10 \\ 0 & 1 & -1 & 4 \end{bmatrix} \xrightarrow{*} \begin{bmatrix} 1 & 1 & -1 & -2 \\ 0 & -1 & -1 & 10 \\ 0 & 0 & -1 & 2 \\ 0 & 1 & -1 & 4 \end{bmatrix}$$

$$\rightarrow \begin{bmatrix} 1 & 1 & -1 & -2 \\ 0 & 1 & 1 & -10 \\ 0 & 0 & -1 & 2 \\ 0 & 0 & -2 & 14 \end{bmatrix} \rightarrow \begin{bmatrix} 1 & 1 & -1 & -2 \\ 0 & 1 & 1 & -10 \\ 0 & 0 & 1 & -2 \\ 0 & 0 & 0 & 10 \end{bmatrix}$$

Two row interchanges were needed (marked with ∗), first rows 1 and 2 and then rows 2 and 3. Hence, as in Theorem 2,

$$P = \begin{bmatrix} 1 & 0 & 0 & 0 \\ 0 & 0 & 1 & 0 \\ 0 & 1 & 0 & 0 \\ 0 & 0 & 0 & 1 \end{bmatrix} \begin{bmatrix} 0 & 1 & 0 & 0 \\ 1 & 0 & 0 & 0 \\ 0 & 0 & 1 & 0 \\ 0 & 0 & 0 & 1 \end{bmatrix} = \begin{bmatrix} 0 & 1 & 0 & 0 \\ 0 & 0 & 1 & 0 \\ 1 & 0 & 0 & 0 \\ 0 & 0 & 0 & 1 \end{bmatrix}$$

If we do these interchanges (in order) to A, the result is PA. Now apply the LU-algorithm to PA:

$$PA = \begin{bmatrix} -1 & -1 & 1 & 2 \\ 2 & 1 & -3 & 6 \\ 0 & 0 & -1 & 2 \\ 0 & 1 & -1 & 4 \end{bmatrix} \to \begin{bmatrix} 1 & 1 & -1 & -2 \\ 0 & -1 & -1 & 10 \\ 0 & 0 & -1 & 2 \\ 0 & 1 & -1 & 4 \end{bmatrix} \to \begin{bmatrix} 1 & 1 & -1 & -2 \\ 0 & 1 & 1 & -10 \\ 0 & 0 & -1 & 2 \\ 0 & 0 & -2 & 14 \end{bmatrix}$$

$$\to \begin{bmatrix} 1 & 1 & -1 & -2 \\ 0 & 1 & 1 & -10 \\ 0 & 0 & 1 & -2 \\ 0 & 0 & 0 & 10 \end{bmatrix} \to \begin{bmatrix} 1 & 1 & -1 & -2 \\ 0 & 1 & 1 & -10 \\ 0 & 0 & 1 & -2 \\ 0 & 0 & 0 & 1 \end{bmatrix} = U$$

Hence, $PA = LU$, where $L = \begin{bmatrix} 1 & 1 & -1 & -2 \\ 0 & 1 & 1 & -10 \\ 0 & 0 & 1 & -2 \\ 0 & 0 & 0 & 1 \end{bmatrix}$ and $U = \begin{bmatrix} -1 & 0 & 0 & 0 \\ 2 & -1 & 0 & 0 \\ 0 & 0 & -1 & 0 \\ 0 & 1 & -2 & 10 \end{bmatrix}$.

Theorem 2 provides an important general factorization theorem for matrices. If A is any $m \times n$ matrix, it asserts that there exists a permutation matrix P and an LU-factorization $PA = LU$. Moreover, it shows that either $P = I$ or $P = P_s \cdots P_2 P_1$, where $P_1, P_2, \ldots, P_s$ are the elementary permutation matrices arising in the reduction of A to row-echelon form. Now observe that $P_i^{-1} = P_i$ for each i (they are elementary row interchanges). Thus, $P^{-1} = P_1 P_2 \cdots P_s$, so the matrix A can be factored as

$$A = P^{-1} LU$$

where P^{-1} is a permutation matrix, L is lower triangular and invertible, and U is a row-echelon matrix. This is called a **PLU-factorization** of A.

The LU-factorization in Theorem 1 is not unique. For example,

$$\begin{bmatrix} 1 & 0 \\ 3 & 2 \end{bmatrix} \begin{bmatrix} 1 & -2 & 3 \\ 0 & 0 & 0 \end{bmatrix} = \begin{bmatrix} 1 & 0 \\ 3 & 1 \end{bmatrix} \begin{bmatrix} 1 & -2 & 3 \\ 0 & 0 & 0 \end{bmatrix}$$

However, it is necessary here that the row-echelon matrix has a row of zeros. Recall that the rank of a matrix A is the number of nonzero rows in any row-echelon matrix U to which A can be carried by row operations. Thus, if A is $m \times n$, the matrix U has no row of zeros if and only if A has rank m.

Theorem 3

Let A be an $m \times n$ matrix that has an LU-factorization

$$A = LU$$

If A has rank m (that is, U has no row of zeros), then L and U are uniquely determined by A.

SECTION 2.7 LU-Factorization

PROOF

Suppose $A = MV$ is another LU-factorization of A, so M is lower triangular and invertible and V is row-echelon. Hence $LU = MV$, and we must show that $L = M$ and $U = V$. We write $N = M^{-1}L$. Then N is lower triangular and invertible (Lemma 1) and $NU = V$, so it suffices to prove that $N = I$. If N is $m \times m$, we use induction on m. The case $m = 1$ is left to the reader. If $m > 1$, observe first that column 1 of V is N times column 1 of U. Thus if either column is zero, so is the other (N is invertible). Hence, we can assume (by deleting zero columns) that the $(1, 1)$-entry is 1 in both U and V.

Now we write $N = \begin{bmatrix} a & 0 \\ X & N_1 \end{bmatrix}$, $U = \begin{bmatrix} 1 & Y \\ 0 & U_1 \end{bmatrix}$, and $V = \begin{bmatrix} 1 & Z \\ 0 & V_1 \end{bmatrix}$ in block form.

Then $NU = V$ becomes $\begin{bmatrix} a & aY \\ X & XY + N_1 U_1 \end{bmatrix} = \begin{bmatrix} 1 & Z \\ 0 & V_1 \end{bmatrix}$. Hence $a = 1$, $Y = Z$, $X = 0$, and $N_1 U_1 = V_1$. But $N_1 U_1 = V_1$ implies $N_1 = I$ by induction, whence $N = I$.

If A is an $m \times m$ invertible matrix, then A has rank m by Theorem 5 Section 2.4. Hence, we get the following important special case of Theorem 3.

Corollary 1

If an invertible matrix A has an LU-factorization $A = LU$, then L and U are uniquely determined by A.

Of course, in this case U is an upper triangular matrix with 1s along the main diagonal.

Proofs of Theorems

PROOF OF THE LU-ALGORITHM

If $c_1, c_2, \ldots, c_r$ are columns of lengths $m, m - 1, \ldots, m - r + 1$, respectively, write $L^{(m)}(c_1, c_2, \ldots, c_r)$ for the lower triangular $m \times m$ matrix obtained from I_m by placing $c_1, c_2, \ldots, c_r$ at the bottom of the first r columns of I_m.

Proceed by induction on n. If $A = 0$ or $n = 1$, it is left to the reader. If $n > 1$, let c_1 denote the leading column of A and let k_1 denote the first column of the $m \times m$ identity matrix. There exist elementary matrices $E_1, \ldots, E_k$ such that, in block form,

$$(E_k \cdots E_2 E_1)A = \begin{bmatrix} 0 & k_1 & \begin{array}{c} X_1 \\ \hline A_1 \end{array} \end{bmatrix} \quad \text{where } (E_k \cdots E_2 E_1)c_1 = k_1.$$

Moreover, each E_j can be taken to be lower triangular (by assumption). Write

$$G = (E_k \cdots E_2 E_1)^{-1} = E_1^{-1} E_2^{-1} \cdots E_k^{-1}$$

Then G is lower triangular, and $GK_1 = c_1$. Also, each E_j (and so each E_j^{-1}) is the result of either multiplying row 1 of I_m by a constant or adding a multiple of row 1 to another row. Hence,

$$G = (E_1^{-1} E_2^{-1} \cdots E_k^{-1})I_m = \begin{bmatrix} c_1 & \begin{array}{c} 0 \\ \hline I_{m-1} \end{array} \end{bmatrix}$$

in block form. Now, by induction, let $A_1 = L_1 U_1$ be an LU-factorization of A_1, where $L_1 = L^{(m-1)}[\mathbf{c}_2, \dots, \mathbf{c}_r]$ and U_1 is row-echelon. Then block multiplication gives

$$G^{-1}A = \left[\, 0 \,\bigg|\, \mathbf{k}_1 \,\bigg|\, \begin{array}{c} X_1 \\ \hline L_1 U_1 \end{array} \,\right] = \left[\begin{array}{c|c} 1 & 0 \\ \hline 0 & L_1 \end{array}\right] \left[\begin{array}{cc|c} 0 & 1 & X_1 \\ \hline 0 & 0 & U_1 \end{array}\right]$$

Hence $A = LU$, where $U = \left[\begin{array}{cc|c} 0 & 1 & X_1 \\ \hline 0 & 0 & U_1 \end{array}\right]$ is row-echelon and

$$L = \left[\, \mathbf{c}_1 \,\bigg|\, \begin{array}{c} 0 \\ \hline I_{m-1} \end{array} \,\right] \left[\begin{array}{c|c} 1 & 0 \\ \hline 0 & L_1 \end{array}\right] = \left[\, \mathbf{c}_1 \,\bigg|\, \begin{array}{c} 0 \\ \hline L \end{array} \,\right] = L^{(m)}[\mathbf{c}_1, \mathbf{c}_2, \dots, \mathbf{c}_r].$$

This completes the proof.

PROOF OF THEOREM 2

Let A be a nonzero $m \times n$ matrix and let $\mathbf{k}_j$ denote column j of I_m. There is a permutation matrix P_1 (where either P_1 is elementary or $P_1 = I_m$) such that the first nonzero column $\mathbf{c}_1$ of $P_1 A$ has a nonzero entry on top. Hence, as in the LU-algorithm,

$$L^{(m)}[\mathbf{c}_1]^{-1} \cdot P_1 \cdot A = \left[\begin{array}{cc|c} 0 & 1 & X_1 \\ \hline 0 & 0 & A_1 \end{array}\right]$$

in block form. Then let P_2 be a permutation matrix (either elementary or I_m) such that

$$P_2 \cdot L^{(m)}[\mathbf{c}_1]^{-1} \cdot P_1 \cdot A = \left[\begin{array}{cc|c} 0 & 1 & X_1 \\ \hline 0 & 0 & A'_1 \end{array}\right]$$

and the first nonzero column $\mathbf{c}_2$ of A'_1 has a nonzero entry on top. Thus,

$$L^{(m)}[\mathbf{k}_1, \mathbf{c}_2]^{-1} \cdot P_2 \cdot L^{(m)}[\mathbf{c}_1]^{-1} \cdot P_1 \cdot A = \left[\begin{array}{cc|c} 0 & 1 & X_1 \\ \hline 0 & 0 & \begin{array}{cc|c} 0 & 1 & X_2 \\ \hline 0 & 0 & A_2 \end{array} \end{array}\right]$$

in block form. Continue to obtain elementary permutation matrices $P_1, P_2, \dots, P_r$ and columns $\mathbf{c}_1, \mathbf{c}_2, \dots, \mathbf{c}_r$ of lengths $m, m-1, \dots,$ such that

$$(L_r P_r L_{r-1} P_{r-1} \cdots L_2 P_2 L_1 P_1) A = U$$

where U is a row-echelon matrix and $L_j = L^{(m)}[\mathbf{k}_1, \dots, \mathbf{k}_{j-1}, \mathbf{c}_j]^{-1}$ for each j, where the notation means the first $j-1$ columns are those of I_m. It is not hard to verify that each L_j has the form $L_j = L^{(m)}[\mathbf{k}_1, \dots, \mathbf{k}_{j-1}, \mathbf{c}'_j]$ where $\mathbf{c}'_j$ is a column of length $m - j + 1$. We now claim that each permutation matrix P_k can be "moved past" each matrix L_j to the right of it, in the sense that

$$P_k L_j = L'_j P_k$$

where $L'_j = L^{(m)}[\mathbf{k}_1, \dots, \mathbf{k}_{j-1}, \mathbf{c}''_j]$ for some column $\mathbf{c}''_j$ of length $m - j + 1$. Given that this is true, we obtain a factorization of the form

$$(L_r L'_{r-1} \cdots L'_2 L'_1)(P_r P_{r-1} \cdots P_2 P_1) A = U$$

If we write $P = P_r P_{r-1} \cdots P_2 P_1$, this shows that PA has an LU-factorization because $L_r L'_{r-1} \cdots L'_2 L'_1$ is lower triangular and invertible. All that remains is to prove the following rather technical result.

SECTION 2.7 LU-Factorization

Lemma 2

Let P_k result from interchanging row k of I_m with a row below it. If $j < k$, let c_j be a column of length $m - j + 1$. Then there is another column c'_j of length $m - j + 1$ such that

$$P_k \cdot L^{(m)}[k_1, \ldots, k_{j-1}, c_j] = L^{(m)}[k_1, \ldots, k_{j-1}, c'_j] \cdot P_k$$

The proof is left as Exercise 11.

EXERCISES 2.7

1. Find an LU-factorization of the following matrices.

   (a) $\begin{bmatrix} 2 & 6 & -2 & 0 & 2 \\ 3 & 9 & -3 & 3 & 1 \\ -1 & -3 & 1 & -3 & 1 \end{bmatrix}$ ◆(b) $\begin{bmatrix} 2 & 4 & 2 \\ 1 & -1 & 3 \\ -1 & 7 & -7 \end{bmatrix}$

   (c) $\begin{bmatrix} 2 & 6 & -2 & 0 & 2 \\ 1 & 5 & -1 & 2 & 5 \\ 3 & 7 & -3 & -2 & 5 \\ -1 & -1 & 1 & 2 & 3 \end{bmatrix}$ ◆(d) $\begin{bmatrix} -1 & -3 & 1 & 0 & -1 \\ 1 & 4 & 1 & 1 & 1 \\ 1 & 2 & -3 & -1 & 1 \\ 0 & -2 & -4 & -2 & 0 \end{bmatrix}$

   (e) $\begin{bmatrix} 2 & 2 & 4 & 6 & 0 & 2 \\ 1 & -1 & 2 & 1 & 3 & 1 \\ -2 & 2 & -4 & -1 & 1 & 6 \\ 0 & 2 & 0 & 3 & 4 & 8 \\ -2 & 4 & -4 & 1 & -2 & 6 \end{bmatrix}$ ◆(f) $\begin{bmatrix} 2 & 2 & -2 & 4 & 2 \\ 1 & -1 & 0 & 2 & 1 \\ 3 & 1 & -2 & 6 & 3 \\ 1 & 3 & -2 & 2 & 1 \end{bmatrix}$

2. Find a permutation matrix P and an LU-factorization of PA if A is:

   (a) $\begin{bmatrix} 0 & 0 & 2 \\ 0 & -1 & 4 \\ 3 & 5 & 1 \end{bmatrix}$ ◆(b) $\begin{bmatrix} 0 & -1 & 2 \\ 0 & 0 & 4 \\ -1 & 2 & 1 \end{bmatrix}$

   (c) $\begin{bmatrix} 0 & -1 & 2 & 1 & 3 \\ -1 & 1 & 3 & 1 & 4 \\ 1 & -1 & -3 & 6 & 2 \\ 2 & -2 & -4 & 1 & 0 \end{bmatrix}$ ◆(d) $\begin{bmatrix} -1 & -2 & 3 & 0 \\ 2 & 4 & -6 & 5 \\ 1 & 1 & -1 & 3 \\ 2 & 5 & -10 & 1 \end{bmatrix}$

3. In each case use the given LU-decomposition of A to solve the system $Ax = b$ by finding y such that $Ly = b$, and then x such that $Ux = y$:

   (a) $A = \begin{bmatrix} 2 & 0 & 0 \\ 0 & -1 & 0 \\ 1 & 1 & 3 \end{bmatrix} \begin{bmatrix} 1 & 0 & 0 & 1 \\ 0 & 0 & 1 & 2 \\ 0 & 0 & 0 & 1 \end{bmatrix}; b = \begin{bmatrix} 1 \\ -1 \\ 2 \end{bmatrix}$

   ◆(b) $A = \begin{bmatrix} 2 & 0 & 0 \\ 1 & 3 & 0 \\ -1 & 2 & 1 \end{bmatrix} \begin{bmatrix} 1 & 1 & 0 & -1 \\ 0 & 1 & 0 & 1 \\ 0 & 0 & 0 & 0 \end{bmatrix}; b = \begin{bmatrix} -2 \\ -1 \\ 1 \end{bmatrix}$

   (c) $A = \begin{bmatrix} -2 & 0 & 0 & 0 \\ 1 & -1 & 0 & 0 \\ -1 & 0 & 2 & 0 \\ 0 & 1 & 0 & 2 \end{bmatrix} \begin{bmatrix} 1 & -1 & 2 & 1 \\ 0 & 1 & 1 & -4 \\ 0 & 0 & 1 & -\frac{1}{2} \\ 0 & 0 & 0 & 1 \end{bmatrix}; b = \begin{bmatrix} 1 \\ -1 \\ 2 \\ 0 \end{bmatrix}$

   ◆(d) $A = \begin{bmatrix} 2 & 0 & 0 & 0 \\ 1 & -1 & 0 & 0 \\ -1 & 1 & 2 & 0 \\ 3 & 0 & 1 & -1 \end{bmatrix} \begin{bmatrix} 1 & -1 & 0 & 1 \\ 0 & 1 & -2 & -1 \\ 0 & 0 & 1 & 1 \\ 0 & 0 & 0 & 0 \end{bmatrix}; b = \begin{bmatrix} 4 \\ -6 \\ 4 \\ 5 \end{bmatrix}$

4. Show that $\begin{bmatrix} 0 & 1 \\ 1 & 0 \end{bmatrix} = LU$ is impossible where L is lower triangular and U is upper triangular.

◆5. Show that we can accomplish any row interchange by using only row operations of other types.

6. (a) Let L and L_1 be invertible lower triangular matrices, and let U and U_1 be invertible upper triangular matrices. Show that $LU = L_1U_1$ if and only if there exists an invertible diagonal matrix D such that $L_1 = LD$ and $U_1 = D^{-1}U$. [Hint: Scrutinize $L^{-1}L_1 = UU_1^{-1}$.]

   ◆(b) Use part (a) to prove Theorem 3 in the case that A is invertible.

◆7. Prove Lemma 1(1). [Hint: Use block multiplication and induction.]

8. Prove Lemma 1(2). [Hint: Use block multiplication and induction.]

9. A triangular matrix is called **unit triangular** if it is square and every main diagonal element is a 1.

   (a) If A can be carried by the gaussian algorithm to row-echelon form using no row interchanges, show that $A = LU$ where L is unit lower triangular and U is upper triangular.

◆(b) Show that the factorization in (a) is unique.

10. Let $c_1, c_2, \ldots, c_r$ be columns of lengths m, $m-1, \ldots, m-r+1$. If k_j denotes column j of I_m, show that $L^{(m)}[c_1, c_2, \ldots, c_r] = L^{(m)}[c_1]\, L^{(m)}[k_1, c_2]\, L^{(m)}[k_1, k_2, c_3] \cdots L^{(m)}[k_1, k_2, \ldots, k_{r-1}, c_r]$. The notation is as in the proof of Theorem 2. [*Hint*: Use induction on m and block multiplication.]

11. Prove Lemma 2. [*Hint*: $P_k^{-1} = P_k$. Write $P_k = \begin{bmatrix} I_k & 0 \\ 0 & P_0 \end{bmatrix}$ in block form where P_0 is an $(m-k) \times (m-k)$ permutation matrix.]

SECTION 2.8 An Application to Input-Output Economic Models[16]

In 1973 Wassily Leontief was awarded the Nobel prize in economics for his work on mathematical models.[17] Roughly speaking, an economic system in this model consists of several industries, each of which produces a product and each of which uses some of the production of the other industries. The following example is typical.

EXAMPLE 1

A primitive society has three basic needs: food, shelter, and clothing. There are thus three industries in the society—the farming, housing, and garment industries—that produce these commodities. Each of these industries consumes a certain proportion of the total output of each commodity according to the following table.

		OUTPUT		
		Farming	Housing	Garment
CONSUMPTION	Farming	0.4	0.2	0.3
	Housing	0.2	0.6	0.4
	Garment	0.4	0.2	0.3

Find the annual prices that each industry must charge for its income to equal its expenditures.

Solution ▶ Let p_1, p_2, and p_3 be the prices charged per year by the farming, housing, and garment industries, respectively, for their total output. To see how these prices are determined, consider the farming industry. It receives p_1 for its production in any year. But it *consumes* products from all these industries in the following amounts (from row 1 of the table): 40% of the food, 20% of the housing, and 30% of the clothing. Hence, the expenditures of the farming industry are $0.4p_1 + 0.2p_2 + 0.3p_3$, so

$$0.4p_1 + 0.2p_2 + 0.3p_3 = p_1$$

A similar analysis of the other two industries leads to the following system of equations.

$$0.4p_1 + 0.2p_2 + 0.3p_3 = p_1$$
$$0.2p_1 + 0.6p_2 + 0.4p_3 = p_2$$
$$0.4p_1 + 0.2p_2 + 0.3p_3 = p_3$$

This has the matrix form $E\mathbf{p} = \mathbf{p}$, where

---

16 The applications in this section and the next are independent and may be taken in any order.
17 See W. W. Leontief, "The world economy of the year 2000," *Scientific American*, Sept. 1980.

$$E = \begin{bmatrix} 0.4 & 0.2 & 0.3 \\ 0.2 & 0.6 & 0.4 \\ 0.4 & 0.2 & 0.3 \end{bmatrix} \quad \text{and} \quad \mathbf{p} = \begin{bmatrix} p_1 \\ p_2 \\ p_3 \end{bmatrix}$$

The equations can be written as the homogeneous system

$$(I - E)\mathbf{p} = \mathbf{0}$$

where I is the 3×3 identity matrix, and the solutions are

$$\mathbf{p} = \begin{bmatrix} 2t \\ 3t \\ 2t \end{bmatrix}$$

where t is a parameter. Thus, the pricing must be such that the total output of the farming industry has the same value as the total output of the garment industry, whereas the total value of the housing industry must be $\frac{3}{2}$ as much.

In general, suppose an economy has n industries, each of which uses some (possibly none) of the production of every industry. We assume first that the economy is **closed** (that is, no product is exported or imported) and that all product is used. Given two industries i and j, let e_{ij} denote the proportion of the total annual output of industry j that is consumed by industry i. Then $E = [e_{ij}]$ is called the **input-output** matrix for the economy. Clearly,

$$0 \leq e_{ij} \leq 1 \quad \text{for all } i \text{ and } j \tag{1}$$

Moreover, all the output from industry j is used by *some* industry (the model is closed), so

$$e_{1j} + e_{2j} + \cdots + e_{ij} = 1 \quad \text{for each } j \tag{2}$$

This condition asserts that each column of E sums to 1. Matrices satisfying conditions 1 and 2 are called **stochastic matrices**.

As in Example 1, let p_i denote the price of the total annual production of industry i. Then p_i is the annual revenue of industry i. On the other hand, industry i spends $e_{i1}p_1 + e_{i2}p_2 + \cdots + e_{in}p_n$ annually for the product it uses ($e_{ij}p_j$ is the cost for product from industry j). The closed economic system is said to be in **equilibrium** if the annual expenditure equals the annual revenue for each industry—that is, if

$$e_{1j}p_1 + e_{2j}p_2 + \cdots + e_{ij}p_n = p_i \quad \text{for each } i = 1, 2, \ldots, n$$

If we write $\mathbf{p} = \begin{bmatrix} p_1 \\ p_2 \\ \vdots \\ p_n \end{bmatrix}$, these equations can be written as the matrix equation

$$E\mathbf{p} = \mathbf{p}$$

This is called the **equilibrium condition**, and the solutions $\mathbf{p}$ are called **equilibrium price structures**. The equilibrium condition can be written as

$$(I - E)\mathbf{p} = \mathbf{0}$$

which is a system of homogeneous equations for $\mathbf{p}$. Moreover, there is always a nontrivial solution $\mathbf{p}$. Indeed, the column sums of $I - E$ are all 0 (because E is stochastic), so the row-echelon form of $I - E$ has a row of zeros. In fact, more is true:

Theorem 1

Let E be any $n \times n$ stochastic matrix. Then there is a nonzero $n \times 1$ matrix $\mathbf{p}$ with nonnegative entries such that $E\mathbf{p} = \mathbf{p}$. If all the entries of E are positive, the matrix $\mathbf{p}$ can be chosen with all entries positive.

Theorem 1 guarantees the existence of an equilibrium price structure for any closed input-output system of the type discussed here. The proof is beyond the scope of this book.[18]

EXAMPLE 2

Find the equilibrium price structures for four industries if the input-output matrix is

$$E = \begin{bmatrix} .6 & .2 & .1 & .1 \\ .3 & .4 & .2 & 0 \\ .1 & .3 & .5 & .2 \\ 0 & .1 & .2 & .7 \end{bmatrix}$$

Find the prices if the total value of business is $1000.

Solution ▶ If $\mathbf{p} = \begin{bmatrix} p_1 \\ p_2 \\ p_3 \\ p_4 \end{bmatrix}$ is the equilibrium price structure, then the equilibrium condition is $E\mathbf{p} = \mathbf{p}$. When we write this as $(I - E)\mathbf{p} = \mathbf{0}$, the methods of Chapter 1 yield the following family of solutions:

$$\mathbf{p} = \begin{bmatrix} 44t \\ 39t \\ 51t \\ 47t \end{bmatrix}$$

where t is a parameter. If we insist that $p_1 + p_2 + p_3 + p_4 = 1000$, then $t = 5.525$ (to four figures). Hence

$$\mathbf{p} = \begin{bmatrix} 243.09 \\ 215.47 \\ 281.76 \\ 259.67 \end{bmatrix}$$

to five figures.

The Open Model

We now assume that there is a demand for products in the **open sector** of the economy, which is the part of the economy other than the producing industries (for example, consumers). Let d_i denote the total value of the demand for product i in the open sector. If p_i and e_{ij} are as before, the value of the annual demand for product i by the producing industries themselves is $e_{i1}p_1 + e_{i2}p_2 + \cdots + e_{in}p_n$, so the total annual revenue p_i of industry i breaks down as follows:

$$p_i = (e_{i1}p_1 + e_{i2}p_2 + \cdots + e_{in}p_n) + d_i \quad \text{for each } i = 1, 2, \ldots, n$$

---

18 The interested reader is referred to P. Lancaster's *Theory of Matrices* (New York: Academic Press, 1969) or to E. Seneta's *Non-negative Matrices* (New York: Wiley, 1973).

SECTION 2.8 An Application to Input-Output Economic Models

The column $\mathbf{d} = \begin{bmatrix} d_1 \\ \vdots \\ d_n \end{bmatrix}$ is called the **demand matrix**, and this gives a matrix equation

$$\mathbf{p} = E\mathbf{p} + \mathbf{d}$$

or

$$(I - E)\mathbf{p} = \mathbf{d} \qquad (*)$$

This is a system of linear equations for $\mathbf{p}$, and we ask for a solution $\mathbf{p}$ with every entry nonnegative. Note that every entry of E is between 0 and 1, but the column sums of E need not equal 1 as in the closed model.

Before proceeding, it is convenient to introduce a useful notation. If $A = [a_{ij}]$ and $B = [b_{ij}]$ are matrices of the same size, we write $A > B$ if $a_{ij} > b_{ij}$ for all i and j, and we write $A \geq B$ if $a_{ij} \geq b_{ij}$ for all i and j. Thus $P \geq 0$ means that every entry of P is nonnegative. Note that $A \geq 0$ and $B \geq 0$ implies that $AB \geq 0$.

Now, given a demand matrix $\mathbf{d} \geq 0$, we look for a production matrix $\mathbf{p} \geq 0$ satisfying equation $(*)$. This certainly exists if $I - E$ is invertible and $(I - E)^{-1} \geq 0$. On the other hand, the fact that $\mathbf{d} \geq 0$ means any solution $\mathbf{p}$ to equation $(*)$ satisfies $\mathbf{p} \geq E\mathbf{p}$. Hence, the following theorem is not too surprising.

Theorem 2

Let $E \geq 0$ be a square matrix. Then $I - E$ is invertible and $(I - E)^{-1} \geq 0$ if and only if there exists a column $\mathbf{p} > 0$ such that $\mathbf{p} > E\mathbf{p}$.

HEURISTIC PROOF

If $(I - E)^{-1} \geq 0$, the existence of $\mathbf{p} > 0$ with $\mathbf{p} > E\mathbf{p}$ is left as Exercise 11. Conversely, suppose such a column $\mathbf{p}$ exists. Observe that

$$(I - E)(I + E + E^2 + \cdots + E^{k-1}) = I - E^k$$

holds for all $k \geq 2$. If we can show that every entry of E^k approaches 0 as k becomes large then, intuitively, the infinite matrix sum

$$U = I + E + E^2 + \cdots$$

exists and $(I - E)U = I$. Since $U \geq 0$, this does it. To show that E^k approaches 0, it suffices to show that $EP < \mu P$ for some number μ with $0 < \mu < 1$ (then $E^k P < \mu^k P$ for all $k \geq 1$ by induction). The existence of μ is left as Exercise 12.

The condition $\mathbf{p} > E\mathbf{p}$ in Theorem 2 has a simple economic interpretation. If $\mathbf{p}$ is a production matrix, entry i of $E\mathbf{p}$ is the total value of all product used by industry i in a year. Hence, the condition $\mathbf{p} > E\mathbf{p}$ means that, for each i, the value of product produced by industry i exceeds the value of the product it uses. In other words, each industry runs at a profit.

EXAMPLE 3

If $E = \begin{bmatrix} 0.6 & 0.2 & 0.3 \\ 0.1 & 0.4 & 0.2 \\ 0.2 & 0.5 & 0.1 \end{bmatrix}$, show that $I - E$ is invertible and $(I - E)^{-1} \geq 0$.

Solution ▶ Use $\mathbf{p} = (3, 2, 2)^T$ in Theorem 2.

If $\mathbf{p}_0 = (1, 1, 1)^T$, the entries of $E\mathbf{p}_0$ are the row sums of E. Hence $\mathbf{p}_0 > E\mathbf{p}_0$ holds if the row sums of E are all less than 1. This proves the first of the following useful facts (the second is Exercise 10).

Corollary 1

Let $E \geq 0$ be a square matrix. In each of the following cases, $I - E$ is invertible and $(I - E)^{-1} \geq 0$:

1. All row sums of E are less than 1.
2. All column sums of E are less than 1.

EXERCISES 2.8

1. Find the possible equilibrium price structures when the input-output matrices are:

   (a) $\begin{bmatrix} 0.1 & 0.2 & 0.3 \\ 0.6 & 0.2 & 0.3 \\ 0.3 & 0.6 & 0.4 \end{bmatrix}$ ♦(b) $\begin{bmatrix} 0.5 & 0 & 0.5 \\ 0.1 & 0.9 & 0.2 \\ 0.4 & 0.1 & 0.3 \end{bmatrix}$

   (c) $\begin{bmatrix} .3 & .1 & .1 & .2 \\ .2 & .3 & .1 & 0 \\ .3 & .3 & .2 & .3 \\ .2 & .3 & .6 & .7 \end{bmatrix}$ ♦(d) $\begin{bmatrix} .5 & 0 & .1 & .1 \\ .2 & .7 & 0 & .1 \\ .1 & .2 & .8 & .2 \\ .2 & .1 & .1 & .6 \end{bmatrix}$

♦2. Three industries A, B, and C are such that all the output of A is used by B, all the output of B is used by C, and all the output of C is used by A. Find the possible equilibrium price structures.

3. Find the possible equilibrium price structures for three industries where the input-output matrix is $\begin{bmatrix} 1 & 0 & 0 \\ 0 & 0 & 1 \\ 0 & 1 & 0 \end{bmatrix}$. Discuss why there are two parameters here.

♦4. Prove Theorem 1 for a 2×2 stochastic matrix E by first writing it in the form
$$E = \begin{bmatrix} a & b \\ 1-a & 1-b \end{bmatrix},$$ where $0 \leq a \leq 1$ and $0 \leq b \leq 1$.

5. If E is an $n \times n$ stochastic matrix and $\mathbf{c}$ is an $n \times 1$ matrix, show that the sum of the entries of $\mathbf{c}$ equals the sum of the entries of the $n \times 1$ matrix $E\mathbf{c}$.

6. Let $W = [1\ 1\ 1\cdots 1]$. Let E and F denote $n \times n$ matrices with nonnegative entries.

   (a) Show that E is a stochastic matrix if and only if $WE = W$.

   (b) Use part (a) to deduce that, if E and F are both stochastic matrices, then EF is also stochastic.

7. Find a 2×2 matrix E with entries between 0 and 1 such that:

   (a) $I - E$ has no inverse.

   ♦(b) $I - E$ has an inverse but not all entries of $(I - E)^{-1}$ are nonnegative.

♦8. If E is a 2×2 matrix with entries between 0 and 1, show that $I - E$ is invertible and $(I - E)^{-1} \geq 0$ if and only if $\text{tr } E < 1 + \det E$. Here, if $E = \begin{bmatrix} a & b \\ c & d \end{bmatrix}$, then $\text{tr } E = a + d$ and $\det E = ad - bc$.

9. In each case show that $I - E$ is invertible and $(I - E)^{-1} \geq 0$.

   (a) $\begin{bmatrix} 0.6 & 0.5 & 0.1 \\ 0.1 & 0.3 & 0.3 \\ 0.2 & 0.1 & 0.4 \end{bmatrix}$ ♦(b) $\begin{bmatrix} 0.7 & 0.1 & 0.3 \\ 0.2 & 0.5 & 0.2 \\ 0.1 & 0.1 & 0.4 \end{bmatrix}$

   (c) $\begin{bmatrix} 0.6 & 0.2 & 0.1 \\ 0.3 & 0.4 & 0.2 \\ 0.2 & 0.5 & 0.1 \end{bmatrix}$ ♦(d) $\begin{bmatrix} 0.8 & 0.1 & 0.1 \\ 0.3 & 0.1 & 0.2 \\ 0.3 & 0.3 & 0.2 \end{bmatrix}$

10. Prove that (1) implies (2) in the Corollary to Theorem 2.

11. If $(I - E)^{-1} \geq 0$, find $\mathbf{p} > 0$ such that $\mathbf{p} > E\mathbf{p}$.

12. If $E\mathbf{p} < \mathbf{p}$ where $E \geq 0$ and $\mathbf{p} > 0$, find a number μ such that $E\mathbf{p} < \mu\mathbf{p}$ and $0 < \mu < 1$.
    [*Hint*: If $E\mathbf{p} = (q_1, ..., q_n)^T$ and $\mathbf{p} = (p_1, ..., p_n)^T$, take any number μ such that
    $$\max\left\{\frac{q_1}{p_1}, ..., \frac{q_n}{p_n}\right\} < \mu < 1.]$$

SECTION 2.9 An Application to Markov Chains

Many natural phenomena progress through various stages and can be in a variety of states at each stage. For example, the weather in a given city progresses day by day and, on any given day, may be sunny or rainy. Here the states are "sun" and "rain," and the weather progresses from one state to another in daily stages. Another example might be a football team: The stages of its evolution are the games it plays, and the possible states are "win," "draw," and "loss."

The general setup is as follows: A "system" evolves through a series of "stages," and at any stage it can be in any one of a finite number of "states." At any given stage, the state to which it will go at the next stage depends on the past and present history of the system—that is, on the sequence of states it has occupied to date.

Definition 2.15   *A **Markov chain** is such an evolving system wherein the state to which it will go next depends only on its present state and does not depend on the earlier history of the system.*[19]

Even in the case of a Markov chain, the state the system will occupy at any stage is determined only in terms of probabilities. In other words, chance plays a role. For example, if a football team wins a particular game, we do not know whether it will win, draw, or lose the next game. On the other hand, we may know that the team tends to persist in winning streaks; for example, if it wins one game it may win the next game $\frac{1}{2}$ of the time, lose $\frac{4}{10}$ of the time, and draw $\frac{1}{10}$ of the time. These fractions are called the **probabilities** of these various possibilities. Similarly, if the team loses, it may lose the next game with probability $\frac{1}{2}$ (that is, half the time), win with probability $\frac{1}{4}$, and draw with probability $\frac{1}{4}$. The probabilities of the various outcomes after a drawn game will also be known.

We shall treat probabilities informally here: *The probability that a given event will occur is the long-run proportion of the time that the event does indeed occur*. Hence, all probabilities are numbers between 0 and 1. A probability of 0 means the event is impossible and never occurs; events with probability 1 are certain to occur.

If a Markov chain is in a particular state, the probabilities that it goes to the various states at the next stage of its evolution are called the **transition probabilities** for the chain, and they are assumed to be known quantities. To motivate the general conditions that follow, consider the following simple example. Here the system is a man, the stages are his successive lunches, and the states are the two restaurants he chooses.

EXAMPLE 1

A man always eats lunch at one of two restaurants, A and B. He never eats at A twice in a row. However, if he eats at B, he is three times as likely to eat at B next time as at A. Initially, he is equally likely to eat at either restaurant.

(a) What is the probability that he eats at A on the third day after the initial one?

(b) What proportion of his lunches does he eat at A?

Solution ▶ The table of transition probabilities follows. The A column indicates that if he eats at A on one day, he never eats there again on the next day and so is certain to go to B.

---
[19] The name honours Andrei Andreyevich Markov (1856–1922) who was a professor at the university in St. Petersburg, Russia.

		Present Lunch	
		A	B
Next Lunch	A	0	0.25
	B	1	0.75

The B column shows that, if he eats at B on one day, he will eat there on the next day $\frac{3}{4}$ of the time and switches to A only $\frac{1}{4}$ of the time.

The restaurant he visits on a given day is not determined. The most that we can expect is to know the probability that he will visit A or B on that day.

Let $\mathbf{s}_m = \begin{bmatrix} s_1^{(m)} \\ s_2^{(m)} \end{bmatrix}$ denote the *state vector* for day m. Here $s_1^{(m)}$ denotes the probability that he eats at A on day m, and $s_2^{(m)}$ is the probability that he eats at B on day m. It is convenient to let $\mathbf{s}_0$ correspond to the initial day. Because he is equally likely to eat at A or B on that initial day, $s_1^{(0)} = 0.5$ and $s_2^{(0)} = 0.5$, so $\mathbf{s}_0 = \begin{bmatrix} 0.5 \\ 0.5 \end{bmatrix}$. Now let

$$P = \begin{bmatrix} 0 & 0.25 \\ 1 & 0.75 \end{bmatrix}$$

denote the *transition matrix*. We claim that the relationship

$$\mathbf{s}_{m+1} = P\mathbf{s}_m$$

holds for all integers $m \geq 0$. This will be derived later; for now, we use it as follows to successively compute $\mathbf{s}_1, \mathbf{s}_2, \mathbf{s}_3, \ldots$.

$$\mathbf{s}_1 = p\mathbf{s}_0 = \begin{bmatrix} 0 & 0.25 \\ 1 & 0.75 \end{bmatrix}\begin{bmatrix} 0.5 \\ 0.5 \end{bmatrix} = \begin{bmatrix} 0.125 \\ 0.875 \end{bmatrix}$$

$$\mathbf{s}_2 = p\mathbf{s}_1 = \begin{bmatrix} 0 & 0.25 \\ 1 & 0.75 \end{bmatrix}\begin{bmatrix} 0.125 \\ 0.875 \end{bmatrix} = \begin{bmatrix} 0.21875 \\ 0.78125 \end{bmatrix}$$

$$\mathbf{s}_3 = p\mathbf{s}_2 = \begin{bmatrix} 0 & 0.25 \\ 1 & 0.75 \end{bmatrix}\begin{bmatrix} 0.21875 \\ 0.78125 \end{bmatrix} = \begin{bmatrix} 0.1953125 \\ 0.8046875 \end{bmatrix}$$

Hence, the probability that his third lunch (after the initial one) is at A is approximately 0.195, whereas the probability that it is at B is 0.805.

If we carry these calculations on, the next state vectors are (to five figures):

$$\mathbf{s}_4 = \begin{bmatrix} 0.20117 \\ 0.79883 \end{bmatrix} \quad \mathbf{s}_5 = \begin{bmatrix} 0.19971 \\ 0.80029 \end{bmatrix}$$

$$\mathbf{s}_6 = \begin{bmatrix} 0.20007 \\ 0.79993 \end{bmatrix} \quad \mathbf{s}_7 = \begin{bmatrix} 0.19998 \\ 0.80002 \end{bmatrix}$$

Moreover, as m increases the entries of $\mathbf{s}_m$ get closer and closer to the corresponding entries of $\begin{bmatrix} 0.2 \\ 0.8 \end{bmatrix}$. Hence, in the long run, he eats 20% of his lunches at A and 80% at B.

Example 1 incorporates most of the essential features of all Markov chains. The general model is as follows: The system evolves through various stages and at each stage can be in exactly one of n distinct states. It progresses through a sequence of states as time goes on. If a Markov chain is in state j at a particular stage of its development, the probability p_{ij} that it goes to state i at the next stage is called the

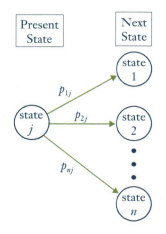

transition probability. The $n \times n$ matrix $P = [p_{ij}]$ is called the **transition matrix** for the Markov chain. The situation is depicted graphically in the diagram.

We make one important assumption about the transition matrix $P = [p_{ij}]$: It does *not* depend on which stage the process is in. This assumption means that the transition probabilities are *independent of time*—that is, they do not change as time goes on. It is this assumption that distinguishes Markov chains in the literature of this subject.

EXAMPLE 2

Suppose the transition matrix of a three-state Markov chain is

$$P = \begin{bmatrix} p_{11} & p_{12} & p_{13} \\ p_{21} & p_{22} & p_{23} \\ p_{31} & p_{32} & p_{33} \end{bmatrix} = \begin{bmatrix} 0.3 & 0.1 & 0.6 \\ 0.5 & 0.9 & 0.2 \\ 0.2 & 0.0 & 0.2 \end{bmatrix} \begin{matrix} 1 \\ 2 \\ 3 \end{matrix} \text{ Next state}$$

$$\begin{matrix} \text{Present state} \\ 1 \quad 2 \quad 3 \end{matrix}$$

If, for example, the system is in state 2, then column 2 lists the probabilities of where it goes next. Thus, the probability is $p_{12} = 0.1$ that it goes from state 2 to state 1, and the probability is $p_{22} = 0.9$ that it goes from state 2 to state 2. The fact that $p_{32} = 0$ means that it is impossible for it to go from state 2 to state 3 at the next stage.

Consider the jth column of the transition matrix P.

$$\begin{bmatrix} p_{1j} \\ p_{2j} \\ \vdots \\ p_{nj} \end{bmatrix}$$

If the system is in state j at some stage of its evolution, the transition probabilities $p_{1j}, p_{2j}, \ldots, p_{nj}$ represent the fraction of the time that the system will move to state 1, state 2, …, state n, respectively, at the next stage. We assume that it has to go to *some* state at each transition, so the sum of these probabilities equals 1:

$$p_{1j} + p_{2j} + \cdots + p_{nj} = 1 \quad \text{for each } j$$

Thus, the columns of P all sum to 1 and the entries of P lie between 0 and 1. Hence P is called a **stochastic matrix**.

As in Example 1, we introduce the following notation: Let $s_i^{(m)}$ denote the probability that the system is in state i after m transitions. The $n \times 1$ matrices

$$\mathbf{s}_m = \begin{bmatrix} s_1^{(m)} \\ s_2^{(m)} \\ \vdots \\ s_n^{(m)} \end{bmatrix} \quad m = 0, 1, 2, \ldots$$

are called the **state vectors** for the Markov chain. Note that the sum of the entries of $\mathbf{s}_m$ must equal 1 because the system must be in *some* state after m transitions. The matrix $\mathbf{s}_0$ is called the **initial state vector** for the Markov chain and is given as part of the data of the particular chain. For example, if the chain has only two states, then an initial vector $\mathbf{s}_0 = \begin{bmatrix} 1 \\ 0 \end{bmatrix}$ means that it started in state 1. If it started in state 2, the initial vector would be $\mathbf{s}_0 = \begin{bmatrix} 0 \\ 1 \end{bmatrix}$. If $\mathbf{s}_0 = \begin{bmatrix} 0.5 \\ 0.5 \end{bmatrix}$, it is equally likely that the system started in state 1 or in state 2.

Theorem 1

Let P be the transition matrix for an n-state Markov chain. If $\mathbf{s}_m$ is the state vector at stage m, then

$$\mathbf{s}_{m+1} = P\mathbf{s}_m.$$

for each $m = 0, 1, 2, \ldots$.

HEURISTIC PROOF

Suppose that the Markov chain has been run N times, each time starting with the same initial state vector. Recall that p_{ij} is the proportion of the time the system goes from state j at some stage to state i at the next stage, whereas $s_i^{(m)}$ is the proportion of the time it is in state i at stage m. Hence

$$s_i^{m+1} N$$

is (approximately) the number of times the system is in state i at stage $m + 1$. We are going to calculate this number another way. The system got to state i at stage $m + 1$ through *some* other state (say state j) at stage m. The number of times it was *in* state j at that stage is (approximately) $s_j^{(m)} N$, so the number of times it got to state i via state j is $p_{ij}(s_j^{(m)} N)$. Summing over j gives the number of times the system is in state i (at stage $m + 1$). This is the number we calculated before, so

$$s_i^{(m+1)} N = p_{i1} s_1^{(m)} N + p_{i2} s_2^{(m)} N + \cdots + p_{in} s_n^{(m)} N$$

Dividing by N gives $s_i^{(m+1)} = p_{i1} s_1^{(m)} + p_{i2} s_2^{(m)} + \cdots + p_{in} s_n^{(m)}$ for each i, and this can be expressed as the matrix equation $\mathbf{s}_{m+1} = P\mathbf{s}_m$.

If the initial probability vector $\mathbf{s}_0$ and the transition matrix P are given, Theorem 1 gives $\mathbf{s}_1, \mathbf{s}_2, \mathbf{s}_3, \ldots$, one after the other, as follows:

$$\mathbf{s}_1 = P\mathbf{s}_0$$
$$\mathbf{s}_2 = P\mathbf{s}_1$$
$$\mathbf{s}_3 = P\mathbf{s}_2$$
$$\vdots$$

Hence, the state vector $\mathbf{s}_m$ is completely determined for each $m = 0, 1, 2, \ldots$ by P and $\mathbf{s}_0$.

EXAMPLE 3

A wolf pack always hunts in one of three regions R_1, R_2, and R_3. Its hunting habits are as follows:

1. If it hunts in some region one day, it is as likely as not to hunt there again the next day.
2. If it hunts in R_1, it never hunts in R_2 the next day.

	R_1	R_2	R_3
R_1	$\frac{1}{2}$	$\frac{1}{4}$	$\frac{1}{4}$
R_2	0	$\frac{1}{2}$	$\frac{1}{4}$
R_3	$\frac{1}{2}$	$\frac{1}{4}$	$\frac{1}{2}$

3. If it hunts in R_2 or R_3, it is equally likely to hunt in each of the other regions the next day.

If the pack hunts in R_1 on Monday, find the probability that it hunts there on Thursday.

Solution ▶ The stages of this process are the successive days; the states are the three regions. The transition matrix P is determined as follows (see the table): The first habit asserts that $p_{11} = p_{22} = p_{33} = \frac{1}{2}$. Now column 1 displays what happens when the pack starts in R_1: It never goes to state 2, so $p_{21} = 0$ and, because the column must sum to 1, $p_{31} = \frac{1}{2}$. Column 2 describes what happens if it starts in R_2: $p_{22} = \frac{1}{2}$ and p_{12} and p_{32} are equal (by habit 3), so $p_{12} = p_{32} = \frac{1}{4}$ because the column sum must equal 1. Column 3 is filled in a similar way.

Now let Monday be the initial stage. Then $\mathbf{s}_0 = \begin{bmatrix} 1 \\ 0 \\ 0 \end{bmatrix}$ because the pack hunts in R_1 on that day. Then $\mathbf{s}_1$, $\mathbf{s}_2$, and $\mathbf{s}_3$ describe Tuesday, Wednesday, and Thursday, respectively, and we compute them using Theorem 1.

$$\mathbf{s}_1 = P\mathbf{s}_0 = \begin{bmatrix} \frac{1}{2} \\ 0 \\ \frac{1}{2} \end{bmatrix} \quad \mathbf{s}_2 = P\mathbf{s}_1 = \begin{bmatrix} \frac{3}{8} \\ \frac{1}{8} \\ \frac{4}{8} \end{bmatrix} \quad \mathbf{s}_3 = P\mathbf{s}_2 = \begin{bmatrix} \frac{11}{32} \\ \frac{6}{32} \\ \frac{15}{32} \end{bmatrix}$$

Hence, the probability that the pack hunts in Region R_1 on Thursday is $\frac{11}{32}$.

Another phenomenon that was observed in Example 1 can be expressed in general terms. The state vectors $\mathbf{s}_0, \mathbf{s}_1, \mathbf{s}_2, \ldots$ were calculated in that example and were found to "approach" $\mathbf{s} = \begin{bmatrix} 0.2 \\ 0.8 \end{bmatrix}$. This means that the first component of $\mathbf{s}_m$ becomes and remains very close to 0.2 as m becomes large, whereas the second component gets close to 0.8 as m increases. When this is the case, we say that $\mathbf{s}_m$ **converges** to $\mathbf{s}$. For large m, then, there is very little error in taking $\mathbf{s}_m = \mathbf{s}$, so the long-term probability that the system is in state 1 is 0.2, whereas the probability that it is in state 2 is 0.8. In Example 1, enough state vectors were computed for the limiting vector S to be apparent. However, there is a better way to do this that works in most cases.

Suppose P is the transition matrix of a Markov chain, and assume that the state vectors $\mathbf{s}_m$ converge to a limiting vector $\mathbf{s}$. Then $\mathbf{s}_m$ is very close to $\mathbf{s}$ for sufficiently large m, so $\mathbf{s}_{m+1}$ is also very close to $\mathbf{s}$. Thus, the equation $\mathbf{s}_{m+1} = P\mathbf{s}_m$ from Theorem 1 is closely approximated by

$$\mathbf{s} = P\mathbf{s}$$

so it is not surprising that $\mathbf{s}$ should be a solution to this matrix equation. Moreover, it is easily solved because it can be written as a system of homogeneous linear equations

$$(I - P)\mathbf{s} = \mathbf{0}$$

with the entries of $\mathbf{s}$ as variables.

In Example 1, where $P = \begin{bmatrix} 0 & 0.25 \\ 1 & 0.75 \end{bmatrix}$, the general solution to $(I - P)\mathbf{s} = \mathbf{0}$ is $\mathbf{s} = \begin{bmatrix} t \\ 4t \end{bmatrix}$, where t is a parameter. But if we insist that the entries of S sum to 1 (as must be true of all state vectors), we find $t = 0.2$ and so $\mathbf{s} = \begin{bmatrix} 0.2 \\ 0.8 \end{bmatrix}$ as before.

All this is predicated on the existence of a limiting vector for the sequence of state vectors of the Markov chain, and such a vector may not always exist. However, it does exist in one commonly occurring situation. A stochastic matrix P is called **regular** if some power P^m of P has every entry greater than zero. The matrix $P = \begin{bmatrix} 0 & 0.25 \\ 1 & 0.75 \end{bmatrix}$ of Example 1 is regular (in this case, each entry of P^2 is positive), and the general theorem is as follows:

Theorem 2

Let P be the transition matrix of a Markov chain and assume that P is regular. Then there is a unique column matrix $\mathbf{s}$ satisfying the following conditions:

1. *$P\mathbf{s} = \mathbf{s}$.*
2. *The entries of $\mathbf{s}$ are positive and sum to 1.*

Moreover, condition 1 can be written as

$$(I - P)\mathbf{s} = \mathbf{0}$$

and so gives a homogeneous system of linear equations for $\mathbf{s}$. Finally, the sequence of state vectors $\mathbf{s}_0, \mathbf{s}_1, \mathbf{s}_2, \ldots$ converges to $\mathbf{s}$ in the sense that if m is large enough, each entry of $\mathbf{s}_m$ is closely approximated by the corresponding entry of $\mathbf{s}$.

This theorem will not be proved here.[20]

If P is the regular transition matrix of a Markov chain, the column $\mathbf{s}$ satisfying conditions 1 and 2 of Theorem 2 is called the **steady-state vector** for the Markov chain. The entries of $\mathbf{s}$ are the long-term probabilities that the chain will be in each of the various states.

EXAMPLE 4

A man eats one of three soups—beef, chicken, and vegetable—each day. He never eats the same soup two days in a row. If he eats beef soup on a certain day, he is equally likely to eat each of the others the next day; if he does not eat beef soup, he is twice as likely to eat it the next day as the alternative.

(a) If he has beef soup one day, what is the probability that he has it again two days later?

(b) What are the long-run probabilities that he eats each of the three soups?

Solution ▶ The states here are B, C, and V, the three soups. The transition matrix P is given in the table. (Recall that, for each state, the corresponding column lists the probabilities for the next state.) If he has beef soup initially, then the initial state vector is

---

[20] The interested reader can find an elementary proof in J. Kemeny, H. Mirkil, J. Snell, and G. Thompson, *Finite Mathematical Structures* (Englewood Cliffs, N.J.: Prentice-Hall, 1958).

$$\mathbf{s}_0 = \begin{bmatrix} 1 \\ 0 \\ 0 \end{bmatrix}$$

Then two days later the state vector is $\mathbf{s}_2$. If P is the transition matrix, then

$$\mathbf{s}_1 = P\mathbf{s}_0 = \tfrac{1}{2}\begin{bmatrix} 0 \\ 1 \\ 1 \end{bmatrix}, \quad \mathbf{s}_2 = P\mathbf{s}_1 = \tfrac{1}{6}\begin{bmatrix} 4 \\ 1 \\ 1 \end{bmatrix}$$

so he eats beef soup two days later with probability $\tfrac{2}{3}$. This answers (a) and also shows that he eats chicken and vegetable soup each with probability $\tfrac{1}{6}$.

To find the long-run probabilities, we must find the steady-state vector $\mathbf{s}$. Theorem 2 applies because P is regular (P^2 has positive entries), so $\mathbf{s}$ satisfies $P\mathbf{s} = \mathbf{s}$. That is, $(I - P)\mathbf{s} = \mathbf{0}$ where

$$I - P = \tfrac{1}{6}\begin{bmatrix} 6 & -4 & -4 \\ -3 & 6 & -2 \\ -3 & -2 & 6 \end{bmatrix}$$

	B	C	V
B	0	$\tfrac{2}{3}$	$\tfrac{2}{3}$
C	$\tfrac{1}{2}$	0	$\tfrac{1}{3}$
V	$\tfrac{1}{2}$	$\tfrac{1}{3}$	0

The solution is $\mathbf{s} = \begin{bmatrix} 4t \\ 3t \\ 3t \end{bmatrix}$, where t is a parameter, and we use $\mathbf{s} = \begin{bmatrix} 0.4 \\ 0.3 \\ 0.3 \end{bmatrix}$ because the entries of S must sum to 1. Hence, in the long run, he eats beef soup 40% of the time and eats chicken soup and vegetable soup each 30% of the time.

EXERCISES 2.9

1. Which of the following stochastic matrices is regular?

   (a) $\begin{bmatrix} 0 & 0 & \tfrac{1}{2} \\ 1 & 0 & \tfrac{1}{2} \\ 0 & 1 & 0 \end{bmatrix}$ ◆(b) $\begin{bmatrix} \tfrac{1}{2} & 0 & \tfrac{1}{3} \\ \tfrac{1}{4} & 1 & \tfrac{1}{3} \\ \tfrac{1}{4} & 0 & \tfrac{1}{3} \end{bmatrix}$

2. In each case find the steady-state vector and, assuming that it starts in state 1, find the probability that it is in state 2 after 3 transitions.

   (a) $\begin{bmatrix} 0.5 & 0.3 \\ 0.5 & 0.7 \end{bmatrix}$ ◆(b) $\begin{bmatrix} \tfrac{1}{2} & 1 \\ \tfrac{1}{2} & 0 \end{bmatrix}$

   (c) $\begin{bmatrix} 0 & \tfrac{1}{2} & \tfrac{1}{4} \\ 1 & 0 & \tfrac{1}{4} \\ 0 & \tfrac{1}{2} & \tfrac{1}{2} \end{bmatrix}$ ◆(d) $\begin{bmatrix} 0.4 & 0.1 & 0.5 \\ 0.2 & 0.6 & 0.2 \\ 0.4 & 0.3 & 0.3 \end{bmatrix}$

   (e) $\begin{bmatrix} 0.8 & 0.0 & 0.2 \\ 0.1 & 0.6 & 0.1 \\ 0.1 & 0.4 & 0.7 \end{bmatrix}$ ◆(f) $\begin{bmatrix} 0.1 & 0.3 & 0.3 \\ 0.3 & 0.1 & 0.6 \\ 0.6 & 0.6 & 0.1 \end{bmatrix}$

3. A fox hunts in three territories A, B, and C. He never hunts in the same territory on two successive days. If he hunts in A, then he hunts in C the next day. If he hunts in B or C, he is twice as likely to hunt in A the next day as in the other territory.

   (a) What proportion of his time does he spend in A, in B, and in C?

   (b) If he hunts in A on Monday (C on Monday), what is the probability that he will hunt in B on Thursday?

4. Assume that there are three social classes—upper, middle, and lower—and that social mobility behaves as follows:

   1. Of the children of upper-class parents, 70% remain upper-class, whereas 10% become middle-class and 20% become lower-class.

   2. Of the children of middle-class parents, 80% remain middle-class, whereas the others are evenly split between the upper class and the lower class.

   3. For the children of lower-class parents, 60% remain lower-class, whereas 30% become middle-class and 10% upper-class.

(a) Find the probability that the grandchild of lower-class parents becomes upper-class.

◆(b) Find the long-term breakdown of society into classes.

5. The prime minister says she will call an election. This gossip is passed from person to person with a probability $p \neq 0$ that the information is passed incorrectly at any stage. Assume that when a person hears the gossip he or she passes it to one person who does not know. Find the long-term probability that a person will hear that there is going to be an election.

◆6. John makes it to work on time one Monday out of four. On other work days his behaviour is as follows: If he is late one day, he is twice as likely to come to work on time the next day as to be late. If he is on time one day, he is as likely to be late as not the next day. Find the probability of his being late and that of his being on time Wednesdays.

7. Suppose you have 1¢ and match coins with a friend. At each match you either win or lose 1¢ with equal probability. If you go broke or ever get 4¢, you quit. Assume your friend never quits. If the states are 0, 1, 2, 3, and 4 representing your wealth, show that the corresponding transition matrix P is not regular. Find the probability that you will go broke after 3 matches.

◆8. A mouse is put into a maze of compartments, as in the diagram. Assume that he always leaves any compartment he enters and that he is equally likely to take any tunnel entry.

(a) If he starts in compartment 1, find the probability that he is in compartment 1 again after 3 moves.

(b) Find the compartment in which he spends most of his time if he is left for a long time.

9. If a stochastic matrix has a 1 on its main diagonal, show that it cannot be regular. Assume it is not 1×1.

10. If $\mathbf{s}_m$ is the stage-m state vector for a Markov chain, show that $\mathbf{s}_{m+k} = P^k \mathbf{s}_m$ holds for all $m \geq 1$ and $k \geq 1$ (where P is the transition matrix).

11. A stochastic matrix is **doubly stochastic** if all the row sums also equal 1. Find the steady-state vector for a doubly stochastic matrix.

◆12. Consider the 2×2 stochastic matrix
$$P = \begin{bmatrix} 1-p & q \\ p & 1-q \end{bmatrix}, \text{ where } 0 < p < 1 \text{ and } 0 < q < 1.$$

(a) Show that $\dfrac{1}{p+q}\begin{bmatrix} q \\ p \end{bmatrix}$ is the steady-state vector for P.

(b) Show that P^m converges to the matrix $\dfrac{1}{p+q}\begin{bmatrix} q & q \\ p & p \end{bmatrix}$ by first verifying inductively that
$$P^m = \frac{1}{p+q}\begin{bmatrix} q & q \\ p & p \end{bmatrix} + \frac{(1-p-q)^m}{p+q}\begin{bmatrix} p & -q \\ -p & q \end{bmatrix}$$
for $m = 1, 2, \ldots$. (It can be shown that the sequence of powers $P, P^2, P^3, \ldots$ of any regular transition matrix converges to the matrix each of whose columns equals the steady-state vector for P.)

SUPPLEMENTARY EXERCISES FOR CHAPTER 2

1. Solve for the matrix X if:

   (a) $PXQ = R$;   (b) $XP = S$;

   where
   $$P = \begin{bmatrix} 1 & 0 \\ 2 & -1 \\ 0 & 3 \end{bmatrix}, Q = \begin{bmatrix} 1 & 1 & -1 \\ 2 & 0 & 3 \end{bmatrix}, R = \begin{bmatrix} -1 & 1 & -4 \\ -4 & 0 & -6 \\ 6 & 6 & -6 \end{bmatrix},$$
   $$S = \begin{bmatrix} 1 & 6 \\ 3 & 1 \end{bmatrix}$$

2. Consider $p(X) = X^3 - 5X^2 + 11X - 4I$.

   (a) If $p(U) = \begin{bmatrix} 1 & 3 \\ -1 & 0 \end{bmatrix}$, compute $p(U^T)$.

   ◆(b) If $p(U) = 0$ where U is $n \times n$, find U^{-1} in terms of U.

3. Show that, if a (possibly nonhomogeneous) system of equations is consistent and has more

variables than equations, then it must have infinitely many solutions. [*Hint*: Use Theorem 2 Section 2.2 and Theorem 1 Section 1.3.]

4. Assume that a system $A\mathbf{x} = \mathbf{b}$ of linear equations has at least two distinct solutions $\mathbf{y}$ and $\mathbf{z}$.

   (a) Show that $\mathbf{x}_k = \mathbf{y} + k(\mathbf{y} - \mathbf{z})$ is a solution for every k.

   ◆(b) Show that $\mathbf{x}_k = \mathbf{x}_m$ implies $k = m$. [*Hint*: See Example 7 Section 2.1.]

   (c) Deduce that $A\mathbf{x} = \mathbf{b}$ has infinitely many solutions.

5. (a) Let A be a 3×3 matrix with all entries on and below the main diagonal zero. Show that $A^3 = 0$.

   (b) Generalize to the $n \times n$ case and prove your answer.

6. Let I_{pq} denote the $n \times n$ matrix with (p, q)-entry equal to 1 and all other entries 0. Show that:

   (a) $I_n = I_{11} + I_{22} + \cdots + I_{nn}$.

   (b) $I_{pq}I_{rs} = \begin{cases} I_{ps} & \text{if } q = r \\ 0 & \text{if } q \neq r \end{cases}$.

   (c) If $A = [a_{ij}]$ is $n \times n$, then $A = \sum_{i=1}^{n}\sum_{j=1}^{n} a_{ij}I_{ij}$.

   ◆(d) If $A = [a_{ij}]$, then $I_{pq}AI_{rs} = a_{qr}I_{ps}$ for all $p, q, r,$ and s.

7. A matrix of the form aI_n, where a is a number, is called an $n \times n$ **scalar matrix**.

   (a) Show that each $n \times n$ scalar matrix commutes with every $n \times n$ matrix.

   ◆(b) Show that A is a scalar matrix if it commutes with every $n \times n$ matrix. [*Hint*: See part (d) of Exercise 6.]

8. Let $M = \begin{bmatrix} A & B \\ C & D \end{bmatrix}$, where $A, B, C,$ and D are all $n \times n$ and each commutes with all the others. If $M^2 = 0$, show that $(A + D)^3 = 0$. [*Hint*: First show that $A^2 = -BC = D^2$ and that $B(A + D) = 0 = C(A + D)$.]

9. If A is 2×2, show that $A^{-1} = A^T$ if and only if
$$A = \begin{bmatrix} \cos\theta & \sin\theta \\ -\sin\theta & \cos\theta \end{bmatrix} \text{ for some } \theta \text{ or}$$
$$A = \begin{bmatrix} \cos\theta & \sin\theta \\ \sin\theta & -\cos\theta \end{bmatrix} \text{ for some } \theta.$$
[*Hint*: If $a^2 + b^2 = 1$, then $a = \cos\theta$, $b = \sin\theta$ for some θ. Use $\cos(\theta - \varphi) = \cos\theta\cos\varphi + \sin\theta\sin\varphi$.]

10. (a) If $A = \begin{bmatrix} 0 & 1 \\ 1 & 0 \end{bmatrix}$, show that $A^2 = I$.

    (b) What is wrong with the following argument? If $A^2 = I$, then $A^2 - I = 0$, so $(A - I)(A + I) = 0$, whence $A = I$ or $A = -I$.

11. Let E and F be elementary matrices obtained from the identity matrix by adding multiples of row k to rows p and q. If $k \neq p$ and $k \neq q$, show that $EF = FE$.

12. If A is a 2×2 real matrix, $A^2 = A$ and $A^T = A$, show that either A is one of
$$\begin{bmatrix} 0 & 0 \\ 0 & 0 \end{bmatrix}, \begin{bmatrix} 1 & 0 \\ 0 & 0 \end{bmatrix}, \begin{bmatrix} 0 & 0 \\ 0 & 1 \end{bmatrix}, \begin{bmatrix} 1 & 0 \\ 0 & 1 \end{bmatrix}, \text{ or } A = \begin{bmatrix} a & b \\ b & 1-a \end{bmatrix}$$
where $a^2 + b^2 = a$, $-\frac{1}{2} \leq b \leq \frac{1}{2}$ and $b \neq 0$.

13. Show that the following are equivalent for matrices P, Q:

    (1) $P, Q,$ and $P + Q$ are all invertible and $(P + Q)^{-1} = P^{-1} + Q^{-1}$.

    (2) P is invertible and $Q = PG$ where $G^2 + G + I = 0$.

3 Determinants and Diagonalization

With each square matrix we can calculate a number, called the determinant of the matrix, which tells us whether or not the matrix is invertible. In fact, determinants can be used to give a formula for the inverse of a matrix. They also arise in calculating certain numbers (called eigenvalues) associated with a matrix. These eigenvalues are essential to a technique called diagonalization that is used in many applications where it is desired to predict the future behaviour of a system. For example, we use it to predict whether a species will become extinct.

Determinants were first studied by Leibnitz in 1696, and the term "determinant" was first used in 1801 by Gauss in his *Disquisitiones Arithmeticae*. Determinants are much older than matrices (which were introduced by Cayley in 1878) and were used extensively in the eighteenth and nineteenth centuries, primarily because of their significance in geometry (see Section 4.4). Although they are somewhat less important today, determinants still play a role in the theory and application of matrix algebra.

SECTION 3.1 The Cofactor Expansion

In Section 2.5 we defined the determinant of a 2×2 matrix $A = \begin{bmatrix} a & b \\ c & d \end{bmatrix}$ as follows:[1]

$$\det A = \begin{vmatrix} a & b \\ c & d \end{vmatrix} = ad - bc$$

and showed (in Example 4) that A has an inverse if and only if $\det A \neq 0$. One objective of this chapter is to do this for *any* square matrix A. There is no difficulty for 1×1 matrices: If $A = [a]$, we define $\det A = \det[a] = a$ and note that A is invertible if and only if $a \neq 0$.

If A is 3×3 and invertible, we look for a suitable definition of $\det A$ by trying to carry A to the identity matrix by row operations. The first column is not zero (A is invertible); suppose the $(1, 1)$-entry a is not zero. Then row operations give

$$A = \begin{bmatrix} a & b & c \\ d & e & f \\ g & h & i \end{bmatrix} \to \begin{bmatrix} a & b & c \\ ad & ae & af \\ ag & ah & ai \end{bmatrix} \to \begin{bmatrix} a & b & c \\ 0 & ae-bd & af-cd \\ 0 & ah-bg & ai-cg \end{bmatrix} = \begin{bmatrix} a & b & c \\ 0 & u & af-cd \\ 0 & v & ai-cg \end{bmatrix}$$

where $u = ae - bd$ and $v = ah - bg$. Since A is invertible, one of u and v is nonzero (by Example 11 Section 2.4); suppose that $u \neq 0$. Then the reduction proceeds

---

[1] Determinants are commonly written $|A| = \det A$ using vertical bars. We will use both notations.

SECTION 3.1 The Cofactor Expansion

$$A \to \begin{bmatrix} a & b & c \\ 0 & u & af - cd \\ 0 & v & ai - cg \end{bmatrix} \to \begin{bmatrix} a & b & c \\ 0 & u & af - cd \\ 0 & uv & u(ai - cg) \end{bmatrix} \to \begin{bmatrix} a & b & c \\ 0 & u & af - cd \\ 0 & 0 & w \end{bmatrix}$$

where $w = u(ai - cg) - v(af - cd) = a(aei + bfg + cdh - ceg - afh - bdi)$. We define

$$\det A = aei + bfg + cdh - ceg - afh - bdi \qquad (*)$$

and observe that $\det A \neq 0$ because $a \det A = w \neq 0$ (is invertible).

To motivate the definition below, collect the terms in $(*)$ involving the entries a, b, and c in row 1 of A:

$$\det A = \begin{vmatrix} a & b & c \\ d & e & f \\ g & h & i \end{vmatrix} = aei + bfg + cdh - ceg - afh - bdi$$

$$= a(ei - fh) - b(di - fg) + c(dh - eg)$$

$$= a \begin{vmatrix} e & f \\ h & i \end{vmatrix} - b \begin{vmatrix} d & f \\ g & i \end{vmatrix} + c \begin{vmatrix} d & e \\ g & h \end{vmatrix}$$

This last expression can be described as follows: To compute the determinant of a 3×3 matrix A, multiply each entry in row 1 by a sign times the determinant of the 2×2 matrix obtained by deleting the row and column of that entry, and add the results. The signs alternate down row 1, starting with $+$. It is this observation that we generalize below.

EXAMPLE 1

$$\det \begin{bmatrix} 2 & 3 & 7 \\ -4 & 0 & 6 \\ 1 & 5 & 0 \end{bmatrix} = 2 \begin{vmatrix} 0 & 6 \\ 5 & 0 \end{vmatrix} - 3 \begin{vmatrix} -4 & 6 \\ 1 & 0 \end{vmatrix} + 7 \begin{vmatrix} -4 & 0 \\ 1 & 5 \end{vmatrix}$$

$$= 2(-30) - 3(-6) + 7(-20)$$

$$= -182.$$

This suggests an inductive method of defining the determinant of any square matrix in terms of determinants of matrices one size smaller. The idea is to define determinants of 3×3 matrices in terms of determinants of 2×2 matrices, then we do 4×4 matrices in terms of 3×3 matrices, and so on.

To describe this, we need some terminology.

Definition 3.1 *Assume that determinants of $(n - 1) \times (n - 1)$ matrices have been defined. Given the $n \times n$ matrix A, let*

A_{ij} *denote the $(n - 1) \times (n - 1)$ matrix obtained from A by deleting row i and column j.*

*Then the (i, j)-**cofactor** $c_{ij}(A)$ is the scalar defined by*

$$c_{ij}(A) = (-1)^{i+j} \det(A_{ij}).$$

*Here $(-1)^{i+j}$ is called the **sign** of the (i, j)-position.*

The sign of a matrix is clearly 1 or -1, and the following diagram is useful for remembering the sign of a position:

$$\begin{bmatrix} + & - & + & - & \cdots \\ - & + & - & + & \cdots \\ + & - & + & - & \cdots \\ - & + & - & + & \cdots \\ \vdots & \vdots & \vdots & \vdots & \end{bmatrix}$$

Note that the signs alternate along each row and column with + in the upper left corner.

EXAMPLE 2

Find the cofactors of positions (1, 2), (3, 1), and (2, 3) in the following matrix.

$$A = \begin{bmatrix} 3 & -1 & 6 \\ 5 & 2 & 7 \\ 8 & 9 & 4 \end{bmatrix}$$

Solution ▶ Here A_{12} is the matrix $\begin{bmatrix} 5 & 7 \\ 8 & 4 \end{bmatrix}$ that remains when row 1 and column 2 are deleted. The sign of position (1, 2) is $(-1)^{1+2} = -1$ (this is also the (1, 2)-entry in the sign diagram), so the (1, 2)-cofactor is

$$c_{12}(A) = (-1)^{1+2}\begin{vmatrix} 5 & 7 \\ 8 & 4 \end{vmatrix} = (-1)(5 \cdot 4 - 7 \cdot 8) = (-1)(-36) = 36$$

Turning to position (3, 1), we find

$$c_{31}(A) = (-1)^{3+1}\det A_{31} = (-1)^{3+1}\begin{vmatrix} -1 & 6 \\ 2 & 7 \end{vmatrix} = (+1)(-7 - 12) = -19$$

Finally, the (2, 3)-cofactor is

$$c_{23}(A) = (-1)^{2+3}\det A_{23} = (-1)^{2+3}\begin{vmatrix} 3 & -1 \\ 8 & 9 \end{vmatrix} = (-1)(27 + 8) = -35$$

Clearly other cofactors can be found—there are nine in all, one for each position in the matrix.

We can now define det A for any square matrix A.

Definition 3.2 *Assume that determinants of $(n-1) \times (n-1)$ matrices have been defined. If $A = [a_{ij}]$ is $n \times n$ define*

$$\det A = a_{11}c_{11}(A) + a_{12}c_{12}(A) + \cdots + a_{1n}c_{1n}(A)$$

*This is called the **cofactor expansion** of det A along row 1.*

It asserts that det A can be computed by multiplying the entries of row 1 by the corresponding cofactors, and adding the results. The astonishing thing is that det A can be computed by taking the cofactor expansion along *any row or column*: Simply multiply each entry of that row or column by the corresponding cofactor and add.

SECTION 3.1 The Cofactor Expansion

> **Theorem 1**
>
> **Cofactor Expansion Theorem**[2]
> The determinant of an $n \times n$ matrix A can be computed by using the cofactor expansion along any row or column of A. That is $\det A$ can be computed by multiplying each entry of the row or column by the corresponding cofactor and adding the results.

The proof will be given in Section 3.6.

EXAMPLE 3

Compute the determinant of $A = \begin{bmatrix} 3 & 4 & 5 \\ 1 & 7 & 2 \\ 9 & 8 & -6 \end{bmatrix}$.

Solution ▶ The cofactor expansion along the first row is as follows:

$$\det A = 3c_{11}(A) + 4c_{12}(A) + 5c_{13}(A)$$
$$= 3\begin{vmatrix} 7 & 2 \\ 8 & -6 \end{vmatrix} - 4\begin{vmatrix} 1 & 2 \\ 9 & -6 \end{vmatrix} + 3\begin{vmatrix} 1 & 7 \\ 9 & 8 \end{vmatrix}$$
$$= 3(-58) - 4(-24) + 5(-55)$$
$$= -353$$

Note that the signs alternate along the row (indeed along *any* row or column). Now we compute $\det A$ by expanding along the first column.

$$\det A = 3c_{11}(A) + 1c_{21}(A) + 9c_{31}(A)$$
$$= 3\begin{vmatrix} 7 & 2 \\ 8 & -6 \end{vmatrix} - \begin{vmatrix} 4 & 5 \\ 8 & -6 \end{vmatrix} + 9\begin{vmatrix} 4 & 5 \\ 7 & 2 \end{vmatrix}$$
$$= 3(-58) - (-64) + 9(-27)$$
$$= -353$$

The reader is invited to verify that $\det A$ can be computed by expanding along any other row or column.

The fact that the cofactor expansion along *any row or column* of a matrix A always gives the same result (the determinant of A) is remarkable, to say the least. The choice of a particular row or column can simplify the calculation.

EXAMPLE 4

Compute $\det A$ where $A = \begin{bmatrix} 3 & 0 & 0 & 0 \\ 5 & 1 & 2 & 0 \\ 2 & 6 & 0 & -1 \\ -6 & 3 & 1 & 0 \end{bmatrix}$.

---

[2] The cofactor expansion is due to Pierre Simon de Laplace (1749–1827), who discovered it in 1772 as part of a study of linear differential equations. Laplace is primarily remembered for his work in astronomy and applied mathematics.

Solution ▶ The first choice we must make is which row or column to use in the cofactor expansion. The expansion involves multiplying entries by cofactors, so the work is minimized when the row or column contains as many zero entries as possible. Row 1 is a best choice in this matrix (column 4 would do as well), and the expansion is

$$\det A = 3c_{11}(A) + 0c_{12}(A) + 0c_{13}(A) + 0c_{14}(A)$$

$$= 3 \begin{vmatrix} 1 & 2 & 0 \\ 6 & 0 & -1 \\ 3 & 1 & 0 \end{vmatrix}$$

This is the first stage of the calculation, and we have succeeded in expressing the determinant of the 4×4 matrix A in terms of the determinant of a 3×3 matrix. The next stage involves this 3×3 matrix. Again, we can use any row or column for the cofactor expansion. The third column is preferred (with two zeros), so

$$\det A = 3\left(0 \begin{vmatrix} 6 & 0 \\ 3 & 1 \end{vmatrix} - (-1) \begin{vmatrix} 1 & 2 \\ 3 & 1 \end{vmatrix} + 0 \begin{vmatrix} 1 & 2 \\ 6 & 0 \end{vmatrix} \right)$$
$$= 3[0 + 1(-5) + 0]$$
$$= -15$$

This completes the calculation.

Computing the determinant of a matrix A can be tedious.[3] For example, if A is a 4×4 matrix, the cofactor expansion along any row or column involves calculating four cofactors, each of which involves the determinant of a 3×3 matrix. And if A is 5×5, the expansion involves five determinants of 4×4 matrices! There is a clear need for some techniques to cut down the work.

The motivation for the method is the observation (see Example 4) that calculating a determinant is simplified a great deal when a row or column consists mostly of zeros. (In fact, when a row or column consists *entirely* of zeros, the determinant is zero—simply expand along that row or column.)

Recall next that one method of *creating* zeros in a matrix is to apply elementary row operations to it. Hence, a natural question to ask is what effect such a row operation has on the determinant of the matrix. It turns out that the effect is easy to determine and that elementary *column* operations can be used in the same way. These observations lead to a technique for evaluating determinants that greatly reduces the labour involved. The necessary information is given in Theorem 2.

Theorem 2

Let A denote an $n \times n$ matrix.

1. If A has a row or column of zeros, $\det A = 0$.
2. If two distinct rows (or columns) of A are interchanged, the determinant of the resulting matrix is $-\det A$.

---

[3] If $A = \begin{bmatrix} a & b & c \\ d & e & f \\ g & h & i \end{bmatrix}$, we can calculate $\det A$ by considering $\begin{bmatrix} a & b & c & a & b \\ d & e & f & d & e \\ g & h & i & g & h \end{bmatrix}$ obtained from A by adjoining columns 1 and 2 on the right. Then $\det A = aei + bfg + cdh - ceg - afh - bdi$, where the positive terms aei, bfg, and cdh are the products down and to the right starting at a, b, and c, and the negative terms ceg, afh, and bdi are the products down and to the left starting at c, a, and b.
Warning: This rule does **not** apply to $n \times n$ matrices where $n > 3$ or $n = 2$.

SECTION 3.1 The Cofactor Expansion

3. If a row (or column) of A is multiplied by a constant u, the determinant of the resulting matrix is $u(\det A)$.
4. If two distinct rows (or columns) of A are identical, $\det A = 0$.
5. If a multiple of one row of A is added to a different row (or if a multiple of a column is added to a different column), the determinant of the resulting matrix is $\det A$.

PROOF

We prove properties 2, 4, and 5 and leave the rest as exercises.

Property 2. If A is $n \times n$, this follows by induction on n. If $n = 2$, the verification is left to the reader. If $n > 2$ and two rows are interchanged, let B denote the resulting matrix. Expand $\det A$ and $\det B$ along a row *other than* the two that were interchanged. The entries in this row are the same for both A and B, but the cofactors in B are the negatives of those in A (by induction) because the corresponding $(n-1) \times (n-1)$ matrices have two rows interchanged. Hence, $\det B = -\det A$, as required. A similar argument works if two columns are interchanged.

Property 4. If two rows of A are equal, let B be the matrix obtained by interchanging them. Then $B = A$, so $\det B = \det A$. But $\det B = -\det A$ by property 2, so $\det A = \det B = 0$. Again, the same argument works for columns.

Property 5. Let B be obtained from $A = [a_{ij}]$ by adding u times row p to row q. Then row q of B is $(a_{q1} + ua_{p1}, a_{q2} + ua_{p2}, \ldots, a_{qn} + ua_{pn})$. The cofactors of these elements in B are the same as in A (they do not involve row q): in symbols, $c_{qj}(B) = c_{qj}(A)$ for each j. Hence, expanding B along row q gives

$$\det B = (a_{q1} + ua_{p1})c_{q1}(A) + (a_{q2} + ua_{p2})c_{q2}(A) + \cdots + (a_{qn} + ua_{pn})c_{qn}(A)$$
$$= [a_{q1}c_{q1}(A) + a_{q2}c_{q2}(A) + \cdots + a_{qn}c_{qn}(A)]$$
$$\quad + u[a_{p1}c_{q1}(A) + a_{p2}c_{q2}(A) + \cdots + a_{pn}c_{qn}(A)]$$
$$= \det A + u \det C$$

where C is the matrix obtained from A by replacing row q by row p (and both expansions are along row q). Because rows p and q of C are equal, $\det C = 0$ by property 4. Hence, $\det B = \det A$, as required. As before, a similar proof holds for columns.

To illustrate Theorem 2, consider the following determinants.

$$\begin{vmatrix} 3 & -1 & 2 \\ 2 & 5 & 1 \\ 0 & 0 & 0 \end{vmatrix} = 0 \qquad \text{(because the last row consists of zeros)}$$

$$\begin{vmatrix} 3 & -1 & 5 \\ 2 & 8 & 7 \\ 1 & 2 & -1 \end{vmatrix} = -\begin{vmatrix} 5 & -1 & 3 \\ 7 & 8 & 2 \\ -1 & 2 & 1 \end{vmatrix} \qquad \text{(because two columns are interchanged)}$$

$$\begin{vmatrix} 8 & 1 & 2 \\ 3 & 0 & 9 \\ 1 & 2 & -1 \end{vmatrix} = 3\begin{vmatrix} 8 & 1 & 2 \\ 1 & 0 & 3 \\ 1 & 2 & -1 \end{vmatrix} \qquad \text{(because the second row of the matrix on the left is 3 times the second row of the matrix on the right)}$$

$$\begin{vmatrix} 2 & 1 & 2 \\ 4 & 0 & 4 \\ 1 & 3 & 1 \end{vmatrix} = 0 \qquad \text{(because two columns are identical)}$$

$$\begin{vmatrix} 2 & 5 & 2 \\ -1 & 2 & 9 \\ 3 & 1 & 1 \end{vmatrix} = \begin{vmatrix} 0 & 9 & 20 \\ -1 & 2 & 9 \\ 3 & 1 & 1 \end{vmatrix}$$ (because twice the second row of the matrix on the left was added to the first row)

The following four examples illustrate how Theorem 2 is used to evaluate determinants.

EXAMPLE 5

Evaluate $\det A$ when $A = \begin{bmatrix} 1 & -1 & 3 \\ 1 & 0 & -1 \\ 2 & 1 & 6 \end{bmatrix}$.

Solution ▶ The matrix does have zero entries, so expansion along (say) the second row would involve somewhat less work. However, a column operation can be used to get a zero in position (2, 3)—namely, add column 1 to column 3. Because this does not change the value of the determinant, we obtain

$$\det A = \begin{vmatrix} 1 & -1 & 3 \\ 1 & 0 & -1 \\ 2 & 1 & 6 \end{vmatrix} = \begin{vmatrix} 1 & -1 & 4 \\ 1 & 0 & 0 \\ 2 & 1 & 8 \end{vmatrix} = -\begin{vmatrix} -1 & 4 \\ 1 & 8 \end{vmatrix} = 12$$

where we expanded the second 3×3 matrix along row 2.

EXAMPLE 6

If $\det \begin{bmatrix} a & b & c \\ p & q & r \\ x & y & z \end{bmatrix} = 6$, evaluate $\det A$ where $A = \begin{bmatrix} a+x & b+y & c+z \\ 3x & 3y & 3z \\ -p & -q & -r \end{bmatrix}$.

Solution ▶ First take common factors out of rows 2 and 3.

$$\det A = 3(-1) \det \begin{bmatrix} a+x & b+y & c+z \\ x & y & z \\ p & q & r \end{bmatrix}$$

Now subtract the second row from the first and interchange the last two rows.

$$\det A = -3 \det \begin{bmatrix} a & b & c \\ x & y & z \\ p & q & r \end{bmatrix} = 3 \det \begin{bmatrix} a & b & c \\ p & q & r \\ x & y & z \end{bmatrix} = 3 \cdot 6 = 18$$

The determinant of a matrix is a sum of products of its entries. In particular, if these entries are polynomials in x, then the determinant itself is a polynomial in x. It is often of interest to determine which values of x make the determinant zero, so it is very useful if the determinant is given in factored form. Theorem 2 can help.

EXAMPLE 7

Find the values of x for which $\det A = 0$, where $A = \begin{bmatrix} 1 & x & x \\ x & 1 & x \\ x & x & 1 \end{bmatrix}$.

Solution ▶ To evaluate det A, first subtract x times row 1 from rows 2 and 3.

$$\det A = \begin{vmatrix} 1 & x & x \\ x & 1 & x \\ x & x & 1 \end{vmatrix} = \begin{vmatrix} 1 & x & x \\ 0 & 1-x^2 & x-x^2 \\ 0 & x-x^2 & 1-x^2 \end{vmatrix} = \begin{vmatrix} 1-x^2 & x-x^2 \\ x-x^2 & 1-x^2 \end{vmatrix}$$

At this stage we could simply evaluate the determinant (the result is $2x^3 - 3x^2 + 1$). But then we would have to factor this polynomial to find the values of x that make it zero. However, this factorization can be obtained directly by first factoring each entry in the determinant and taking a common factor of $(1-x)$ from each row.

$$\det A = \begin{vmatrix} (1-x)(1+x) & x(1-x) \\ x(1-x) & (1-x)(1+x) \end{vmatrix} = (1-x)^2 \begin{vmatrix} 1+x & x \\ x & 1+x \end{vmatrix}$$
$$= (1-x)^2 (2x+1)$$

Hence, det $A = 0$ means $(1-x)^2(2x+1) = 0$, that is $x = 1$ or $x = -\frac{1}{2}$.

EXAMPLE 8

If a_1, a_2, and a_3 are given show that

$$\det \begin{bmatrix} 1 & a_1 & a_1^2 \\ 1 & a_2 & a_2^2 \\ 1 & a_3 & a_3^2 \end{bmatrix} = (a_3 - a_1)(a_3 - a_2)(a_2 - a_1)$$

Solution ▶ Begin by subtracting row 1 from rows 2 and 3, and then expand along column 1:

$$\det \begin{bmatrix} 1 & a_1 & a_1^2 \\ 1 & a_2 & a_2^2 \\ 1 & a_3 & a_3^2 \end{bmatrix} = \det \begin{bmatrix} 1 & a_1 & a_1^2 \\ 0 & a_2 - a_1 & a_2^2 - a_1^2 \\ 0 & a_3 - a_1 & a_3^2 - a_1^2 \end{bmatrix} = \det \begin{bmatrix} a_2 - a_1 & a_2^2 - a_1^2 \\ a_3 - a_1 & a_3^2 - a_1^2 \end{bmatrix}$$

Now $(a_2 - a_1)$ and $(a_3 - a_1)$ are common factors in rows 1 and 2, respectively, so

$$\det \begin{bmatrix} 1 & a_1 & a_1^2 \\ 1 & a_2 & a_2^2 \\ 1 & a_3 & a_3^2 \end{bmatrix} = (a_2 - a_1)(a_3 - a_1) \det \begin{bmatrix} 1 & a_2 + a_1 \\ 1 & a_3 + a_1 \end{bmatrix}$$
$$= (a_2 - a_1)(a_3 - a_1)(a_3 - a_2)$$

The matrix in Example 8 is called a Vandermonde matrix, and the formula for its determinant can be generalized to the $n \times n$ case (see Theorem 7 Section 3.2).

If A is an $n \times n$ matrix, forming uA means multiplying *every* row of A by u. Applying property 3 of Theorem 2, we can take the common factor u out of each row and so obtain the following useful result.

Theorem 3

If A is an $n \times n$ matrix, then $\det(uA) = u^n \det A$ for any number u.

The next example displays a type of matrix whose determinant is easy to compute.

EXAMPLE 9

Evaluate $\det A$ if $A = \begin{bmatrix} a & 0 & 0 & 0 \\ u & b & 0 & 0 \\ v & w & c & 0 \\ x & y & z & d \end{bmatrix}$.

Solution ▶ Expand along row 1 to get $\det A = a \begin{vmatrix} b & 0 & 0 \\ w & c & 0 \\ y & z & d \end{vmatrix}$. Now expand this along the top row to get $\det A = ab \begin{vmatrix} c & 0 \\ z & d \end{vmatrix} = abcd$, the product of the main diagonal entries.

A square matrix is called a **lower triangular matrix** if all entries above the main diagonal are zero (as in Example 9). Similarly, an **upper triangular matrix** is one for which all entries below the main diagonal are zero. A **triangular matrix** is one that is either upper or lower triangular. Theorem 4 gives an easy rule for calculating the determinant of any triangular matrix. The proof is like the solution to Example 9.

Theorem 4

If A is a square triangular matrix, then $\det A$ is the product of the entries on the main diagonal.

Theorem 4 is useful in computer calculations because it is a routine matter to carry a matrix to triangular form using row operations.

Block matrices such as those in the next theorem arise frequently in practice, and the theorem gives an easy method for computing their determinants. This dovetails with Example 11 Section 2.4.

Theorem 5

Consider matrices $\begin{bmatrix} A & X \\ 0 & B \end{bmatrix}$ and $\begin{bmatrix} A & 0 \\ Y & B \end{bmatrix}$ in block form, where A and B are square matrices. Then

$$\det \begin{bmatrix} A & X \\ 0 & B \end{bmatrix} = \det A \det B \quad \text{and} \quad \det \begin{bmatrix} A & 0 \\ Y & B \end{bmatrix} = \det A \det B$$

PROOF

Write $T = \begin{bmatrix} A & X \\ 0 & B \end{bmatrix}$ and proceed by induction on k where A is $k \times k$. If $k = 1$, it is the Laplace expansion along column 1. In general let $S_i(T)$ denote the matrix obtained from T by deleting row i and column 1. Then the cofactor expansion of $\det T$ along the first column is

$$\det T = a_{11}\det(S_1(T)) - a_{21}\det(S_2(T)) + \cdots \pm a_{k1}\det(S_k(T)) \qquad (*)$$

SECTION 3.1 The Cofactor Expansion

where $a_{11}, a_{21}, \ldots, a_{k1}$ are the entries in the first column of A. But
$$S_i(T) = \begin{bmatrix} S_i(A) & X_i \\ 0 & B \end{bmatrix}$$ for each $i = 1, 2, \ldots, k$, so $\det(S_i(T)) = \det(S_i(A)) \cdot \det B$ by induction. Hence, equation (∗) becomes

$$\det T = \{a_{11}\det(S_1(T)) - a_{21}\det(S_2(T)) + \cdots \pm a_{k1}\det(S_k(T))\} \det B$$
$$= \{\det A\} \det B$$

as required. The lower triangular case is similar.

EXAMPLE 10

$$\det \begin{bmatrix} 2 & 3 & 1 & 3 \\ 1 & -2 & -1 & 1 \\ 0 & 1 & 0 & 1 \\ 0 & 4 & 0 & 1 \end{bmatrix} = -\begin{vmatrix} 2 & 1 & 3 & 3 \\ 1 & -1 & -2 & 1 \\ 0 & 0 & 1 & 1 \\ 0 & 0 & 4 & 1 \end{vmatrix} = -\begin{vmatrix} 2 & 1 \\ 1 & -1 \end{vmatrix} \begin{vmatrix} 1 & 1 \\ 4 & 1 \end{vmatrix} = -(-3)(-3) = -9$$

The next result shows that $\det A$ is a linear transformation when regarded as a function of a fixed column of A. The proof is Exercise 21.

Theorem 6

Given columns $c_1, \ldots, c_{j-1}, c_{j+1}, \ldots, c_n$ in $\mathbb{R}^n$, define $T : \mathbb{R}^n \to \mathbb{R}$ by

$$T(\mathbf{x}) = \det[c_1 \ \cdots \ c_{j-1} \ \mathbf{x} \ c_{j+1} \ \cdots \ c_n] \text{ for all } \mathbf{x} \text{ in } \mathbb{R}^n.$$

Then, for all $\mathbf{x}$ and $\mathbf{y}$ in $\mathbb{R}^n$ and all a in $\mathbb{R}$,

$$T(\mathbf{x} + \mathbf{y}) = T(\mathbf{x}) + T(\mathbf{y}) \text{ and } T(a\mathbf{x}) = aT(\mathbf{x})$$

EXERCISES 3.1

1. Compute the determinants of the following matrices.

   (a) $\begin{bmatrix} 2 & -1 \\ 3 & 2 \end{bmatrix}$

   ◆(b) $\begin{bmatrix} 6 & 9 \\ 8 & 12 \end{bmatrix}$

   (c) $\begin{bmatrix} a^2 & ab \\ ab & b^2 \end{bmatrix}$

   ◆(d) $\begin{bmatrix} a+1 & a \\ a & a-1 \end{bmatrix}$

   (e) $\begin{bmatrix} \cos\theta & -\sin\theta \\ \sin\theta & \cos\theta \end{bmatrix}$

   ◆(f) $\begin{bmatrix} 2 & 0 & -3 \\ 1 & 2 & 5 \\ 0 & 3 & 0 \end{bmatrix}$

   (g) $\begin{bmatrix} 1 & 2 & 3 \\ 4 & 5 & 6 \\ 7 & 8 & 9 \end{bmatrix}$

   ◆(h) $\begin{bmatrix} 0 & a & 0 \\ b & c & d \\ 0 & e & 0 \end{bmatrix}$

   (i) $\begin{bmatrix} 1 & b & c \\ b & c & 1 \\ c & 1 & b \end{bmatrix}$

   ◆(j) $\begin{bmatrix} 0 & a & b \\ a & 0 & c \\ b & c & 0 \end{bmatrix}$

   (k) $\begin{bmatrix} 0 & 1 & -1 & 0 \\ 3 & 0 & 0 & 2 \\ 0 & 1 & 2 & 1 \\ 5 & 0 & 0 & 7 \end{bmatrix}$

   ◆(l) $\begin{bmatrix} 1 & 0 & 3 & 1 \\ 2 & 2 & 6 & 0 \\ -1 & 0 & -3 & 1 \\ 4 & 1 & 12 & 0 \end{bmatrix}$

   (m) $\begin{bmatrix} 3 & 1 & -5 & 2 \\ 1 & 3 & 0 & 1 \\ 1 & 0 & 5 & 2 \\ 1 & 1 & 2 & -1 \end{bmatrix}$

   ◆(n) $\begin{bmatrix} 4 & -1 & 3 & -1 \\ 3 & 1 & 0 & 2 \\ 0 & 1 & 2 & 2 \\ 1 & 2 & -1 & 1 \end{bmatrix}$

   (o) $\begin{bmatrix} 1 & -1 & 5 & 5 \\ 3 & 1 & 2 & 4 \\ -1 & -3 & 8 & 0 \\ 1 & 1 & 2 & -1 \end{bmatrix}$

   ◆(p) $\begin{bmatrix} 0 & 0 & 0 & a \\ 0 & 0 & b & p \\ 0 & c & q & k \\ d & s & t & u \end{bmatrix}$

2. Show that $\det A = 0$ if A has a row or column consisting of zeros.

3. Show that the sign of the position in the last row and the last column of A is always $+1$.

4. Show that $\det I = 1$ for any identity matrix I.

5. Evaluate the determinant of each matrix by reducing it to upper triangular form.

  (a) $\begin{bmatrix} 1 & -1 & 2 \\ 3 & 1 & 1 \\ 2 & -1 & 3 \end{bmatrix}$
  ◆(b) $\begin{bmatrix} -1 & 3 & 1 \\ 2 & 5 & 3 \\ 1 & -2 & 1 \end{bmatrix}$

  (c) $\begin{bmatrix} -1 & -1 & 1 & 0 \\ 2 & 1 & 1 & 3 \\ 0 & 1 & 1 & 2 \\ 1 & 3 & -1 & 2 \end{bmatrix}$
  ◆(d) $\begin{bmatrix} 2 & 3 & 1 & 1 \\ 0 & 2 & -1 & 3 \\ 0 & 5 & 1 & 1 \\ 1 & 1 & 2 & 5 \end{bmatrix}$

6. Evaluate by cursory inspection:

  (a) $\det \begin{bmatrix} a & b & c \\ a+1 & b+1 & c+1 \\ a-1 & b-1 & c-1 \end{bmatrix}$

  ◆(b) $\det \begin{bmatrix} a & b & c \\ a+b & 2b & c+b \\ 2 & 2 & 2 \end{bmatrix}$

7. If $\det \begin{bmatrix} a & b & c \\ p & q & r \\ x & y & z \end{bmatrix} = -1$, compute:

  (a) $\det \begin{bmatrix} -x & -y & -z \\ 3p+a & 3q+b & 3r+c \\ 2p & 2q & 2r \end{bmatrix}$

  ◆(b) $\det \begin{bmatrix} -2a & -2b & -2c \\ 2p+x & 2q+y & 2r+z \\ 3x & 3y & 3z \end{bmatrix}$

8. Show that:

  (a) $\det \begin{bmatrix} p+x & q+y & r+z \\ a+x & b+y & c+z \\ a+p & b+q & c+r \end{bmatrix} = 2 \det \begin{bmatrix} a & b & c \\ p & q & r \\ x & y & z \end{bmatrix}$

  ◆(b) $\det \begin{bmatrix} 2a+p & 2b+q & 2c+r \\ 2p+x & 2q+y & 2r+z \\ 2x+a & 2y+b & 2z+c \end{bmatrix} = 9 \det \begin{bmatrix} a & b & c \\ p & q & r \\ x & y & z \end{bmatrix}$

9. In each case either prove the statement or give an example showing that it is false:

  (a) $\det(A + B) = \det A + \det B$.

  ◆(b) If $\det A = 0$, then A has two equal rows.

  (c) If A is 2×2, then $\det(A^T) = \det A$.

  ◆(d) If R is the reduced row-echelon form of A, then $\det A = \det R$.

  (e) If A is 2×2, then $\det(7A) = 49 \det A$.

  ◆(f) $\det(A^T) = -\det A$.

  (g) $\det(-A) = -\det A$.

  ◆(h) If $\det A = \det B$ where A and B are the same size, then $A = B$.

10. Compute the determinant of each matrix, using Theorem 5.

  (a) $\begin{bmatrix} 1 & -1 & 2 & 0 & -2 \\ 0 & 1 & 0 & 4 & 1 \\ 1 & 1 & 5 & 0 & 0 \\ 0 & 0 & 3 & -1 \\ 0 & 0 & 0 & 1 & 1 \end{bmatrix}$
  ◆(b) $\begin{bmatrix} 1 & 2 & 0 & 3 & 0 \\ -1 & 3 & 1 & 4 & 0 \\ 0 & 0 & 2 & 1 & 1 \\ 0 & 0 & -1 & 0 & 2 \\ 0 & 0 & 3 & 0 & 1 \end{bmatrix}$

11. If $\det A = 2$, $\det B = -1$, and $\det C = 3$, find:

  (a) $\det \begin{bmatrix} A & X & Y \\ 0 & B & Z \\ 0 & 0 & C \end{bmatrix}$
  ◆(b) $\det \begin{bmatrix} A & 0 & 0 \\ X & B & 0 \\ Y & Z & C \end{bmatrix}$

  (c) $\det \begin{bmatrix} A & X & Y \\ 0 & B & 0 \\ 0 & Z & C \end{bmatrix}$
  ◆(d) $\det \begin{bmatrix} A & X & 0 \\ 0 & B & 0 \\ Y & Z & C \end{bmatrix}$

12. If A has three columns with only the top two entries nonzero, show that $\det A = 0$.

13. (a) Find $\det A$ if A is 3×3 and $\det(2A) = 6$.

  (b) Under what conditions is $\det(-A) = \det A$?

14. Evaluate by first adding all other rows to the first row.

  (a) $\det \begin{bmatrix} x-1 & 2 & 3 \\ 2 & -3 & x-2 \\ -2 & x & -2 \end{bmatrix}$

  ◆(b) $\det \begin{bmatrix} x-1 & -3 & 1 \\ 2 & -1 & x-1 \\ -3 & x+2 & -2 \end{bmatrix}$

15. (a) Find b if $\det \begin{bmatrix} 5 & -1 & x \\ 2 & 6 & y \\ -5 & 4 & z \end{bmatrix} = ax + by + cz$.

  ◆(b) Find c if $\det \begin{bmatrix} 2 & x & -1 \\ 1 & y & 3 \\ -3 & z & 4 \end{bmatrix} = ax + by + cz$.

16. Find the real numbers x and y such that $\det A = 0$ if:

  (a) $A = \begin{bmatrix} 0 & x & y \\ y & 0 & x \\ x & y & 0 \end{bmatrix}$
  ◆(b) $A = \begin{bmatrix} 1 & x & x \\ -x & -2 & x \\ -x & -x & -3 \end{bmatrix}$

(c) $A = \begin{bmatrix} 1 & x & x^2 & x^3 \\ x & x^2 & x^3 & 1 \\ x^2 & x^3 & 1 & x \\ x^3 & 1 & x & x^2 \end{bmatrix}$ ♦(d) $A = \begin{bmatrix} x & y & 0 & 0 \\ 0 & x & y & 0 \\ 0 & 0 & x & y \\ y & 0 & 0 & x \end{bmatrix}$

17. Show that $\det \begin{bmatrix} 0 & 1 & 1 & 1 \\ 1 & 0 & x & x \\ 1 & x & 0 & x \\ 1 & x & x & 0 \end{bmatrix} = -3x^2$.

18. Show that $\det \begin{bmatrix} 1 & x & x^2 & x^3 \\ a & 1 & x & x^2 \\ p & b & 1 & x \\ q & r & c & 1 \end{bmatrix}$
    $= (1 - ax)(1 - bx)(1 - cx)$.

19. Given the polynomial $p(x) = a + bx + cx^2 + dx^3 + x^4$,
    the matrix $C = \begin{bmatrix} 0 & 1 & 0 & 0 \\ 0 & 0 & 1 & 0 \\ 0 & 0 & 0 & 1 \\ -a & -b & -c & -d \end{bmatrix}$
    is called the **companion matrix** of $p(x)$.
    Show that $\det(xI - C) = p(x)$.

20. Show that $\det \begin{bmatrix} a+x & b+x & c+x \\ b+x & c+x & a+x \\ c+x & a+x & b+x \end{bmatrix}$
    $= (a + b + c + 3x)[(ab + ac + bc) - (a^2 + b^2 + c^2)]$.

21. Prove Theorem 6. [*Hint*: Expand the determinant along column j.]

22. Show that
    $\det \begin{bmatrix} 0 & 0 & \cdots & 0 & a_1 \\ 0 & 0 & \cdots & a_2 & * \\ \vdots & \vdots & & \vdots & \vdots \\ 0 & a_{n-1} & \cdots & * & * \\ a_n & * & \cdots & * & * \end{bmatrix} = (-1)^k a_1 a_2 \cdots a_n$

where either $n = 2k$ or $n = 2k + 1$, and *-entries are arbitrary.

23. By expanding along the first column, show that:
    $\det \begin{bmatrix} 1 & 1 & 0 & 0 & \cdots & 0 & 0 \\ 0 & 1 & 1 & 0 & \cdots & 0 & 0 \\ 0 & 0 & 1 & 1 & \cdots & 0 & 0 \\ \vdots & \vdots & \vdots & \vdots & & \vdots & \vdots \\ 0 & 0 & 0 & 0 & \cdots & 1 & 1 \\ 1 & 0 & 0 & 0 & \cdots & 0 & 1 \end{bmatrix} = 1 + (-1)^{n+1}$
    if the matrix is $n \times n$, $n \geq 2$.

♦24. Form matrix B from a matrix A by writing the columns of A in reverse order. Express $\det B$ in terms of $\det A$.

25. Prove property 3 of Theorem 2 by expanding along the row (or column) in question.

26. Show that the line through two distinct points (x_1, y_1) and (x_2, y_2) in the plane has equation
    $\det \begin{bmatrix} x & y & 1 \\ x_1 & y_1 & 1 \\ x_2 & y_2 & 1 \end{bmatrix} = 0$.

27. Let A be an $n \times n$ matrix. Given a polynomial $p(x) = a_0 + a_1 x + \cdots + a_m x^m$, we write
    $p(A) = a_0 I + a_1 A + \cdots + a_m A^m$.

    For example, if $p(x) = 2 - 3x + 5x^2$, then $p(A) = 2I - 3A + 5A^2$. The *characteristic polynomial* of A is defined to be $c_A(x) = \det[xI - A]$, and the Cayley-Hamilton theorem asserts that $c_A(A) = 0$ for any matrix A.

    (a) Verify the theorem for

    (i) $A = \begin{bmatrix} 3 & 2 \\ 1 & -1 \end{bmatrix}$ and (ii) $A = \begin{bmatrix} 1 & -1 & 1 \\ 0 & 1 & 0 \\ 8 & 2 & 2 \end{bmatrix}$.

    (b) Prove the theorem for $A = \begin{bmatrix} a & b \\ c & d \end{bmatrix}$.

SECTION 3.2 Determinants and Matrix Inverses

In this section, several theorems about determinants are derived. One consequence of these theorems is that a square matrix A is invertible if and only if $\det A \neq 0$. Moreover, determinants are used to give a formula for A^{-1} which, in turn, yields a formula (called Cramer's rule) for the solution of any system of linear equations with an invertible coefficient matrix.

We begin with a remarkable theorem (due to Cauchy in 1812) about the determinant of a product of matrices. The proof is given at the end of this section.

> ### Theorem 1
>
> **Product Theorem**
> If A and B are $n \times n$ matrices, then $\det(AB) = \det A \det B$.

The complexity of matrix multiplication makes the product theorem quite unexpected. Here is an example where it reveals an important numerical identity.

> **EXAMPLE 1**
>
> If $A = \begin{bmatrix} a & b \\ -b & a \end{bmatrix}$ and $B = \begin{bmatrix} c & d \\ -d & c \end{bmatrix}$, then $AB = \begin{bmatrix} ac - bd & ad + bc \\ -(ad + bc) & ac - bd \end{bmatrix}$.
> Hence $\det A \det B = \det(AB)$ gives the identity
> $$(a^2 + b^2)(c^2 + d^2) = (ac - bd)^2 + (ad + bc)^2$$

Theorem 1 extends easily to $\det(ABC) = \det A \det B \det C$. In fact, induction gives
$$\det(A_1 A_2 \cdots A_{k-1} A_k) = \det A_1 \det A_2 \cdots \det A_{k-1} \det A_k$$
for any square matrices $A_1, \ldots, A_k$ of the same size. In particular, if each $A_i = A$, we obtain
$$\det(A^k) = (\det A)^k \quad \text{for any } k \geq 1$$

We can now give the invertibility condition.

> ### Theorem 2
>
> An $n \times n$ matrix A is invertible if and only if $\det A \neq 0$. When this is the case,
> $$\det(A^{-1}) = \frac{1}{\det A}.$$

> **PROOF**
>
> If A is invertible, then $AA^{-1} = I$; so the product theorem gives
> $$1 = \det I = \det(AA^{-1}) = \det A \det A^{-1}$$
> Hence, $\det A \neq 0$ and also $\det A^{-1} = \frac{1}{\det A}$.
>
> Conversely, if $\det A \neq 0$, we show that A can be carried to I by elementary row operations (and invoke Theorem 5 Section 2.4). Certainly, A can be carried to its reduced row-echelon form R, so $R = E_k \cdots E_2 E_1 A$ where the E_i are elementary matrices (Theorem 1 Section 2.5). Hence the product theorem gives
> $$\det R = \det E_k \cdots \det E_2 \det E_1 \det A$$
> Since $\det E \neq 0$ for all elementary matrices E, this shows $\det R \neq 0$. In particular, R has no row of zeros, so $R = I$ because R is square and reduced row-echelon. This is what we wanted.

SECTION 3.2 Determinants and Matrix Inverses

EXAMPLE 2

For which values of c does $A = \begin{bmatrix} 1 & 0 & -c \\ -1 & 3 & 1 \\ 0 & 2c & -4 \end{bmatrix}$ have an inverse?

Solution ▶ Compute det A by first adding c times column 1 to column 3 and then expanding along row 1.

$$\det A = \det \begin{bmatrix} 1 & 0 & -c \\ -1 & 3 & 1 \\ 0 & 2c & -4 \end{bmatrix} = \det \begin{bmatrix} 1 & 0 & 0 \\ -1 & 3 & 1-c \\ 0 & 2c & -4 \end{bmatrix} = 2(c+2)(c-3).$$

Hence, det $A = 0$ if $c = -2$ or $c = 3$, and A has an inverse if $c \neq -2$ and $c \neq 3$.

EXAMPLE 3

If a product $A_1 A_2 \cdots A_k$ of square matrices is invertible, show that each A_i is invertible.

Solution ▶ We have $A_1 A_2 \cdots A_k = \det(A_1 A_2 \cdots A_k)$ by the product theorem, and $\det(A_1 A_2 \cdots A_k) \neq 0$ by Theorem 2 because $A_1 A_2 \cdots A_k$ is invertible. Hence

$$\det A_1 \det A_2 \cdots \det A_k \neq 0,$$

so det $A_i \neq 0$ for each i. This shows that each A_i is invertible, again by Theorem 2.

Theorem 3

If A is any square matrix, $\det A^T = \det A$.

PROOF

Consider first the case of an elementary matrix E. If E is of type I or II, then $E^T = E$; so certainly $\det E^T = \det E$. If E is of type III, then E^T is also of type III; so $\det E^T = 1 = \det E$ by Theorem 2 Section 3.1. Hence, $\det E^T = \det E$ for every elementary matrix E.

Now let A be any square matrix. If A is not invertible, then neither is A^T; so $\det A^T = 0 = \det A$ by Theorem 2. On the other hand, if A is invertible, then $A = E_k \cdots E_2 E_1$, where the E_i are elementary matrices (Theorem 2 Section 2.5). Hence, $A^T = E_1^T E_2^T \cdots E_k^T$ so the product theorem gives

$$\det A^T = \det E_1^T \det E_2^T \cdots \det E_k^T = \det E_1 \det E_2 \cdots \det E_k$$
$$= \det E_k \cdots \det E_2 \det E_1$$
$$= \det A$$

This completes the proof.

EXAMPLE 4

If $\det A = 2$ and $\det B = 5$, calculate $\det(A^3 B^{-1} A^T B^2)$.

Solution ▶ We use several of the facts just derived.
$$\det(A^3 B^{-1} A^T B^2) = \det(A^3)\det(B^{-1})\det(A^T)\det(B^2)$$
$$= (\det A)^3 \frac{1}{\det B} \det A (\det B)^2$$
$$= 2^3 \cdot \tfrac{1}{5} \cdot 2 \cdot 5^2$$
$$= 80$$

EXAMPLE 5

A square matrix is called **orthogonal** if $A^{-1} = A^T$. What are the possible values of $\det A$ if A is orthogonal?

Solution ▶ If A is orthogonal, we have $I = AA^T$. Take determinants to obtain $1 = \det I = \det(AA^T) = \det A \det A^T = (\det A)^2$. Since $\det A$ is a number, this means $\det A = \pm 1$.

Hence Theorems 4 and 5 of Section 2.6 imply that rotation about the origin and reflection about a line through the origin in $\mathbb{R}^2$ have orthogonal matrices with determinants 1 and -1 respectively. In fact they are the *only* such transformations of $\mathbb{R}^2$. We have more to say about this in Section 8.2.

Adjugates

In Section 2.4 we defined the adjugate of a 2×2 matrix $A = \begin{bmatrix} a & b \\ c & d \end{bmatrix}$ to be $\text{adj}(A) = \begin{bmatrix} d & -b \\ -c & a \end{bmatrix}$. Then we verified that $A(\text{adj } A) = (\det A)I = (\text{adj } A)A$ and hence that, if $\det A \neq 0$, $A^{-1} = \frac{1}{\det A} \text{adj } A$. We are now able to define the adjugate of an arbitrary square matrix and to show that this formula for the inverse remains valid (when the inverse exists).

Recall that the (i, j)-cofactor $c_{ij}(A)$ of a square matrix A is a number defined for each position (i, j) in the matrix. If A is a square matrix, the **cofactor matrix of** A is defined to be the matrix $[c_{ij}(A)]$ whose (i, j)-entry is the (i, j)-cofactor of A.

Definition 3.3   The **adjugate**[4] *of* A, *denoted* $\text{adj}(A)$, *is the transpose of this cofactor matrix; in symbols,*
$$\text{adj}(A) = [c_{ij}(A)]^T$$

This agrees with the earlier definition for a 2×2 matrix A as the reader can verify.

---

[4] This is also called the classical adjoint of A, but the term "adjoint" has another meaning.

EXAMPLE 6

Compute the adjugate of $A = \begin{bmatrix} 1 & 3 & -2 \\ 0 & 1 & 5 \\ -2 & -6 & 7 \end{bmatrix}$ and calculate $A(\text{adj } A)$ and $(\text{adj } A)A$.

Solution ▶ We first find the cofactor matrix.

$$\begin{bmatrix} c_{11}(A) & c_{12}(A) & c_{13}(A) \\ c_{21}(A) & c_{22}(A) & c_{23}(A) \\ c_{31}(A) & c_{32}(A) & c_{33}(A) \end{bmatrix} = \begin{bmatrix} \begin{vmatrix} 1 & 5 \\ -6 & 7 \end{vmatrix} & -\begin{vmatrix} 0 & 5 \\ -2 & 7 \end{vmatrix} & \begin{vmatrix} 0 & 1 \\ -2 & -6 \end{vmatrix} \\ -\begin{vmatrix} 3 & -2 \\ -6 & 7 \end{vmatrix} & \begin{vmatrix} 1 & -2 \\ -2 & 7 \end{vmatrix} & -\begin{vmatrix} 1 & 3 \\ -2 & -6 \end{vmatrix} \\ \begin{vmatrix} 3 & -2 \\ 1 & 5 \end{vmatrix} & -\begin{vmatrix} 1 & -2 \\ 0 & 5 \end{vmatrix} & \begin{vmatrix} 1 & 3 \\ 0 & 1 \end{vmatrix} \end{bmatrix}$$

$$= \begin{bmatrix} 37 & -10 & 2 \\ -9 & 3 & 0 \\ 17 & -5 & 1 \end{bmatrix}$$

Then the adjugate of A is the transpose of this cofactor matrix.

$$\text{adj } A = \begin{bmatrix} 37 & -10 & 2 \\ -9 & 3 & 0 \\ 17 & -5 & 1 \end{bmatrix}^T = \begin{bmatrix} 37 & -9 & 17 \\ -10 & 3 & -5 \\ 2 & 0 & 1 \end{bmatrix}$$

The computation of $A(\text{adj } A)$ gives

$$A(\text{adj } A) = \begin{bmatrix} 1 & 3 & -2 \\ 0 & 1 & 5 \\ -2 & -6 & 7 \end{bmatrix} \begin{bmatrix} 37 & -9 & 17 \\ -10 & 3 & -5 \\ 2 & 0 & 1 \end{bmatrix} = \begin{bmatrix} 3 & 0 & 0 \\ 0 & 3 & 0 \\ 0 & 0 & 3 \end{bmatrix} = 3I$$

and the reader can verify that also $(\text{adj } A)A = 3I$. Hence, analogy with the 2×2 case would indicate that $\det A = 3$; this is, in fact, the case.

The relationship $A(\text{adj } A) = (\det A)I$ holds for any square matrix A. To see why this is so, consider the general 3×3 case. Writing $c_{ij}(A) = c_{ij}$ for short, we have

$$\text{adj } A = \begin{bmatrix} c_{11} & c_{12} & c_{13} \\ c_{21} & c_{22} & c_{23} \\ c_{31} & c_{32} & c_{33} \end{bmatrix}^T = \begin{bmatrix} c_{11} & c_{21} & c_{31} \\ c_{12} & c_{22} & c_{32} \\ c_{13} & c_{23} & c_{33} \end{bmatrix}$$

If $A = [a_{ij}]$ in the usual notation, we are to verify that $A(\text{adj } A) = (\det A)I$. That is,

$$A(\text{adj } A) = \begin{bmatrix} a_{11} & a_{12} & a_{13} \\ a_{21} & a_{22} & a_{23} \\ a_{31} & a_{32} & a_{33} \end{bmatrix} \begin{bmatrix} c_{11} & c_{21} & c_{31} \\ c_{12} & c_{22} & c_{32} \\ c_{13} & c_{23} & c_{33} \end{bmatrix} = \begin{bmatrix} \det A & 0 & 0 \\ 0 & \det A & 0 \\ 0 & 0 & \det A \end{bmatrix}$$

Consider the $(1, 1)$-entry in the product. It is given by $a_{11}c_{11} + a_{12}c_{12} + a_{13}c_{13}$, and this is just the cofactor expansion of $\det A$ along the first row of A. Similarly, the $(2, 2)$-entry and the $(3, 3)$-entry are the cofactor expansions of $\det A$ along rows 2 and 3, respectively.

So it remains to be seen why the off-diagonal elements in the matrix product $A(\text{adj } A)$ are all zero. Consider the $(1, 2)$-entry of the product. It is given by $a_{11}c_{21} + a_{12}c_{22} + a_{13}c_{23}$. This *looks* like the cofactor expansion of the determinant of *some* matrix. To see which, observe that c_{21}, c_{22}, and c_{23} are all computed by *deleting* row 2 of A (and one of the columns), so they remain the same if row 2 of A is changed. In particular, if row 2 of A is replaced by row 1, we obtain

$$a_{11}c_{21} + a_{12}c_{22} + a_{13}c_{23} = \det \begin{bmatrix} a_{11} & a_{12} & a_{13} \\ a_{11} & a_{12} & a_{13} \\ a_{31} & a_{32} & a_{33} \end{bmatrix} = 0$$

where the expansion is along row 2 and where the determinant is zero because two rows are identical. A similar argument shows that the other off-diagonal entries are zero.

This argument works in general and yields the first part of Theorem 4. The second assertion follows from the first by multiplying through by the scalar $\frac{1}{\det A}$.

Theorem 4

Adjugate Formula
If A is any square matrix, then

$$A(\text{adj } A) = (\det A)I = (\text{adj } A)A$$

In particular, if $\det A \neq 0$, the inverse of A is given by

$$A^{-1} = \frac{1}{\det A} \text{adj } A$$

It is important to note that this theorem is *not* an efficient way to find the inverse of the matrix A. For example, if A were 10×10, the calculation of adj A would require computing $10^2 = 100$ determinants of 9×9 matrices! On the other hand, the matrix inversion algorithm would find A^{-1} with about the same effort as finding $\det A$. Clearly, Theorem 4 is not a *practical* result: its virtue is that it gives a formula for A^{-1} that is useful for *theoretical* purposes.

EXAMPLE 7

Find the $(2, 3)$-entry of A^{-1} if $A = \begin{bmatrix} 2 & 1 & 3 \\ 5 & -7 & 1 \\ 3 & 0 & -6 \end{bmatrix}$.

Solution ▶ First compute $\det A = \begin{vmatrix} 2 & 1 & 3 \\ 5 & -7 & 1 \\ 3 & 0 & -6 \end{vmatrix} = \begin{vmatrix} 2 & 1 & 7 \\ 5 & -7 & 11 \\ 3 & 0 & 0 \end{vmatrix} = 3 \begin{vmatrix} 1 & 7 \\ -7 & 11 \end{vmatrix} = 180.$

Since $A^{-1} = \frac{1}{\det A} \text{adj } A = \frac{1}{180}[c_{ij}(A)]^T$, the $(2, 3)$-entry of A^{-1} is the $(3, 2)$-entry of the matrix $\frac{1}{180}[c_{ij}(A)]$; that is, it equals $\frac{1}{180}c_{32}(A) = \frac{1}{180}\left(-\begin{vmatrix} 2 & 3 \\ 5 & 1 \end{vmatrix}\right) = \frac{13}{180}$.

EXAMPLE 8

If A is $n \times n$, $n \geq 2$, show that $\det(\text{adj } A) = (\det A)^{n-1}$.

Solution ▶ Write $d = \det A$; we must show that $\det(\text{adj } A) = d^{n-1}$. We have $A(\text{adj } A) = dI$ by Theorem 4, so taking determinants gives $d \det(\text{adj } A) = d^n$. Hence we are done if $d \neq 0$. Assume $d = 0$; we must show that $\det(\text{adj } A) = 0$, that is, adj A is not invertible. If $A \neq 0$, this follows from $A(\text{adj } A) = dI = 0$; if $A = 0$, it follows because then adj $A = 0$.

Cramer's Rule

Theorem 4 has a nice application to linear equations. Suppose
$$A\mathbf{x} = \mathbf{b}$$
is a system of n equations in n variables $x_1, x_2, \ldots, x_n$. Here A is the $n \times n$ coefficient matrix, and $\mathbf{x}$ and $\mathbf{b}$ are the columns

$$\mathbf{x} = \begin{bmatrix} x_1 \\ x_2 \\ \vdots \\ x_n \end{bmatrix} \quad \text{and} \quad \mathbf{b} = \begin{bmatrix} b_1 \\ b_2 \\ \vdots \\ b_n \end{bmatrix}$$

of variables and constants, respectively. If $\det A \neq 0$, we left multiply by A^{-1} to obtain the solution $\mathbf{x} = A^{-1}\mathbf{b}$. When we use the adjugate formula, this becomes

$$\begin{bmatrix} x_1 \\ x_2 \\ \vdots \\ x_n \end{bmatrix} = \frac{1}{\det A}(\text{adj } A)\mathbf{b}$$

$$= \frac{1}{\det A} \begin{bmatrix} c_{11}(A) & c_{21}(A) & \cdots & c_{n1}(A) \\ c_{12}(A) & c_{22}(A) & \cdots & c_{n2}(A) \\ \vdots & \vdots & & \vdots \\ c_{1n}(A) & c_{2n}(A) & \cdots & c_{nn}(A) \end{bmatrix} \begin{bmatrix} b_1 \\ b_2 \\ \vdots \\ b_n \end{bmatrix}$$

Hence, the variables $x_1, x_2, \ldots, x_n$ are given by

$$x_1 = \frac{1}{\det A}[b_1 c_{11}(A) + b_2 c_{21}(A) + \cdots + b_n c_{n1}(A)]$$
$$x_2 = \frac{1}{\det A}[b_1 c_{12}(A) + b_2 c_{22}(A) + \cdots + b_n c_{n2}(A)]$$
$$\vdots \qquad \vdots$$
$$x_n = \frac{1}{\det A}[b_1 c_{1n}(A) + b_2 c_{2n}(A) + \cdots + b_n c_{nn}(A)]$$

Now the quantity $b_1 c_{11}(A) + b_2 c_{21}(A) + \cdots + b_n c_{n1}(A)$ occurring in the formula for x_1 looks like the cofactor expansion of the determinant of a matrix. The cofactors involved are $c_{11}(A), c_{21}(A), \ldots, c_{n1}(A)$, corresponding to the first column of A. If A_1 is obtained from A by replacing the first column of A by $\mathbf{b}$, then $c_{i1}(A_1) = c_{i1}(A)$ for each i because column 1 is deleted when computing them. Hence, expanding $\det(A_1)$ by the first column gives

$$\det A_1 = b_1 c_{11}(A_1) + b_2 c_{21}(A_1) + \cdots + b_n c_{n1}(A_1)$$
$$= b_1 c_{11}(A) + b_2 c_{21}(A) + \cdots + b_n c_{n1}(A)$$
$$= (\det A)x_1$$

Hence, $x_1 = \dfrac{\det A_1}{\det A}$, and similar results hold for the other variables.

Theorem 5

Cramer's Rule[5]

If A is an invertible $n \times n$ matrix, the solution to the system
$$A\mathbf{x} = \mathbf{b}$$
of n equations in the variables $x_1, x_2, \ldots, x_n$ is given by

$$x_1 = \frac{\det A_1}{\det A}, \quad x_2 = \frac{\det A_2}{\det A}, \quad \ldots, \quad x_n = \frac{\det A_n}{\det A}$$

where, for each k, A_k is the matrix obtained from A by replacing column k by $\mathbf{b}$.

---

5   Gabriel Cramer (1704–1752) was a Swiss mathematician who wrote an introductory work on algebraic curves. He popularized the rule that bears his name, but the idea was known earlier.

EXAMPLE 9

Find x_1, given the following system of equations.

$$5x_1 + x_2 - x_3 = 4$$
$$9x_1 + x_2 - x_3 = 1$$
$$x_1 - x_2 + 5x_3 = 2$$

Solution ▶ Compute the determinants of the coefficient matrix A and the matrix A_1 obtained from it by replacing the first column by the column of constants.

$$\det A = \det \begin{bmatrix} 5 & 1 & -1 \\ 9 & 1 & -1 \\ 1 & -1 & 5 \end{bmatrix} = -16$$

$$\det A_1 = \det \begin{bmatrix} 4 & 1 & -1 \\ 1 & 1 & -1 \\ 2 & -1 & 5 \end{bmatrix} = 12$$

Hence, $x_1 = \dfrac{\det A_1}{\det A} = -\dfrac{3}{4}$ by Cramer's rule.

Cramer's rule is *not* an efficient way to solve linear systems or invert matrices. True, it enabled us to calculate x_1 here without computing x_2 or x_3. Although this might seem an advantage, the truth of the matter is that, for large systems of equations, the number of computations needed to find *all* the variables by the gaussian algorithm is comparable to the number required to find *one* of the determinants involved in Cramer's rule. Furthermore, the algorithm works when the matrix of the system is not invertible and even when the coefficient matrix is not square. Like the adjugate formula, then, Cramer's rule is *not* a practical numerical technique; its virtue is theoretical.

Polynomial Interpolation

EXAMPLE 10

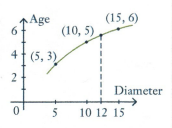

A forester wants to estimate the age (in years) of a tree by measuring the diameter of the trunk (in cm). She obtains the following data:

	Tree 1	Tree 2	Tree 3
Trunk Diameter	5	10	15
Age	3	5	6

Estimate the age of a tree with a trunk diameter of 12 cm.

Solution ▶ The forester decides to "fit" a quadratic polynomial

$$p(x) = r_0 + r_1 x + r_2 x^2$$

to the data, that is choose the coefficients r_0, r_1, and r_2 so that $p(5) = 3$, $p(10) = 5$, and $p(15) = 6$, and then use $p(12)$ as the estimate. These conditions give three linear equations:

$$r_0 + 5r_1 + 25r_2 = 3$$
$$r_0 + 10r_1 + 100r_2 = 5$$
$$r_0 + 15r_1 + 225r_2 = 6$$

The (unique) solution is $r_0 = 0$, $r_1 = \frac{7}{10}$, and $r_2 = -\frac{1}{50}$, so
$$p(x) = \frac{7}{10}x - \frac{1}{50}x^2 = \frac{1}{50}x(35 - x).$$
Hence the estimate is $p(12) = 5.52$.

As in Example 10, it often happens that two variables x and y are related but the actual functional form $y = f(x)$ of the relationship is unknown. Suppose that for certain values $x_1, x_2, \ldots, x_n$ of x the corresponding values $y_1, y_2, \ldots, y_n$ are known (say from experimental measurements). One way to estimate the value of y corresponding to some other value a of x is to find a polynomial[6]

$$p(x) = r_0 + r_1 x + r_2 x + \cdots + r_{n-1} x^{n-1}$$

that "fits" the data, that is $p(x_i) = y_i$ holds for each $i = 1, 2, \ldots, n$. Then the estimate for y is $p(a)$. As we will see, such a polynomial always exists if the x_i are distinct.

The conditions that $p(x_i) = y_i$ are

$$\begin{aligned} r_0 + r_1 x_1 + r_2 x_1^2 + \cdots + r_{n-1} x_1^{n-2} &= y_1 \\ r_0 + r_1 x_2 + r_2 x_2^2 + \cdots + r_{n-1} x_2^{n-2} &= y_2 \\ \vdots \quad \vdots \quad \vdots \quad \quad \vdots \quad \quad \vdots \\ r_0 + r_1 x_n + r_2 x_n^2 + \cdots + r_{n-1} x_n^{n-2} &= y_n \end{aligned}$$

In matrix form, this is

$$\begin{bmatrix} 1 & x_1 & x_1^2 & \cdots & x_1^{n-1} \\ 1 & x_2 & x_2^2 & \cdots & x_2^{n-1} \\ \vdots & \vdots & \vdots & & \vdots \\ 1 & x_n & x_n^2 & \cdots & x_n^{n-1} \end{bmatrix} \begin{bmatrix} r_0 \\ r_1 \\ \vdots \\ r_{n-1} \end{bmatrix} = \begin{bmatrix} y_1 \\ y_2 \\ \vdots \\ y_n \end{bmatrix} \qquad (*)$$

It can be shown (see Theorem 7) that the determinant of the coefficient matrix equals the product of all terms $(x_i - x_j)$ with $i > j$ and so is nonzero (because the x_i are distinct). Hence the equations have a unique solution $r_0, r_1, \ldots, r_{n-1}$. This proves

Theorem 6

Let n data pairs $(x_1, y_1), (x_2, y_2), \ldots, (x_n, y_n)$ be given, and assume that the x_i are distinct. Then there exists a unique polynomial

$$p(x) = r_0 + r_1 x + r_2 x^2 + \cdots + r_{n-1} x^{n-1}$$

such that $p(x_i) = y_i$ for each $i = 1, 2, \ldots, n$.

The polynomial in Theorem 6 is called the **interpolating polynomial** for the data.

We conclude by evaluating the determinant of the coefficient matrix in $(*)$.

If $a_1, a_2, \ldots, a_n$ are numbers, the determinant

---

[6] A **polynomial** is an expression of the form $a_0 + a_1 x + a_2 x^2 + \cdots + a_n x^n$ where the a_i are numbers and x is a variable. If $a_n \neq 0$, the integer n is called the degree of the polynomial, and a_n is called the leading coefficient. See Appendix D.

$$\det \begin{bmatrix} 1 & a_1 & a_1^2 & \cdots & a_1^{n-1} \\ 1 & a_2 & a_2^2 & \cdots & a_2^{n-1} \\ 1 & a_3 & a_3^2 & \cdots & a_3^{n-1} \\ \vdots & \vdots & \vdots & & \vdots \\ 1 & a_n & a_n^2 & \cdots & a_n^{n-1} \end{bmatrix}$$

is called a **Vandermonde determinant**.[7] There is a simple formula for this determinant. If $n = 2$, it equals $(a_2 - a_1)$; if $n = 3$, it is $(a_3 - a_2)(a_3 - a_1)(a_2 - a_1)$ by Example 8 Section 3.1. The general result is the product

$$\prod_{1 \leq j < i \leq n}(a_i - a_j)$$

of all factors $(a_i - a_j)$ where $1 \leq j < i \leq n$. For example, if $n = 4$, it is

$$(a_4 - a_3)(a_4 - a_2)(a_4 - a_1)(a_3 - a_2)(a_3 - a_1)(a_2 - a_1).$$

Theorem 7

Let $a_1, a_2, \ldots, a_n$ be numbers where $n \geq 2$. Then the corresponding Vandermonde determinant is given by

$$\det \begin{bmatrix} 1 & a_1 & a_1^2 & \cdots & a_1^{n-1} \\ 1 & a_2 & a_2^2 & \cdots & a_2^{n-1} \\ 1 & a_3 & a_3^2 & \cdots & a_3^{n-1} \\ \vdots & \vdots & \vdots & & \vdots \\ 1 & a_n & a_n^2 & \cdots & a_n^{n-1} \end{bmatrix} = \prod_{1 \leq j < i \leq n}(a_i - a_j)$$

PROOF

We may assume that the a_i are distinct; otherwise both sides are zero. We proceed by induction on $n \geq 2$; we have it for $n = 2, 3$. So assume it holds for $n - 1$. The trick is to replace a_n by a variable x, and consider the determinant

$$p(x) = \det \begin{bmatrix} 1 & a_1 & a_1^2 & \cdots & a_1^{n-1} \\ 1 & a_2 & a_2^2 & \cdots & a_2^{n-1} \\ \vdots & \vdots & \vdots & & \vdots \\ 1 & a_{n-1} & a_{n-1}^2 & \cdots & a_{n-1}^{n-1} \\ 1 & x & x^2 & \cdots & x^{n-1} \end{bmatrix}.$$

Then $p(x)$ is a polynomial of degree at most $n - 1$ (expand along the last row), and $p(a_i) = 0$ for $i = 1, 2, \ldots, n - 1$ because in each case there are two identical rows in the determinant. In particular, $p(a_1) = 0$, so we have $p(x) = (x - a_1)p_1(x)$ by the factor theorem (see Appendix D). Since $a_2 \neq a_1$, we obtain $p_1(a_2) = 0$, and so $p_1(x) = (x - a_2)p_2(x)$. Thus $p(x) = (x - a_1)(x - a_2)p_2(x)$. As the a_i are distinct, this process continues to obtain

$$p(x) = (x - a_1)(x - a_2) \cdots (x - a_{n-1})d \qquad (**)$$

---

[7] Alexandre Théophile Vandermonde (1735–1796) was a French mathematician who made contributions to the theory of equations.

SECTION 3.2 Determinants and Matrix Inverses

where d is the coefficient of x^{n-1} in $p(x)$. By the cofactor expansion of $p(x)$ along the last row we get

$$d = (-1)^{n+n} \det \begin{bmatrix} 1 & a_1 & a_1^2 & \cdots & a_1^{n-2} \\ 1 & a_2 & a_2^2 & \cdots & a_2^{n-2} \\ \vdots & \vdots & \vdots & & \vdots \\ 1 & a_{n-1} & a_{n-1}^2 & \cdots\cdots & a_{n-1}^{n-2} \end{bmatrix}.$$

Because $(-1)^{n+n} = 1$, the induction hypothesis shows that d is the product of all factors $(a_i - a_j)$ where $1 \leq j < i \leq n - 1$. The result now follows from (**) by substituting a_n for x in $p(x)$.

PROOF OF THEOREM 1

If A and B are $n \times n$ matrices we must show that

$$\det(AB) = \det A \det B. \tag{*}$$

Recall that if E is an elementary matrix obtained by doing one row operation to I_n, then doing that operation to a matrix C (Lemma 1 Section 2.5) results in EC. By looking at the three types of elementary matrices separately, Theorem 2 Section 3.1 shows that

$$\det(EC) = \det E \det C \quad \text{for any matrix } C. \tag{**}$$

Thus if $E_1, E_2, \ldots, E_k$ are all elementary matrices, it follows by induction that

$$\det(E_k \cdots E_2 E_1 C) = \det E_k \cdots \det E_2 \det E_1 \det C \quad \text{for any matrix } C. \tag{***}$$

Lemma. If A has no inverse, then $\det A = 0$.

Proof. Let $A \to R$ where R is reduced row-echelon, say $E_n \cdots E_2 E_1 A = R$. Then R has a row of zeros by Theorem 5(4) Section 2.4, and hence $\det R = 0$. But then (***) gives $\det A = 0$ because $\det E \neq 0$ for any elementary matrix E. This proves the Lemma.

Now we can prove (*) by considering two cases.

Case 1. A has no inverse. Then AB also has no inverse (otherwise $A[B(AB)^{-1}] = I$ so A is invertible by the Corollary 2 to Theorem 5 Section 2.4). Hence the above Lemma (twice) gives

$$\det(AB) = 0 = 0 \det B = \det A \det B.$$

proving (*) in this case.

Case 2. A has an inverse. Then A is a product of elementary matrices by Theorem 2 Section 2.5, say $A = E_1 E_2 \ldots E_k$. Then (***) with $C = I$ gives

$$\det A = \det(E_1 E_2 \cdots E_k) = \det E_1 \det E_2 \cdots \det E_k.$$

But then (***) with $C = B$ gives

$$\det(AB) = \det[(E_1 E_2 \cdots E_k)B] = \det E_1 \det E_2 \cdots \det E_k \det B = \det A \det B,$$

and (*) holds in this case too.

EXERCISES 3.2

1. Find the adjugate of each of the following matrices.

   (a) $\begin{bmatrix} 5 & 1 & 3 \\ -1 & 2 & 3 \\ 1 & 4 & 8 \end{bmatrix}$
   ◆(b) $\begin{bmatrix} 1 & -1 & 2 \\ 3 & 1 & 0 \\ 0 & -1 & 1 \end{bmatrix}$

   (c) $\begin{bmatrix} 1 & 0 & -1 \\ -1 & 1 & 0 \\ 0 & -1 & 1 \end{bmatrix}$
   ◆(d) $\frac{1}{3}\begin{bmatrix} -1 & 2 & 2 \\ 2 & -1 & 2 \\ 2 & 2 & -1 \end{bmatrix}$

2. Use determinants to find which real values of c make each of the following matrices invertible.

   (a) $\begin{bmatrix} 1 & 0 & 3 \\ 3 & -4 & c \\ 2 & 5 & 8 \end{bmatrix}$
   ◆(b) $\begin{bmatrix} 0 & c & -c \\ -1 & 2 & 1 \\ c & -c & c \end{bmatrix}$

   (c) $\begin{bmatrix} c & 1 & 0 \\ 0 & 2 & c \\ -1 & c & 5 \end{bmatrix}$
   ◆(d) $\begin{bmatrix} 4 & c & 3 \\ c & 2 & c \\ 5 & c & 4 \end{bmatrix}$

   (e) $\begin{bmatrix} 1 & 2 & -1 \\ 0 & -1 & c \\ 2 & c & 1 \end{bmatrix}$
   ◆(f) $\begin{bmatrix} 1 & c & -1 \\ c & 1 & 1 \\ 0 & 1 & c \end{bmatrix}$

3. Let A, B, and C denote $n \times n$ matrices and assume that $\det A = -1$, $\det B = 2$, and $\det C = 3$. Evaluate:

   (a) $\det(A^3 BC^T B^{-1})$
   ◆(b) $\det(B^2 C^{-1} AB^{-1} C^T)$

4. Let A and B be invertible $n \times n$ matrices. Evaluate:

   (a) $\det(B^{-1} AB)$
   ◆(b) $\det(A^{-1} B^{-1} AB)$

5. If A is 3×3 and $\det(2A^{-1}) = -4 = \det(A^3(B^{-1})^T)$, find $\det A$ and $\det B$.

6. Let $A = \begin{bmatrix} a & b & c \\ p & q & r \\ u & v & w \end{bmatrix}$ and assume that $\det A = 3$. Compute:

   (a) $\det(2B^{-1})$ where $B = \begin{bmatrix} 4u & 2a & -p \\ 4v & 2b & -q \\ 4w & 2c & -r \end{bmatrix}$

   ◆(b) $\det(2C^{-1})$ where $C = \begin{bmatrix} 2p & -a+u & 3u \\ 2q & -b+v & 3v \\ 2r & -c+w & 3w \end{bmatrix}$

7. If $\det\begin{bmatrix} a & b \\ c & d \end{bmatrix} = -2$, calculate:

   (a) $\det\begin{bmatrix} 2 & -2 & 0 \\ c+1 & -1 & 2a \\ d-2 & 2 & 2b \end{bmatrix}$

   ◆(b) $\det\begin{bmatrix} 2b & 0 & 4d \\ 1 & 2 & -2 \\ a+1 & 2 & 2(c-1) \end{bmatrix}$

   (c) $\det(3A^{-1})$ where $A = \begin{bmatrix} 3c & a+c \\ 3d & b+d \end{bmatrix}$

8. Solve each of the following by Cramer's rule:

   (a) $2x + y = 1$
       $3x + 7y = -2$
   ◆(b) $3x + 4y = 9$
       $2x - y = -1$

   (c) $5x + y - z = -7$
       $2x - y - 2z = 6$
       $3x + 2z = -7$
   ◆(d) $4x - y + 3z = 1$
       $6x + 2y - z = 0$
       $3x + 3y + 2z = -1$

9. Use Theorem 4 to find the (2, 3)-entry of A^{-1} if:

   (a) $A = \begin{bmatrix} 3 & 2 & 1 \\ 1 & 1 & 2 \\ -1 & 2 & 1 \end{bmatrix}$
   ◆(b) $A = \begin{bmatrix} 1 & 2 & -1 \\ 3 & 1 & 1 \\ 0 & 4 & 7 \end{bmatrix}$

10. Explain what can be said about $\det A$ if:

    (a) $A^2 = A$
    ◆(b) $A^2 = I$
    (c) $A^3 = A$
    ◆(d) $PA = P$ and P is invertible
    (e) $A^2 = uA$ and A is $n \times n$
    ◆(f) $A = -A^T$ and A is $n \times n$
    (g) $A^2 + I = 0$ and A is $n \times n$

11. Let A be $n \times n$. Show that $uA = (uI)A$, and use this with Theorem 1 to deduce the result in Theorem 3 Section 3.1: $\det(uA) = u^n \det A$.

12. If A and B are $n \times n$ matrices, $AB = -BA$, and n is odd, show that either A or B has no inverse.

13. Show that $\det AB = \det BA$ holds for any two $n \times n$ matrices A and B.

14. If $A^k = 0$ for some $k \geq 1$, show that A is not invertible.

◆15. If $A^{-1} = A^T$, describe the cofactor matrix of A in terms of A.

16. Show that no 3×3 matrix A exists such that $A^2 + I = 0$. Find a 2×2 matrix A with this property.

17. Show that $\det(A + B^T) = \det(A^T + B)$ for any $n \times n$ matrices A and B.

18. Let A and B be invertible $n \times n$ matrices. Show that $\det A = \det B$ if and only if $A = UB$ where U is a matrix with $\det U = 1$.

♦19. For each of the matrices in Exercise 2, find the inverse for those values of c for which it exists.

20. In each case either prove the statement or give an example showing that it is false:

    (a) If adj A exists, then A is invertible.
    ♦(b) If A is invertible and adj $A = A^{-1}$, then $\det A = 1$.
    (c) $\det(AB) = \det(B^T A)$.
    ♦(d) If $\det A \neq 0$ and $AB = AC$, then $B = C$.
    (e) If $A^T = -A$, then $\det A = -1$.
    ♦(f) If adj $A = 0$, then $A = 0$.
    (g) If A is invertible, then adj A is invertible.
    ♦(h) If A has a row of zeros, so also does adj A.
    (i) $\det(A^T A) > 0$.
    ♦(j) $\det(I + A) = 1 + \det A$.
    (k) If AB is invertible, then A and B are invertible.
    ♦(l) If $\det A = 1$, then adj $A = A$.

21. If A is 2×2 and $\det A = 0$, show that one column of A is a scalar multiple of the other. [*Hint*: Definition 2.5 and Theorem 5(2) Section 2.4.]

22. Find a polynomial $p(x)$ of degree 2 such that:
    (a) $p(0) = 2, p(1) = 3, p(3) = 8$
    ♦(b) $p(0) = 5, p(1) = 3, p(2) = 5$

23. Find a polynomial $p(x)$ of degree 3 such that:
    (a) $p(0) = p(1) = 1, p(-1) = 4, p(2) = -5$
    ♦(b) $p(0) = p(1) = 1, p(-1) = 2, p(-2) = -3$

24. Given the following data pairs, find the interpolating polynomial of degree 3 and estimate the value of y corresponding to $x = 1.5$.
    (a) $(0, 1), (1, 2), (2, 5), (3, 10)$
    ♦(b) $(0, 1), (1, 1.49), (2, -0.42), (3, -11.33)$
    (c) $(0, 2), (1, 2.03), (2, -0.40), (-1, 0.89)$

25. If $A = \begin{bmatrix} 1 & a & b \\ -a & 1 & c \\ -b & -c & 1 \end{bmatrix}$ show that $\det A = 1 + a^2 + b^2 + c^2$. Hence, find A^{-1} for any a, b, and c.

26. (a) Show that $A = \begin{bmatrix} a & p & q \\ 0 & b & r \\ 0 & 0 & c \end{bmatrix}$ has an inverse if and only if $abc \neq 0$, and find A^{-1} in that case.
    ♦(b) Show that if an upper triangular matrix is invertible, the inverse is also upper triangular.

27. Let A be a matrix each of whose entries are integers. Show that each of the following conditions implies the other.
    (1) A is invertible and A^{-1} has integer entries.
    (2) $\det A = 1$ or -1.

♦28. If $A^{-1} = \begin{bmatrix} 3 & 0 & 1 \\ 0 & 2 & 3 \\ 3 & 1 & -1 \end{bmatrix}$, find adj A.

29. If A is 3×3 and $\det A = 2$, find $\det(A^{-1} + 4 \text{ adj } A)$.

30. Show that $\det \begin{bmatrix} 0 & A \\ B & X \end{bmatrix} = \det A \det B$ when A and B are 2×2. What if A and B are 3×3? [*Hint*: Block multiply by $\begin{bmatrix} 0 & I \\ I & 0 \end{bmatrix}$.]

31. Let A be $n \times n$, $n \geq 2$, and assume one column of A consists of zeros. Find the possible values of rank(adj A).

32. If A is 3×3 and invertible, compute $\det(-A^2 (\text{adj } A)^{-1})$.

33. Show that $\text{adj}(uA) = u^{n-1} \text{adj } A$ for all $n \times n$ matrices A.

34. Let A and B denote invertible $n \times n$ matrices. Show that:
    (a) $\text{adj}(\text{adj } A) = (\det A)^{n-2} A$ (here $n \geq 2$) [*Hint*: See Example 8.]
    ♦(b) $\text{adj}(A^{-1}) = (\text{adj } A)^{-1}$
    (c) $\text{adj}(A^T) = (\text{adj } A)^T$
    ♦(d) $\text{adj}(AB) = (\text{adj } B)(\text{adj } A)$ [*Hint*: Show that $AB \text{ adj}(AB) = AB \text{ adj } B \text{ adj } A$.]

SECTION 3.3 Diagonalization and Eigenvalues

The world is filled with examples of systems that evolve in time—the weather in a region, the economy of a nation, the diversity of an ecosystem, etc. Describing such systems is difficult in general and various methods have been developed in special cases. In this section we describe one such method, called *diagonalization*, which is one of the most important techniques in linear algebra. A very fertile example of this procedure is in modelling the growth of the population of an animal species. This has attracted more attention in recent years with the ever increasing awareness that many species are endangered. To motivate the technique, we begin by setting up a simple model of a bird population in which we make assumptions about survival and reproduction rates.

EXAMPLE 1

Consider the evolution of the population of a species of birds. Because the number of males and females are nearly equal, we count only females. We assume that each female remains a juvenile for one year and then becomes an adult, and that only adults have offspring. We make three assumptions about reproduction and survival rates:

1. The number of juvenile females hatched in any year is twice the number of adult females alive the year before (we say the **reproduction rate** is 2).
2. Half of the adult females in any year survive to the next year (the **adult survival rate** is $\frac{1}{2}$).
3. One quarter of the juvenile females in any year survive into adulthood (the **juvenile survival rate** is $\frac{1}{4}$).

If there were 100 adult females and 40 juvenile females alive initially, compute the population of females k years later.

Solution ▶ Let a_k and j_k denote, respectively, the number of adult and juvenile females after k years, so that the total female population is the sum $a_k + j_k$. Assumption 1 shows that $j_{k+1} = 2a_k$, while assumptions 2 and 3 show that $a_{k+1} = \frac{1}{2}a_k + \frac{1}{4}j_k$. Hence the numbers a_k and j_k in successive years are related by the following equations:

$$a_{k+1} = \tfrac{1}{2}a_k + \tfrac{1}{4}j_k$$
$$j_{k+1} = 2a_k$$

If we write $\mathbf{v}_k = \begin{bmatrix} a_k \\ j_k \end{bmatrix}$ and $A = \begin{bmatrix} \frac{1}{2} & \frac{1}{4} \\ 2 & 0 \end{bmatrix}$, these equations take the matrix form

$$\mathbf{v}_{k+1} = A\mathbf{v}_k, \text{ for each } k = 0, 1, 2, \ldots$$

Taking $k = 0$ gives $\mathbf{v}_1 = A\mathbf{v}_0$, then taking $k = 1$ gives $\mathbf{v}_2 = A\mathbf{v}_1 = A^2\mathbf{v}_0$, and taking $k = 2$ gives $\mathbf{v}_3 = A\mathbf{v}_2 = A^3\mathbf{v}_0$. Continuing in this way, we get

$$\mathbf{v}_k = A^k\mathbf{v}_0 \quad \text{for each } k = 0, 1, 2, \ldots.$$

Since $\mathbf{v}_0 = \begin{bmatrix} a_0 \\ j_0 \end{bmatrix} = \begin{bmatrix} 100 \\ 40 \end{bmatrix}$ is known, finding the population profile $\mathbf{v}_k$ amounts to computing A^k for all $k \geq 0$. We will complete this calculation in Example 12 after some new techniques have been developed.

SECTION 3.3 *Diagonalization and Eigenvalues*

Let A be a fixed $n \times n$ matrix. A sequence $\mathbf{v}_0, \mathbf{v}_1, \mathbf{v}_2, \ldots$ of column vectors in $\mathbb{R}^n$ is called a **linear dynamical system**[8] if $\mathbf{v}_0$ is known and the other $\mathbf{v}_k$ are determined (as in Example 1) by the conditions

$$\mathbf{v}_{k+1} = A\mathbf{v}_k \quad \text{for each } k = 0, 1, 2, \ldots.$$

These conditions are called a **matrix recurrence** for the vectors $\mathbf{v}_k$. As in Example 1, they imply that

$$\mathbf{v}_k = A^k \mathbf{v}_0 \quad \text{for all } k \geq 0,$$

so finding the columns $\mathbf{v}_k$ amounts to calculating A^k for $k \geq 0$.

Direct computation of the powers A^k of a square matrix A can be time-consuming, so we adopt an indirect method that is commonly used. The idea is to first **diagonalize** the matrix A, that is, to find an invertible matrix P such that

$$P^{-1}AP = D \quad \text{is a diagonal matrix} \tag{$*$}$$

This works because the powers D^k of the diagonal matrix D are easy to compute, and ($*$) enables us to compute powers A^k of the matrix A in terms of powers D^k of D. Indeed, we can solve ($*$) for A to get $A = PDP^{-1}$. Squaring this gives

$$A^2 = (PDP^{-1})(PDP^{-1}) = PD^2P^{-1}$$

Using this we can compute A^3 as follows:

$$A^3 = AA^2 = (PDP^{-1})(PD^2P^{-1}) = PD^3P^{-1}$$

Continuing in this way we obtain Theorem 1 (even if D is not diagonal).

Theorem 1

If $A = PDP^{-1}$ then $A^k = PD^kP^{-1}$ for each $k = 1, 2, \ldots$.

Hence computing A^k comes down to finding an invertible matrix P as in equation ($*$). To do this it is necessary to first compute certain numbers (called eigenvalues) associated with the matrix A.

Eigenvalues and Eigenvectors

Definition 3.4

If A is an $n \times n$ matrix, a number λ is called an **eigenvalue** of A if

$$A\mathbf{x} = \lambda\mathbf{x} \quad \text{for some column } \mathbf{x} \neq \mathbf{0} \text{ in } \mathbb{R}^n$$

In this case, $\mathbf{x}$ is called an **eigenvector** of A corresponding to the eigenvalue λ, or a λ-**eigenvector** for short.

EXAMPLE 2

If $A = \begin{bmatrix} 3 & 5 \\ 1 & -1 \end{bmatrix}$ and $\mathbf{x} = \begin{bmatrix} 5 \\ 1 \end{bmatrix}$, then $A\mathbf{x} = 4\mathbf{x}$ so $\lambda = 4$ is an eigenvalue of A with corresponding eigenvector $\mathbf{x}$.

---

[8] More precisely, this is *a linear discrete* dynamical system. Many models regard $\mathbf{v}_t$ as a continuous function of the time t, and replace our condition between $\mathbf{v}_{k+1}$ and $A\mathbf{v}_k$ with a differential relationship viewed as functions of time.

The matrix A in Example 2 has another eigenvalue in addition to $\lambda = 4$. To find it, we develop a general procedure for *any* $n \times n$ matrix A.

By definition a number λ is an eigenvalue of the $n \times n$ matrix A if and only if $A\mathbf{x} = \lambda\mathbf{x}$ for some column $\mathbf{x} \neq \mathbf{0}$. This is equivalent to asking that the homogeneous system

$$(\lambda I - A)\mathbf{x} = \mathbf{0}$$

of linear equations has a nontrivial solution $\mathbf{x} \neq \mathbf{0}$. By Theorem 5 Section 2.4 this happens if and only if the matrix $\lambda I - A$ is not invertible and this, in turn, holds if and only if the determinant of the coefficient matrix is zero:

$$\det(\lambda I - A) = 0$$

This last condition prompts the following definition:

Definition 3.5 *If A is an $n \times n$ matrix, the **characteristic polynomial** $c_A(x)$ of A is defined by*

$$c_A(x) = \det(xI - A)$$

Note that $c_A(x)$ is indeed a polynomial in the variable x, and it has degree n when A is an $n \times n$ matrix (this is illustrated in the examples below). The above discussion shows that a number λ is an eigenvalue of A if and only if $c_A(\lambda) = 0$, that is if and only if λ is a **root** of the characteristic polynomial $c_A(x)$. We record these observations in

Theorem 2

Let A be an $n \times n$ matrix.
1. The eigenvalues λ of A are the roots of the characteristic polynomial $c_A(x)$ of A.
2. The λ-eigenvectors $\mathbf{x}$ are the nonzero solutions to the homogeneous system

$$(\lambda I - A)\mathbf{x} = \mathbf{0}$$

of linear equations with $\lambda I - A$ as coefficient matrix.

In practice, solving the equations in part 2 of Theorem 2 is a routine application of gaussian elimination, but finding the eigenvalues can be difficult, often requiring computers (see Section 8.5). For now, the examples and exercises will be constructed so that the roots of the characteristic polynomials are relatively easy to find (usually integers). However, the reader should not be misled by this into thinking that eigenvalues are so easily obtained for the matrices that occur in practical applications!

EXAMPLE 3

Find the characteristic polynomial of the matrix $A = \begin{bmatrix} 3 & 5 \\ 1 & -1 \end{bmatrix}$ discussed in Example 2, and then find all the eigenvalues and their eigenvectors.

Solution ▶ Since $xI - A = \begin{bmatrix} x & 0 \\ 0 & x \end{bmatrix} - \begin{bmatrix} 3 & 5 \\ 1 & -1 \end{bmatrix} = \begin{bmatrix} x-3 & -5 \\ -1 & x+1 \end{bmatrix}$, we get

$$c_A(x) = \det\begin{bmatrix} x-3 & -5 \\ -1 & x+1 \end{bmatrix} = x^2 - 2x - 8 = (x-4)(x+2)$$

SECTION 3.3 Diagonalization and Eigenvalues

Hence, the roots of $c_A(x)$ are $\lambda_1 = 4$ and $\lambda_2 = -2$, so these are the eigenvalues of A. Note that $\lambda_1 = 4$ was the eigenvalue mentioned in Example 2, but we have found a new one: $\lambda_2 = -2$.

To find the eigenvectors corresponding to $\lambda_2 = -2$, observe that in this case

$$(\lambda_2 I - A)\mathbf{x} = \begin{bmatrix} \lambda_2 - 3 & -5 \\ -1 & \lambda_2 + 1 \end{bmatrix} = \begin{bmatrix} -5 & -5 \\ -1 & -1 \end{bmatrix}$$

so the general solution to $(\lambda_2 I - A)\mathbf{x} = \mathbf{0}$ is $\mathbf{x} = t\begin{bmatrix} -1 \\ 1 \end{bmatrix}$ where t is an arbitrary real number. Hence, the eigenvectors $\mathbf{x}$ corresponding to λ_2 are $\mathbf{x} = t\begin{bmatrix} -1 \\ 1 \end{bmatrix}$ where $t \neq 0$ is arbitrary. Similarly, $\lambda_1 = 4$ gives rise to the eigenvectors $\mathbf{x} = t\begin{bmatrix} 5 \\ 1 \end{bmatrix}$, $t \neq 0$, which includes the observation in Example 2.

Note that a square matrix A has *many* eigenvectors associated with any given eigenvalue λ. In fact *every* nonzero solution $\mathbf{x}$ of $(\lambda I - A)\mathbf{x} = \mathbf{0}$ is an eigenvector. Recall that these solutions are all linear combinations of certain basic solutions determined by the gaussian algorithm (see Theorem 2 Section 1.3). Observe that any nonzero multiple of an eigenvector is again an eigenvector,[9] and such multiples are often more convenient.[10] Any set of nonzero multiples of the basic solutions of $(\lambda I - A)\mathbf{x} = \mathbf{0}$ will be called a set of **basic eigenvectors** corresponding to λ.

EXAMPLE 4

Find the characteristic polynomial, eigenvalues, and basic eigenvectors for
$$A = \begin{bmatrix} 2 & 0 & 0 \\ 1 & 2 & -1 \\ 1 & 3 & -2 \end{bmatrix}.$$

Solution ▶ Here the characteristic polynomial is given by

$$c_A(x) = \det \begin{bmatrix} x-2 & 0 & 0 \\ -1 & x-2 & 1 \\ -1 & -3 & x+2 \end{bmatrix} = (x-2)(x-1)(x+1)$$

so the eigenvalues are $\lambda_1 = 2$, $\lambda_2 = 1$, and $\lambda_3 = -1$. To find all eigenvectors for $\lambda_1 = 2$, compute

$$\lambda_1 I - A = \begin{bmatrix} \lambda_1 - 2 & 0 & 0 \\ -1 & \lambda_1 - 2 & 1 \\ -1 & -3 & \lambda_1 + 2 \end{bmatrix} = \begin{bmatrix} 0 & 0 & 0 \\ -1 & 0 & 1 \\ -1 & -3 & 4 \end{bmatrix}$$

---

9 In fact, any nonzero linear combination of λ-eigenvectors is again a λ-eigenvector.
10 Allowing nonzero multiples helps eliminate round-off error when the eigenvectors involve fractions.

We want the (nonzero) solutions to $(\lambda_1 I - A)\mathbf{x} = \mathbf{0}$. The augmented matrix becomes

$$\begin{bmatrix} 0 & 0 & 0 & | & 0 \\ -1 & 0 & 1 & | & 0 \\ -1 & -3 & 4 & | & 0 \end{bmatrix} \rightarrow \begin{bmatrix} 1 & 0 & -1 & | & 0 \\ 0 & 1 & -1 & | & 0 \\ 0 & 0 & 0 & | & 0 \end{bmatrix}$$

using row operations. Hence, the general solution $\mathbf{x}$ to $(\lambda_1 I - A)\mathbf{x} = \mathbf{0}$ is $\mathbf{x} = t \begin{bmatrix} 1 \\ 1 \\ 1 \end{bmatrix}$, where t is arbitrary, so we can use $\mathbf{x}_1 = \begin{bmatrix} 1 \\ 1 \\ 1 \end{bmatrix}$ as the basic eigenvector corresponding to $\lambda_1 = 2$. As the reader can verify, the gaussian algorithm gives basic eigenvectors $\mathbf{x}_2 = \begin{bmatrix} 0 \\ 1 \\ 1 \end{bmatrix}$ and $\mathbf{x}_3 = \begin{bmatrix} 0 \\ \frac{1}{3} \\ 1 \end{bmatrix}$ corresponding to $\lambda_2 = 1$ and $\lambda_3 = -1$, respectively. Note that to eliminate fractions, we could instead use $3\mathbf{x}_3 = \begin{bmatrix} 0 \\ 1 \\ 3 \end{bmatrix}$ as the basic λ_3-eigenvector.

EXAMPLE 5

If A is a square matrix, show that A and A^T have the same characteristic polynomial, and hence the same eigenvalues.

Solution ▶ We use the fact that $xI - A^T = (xI - A)^T$. Then

$$c_{A^T}(x) = \det(xI - A^T) = \det[(xI - A)^T] = \det(xI - A) = c_A(x)$$

by Theorem 3 Section 3.2. Hence $c_{A^T}(x)$ and $c_A(x)$ have the same roots, and so A^T and A have the same eigenvalues (by Theorem 2).

The eigenvalues of a matrix need not be distinct. For example, if $A = \begin{bmatrix} 1 & 1 \\ 0 & 1 \end{bmatrix}$ the characteristic polynomial is $(x - 1)^2$ so the eigenvalue 1 occurs twice. Furthermore, eigenvalues are usually not computed as the roots of the characteristic polynomial. There are iterative, numerical methods (for example the QR-algorithm in Section 8.5) that are much more efficient for large matrices.

A-Invariance

If A is a 2×2 matrix, we can describe the eigenvectors of A geometrically using the following concept. A line L through the origin in $\mathbb{R}^2$ is called A-**invariant** if $A\mathbf{x}$ is in L whenever $\mathbf{x}$ is in L. If we think of A as a linear transformation $\mathbb{R}^2 \to \mathbb{R}^2$, this asks that A carries L into itself, that is the image $A\mathbf{x}$ of each vector $\mathbf{x}$ in L is again in L.

EXAMPLE 6

The x axis $L = \left\{ \begin{bmatrix} x \\ 0 \end{bmatrix} \mid x \text{ in } \mathbb{R} \right\}$ is A-invariant for any matrix of the form $A = \begin{bmatrix} a & b \\ 0 & c \end{bmatrix}$ because $\begin{bmatrix} a & b \\ 0 & c \end{bmatrix}\begin{bmatrix} x \\ 0 \end{bmatrix} = \begin{bmatrix} ax \\ 0 \end{bmatrix}$ is L for all $\mathbf{x} = \begin{bmatrix} x \\ 0 \end{bmatrix}$ in L.

SECTION 3.3 Diagonalization and Eigenvalues

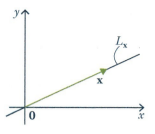

To see the connection with eigenvectors, let $\mathbf{x} \neq \mathbf{0}$ be any nonzero vector in $\mathbb{R}^2$ and let $L_\mathbf{x}$ denote the unique line through the origin containing $\mathbf{x}$ (see the diagram). By the definition of scalar multiplication in Section 2.6, we see that $L_\mathbf{x}$ consists of all scalar multiples of $\mathbf{x}$, that is

$$L_\mathbf{x} = \mathbb{R}\mathbf{x} = \{t\mathbf{x} \mid t \text{ in } \mathbb{R}\}.$$

Now suppose that $\mathbf{x}$ is an eigenvector of A, say $A\mathbf{x} = \lambda\mathbf{x}$ for some λ in $\mathbb{R}$. Then if $t\mathbf{x}$ is in $L_\mathbf{x}$ then

$$A(t\mathbf{x}) = t(A\mathbf{x}) = t(\lambda\mathbf{x}) = (t\lambda)\mathbf{x} \text{ is again in } L_\mathbf{x}.$$

That is, $L_\mathbf{x}$ is A-invariant. On the other hand, if $L_\mathbf{x}$ is A-invariant then $A\mathbf{x}$ is in $L_\mathbf{x}$ (since $\mathbf{x}$ is in $L_\mathbf{x}$). Hence $A\mathbf{x} = t\mathbf{x}$ for some t in $\mathbb{R}$, so $\mathbf{x}$ is an eigenvector for A (with eigenvalue t). This proves:

Theorem 3

Let A be a 2×2 matrix, let $\mathbf{x} \neq \mathbf{0}$ be a vector in $\mathbb{R}^2$, and let $L_\mathbf{x}$ be the line through the origin in $\mathbb{R}^2$ containing $\mathbf{x}$. Then

$$\mathbf{x} \text{ is an eigenvector of } A \quad \text{if and only if} \quad L_\mathbf{x} \text{ is } A\text{-invariant}.$$

EXAMPLE 7

1. If θ is not a multiple of π, show that $A = \begin{bmatrix} \cos\theta & -\sin\theta \\ \sin\theta & \cos\theta \end{bmatrix}$ has no real eigenvalue.
2. If m is real show that $B = \dfrac{1}{1+m^2}\begin{bmatrix} 1-m^2 & 2m \\ 2m & m^2-1 \end{bmatrix}$ has a 1 as an eigenvalue.

Solution ▶

(1) A induces rotation about the origin through the angle θ (Theorem 4 Section 2.6). Since θ is not a multiple of π, this shows that no line through the origin is A-invariant. Hence A has no eigenvector by Theorem 3, and so has no eigenvalue.

(2) B induces reflection Q_m in the line through the origin with slope m by Theorem 5 Section 2.6. If $\mathbf{x}$ is any nonzero point on this line then it is clear that $Q_m\mathbf{x} = \mathbf{x}$, that is $Q_m\mathbf{x} = 1\mathbf{x}$. Hence 1 is an eigenvalue (with eigenvector $\mathbf{x}$).

If $\theta = \frac{\pi}{2}$ in Example 7(1), then $A = \begin{bmatrix} 0 & -1 \\ 1 & 0 \end{bmatrix}$, so $c_A(x) = x^2 + 1$. This polynomial has no root in $\mathbb{R}$, so A has no (real) eigenvalue, and hence no eigenvector. In fact its eigenvalues are the complex numbers i and $-i$, with corresponding eigenvectors $\begin{bmatrix} 1 \\ -i \end{bmatrix}$ and $\begin{bmatrix} 1 \\ i \end{bmatrix}$. In other words, A *has* eigenvalues and eigenvectors, just not real ones.

Note that *every* polynomial has complex roots,[11] so every matrix has complex eigenvalues. While these eigenvalues may very well be real, this suggests that we really should be doing linear algebra over the complex numbers. Indeed, everything

---
[11] This is called the *Fundamental Theorem of Algebra* and was first proved by Gauss in his doctoral dissertation.

we have done (gaussian elimination, matrix algebra, determinants, etc.) works if all the scalars are complex.

Diagonalization

An $n \times n$ matrix D is called a **diagonal matrix** if all its entries off the main diagonal are zero, that is if D has the form

$$D = \begin{bmatrix} \lambda_1 & 0 & \cdots & 0 \\ 0 & \lambda_2 & \cdots & 0 \\ \vdots & \vdots & \ddots & \vdots \\ 0 & 0 & \cdots & \lambda_n \end{bmatrix} = \text{diag}(\lambda_1, \lambda_2, \ldots, \lambda_n)$$

where $\lambda_1, \lambda_2, \ldots, \lambda_n$ are numbers. Calculations with diagonal matrices are very easy. Indeed, if $D = \text{diag}(\lambda_1, \lambda_2, \ldots, \lambda_n)$ and $E = \text{diag}(\mu_1, \mu_2, \ldots, \mu_n)$ are two diagonal matrices, their product DE and sum $D + E$ are again diagonal, and are obtained by doing the same operations to corresponding diagonal elements:

$$DE = \text{diag}(\lambda_1\mu_1, \lambda_2\mu_2, \ldots, \lambda_n\mu_n)$$
$$D + E = \text{diag}(\lambda_1 + \mu_1, \lambda_2 + \mu_2, \ldots, \lambda_n + \mu_n)$$

Because of the simplicity of these formulas, and with an eye on Theorem 1 and the discussion preceding it, we make another definition:

Definition 3.6

An $n \times n$ matrix A is called **diagonalizable** if

$$P^{-1}AP \quad \text{is diagonal for some invertible } n \times n \text{ matrix } P$$

Here the invertible matrix P is called a **diagonalizing matrix** for A.

To discover when such a matrix P exists, we let $\mathbf{x}_1, \mathbf{x}_2, \ldots, \mathbf{x}_n$ denote the columns of P and look for ways to determine when such $\mathbf{x}_i$ exist and how to compute them. To this end, write P in terms of its columns as follows:

$$P = [\mathbf{x}_1, \mathbf{x}_2, \ldots, \mathbf{x}_n]$$

Observe that $P^{-1}AP = D$ for some diagonal matrix D holds if and only if

$$AP = PD$$

If we write $D = \text{diag}(\lambda_1, \lambda_2, \ldots, \lambda_n)$, where the λ_i are numbers to be determined, the equation $AP = PD$ becomes

$$A[\mathbf{x}_1, \mathbf{x}_2, \ldots, \mathbf{x}_n] = [\mathbf{x}_1, \mathbf{x}_2, \ldots, \mathbf{x}_n] \begin{bmatrix} \lambda_1 & 0 & \cdots & 0 \\ 0 & \lambda_2 & \cdots & 0 \\ \vdots & \vdots & \ddots & \vdots \\ 0 & 0 & \cdots & \lambda_n \end{bmatrix}$$

By the definition of matrix multiplication, each side simplifies as follows

$$[A\mathbf{x}_1 \ A\mathbf{x}_2 \ \cdots \ A\mathbf{x}_n] = [\lambda_1\mathbf{x}_1 \ \lambda_2\mathbf{x}_2 \ \cdots \ \lambda_n\mathbf{x}_n]$$

Comparing columns shows that $A\mathbf{x}_i = \lambda_i\mathbf{x}_i$ for each i, so

$$P^{-1}AP = D \quad \text{if and only if} \quad A\mathbf{x}_i = \lambda_i\mathbf{x}_i \text{ for each } i.$$

In other words, $P^{-1}AP = D$ holds if and only if the diagonal entries of D are eigenvalues of A and the columns of P are corresponding eigenvectors. This proves the following fundamental result.

SECTION 3.3 Diagonalization and Eigenvalues

Theorem 4

Let A be an $n \times n$ matrix.

1. A is diagonalizable if and only if it has eigenvectors $\mathbf{x}_1, \mathbf{x}_2, \ldots, \mathbf{x}_n$ such that the matrix $P = [\mathbf{x}_1 \ \mathbf{x}_2 \ \cdots \ \mathbf{x}_n]$ is invertible.
2. When this is the case, $P^{-1}AP = \operatorname{diag}(\lambda_1, \lambda_2, \ldots, \lambda_n)$ where, for each i, λ_i is the eigenvalue of A corresponding to $\mathbf{x}_i$.

EXAMPLE 8

Diagonalize the matrix $A = \begin{bmatrix} 2 & 0 & 0 \\ 1 & 2 & -1 \\ 1 & 3 & -2 \end{bmatrix}$ in Example 4.

Solution ▶ By Example 4, the eigenvalues of A are $\lambda_1 = 2$, $\lambda_2 = 1$, and $\lambda_3 = -1$, with corresponding basic eigenvectors $\mathbf{x}_1 = \begin{bmatrix} 1 \\ 1 \\ 1 \end{bmatrix}$, $\mathbf{x}_2 = \begin{bmatrix} 0 \\ 1 \\ 1 \end{bmatrix}$, and $\mathbf{x}_3 = \begin{bmatrix} 0 \\ 1 \\ 3 \end{bmatrix}$, respectively. Since the matrix $P = [\mathbf{x}_1 \ \mathbf{x}_2 \ \cdots \ \mathbf{x}_n] = \begin{bmatrix} 1 & 0 & 0 \\ 1 & 1 & 1 \\ 1 & 1 & 3 \end{bmatrix}$ is invertible, Theorem 4 guarantees that $P^{-1}AP = \begin{bmatrix} \lambda_1 & 0 & 0 \\ 0 & \lambda_2 & 0 \\ 0 & 0 & \lambda_3 \end{bmatrix} = \begin{bmatrix} 2 & 0 & 0 \\ 0 & 1 & 0 \\ 0 & 0 & -1 \end{bmatrix} = D$.

The reader can verify this directly—easier to check $AP = PD$.

In Example 8, suppose we let $Q = [\mathbf{x}_2 \ \mathbf{x}_1 \ \mathbf{x}_3]$ be the matrix formed from the eigenvectors $\mathbf{x}_1$, $\mathbf{x}_2$, and $\mathbf{x}_3$ of A, but in a *different order* than that used to form P. Then $Q^{-1}AQ = \operatorname{diag}(\lambda_2, \lambda_1, \lambda_3)$ is diagonal by Theorem 4, but the eigenvalues are in the *new* order. Hence we can choose the diagonalizing matrix P so that the eigenvalues λ_i appear in any order we want along the main diagonal of D.

In every example above each eigenvalue has had only one basic eigenvector. Here is a diagonalizable matrix where this is not the case.

EXAMPLE 9

Diagonalize the matrix $A = \begin{bmatrix} 0 & 1 & 1 \\ 1 & 0 & 1 \\ 1 & 1 & 0 \end{bmatrix}$.

Solution ▶ To compute the characteristic polynomial of A first add rows 2 and 3 of $xI - A$ to row 1:

$$c_A(x) = \det\begin{bmatrix} x & -1 & -1 \\ -1 & x & -1 \\ -1 & -1 & x \end{bmatrix} = \det\begin{bmatrix} x-2 & x-2 & x-2 \\ -1 & x & -1 \\ -1 & -1 & x \end{bmatrix}$$

$$= \det\begin{bmatrix} x-2 & 0 & 0 \\ -1 & x+1 & 0 \\ -1 & 0 & x+1 \end{bmatrix} = (x-2)(x+1)^2$$

Hence the eigenvalues are $\lambda_1 = 2$ and $\lambda_2 = -1$, with λ_2 repeated twice (we say that λ_2 has *multiplicity* two). However, A *is* diagonalizable. For $\lambda_1 = 2$, the system of equations $(\lambda_1 I - A)\mathbf{x} = \mathbf{0}$ has general solution $\mathbf{x} = t\begin{bmatrix} 1 \\ 1 \\ 1 \end{bmatrix}$ as the reader can verify, so a basic λ_1-eigenvector is $\mathbf{x}_1 = \begin{bmatrix} 1 \\ 1 \\ 1 \end{bmatrix}$.

Turning to the repeated eigenvalue $\lambda_2 = -1$, we must solve $(\lambda_2 I - A)\mathbf{x} = \mathbf{0}$. By gaussian elimination, the general solution is $\mathbf{x} = s\begin{bmatrix} -1 \\ 1 \\ 0 \end{bmatrix} + t\begin{bmatrix} -1 \\ 0 \\ 1 \end{bmatrix}$ where s and t are arbitrary. Hence the gaussian algorithm produces *two* basic λ_2-eigenvectors $\mathbf{x}_2 = \begin{bmatrix} -1 \\ 1 \\ 0 \end{bmatrix}$ and $\mathbf{y}_2 = \begin{bmatrix} -1 \\ 0 \\ 1 \end{bmatrix}$. If we take $P = [\mathbf{x}_1 \ \mathbf{x}_2 \ \mathbf{y}_2] = \begin{bmatrix} 1 & -1 & -1 \\ 1 & 1 & 0 \\ 1 & 0 & 1 \end{bmatrix}$ we find that P is invertible. Hence $P^{-1}AP = \text{diag}(2, -1, -1)$ by Theorem 4.

Example 9 typifies every diagonalizable matrix. To describe the general case, we need some terminology.

Definition 3.7 *An eigenvalue λ of a square matrix A is said to have **multiplicity** m if it occurs m times as a root of the characteristic polynomial $c_A(x)$.*

Thus, for example, the eigenvalue $\lambda_2 = -1$ in Example 9 has multiplicity 2. In that example the gaussian algorithm yields two basic λ_2-eigenvectors, the same number as the multiplicity. This works in general.

Theorem 5

A square matrix A is diagonalizable if and only if every eigenvalue λ of multiplicity m yields exactly m basic eigenvectors; that is, if and only if the general solution of the system $(\lambda I - A)\mathbf{x} = \mathbf{0}$ has exactly m parameters.

One case of Theorem 5 deserves mention.

Theorem 6

An $n \times n$ matrix with n distinct eigenvalues is diagonalizable.

SECTION 3.3 Diagonalization and Eigenvalues

The proofs of Theorems 5 and 6 require more advanced techniques and are given in Chapter 5. The following procedure summarizes the method.

Diagonalization Algorithm

To diagonalize an $n \times n$ matrix A:

Step 1. Find the distinct eigenvalues λ of A.

Step 2. Compute the basic eigenvectors corresponding to each of these eigenvalues λ as basic solutions of the homogeneous system $(\lambda I - A)\mathbf{x} = \mathbf{0}$.

Step 3. The matrix A is diagonalizable if and only if there are n basic eigenvectors in all.

Step 4. If A is diagonalizable, the $n \times n$ matrix P with these basic eigenvectors as its columns is a diagonalizing matrix for A, that is, P is invertible and $P^{-1}AP$ is diagonal.

The diagonalization algorithm is valid even if the eigenvalues are nonreal complex numbers. In this case the eigenvectors will also have complex entries, but we will not pursue this here.

EXAMPLE 10

Show that $A = \begin{bmatrix} 1 & 1 \\ 0 & 1 \end{bmatrix}$ is not diagonalizable.

Solution 1 ▶ The characteristic polynomial is $c_A(x) = (x-1)^2$, so A has only one eigenvalue $\lambda_1 = 1$ of multiplicity 2. But the system of equations $(\lambda_1 I - A)\mathbf{x} = \mathbf{0}$ has general solution $t\begin{bmatrix} 1 \\ 0 \end{bmatrix}$, so there is only one parameter, and so only one basic eigenvector $\begin{bmatrix} 1 \\ 2 \end{bmatrix}$. Hence A is not diagonalizable.

Solution 2 ▶ We have $c_A(x) = (x-1)^2$ so the only eigenvalue of A is $\lambda = 1$. Hence, if A were diagonalizable, Theorem 4 would give $P^{-1}AP = \begin{bmatrix} 1 & 0 \\ 0 & 1 \end{bmatrix} = I$ for some invertible matrix P. But then $A = PIP^{-1} = I$, which is not the case. So A cannot be diagonalizable.

Diagonalizable matrices share many properties of their eigenvalues. The following example illustrates why.

EXAMPLE 11

If $\lambda^3 = 5\lambda$ for every eigenvalue of the diagonalizable matrix A, show that $A^3 = 5A$.

Solution ▶ Let $P^{-1}AP = D = \text{diag}(\lambda_1, \ldots, \lambda_n)$. Because $\lambda_i^3 = 5\lambda_i$ for each i, we obtain

$$D^3 = \text{diag}(\lambda_1^3, \ldots, \lambda_n^3) = \text{diag}(5\lambda_1, \ldots, 5\lambda_n) = 5D$$

Hence $A^3 = (PDP^{-1})^3 = PD^3P^{-1} = P(5D)P^{-1} = 5(PDP^{-1}) = 5A$ using Theorem 1. This is what we wanted.

If $p(x)$ is any polynomial and $p(\lambda) = 0$ for every eigenvalue of the diagonalizable matrix A, an argument similar to that in Example 11 shows that $p(A) = 0$. Thus Example 11 deals with the case $p(x) = x^3 - 5x$. In general, $p(A)$ is called the *evaluation* of the polynomial $p(x)$ at the matrix A. For example, if $p(x) = 2x^3 - 3x + 5$, then $p(A) = 2A^3 - 3A + 5I$—note the use of the identity matrix.

In particular, if $c_A(x)$ denotes the characteristic polynomial of A, we certainly have $c_A(\lambda) = 0$ for each eigenvalue λ of A (Theorem 2). Hence $c_A(A) = 0$ for every diagonalizable matrix A. This is, in fact, true for *any* square matrix, diagonalizable or not, and the general result is called the Cayley-Hamilton theorem. It is proved in Section 8.6 and again in Section 9.4.

Linear Dynamical Systems

We began Section 3.3 with an example from ecology which models the evolution of the population of a species of birds as time goes on. As promised, we now complete the example—Example 12 below.

The bird population was described by computing the female population profile $\mathbf{v}_k = \begin{bmatrix} a_k \\ j_k \end{bmatrix}$ of the species, where a_k and j_k represent the number of adult and juvenile females present k years after the initial values a_0 and j_0 were observed. The model assumes that these numbers are related by the following equations:

$$a_{k+1} = \tfrac{1}{2}a_k + \tfrac{1}{4}j_k$$
$$j_{k+1} = 2a_k$$

If we write $A = \begin{bmatrix} \tfrac{1}{2} & \tfrac{1}{4} \\ 2 & 0 \end{bmatrix}$, the columns $\mathbf{v}_k$ satisfy $\mathbf{v}_{k+1} = A\mathbf{v}_k$ for each $k = 0, 1, 2, \ldots$. Hence $\mathbf{v}_k = A^k\mathbf{v}_0$ for each $k = 1, 2, \ldots$. We can now use our diagonalization techniques to determine the population profile $\mathbf{v}_k$ for all values of k in terms of the initial values.

EXAMPLE 12

Assuming that the initial values were $a_0 = 100$ adult females and $j_0 = 40$ juvenile females, compute a_k and j_k for $k = 1, 2, \ldots$.

SECTION 3.3 Diagonalization and Eigenvalues

Solution ▶ The characteristic polynomial of the matrix $A = \begin{bmatrix} \frac{1}{2} & \frac{1}{4} \\ 2 & 0 \end{bmatrix}$ is $c_A(x) = x^2 - \frac{1}{2}x - \frac{1}{2} = (x-1)(x+\frac{1}{2})$, so the eigenvalues are $\lambda_1 = 1$ and $\lambda_2 = -\frac{1}{2}$ and gaussian elimination gives corresponding basic eigenvectors $\begin{bmatrix} \frac{1}{2} \\ 1 \end{bmatrix}$ and $\begin{bmatrix} -\frac{1}{4} \\ 1 \end{bmatrix}$. For convenience, we can use multiples $\mathbf{x}_1 = \begin{bmatrix} 1 \\ 2 \end{bmatrix}$ and $\mathbf{x}_2 = \begin{bmatrix} -1 \\ 4 \end{bmatrix}$ respectively. Hence a diagonalizing matrix is $P = \begin{bmatrix} 1 & -1 \\ 2 & 4 \end{bmatrix}$ and we obtain

$$P^{-1}AP = D \quad \text{where } D = \begin{bmatrix} 1 & 0 \\ 0 & -\frac{1}{2} \end{bmatrix}.$$

This gives $A = PDP^{-1}$ so, for each $k \geq 0$, we can compute A^k explicitly:

$$A^k = PD^kP^{-1} = \begin{bmatrix} 1 & -1 \\ 2 & 4 \end{bmatrix}\begin{bmatrix} 1 & 0 \\ 0 & (-\frac{1}{2})^k \end{bmatrix}\frac{1}{6}\begin{bmatrix} 4 & 1 \\ -2 & 4 \end{bmatrix}$$

$$= \frac{1}{6}\begin{bmatrix} 4 + 2(-\frac{1}{2})^k & 1 - (-\frac{1}{2})^k \\ 8 - 8(-\frac{1}{2})^k & 2 + 4(-\frac{1}{2})^k \end{bmatrix}$$

Hence we obtain

$$\begin{bmatrix} a_k \\ j_k \end{bmatrix} = \mathbf{v}_k = A^k \mathbf{v}_0 = \frac{1}{6}\begin{bmatrix} 4 + 2(-\frac{1}{2})^k & 1 - (-\frac{1}{2})^k \\ 8 - 8(-\frac{1}{2})^k & 2 + 4(-\frac{1}{2})^k \end{bmatrix}\begin{bmatrix} 100 \\ 40 \end{bmatrix}$$

$$= \frac{1}{6}\begin{bmatrix} 440 + 160(-\frac{1}{2})^k \\ 880 - 640(-\frac{1}{2})^k \end{bmatrix}.$$

Equating top and bottom entries, we obtain exact formulas for a_k and j_k:

$$a_k = \frac{220}{3} + \frac{80}{3}\left(-\frac{1}{2}\right)^k \quad \text{and} \quad j_k = \frac{440}{3} + \frac{320}{3}\left(-\frac{1}{2}\right)^k \quad \text{for } k = 1, 2, \ldots.$$

In practice, the exact values of a_k and j_k are not usually required. What is needed is a measure of how these numbers behave for large values of k. This is easy to obtain here. Since $\left(-\frac{1}{2}\right)^k$ is nearly zero for large k, we have the following approximate values

$$a_k \approx \frac{220}{3} \quad \text{and } j_k \approx \frac{440}{3} \quad \text{if } k \text{ is large.}$$

Hence, in the long term, the female population stabilizes with approximately twice as many juveniles as adults.

Definition 3.8 *If A is an $n \times n$ matrix, a sequence $\mathbf{v}_0, \mathbf{v}_1, \mathbf{v}_2, \ldots$ of columns in $\mathbb{R}^n$ is called a **linear dynamical system** if $\mathbf{v}_0$ is specified and $\mathbf{v}_1, \mathbf{v}_2, \ldots$ are given by the matrix recurrence $\mathbf{v}_{k+1} = A\mathbf{v}_k$ for each $k \geq 0$.*

As before, we obtain

$$\mathbf{v}_k = A^k \mathbf{v}_0 \text{ for each } k = 1, 2, \ldots \quad (*)$$

Hence the columns $\mathbf{v}_k$ are determined by the powers A^k of the matrix A and, as we have seen, these powers can be efficiently computed if A is diagonalizable. In fact $(*)$ can be used to give a nice "formula" for the columns $\mathbf{v}_k$ in this case.

Assume that A is diagonalizable with eigenvalues $\lambda_1, \lambda_2, \ldots, \lambda_n$ and corresponding basic eigenvectors $\mathbf{x}_1, \mathbf{x}_2, \ldots, \mathbf{x}_n$. If $P = [\mathbf{x}_1 \ \mathbf{x}_2 \ \cdots \ \mathbf{x}_n]$ is a diagonalizing matrix with the $\mathbf{x}_i$ as columns, then P is invertible and

$$P^{-1}AP = D = \text{diag}(\lambda_1, \lambda_2, \ldots, \lambda_n)$$

by Theorem 4. Hence $A = PDP^{-1}$ so (*) and Theorem 1 give

$$\mathbf{v}_k = A^k \mathbf{v}_0 = (PDP^{-1})^k \mathbf{v}_0 = (PD^k P^{-1})\mathbf{v}_0 = PD^k(P^{-1}\mathbf{v}_0)$$

for each $k = 1, 2, \ldots$. For convenience, we denote the column $P^{-1}\mathbf{v}_0$ arising here as follows:

$$\mathbf{b} = P^{-1}\mathbf{v}_0 = \begin{bmatrix} b_1 \\ b_2 \\ \vdots \\ b_n \end{bmatrix}$$

Then matrix multiplication gives

$$\begin{aligned}
\mathbf{v}_k &= PD^k(P^{-1}\mathbf{v}_0) \\
&= [\mathbf{x}_1 \ \mathbf{x}_2 \ \cdots \ \mathbf{x}_n] \begin{bmatrix} \lambda_1^k & 0 & \cdots & 0 \\ 0 & \lambda_2^k & \cdots & 0 \\ \vdots & \vdots & \ddots & \vdots \\ 0 & 0 & \cdots & \lambda_n^k \end{bmatrix} \begin{bmatrix} b_1 \\ b_2 \\ \vdots \\ b_n \end{bmatrix} \\
&= [\mathbf{x}_1 \ \mathbf{x}_2 \ \cdots \ \mathbf{x}_n] \begin{bmatrix} b_1 \lambda_1^k \\ b_2 \lambda_2^k \\ \vdots \\ b_n \lambda_n^k \end{bmatrix} \\
&= b_1 \lambda_1^k \mathbf{x}_1 + b_2 \lambda_2^k \mathbf{x}_2 + \cdots + b_n \lambda_n^k \mathbf{x}_n
\end{aligned} \quad (**)$$

for each $k \geq 0$. This is a useful **exact formula** for the columns $\mathbf{v}_k$. Note that, in particular, $\mathbf{v}_0 = b_1\mathbf{x}_1 + b_2\mathbf{x}_2 + \cdots + b_n\mathbf{x}_n$.

However, such an exact formula for $\mathbf{v}_k$ is often not required in practice; all that is needed is to *estimate* $\mathbf{v}_k$ for large values of k (as was done in Example 12). This can be easily done if A has a largest eigenvalue. An eigenvalue λ of a matrix A is called a **dominant eigenvalue** of A if it has multiplicity 1 and

$$|\lambda| > |\mu| \quad \text{for all eigenvalues } \mu \neq \lambda$$

where $|\lambda|$ denotes the absolute value of the number λ. For example, $\lambda_1 = 1$ is dominant in Example 12.

Returning to the above discussion, suppose that A *has* a dominant eigenvalue. By choosing the order in which the columns $\mathbf{x}_i$ are placed in P, we may assume that λ_1 is dominant among the eigenvalues $\lambda_1, \lambda_2, \ldots, \lambda_n$ of A (see the discussion following Example 8). Now recall the exact expression for V_k in (**) above:

$$\mathbf{v}_k = b_1 \lambda_1^k \mathbf{x}_1 + b_2 \lambda_2^k \mathbf{x}_2 + \cdots + b_n \lambda_n^k \mathbf{x}_n$$

Take λ_1^k out as a common factor in this equation to get

$$\mathbf{v}_k = \lambda_1^k \left[b_1 \mathbf{x}_1 + b_2 \left(\frac{\lambda_2}{\lambda_1}\right)^k \mathbf{x}_2 + \cdots + b_n \left(\frac{\lambda_n}{\lambda_1}\right)^k \mathbf{x}_n \right]$$

for each $k \geq 0$. Since λ_1 is dominant, we have $|\lambda_i| < |\lambda_1|$ for each $i \geq 2$, so each of the numbers $(\lambda_i/\lambda_1)^k$ become small in absolute value as k increases. Hence $\mathbf{v}_k$ is approximately equal to the first term $\lambda_1^k b_1 \mathbf{x}_1$, and we write this as $\mathbf{v}_k \approx \lambda_1^k b_1 \mathbf{x}_1$. These observations are summarized in the following theorem (together with the above exact formula for $\mathbf{v}_k$).

SECTION 3.3 Diagonalization and Eigenvalues

Theorem 7

Consider the dynamical system $\mathbf{v}_0, \mathbf{v}_1, \mathbf{v}_2, \ldots$ with matrix recurrence

$$\mathbf{v}_{k+1} = A\mathbf{v}_k \quad \text{for } k \geq 0$$

where A and $\mathbf{v}_0$ are given. Assume that A is a diagonalizable $n \times n$ matrix with eigenvalues $\lambda_1, \lambda_2, \ldots, \lambda_n$ and corresponding basic eigenvectors $\mathbf{x}_1, \mathbf{x}_2, \ldots, \mathbf{x}_n$, and let $P = [\mathbf{x}_1 \ \mathbf{x}_2 \ \cdots \ \mathbf{x}_n]$ be the diagonalizing matrix. Then an exact formula for $\mathbf{v}_k$ is

$$\mathbf{v}_k = b_1\lambda_1^k\mathbf{x}_1 + b_2\lambda_2^k\mathbf{x}_2 + \cdots + b_n\lambda_n^k\mathbf{x}_n \quad \text{for each } k \geq 0$$

where the coefficients b_i come from

$$\mathbf{b} = P^{-1}\mathbf{v}_0 = \begin{bmatrix} b_1 \\ b_2 \\ \vdots \\ b_n \end{bmatrix}.$$

Moreover, if A has dominant eigenvalue λ_1,[12] then $\mathbf{v}_k$ is approximated by

$$\mathbf{v}_k = b_1\lambda_1^k\mathbf{x}_1 \quad \text{for sufficiently large } k.$$

EXAMPLE 13

Returning to Example 12, we see that $\lambda_1 = 1$ is the dominant eigenvalue, with eigenvector $\mathbf{x}_1 = \begin{bmatrix} 1 \\ 2 \end{bmatrix}$. Here $P = \begin{bmatrix} 1 & -1 \\ 2 & 4 \end{bmatrix}$ and $\mathbf{v}_0 = \begin{bmatrix} 100 \\ 40 \end{bmatrix}$, so $P^{-1}\mathbf{v}_0 = \frac{1}{3}\begin{bmatrix} 220 \\ -80 \end{bmatrix}$.
Hence $b_1 = \frac{220}{3}$ in the notation of Theorem 7, so

$$\begin{bmatrix} a_k \\ j_k \end{bmatrix} = \mathbf{v}_k \approx b_1\lambda_1^k\mathbf{x}_1 = \frac{220}{3}1^k\begin{bmatrix} 1 \\ 2 \end{bmatrix}$$

where k is large. Hence $a_k \approx \frac{220}{3}$ and $j_k \approx \frac{440}{3}$, as in Example 12.

This next example uses Theorem 7 to solve a "linear recurrence." See also Section 3.4.

EXAMPLE 14

Suppose a sequence $x_0, x_1, x_2, \ldots$ is determined by insisting that

$$x_0 = 1, \ x_1 = -1, \text{ and } x_{k+2} = 2x_k - x_{k+1} \text{ for every } k \geq 0.$$

Find a formula for x_k in terms of k.

Solution ▶ Using the linear recurrence $x_{k+2} = 2x_k - x_{k+1}$ repeatedly gives

$$x_2 = 2x_0 - x_1 = 3, \quad x_3 = 2x_1 - x_2 = 5, \quad x_4 = 11, \quad x_5 = 21, \ldots$$

so the x_i are determined but no pattern is apparent. The idea is to find

$$\mathbf{v}_k = \begin{bmatrix} x_k \\ x_{k+1} \end{bmatrix}$$ for each k instead, and then retrieve x_k as the top component

of $\mathbf{v}_k$. The reason this works is that the linear recurrence guarantees that these $\mathbf{v}_k$ are a dynamical system:

---

[12] Similar results can be found in other situations. If for example, eigenvalues λ_1 and λ_2 (possibly equal) satisfy $|\lambda_1| = |\lambda_2| > |\lambda_i|$ for all $i > 2$, then we obtain $\mathbf{v}_k \approx b_1\lambda_1^k\mathbf{x}_1 + b_2\lambda_2^k\mathbf{x}_2$ for large k.

$$\mathbf{v}_{k+1} = \begin{bmatrix} x_{k+1} \\ x_{k+2} \end{bmatrix} = \begin{bmatrix} x_{k+1} \\ 2x_k - x_{k+1} \end{bmatrix} = A\mathbf{v}_k \text{ where } A = \begin{bmatrix} 0 & 1 \\ 2 & -1 \end{bmatrix}.$$

The eigenvalues of A are $\lambda_1 = -2$ and $\lambda_2 = 1$ with eigenvectors $\mathbf{x}_1 = \begin{bmatrix} 1 \\ -2 \end{bmatrix}$ and $\mathbf{x}_2 = \begin{bmatrix} 1 \\ 1 \end{bmatrix}$, so the diagonalizing matrix is $P = \begin{bmatrix} 1 & 1 \\ -2 & 1 \end{bmatrix}$.

Moreover, $\mathbf{b} = P_0^{-1}\mathbf{v}_0 = \frac{1}{3}\begin{bmatrix} 2 \\ 1 \end{bmatrix}$ so the exact formula for $\mathbf{v}_k$ is

$$\begin{bmatrix} x_k \\ x_{k+1} \end{bmatrix} = \mathbf{v}_k = b_1\lambda_1^k \mathbf{x}_1 + b_2\lambda_2^k \mathbf{x}_2 = \frac{2}{3}(-2)^k \begin{bmatrix} 1 \\ -2 \end{bmatrix} + \frac{1}{3}1^k \begin{bmatrix} 1 \\ 1 \end{bmatrix}.$$

Equating top entries gives the desired formula for x_k:

$$x_k = \frac{1}{3}\left[2(-2)^k + 1\right] \text{ for all } k = 0, 1, 2, \ldots.$$

The reader should check this for the first few values of k.

Graphical Description of Dynamical Systems

If a dynamical system $\mathbf{v}_{k+1} = A\mathbf{v}_k$ is given, the sequence $\mathbf{v}_0, \mathbf{v}_1, \mathbf{v}_2, \ldots$ is called the **trajectory** of the system starting at $\mathbf{v}_0$. It is instructive to obtain a graphical plot of the system by writing $\mathbf{v}_k = \begin{bmatrix} x_k \\ y_k \end{bmatrix}$ and plotting the successive values as points in the plane, identifying $\mathbf{v}_k$ with the point (x_k, y_k) in the plane. We give several examples which illustrate properties of dynamical systems. For ease of calculation we assume that the matrix A is simple, usually diagonal.

EXAMPLE 15

Let $A = \begin{bmatrix} \frac{1}{2} & 0 \\ 0 & \frac{1}{3} \end{bmatrix}$. Then the eigenvalues are $\frac{1}{2}$ and $\frac{1}{3}$, with corresponding eigenvectors $\mathbf{x}_1 = \begin{bmatrix} 1 \\ 0 \end{bmatrix}$ and $\mathbf{x}_2 = \begin{bmatrix} 0 \\ 1 \end{bmatrix}$. The exact formula is

$$\mathbf{v}_k = b_1(\tfrac{1}{2})^k \begin{bmatrix} 1 \\ 0 \end{bmatrix} + b_2(\tfrac{1}{3})^k \begin{bmatrix} 0 \\ 1 \end{bmatrix}$$

for $k = 0, 1, 2, \ldots$ by Theorem 7, where the coefficients b_1 and b_2 depend on the initial point $\mathbf{v}_0$. Several trajectories are plotted in the diagram and, for each choice of $\mathbf{v}_0$, the trajectories converge toward the origin because both eigenvalues are less than 1 in absolute value. For this reason, the origin is called an **attractor** for the system.

EXAMPLE 16

Let $A = \begin{bmatrix} \frac{3}{2} & 0 \\ 0 & \frac{4}{3} \end{bmatrix}$. Here the eigenvalues are $\frac{3}{2}$ and $\frac{4}{3}$, with corresponding eigenvectors $\mathbf{x}_1 = \begin{bmatrix} 1 \\ 0 \end{bmatrix}$ and $\mathbf{x}_2 = \begin{bmatrix} 0 \\ 1 \end{bmatrix}$ as before. The exact formula is

$$\mathbf{v}_k = b_1(\tfrac{3}{2})^k \begin{bmatrix} 1 \\ 0 \end{bmatrix} + b_2(\tfrac{4}{3})^k \begin{bmatrix} 0 \\ 1 \end{bmatrix}$$

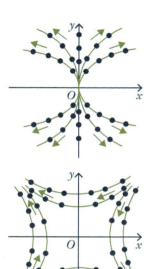

for $k = 0, 1, 2, \ldots$. Since both eigenvalues are greater than 1 in absolute value, the trajectories diverge away from the origin for every choice of initial point V_0. For this reason, the origin is called a **repellor** for the system.[13]

EXAMPLE 17

Let $A = \begin{bmatrix} 1 & -\frac{1}{2} \\ -\frac{1}{2} & 1 \end{bmatrix}$. Now the eigenvalues are $\frac{3}{2}$ and $\frac{1}{2}$, with corresponding eigenvectors $\mathbf{x}_1 = \begin{bmatrix} -1 \\ 1 \end{bmatrix}$ and $\mathbf{x}_2 = \begin{bmatrix} 1 \\ 1 \end{bmatrix}$. The exact formula is

$$\mathbf{v}_k = b_1 (\tfrac{3}{2})^k \begin{bmatrix} -1 \\ 1 \end{bmatrix} + b_2 (\tfrac{1}{2})^k \begin{bmatrix} 1 \\ 1 \end{bmatrix}$$

for $k = 0, 1, 2, \ldots$. In this case $\frac{3}{2}$ is the dominate eigenvalue so, if $b_1 \neq 0$, we have $\mathbf{v}_k \approx b_1 (\tfrac{3}{2})^k \begin{bmatrix} -1 \\ 1 \end{bmatrix}$ for large k and $\mathbf{v}_k$ is approaching the line $y = -x$. However, if $b_1 = 0$, then $\mathbf{v}_k = b_2 (\tfrac{1}{2})^k \begin{bmatrix} 1 \\ 1 \end{bmatrix}$ and so approaches the origin along the line $y = x$. In general the trajectories appear as in the diagram, and the origin is called a **saddle point** for the dynamical system in this case.

EXAMPLE 18

Let $A = \begin{bmatrix} 0 & \frac{1}{2} \\ -\frac{1}{2} & 0 \end{bmatrix}$. Now the characteristic polynomial is $c_A(x) = x^2 + \frac{1}{4}$, so the eigenvalues are the complex numbers $\frac{i}{2}$ and $-\frac{i}{2}$ where $i^2 = -1$. Hence A is not diagonalizable as a real matrix. However, the trajectories are not difficult to describe. If we start with $\mathbf{v}_0 = \begin{bmatrix} 1 \\ 1 \end{bmatrix}$, then the trajectory begins as

$$\mathbf{v}_1 = \begin{bmatrix} \frac{1}{2} \\ -\frac{1}{2} \end{bmatrix}, \quad \mathbf{v}_2 = \begin{bmatrix} -\frac{1}{4} \\ -\frac{1}{4} \end{bmatrix}, \quad \mathbf{v}_3 = \begin{bmatrix} -\frac{1}{8} \\ \frac{1}{8} \end{bmatrix}, \quad \mathbf{v}_4 = \begin{bmatrix} \frac{1}{16} \\ \frac{1}{16} \end{bmatrix}, \quad \mathbf{v}_5 = \begin{bmatrix} \frac{1}{32} \\ -\frac{1}{32} \end{bmatrix}, \quad \mathbf{v}_6 = \begin{bmatrix} -\frac{1}{64} \\ -\frac{1}{64} \end{bmatrix}, \ldots$$

Five of these points are plotted in the diagram. Here each trajectory spirals in toward the origin, so the origin is an attractor. Note that the two (complex) eigenvalues have absolute value less than 1 here. If they had absolute value greater than 1, the trajectories would spiral out from the origin.

Google PageRank

Dominant eigenvalues are useful to the Google search engine for finding information on the Web. If an information query comes in from a client, Google has a sophisticated method of establishing the "relevance" of each site to that query. When the relevant sites have been determined, they are placed in order of importance using a ranking of *all* sites called the PageRank. The relevant sites with the highest PageRank are the ones presented to the client. It is the construction of the PageRank that is our interest here.

---

[13] In fact, $P = I$ here, so $\mathbf{v}_0 = \begin{bmatrix} b_1 \\ b_2 \end{bmatrix}$

The Web contains many links from one site to another. Google interprets a link from site j to site i as a "vote" for the importance of site i. Hence if site i has more links to it than does site j, then i is regarded as more "important" and assigned a higher PageRank. One way to look at this is to view the sites as vertices in a huge directed graph (see Section 2.2). Then if site j links to site i there is an edge from j to i, and hence the (i,j)-entry is a 1 in the associated adjacency matrix (called the *connectivity* matrix in this context). Thus a large number of 1s in row i of this matrix is a measure of the PageRank of site i.[14]

However this does not take into account the PageRank of the sites that link to i. Intuitively, the higher the rank of these sites, the higher the rank of site i. One approach is to compute a dominant eigenvector $\mathbf{x}$ for the connectivity matrix. In most cases the entries of $\mathbf{x}$ can be chosen to be positive with sum 1. Each site corresponds to an entry of $\mathbf{x}$, so the sum of the entries of sites linking to a given site i is a measure of the rank of site i. In fact, Google chooses the PageRank of a site so that it is proportional to this sum.[15]

EXERCISES 3.3

1. In each case find the characteristic polynomial, eigenvalues, eigenvectors, and (if possible) an invertible matrix P such that $P^{-1}AP$ is diagonal.

   (a) $A = \begin{bmatrix} 1 & 2 \\ 3 & 2 \end{bmatrix}$
   ◆(b) $A = \begin{bmatrix} 2 & -4 \\ -1 & -1 \end{bmatrix}$

   (c) $A = \begin{bmatrix} 7 & 0 & -4 \\ 0 & 5 & 0 \\ 5 & 0 & -2 \end{bmatrix}$
   ◆(d) $A = \begin{bmatrix} 1 & 1 & -3 \\ 2 & 0 & 6 \\ 1 & -1 & 5 \end{bmatrix}$

   (e) $A = \begin{bmatrix} 1 & -2 & 3 \\ 2 & 6 & -6 \\ 1 & 2 & -1 \end{bmatrix}$
   ◆(f) $A = \begin{bmatrix} 0 & 1 & 0 \\ 3 & 0 & 1 \\ 2 & 0 & 0 \end{bmatrix}$

   (g) $A = \begin{bmatrix} 3 & 1 & 1 \\ -4 & -2 & -5 \\ 2 & 2 & 5 \end{bmatrix}$
   ◆(h) $A = \begin{bmatrix} 2 & 1 & 1 \\ 0 & 1 & 0 \\ 1 & -1 & 2 \end{bmatrix}$

   (i) $A = \begin{bmatrix} \lambda & 0 & 0 \\ 0 & \lambda & 0 \\ 0 & 0 & \mu \end{bmatrix}, \lambda \neq \mu$

2. Consider a linear dynamical system $\mathbf{v}_{k+1} = A\mathbf{v}_k$ for $k \geq 0$. In each case approximate $\mathbf{v}_k$ using Theorem 7.

   (a) $A = \begin{bmatrix} 2 & 1 \\ 4 & -1 \end{bmatrix}, \mathbf{v}_0 = \begin{bmatrix} 1 \\ 2 \end{bmatrix}$

   ◆(b) $A = \begin{bmatrix} 3 & -2 \\ 2 & -2 \end{bmatrix}, \mathbf{v}_0 = \begin{bmatrix} 3 \\ -1 \end{bmatrix}$

   (c) $A = \begin{bmatrix} 1 & 0 & 0 \\ 1 & 2 & 3 \\ 1 & 4 & 1 \end{bmatrix}, \mathbf{v}_0 = \begin{bmatrix} 1 \\ 1 \\ 1 \end{bmatrix}$

   ◆(d) $A = \begin{bmatrix} 1 & 3 & 2 \\ -1 & 2 & 1 \\ 4 & -1 & -1 \end{bmatrix}, \mathbf{v}_0 = \begin{bmatrix} 2 \\ 0 \\ 1 \end{bmatrix}$

3. Show that A has $\lambda = 0$ as an eigenvalue if and only if A is not invertible.

◆4. Let A denote an $n \times n$ matrix and put $A_1 = A - \alpha I$, α in $\mathbb{R}$. Show that λ is an eigenvalue of A if and only if $\lambda - \alpha$ is an eigenvalue of A_1. (Hence, the eigenvalues of A_1 are just those of A "shifted" by α.) How do the eigenvectors compare?

5. Show that the eigenvalues of $\begin{bmatrix} \cos\theta & -\sin\theta \\ \sin\theta & \cos\theta \end{bmatrix}$ are $e^{i\theta}$ and $e^{-i\theta}$. (See Appendix A.)

6. Find the characteristic polynomial of the $n \times n$ identity matrix I. Show that I has exactly one eigenvalue and find the eigenvectors.

7. Given $A = \begin{bmatrix} a & b \\ c & d \end{bmatrix}$, show that:

   (a) $c_A(x) = x^2 - \text{tr } Ax + \det A$, where $\text{tr } A = a + d$ is called the **trace** of A.

---

14 For more on PageRank, visit http://www.google.com/technology/.

15 See the articles "Searching the web with eigenvectors" by Herbert S. Wilf, UMAP Journal 23(2), 2002, pages 101–103, and "The worlds largest matrix computation: Google's PageRank is an eigenvector of a matrix of order 2.7 billion" by Cleve Moler, Matlab News and Notes, October 2002, pages 12–13.

(b) The eigenvalues are
$\frac{1}{2}\left[(a+d) \pm \sqrt{(a-d)^2 + 4bc}\right]$

8. In each case, find $P^{-1}AP$ and then compute A^n.

   (a) $A = \begin{bmatrix} 6 & -5 \\ 2 & -1 \end{bmatrix}, P = \begin{bmatrix} 1 & 5 \\ 1 & 2 \end{bmatrix}$

   ◆(b) $A = \begin{bmatrix} -7 & -12 \\ 6 & 10 \end{bmatrix}, P = \begin{bmatrix} -3 & 4 \\ 2 & -3 \end{bmatrix}$

   [Hint: $(PDP^{-1})^n = PD^nP^{-1}$ for each $n = 1, 2, \ldots$]

◆9. (a) If $A = \begin{bmatrix} 1 & 3 \\ 0 & 2 \end{bmatrix}$ and $B = \begin{bmatrix} 2 & 0 \\ 0 & 1 \end{bmatrix}$, verify that A and B are diagonalizable, but AB is not.

   ◆(b) If $D = \begin{bmatrix} 1 & 0 \\ 0 & -1 \end{bmatrix}$, find a diagonalizable matrix A such that $D + A$ is not diagonalizable.

10. If A is an $n \times n$ matrix, show that A is diagonalizable if and only if A^T is diagonalizable.

11. If A is diagonalizable, show that each of the following is also diagonalizable.

    (a) $A^n, n \geq 1$ ◆(b) kA, k any scalar.

    (c) $p(A)$, $p(x)$ any polynomial (Theorem 1)

    ◆(d) $U^{-1}AU$ for any invertible matrix U.

    (e) $kI + A$ for any scalar k.

◆12. Give an example of two diagonalizable matrices A and B whose sum $A + B$ is not diagonalizable.

13. If A is diagonalizable and 1 and -1 are the only eigenvalues, show that $A^{-1} = A$.

◆14. If A is diagonalizable and 0 and 1 are the only eigenvalues, show that $A^2 = A$.

15. If A is diagonalizable and $\lambda \geq 0$ for each eigenvalue of A, show that $A = B^2$ for some matrix B.

16. If $P^{-1}AP$ and $P^{-1}BP$ are both diagonal, show that $AB = BA$. [Hint: Diagonal matrices commute.]

17. A square matrix A is called **nilpotent** if $A^n = 0$ for some $n \geq 1$. Find all nilpotent diagonalizable matrices. [Hint: Theorem 1.]

18. Let A be any $n \times n$ matrix and $r \neq 0$ a real number.

    (a) Show that the eigenvalues of rA are precisely the numbers $r\lambda$, where λ is an eigenvalue of A.

    ◆(b) Show that $c_{rA}(x) = r^n c_A(\frac{x}{r})$.

19. (a) If all rows of A have the same sum s, show that s is an eigenvalue.

    (b) If all columns of A have the same sum s, show that s is an eigenvalue.

20. Let A be an invertible $n \times n$ matrix.

    (a) Show that the eigenvalues of A are nonzero.

    ◆(b) Show that the eigenvalues of A^{-1} are precisely the numbers $1/\lambda$, where λ is an eigenvalue of A.

    (c) Show that $c_{A^{-1}}(x) = \frac{(-x)^n}{\det A} c_A(\frac{1}{x})$.

21. Suppose λ is an eigenvalue of a square matrix A with eigenvector $\mathbf{x} \neq \mathbf{0}$.

    (a) Show that λ^2 is an eigenvalue of A^2 (with the same $\mathbf{x}$).

    ◆(b) Show that $\lambda^3 - 2\lambda + 3$ is an eigenvalue of $A^3 - 2A + 3I$.

    (c) Show that $p(\lambda)$ is an eigenvalue of $p(A)$ for any nonzero polynomial $p(x)$.

22. If A is an $n \times n$ matrix, show that $c_{A^2}(x^2) = (-1)^n c_A(x) c_A(-x)$.

23. An $n \times n$ matrix A is called nilpotent if $A^m = 0$ for some $m \geq 1$.

    (a) Show that every triangular matrix with zeros on the main diagonal is nilpotent.

    ◆(b) If A is nilpotent, show that $\lambda = 0$ is the only eigenvalue (even complex) of A.

    (c) Deduce that $c_A(x) = x^n$, if A is $n \times n$ and nilpotent.

24. Let A be diagonalizable with real eigenvalues and assume that $A^m = I$ for some $m \geq 1$.

    ◆(a) Show that $A^2 = I$.

    (b) If m is odd, show that $A = I$.

    [Hint: Theorem 3 Appendix A.]

25. Let $A^2 = I$, and assume that $A \neq I$ and $A \neq -I$.

    (a) Show that the only eigenvalues of A are $\lambda = 1$ and $\lambda = -1$.

    (b) Show that A is diagonalizable. [Hint: Verify that $A(A + I) = A + I$ and

$A(A - I) = -(A - I)$, and then look at nonzero columns of $A + I$ and of $A - I$.]

(c) If $Q_m : \mathbb{R}^2 \to \mathbb{R}^2$ is reflection in the line $y = mx$ where $m \neq 0$, use (b) to show that the matrix of Q_m is diagonalizable for each m.

(d) Now prove (c) geometrically using Theorem 3.

26. Let $A = \begin{bmatrix} 2 & 3 & -3 \\ 1 & 0 & -1 \\ 1 & 1 & -2 \end{bmatrix}$ and $B = \begin{bmatrix} 0 & 1 & 0 \\ 3 & 0 & 1 \\ 2 & 0 & 0 \end{bmatrix}$. Show that $c_A(x) = c_B(x) = (x + 1)^2(x - 2)$, but A is diagonalizable and B is not.

◆27. (a) Show that the only diagonalizable matrix A that has only one eigenvalue λ is the scalar matrix $A = \lambda I$.

(b) Is $\begin{bmatrix} 3 & -2 \\ 2 & -1 \end{bmatrix}$ diagonalizable?

28. Characterize the diagonalizable $n \times n$ matrices A such that $A^2 - 3A + 2I = 0$ in terms of their eigenvalues. [*Hint*: Theorem 1.]

29. Let $A = \begin{bmatrix} B & 0 \\ 0 & C \end{bmatrix}$ where B and C are square matrices.

(a) If B and C are diagonalizable via Q and R (that is, $Q^{-1}BQ$ and $R^{-1}CR$ are diagonal), show that A is diagonalizable via $\begin{bmatrix} Q & 0 \\ 0 & R \end{bmatrix}$.

(b) Use (a) to diagonalize A if $B = \begin{bmatrix} 5 & 3 \\ 3 & 5 \end{bmatrix}$ and $C = \begin{bmatrix} 7 & -1 \\ -1 & 7 \end{bmatrix}$.

30. Let $A = \begin{bmatrix} B & 0 \\ 0 & C \end{bmatrix}$, where B and C are square matrices.

(a) Show that $c_A(x) = c_B(x)c_C(x)$.

(b) If $\mathbf{x}$ and $\mathbf{y}$ are eigenvectors of B and C, respectively, show that $\begin{bmatrix} \mathbf{x} \\ 0 \end{bmatrix}$ and $\begin{bmatrix} 0 \\ \mathbf{y} \end{bmatrix}$ are eigenvectors of A, and show how every eigenvector of A arises from such eigenvectors.

31. Referring to the model in Example 1, determine if the population stabilizes, becomes extinct, or becomes large in each case. Denote the adult and juvenile survival rates as A and J, and the reproduction rate as R.

	R	A	J
(a)	2	$\frac{1}{2}$	$\frac{1}{2}$
◆(b)	3	$\frac{1}{4}$	$\frac{1}{4}$
(c)	2	$\frac{1}{4}$	$\frac{1}{3}$
◆(d)	3	$\frac{3}{5}$	$\frac{1}{5}$

32. In the model of Example 1, does the final outcome depend on the initial population of adult and juvenile females? Support your answer.

33. In Example 1, keep the same reproduction rate of 2 and the same adult survival rate of $\frac{1}{2}$, but suppose that the juvenile survival rate is ρ. Determine which values of ρ cause the population to become extinct or to become large.

◆34. In Example 1, let the juvenile survival rate be $\frac{2}{5}$, and let the reproduction rate be 2. What values of the adult survival rate α will ensure that the population stabilizes?

SECTION 3.4 An Application to Linear Recurrences

It often happens that a problem can be solved by finding a sequence of numbers x_0, $x_1, x_2, \ldots$ where the first few are known, and subsequent numbers are given in terms of earlier ones. Here is a combinatorial example where the object is to count the number of ways to do something.

EXAMPLE 1

An urban planner wants to determine the number x_k of ways that a row of k parking spaces can be filled with cars and trucks if trucks take up two spaces each. Find the first few values of x_k.

SECTION 3.4 An Application to Linear Recurrences

Solution ▶ Clearly, $x_0 = 1$ and $x_1 = 1$, while $x_2 = 2$ since there can be two cars or one truck. We have $x_3 = 3$ (the 3 configurations are ccc, cT, and Tc) and $x_4 = 5$ ($cccc$, ccT, cTc, Tcc, and TT). The key to this method is to find a way to express each subsequent x_k in terms of earlier values. In this case we claim that

$$x_{k+2} = x_k + x_{k+1} \quad \text{for every } k \geq 0 \qquad (*)$$

Indeed, every way to fill $k + 2$ spaces falls into one of two categories: Either a car is parked in the first space (and the remaining $k + 1$ spaces are filled in x_{k+1} ways), or a truck is parked in the first two spaces (with the other k spaces filled in x_k ways). Hence, there are $x_{k+1} + x_k$ ways to fill the $k + 2$ spaces. This is $(*)$.

The recurrence $(*)$ determines x_k for every $k \geq 2$ since x_0 and x_1 are given. In fact, the first few values are

$$\begin{aligned} x_0 &= 1 \\ x_1 &= 1 \\ x_2 &= x_0 + x_1 = 2 \\ x_3 &= x_1 + x_2 = 3 \\ x_4 &= x_2 + x_3 = 5 \\ x_5 &= x_3 + x_4 = 8 \\ &\vdots \end{aligned}$$

Clearly, we can find x_k for any value of k, but one wishes for a "formula" for x_k as a function of k. It turns out that such a formula can be found using diagonalization. We will return to this example later.

A sequence $x_0, x_1, x_2, \ldots$ of numbers is said to be given **recursively** if each number in the sequence is completely determined by those that come before it. Such sequences arise frequently in mathematics and computer science, and also occur in other parts of science. The formula $x_{k+2} = x_{k+1} + x_k$ in Example 1 is an example of a **linear recurrence relation** of length 2 because x_{k+2} is the sum of the two preceding terms x_{k+1} and x_k; in general, the **length** is m if x_{k+m} is a sum of multiples of $x_k, x_{k+1}, \ldots, x_{k+m-1}$.

The simplest linear recursive sequences are of length 1, that is x_{k+1} is a fixed multiple of x_k for each k, say $x_{k+1} = ax_k$. If x_0 is specified, then $x_1 = ax_0$, $x_2 = ax_1 = a^2 x_0$, and $x_3 = ax_2 = a^3 x_0, \ldots$. Continuing, we obtain $x_k = a^k x_0$ for each $k \geq 0$, which is an explicit formula for x_k as a function of k (when x_0 is given).

Such formulas are not always so easy to find for all choices of the initial values. Here is an example where diagonalization helps.

EXAMPLE 2

Suppose the numbers $x_0, x_1, x_2, \ldots$ are given by the linear recurrence relation

$$x_{k+2} = x_{k+1} + 6x_k \quad \text{for } k \geq 0$$

where x_0 and x_1 are specified. Find a formula for x_k when $x_0 = 1$ and $x_1 = 3$, and also when $x_0 = 1$ and $x_1 = 1$.

Solution ▶ If $x_0 = 1$ and $x_1 = 3$, then $x_2 = x_1 + 6x_0 = 9$, $x_3 = x_2 + 6x_1 = 27$, $x_4 = x_3 + 6x_2 = 81$, and it is apparent that

$x_k = 3^k$ for $k = 0, 1, 2, 3$, and 4.

This formula holds for all k because it is true for $k = 0$ and $k = 1$, and it satisfies the recurrence $x_{k+2} = x_{k+1} + 6x_k$ for each k as is readily checked.

However, if we begin instead with $x_0 = 1$ and $x_1 = 1$, the sequence continues $x_2 = 7$, $x_3 = 13$, $x_4 = 55$, $x_5 = 133$, ... In this case, the sequence is uniquely determined but no formula is apparent. Nonetheless, a simple device transforms the recurrence into a matrix recurrence to which our diagonalization techniques apply.

The idea is to compute the sequence $\mathbf{v}_0, \mathbf{v}_1, \mathbf{v}_2, \ldots$ of columns instead of the numbers $x_0, x_1, x_2, \ldots$, where

$$\mathbf{v}_k = \begin{bmatrix} x_k \\ x_{k+1} \end{bmatrix} \text{ for each } k \geq 0$$

Then $\mathbf{v}_0 = \begin{bmatrix} x_0 \\ x_1 \end{bmatrix} = \begin{bmatrix} 1 \\ 1 \end{bmatrix}$ is specified, and the numerical recurrence $x_{k+2} = x_{k+1} + 6x_k$ transforms into a matrix recurrence as follows:

$$\mathbf{v}_{k+1} = \begin{bmatrix} x_{k+1} \\ x_{k+2} \end{bmatrix} = \begin{bmatrix} x_{k+1} \\ 6x_k + x_{k+1} \end{bmatrix} = \begin{bmatrix} 0 & 1 \\ 6 & 1 \end{bmatrix} \begin{bmatrix} x_k \\ x_{k+1} \end{bmatrix} = A\mathbf{v}_k$$

where $A = \begin{bmatrix} 0 & 1 \\ 6 & 1 \end{bmatrix}$. Thus these columns $\mathbf{v}_k$ are a linear dynamical system, so Theorem 7 Section 3.3 applies provided the matrix A is diagonalizable.

We have $c_A(x) = (x - 3)(x + 2)$ so the eigenvalues are $\lambda_1 = 3$ and $\lambda_2 = -2$ with corresponding eigenvectors $\mathbf{x}_1 = \begin{bmatrix} 1 \\ 3 \end{bmatrix}$ and $\mathbf{x}_2 = \begin{bmatrix} -1 \\ 2 \end{bmatrix}$, as the reader can check.

Since $P = [\mathbf{x}_1 \; \mathbf{x}_2] = \begin{bmatrix} 1 & -1 \\ 3 & 2 \end{bmatrix}$ is invertible, it is a diagonalizing matrix for A. The coefficients b_i in Theorem 7 Section 3.3 are given by $\begin{bmatrix} b_1 \\ b_2 \end{bmatrix} = P^{-1}\mathbf{v}_0 = \begin{bmatrix} \frac{3}{5} \\ \frac{-2}{5} \end{bmatrix}$, so that the theorem gives

$$\begin{bmatrix} x_k \\ x_{k+1} \end{bmatrix} = \mathbf{v}_k = b_1 \lambda_1^k \mathbf{x}_1 + b_2 \lambda_2^k \mathbf{x}_2 = \tfrac{3}{5} 3^k \begin{bmatrix} 1 \\ 3 \end{bmatrix} + \tfrac{-2}{5}(-2)^k \begin{bmatrix} -1 \\ 2 \end{bmatrix}$$

Equating top entries yields

$$x_k = \tfrac{1}{5}\left[3^{k+1} - (-2)^{k+1}\right] \text{ for } k \geq 0$$

This gives $x_0 = 1 = x_1$, and it satisfies the recurrence $x_{k+2} = x_{k+1} + 6x_k$ as is easily verified. Hence, it is the desired formula for the x_k.

Returning to Example 1, these methods give an exact formula and a good approximation for the numbers x_k in that problem.

EXAMPLE 3

In Example 1, an urban planner wants to determine x_k, the number of ways that a row of k parking spaces can be filled with cars and trucks if trucks take up two spaces each. Find a formula for x_k and estimate it for large k.

Solution ▶ We saw in Example 1 that the numbers x_k satisfy a linear recurrence
$$x_{k+2} = x_k + x_{k+1} \quad \text{for every } k \geq 0$$
If we write $\mathbf{v}_k = \begin{bmatrix} x_k \\ x_{k+1} \end{bmatrix}$ as before, this recurrence becomes a matrix recurrence for the $\mathbf{v}_k$:
$$\mathbf{v}_{k+1} = \begin{bmatrix} x_{k+1} \\ x_{k+2} \end{bmatrix} = \begin{bmatrix} x_{k+1} \\ x_k + x_{k+1} \end{bmatrix} = \begin{bmatrix} 0 & 1 \\ 1 & 1 \end{bmatrix} \begin{bmatrix} x_k \\ x_{k+1} \end{bmatrix} = A\mathbf{v}_k$$
for all $k \geq 0$ where $A = \begin{bmatrix} 0 & 1 \\ 1 & 1 \end{bmatrix}$. Moreover, A is diagonalizable here. The characteristic polynomial is $c_A(x) = x^2 - x - 1$ with roots $\frac{1}{2}[1 \pm \sqrt{5}]$ by the quadratic formula, so A has eigenvalues
$$\lambda_1 = \tfrac{1}{2}[1 + \sqrt{5}] \quad \text{and} \quad \lambda_1 = \tfrac{1}{2}[1 - \sqrt{5}]$$
Corresponding eigenvectors are $\mathbf{x}_1 = \begin{bmatrix} 1 \\ \lambda_1 \end{bmatrix}$ and $\mathbf{x}_2 = \begin{bmatrix} 1 \\ \lambda_2 \end{bmatrix}$ respectively as the reader can verify. As the matrix $P = [\mathbf{x}_1 \ \mathbf{x}_2] = \begin{bmatrix} 1 & 1 \\ \lambda_1 & \lambda_2 \end{bmatrix}$ is invertible, it is a diagonalizing matrix for A. We compute the coefficients b_1 and b_2 (in Theorem 7 Section 3.3) as follows:
$$\begin{bmatrix} b_1 \\ b_2 \end{bmatrix} = P^{-1}\mathbf{v}_0 = \tfrac{1}{-\sqrt{5}} \begin{bmatrix} \lambda_2 & -1 \\ -\lambda_1 & 1 \end{bmatrix} \begin{bmatrix} 1 \\ 1 \end{bmatrix} = \tfrac{1}{\sqrt{5}} \begin{bmatrix} \lambda_1 \\ -\lambda_2 \end{bmatrix}$$
where we used the fact that $\lambda_1 + \lambda_2 = 1$. Thus Theorem 7 Section 3.3 gives
$$\begin{bmatrix} x_k \\ x_{k+1} \end{bmatrix} = \mathbf{v}_k = b_1 \lambda_1^k \mathbf{x}_1 + b_2 \lambda_2^k \mathbf{x}_2 = \tfrac{\lambda_1}{\sqrt{5}} \lambda_1^k \begin{bmatrix} 1 \\ \lambda_1 \end{bmatrix} - \tfrac{\lambda_2}{\sqrt{5}} \lambda_2^k \begin{bmatrix} 1 \\ \lambda_2 \end{bmatrix}$$
Comparing top entries gives an exact formula for the numbers x_k:
$$x_k = \tfrac{1}{\sqrt{5}} \left[\lambda_1^{k+1} - \lambda_2^{k+1} \right] \quad \text{for } k \geq 0$$
Finally, observe that λ_1 is dominant here (in fact, $\lambda_1 = 1.618$ and $\lambda_2 = -0.618$ to three decimal places) so λ_2^{k+1} is negligible compared with λ_1^{k+1} if k is large. Thus,
$$x_k \approx \tfrac{1}{\sqrt{5}} \lambda_1^{k+1} \quad \text{for each } k \geq 0$$
This is a good approximation, even for as small a value as $k = 12$. Indeed, repeated use of the recurrence $x_{k+2} = x_k + x_{k+1}$ gives the exact value $x_{12} = 233$, while the approximation is $x_{12} \approx \frac{(1.618)^{13}}{\sqrt{5}} = 232.94$.

The sequence $x_0, x_1, x_2, \ldots$ in Example 3 was first discussed in 1202 by Leonardo Pisano of Pisa, also known as Fibonacci,[16] and is now called the **Fibonacci sequence**. It is completely determined by the conditions $x_0 = 1$, $x_1 = 1$ and the recurrence $x_{k+2} = x_k + x_{k+1}$ for each $k \geq 0$. These numbers have been studied for centuries and have many interesting properties (there is even a journal, the *Fibonacci Quarterly*, devoted exclusively to them). For example, biologists have discovered that the arrangement of leaves around the stems of some plants follow a Fibonacci pattern. The formula $x_k = \frac{1}{\sqrt{5}} \left[\lambda_1^{k+1} - \lambda_2^{k+1} \right]$ in Example 3 is called the **Binet**

---

[16] The problem Fibonacci discussed was: "How many pairs of rabbits will be produced in a year, beginning with a single pair, if in every month each pair brings forth a new pair that becomes productive from the second month on? Assume no pairs die." The number of pairs satisfies the Fibonacci recurrence.

formula. It is remarkable in that the x_k are integers but λ_1 and λ_2 are not. This phenomenon can occur even if the eigenvalues λ_i are nonreal complex numbers.

We conclude with an example showing that *nonlinear* recurrences can be very complicated.

EXAMPLE 4

Suppose a sequence $x_0, x_1, x_2, \ldots$ satisfies the following recurrence:

$$x_{k+1} = \begin{cases} \frac{1}{2} x_k & \text{if } x_k \text{ is even} \\ 3x_k + 1 & \text{if } x_k \text{ is odd} \end{cases}$$

If $x_0 = 1$, the sequence is 1, 4, 2, 1, 4, 2, 1, ... and so continues to cycle indefinitely. The same thing happens if $x_0 = 7$. Then the sequence is

7, 22, 11, 34, 17, 52, 26, 13, 40, 20, 10, 5, 16, 8, 4, 2, 1, ...

and it again cycles. However, it is not known whether every choice of x_0 will lead eventually to 1. It is quite possible that, for some x_0, the sequence will continue to produce different values indefinitely, or will repeat a value and cycle without reaching 1. No one knows for sure.

EXERCISES 3.4

1. Solve the following linear recurrences.

   (a) $x_{k+2} = 3x_k + 2x_{k+1}$, where $x_0 = 1$ and $x_1 = 1$.

   ◆(b) $x_{k+2} = 2x_k - x_{k+1}$, where $x_0 = 1$ and $x_1 = 2$.

   (c) $x_{k+2} = 2x_k + x_{k+1}$, where $x_0 = 0$ and $x_1 = 1$.

   ◆(d) $x_{k+2} = 6x_k - x_{k+1}$, where $x_0 = 1$ and $x_1 = 1$.

2. Solve the following linear recurrences.

   (a) $x_{k+3} = 6x_{k+2} - 11x_{k+1} + 6x_k$, where $x_0 = 1$, $x_1 = 0$, and $x_2 = 1$.

   ◆(b) $x_{k+3} = -2x_{k+2} + x_{k+1} + 2x_k$, where $x_0 = 1$, $x_1 = 0$, and $x_2 = 1$.

   [*Hint*: Use $\mathbf{v}_k = \begin{bmatrix} x_k \\ x_{k+1} \\ x_{k+2} \end{bmatrix}$.]

3. In Example 1 suppose busses are also allowed to park, and let x_k denote the number of ways a row of k parking spaces can be filled with cars, trucks, and busses.

   (a) If trucks and busses take up 2 and 3 spaces respectively, show that $x_{k+3} = x_k + x_{k+1} + x_{k+2}$ for each k, and use this recurrence to compute x_{10}. [*Hint*: The eigenvalues are of little use.]

   ◆(b) If busses take up 4 spaces, find a recurrence for the x_k and compute x_{10}.

4. A man must climb a flight of k steps. He always takes one or two steps at a time. Thus he can climb 3 steps in the following ways: 1, 1, 1; 1, 2; or 2, 1. Find s_k, the number of ways he can climb the flight of k steps. [*Hint*: Fibonacci.]

◆5. How many "words" of k letters can be made from the letters {a, b} if there are no adjacent a's?

6. How many sequences of k flips of a coin are there with no *HH*?

◆7. Find x_k, the number of ways to make a stack of k poker chips if only red, blue, and gold chips are used and no two gold chips are adjacent. [*Hint*: Show that $x_{k+2} = 2x_{k+1} + 2x_k$ by considering how many stacks have a red, blue, or gold chip on top.]

8. A nuclear reactor contains α- and β-particles. In every second each α-particle splits into three β-particles, and each β-particle splits into an α-particle and two β-particles. If there is a single α-particle in the reactor at time $t = 0$, how many α-particles are there at $t = 20$ seconds? [*Hint*: Let x_k and y_k denote the number of α- and β-particles at time $t = k$ seconds. Find x_{k+1} and y_{k+1} in terms of x_k and y_k.]

9. The annual yield of wheat in a certain country has been found to equal the average of the yield in the previous two years. If the yields in 1990 and 1991 were 10 and 12 million tons respectively, find a formula for the yield k years after 1990. What is the long-term average yield?

10. Find the general solution to the recurrence $x_{k+1} = rx_k + c$ where r and c are constants. [*Hint*: Consider the cases $r = 1$ and $r \neq 1$ separately. If $r \neq 1$, you will need the identity $1 + r + r^2 + \cdots + r^{n-1} = \frac{1-r^n}{1-r}$ for $n \geq 1$.]

11. Consider the length 3 recurrence $x_{k+3} = ax_k + bx_{k+1} + cx_{k+2}$.

   (a) If $\mathbf{v}_k = \begin{bmatrix} x_k \\ x_{k+1} \\ x_{k+2} \end{bmatrix}$ and $A = \begin{bmatrix} 0 & 1 & 0 \\ 0 & 0 & 1 \\ a & b & c \end{bmatrix}$, show that $\mathbf{v}_{k+1} = A\mathbf{v}_k$.

   ◆(b) If λ is any eigenvalue of A, show that
   $$\mathbf{x} = \begin{bmatrix} 1 \\ \lambda \\ \lambda^2 \end{bmatrix}$$ is a λ-eigenvector.

   [*Hint*: Show directly that $A\mathbf{x} = \lambda\mathbf{x}$.]

   (c) Generalize (a) and (b) to a recurrence $x_{k+4} = ax_k + bx_{k+1} + cx_{k+2} + dx_{k+3}$ of length 4.

12. Consider the recurrence $x_{k+2} = ax_{k+1} + bx_k + c$ where c may not be zero.

   (a) If $a + b \neq 1$ show that p can be found such that, if we set $y_k = x_k + p$, then $y_{k+2} = ay_{k+1} + by_k$. [Hence, the sequence x_k can be found provided y_k can be found by the methods of this section (or otherwise).]

   ◆(b) Use (a) to solve the recurrence $x_{k+2} = x_{k+1} + 6x_k + 5$ where $x_0 = 1$ and $x_1 = 1$.

13. Consider the recurrence
$$x_{k+2} = ax_{k+1} + bx_k + c(k) \quad (*)$$
where $c(k)$ is a function of k, and consider the related recurrence
$$x_{k+2} = ax_{k+1} + bx_k \quad (**)$$
Suppose that $x_k = p_k$ is a particular solution of $(*)$.

   ◆(a) If q_k is any solution of $(**)$, show that $q_k + p_k$ is a solution of $(*)$.

   (b) Show that every solution of $(*)$ arises as in (a) as the sum of a solution of $(**)$ plus the particular solution p_k of $(*)$.

SECTION 3.5 An Application to Systems of Differential Equations

A function f of a real variable is said to be **differentiable** if its derivative exists and, in this case, we let f' denote the derivative. If f and g are differentiable functions, a system

$$f' = 3f + 5g$$
$$g' = -f + 2g$$

is called a *system of first order differential equations*, or a *differential system* for short. Solving many practical problems often comes down to finding sets of functions that satisfy such a system (often involving more than two functions). In this section we show how diagonalization can help. Of course an acquaintance with calculus is required.

The Exponential Function

The simplest differential system is the following single equation:

$$f' = af \quad \text{where } a \text{ is a constant.} \quad (*)$$

It is easily verified that $f(x) = e^{ax}$ is one solution; in fact, equation $(*)$ is simple enough for us to find *all* solutions. Suppose that f is any solution, so that

$f'(x) = af(x)$ for all x. Consider the new function g given by $g(x) = f(x)e^{-ax}$. Then the product rule of differentiation gives

$$g'(x) = f(x)[-ae^{-ax}] + f'(x)e^{-ax}$$
$$= -af(x)e^{-ax} + [af(x)]e^{-ax}$$
$$= 0$$

for all x. Hence the function $g(x)$ has zero derivative and so must be a constant, say $g(x) = c$. Thus $c = g(x) = f(x)e^{-ax}$, that is

$$f(x) = ce^{ax}.$$

In other words, every solution $f(x)$ of $(*)$ is just a scalar multiple of e^{ax}. Since every such scalar multiple is easily seen to be a solution of $(*)$, we have proved

Theorem 1

The set of solutions to $f' = af$ is $\{ce^{ax} \mid c \text{ any constant}\} = \mathbb{R}e^{ax}$.

Remarkably, this result together with diagonalization enables us to solve a wide variety of differential systems.

EXAMPLE 1

Assume that the number $n(t)$ of bacteria in a culture at time t has the property that the rate of change of n is proportional to n itself. If there are n_0 bacteria present when $t = 0$, find the number at time t.

Solution ▶ Let k denote the proportionality constant. The rate of change of $n(t)$ is its time-derivative $n'(t)$, so the given relationship is $n'(t) = kn(t)$. Thus Theorem 1 shows that all solutions n are given by $n(t) = ce^{kt}$, where c is a constant. In this case, the constant c is determined by the requirement that there be n_0 bacteria present when $t = 0$. Hence $n_0 = n(0) = ce^{k0} = c$, so

$$n(t) = n_0 e^{kt}$$

gives the number at time t. Of course the constant k depends on the strain of bacteria.

The condition that $n(0) = n_0$ in Example 1 is called an **initial condition** or a **boundary condition** and serves to select one solution from the available solutions.

General Differential Systems

Solving a variety of problems, particularly in science and engineering, comes down to solving a system of linear differential equations. Diagonalization enters into this as follows. The general problem is to find differentiable functions $f_1, f_2, \ldots, f_n$ that satisfy a system of equations of the form

SECTION 3.5   An Application to Systems of Differential Equations

$$f_1' = a_{11}f_1 + a_{12}f_2 + \cdots + a_{1n}f_n$$
$$f_2' = a_{21}f_1 + a_{22}f_2 + \cdots + a_{2n}f_n$$
$$\vdots \quad \vdots \quad \vdots \quad \vdots$$
$$f_n' = a_{n1}f_1 + a_{n2}f_2 + \cdots + a_{nn}f_n$$

where the a_{ij} are constants. This is called a **linear system of differential equations** or simply a **differential system**. The first step is to put it in matrix form. Write

$$\mathbf{f}' = \begin{bmatrix} f_1 \\ f_2 \\ \vdots \\ f_n \end{bmatrix} \quad \mathbf{f}' = \begin{bmatrix} f_1' \\ f_2' \\ \vdots \\ f_n' \end{bmatrix} \quad A = \begin{bmatrix} a_{11} & a_{12} & \cdots & a_{1n} \\ a_{21} & a_{22} & \cdots & a_{2n} \\ \vdots & \vdots & & \vdots \\ a_{n1} & a_{n2} & \cdots & a_{nn} \end{bmatrix}$$

Then the system can be written compactly using matrix multiplication:

$$\mathbf{f}' = A\mathbf{f}$$

Hence, given the matrix A, the problem is to find a column $\mathbf{f}$ of differentiable functions that satisfies this condition. This can be done if A is diagonalizable. Here is an example.

EXAMPLE 2

Find a solution to the system

$$f_1' = f_1 + 3f_2$$
$$f_2' = 2f_1 + 2f_2$$

that satisfies $f_1(0) = 0$, $f_2(0) = 5$.

Solution ▶ This is $\mathbf{f}' = A\mathbf{f}$, where $\mathbf{f} = \begin{bmatrix} f_1 \\ f_2 \end{bmatrix}$ and $A = \begin{bmatrix} 1 & 3 \\ 2 & 2 \end{bmatrix}$. The reader can verify that $c_A(x) = (x - 4)(x + 1)$, and that $\mathbf{x}_1 = \begin{bmatrix} 1 \\ 1 \end{bmatrix}$ and $\mathbf{x}_2 = \begin{bmatrix} 3 \\ -2 \end{bmatrix}$ are eigenvectors corresponding to the eigenvalues 4 and -1, respectively. Hence the diagonalization algorithm gives $P^{-1}AP = \begin{bmatrix} 4 & 0 \\ 0 & -1 \end{bmatrix}$, where $P = [\mathbf{x}_1 \ \mathbf{x}_2] = \begin{bmatrix} 1 & 3 \\ 1 & -2 \end{bmatrix}$. Now consider new functions g_1 and g_2 given by $\mathbf{f} = P\mathbf{g}$ (equivalently, $\mathbf{g} = P^{-1}\mathbf{f}$), where $\mathbf{g} = \begin{bmatrix} g_1 \\ g_2 \end{bmatrix}$. Then

$$\begin{bmatrix} f_1 \\ f_2 \end{bmatrix} = \begin{bmatrix} 1 & 3 \\ 1 & -2 \end{bmatrix} \begin{bmatrix} g_1 \\ g_2 \end{bmatrix} \quad \text{that is,} \quad \begin{matrix} f_1 = g_1 + 3g_2 \\ f_2 = g_1 - 2g_2 \end{matrix}$$

Hence $f_1' = g_1' + 3g_2'$ and $f_2' = g_1' - 2g_2'$ so that

$$\mathbf{f}' = \begin{bmatrix} f_1' \\ f_2' \end{bmatrix} = \begin{bmatrix} 1 & 3 \\ 1 & -2 \end{bmatrix} \begin{bmatrix} g_1' \\ g_2' \end{bmatrix} = P\mathbf{g}'$$

If this is substituted in $\mathbf{f}' = A\mathbf{f}$, the result is $P\mathbf{g}' = AP\mathbf{g}$, whence

$$\mathbf{g}' = P^{-1}AP\mathbf{g}$$

But this means that

$$\begin{bmatrix} g_1' \\ g_2' \end{bmatrix} = \begin{bmatrix} 4 & 0 \\ 0 & -1 \end{bmatrix} \begin{bmatrix} g_1 \\ g_2 \end{bmatrix}, \quad \text{so} \quad \begin{matrix} g_1' = 4g_1 \\ g_2' = -g_2 \end{matrix}$$

Hence Theorem 1 gives $g_1(x) = ce^{4x}$, $g_2(x) = de^{-x}$, where c and d are constants. Finally, then,

$$\begin{bmatrix} f_1(x) \\ f_2(x) \end{bmatrix} = P \begin{bmatrix} g_1(x) \\ g_2(x) \end{bmatrix} = \begin{bmatrix} 1 & 3 \\ 1 & -2 \end{bmatrix} \begin{bmatrix} ce^{4x} \\ de^{-x} \end{bmatrix} = \begin{bmatrix} ce^{4x} + 3de^{-x} \\ ce^{4x} - 2de^{-x} \end{bmatrix}$$

so the *general solution* is

$$\begin{matrix} f_1(x) = ce^{4x} + 3de^{-x} \\ f_2(x) = ce^{4x} - 2de^{-x} \end{matrix} \quad c \text{ and } d \text{ constants.}$$

It is worth observing that this can be written in matrix form as

$$\begin{bmatrix} f_1(x) \\ f_2(x) \end{bmatrix} = c \begin{bmatrix} 1 \\ 1 \end{bmatrix} e^{4x} + d \begin{bmatrix} 3 \\ -2 \end{bmatrix} e^{-x}$$

That is,

$$\mathbf{f}(x) = c\mathbf{x}_1 e^{4x} + d\mathbf{x}_2 e^{-x}$$

This form of the solution works more generally, as will be shown.

Finally, the requirement that $f_1(0) = 0$ and $f_2(0) = 5$ in this example determines the constants c and d:

$$0 = f_1(0) = ce^0 + 3de^0 = c + 3d$$
$$5 = f_2(0) = ce^0 - 2de^0 = c - 2d$$

These equations give $c = 3$ and $d = -1$, so

$$\begin{matrix} f_1(x) = 3e^{4x} - 3e^{-x} \\ f_2(x) = 3e^{4x} + 2e^{-x} \end{matrix}$$

satisfy all the requirements.

The technique in this example works in general.

Theorem 2

Consider a linear system

$$\mathbf{f}' = A\mathbf{f}$$

of differential equations, where A is an $n \times n$ diagonalizable matrix. Let $P^{-1}AP$ be diagonal, where P is given in terms of its columns

$$P = [\mathbf{x}_1, \mathbf{x}_2, \ldots, \mathbf{x}_n]$$

and $\{\mathbf{x}_1, \mathbf{x}_2, \ldots, \mathbf{x}_n\}$ are eigenvectors of A. If $\mathbf{x}_i$ corresponds to the eigenvalue λ_i for each i, then every solution $\mathbf{f}$ of $\mathbf{f}' = A\mathbf{f}$ has the form

$$\mathbf{f}(x) = c_1 \mathbf{x}_1 e^{\lambda_1 x} + c_2 \mathbf{x}_2 e^{\lambda_2 x} + \cdots + c_n \mathbf{x}_2 e^{\lambda_n x}$$

where $c_1, c_2, \ldots, c_n$ are arbitrary constants.

SECTION 3.5 An Application to Systems of Differential Equations

PROOF

By Theorem 4 Section 3.3, the matrix $P = [\mathbf{x}_1, \mathbf{x}_2, \ldots, \mathbf{x}_n]$ is invertible and

$$P^{-1}AP = \begin{bmatrix} \lambda_1 & 0 & \cdots & 0 \\ 0 & \lambda_2 & \cdots & 0 \\ \vdots & \vdots & & \vdots \\ 0 & 0 & \cdots & \lambda_n \end{bmatrix}.$$

As in Example 2, write $\mathbf{f} = \begin{bmatrix} f_1 \\ f_2 \\ \vdots \\ f_n \end{bmatrix}$ and define $\mathbf{g} = \begin{bmatrix} g_1 \\ g_2 \\ \vdots \\ g_n \end{bmatrix}$ by $\mathbf{g} = P^{-1}\mathbf{f}$; equivalently, $\mathbf{f} = P\mathbf{g}$. If $P = [p_{ij}]$, this gives

$$f_i = p_{i1}g_1 + p_{i2}g_2 + \cdots + p_{in}g_n.$$

Since the p_{ij} are constants, differentiation preserves this relationship:

$$f_i' = p_{i1}g_1' + p_{i2}g_2' + \cdots + p_{in}g_n'.$$

so $\mathbf{f}' = P\mathbf{g}'$. Substituting this into $\mathbf{f}' = A\mathbf{f}$ gives $P\mathbf{g}' = AP\mathbf{g}$. But then multiplication by P^{-1} gives $\mathbf{g}' = P^{-1}AP\mathbf{g}$, so the original system of equations $\mathbf{f}' = A\mathbf{f}$ for $\mathbf{f}$ becomes much simpler in terms of $\mathbf{g}$:

$$\begin{bmatrix} g_1' \\ g_2' \\ \vdots \\ g_n' \end{bmatrix} = \begin{bmatrix} \lambda_1 & 0 & \cdots & 0 \\ 0 & \lambda_2 & \cdots & 0 \\ \vdots & \vdots & & \vdots \\ 0 & 0 & \cdots & \lambda_n \end{bmatrix} \begin{bmatrix} g_1 \\ g_2 \\ \vdots \\ g_n \end{bmatrix}$$

Hence $g_i' = \lambda_i g_i$ holds for each i, and Theorem 1 implies that the only solutions are

$$g_i(x) = c_i e^{\lambda_i x} \quad c_i \text{ some constant.}$$

Then the relationship $\mathbf{f} = P\mathbf{g}$ gives the functions $f_1, f_2, \ldots, f_n$ as follows:

$$\mathbf{f}(x) = [\mathbf{x}_1, \mathbf{x}_2, \ldots, \mathbf{x}_n] \begin{bmatrix} c_1 e^{\lambda_1 x} \\ c_2 e^{\lambda_2 x} \\ \vdots \\ c_n e^{\lambda_n x} \end{bmatrix} = c_1 \mathbf{x}_1 e^{\lambda_1 x} + c_2 \mathbf{x}_2 e^{\lambda_2 x} + \cdots + c_n \mathbf{x}_2 e^{\lambda_n x}$$

This is what we wanted.

The theorem shows that *every* solution to $\mathbf{f}' = A\mathbf{f}$ is a linear combination

$$\mathbf{f}(x) = c_1 \mathbf{x}_1 e^{\lambda_1 x} + c_2 \mathbf{x}_2 e^{\lambda_2 x} + \cdots + c_n \mathbf{x}_2 e^{\lambda_n x}$$

where the coefficients c_i are arbitrary. Hence this is called the **general solution** to the system of differential equations. In most cases the solution functions $f_i(x)$ are required to satisfy boundary conditions, often of the form $f_i(a) = b_i$, where $a, b_1, \ldots, b_n$ are prescribed numbers. These conditions determine the constants c_i. The following example illustrates this and displays a situation where one eigenvalue has multiplicity greater than 1.

EXAMPLE 3

Find the general solution to the system

$$\begin{aligned} f_1' &= 5f_1 + 8f_2 + 16f_3 \\ f_2' &= 4f_1 + f_2 + 8f_3 \\ f_3' &= -4f_1 - 4f_2 - 11f_3 \end{aligned}$$

Then find a solution satisfying the boundary conditions $f_1(0) = f_2(0) = f_3(0) = 1$.

Solution ▶ The system has the form $\mathbf{f}' = A\mathbf{f}$, where $A = \begin{bmatrix} 5 & 8 & 16 \\ 4 & 1 & 8 \\ -4 & -4 & -11 \end{bmatrix}$. Then $c_A(x) = (x+3)^2(x-1)$ and eigenvectors corresponding to the eigenvalues -3, -3, and 1 are, respectively,

$$\mathbf{x}_1 = \begin{bmatrix} -1 \\ 1 \\ 0 \end{bmatrix} \quad \mathbf{x}_2 = \begin{bmatrix} -2 \\ 0 \\ 1 \end{bmatrix} \quad \mathbf{x}_3 = \begin{bmatrix} 2 \\ 1 \\ -1 \end{bmatrix}$$

Hence, by Theorem 2, the general solution is

$$\mathbf{f}(x) = c_1 \begin{bmatrix} -1 \\ 1 \\ 0 \end{bmatrix} e^{-3x} + c_2 \begin{bmatrix} -2 \\ 0 \\ 1 \end{bmatrix} e^{-3x} + c_3 \begin{bmatrix} 2 \\ 1 \\ -1 \end{bmatrix} e^x, \quad c_i \text{ constants.}$$

The boundary conditions $f_1(0) = f_2(0) = f_3(0) = 1$ determine the constants c_i.

$$\begin{bmatrix} 1 \\ 1 \\ 1 \end{bmatrix} = \mathbf{f}(0) = c_1 \begin{bmatrix} -1 \\ 1 \\ 0 \end{bmatrix} + c_2 \begin{bmatrix} -2 \\ 0 \\ 1 \end{bmatrix} + c_3 \begin{bmatrix} 2 \\ 1 \\ -1 \end{bmatrix}$$

$$= \begin{bmatrix} -1 & -2 & 2 \\ 1 & 0 & 1 \\ 0 & 1 & -1 \end{bmatrix} \begin{bmatrix} c_1 \\ c_2 \\ c_3 \end{bmatrix}$$

The solution is $c_1 = -3$, $c_2 = 5$, $c_3 = 4$, so the required specific solution is

$$\begin{aligned} f_1(x) &= -7e^{-3x} + 8e^x \\ f_2(x) &= -3e^{-3x} + 4e^x \\ f_3(x) &= 5e^{-3x} - 4e^x \end{aligned}$$

EXERCISES 3.5

1. Use Theorem 1 to find the general solution to each of the following systems. Then find a specific solution satisfying the given boundary condition.

   (a) $f_1' = 2f_1 + 4f_2$, $f_1(0) = 0$
   $f_2' = 3f_1 + 3f_2$, $f_2(0) = 1$

   ◆(b) $f_1' = -f_1 + 5f_2$, $f_1(0) = 1$
   $f_2' = f_1 + 3f_2$, $f_2(0) = -1$

   (c) $f_1' = 4f_2 + 4f_3$
   $f_2' = f_1 + f_2 - 2f_3$
   $f_3' = -f_1 + f_2 + 4f_3$
   $f_1(0) = f_2(0) = f_3(0) = 1$

   ◆(d) $f_1' = 2f_1 + f_2 + 2f_3$
   $f_2' = 2f_1 + 2f_2 - 2f_3$
   $f_3' = 3f_1 + f_2 + f_3$
   $f_1(0) = f_2(0) = f_3(0) = 1$

2. Show that the solution to $f' = af$ satisfying $f(x_0) = k$ is $f(x) = ke^{a(x-x_0)}$.

3. A radioactive element decays at a rate proportional to the amount present. Suppose an initial mass of 10 g decays to 8 g in 3 hours.

   (a) Find the mass t hours later.

   ◆(b) Find the half-life of the element—the time taken to decay to half its mass.

4. The population $N(t)$ of a region at time t increases at a rate proportional to the population. If the population doubles every 5 years and is 3 million initially, find $N(t)$.

5. Let A be an invertible diagonalizable $n \times n$ matrix and let $\mathbf{b}$ be an n-column of constant functions. We can solve the system $\mathbf{f}' = A\mathbf{f} + \mathbf{b}$ as follows:

   ◆(a) If $\mathbf{g}$ satisfies $\mathbf{g}' = A\mathbf{g}$ (using Theorem 2), show that $\mathbf{f} = \mathbf{g} - A^{-1}\mathbf{b}$ is a solution to $\mathbf{f}' = A\mathbf{f} + \mathbf{b}$.

   (b) Show that every solution to $\mathbf{f}' = A\mathbf{f} + \mathbf{b}$ arises as in (a) for some solution $\mathbf{g}$ to $\mathbf{g}' = A\mathbf{g}$.

6. Denote the second derivative of f by $f'' = (f')'$. Consider the second order differential equation
$$f'' - a_1 f' - a_2 f = 0, \; a_1 \text{ and } a_2 \text{ real numbers.} \quad (*)$$

   (a) If f is a solution to $(*)$ let $f_1 = f$ and $f_2 = f' - a_1 f$. Show that
$$\begin{cases} f_1' = a_1 f_1 + f_2 \\ f_2' = a_2 f_1 \end{cases}, \text{ that is } \begin{bmatrix} f_1' \\ f_2' \end{bmatrix} = \begin{bmatrix} a_1 & 1 \\ a_2 & 0 \end{bmatrix} \begin{bmatrix} f_1 \\ f_2 \end{bmatrix}.$$

   ◆(b) Conversely, if $\begin{bmatrix} f_1 \\ f_2 \end{bmatrix}$ is a solution to the system in (a), show that f_1 is a solution to $(*)$.

7. Writing $f''' = (f'')'$, consider the third order differential equation
$$f''' - a_1 f'' - a_2 f' - a_3 f = 0$$
where a_1, a_2, and a_3 are real numbers. Let $f_1 = f$, $f_2 = f' - a_1 f$ and $f_3 = f'' - a_1 f' - a_2 f''$.

   (a) Show that $\begin{bmatrix} f_1 \\ f_2 \\ f_3 \end{bmatrix}$ is a solution to the system
$$\begin{cases} f_1' = a_1 f_1 + f_2 \\ f_2' = a_2 f_1 + f_3 \\ f_3' = a_3 f_1 \end{cases}, \text{ that is } \begin{bmatrix} f_1' \\ f_2' \\ f_3' \end{bmatrix} = \begin{bmatrix} a_1 & 1 & 0 \\ a_2 & 0 & 1 \\ a_3 & 0 & 0 \end{bmatrix} \begin{bmatrix} f_1 \\ f_2 \\ f_3 \end{bmatrix}.$$

   (b) Show further that if $\begin{bmatrix} f_1 \\ f_2 \\ f_3 \end{bmatrix}$ is any solution to this system, then $f = f_1$ is a solution to $(*)$.
   Remark. A similar construction casts every linear differential equation of order n (with constant coefficients) as an $n \times n$ linear system of first order equations. However, the matrix need not be diagonalizable, so other methods have been developed.

SECTION 3.6 Proof of the Cofactor Expansion Theorem

Recall that our definition of the term *determinant* is inductive: The determinant of any 1×1 matrix is defined first; then it is used to define the determinants of 2×2 matrices. Then that is used for the 3×3 case, and so on. The case of a 1×1 matrix $[a]$ poses no problem. We simply define

$$\det[a] = a$$

as in Section 3.1. Given an $n \times n$ matrix A, define A_{ij} to be the $(n-1) \times (n-1)$ matrix obtained from A by deleting row i and column j. Now assume that the determinant of any $(n-1) \times (n-1)$ matrix has been defined. Then the determinant of A is *defined* to be

$$\det A = a_{11} \det A_{11} - a_{21} \det A_{21} + \cdots + (-1)^{n+1} a_{n1} \det A_{n1}$$
$$= \sum_{i=1}^{n} (-1)^{i+1} a_{i1} \det A_{i1}$$

where summation notation has been introduced for convenience.[17] Observe that, in the terminology of Section 3.1, this is just the cofactor expansion of $\det A$ along the first column, and that $(-1)^{i+j} \det A_{ij}$ is the (i, j)-cofactor (previously denoted

---

17 Summation notation is a convenient shorthand way to write sums of similar expressions. For example $a_1 + a_2 + a_3 + a_4 = \sum_{h=1}^{4} a_h$, $a_5 b_5 + a_6 b_6 + a_7 b_7 + a_8 b_8 = \sum_{k=5}^{8} a_k b_k$, and $1^2 + 2^2 + 3^2 + 4^2 + 5^2 = \sum_{j=1}^{5} j^2$.

as $c_{ij}(A)$).[18] To illustrate the definition, consider the 2×2 matrix $A = \begin{bmatrix} a_{11} & a_{12} \\ a_{21} & a_{22} \end{bmatrix}$. Then the definition gives

$$\det \begin{bmatrix} a_{11} & a_{12} \\ a_{21} & a_{22} \end{bmatrix} = a_{11} \det[a_{22}] - a_{21} \det[a_{12}] = a_{11}a_{22} - a_{21}a_{12}$$

and this is the same as the definition in Section 3.1.

Of course, the task now is to use this definition to *prove* that the cofactor expansion along *any* row or column yields $\det A$ (this is Theorem 1 Section 3.1). The proof proceeds by first establishing the properties of determinants stated in Theorem 2 Section 3.1, but for *rows* only (see Lemma 2). This being done, the full proof of Theorem 1 Section 3.1 is not difficult. The proof of Lemma 2 requires the following preliminary result.

Lemma 1

Let A, B, and C be $n \times n$ matrices that are identical except that the pth row of A is the sum of the pth rows of B and C. Then

$$\det A = \det B + \det C$$

PROOF

We proceed by induction on n, the cases $n = 1$ and $n = 2$ being easily checked. Consider a_{i1} and A_{i1}:

Case 1: If $i \neq p$,

$$a_{i1} = b_{i1} = c_{i1} \quad \text{and} \quad \det A_{i1} = \det B_{i1} = \det C_{i1}$$

by induction because A_{i1}, B_{i1}, C_{i1} are identical except that one row of A_{i1} is the sum of the corresponding rows of B_{i1} and C_{i1}.

Case 2: If $i = p$,

$$a_{p1} = b_{p1} + c_{p1} \quad \text{and} \quad A_{p1} = B_{p1} = C_{p1}$$

Now write out the defining sum for $\det A$, splitting off the pth term for special attention.

$$\det A = \sum_{i \neq p} a_{i1}(-1)^{i+1} \det A_{i1} + a_{p1}(-1)^{p+1} \det A_{p1}$$

$$= \sum_{i \neq p} a_{i1}(-1)^{i+1} [\det B_{i1} + \det B_{i1}] + (b_{p1} + c_{p1})(-1)^{p+1} \det A_{p1}$$

where $\det A_{i1} = \det B_{i1} + \det C_{i1}$ by induction. But the terms here involving B_{i1} and b_{p1} add up to $\det B$ because $a_{i1} = b_{i1}$ if $i \neq p$ and $A_{p1} = B_{p1}$. Similarly, the terms involving C_{i1} and c_{p1} add up to $\det C$. Hence $\det A = \det B + \det C$, as required.

---

[18] Note that we used the expansion along *row* 1 at the beginning of Section 3.1. The column 1 expansion definition is more convenient here.

SECTION 3.6 Proof of the Cofactor Expansion Theorem

Lemma 2

Let $A = [a_{ij}]$ denote an $n \times n$ matrix.

1. If $B = [b_{ij}]$ is formed from A by multiplying a row of A by a number u, then $\det B = u \det A$.
2. If A contains a row of zeros, then $\det A = 0$.
3. If $B = [b_{ij}]$ is formed by interchanging two rows of A, then $\det B = -\det A$.
4. If A contains two identical rows, then $\det A = 0$.
5. If $B = [b_{ij}]$ is formed by adding a multiple of one row of A to a different row, then $\det B = \det A$.

PROOF

For later reference the defining sums for $\det A$ and $\det B$ are as follows:

$$\det A = \sum_{i=1}^{n} a_{i1}(-1)^{i+1} \det A_{i1} \qquad (*)$$

$$\det B = \sum_{i=1}^{n} b_{i1}(-1)^{i+1} \det B_{i1} \qquad (**)$$

Property 1. The proof is by induction on n, the cases $n = 1$ and $n = 2$ being easily verified. Consider the ith term in the sum $(**)$ for $\det B$ where B is the result of multiplying row p of A by u.

(a) If $i \neq p$, then $b_{i1} = a_{i1}$ and $\det B_{i1} = u \det A_{i1}$ by induction because B_{i1} comes from A_{i1} by multiplying a row by u.

(b) If $i = p$, then $b_{p1} = ua_{p1}$ and $B_{p1} = A_{p1}$.

In either case, each term in equation $(**)$ is u times the corresponding term in equation $(*)$, so it is clear that $\det B = u \det A$.

Property 2. This is clear by property 1 because the row of zeros has a common factor $u = 0$.

Property 3. Observe first that it suffices to prove property 3 for interchanges of adjacent rows. (Rows p and q ($q > p$) can be interchanged by carrying out $2(q - p) - 1$ adjacent changes, which results in an *odd* number of sign changes in the determinant.) So suppose that rows p and $p + 1$ of A are interchanged to obtain B. Again consider the ith term in $(**)$.

(a) If $i \neq p$ and $i \neq p + 1$, then $b_{i1} = a_{i1}$ and $\det B_{i1} = -\det A_{i1}$ by induction because B_{i1} results from interchanging adjacent rows in A_{i1}. Hence the ith term in $(**)$ is the negative of the ith term in $(*)$. Hence $\det B = -\det A$ in this case.

(b) If $i = p$ or $i = p + 1$, then $b_{p1} = a_{p+1,1}$ and $B_{p1} = A_{p+1,1}$, whereas $b_{p+1,1} = a_{p1}$ and $B_{p+1,1} = A_{p1}$. Hence terms p and $p + 1$ in $(**)$ are

$$b_{p1}(-1)^{p+1} \det B_{p1} = -a_{p+1,1}(-1)^{(p+1)+1} \det(A_{p+1,1})$$
$$b_{p+1,1}(-1)^{(p+1)+1} \det(B_{p+1,1}) = -a_{p1}(-1)^{p+1} \det A_{p1}$$

This means that terms p and $p + 1$ in $(**)$ are the same as these terms in $(*)$, except that the order is reversed and the signs are changed. Thus the sum $(**)$ is the negative of the sum $(*)$; that is, $\det B = -\det A$.

Property 4. If rows p and q in A are identical, let B be obtained from A by interchanging these rows. Then $B = A$ so $\det A = \det B$. But $\det B = -\det A$ by property 3 so $\det A = -\det A$. This implies that $\det A = 0$.

Property 5. Suppose B results from adding u times row q of A to row p. Then Lemma 1 applies to B to show that $\det B = \det A + \det C$, where C is obtained from A by replacing row p by u times row q. It now follows from properties 1 and 4 that $\det C = 0$ so $\det B = \det A$, as asserted.

These facts are enough to enable us to prove Theorem 1 Section 3.1. For convenience, it is restated here in the notation of the foregoing lemmas. The only difference between the notations is that the (i, j)-cofactor of an $n \times n$ matrix A was denoted earlier by

$$c_{ij}(A) = (-1)^{i+j} \det A_{ij}$$

Theorem 1

If $A = [a_{ij}]$ is an $n \times n$ matrix, then

1. $\det A = \sum_{i=1}^{n} a_{ij}(-1)^{i+j} \det A_{ij}$ (cofactor expansion along column j).

2. $\det A = \sum_{j=1}^{n} a_{ij}(-1)^{i+j} \det A_{ij}$ (cofactor expansion along row i).

Here A_{ij} denotes the matrix obtained from A by deleting row i and column j.

PROOF

Lemma 2 establishes the truth of Theorem 2 Section 3.1 for *rows*. With this information, the arguments in Section 3.2 proceed exactly as written to establish that $\det A = \det A^T$ holds for any $n \times n$ matrix A. Now suppose B is obtained from A by interchanging two columns. Then B^T is obtained from A^T by interchanging two rows so, by property 3 of Lemma 2,

$$\det B = \det B^T = -\det A^T = -\det A$$

Hence property 3 of Lemma 2 holds for *columns* too.

This enables us to prove the cofactor expansion for columns. Given an $n \times n$ matrix $A = [a_{ij}]$, let $B = [b_{ij}]$ be obtained by moving column j to the left side, using $j - 1$ interchanges of adjacent columns. Then $\det B = (-1)^{j-1} \det A$ and, because $B_{i1} = A_{ij}$ and $b_{i1} = a_{ij}$ for all i, we obtain

$$\det A = (-1)^{j-1} \det B = (-1)^{j-1} \sum_{i=1}^{n} b_{i1}(-1)^{i+1} \det B_{i1}$$
$$= \sum_{i=1}^{n} a_{ij}(-1)^{i+j} \det A_{ij}$$

This is the cofactor expansion of $\det A$ along column j.

SECTION 3.6 Proof of the Cofactor Expansion Theorem

Finally, to prove the row expansion, write $B = A^T$. Then $B_{ij} = (A_{ij}^T)$ and $b_{ij} = a_{ji}$ for all i and j. Expanding $\det B$ along column j gives

$$\det A = \det A^T = \det B = \sum_{i=1}^{n} b_{ij}(-1)^{i+j} \det B_{ij}$$

$$= \sum_{i=1}^{n} a_{ji}(-1)^{j+i} \det \left[(A_{ji}^T)\right] = \sum_{i=1}^{n} a_{ji}(-1)^{j+i} \det A_{ji}$$

This is the required expansion of $\det A$ along row j.

EXERCISES 3.6

1. Prove Lemma 1 for columns.

◆2. Verify that interchanging rows p and q ($q > p$) can be accomplished using $2(q - p) - 1$ adjacent interchanges.

3. If u is a number and A is an $n \times n$ matrix, prove that $\det(uA) = u^n \det A$ by induction on n, using only the definition of $\det A$.

SUPPLEMENTARY EXERCISES FOR CHAPTER 3

1. Show that
$$\det \begin{bmatrix} a + px & b + qx & c + rx \\ p + ux & q + vx & r + wx \\ u + ax & v + bx & w + cx \end{bmatrix}$$
$$= (1 + x^3)\det \begin{bmatrix} a & b & c \\ p & q & r \\ u & v & w \end{bmatrix}$$

2. (a) Show that $(A_{ij})^T = (A^T)_{ji}$ for all i, j, and all square matrices A.

   ◆(b) Use (a) to prove that $\det A^T = \det A$.
   [*Hint*: Induction on n where A is $n \times n$.]

3. Show that $\det \begin{bmatrix} 0 & I_n \\ I_m & 0 \end{bmatrix} = (-1)^{nm}$ for all $n \geq 1$ and $m \geq 1$.

4. Show that
$$\det \begin{bmatrix} 1 & a & a^3 \\ 1 & b & b^3 \\ 1 & c & c^3 \end{bmatrix} = (b - a)(c - a)(c - b)(a + b + c).$$

5. Let $A = \begin{bmatrix} R_1 \\ R_2 \end{bmatrix}$ be a 2×2 matrix with rows R_1 and R_2. If $\det A = 5$, find $\det B$ where
$$B = \begin{bmatrix} 3R_1 + 2R_3 \\ 2R_1 + 5R_2 \end{bmatrix}.$$

6. Let $A = \begin{bmatrix} 3 & -4 \\ 2 & -3 \end{bmatrix}$, and let $\mathbf{v}_k = A^k \mathbf{v}_0$ for each $k \geq 0$.
   (a) Show that A has no dominant eigenvalue.
   (b) Find $\mathbf{v}_k$ if $\mathbf{v}_0$ equals: (i) $\begin{bmatrix} 1 \\ 1 \end{bmatrix}$ (ii) $\begin{bmatrix} 2 \\ 1 \end{bmatrix}$
   (iii) $\begin{bmatrix} x \\ y \end{bmatrix} \neq \begin{bmatrix} 1 \\ 1 \end{bmatrix}$ or $\begin{bmatrix} 2 \\ 1 \end{bmatrix}$

4 Vector Geometry

SECTION 4.1 Vectors and Lines

In this chapter we study the geometry of 3-dimensional space. We view a point in 3-space as an arrow from the origin to that point. Doing so provides a "picture" of the point that is truly worth a thousand words. We used this idea earlier, in Section 2.6, to describe rotations, reflections, and projections of the plane $\mathbb{R}^2$. We now apply the same techniques to 3-space to examine similar transformations of $\mathbb{R}^3$. Moreover, the method enables us to completely describe all lines and planes in space.

Vectors in $\mathbb{R}^3$

Introduce a coordinate system in 3-dimensional space in the usual way. First choose a point O called the *origin*, then choose three mutually perpendicular lines through O, called the x, y, and z *axes*, and establish a number scale on each axis with zero at the origin. Given a point P in 3-space we associate three numbers x, y, and z with P, as described in Figure 1. These numbers are called the *coordinates* of P, and we denote the point as (x, y, z), or $P(x, y, z)$ to emphasize the label P. The result is called a *cartesian*[1] coordinate system for 3-space, and the resulting description of 3-space is called *cartesian geometry*.

As in the plane, we introduce vectors by identifying each point $P(x, y, z)$ with the vector $\mathbf{v} = \begin{bmatrix} x \\ y \\ z \end{bmatrix}$ in $\mathbb{R}^3$, represented by the **arrow** from the origin to P as in Figure 1. Informally, we say that the point *P has vector* $\mathbf{v}$, and that vector $\mathbf{v}$ *has point P*. In this way 3-space is identified with $\mathbb{R}^3$, and this identification will be made throughout this chapter, often without comment. In particular, the terms "vector" and "point" are interchangeable.[2] The resulting description of 3-space is called **vector geometry**. Note that the origin is $\mathbf{0} = \begin{bmatrix} 0 \\ 0 \\ 0 \end{bmatrix}$.

FIGURE 1

Length and Direction

We are going to discuss two fundamental geometric properties of vectors in $\mathbb{R}^3$: length and direction. First, if $\mathbf{v}$ is a vector with point P, the **length** $\|\mathbf{v}\|$ of vector

---

[1] Named after René Descartes who introduced the idea in 1637.
[2] Recall that we defined $\mathbb{R}^n$ as the set of all ordered n-tuples of real numbers, and reserved the right to denote them as rows or as columns.

v is defined to be the distance from the origin to P, that is the length of the arrow representing **v**. The following properties of length will be used frequently.

Theorem 1

Let $\mathbf{v} = \begin{bmatrix} x \\ y \\ z \end{bmatrix}$ be a vector.

(1) $\|\mathbf{v}\| = \sqrt{x^2 + y^2 + z^2}$.[3]
(2) $\mathbf{v} = \mathbf{0}$ if and only if $\|\mathbf{v}\| = 0$
(3) $\|a\mathbf{v}\| = |a|\,\|\mathbf{v}\|$ for all scalars a.[4]

PROOF

Let v have point $P = (x, y, z)$.

(1) In Figure 2, $\|\mathbf{v}\|$ is the hypotenuse of the right triangle OQP, and so $\|\mathbf{v}\|^2 = h^2 + z^2$ by Pythagoras' theorem.[5] But h is the hypotenuse of the right triangle ORQ, so $h^2 = x^2 + y^2$. Now (1) follows by eliminating h^2 and taking positive square roots.

(2) If $\|\mathbf{v}\| = 0$, then $x^2 + y^2 + z^2 = 0$ by (1). Because squares of real numbers are nonnegative, it follows that $x = y = z = 0$, and hence that $\mathbf{v} = \mathbf{0}$. The converse is because $\|\mathbf{0}\| = 0$.

(3) We have $a\mathbf{v} = (ax, ay, az)$ so (1) gives $\|a\mathbf{v}\|^2 = (ax)^2 + (ay)^2 + (az)^2 = a^2\|\mathbf{v}\|^2$. Hence $\|a\mathbf{v}\| = \sqrt{a^2}\,\|\mathbf{v}\|$, and we are done because $\sqrt{a^2} = |a|$ for any real number a.

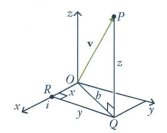

FIGURE 2

Of course the $\mathbb{R}^2$-version of Theorem 1 also holds.

EXAMPLE 1

If $\mathbf{v} = \begin{bmatrix} 2 \\ -1 \\ 3 \end{bmatrix}$ then $\|\mathbf{v}\| = \sqrt{4 + 1 + 9} = \sqrt{14}$. Similarly if $\mathbf{v} = \begin{bmatrix} 3 \\ -4 \end{bmatrix}$ in 2-space then $\|\mathbf{v}\| = \sqrt{9 + 16} = 5$.

When we view two nonzero vectors as arrows emanating from the origin, it is clear geometrically what we mean by saying that they have the same or opposite **direction**. This leads to a fundamental new description of vectors.

---

[3] When we write $\sqrt{p}$ we mean the positive square root of p.
[4] Recall that the absolute value $|a|$ of a real number is defined by $|a| = \begin{cases} a & \text{if } a \geq 0 \\ -a & \text{if } a < 0 \end{cases}$.
[5] Pythagoras' theorem states that if a and b are sides of right triangle with hypotenuse c, then $a^2 + b^2 = c^2$. A proof is given at the end of this section.

Theorem 2

Let $\mathbf{v} \neq \mathbf{0}$ and $\mathbf{w} \neq \mathbf{0}$ be vectors in $\mathbb{R}^3$. Then $\mathbf{v} = \mathbf{w}$ as matrices if and only if $\mathbf{v}$ and $\mathbf{w}$ have the same direction and the same length.[6]

PROOF

If $\mathbf{v} = \mathbf{w}$, they clearly have the same direction and length. Conversely, let $\mathbf{v}$ and $\mathbf{w}$ be vectors with points $P(x, y, z)$ and $Q(x_1, y_1, z_1)$ respectively. If $\mathbf{v}$ and $\mathbf{w}$ have the same length and direction then, geometrically, P and Q must be the same point (see Figure 3). Hence $x = x_1, y = y_1,$ and $z = z_1$, that is $\mathbf{v} = \begin{bmatrix} x \\ y \\ z \end{bmatrix} = \begin{bmatrix} x_1 \\ y_1 \\ z_1 \end{bmatrix} = \mathbf{w}$.

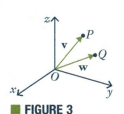

FIGURE 3

A characterization of a vector in terms of its length and direction only is called an **intrinsic** description of the vector. The point to note is that such a description does *not* depend on the choice of coordinate system in $\mathbb{R}^3$. Such descriptions are important in applications because physical laws are often stated in terms of vectors, and these laws cannot depend on the particular coordinate system used to describe the situation.

Geometric Vectors

If A and B are distinct points in space, the arrow from A to B has length and direction. Hence:

Definition 4.1

Suppose that A and B are any two points in $\mathbb{R}^3$. In Figure 4 the line segment from A to B is denoted $\overrightarrow{AB}$ and is called the **geometric vector** from A to B. Point A is called the **tail** of $\overrightarrow{AB}$, B is called the **tip** of $\overrightarrow{AB}$, and the **length** of $\overrightarrow{AB}$ is denoted $\|\overrightarrow{AB}\|$.

Note that if $\mathbf{v}$ is any vector in $\mathbb{R}^3$ with point P then $\mathbf{v} = \overrightarrow{OP}$ is itself a geometric vector where O is the origin. Referring to $\overrightarrow{AB}$ as a "vector" seems justified by Theorem 2 because it has a direction (from A to B) and a length $\|\overrightarrow{AB}\|$. However there appears to be a problem because two geometric vectors can have the same length and direction even if the tips and tails are different. For example $\overrightarrow{AB}$ and $\overrightarrow{PQ}$ in Figure 5 have the same length $\sqrt{5}$ and the same direction (1 unit left and 2 units up) so, by Theorem 2, they are the same vector! The best way to understand this apparent paradox is to see $\overrightarrow{AB}$ and $\overrightarrow{PQ}$ as different *representations* of the same underlying vector $\begin{bmatrix} -1 \\ 2 \end{bmatrix}$.[7] Once it is clarified, this phenomenon is a great benefit because, thanks to Theorem 2, it means that the same geometric vector can be positioned anywhere in space; what is important is the length and direction, not the location of the tip and tail. This ability to move geometric vectors about is very useful as we shall soon see.

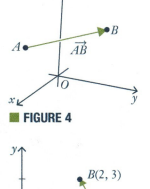

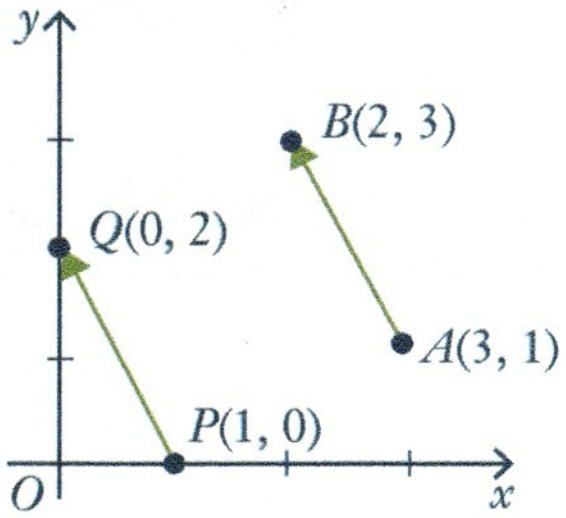

FIGURE 4

FIGURE 5

---

[6] It is Theorem 2 that gives vectors their power in science and engineering because many physical quantities are determined by their length and magnitude (and are called **vector quantities**). For example, saying that an airplane is flying at 200 km/h does not describe where it is going; the direction must also be specified. The speed and direction comprise the **velocity** of the airplane, a vector quantity.

[7] Fractions provide another example of quantities that can be the same but *look* different. For example $\frac{6}{9}$ and $\frac{14}{21}$ certainly appear different, but they are equal fractions—both equal $\frac{2}{3}$ in "lowest terms".

The Parallelogram Law

FIGURE 6

We now give an intrinsic description of the sum of two vectors **v** and **w** in $\mathbb{R}^3$, that is a description that depends only on the lengths and directions of **v** and **w** and not on the choice of coordinate system. Using Theorem 2 we can think of these vectors as having a common tail A. If their tips are P and Q respectively, then they both lie in a plane $\mathcal{P}$ containing A, P, and Q, as shown in Figure 6. The vectors **v** and **w** create a parallelogram[8] in $\mathcal{P}$, shaded in Figure 6, called the parallelogram **determined** by **v** and **w**.

If we now choose a coordinate system in the plane $\mathcal{P}$ with A as origin, then the parallelogram law in the plane (Section 2.6) shows that their sum $\mathbf{v} + \mathbf{w}$ is the diagonal of the parallelogram they determine with tail A. This is an intrinsic description of the sum $\mathbf{v} + \mathbf{w}$ because it makes no reference to coordinates. This discussion proves:

The Parallelogram Law

*In the parallelogram determined by two vectors **v** and **w**, the vector $\mathbf{v} + \mathbf{w}$ is the diagonal with the same tail as **v** and **w**.*

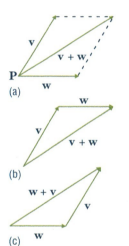

(a)
(b)
(c)

FIGURE 7

Because a vector can be positioned with its tail at any point, the parallelogram law leads to another way to view vector addition. In Figure 7(a) the sum $\mathbf{v} + \mathbf{w}$ of two vectors **v** and **w** is shown as given by the parallelogram law. If **w** is moved so its tail coincides with the tip of **v** (Figure 7(b)) then the sum $\mathbf{v} + \mathbf{w}$ is seen as "first **v** and then **w**. Similarly, moving the tail of **v** to the tip of **w** shows in Figure 7(c) that $\mathbf{v} + \mathbf{w}$ is "first **w** and then **v**." This will be referred to as the **tip-to-tail rule**, and it gives a graphic illustration of why $\mathbf{v} + \mathbf{w} = \mathbf{w} + \mathbf{v}$.

Since $\overrightarrow{AB}$ denotes the vector from a point A to a point B, the tip-to-tail rule takes the easily remembered form

$$\overrightarrow{AB} + \overrightarrow{BC} = \overrightarrow{AC}$$

for any points A, B, and C. The next example uses this to derive a theorem in geometry without using coordinates.

EXAMPLE 2

Show that the diagonals of a parallelogram bisect each other.

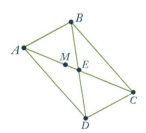

Solution ▶ Let the parallelogram have vertices A, B, C, and D, as shown; let E denote the intersection of the two diagonals; and let M denote the midpoint of diagonal AC. We must show that $M = E$ and that this is the midpoint of diagonal BD. This is accomplished by showing that $\overrightarrow{BM} = \overrightarrow{MD}$. (Then the fact that these vectors have the same direction means that $M = E$, and the fact that they have the same length means that $M = E$ is the midpoint of BD.) Now $\overrightarrow{AM} = \overrightarrow{MC}$ because M is the midpoint of AC, and $\overrightarrow{BA} = \overrightarrow{CD}$ because the figure is a parallelogram. Hence

$$\overrightarrow{BM} = \overrightarrow{BA} + \overrightarrow{AM} = \overrightarrow{CD} + \overrightarrow{MC} = \overrightarrow{MC} + \overrightarrow{CD} = \overrightarrow{MD}$$

where the first and last equalities use the tip-to-tail rule of vector addition.

---

[8] Recall that a parallelogram is a four-sided figure whose opposite sides are parallel and of equal length.

FIGURE 8

One reason for the importance of the tip-to-tail rule is that it means two or more vectors can be added by placing them tip-to-tail in sequence. This gives a useful "picture" of the sum of several vectors, and is illustrated for three vectors in Figure 8 where $\mathbf{u} + \mathbf{v} + \mathbf{w}$ is viewed as first $\mathbf{u}$, then $\mathbf{v}$, then $\mathbf{w}$.

There is a simple geometrical way to visualize the (matrix) **difference** $\mathbf{v} - \mathbf{w}$ of two vectors. If $\mathbf{v}$ and $\mathbf{w}$ are positioned so that they have a common tail A (see Figure 9), and if B and C are their respective tips, then the tip-to-tail rule gives $\mathbf{w} + \overrightarrow{CB} = \mathbf{v}$. Hence $\mathbf{v} - \mathbf{w} = \overrightarrow{CB}$ is the vector from the tip of $\mathbf{w}$ to the tip of $\mathbf{v}$. Thus both $\mathbf{v} - \mathbf{w}$ and $\mathbf{v} + \mathbf{w}$ appear as diagonals in the parallelogram determined by $\mathbf{v}$ and $\mathbf{w}$ (see Figure 9). We record this for reference.

Theorem 3

If $\mathbf{v}$ and $\mathbf{w}$ have a common tail, then $\mathbf{v} - \mathbf{w}$ is the vector from the tip of $\mathbf{w}$ to the tip of $\mathbf{v}$.

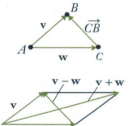

FIGURE 9

One of the most useful applications of vector subtraction is that it gives a simple formula for the vector from one point to another, and for the distance between the points.

Theorem 4

Let $P_1(x_1, y_1, z_1)$ and $P_2(x_2, y_2, z_2)$ be two points. Then:

1. $\overrightarrow{P_1P_2} = \begin{bmatrix} x_2 - x_1 \\ y_2 - y_1 \\ z_2 - z_1 \end{bmatrix}$.

2. The distance between P_1 and P_2 is $\sqrt{(x_2 - x_1)^2 + (y_2 - y_1)^2 + (z_2 - z_1)^2}$.

PROOF

If O is the origin, write $\mathbf{v}_1 = \overrightarrow{OP_1} = \begin{bmatrix} x_1 \\ y_1 \\ z_1 \end{bmatrix}$ and $\mathbf{v}_2 = \overrightarrow{OP_2} = \begin{bmatrix} x_2 \\ y_2 \\ z_2 \end{bmatrix}$ as in Figure 10. Then Theorem 3 gives $\overrightarrow{P_1P_2} = \mathbf{v}_2 - \mathbf{v}_1$, and (1) follows. But the distance between P_1 and P_2 is $\|\overrightarrow{P_1P_2}\|$, so (2) follows from (1) and Theorem 1.

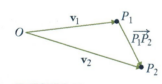

FIGURE 10

Of course the $\mathbb{R}^2$-version of Theorem 4 is also valid: If $P_1(x_1, y_1)$ and $P_2(x_2, y_2)$ are points in $\mathbb{R}^2$, then $\overrightarrow{P_1P_2} = \begin{bmatrix} x_2 - x_1 \\ y_2 - y_1 \end{bmatrix}$, and the distance between P_1 and P_2 is $\sqrt{(x_2 - x_1)^2 + (y_2 - y_1)^2}$.

EXAMPLE 3

The distance between $P_1(2, -1, 3)$ and $P_2(1, 1, 4)$ is $\sqrt{(-1)^2 + (2)^2 + (1)^2} = \sqrt{6}$, and the vector from P_1 to P_2 is $\overrightarrow{P_1P_2} = \begin{bmatrix} -1 \\ 2 \\ 1 \end{bmatrix}$.

SECTION 4.1　Vectors and Lines

As for the parallelogram law, the intrinsic rule for finding the length and direction of a scalar multiple of a vector in $\mathbb{R}^3$ follows easily from the same situation in $\mathbb{R}^2$.

Scalar Multiplication

Scalar Multiple Law
If a is a real number and $\mathbf{v} \neq \mathbf{0}$ is a vector then:

(1) The length of $a\mathbf{v}$ is $\|a\mathbf{v}\| = |a|\|\mathbf{v}\|$.

(2) If $a\mathbf{v} \neq \mathbf{0}$,[9] the direction of $a\mathbf{v}$ is $\begin{cases} \text{the same as } \mathbf{v} \text{ if } a > 0, \\ \text{opposite to } \mathbf{v} \text{ if } a < 0. \end{cases}$

PROOF

(1) This part of Theorem 1.

(2) Let O denote the origin in $\mathbb{R}^3$, let $\mathbf{v}$ have point P, and choose any plane containing O and P. If we set up a coordinate system in this plane with O as origin, then $\mathbf{v} = \overrightarrow{OP}$ so the result in (2) follows from the scalar multiple law in the plane (Section 2.6).

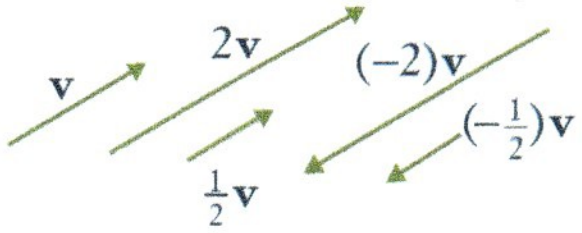

FIGURE 11

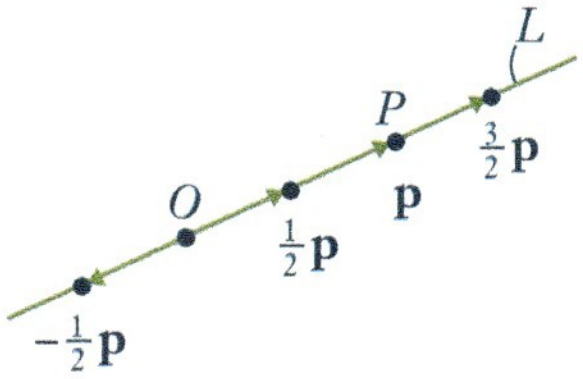

Figure 11 gives several examples of scalar multiples of a vector $\mathbf{v}$.

Consider a line L through the origin, let P be any point on L other than the origin O, and let $\mathbf{p} = \overrightarrow{OP}$. If $t \neq 0$, then $t\mathbf{p}$ is a point on L because it has direction the same or opposite as that of $\mathbf{p}$. Moreover $t > 0$ or $t < 0$ according as the point $t\mathbf{p}$ lies on the same or opposite side of the origin as P. This is illustrated in Figure 12.

A vector $\mathbf{u}$ is called a **unit vector** if $\|\mathbf{u}\| = 1$. Then $\mathbf{i} = \begin{bmatrix} 1 \\ 0 \\ 0 \end{bmatrix}, \mathbf{j} = \begin{bmatrix} 0 \\ 1 \\ 0 \end{bmatrix}$, and $\mathbf{k} = \begin{bmatrix} 0 \\ 0 \\ 1 \end{bmatrix}$ are unit vectors, called the **coordinate** vectors. We discuss them in more detail in Section 4.2.

FIGURE 12

EXAMPLE 4

If $\mathbf{v} \neq \mathbf{0}$ show that $\dfrac{1}{\|\mathbf{v}\|}\mathbf{v}$ is the unique unit vector in the same direction as $\mathbf{v}$.

Solution ▶ The vectors in the same direction as $\mathbf{v}$ are the scalar multiples $a\mathbf{v}$ where $a > 0$. But $\|a\mathbf{v}\| = |a|\|\mathbf{v}\| = a\|\mathbf{v}\|$ when $a > 0$, so $a\mathbf{v}$ is a unit vector if and only if $a = \dfrac{1}{\|\mathbf{v}\|}$.

The next example shows how to find the coordinates of a point on the line segment between two given points. The technique is important and will be used again below.

---

[9] Since the zero vector has no direction, we deal only with the case $a\mathbf{v} \neq \mathbf{0}$.

EXAMPLE 5

Let $\mathbf{p}_1$ and $\mathbf{p}_2$ be the vectors of two points P_1 and P_2. If M is the point one third the way from P_1 to P_2, show that the vector $\mathbf{m}$ of M is given by

$$\mathbf{m} = \tfrac{2}{3}\mathbf{p}_1 + \tfrac{1}{3}\mathbf{p}_2$$

Conclude that if $P_1 = P_1(x_1, y_1, z_1)$ and $P_2 = P_2(x_2, y_2, z_2)$, then M has coordinates

$$M = M(\tfrac{2}{3}x_1 + \tfrac{1}{3}x_2, \tfrac{2}{3}y_1 + \tfrac{1}{3}y_2, \tfrac{2}{3}z_1 + \tfrac{1}{3}z_2).$$

Solution ▶ The vectors $\mathbf{p}_1$, $\mathbf{p}_2$, and $\mathbf{m}$ are shown in the diagram. We have $\overrightarrow{P_1M} = \tfrac{1}{3}\overrightarrow{P_1P_2}$ because $\overrightarrow{P_1M}$ is in the same direction as $\overrightarrow{P_1P_2}$ and $\tfrac{1}{3}$ as long. By Theorem 3 we have $\overrightarrow{P_1P_2} = \mathbf{p}_2 - \mathbf{p}_1$, so tip-to-tail addition gives

$$\mathbf{m} = \mathbf{p}_1 + \overrightarrow{P_1M} = \mathbf{p}_1 + \tfrac{1}{3}(\mathbf{p}_2 - \mathbf{p}_1) = \tfrac{2}{3}\mathbf{p}_1 + \tfrac{1}{3}\mathbf{p}_2$$

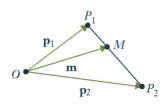

as required. For the coordinates, we have $\mathbf{p}_1 = \begin{bmatrix} x_1 \\ y_1 \\ z_1 \end{bmatrix}$ and $\mathbf{p}_2 = \begin{bmatrix} x_2 \\ y_2 \\ z_2 \end{bmatrix}$, so

$$\mathbf{m} = \tfrac{2}{3}\begin{bmatrix} x_1 \\ y_1 \\ z_1 \end{bmatrix} + \tfrac{1}{3}\begin{bmatrix} x_2 \\ y_2 \\ z_2 \end{bmatrix} = \begin{bmatrix} \tfrac{2}{3}x_1 + \tfrac{1}{3}x_2 \\ \tfrac{2}{3}y_1 + \tfrac{1}{3}y_2 \\ \tfrac{2}{3}z_1 + \tfrac{1}{3}z_2 \end{bmatrix}$$

by matrix addition. The last statement follows.

Note that in Example 5 $\mathbf{m} = \tfrac{2}{3}\mathbf{p}_1 + \tfrac{1}{3}\mathbf{p}_2$ is a "weighted average" of $\mathbf{p}_1$ and $\mathbf{p}_2$ with more weight on $\mathbf{p}_1$ because $\mathbf{m}$ is closer to $\mathbf{p}_1$.

The point M halfway between points P_1 and P_2 is called the **midpoint** between these points. In the same way, the vector $\mathbf{m}$ of M is

$$\mathbf{m} = \tfrac{1}{2}\mathbf{p}_1 + \tfrac{1}{2}\mathbf{p}_2 = \tfrac{1}{2}(\mathbf{p}_1 + \mathbf{p}_2)$$

as the reader can verify, so $\mathbf{m}$ is the "average" of $\mathbf{p}_1$ and $\mathbf{p}_2$ in this case.

EXAMPLE 6

Show that the midpoints of the four sides of any quadrilateral are the vertices of a parallelogram. Here a quadrilateral is any figure with four vertices and straight sides.

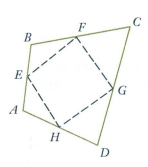

Solution ▶ Suppose that the vertices of the quadrilateral are A, B, C, and D (in that order) and that E, F, G, and H are the midpoints of the sides as shown in the diagram. It suffices to show $\overrightarrow{EF} = \overrightarrow{HG}$ (because then sides EF and HG are parallel and of equal length). Now the fact that E is the midpoint of AB means that $\overrightarrow{EB} = \tfrac{1}{2}\overrightarrow{AB}$. Similarly, $\overrightarrow{BF} = \tfrac{1}{2}\overrightarrow{BC}$, so

$$\overrightarrow{EF} = \overrightarrow{EB} + \overrightarrow{BF} = \tfrac{1}{2}\overrightarrow{AB} + \tfrac{1}{2}\overrightarrow{BC} = \tfrac{1}{2}(\overrightarrow{AB} + \overrightarrow{BC}) = \tfrac{1}{2}\overrightarrow{AC}$$

A similar argument shows that $\overrightarrow{HG} = \tfrac{1}{2}\overrightarrow{AC}$ too, so $\overrightarrow{EF} = \overrightarrow{HG}$ as required.

SECTION 4.1 Vectors and Lines

Definition 4.2 *Two nonzero vectors are called **parallel** if they have the same or opposite direction.*

Many geometrical propositions involve this notion, so the following theorem will be referred to repeatedly.

Theorem 5

*Two nonzero vectors **v** and **w** are parallel if and only if one is a scalar multiple of the other.*

PROOF

If one of them is a scalar multiple of the other, they are parallel by the scalar multiple law.

Conversely, assume that **v** and **w** are parallel and write $d = \frac{\|\mathbf{v}\|}{\|\mathbf{w}\|}$ for convenience. Then **v** and **w** have the same or opposite direction. If they have the same direction we show that $\mathbf{v} = d\mathbf{w}$ by showing that **v** and $d\mathbf{w}$ have the same length and direction. In fact, $\|d\mathbf{w}\| = |d|\, \|\mathbf{w}\| = \|\mathbf{v}\|$ by Theorem 1; as to the direction, $d\mathbf{w}$ and **w** have the same direction because $d > 0$, and this is the direction of **v** by assumption. Hence $\mathbf{v} = d\mathbf{w}$ in this case by Theorem 2. In the other case, **v** and **w** have opposite direction and a similar argument shows that $\mathbf{v} = -d\mathbf{w}$. We leave the details to the reader.

EXAMPLE 7

Given points $P(2, -1, 4)$, $Q(3, -1, 3)$, $A(0, 2, 1)$, and $B(1, 3, 0)$, determine if $\overrightarrow{PQ}$ and $\overrightarrow{AB}$ are parallel.

Solution ▶ By Theorem 3, $\overrightarrow{PQ} = (1, 0, -1)$ and $\overrightarrow{AB} = (1, 1, -1)$. If $\overrightarrow{PQ} = t\overrightarrow{AB}$ then $(1, 0, -1) = (t, t, -t)$, so $1 = t$ and $0 = t$, which is impossible. Hence $\overrightarrow{PQ}$ is *not* a scalar multiple of $\overrightarrow{AB}$, so these vectors are not parallel by Theorem 5.

Lines in Space

These vector techniques can be used to give a very simple way of describing straight lines in space. In order to do this, we first need a way to specify the orientation of such a line, much as the slope does in the plane.

Definition 4.3 *With this in mind, we call a nonzero vector $\mathbf{d} \neq \mathbf{0}$ a **direction vector** for the line if it is parallel to $\overrightarrow{AB}$ for some pair of distinct points A and B on the line.*

Of course it is then parallel to $\overrightarrow{CD}$ for *any* distinct points C and D on the line. In particular, any nonzero scalar multiple of **d** will also serve as a direction vector of the line.

We use the fact that there is exactly one line that passes through a particular point $P_0(x_0, y_0, z_0)$ and has a given direction vector $\mathbf{d} = \begin{bmatrix} a \\ b \\ c \end{bmatrix}$. We want to describe this line by giving a condition on x, y, and z that the point $P(x, y, z)$ lies on

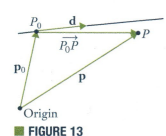
FIGURE 13

this line. Let $\mathbf{p}_0 = \begin{bmatrix} x_0 \\ y_0 \\ z_0 \end{bmatrix}$ and $\mathbf{p} = \begin{bmatrix} x \\ y \\ z \end{bmatrix}$ denote the vectors of P_0 and P, respectively (see Figure 13). Then

$$\mathbf{p} = \mathbf{p}_0 + \overrightarrow{P_0P}$$

Hence P lies on the line if and only if $\overrightarrow{P_0P}$ is parallel to $\mathbf{d}$—that is, if and only if $\overrightarrow{P_0P} = t\mathbf{d}$ for some scalar t by Theorem 5. Thus $\mathbf{p}$ is the vector of a point on the line if and only if $\mathbf{p} = \mathbf{p}_0 + t\mathbf{d}$ for some scalar t. This discussion is summed up as follows.

Vector Equation of a Line

The line parallel to $\mathbf{d} \neq \mathbf{0}$ through the point with vector $\mathbf{p}_0$ is given by

$$\mathbf{p} = \mathbf{p}_0 + t\mathbf{d} \quad t \text{ any scalar}$$

In other words, the point $\mathbf{p}$ is on this line if and only if a real number t exists such that $\mathbf{p} = \mathbf{p}_0 + t\mathbf{d}$.

In component form the vector equation becomes

$$\begin{bmatrix} x \\ y \\ z \end{bmatrix} = \begin{bmatrix} x_0 \\ y_0 \\ z_0 \end{bmatrix} + t \begin{bmatrix} a \\ b \\ c \end{bmatrix}$$

Equating components gives a different description of the line.

Parametric Equations of a Line

The line through $P_0(x_0, y_0, z_0)$ with direction vector $\mathbf{d} = \begin{bmatrix} a \\ b \\ c \end{bmatrix} \neq \mathbf{0}$ is given by

$$\begin{aligned} x &= x_0 + ta \\ y &= y_0 + tb \quad t \text{ any scalar} \\ z &= z_0 + tc \end{aligned}$$

In other words, the point $P(x, y, z)$ is on this line if and only if a real number t exists such that $x = x_0 + ta$, $y = y_0 + tb$, and $z = z_0 + tc$.

EXAMPLE 8

Find the equations of the line through the points $P_0(2, 0, 1)$ and $P_1(4, -1, 1)$.

Solution ▶ Let $\mathbf{d} = \overrightarrow{P_0P_1} = \begin{bmatrix} 2 \\ 1 \\ 0 \end{bmatrix}$ denote the vector from P_0 to P_1. Then $\mathbf{d}$ is parallel to the line (P_0 and P_1 are *on* the line), so $\mathbf{d}$ serves as a direction vector for the line. Using P_0 as the point on the line leads to the parametric equations

$$\begin{aligned} x &= 2 + 2t \\ y &= -t \quad t \text{ a parameter} \\ z &= 1 \end{aligned}$$

Note that if P_1 is used (rather than P_0), the equations are

SECTION 4.1 Vectors and Lines

$$x = 4 + 2s$$
$$y = -1 - s \quad s \text{ a parameter}$$
$$z = 1$$

These are different from the preceding equations, but this is merely the result of a change of parameter. In fact, $s = t - 1$.

EXAMPLE 9

Find the equations of the line through $P_0(3, -1, 2)$ parallel to the line with equations

$$x = -1 + 2t$$
$$y = 1 + t$$
$$z = -3 + 4t$$

Solution ▶ The coefficients of t give a direction vector $\mathbf{d} = \begin{bmatrix} 2 \\ 1 \\ 4 \end{bmatrix}$ of the given line. Because the line we seek is parallel to this line, $\mathbf{d}$ also serves as a direction vector for the new line. It passes through P_0, so the parametric equations are

$$x = 3 + 2t$$
$$y = -1 + t$$
$$z = 2 + 4t$$

EXAMPLE 10

Determine whether the following lines intersect and, if so, find the point of intersection.

$$\begin{array}{ll} x = 1 - 3t & x = -1 + s \\ y = 2 + 5t & y = 3 - 4s \\ z = 1 + t & z = 1 - s \end{array}$$

Solution ▶ Suppose $\mathbf{p} = P(x, y, z)$ lies on both lines. Then

$$\begin{bmatrix} 1 - 3t \\ 2 + 5t \\ 1 + t \end{bmatrix} = \begin{bmatrix} x \\ y \\ z \end{bmatrix} = \begin{bmatrix} -1 + s \\ 3 - 4s \\ 1 - s \end{bmatrix} \text{ for some } t \text{ and } s,$$

where the first (second) equation is because P lies on the first (second) line. Hence the lines intersect if and only if the three equations

$$1 - 3t = -1 + s$$
$$2 + 5t = 3 - 4s$$
$$1 + t = 1 - s$$

have a solution. In this case, $t = 1$ and $s = -1$ satisfy all three equations, so the lines *do* intersect and the point of intersection is

$$\mathbf{p} = \begin{bmatrix} 1 - 3t \\ 2 + 5t \\ 1 + t \end{bmatrix} = \begin{bmatrix} -2 \\ 7 \\ 2 \end{bmatrix}$$

using $t = 1$. Of course, this point can also be found from $\mathbf{p} = \begin{bmatrix} -1 + s \\ 3 - 4s \\ 1 - s \end{bmatrix}$ using $s = -1$.

EXAMPLE 11

Show that the line through $P_0(x_0, y_0)$ with slope m has direction vector $\mathbf{d} = \begin{bmatrix} 1 \\ m \end{bmatrix}$ and equation $y - y_0 = m(x - x_0)$. This equation is called the *point-slope* formula.

Solution ▶ Let $P_1(x_1, y_1)$ be the point on the line one unit to the right of P_0 (see the diagram). Hence $x_1 = x_0 + 1$. Then $\mathbf{d} = \overrightarrow{P_0P_1}$ serves as direction vector of the line, and $\mathbf{d} = \begin{bmatrix} x_1 - x_0 \\ y_1 - y_0 \end{bmatrix} = \begin{bmatrix} 1 \\ y_1 - y_0 \end{bmatrix}$. But the slope m can be computed as follows:

$$m = \frac{y_1 - y_0}{x_1 - x_0} = \frac{y_1 - y_0}{1} = y_1 - y_0$$

Hence $\mathbf{d} = \begin{bmatrix} 1 \\ m \end{bmatrix}$ and the parametric equations are $x = x_0 + t, y = y_0 + mt$. Eliminating t gives $y - y_0 = mt = m(x - x_0)$, as asserted.

Note that the vertical line through $P_0(x_0, y_0)$ has a direction vector $\mathbf{d} = \begin{bmatrix} 0 \\ 1 \end{bmatrix}$ that is *not* of the form $\begin{bmatrix} 1 \\ m \end{bmatrix}$ for any m. This result confirms that the notion of slope makes no sense in this case. However, the vector method gives parametric equations for the line:

$$x = x_0$$
$$y = y_0 + t$$

Because y is arbitrary here (t is arbitrary), this is usually written simply as $x = x_0$.

Pythagoras' Theorem

The pythagorean theorem was known earlier, but Pythagoras (c. 550 B.C.) is credited with giving the first rigorous, logical, deductive proof of the result. The proof we give depends on a basic property of similar triangles: ratios of corresponding sides are equal.

Theorem 6

Pythagoras' Theorem
Given a right-angled triangle with hypotenuse c and sides a and b, then $a^2 + b^2 = c^2$.

PROOF

Let A, B, and C be the vertices of the triangle as in Figure 14. Draw a perpendicular from C to the point D on the hypotenuse, and let p and q be the lengths of BD and DA respectively. Then DBC and CBA are similar triangles so $\frac{p}{a} = \frac{a}{c}$.

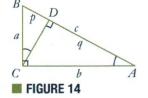
FIGURE 14

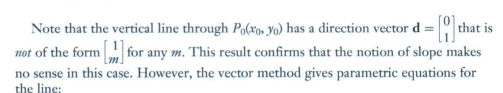

SECTION 4.1 Vectors and Lines

This means $a^2 = pc$. In the same way, the similarity of DCA and CBA gives $\frac{q}{b} = \frac{b}{c}$, whence $b^2 = qc$. But then

$$a^2 + b^2 = pc + qc = (p + q)c = c^2$$

because $p + q = c$. This proves Pythagoras' theorem.

EXERCISES 4.1

1. Compute $\|\mathbf{v}\|$ if $\mathbf{v}$ equals:

   (a) $\begin{bmatrix} 2 \\ -1 \\ 2 \end{bmatrix}$
   ◆(b) $\begin{bmatrix} 1 \\ -1 \\ 2 \end{bmatrix}$

   (c) $\begin{bmatrix} 1 \\ 0 \\ -1 \end{bmatrix}$
   ◆(d) $\begin{bmatrix} -1 \\ 0 \\ 2 \end{bmatrix}$

   (e) $2\begin{bmatrix} 1 \\ -1 \\ 2 \end{bmatrix}$
   ◆(f) $-3\begin{bmatrix} 1 \\ 1 \\ 2 \end{bmatrix}$

2. Find a unit vector in the direction of:

   (a) $\begin{bmatrix} 7 \\ -1 \\ 5 \end{bmatrix}$
   ◆(b) $\begin{bmatrix} -2 \\ -1 \\ 2 \end{bmatrix}$

3. (a) Find a unit vector in the direction from $\begin{bmatrix} 3 \\ -1 \\ 4 \end{bmatrix}$ to $\begin{bmatrix} 1 \\ 3 \\ 5 \end{bmatrix}$.

   (b) If $\mathbf{u} \neq \mathbf{0}$, for which values of a is $a\mathbf{u}$ a unit vector?

4. Find the distance between the following pairs of points.

   (a) $\begin{bmatrix} 3 \\ -1 \\ 0 \end{bmatrix}$ and $\begin{bmatrix} 2 \\ -1 \\ 1 \end{bmatrix}$
   ◆(b) $\begin{bmatrix} 2 \\ -1 \\ 2 \end{bmatrix}$ and $\begin{bmatrix} 2 \\ 0 \\ 1 \end{bmatrix}$

   (c) $\begin{bmatrix} -3 \\ 5 \\ 2 \end{bmatrix}$ and $\begin{bmatrix} 1 \\ 3 \\ 3 \end{bmatrix}$
   ◆(d) $\begin{bmatrix} 4 \\ 0 \\ -2 \end{bmatrix}$ and $\begin{bmatrix} 3 \\ 2 \\ 0 \end{bmatrix}$

5. Use vectors to show that the line joining the midpoints of two sides of a triangle is parallel to the third side and half as long.

6. Let A, B, and C denote the three vertices of a triangle.

   (a) If E is the midpoint of side BC, show that
   $$\overrightarrow{AE} = \tfrac{1}{2}(\overrightarrow{AB} + \overrightarrow{AC}).$$

   ◆(b) If F is the midpoint of side AC, show that
   $$\overrightarrow{FE} = \tfrac{1}{2}\overrightarrow{AB}.$$

7. Determine whether $\mathbf{u}$ and $\mathbf{v}$ are parallel in each of the following cases.

   (a) $\mathbf{u} = \begin{bmatrix} -3 \\ -6 \\ 3 \end{bmatrix}$; $\mathbf{v} = \begin{bmatrix} 5 \\ 10 \\ -5 \end{bmatrix}$

   ◆(b) $\mathbf{u} = \begin{bmatrix} 3 \\ -6 \\ 3 \end{bmatrix}$; $\mathbf{v} = \begin{bmatrix} -1 \\ 2 \\ -1 \end{bmatrix}$

   (c) $\mathbf{u} = \begin{bmatrix} 1 \\ 0 \\ 1 \end{bmatrix}$; $\mathbf{v} = \begin{bmatrix} -1 \\ 0 \\ 1 \end{bmatrix}$

   ◆(d) $\mathbf{u} = \begin{bmatrix} 2 \\ 0 \\ -1 \end{bmatrix}$; $\mathbf{v} = \begin{bmatrix} -8 \\ 0 \\ 4 \end{bmatrix}$

8. Let $\mathbf{p}$ and $\mathbf{q}$ be the vectors of points P and Q, respectively, and let R be the point whose vector is $\mathbf{p} + \mathbf{q}$. Express the following in terms of $\mathbf{p}$ and $\mathbf{q}$.

   (a) $\overrightarrow{QP}$
   ◆(b) $\overrightarrow{QR}$

   (c) $\overrightarrow{RP}$
   ◆(d) $\overrightarrow{RO}$ where O is the origin

9. In each case, find $\overrightarrow{PQ}$ and $\|\overrightarrow{PQ}\|$.

   (a) $P(1, -1, 3)$, $Q(3, 1, 0)$
   ◆(b) $P(2, 0, 1)$, $Q(1, -1, 6)$
   (c) $P(1, 0, 1)$, $Q(1, 0, -3)$
   ◆(d) $P(1, -1, 2)$, $Q(1, -1, 2)$
   (e) $P(1, 0, -3)$, $Q(-1, 0, 3)$
   ◆(f) $P(3, -1, 6)$, $Q(1, 1, 4)$

10. In each case, find a point Q such that $\overrightarrow{PQ}$ has (i) the same direction as $\mathbf{v}$; (ii) the opposite direction to $\mathbf{v}$.

    (a) $P(-1, 2, 2)$, $\mathbf{v} = \begin{bmatrix} 1 \\ 3 \\ 1 \end{bmatrix}$

    ◆(b) $P(3, 0, -1)$, $\mathbf{v} = \begin{bmatrix} 2 \\ -1 \\ 3 \end{bmatrix}$

11. Let $\mathbf{u} = \begin{bmatrix} 3 \\ -1 \\ 0 \end{bmatrix}$, $\mathbf{v} = \begin{bmatrix} 4 \\ 0 \\ 1 \end{bmatrix}$, and $\mathbf{w} = \begin{bmatrix} -1 \\ 1 \\ 5 \end{bmatrix}$.

    In each case, find $\mathbf{x}$ such that:

    (a) $3(2\mathbf{u} + \mathbf{x}) + \mathbf{w} = 2\mathbf{x} - \mathbf{v}$

    ◆(b) $2(3\mathbf{v} - \mathbf{x}) = 5\mathbf{w} + \mathbf{u} - 3\mathbf{x}$

12. Let $\mathbf{u} = \begin{bmatrix} 1 \\ 1 \\ 2 \end{bmatrix}$, $\mathbf{v} = \begin{bmatrix} 0 \\ 1 \\ 2 \end{bmatrix}$, and $\mathbf{w} = \begin{bmatrix} 1 \\ 0 \\ -1 \end{bmatrix}$. In each case, find numbers a, b, and c such that $\mathbf{x} = a\mathbf{u} + b\mathbf{v} + c\mathbf{w}$.

    (a) $\mathbf{x} = \begin{bmatrix} 2 \\ -1 \\ 6 \end{bmatrix}$ ◆(b) $\mathbf{x} = \begin{bmatrix} 1 \\ 3 \\ 0 \end{bmatrix}$

13. Let $\mathbf{u} = \begin{bmatrix} 3 \\ -1 \\ 0 \end{bmatrix}$, $\mathbf{v} = \begin{bmatrix} 4 \\ 0 \\ 1 \end{bmatrix}$, and $\mathbf{z} = \begin{bmatrix} 1 \\ 1 \\ 1 \end{bmatrix}$. In each case, show that there are no numbers a, b, and c such that:

    (a) $a\mathbf{u} + b\mathbf{v} + c\mathbf{z} = \begin{bmatrix} 1 \\ 2 \\ 1 \end{bmatrix}$

    (b) $a\mathbf{u} + b\mathbf{v} + c\mathbf{z} = \begin{bmatrix} 5 \\ 6 \\ -1 \end{bmatrix}$

14. Let $P_1 = P_1(2, 1, -2)$ and $P_2 = P_2(1, -2, 0)$. Find the coordinates of the point P:

    (a) $\frac{1}{5}$ the way from P_1 to P_2

    ◆(b) $\frac{1}{4}$ the way from P_2 to P_1

15. Find the two points trisecting the segment between $P(2, 3, 5)$ and $Q(8, -6, 2)$.

16. Let $P_1 = P_1(x_1, y_1, z_1)$ and $P_2 = P_2(x_2, y_2, z_2)$ be two points with vectors $\mathbf{p}_1$ and $\mathbf{p}_2$, respectively. If r and s are positive integers, show that the point P lying $\frac{r}{r+s}$ the way from P_1 to P_2 has vector

    $$\mathbf{p} = \left(\frac{s}{r+s}\right)\mathbf{p}_1 + \left(\frac{r}{r+s}\right)\mathbf{p}_2.$$

17. In each case, find the point Q:

    (a) $\overrightarrow{PQ} = \begin{bmatrix} 2 \\ 0 \\ -3 \end{bmatrix}$ and $P = P(2, -3, 1)$

    ◆(b) $\overrightarrow{PQ} = \begin{bmatrix} -1 \\ 4 \\ 7 \end{bmatrix}$ and $P = P(1, 3, -4)$

18. Let $\mathbf{u} = \begin{bmatrix} 2 \\ 0 \\ -4 \end{bmatrix}$ and $\mathbf{v} = \begin{bmatrix} 2 \\ 1 \\ -2 \end{bmatrix}$. In each case find $\mathbf{x}$:

    (a) $2\mathbf{u} - \|\mathbf{v}\|\mathbf{v} = \frac{3}{2}(\mathbf{u} - 2\mathbf{x})$

    ◆(b) $3\mathbf{u} + 7\mathbf{v} = \|\mathbf{u}\|^2(2\mathbf{x} + \mathbf{v})$

19. Find all vectors $\mathbf{u}$ that are parallel to $\mathbf{v} = \begin{bmatrix} 3 \\ -2 \\ 1 \end{bmatrix}$ and satisfy $\|\mathbf{u}\| = 3\|\mathbf{v}\|$.

20. Let P, Q, and R be the vertices of a parallelogram with adjacent sides PQ and PR. In each case, find the other vertex S.

    (a) $P(3, -1, -1)$, $Q(1, -2, 0)$, $R(1, -1, 2)$

    ◆(b) $P(2, 0, -1)$, $Q(-2, 4, 1)$, $R(3, -1, 0)$

21. In each case either prove the statement or give an example showing that it is false.

    (a) The zero vector $\mathbf{0}$ is the only vector of length 0.

    ◆(b) If $\|\mathbf{v} - \mathbf{w}\| = 0$, then $\mathbf{v} = \mathbf{w}$.

    (c) If $\mathbf{v} = -\mathbf{v}$, then $\mathbf{v} = \mathbf{0}$.

    ◆(d) If $\|\mathbf{v}\| = \|\mathbf{w}\|$, then $\mathbf{v} = \mathbf{w}$.

    (e) If $\|\mathbf{v}\| = \|\mathbf{w}\|$, then $\mathbf{v} = \pm\mathbf{w}$.

    ◆(f) If $\mathbf{v} = t\mathbf{w}$ for some scalar t, then $\mathbf{v}$ and $\mathbf{w}$ have the same direction.

    (g) If $\mathbf{v}$, $\mathbf{w}$, and $\mathbf{v} + \mathbf{w}$ are nonzero, and $\mathbf{v}$ and $\mathbf{v} + \mathbf{w}$ parallel, then $\mathbf{v}$ and $\mathbf{w}$ are parallel.

    ◆(h) $\|-5\mathbf{v}\| = -5\|\mathbf{v}\|$, for all $\mathbf{v}$.

    (i) If $\|\mathbf{v}\| = \|2\mathbf{v}\|$, then $\mathbf{v} = \mathbf{0}$.

    ◆(j) $\|\mathbf{v} + \mathbf{w}\| = \|\mathbf{v}\| + \|\mathbf{w}\|$, for all $\mathbf{v}$ and $\mathbf{w}$.

22. Find the vector and parametric equations of the following lines.

    (a) The line parallel to $\begin{bmatrix} 2 \\ -1 \\ 0 \end{bmatrix}$ and passing through $P(1, -1, 3)$.

    ◆(b) The line passing through $P(3, -1, 4)$ and $Q(1, 0, -1)$.

    (c) The line passing through $P(3, -1, 4)$ and $Q(3, -1, 5)$.

    ◆(d) The line parallel to $\begin{bmatrix} 1 \\ 1 \\ 1 \end{bmatrix}$ and passing through $P(1, 1, 1)$.

(e) The line passing through $P(1, 0, -3)$ and parallel to the line with parametric equations $x = -1 + 2t, y = 2 - t,$ and $z = 3 + 3t$.

◆(f) The line passing through $P(2, -1, 1)$ and parallel to the line with parametric equations $x = 2 - t, y = 1,$ and $z = t$.

(g) The lines through $P(1, 0, 1)$ that meet the line with vector equation $\mathbf{p} = \begin{bmatrix} 1 \\ 2 \\ 0 \end{bmatrix} + t \begin{bmatrix} 2 \\ -1 \\ 2 \end{bmatrix}$ at points at distance 3 from $P_0(1, 2, 0)$.

23. In each case, verify that the points P and Q lie on the line.

(a) $x = 3 - 4t$ $P(-1, 3, 0), Q(11, 0, 3)$
$y = 2 + t$
$z = 1 - t$

◆(b) $x = 4 - t$ $P(2, 3, -3), Q(-1, 3, -9)$
$y = 3$
$z = 1 - 2t$

24. Find the point of intersection (if any) of the following pairs of lines.

(a) $x = 3 + t$ $x = 4 + 2s$
$y = 1 - 2t$ $y = 6 + 3s$
$z = 3 + 3t$ $z = 1 + s$

◆(b) $x = 1 - t$ $x = 2s$
$y = 2 + 2t$ $y = 1 + s$
$z = -1 + 3t$ $z = 3$

(c) $\begin{bmatrix} x \\ y \\ z \end{bmatrix} = \begin{bmatrix} 3 \\ -1 \\ 2 \end{bmatrix} + t \begin{bmatrix} 1 \\ 1 \\ -1 \end{bmatrix}$

$\begin{bmatrix} x \\ y \\ z \end{bmatrix} = \begin{bmatrix} 1 \\ 1 \\ -2 \end{bmatrix} + s \begin{bmatrix} 2 \\ 0 \\ 3 \end{bmatrix}$

◆(d) $\begin{bmatrix} x \\ y \\ z \end{bmatrix} = \begin{bmatrix} 4 \\ -1 \\ 5 \end{bmatrix} + t \begin{bmatrix} 1 \\ 0 \\ 1 \end{bmatrix}$

$\begin{bmatrix} x \\ y \\ z \end{bmatrix} = \begin{bmatrix} 2 \\ -7 \\ 12 \end{bmatrix} + s \begin{bmatrix} 0 \\ -2 \\ 3 \end{bmatrix}$

25. Show that if a line passes through the origin, the vectors of points on the line are all scalar multiples of some fixed nonzero vector.

26. Show that every line parallel to the z axis has parametric equations $x = x_0, y = y_0, z = t$ for some fixed numbers x_0 and y_0.

27. Let $\mathbf{d} = \begin{bmatrix} a \\ b \\ c \end{bmatrix}$ be a vector where a, b, and c are *all* nonzero. Show that the equations of the line through $P_0(x_0, y_0, z_0)$ with direction vector $\mathbf{d}$ can be written in the form
$$\frac{x - x_0}{a} = \frac{y - y_0}{b} = \frac{z - z_0}{c}$$
This is called the **symmetric form** of the equations.

28. A parallelogram has sides AB, BC, CD, and DA. Given $A(1, -1, 2)$, $C(2, 1, 0)$, and the midpoint $M(1, 0, -3)$ of AB, find $\overrightarrow{BD}$.

◆29. Find all points C on the line through $A(1, -1, 2)$ and $B = (2, 0, 1)$ such that $\|\overrightarrow{AC}\| = 2\|\overrightarrow{BC}\|$.

30. Let A, B, C, D, E, and F be the vertices of a regular hexagon, taken in order. Show that $\overrightarrow{AB} + \overrightarrow{AC} + \overrightarrow{AD} + \overrightarrow{AE} + \overrightarrow{AF} = 3\overrightarrow{AD}$.

31. (a) Let $P_1, P_2, P_3, P_4, P_5,$ and P_6 be six points equally spaced on a circle with centre C. Show that
$$\overrightarrow{CP_1} + \overrightarrow{CP_2} + \overrightarrow{CP_3} + \overrightarrow{CP_4} + \overrightarrow{CP_5} + \overrightarrow{CP_6} = \mathbf{0}.$$

◆(b) Show that the conclusion in part (a) holds for any *even* set of points evenly spaced on the circle.

(c) Show that the conclusion in part (a) holds for *three* points.

(d) Do you think it works for *any* finite set of points evenly spaced around the circle?

32. Consider a quadrilateral with vertices A, B, C, and D in order (as shown in the diagram).

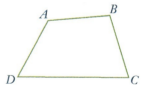

If the diagonals AC and BD bisect each other, show that the quadrilateral is a parallelogram. (This is the converse of Example 2.) [*Hint*: Let E be the intersection of the diagonals. Show that $\overrightarrow{AB} = \overrightarrow{DC}$ by writing $\overrightarrow{AB} = \overrightarrow{AE} + \overrightarrow{EB}$.]

◆33. Consider the parallelogram $ABCD$ (see diagram), and let E be the midpoint of side AD.

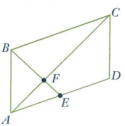

Show that BE and AC trisect each other; that is, show that the intersection point is one-third of the way from E to B and from A to C. [*Hint*: If F is one-third of the way from A to C, show that $2\overrightarrow{EF} = \overrightarrow{FB}$ and argue as in Example 2.]

34. The line from a vertex of a triangle to the midpoint of the opposite side is called a **median** of the triangle. If the vertices of a triangle have vectors $\mathbf{u}$, $\mathbf{v}$, and $\mathbf{w}$, show that the point on each median that is $\frac{1}{3}$ the way from the midpoint to the vertex has vector $\frac{1}{3}(\mathbf{u} + \mathbf{v} + \mathbf{w})$. Conclude that the point C with vector $\frac{1}{3}(\mathbf{u} + \mathbf{v} + \mathbf{w})$ lies on all three medians. This point C is called the **centroid** of the triangle.

35. Given four noncoplanar points in space, the figure with these points as vertices is called a **tetrahedron**. The line from a vertex through the centroid (see previous exercise) of the triangle formed by the remaining vertices is called a **median** of the tetrahedron. If $\mathbf{u}$, $\mathbf{v}$, $\mathbf{w}$, and $\mathbf{x}$ are the vectors of the four vertices, show that the point on a median one-fourth the way from the centroid to the vertex has vector $\frac{1}{4}(\mathbf{u} + \mathbf{v} + \mathbf{w} + \mathbf{x})$. Conclude that the four medians are concurrent.

SECTION 4.2 Projections and Planes

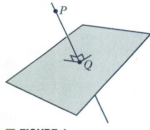

FIGURE 1

Any student of geometry soon realizes that the notion of perpendicular lines is fundamental. As an illustration, suppose a point P and a plane are given and it is desired to find the point Q that lies in the plane and is closest to P, as shown in Figure 1. Clearly, what is required is to find the line through P that is perpendicular to the plane and then to obtain Q as the point of intersection of this line with the plane. Finding the line *perpendicular* to the plane requires a way to determine when two vectors are perpendicular. This can be done using the idea of the dot product of two vectors.

The Dot Product and Angles

Definition 4.4   Given vectors $\mathbf{v} = \begin{bmatrix} x_1 \\ y_1 \\ z_1 \end{bmatrix}$ and $\mathbf{w} = \begin{bmatrix} x_2 \\ y_2 \\ z_2 \end{bmatrix}$, their **dot product** $\mathbf{v} \cdot \mathbf{w}$ is a number defined

$$\mathbf{v} \cdot \mathbf{w} = x_1 x_2 + y_1 y_2 + z_1 z_2 = \mathbf{v}^T \mathbf{w}$$

Because $\mathbf{v} \cdot \mathbf{w}$ is a number, it is sometimes called the **scalar product** of $\mathbf{v}$ and $\mathbf{w}$.[10]

EXAMPLE 1

If $\mathbf{v} = \begin{bmatrix} 2 \\ -1 \\ 3 \end{bmatrix}$ and $\mathbf{w} = \begin{bmatrix} 1 \\ 4 \\ -1 \end{bmatrix}$, then $\mathbf{v} \cdot \mathbf{w} = 2 \cdot 1 + (-1) \cdot 4 + 3 \cdot (-1) = -5$.

The next theorem lists several basic properties of the dot product.

---

10 Similarly, if $\mathbf{v} = \begin{bmatrix} x_1 \\ y_1 \end{bmatrix}$ and $\mathbf{w} = \begin{bmatrix} x_2 \\ y_2 \end{bmatrix}$ in $\mathbb{R}^2$, then $\mathbf{v} \cdot \mathbf{w} = x_1 x_2 + y_1 y_2$.

SECTION 4.2 Projections and Planes

Theorem 1

Let **u**, **v**, *and* **w** *denote vectors in* $\mathbb{R}^3$ *(or* $\mathbb{R}^2$*).*

1. $\mathbf{v} \cdot \mathbf{w}$ *is a real number.*
2. $\mathbf{v} \cdot \mathbf{w} = \mathbf{w} \cdot \mathbf{v}$.
3. $\mathbf{v} \cdot \mathbf{0} = 0 = \mathbf{0} \cdot \mathbf{v}$.
4. $\mathbf{v} \cdot \mathbf{v} = \|\mathbf{v}\|^2$.
5. $(k\mathbf{v}) \cdot \mathbf{w} = k(\mathbf{w} \cdot \mathbf{v}) = \mathbf{v} \cdot (k\mathbf{w})$ *for all scalars* k.
6. $\mathbf{u} \cdot (\mathbf{v} \pm \mathbf{w}) = \mathbf{u} \cdot \mathbf{v} \pm \mathbf{u} \cdot \mathbf{w}$

PROOF

(1), (2), and (3) are easily verified, and (4) comes from Theorem 1 Section 4.1. The rest are properties of matrix arithmetic (because $\mathbf{w} \cdot \mathbf{v} = \mathbf{v}^T \mathbf{w}$, and are left to the reader.

The properties in Theorem 1 enable us to do calculations like
$$3\mathbf{u} \cdot (2\mathbf{v} - 3\mathbf{w} + 4\mathbf{z}) = 6(\mathbf{u} \cdot \mathbf{v}) - 9(\mathbf{u} \cdot \mathbf{w}) + 12(\mathbf{u} \cdot \mathbf{z})$$
and such computations will be used without comment below. Here is an example.

EXAMPLE 2

Verify that $\|\mathbf{v} - 3\mathbf{w}\|^2 = 1$ when $\|\mathbf{v}\| = 2$, $\|\mathbf{w}\| = 1$, and $\mathbf{v} \cdot \mathbf{w} = 2$.

Solution ▶ We apply Theorem 1 several times:
$$\begin{aligned}\|\mathbf{v} - 3\mathbf{w}\|^2 &= (\mathbf{v} - 3\mathbf{w}) \cdot (\mathbf{v} - 3\mathbf{w}) \\ &= \mathbf{v} \cdot (\mathbf{v} - 3\mathbf{w}) - 3\mathbf{w} \cdot (\mathbf{v} - 3\mathbf{w}) \\ &= \mathbf{v} \cdot \mathbf{v} - 3(\mathbf{v} \cdot \mathbf{w}) - 3(\mathbf{w} \cdot \mathbf{v}) + 9(\mathbf{w} \cdot \mathbf{w}) \\ &= \|\mathbf{v}\|^2 - 6(\mathbf{v} \cdot \mathbf{w}) + 9\|\mathbf{v}\|^2 \\ &= 4 - 12 + 9 = 1.\end{aligned}$$

There is an intrinsic description of the dot product of two nonzero vectors in $\mathbb{R}^3$. To understand it we require the following result from trigonometry.

Law of Cosines

If a triangle has sides a, b, *and* c, *and if* θ *is the interior angle opposite* c *then*
$$c^2 = a^2 + b^2 - 2ab \cos \theta.$$

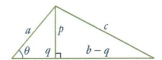

FIGURE 2

> **PROOF**
>
> We prove it when is θ acute, that is $0 \leq \theta < \frac{\pi}{2}$; the obtuse case is similar. In Figure 2 we have $p = a \sin \theta$ and $q = a \cos \theta$. Hence Pythagoras' theorem gives
> $$c^2 = p^2 + (b-q)^2 = a^2 \sin^2 \theta + (b - a\cos\theta)^2$$
> $$= a^2(\sin^2 \theta + \cos^2 \theta) + b^2 - 2ab \cos \theta.$$
> The law of cosines follows because $\sin^2 \theta + \cos^2 \theta = 1$ for any angle θ.

Note that the law of cosines reduces to Pythagoras' theorem if θ is a right angle (because $\cos \frac{\pi}{2} = 0$).

Now let $\mathbf{v}$ and $\mathbf{w}$ be nonzero vectors positioned with a common tail as in Figure 3. Then they determine a unique angle θ in the range

$$0 \leq \theta \leq \pi$$

This angle θ will be called the **angle between** $\mathbf{v}$ and $\mathbf{w}$. Figure 2 illustrates when θ is acute (less than $\frac{\pi}{2}$) and obtuse (greater than $\frac{\pi}{2}$). Clearly $\mathbf{v}$ and $\mathbf{w}$ are parallel if θ is either 0 or π. Note that we do not define the angle between $\mathbf{v}$ and $\mathbf{w}$ if one of these vectors is $\mathbf{0}$.

The next result gives an easy way to compute the angle between two nonzero vectors using the dot product.

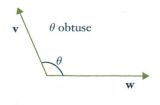

FIGURE 3

> **Theorem 2**
>
> Let $\mathbf{v}$ and $\mathbf{w}$ be nonzero vectors. If θ is the angle between $\mathbf{v}$ and $\mathbf{w}$, then
> $$\mathbf{v} \cdot \mathbf{w} = \|\mathbf{v}\|\|\mathbf{w}\|\cos \theta$$

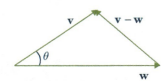

FIGURE 4

> **PROOF**
>
> We calculate $\|\mathbf{v} - \mathbf{w}\|^2$ in two ways. First apply the law of cosines to the triangle in Figure 4 to obtain:
> $$\|\mathbf{v} - \mathbf{w}\|^2 = \|\mathbf{v}\|^2 + \|\mathbf{w}\|^2 - 2\|\mathbf{v}\|\|\mathbf{w}\|\cos \theta$$
> On the other hand, we use Theorem 1:
> $$\|\mathbf{v} - \mathbf{w}\|^2 = (\mathbf{v} - \mathbf{w}) \cdot (\mathbf{v} - \mathbf{w})$$
> $$= \mathbf{v} \cdot \mathbf{v} - \mathbf{v} \cdot \mathbf{w} - \mathbf{w} \cdot \mathbf{v} + \mathbf{w} \cdot \mathbf{w}$$
> $$= \|\mathbf{v}\|^2 - 2(\mathbf{v} \cdot \mathbf{w}) + \|\mathbf{w}\|^2$$
> Comparing these we see that $-2\|\mathbf{v}\|\|\mathbf{w}\|\cos \theta = -2(\mathbf{v} \cdot \mathbf{w})$, and the result follows.

If $\mathbf{v}$ and $\mathbf{w}$ are nonzero vectors, Theorem 2 gives an intrinsic description of $\mathbf{v} \cdot \mathbf{w}$ because $\|\mathbf{v}\|$, $\|\mathbf{w}\|$, and the angle θ between $\mathbf{v}$ and $\mathbf{w}$ do not depend on the choice of coordinate system. Moreover, since $\|\mathbf{v}\|$ and $\|\mathbf{v}\|$ are nonzero ($\mathbf{v}$ and $\mathbf{w}$ are nonzero vectors), it gives a formula for the cosine of the angle θ:

$$\cos \theta = \frac{\mathbf{v} \cdot \mathbf{w}}{\|\mathbf{v}\|\|\mathbf{w}\|} \quad (*)$$

Since $0 \leq \theta \leq \pi$, this can be used to find θ.

SECTION 4.2 Projections and Planes

EXAMPLE 3

Compute the angle between $\mathbf{u} = \begin{bmatrix} -1 \\ 1 \\ 2 \end{bmatrix}$ and $\mathbf{v} = \begin{bmatrix} 2 \\ 1 \\ -1 \end{bmatrix}$.

Solution ▶ Compute $\cos \theta = \dfrac{\mathbf{v} \cdot \mathbf{w}}{\|\mathbf{v}\|\|\mathbf{w}\|} = \dfrac{-2+1-2}{\sqrt{6}\sqrt{6}} = -\dfrac{1}{2}$. Now recall that $\cos \theta$ and $\sin \theta$ are defined so that $(\cos \theta, \sin \theta)$ is the point on the unit circle determined by the angle θ (drawn counterclockwise, starting from the positive x axis). In the present case, we know that $\cos \theta = -\dfrac{1}{2}$ and that $0 \leq \theta \leq \pi$. Because $\cos \dfrac{\pi}{3} = \dfrac{1}{2}$, it follows that $\theta = \dfrac{2\pi}{3}$ (see the diagram).

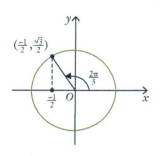

If $\mathbf{v}$ and $\mathbf{w}$ are nonzero, (∗) shows that $\cos \theta$ has the same sign as $\mathbf{v} \cdot \mathbf{w}$, so

$\mathbf{v} \cdot \mathbf{w} > 0$ if and only if θ is acute $(0 \leq \theta < \dfrac{\pi}{2})$
$\mathbf{v} \cdot \mathbf{w} < 0$ if and only if θ is obtuse $(\dfrac{\pi}{2} < \theta \leq 0)$
$\mathbf{v} \cdot \mathbf{w} = 0$ if and only if $\theta = \dfrac{\pi}{2}$

In this last case, the (nonzero) vectors are perpendicular. The following terminology is used in linear algebra:

Definition 4.5 *Two vectors $\mathbf{v}$ and $\mathbf{w}$ are said to be **orthogonal** if $\mathbf{v} = \mathbf{0}$ or $\mathbf{w} = \mathbf{0}$ or the angle between them is $\dfrac{\pi}{2}$.*

Since $\mathbf{v} \cdot \mathbf{w} = 0$ if either $\mathbf{v} = \mathbf{0}$ or $\mathbf{w} = \mathbf{0}$, we have the following theorem:

Theorem 3

Two vectors $\mathbf{v}$ and $\mathbf{w}$ are orthogonal if and only if $\mathbf{v} \cdot \mathbf{w} = 0$.

EXAMPLE 4

Show that the points $P(3, -1, 1)$, $Q(4, 1, 4)$, and $R(6, 0, 4)$ are the vertices of a right triangle.

Solution ▶ The vectors along the sides of the triangle are

$$\overrightarrow{PQ} = \begin{bmatrix} 1 \\ 2 \\ 3 \end{bmatrix}, \quad \overrightarrow{PR} = \begin{bmatrix} 3 \\ 1 \\ 3 \end{bmatrix}, \text{ and } \overrightarrow{QR} = \begin{bmatrix} 2 \\ -1 \\ 0 \end{bmatrix}$$

Evidently $\overrightarrow{PQ} \cdot \overrightarrow{QR} = 2 - 2 + 0 = 0$, so $\overrightarrow{PQ}$ and $\overrightarrow{QR}$ are orthogonal vectors. This means sides PQ and QR are perpendicular—that is, the angle at Q is a right angle.

Example 5 demonstrates how the dot product can be used to verify geometrical theorems involving perpendicular lines.

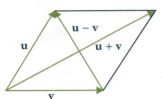

EXAMPLE 5

A parallelogram with sides of equal length is called a **rhombus**. Show that the diagonals of a rhombus are perpendicular.

Solution ▶ Let $\mathbf{u}$ and $\mathbf{v}$ denote vectors along two adjacent sides of a rhombus, as shown in the diagram. Then the diagonals are $\mathbf{u} - \mathbf{v}$ and $\mathbf{u} + \mathbf{v}$, and we compute

$$(\mathbf{u} - \mathbf{v}) \cdot (\mathbf{u} + \mathbf{v}) = \mathbf{u} \cdot (\mathbf{u} + \mathbf{v}) - \mathbf{v} \cdot (\mathbf{u} + \mathbf{v})$$
$$= \mathbf{u} \cdot \mathbf{u} + \mathbf{u} \cdot \mathbf{v} - \mathbf{v} \cdot \mathbf{u} - \mathbf{v} \cdot \mathbf{v}$$
$$= \|\mathbf{u}\|^2 - \|\mathbf{v}\|^2$$
$$= 0$$

because $\|\mathbf{u}\| = \|\mathbf{v}\|$ (it is a rhombus). Hence $\mathbf{u} - \mathbf{v}$ and $\mathbf{u} + \mathbf{v}$ are orthogonal.

Projections

In applications of vectors, it is frequently useful to write a vector as the sum of two orthogonal vectors. Here is an example.

EXAMPLE 6

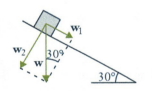

Suppose a ten-kilogram block is placed on a flat surface inclined 30° to the horizontal as in the diagram. Neglecting friction, how much force is required to keep the block from sliding down the surface?

Solution ▶ Let $\mathbf{w}$ denote the weight (force due to gravity) exerted on the block. Then $\|\mathbf{w}\| = 10$ kilograms and the direction of $\mathbf{w}$ is vertically down as in the diagram. The idea is to write $\mathbf{w}$ as a sum $\mathbf{w} = \mathbf{w}_1 + \mathbf{w}_2$ where $\mathbf{w}_1$ is parallel to the inclined surface and $\mathbf{w}_2$ is perpendicular to the surface. Since there is no friction, the force required is $-\mathbf{w}_1$ because the force $\mathbf{w}_2$ has no effect parallel to the surface. As the angle between $\mathbf{w}$ and $\mathbf{w}_2$ is 30° in the diagram, we have $\frac{\|\mathbf{w}_1\|}{\|\mathbf{w}\|} = \sin 30° = \frac{1}{2}$. Hence $\|\mathbf{w}_1\| = \frac{1}{2}\|\mathbf{w}\| = \frac{1}{2}10 = 5$. Thus the required force has a magnitude of 5 kilograms weight directed up the surface.

If a nonzero vector $\mathbf{d}$ is specified, the key idea in Example 6 is to be able to write an arbitrary vector $\mathbf{u}$ as a sum of two vectors,

$$\mathbf{u} = \mathbf{u}_1 + \mathbf{u}_2$$

where $\mathbf{u}_1$ is parallel to $\mathbf{d}$ and $\mathbf{u}_2 = \mathbf{u} - \mathbf{u}_1$ is orthogonal to $\mathbf{d}$. Suppose that $\mathbf{u}$ and $\mathbf{d} \neq \mathbf{0}$ emanate from a common tail Q (see Figure 5). Let P be the tip of $\mathbf{u}$, and let P_1 denote the foot of the perpendicular from P to the line through Q parallel to $\mathbf{d}$. Then $\mathbf{u}_1 = \overrightarrow{QP_1}$ has the required properties:

1. $\mathbf{u}_1$ is parallel to $\mathbf{d}$.
2. $\mathbf{u}_2 = \mathbf{u} - \mathbf{u}_1$ is orthogonal to $\mathbf{d}$.
3. $\mathbf{u} = \mathbf{u}_1 + \mathbf{u}_2$.

■ FIGURE 5

Definition 4.6 The vector $\mathbf{u}_1 = \overrightarrow{QP_1}$ in Figure 5 is called **the projection of u on d**. *It is denoted*

$$\mathbf{u}_1 = \text{proj}_\mathbf{d}\, \mathbf{u}$$

SECTION 4.2 Projections and Planes

In Figure 5(a) the vector $u_1 = \text{proj}_d\ u$ has the same direction as d; however, u_1 and d have opposite directions if the angle between u and d is greater than $\frac{\pi}{2}$. (Figure 5(b)). Note that the projection $u_1 = \text{proj}_d\ u$ is zero if and only if u and d are orthogonal.

Calculating the projection of u on $d \neq 0$ is remarkably easy.

Theorem 4

Let u and $d \neq 0$ be vectors.

1. The projection of u on d is given by $\text{proj}_d\ u = \dfrac{u \cdot d}{\|d\|^2} d$.
2. The vector $u - \text{proj}_d\ u$ is orthogonal to d.

PROOF

The vector $u_1 = \text{proj}_d\ u$ is parallel to d and so has the form $u_1 = td$ for some scalar t. The requirement that $u - u_1$ and d are orthogonal determines t. In fact, it means that $(u - u_1) \cdot d = 0$ by Theorem 3. If $u_1 = td$ is substituted here, the condition is

$$0 = (u - td) \cdot d = u \cdot d - t(d \cdot d) = u \cdot d - t\|d\|^2$$

It follows that $t = \dfrac{u \cdot d}{\|d\|^2}$, where the assumption that $d \neq 0$ guarantees that $\|d\|^2 \neq 0$.

EXAMPLE 7

Find the projection of $u = \begin{bmatrix} 2 \\ -3 \\ 1 \end{bmatrix}$ on $d = \begin{bmatrix} 1 \\ -1 \\ 3 \end{bmatrix}$ and express $u = u_1 + u_2$ where u_1 is parallel to d and u_2 is orthogonal to d.

Solution ▶ The projection u_1 of u on d is

$$u_1 = \text{proj}_d\ u = \frac{u \cdot d}{\|d\|^2} d = \frac{2 + 3 + 3}{1^2 + (-1)^2 + 3^2} \begin{bmatrix} 1 \\ -1 \\ 3 \end{bmatrix} = \frac{8}{11} \begin{bmatrix} 1 \\ -1 \\ 3 \end{bmatrix}$$

Hence $u_2 = u - u_1 = \frac{1}{11} \begin{bmatrix} 14 \\ -25 \\ -13 \end{bmatrix}$, and this is orthogonal to d by Theorem 4 (alternatively, observe that $d \cdot u_2 = 0$). Since $u = u_1 + u_2$, we are done.

EXAMPLE 8

Find the shortest distance (see diagram) from the point $P(1, 3, -2)$ to the line through $P_0(2, 0, -1)$ with direction vector $d = \begin{bmatrix} 1 \\ -1 \\ 0 \end{bmatrix}$. Also find the point Q that lies on the line and is closest to P.

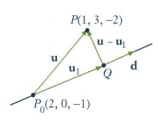

Solution ▶ Let $\mathbf{u} = \begin{bmatrix} 1 \\ 3 \\ -2 \end{bmatrix} - \begin{bmatrix} 2 \\ 0 \\ -1 \end{bmatrix} = \begin{bmatrix} -1 \\ 3 \\ -1 \end{bmatrix}$ denote the vector from P_0 to P, and let $\mathbf{u}_1$ denote the projection of $\mathbf{u}$ on $\mathbf{d}$. Thus

$$\mathbf{u}_1 = \frac{\mathbf{u} \cdot \mathbf{d}}{\|\mathbf{d}\|^2}\mathbf{d} = \frac{-1-3+0}{1^2+(-1)^2+0^2}\mathbf{d} = -2\mathbf{d} = \begin{bmatrix} -2 \\ 2 \\ 0 \end{bmatrix}$$

by Theorem 4. We see geometrically that the point Q on the line is closest to P, so the distance is

$$\|\overrightarrow{QP}\| = \|\mathbf{u} - \mathbf{u}_1\| = \left\| \begin{bmatrix} 1 \\ 1 \\ -1 \end{bmatrix} \right\| = \sqrt{3}$$

To find the coordinates of Q, let $\mathbf{p}_0$ and $\mathbf{q}$ denote the vectors of P_0 and Q, respectively. Then $\mathbf{p}_0 = \begin{bmatrix} 2 \\ 0 \\ -1 \end{bmatrix}$ and $\mathbf{q} = \mathbf{p}_0 + \mathbf{u}_1 = \begin{bmatrix} 0 \\ 2 \\ -1 \end{bmatrix}$.

Hence $Q(0, 2, -1)$ is the required point. It can be checked that the distance from Q to P is $\sqrt{3}$, as expected.

Planes

It is evident geometrically that among all planes that are perpendicular to a given straight line there is exactly one containing any given point. This fact can be used to give a very simple description of a plane. To do this, it is necessary to introduce the following notion:

Definition 4.7 *A nonzero vector $\mathbf{n}$ is called a **normal** for a plane if it is orthogonal to every vector in the plane.*

For example, the coordinate vector $\mathbf{k}$ is a normal for the x-y plane.

Given a point $P_0 = P_0(x_0, y_0, z_0)$ and a nonzero vector $\mathbf{n}$, there is a unique plane through P_0 with normal $\mathbf{n}$, shaded in Figure 6. A point $P = P(x, y, z)$ lies on this plane if and only if the vector $\overrightarrow{P_0P}$ is orthogonal to $\mathbf{n}$—that is, if and only if $\mathbf{n} \cdot \overrightarrow{P_0P} = 0$. Because $\overrightarrow{P_0P} = \begin{bmatrix} x - x_0 \\ y - y_0 \\ z - z_0 \end{bmatrix}$ this gives the following result:

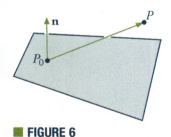

FIGURE 6

Scalar Equation of a Plane

The plane through $P_0(x_0, y_0, z_0)$ with normal $\mathbf{n} = \begin{bmatrix} a \\ b \\ c \end{bmatrix} \neq \mathbf{0}$ as a normal vector is given by

$$a(x - x_0) + b(y - y_0) + c(z - z_0) = 0$$

In other words, a point $P(x, y, z)$ is on this plane if and only if x, y, and z satisfy this equation.

SECTION 4.2 Projections and Planes

EXAMPLE 9

Find an equation of the plane through $P_0(1, -1, 3)$ with $\mathbf{n} = \begin{bmatrix} 3 \\ -1 \\ 2 \end{bmatrix}$ as normal.

Solution ▶ Here the general scalar equation becomes
$$3(x - 1) - (y + 1) + 2(z - 3) = 0$$
This simplifies to $3x - y + 2z = 10$.

If we write $d = ax_0 + by_0 + cz_0$, the scalar equation shows that every plane with normal $\mathbf{n} = \begin{bmatrix} a \\ b \\ c \end{bmatrix}$ has a linear equation of the form

$$ax + by + cz = d \qquad (*)$$

for some constant d. Conversely, the graph of this equation is a plane with $\mathbf{n} = \begin{bmatrix} a \\ b \\ c \end{bmatrix}$ as a normal vector (assuming that a, b, and c are not all zero).

EXAMPLE 10

Find an equation of the plane through $P_0(3, -1, 2)$ that is parallel to the plane with equation $2x - 3y = 6$.

Solution ▶ The plane with equation $2x - 3y = 6$ has normal $\mathbf{n} = \begin{bmatrix} 2 \\ -3 \\ 0 \end{bmatrix}$. Because the two planes are parallel, $\mathbf{n}$ serves as a normal for the plane we seek, so the equation is $2x - 3y = d$ for some d by equation (*). Insisting that $P_0(3, -1, 2)$ lies on the plane determines d; that is, $d = 2 \cdot 3 - 3(-1) = 9$. Hence, the equation is $2x - 3y = 9$.

Consider points $P_0(x_0, y_0, z_0)$ and $P(x, y, z)$ with vectors $\mathbf{p}_0 = \begin{bmatrix} x_0 \\ y_0 \\ z_0 \end{bmatrix}$ and $\mathbf{p} = \begin{bmatrix} x \\ y \\ z \end{bmatrix}$.

Given a nonzero vector $\mathbf{n}$, the scalar equation of the plane through $P_0(x_0, y_0, z_0)$ with normal $\mathbf{n} = \begin{bmatrix} a \\ b \\ c \end{bmatrix}$ takes the vector form:

Vector Equation of a Plane

The plane with normal $\mathbf{n} \neq \mathbf{0}$ through the point with vector $\mathbf{p}_0$ is given by
$$\mathbf{n} \cdot (\mathbf{p} - \mathbf{p}_0) = 0$$
In other words, the point with vector $\mathbf{p}$ is on the plane if and only if $\mathbf{p}$ satisfies this condition.

Moreover, equation (*) translates as follows:

Every plane with normal $\mathbf{n}$ has vector equation $\mathbf{n} \cdot \mathbf{p} = d$ for some number d.

This is useful in the second solution of Example 11.

EXAMPLE 11

Find the shortest distance from the point $P(2, 1, -3)$ to the plane with equation $3x - y + 4z = 1$. Also find the point Q on this plane closest to P.

Solution 1 ▶ The plane in question has normal $\mathbf{n} = \begin{bmatrix} 3 \\ -1 \\ 4 \end{bmatrix}$. Choose any point P_0 on the plane—say $P_0(0, -1, 0)$—and let $Q(x, y, z)$ be the point on the plane closest to P (see the diagram). The vector from P_0 to P is $\mathbf{u} = \begin{bmatrix} 2 \\ 2 \\ -3 \end{bmatrix}$. Now erect $\mathbf{n}$ with its tail at P_0. Then $\overrightarrow{QP} = \mathbf{u}_1$ and $\mathbf{u}_1$ is the projection of $\mathbf{u}$ on $\mathbf{n}$:

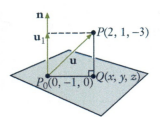

$$\mathbf{u}_1 = \frac{\mathbf{n} \cdot \mathbf{u}}{\|\mathbf{n}\|^2} \mathbf{n} = \frac{-8}{26} \begin{bmatrix} 3 \\ -1 \\ 4 \end{bmatrix} = \frac{-4}{13} \begin{bmatrix} 3 \\ -1 \\ 4 \end{bmatrix}$$

Hence the distance is $\|\overrightarrow{QP}\| = \|\mathbf{u}_1\| = \frac{4\sqrt{26}}{13}$. To calculate the point Q, let $\mathbf{q} = \begin{bmatrix} x \\ y \\ z \end{bmatrix}$ and $\mathbf{p}_0 = \begin{bmatrix} 0 \\ -1 \\ 0 \end{bmatrix}$ be the vectors of Q and P_0. Then

$$\mathbf{q} = \mathbf{p}_0 + \mathbf{u} - \mathbf{u}_1 = \begin{bmatrix} 0 \\ -1 \\ 0 \end{bmatrix} + \begin{bmatrix} 2 \\ 2 \\ -3 \end{bmatrix} + \frac{4}{13} \begin{bmatrix} 3 \\ -1 \\ 4 \end{bmatrix} = \begin{bmatrix} \frac{38}{13} \\ \frac{9}{13} \\ \frac{-23}{13} \end{bmatrix}$$

This gives the coordinates of $Q(\frac{38}{13}, \frac{9}{13}, \frac{-23}{13})$.

Solution 2 ▶ Let $\mathbf{q} = \begin{bmatrix} x \\ y \\ z \end{bmatrix}$ and $\mathbf{p} = \begin{bmatrix} 2 \\ 1 \\ -3 \end{bmatrix}$ be the vectors of Q and P. Then Q is on the line through P with direction vector $\mathbf{n}$, so $\mathbf{q} = \mathbf{p} + t\mathbf{n}$ for some scalar t. In addition, Q lies on the plane, so $\mathbf{n} \cdot \mathbf{q} = 1$. This determines t:

$$1 = \mathbf{n} \cdot \mathbf{q} = \mathbf{n} \cdot (\mathbf{p} + t\mathbf{n}) = \mathbf{n} \cdot \mathbf{p} + t\|\mathbf{n}\|^2 = -7 + t(26)$$

This gives $t = \frac{8}{26} = \frac{4}{13}$, so

$$\begin{bmatrix} x \\ y \\ z \end{bmatrix} = \mathbf{q} = \mathbf{p} + t\mathbf{n} = \begin{bmatrix} 2 \\ 1 \\ -3 \end{bmatrix} + \frac{4}{13} \begin{bmatrix} 3 \\ -1 \\ 4 \end{bmatrix} = \frac{1}{13} \begin{bmatrix} 38 \\ 9 \\ -23 \end{bmatrix}$$

as before. This determines Q (in the diagram), and the reader can verify that the required distance is $\|\overrightarrow{QP}\| = \frac{4}{13}\sqrt{26}$, as before.

The Cross Product

If P, Q, and R are three distinct points in $\mathbb{R}^3$ that are not all on some line, it is clear geometrically that there is a unique plane containing all three. The vectors $\overrightarrow{PQ}$ and $\overrightarrow{PR}$ both lie in this plane, so finding a normal amounts to finding a nonzero vector orthogonal to both $\overrightarrow{PQ}$ and $\overrightarrow{PR}$. The cross product provides a systematic way to do this.

SECTION 4.2 Projections and Planes

Definition 4.8 Given vectors $\mathbf{v}_1 = \begin{bmatrix} x_1 \\ y_1 \\ z_1 \end{bmatrix}$ and $\mathbf{v}_2 = \begin{bmatrix} x_2 \\ y_2 \\ z_2 \end{bmatrix}$, define the **cross product** $\mathbf{v}_1 \times \mathbf{v}_2$ by

$$\mathbf{v}_1 \times \mathbf{v}_2 = \begin{bmatrix} y_1 z_2 - z_1 y_2 \\ -(x_1 z_2 - z_1 x_2) \\ x_1 y_2 - y_1 x_2 \end{bmatrix}.$$

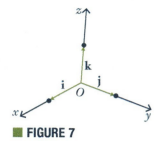

■ FIGURE 7

(Because it is a vector, $\mathbf{v}_1 \times \mathbf{v}_2$ is often called the **vector product**.) There is an easy way to remember this definition using the **coordinate vectors**:

$$\mathbf{i} = \begin{bmatrix} 1 \\ 0 \\ 0 \end{bmatrix}, \mathbf{j} = \begin{bmatrix} 0 \\ 1 \\ 0 \end{bmatrix}, \text{ and } \mathbf{k} = \begin{bmatrix} 0 \\ 0 \\ 1 \end{bmatrix}$$

They are vectors of length 1 pointing along the positive x, y, and z axes, respectively, as in Figure 7. The reason for the name is that any vector can be written as

$$\begin{bmatrix} x \\ y \\ z \end{bmatrix} = x\mathbf{i} + y\mathbf{j} + z\mathbf{k}.$$

With this, the cross product can be described as follows:

Determinant Form of the Cross Product

If $\mathbf{v}_1 = \begin{bmatrix} x_1 \\ y_1 \\ z_1 \end{bmatrix}$ and $\mathbf{v}_2 = \begin{bmatrix} x_2 \\ y_2 \\ z_2 \end{bmatrix}$ are two vectors, then

$$\mathbf{v}_1 \times \mathbf{v}_2 = \det \begin{bmatrix} \mathbf{i} & x_1 & x_2 \\ \mathbf{j} & y_1 & y_2 \\ \mathbf{k} & z_1 & z_2 \end{bmatrix} = \begin{vmatrix} y_1 & y_2 \\ z_1 & z_2 \end{vmatrix} \mathbf{i} - \begin{vmatrix} x_1 & x_2 \\ z_1 & z_2 \end{vmatrix} \mathbf{j} + \begin{vmatrix} x_1 & x_2 \\ y_1 & y_2 \end{vmatrix} \mathbf{k}$$

where the determinant is expanded along the first column.

EXAMPLE 12

If $\mathbf{v} = \begin{bmatrix} 2 \\ -1 \\ 4 \end{bmatrix}$ and $\mathbf{w} = \begin{bmatrix} 1 \\ 3 \\ 7 \end{bmatrix}$, then

$$\mathbf{v}_1 \times \mathbf{v}_2 = \det \begin{bmatrix} \mathbf{i} & 2 & 1 \\ \mathbf{j} & -1 & 3 \\ \mathbf{k} & 4 & 7 \end{bmatrix} = \begin{vmatrix} -1 & 3 \\ 4 & 7 \end{vmatrix} \mathbf{i} - \begin{vmatrix} 2 & 1 \\ 4 & 7 \end{vmatrix} \mathbf{j} + \begin{vmatrix} 2 & 1 \\ -1 & 3 \end{vmatrix} \mathbf{k}$$

$$= -19\mathbf{i} - 10\mathbf{j} + 7\mathbf{k}$$

$$= \begin{bmatrix} -19 \\ -10 \\ 7 \end{bmatrix}$$

Observe that $\mathbf{v} \times \mathbf{w}$ is orthogonal to both $\mathbf{v}$ and $\mathbf{w}$ in Example 12. This holds in general as can be verified directly by computing $\mathbf{v} \cdot (\mathbf{v} \times \mathbf{w})$ and $\mathbf{w} \cdot (\mathbf{v} \times \mathbf{w})$, and is recorded as the first part of the following theorem. It will follow from a more general result which, together with the second part, will be proved in Section 4.3 where a more detailed study of the cross product will be undertaken.

Theorem 5

Let **v** and **w** be vectors in $\mathbb{R}^3$.

1. $\mathbf{v} \times \mathbf{w}$ is a vector orthogonal to both **v** and **w**.
2. If **v** and **w** are nonzero, then $\mathbf{v} \times \mathbf{w} = \mathbf{0}$ if and only if **v** and **w** are parallel.

It is interesting to contrast Theorem 5(2) with the assertion (in Theorem 3) that
$$\mathbf{v} \cdot \mathbf{w} = 0 \quad \text{if and only if } \mathbf{v} \text{ and } \mathbf{w} \text{ are orthogonal.}$$

EXAMPLE 13

Find the equation of the plane through $P(1, 3, -2)$, $Q(1, 1, 5)$, and $R(2, -2, 3)$.

Solution ▶ The vectors $\overrightarrow{PQ} = \begin{bmatrix} 0 \\ -2 \\ 7 \end{bmatrix}$ and $\overrightarrow{PR} = \begin{bmatrix} 1 \\ -5 \\ 5 \end{bmatrix}$ lie in the plane, so

$$\overrightarrow{PQ} \times \overrightarrow{PR} = \det \begin{bmatrix} \mathbf{i} & 0 & 1 \\ \mathbf{j} & -2 & -5 \\ \mathbf{k} & 7 & 5 \end{bmatrix} = 25\mathbf{i} + 7\mathbf{j} + 2\mathbf{k} = \begin{bmatrix} 25 \\ 7 \\ 2 \end{bmatrix}$$

is a normal for the plane (being orthogonal to both $\overrightarrow{PQ}$ and $\overrightarrow{PR}$). Hence the plane has equation

$$25x + 7y + 2z = d \quad \text{for some number } d.$$

Since $P(1, 3, -2)$ lies in the plane we have $25 \cdot 1 + 7 \cdot 3 + 2(-2) = d$. Hence $d = 42$ and the equation is $25x + 7y + 2z = 42$. Incidentally, the same equation is obtained (verify) if $\overrightarrow{QP}$ and $\overrightarrow{QR}$, or $\overrightarrow{RP}$ and $\overrightarrow{RQ}$, are used as the vectors in the plane.

EXAMPLE 14

Find the shortest distance between the nonparallel lines

$$\begin{bmatrix} x \\ y \\ z \end{bmatrix} = \begin{bmatrix} 1 \\ 0 \\ -1 \end{bmatrix} + t \begin{bmatrix} 2 \\ 0 \\ 1 \end{bmatrix} \quad \text{and} \quad \begin{bmatrix} x \\ y \\ z \end{bmatrix} = \begin{bmatrix} 3 \\ 1 \\ 0 \end{bmatrix} + s \begin{bmatrix} 1 \\ 1 \\ -1 \end{bmatrix}$$

Then find the points A and B on the lines that are closest together.

Solution ▶ Direction vectors for the two lines are $\mathbf{d}_1 = \begin{bmatrix} 2 \\ 0 \\ 1 \end{bmatrix}$ and $\mathbf{d}_2 = \begin{bmatrix} 1 \\ 1 \\ -1 \end{bmatrix}$, so

$$\mathbf{n} = \mathbf{d}_1 \times \mathbf{d}_2 = \det \begin{bmatrix} \mathbf{i} & 2 & 1 \\ \mathbf{j} & 0 & 1 \\ \mathbf{k} & 1 & -1 \end{bmatrix} = \begin{bmatrix} -1 \\ 3 \\ 2 \end{bmatrix}$$

SECTION 4.2 Projections and Planes

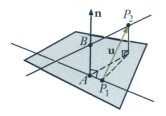

is perpendicular to both lines. Consider the plane shaded in the diagram containing the first line with **n** as normal. This plane contains $P_1(1, 0, -1)$ and is parallel to the second line. Because $P_2(3, 1, 0)$ is on the second line, the distance in question is just the shortest distance between $P_2(3, 1, 0)$ and this plane. The vector **u** from P_1 to P_2 is $\mathbf{u} = \overrightarrow{P_1P_2} = \begin{bmatrix} 2 \\ 1 \\ 1 \end{bmatrix}$ and so, as in Example 11, the distance is the length of the projection of **u** on **n**.

$$\text{distance} = \left\| \frac{\mathbf{u} \cdot \mathbf{n}}{\|\mathbf{n}\|^2} \mathbf{n} \right\| = \frac{|\mathbf{u} \cdot \mathbf{n}|}{\|\mathbf{n}\|} = \frac{3}{\sqrt{14}} = \frac{3\sqrt{14}}{14}$$

Note that it is necessary that $\mathbf{n} = \mathbf{d}_1 \times \mathbf{d}_2$ be nonzero for this calculation to be possible. As is shown later (Theorem 4 Section 4.3), this is guaranteed by the fact that $\mathbf{d}_1$ and $\mathbf{d}_2$ are *not* parallel.

The points A and B have coordinates $A(1 + 2t, 0, t - 1)$ and $B(3 + s, 1 + s, -s)$ for some s and t, so $\overrightarrow{AB} = \begin{bmatrix} 2 + s - 2t \\ 1 + s \\ 1 - s - t \end{bmatrix}$. This vector is orthogonal to both $\mathbf{d}_1$ and $\mathbf{d}_2$, and the conditions $\overrightarrow{AB} \cdot \mathbf{d}_1 = 0$ and $\overrightarrow{AB} \cdot \mathbf{d}_2 = 0$ give equations $5t - s = 5$ and $t - 3s = 2$. The solution is $s = \frac{-5}{14}$ and $t = \frac{13}{14}$, so the points are $A\left(\frac{40}{14}, 0, \frac{-1}{14}\right)$ and $B\left(\frac{37}{14}, \frac{9}{14}, \frac{5}{14}\right)$. We have $\|\overrightarrow{AB}\| = \frac{3\sqrt{14}}{14}$, as before.

EXERCISES 4.2

1. Compute $\mathbf{u} \cdot \mathbf{v}$ where:

   (a) $\mathbf{u} = \begin{bmatrix} 2 \\ -1 \\ 3 \end{bmatrix}, \mathbf{v} = \begin{bmatrix} -1 \\ 1 \\ 1 \end{bmatrix}$

   ◆(b) $\mathbf{u} = \begin{bmatrix} 1 \\ 2 \\ -1 \end{bmatrix}, \mathbf{v} = \mathbf{u}$

   (c) $\mathbf{u} = \begin{bmatrix} 1 \\ 1 \\ -3 \end{bmatrix}, \mathbf{v} = \begin{bmatrix} 2 \\ -1 \\ 1 \end{bmatrix}$

   ◆(d) $\mathbf{u} = \begin{bmatrix} 3 \\ -1 \\ 5 \end{bmatrix}, \mathbf{v} = \begin{bmatrix} 6 \\ -7 \\ -5 \end{bmatrix}$

   (e) $\mathbf{u} = \begin{bmatrix} x \\ y \\ z \end{bmatrix}, \mathbf{v} = \begin{bmatrix} a \\ b \\ c \end{bmatrix}$

   ◆(f) $\mathbf{u} = \begin{bmatrix} a \\ b \\ c \end{bmatrix}, \mathbf{v} = \mathbf{0}$

2. Find the angle between the following pairs of vectors.

   (a) $\mathbf{u} = \begin{bmatrix} 1 \\ 0 \\ 3 \end{bmatrix}, \mathbf{v} = \begin{bmatrix} 2 \\ 0 \\ 1 \end{bmatrix}$ ◆(b) $\mathbf{u} = \begin{bmatrix} 3 \\ -1 \\ 0 \end{bmatrix}, \mathbf{v} = \begin{bmatrix} -6 \\ 2 \\ 0 \end{bmatrix}$

   (c) $\mathbf{u} = \begin{bmatrix} 7 \\ -1 \\ 3 \end{bmatrix}, \mathbf{v} = \begin{bmatrix} 1 \\ 4 \\ -1 \end{bmatrix}$

   ◆(d) $\mathbf{u} = \begin{bmatrix} 2 \\ 1 \\ -1 \end{bmatrix}, \mathbf{v} = \begin{bmatrix} 3 \\ 6 \\ 3 \end{bmatrix}$ (e) $\mathbf{u} = \begin{bmatrix} 1 \\ -1 \\ 0 \end{bmatrix}, \mathbf{v} = \begin{bmatrix} 0 \\ 1 \\ 1 \end{bmatrix}$

   ◆(f) $\mathbf{u} = \begin{bmatrix} 0 \\ 3 \\ 4 \end{bmatrix}, \mathbf{v} = \begin{bmatrix} 5\sqrt{2} \\ -7 \\ -1 \end{bmatrix}$

3. Find all real numbers x such that:

   (a) $\begin{bmatrix} 2 \\ -1 \\ 3 \end{bmatrix}$ and $\begin{bmatrix} x \\ -2 \\ 1 \end{bmatrix}$ are orthogonal.

   ◆(b) $\begin{bmatrix} 2 \\ -1 \\ 1 \end{bmatrix}$ and $\begin{bmatrix} 1 \\ x \\ 2 \end{bmatrix}$ are at an angle of $\frac{\pi}{3}$.

4. Find all vectors $\mathbf{v} = \begin{bmatrix} x \\ y \\ z \end{bmatrix}$ orthogonal to both:

(a) $\mathbf{u}_1 = \begin{bmatrix} -1 \\ -3 \\ 2 \end{bmatrix}, \mathbf{u}_2 = \begin{bmatrix} 0 \\ 1 \\ 1 \end{bmatrix}$

◆(b) $\mathbf{u}_1 = \begin{bmatrix} 3 \\ -1 \\ 2 \end{bmatrix}, \mathbf{u}_2 = \begin{bmatrix} 2 \\ 0 \\ 1 \end{bmatrix}$

(c) $\mathbf{u}_1 = \begin{bmatrix} 2 \\ 0 \\ -1 \end{bmatrix}, \mathbf{u}_2 = \begin{bmatrix} -4 \\ 0 \\ 2 \end{bmatrix}$

◆(d) $\mathbf{u}_1 = \begin{bmatrix} 2 \\ -1 \\ 3 \end{bmatrix}, \mathbf{u}_2 = \begin{bmatrix} 0 \\ 0 \\ 0 \end{bmatrix}$

5. Find two orthogonal vectors that are both orthogonal to $\mathbf{v} = \begin{bmatrix} 1 \\ 2 \\ 0 \end{bmatrix}$.

6. Consider the triangle with vertices $P(2, 0, -3)$, $Q(5, -2, 1)$, and $R(7, 5, 3)$.

   (a) Show that it is a right-angled triangle.

   ◆(b) Find the lengths of the three sides and verify the Pythagorean theorem.

7. Show that the triangle with vertices $A(4, -7, 9)$, $B(6, 4, 4)$, and $C(7, 10, -6)$ is not a right-angled triangle.

8. Find the three internal angles of the triangle with vertices:

   (a) $A(3, 1, -2), B(3, 0, -1)$, and $C(5, 2, -1)$

   ◆(b) $A(3, 1, -2), B(5, 2, -1)$, and $C(4, 3, -3)$

9. Show that the line through $P_0(3, 1, 4)$ and $P_1(2, 1, 3)$ is perpendicular to the line through $P_2(1, -1, 2)$ and $P_3(0, 5, 3)$.

10. In each case, compute the projection of $\mathbf{u}$ on $\mathbf{v}$.

    (a) $\mathbf{u} = \begin{bmatrix} 5 \\ 7 \\ 1 \end{bmatrix}, \mathbf{v} = \begin{bmatrix} 2 \\ -1 \\ 3 \end{bmatrix}$

    ◆(b) $\mathbf{u} = \begin{bmatrix} 3 \\ -2 \\ 1 \end{bmatrix}, \mathbf{v} = \begin{bmatrix} 4 \\ 1 \\ 1 \end{bmatrix}$

    c) $\mathbf{u} = \begin{bmatrix} 1 \\ -1 \\ 2 \end{bmatrix}, \mathbf{v} = \begin{bmatrix} 3 \\ -1 \\ 1 \end{bmatrix}$

    ◆(d) $\mathbf{u} = \begin{bmatrix} 3 \\ -2 \\ -1 \end{bmatrix}, \mathbf{v} = \begin{bmatrix} -6 \\ 4 \\ 2 \end{bmatrix}$

11. In each case, write $\mathbf{u} = \mathbf{u}_1 + \mathbf{u}_2$, where $\mathbf{u}_1$ is parallel to $\mathbf{v}$ and $\mathbf{u}_2$ is orthogonal to $\mathbf{v}$.

    (a) $\mathbf{u} = \begin{bmatrix} 2 \\ -1 \\ 1 \end{bmatrix}, \mathbf{v} = \begin{bmatrix} 1 \\ -1 \\ 3 \end{bmatrix}$

    ◆(b) $\mathbf{u} = \begin{bmatrix} 3 \\ 1 \\ 0 \end{bmatrix}, \mathbf{v} = \begin{bmatrix} -2 \\ 1 \\ 4 \end{bmatrix}$

    (c) $\mathbf{u} = \begin{bmatrix} 2 \\ -1 \\ 0 \end{bmatrix}, \mathbf{v} = \begin{bmatrix} 3 \\ 1 \\ -1 \end{bmatrix}$

    ◆(d) $\mathbf{u} = \begin{bmatrix} 3 \\ -2 \\ 1 \end{bmatrix}, \mathbf{v} = \begin{bmatrix} -6 \\ 4 \\ -1 \end{bmatrix}$

12. Calculate the distance from the point P to the line in each case and find the point Q on the line closest to P.

    (a) $P(3, 2, -1)$ line: $\begin{bmatrix} x \\ y \\ z \end{bmatrix} = \begin{bmatrix} 2 \\ 1 \\ 3 \end{bmatrix} + t\begin{bmatrix} 3 \\ -1 \\ -2 \end{bmatrix}$

    ◆(b) $P(1, -1, 3)$ line: $\begin{bmatrix} x \\ y \\ z \end{bmatrix} = \begin{bmatrix} 1 \\ 0 \\ -1 \end{bmatrix} + t\begin{bmatrix} 3 \\ 1 \\ 4 \end{bmatrix}$

13. Compute $\mathbf{u} \times \mathbf{v}$ where:

    (a) $\mathbf{u} = \begin{bmatrix} 1 \\ 2 \\ 3 \end{bmatrix}, \mathbf{v} = \begin{bmatrix} 1 \\ 1 \\ 2 \end{bmatrix}$

    ◆(b) $\mathbf{u} = \begin{bmatrix} 3 \\ -1 \\ 0 \end{bmatrix}, \mathbf{v} = \begin{bmatrix} -6 \\ 2 \\ 0 \end{bmatrix}$

    (c) $\mathbf{u} = \begin{bmatrix} 3 \\ -2 \\ 1 \end{bmatrix}, \mathbf{v} = \begin{bmatrix} 1 \\ 1 \\ -1 \end{bmatrix}$

    ◆(d) $\mathbf{u} = \begin{bmatrix} 2 \\ 0 \\ -1 \end{bmatrix}, \mathbf{v} = \begin{bmatrix} 1 \\ 4 \\ 7 \end{bmatrix}$

14. Find an equation of each of the following planes.

    (a) Passing through $A(2, 1, 3), B(3, -1, 5)$, and $C(1, 2, -3)$.

    ◆(b) Passing through $A(1, -1, 6), B(0, 0, 1)$, and $C(4, 7, -11)$.

    (c) Passing through $P(2, -3, 5)$ and parallel to the plane with equation $3x - 2y - z = 0$.

    ◆(d) Passing through $P(3, 0, -1)$ and parallel to the plane with equation $2x - y + z = 3$.

(e) Containing $P(3, 0, -1)$ and the line
$$\begin{bmatrix} x \\ y \\ z \end{bmatrix} = \begin{bmatrix} 0 \\ 0 \\ 2 \end{bmatrix} + t \begin{bmatrix} 1 \\ 0 \\ 1 \end{bmatrix}.$$

◆(f) Containing $P(2, 1, 0)$ and the line
$$\begin{bmatrix} x \\ y \\ z \end{bmatrix} = \begin{bmatrix} 3 \\ -1 \\ 2 \end{bmatrix} + t \begin{bmatrix} 1 \\ 0 \\ -1 \end{bmatrix}.$$

(g) Containing the lines
$$\begin{bmatrix} x \\ y \\ z \end{bmatrix} = \begin{bmatrix} 1 \\ -1 \\ 2 \end{bmatrix} + t \begin{bmatrix} 1 \\ 1 \\ 1 \end{bmatrix} \text{ and } \begin{bmatrix} x \\ y \\ z \end{bmatrix} = \begin{bmatrix} 0 \\ 0 \\ 2 \end{bmatrix} + t \begin{bmatrix} 1 \\ -1 \\ 0 \end{bmatrix}.$$

◆(h) Containing the lines $\begin{bmatrix} x \\ y \\ z \end{bmatrix} = \begin{bmatrix} 3 \\ 1 \\ 0 \end{bmatrix} + t \begin{bmatrix} 1 \\ -1 \\ 3 \end{bmatrix}$

and $\begin{bmatrix} x \\ y \\ z \end{bmatrix} = \begin{bmatrix} 0 \\ -2 \\ 5 \end{bmatrix} + t \begin{bmatrix} 2 \\ 1 \\ -1 \end{bmatrix}.$

(i) Each point of which is equidistant from $P(2, -1, 3)$ and $Q(1, 1, -1)$.

◆(j) Each point of which is equidistant from $P(0, 1, -1)$ and $Q(2, -1, -3)$.

15. In each case, find a vector equation of the line.

   (a) Passing through $P(3, -1, 4)$ and perpendicular to the plane $3x - 2y - z = 0$.

   ◆(b) Passing through $P(2, -1, 3)$ and perpendicular to the plane $2x + y = 1$.

   (c) Passing through $P(0, 0, 0)$ and perpendicular to the lines
$$\begin{bmatrix} x \\ y \\ z \end{bmatrix} = \begin{bmatrix} 1 \\ 1 \\ 0 \end{bmatrix} + t \begin{bmatrix} 2 \\ 0 \\ -1 \end{bmatrix} \text{ and } \begin{bmatrix} x \\ y \\ z \end{bmatrix} = \begin{bmatrix} 2 \\ 1 \\ -3 \end{bmatrix} + t \begin{bmatrix} 1 \\ -1 \\ 5 \end{bmatrix}.$$

   ◆(d) Passing through $P(1, 1, -1)$, and perpendicular to the lines
$$\begin{bmatrix} x \\ y \\ z \end{bmatrix} = \begin{bmatrix} 2 \\ 0 \\ 1 \end{bmatrix} + t \begin{bmatrix} 1 \\ 1 \\ -2 \end{bmatrix} \text{ and } \begin{bmatrix} x \\ y \\ z \end{bmatrix} = \begin{bmatrix} 5 \\ 5 \\ -2 \end{bmatrix} + t \begin{bmatrix} 1 \\ 2 \\ -3 \end{bmatrix}.$$

   (e) Passing through $P(2, 1, -1)$, intersecting the line $\begin{bmatrix} x \\ y \\ z \end{bmatrix} = \begin{bmatrix} 1 \\ 2 \\ -1 \end{bmatrix} + t \begin{bmatrix} 3 \\ 0 \\ 1 \end{bmatrix}$, and perpendicular to that line.

   ◆(f) Passing through $P(1, 1, 2)$, intersecting the line $\begin{bmatrix} x \\ y \\ z \end{bmatrix} = \begin{bmatrix} 2 \\ 1 \\ 0 \end{bmatrix} + t \begin{bmatrix} 1 \\ 1 \\ 1 \end{bmatrix}$, and perpendicular to that line.

16. In each case, find the shortest distance from the point P to the plane and find the point Q on the plane closest to P.

   (a) $P(2, 3, 0)$; plane with equation $5x + y + z = 1$.

   ◆(b) $P(3, 1, -1)$; plane with equation $2x + y - z = 6$.

17. (a) Does the line through $P(1, 2, -3)$ with direction vector $\mathbf{d} = \begin{bmatrix} 1 \\ 2 \\ -3 \end{bmatrix}$ lie in the plane $2x - y - z = 3$? Explain.

   ◆(b) Does the plane through $P(4, 0, 5)$, $Q(2, 2, 1)$, and $R(1, -1, 2)$ pass through the origin? Explain.

18. Show that every plane containing $P(1, 2, -1)$ and $Q(2, 0, 1)$ must also contain $R(-1, 6, -5)$.

19. Find the equations of the line of intersection of the following planes.

   (a) $2x - 3y + 2z = 5$ and $x + 2y - z = 4$.

   ◆(b) $3x + y - 2z = 1$ and $x + y + z = 5$.

20. In each case, find all points of intersection of the given plane and the line $\begin{bmatrix} x \\ y \\ z \end{bmatrix} = \begin{bmatrix} 1 \\ -2 \\ 3 \end{bmatrix} + t \begin{bmatrix} 2 \\ 5 \\ -1 \end{bmatrix}$.

   (a) $x - 3y + 2z = 4$ ◆(b) $2x - y - z = 5$
   (c) $3x - y + z = 8$ ◆(d) $-x - 4y - 3z = 6$

21. Find the equation of *all* planes:

   (a) Perpendicular to the line $\begin{bmatrix} x \\ y \\ z \end{bmatrix} = \begin{bmatrix} 2 \\ -1 \\ 3 \end{bmatrix} + t \begin{bmatrix} 2 \\ 1 \\ 3 \end{bmatrix}$.

   ◆(b) Perpendicular to the line $\begin{bmatrix} x \\ y \\ z \end{bmatrix} = \begin{bmatrix} 1 \\ 0 \\ -1 \end{bmatrix} + t \begin{bmatrix} 3 \\ 0 \\ 2 \end{bmatrix}$.

   (c) Containing the origin.

   ◆(d) Containing $P(3, 2, -4)$.

   (e) Containing $P(1, 1, -1)$ and $Q(0, 1, 1)$.

   ◆(f) Containing $P(2, -1, 1)$ and $Q(1, 0, 0)$.

   (g) Containing the line $\begin{bmatrix} x \\ y \\ z \end{bmatrix} = \begin{bmatrix} 2 \\ 1 \\ 0 \end{bmatrix} + t \begin{bmatrix} 1 \\ -1 \\ 0 \end{bmatrix}$.

   ◆(h) Containing the line $\begin{bmatrix} x \\ y \\ z \end{bmatrix} = \begin{bmatrix} 3 \\ 0 \\ 2 \end{bmatrix} + t \begin{bmatrix} 1 \\ -2 \\ -1 \end{bmatrix}$.

22. If a plane contains two distinct points P_1 and P_2, show that it contains every point on the line through P_1 and P_2.

23. Find the shortest distance between the following pairs of parallel lines.

(a) $\begin{bmatrix} x \\ y \\ z \end{bmatrix} = \begin{bmatrix} 2 \\ -1 \\ 3 \end{bmatrix} + t \begin{bmatrix} 1 \\ -1 \\ 4 \end{bmatrix}$; $\begin{bmatrix} x \\ y \\ z \end{bmatrix} = \begin{bmatrix} 1 \\ 0 \\ 1 \end{bmatrix} + t \begin{bmatrix} 1 \\ -1 \\ 4 \end{bmatrix}$

◆(b) $\begin{bmatrix} x \\ y \\ z \end{bmatrix} = \begin{bmatrix} 3 \\ 0 \\ 2 \end{bmatrix} + t \begin{bmatrix} 3 \\ 1 \\ 0 \end{bmatrix}$; $\begin{bmatrix} x \\ y \\ z \end{bmatrix} = \begin{bmatrix} -1 \\ 2 \\ 2 \end{bmatrix} + t \begin{bmatrix} 3 \\ 1 \\ 0 \end{bmatrix}$

24. Find the shortest distance between the following pairs of nonparallel lines and find the points on the lines that are closest together.

(a) $\begin{bmatrix} x \\ y \\ z \end{bmatrix} = \begin{bmatrix} 3 \\ 0 \\ 1 \end{bmatrix} + s \begin{bmatrix} 2 \\ 1 \\ -3 \end{bmatrix}$; $\begin{bmatrix} x \\ y \\ z \end{bmatrix} = \begin{bmatrix} 1 \\ 1 \\ -1 \end{bmatrix} + t \begin{bmatrix} 1 \\ 0 \\ 1 \end{bmatrix}$

◆(b) $\begin{bmatrix} x \\ y \\ z \end{bmatrix} = \begin{bmatrix} 1 \\ -1 \\ 0 \end{bmatrix} + s \begin{bmatrix} 1 \\ 1 \\ 1 \end{bmatrix}$; $\begin{bmatrix} x \\ y \\ z \end{bmatrix} = \begin{bmatrix} 2 \\ -1 \\ 3 \end{bmatrix} + t \begin{bmatrix} 3 \\ 1 \\ 0 \end{bmatrix}$

(c) $\begin{bmatrix} x \\ y \\ z \end{bmatrix} = \begin{bmatrix} 3 \\ 1 \\ -1 \end{bmatrix} + s \begin{bmatrix} 1 \\ 1 \\ -1 \end{bmatrix}$; $\begin{bmatrix} x \\ y \\ z \end{bmatrix} = \begin{bmatrix} 1 \\ 2 \\ 0 \end{bmatrix} + t \begin{bmatrix} 1 \\ 0 \\ 2 \end{bmatrix}$

◆(d) $\begin{bmatrix} x \\ y \\ z \end{bmatrix} = \begin{bmatrix} 1 \\ 2 \\ 3 \end{bmatrix} + s \begin{bmatrix} 2 \\ 0 \\ -1 \end{bmatrix}$; $\begin{bmatrix} x \\ y \\ z \end{bmatrix} = \begin{bmatrix} 3 \\ -1 \\ 0 \end{bmatrix} + t \begin{bmatrix} 1 \\ 1 \\ 0 \end{bmatrix}$

25. Show that two lines in the plane with slopes m_1 and m_2 are perpendicular if and only if $m_1 m_2 = -1$. [*Hint*: Example 11 Section 4.1.]

26. (a) Show that, of the four diagonals of a cube, no pair is perpendicular.

    ◆(b) Show that each diagonal is perpendicular to the face diagonals it does not meet.

27. Given a rectangular solid with sides of lengths 1, 1, and $\sqrt{2}$, find the angle between a diagonal and one of the longest sides.

◆28. Consider a rectangular solid with sides of lengths a, b, and c. Show that it has two orthogonal diagonals if and only if the sum of two of a^2, b^2, and c^2 equals the third.

29. Let A, B, and $C(2, -1, 1)$ be the vertices of a triangle where $\overrightarrow{AB}$ is parallel to $\begin{bmatrix} 1 \\ -1 \\ 1 \end{bmatrix}$, $\overrightarrow{AC}$ is parallel to $\begin{bmatrix} 2 \\ 0 \\ -1 \end{bmatrix}$, and angle $C = 90°$. Find the equation of the line through B and C.

30. If the diagonals of a parallelogram have equal length, show that the parallelogram is a rectangle.

31. Given $\mathbf{v} = \begin{bmatrix} x \\ y \\ z \end{bmatrix}$ in component form, show that the projections of $\mathbf{v}$ on $\mathbf{i}$, $\mathbf{j}$, and $\mathbf{k}$ are $x\mathbf{i}$, $y\mathbf{j}$, and $z\mathbf{k}$, respectively.

32. (a) Can $\mathbf{u} \cdot \mathbf{v} = -7$ if $\|\mathbf{u}\| = 3$ and $\|\mathbf{v}\| = 2$? Defend your answer.

    (b) Find $\mathbf{u} \cdot \mathbf{v}$ if $\mathbf{u} = \begin{bmatrix} 2 \\ -1 \\ 2 \end{bmatrix}$, $\|\mathbf{v}\| = 6$, and the angle between $\mathbf{u}$ and $\mathbf{v}$ is $\frac{2\pi}{3}$.

33. Show that $(\mathbf{u} + \mathbf{v}) \cdot (\mathbf{u} - \mathbf{v}) = \|\mathbf{u}\|^2 - \|\mathbf{v}\|^2$ for any vectors $\mathbf{u}$ and $\mathbf{v}$.

34. (a) Show that $\|\mathbf{u} + \mathbf{v}\|^2 + \|\mathbf{u} - \mathbf{v}\|^2 = 2(\|\mathbf{u}\|^2 + \|\mathbf{v}\|^2)$ for any vectors $\mathbf{u}$ and $\mathbf{v}$.

    ◆(b) What does this say about parallelograms?

35. Show that if the diagonals of a parallelogram are perpendicular, it is necessarily a rhombus. [*Hint*: Example 5.]

36. Let A and B be the end points of a diameter of a circle (see the diagram). If C is any point on the circle, show that AC and BC are perpendicular. [*Hint*: Express $\overrightarrow{AC}$ and $\overrightarrow{BC}$ in terms of $\mathbf{u} = \overrightarrow{OA}$ and $\mathbf{v} = \overrightarrow{OC}$, where O is the centre.]

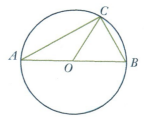

37. Show that $\mathbf{u}$ and $\mathbf{v}$ are orthogonal, if and only if $\|\mathbf{u} + \mathbf{v}\|^2 = \|\mathbf{u}\|^2 + \|\mathbf{v}\|^2$.

38. Let $\mathbf{u}$, $\mathbf{v}$, and $\mathbf{w}$ be pairwise orthogonal vectors.

    (a) Show that $\|\mathbf{u} + \mathbf{v} + \mathbf{w}\|^2 = \|\mathbf{u}\|^2 + \|\mathbf{v}\|^2 + \|\mathbf{w}\|^2$.

    ◆(b) If $\mathbf{u}$, $\mathbf{v}$, and $\mathbf{w}$ are all the same length, show that they all make the same angle with $\mathbf{u} + \mathbf{v} + \mathbf{w}$.

39. (a) Show that $\mathbf{n} = \begin{bmatrix} a \\ b \end{bmatrix}$ is orthogonal to every vector along the line $ax + by + c = 0$.

◆(b) Show that the shortest distance from $P_0(x_0, y_0)$ to the line is $\dfrac{|ax_0 + by_0 + c|}{\sqrt{a^2 + b^2}}$.

[*Hint*: If P_1 is on the line, project $\mathbf{u} = \overrightarrow{P_1 P_0}$ on $\mathbf{n}$.]

40. Assume $\mathbf{u}$ and $\mathbf{v}$ are nonzero vectors that are not parallel. Show that $\mathbf{w} = \|\mathbf{u}\|\mathbf{v} + \|\mathbf{v}\|\mathbf{u}$ is a nonzero vector that bisects the angle between $\mathbf{u}$ and $\mathbf{v}$.

41. Let α, β, and γ be the angles a vector $\mathbf{v} \neq \mathbf{0}$ makes with the positive x, y, and z axes, respectively. Then $\cos \alpha$, $\cos \beta$, and $\cos \gamma$ are called the **direction cosines** of the vector $\mathbf{v}$.

(a) If $\mathbf{v} = \begin{bmatrix} a \\ b \\ c \end{bmatrix}$, show that $\cos \alpha = \dfrac{a}{\|\mathbf{v}\|}$, $\cos \beta = \dfrac{b}{\|\mathbf{v}\|}$, and $\cos \gamma = \dfrac{c}{\|\mathbf{v}\|}$.

◆(b) Show that $\cos^2 \alpha + \cos^2 \beta + \cos^2 \gamma = 1$.

42. Let $\mathbf{v} \neq \mathbf{0}$ be any nonzero vector and suppose that a vector $\mathbf{u}$ can be written as $\mathbf{u} = \mathbf{p} + \mathbf{q}$, where $\mathbf{p}$ is parallel to $\mathbf{v}$ and $\mathbf{q}$ is orthogonal to $\mathbf{v}$. Show that $\mathbf{p}$ must equal the projection of $\mathbf{u}$ on $\mathbf{v}$. [*Hint*: Argue as in the proof of Theorem 4.]

43. Let $\mathbf{v} \neq \mathbf{0}$ be a nonzero vector and let $a \neq 0$ be a scalar. If $\mathbf{u}$ is any vector, show that the projection of $\mathbf{u}$ on $\mathbf{v}$ equals the projection of $\mathbf{u}$ on $a\mathbf{v}$.

44. (a) Show that the **Cauchy-Schwarz inequality** $|\mathbf{u} \cdot \mathbf{v}| \leq \|\mathbf{u}\|\|\mathbf{v}\|$ holds for all vectors $\mathbf{u}$ and $\mathbf{v}$. [*Hint*: $|\cos \theta| \leq 1$ for all angles θ.]

(b) Show that $|\mathbf{u} \cdot \mathbf{v}| = \|\mathbf{u}\|\|\mathbf{v}\|$ if and only if $\mathbf{u}$ and $\mathbf{v}$ are parallel. [*Hint*: When is $\cos \theta = \pm 1$?]

(c) Show that
$$|x_1 x_2 + y_1 y_2 + z_1 z_2| \leq \sqrt{x_1^2 + y_1^2 + z_1^2} \sqrt{x_2^2 + y_2^2 + z_2^2}$$
holds for all numbers x_1, x_2, y_1, y_2, z_1, and z_2.

◆(d) Show that $|xy + yz + zx| \leq x^2 + y^2 + z^2$ for all x, y, and z.

(e) Show that $(x + y + z)^2 \leq 3(x^2 + y^2 + z^2)$ holds for all x, y, and z.

45. Prove that the **triangle inequality** $\|\mathbf{u} \cdot \mathbf{v}\| \leq \|\mathbf{u}\| + \|\mathbf{v}\|$ holds for all vectors $\mathbf{u}$ and $\mathbf{v}$. [*Hint*: Consider the triangle with $\mathbf{u}$ and $\mathbf{v}$ as two sides.]

SECTION 4.3 More on the Cross Product

The cross product $\mathbf{v} \times \mathbf{w}$ of two $\mathbb{R}^3$-vectors $\mathbf{v} = \begin{bmatrix} x_1 \\ y_1 \\ z_1 \end{bmatrix}$ and $\mathbf{w} = \begin{bmatrix} x_2 \\ y_2 \\ z_2 \end{bmatrix}$ was defined in Section 4.2 where we observed that it can be best remembered using a determinant:

$$\mathbf{v} \times \mathbf{w} = \det \begin{bmatrix} \mathbf{i} & x_1 & x_2 \\ \mathbf{j} & y_1 & y_2 \\ \mathbf{k} & z_1 & z_2 \end{bmatrix} = \begin{vmatrix} y_1 & y_2 \\ z_1 & z_2 \end{vmatrix} \mathbf{i} - \begin{vmatrix} x_1 & x_2 \\ z_1 & z_2 \end{vmatrix} \mathbf{j} + \begin{vmatrix} x_1 & x_2 \\ y_1 & y_2 \end{vmatrix} \mathbf{k} \qquad (*)$$

Here $\mathbf{i} = \begin{bmatrix} 1 \\ 0 \\ 0 \end{bmatrix}$, $\mathbf{j} = \begin{bmatrix} 0 \\ 1 \\ 0 \end{bmatrix}$, and $\mathbf{k} = \begin{bmatrix} 1 \\ 0 \\ 0 \end{bmatrix}$ are the coordinate vectors, and the determinant is expanded along the first column. We observed (but did not prove) in Theorem 5 Section 4.2 that $\mathbf{v} \times \mathbf{w}$ is orthogonal to both $\mathbf{v}$ and $\mathbf{w}$. This follows easily from the next result.

Theorem 1

If $\mathbf{u} = \begin{bmatrix} x_0 \\ y_0 \\ z_0 \end{bmatrix}$, $\mathbf{v} = \begin{bmatrix} x_1 \\ y_1 \\ z_1 \end{bmatrix}$, and $\mathbf{w} = \begin{bmatrix} x_2 \\ y_2 \\ z_2 \end{bmatrix}$, then $\mathbf{u} \cdot (\mathbf{v} \times \mathbf{w}) = \det \begin{bmatrix} x_0 & x_1 & x_2 \\ y_0 & y_1 & y_2 \\ z_0 & z_1 & z_2 \end{bmatrix}$.

PROOF

Recall that $\mathbf{u} \cdot (\mathbf{v} \times \mathbf{w})$ is computed by multiplying corresponding components of $\mathbf{u}$ and $\mathbf{v} \times \mathbf{w}$ and then adding. Using (*), the result is:

$$\mathbf{u} \cdot (\mathbf{v} \times \mathbf{w}) = x_0 \left(\begin{vmatrix} y_1 & y_2 \\ z_1 & z_2 \end{vmatrix} \right) + y_0 \left(-\begin{vmatrix} x_1 & x_2 \\ z_1 & z_2 \end{vmatrix} \right) + z_0 \left(\begin{vmatrix} x_1 & x_2 \\ y_1 & y_2 \end{vmatrix} \right) = \det \begin{bmatrix} x_0 & x_1 & x_2 \\ y_0 & y_1 & y_2 \\ z_0 & z_1 & z_2 \end{bmatrix}$$

where the last determinant is expanded along column 1.

The result in Theorem 1 can be succinctly stated as follows: If $\mathbf{u}$, $\mathbf{v}$, and $\mathbf{w}$ are three vectors in $\mathbb{R}^3$, then

$$\mathbf{u} \cdot (\mathbf{v} \times \mathbf{w}) = \det[\mathbf{u} \, \mathbf{v} \, \mathbf{w}]$$

where $[\mathbf{u} \, \mathbf{v} \, \mathbf{w}]$ denotes the matrix with $\mathbf{u}$, $\mathbf{v}$, and $\mathbf{w}$ as its columns. Now it is clear that $\mathbf{v} \times \mathbf{w}$ is orthogonal to both $\mathbf{v}$ and $\mathbf{w}$ because the determinant of a matrix is zero if two columns are identical.

Because of (*) and Theorem 1, several of the following properties of the cross product follow from properties of determinants (they can also be verified directly).

Theorem 2

Let $\mathbf{u}$, $\mathbf{v}$, and $\mathbf{w}$ denote arbitrary vectors in $\mathbb{R}^3$.

1. $\mathbf{u} \times \mathbf{v}$ is a vector.
2. $\mathbf{u} \times \mathbf{v}$ is orthogonal to both $\mathbf{u}$ and $\mathbf{v}$.
3. $\mathbf{u} \times \mathbf{0} = \mathbf{0} = \mathbf{0} \times \mathbf{u}$.
4. $\mathbf{u} \times \mathbf{u} = \mathbf{0}$.
5. $\mathbf{u} \times \mathbf{v} = -(\mathbf{v} \times \mathbf{u})$.
6. $(k\mathbf{u}) \times \mathbf{v} = k(\mathbf{u} \times \mathbf{v}) = \mathbf{u} \times (k\mathbf{v})$ for any scalar k.
7. $\mathbf{u} \times (\mathbf{v} + \mathbf{w}) = (\mathbf{u} \times \mathbf{v}) + (\mathbf{u} \times \mathbf{w})$.
8. $(\mathbf{v} + \mathbf{w}) \times \mathbf{u} = (\mathbf{v} \times \mathbf{u}) + (\mathbf{w} \times \mathbf{u})$.

PROOF

(1) is clear; (2) follows from Theorem 1; and (3) and (4) follow because the determinant of a matrix is zero if one column is zero or if two columns are identical. If two columns are interchanged, the determinant changes sign, and this proves (5). The proofs of (6), (7), and (8) are left as Exercise 15.

SECTION 4.3 More on the Cross Product

We now come to a fundamental relationship between the dot and cross products.

Joseph Louis Lagrange.
Photo © Corbis.

Theorem 3

Lagrange Identity[11]
If $\mathbf{u}$ and $\mathbf{v}$ are any two vectors in $\mathbb{R}^3$, then
$$\|\mathbf{u} \times \mathbf{v}\|^2 = \|\mathbf{u}\|^2 \|\mathbf{v}\|^2 - (\mathbf{u} \cdot \mathbf{v})^2$$

PROOF

Given $\mathbf{u}$ and $\mathbf{v}$, introduce a coordinate system and write $\mathbf{u} = \begin{bmatrix} x_1 \\ y_1 \\ z_1 \end{bmatrix}$ and $\mathbf{v} = \begin{bmatrix} x_2 \\ y_2 \\ z_2 \end{bmatrix}$ in component form. Then all the terms in the identity can be computed in terms of the components. The detailed proof is left as Exercise 14.

An expression for the magnitude of the vector $\mathbf{u} \times \mathbf{v}$ can be easily obtained from the Lagrange identity. If θ is the angle between $\mathbf{u}$ and $\mathbf{v}$, substituting $\mathbf{u} \cdot \mathbf{v} = \|\mathbf{u}\| \|\mathbf{v}\| \cos \theta$ into the Lagrange identity gives
$$\|\mathbf{u} \times \mathbf{v}\|^2 = \|\mathbf{u}\|^2 \|\mathbf{v}\|^2 - \|\mathbf{u}\|^2 \|\mathbf{v}\|^2 \cos^2 \theta = \|\mathbf{u}\|^2 \|\mathbf{v}\|^2 \sin^2 \theta$$
using the fact that $1 - \cos^2 \theta = \sin^2 \theta$. But $\sin \theta$ is nonnegative on the range $0 \leq \theta \leq \pi$, so taking the positive square root of both sides gives
$$\|\mathbf{u} \times \mathbf{v}\| = \|\mathbf{u}\| \|\mathbf{v}\| \sin \theta$$

This expression for $\|\mathbf{u} \times \mathbf{v}\|$ makes no reference to a coordinate system and, moreover, it has a nice geometrical interpretation. The parallelogram determined by the vectors $\mathbf{u}$ and $\mathbf{v}$ has base length $\|\mathbf{v}\|$ and altitude $\|\mathbf{u}\| \sin \theta$ (see Figure 1). Hence the area of the parallelogram formed by $\mathbf{u}$ and $\mathbf{v}$ is
$$(\|\mathbf{u}\| \sin \theta) \|\mathbf{v}\| = \|\mathbf{u} \times \mathbf{v}\|$$
This proves the first part of Theorem 4.

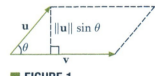

FIGURE 1

Theorem 4

If $\mathbf{u}$ and $\mathbf{v}$ are two nonzero vectors and θ is the angle between $\mathbf{u}$ and $\mathbf{v}$, then
1. $\|\mathbf{u} \times \mathbf{v}\| = \|\mathbf{u}\| \|\mathbf{v}\| \sin \theta$ = area of the parallelogram determined by $\mathbf{u}$ and $\mathbf{v}$.
2. $\mathbf{u}$ and $\mathbf{v}$ are parallel if and only if $\mathbf{u} \times \mathbf{v} = \mathbf{0}$.

---

[11] Joseph Louis Lagrange (1736–1813) was born in Italy and spent his early years in Turin. At the age of 19 he solved a famous problem by inventing an entirely new method, known today as the calculus of variations, and went on to become one of the greatest mathematicians of all time. His work brought a new level of rigour to analysis and his *Mécanique Analytique* is a masterpiece in which he introduced methods still in use. In 1766 he was appointed to the Berlin Academy by Frederik the Great who asserted that the "greatest mathematician in Europe" should be at the court of the "greatest king in Europe." After the death of Frederick, Lagrange went to Paris at the invitation of Louis XVI. He remained there throughout the revolution and was made a count by Napoleon.

PROOF OF (2)

By (1), $\mathbf{u} \times \mathbf{v} = \mathbf{0}$ if and only if the area of the parallelogram is zero. By Figure 1 the area vanishes if and only if $\mathbf{u}$ and $\mathbf{v}$ have the same or opposite direction—that is, if and only if they are parallel.

EXAMPLE 1

Find the area of the triangle with vertices $P(2, 1, 0)$, $Q(3, -1, 1)$, and $R(1, 0, 1)$.

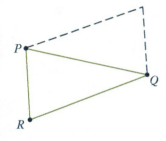

Solution ▶ We have $\overrightarrow{RP} = \begin{bmatrix} 1 \\ 1 \\ -1 \end{bmatrix}$ and $\overrightarrow{RQ} = \begin{bmatrix} 2 \\ -1 \\ 0 \end{bmatrix}$. The area of the triangle is half the area of the parallelogram (see the diagram), and so equals $\frac{1}{2}\|\overrightarrow{RP} \times \overrightarrow{RQ}\|$. We have

$$\overrightarrow{RP} \times \overrightarrow{RQ} = \det \begin{bmatrix} \mathbf{i} & 1 & 2 \\ \mathbf{j} & 1 & -1 \\ \mathbf{k} & -1 & 0 \end{bmatrix} = \begin{bmatrix} -1 \\ -2 \\ -3 \end{bmatrix},$$

so the area of the triangle is $\frac{1}{2}\|\overrightarrow{RP} \times \overrightarrow{RQ}\| = \frac{1}{2}\sqrt{1 + 4 + 9} = \frac{1}{2}\sqrt{14}$.

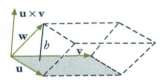

FIGURE 2

If three vectors $\mathbf{u}$, $\mathbf{v}$, and $\mathbf{w}$ are given, they determine a "squashed" rectangular solid called a **parallelepiped** (Figure 2), and it is often useful to be able to find the volume of such a solid. The base of the solid is the parallelogram determined by $\mathbf{u}$ and $\mathbf{v}$, so it has area $A = \|\mathbf{u} \times \mathbf{v}\|$ by Theorem 4. The height of the solid is the length h of the projection of $\mathbf{w}$ on $\mathbf{u} \times \mathbf{v}$. Hence

$$h = \left|\frac{\mathbf{w} \cdot (\mathbf{u} \times \mathbf{v})}{\|\mathbf{u} \times \mathbf{v}\|^2}\right| \|\mathbf{u} \times \mathbf{v}\| = \frac{|\mathbf{w} \cdot (\mathbf{u} \times \mathbf{v})|}{\|\mathbf{u} \times \mathbf{v}\|} = \frac{|\mathbf{w} \cdot (\mathbf{u} \times \mathbf{v})|}{A}$$

Thus the volume of the parallelepiped is $hA = |\mathbf{w} \cdot (\mathbf{u} \times \mathbf{v})|$. This proves

Theorem 5

The volume of the parallelepiped determined by three vectors $\mathbf{w}$, $\mathbf{u}$, and $\mathbf{v}$ (Figure 2) is given by $|\mathbf{w} \cdot (\mathbf{u} \times \mathbf{v})|$.

EXAMPLE 2

Find the volume of the parallelepiped determined by the vectors

$$\mathbf{w} = \begin{bmatrix} 1 \\ 2 \\ -1 \end{bmatrix}, \mathbf{u} = \begin{bmatrix} 1 \\ 1 \\ 0 \end{bmatrix}, \text{ and } \mathbf{v} = \begin{bmatrix} -2 \\ 0 \\ 1 \end{bmatrix}.$$

Solution ▶ By Theorem 1, $\mathbf{w} \cdot (\mathbf{u} \times \mathbf{v}) = \det \begin{bmatrix} 1 & 1 & -2 \\ 2 & 1 & 0 \\ -1 & 0 & 1 \end{bmatrix} = -3$.

Hence the volume is $|\mathbf{w} \cdot (\mathbf{u} \times \mathbf{v})| = |-3| = 3$ by Theorem 5.

SECTION 4.3 More on the Cross Product

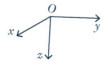

Left-hand system

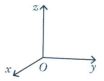

Right-hand system

FIGURE 3

We can now give an intrinsic description of the cross product $\mathbf{u} \times \mathbf{v}$. Its magnitude $\|\mathbf{u} \times \mathbf{v}\| = \|\mathbf{u}\|\|\mathbf{v}\| \sin \theta$ is coordinate-free. If $\mathbf{u} \times \mathbf{v} \neq \mathbf{0}$, its direction is very nearly determined by the fact that it is orthogonal to both $\mathbf{u}$ and $\mathbf{v}$ and so points along the line normal to the plane determined by $\mathbf{u}$ and $\mathbf{v}$. It remains only to decide which of the two possible directions is correct.

Before this can be done, the basic issue of how coordinates are assigned must be clarified. When coordinate axes are chosen in space, the procedure is as follows: An origin is selected, two perpendicular lines (the x and y axes) are chosen through the origin, and a positive direction on each of these axes is selected quite arbitrarily. Then the line through the origin normal to this x-y plane is called the z axis, but there is a choice of which direction on this axis is the positive one. The two possibilities are shown in Figure 3, and it is a standard convention that cartesian coordinates are always **right-hand coordinate systems**. The reason for this terminology is that, in such a system, if the z axis is grasped in the right hand with the thumb pointing in the positive z direction, then the fingers curl around from the positive x axis to the positive y axis (through a right angle).

Suppose now that $\mathbf{u}$ and $\mathbf{v}$ are given and that θ is the angle between them (so $0 \leq \theta \leq \pi$). Then the direction of $\|\mathbf{u} \times \mathbf{v}\|$ is given by the right-hand rule.

Right-hand Rule

If the vector $\mathbf{u} \times \mathbf{v}$ is grasped in the right hand and the fingers curl around from $\mathbf{u}$ to $\mathbf{v}$ through the angle θ, the thumb points in the direction for $\mathbf{u} \times \mathbf{v}$.

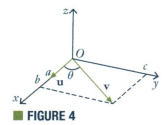

FIGURE 4

To indicate why this is true, introduce coordinates in $\mathbb{R}^3$ as follows: Let $\mathbf{u}$ and $\mathbf{v}$ have a common tail O, choose the origin at O, choose the x axis so that $\mathbf{u}$ points in the positive x direction, and then choose the y axis so that $\mathbf{v}$ is in the x-y plane and the positive y axis is on the same side of the x axis as $\mathbf{v}$. Then, in this system, $\mathbf{u}$ and $\mathbf{v}$ have component form $\mathbf{u} = \begin{bmatrix} a \\ 0 \\ 0 \end{bmatrix}$ and $\mathbf{v} = \begin{bmatrix} b \\ c \\ 0 \end{bmatrix}$ where $a > 0$ and $c > 0$. The situation is depicted in Figure 4. The right-hand rule asserts that $\mathbf{u} \times \mathbf{v}$ should point in the positive z direction. But our definition of $\mathbf{u} \times \mathbf{v}$ gives

$$\mathbf{u} \times \mathbf{v} = \det \begin{bmatrix} \mathbf{i} & a & b \\ \mathbf{j} & 0 & c \\ \mathbf{k} & 0 & 0 \end{bmatrix} = \begin{bmatrix} 0 \\ 0 \\ ac \end{bmatrix} = (ac)\mathbf{k}$$

and $(ac)\mathbf{k}$ has the positive z direction because $ac > 0$.

EXERCISES 4.3

1. If $\mathbf{i}$, $\mathbf{j}$, and $\mathbf{k}$ are the coordinate vectors, verify that $\mathbf{i} \times \mathbf{j} = \mathbf{k}$, $\mathbf{j} \times \mathbf{k} = \mathbf{i}$, and $\mathbf{k} \times \mathbf{i} = \mathbf{j}$.

2. Show that $\mathbf{u} \times (\mathbf{v} \times \mathbf{w})$ need not equal $(\mathbf{u} \times \mathbf{v}) \times \mathbf{w}$ by calculating both when
$\mathbf{u} = \begin{bmatrix} 1 \\ 1 \\ 1 \end{bmatrix}$, $\mathbf{v} = \begin{bmatrix} 1 \\ 1 \\ 0 \end{bmatrix}$, and $\mathbf{w} = \begin{bmatrix} 0 \\ 0 \\ 1 \end{bmatrix}$.

3. Find two unit vectors orthogonal to both $\mathbf{u}$ and $\mathbf{v}$ if:

   (a) $\mathbf{u} = \begin{bmatrix} 1 \\ 2 \\ 2 \end{bmatrix}$, $\mathbf{v} = \begin{bmatrix} 2 \\ -1 \\ 2 \end{bmatrix}$ ◆(b) $\mathbf{u} = \begin{bmatrix} 1 \\ 2 \\ -1 \end{bmatrix}$, $\mathbf{v} = \begin{bmatrix} 3 \\ 1 \\ 2 \end{bmatrix}$

4. Find the area of the triangle with the following vertices.

   (a) $A(3, -1, 2)$, $B(1, 1, 0)$, and $C(1, 2, -1)$

   ◆(b) $A(3, 0, 1)$, $B(5, 1, 0)$, and $C(7, 2, -1)$

   (c) $A(1, 1, -1)$, $B(2, 0, 1)$, and $C(1, -1, 3)$

   ◆(d) $A(3, -1, 1)$, $B(4, 1, 0)$, and $C(2, -3, 0)$

5. Find the volume of the parallelepiped determined by $\mathbf{w}$, $\mathbf{u}$, and $\mathbf{v}$ when:

   (a) $\mathbf{w} = \begin{bmatrix} 2 \\ 1 \\ 1 \end{bmatrix}$, $\mathbf{v} = \begin{bmatrix} 1 \\ 0 \\ 2 \end{bmatrix}$, and $\mathbf{u} = \begin{bmatrix} 2 \\ 1 \\ -1 \end{bmatrix}$

   ◆(b) $\mathbf{w} = \begin{bmatrix} 1 \\ 0 \\ 3 \end{bmatrix}$, $\mathbf{v} = \begin{bmatrix} 2 \\ 1 \\ -3 \end{bmatrix}$, and $\mathbf{u} = \begin{bmatrix} 1 \\ 1 \\ 1 \end{bmatrix}$

6. Let P_0 be a point with vector $\mathbf{p}_0$, and let $ax + by + cz = d$ be the equation of a plane with normal $\mathbf{n} = \begin{bmatrix} a \\ b \\ c \end{bmatrix}$.

   (a) Show that the point on the plane closest to P_0 has vector $\mathbf{p}$ given by
   $$\mathbf{p} = \mathbf{p}_0 + \frac{d - (\mathbf{p}_0 \cdot \mathbf{n})}{\|\mathbf{n}\|^2} \mathbf{n}.$$
   [Hint: $\mathbf{p} = \mathbf{p}_0 + t\mathbf{n}$ for some t, and $\mathbf{p} \cdot \mathbf{n} = d$.]

   ◆(b) Show that the shortest distance from P_0 to the plane is $\dfrac{|d - (\mathbf{p}_0 \cdot \mathbf{n})|}{\|\mathbf{n}\|}$.

   (c) Let P_0' denote the reflection of P_0 in the plane—that is, the point on the opposite side of the plane such that the line through P_0 and P_0' is perpendicular to the plane. Show that $\mathbf{p}_0 + 2\dfrac{d - (\mathbf{p}_0 \cdot \mathbf{n})}{\|\mathbf{n}\|^2}\mathbf{n}$ is the vector of P_0'.

7. Simplify $(a\mathbf{u} + b\mathbf{v}) \times (c\mathbf{u} + d\mathbf{v})$.

8. Show that the shortest distance from a point P to the line through P_0 with direction vector $\mathbf{d}$ is $\dfrac{\|\overrightarrow{P_0P} \times \mathbf{d}\|}{\|\mathbf{d}\|}$.

9. Let $\mathbf{u}$ and $\mathbf{v}$ be nonzero, nonorthogonal vectors. If θ is the angle between them, show that $\tan \theta = \dfrac{\|\mathbf{u} \times \mathbf{v}\|}{\mathbf{u} \cdot \mathbf{v}}$.

◆10. Show that points A, B, and C are all on one line if and only if $\overrightarrow{AB} \times \overrightarrow{AC} = \mathbf{0}$.

11. Show that points A, B, C, and D are all on one plane if and only if $\overrightarrow{AB} \cdot (\overrightarrow{AB} \times \overrightarrow{AC}) = 0$.

◆12. Use Theorem 5 to confirm that, if $\mathbf{u}$, $\mathbf{v}$, and $\mathbf{w}$ are mutually perpendicular, the (rectangular) parallelepiped they determine has volume $\|\mathbf{u}\|\|\mathbf{v}\|\|\mathbf{w}\|$.

13. Show that the volume of the parallelepiped determined by $\mathbf{u}$, $\mathbf{v}$, and $\mathbf{u} \times \mathbf{v}$ is $\|\mathbf{u} \times \mathbf{v}\|^2$.

14. Complete the proof of Theorem 3.

15. Prove the following properties in Theorem 2.

   (a) Property 6   ◆(b) Property 7

   (c) Property 8

16. (a) Show that
   $$\mathbf{w} \cdot (\mathbf{u} \times \mathbf{v}) = \mathbf{u} \cdot (\mathbf{v} \times \mathbf{w}) = \mathbf{v} \times (\mathbf{w} \times \mathbf{u})$$
   holds for all vectors $\mathbf{w}$, $\mathbf{u}$, and $\mathbf{v}$.

   ◆(b) Show that $\mathbf{v} - \mathbf{w}$ and $(\mathbf{u} \times \mathbf{v}) + (\mathbf{v} \times \mathbf{w}) + (\mathbf{w} \times \mathbf{u})$ are orthogonal.

17. Show that $\mathbf{u} \times (\mathbf{v} \times \mathbf{w}) = (\mathbf{u} \cdot \mathbf{w})\mathbf{v} - (\mathbf{u} \times \mathbf{v})\mathbf{w}$.
   [Hint: First do it for $\mathbf{u} = \mathbf{i}$, $\mathbf{j}$, and $\mathbf{k}$; then write $\mathbf{u} = x\mathbf{i} + y\mathbf{j} + z\mathbf{k}$ and use Theorem 2.]

18. Prove the **Jacobi identity**:
   $\mathbf{u} \times (\mathbf{v} \times \mathbf{w}) + \mathbf{v} \times (\mathbf{w} \times \mathbf{u}) + \mathbf{w} \times (\mathbf{u} \times \mathbf{v}) = \mathbf{0}$.
   [Hint: The preceding exercise.]

19. Show that
   $(\mathbf{u} \times \mathbf{v}) \cdot (\mathbf{w} \times \mathbf{z}) = \det\begin{bmatrix} \mathbf{u} \cdot \mathbf{w} & \mathbf{u} \cdot \mathbf{z} \\ \mathbf{v} \cdot \mathbf{w} & \mathbf{v} \cdot \mathbf{z} \end{bmatrix}$.
   [Hint: Exercises 16 and 17.]

20. Let P, Q, R, and S be four points, not all on one plane, as in the diagram. Show that the volume of the pyramid they determine is
   $\frac{1}{6}|\overrightarrow{PQ} \cdot (\overrightarrow{PR} \times \overrightarrow{PS})|$.

   [Hint: The volume of a cone with base area A and height h as in the diagram below right is $\frac{1}{3}Ah$.]

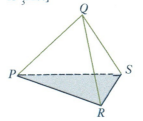
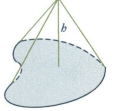

21. Consider a triangle with vertices A, B, and C, as in the diagram below. Let α, β, and γ denote the angles at A, B, and C, respectively, and let a, b, and c denote the lengths of the sides opposite A, B, and C, respectively. Write $\mathbf{u} = \overrightarrow{AB}$, $\mathbf{v} = \overrightarrow{BC}$, and $\mathbf{w} = \overrightarrow{CA}$.

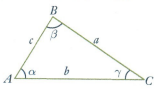

(a) Deduce that $\mathbf{u} + \mathbf{v} + \mathbf{w} = \mathbf{0}$.

(b) Show that $\mathbf{u} \times \mathbf{v} = \mathbf{w} \times \mathbf{u} = \mathbf{v} \times \mathbf{w}$. [*Hint:* Compute $\mathbf{u} \times (\mathbf{u} + \mathbf{v} + \mathbf{w})$ and $\mathbf{v} \times (\mathbf{u} + \mathbf{v} + \mathbf{w})$.]

(c) Deduce the **law of sines**:
$$\frac{\sin \alpha}{a} = \frac{\sin \beta}{b} = \frac{\sin \gamma}{c}$$

◆22. Show that the (shortest) distance between two planes $\mathbf{n} \cdot \mathbf{p} = d_1$ and $\mathbf{n} \cdot \mathbf{p} = d_2$ with $\mathbf{n}$ as normal is $\dfrac{|d_2 - d_1|}{\|\mathbf{n}\|}$.

23. Let A and B be points other than the origin, and let $\mathbf{a}$ and $\mathbf{b}$ be their vectors. If $\mathbf{a}$ and $\mathbf{b}$ are not parallel, show that the plane through A, B, and the origin is given by
$$\left\{P(x, y, z) \bigg| \begin{bmatrix} x \\ y \\ z \end{bmatrix} = s\mathbf{a} + t\mathbf{b} \text{ for some } s \text{ and } t\right\}.$$

24. Let A be a 2×3 matrix of rank 2 with rows $\mathbf{r}_1$ and $\mathbf{r}_2$. Show that $P = \{XA | X = [x\ y]; x, y \text{ arbitrary}\}$ is the plane through the origin with normal $\mathbf{r}_1 \times \mathbf{r}_2$.

25. Given the cube with vertices $P(x, y, z)$, where each of x, y, and z is either 0 or 2, consider the plane perpendicular to the diagonal through $P(0, 0, 0)$ and $P(2, 2, 2)$ and bisecting it.

(a) Show that the plane meets six of the edges of the cube and bisects them.

(b) Show that the six points in (a) are the vertices of a regular hexagon.

SECTION 4.4 Linear Operators on $\mathbb{R}^3$

Recall that a transformation $T: \mathbb{R}^n \to \mathbb{R}^m$ is called *linear* if $T(\mathbf{x} + \mathbf{y}) = T(\mathbf{x}) + T(\mathbf{y})$ and $T(a\mathbf{x}) = aT(\mathbf{x})$ holds for all $\mathbf{x}$ and $\mathbf{y}$ in $\mathbb{R}^n$ and all scalars a. In this case we showed (in Theorem 2 Section 2.6) that there exists an $m \times n$ matrix A such that $T(\mathbf{x}) = A\mathbf{x}$ for all $\mathbf{x}$ in $\mathbb{R}^n$, and we say that T is the **matrix transformation induced** by A.

Definition 4.9 *A linear transformation*
$$T: \mathbb{R}^n \to \mathbb{R}^n$$
is called a **linear operator** *on* $\mathbb{R}^n$.

In Section 2.6 we investigated three important linear operators on $\mathbb{R}^2$: rotations about the origin, reflections in a line through the origin, and projections on this line.

In this section we investigate the analogous operators on $\mathbb{R}^3$: Rotations about a line through the origin, reflections in a plane through the origin, and projections onto a plane or line through the origin in $\mathbb{R}^3$. In every case we show that the operator is linear, and we find the matrices of all the reflections and projections.

To do this we must prove that these reflections, projections, and rotations are actually *linear* operators on $\mathbb{R}^3$. In the case of reflections and rotations, it is convenient to examine a more general situation. A transformation $T: \mathbb{R}^3 \to \mathbb{R}^3$ is said to be **distance preserving** if the distance between $T(\mathbf{v})$ and $T(\mathbf{w})$ is the same as the distance between $\mathbf{v}$ and $\mathbf{w}$ for all $\mathbf{v}$ and $\mathbf{w}$ in $\mathbb{R}^3$; that is,

$$\|T(\mathbf{v}) - T(\mathbf{w})\| = \|\mathbf{v} - \mathbf{w}\| \text{ for all } \mathbf{v} \text{ and } \mathbf{w} \text{ in } \mathbb{R}^3. \quad (*)$$

Clearly reflections and rotations are distance preserving, and both carry $\mathbf{0}$ to $\mathbf{0}$, so the following theorem shows that they are both linear.

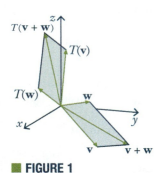

FIGURE 1

Theorem 1

If $T : \mathbb{R}^3 \to \mathbb{R}^3$ is distance preserving, and if $T(\mathbf{0}) = \mathbf{0}$, then T is linear.

PROOF

Since $T(\mathbf{0}) = \mathbf{0}$, taking $\mathbf{w} = \mathbf{0}$ in (*) shows that $\|T(\mathbf{v})\| = \|\mathbf{v}\|$ for all $\mathbf{v}$ in $\mathbb{R}^3$, that is T preserves length. Also, $\|T(\mathbf{v}) - T(\mathbf{w})\|^2 = \|\mathbf{v} - \mathbf{w}\|^2$ by (*). Since $\|\mathbf{v} - \mathbf{w}\|^2 = \|\mathbf{v}\|^2 - 2\mathbf{v} \cdot \mathbf{w} + \|\mathbf{w}\|^2$ always holds, it follows that $T(\mathbf{v}) \cdot T(\mathbf{w}) = \mathbf{v} \cdot \mathbf{w}$ for all $\mathbf{v}$ and $\mathbf{w}$. Hence (by Theorem 2 Section 4.2) the angle between $T(\mathbf{v})$ and $T(\mathbf{w})$ is the same as the angle between $\mathbf{v}$ and $\mathbf{w}$ for all (nonzero) vectors $\mathbf{v}$ and $\mathbf{w}$ in $\mathbb{R}^3$.

With this we can show that T is linear. Given nonzero vectors $\mathbf{v}$ and $\mathbf{w}$ in $\mathbb{R}^3$, the vector $\mathbf{v} + \mathbf{w}$ is the diagonal of the parallelogram determined by $\mathbf{v}$ and $\mathbf{w}$. By the preceding paragraph, the effect of T is to carry this *entire parallelogram* to the parallelogram determined by $T(\mathbf{v})$ and $T(\mathbf{w})$, with diagonal $T(\mathbf{v} + \mathbf{w})$. But this diagonal is $T(\mathbf{v}) + T(\mathbf{w})$ by the parallelogram law (see Figure 1).

In other words, $T(\mathbf{v} + \mathbf{w}) = T(\mathbf{v}) + T(\mathbf{w})$. A similar argument shows that $T(a\mathbf{v}) = aT(\mathbf{v})$ for all scalars a, proving that T is indeed linear.

Distance-preserving linear operators are called **isometries**, and we return to them in Section 10.4.

Reflections and Projections

In Section 2.6 we studied the reflection $Q_m : \mathbb{R}^2 \to \mathbb{R}^2$ in the line $y = mx$ and projection $P_m : \mathbb{R}^2 \to \mathbb{R}^2$ on the same line. We found (in Theorems 5 and 6, Section 2.6) that they are both linear and

$$Q_m \text{ has matrix } \frac{1}{1+m^2}\begin{bmatrix} 1-m^2 & 2m \\ 2m & m^2-1 \end{bmatrix} \quad \text{and} \quad P_m \text{ has matrix } \frac{1}{1+m^2}\begin{bmatrix} 1 & m \\ m & m^2 \end{bmatrix}.$$

We now look at the analogues in $\mathbb{R}^3$.

Let L denote a line through the origin in $\mathbb{R}^3$. Given a vector $\mathbf{v}$ in $\mathbb{R}^3$, the reflection $Q_L(\mathbf{v})$ of $\mathbf{v}$ in L and the projection $P_L(\mathbf{v})$ of $\mathbf{v}$ on L are defined in Figure 2. In the same figure, we see that

$$P_L(\mathbf{v}) = \mathbf{v} + \tfrac{1}{2}[Q_L(\mathbf{v}) - \mathbf{v}] = \tfrac{1}{2}[Q_L(\mathbf{v}) + \mathbf{v}] \qquad (**)$$

so the fact that Q_L is linear (by Theorem 1) shows that P_L is also linear.[12] However,

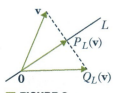

FIGURE 2

Theorem 4 Section 4.2 gives us the matrix of P_L directly. In fact, if $\mathbf{d} = \begin{bmatrix} a \\ b \\ c \end{bmatrix} \neq \mathbf{0}$ is a direction vector for L, and we write $\mathbf{v} = \begin{bmatrix} x \\ y \\ z \end{bmatrix}$, then

$$P_L(\mathbf{v}) = \frac{\mathbf{v} \cdot \mathbf{d}}{\|\mathbf{d}\|^2}\mathbf{d} = \frac{ax+by+cz}{a^2+b^2+c^2}\begin{bmatrix} a \\ b \\ c \end{bmatrix} = \frac{1}{a^2+b^2+c^2}\begin{bmatrix} a^2 & ab & ac \\ ab & b^2 & bc \\ ac & bc & c^2 \end{bmatrix}\begin{bmatrix} x \\ y \\ z \end{bmatrix}$$

as the reader can verify. Note that this shows directly that P_L is a matrix transformation and so gives another proof that it is linear.

---

12 Note that Theorem 1 does *not* apply to P_L since it does not preserve distance.

SECTION 4.4 Linear Operators on $\mathbb{R}^3$

Theorem 2

Let L denote the line through the origin in $\mathbb{R}^3$ with direction vector $\mathbf{d} = \begin{bmatrix} a \\ b \\ c \end{bmatrix} \neq \mathbf{0}$. Then P_L and Q_L are both linear and

$$P_L \text{ has matrix } \frac{1}{a^2 + b^2 + c^2} \begin{bmatrix} a^2 & ab & ac \\ ab & b^2 & bc \\ ac & bc & c^2 \end{bmatrix},$$

$$Q_L \text{ has matrix } \frac{1}{a^2 + b^2 + c^2} \begin{bmatrix} a^2 - b^2 - c^2 & 2ab & 2ac \\ 2ab & b^2 - a^2 - c^2 & 2bc \\ 2ac & 2bc & c^2 - a^2 - b^2 \end{bmatrix}.$$

PROOF

It remains to find the matrix of Q_L. But (**) implies that $Q_L(\mathbf{v}) = 2P_L(\mathbf{v}) - \mathbf{v}$ for each $\mathbf{v}$ in $\mathbb{R}^3$, so if $\mathbf{v} = \begin{bmatrix} x \\ y \\ z \end{bmatrix}$ we obtain (with some matrix arithmetic):

$$Q_L(\mathbf{v}) = \left\{ \frac{2}{a^2 + b^2 + c^2} \begin{bmatrix} a^2 & ab & ac \\ ab & b^2 & bc \\ ac & bc & c^2 \end{bmatrix} - \begin{bmatrix} 1 & 0 & 0 \\ 0 & 1 & 0 \\ 0 & 0 & 1 \end{bmatrix} \right\} \begin{bmatrix} x \\ y \\ z \end{bmatrix}$$

$$= \frac{1}{a^2 + b^2 + c^2} \begin{bmatrix} a^2 - b^2 - c^2 & 2ab & 2ac \\ 2ab & b^2 - a^2 - c^2 & 2bc \\ 2ac & 2bc & c^2 - a^2 - b^2 \end{bmatrix} \begin{bmatrix} x \\ y \\ z \end{bmatrix}$$

as required.

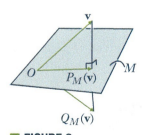

FIGURE 3

In $\mathbb{R}^3$ we can reflect in planes as well as lines. Let M denote a plane through the origin in $\mathbb{R}^3$. Given a vector $\mathbf{v}$ in $\mathbb{R}^3$, the reflection $Q_M(\mathbf{v})$ of $\mathbf{v}$ in M and the projection $P_M(\mathbf{v})$ of $\mathbf{v}$ on M are defined in Figure 3. As above, we have

$$P_M(\mathbf{v}) = \mathbf{v} + \tfrac{1}{2}[Q_M(\mathbf{v}) - \mathbf{v}] = \tfrac{1}{2}[Q_M(\mathbf{v}) + \mathbf{v}]$$

so the fact that Q_M is linear (again by Theorem 1) shows that P_M is also linear. Again we can obtain the matrix directly. If $\mathbf{n}$ is a normal for the plane M, then Figure 3 shows that

$$P_M(\mathbf{v}) = \mathbf{v} - \text{proj}_\mathbf{n}(\mathbf{v}) = \mathbf{v} - \frac{\mathbf{v} \cdot \mathbf{n}}{\|\mathbf{n}\|^2} \mathbf{n} \text{ for all vectors } \mathbf{v}.$$

If $\mathbf{n} = \begin{bmatrix} a \\ b \\ c \end{bmatrix} \neq \mathbf{0}$ and $\mathbf{v} = \begin{bmatrix} x \\ y \\ z \end{bmatrix}$, a computation like the above gives

$$P_M(\mathbf{v}) = \left\{ \begin{bmatrix} 1 & 0 & 0 \\ 0 & 1 & 0 \\ 0 & 0 & 1 \end{bmatrix} \begin{bmatrix} x \\ y \\ z \end{bmatrix} - \frac{ax + by + cz}{a^2 + b^2 + c^2} \begin{bmatrix} a \\ b \\ c \end{bmatrix} \right\} = \frac{1}{a^2 + b^2 + c^2} \begin{bmatrix} b^2 + c^2 & -ab & -ac \\ -ab & a^2 + c^2 & -bc \\ -ac & -bc & b^2 + c^2 \end{bmatrix} \begin{bmatrix} x \\ y \\ z \end{bmatrix}.$$

Chapter 4   Vector Geometry

This proves the first part of

> **Theorem 3**
>
> Let M denote the plane through the origin in $\mathbb{R}^3$ with normal $\mathbf{n} = \begin{bmatrix} a \\ b \\ c \end{bmatrix} \neq \mathbf{0}$. Then P_M and Q_M are both linear and
>
> $$P_M \text{ has matrix } \frac{1}{a^2+b^2+c^2} \begin{bmatrix} b^2+c^2 & -ab & -ac \\ -ab & a^2+c^2 & -bc \\ -ac & -bc & a^2+b^2 \end{bmatrix},$$
>
> $$Q_M \text{ has matrix } \frac{1}{a^2+b^2+c^2} \begin{bmatrix} b^2+c^2-a^2 & -2ab & -2ac \\ -2ab & a^2+c^2-b^2 & -2bc \\ -2ac & -2bc & a^2+b^2-c^2 \end{bmatrix}.$$

> **PROOF**
>
> It remains to compute the matrix of Q_M. Since $Q_M(\mathbf{v}) = 2P_M(\mathbf{v}) - \mathbf{v}$ for each $\mathbf{v}$ in $\mathbb{R}^3$, the computation is similar to the above and is left as an exercise for the reader.

Rotations

In Section 2.6 we studied the rotation $R_\theta : \mathbb{R}^2 \to \mathbb{R}^2$ counterclockwise about the origin through the angle θ. Moreover, we showed in Theorem 4 Section 2.6 that R_θ is linear and has matrix $\begin{bmatrix} \cos\theta & -\sin\theta \\ \sin\theta & \cos\theta \end{bmatrix}$. One extension of this is given in the following example.

> **EXAMPLE 1**
>
> Let $R_{z,\theta} : \mathbb{R}^3 \to \mathbb{R}^3$ denote rotation of $\mathbb{R}^3$ about the z axis through an angle θ from the positive x axis toward the positive y axis. Show that $R_{z,\theta}$ is linear and find its matrix.
>
> **Solution** ▶ First R is distance preserving and so is linear by Theorem 1. Hence we apply Theorem 2 Section 2.6 to obtain the matrix of $R_{z,\theta}$.
>
> Let $\mathbf{i} = \begin{bmatrix} 1 \\ 0 \\ 0 \end{bmatrix}, \mathbf{j} = \begin{bmatrix} 0 \\ 1 \\ 0 \end{bmatrix}$, and $\mathbf{k} = \begin{bmatrix} 0 \\ 0 \\ 1 \end{bmatrix}$ denote the standard basis of $\mathbb{R}^3$; we must find $R_{z,\theta}(\mathbf{i}), R_{z,\theta}(\mathbf{j})$, and $R_{z,\theta}(\mathbf{k})$. Clearly $R_{z,\theta}(\mathbf{k}) = \mathbf{k}$. The effect of $R_{z,\theta}$ on the x-y plane is to rotate it counterclockwise through the angle θ. Hence Figure 4 gives
>
> $$R_{z,\theta}(\mathbf{i}) = \begin{bmatrix} \cos\theta \\ \sin\theta \\ 0 \end{bmatrix}, \quad R_{z,\theta}(\mathbf{j}) = \begin{bmatrix} -\sin\theta \\ \cos\theta \\ 0 \end{bmatrix}$$

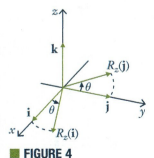

FIGURE 4

so, by Theorem 2 Section 2.6, $R_{z,\theta}$ has matrix

$$[R_{z,\theta}(\mathbf{i})\ R_{z,\theta}(\mathbf{j})\ R_{z,\theta}(\mathbf{k})] = \begin{bmatrix} \cos\theta & -\sin\theta & 0 \\ \sin\theta & \cos\theta & 0 \\ 0 & 0 & 1 \end{bmatrix}.$$

Example 1 begs to be generalized. Given a line L through the origin in $\mathbb{R}^3$, every rotation about L through a fixed angle is clearly distance preserving, and so is a linear operator by Theorem 1. However, giving a precise description of the matrix of this rotation is not easy and will have to wait until more techniques are available.

Transformations of Areas and Volumes

Let $\mathbf{v}$ be a nonzero vector in $\mathbb{R}^3$. Each vector in the same direction as $\mathbf{v}$ whose length is a fraction s of the length of $\mathbf{v}$ has the form $s\mathbf{v}$ (see Figure 5). With this, scrutiny of Figure 6 shows that a vector $\mathbf{u}$ is in the parallelogram determined by $\mathbf{v}$ and $\mathbf{w}$ if and only if it has the form $\mathbf{u} = s\mathbf{v} + t\mathbf{w}$ where $0 \leq s \leq 1$ and $0 \leq t \leq 1$. But then, if $T: \mathbb{R}^3 \to \mathbb{R}^3$ is a linear transformation, we have

$$T(s\mathbf{v} + t\mathbf{w}) = T(s\mathbf{v}) + T(t\mathbf{w}) = sT(\mathbf{v}) + tT(\mathbf{w}).$$

Hence $T(s\mathbf{v} + t\mathbf{w})$ is in the parallelogram determined by $T(\mathbf{v})$ and $T(\mathbf{w})$. Conversely, every vector in this parallelogram has the form $T(s\mathbf{v} + t\mathbf{w})$ where $s\mathbf{v} + t\mathbf{w}$ is in the parallelogram determined by $\mathbf{v}$ and $\mathbf{w}$. For this reason, the parallelogram determined by $T(\mathbf{v})$ and $T(\mathbf{w})$ is called the **image** of the parallelogram determined by $\mathbf{v}$ and $\mathbf{w}$. We record this discussion as:

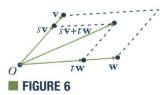

FIGURE 5

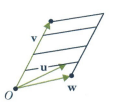

FIGURE 6

Theorem 4

If $T: \mathbb{R}^3 \to \mathbb{R}^3$ (or $\mathbb{R}^2 \to \mathbb{R}^2$) is a linear operator, the image of the parallelogram determined by vectors $\mathbf{v}$ and $\mathbf{w}$ is the parallelogram determined by $T(\mathbf{v})$ and $T(\mathbf{w})$.

This result is illustrated in Figure 7, and was used in Examples 15 and 16 Section 2.2 to reveal the effect of expansion and shear transformations.

Now we are interested in the effect of a linear transformation $T: \mathbb{R}^3 \to \mathbb{R}^3$ on the parallelepiped determined by three vectors $\mathbf{u}$, $\mathbf{v}$, and $\mathbf{w}$ in $\mathbb{R}^3$ (see the discussion preceding Theorem 5 Section 4.3). If T has matrix A, Theorem 4 shows that this parallelepiped is carried to the parallelepiped determined by $T(\mathbf{u}) = A\mathbf{u}$, $T(\mathbf{v}) = A\mathbf{v}$, and $T(\mathbf{w}) = A\mathbf{w}$. In particular, we want to discover how the volume changes, and it turns out to be closely related to the determinant of the matrix A.

Theorem 5

Let $\text{vol}(\mathbf{u}, \mathbf{v}, \mathbf{w})$ denote the volume of the parallelepiped determined by three vectors $\mathbf{u}$, $\mathbf{v}$, and $\mathbf{w}$ in $\mathbb{R}^3$, and let $\text{area}(\mathbf{p}, \mathbf{q})$ denote the area of the parallelogram determined by two vectors $\mathbf{p}$ and $\mathbf{q}$ in $\mathbb{R}^2$. Then:

1. If A is a 3×3 matrix, then $\text{vol}(A\mathbf{u}, A\mathbf{v}, A\mathbf{w}) = |\det(A)| \cdot \text{vol}(\mathbf{u}, \mathbf{v}, \mathbf{w})$.
2. If A is a 2×2 matrix, then $\text{area}(A\mathbf{p}, A\mathbf{q}) = |\det(A)| \cdot \text{area}(\mathbf{p}, \mathbf{q})$.

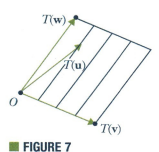

FIGURE 7

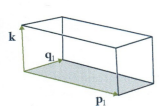

PROOF

1. Let $[\mathbf{u}\ \mathbf{v}\ \mathbf{w}]$ denote the 3×3 matrix with columns $\mathbf{u}$, $\mathbf{v}$, and $\mathbf{w}$. Then
$$\text{vol}(A\mathbf{u}, A\mathbf{v}, A\mathbf{w}) = |A\mathbf{u} \cdot (A\mathbf{v} \times A\mathbf{w})|$$
by Theorem 5 Section 4.3. Now apply Theorem 1 Section 4.3 twice to get
$$A\mathbf{u} \cdot (A\mathbf{v} \times A\mathbf{w}) = \det[A\mathbf{u}\ A\mathbf{v}\ A\mathbf{w}] = \det\{A[\mathbf{u}\ \mathbf{v}\ \mathbf{w}]\}$$
$$= \det(A) \det[\mathbf{u}\ \mathbf{v}\ \mathbf{w}]$$
$$= \det(A)(\mathbf{u} \cdot (\mathbf{v} \times \mathbf{w}))$$
where we used Definition 2.9 and the product theorem for determinants. Finally (1) follows from Theorem 5 Section 4.3 by taking absolute values.

2. Given $\mathbf{p} = \begin{bmatrix} x \\ y \end{bmatrix}$ in $\mathbb{R}^2$, write $\mathbf{p}_1 = \begin{bmatrix} x \\ y \\ 0 \end{bmatrix}$ in $\mathbb{R}^3$. By the diagram,
$$\text{area}(\mathbf{p}, \mathbf{q}) = \text{vol}(\mathbf{p}_1, \mathbf{q}_1, \mathbf{k})$$ where $\mathbf{k}$ is the (length 1) coordinate vector along the z axis. If A is a 2×2 matrix, write $A_1 = \begin{bmatrix} A & 0 \\ 0 & 1 \end{bmatrix}$ in block form, and observe that $(A\mathbf{v})_1 = (A_1 \mathbf{v}_1)$ for all $\mathbf{v}$ in $\mathbb{R}^2$ and $A_1 \mathbf{k} = \mathbf{k}$. Hence part (1) if this theorem shows
$$\text{area}(A\mathbf{p}, A\mathbf{q}) = \text{vol}(A_1\mathbf{p}_1, A_1\mathbf{q}_1, A_1\mathbf{k})$$
$$= |\det(A_1)|\text{vol}(\mathbf{p}_1, \mathbf{q}_1, \mathbf{k})$$
$$= |\det(A)|\text{area}(\mathbf{p}, \mathbf{q})$$
as required.

Define the **unit square** and **unit cube** to be the square and cube corresponding to the coordinate vectors in $\mathbb{R}^2$ and $\mathbb{R}^3$, respectively. Then Theorem 5 gives a geometrical meaning to the determinant of a matrix A:

- If A is a 2×2 matrix, then $|\det(A)|$ is the area of the image of the unit square under multiplication by A;
- If A is a 3×3 matrix, then $|\det(A)|$ is the volume of the image of the unit cube under multiplication by A.

These results, together with the importance of areas and volumes in geometry, were among the reasons for the initial development of determinants.

EXERCISES 4.4

1. In each case show that that T is either projection on a line, reflection in a line, or rotation through an angle, and find the line or angle.

   (a) $T\begin{bmatrix} x \\ y \end{bmatrix} = \frac{1}{5}\begin{bmatrix} x + 2y \\ 2x + 4y \end{bmatrix}$ ◆(b) $T\begin{bmatrix} x \\ y \end{bmatrix} = \frac{1}{2}\begin{bmatrix} x - y \\ y - x \end{bmatrix}$

   (c) $T\begin{bmatrix} x \\ y \end{bmatrix} = \frac{1}{\sqrt{2}}\begin{bmatrix} -x - y \\ x - y \end{bmatrix}$ ◆(d) $T\begin{bmatrix} x \\ y \end{bmatrix} = \frac{1}{5}\begin{bmatrix} -3x + 4y \\ 4x + 3y \end{bmatrix}$

   (e) $T\begin{bmatrix} x \\ y \end{bmatrix} = \begin{bmatrix} -y \\ -x \end{bmatrix}$ ◆(f) $T\begin{bmatrix} x \\ y \end{bmatrix} = \frac{1}{2}\begin{bmatrix} x - \sqrt{3}y \\ \sqrt{3}x + y \end{bmatrix}$

2. Determine the effect of the following transformations.

   (a) Rotation through $\frac{\pi}{2}$, followed by projection on the y axis, followed by reflection in the line $y = x$.

   ◆(b) Projection on the line $y = x$ followed by projection on the line $y = -x$.

   (c) Projection on the x axis followed by reflection in the line $y = x$.

3. In each case solve the problem by finding the matrix of the operator.

   (a) Find the projection of $\mathbf{v} = \begin{bmatrix} 1 \\ -2 \\ 3 \end{bmatrix}$ on the plane with equation $3x - 5y + 2z = 0$.

   ◆(b) Find the projection of $\mathbf{v} = \begin{bmatrix} 0 \\ 1 \\ -3 \end{bmatrix}$ on the plane with equation $2x - y + 4z = 0$.

   (c) Find the reflection of $\mathbf{v} = \begin{bmatrix} 1 \\ -2 \\ 3 \end{bmatrix}$ in the plane with equation $x - y + 3z = 0$.

   ◆(d) Find the reflection of $\mathbf{v} = \begin{bmatrix} 0 \\ 1 \\ -3 \end{bmatrix}$ in the plane with equation $2x + y - 5z = 0$.

   (e) Find the reflection of $\mathbf{v} = \begin{bmatrix} 2 \\ 5 \\ -1 \end{bmatrix}$ in the line with equation $\begin{bmatrix} x \\ y \\ z \end{bmatrix} = t \begin{bmatrix} 1 \\ 1 \\ -2 \end{bmatrix}$.

   ◆(f) Find the projection of $\mathbf{v} = \begin{bmatrix} 1 \\ -1 \\ 7 \end{bmatrix}$ on the line with equation $\begin{bmatrix} x \\ y \\ z \end{bmatrix} = t \begin{bmatrix} 3 \\ 0 \\ 4 \end{bmatrix}$.

   (g) Find the projection of $\mathbf{v} = \begin{bmatrix} 1 \\ 1 \\ -3 \end{bmatrix}$ on the line with equation $\begin{bmatrix} x \\ y \\ z \end{bmatrix} = t \begin{bmatrix} 2 \\ 0 \\ -3 \end{bmatrix}$.

   ◆(h) Find the reflection of $\mathbf{v} = \begin{bmatrix} 2 \\ -5 \\ 0 \end{bmatrix}$ in the line with equation $\begin{bmatrix} x \\ y \\ z \end{bmatrix} = t \begin{bmatrix} 1 \\ 1 \\ -3 \end{bmatrix}$.

4. (a) Find the rotation of $\mathbf{v} = \begin{bmatrix} 2 \\ 3 \\ -1 \end{bmatrix}$ about the z axis through $\theta = \frac{\pi}{4}$.

   ◆(b) Find the rotation of $\mathbf{v} = \begin{bmatrix} 1 \\ 0 \\ 3 \end{bmatrix}$ about the z axis through $\theta = \frac{\pi}{6}$.

5. Find the matrix of the rotation in $\mathbb{R}^3$ about the x axis through the angle θ (from the positive y axis to the positive z axis).

◆6. Find the matrix of the rotation about the y axis through the angle θ (from the positive x axis to the positive z axis).

7. If A is 3×3, show that the image of the line in $\mathbb{R}^3$ through $\mathbf{p}_0$ with direction vector $\mathbf{d}$ is the line through $A\mathbf{p}_0$ with direction vector $A\mathbf{d}$, assuming that $A\mathbf{d} \neq \mathbf{0}$. What happens if $A\mathbf{d} = \mathbf{0}$?

8. If A is 3×3 and invertible, show that the image of the plane through the origin with normal $\mathbf{n}$ is the plane through the origin with normal $\mathbf{n}_1 = B\mathbf{n}$ where $B = (A^{-1})^T$. [Hint: Use the fact that $\mathbf{v} \cdot \mathbf{w} = \mathbf{v}^T\mathbf{w}$ to show that $\mathbf{n}_1 \cdot (A\mathbf{p}) = \mathbf{n} \cdot \mathbf{p}$ for each $\mathbf{p}$ in $\mathbb{R}^3$.]

9. Let L be the line through the origin in $\mathbb{R}^2$ with direction vector $\mathbf{d} = \begin{bmatrix} a \\ b \end{bmatrix} \neq \mathbf{0}$.

   ◆(a) If P_L denotes projection on L, show that P_L has matrix $\dfrac{1}{a^2 + b^2} \begin{bmatrix} a^2 & ab \\ ab & b^2 \end{bmatrix}$.

   (b) If Q_L denotes reflection in L, show that Q_L has matrix $\dfrac{1}{a^2 + b^2} \begin{bmatrix} a^2 - b^2 & 2ab \\ 2ab & b^2 - a^2 \end{bmatrix}$.

10. Let $\mathbf{n}$ be a nonzero vector in $\mathbb{R}^3$, let L be the line through the origin with direction vector $\mathbf{n}$, and let M be the plane through the origin with normal $\mathbf{n}$. Show that $P_L(\mathbf{v}) = Q_L(\mathbf{v}) + P_M(\mathbf{v})$ for all $\mathbf{v}$ in $\mathbb{R}^3$. [In this case, we say that $P_L = Q_L + P_M$.]

11. If M is the plane through the origin in $\mathbb{R}^3$ with normal $\mathbf{n} = \begin{bmatrix} a \\ b \\ c \end{bmatrix}$, show that Q_M has matrix

$$\frac{1}{a^2 + b^2 + c^2} \begin{bmatrix} b^2 + c^2 - a^2 & -2ab & -2ac \\ -2ab & a^2 + c^2 - b^2 & -2bc \\ -2ac & -2bc & a^2 + b^2 - c^2 \end{bmatrix}$$

SECTION 4.5 An Application to Computer Graphics

Computer graphics deals with images displayed on a computer screen, and so arises in a variety of applications, ranging from word processors, to *Star Wars* animations, to video games, to wire-frame images of an airplane. These images consist of a number of points on the screen, together with instructions on how to fill in areas bounded by lines and curves. Often curves are approximated by a set of short straight-line segments, so that the curve is specified by a series of points on the screen at the end of these segments. Matrix transformations are important here because matrix images of straight line segments are again line segments.[13] Note that a colour image requires that three images are sent, one to each of the red, green, and blue phosphorus dots on the screen, in varying intensities.

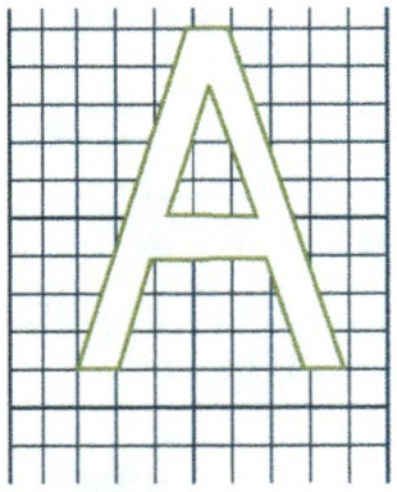

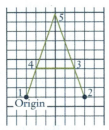

■ FIGURE 1

Consider displaying the letter A. In reality, it is depicted on the screen, as in Figure 1, by specifying the coordinates of the 11 corners and filling in the interior. For simplicity, we will disregard the thickness of the letter, so we require only five coordinates as in Figure 2. This simplified letter can then be stored as a data matrix

$$D = \begin{bmatrix} \text{Vertex} & 1 & 2 & 3 & 4 & 5 \\ & 0 & 6 & 5 & 1 & 3 \\ & 0 & 0 & 3 & 3 & 9 \end{bmatrix}$$

where the columns are the coordinates of the vertices in order. Then if we want to transform the letter by a 2×2 matrix A, we left-multiply this data matrix by A (the effect is to multiply each column by A and so transform each vertex).

■ FIGURE 2

For example, we can slant the letter to the right by multiplying by an x-shear matrix $A = \begin{bmatrix} 1 & 0.2 \\ 0 & 1 \end{bmatrix}$—see Section 2.2. The result is the letter with data matrix

$$AD = \begin{bmatrix} 1 & 0.2 \\ 0 & 1 \end{bmatrix} \begin{bmatrix} 0 & 6 & 5 & 1 & 3 \\ 0 & 0 & 3 & 3 & 9 \end{bmatrix} = \begin{bmatrix} 0 & 6 & 5.6 & 1.6 & 4.8 \\ 0 & 0 & 3 & 3 & 9 \end{bmatrix}$$

which is shown in Figure 3. If we want to make this slanted matrix narrower, we can now apply an x-scale matrix $B = \begin{bmatrix} 0.8 & 0 \\ 0 & 1 \end{bmatrix}$ that shrinks the x-coordinate by 0.8. The result is the composite transformation

$$BAD = \begin{bmatrix} 0.8 & 0 \\ 0 & 1 \end{bmatrix} \begin{bmatrix} 1 & 0.2 \\ 0 & 1 \end{bmatrix} \begin{bmatrix} 0 & 6 & 5 & 1 & 3 \\ 0 & 0 & 3 & 3 & 9 \end{bmatrix} = \begin{bmatrix} 0 & 4.8 & 4.48 & 1.28 & 3.84 \\ 0 & 0 & 3 & 3 & 9 \end{bmatrix}$$

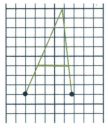

■ FIGURE 3

which is drawn in Figure 4.

On the other hand, we can rotate the letter about the origin through $\frac{\pi}{6}$ (or 30°) by multiplying by the matrix $R_{\frac{\pi}{6}} = \begin{bmatrix} \cos(\frac{\pi}{6}) & -\sin(\frac{\pi}{6}) \\ \sin(\frac{\pi}{6}) & \cos(\frac{\pi}{6}) \end{bmatrix} = \begin{bmatrix} 0.866 & -0.5 \\ 0.5 & 0.866 \end{bmatrix}$.

This gives

$$R_{\frac{\pi}{6}}D = \begin{bmatrix} 0.866 & -0.5 \\ 0.5 & 0.866 \end{bmatrix} \begin{bmatrix} 0 & 6 & 5 & 1 & 3 \\ 0 & 0 & 3 & 3 & 9 \end{bmatrix} = \begin{bmatrix} 0 & 5.196 & 2.83 & -0.634 & -1.902 \\ 0 & 3 & 5.098 & 3.098 & 9.294 \end{bmatrix}$$

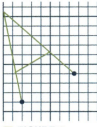

■ FIGURE 4

and is plotted in Figure 5.

This poses a problem: How do we rotate at a point other than the origin? It turns out that we can do this when we have solved another more basic problem. It is clearly important to be able to translate a screen image by a fixed vector $\mathbf{w}$, that is apply the transformation $T_{\mathbf{w}} : \mathbb{R}^2 \to \mathbb{R}^2$ given by $T_{\mathbf{w}}(\mathbf{v}) = \mathbf{v} + \mathbf{w}$ for all $\mathbf{v}$ in $\mathbb{R}^2$. The problem is that these translations are not matrix transformations $\mathbb{R}^2 \to \mathbb{R}^2$ because they do not carry $\mathbf{0}$ to $\mathbf{0}$ (unless $\mathbf{w} = \mathbf{0}$). However, there is a clever way around this.

■ FIGURE 5

---

13 If $\mathbf{v}_0$ and $\mathbf{v}_1$ are vectors, the vector from $\mathbf{v}_0$ to $\mathbf{v}_1$ is $\mathbf{d} = \mathbf{v}_1 - \mathbf{v}_0$. So a vector $\mathbf{v}$ lies on the line segment between $\mathbf{v}_0$ and $\mathbf{v}_1$ if and only if $\mathbf{v} = \mathbf{v}_0 + t\mathbf{d}$ for some number t in the range $0 \leq t \leq 1$. Thus the image of this segment is the set of vectors $A\mathbf{v} = A\mathbf{v}_0 + tA\mathbf{d}$ with $0 \leq t \leq 1$, that is the image is the segment between $A\mathbf{v}_0$ and $A\mathbf{v}_1$.

SECTION 4.5 An Application to Computer Graphics

The idea is to represent a point $\mathbf{v} = \begin{bmatrix} x \\ y \end{bmatrix}$ as a 3×1 column $\begin{bmatrix} x \\ y \\ 1 \end{bmatrix}$, called the **homogeneous coordinates** of $\mathbf{v}$. Then translation by $\mathbf{w} = \begin{bmatrix} p \\ q \end{bmatrix}$ can be achieved by multiplying by a 3×3 matrix:

$$\begin{bmatrix} 1 & 0 & p \\ 0 & 1 & q \\ 0 & 0 & 1 \end{bmatrix} \begin{bmatrix} x \\ y \\ 1 \end{bmatrix} = \begin{bmatrix} x+p \\ y+q \\ 1 \end{bmatrix} = \begin{bmatrix} T_{\mathbf{w}}(\mathbf{v}) \\ 1 \end{bmatrix}$$

Thus, by using homogeneous coordinates we can implement the translation $T_{\mathbf{w}}$ in the top two coordinates. On the other hand, the matrix transformation induced by $A = \begin{bmatrix} a & b \\ c & d \end{bmatrix}$ is also given by a 3×3 matrix:

$$\begin{bmatrix} a & b & 0 \\ c & d & 0 \\ 0 & 0 & 1 \end{bmatrix} \begin{bmatrix} x \\ y \\ 1 \end{bmatrix} = \begin{bmatrix} ax+by \\ cx+dy \\ 1 \end{bmatrix} = \begin{bmatrix} A\mathbf{v} \\ 1 \end{bmatrix}$$

So everything can be accomplished at the expense of using 3×3 matrices and homogeneous coordinates.

EXAMPLE 1

Rotate the letter A in Figure 2 through $\frac{\pi}{6}$ about the point $\begin{bmatrix} 4 \\ 5 \end{bmatrix}$.

Solution ▶ Using homogenous coordinates for the vertices of the letter results in a data matrix with three rows:

$$K_d = \begin{bmatrix} 0 & 6 & 5 & 1 & 3 \\ 0 & 0 & 3 & 3 & 9 \\ 1 & 1 & 1 & 1 & 1 \end{bmatrix}$$

If we write $\mathbf{w} = \begin{bmatrix} 4 \\ 5 \end{bmatrix}$, the idea is to use a composite of transformations: First translate the letter by $-\mathbf{w}$ so that the point $\mathbf{w}$ moves to the origin, then rotate this translated letter, and then translate it by $\mathbf{w}$ back to its original position. The matrix arithmetic is as follows (remember the order of composition!):

$$\begin{bmatrix} 1 & 0 & 4 \\ 0 & 1 & 5 \\ 0 & 0 & 1 \end{bmatrix} \begin{bmatrix} 0.866 & -0.5 & 0 \\ 0.5 & 0.866 & 0 \\ 0 & 0 & 1 \end{bmatrix} \begin{bmatrix} 1 & 0 & -4 \\ 0 & 1 & -5 \\ 0 & 0 & 1 \end{bmatrix} \begin{bmatrix} 0 & 6 & 5 & 1 & 3 \\ 0 & 0 & 3 & 3 & 9 \\ 1 & 1 & 1 & 1 & 1 \end{bmatrix}$$

$$= \begin{bmatrix} 3.036 & 8.232 & 5.866 & 2.402 & 1.134 \\ -1.33 & 1.67 & 3.768 & 1.768 & 7.964 \\ 1 & 1 & 1 & 1 & 1 \end{bmatrix}$$

This is plotted in Figure 6.

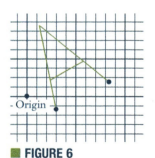

FIGURE 6

This discussion merely touches the surface of computer graphics, and the reader is referred to specialized books on the subject. Realistic graphic rendering requires an enormous number of matrix calculations. In fact, matrix multiplication algorithms are now embedded in microchip circuits, and can perform over 100 million matrix multiplications per second. This is particularly important in the field of three-dimensional graphics where the homogeneous coordinates have four components and 4×4 matrices are required.

EXERCISES 4.5

1. Consider the letter A described in Figure 2. Find the data matrix for the letter obtained by:
   (a) Rotating the letter through $\frac{\pi}{4}$ about the origin.
   ◆(b) Rotating the letter through $\frac{\pi}{4}$ about the point $\begin{bmatrix} 1 \\ 2 \end{bmatrix}$.

2. Find the matrix for turning the letter A in Figure 2 upside-down in place.

3. Find the 3×3 matrix for reflecting in the line $y = mx + b$. Use $\begin{bmatrix} 1 \\ m \end{bmatrix}$ as direction vector for the line.

4. Find the 3×3 matrix for rotating through the angle θ about the point $P(a, b)$.

5. Find the reflection of the point P in the line $y = 1 + 2x$ in $\mathbb{R}^2$ if:
   (a) $P = P(1, 1)$
   ◆(b) $P = P(1, 4)$
   (c) What about $P = P(1, 3)$? Explain. [Hint: Example 1 and Section 4.4.]

SUPPLEMENTARY EXERCISES FOR CHAPTER 4

1. Suppose that $\mathbf{u}$ and $\mathbf{v}$ are nonzero vectors. If $\mathbf{u}$ and $\mathbf{v}$ are not parallel, and $a\mathbf{u} + b\mathbf{v} = a_1\mathbf{u} + b_1\mathbf{v}$, show that $a = a_1$ and $b = b_1$.

2. Consider a triangle with vertices A, B, and C. Let E and F be the midpoints of sides AB and AC, respectively, and let the medians EC and FB meet at O. Write $\overrightarrow{EO} = s\overrightarrow{EC}$ and $\overrightarrow{FO} = t\overrightarrow{FB}$, where s and t are scalars. Show that $s = t = \frac{1}{3}$ by expressing $\overrightarrow{AO}$ two ways in the form $a\overrightarrow{EO} + b\overrightarrow{AC}$, and applying Exercise 1. Conclude that the medians of a triangle meet at the point on each that is one-third of the way from the midpoint to the vertex (and so are concurrent).

3. A river flows at 1 km/h and a swimmer moves at 2 km/h (relative to the water). At what angle must he swim to go straight across? What is his resulting speed?

◆4. A wind is blowing from the south at 75 knots, and an airplane flies heading east at 100 knots. Find the resulting velocity of the airplane.

5. An airplane pilot flies at 300 km/h in a direction 30° south of east. The wind is blowing from the south at 150 km/h.
   (a) Find the resulting direction and speed of the airplane.
   (b) Find the speed of the airplane if the wind is from the west (at 150 km/h).

◆6. A rescue boat has a top speed of 13 knots. The captain wants to go due east as fast as possible in water with a current of 5 knots due south. Find the velocity vector $\mathbf{v} = (x, y)$ that she must achieve, assuming the $\mathbf{x}$ and $\mathbf{y}$ axes point east and north, respectively, and find her resulting speed.

7. A boat goes 12 knots heading north. The current is 5 knots from the west. In what direction does the boat actually move and at what speed?

8. Show that the distance from a point A (with vector $\mathbf{a}$) to the plane with vector equation $\mathbf{n} \cdot \mathbf{p} = d$ is $\frac{1}{\|\mathbf{n}\|}|\mathbf{n} \cdot \mathbf{a} - d|$.

9. If two distinct points lie in a plane, show that the line through these points is contained in the plane.

10. The line through a vertex of a triangle, perpendicular to the opposite side, is called an **altitude** of the triangle. Show that the three altitudes of any triangle are concurrent. (The intersection of the altitudes is called the **orthocentre** of the triangle.) [Hint: If P is the intersection of two of the altitudes, show that the line through P and the remaining vertex is perpendicular to the remaining side.]

5 The Vector Space $\mathbb{R}^n$

SECTION 5.1 Subspaces and Spanning

In Section 2.2 we introduced the set $\mathbb{R}^n$ of all n-tuples (called *vectors*), and began our investigation of the matrix transformations $\mathbb{R}^n \to \mathbb{R}^m$ given by matrix multiplication by an $m \times n$ matrix. Particular attention was paid to the euclidean plane $\mathbb{R}^2$ where certain simple geometric transformations were seen to be matrix transformations. Then in Section 2.6 we introduced linear transformations, showed that they are all matrix transformations, and found the matrices of rotations and reflections in $\mathbb{R}^2$. We returned to this in Section 4.4 where we showed that projections, reflections, and rotations of $\mathbb{R}^2$ and $\mathbb{R}^3$ were all linear, and where we related areas and volumes to determinants.

In this chapter we investigate $\mathbb{R}^n$ in full generality, and introduce some of the most important concepts and methods in linear algebra. The n-tuples in $\mathbb{R}^n$ will continue to be denoted **x**, **y**, and so on, and will be written as rows or columns depending on the context.

Subspaces of $\mathbb{R}^n$

Definition 5.1   *A set[1] U of vectors in $\mathbb{R}^n$ is called a **subspace** of $\mathbb{R}^n$ if it satisfies the following properties:*

  S1. *The zero vector **0** is in U.*

  S2. *If **x** and **y** are in U, then **x** + **y** is also in U.*

  S3. *If **x** is in U, then $a\mathbf{x}$ is in U for every real number a.*

We say that the subset U is **closed under addition** if S2 holds, and that U is **closed under scalar multiplication** if S3 holds.

Clearly $\mathbb{R}^n$ is a subspace of itself. The set $U = \{\mathbf{0}\}$, consisting of only the zero vector, is also a subspace because $\mathbf{0} + \mathbf{0} = \mathbf{0}$ and $a\mathbf{0} = \mathbf{0}$ for each a in $\mathbb{R}$; it is called the **zero subspace**. Any subspace of $\mathbb{R}^n$ other than $\{\mathbf{0}\}$ or $\mathbb{R}^n$ is called a **proper** subspace.

---

[1] We use the language of sets. Informally, a **set** X is a collection of objects, called the **elements** of the set. The fact that x is an element of X is denoted $x \in X$. Two sets X and Y are called equal (written $X = Y$) if they have the same elements. If every element of X is in the set Y, we say that X is a **subset** of Y, and write $X \subseteq Y$. Hence $X \subseteq Y$ and $Y \subseteq X$ both hold if and only if $X = Y$.

We saw in Section 4.2 that every plane M through the origin in $\mathbb{R}^3$ has equation $ax + by + cz = 0$ where a, b, and c are not all zero. Here $\mathbf{n} = \begin{bmatrix} a \\ b \\ c \end{bmatrix}$ is a normal for the plane and

$$M = \{\mathbf{v} \text{ in } \mathbb{R}^3 \mid \mathbf{n} \cdot \mathbf{v} = 0\}$$

where $\mathbf{v} = \begin{bmatrix} x \\ y \\ z \end{bmatrix}$ and $\mathbf{n} \cdot \mathbf{v}$ denotes the dot product introduced in Section 2.2 (see the

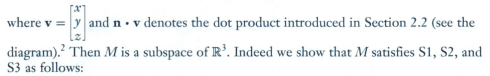

diagram).[2] Then M is a subspace of $\mathbb{R}^3$. Indeed we show that M satisfies S1, S2, and S3 as follows:

S1. $\mathbf{0}$ is in M because $\mathbf{n} \cdot \mathbf{0} = 0$;

S2. If $\mathbf{v}$ and $\mathbf{v}_1$ are in M, then $\mathbf{n} \cdot (\mathbf{v} + \mathbf{v}_1) = \mathbf{n} \cdot \mathbf{v} + \mathbf{n} \cdot \mathbf{v}_1 = 0 + 0 = 0$, so $\mathbf{v} + \mathbf{v}_1$ is in M;

S3. If $\mathbf{v}$ is in M, then $\mathbf{n} \cdot (a\mathbf{v}) = a(\mathbf{n} \cdot \mathbf{v}) = a(0) = 0$, so $a\mathbf{v}$ is in M.

This proves the first part of

EXAMPLE 1

Planes and lines through the origin in $\mathbb{R}^3$ are all subspaces of $\mathbb{R}^3$.

Solution ▶ We dealt with planes above. If L is a line through the origin with direction vector $\mathbf{d}$, then $L = \{t\mathbf{d} \mid t \text{ in } \mathbb{R}\}$ (see the diagram). We leave it as an exercise to verify that L satisfies S1, S2, and S3.

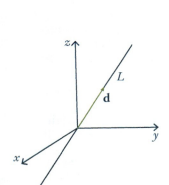

Example 1 shows that lines through the origin in $\mathbb{R}^2$ are subspaces; in fact, they are the *only* proper subspaces of $\mathbb{R}^2$ (Exercise 24). Indeed, we shall see in Example 14 Section 5.2 that lines and planes through the origin in $\mathbb{R}^3$ are the only proper subspaces of $\mathbb{R}^3$. Thus the geometry of lines and planes through the origin is captured by the subspace concept. (Note that *every* line or plane is just a translation of one of these.)

Subspaces can also be used to describe important features of an $m \times n$ matrix A. The **null space** of A, denoted null A, and the **image space** of A, denoted im A, are defined by

$$\text{null } A = \{\mathbf{x} \text{ in } \mathbb{R}^n \mid A\mathbf{x} = \mathbf{0}\} \quad \text{and} \quad \text{im } A = \{A\mathbf{x} \mid \mathbf{x} \text{ in } \mathbb{R}^n\}$$

In the language of Chapter 2, null A consists of all solutions $\mathbf{x}$ in $\mathbb{R}^n$ of the homogeneous system $A\mathbf{x} = \mathbf{0}$, and im A is the set of all vectors $\mathbf{y}$ in $\mathbb{R}^m$ such that $A\mathbf{x} = \mathbf{y}$ *has* a solution $\mathbf{x}$. Note that $\mathbf{x}$ is in null A if it satisfies the *condition* $A\mathbf{x} = \mathbf{0}$, while im A consists of vectors of the *form* $A\mathbf{x}$ for some $\mathbf{x}$ in $\mathbb{R}^n$. These two ways to describe subsets occur frequently.

EXAMPLE 2

If A is an $m \times n$ matrix, then:

1. null A is a subspace of $\mathbb{R}^n$.
2. im A is a subspace of $\mathbb{R}^m$.

---

[2] We are using set notation here. In general $\{q \mid p\}$ means the set of all objects q with property p.

Solution ▶

1. The zero vector $\mathbf{0}$ in $\mathbb{R}^n$ lies in null A because $A\mathbf{0} = \mathbf{0}$.[3] If $\mathbf{x}$ and $\mathbf{x}_1$ are in null A, then $\mathbf{x} + \mathbf{x}_1$ and $a\mathbf{x}$ are in null A because they satisfy the required condition:
$$A(\mathbf{x} + \mathbf{x}_1) = A\mathbf{x} + A\mathbf{x}_1 = \mathbf{0} + \mathbf{0} = \mathbf{0} \quad \text{and} \quad A(a\mathbf{x}) = a(A\mathbf{x}) = a\mathbf{0} = \mathbf{0}$$
Hence null A satisfies S1, S2, and S3, and so is a subspace of $\mathbb{R}^n$.

2. The zero vector $\mathbf{0}$ in $\mathbb{R}^m$ lies in im A because $\mathbf{0} = A\mathbf{0}$. Suppose that $\mathbf{y}$ and $\mathbf{y}_1$ are in im A, say $\mathbf{y} = A\mathbf{x}$ and $\mathbf{y}_1 = A\mathbf{x}_1$ where $\mathbf{x}$ and $\mathbf{x}_1$ are in $\mathbb{R}^n$. Then
$$\mathbf{y} + \mathbf{y}_1 = A\mathbf{x} + A\mathbf{x}_1 = A(\mathbf{x} + \mathbf{x}_1) \quad \text{and} \quad a\mathbf{y} = a(A\mathbf{x}) = A(a\mathbf{x})$$
show that $\mathbf{y} + \mathbf{y}_1$ and $a\mathbf{y}$ are both in im A (they have the required form). Hence im A is a subspace of $\mathbb{R}^m$.

There are other important subspaces associated with a matrix A that clarify basic properties of A. If A is an $n \times n$ matrix and λ is any number, let
$$E_\lambda(A) = \{\mathbf{x} \text{ in } \mathbb{R}^n \mid A\mathbf{x} = \lambda\mathbf{x}\}.$$
A vector $\mathbf{x}$ is in $E_\lambda(A)$ if and only if $(\lambda I - A)\mathbf{x} = \mathbf{0}$, so Example 2 gives:

EXAMPLE 3

$E_\lambda(A) = \text{null}(\lambda I - A)$ is a subspace of $\mathbb{R}^n$ for each $n \times n$ matrix A and number λ.

$E_\lambda(A)$ is called the **eigenspace** of A corresponding to λ. The reason for the name is that, in the terminology of Section 3.3, λ is an **eigenvalue** of A if $E_\lambda(A) \neq \{\mathbf{0}\}$. In this case the nonzero vectors in $E_\lambda(A)$ are called the **eigenvectors** of A corresponding to λ.

The reader should not get the impression that *every* subset of $\mathbb{R}^n$ is a subspace. For example:
$$U_1 = \left\{ \begin{bmatrix} x \\ y \end{bmatrix} \mid x \geq 0 \right\} \text{ satisfies S1 and S2, but not S3;}$$
$$U_2 = \left\{ \begin{bmatrix} x \\ y \end{bmatrix} \mid x^2 = y^2 \right\} \text{ satisfies S1 and S3, but not S2;}$$
Hence neither U_1 nor U_2 is a subspace of $\mathbb{R}^2$. (However, see Exercise 20.)

Spanning Sets

Let $\mathbf{v}$ and $\mathbf{w}$ be two nonzero, nonparallel vectors in $\mathbb{R}^3$ with their tails at the origin. The plane M through the origin containing these vectors is described in Section 4.2 by saying that $\mathbf{n} = \mathbf{v} \times \mathbf{w}$ is a *normal* for M, and that M consists of all vectors $\mathbf{p}$ such that $\mathbf{n} \cdot \mathbf{p} = 0$.[4] While this is a very useful way to look at planes, there is another approach that is at least as useful in $\mathbb{R}^3$ and, more importantly, works for all subspaces of $\mathbb{R}^n$ for any $n \geq 1$.

---

3 We are using **0** to represent the zero vector in both $\mathbb{R}^m$ and $\mathbb{R}^n$. This abuse of notation is common and causes no confusion once everybody knows what is going on.

4 The vector $\mathbf{n} = \mathbf{v} \times \mathbf{w}$ is nonzero because $\mathbf{v}$ and $\mathbf{w}$ are not parallel.

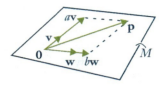

The idea is as follows: Observe that, by the diagram, a vector **p** is in M if and only if it has the form

$$\mathbf{p} = a\mathbf{v} + b\mathbf{w}$$

for certain real numbers a and b (we say that **p** is a *linear combination* of **v** and **w**). Hence we can describe M as

$$M = \{a\mathbf{x} + b\mathbf{w} \mid a, b \text{ in } \mathbb{R}\}.^5$$

and we say that $\{\mathbf{v}, \mathbf{w}\}$ is a *spanning set* for M. It is this notion of a spanning set that provides a way to describe all subspaces of $\mathbb{R}^n$.

As in Section 1.3, given vectors $\mathbf{x}_1, \mathbf{x}_2, ..., \mathbf{x}_k$ in $\mathbb{R}^n$, a vector of the form

$$t_1\mathbf{x}_1 + t_2\mathbf{x}_2 + \cdots + t_k\mathbf{x}_k \quad \text{where the } t_i \text{ are scalars}$$

is called a **linear combination** of the $\mathbf{x}_i$, and t_i is called the **coefficient** of $\mathbf{x}_i$ in the linear combination.

Definition 5.2

*The set of all such linear combinations is called the **span** of the $\mathbf{x}_i$ and is denoted*

$$\text{span}\{\mathbf{x}_1, \mathbf{x}_2, ..., \mathbf{x}_k\} = \{t_1\mathbf{x}_1 + t_2\mathbf{x}_2 + \cdots + t_k\mathbf{x}_k \mid t_i \text{ in } \mathbb{R}\}.$$

*If $V = \text{span}\{\mathbf{x}_1, \mathbf{x}_2, ..., \mathbf{x}_k\}$, we say that V is **spanned** by the vectors $\mathbf{x}_1, \mathbf{x}_2, ..., \mathbf{x}_k$, and that the vectors $\mathbf{x}_1, \mathbf{x}_2, ..., \mathbf{x}_k$ **span** the space V.*

Two examples:

$$\text{span}\{\mathbf{x}\} = \{t\mathbf{x} \mid t \text{ in } \mathbb{R}\},$$

which we write as $\text{span}\{\mathbf{x}\} = \mathbb{R}\mathbf{x}$ for simplicity.

$$\text{span}\{\mathbf{x}, \mathbf{y}\} = \{r\mathbf{x} + s\mathbf{y} \mid r, s \text{ in } \mathbb{R}\}.$$

In particular, the above discussion shows that, if **v** and **w** are two nonzero, nonparallel vectors in $\mathbb{R}^3$, then

$$M = \text{span}\{\mathbf{v}, \mathbf{w}\}$$

is the plane in $\mathbb{R}^3$ containing **v** and **w**. Moreover, if **d** is any nonzero vector in $\mathbb{R}^3$ (or $\mathbb{R}^2$), then

$$L = \text{span}\{\mathbf{v}\} = \{t\mathbf{d} \mid t \text{ in } \mathbb{R}\} = \mathbb{R}\mathbf{d}$$

is the line with direction vector **d** (see also Lemma 1 Section 3.3). Hence lines and planes can both be described in terms of spanning sets.

EXAMPLE 4

Let $\mathbf{x} = (2, -1, 2, 1)$ and $\mathbf{y} = (3, 4, -1, 1)$ in $\mathbb{R}^4$. Determine whether $\mathbf{p} = (0, -11, 8, 1)$ or $\mathbf{q} = (2, 3, 1, 2)$ are in $U = \text{span}\{\mathbf{x}, \mathbf{y}\}$.

Solution ▶ The vector **p** is in U if and only if $\mathbf{p} = s\mathbf{x} + t\mathbf{y}$ for scalars s and t. Equating components gives equations

$$2s + 3t = 0, \quad -s + 4t = -11, \quad 2s - t = 8, \quad \text{and} \quad s + t = 1.$$

This linear system has solution $s = 3$ and $t = -2$, so **p** is in U. On the other hand, asking that $\mathbf{q} = s\mathbf{x} + t\mathbf{y}$ leads to equations

$$2s + 3t = 2, \quad -s + 4t = 3, \quad 2s - t = 1, \quad \text{and} \quad s + t = 2$$

and this system has *no* solution. So **q** does *not* lie in U.

---

5 In particular, this implies that any vector **p** orthogonal to $\mathbf{v} \times \mathbf{w}$ must be a linear combination $\mathbf{p} = a\mathbf{v} + b\mathbf{w}$ of **v** and **w** for some a and b. Can you prove this directly?

SECTION 5.1 Subspaces and Spanning

Theorem 1

Let $U = \text{span}\{\mathbf{x}_1, \mathbf{x}_2, \ldots, \mathbf{x}_k\}$ in $\mathbb{R}^n$. Then:
1. U is a subspace of $\mathbb{R}^n$ containing each X_i.
2. If W is a subspace of $\mathbb{R}^n$ and each X_i is in W, then $U \subseteq W$.

PROOF

Write $U = \text{span}\{\mathbf{x}_1, \mathbf{x}_2, \ldots, \mathbf{x}_k\}$ for convenience.

1. The zero vector $\mathbf{0}$ is in U because $\mathbf{0} = 0\mathbf{x}_1 + 0\mathbf{x}_2 + \cdots + 0\mathbf{x}_k$ is a linear combination of the $\mathbf{x}_i$. If $\mathbf{x} = t_1\mathbf{x}_1 + t_2\mathbf{x}_2 + \cdots + t_k\mathbf{x}_k$ and $\mathbf{y} = s_1\mathbf{x}_1 + s_2\mathbf{x}_2 + \cdots + s_k\mathbf{x}_k$ are in U, then $\mathbf{x} + \mathbf{y}$ and $a\mathbf{x}$ are in U because

$$\mathbf{x} + \mathbf{y} = (t_1 + s_1)\mathbf{x}_1 + (t_2 + s_2)\mathbf{x}_2 + \cdots + (t_k + s_k)\mathbf{x}_1, \text{ and}$$
$$a\mathbf{x} = (at_1)\mathbf{x}_1 + (at_2)\mathbf{x}_2 + \cdots + (at_k)\mathbf{x}_1.$$

Hence S1, S2, and S3 are satisfied for U, proving (1).

2. Let $\mathbf{x} = t_1\mathbf{x}_1 + t_2\mathbf{x}_2 + \cdots + t_k\mathbf{x}_k$ where the t_i are scalars and each $\mathbf{x}_i$ is in W. Then each $t_i\mathbf{x}_i$ is in W because W satisfies S3. But then $\mathbf{x}$ is in W because W satisfies S2 (verify). This proves (2).

Condition (2) in Theorem 1 can be expressed by saying that $\text{span}\{\mathbf{x}_1, \mathbf{x}_2, \ldots, \mathbf{x}_k\}$ is the *smallest* subspace of $\mathbb{R}^n$ that contains each $\mathbf{x}_i$. This is useful for showing that two subspaces U and W are equal, since this amounts to showing that both $U \subseteq W$ and $W \subseteq U$. Here is an example of how it is used.

EXAMPLE 5

If $\mathbf{x}$ and $\mathbf{y}$ are in $\mathbb{R}^n$, show that $\text{span}\{\mathbf{x}, \mathbf{y}\} = \text{span}\{\mathbf{x} + \mathbf{y}, \mathbf{x} - \mathbf{y}\}$.

Solution ▶ Since both $\mathbf{x} + \mathbf{y}$ and $\mathbf{x} - \mathbf{y}$ are in $\text{span}\{\mathbf{x}, \mathbf{y}\}$, Theorem 1 gives

$$\text{span}\{\mathbf{x} + \mathbf{y}, \mathbf{x} - \mathbf{y}\} \subseteq \text{span}\{\mathbf{x}, \mathbf{y}\}.$$

But $\mathbf{x} = \frac{1}{2}(\mathbf{x} + \mathbf{y}) + \frac{1}{2}(\mathbf{x} - \mathbf{y})$ and $\mathbf{y} = \frac{1}{2}(\mathbf{x} + \mathbf{y}) - \frac{1}{2}(\mathbf{x} - \mathbf{y})$ are both in $\text{span}\{\mathbf{x} + \mathbf{y}, \mathbf{x} - \mathbf{y}\}$, so

$$\text{span}\{\mathbf{x}, \mathbf{y}\} \subseteq \text{span}\{\mathbf{x} + \mathbf{y}, \mathbf{x} - \mathbf{y}\}$$

again by Theorem 1. Thus $\text{span}\{\mathbf{x}, \mathbf{y}\} = \text{span}\{\mathbf{x} + \mathbf{y}, \mathbf{x} - \mathbf{y}\}$, as desired.

It turns out that many important subspaces are best described by giving a spanning set. Here are three examples, beginning with an important spanning set for $\mathbb{R}^n$ itself. Column j of the $n \times n$ identity matrix I_n is denoted $\mathbf{e}_j$ and called the *j*th **coordinate vector** in $\mathbb{R}^n$, and the set $\{\mathbf{e}_1, \mathbf{e}_2, \ldots, \mathbf{e}_n\}$ is called the **standard basis** of $\mathbb{R}^n$. If $\mathbf{x} = \begin{bmatrix} x_1 \\ x_2 \\ \vdots \\ x_n \end{bmatrix}$ is any vector in $\mathbb{R}^n$, then $\mathbf{x} = x_1\mathbf{e}_1 + x_2\mathbf{e}_2 + \cdots + x_n\mathbf{e}_n$, as the reader can verify. This proves:

EXAMPLE 6

$\mathbb{R}^n = \text{span}\{e_1, e_2, \ldots, e_n\}$ where $e_1, e_2, \ldots, e_n$ are the columns of I_n.

If A is an $m \times n$ matrix A, the next two examples show that it is a routine matter to find spanning sets for null A and im A.

EXAMPLE 7

Given an $m \times n$ matrix A, let $x_1, x_2, \ldots, x_k$ denote the basic solutions to the system $Ax = 0$ given by the gaussian algorithm. Then

$$\text{null } A = \text{span}\{x_1, x_2, \ldots, x_k\}.$$

Solution ▶ If x is in null A, then $Ax = 0$ so Theorem 2 Section 1.3 shows that x is a linear combination of the basic solutions; that is, null $A \subseteq \text{span}\{x_1, x_2, \ldots, x_k\}$. On the other hand, if x is in span$\{x_1, x_2, \ldots, x_k\}$, then $x = t_1 x_1 + t_2 x_2 + \cdots + t_k x_k$ for scalars t_i, so

$$Ax = t_1 A x_1 + t_2 A x_2 + \cdots + t_k A x_k = t_1 0 + t_2 0 + \cdots + t_k 0 = 0.$$

This shows that x is in null A, and hence that span$\{x_1, x_2, \ldots, x_k\} \subseteq$ null A. Thus we have equality.

EXAMPLE 8

Let $c_1, c_2, \ldots, c_n$ denote the columns of the $m \times n$ matrix A. Then

$$\text{im } A = \text{span}\{c_1, c_2, \ldots, c_n\}.$$

Solution ▶ If $\{e_1, e_2, \ldots, e_n\}$ is the standard basis of $\mathbb{R}^n$, observe that

$$[Ae_1 \ Ae_2 \ \cdots \ Ae_n] = A[e_1 \ e_2 \ \cdots \ e_n] = AI_n = A = [c_1 \ c_2 \ \cdots \ c_n].$$

Hence $c_i = Ae_i$ is in im A for each i, so span$\{c_1, c_2, \ldots, c_n\} \subseteq$ im A.

Conversely, let y be in im A, say $y = Ax$ for some x in $\mathbb{R}^n$. If $x = \begin{bmatrix} x_1 \\ x_2 \\ \vdots \\ x_n \end{bmatrix}$, then Definition 2.5 gives

$$y = Ax = x_1 c_1 + x_2 c_2 + \cdots + x_n c_n \text{ is in span}\{c_1, c_2, \ldots, c_n\}.$$

This shows that im $A \subseteq \text{span}\{c_1, c_2, \ldots, c_n\}$, and the result follows.

EXERCISES 5.1

We often write vectors in $\mathbb{R}^n$ as rows.

1. In each case determine whether U is a subspace of $\mathbb{R}^3$. Support your answer.

    (a) $U = \{(1, s, t) \mid s \text{ and } t \text{ in } \mathbb{R}\}$.

    ◆(b) $U = \{(0, s, t) \mid s \text{ and } t \text{ in } \mathbb{R}\}$.

    (c) $U = \{(r, s, t) \mid r, s, \text{ and } t \text{ in } \mathbb{R}, -r + 3s + 2t = 0\}$.

    ◆(d) $U = \{(r, 3s, r - 2) \mid r \text{ and } s \text{ in } \mathbb{R}\}$.

    (e) $U = \{(r, 0, s) \mid r^2 + s^2 = 0, r \text{ and } s \text{ in } \mathbb{R}\}$.

    ◆(f) $U = \{(2r, -s^2, t) \mid r, s, \text{ and } t \text{ in } \mathbb{R}\}$.

2. In each case determine if **x** lies in $U = \text{span}\{\mathbf{y}, \mathbf{z}\}$. If **x** is in U, write it as a linear combination of **y** and **z**; if **x** is not in U, show why not.

   (a) $\mathbf{x} = (2, -1, 0, 1)$, $\mathbf{y} = (1, 0, 0, 1)$, and $\mathbf{z} = (0, 1, 0, 1)$.

   ◆(b) $\mathbf{x} = (1, 2, 15, 11)$, $\mathbf{y} = (2, -1, 0, 2)$, and $\mathbf{z} = (1, -1, -3, 1)$.

   (c) $\mathbf{x} = (8, 3, -13, 20)$, $\mathbf{y} = (2, 1, -3, 5)$, and $\mathbf{z} = (-1, 0, 2, -3)$.

   ◆(d) $\mathbf{x} = (2, 5, 8, 3)$, $\mathbf{y} = (2, -1, 0, 5)$, and $\mathbf{z} = (-1, 2, 2, -3)$.

3. In each case determine if the given vectors span $\mathbb{R}^4$. Support your answer.

   (a) $\{(1, 1, 1, 1), (0, 1, 1, 1), (0, 0, 1, 1), (0, 0, 0, 1)\}$.

   ◆(b) $\{(1, 3, -5, 0), (-2, 1, 0, 0), (0, 2, 1, -1), (1, -4, 5, 0)\}$.

4. Is it possible that $\{(1, 2, 0), (2, 0, 3)\}$ can span the subspace $U = \{(r, s, 0) \mid r \text{ and } s \text{ in } \mathbb{R}\}$? Defend your answer.

5. Give a spanning set for the zero subspace $\{\mathbf{0}\}$ of $\mathbb{R}^n$.

6. Is $\mathbb{R}^2$ a subspace of $\mathbb{R}^3$? Defend your answer.

7. If $U = \text{span}\{\mathbf{x}, \mathbf{y}, \mathbf{z}\}$ in $\mathbb{R}^n$, show that $U = \text{span}\{\mathbf{x} + t\mathbf{z}, \mathbf{y}, \mathbf{z}\}$ for every t in $\mathbb{R}$.

8. If $U = \text{span}\{\mathbf{x}, \mathbf{y}, \mathbf{z}\}$ in $\mathbb{R}^n$, show that $U = \text{span}\{\mathbf{x} + \mathbf{y}, \mathbf{y} + \mathbf{z}, \mathbf{z} + \mathbf{x}\}$.

9. If $a \neq 0$ is a scalar, show that $\text{span}\{a\mathbf{x}\} = \text{span}\{\mathbf{x}\}$ for every vector **x** in $\mathbb{R}^n$.

◆10. If $a_1, a_2, \ldots, a_k$ are nonzero scalars, show that $\text{span}\{a_1\mathbf{x}_1, a_2\mathbf{x}_2, \ldots, a_k\mathbf{x}_k\} = \text{span}\{\mathbf{x}_1, \mathbf{x}_2, \ldots, \mathbf{x}_k\}$ for any vectors $\mathbf{x}_i$ in $\mathbb{R}^n$.

11. If $\mathbf{x} \neq \mathbf{0}$ in $\mathbb{R}^n$, determine all subspaces of $\text{span}\{\mathbf{x}\}$.

◆12. Suppose that $U = \text{span}\{\mathbf{x}_1, \mathbf{x}_2, \ldots, \mathbf{x}_k\}$ where each $\mathbf{x}_i$ is in $\mathbb{R}^n$. If A is an $m \times n$ matrix and $A\mathbf{x}_i = \mathbf{0}$ for each i, show that $A\mathbf{y} = \mathbf{0}$ for every vector **y** in U.

13. If A is an $m \times n$ matrix, show that, for each invertible $m \times m$ matrix U, $\text{null}(A) = \text{null}(UA)$.

14. If A is an $m \times n$ matrix, show that, for each invertible $n \times n$ matrix V, $\text{im}(A) = \text{im}(AV)$.

15. Let U be a subspace of $\mathbb{R}^n$, and let **x** be a vector in $\mathbb{R}^n$.

   (a) If $a\mathbf{x}$ is in U where $a \neq 0$ is a number, show that **x** is in U.

   ◆(b) If **y** and $\mathbf{x} + \mathbf{y}$ are in U where **y** is a vector in $\mathbb{R}^n$, show that **x** is in U.

16. In each case either show that the statement is true or give an example showing that it is false.

   (a) If $U \neq \mathbb{R}^n$ is a subspace of $\mathbb{R}^n$ and $\mathbf{x} + \mathbf{y}$ is in U, then **x** and **y** are both in U.

   ◆(b) If U is a subspace of $\mathbb{R}^n$ and $r\mathbf{x}$ is in U for all r in $\mathbb{R}$, then **x** is in U.

   (c) If U is a subspace of $\mathbb{R}^n$ and **x** is in U, then $-\mathbf{x}$ is also in U.

   ◆(d) If **x** is in U and $U = \text{span}\{\mathbf{y}, \mathbf{z}\}$, then $U = \text{span}\{\mathbf{x}, \mathbf{y}, \mathbf{z}\}$.

   (e) The empty set of vectors in $\mathbb{R}^n$ is a subspace of $\mathbb{R}^n$.

   ◆(f) $\begin{bmatrix} 0 \\ 1 \end{bmatrix}$ is in $\text{span}\left\{\begin{bmatrix} 1 \\ 0 \end{bmatrix}, \begin{bmatrix} 2 \\ 0 \end{bmatrix}\right\}$.

17. (a) If A and B are $m \times n$ matrices, show that $U = \{\mathbf{x} \text{ in } \mathbb{R}^n \mid A\mathbf{x} = B\mathbf{x}\}$ is a subspace of $\mathbb{R}^n$.

   (b) What if A is $m \times n$, B is $k \times n$, and $m \neq k$?

18. Suppose that $\mathbf{x}_1, \mathbf{x}_2, \ldots, \mathbf{x}_k$ are vectors in $\mathbb{R}^n$. If $\mathbf{y} = a_1\mathbf{x}_1 + a_2\mathbf{x}_2 + \cdots + a_k\mathbf{x}_k$ where $a_1 \neq 0$, show that $\text{span}\{\mathbf{x}_1, \mathbf{x}_2, \ldots, \mathbf{x}_k\} = \text{span}\{\mathbf{y}_1, \mathbf{x}_2, \ldots, \mathbf{x}_k\}$.

19. If $U \neq \{\mathbf{0}\}$ is a subspace of $\mathbb{R}$, show that $U = \mathbb{R}$.

◆20. Let U be a nonempty subset of $\mathbb{R}^n$. Show that U is a subspace if and only if S2 and S3 hold.

21. If S and T are nonempty sets of vectors in $\mathbb{R}^n$, and if $S \subseteq T$, show that $\text{span}\{S\} \subseteq \text{span}\{T\}$.

22. Let U and W be subspaces of $\mathbb{R}^n$. Define their **intersection** $U \cap W$ and their **sum** $U + W$ as follows:

   $U \cap W = \{\mathbf{x} \text{ in } \mathbb{R}^n \mid \mathbf{x} \text{ belongs to both } U \text{ and } W\}$.

   $U + W = \{\mathbf{x} \text{ in } \mathbb{R}^n \mid \mathbf{x} \text{ is a sum of a vector in } U \text{ and a vector in } W\}$.

   (a) Show that $U \cap W$ is a subspace of $\mathbb{R}^n$.

   ◆(b) Show that $U + W$ is a subspace of $\mathbb{R}^n$.

23. Let P denote an invertible $n \times n$ matrix. If λ is a number, show that $E_\lambda(PAP^{-1}) = \{P\mathbf{x} \mid \mathbf{x} \text{ is in } E_\lambda(A)\}$ for each $n \times n$ matrix A.

24. Show that every proper subspace U of $\mathbb{R}^2$ is a line through the origin. [*Hint*: If **d** is a nonzero vector in U, let $L = \mathbb{R}\mathbf{d} = \{r\mathbf{d} \mid r \text{ in } \mathbb{R}\}$ denote the line with direction vector **d**. If **u** is in U but not in L, argue geometrically that every vector **v** in $\mathbb{R}^2$ is a linear combination of **u** and **d**.]

SECTION 5.2 Independence and Dimension

Some spanning sets are better than others. If $U = \text{span}\{\mathbf{x}_1, \mathbf{x}_2, \ldots, \mathbf{x}_k\}$ is a subspace of $\mathbb{R}^n$, then every vector in U can be written as a linear combination of the $\mathbf{x}_i$ in at least one way. Our interest here is in spanning sets where each vector in U has a *exactly one* representation as a linear combination of these vectors.

Linear Independence

Given $\mathbf{x}_1, \mathbf{x}_2, \ldots, \mathbf{x}_k$ in $\mathbb{R}^n$, suppose that two linear combinations are equal:

$$r_1\mathbf{x}_1 + r_2\mathbf{x}_2 + \cdots + r_k\mathbf{x}_k = s_1\mathbf{x}_1 + s_2\mathbf{x}_2 + \cdots + s_k\mathbf{x}_k.$$

We are looking for a condition on the set $\{\mathbf{x}_1, \mathbf{x}_2, \ldots, \mathbf{x}_k\}$ of vectors that guarantees that this representation is *unique*; that is, $r_i = s_i$ for each i. Taking all terms to the left side gives

$$(r_1 - s_1)\mathbf{x}_1 + (r_2 - s_2)\mathbf{x}_2 + \cdots + (r_k - s_k)\mathbf{x}_k = \mathbf{0}.$$

so the required condition is that this equation forces all the coefficients $r_i - s_i$ to be zero.

Definition 5.3   With this in mind, we call a set $\{\mathbf{x}_1, \mathbf{x}_2, \ldots, \mathbf{x}_k\}$ of vectors **linearly independent** (or simply **independent**) if it satisfies the following condition:

$$\text{If } t_1\mathbf{x}_1 + t_2\mathbf{x}_2 + \cdots + t_k\mathbf{x}_k = \mathbf{0} \quad \text{then} \quad t_1 = t_2 = \cdots = t_k = 0.$$

We record the result of the above discussion for reference.

Theorem 1

If $\{\mathbf{x}_1, \mathbf{x}_2, \ldots, \mathbf{x}_k\}$ is an independent set of vectors in $\mathbb{R}^n$, then every vector in $\text{span}\{\mathbf{x}_1, \mathbf{x}_2, \ldots, \mathbf{x}_k\}$ has a **unique** representation as a linear combination of the $\mathbf{x}_i$.

It is useful to state the definition of independence in different language. Let us say that a linear combination **vanishes** if it equals the zero vector, and call a linear combination **trivial** if every coefficient is zero. Then the definition of independence can be compactly stated as follows:

> A set of vectors is independent if and only if the only linear combination that vanishes is the trivial one.

Hence we have a procedure for checking that a set of vectors is independent:

Independence Test

To verify that a set $\{\mathbf{x}_1, \mathbf{x}_2, \ldots, \mathbf{x}_k\}$ of vectors in $\mathbb{R}^n$ is independent, proceed as follows:

1. Set a linear combination equal to zero: $t_1\mathbf{x}_1 + t_2\mathbf{x}_2 + \cdots + t_k\mathbf{x}_k = \mathbf{0}$.
2. Show that $t_i = 0$ for each i (that is, the linear combination is trivial).

Of course, if some nontrivial linear combination vanishes, the vectors are *not independent*.

SECTION 5.2 Independence and Dimension

EXAMPLE 1

Determine whether $\{(1, 0, -2, 5), (2, 1, 0, -1), (1, 1, 2, 1)\}$ is independent in $\mathbb{R}^4$.

Solution ▶ Suppose a linear combination vanishes:
$$r(1, 0, -2, 5) + s(2, 1, 0, -1) + t(1, 1, 2, 1) = (0, 0, 0, 0).$$
Equating corresponding entries gives a system of four equations:
$$r + 2s + t = 0, \quad s + t = 0, \quad -2r + 2t = 0, \quad \text{and} \quad 5r - s + t = 0.$$
The only solution is the trivial one $r = s = t = 0$ (verify), so these vectors are independent by the independence test.

EXAMPLE 2

Show that the standard basis $\{\mathbf{e}_1, \mathbf{e}_2, ..., \mathbf{e}_k\}$ of $\mathbb{R}^n$ is independent.

Solution ▶ The components of $t_1\mathbf{e}_1 + t_2\mathbf{e}_2 + \cdots + t_n\mathbf{e}_n$ are $t_1, t_2, ..., t_n$ (see the discussion preceding Example 6 Section 5.1) So the linear combination vanishes if and only if each $t_i = 0$. Hence the independence test applies.

EXAMPLE 3

If $\{\mathbf{x}, \mathbf{y}\}$ is independent, show that $\{2\mathbf{x} + 3\mathbf{y}, \mathbf{x} - 5\mathbf{y}\}$ is also independent.

Solution ▶ If $s(2\mathbf{x} + 3\mathbf{y}) + t(\mathbf{x} - 5\mathbf{y}) = \mathbf{0}$, collect terms to get $(2s + t)\mathbf{x} + (3s - 5t)\mathbf{y} = \mathbf{0}$. Since $\{\mathbf{x}, \mathbf{y}\}$ is independent this combination must be trivial; that is, $2s + t = 0$ and $3s - 5t = 0$. These equations have only the trivial solution $s = t = 0$, as required.

EXAMPLE 4

Show that the zero vector in $\mathbb{R}^n$ does not belong to any independent set.

Solution ▶ No set $\{\mathbf{0}, \mathbf{x}_1, \mathbf{x}_2, ..., \mathbf{x}_k\}$ of vectors is independent because we have a vanishing, nontrivial linear combination $1 \cdot \mathbf{0} + 0\mathbf{x}_1 + 0\mathbf{x}_2 + \cdots + 0\mathbf{x}_k = \mathbf{0}$.

EXAMPLE 5

Given $\mathbf{x}$ in $\mathbb{R}^n$, show that $\{\mathbf{x}\}$ is independent if and only if $\mathbf{x} \neq \mathbf{0}$.

Solution ▶ A vanishing linear combination from $\{\mathbf{x}\}$ takes the form $t\mathbf{x} = \mathbf{0}$, t in $\mathbb{R}$. This implies that $t = 0$ because $\mathbf{x} \neq \mathbf{0}$.

The next example will be needed later.

EXAMPLE 6

Show that the nonzero rows of a row-echelon matrix R are independent.

Solution ▶ We illustrate the case with 3 leading 1s; the general case is analogous. Suppose R has the form $R = \begin{bmatrix} 0 & 1 & * & * & * & * \\ 0 & 0 & 0 & 1 & * & * \\ 0 & 0 & 0 & 0 & 1 & * \\ 0 & 0 & 0 & 0 & 0 & 0 \end{bmatrix}$ where $*$ indicates a nonspecified number. Let R_1, R_2, and R_3 denote the nonzero rows of R. If $t_1 R_1 + t_2 R_2 + t_3 R_3 = 0$ we show that $t_1 = 0$, then $t_2 = 0$, and finally $t_3 = 0$. The condition $t_1 R_1 + t_2 R_2 + t_3 R_3 = 0$ becomes

$$(0, t_1, *, *, *, *) + (0, 0, 0, t_2, *, *) + (0, 0, 0, 0, t_3, *) = (0, 0, 0, 0, 0, 0).$$

Equating second entries show that $t_1 = 0$, so the condition becomes $t_2 R_2 + t_3 R_3 = 0$. Now the same argument shows that $t_2 = 0$. Finally, this gives $t_3 R_3 = 0$ and we obtain $t_3 = 0$.

A set of vectors in $\mathbb{R}^n$ is called **linearly dependent** (or simply **dependent**) if it is *not* linearly independent, equivalently if some nontrivial linear combination vanishes.

EXAMPLE 7

If **v** and **w** are nonzero vectors in $\mathbb{R}^3$, show that $\{\mathbf{v}, \mathbf{w}\}$ is dependent if and only if **v** and **w** are parallel.

Solution ▶ If **v** and **w** are parallel, then one is a scalar multiple of the other (Theorem 4 Section 4.1), say $\mathbf{v} = a\mathbf{w}$ for some scalar a. Then the nontrivial linear combination $\mathbf{v} - a\mathbf{w} = \mathbf{0}$ vanishes, so $\{\mathbf{v}, \mathbf{w}\}$ is dependent.

Conversely, if $\{\mathbf{v}, \mathbf{w}\}$ is dependent, let $s\mathbf{v} + t\mathbf{w} = \mathbf{0}$ be nontrivial, say $s \neq 0$. Then $\mathbf{v} = -\frac{t}{s}\mathbf{w}$, so **v** and **w** are parallel (by Theorem 4 Section 4.1). A similar argument works if $t \neq 0$.

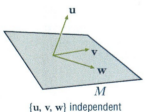

$\{\mathbf{u}, \mathbf{v}, \mathbf{w}\}$ independent

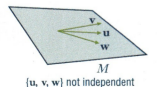

$\{\mathbf{u}, \mathbf{v}, \mathbf{w}\}$ not independent

With this we can give a geometric description of what it means for a set $\{\mathbf{u}, \mathbf{v}, \mathbf{w}\}$ in $\mathbb{R}^3$ to be independent. Note that this requirement means that $\{\mathbf{v}, \mathbf{w}\}$ is also independent ($a\mathbf{v} + b\mathbf{w} = \mathbf{0}$ means that $0\mathbf{u} + a\mathbf{v} + b\mathbf{w} = \mathbf{0}$), so $M = \text{span}\{\mathbf{v}, \mathbf{w}\}$ is the plane containing **v**, **w**, and **0** (see the discussion preceding Example 4 Section 5.1). So we assume that $\{\mathbf{v}, \mathbf{w}\}$ is independent in the following example.

EXAMPLE 8

Let **u**, **v**, and **w** be nonzero vectors in $\mathbb{R}^3$ where $\{\mathbf{v}, \mathbf{w}\}$ independent. Show that $\{\mathbf{u}, \mathbf{v}, \mathbf{w}\}$ is independent if and only if **u** is not in the plane $M = \text{span}\{\mathbf{v}, \mathbf{w}\}$. This is illustrated in the diagrams.

SECTION 5.2 *Independence and Dimension*

> **Solution** ▶ If $\{\mathbf{u}, \mathbf{v}, \mathbf{w}\}$ is independent, suppose $\mathbf{u}$ is in the plane $M = \text{span}\{\mathbf{v}, \mathbf{w}\}$, say $\mathbf{u} = a\mathbf{v} + b\mathbf{w}$, where a and b are in $\mathbb{R}$. Then $1\mathbf{u} - a\mathbf{v} - b\mathbf{w} = \mathbf{0}$, contradicting the independence of $\{\mathbf{u}, \mathbf{v}, \mathbf{w}\}$.
>
> On the other hand, suppose that $\mathbf{u}$ is not in M; we must show that $\{\mathbf{u}, \mathbf{v}, \mathbf{w}\}$ is independent. If $r\mathbf{u} + s\mathbf{v} + t\mathbf{w} = \mathbf{0}$ where r, s, and t are in $\mathbb{R}^3$, then $r = 0$ since otherwise $\mathbf{u} = \frac{-s}{r}\mathbf{v} + \frac{-t}{r}\mathbf{w}$ is in M. But then $s\mathbf{v} + t\mathbf{w} = \mathbf{0}$, so $s = t = 0$ by our assumption. This shows that $\{\mathbf{u}, \mathbf{v}, \mathbf{w}\}$ is independent, as required.

By Theorem 5 Section 2.4, the following conditions are equivalent for an $n \times n$ matrix A:

1. A is invertible.
2. If $A\mathbf{x} = \mathbf{0}$ where $\mathbf{x}$ is in $\mathbb{R}^n$, then $\mathbf{x} = \mathbf{0}$.
3. $A\mathbf{x} = \mathbf{b}$ has a solution $\mathbf{x}$ for every vector $\mathbf{b}$ in $\mathbb{R}^n$.

While condition 1 makes no sense if A is not square, conditions 2 and 3 are meaningful for any matrix A and, in fact, are related to independence and spanning.

Indeed, if $\mathbf{c}_1, \mathbf{c}_2, \ldots, \mathbf{c}_n$ are the columns of A, and if we write $\mathbf{x} = \begin{bmatrix} x_1 \\ x_2 \\ \vdots \\ x_n \end{bmatrix}$, then

$$A\mathbf{x} = x_1\mathbf{c}_1 + x_2\mathbf{c}_2 + \cdots + x_n\mathbf{c}_n$$

by Definition 2.5. Hence the definitions of independence and spanning show, respectively, that condition 2 is equivalent to the independence of $\{\mathbf{c}_1, \mathbf{c}_2, \ldots, \mathbf{c}_n\}$ and condition 3 is equivalent to the requirement that $\text{span}\{\mathbf{c}_1, \mathbf{c}_2, \ldots, \mathbf{c}_n\} = \mathbb{R}^m$. This discussion is summarized in the following theorem:

Theorem 2

If A is an $m \times n$ matrix, let $\{\mathbf{c}_1, \mathbf{c}_2, \ldots, \mathbf{c}_n\}$ denote the columns of A.

1. $\{\mathbf{c}_1, \mathbf{c}_2, \ldots, \mathbf{c}_n\}$ is independent in $\mathbb{R}^m$ if and only if $A\mathbf{x} = \mathbf{0}$, $\mathbf{x}$ in $\mathbb{R}^n$, implies $\mathbf{x} = \mathbf{0}$.
2. $\mathbb{R}^m = \text{span}\{\mathbf{c}_1, \mathbf{c}_2, \ldots, \mathbf{c}_n\}$ if and only if $A\mathbf{x} = \mathbf{b}$ has a solution $\mathbf{x}$ for every vector $\mathbf{b}$ in $\mathbb{R}^m$.

For a *square* matrix A, Theorem 2 characterizes the invertibility of A in terms of the spanning and independence of its columns (see the discussion preceding Theorem 2). It is important to be able to discuss these notions for *rows*. If $\mathbf{x}_1, \mathbf{x}_2, \ldots, \mathbf{x}_k$ are $1 \times n$ rows, we define $\text{span}\{\mathbf{x}_1, \mathbf{x}_2, \ldots, \mathbf{x}_k\}$ to be the set of all linear combinations of the $\mathbf{x}_i$ (as matrices), and we say that $\{\mathbf{x}_1, \mathbf{x}_2, \ldots, \mathbf{x}_k\}$ is linearly independent if the only vanishing linear combination is the trivial one (that is, if $\{\mathbf{x}_1^T, \mathbf{x}_2^T, \ldots, \mathbf{x}_k^T\}$ is independent in $\mathbb{R}^n$, as the reader can verify).[6]

Theorem 3

The following are equivalent for an $n \times n$ matrix A:

1. A is invertible.

---

[6] It is best to view columns and rows as just two different *notations* for ordered n-tuples. This discussion will become redundant in Chapter 6 where we define the general notion of a vector space.

2. The columns of A are linearly independent.
3. The columns of A span $\mathbb{R}^n$.
4. The rows of A are linearly independent.
5. The rows of A span the set of all $1 \times n$ rows.

PROOF

Let $\mathbf{c}_1, \mathbf{c}_2, \ldots, \mathbf{c}_n$ denote the columns of A.

(1) $\Leftrightarrow$ (2). By Theorem 5 Section 2.4, A is invertible if and only if $A\mathbf{x} = \mathbf{0}$ implies $\mathbf{x} = \mathbf{0}$; this holds if and only if $\{\mathbf{c}_1, \mathbf{c}_2, \ldots, \mathbf{c}_n\}$ is independent by Theorem 2.

(1) $\Leftrightarrow$ (3). Again by Theorem 5 Section 2.4, A is invertible if and only if $A\mathbf{x} = \mathbf{b}$ has a solution for every column B in $\mathbb{R}^n$; this holds if and only if span$\{\mathbf{c}_1, \mathbf{c}_2, \ldots, \mathbf{c}_n\} = \mathbb{R}^n$ by Theorem 2.

(1) $\Leftrightarrow$ (4). The matrix A is invertible if and only if A^T is invertible (by the Corollary to Theorem 4 Section 2.4); this in turn holds if and only if A^T has independent columns (by (1) $\Leftrightarrow$ (2)); finally, this last statement holds if and only if A has independent rows (because the rows of A are the transposes of the columns of A^T).

(1) $\Leftrightarrow$ (5). The proof is similar to (1) $\Leftrightarrow$ (4).

EXAMPLE 9

Show that $S = \{(2, -2, 5), (-3, 1, 1), (2, 7, -4)\}$ is independent in $\mathbb{R}^3$.

Solution ▶ Consider the matrix $A = \begin{bmatrix} 2 & -2 & 5 \\ -3 & 1 & 1 \\ 2 & 7 & -4 \end{bmatrix}$ with the vectors in S as its rows. A routine computation shows that det $A = -117 \neq 0$, so A is invertible. Hence S is independent by Theorem 3. Note that Theorem 3 also shows that $\mathbb{R}^3 =$ span S.

Dimension

It is common geometrical language to say that $\mathbb{R}^3$ is 3-dimensional, that planes are 2-dimensional and that lines are 1-dimensional. The next theorem is a basic tool for clarifying this idea of "dimension". Its importance is difficult to exaggerate.

Theorem 4

Fundamental Theorem
Let U be a subspace of $\mathbb{R}^n$. If U is spanned by m vectors, and if U contains k linearly independent vectors, then $k \leq m$.

This proof is given in Theorem 2 Section 6.3 in much greater generality.

SECTION 5.2 *Independence and Dimension*

Definition 5.4   *If U is a subspace of $\mathbb{R}^n$, a set $\{\mathbf{x}_1, \mathbf{x}_2, ..., \mathbf{x}_m\}$ of vectors in U is called a **basis** of U if it satisfies the following two conditions:*

1. *$\{\mathbf{x}_1, \mathbf{x}_2, ..., \mathbf{x}_m\}$ is linearly independent.*
2. *$U = \text{span}\{\mathbf{x}_1, \mathbf{x}_2, ..., \mathbf{x}_m\}$.*

The most remarkable result about bases[7] is:

Theorem 5

Invariance Theorem

If $\{\mathbf{x}_1, \mathbf{x}_2, ..., \mathbf{x}_m\}$ and $\{\mathbf{y}_1, \mathbf{y}_2, ..., \mathbf{y}_k\}$ are bases of a subspace U of $\mathbb{R}^n$, then $m = k$.

PROOF

We have $k \leq m$ by the fundamental theorem because $\{\mathbf{x}_1, \mathbf{x}_2, ..., \mathbf{x}_m\}$ spans U, and $\{\mathbf{y}_1, \mathbf{y}_2, ..., \mathbf{y}_k\}$ is independent. Similarly, by interchanging **x**s and **y**s we get $m \leq k$. Hence $m = k$.

The invariance theorem guarantees that there is no ambiguity in the following definition:

Definition 5.5   *If U is a subspace of $\mathbb{R}^n$ and $\{\mathbf{x}_1, \mathbf{x}_2, ..., \mathbf{x}_m\}$ is any basis of U, the number, m, of vectors in the basis is called the **dimension** of U, denoted*

$$\dim U = m.$$

The importance of the invariance theorem is that the dimension of U can be determined by counting the number of vectors in *any* basis.[8]

Let $\{\mathbf{e}_1, \mathbf{e}_2, ..., \mathbf{e}_n\}$ denote the standard basis of $\mathbb{R}^n$, that is the set of columns of the identity matrix. Then $\mathbb{R}^n = \text{span}\{\mathbf{e}_1, \mathbf{e}_2, ..., \mathbf{e}_n\}$ by Example 6 Section 5.1, and $\{\mathbf{e}_1, \mathbf{e}_2, ..., \mathbf{e}_n\}$ is independent by Example 2. Hence it is indeed a basis of $\mathbb{R}^n$ in the present terminology, and we have

EXAMPLE 10

$\dim(\mathbb{R}^n) = n$ and $\{\mathbf{e}_1, \mathbf{e}_2, ..., \mathbf{e}_n\}$ is a basis.

This agrees with our geometric sense that $\mathbb{R}^2$ is two-dimensional and $\mathbb{R}^3$ is three-dimensional. It also says that $\mathbb{R}^1 = \mathbb{R}$ is one-dimensional, and $\{1\}$ is a basis. Returning to subspaces of $\mathbb{R}^n$, we define

$$\dim \{\mathbf{0}\} = 0.$$

This amounts to saying $\{\mathbf{0}\}$ has a basis containing *no* vectors. This makes sense because $\mathbf{0}$ cannot belong to *any* independent set (Example 4).

---

[7] The plural of "basis" is "bases".
[8] We will show in Theorem 6 that every subspace of $\mathbb{R}^n$ does indeed *have* a basis.

EXAMPLE 11

Let $U = \left\{ \begin{bmatrix} r \\ s \\ r \end{bmatrix} \mid r, s \text{ in } \mathbb{R} \right\}$. Show that U is a subspace of $\mathbb{R}^3$, find a basis, and calculate dim U.

Solution ▶ Clearly, $\begin{bmatrix} r \\ s \\ r \end{bmatrix} = r\mathbf{u} + s\mathbf{v}$ where $\mathbf{u} = \begin{bmatrix} 1 \\ 0 \\ 1 \end{bmatrix}$ and $\mathbf{v} = \begin{bmatrix} 0 \\ 1 \\ 0 \end{bmatrix}$. It follows that $U = \text{span}\{\mathbf{u}, \mathbf{v}\}$, and hence that U is a subspace of $\mathbb{R}^3$. Moreover, if $r\mathbf{u} + s\mathbf{v} = \mathbf{0}$, then $\begin{bmatrix} r \\ s \\ r \end{bmatrix} = \begin{bmatrix} 0 \\ 0 \\ 0 \end{bmatrix}$ so $r = s = 0$. Hence $\{\mathbf{u}, \mathbf{v}\}$ is independent, and so a basis of U. This means dim $U = 2$.

EXAMPLE 12

Let $B = \{\mathbf{x}_1, \mathbf{x}_2, \ldots, \mathbf{x}_n\}$ be a basis of $\mathbb{R}^n$. If A is an invertible $n \times n$ matrix, then $D = \{A\mathbf{x}_1, A\mathbf{x}_2, \ldots, A\mathbf{x}_n\}$ is also a basis of $\mathbb{R}^n$.

Solution ▶ Let $\mathbf{x}$ be a vector in $\mathbb{R}^n$. Then $A^{-1}\mathbf{x}$ is in $\mathbb{R}^n$ so, since B is a basis, we have $A^{-1}\mathbf{x} = t_1\mathbf{x}_1 + t_2\mathbf{x}_2 + \cdots + t_n\mathbf{x}_n$ for t_i in $\mathbb{R}$. Left multiplication by A gives $\mathbf{x} = t_1(A\mathbf{x}_1) + t_2(A\mathbf{x}_2) + \cdots + t_n(A\mathbf{x}_n)$, and it follows that D spans $\mathbb{R}^n$. To show independence, let $s_1(A\mathbf{x}_1) + s_2(A\mathbf{x}_2) + \cdots + s_n(A\mathbf{x}_n) = \mathbf{0}$, where the s_i are in $\mathbb{R}$. Then $A(s_1\mathbf{x}_1 + s_2\mathbf{x}_2 + \cdots + s_n\mathbf{x}_n) = \mathbf{0}$ so left multiplication by A^{-1} gives $s_1\mathbf{x}_1 + s_2\mathbf{x}_2 + \cdots + s_n\mathbf{x}_n = \mathbf{0}$. Now the independence of B shows that each $s_i = 0$, and so proves the independence of D. Hence D is a basis of $\mathbb{R}^n$.

While we have found bases in many subspaces of $\mathbb{R}^n$, we have not yet shown that *every* subspace *has* a basis. This is part of the next theorem, the proof of which is deferred to Section 6.4 where it will be proved in more generality.

Theorem 6

Let $U \neq \{\mathbf{0}\}$ be a subspace of $\mathbb{R}^n$. Then:
1. U has a basis and dim $U \leq n$.
2. Any independent set in U can be enlarged (by adding vectors from the standard basis) to a basis of U.
3. Any spanning set for U can be cut down (by deleting vectors) to a basis of U.

EXAMPLE 13

Find a basis of $\mathbb{R}^4$ containing $S = \{\mathbf{u}, \mathbf{v}\}$ where $\mathbf{u} = (0, 1, 2, 3)$ and $\mathbf{v} = (2, -1, 0, 1)$.

Solution ▶ By Theorem 6 we can find such a basis by adding vectors from the standard basis of $\mathbb{R}^4$ to S. If we try $\mathbf{e}_1 = (1, 0, 0, 0)$, we find easily that $\{\mathbf{e}_1, \mathbf{u}, \mathbf{v}\}$ is independent. Now add another vector from the standard basis, say $\mathbf{e}_2$.

SECTION 5.2 Independence and Dimension

Again we find that $B = \{\mathbf{e}_1, \mathbf{e}_2, \mathbf{u}, \mathbf{v}\}$ is independent. Since B has $4 = \dim \mathbb{R}^4$ vectors, then B must span $\mathbb{R}^4$ by Theorem 7 below (or simply verify it directly). Hence B is a basis of $\mathbb{R}^4$.

Theorem 6 has a number of useful consequences. Here is the first.

Theorem 7

Let U be a subspace of $\mathbb{R}^n$ where $\dim U = m$ and let $B = \{\mathbf{x}_1, \mathbf{x}_2, \ldots, \mathbf{x}_m\}$ be a set of m vectors in U. Then B is independent if and only if B spans U.

PROOF

Suppose B is independent. If B does not span U then, by Theorem 6, B can be enlarged to a basis of U containing more than m vectors. This contradicts the invariance theorem because $\dim U = m$, so B spans U. Conversely, if B spans U but is not independent, then B can be cut down to a basis of U containing fewer than m vectors, again a contradiction. So B is independent, as required.

As we saw in Example 13, Theorem 7 is a "labour-saving" result. It asserts that, given a subspace U of dimension m and a set B of exactly m vectors in U, to prove that B is a basis of U it suffices to show either that B spans U or that B is independent. It is not necessary to verify both properties.

Theorem 8

Let $U \subseteq W$ be subspaces of $\mathbb{R}^n$. Then:

1. $\dim U \leq \dim W$.
2. If $\dim U = \dim W$, then $U = W$.

PROOF

Write $\dim W = k$, and let B be a basis of U.

1. If $\dim U > k$, then B is an independent set in W containing more than k vectors, contradicting the fundamental theorem. So $\dim U \leq k = \dim W$.

2. If $\dim U = k$, then B is an independent set in W containing $k = \dim W$ vectors, so B spans W by Theorem 7. Hence $W = \operatorname{span} B = U$, proving (2).

It follows from Theorem 8 that if U is a subspace of $\mathbb{R}^n$, then $\dim U$ is one of the integers $0, 1, 2, \ldots, n$, and that:

$$\dim U = 0 \quad \text{if and only if } U = \{\mathbf{0}\},$$
$$\dim U = n \quad \text{if and only if } U = \mathbb{R}^n$$

The other subspaces are called **proper**. The following example uses Theorem 8 to show that the proper subspaces of $\mathbb{R}^2$ are the lines through the origin, while the proper subspaces of $\mathbb{R}^3$ are the lines and planes through the origin.

EXAMPLE 14

1. If U is a subspace of $\mathbb{R}^2$ or $\mathbb{R}^3$, then dim $U = 1$ if and only if U is a line through the origin.
2. If U is a subspace of $\mathbb{R}^3$, then dim $U = 2$ if and only if U is a plane through the origin.

PROOF

1. Since dim $U = 1$, let $\{\mathbf{u}\}$ be a basis of U. Then $U = \text{span}\{\mathbf{u}\} = \{t\mathbf{u} \mid t \text{ in } \mathbb{R}\}$, so U is the line through the origin with direction vector $\mathbf{u}$. Conversely each line L with direction vector $\mathbf{d} \neq \mathbf{0}$ has the form $L = \{t\mathbf{d} \mid t \text{ in } \mathbb{R}\}$. Hence $\{\mathbf{d}\}$ is a basis of U, so U has dimension 1.

2. If $U \subseteq \mathbb{R}^3$ has dimension 2, let $\{\mathbf{v}, \mathbf{w}\}$ be a basis of U. Then $\mathbf{v}$ and $\mathbf{w}$ are not parallel (by Example 7) so $\mathbf{n} = \mathbf{v} \times \mathbf{w} \neq \mathbf{0}$. Let $P = \{\mathbf{x} \text{ in } \mathbb{R}^3 \mid \mathbf{n} \cdot \mathbf{x} = 0\}$ denote the plane through the origin with normal $\mathbf{n}$. Then P is a subspace of $\mathbb{R}^3$ (Example 1 Section 5.1) and both $\mathbf{v}$ and $\mathbf{w}$ lie in P (they are orthogonal to $\mathbf{n}$), so $U = \text{span}\{\mathbf{v}, \mathbf{w}\} \subseteq P$ by Theorem 1 Section 5.1. Hence

$$U \subseteq P \subseteq \mathbb{R}^3.$$

Since dim $U = 2$ and $\dim(\mathbb{R}^3) = 3$, it follows from Theorem 8 that dim $P = 2$ or 3, whence $P = U$ or $\mathbb{R}^3$. But $P \neq \mathbb{R}^3$ (for example, $\mathbf{n}$ is not in P) and so $U = P$ is a plane through the origin.

Conversely, if U is a plane through the origin, then dim $U = 0, 1, 2,$ or 3 by Theorem 8. But dim $U \neq 0$ or 3 because $U \neq \{\mathbf{0}\}$ and $U \neq \mathbb{R}^3$, and dim $U \neq 1$ by (1). So dim $U = 2$.

Note that this proof shows that if $\mathbf{v}$ and $\mathbf{w}$ are nonzero, nonparallel vectors in $\mathbb{R}^3$, then span$\{\mathbf{v}, \mathbf{w}\}$ is the plane with normal $\mathbf{n} = \mathbf{v} \times \mathbf{w}$. We gave a geometrical verification of this fact in Section 5.1.

EXERCISES 5.2

In Exercises 1–6 we write vectors $\mathbb{R}^n$ as rows.

1. Which of the following subsets are independent? Support your answer.

   (a) $\{(1, -1, 0), (3, 2, -1), (3, 5, -2)\}$ in $\mathbb{R}^3$.

   ◆(b) $\{(1, 1, 1), (1, -1, 1), (0, 0, 1)\}$ in $\mathbb{R}^3$.

   (c) $\{(1, -1, 1, -1), (2, 0, 1, 0), (0, -2, 1, -2)\}$ in $\mathbb{R}^4$.

   ◆(d) $\{(1, 1, 0, 0), (1, 0, 1, 0), (0, 0, 1, 1), (0, 1, 0, 1)\}$ in $\mathbb{R}^4$.

2. Let $\{\mathbf{x}, \mathbf{y}, \mathbf{z}, \mathbf{w}\}$ be an independent set in $\mathbb{R}^n$. Which of the following sets is independent? Support your answer.

   (a) $\{\mathbf{x} - \mathbf{y}, \mathbf{y} - \mathbf{z}, \mathbf{z} - \mathbf{x}\}$

   ◆(b) $\{\mathbf{x} + \mathbf{y}, \mathbf{y} + \mathbf{z}, \mathbf{z} + \mathbf{x}\}$

   (c) $\{\mathbf{x} - \mathbf{y}, \mathbf{y} - \mathbf{z}, \mathbf{z} - \mathbf{w}, \mathbf{w} - \mathbf{x}\}$

   ◆(d) $\{\mathbf{x} + \mathbf{y}, \mathbf{y} + \mathbf{z}, \mathbf{z} + \mathbf{w}, \mathbf{w} + \mathbf{x}\}$

3. Find a basis and calculate the dimension of the following subspaces of $\mathbb{R}^4$.

   (a) span$\{(1, -1, 2, 0), (2, 3, 0, 3), (1, 9, -6, 6)\}$.

   ◆(b) span$\{(2, 1, 0, -1), (-1, 1, 1, 1), (2, 7, 4, 1)\}$.

   (c) span$\{(-1, 2, 1, 0), (2, 0, 3, -1), (4, 4, 11, -3), (3, -2, 2, -1)\}$.

   ◆(d) span$\{(-2, 0, 3, 1), (1, 2, -1, 0), (-2, 8, 5, 3), (-1, 2, 2, 1)\}$.

4. Find a basis and calculate the dimension of the following subspaces of $\mathbb{R}^4$.

(a) $U = \left\{ \begin{bmatrix} a \\ a+b \\ a-b \\ b \end{bmatrix} \,\middle|\, a \text{ and } b \text{ in } \mathbb{R} \right\}$.

(b) $U = \left\{ \begin{bmatrix} a+b \\ a-b \\ b \\ a \end{bmatrix} \,\middle|\, a \text{ and } b \text{ in } \mathbb{R} \right\}$.

(c) $U = \left\{ \begin{bmatrix} a \\ b \\ c+a \\ c \end{bmatrix} \,\middle|\, a, b, \text{ and } c \text{ in } \mathbb{R} \right\}$.

(d) $U = \left\{ \begin{bmatrix} a-b \\ b+c \\ a \\ b+c \end{bmatrix} \,\middle|\, a, b, \text{ and } c \text{ in } \mathbb{R} \right\}$.

(e) $U = \left\{ \begin{bmatrix} a \\ b \\ c \\ d \end{bmatrix} \,\middle|\, a+b-c+d=0 \text{ in } \mathbb{R} \right\}$.

(f) $U = \left\{ \begin{bmatrix} a \\ b \\ c \\ d \end{bmatrix} \,\middle|\, a+b=c+d \text{ in } \mathbb{R} \right\}$.

5. Suppose that $\{\mathbf{x}, \mathbf{y}, \mathbf{z}, \mathbf{w}\}$ is a basis of $\mathbb{R}^4$. Show that:

(a) $\{\mathbf{x} + a\mathbf{w}, \mathbf{y}, \mathbf{z}, \mathbf{w}\}$ is also a basis of $\mathbb{R}^4$ for any choice of the scalar a.

(b) $\{\mathbf{x} + \mathbf{w}, \mathbf{y} + \mathbf{w}, \mathbf{z} + \mathbf{w}, \mathbf{w}\}$ is also a basis of $\mathbb{R}^4$.

(c) $\{\mathbf{x}, \mathbf{x} + \mathbf{y}, \mathbf{x} + \mathbf{y} + \mathbf{z}, \mathbf{x} + \mathbf{y} + \mathbf{z} + \mathbf{w}\}$ is also a basis of $\mathbb{R}^4$.

6. Use Theorem 3 to determine if the following sets of vectors are a basis of the indicated space.

(a) $\{(3, -1), (2, 2)\}$ in $\mathbb{R}^2$.

(b) $\{(1, 1, -1), (1, -1, 1), (0, 0, 1)\}$ in $\mathbb{R}^3$.

(c) $\{(-1, 1, -1), (1, -1, 2), (0, 0, 1)\}$ in $\mathbb{R}^3$.

(d) $\{(5, 2, -1), (1, 0, 1), (3, -1, 0)\}$ in $\mathbb{R}^3$.

(e) $\{(2, 1, -1, 3), (1, 1, 0, 2), (0, 1, 0, -3), (-1, 2, 3, 1)\}$ in $\mathbb{R}^4$.

(f) $\{(1, 0, -2, 5), (4, 4, -3, 2), (0, 1, 0, -3), (1, 3, 3, -10)\}$ in $\mathbb{R}^4$.

7. In each case show that the statement is true or give an example showing that it is false.

(a) If $\{\mathbf{x}, \mathbf{y}\}$ is independent, then $\{\mathbf{x}, \mathbf{y}, \mathbf{x} + \mathbf{y}\}$ is independent.

(b) If $\{\mathbf{x}, \mathbf{y}, \mathbf{z}\}$ is independent, then $\{\mathbf{y}, \mathbf{z}\}$ is independent.

(c) If $\{\mathbf{y}, \mathbf{z}\}$ is dependent, then $\{\mathbf{x}, \mathbf{y}, \mathbf{z}\}$ is dependent for any $\mathbf{x}$.

(d) If all of $\mathbf{x}_1, \mathbf{x}_2, \ldots, \mathbf{x}_k$ are nonzero, then $\{\mathbf{x}_1, \mathbf{x}_2, \ldots, \mathbf{x}_k\}$ is independent.

(e) If one of $\mathbf{x}_1, \mathbf{x}_2, \ldots, \mathbf{x}_k$ is zero, then $\{\mathbf{x}_1, \mathbf{x}_2, \ldots, \mathbf{x}_k\}$ is dependent.

(f) If $a\mathbf{x} + b\mathbf{y} + c\mathbf{z} = \mathbf{0}$, then $\{\mathbf{x}, \mathbf{y}, \mathbf{z}\}$ is independent.

(g) If $\{\mathbf{x}, \mathbf{y}, \mathbf{z}\}$ is independent, then $a\mathbf{x} + b\mathbf{y} + c\mathbf{z} = \mathbf{0}$ for some a, b, and c in $\mathbb{R}$.

(h) If $\{\mathbf{x}_1, \mathbf{x}_2, \ldots, \mathbf{x}_k\}$ is dependent, then $t_1\mathbf{x}_1 + t_2\mathbf{x}_2 + \cdots + t_k\mathbf{x}_k = \mathbf{0}$ for some numbers t_i in $\mathbb{R}$ not all zero.

(i) If $\{\mathbf{x}_1, \mathbf{x}_2, \ldots, \mathbf{x}_k\}$ is independent, then $t_1\mathbf{x}_1 + t_2\mathbf{x}_2 + \cdots + t_k\mathbf{x}_k = \mathbf{0}$ for some t_i in $\mathbb{R}$.

8. If A is an $n \times n$ matrix, show that $\det A = 0$ if and only if some column of A is a linear combination of the other columns.

9. Let $\{\mathbf{x}, \mathbf{y}, \mathbf{z}\}$ be a linearly independent set in $\mathbb{R}^4$. Show that $\{\mathbf{x}, \mathbf{y}, \mathbf{z}, \mathbf{e}_k\}$ is a basis of $\mathbb{R}^4$ for some $\mathbf{e}_k$ in the standard basis $\{\mathbf{e}_1, \mathbf{e}_2, \mathbf{e}_3, \mathbf{e}_4\}$.

10. If $\{\mathbf{x}_1, \mathbf{x}_2, \mathbf{x}_3, \mathbf{x}_4, \mathbf{x}_5, \mathbf{x}_6\}$ is an independent set of vectors, show that the subset $\{\mathbf{x}_2, \mathbf{x}_3, \mathbf{x}_5\}$ is also independent.

11. Let A be any $m \times n$ matrix, and let $\mathbf{b}_1, \mathbf{b}_2, \mathbf{b}_3, \ldots, \mathbf{b}_k$ be columns in $\mathbb{R}^m$ such that the system $A\mathbf{x} = \mathbf{b}_i$ has a solution $\mathbf{x}_i$ for each i. If $\{\mathbf{b}_1, \mathbf{b}_2, \mathbf{b}_3, \ldots, \mathbf{b}_k\}$ is independent in $\mathbb{R}^m$, show that $\{\mathbf{x}_1, \mathbf{x}_2, \mathbf{x}_3, \ldots, \mathbf{x}_k\}$ is independent in $\mathbb{R}^n$.

12. If $\{\mathbf{x}_1, \mathbf{x}_2, \mathbf{x}_3, \ldots, \mathbf{x}_k\}$ is independent, show that $\{\mathbf{x}_1, \mathbf{x}_1 + \mathbf{x}_2, \mathbf{x}_1 + \mathbf{x}_2 + \mathbf{x}_3, \ldots, \mathbf{x}_1 + \mathbf{x}_2 + \cdots + \mathbf{x}_k\}$ is also independent.

13. If $\{\mathbf{y}, \mathbf{x}_1, \mathbf{x}_2, \mathbf{x}_3, \ldots, \mathbf{x}_k\}$ is independent, show that $\{\mathbf{y} + \mathbf{x}_1, \mathbf{y} + \mathbf{x}_2, \mathbf{y} + \mathbf{x}_3, \ldots, \mathbf{y} + \mathbf{x}_k\}$ is also independent.

14. If $\{\mathbf{x}_1, \mathbf{x}_2, \ldots, \mathbf{x}_k\}$ is independent in $\mathbb{R}^n$, and if $\mathbf{y}$ is not in $\text{span}\{\mathbf{x}_1, \mathbf{x}_2, \ldots, \mathbf{x}_k\}$, show that $\{\mathbf{x}_1, \mathbf{x}_2, \ldots, \mathbf{x}_k, \mathbf{y}\}$ is independent.

15. If A and B are matrices and the columns of AB are independent, show that the columns of B are independent.

16. Suppose that $\{\mathbf{x}, \mathbf{y}\}$ is a basis of $\mathbb{R}^2$, and let
$A = \begin{bmatrix} a & b \\ c & d \end{bmatrix}$.

   (a) If A is invertible, show that $\{a\mathbf{x} + b\mathbf{y}, c\mathbf{x} + d\mathbf{y}\}$ is a basis of $\mathbb{R}^2$.

   ◆(b) If $\{a\mathbf{x} + b\mathbf{y}, c\mathbf{x} + d\mathbf{y}\}$ is a basis of $\mathbb{R}^2$, show that A is invertible.

17. Let A denote an $m \times n$ matrix.

   (a) Show that null A = null(UA) for every invertible $m \times m$ matrix U.

   ◆(b) Show that dim(null A) = dim(null(AV)) for every invertible $n \times n$ matrix V. [*Hint*: If $\{\mathbf{x}_1, \mathbf{x}_2, \ldots, \mathbf{x}_k\}$ is a basis of null A, show that $\{V^{-1}\mathbf{x}_1, V^{-1}\mathbf{x}_2, \ldots, V^{-1}\mathbf{x}_k\}$ is a basis of null(AV).]

18. Let A denote an $m \times n$ matrix.

   (a) Show that im A = im(AV) for every invertible $n \times n$ matrix V.

   (b) Show that dim(im A) = dim(im(UA)) for every invertible $m \times m$ matrix U. [*Hint*: If $\{\mathbf{y}_1, \mathbf{y}_2, \ldots, \mathbf{y}_k\}$ is a basis of im(UA), show that $\{U^{-1}\mathbf{y}_1, U^{-1}\mathbf{y}_2, \ldots, U^{-1}\mathbf{y}_k\}$ is a basis of im A.]

19. Let U and W denote subspaces of $\mathbb{R}^n$, and assume that $U \subseteq W$. If dim $U = n - 1$, show that either $W = U$ or $W = \mathbb{R}^n$.

◆20. Let U and W denote subspaces of $\mathbb{R}^n$, and assume that $U \subseteq W$. If dim $W = 1$, show that either $U = \{\mathbf{0}\}$ or $U = W$.

SECTION 5.3 Orthogonality

Length and orthogonality are basic concepts in geometry and, in $\mathbb{R}^2$ and $\mathbb{R}^3$, they both can be defined using the dot product. In this section we extend the dot product to vectors in $\mathbb{R}^n$, and so endow $\mathbb{R}^n$ with euclidean geometry. We then introduce the idea of an orthogonal basis—one of the most useful concepts in linear algebra, and begin exploring some of its applications.

Dot Product, Length, and Distance

If $\mathbf{x} = (x_1, x_2, \ldots, x_n)$ and $\mathbf{y} = (y_1, y_2, \ldots, y_n)$ are two n-tuples in $\mathbb{R}^n$, recall that their **dot product** was defined in Section 2.2 as follows:

$$\mathbf{x} \cdot \mathbf{y} = x_1 y_1 + x_2 y_2 + \cdots + x_n y_n.$$

Observe that if $\mathbf{x}$ and $\mathbf{y}$ are written as columns then $\mathbf{x} \cdot \mathbf{y} = \mathbf{x}^T \mathbf{y}$ is a matrix product (and $\mathbf{x} \cdot \mathbf{y} = \mathbf{x}\mathbf{y}^T$ if they are written as rows). Here $\mathbf{x} \cdot \mathbf{y}$ is a 1×1 matrix, which we take to be a number.

Definition 5.6 *As in $\mathbb{R}^3$, the **length** $\|\mathbf{x}\|$ of the vector is defined by*

$$\|\mathbf{x}\| = \sqrt{\mathbf{x} \cdot \mathbf{x}} = \sqrt{x_1^2 + x_2^2 + \cdots + x_n^2}$$

Where $\sqrt{(\ )}$ indicates the positive square root.

A vector $\mathbf{x}$ of length 1 is called a **unit vector**. If $\mathbf{x} \neq \mathbf{0}$, then $\|\mathbf{x}\| \neq 0$ and it follows easily that $\frac{1}{\|\mathbf{x}\|}\mathbf{x}$ is a unit vector (see Theorem 6 below), a fact that we shall use later.

EXAMPLE 1

If $\mathbf{x} = (1, -1, -3, 1)$ and $\mathbf{y} = (2, 1, 1, 0)$ in $\mathbb{R}^4$, then $\mathbf{x} \cdot \mathbf{y} = 2 - 1 - 3 + 0 = -2$ and $\|\mathbf{x}\| = \sqrt{1 + 1 + 9 + 1} = \sqrt{12} = 2\sqrt{3}$. Hence $\frac{1}{2\sqrt{3}}\mathbf{x}$ is a unit vector; similarly $\frac{1}{\sqrt{6}}\mathbf{y}$ is a unit vector.

SECTION 5.3 Orthogonality

These definitions agree with those in $\mathbb{R}^2$ and $\mathbb{R}^3$, and many properties carry over to $\mathbb{R}^n$:

Theorem 1

Let $\mathbf{x}$, $\mathbf{y}$, and $\mathbf{z}$ denote vectors in $\mathbb{R}^n$. Then:
1. $\mathbf{x} \cdot \mathbf{y} = \mathbf{y} \cdot \mathbf{x}$.
2. $\mathbf{x} \cdot (\mathbf{y} + \mathbf{z}) = \mathbf{x} \cdot \mathbf{y} + \mathbf{x} \cdot \mathbf{z}$.
3. $(a\mathbf{x}) \cdot \mathbf{y} = a(\mathbf{x} \cdot \mathbf{y}) = \mathbf{x} \cdot (a\mathbf{y})$ for all scalars a.
4. $\|\mathbf{x}\|^2 = \mathbf{x} \cdot \mathbf{x}$.
5. $\|\mathbf{x}\| \geq 0$, and $\|\mathbf{x}\| = 0$ if and only if $\mathbf{x} = \mathbf{0}$.
6. $\|a\mathbf{x}\| = |a|\,\|\mathbf{x}\|$ for all scalars a.

PROOF

(1), (2), and (3) follow from matrix arithmetic because $\mathbf{x} \cdot \mathbf{y} = \mathbf{x}^T\mathbf{y}$; (4) is clear from the definition; and (6) is a routine verification since $|a| = \sqrt{a^2}$. If $\mathbf{x} = (x_1, x_2, \ldots, x_n)$, then $\|\mathbf{x}\| = \sqrt{x_1^2 + x_2^2 + \cdots + x_n^2}$, so $\|\mathbf{x}\| = 0$ if and only if $x_1^2 + x_2^2 + \cdots + x_n^2 = 0$. Since each x_i is a real number this happens if and only if $x_i = 0$ for each i; that is, if and only if $\mathbf{x} = \mathbf{0}$. This proves (5).

Because of Theorem 1, computations with dot products in $\mathbb{R}^n$ are similar to those in $\mathbb{R}^3$. In particular, the dot product

$$(\mathbf{x}_1 + \mathbf{x}_2 + \cdots + \mathbf{x}_m) \cdot (\mathbf{y}_1 + \mathbf{y}_2 + \cdots + \mathbf{y}_k)$$

equals the sum of mk terms, $\mathbf{x}_i \cdot \mathbf{y}_j$, one for each choice of i and j. For example:

$$(3\mathbf{x} - 4\mathbf{y}) \cdot (7\mathbf{x} + 2\mathbf{y}) = 21(\mathbf{x} \cdot \mathbf{x}) + 6(\mathbf{x} \cdot \mathbf{y}) - 28(\mathbf{y} \cdot \mathbf{x}) - 8(\mathbf{y} \cdot \mathbf{y})$$
$$= 21\|\mathbf{x}\|^2 - 22(\mathbf{x} \cdot \mathbf{y}) - 8\|\mathbf{y}\|^2$$

holds for all vectors $\mathbf{x}$ and $\mathbf{y}$.

EXAMPLE 2

Show that $\|\mathbf{x} + \mathbf{y}\|^2 = \|\mathbf{x}\|^2 + 2(\mathbf{x} \cdot \mathbf{y}) + \|\mathbf{y}\|^2$ for any $\mathbf{x}$ and $\mathbf{y}$ in $\mathbb{R}^n$.

Solution ▶ Using Theorem 1 several times:

$$\|\mathbf{x} + \mathbf{y}\|^2 = (\mathbf{x} + \mathbf{y}) \cdot (\mathbf{x} + \mathbf{y}) = \mathbf{x} \cdot \mathbf{x} + \mathbf{x} \cdot \mathbf{y} + \mathbf{y} \cdot \mathbf{x} + \mathbf{y} \cdot \mathbf{y}$$
$$= \|\mathbf{x}\|^2 + 2(\mathbf{x} \cdot \mathbf{y}) + \|\mathbf{y}\|^2$$

EXAMPLE 3

Suppose that $\mathbb{R}^n = \text{span}\{\mathbf{f}_1, \mathbf{f}_2, \ldots, \mathbf{f}_k\}$ for some vectors $\mathbf{f}_i$. If $\mathbf{x} \cdot \mathbf{f}_i = 0$ for each i where $\mathbf{x}$ is in $\mathbb{R}^n$, show that $\mathbf{x} = \mathbf{0}$.

Solution ▶ We show $\mathbf{x} = \mathbf{0}$ by showing that $\|\mathbf{x}\| = 0$ and using (5) of Theorem 1. Since the $\mathbf{f}_i$ span $\mathbb{R}^n$, write $\mathbf{x} = t_1\mathbf{f}_1 + t_2\mathbf{f}_2 + \cdots + t_k\mathbf{f}_k$ where the t_i are in $\mathbb{R}$. Then

$$\begin{aligned}\|\mathbf{x}\|^2 &= \mathbf{x} \cdot \mathbf{x} = \mathbf{x} \cdot (t_1\mathbf{f}_1 + t_2\mathbf{f}_2 + \cdots + t_k\mathbf{f}_k) \\ &= t_1(\mathbf{x} \cdot \mathbf{f}_1) + t_2(\mathbf{x} \cdot \mathbf{f}_2) + \cdots + t_k(\mathbf{x} \cdot \mathbf{f}_k) \\ &= t_1(0) + t_2(0) + \cdots + t_k(0) \\ &= 0.\end{aligned}$$

We saw in Section 4.2 that if $\mathbf{u}$ and $\mathbf{v}$ are nonzero vectors in $\mathbb{R}^3$, then $\dfrac{\mathbf{u} \cdot \mathbf{v}}{\|\mathbf{u}\|\|\mathbf{v}\|} = \cos\theta$ where θ is the angle between $\mathbf{u}$ and $\mathbf{v}$. Since $|\cos\theta| \leq 1$ for any angle θ, this shows that $|\mathbf{u} \cdot \mathbf{v}| \leq \|\mathbf{u}\|\|\mathbf{v}\|$. In this form the result holds in $\mathbb{R}^n$.

Theorem 2

Cauchy Inequality[9]

If $\mathbf{x}$ and $\mathbf{y}$ are vectors in $\mathbb{R}^n$, then

$$|\mathbf{x} \cdot \mathbf{y}| \leq \|\mathbf{x}\|\|\mathbf{y}\|.$$

Moreover $|\mathbf{x} \cdot \mathbf{y}| = \|\mathbf{x}\|\|\mathbf{y}\|$ if and only if one of $\mathbf{x}$ and $\mathbf{y}$ is a multiple of the other.

Augustin Louis Cauchy.
Photo © Corbis.

PROOF

The inequality holds if $\mathbf{x} = \mathbf{0}$ or $\mathbf{y} = \mathbf{0}$ (in fact it is equality). Otherwise, write $\|\mathbf{x}\| = a > 0$ and $\|\mathbf{y}\| = b > 0$ for convenience. A computation like that preceding Example 2 gives

$$\|b\mathbf{x} - a\mathbf{y}\|^2 = 2ab(ab - \mathbf{x} \cdot \mathbf{y}) \quad \text{and} \quad \|b\mathbf{x} - a\mathbf{y}\|^2 = 2ab(ab + \mathbf{x} \cdot \mathbf{y}). \qquad (*)$$

It follows that $ab - \mathbf{x} \cdot \mathbf{y} \geq 0$ and $ab + \mathbf{x} \cdot \mathbf{y} \geq 0$, and hence that $-ab \leq \mathbf{x} \cdot \mathbf{y} \leq ab$. Hence $|\mathbf{x} \cdot \mathbf{y}| \leq ab = \|\mathbf{x}\|\|\mathbf{y}\|$, proving the Cauchy inequality.

If equality holds, then $|\mathbf{x} \cdot \mathbf{y}| = ab$, so $\mathbf{x} \cdot \mathbf{y} = ab$ or $\mathbf{x} \cdot \mathbf{y} = -ab$. Hence $(*)$ shows that $b\mathbf{x} - a\mathbf{y} = 0$ or $b\mathbf{x} + a\mathbf{y} = 0$, so one of $\mathbf{x}$ and $\mathbf{y}$ is a multiple of the other (even if $a = 0$ or $b = 0$).

The Cauchy inequality is equivalent to $(\mathbf{x} \cdot \mathbf{y})^2 \leq \|\mathbf{x}\|^2\|\mathbf{y}\|^2$. In $\mathbb{R}^5$ this becomes

$$(x_1y_1 + x_2y_2 + x_3y_3 + x_4y_4 + x_5y_5)^2 \\ \leq (x_1^2 + x_2^2 + x_3^2 + x_4^2 + x_5^2)(y_1^2 + y_2^2 + y_3^2 + y_4^2 + y_5^2)$$

for all x_i and y_i in $\mathbb{R}$.

There is an important consequence of the Cauchy inequality. Given $\mathbf{x}$ and $\mathbf{y}$ in $\mathbb{R}^n$, use Example 2 and the fact that $\mathbf{x} \cdot \mathbf{y} \leq \|\mathbf{x}\|\|\mathbf{y}\|$ to compute

$$\|\mathbf{x} + \mathbf{y}\|^2 = \|\mathbf{x}\|^2 + 2(\mathbf{x} \cdot \mathbf{y}) + \|\mathbf{y}\|^2 \leq \|\mathbf{x}\|^2 + 2\|\mathbf{x}\|\|\mathbf{y}\| + \|\mathbf{y}\|^2 = (\|\mathbf{x} + \mathbf{y}\|)^2.$$

Taking positive square roots gives:

---

[9] Augustin Louis Cauchy (1789–1857) was born in Paris and became a professor at the École Polytechnique at the age of 26. He was one of the great mathematicians, producing more than 700 papers, and is best remembered for his work in analysis in which he established new standards of rigour and founded the theory of functions of a complex variable. He was a devout Catholic with a long-term interest in charitable work, and he was a royalist, following King Charles X into exile in Prague after he was deposed in 1830. Theorem 2 first appeared in his 1812 memoir on determinants.

Corollary 1

Triangle Inequality
If **x** and **y** are vectors in $\mathbb{R}^n$, then $\|\mathbf{x} + \mathbf{y}\| \leq \|\mathbf{x}\| + \|\mathbf{y}\|$.

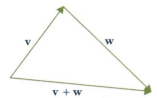

The reason for the name comes from the observation that in $\mathbb{R}^3$ the inequality asserts that the sum of the lengths of two sides of a triangle is not less than the length of the third side. This is illustrated in the first diagram.

Definition 5.7 If **x** and **y** are two vectors in $\mathbb{R}^n$, we define the **distance** $d(\mathbf{x}, \mathbf{y})$ between **x** and **y** by

$$d(\mathbf{x}, \mathbf{y}) = \|\mathbf{x} - \mathbf{y}\|$$

The motivation again comes from $\mathbb{R}^3$ as is clear in the second diagram. This distance function has all the intuitive properties of distance in $\mathbb{R}^3$, including another version of the triangle inequality.

Theorem 3

If **x**, **y**, and **z** are three vectors in $\mathbb{R}^n$ we have:
1. $d(\mathbf{x}, \mathbf{y}) \geq 0$ for all **x** and **y**.
2. $d(\mathbf{x}, \mathbf{y}) = 0$ if and only if $\mathbf{x} = \mathbf{y}$.
3. $d(\mathbf{x}, \mathbf{y}) = d(\mathbf{y}, \mathbf{x})$.
4. $d(\mathbf{x}, \mathbf{z}) \leq d(\mathbf{x}, \mathbf{y}) + d(\mathbf{y}, \mathbf{z})$. Triangle inequality.

PROOF

(1) and (2) restate part (5) of Theorem 1 because $d(\mathbf{x}, \mathbf{y}) = \|\mathbf{x} - \mathbf{y}\|$, and (3) follows because $\|\mathbf{u}\| = \|-\mathbf{u}\|$ for every vector **u** in $\mathbb{R}^n$. To prove (4) use the Corollary to Theorem 2:

$$d(\mathbf{x}, \mathbf{z}) = \|\mathbf{x} - \mathbf{z}\| = \|(\mathbf{x} - \mathbf{y}) + (\mathbf{y} - \mathbf{z})\|$$
$$\leq \|(\mathbf{x} - \mathbf{y})\| + \|(\mathbf{y} - \mathbf{z})\| = d(\mathbf{x}, \mathbf{y}) + d(\mathbf{y}, \mathbf{z})$$

Orthogonal Sets and the Expansion Theorem

Definition 5.8 We say that two vectors **x** and **y** in $\mathbb{R}^n$ are **orthogonal** if $\mathbf{x} \cdot \mathbf{y} = 0$, extending the terminology in $\mathbb{R}^3$ (See Theorem 3 Section 4.2). More generally, a set $\{\mathbf{x}_1, \mathbf{x}_2, ..., \mathbf{x}_k\}$ of vectors in $\mathbb{R}^n$ is called an **orthogonal set** if

$$\mathbf{x}_i \cdot \mathbf{x}_j = 0 \text{ for all } i \neq j \quad \text{and} \quad \mathbf{x}_i \neq \mathbf{0} \text{ for all } i.^{10}$$

Note that $\{\mathbf{x}\}$ is an orthogonal set if $\mathbf{x} \neq \mathbf{0}$. A set $\{\mathbf{x}_1, \mathbf{x}_2, ..., \mathbf{x}_k\}$ of vectors in $\mathbb{R}^n$ is called **orthonormal** if it is orthogonal and, in addition, each $\mathbf{x}_i$ is a unit vector:

$$\|\mathbf{x}_i\| = 1 \text{ for each } i.$$

---

[10] The reason for insisting that orthogonal sets consist of *nonzero* vectors is that we will be primarily concerned with orthogonal bases.

EXAMPLE 4

The standard basis $\{e_1, e_2, ..., e_n\}$ is an orthonormal set in $\mathbb{R}^n$.

The routine verification is left to the reader, as is the proof of:

EXAMPLE 5

If $\{x_1, x_2, ..., x_k\}$ is orthogonal, so also is $\{a_1 x_1, a_2 x_2, ..., a_k x_k\}$ for any nonzero scalars a_i.

If $\mathbf{x} \neq \mathbf{0}$, it follows from item (6) of Theorem 1 that $\frac{1}{\|\mathbf{x}\|}\mathbf{x}$ is a unit vector, that is it has length 1.

Definition 5.9 Hence if $\{x_1, x_2, ..., x_k\}$ is an orthogonal set, then $\left\{\frac{1}{\|x_1\|}x_1, \frac{1}{\|x_2\|}x_2, ..., \frac{1}{\|x_k\|}x_k\right\}$ is an orthonormal set, and we say that it is the result of **normalizing** the orthogonal set $\{x_1, x_2, ..., x_k\}$.

EXAMPLE 6

If $\mathbf{f}_1 = \begin{bmatrix} 1 \\ 1 \\ 1 \\ -1 \end{bmatrix}$, $\mathbf{f}_2 = \begin{bmatrix} 1 \\ 0 \\ 1 \\ 2 \end{bmatrix}$, $\mathbf{f}_3 = \begin{bmatrix} -1 \\ 0 \\ 1 \\ 0 \end{bmatrix}$, and $\mathbf{f}_4 = \begin{bmatrix} -1 \\ 3 \\ -1 \\ 1 \end{bmatrix}$ then $\{\mathbf{f}_1, \mathbf{f}_2, \mathbf{f}_3, \mathbf{f}_4\}$ is an orthogonal set in $\mathbb{R}^4$ as is easily verified. After normalizing, the corresponding orthonormal set is $\{\frac{1}{2}\mathbf{f}_1, \frac{1}{\sqrt{6}}\mathbf{f}_2, \frac{1}{\sqrt{2}}\mathbf{f}_3, \frac{1}{2\sqrt{3}}\mathbf{f}_4\}$.

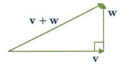

The most important result about orthogonality is Pythagoras' theorem. Given orthogonal vectors $\mathbf{v}$ and $\mathbf{w}$ in $\mathbb{R}^3$, it asserts that $\|\mathbf{v} + \mathbf{w}\|^2 = \|\mathbf{v}\|^2 + \|\mathbf{w}\|^2$ as in the diagram. In this form the result holds for any orthogonal set in $\mathbb{R}^n$.

Theorem 4

Pythagoras' Theorem
If $\{x_1, x_2, ..., x_k\}$ is a orthogonal set in $\mathbb{R}^n$, then
$$\|x_1 + x_2 + \cdots + x_k\|^2 = \|x_1\|^2 + \|x_2\|^2 + \cdots + \|x_k\|^2.$$

PROOF

The fact that $\mathbf{x}_i \cdot \mathbf{x}_j = 0$ whenever $i \neq j$ gives
$$\begin{aligned} \|x_1 + x_2 + \cdots + x_k\|^2 &= (x_1 + x_2 + \cdots + x_k) \cdot (x_1 + x_2 + \cdots + x_k) \\ &= (x_1 \cdot x_1 + x_2 \cdot x_2 + \cdots + x_k \cdot x_k) + \sum_{i \neq j} x_i \cdot x_j \\ &= \|x_1\|^2 + \|x_2\|^2 + \cdots + \|x_k\|^2 + 0. \end{aligned}$$

This is what we wanted.

SECTION 5.3 Orthogonality

If **v** and **w** are orthogonal, nonzero vectors in $\mathbb{R}^3$, then they are certainly not parallel, and so are linearly independent by Example 7 Section 5.2. The next theorem gives a far-reaching extension of this observation.

Theorem 5

Every orthogonal set in $\mathbb{R}^n$ is linearly independent.

PROOF

Let $\{\mathbf{x}_1, \mathbf{x}_2, \ldots, \mathbf{x}_k\}$ be an orthogonal set in $\mathbb{R}^n$ and suppose a linear combination vanishes: $t_1\mathbf{x}_1 + t_2\mathbf{x}_2 + \cdots + t_k\mathbf{x}_k = \mathbf{0}$. Then

$$\begin{aligned} 0 = \mathbf{x}_1 \cdot \mathbf{0} &= \mathbf{x}_1 \cdot (t_1\mathbf{x}_1 + t_2\mathbf{x}_2 + \cdots + t_k\mathbf{x}_k) \\ &= t_1(\mathbf{x}_1 \cdot \mathbf{x}_1) + t_2(\mathbf{x}_1 \cdot \mathbf{x}_2) + \cdots + t_k(\mathbf{x}_1 \cdot \mathbf{x}_k) \\ &= t_1\|\mathbf{x}_1\|^2 + t_2(0) + \cdots + t_k(0) \\ &= t_1\|\mathbf{x}_1\|^2 \end{aligned}$$

Since $\|\mathbf{x}_1\|^2 \neq 0$, this implies that $t_1 = 0$. Similarly $t_i = 0$ for each i.

Theorem 5 suggests considering orthogonal bases for $\mathbb{R}^n$, that is orthogonal sets that span $\mathbb{R}^n$. These turn out to be the best bases in the sense that, when expanding a vector as a linear combination of the basis vectors, there are explicit formulas for the coefficients.

Theorem 6

Expansion Theorem

Let $\{\mathbf{f}_1, \mathbf{f}_2, \ldots, \mathbf{f}_m\}$ be an orthogonal basis of a subspace U of $\mathbb{R}^n$. If $\mathbf{x}$ is any vector in U, we have

$$\mathbf{x} = \left(\frac{\mathbf{x} \cdot \mathbf{f}_1}{\|\mathbf{f}_1\|^2}\right)\mathbf{f}_1 + \left(\frac{\mathbf{x} \cdot \mathbf{f}_2}{\|\mathbf{f}_2\|^2}\right)\mathbf{f}_2 + \cdots + \left(\frac{\mathbf{x} \cdot \mathbf{f}_m}{\|\mathbf{f}_m\|^2}\right)\mathbf{f}_m.$$

PROOF

Since $\{\mathbf{f}_1, \mathbf{f}_2, \ldots, \mathbf{f}_m\}$ spans U, we have $\mathbf{x} = t_1\mathbf{f}_1 + t_2\mathbf{f}_2 + \cdots + t_m\mathbf{f}_m$ where the t_i are scalars. To find t_1 we take the dot product of both sides with $\mathbf{f}_1$:

$$\begin{aligned} \mathbf{x} \cdot \mathbf{f}_1 &= (t_1\mathbf{f}_1 + t_2\mathbf{f}_2 + \cdots + t_m\mathbf{f}_m) \cdot \mathbf{f}_1 \\ &= t_1(\mathbf{f}_1 \cdot \mathbf{f}_1) + t_2(\mathbf{f}_2 \cdot \mathbf{f}_1) + \cdots + t_m(\mathbf{f}_m \cdot \mathbf{f}_1) \\ &= t_1\|\mathbf{f}_1\|^2 + t_2(0) + \cdots + t_m(0) \\ &= t_1\|\mathbf{f}_1\|^2 \end{aligned}$$

Since $\mathbf{f}_1 \neq \mathbf{0}$, this gives $t_1 = \dfrac{\mathbf{x} \cdot \mathbf{f}_1}{\|\mathbf{f}_1\|^2}$. Similarly, $t_i = \dfrac{\mathbf{x} \cdot \mathbf{f}_i}{\|\mathbf{f}_i\|^2}$ for each i.

The expansion in Theorem 6 of $\mathbf{x}$ as a linear combination of the orthogonal basis $\{\mathbf{f}_1, \mathbf{f}_2, \ldots, \mathbf{f}_m\}$ is called the **Fourier expansion** of $\mathbf{x}$, and the coefficients $t_1 = \dfrac{\mathbf{x} \cdot \mathbf{f}_i}{\|\mathbf{f}_i\|^2}$ are called the **Fourier coefficients**. Note that if $\{\mathbf{f}_1, \mathbf{f}_2, \ldots, \mathbf{f}_m\}$ is actually

EXAMPLE 7

Expand $\mathbf{x} = (a, b, c, d)$ as a linear combination of the orthogonal basis $\{\mathbf{f}_1, \mathbf{f}_2, \mathbf{f}_3, \mathbf{f}_4\}$ of $\mathbb{R}^4$ given in Example 6.

Solution ▶ We have $\mathbf{f}_1 = (1, 1, 1, -1)$, $\mathbf{f}_2 = (1, 0, 1, 2)$, $\mathbf{f}_3 = (-1, 0, 1, 0)$, and $\mathbf{f}_4 = (-1, 3, -1, 1)$ so the Fourier coefficients are

$$t_1 = \frac{\mathbf{x} \cdot \mathbf{f}_1}{\|\mathbf{f}_1\|^2} = \tfrac{1}{4}(a + b + c + d) \qquad t_3 = \frac{\mathbf{x} \cdot \mathbf{f}_3}{\|\mathbf{f}_3\|^2} = \tfrac{1}{2}(-a + c)$$

$$t_2 = \frac{\mathbf{x} \cdot \mathbf{f}_2}{\|\mathbf{f}_2\|^2} = \tfrac{1}{6}(a + c + 2d) \qquad t_4 = \frac{\mathbf{x} \cdot \mathbf{f}_4}{\|\mathbf{f}_4\|^2} = \tfrac{1}{12}(-a + 3b - c + d)$$

The reader can verify that indeed $\mathbf{x} = t_1\mathbf{f}_1 + t_2\mathbf{f}_2 + t_3\mathbf{f}_3 + t_4\mathbf{f}_4$.

A natural question arises here: Does every subspace U of $\mathbb{R}^n$ *have* an orthogonal basis? The answer is "yes"; in fact, there is a systematic procedure, called the Gram-Schmidt algorithm, for turning any basis of U into an orthogonal one. This leads to a definition of the projection onto a subspace U that generalizes the projection along a vector used in $\mathbb{R}^2$ and $\mathbb{R}^3$. All this is discussed in Section 8.1.

EXERCISES 5.3

We often write vectors in $\mathbb{R}^n$ as row n-tuples.

1. Obtain orthonormal bases of $\mathbb{R}^3$ by normalizing the following.

   (a) $\{(1, -1, 2), (0, 2, 1), (5, 1, -2)\}$

   ◆(b) $\{(1, 1, 1), (4, 1, -5), (2, -3, 1)\}$

2. In each case, show that the set of vectors is orthogonal in $\mathbb{R}^4$.

   (a) $\{(1, -1, 2, 5), (4, 1, 1, -1), (-7, 28, 5, 5)\}$

   (b) $\{(2, -1, 4, 5), (0, -1, 1, -1), (0, 3, 2, -1)\}$

3. In each case, show that B is an orthogonal basis of $\mathbb{R}^3$ and use Theorem 6 to expand $\mathbf{x} = (a, b, c)$ as a linear combination of the basis vectors.

   (a) $B = \{(1, -1, 3), (-2, 1, 1), (4, 7, 1)\}$

   ◆(b) $B = \{(1, 0, -1), (1, 4, 1), (2, -1, 2)\}$

   (c) $B = \{(1, 2, 3), (-1, -1, 1), (5, -4, 1)\}$

   ◆(d) $B = \{(1, 1, 1), (1, -1, 0), (1, 1, -2)\}$

4. In each case, write $\mathbf{x}$ as a linear combination of the orthogonal basis of the subspace U.

   (a) $\mathbf{x} = (13, -20, 15)$; $U = \text{span}\{(1, -2, 3), (-1, 1, 1)\}$

   ◆(b) $\mathbf{x} = (14, 1, -8, 5)$; $U = \text{span}\{(2, -1, 0, 3), (2, 1, -2, -1)\}$

5. In each case, find all (a, b, c, d) in $\mathbb{R}^4$ such that the given set is orthogonal.

   (a) $\{(1, 2, 1, 0), (1, -1, 1, 3), (2, -1, 0, -1), (a, b, c, d)\}$

   ◆(b) $\{(1, 0, -1, 1), (2, 1, 1, -1), (1, -3, 1, 0), (a, b, c, d)\}$

6. If $\|\mathbf{x}\| = 3$, $\|\mathbf{y}\| = 1$, and $\mathbf{x} \cdot \mathbf{y} = -2$, compute:

   (a) $\|3\mathbf{x} - 5\mathbf{y}\|$ ◆(b) $\|2\mathbf{x} + 7\mathbf{y}\|$

   (c) $(3\mathbf{x} - \mathbf{y}) \cdot (2\mathbf{y} - \mathbf{x})$

   ◆(d) $(\mathbf{x} - 2\mathbf{y}) \cdot (3\mathbf{x} + 5\mathbf{y})$

7. In each case either show that the statement is true or give an example showing that it is false.

   (a) Every independent set in $\mathbb{R}^n$ is orthogonal.

   ◆(b) If $\{\mathbf{x}, \mathbf{y}\}$ is an orthogonal set in $\mathbb{R}^n$, then $\{\mathbf{x}, \mathbf{x} + \mathbf{y}\}$ is also orthogonal.

(c) If {**x**, **y**} and {**z**, **w**} are both orthogonal in $\mathbb{R}^n$, then {**x**, **y**, **z**, **w**} is also orthogonal.

◆(d) If {**x**₁, **x**₂} and {**y**₁, **y**₂, **y**₃} are both orthogonal and $\mathbf{x}_i \cdot \mathbf{y}_j = 0$ for all i and j, then {**x**₁, **x**₂, **y**₁, **y**₂, **y**₃} is orthogonal.

(e) If {**x**₁, **x**₂, ..., **x**ₙ} is orthogonal in $\mathbb{R}^n$, then $\mathbb{R}^n = \text{span}\{\mathbf{x}_1, \mathbf{x}_2, ..., \mathbf{x}_n\}$.

◆(f) If $\mathbf{x} \neq \mathbf{0}$ in $\mathbb{R}^n$, then {**x**} is an orthogonal set.

8. Let **v** denote a nonzero vector in $\mathbb{R}^n$.

   (a) Show that $P = \{\mathbf{x} \text{ in } \mathbb{R}^n \mid \mathbf{x} \cdot \mathbf{v} = 0\}$ is a subspace of $\mathbb{R}^n$.

   (b) Show that $\mathbb{R}\mathbf{v} = \{t\mathbf{v} \mid t \text{ in } \mathbb{R}\}$ is a subspace of $\mathbb{R}^n$.

   (c) Describe P and $\mathbb{R}\mathbf{v}$ geometrically when $n = 3$.

◆9. If A is an $m \times n$ matrix with orthonormal columns, show that $A^T A = I_n$.
   [*Hint*: If $\mathbf{c}_1, \mathbf{c}_2, ..., \mathbf{c}_n$ are the columns of A, show that column j of $A^T A$ has entries $\mathbf{c}_1 \cdot \mathbf{c}_j, \mathbf{c}_2 \cdot \mathbf{c}_j, ..., \mathbf{c}_n \cdot \mathbf{c}_j$].

10. Use the Cauchy inequality to show that $\sqrt{xy} \leq \frac{1}{2}(x + y)$ for all $x \geq 0$ and $y \geq 0$. Here $\sqrt{xy}$ and $\frac{1}{2}(x + y)$ are called, respectively, the *geometric mean* and *arithmetic mean* of x and y.
    [*Hint*: Use $\mathbf{x} = \begin{bmatrix} \sqrt{x} \\ \sqrt{y} \end{bmatrix}$ and $\mathbf{y} = \begin{bmatrix} \sqrt{y} \\ \sqrt{x} \end{bmatrix}$.]

11. Use the Cauchy inequality to prove that:

    (a) $(r_1 + r_2 + \cdots + r_n)^2 \leq n(r_1^2 + r_2^2 + \cdots + r_n^2)$ for all r_i in $\mathbb{R}$ and all $n \geq 1$.

◆(b) $r_1 r_2 + r_1 r_3 + r_2 r_3 \leq r_1^2 + r_2^2 + r_3^2$ for all $r_1, r_2,$ and r_3 in $\mathbb{R}$. [*Hint*: See part (a).]

12. (a) Show that **x** and **y** are orthogonal in $\mathbb{R}^n$ if and only if $\|\mathbf{x} + \mathbf{y}\| = \|\mathbf{x} - \mathbf{y}\|$.

◆(b) Show that $\mathbf{x} + \mathbf{y}$ and $\mathbf{x} - \mathbf{y}$ are orthogonal in $\mathbb{R}^n$ if and only if $\|\mathbf{x}\| = \|\mathbf{y}\|$.

13. (a) Show that $\|\mathbf{x} + \mathbf{y}\|^2 = \|\mathbf{x}\|^2 + \|\mathbf{y}\|^2$ if and only if **x** is orthogonal to **y**.

    (b) If $\mathbf{x} = \begin{bmatrix} 1 \\ 1 \end{bmatrix}, \mathbf{y} = \begin{bmatrix} 1 \\ 0 \end{bmatrix}$ and $\mathbf{z} = \begin{bmatrix} -2 \\ 3 \end{bmatrix}$, show that $\|\mathbf{x} + \mathbf{y} + \mathbf{z}\|^2 = \|\mathbf{x}\|^2 + \|\mathbf{y}\|^2 + \|\mathbf{z}\|^2$ but $\mathbf{x} \cdot \mathbf{y} \neq 0, \mathbf{x} \cdot \mathbf{z} \neq 0,$ and $\mathbf{y} \cdot \mathbf{z} \neq 0$.

14. (a) Show that $\mathbf{x} \cdot \mathbf{y} = \frac{1}{4}[\|\mathbf{x} + \mathbf{y}\|^2 - \|\mathbf{x} - \mathbf{y}\|^2]$ for all **x**, **y** in $\mathbb{R}^n$.

    (b) Show that $\|\mathbf{x}\|^2 + \|\mathbf{y}\|^2 = \frac{1}{2}\|\mathbf{x} + \mathbf{y}\|^2 + \|\mathbf{x} - \mathbf{y}\|^2$ for all **x**, **y** in $\mathbb{R}^n$.

◆15. If A is $n \times n$, show that every eigenvalue of $A^T A$ is nonnegative. [*Hint*: Compute $\|A\mathbf{x}\|^2$ where **x** is an eigenvector.]

16. If $\mathbb{R}^n = \text{span}\{\mathbf{x}_1, ..., \mathbf{x}_m\}$ and $\mathbf{x} \cdot \mathbf{x}_i = 0$ for all i, show that $\mathbf{x} = \mathbf{0}$. [*Hint*: Show $\|\mathbf{x}\| = 0$.]

17. If $\mathbb{R}^n = \text{span}\{\mathbf{x}_1, ..., \mathbf{x}_m\}$ and $\mathbf{x} \cdot \mathbf{x}_i = \mathbf{y} \cdot \mathbf{x}_i$ for all i, show that $\mathbf{x} = \mathbf{y}$. [*Hint*: Preceding Exercise.]

18. Let {**e**₁, ..., **e**ₙ} be an orthogonal basis of $\mathbb{R}^n$. Given **x** and **y** in $\mathbb{R}^n$, show that
$$\mathbf{x} \cdot \mathbf{y} = \frac{(\mathbf{x} \cdot \mathbf{e}_1)(\mathbf{y} \cdot \mathbf{e}_1)}{\|\mathbf{e}_1\|^2} + \cdots + \frac{(\mathbf{x} \cdot \mathbf{e}_n)(\mathbf{y} \cdot \mathbf{e}_n)}{\|\mathbf{e}_n\|^2}.$$

SECTION 5.4 Rank of a Matrix

In this section we use the concept of dimension to clarify the definition of the rank of a matrix given in Section 1.2, and to study its properties. This requires that we deal with rows and columns in the same way. While it has been our custom to write the n-tuples in $\mathbb{R}^n$ as columns, in this section we will frequently write them as rows. Subspaces, independence, spanning, and dimension are defined for rows using matrix operations, just as for columns. If A is an $m \times n$ matrix, we define:

Definition 5.10 The **column space**, col A, of A is the subspace of $\mathbb{R}^m$ spanned by the columns of A. The **row space**, row A, of A is the subspace of $\mathbb{R}^n$ spanned by the rows of A.

Much of what we do in this section involves these subspaces. We begin with:

Lemma 1

Let A and B denote $m \times n$ matrices.
1. If $A \to B$ by elementary row operations, then row A = row B.
2. If $A \to B$ by elementary column operations, then col A = col B.

PROOF

We prove (1); the proof of (2) is analogous. It is enough to do it in the case when $A \to B$ by a single row operation. Let $R_1, R_2, \ldots, R_m$ denote the rows of A. The row operation $A \to B$ either interchanges two rows, multiplies a row by a nonzero constant, or adds a multiple of a row to a different row. We leave the first two cases to the reader. In the last case, suppose that a times row p is added to row q where $p < q$. Then the rows of B are $R_1, \ldots, R_p, \ldots, R_q + aR_p, \ldots, R_m$, and Theorem 1 Section 5.1 shows that

$$\text{span}\{R_1, \ldots, R_p, \ldots, R_q, \ldots, R_m\} = \text{span}\{R_1, \ldots, R_p, \ldots, R_q + aR_p, \ldots, R_m\}.$$

That is, row A = row B.

If A is any matrix, we can carry $A \to R$ by elementary row operations where R is a row-echelon matrix. Hence row A = row R by Lemma 1; so the first part of the following result is of interest.

Lemma 2

If R is a row-echelon matrix, then
1. The nonzero rows of R are a basis of row R.
2. The columns of R containing leading ones are a basis of col R.

PROOF

The rows of R are independent by Example 6 Section 5.2, and they span row R by definition. This proves *1*.

Let $\mathbf{c}_{j_1}, \mathbf{c}_{j_2}, \ldots, \mathbf{c}_{j_r}$ denote the columns of R containing leading 1s. Then $\{\mathbf{c}_{j_1}, \mathbf{c}_{j_2}, \ldots, \mathbf{c}_{j_r}\}$ is independent because the leading 1s are in different rows (and have zeros below and to the left of them). Let U denote the subspace of all columns in $\mathbb{R}^m$ in which the last $m - r$ entries are zero. Then $\dim U = r$ (it is just $\mathbb{R}^r$ with extra zeros). Hence the independent set $\{\mathbf{c}_{j_1}, \mathbf{c}_{j_2}, \ldots, \mathbf{c}_{j_r}\}$ is a basis of U by Theorem 7 Section 5.2. Since each $\mathbf{c}_{j_i}$ is in col R, it follows that col $R = U$, proving (2).

With Lemma 2 we can fill a gap in the definition of the rank of a matrix given in Chapter 1. Let A be any matrix and suppose A is carried to some row-echelon matrix R by row operations. Note that R is not unique. In Section 1.2 we defined the **rank** of A, denoted rank A, to be the number of leading 1s in R, that is the number of nonzero rows of R. The fact that this number does not depend on the choice of R was not proved in Section 1.2. However part 1 of Lemma 2 shows that

SECTION 5.4 Rank of a Matrix

$$\text{rank } A = \dim(\text{row } A)$$

and hence that rank A is independent of R.

Lemma 2 can be used to find bases of subspaces of $\mathbb{R}^n$ (written as rows). Here is an example.

EXAMPLE 1

Find a basis of $U = \text{span}\{(1, 1, 2, 3), (2, 4, 1, 0), (1, 5, -4, -9)\}$.

Solution ▶ U is the row space of $\begin{bmatrix} 1 & 1 & 2 & 3 \\ 2 & 4 & 1 & 0 \\ 1 & 5 & -4 & -9 \end{bmatrix}$. This matrix has row-echelon form $\begin{bmatrix} 1 & 1 & 2 & 3 \\ 0 & 1 & -\frac{3}{2} & -3 \\ 0 & 0 & 0 & 0 \end{bmatrix}$, so $\{(1, 1, 2, 3), (0, 1, -\frac{3}{2}, -3)\}$ is basis of U by Lemma 2.

Note that $\{(1, 1, 2, 3), (0, 2, -3, -6)\}$ is another basis that avoids fractions.

Lemmas 1 and 2 are enough to prove the following fundamental theorem.

Theorem 1

Let A denote any $m \times n$ matrix of rank r. Then

$$\dim(\text{col } A) = \dim(\text{row } A) = r.$$

Moreover, if A is carried to a row-echelon matrix R by row operations, then

1. The r nonzero rows of R are a basis of row A.
2. If the leading 1s lie in columns $j_1, j_2, \ldots, j_r$ of R, then columns $j_1, j_2, \ldots, j_r$ of A are a basis of col A.

PROOF

We have row A = row R by Lemma 1, so (1) follows from Lemma 2. Moreover, $R = UA$ for some invertible matrix U by Theorem 1 Section 2.5. Now write $A = [\mathbf{c}_1 \ \mathbf{c}_2 \ \cdots \ \mathbf{c}_n]$ where $\mathbf{c}_1, \mathbf{c}_2, \ldots, \mathbf{c}_n$ are the columns of A. Then

$$R = UA = U[\mathbf{c}_1 \ \mathbf{c}_2 \ \cdots \ \mathbf{c}_n] = [U\mathbf{c}_1 \ U\mathbf{c}_2 \ \cdots \ U\mathbf{c}_n].$$

Thus, in the notation of (2), the set $B = \{U\mathbf{c}_{j_1}, U\mathbf{c}_{j_2}, \ldots, U\mathbf{c}_{j_r}\}$ is a basis of col R by Lemma 2. So, to prove (2) and the fact that $\dim(\text{col } A) = r$, it is enough to show that $D = \{\mathbf{c}_{j_1}, \mathbf{c}_{j_2}, \ldots, \mathbf{c}_{j_r}\}$ is a basis of col A. First, D is linearly independent because U is invertible (verify), so we show that, for each j, column $\mathbf{c}_j$ is a linear combination of the $\mathbf{c}_{j_i}$. But $U\mathbf{c}_j$ is column j of R, and so is a linear combination of the $U\mathbf{c}_{j_i}$, say $U\mathbf{c}_j = a_1 U\mathbf{c}_{j_1} + a_2 U\mathbf{c}_{j_2} + \cdots + a_r U\mathbf{c}_{j_r}$ where each a_i is a real number.

Since U is invertible, it follows that $\mathbf{c}_j = a_1 \mathbf{c}_{j_1} + a_2 \mathbf{c}_{j_2} + \cdots + a_r \mathbf{c}_{j_r}$ and the proof is complete.

EXAMPLE 2

Compute the rank of $A = \begin{bmatrix} 1 & 2 & 2 & -1 \\ 3 & 6 & 5 & 0 \\ 1 & 2 & 1 & 2 \end{bmatrix}$ and find bases for row A and col A.

Solution ▶ The reduction of A to row-echelon form is as follows:

$$\begin{bmatrix} 1 & 2 & 2 & -1 \\ 3 & 6 & 5 & 0 \\ 1 & 2 & 1 & 2 \end{bmatrix} \to \begin{bmatrix} 1 & 2 & 2 & -1 \\ 0 & 0 & -1 & 3 \\ 0 & 0 & -1 & 3 \end{bmatrix} \to \begin{bmatrix} 1 & 2 & 2 & -1 \\ 0 & 0 & 1 & -3 \\ 0 & 0 & 0 & 0 \end{bmatrix}$$

Hence rank $A = 2$, and $\{[1\ 2\ 2\ -1], [0\ 0\ 1\ -3]\}$ is a basis of row A by Lemma 2. Since the leading 1s are in columns 1 and 3 of the row-echelon matrix, Theorem 1 shows that columns 1 and 3 of A are a basis $\left\{ \begin{bmatrix} 1 \\ 3 \\ 1 \end{bmatrix}, \begin{bmatrix} 2 \\ 5 \\ 1 \end{bmatrix} \right\}$ of col A.

Theorem 1 has several important consequences. The first, Corollary 1 below, follows because the rows of A are independent (respectively span row A) if and only if their transposes are independent (respectively span col A).

Corollary 1

If A is any matrix, then rank A = rank(A^T).

If A is an $m \times n$ matrix, we have col $A \subseteq \mathbb{R}^m$ and row $A \subseteq \mathbb{R}^n$. Hence Theorem 8 Section 5.2 shows that dim(col A) $\leq$ dim($\mathbb{R}^m$) = m and dim(row A) $\leq$ dim($\mathbb{R}^n$) = n. Thus Theorem 1 gives:

Corollary 2

If A is an $m \times n$ matrix, then rank $A \leq m$ and rank $A \leq n$.

Corollary 3

Rank A = rank(UA) = rank(AV) whenever U and V are invertible.

PROOF

Lemma 1 gives rank A = rank(UA). Using this and Corollary 1 we get
$$\text{rank}(AV) = \text{rank}(AV)^T = \text{rank}(V^T A^T) = \text{rank}(A^T) = \text{rank } A.$$

The next corollary requires a preliminary lemma.

SECTION 5.4 Rank of a Matrix

Lemma 3

Let A, U, and V be matrices of sizes $m \times n$, $p \times m$, and $n \times q$ respectively.
(1) col$(AV) \subseteq$ col A, with equality if V is (square and) invertible.
(2) row$(UA) \subseteq$ row A, with equality if U is (square and) invertible.

PROOF

For (1), write $V = [\mathbf{v}_1, \mathbf{v}_2, \ldots, \mathbf{v}_q]$ where $\mathbf{v}_j$ is column j of V. Then we have $AV = [A\mathbf{v}_1, A\mathbf{v}_2, \ldots, A\mathbf{v}_q]$, and each $A\mathbf{v}_j$ is in col A by Definition 1 Section 2.2. It follows that col$(AV) \subseteq$ col A. If V is invertible, we obtain col $A = $ col$[(AV)V^{-1}] \subseteq$ col(AV) in the same way. This proves (1).

As to (2), we have col$[(UA)^T] = $ col$(A^T U^T) \subseteq$ col(A^T) by (1), from which row$(UA) \subseteq$ row A. If U is invertible, this is equality as in the proof of (1).

Corollary 4

If A is $m \times n$ and B is $n \times m$, then rank $AB \leq$ rank A and rank $AB \leq$ rank B.

PROOF

By Lemma 3, col$(AB) \subseteq$ col A and row$(BA) \subseteq$ row A, so Theorem 1 applies.

In Section 5.1 we discussed two other subspaces associated with an $m \times n$ matrix A: the null space null(A) and the image space im(A)

$$\text{null}(A) = \{\mathbf{x} \text{ in } \mathbb{R}^n \mid A\mathbf{x} = \mathbf{0}\} \quad \text{and} \quad \text{im}(A) = \{A\mathbf{x} \mid \mathbf{x} \text{ in } \mathbb{R}^n\}.$$

Using rank, there are simple ways to find bases of these spaces. If A has rank r, we have im$(A) = $ col(A) by Example 8 Section 5.1, so dim[im(A)] = dim[col(A)] = r. Hence Theorem 1 provides a method of finding a basis of im(A). This is recorded as part (2) of the following theorem.

Theorem 2

Let A denote an $m \times n$ matrix of rank r. Then
(1) The $n - r$ basic solutions to the system $A\mathbf{x} = \mathbf{0}$ provided by the gaussian algorithm are a basis of null(A), so dim[null(A)] = $n - r$.
(2) Theorem 1 provides a basis of im$(A) = $ col(A), and dim[im(A)] = r.

PROOF

It remains to prove (1). We already know (Theorem 1 Section 2.2) that null(A) is spanned by the $n - r$ basic solutions of $A\mathbf{x} = \mathbf{0}$. Hence using Theorem 7 Section 5.2, it suffices to show that dim[null(A)] = $n - r$. So let $\{\mathbf{x}_1, \ldots, \mathbf{x}_k\}$ be a basis of null(A), and extend it to a basis $\{\mathbf{x}_1, \ldots, \mathbf{x}_k, \mathbf{x}_{k+1}, \ldots, \mathbf{x}_n\}$ of $\mathbb{R}^n$ (by Theorem 6 Section 5.2). It is enough to show that $\{A\mathbf{x}_{k+1}, \ldots, A\mathbf{x}_n\}$ is a basis of im(A); then $n - k = r$ by the above and so $k = n - r$ as required.

Spanning. Choose $A\mathbf{x}$ in $\text{im}(A)$, $\mathbf{x}$ in $\mathbb{R}^n$, and write
$\mathbf{x} = a_1\mathbf{x}_1 + \cdots + a_k\mathbf{x}_k + a_{k+1}\mathbf{x}_{k+1} + \cdots + a_n\mathbf{x}_n$ where the a_i are in $\mathbb{R}$.
Then $A\mathbf{x} = a_{k+1}A\mathbf{x}_{k+1} + \cdots + a_n A\mathbf{x}_n$ because $\{\mathbf{x}_1, \ldots, \mathbf{x}_k\} \subseteq \text{null}(A)$.

Independence. Let $t_{k+1}A\mathbf{x}_{k+1} + \cdots + t_n A\mathbf{x}_n = \mathbf{0}$, t_i in $\mathbb{R}$. Then $t_{k+1}\mathbf{x}_{k+1} + \cdots + t_n\mathbf{x}_n$
is in null A, so $t_{k+1}\mathbf{x}_{k+1} + \cdots + t_n\mathbf{x}_n = t_1\mathbf{x}_1 + \cdots + t_k\mathbf{x}_k$ for some $t_1, \ldots, t_k$ in $\mathbb{R}$.
But then the independence of the $\mathbf{x}_i$ shows that $t_i = 0$ for every i.

EXAMPLE 3

If $A = \begin{bmatrix} 1 & -2 & 1 & 1 \\ -1 & 2 & 0 & 1 \\ 2 & -4 & 1 & 0 \end{bmatrix}$, find bases of null($A$) and im($A$), and so find their dimensions.

Solution ▶ If $\mathbf{x}$ is in null(A), then $A\mathbf{x} = \mathbf{0}$, so $\mathbf{x}$ is given by solving the system $A\mathbf{x} = \mathbf{0}$. The reduction of the augmented matrix to reduced form is

$$\begin{bmatrix} 1 & -2 & 1 & 1 & | & 0 \\ -1 & 2 & 0 & 1 & | & 0 \\ 2 & -4 & 1 & 0 & | & 0 \end{bmatrix} \rightarrow \begin{bmatrix} 1 & -2 & 0 & -1 & | & 0 \\ 0 & 0 & 1 & 2 & | & 0 \\ 0 & 0 & 0 & 0 & | & 0 \end{bmatrix}$$

Hence $r = \text{rank}(A) = 2$. Here, $\text{im}(A) = \text{col}(A)$ has basis $\left\{ \begin{bmatrix} 1 \\ -1 \\ 2 \end{bmatrix}, \begin{bmatrix} 1 \\ 0 \\ 1 \end{bmatrix} \right\}$ by Theorem 1 because the leading 1s are in columns 1 and 3. In particular, $\dim[\text{im}(A)] = 2 = r$ as in Theorem 2.

Turning to null(A), we use gaussian elimination. The leading variables are x_1 and x_3, so the nonleading variables become parameters: $x_2 = s$ and $x_4 = t$. It follows from the reduced matrix that $x_1 = 2s + t$ and $x_3 = -2t$, so the general solution is

$$\mathbf{x} = \begin{bmatrix} x_1 \\ x_2 \\ x_3 \\ x_4 \end{bmatrix} = \begin{bmatrix} 2s + t \\ s \\ -2t \\ t \end{bmatrix} = s\mathbf{x}_1 + t\mathbf{x}_2 \text{ where } \mathbf{x}_1 = \begin{bmatrix} 2 \\ 1 \\ 0 \\ 0 \end{bmatrix}, \text{ and } \mathbf{x}_2 = \begin{bmatrix} 1 \\ 0 \\ -2 \\ 1 \end{bmatrix}.$$

Hence null(A). But $\mathbf{x}_1$ and $\mathbf{x}_2$ *are* solutions (basic), so

$$\text{null}(A) = \text{span}\{\mathbf{x}_1, \mathbf{x}_2\}$$

However Theorem 2 asserts that $\{\mathbf{x}_1, \mathbf{x}_2\}$ is a *basis* of null(A). (In fact it is easy to verify directly that $\{\mathbf{x}_1, \mathbf{x}_2\}$ is independent in this case.) In particular, $\dim[\text{null}(A)] = 2 = n - r$, as Theorem 2 asserts.

Let A be an $m \times n$ matrix. Corollary 2 of the Theorem 1 asserts that rank $A \leq m$ and rank $A \leq n$, and it is natural to ask when these extreme cases arise. If $\mathbf{c}_1, \mathbf{c}_2, \ldots, \mathbf{c}_n$ are the columns of A, Theorem 2 Section 5.2 shows that $\{\mathbf{c}_1, \mathbf{c}_2, \ldots, \mathbf{c}_n\}$ spans $\mathbb{R}^m$ if and only if the system $A\mathbf{x} = \mathbf{b}$ is consistent for every $\mathbf{b}$ in $\mathbb{R}^m$, and that $\{\mathbf{c}_1, \mathbf{c}_2, \ldots, \mathbf{c}_n\}$ is independent if and only if $A\mathbf{x} = \mathbf{0}$, $\mathbf{x}$ in $\mathbb{R}^n$, implies $\mathbf{x} = \mathbf{0}$. The next two useful theorems improve on both these results, and relate them to when the rank of A is n or m.

SECTION 5.4 Rank of a Matrix

Theorem 3

The following are equivalent for an $m \times n$ matrix A:

1. rank $A = n$.
2. The rows of A span $\mathbb{R}^n$.
3. The columns of A are linearly independent in $\mathbb{R}^m$.
4. The $n \times n$ matrix $A^T A$ is invertible.
5. $CA = I_n$ for some $n \times m$ matrix C.
6. If $A\mathbf{x} = \mathbf{0}$, $\mathbf{x}$ in $\mathbb{R}^n$, then $\mathbf{x} = \mathbf{0}$.

PROOF

(1) $\Rightarrow$ (2). We have row $A \subseteq \mathbb{R}^n$, and dim(row A) $= n$ by (1), so row $A = \mathbb{R}^n$ by Theorem 8 Section 5.2. This is (2).

(2) $\Rightarrow$ (3). By (2), row $A = \mathbb{R}^n$, so rank $A = n$. This means dim(col A) $= n$. Since the n columns of A span col A, they are independent by Theorem 7 Section 5.2.

(3) $\Rightarrow$ (4). If $(A^T A)\mathbf{x} = \mathbf{0}$, $\mathbf{x}$ in $\mathbb{R}^n$, we show that $\mathbf{x} = \mathbf{0}$ (Theorem 5 Section 2.4). We have
$$\|A\mathbf{x}\|^2 = (A\mathbf{x})^T A\mathbf{x} = \mathbf{x}^T A^T A\mathbf{x} = \mathbf{x}^T \mathbf{0} = \mathbf{0}.$$
Hence $A\mathbf{x} = \mathbf{0}$, so $\mathbf{x} = \mathbf{0}$ by (3) and Theorem 2 Section 5.2.

(4) $\Rightarrow$ (5). Given (4), take $C = (A^T A)^{-1} A^T$.

(5) $\Rightarrow$ (6). If $A\mathbf{x} = \mathbf{0}$, then left multiplication by C (from (5)) gives $\mathbf{x} = \mathbf{0}$.

(6) $\Rightarrow$ (1). Given (6), the columns of A are independent by Theorem 2 Section 5.2. Hence dim(col A) $= n$, and (1) follows.

Theorem 4

The following are equivalent for an $m \times n$ matrix A:

1. rank $A = m$.
2. The columns of A span $\mathbb{R}^m$.
3. The rows of A are linearly independent in $\mathbb{R}^n$.
4. The $m \times m$ matrix AA^T is invertible.
5. $AC = I_m$ for some $n \times m$ matrix C.
6. The system $A\mathbf{x} = \mathbf{b}$ is consistent for every $\mathbf{b}$ in $\mathbb{R}^m$.

PROOF

(1) $\Rightarrow$ (2). By (1), dim(col A) $= m$, so col $A = \mathbb{R}^m$ by Theorem 8 Section 5.2.

(2) $\Rightarrow$ (3). By (2), col $A = \mathbb{R}^m$, so rank $A = m$. This means dim(row A) $= m$. Since the m rows of A span row A, they are independent by Theorem 7 Section 5.2.

(3) ⇒ (4). We have rank $A = m$ by (3), so the $n \times m$ matrix A^T has rank m. Hence applying Theorem 3 to A^T in place of A shows that $(A^T)^T A^T$ is invertible, proving (4).

(4) ⇒ (5). Given (4), take $C = A^T(AA^T)^{-1}$ in (5).

(5) ⇒ (6). Comparing columns in $AC = I_m$ gives $A\mathbf{c}_j = \mathbf{e}_j$ for each j, where $\mathbf{c}_j$ and $\mathbf{e}_j$ denote column j of C and I_m respectively. Given $\mathbf{b}$ in $\mathbb{R}^m$, write $\mathbf{b} = \sum_{j=1}^{m} r_j \mathbf{e}_j$, r_j in $\mathbb{R}$. Then $A\mathbf{x} = \mathbf{b}$ holds with $\mathbf{x} = \sum_{j=1}^{m} r_j \mathbf{c}_j$, as the reader can verify.

(6) ⇒ (1). Given (6), the columns of A span $\mathbb{R}^m$ by Theorem 2 Section 5.2. Thus col $A = \mathbb{R}^m$ and (1) follows.

EXAMPLE 4

Show that $\begin{bmatrix} 3 & x+y+z \\ x+y+z & x^2+y^2+z^2 \end{bmatrix}$ is invertible if x, y, and z are not all equal.

Solution ▸ The given matrix has the form $A^T A$ where $a = \begin{bmatrix} 1 & x \\ 1 & y \\ 1 & z \end{bmatrix}$ has independent columns because x, y, and z are not all equal (verify). Hence Theorem 3 applies.

Theorems 4 and 5 relate several important properties of an $m \times n$ matrix A to the invertibility of the square, symmetric matrices $A^T A$ and AA^T. In fact, even if the columns of A are not independent or do not span $\mathbb{R}^m$, the matrices $A^T A$ and AA^T are both symmetric and, as such, have real eigenvalues as we shall see. We return to this in Chapter 7.

EXERCISES 5.4

1. In each case find bases for the row and column spaces of A and determine the rank of A.

(a) $A = \begin{bmatrix} 2 & -4 & 6 & 8 \\ 2 & -1 & 3 & 2 \\ 4 & -5 & 9 & 10 \\ 0 & -1 & 1 & 2 \end{bmatrix}$ ◆(b) $A = \begin{bmatrix} 2 & -1 & 1 \\ -2 & 1 & 1 \\ 4 & -2 & 3 \\ -6 & 3 & 0 \end{bmatrix}$

(c) $A = \begin{bmatrix} 1 & -1 & 5 & -2 & 2 \\ 2 & -2 & -2 & 5 & 1 \\ 0 & 0 & -12 & 9 & -3 \\ -1 & 1 & 7 & -7 & 1 \end{bmatrix}$

◆(d) $A = \begin{bmatrix} 1 & 2 & -1 & 3 \\ -3 & -6 & 3 & -2 \end{bmatrix}$

2. In each case find a basis of the subspace U.

(a) $U = \text{span}\{(1, -1, 0, 3), (2, 1, 5, 1), (4, -2, 5, 7)\}$

◆(b) $U = \text{span}\{(1, -1, 2, 5, 1), (3, 1, 4, 2, 7), (1, 1, 0, 0, 0), (5, 1, 6, 7, 8)\}$

(c) $U = \text{span}\left\{ \begin{bmatrix} 1 \\ 1 \\ 0 \\ 0 \end{bmatrix}, \begin{bmatrix} 0 \\ 0 \\ 1 \\ 1 \end{bmatrix}, \begin{bmatrix} 1 \\ 0 \\ 1 \\ 0 \end{bmatrix}, \begin{bmatrix} 0 \\ 1 \\ 0 \\ 1 \end{bmatrix} \right\}$

◆(d) $U = \text{span}\left\{ \begin{bmatrix} 1 \\ 5 \\ -6 \end{bmatrix}, \begin{bmatrix} 2 \\ 6 \\ -8 \end{bmatrix}, \begin{bmatrix} 3 \\ 7 \\ -10 \end{bmatrix}, \begin{bmatrix} 4 \\ 8 \\ 12 \end{bmatrix} \right\}$

3. (a) Can a 3×4 matrix have independent columns? Independent rows? Explain.

◆(b) If A is 4×3 and rank $A = 2$, can A have independent columns? Independent rows? Explain.

(c) If A is an $m \times n$ matrix and rank $A = m$, show that $m \leq n$.

◆(d) Can a nonsquare matrix have its rows independent and its columns independent? Explain.

(e) Can the null space of a 3×6 matrix have dimension 2? Explain.

◆(f) Suppose that A is 5×4 and $\text{null}(A) = \mathbb{R}\mathbf{x}$ for some column $\mathbf{x} \neq \mathbf{0}$. Can $\dim(\text{im } A) = 2$?

◆4. If A is $m \times n$ show that $\text{col}(A) = \{A\mathbf{x} \mid \mathbf{x} \text{ in } \mathbb{R}^n\}$.

5. If A is $m \times n$ and B is $n \times m$, show that $AB = 0$ if and only if $\text{col } B \subseteq \text{null } A$.

6. Show that the rank does not change when an elementary row or column operation is performed on a matrix.

7. In each case find a basis of the null space of A. Then compute rank A and verify (1) of Theorem 2.

(a) $A = \begin{bmatrix} 3 & 1 & 1 \\ 2 & 0 & 1 \\ 4 & 2 & 1 \\ 1 & -1 & 1 \end{bmatrix}$

◆(b) $A = \begin{bmatrix} 3 & 5 & 5 & 2 & 0 \\ 1 & 0 & 2 & 2 & 1 \\ 1 & 1 & 1 & -2 & -2 \\ -2 & 0 & -4 & -4 & -2 \end{bmatrix}$

8. Let $A = \mathbf{c}R$ where $\mathbf{c} \neq \mathbf{0}$ is a column in $\mathbb{R}^m$ and $\mathbf{r} \neq \mathbf{0}$ is a row in $\mathbb{R}^n$.

(a) Show that $\text{col } A = \text{span}\{\mathbf{c}\}$ and $\text{row } A = \text{span}\{\mathbf{r}\}$.

◆(b) Find $\dim(\text{null } A)$.

(c) Show that $\text{null } A = \text{null } \mathbf{r}$.

9. Let A be $m \times n$ with columns $\mathbf{c}_1, \mathbf{c}_2, \ldots, \mathbf{c}_n$.

(a) If $\{\mathbf{c}_1, \ldots, \mathbf{c}_n\}$ is independent, show $\text{null } A = \{\mathbf{0}\}$.

◆(b) If $\text{null } A = \{\mathbf{0}\}$, show that $\{\mathbf{c}_1, \ldots, \mathbf{c}_n\}$ is independent.

10. Let A be an $n \times n$ matrix.

(a) Show that $A^2 = 0$ if and only if $\text{col } A \subseteq \text{null } A$.

◆(b) Conclude that if $A^2 = 0$, then rank $A \leq \frac{n}{2}$.

(c) Find a matrix A for which $\text{col } A = \text{null } A$.

11. Let B be $m \times n$ and let AB be $k \times n$. If rank $B = \text{rank}(AB)$, show that $\text{null } B = \text{null}(AB)$. [Hint: Theorem 1.]

◆12. Give a careful argument why $\text{rank}(A^T) = \text{rank } A$.

13. Let A be an $m \times n$ matrix with columns $\mathbf{c}_1, \mathbf{c}_2, \ldots, \mathbf{c}_n$. If rank $A = n$, show that $\{A^T\mathbf{c}_1, A^T\mathbf{c}_2, \ldots, A^T\mathbf{c}_n\}$ is a basis of $\mathbb{R}^n$.

14. If A is $m \times n$ and $\mathbf{b}$ is $m \times 1$, show that $\mathbf{b}$ lies in the column space of A if and only if $\text{rank}[A \ \mathbf{b}] = \text{rank } A$.

15. (a) Show that $A\mathbf{x} = \mathbf{b}$ has a solution if and only if rank $A = \text{rank}[A \ \mathbf{b}]$. [Hint: Exercises 12 and 14.]

◆(b) If $A\mathbf{x} = \mathbf{b}$ has no solution, show that $\text{rank}[A \ \mathbf{b}] = 1 + \text{rank } A$.

16. Let X be a $k \times m$ matrix. If I is the $m \times m$ identity matrix, show that $I + X^TX$ is invertible. [Hint: $I + X^TX = A^TA$ where $A = \begin{bmatrix} I \\ X \end{bmatrix}$ in block form.]

17. If A is $m \times n$ of rank r, show that A can be factored as $A = PQ$ where P is $m \times r$ with r independent columns, and Q is $r \times n$ with r independent rows. [Hint: Let $UAV = \begin{bmatrix} I_r & 0 \\ 0 & 0 \end{bmatrix}$ by Theorem 3, Section 2.5, and write $U^{-1} = \begin{bmatrix} U_1 & U_2 \\ U_3 & U_4 \end{bmatrix}$ and $V^{-1} = \begin{bmatrix} V_1 & V_2 \\ V_3 & V_4 \end{bmatrix}$ in block form, where U_1 and V_1 are $r \times r$.]

18. (a) Show that if A and B have independent columns, so does AB.

(b) Show that if A and B have independent rows, so does AB.

19. A matrix obtained from A by deleting rows and columns is called a **submatrix** of A. If A has an invertible $k \times k$ submatrix, show that rank $A \geq k$. [Hint: Show that row and column operations carry $A \to \begin{bmatrix} I_k & P \\ 0 & Q \end{bmatrix}$ in block form.] *Remark*: It can be shown that rank A is the largest integer r such that A has an invertible $r \times r$ submatrix.

SECTION 5.5   Similarity and Diagonalization

In Section 3.3 we studied diagonalization of a square matrix A, and found important applications (for example to linear dynamical systems). We can now utilize the concepts of subspace, basis, and dimension to clarify the diagonalization process, reveal some new results, and prove some theorems which could not be demonstrated in Section 3.3.

Before proceeding, we introduce a notion that simplifies the discussion of diagonalization, and is used throughout the book.

Similar Matrices

Definition 5.11 *If A and B are $n \times n$ matrices, we say that A and B are* **similar**, *and write $A \sim B$, if $B = P^{-1}AP$ for some invertible matrix P.*

Note that $A \sim B$ if and only if $B = QAQ^{-1}$ where Q is invertible (write $P^{-1} = Q$). The language of similarity is used throughout linear algebra. For example, a matrix A is diagonalizable if and only if it is similar to a diagonal matrix.

If $A \sim B$, then necessarily $B \sim A$. To see why, suppose that $B = P^{-1}AP$. Then $A = PBP^{-1} = Q^{-1}BQ$ where $Q = P^{-1}$ is invertible. This proves the second of the following properties of similarity (the others are left as an exercise):

1. $A \sim A$ for all square matrices A.
2. If $A \sim B$, then $B \sim A$. $(*)$
3. If $A \sim B$ and $B \sim C$, then $A \sim C$.

These properties are often expressed by saying that the similarity relation $\sim$ is an **equivalence relation** on the set of $n \times n$ matrices. Here is an example showing how these properties are used.

EXAMPLE 1

If A is similar to B and either A or B is diagonalizable, show that the other is also diagonalizable.

Solution ▶ We have $A \sim B$. Suppose that A is diagonalizable, say $A \sim D$ where D is diagonal. Since $B \sim A$ by (2) of $(*)$, we have $B \sim A$ and $A \sim D$. Hence $B \sim D$ by (3) of $(*)$, so B is diagonalizable too. An analogous argument works if we assume instead that B is diagonalizable.

Similarity is compatible with inverses, transposes, and powers:

If $A \sim B$ then $A^{-1} \sim B^{-1}, A^T \sim B^T$, and $A^k \sim B^k$ for all integers $k \geq 1$.

The proofs are routine matrix computations using Theorem 1 Section 3.3. Thus, for example, if A is diagonalizable, so also are A^T, A^{-1} (if it exists), and A^k (for each $k \geq 1$). Indeed, if $A \sim D$ where D is a diagonal matrix, we obtain $A^T \sim D^T$, $A^{-1} \sim D^{-1}$, and $A^k \sim D^k$, and each of the matrices D^T, D^{-1}, and D^k is diagonal.

We pause to introduce a simple matrix function that will be referred to later.

SECTION 5.5 Similarity and Diagonalization

Definition 5.12

*The **trace** tr A of an $n \times n$ matrix A is defined to be the sum of the main diagonal elements of A.*

In other words:

$$\text{If } A = [a_{ij}], \text{ then tr } A = a_{11} + a_{22} + \cdots + a_{nn}.$$

It is evident that $\text{tr}(A + B) = \text{tr } A + \text{tr } B$ and that $\text{tr}(cA) = c \text{ tr } A$ holds for all $n \times n$ matrices A and B and all scalars c. The following fact is more surprising.

Lemma 1

Let A and B be $n \times n$ matrices. Then $\text{tr}(AB) = \text{tr}(BA)$.

PROOF

Write $A = [a_{ij}]$ and $B = [b_{ij}]$. For each i, the (i, i)-entry d_i of the matrix AB is $d_i = a_{i1}b_{1i} + a_{i2}b_{2i} + \cdots + a_{in}b_{ni} = \sum_j a_{ij}b_{ji}$. Hence

$$\text{tr}(AB) = d_1 + d_2 + \cdots + d_n = \sum_i d_i = \sum_i \left(\sum_j a_{ij}b_{ji} \right).$$

Similarly we have $\text{tr}(BA) = \sum_i \left(\sum_j b_{ij}a_{ji} \right)$. Since these two double sums are the same, Lemma 1 is proved.

As the name indicates, similar matrices share many properties, some of which are collected in the next theorem for reference.

Theorem 1

If A and B are similar $n \times n$ matrices, then A and B have the same determinant, rank, trace, characteristic polynomial, and eigenvalues.

PROOF

Let $B = P^{-1}AP$ for some invertible matrix P. Then we have

$$\det B = \det(P^{-1}) \det A \det P = \det A \quad \text{because} \quad \det(P^{-1}) = 1/\det P.$$

Similarly, rank $B = \text{rank}(P^{-1}AP) = \text{rank } A$ by Corollary 3 of Theorem 1 Section 5.4. Next Lemma 1 gives

$$\text{tr}(P^{-1}AP) = \text{tr}[P^{-1}(AP)] = \text{tr}[(AP)P^{-1}] = \text{tr } A.$$

As to the characteristic polynomial,

$$c_B(x) = \det(xI - B) = \det\{x(P^{-1}IP) - P^{-1}AP\}$$
$$= \det\{P^{-1}(xI - A)P\}$$
$$= \det(xI - A)$$
$$= c_A(x).$$

Finally, this shows that A and B have the same eigenvalues because the eigenvalues of a matrix are the roots of its characteristic polynomial.

EXAMPLE 2

Sharing the five properties in Theorem 1 does not guarantee that two matrices are similar. The matrices $A = \begin{bmatrix} 1 & 1 \\ 0 & 1 \end{bmatrix}$ and $I = \begin{bmatrix} 1 & 0 \\ 0 & 1 \end{bmatrix}$ have the same determinant, rank, trace, characteristic polynomial, and eigenvalues, but they are not similar because $P^{-1}IP = I$ for any invertible matrix P.

Diagonalization Revisited

Recall that a square matrix A is **diagonalizable** if there exists an invertible matrix P such that $P^{-1}AP = D$ is a diagonal matrix, that is if A is similar to a diagonal matrix D. Unfortunately, not all matrices are diagonalizable, for example $\begin{bmatrix} 1 & 1 \\ 0 & 1 \end{bmatrix}$ (see Example 10 Section 3.3). Determining whether A is diagonalizable is closely related to the eigenvalues and eigenvectors of A. Recall that a number λ is called an **eigenvalue** of A if $A\mathbf{x} = \lambda\mathbf{x}$ for some nonzero column $\mathbf{x}$ in $\mathbb{R}^n$, and any such nonzero vector $\mathbf{x}$ is called an **eigenvector** of A corresponding to λ (or simply a λ-eigenvector of A). The eigenvalues and eigenvectors of A are closely related to the **characteristic polynomial** $c_A(x)$ of A, defined by

$$c_A(x) = \det(xI - A).$$

If A is $n \times n$ this is a polynomial of degree n, and its relationship to the eigenvalues is given in the following theorem (a repeat of Theorem 2 Section 3.3).

Theorem 2

Let A be an $n \times n$ matrix.
1. The eigenvalues λ of A are the roots of the characteristic polynomial $c_A(x)$ of A.
2. The λ-eigenvectors $\mathbf{x}$ are the nonzero solutions to the homogeneous system

$$(\lambda I - A)\mathbf{x} = \mathbf{0}$$

of linear equations with $\lambda I - A$ as coefficient matrix.

EXAMPLE 3

Show that the eigenvalues of a triangular matrix are the main diagonal entries.

Solution ▶ Assume that A is triangular. Then the matrix $xI - A$ is also triangular and has diagonal entries $(x - a_{11}), (x - a_{22}), \ldots, (x - a_{nn})$ where $A = [a_{ij}]$. Hence Theorem 4 Section 3.1 gives

$$c_A(x) = (x - a_{11})(x - a_{22})\cdots(x - a_{nn})$$

and the result follows because the eigenvalues are the roots of $c_A(x)$.

Theorem 4 Section 3.3 asserts (in part) that an $n \times n$ matrix A is diagonalizable if and only if it has n eigenvectors $\mathbf{x}_1, \ldots, \mathbf{x}_n$ such that the matrix $P = [\mathbf{x}_1 \cdots \mathbf{x}_n]$ with the $\mathbf{x}_i$ as columns is invertible. This is equivalent to requiring that $\{\mathbf{x}_1, \ldots, \mathbf{x}_n\}$ is a basis of $\mathbb{R}^n$ consisting of eigenvectors of A. Hence we can restate Theorem 4 Section 3.3 as follows:

SECTION 5.5 Similarity and Diagonalization

Theorem 3

Let A be an $n \times n$ matrix.

1. A is diagonalizable if and only if $\mathbb{R}^n$ has a basis $\{\mathbf{x}_1, \mathbf{x}_2, \ldots, \mathbf{x}_n\}$ consisting of eigenvectors of A.
2. When this is the case, the matrix $P = [\mathbf{x}_1 \ \mathbf{x}_2 \ \cdots \ \mathbf{x}_n]$ is invertible and $P^{-1}AP = \text{diag}(\lambda_1, \lambda_2, \ldots, \lambda_n)$ where, for each i, λ_i is the eigenvalue of A corresponding to $\mathbf{x}_i$.

The next result is a basic tool for determining when a matrix is diagonalizable. It reveals an important connection between eigenvalues and linear independence: Eigenvectors corresponding to distinct eigenvalues are necessarily linearly independent.

Theorem 4

Let $\mathbf{x}_1, \mathbf{x}_2, \ldots, \mathbf{x}_k$ be eigenvectors corresponding to distinct eigenvalues $\lambda_1, \lambda_2, \ldots, \lambda_k$ of an $n \times n$ matrix A. Then $\{\mathbf{x}_1, \mathbf{x}_2, \ldots, \mathbf{x}_k\}$ is a linearly independent set.

PROOF

We use induction on k. If $k = 1$, then $\{\mathbf{x}_1\}$ is independent because $\mathbf{x}_1 \neq \mathbf{0}$. In general, suppose the theorem is true for some $k \geq 1$. Given eigenvectors $\{\mathbf{x}_1, \mathbf{x}_2, \ldots, \mathbf{x}_{k+1}\}$, suppose a linear combination vanishes:

$$t_1 \mathbf{x}_1 + t_2 \mathbf{x}_2 + \cdots + t_{k+1} \mathbf{x}_{k+1} = \mathbf{0}. \qquad (*)$$

We must show that each $t_i = 0$. Left multiply $(*)$ by A and use the fact that $A\mathbf{x}_i = \lambda_i \mathbf{x}_i$ to get

$$t_1 \lambda_1 \mathbf{x}_1 + t_2 \lambda_2 \mathbf{x}_2 + \cdots + t_{k+1} \lambda_{k+1} \mathbf{x}_{k+1} = \mathbf{0}. \qquad (**)$$

If we multiply $(*)$ by λ_1 and subtract the result from $(**)$, the first terms cancel and we obtain

$$t_2(\lambda_2 - \lambda_1)\mathbf{x}_2 + t_3(\lambda_3 - \lambda_1)\mathbf{x}_3 + \cdots + t_{k+1}(\lambda_{k+1} - \lambda_1)\mathbf{x}_{k+1} = \mathbf{0}.$$

Since $\mathbf{x}_2, \mathbf{x}_3, \ldots, \mathbf{x}_{k+1}$ correspond to distinct eigenvalues $\lambda_2, \lambda_3, \ldots, \lambda_{k+1}$, the set $\{\mathbf{x}_2, \mathbf{x}_3, \ldots, \mathbf{x}_{k+1}\}$ is independent by the induction hypothesis. Hence,

$$t_2(\lambda_2 - \lambda_1) = 0, \quad t_3(\lambda_3 - \lambda_1) = 0, \quad \ldots, \quad t_{k+1}(\lambda_{k+1} - \lambda_1) = 0,$$

and so $t_2 = t_3 = \cdots = t_{k+1} = 0$ because the λ_i are distinct. Hence $(*)$ becomes $t_1 \mathbf{x}_1 = \mathbf{0}$, which implies that $t_1 = 0$ because $\mathbf{x}_1 \neq \mathbf{0}$. This is what we wanted.

Theorem 4 will be applied several times; we begin by using it to give a useful condition for when a matrix is diagonalizable.

Theorem 5

If A is an $n \times n$ matrix with n distinct eigenvalues, then A is diagonalizable.

PROOF

Choose one eigenvector for each of the n distinct eigenvalues. Then these eigenvectors are independent by Theorem 4, and so are a basis of $\mathbb{R}^n$ by Theorem 7 Section 5.2. Now use Theorem 3.

EXAMPLE 4

Show that $A = \begin{bmatrix} 1 & 0 & 0 \\ 1 & 2 & 3 \\ -1 & 1 & 0 \end{bmatrix}$ is diagonalizable.

Solution ▶ A routine computation shows that $c_A(x) = (x-1)(x-3)(x+1)$ and so has distinct eigenvalues 1, 3, and -1. Hence Theorem 5 applies.

However, a matrix can have multiple eigenvalues as we saw in Section 3.3. To deal with this situation, we prove an important lemma which formalizes a technique that is basic to diagonalization, and which will be used three times below.

Lemma 2

Let $\{\mathbf{x}_1, \mathbf{x}_2, \ldots, \mathbf{x}_k\}$ be a linearly independent set of eigenvectors of an $n \times n$ matrix A, extend it to a basis $\{\mathbf{x}_1, \mathbf{x}_2, \ldots, \mathbf{x}_k, \ldots, \mathbf{x}_n\}$ of $\mathbb{R}^n$, and let

$$P = [\mathbf{x}_1 \ \mathbf{x}_2 \ \cdots \ \mathbf{x}_n]$$

be the (invertible) $n \times n$ matrix with the $\mathbf{x}_i$ as its columns. If $\lambda_1, \lambda_2, \ldots, \lambda_k$ are the (not necessarily distinct) eigenvalues of A corresponding to $\mathbf{x}_1, \mathbf{x}_2, \ldots, \mathbf{x}_k$ respectively, then $P^{-1}AP$ has block form

$$P^{-1}AP = \begin{bmatrix} \text{diag}(\lambda_1, \lambda_2, \ldots, \lambda_k) & B \\ 0 & A_1 \end{bmatrix}$$

where B has size $k \times (n-k)$ and A_1 has size $(n-k) \times (n-k)$.

PROOF

If $\{\mathbf{e}_1, \mathbf{e}_2, \ldots, \mathbf{e}_n\}$ is the standard basis of $\mathbb{R}^n$, then

$$[\mathbf{e}_1 \ \mathbf{e}_2 \ \ldots \ \mathbf{e}_n] = I_n = P^{-1}P = P^{-1}[\mathbf{x}_1 \ \mathbf{x}_2 \ \cdots \ \mathbf{x}_n]$$
$$= [P^{-1}\mathbf{x}_1 \ P^{-1}\mathbf{x}_2 \ \cdots \ P^{-1}\mathbf{x}_n]$$

Comparing columns, we have $P^{-1}\mathbf{x}_i = \mathbf{e}_i$ for each $1 \leq i \leq n$. On the other hand, observe that

$$P^{-1}AP = P^{-1}A[\mathbf{x}_1 \ \mathbf{x}_2 \ \cdots \ \mathbf{x}_n] = [(P^{-1}A)\mathbf{x}_1 \ (P^{-1}A)\mathbf{x}_2 \ \cdots \ (P^{-1}A)\mathbf{x}_n].$$

Hence, if $1 \leq i \leq k$, column i of $P^{-1}AP$ is

$$(P^{-1}A)\mathbf{x}_i = P^{-1}(\lambda_i \mathbf{x}_i) = \lambda_i(P^{-1}\mathbf{x}_1) = \lambda_i \mathbf{e}_i.$$

This describes the first k columns of $P^{-1}AP$, and Lemma 2 follows.

Note that Lemma 2 (with $k = n$) shows that an $n \times n$ matrix A is diagonalizable if $\mathbb{R}^n$ has a basis of eigenvectors of A, as in (1) of Theorem 3.

SECTION 5.5 Similarity and Diagonalization

Definition 5.13 *If λ is an eigenvalue of an $n \times n$ matrix A, define the **eigenspace** of A corresponding to λ by*

$$E_\lambda(A) = \{\mathbf{x} \text{ in } \mathbb{R}^n \mid A\mathbf{x} = \lambda\mathbf{x}\}.$$

This is a subspace of $\mathbb{R}^n$ and the eigenvectors corresponding to λ are just the nonzero vectors in $E_\lambda(A)$. In fact $E_\lambda(A)$ is the null space of the matrix $(\lambda I - A)$:

$$E_\lambda(A) = \{\mathbf{x} \mid (\lambda I - A)\mathbf{x} = \mathbf{0}\} = \text{null}(\lambda I - A).$$

Hence, by Theorem 2 Section 5.4, the basic solutions of the homogeneous system $(\lambda I - A)\mathbf{x} = \mathbf{0}$ given by the gaussian algorithm form a basis for $E_\lambda(A)$. In particular

$$\dim E_\lambda(A) \text{ is the number of basic solutions } \mathbf{x} \text{ of } (\lambda I - A)\mathbf{x} = \mathbf{0}. \quad (***)$$

Now recall (Definition 3.7) that the **multiplicity**[11] of an eigenvalue λ of A is the number of times λ occurs as a root of the characteristic polynomial $c_A(x)$ of A. In other words, the multiplicity of λ is the largest integer $m \geq 1$ such that

$$c_A(x) = (x - \lambda)^m g(x)$$

for some polynomial $g(x)$. Because of $(***)$, the assertion (without proof) in Theorem 5 Section 3.3 can be stated as follows: A square matrix is diagonalizable if and only if the multiplicity of each eigenvalue λ equals $\dim[E_\lambda(A)]$. We are going to prove this, and the proof requires the following result which is valid for *any* square matrix, diagonalizable or not.

Lemma 3

Let λ be an eigenvalue of multiplicity m of a square matrix A. Then $\dim[E_\lambda(A)] \leq m$.

PROOF

Write $\dim[E_\lambda(A)] = d$. It suffices to show that $c_A(x) = (x - \lambda)^d g(x)$ for some polynomial $g(x)$, because m is the highest power of $(x - \lambda)$ that divides $c_A(x)$. To this end, let $\{\mathbf{x}_1, \mathbf{x}_2, \ldots, \mathbf{x}_d\}$ be a basis of $E_\lambda(A)$. Then Lemma 2 shows that an invertible $n \times n$ matrix P exists such that

$$P^{-1}AP = \begin{bmatrix} \lambda I_d & B \\ 0 & A_1 \end{bmatrix}$$

in block form, where I_d denotes the $d \times d$ identity matrix. Now write $A' = P^{-1}AP$ and observe that $c_{A'}(x) = c_A(x)$ by Theorem 1. But Theorem 5 Section 3.1 gives

$$c_A(x) = c_{A'}(x) = \det(xI_n - A') = \det\begin{bmatrix} (x - \lambda)I_d & -B \\ 0 & xI_{n-d} - A_1 \end{bmatrix}$$

$$= \det[(x - \lambda)I_d] \det[(xI_{n-d} - A_1)]$$
$$= (x - \lambda)^d g(x).$$

where $g(x) = c_{A_1}(x)$. This is what we wanted.

It is impossible to ignore the question when equality holds in Lemma 3 for each eigenvalue λ. It turns out that this characterizes the diagonalizable $n \times n$ matrices A for which $c_A(x)$ **factors completely** over $\mathbb{R}$. By this we mean that $c_A(x) = (x - \lambda_1)(x - \lambda_2)\cdots(x - \lambda_n)$, where the λ_i are *real* numbers (not necessarily

---

[11] This is often called the *algebraic* multiplicity of λ.

distinct); in other words, every eigenvalue of A is real. This need not happen (consider $A = \begin{bmatrix} 0 & -1 \\ 1 & 0 \end{bmatrix}$), and we investigate the general case below.

Theorem 6

The following are equivalent for a square matrix A for which $c_A(x)$ factors completely.
1. *A is diagonalizable.*
2. *$\dim[E_\lambda(A)]$ equals the multiplicity of λ for every eigenvalue λ of the matrix A.*

PROOF

Let A be $n \times n$ and let $\lambda_1, \lambda_2, \ldots, \lambda_k$ be the distinct eigenvalues of A. For each i, let m_i denote the multiplicity of λ_i and write $d_i = \dim[E_{\lambda_i}(A)]$. Then

$$c_A(x) = (x - \lambda_1)^{m_1}(x - \lambda_2)^{m_2}\cdots(x - \lambda_n)^{m_k}$$

so $m_1 + \cdots + m_k = n$ because $c_A(x)$ has degree n. Moreover, $d_i \leq m_i$ for each i by Lemma 3.

(1) $\Rightarrow$ (2). By (1), $\mathbb{R}^n$ has a basis of n eigenvectors of A, so let t_i of them lie in $E_{\lambda_i}(A)$ for each i. Since the subspace spanned by these t_i eigenvectors has dimension t_i, we have $t_i \leq d_i$ for each i by Theorem 4 Section 5.2. Hence

$$n = t_1 + \cdots + t_k \leq d_1 + \cdots + d_k \leq m_1 + \cdots + m_k = n.$$

It follows that $d_1 + \cdots + d_k = m_1 + \cdots + m_k$ so, since $d_i \leq m_i$ for each i, we must have $d_i = m_i$. This is (2).

(2) $\Rightarrow$ (1). Let B_i denote a basis of $E_{\lambda_i}(A)$ for each i, and let $B = B_1 \cup \cdots \cup B_k$. Since each B_i contains m_i vectors by (2), and since the B_i are pairwise disjoint (the λ_i are distinct), it follows that B contains n vectors. So it suffices to show that B is linearly independent (then B is a basis of $\mathbb{R}^n$). Suppose a linear combination of the vectors in B vanishes, and let $\mathbf{y}_i$ denote the sum of all terms that come from B_i. Then $\mathbf{y}_i$ lies in $E_{\lambda_i}(A)$ for each i, so the nonzero $\mathbf{y}_i$ are independent by Theorem 4 (as the λ_i are distinct). Since the sum of the $\mathbf{y}_i$ is zero, it follows that $\mathbf{y}_i = \mathbf{0}$ for each i. Hence all coefficients of terms in $\mathbf{y}_i$ are zero (because B_i is independent). Since this holds for each i, it shows that B is independent.

EXAMPLE 5

If $A = \begin{bmatrix} 5 & 8 & 16 \\ 4 & 1 & 8 \\ -4 & -4 & -11 \end{bmatrix}$ and $B = \begin{bmatrix} 2 & 1 & 1 \\ 2 & 1 & -2 \\ -1 & 0 & -2 \end{bmatrix}$, show that A is diagonalizable but B is not.

Solution ▶ We have $c_A(x) = (x + 3)^2(x - 1)$ so the eigenvalues are $\lambda_1 = -3$ and $\lambda_2 = 1$. The corresponding eigenspaces are $E_{\lambda_1}(A) = \text{span}\{\mathbf{x}_1, \mathbf{x}_2\}$ and $E_{\lambda_2}(A) = \text{span}\{\mathbf{x}_3\}$ where

$$\mathbf{x}_1 = \begin{bmatrix} -1 \\ 1 \\ 0 \end{bmatrix}, \quad \mathbf{x}_2 = \begin{bmatrix} -2 \\ 0 \\ 1 \end{bmatrix}, \quad \mathbf{x}_3 = \begin{bmatrix} 2 \\ 1 \\ -1 \end{bmatrix}$$

as the reader can verify. Since $\{\mathbf{x}_1, \mathbf{x}_2\}$ is independent, we have $\dim(E_{\lambda_1}(A)) = 2$ which is the multiplicity of λ_1. Similarly, $\dim(E_{\lambda_2}(A)) = 1$ equals the multiplicity of λ_2. Hence A is diagonalizable by Theorem 6, and a diagonalizing matrix is $P = [\mathbf{x}_1 \ \mathbf{x}_2 \ \mathbf{x}_3]$.

Turning to B, $c_B(x) = (x + 1)^2(x - 3)$ so the eigenvalues are $\lambda_1 = -1$ and $\lambda_2 = 3$. The corresponding eigenspaces are $E_\lambda 1(B) = \text{span}\{\mathbf{y}_1\}$ and $E_\lambda 2(B) = \text{span}\{\mathbf{y}_2\}$ where

$$\mathbf{y}_1 = \begin{bmatrix} -1 \\ 2 \\ 1 \end{bmatrix}, \quad \mathbf{y}_2 = \begin{bmatrix} 5 \\ 6 \\ -1 \end{bmatrix}.$$

Here $\dim(E_{\lambda_1}(B)) = 1$ is *smaller* than the multiplicity of λ_1, so the matrix B is *not* diagonalizable, again by Theorem 6. The fact that $\dim(E_{\lambda_1}(B)) = 1$ means that there is no possibility of finding *three* linearly independent eigenvectors.

Complex Eigenvalues

All the matrices we have considered have had real eigenvalues. But this need not be the case: The matrix $A = \begin{bmatrix} 0 & -1 \\ 1 & 0 \end{bmatrix}$ has characteristic polynomial $c_A(x) = x^2 + 1$ which has no real roots. Nonetheless, this matrix is diagonalizable; the only difference is that we must use a larger set of scalars, the complex numbers. The basic properties of these numbers are outlined in Appendix A.

Indeed, nearly everything we have done for real matrices can be done for complex matrices. The methods are the same; the only difference is that the arithmetic is carried out with complex numbers rather than real ones. For example, the gaussian algorithm works in exactly the same way to solve systems of linear equations with complex coefficients, matrix multiplication is defined the same way, and the matrix inversion algorithm works in the same way.

But the complex numbers are better than the real numbers in one respect: While there are polynomials like $x^2 + 1$ with real coefficients that have no real root, this problem does not arise with the complex numbers: *Every* nonconstant polynomial with complex coefficients has a complex root, and hence factors completely as a product of linear factors. This fact is known as the fundamental theorem of algebra.[12]

EXAMPLE 6

Diagonalize the matrix $A = \begin{bmatrix} 0 & -1 \\ 1 & 0 \end{bmatrix}$.

Solution ▶ The characteristic polynomial of A is

$$c_A(x) = \det(xI - A) = x^2 + 1 = (x - i)(x + i)$$

where $i^2 = -1$. Hence the eigenvalues are $\lambda_1 = i$ and $\lambda_2 = -i$, with corresponding eigenvectors $\mathbf{x}_1 = \begin{bmatrix} 1 \\ -i \end{bmatrix}$ and $\mathbf{x}_2 = \begin{bmatrix} 1 \\ i \end{bmatrix}$. Hence A is diagonalizable by the complex version of Theorem 5, and the complex version of Theorem 3 shows that $P = [\mathbf{x}_1 \ \mathbf{x}_2] = \begin{bmatrix} 1 & 1 \\ -i & i \end{bmatrix}$ is invertible and $P^{-1}AP = \begin{bmatrix} \lambda_1 & 0 \\ 0 & \lambda_2 \end{bmatrix} = \begin{bmatrix} i & 0 \\ 0 & -i \end{bmatrix}$. Of course, this can be checked directly.

We shall return to complex linear algebra in Section 8.6.

---

12 This was a famous open problem in 1799 when Gauss solved it at the age of 22 in his Ph.D. dissertation.

Symmetric Matrices[13]

On the other hand, many of the applications of linear algebra involve a real matrix A and, while A will have complex eigenvalues by the fundamental theorem of algebra, it is always of interest to know when the eigenvalues are, in fact, real. While this can happen in a variety of ways, it turns out to hold whenever A is symmetric. This important theorem will be used extensively later. Surprisingly, the theory of *complex* eigenvalues can be used to prove this useful result about *real* eigenvalues.

Let $\bar{z}$ denote the conjugate of a complex number z. If A is a complex matrix, the **conjugate matrix** $\bar{A}$ is defined to be the matrix obtained from A by conjugating every entry. Thus, if $A = [z_{ij}]$, then $\bar{A} = [\bar{z}_{ij}]$. For example,

$$\text{If } A = \begin{bmatrix} -i+2 & 5 \\ i & 3+4i \end{bmatrix} \text{ then } \bar{A} = \begin{bmatrix} i+2 & 5 \\ -i & 3-4i \end{bmatrix}$$

Recall that $\overline{z+w} = \bar{z} + \bar{w}$ and $\overline{zw} = \bar{z}\,\bar{w}$ hold for all complex numbers z and w. It follows that if A and B are two complex matrices, then

$$\overline{A+B} = \bar{A} + \bar{B}, \quad \overline{AB} = \bar{A}\,\bar{B} \quad \text{and} \quad \overline{\lambda A} = \bar{\lambda}\,\bar{A}$$

hold for all complex scalars λ. These facts are used in the proof of the following theorem.

Theorem 7

Let A be a symmetric real matrix. If λ is any complex eigenvalue of A, then λ is real.[14]

PROOF

Observe that $\bar{A} = A$ because A is real. If λ is an eigenvalue of A, we show that λ is real by showing that $\bar{\lambda} = \lambda$. Let $\mathbf{x}$ be a (possibly complex) eigenvector corresponding to λ, so that $\mathbf{x} \neq \mathbf{0}$ and $A\mathbf{x} = \lambda\mathbf{x}$. Define $c = \mathbf{x}^T \bar{\mathbf{x}}$.

If we write $\mathbf{x} = (z_1, z_2, \ldots, z_n)$ where the z_i are complex numbers, we have

$$c = \mathbf{x}^T \bar{\mathbf{x}} = z_1 \bar{z}_1 + z_2 \bar{z}_2 + \cdots + z_n \bar{z}_n = |\bar{z}_1|^2 + |\bar{z}_2|^2 + \cdots + |\bar{z}_n|^2.$$

Thus c is a real number, and $c > 0$ because at least one of the $z_i \neq 0$ (as $\mathbf{x} \neq \mathbf{0}$). We show that $\bar{\lambda} = \lambda$ by verifying that $\lambda c = \bar{\lambda} c$. We have

$$\lambda c = \lambda(\mathbf{x}^T \bar{\mathbf{x}}) = (\lambda \mathbf{x})^T \bar{\mathbf{x}} = (A\mathbf{x})^T \bar{\mathbf{x}} = \mathbf{x}^T A^T \bar{\mathbf{x}}.$$

At this point we use the hypothesis that A is symmetric and real. This means $A^T = A = \bar{A}$, so we continue the calculation:

$$\lambda c = \mathbf{x}^T A^T \bar{\mathbf{x}} = \mathbf{x}^T (\bar{A}\,\bar{\mathbf{x}}) = \mathbf{x}^T (\overline{A\mathbf{x}}) = \mathbf{x}^T (\overline{\lambda \mathbf{x}})$$
$$= \mathbf{x}^T (\bar{\lambda}\,\bar{\mathbf{x}})$$
$$= \bar{\lambda} \mathbf{x}^T \bar{\mathbf{x}}$$
$$= \bar{\lambda} c$$

as required.

The technique in the proof of Theorem 7 will be used again when we return to complex linear algebra in Section 8.6.

---

[13] This discussion uses complex conjugation and absolute value. These topics are discussed in Appendix A.
[14] This theorem was first proved in 1829 by the great French mathematician Augustin Louis Cauchy (1789–1857).

SECTION 5.5 Similarity and Diagonalization

EXAMPLE 7

Verify Theorem 7 for every real, symmetric 2×2 matrix A.

Solution ▶ If $A = \begin{bmatrix} a & b \\ b & c \end{bmatrix}$ we have $c_A(x) = x^2 - (a + c)x + (ac - b^2)$, so the eigenvalues are given by $\lambda = \frac{1}{2}\left[(a + c) \pm \sqrt{(a + c)^2 - 4(ac - b^2)}\right]$. But here

$$(a + c)^2 - 4(ac - b^2) = (a - c)^2 + 4b^2 \geq 0$$

for any choice of a, b, and c. Hence, the eigenvalues are real numbers.

EXERCISES 5.5

1. By computing the trace, determinant, and rank, show that A and B are *not* similar in each case.

   (a) $A = \begin{bmatrix} 1 & 2 \\ 2 & 1 \end{bmatrix}$, $B = \begin{bmatrix} 1 & 1 \\ -1 & 1 \end{bmatrix}$

   ◆(b) $A = \begin{bmatrix} 3 & 1 \\ 2 & -1 \end{bmatrix}$, $B = \begin{bmatrix} 1 & 1 \\ 2 & 1 \end{bmatrix}$

   (c) $A = \begin{bmatrix} 2 & 1 \\ 1 & -1 \end{bmatrix}$, $B = \begin{bmatrix} 3 & 0 \\ 1 & -1 \end{bmatrix}$

   ◆(d) $A = \begin{bmatrix} 3 & 1 \\ -1 & 2 \end{bmatrix}$, $B = \begin{bmatrix} 2 & -1 \\ 3 & 2 \end{bmatrix}$

   (e) $A = \begin{bmatrix} 2 & 1 & 1 \\ 1 & 0 & 1 \\ 1 & 1 & 0 \end{bmatrix}$, $B = \begin{bmatrix} 1 & -2 & 1 \\ -2 & 4 & -2 \\ -3 & 6 & -3 \end{bmatrix}$

   ◆(f) $A = \begin{bmatrix} 1 & 2 & -3 \\ 1 & -1 & 2 \\ 0 & 3 & -5 \end{bmatrix}$, $B = \begin{bmatrix} -2 & 1 & 3 \\ 6 & -3 & -9 \\ 0 & 0 & 0 \end{bmatrix}$

2. Show that $\begin{bmatrix} 1 & 2 & -1 & 0 \\ 2 & 0 & 1 & 1 \\ 1 & 1 & 0 & -1 \\ 4 & 3 & 0 & 0 \end{bmatrix}$ and $\begin{bmatrix} 1 & -1 & 3 & 0 \\ -1 & 0 & 1 & 1 \\ 0 & -1 & 4 & 1 \\ 5 & -1 & -1 & -4 \end{bmatrix}$ are *not* similar.

3. If $A \sim B$, show that:

   (a) $A^T \sim B^T$ ◆(b) $A^{-1} \sim B^{-1}$

   (c) $rA \sim rB$ for r in $\mathbb{R}$ (d) $A^n \sim B^n$ for $n \geq 1$

4. In each case, decide whether the matrix A is diagonalizable. If so, find P such that $P^{-1}AP$ is diagonal.

   (a) $\begin{bmatrix} 1 & 0 & 0 \\ 1 & 2 & 1 \\ 0 & 0 & 1 \end{bmatrix}$ ◆(b) $\begin{bmatrix} 3 & 0 & 6 \\ 0 & -3 & 0 \\ 5 & 0 & 2 \end{bmatrix}$

   (c) $\begin{bmatrix} 3 & 1 & 6 \\ 2 & 1 & 0 \\ -1 & 0 & -3 \end{bmatrix}$ ◆(d) $\begin{bmatrix} 4 & 0 & 0 \\ 0 & 2 & 2 \\ 2 & 3 & 1 \end{bmatrix}$

5. If A is invertible, show that AB is similar to BA for all B.

6. Show that the only matrix similar to a scalar matrix $A = rI$, r in $\mathbb{R}$, is A itself.

7. Let λ be an eigenvalue of A with corresponding eigenvector $\mathbf{x}$. If $B = P^{-1}AP$ is similar to A, show that $P^{-1}\mathbf{x}$ is an eigenvector of B corresponding to λ.

8. If $A \sim B$ and A has any of the following properties, show that B has the same property.

   (a) Idempotent, that is $A^2 = A$.

   ◆(b) Nilpotent, that is $A^k = 0$ for some $k \geq 1$.

   (c) Invertible.

9. Let A denote an $n \times n$ upper triangular matrix.

   (a) If all the main diagonal entries of A are distinct, show that A is diagonalizable.

   ◆(b) If all the main diagonal entries of A are equal, show that A is diagonalizable only if it is *already* diagonal.

   (c) Show that $\begin{bmatrix} 1 & 0 & 1 \\ 0 & 1 & 0 \\ 0 & 0 & 2 \end{bmatrix}$ is diagonalizable but that $\begin{bmatrix} 1 & 1 & 0 \\ 0 & 1 & 0 \\ 0 & 0 & 2 \end{bmatrix}$ is not diagonalizable.

10. Let A be a diagonalizable $n \times n$ matrix with eigenvalues $\lambda_1, \lambda_2, \ldots, \lambda_n$ (including multiplicities). Show that:

   (a) $\det A = \lambda_1 \lambda_2 \cdots \lambda_n$

   ◆(b) $\operatorname{tr} A = \lambda_1 + \lambda_2 + \cdots + \lambda_n$

11. Given a polynomial $p(x) = r_0 + r_1 x + \cdots + r_n x^n$ and a square matrix A, the matrix $p(A) = r_0 I + r_1 A + \cdots + r_n A^n$ is called the **evaluation** of $p(x)$ at A. Let $B = P^{-1}AP$. Show that $p(B) = P^{-1}p(A)P$ for all polynomials $p(x)$.

12. Let P be an invertible $n \times n$ matrix. If A is any $n \times n$ matrix, write $T_P(A) = P^{-1}AP$. Verify that:

   (a) $T_P(I) = I$

   ◆(b) $T_P(AB) = T_P(A)T_P(B)$

   (c) $T_P(A + B) = T_P(A) + T_P(B)$

   (d) $T_P(rA) = rT_P(A)$

   (e) $T_P(A^k) = [T_P(A)]^k$ for $k \geq 1$

   (f) If A is invertible, $T_P(A^{-1}) = [T_P(A)]^{-1}$.

   (g) If Q is invertible, $T_Q[T_P(A)] = T_{PQ}(A)$.

13. (a) Show that two diagonalizable matrices are similar if and only if they have the same eigenvalues with the same multiplicities.

   ◆(b) If A is diagonalizable, show that $A \sim A^T$.

   (c) Show that $A \sim A^T$ if $A = \begin{bmatrix} 1 & 1 \\ 0 & 1 \end{bmatrix}$.

14. If A is 2×2 and diagonalizable, show that $C(A) = \{X \mid XA = AX\}$ has dimension 2 or 4. [*Hint*: If $P^{-1}AP = D$, show that X is in $C(A)$ if and only if $P^{-1}XP$ is in $C(D)$.]

15. If A is diagonalizable and $p(x)$ is a polynomial such that $p(\lambda) = 0$ for all eigenvalues λ of A, show that $p(A) = 0$ (see Example 9 Section 3.3). In particular, show $c_A(A) = 0$. [*Remark*: $c_A(A) = 0$ for *all* square matrices A—this is the Cayley-Hamilton theorem (see Theorem 2 Section 9.4).]

16. Let A be $n \times n$ with n distinct real eigenvalues. If $AC = CA$, show that C is diagonalizable.

17. Let $A = \begin{bmatrix} 0 & a & b \\ a & 0 & c \\ b & c & 0 \end{bmatrix}$ and $B = \begin{bmatrix} c & a & b \\ a & b & c \\ b & c & a \end{bmatrix}$.

   (a) Show that $x^3 - (a^2 + b^2 + c^2)x - 2abc$ has real roots by considering A.

   ◆(b) Show that $a^2 + b^2 + c^2 \geq ab + ac + bc$ by considering B.

18. Assume the 2×2 matrix A is similar to an upper triangular matrix. If $\operatorname{tr} A = 0 = \operatorname{tr} A^2$, show that $A^2 = 0$.

19. Show that A is similar to A^T for all 2×2 matrices A. [*Hint*: Let $A = \begin{bmatrix} a & b \\ c & d \end{bmatrix}$. If $c = 0$, treat the cases $b = 0$ and $b \neq 0$ separately. If $c \neq 0$, reduce to the case $c = 1$ using Exercise 12(d).]

20. Refer to Section 3.4 on linear recurrences. Assume that the sequence $x_0, x_1, x_2, \ldots$ satisfies
$$x_{n+k} = r_0 x_n + r_1 x_{n+1} + \cdots + r_{k-1} x_{n+k-1}$$
for all $n \geq 0$. Define
$$A = \begin{bmatrix} 0 & 1 & 0 & \cdots & 0 \\ 0 & 0 & 1 & \cdots & 0 \\ \vdots & \vdots & \vdots & & \vdots \\ 0 & 0 & 0 & \cdots & 1 \\ r_0 & r_1 & r_2 & \cdots & r_{k-1} \end{bmatrix}, \quad V_n = \begin{bmatrix} x_n \\ x_{n+1} \\ \vdots \\ x_{n+k-1} \end{bmatrix}.$$
Then show that:

   (a) $V_n = A^n V_0$ for all n.

   (b) $c_A(x) = x^k - r_{k-1} x^{k-1} - \cdots - r_1 x - r_0$.

   (c) If λ is an eigenvalue of A, the eigenspace E_λ has dimension 1, and $\mathbf{x} = (1, \lambda, \lambda^2, \ldots, \lambda^{k-1})^T$ is an eigenvector. [*Hint*: Use $c_A(\lambda) = 0$ to show that $E_\lambda = \mathbb{R}\mathbf{x}$.]

   (d) A is diagonalizable if and only if the eigenvalues of A are distinct. [*Hint*: See part (c) and Theorem 4.]

   (e) If $\lambda_1, \lambda_2, \ldots, \lambda_k$ are distinct real eigenvalues, there exist constants $t_1, t_2, \ldots, t_k$ such that $x_n = t_1 \lambda_1^n + \cdots + t_k \lambda_k^n$ holds for all n. [*Hint*: If D is diagonal with $\lambda_1, \lambda_2, \ldots, \lambda_k$ as the main diagonal entries, show that $A^n = PD^n P^{-1}$ has entries that are linear combinations of $\lambda_1^n, \lambda_2^n, \ldots, \lambda_k^n$.]

SECTION 5.6 Best Approximation and Least Squares

Often an exact solution to a problem in applied mathematics is difficult to obtain. However, it is usually just as useful to find arbitrarily close approximations to a solution. In particular, finding "linear approximations" is a potent technique in applied mathematics. One basic case is the situation where a system of linear equations has no solution, and it is desirable to find a "best approximation" to a solution to the system. In this section best approximations are defined and a method for finding them is described. The result is then applied to "least squares" approximation of data.

Suppose A is an $m \times n$ matrix and $\mathbf{b}$ is a column in $\mathbb{R}^m$, and consider the system

$$A\mathbf{x} = \mathbf{b}$$

of m linear equations in n variables. This need not have a solution. However, given any column $\mathbf{z}$ in $\mathbb{R}^n$, the distance $\|\mathbf{b} - A\mathbf{z}\|$ is a measure of how far $A\mathbf{z}$ is from $\mathbf{b}$. Hence it is natural to ask whether there is a column $\mathbf{z}$ in $\mathbb{R}^n$ that is as close as possible to a solution in the sense that

$$\|\mathbf{b} - A\mathbf{z}\|$$

is the minimum value of $\|\mathbf{b} - A\mathbf{x}\|$ as $\mathbf{x}$ ranges over all columns in $\mathbb{R}^n$.

The answer is "yes", and to describe it define

$$U = \{A\mathbf{x} \mid \mathbf{x} \text{ lies in } \mathbb{R}^n\}.$$

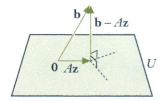

This is a subspace of $\mathbb{R}^n$ (verify) and we want a vector $A\mathbf{z}$ in U as close as possible to $\mathbf{b}$. That there is such a vector is clear geometrically if $n = 3$ by the diagram. In general such a vector $A\mathbf{z}$ exists by a general result called the *projection theorem* that will be proved in Chapter 8 (Theorem 3 Section 8.1). Moreover, the projection theorem gives a simple way to compute $\mathbf{z}$ because it also shows that the vector $\mathbf{b} - A\mathbf{z}$ is *orthogonal* to every vector $A\mathbf{x}$ in U. Thus, for all $\mathbf{x}$ in $\mathbb{R}^n$,

$$0 = (A\mathbf{x}) \cdot (\mathbf{b} - A\mathbf{z}) = (A\mathbf{x})^T(\mathbf{b} - A\mathbf{z}) = \mathbf{x}^T A^T(\mathbf{b} - A\mathbf{z})$$
$$= \mathbf{x} \cdot [A^T(\mathbf{b} - A\mathbf{z})]$$

In other words, the vector $A^T(\mathbf{b} - A\mathbf{z})$ in $\mathbb{R}^n$ is orthogonal to *every* vector in $\mathbb{R}^n$ and so must be zero (being orthogonal to itself). Hence $\mathbf{z}$ satisfies

$$(A^T A)\mathbf{z} = A^T \mathbf{b}.$$

Definition 5.14 *This is a system of linear equations called the **normal equations** for $\mathbf{z}$.*

Note that this system can have more than one solution (see Exercise 5). However, the $n \times n$ matrix $A^T A$ is invertible if (and only if) the columns of A are linearly independent (Theorem 3 Section 5.4); so, in this case, $\mathbf{z}$ is uniquely determined and is given explicitly by $\mathbf{z} = (A^T A)^{-1} A^T \mathbf{b}$. However, the most efficient way to find $\mathbf{z}$ is to apply gaussian elimination to the normal equations.

This discussion is summarized in the following theorem.

Theorem 1

Best Approximation Theorem
Let A be an $m \times n$ matrix, let $\mathbf{b}$ be any column in $\mathbb{R}^m$, and consider the system

$$A\mathbf{x} = \mathbf{b}$$

of m equations in n variables.

(1) Any solution $\mathbf{z}$ to the normal equations
$$(A^TA)\mathbf{z} = A^T\mathbf{b}$$
is a best approximation to a solution to $A\mathbf{x} = \mathbf{b}$ in the sense that $\|\mathbf{b} - A\mathbf{z}\|$ is the minimum value of $\|\mathbf{b} - A\mathbf{x}\|$ as $\mathbf{x}$ ranges over all columns in $\mathbb{R}^n$.

(2) If the columns of A are linearly independent, then A^TA is invertible and $\mathbf{z}$ is given uniquely by $\mathbf{z} = (A^TA)^{-1}A^T\mathbf{b}$.

We note in passing that if A is $n \times n$ and invertible, then
$$\mathbf{z} = (A^TA)^{-1}A^T\mathbf{b} = A^{-1}\mathbf{b}$$
is the solution to the system of equations, and $\|\mathbf{b} - A\mathbf{z}\| = 0$. Hence if A has independent columns, then $(A^TA)^{-1}A^T$ is playing the role of the inverse of the nonsquare matrix A. The matrix $A^T(AA^T)^{-1}$ plays a similar role when the rows of A are linearly independent. These are both special cases of the **generalized inverse** of a matrix A (see Exercise 14). However, we shall not pursue this topic here.

EXAMPLE 1

The system of linear equations
$$\begin{aligned} 3x - y &= 4 \\ x + 2y &= 0 \\ 2x + y &= 1 \end{aligned}$$
has no solution. Find the vector $\mathbf{z} = \begin{bmatrix} x_0 \\ y_0 \end{bmatrix}$ that best approximates a solution.

Solution ▶ In this case,
$$A = \begin{bmatrix} 3 & -1 \\ 1 & 2 \\ 2 & 1 \end{bmatrix}, \text{ so } A^TA = \begin{bmatrix} 3 & 1 & 2 \\ -1 & 2 & 1 \end{bmatrix} \begin{bmatrix} 3 & -1 \\ 1 & 2 \\ 2 & 1 \end{bmatrix} = \begin{bmatrix} 14 & 1 \\ 1 & 6 \end{bmatrix}$$
is invertible. The normal equations $(A^TA)\mathbf{z} = A^T\mathbf{b}$ are
$$\begin{bmatrix} 14 & 1 \\ 1 & 6 \end{bmatrix} \mathbf{z} = \begin{bmatrix} 14 \\ -3 \end{bmatrix}, \text{ so } \mathbf{z} = \tfrac{1}{83}\begin{bmatrix} 87 \\ -56 \end{bmatrix}.$$

Thus $x_0 = \frac{87}{83}$ and $y_0 = \frac{-56}{83}$. With these values of x and y, the left sides of the equations are, approximately,
$$\begin{aligned} 3x_0 - y_0 &= \tfrac{317}{83} = 3.82 \\ x_0 + 2y_0 &= \tfrac{-25}{83} = -0.30 \\ 2x_0 + y_0 &= \tfrac{118}{83} = 1.42 \end{aligned}$$

This is as close as possible to a solution.

EXAMPLE 2

The average number g of goals per game scored by a hockey player seems to be related linearly to two factors: the number x_1 of years of experience and the number x_2 of goals in the preceding 10 games. The data on the following page were collected on four players. Find the linear function $g = a_0 + a_1x_1 + a_2x_2$ that best fits these data.

SECTION 5.6 Best Approximation and Least Squares

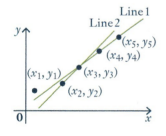

g	x_1	x_2
0.8	5	3
0.8	3	4
0.6	1	5
0.4	2	1

Solution ▶ If the relationship is given by $g = r_0 + r_1 x_1 + r_2 x_2$, then the data can be described as follows:

$$\begin{bmatrix} 1 & 5 & 3 \\ 1 & 3 & 4 \\ 1 & 1 & 5 \\ 1 & 2 & 1 \end{bmatrix} \begin{bmatrix} r_0 \\ r_1 \\ r_2 \end{bmatrix} = \begin{bmatrix} 0.8 \\ 0.8 \\ 0.6 \\ 0.4 \end{bmatrix}$$

Using the notation in Theorem 1, we get

$$\mathbf{z} = (A^T A)^{-1} A^T \mathbf{b}$$

$$= \tfrac{1}{42} \begin{bmatrix} 119 & -17 & -19 \\ -17 & 5 & 1 \\ -19 & 1 & 5 \end{bmatrix} \begin{bmatrix} 1 & 1 & 1 & 1 \\ 5 & 3 & 1 & 2 \\ 3 & 4 & 5 & 1 \end{bmatrix} \begin{bmatrix} 0.8 \\ 0.8 \\ 0.6 \\ 0.4 \end{bmatrix} = \begin{bmatrix} 0.14 \\ 0.09 \\ 0.08 \end{bmatrix}$$

Hence the best-fitting function is $g = 0.14 + 0.09 x_1 + 0.08 x_2$. The amount of computation would have been reduced if the normal equations had been constructed and then solved by gaussian elimination.

Least Squares Approximation

In many scientific investigations, data are collected that relate two variables. For example, if x is the number of dollars spent on advertising by a manufacturer and y is the value of sales in the region in question, the manufacturer could generate data by spending $x_1, x_2, \dots, x_n$ dollars at different times and measuring the corresponding sales values $y_1, y_2, \dots, y_n$.

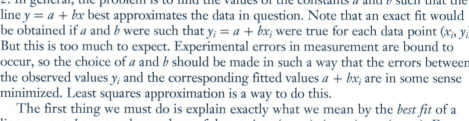

Suppose it is known that a linear relationship exists between the variables x and y—in other words, that $y = a + bx$ for some constants a and b. If the data are plotted, the points $(x_1, y_1), (x_2, y_2), \dots, (x_n, y_n)$ may appear to lie on a straight line and estimating a and b requires finding the "best-fitting" line through these data points. For example, if five data points occur as shown in the diagram, line 1 is clearly a better fit than line 2. In general, the problem is to find the values of the constants a and b such that the line $y = a + bx$ best approximates the data in question. Note that an exact fit would be obtained if a and b were such that $y_i = a + bx_i$ were true for each data point (x_i, y_i). But this is too much to expect. Experimental errors in measurement are bound to occur, so the choice of a and b should be made in such a way that the errors between the observed values y_i and the corresponding fitted values $a + bx_i$ are in some sense minimized. Least squares approximation is a way to do this.

The first thing we must do is explain exactly what we mean by the *best fit* of a line $y = a + bx$ to an observed set of data points $(x_1, y_1), (x_2, y_2), \dots, (x_n, y_n)$. For convenience, write the linear function $r_0 + r_1 x$ as

$$f(x) = r_0 + r_1 x$$

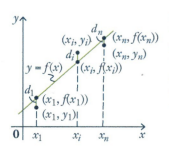

so that the fitted points (on the line) have coordinates $(x_1, f(x_1)), \dots, (x_n, f(x_n))$. The second diagram is a sketch of what the line $y = f(x)$ might look like. For each i the observed data point (x_i, y_i) and the fitted point $(x_i, f(x_i))$ need not be the same, and the distance d_i between them measures how far the line misses the observed point. For this reason d_i is often called the **error** at x_i, and a natural measure of how close the line $y = f(x)$ is to the observed data points is the sum $d_1 + d_2 + \cdots + d_n$ of all these errors. However, it turns out to be better to use the sum of squares

$$S = d_1^2 + d_2^2 + \cdots + d_n^2$$

as the measure of error, and the line $y = f(x)$ is to be chosen so as to make this sum as small as possible. This line is said to be the **least squares approximating line** for the data points $(x_1, y_1), (x_2, y_2), \ldots, (x_n, y_n)$.

The square of the error d_i is given by $d_i^2 = [y_i - f(x_i)]^2$ for each i, so the quantity S to be minimized is the sum:

$$S = [y_1 - f(x_1)]^2 + [y_2 - f(x_2)]^2 + \cdots + [y_n - f(x_n)]^2.$$

Note that all the numbers x_i and y_i are *given* here; what is required is that the *function f* be chosen in such a way as to minimize S. Because $f(x) = r_0 + r_1 x$, this amounts to choosing r_0 and r_1 to minimize S. This problem can be solved using Theorem 1. The following notation is convenient.

$$\mathbf{x} = \begin{bmatrix} x_1 \\ x_2 \\ \vdots \\ x_n \end{bmatrix} \quad \mathbf{y} = \begin{bmatrix} y_1 \\ y_2 \\ \vdots \\ y_n \end{bmatrix} \quad \text{and} \quad f(\mathbf{x}) = \begin{bmatrix} f(x_1) \\ f(x_2) \\ \vdots \\ f(x_n) \end{bmatrix} = \begin{bmatrix} r_0 + r_1 x_1 \\ r_0 + r_1 x_2 \\ \vdots \\ r_0 + r_1 x_n \end{bmatrix}$$

Then the problem takes the following form: Choose r_0 and r_1 such that

$$S = [y_1 - f(x_1)]^2 + [y_2 - f(x_2)]^2 + \cdots + [y_n - f(x_n)]^2 = \|\mathbf{y} - f(\mathbf{x})\|^2$$

is as small as possible. Now write

$$M = \begin{bmatrix} 1 & x_1 \\ 1 & x_2 \\ \vdots & \vdots \\ 1 & x_n \end{bmatrix} \quad \text{and} \quad \mathbf{r} = \begin{bmatrix} r_0 \\ r_1 \end{bmatrix}.$$

Then $M\mathbf{r} = f(\mathbf{x})$, so we are looking for a column $\mathbf{r} = \begin{bmatrix} r_0 \\ r_1 \end{bmatrix}$ such that $\|\mathbf{y} - M\mathbf{r}\|^2$ is as small as possible. In other words, we are looking for a best approximation $\mathbf{z}$ to the system $M\mathbf{r} = \mathbf{y}$. Hence Theorem 1 applies directly, and we have

Theorem 2

Suppose that n data points $(x_1, y_1), (x_2, y_2), \ldots, (x_n, y_n)$ are given, where at least two of $x_1, x_2, \ldots, x_n$ are distinct. Put

$$\mathbf{y} = \begin{bmatrix} y_1 \\ y_2 \\ \vdots \\ y_n \end{bmatrix} \quad M = \begin{bmatrix} 1 & x_1 \\ 1 & x_2 \\ \vdots & \vdots \\ 1 & x_n \end{bmatrix}$$

Then the least squares approximating line for these data points has equation

$$y = z_0 + z_1 x$$

where $\mathbf{z} = \begin{bmatrix} z_0 \\ z_1 \end{bmatrix}$ is found by gaussian elimination from the normal equations

$$(M^T M)\mathbf{z} = M^T \mathbf{y}.$$

The condition that at least two of $x_1, x_2, \ldots, x_n$ are distinct ensures that $M^T M$ is an invertible matrix, so $\mathbf{z}$ is unique:

$$\mathbf{z} = (M^T M)^{-1} M^T \mathbf{y}.$$

SECTION 5.6 Best Approximation and Least Squares

EXAMPLE 3

Let data points $(x_1, y_1), (x_2, y_2), \ldots, (x_5, y_5)$ be given as in the accompanying table. Find the least squares approximating line for these data.

x	y
1	1
3	2
4	3
6	4
7	5

Solution ▶ In this case we have

$$M^T M = \begin{bmatrix} 1 & 1 & \cdots & 1 \\ x_1 & x_2 & \cdots & x_5 \end{bmatrix} \begin{bmatrix} 1 & x_1 \\ 1 & x_2 \\ \vdots & \vdots \\ 1 & x_5 \end{bmatrix}$$

$$= \begin{bmatrix} 5 & x_1 + \cdots + x_5 \\ x_1 + \cdots + x_5 & x_1^2 + \cdots + x_5^2 \end{bmatrix} = \begin{bmatrix} 5 & 21 \\ 21 & 111 \end{bmatrix},$$

and $M^T \mathbf{y} = \begin{bmatrix} 1 & 1 & \cdots & 1 \\ x_1 & x_2 & \cdots & x_5 \end{bmatrix} \begin{bmatrix} y_1 \\ y_2 \\ \vdots \\ y_5 \end{bmatrix}$

$$= \begin{bmatrix} y_1 + y_2 + \cdots + y_5 \\ x_1 y_1 + x_2 y_2 + \cdots + x_5 y_5 \end{bmatrix} = \begin{bmatrix} 15 \\ 78 \end{bmatrix},$$

so the normal equations $(M^T M)\mathbf{z} = M^T \mathbf{y}$ for $\mathbf{z} = \begin{bmatrix} z_0 \\ z_1 \end{bmatrix}$ become

$$\begin{bmatrix} 5 & 21 \\ 21 & 111 \end{bmatrix} \begin{bmatrix} z_0 \\ z_1 \end{bmatrix} = \begin{bmatrix} 15 \\ 78 \end{bmatrix}.$$

The solution (using gaussian elimination) is $\mathbf{z} = \begin{bmatrix} z_0 \\ z_1 \end{bmatrix} = \begin{bmatrix} 0.24 \\ 0.66 \end{bmatrix}$ to two decimal places, so the least squares approximating line for these data is $y = 0.24 + 0.66x$. Note that $M^T M$ is indeed invertible here (the determinant is 114), and the exact solution is

$$\mathbf{z} = (M^T M)^{-1} M^T \mathbf{y} = \tfrac{1}{114} \begin{bmatrix} 111 & -21 \\ -21 & 5 \end{bmatrix} \begin{bmatrix} 15 \\ 78 \end{bmatrix} = \tfrac{1}{114} \begin{bmatrix} 27 \\ 75 \end{bmatrix} = \tfrac{1}{38} \begin{bmatrix} 9 \\ 25 \end{bmatrix}.$$

Least Squares Approximating Polynomials

Suppose now that, rather than a straight line, we want to find a polynomial

$$y = f(x) = r_0 + r_1 x + r_2 x^2 + \cdots + r_m x^m$$

of degree m that best approximates the data pairs $(x_1, y_1), (x_2, y_2), \ldots, (x_n, y_n)$. As before, write

$$\mathbf{x} = \begin{bmatrix} x_1 \\ x_2 \\ \vdots \\ x_n \end{bmatrix} \quad \mathbf{y} = \begin{bmatrix} y_1 \\ y_2 \\ \vdots \\ y_n \end{bmatrix} \quad \text{and} \quad f(\mathbf{x}) = \begin{bmatrix} f(x_1) \\ f(x_2) \\ \vdots \\ f(x_n) \end{bmatrix}$$

For each x_i we have two values of the variable y, the observed value y_i, and the computed value $f(x_i)$. The problem is to choose $f(x)$—that is, choose $r_0, r_1, \ldots, r_m$—such that the $f(x_i)$ are as close as possible to the y_i. Again we define "as close as possible" by the least squares condition: We choose the r_i such that

$$\|\mathbf{y} - f(\mathbf{x})\|^2 = [y_1 - f(x_1)]^2 + [y_2 - f(x_2)]^2 + \cdots + [y_n - f(x_n)]^2$$

is as small as possible.

Definition 5.15 *A polynomial $f(x)$ satisfying this condition is called a* **least squares approximating polynomial** *of degree m for the given data pairs.*

If we write

$$M = \begin{bmatrix} 1 & x_1 & x_1^2 & \cdots & x_1^m \\ 1 & x_2 & x_2^2 & \cdots & x_2^m \\ \vdots & \vdots & \vdots & & \vdots \\ 1 & x_n & x_n^2 & \cdots & x_n^m \end{bmatrix} \quad \text{and} \quad \mathbf{r} = \begin{bmatrix} r_0 \\ r_1 \\ \vdots \\ r_m \end{bmatrix}$$

we see that $f(\mathbf{x}) = M\mathbf{r}$. Hence we want to find R such that $\|\mathbf{y} - M\mathbf{r}\|^2$ is as small as possible; that is, we want a best approximation $\mathbf{z}$ to the system $M\mathbf{r} = \mathbf{y}$. Theorem 1 gives the first part of Theorem 3.

Theorem 3

Let n data pairs $(x_1, y_1), (x_2, y_2), \ldots, (x_n, y_n)$ be given, and write

$$\mathbf{y} = \begin{bmatrix} y_1 \\ y_2 \\ \vdots \\ y_n \end{bmatrix} \quad M = \begin{bmatrix} 1 & x_1 & x_1^2 & \cdots & x_1^m \\ 1 & x_2 & x_2^2 & \cdots & x_2^m \\ \vdots & \vdots & \vdots & & \vdots \\ 1 & x_n & x_n^2 & \cdots & x_n^m \end{bmatrix} \quad \mathbf{z} = \begin{bmatrix} z_0 \\ z_1 \\ \vdots \\ z_m \end{bmatrix}$$

1. If $\mathbf{z}$ is any solution to the normal equations

$$(M^T M)\mathbf{z} = M^T \mathbf{y}$$

then the polynomial

$$z_0 + z_1 x + z_2 x^2 + \cdots + z_m x^m$$

is a least squares approximating polynomial of degree m for the given data pairs.

2. If at least $m + 1$ of the numbers $x_1, x_2, \ldots, x_n$ are distinct (so $n \geq m + 1$), the matrix $M^T M$ is invertible and $\mathbf{z}$ is uniquely determined by

$$\mathbf{z} = (M^T M)^{-1} M^T \mathbf{y}$$

PROOF

It remains to prove (2), and for that we show that the columns of M are linearly independent (Theorem 3 Section 5.4). Suppose a linear combination of the columns vanishes:

$$r_0 \begin{bmatrix} 1 \\ 1 \\ \vdots \\ 1 \end{bmatrix} + r_1 \begin{bmatrix} x_1 \\ x_2 \\ \vdots \\ x_n \end{bmatrix} + \cdots + r_m \begin{bmatrix} x_1^m \\ x_2^m \\ \vdots \\ x_n^m \end{bmatrix} = \begin{bmatrix} 0 \\ 0 \\ \vdots \\ 0 \end{bmatrix}$$

If we write $q(x) = r_0 + r_1 x + \cdots + r_m x^m$, equating coefficients shows that $q(x_1) = q(x_2) = \cdots = q(x_n) = 0$. Hence $q(x)$ is a polynomial of degree m with at least $m + 1$ distinct roots, so $q(x)$ must be the zero polynomial (see Appendix D or Theorem 4 Section 6.5). Thus $r_0 = r_1 = \cdots = r_m = 0$ as required.

SECTION 5.6 Best Approximation and Least Squares

EXAMPLE 4

Find the least squares approximating quadratic $y = z_0 + z_1 x + z_2 x^2$ for the following data points.

$$(-3, 3), (-1, 1), (0, 1), (1, 2), (3, 4)$$

Solution ▶ This is an instance of Theorem 3 with $m = 2$. Here

$$\mathbf{y} = \begin{bmatrix} 3 \\ 1 \\ 1 \\ 2 \\ 4 \end{bmatrix} \quad M = \begin{bmatrix} 1 & -3 & 9 \\ 1 & -1 & 1 \\ 1 & 0 & 0 \\ 1 & 1 & 1 \\ 1 & 3 & 9 \end{bmatrix}$$

Hence,

$$M^T M = \begin{bmatrix} 1 & 1 & 1 & 1 & 1 \\ -3 & -1 & 0 & 1 & 3 \\ 9 & 1 & 0 & 1 & 9 \end{bmatrix} \begin{bmatrix} 1 & -3 & 9 \\ 1 & -1 & 1 \\ 1 & 0 & 0 \\ 1 & 1 & 1 \\ 1 & 3 & 9 \end{bmatrix} = \begin{bmatrix} 5 & 0 & 20 \\ 0 & 20 & 0 \\ 20 & 0 & 164 \end{bmatrix}$$

$$M^T \mathbf{y} = \begin{bmatrix} 1 & 1 & 1 & 1 & 1 \\ -3 & -1 & 0 & 1 & 3 \\ 9 & 1 & 0 & 1 & 9 \end{bmatrix} \begin{bmatrix} 3 \\ 1 \\ 1 \\ 2 \\ 4 \end{bmatrix} = \begin{bmatrix} 11 \\ 4 \\ 66 \end{bmatrix}$$

The normal equations for $\mathbf{z}$ are

$$\begin{bmatrix} 5 & 0 & 20 \\ 0 & 20 & 0 \\ 20 & 0 & 164 \end{bmatrix} \mathbf{z} = \begin{bmatrix} 11 \\ 4 \\ 66 \end{bmatrix} \quad \text{whence } \mathbf{z} = \begin{bmatrix} 1.15 \\ 0.20 \\ 0.26 \end{bmatrix}$$

This means that the least squares approximating quadratic for these data is $y = 1.15 + 0.20x + 0.26x^2$.

Other Functions

There is an extension of Theorem 3 that should be mentioned. Given data pairs $(x_1, y_1), (x_2, y_2), \ldots, (x_n, y_n)$, that theorem shows how to find a polynomial

$$f(x) = r_0 + r_1 x + \cdots + r_m x^m$$

such that $\|\mathbf{y} - f(\mathbf{x})\|^2$ is as small as possible, where $\mathbf{x}$ and $f(\mathbf{x})$ are as before. Choosing the appropriate polynomial $f(x)$ amounts to choosing the coefficients $r_0, r_1, \ldots, r_m$, and Theorem 3 gives a formula for the optimal choices. Here $f(x)$ is a linear combination of the functions $1, x, x^2, \ldots, x^m$ where the r_i are the coefficients, and this suggests applying the method to other functions. If $f_0(x), f_1(x), \ldots, f_m(x)$ are given functions, write

$$f(x) = r_0 f_0(x) + r_1 f_1(x) + \cdots + r_m f_m(x)$$

where the r_i are real numbers. Then the more general question is whether $r_0, r_1, \ldots, r_m$ can be found such that $\|\mathbf{y} - f(\mathbf{x})\|^2$ is as small as possible where

$$f(\mathbf{x}) = \begin{bmatrix} f(x_1) \\ f(x_2) \\ \vdots \\ f(x_m) \end{bmatrix}$$

Such a function $f(\mathbf{x})$ is called a **least squares best approximation** for these data pairs of the form $r_0 f_0(x) + r_1 f_1(x) + \cdots + r_m f_m(x)$, r_i in $\mathbb{R}$. The proof of Theorem 3 goes through to prove

Theorem 4

Let n data pairs $(x_1, y_1), (x_2, y_2), \ldots, (x_n, y_n)$ be given, and suppose that $m + 1$ functions $f_0(x), f_1(x), \ldots, f_m(x)$ are specified. Write

$$\mathbf{y} = \begin{bmatrix} y_1 \\ y_2 \\ \vdots \\ y_n \end{bmatrix} \quad M = \begin{bmatrix} f_0(x_1) & f_1(x_1) & \cdots & f_m(x_1) \\ f_0(x_2) & f_1(x_2) & \cdots & f_m(x_2) \\ \vdots & \vdots & & \vdots \\ f_0(x_n) & f_1(x_n) & \cdots & f_m(x_n) \end{bmatrix} \quad \mathbf{z} = \begin{bmatrix} z_1 \\ z_2 \\ \vdots \\ z_m \end{bmatrix}$$

(1) If $\mathbf{z}$ is any solution to the normal equations

$$(M^T M)\mathbf{z} = M^T \mathbf{y},$$

then the function

$$z_0 f_0(x) + z_1 f_1(x) + \cdots + z_m f_m(x)$$

is the best approximation for these data among all functions of the form $r_0 f_0(x) + r_1 f_1(x) + \cdots + r_m f_m(x)$ where the r_i are in $\mathbb{R}$.

(2) If $M^T M$ is invertible (that is, if $\text{rank}(M) = m + 1$), then $\mathbf{z}$ is uniquely determined; in fact, $\mathbf{z} = (M^T M)^{-1}(M^T \mathbf{y})$.

Clearly Theorem 4 contains Theorem 3 as a special case, but there is no simple test in general for whether $M^T M$ is invertible. Conditions for this to hold depend on the choice of the functions $f_0(x), f_1(x), \ldots, f_m(x)$.

EXAMPLE 5

Given the data pairs $(-1, 0)$, $(0, 1)$, and $(1, 4)$, find the least squares approximating function of the form $r_0 x + r_1 2^x$.

Solution ▶ The functions are $f_0(x) = x$ and $f_1(x) = 2^x$, so the matrix M is

$$M = \begin{bmatrix} f_0(x_1) & f_1(x_1) \\ f_0(x_2) & f_1(x_2) \\ f_0(x_3) & f_1(x_3) \end{bmatrix} = \begin{bmatrix} -1 & 2^{-1} \\ 0 & 2^0 \\ 1 & 2^1 \end{bmatrix} = \tfrac{1}{2} \begin{bmatrix} -2 & 1 \\ 0 & 2 \\ 2 & 4 \end{bmatrix}$$

In this case $M^T M = \tfrac{1}{4}\begin{bmatrix} 8 & 6 \\ 6 & 21 \end{bmatrix}$ is invertible, so the normal equations

$$\frac{1}{4}\begin{bmatrix} 8 & 6 \\ 6 & 21 \end{bmatrix} \mathbf{z} = \begin{bmatrix} 4 \\ 9 \end{bmatrix} \text{ have a unique solution } \mathbf{z} = \frac{1}{11}\begin{bmatrix} 10 \\ 16 \end{bmatrix}$$

Hence the best-fitting function of the form $r_0 x + r_1 2^x$ is $\overline{f}(x) = \frac{10}{11}x + \frac{16}{11}2^x$.

Note that $\overline{f}(\mathbf{x}) = \begin{bmatrix} \overline{f}(-1) \\ \overline{f}(0) \\ \overline{f}(1) \end{bmatrix} = \begin{bmatrix} \frac{-2}{11} \\ \frac{16}{11} \\ \frac{42}{11} \end{bmatrix}$, compared with $\mathbf{y} = \begin{bmatrix} 0 \\ 1 \\ 4 \end{bmatrix}$.

EXERCISES 5.6

1. Find the best approximation to a solution of each of the following systems of equations.

   (a) $\quad x + y - z = 5$
   $\quad\;\; 2x - y + 6z = 1$
   $\quad\;\; 3x + 2y - z = 6$
   $\quad\;\; -x + 4y + z = 0$

   ◆(b) $\;\; 3x + y + z = 6$
   $\quad\;\; 2x + 3y - z = 1$
   $\quad\;\; 2x - y + z = 0$
   $\quad\;\; 3x - 3y + 3z = 8$

2. Find the least squares approximating line $y = z_0 + z_1 x$ for each of the following sets of data points.

   (a) $(1, 1), (3, 2), (4, 3), (6, 4)$

   ◆(b) $(2, 4), (4, 3), (7, 2), (8, 1)$

   (c) $(-1, -1), (0, 1), (1, 2), (2, 4), (3, 6)$

   ◆(d) $(-2, 3), (-1, 1), (0, 0), (1, -2), (2, -4)$

3. Find the least squares approximating quadratic $y = z_0 + z_1 x + z_2 x^2$ for each of the following sets of data points.

   (a) $(0, 1), (2, 2), (3, 3), (4, 5)$

   ◆(b) $(-2, 1), (0, 0), (3, 2), (4, 3)$

4. Find a least squares approximating function of the form $r_0 x + r_1 x^2 + r_2 2^x$ for each of the following sets of data pairs.

   (a) $(-1, 1), (0, 3), (1, 1), (2, 0)$

   ◆(b) $(0, 1), (1, 1), (2, 5), (3, 10)$

5. Find the least squares approximating function of the form $r_0 + r_1 x^2 + r_2 \sin\frac{\pi x}{2}$ for each of the following sets of data pairs.

   (a) $(0, 3), (1, 0), (1, -1), (-1, 2)$

   ◆(b) $(-1, \frac{1}{2}), (0, 1), (2, 5), (3, 9)$

6. If M is a square invertible matrix, show that $\mathbf{z} = M^{-1}\mathbf{y}$ (in the notation of Theorem 3).

◆7. Newton's laws of motion imply that an object dropped from rest at a height of 100 metres will be at a height $s = 100 - \frac{1}{2}gt^2$ metres t seconds later, where g is a constant called the acceleration due to gravity. The values of s and t given in the table are observed. Write $x = t^2$, find the least squares approximating line $s = a + bx$ for these data, and use b to estimate g.

Then find the least squares approximating quadratic $s = a_0 + a_1 t + a_2 t^2$ and use the value of a_2 to estimate g.

t	1	2	3
s	95	80	56

8. A naturalist measured the heights y_i (in metres) of several spruce trees with trunk diameters x_i (in centimetres). The data are as given in the table. Find the least squares approximating line for these data and use it to estimate the height of a spruce tree with a trunk of diameter 10 cm.

x_i	5	7	8	12	13	16
y_i	2	3.3	4	7.3	7.9	10.1

◆9. The yield y of wheat in bushels per acre appears to be a linear function of the number of days x_1 of sunshine, the number of inches x_2 of rain, and the number of pounds x_3 of fertilizer applied per acre. Find the best fit to the data in the table by an equation of the form $y = r_0 + r_1 x_1 + r_2 x_2 + r_3 x_3$. [Hint: If a calculator for inverting $A^T A$ is not available, the inverse is given in the answer.]

y	x_1	x_2	x_3
28	50	18	10
30	40	20	16
21	35	14	10
23	40	12	12
23	30	16	14

10. (a) Use $m = 0$ in Theorem 3 to show that the best-fitting horizontal line $y = a_0$ through the data points $(x_1, y_1), \ldots, (x_m, y_n)$ is $y = \frac{1}{n}(y_1 + y_2 + \cdots + y_n)$, the average of the y coordinates.

♦(b) Deduce the conclusion in (a) without using Theorem 3.

11. Assume $n = m + 1$ in Theorem 3 (so M is square). If the x_i are distinct, use Theorem 6 Section 3.2 to show that M is invertible. Deduce that $\mathbf{z} = M^{-1}\mathbf{y}$ and that the least squares polynomial is the interpolating polynomial (Theorem 6 Section 3.2) and actually passes through all the data points.

12. Let A be any $m \times n$ matrix and write $K = \{\mathbf{x} \mid A^T A \mathbf{x} = \mathbf{0}\}$. Let B be an m-column. Show that, if $\mathbf{z}$ is an n-column such that $\|\mathbf{b} - A\mathbf{z}\|$ is minimal, then *all* such vectors have the form $\mathbf{z} + \mathbf{x}$ for some $\mathbf{x}$ in K. [*Hint:* $\|\mathbf{b} - A\mathbf{y}\|$ is minimal if and only if $A^T A \mathbf{y} = A^T \mathbf{b}$.]

13. Given the situation in Theorem 4, write
$$f(x) = r_0 p_0(x) + r_1 p_1(x) + \cdots + r_m p_m(x)$$
Suppose that $f(x)$ has at most k roots for any choice of the coefficients $r_0, r_1, \ldots, r_m$, not all zero.

(a) Show that $M^T M$ is invertible if at least $k + 1$ of the x_i are distinct.

♦(b) If at least two of the x_i are distinct, show that there is always a best approximation of the form $r_0 + r_1 e^x$.

(c) If at least three of the x_i are distinct, show that there is always a best approximation of the form $r_0 + r_1 x + r_2 e^x$. [Calculus is needed.]

14. If A is an $m \times n$ matrix, it can be proved that there exists a unique $n \times m$ matrix $A^\#$ satisfying the following four conditions: $AA^\# A = A$; $A^\# A A^\# = A^\#$; $AA^\#$ and $A^\# A$ are symmetric. The matrix $A^\#$ is called the **generalized inverse** of A, or the **Moore-Penrose** inverse.

(a) If A is square and invertible, show that $A^\# = A^{-1}$.

(b) If rank $A = m$, show that $A^\# = A^T(AA^T)^{-1}$.

(c) If rank $A = n$, show that $A^\# = (A^T A)^{-1} A^T$.

SECTION 5.7 An Application to Correlation and Variance

Suppose the heights $h_1, h_2, \ldots, h_n$ of n men are measured. Such a data set is called a **sample** of the heights of all the men in the population under study, and various questions are often asked about such a sample: What is the average height in the sample? How much variation is there in the sample heights, and how can it be measured? What can be inferred from the sample about the heights of all men in the population? How do these heights compare to heights of men in neighbouring countries? Does the prevalence of smoking affect the height of a man?

The analysis of samples, and of inferences that can be drawn from them, is a subject called *mathematical statistics*, and an extensive body of information has been developed to answer many such questions. In this section we will describe a few ways that linear algebra can be used.

It is convenient to represent a sample $\{x_1, x_2, \ldots, x_n\}$ as a **sample vector**[15] $\mathbf{x} = [x_1 \; x_2 \; \cdots \; x_n]$ in $\mathbb{R}^n$. This being done, the dot product in $\mathbb{R}^n$ provides a convenient tool to study the sample and describe some of the statistical concepts

---

15 We write vectors in $\mathbb{R}^n$ as row matrices, for convenience.

SECTION 5.7 An Application to Correlation and Variance

related to it. The most widely known statistic for describing a data set is the **sample mean** $\bar{x}$ defined by[16]

$$\bar{x} = \tfrac{1}{n}(x_1 + x_2 + \cdots + x_n) = \tfrac{1}{n}\sum_{i=1}^{n} x_i.$$

The mean $\bar{x}$ is "typical" of the sample values x_i, but may not itself be one of them. The number $x_i - \bar{x}$ is called the **deviation** of x_i from the mean $\bar{x}$. The deviation is positive if $x_i > \bar{x}$ and it is negative if $x_i < \bar{x}$. Moreover, the sum of these deviations is zero:

$$\sum_{i=1}^{n}(x_i - \bar{x}) = \left(\sum_{i=1}^{n} x_i\right) - n\bar{x} = n\bar{x} - n\bar{x} = 0. \qquad (*)$$

This is described by saying that the sample mean $\bar{x}$ is *central* to the sample values x_i.

If the mean $\bar{x}$ is subtracted from each data value x_i, the resulting data $x_i - \bar{x}$ are said to be **centred**. The corresponding data vector is

$$\mathbf{x}_c = [x_1 - \bar{x} \quad x_2 - \bar{x} \quad \cdots \quad x_n - \bar{x}]$$

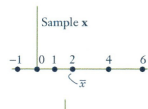

Sample x

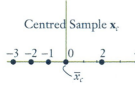

Centred Sample $\mathbf{x}_c$

and (*) shows that the mean $\bar{x}_c = 0$. For example, the sample $\mathbf{x} = [-1\ 0\ 1\ 4\ 6]$ is plotted in the first diagram. The mean is $\bar{x} = 2$, and the centred sample $\mathbf{x}_c = [-3\ -2\ -1\ 2\ 4]$ is also plotted. Thus, the effect of centring is to shift the data by an amount $\bar{x}$ (to the left if $\bar{x}$ is positive) so that the mean moves to 0.

Another question that arises about samples is how much variability there is in the sample $\mathbf{x} = [x_1\ x_2\ \cdots\ x_n]$; that is, how widely are the data "spread out" around the sample mean $\bar{x}$. A natural measure of variability would be the sum of the deviations of the x_i about the mean, but this sum is zero by (*); these deviations cancel out. To avoid this cancellation, statisticians use the *squares* $(x_i - \bar{x})^2$ of the deviations as a measure of variability. More precisely, they compute a statistic called the **sample variance** s_x^2, defined[17] as follows:

$$s_x^2 = \tfrac{1}{n-1}[(x_1 - \bar{x})^2 + (x_2 - \bar{x})^2 + \cdots + (x_n - \bar{x})^2] = \tfrac{1}{n-1}\sum_{i=1}^{n}(x_i - \bar{x})^2.$$

The sample variance will be large if there are many x_i at a large distance from the mean $\bar{x}$, and it will be small if all the x_i are tightly clustered about the mean. The variance is clearly nonnegative (hence the notation s_x^2), and the square root s_x of the variance is called the **sample standard deviation**.

The sample mean and variance can be conveniently described using the dot product. Let

$$\mathbf{1} = [1\ 1\ \cdots\ 1]$$

denote the row with every entry equal to 1. If $\mathbf{x} = [x_1\ x_2\ \cdots\ x_n]$, then $\mathbf{x} \cdot \mathbf{1} = x_1 + x_2 + \cdots + x_n$, so the sample mean is given by the formula

$$\bar{x} = \tfrac{1}{n}(\mathbf{x} \cdot \mathbf{1}).$$

Moreover, remembering that $\bar{x}$ is a scalar, we have $\bar{x}\mathbf{1} = [\bar{x}\ \bar{x}\ \cdots\ \bar{x}]$, so the centred sample vector $\mathbf{x}_c$ is given by

$$\mathbf{x}_c = \mathbf{x} - \bar{x}\mathbf{1} = [x_1 - \bar{x} \quad x_2 - \bar{x} \quad \cdots \quad x_n - \bar{x}].$$

Thus we obtain a formula for the sample variance:

$$s_x^2 = \tfrac{1}{n-1}\|\mathbf{x}_c\|^2 = \tfrac{1}{n-1}\|\mathbf{x} - \bar{x}\mathbf{1}\|^2.$$

Linear algebra is also useful for comparing two different samples. To illustrate how, consider two examples.

---

[16] The mean is often called the "average" of the sample values x_i, but statisticians use the term "mean".

[17] Since there are n sample values, it seems more natural to divide by n here, rather than by $n - 1$. The reason for using $n - 1$ is that then the sample variance s_x^2 provides a better estimate of the variance of the entire population from which the sample was drawn.

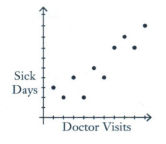
Doctor Visits

The following table represents the number of sick days at work per year and the yearly number of visits to a physician for 10 individuals.

Individual	1	2	3	4	5	6	7	8	9	10
Doctor visits	2	6	8	1	5	10	3	9	7	4
Sick days	2	4	8	3	5	9	4	7	7	2

The data are plotted in the **scatter diagram** where it is evident that, roughly speaking, the more visits to the doctor the more sick days. This is an example of a *positive correlation* between sick days and doctor visits.

Now consider the following table representing the daily doses of vitamin C and the number of sick days.

Individual	1	2	3	4	5	6	7	8	9	10
Vitamin C	1	5	7	0	4	9	2	8	6	3
Sick days	5	2	2	6	2	1	4	3	2	5

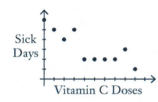
Vitamin C Doses

The scatter diagram is plotted as shown and it appears that the more vitamin C taken, the fewer sick days. In this case there is a *negative correlation* between daily vitamin C and sick days.

In both these situations, we have **paired samples**, that is observations of two variables are made for ten individuals: doctor visits and sick days in the first case; daily vitamin C and sick days in the second case. The scatter diagrams point to a relationship between these variables, and there is a way to use the sample to compute a number, called the correlation coefficient, that measures the degree to which the variables are associated.

To motivate the definition of the correlation coefficient, suppose two paired samples $\mathbf{x} = [x_1 \; x_2 \; \cdots \; x_n]$, and $\mathbf{y} = [y_1 \; y_2 \; \cdots \; y_n]$ are given and consider the centred samples

$$\mathbf{x}_c = [x_1 - \bar{x} \quad x_2 - \bar{x} \quad \cdots \quad x_n - \bar{x}] \quad \text{and} \quad \mathbf{y}_c = [y_1 - \bar{y} \quad y_2 - \bar{y} \quad \cdots \quad y_n - \bar{y}]$$

If x_k is large among the x_i's, then the deviation $x_k - \bar{x}$ will be positive; and $x_k - \bar{x}$ will be negative if x_k is small among the x_i's. The situation is similar for $\mathbf{y}$, and the following table displays the sign of the quantity $(x_i - \bar{x})(y_k - \bar{y})$ in all four cases:

Sign of $(x_i - \bar{x})(y_k - \bar{y})$:

	x_i large	x_i small
y_i large	positive	negative
y_i small	negative	positive

Intuitively, if $\mathbf{x}$ and $\mathbf{y}$ are positively correlated, then two things happen:

1. Large values of the x_i tend to be associated with large values of the y_i, and
2. Small values of the x_i tend to be associated with small values of the y_i.

It follows from the table that, if $\mathbf{x}$ and $\mathbf{y}$ are positively correlated, then the dot product

$$\mathbf{x}_c \cdot \mathbf{y}_c = \sum_{i=1}^{n}(x_i - \bar{x})(y_i - \bar{y})$$

is positive. Similarly $\mathbf{x}_c \cdot \mathbf{y}_c$ is negative if $\mathbf{x}$ and $\mathbf{y}$ are negatively correlated. With this in mind, the **sample correlation coefficient**[18] r is defined by

---

[18] The idea of using a single number to measure the degree of relationship between different variables was pioneered by Francis Galton (1822–1911). He was studying the degree to which characteristics of an offspring relate to those of its parents. The idea was refined by Karl Pearson (1857–1936) and r is often referred to as the Pearson correlation coefficient.

SECTION 5.7 An Application to Correlation and Variance

$$r = r(\mathbf{x}, \mathbf{y}) = \frac{\mathbf{x}_c \cdot \mathbf{y}_c}{\|\mathbf{x}_c\| \|\mathbf{y}_c\|}.$$

Bearing the situation in $\mathbb{R}^3$ in mind, r is the cosine of the "angle" between the vectors $\mathbf{x}_c$ and $\mathbf{y}_c$, and so we would expect it to lie between -1 and 1. Moreover, we would expect r to be near 1 (or -1) if these vectors were pointing in the same (opposite) direction, that is the "angle" is near zero (or π).

This is confirmed by Theorem 1 below, and it is also borne out in the examples above. If we compute the correlation between sick days and visits to the physician (in the first scatter diagram above) the result is $r = 0.90$ as expected. On the other hand, the correlation between daily vitamin C doses and sick days (second scatter diagram) is $r = -0.84$.

However, a word of caution is in order here. We *cannot* conclude from the second example that taking more vitamin C will reduce the number of sick days at work. The (negative) correlation may arise because of some third factor that is related to both variables. For example, case it may be that less healthy people are inclined to take more vitamin C. Correlation does *not* imply causation. Similarly, the correlation between sick days and visits to the doctor does not mean that having many sick days *causes* more visits to the doctor. A correlation between two variables may point to the existence of other underlying factors, but it does not necessarily mean that there is a causality relationship between the variables.

Our discussion of the dot product in $\mathbb{R}^n$ provides the basic properties of the correlation coefficient:

Theorem 1

Let $\mathbf{x} = [x_1 \ x_2 \ \cdots \ x_n]$ and $\mathbf{y} = [y_1 \ y_2 \ \cdots \ y_n]$ be (nonzero) paired samples, and let $r = r(\mathbf{x}, \mathbf{y})$ denote the correlation coefficient. Then:

1. $-1 \leq r \leq 1$.
2. $r = 1$ if and only if there exist a and $b > 0$ such that $y_i = a + bx_i$ for each i.
3. $r = -1$ if and only if there exist a and $b < 0$ such that $y_i = a + bx_i$ for each i.

PROOF

The Cauchy inequality (Theorem 2 Section 5.3) proves (1), and also shows that $r = \pm 1$ if and only if one of $\mathbf{x}_c$ and $\mathbf{y}_c$ is a scalar multiple of the other. This in turn holds if and only if $\mathbf{y}_c = b\mathbf{x}_c$ for some $b \neq 0$, and it is easy to verify that $r = 1$ when $b > 0$ and $r = -1$ when $b < 0$.

Finally, $\mathbf{y}_c = b\mathbf{x}_c$ means $y_1 - \bar{y} = b(x_1 - \bar{x})$ for each i; that is, $y_i = a + bx_i$ where $a = \bar{y} - b\bar{x}$. Conversely, if $y_i = a + bx_i$, then $\bar{y} = a + b\bar{x}$ (verify), so $y_1 - \bar{y} = (a + bx_i) - (a + b\bar{x}) = b(x_1 - \bar{x})$ for each i. In other words, $\mathbf{y}_c = b\mathbf{x}_c$. This completes the proof.

Properties (2) and (3) in Theorem 1 show that $r(\mathbf{x}, \mathbf{y}) = 1$ means that there is a linear relation with *positive* slope between the paired data (so large x values are paired with large y values). Similarly, $r(\mathbf{x}, \mathbf{y}) = -1$ means that there is a linear relation with *negative* slope between the paired data (so small x values are paired with small y values). This is borne out in the two scatter diagrams above.

We conclude by using the dot product to derive some useful formulas for computing variances and correlation coefficients. Given samples $\mathbf{x} = [x_1\ x_2\ \cdots\ x_n]$, and $\mathbf{y} = [y_1\ y_2\ \cdots\ y_n]$, the key observation is the following formula:

$$\mathbf{x}_c \cdot \mathbf{y}_c = \mathbf{x} \cdot \mathbf{y} - n\bar{x}\bar{y}. \qquad (**)$$

Indeed, remembering that $\bar{x}$ and $\bar{y}$ are scalars:

$$\begin{aligned}
\mathbf{x}_c \cdot \mathbf{y}_c &= (\mathbf{x} - \bar{x}\mathbf{1}) \cdot (\mathbf{y} - \bar{y}\mathbf{1}) \\
&= \mathbf{x} \cdot \mathbf{y} - \mathbf{x} \cdot (\bar{y}\mathbf{1}) - (\bar{x}\mathbf{1}) \cdot \mathbf{y} + (\bar{x}\mathbf{1}) \cdot (\bar{y}\mathbf{1}) \\
&= \mathbf{x} \cdot \mathbf{y} - \bar{y}(\mathbf{x} \cdot \mathbf{1}) - \bar{x}(\mathbf{1} \cdot \mathbf{y}) + \bar{x}\bar{y}(\mathbf{1} \cdot \mathbf{1}) \\
&= \mathbf{x} \cdot \mathbf{y} - \bar{y}(n\bar{x}) - \bar{x}(n\bar{y}) + \bar{x}\bar{y}(n) \\
&= \mathbf{x} \cdot \mathbf{y} - n\bar{x}\bar{y}.
\end{aligned}$$

Taking $\mathbf{y} = \mathbf{x}$ in $(**)$ gives a formula for the variance $s_x^2 = \frac{1}{n-1}\|\mathbf{x}_c\|^2$ of $\mathbf{x}$.

Variance Formula

If x is a sample vector, then $s_x^2 = \frac{1}{n-1}(\|\mathbf{x}_c\|^2 - n\bar{x}^2)$.

We also get a convenient formula for the correlation coefficient,

$$r = r(\mathbf{x}, \mathbf{y}) = \frac{\mathbf{x}_c \cdot \mathbf{y}_c}{\|\mathbf{x}_c\|\,\|\mathbf{y}_c\|}.$$

Moreover, $(**)$ and the fact that $s_x^2 = \frac{1}{n-1}\|\mathbf{x}_c\|^2$ give:

Correlation Formula

If $\mathbf{x}$ and $\mathbf{y}$ are sample vectors, then

$$r = r(\mathbf{x}, \mathbf{y}) = \frac{\mathbf{x} \cdot \mathbf{y} - n\bar{x}\bar{y}}{(n-1)s_x s_y}.$$

Finally, we give a method that simplifies the computations of variances and correlations.

Data Scaling

Let $\mathbf{x} = [x_1\ x_2\ \cdots\ x_n]$ and $\mathbf{y} = [y_1\ y_2\ \cdots\ y_n]$ be sample vectors. Given constants a, b, c, and d, consider new samples $\mathbf{z} = [z_1\ z_2\ \cdots\ z_n]$ and $\mathbf{w} = [w_1\ w_2\ \cdots\ w_n]$ where $z_i = a + bx_i$ for each i and $w_i = c + dy_i$ for each i. Then:

(a) $\bar{z} = a + b\bar{x}$.

(b) $s_z^2 = b^2 s_x^2$, so $s_z = |b|s_x$.

(c) If b and d have the same sign, then $r(\mathbf{x}, \mathbf{y}) = r(\mathbf{z}, \mathbf{w})$.

The verification is left as an exercise.

For example, if $\mathbf{x} = [101\ 98\ 103\ 99\ 100\ 97]$, subtracting 100 yields $\mathbf{z} = [1\ -2\ 3\ -1\ 0\ -3]$. A routine calculation shows that $\bar{z} = -\frac{1}{3}$ and $s_z^2 = \frac{14}{3}$, so $\bar{x} = 100 - \frac{1}{3} = 99.67$, and $s_x^2 = \frac{14}{3} = 4.67$.

EXERCISES 5.7

1. The following table gives IQ scores for 10 fathers and their eldest sons. Calculate the means, the variances, and the correlation coefficient r. (The data scaling formula is useful.)

	1	2	3	4	5	6	7	8	9	10
Father's IQ	140	131	120	115	110	106	100	95	91	86
Son's IQ	130	138	110	99	109	120	105	99	100	94

◆2. The following table gives the number of years of education and the annual income (in thousands) of 10 individuals. Find the means, the variances, and the correlation coefficient. (Again the data scaling formula is useful.)

Individual	1	2	3	4	5	6	7	8	9	10
Years of education	12	16	13	18	19	12	18	19	12	14
Yearly income (1000's)	31	48	35	28	55	40	39	60	32	35

3. If $\mathbf{x}$ is a sample vector, and $\mathbf{x}_c$ is the centred sample, show that $\bar{x}_c = 0$ and the standard deviation of $\mathbf{x}_c$ is s_x.

4. Prove the data scaling formulas found on page 292: (a), ◆(b), and (c).

SUPPLEMENTARY EXERCISES FOR CHAPTER 5

1. In each case either show that the statement is true or give an example showing that it is false. Throughout, $\mathbf{x}, \mathbf{y}, \mathbf{z}, \mathbf{x}_1, \mathbf{x}_2, \ldots, \mathbf{x}_n$ denote vectors in $\mathbb{R}^n$.

   (a) If U is a subspace of $\mathbb{R}^n$ and $\mathbf{x} + \mathbf{y}$ is in U, then $\mathbf{x}$ and $\mathbf{y}$ are both in U.

   ◆(b) If U is a subspace of $\mathbb{R}^n$ and $r\mathbf{x}$ is in U, then $\mathbf{x}$ is in U.

   (c) If U is a nonempty set and $s\mathbf{x} + t\mathbf{y}$ is in U for any s and t whenever $\mathbf{x}$ and $\mathbf{y}$ are in U, then U is a subspace.

   ◆(d) If U is a subspace of $\mathbb{R}^n$ and $\mathbf{x}$ is in U, then $-\mathbf{x}$ is in U.

   (e) If $\{\mathbf{x}, \mathbf{y}\}$ is independent, then $\{\mathbf{x}, \mathbf{y}, \mathbf{x} + \mathbf{y}\}$ is independent.

   ◆(f) If $\{\mathbf{x}, \mathbf{y}, \mathbf{z}\}$ is independent, then $\{\mathbf{x}, \mathbf{y}\}$ is independent.

   (g) If $\{\mathbf{x}, \mathbf{y}\}$ is not independent, then $\{\mathbf{x}, \mathbf{y}, \mathbf{z}\}$ is not independent.

   ◆(h) If all of $\mathbf{x}_1, \mathbf{x}_2, \ldots, \mathbf{x}_n$ are nonzero, then $\{\mathbf{x}_1, \mathbf{x}_2, \ldots, \mathbf{x}_n\}$ is independent.

   (i) If one of $\mathbf{x}_1, \mathbf{x}_2, \ldots, \mathbf{x}_n$ is zero, then $\{\mathbf{x}_1, \mathbf{x}_2, \ldots, \mathbf{x}_n\}$ is not independent.

   ◆(j) If $a\mathbf{x} + b\mathbf{y} + c\mathbf{z} = \mathbf{0}$ where a, b, and c are in $\mathbb{R}$, then $\{\mathbf{x}, \mathbf{y}, \mathbf{z}\}$ is independent.

   (k) If $\{\mathbf{x}, \mathbf{y}, \mathbf{z}\}$ is independent, then $a\mathbf{x} + b\mathbf{y} + c\mathbf{z} = \mathbf{0}$ for some a, b, and c in $\mathbb{R}$.

   ◆(l) If $\{\mathbf{x}_1, \mathbf{x}_2, \ldots, \mathbf{x}_n\}$ is not independent, then $t_1\mathbf{x}_1 + t_2\mathbf{x}_2 + \cdots + t_n\mathbf{x}_n = \mathbf{0}$ for t_i in $\mathbb{R}$ not all zero.

   (m) If $\{\mathbf{x}_1, \mathbf{x}_2, \ldots, \mathbf{x}_n\}$ is independent, then $t_1\mathbf{x}_1 + t_2\mathbf{x}_2 + \cdots + t_n\mathbf{x}_n = \mathbf{0}$ for some t_i in $\mathbb{R}$.

   ◆(n) Every set of four non-zero vectors in $\mathbb{R}^4$ is a basis.

   (o) No basis of $\mathbb{R}^3$ can contain a vector with a component 0.

   ◆(p) $\mathbb{R}^3$ has a basis of the form $\{\mathbf{x}, \mathbf{x} + \mathbf{y}, \mathbf{y}\}$ where $\mathbf{x}$ and $\mathbf{y}$ are vectors.

   (q) Every basis of $\mathbb{R}^5$ contains one column of I_5.

   ◆(r) Every nonempty subset of a basis of $\mathbb{R}^3$ is again a basis of $\mathbb{R}^3$.

   (s) If $\{\mathbf{x}_1, \mathbf{x}_2, \mathbf{x}_3, \mathbf{x}_4\}$ and $\{\mathbf{y}_1, \mathbf{y}_2, \mathbf{y}_3, \mathbf{y}_4\}$ are bases of $\mathbb{R}^4$, then $\{\mathbf{x}_1 + \mathbf{y}_1, \mathbf{x}_2 + \mathbf{y}_2, \mathbf{x}_3 + \mathbf{y}_3, \mathbf{x}_4 + \mathbf{y}_4\}$ is also a basis of $\mathbb{R}^4$.

Vector Spaces

6

In this chapter we introduce vector spaces in full generality. The reader will notice some similarity with the discussion of the space $\mathbb{R}^n$ in Chapter 5. In fact much of the present material has been developed in that context, and there is some repetition. However, Chapter 6 deals with the notion of an *abstract* vector space, a concept that will be new to most readers. It turns out that there are many systems in which a natural addition and scalar multiplication are defined and satisfy the usual rules familiar from $\mathbb{R}^n$. The study of abstract vector spaces is a way to deal with all these examples simultaneously. The new aspect is that we are dealing with an abstract system in which *all we know* about the vectors is that they are objects that can be added and multiplied by a scalar and satisfy rules familiar from $\mathbb{R}^n$.

The novel thing is the *abstraction*. Getting used to this new conceptual level is facilitated by the work done in Chapter 5: First, the vector manipulations are familiar, giving the reader more time to become accustomed to the abstract setting; and, second, the mental images developed in the concrete setting of $\mathbb{R}^n$ serve as an aid to doing many of the exercises in Chapter 6.

The concept of a vector space was first introduced in 1844 by the German mathematician Hermann Grassmann (1809–1877), but his work did not receive the attention it deserved. It was not until 1888 that the Italian mathematician Guiseppe Peano (1858–1932) clarified Grassmann's work in his book *Calcolo Geometrico* and gave the vector space axioms in their present form. Vector spaces became established with the work of the Polish mathematician Stephan Banach (1892–1945), and the idea was finally accepted in 1918 when Hermann Weyl (1885–1955) used it in his widely read book *Raum-Zeit-Materie* ("Space-Time-Matter"), an introduction to the general theory of relativity.

SECTION 6.1 Examples and Basic Properties

Many mathematical entities have the property that they can be added and multiplied by a number. Numbers themselves have this property, as do $m \times n$ matrices: The sum of two such matrices is again $m \times n$ as is any scalar multiple of such a matrix. Polynomials are another familiar example, as are the geometric vectors in Chapter 4. It turns out that there are many other types of mathematical objects that can be added and multiplied by a scalar, and the general study of such systems is introduced in this chapter. Remarkably, much of what we could say in Chapter 5 about the dimension of subspaces in $\mathbb{R}^n$ can be formulated in this generality.

SECTION 6.1 Examples and Basic Properties

Definition 6.1

*A **vector space** consists of a nonempty set V of objects (called **vectors**) that can be added, that can be multiplied by a real number (called a **scalar** in this context), and for which certain axioms hold.[1] If $\mathbf{v}$ and $\mathbf{w}$ are two vectors in V, their sum is expressed as $\mathbf{v} + \mathbf{w}$, and the scalar product of $\mathbf{v}$ by a real number a is denoted as $a\mathbf{v}$. These operations are called **vector addition** and **scalar multiplication**, respectively, and the following axioms are assumed to hold.*

Axioms for vector addition

A1. If $\mathbf{u}$ and $\mathbf{v}$ are in V, then $\mathbf{u} + \mathbf{v}$ is in V.

A2. $\mathbf{u} + \mathbf{v} = \mathbf{v} + \mathbf{u}$ for all $\mathbf{u}$ and $\mathbf{v}$ in V.

A3. $\mathbf{u} + (\mathbf{v} + \mathbf{w}) = (\mathbf{u} + \mathbf{v}) + \mathbf{w}$ for all $\mathbf{u}, \mathbf{v},$ and $\mathbf{w}$ in V.

A4. An element $\mathbf{0}$ in V exists such that $\mathbf{v} + \mathbf{0} = \mathbf{v} = \mathbf{0} + \mathbf{v}$ for every $\mathbf{v}$ in V.

A5. For each $\mathbf{v}$ in V, an element $-\mathbf{v}$ in V exists such that $-\mathbf{v} + \mathbf{v} = \mathbf{0}$ and $\mathbf{v} + (-\mathbf{v}) = \mathbf{0}$.

Axioms for scalar multiplication

S1. If $\mathbf{v}$ is in V, then $a\mathbf{v}$ is in V for all a in $\mathbb{R}$.

S2. $a(\mathbf{v} + \mathbf{w}) = a\mathbf{v} + a\mathbf{w}$ for all $\mathbf{v}$ and $\mathbf{w}$ in V and all a in $\mathbb{R}$.

S3. $(a + b)\mathbf{v} = a\mathbf{v} + b\mathbf{v}$ for all $\mathbf{v}$ in V and all a and b in $\mathbb{R}$.

S4. $a(b\mathbf{v}) = (ab)\mathbf{v}$ for all $\mathbf{v}$ in V and all a and b in $\mathbb{R}$.

S5. $1\mathbf{v} = \mathbf{v}$ for all $\mathbf{v}$ in V.

The content of axioms A1 and S1 is described by saying that V is **closed** under vector addition and scalar multiplication. The element $\mathbf{0}$ in axiom A4 is called the **zero vector**, and the vector $-\mathbf{v}$ in axiom A5 is called the **negative** of $\mathbf{v}$.

The rules of matrix arithmetic, when applied to $\mathbb{R}^n$, give

EXAMPLE 1

$\mathbb{R}^n$ is a vector space using matrix addition and scalar multiplication.[2]

It is important to realize that, in a general vector space, the vectors need not be n-tuples as in $\mathbb{R}^n$. They can be any kind of objects at all as long as the addition and scalar multiplication are defined and the axioms are satisfied. The following examples illustrate the diversity of the concept.

The space $\mathbb{R}^n$ consists of special types of matrices. More generally, let $\mathbf{M}_{mn}$ denote the set of all $m \times n$ matrices with real entries. Then Theorem 1 Section 2.1 gives:

EXAMPLE 2

The set $\mathbf{M}_{mn}$ of all $m \times n$ matrices is a vector space using matrix addition and scalar multiplication. The zero element in this vector space is the zero matrix of size $m \times n$, and the vector space negative of a matrix (required by axiom A5) is the usual matrix negative discussed in Section 2.1. Note that $\mathbf{M}_{mn}$ is just $\mathbb{R}^{mn}$ in different notation.

---

[1] The scalars will usually be real numbers, but they could be complex numbers, or elements of an algebraic system called a field. Another example is the field $\mathbb{Q}$ of rational numbers. We will look briefly at finite fields in Section 8.7.

[2] We will usually write the vectors in $\mathbb{R}^n$ as n-tuples. However, if it is convenient, we will sometimes denote them as rows or columns.

In Chapter 5 we identified many important subspaces of $\mathbb{R}^n$ such as im A and null A for a matrix A. These are all vector spaces.

EXAMPLE 3

Show that every subspace of $\mathbb{R}^n$ is a vector space in its own right using the addition and scalar multiplication of $\mathbb{R}^n$.

Solution ▶ Axioms A1 and S1 are two of the defining conditions for a subspace U of $\mathbb{R}^n$ (see Section 5.1). The other eight axioms for a vector space are inherited from $\mathbb{R}^n$. For example, if $\mathbf{x}$ and $\mathbf{y}$ are in U and a is a scalar, then $a(\mathbf{x} + \mathbf{y}) = a\mathbf{x} + a\mathbf{y}$ because $\mathbf{x}$ and $\mathbf{y}$ are in $\mathbb{R}^n$. This shows that axiom S2 holds for U; similarly, the other axioms also hold for U.

EXAMPLE 4

Let V denote the set of all ordered pairs (x, y) and define addition in V as in $\mathbb{R}^2$. However, define a new scalar multiplication in V by

$$a(x, y) = (ay, ax)$$

Determine if V is a vector space with these operations.

Solution ▶ Axioms A1 to A5 are valid for V because they hold for matrices. Also $a(x, y) = (ay, ax)$ is again in V, so axiom S1 holds. To verify axiom S2, let $\mathbf{v} = (x, y)$ and $\mathbf{w} = (x_1, y_1)$ be typical elements in V and compute

$$a(\mathbf{v} + \mathbf{w}) = a(x + x_1, y + y_1) = (a(y + y_1), a(x + x_1))$$
$$a\mathbf{v} + a\mathbf{w} = (ay, ax) + (ay_1, ax_1) = (ay + ay_1, ax + ax_1)$$

Because these are equal, axiom S2 holds. Similarly, the reader can verify that axiom S3 holds. However, axiom S4 fails because

$$a(b(x, y)) = a(by, bx) = (abx, aby)$$

need not equal $ab(x, y) = (aby, abx)$. Hence, V is *not* a vector space. (In fact, axiom S5 also fails.)

Sets of polynomials provide another important source of examples of vector spaces, so we review some basic facts. A **polynomial** in an indeterminate x is an expression

$$p(x) = a_0 + a_1 x + a_2 x^2 + \cdots + a_n x^n$$

where $a_0, a_1, a_2, \ldots, a_n$ are real numbers called the **coefficients** of the polynomial. If all the coefficients are zero, the polynomial is called the **zero polynomial** and is denoted simply as 0. If $p(x) \neq 0$, the highest power of x with a nonzero coefficient is called the **degree** of $p(x)$ denoted as deg $p(x)$. The coefficient itself is called the **leading coefficient** of $p(x)$. Hence deg$(3 + 5x) = 1$, deg$(1 + x + x^2) = 2$, and deg$(4) = 0$. (The degree of the zero polynomial is not defined.)

Let **P** denote the set of all polynomials and suppose that

$$p(x) = a_0 + a_1 x + a_2 x^2 + \cdots$$
$$q(x) = b_0 + b_1 x + b_2 x^2 + \cdots$$

SECTION 6.1 Examples and Basic Properties

are two polynomials in **P** (possibly of different degrees). Then $p(x)$ and $q(x)$ are called **equal** [written $p(x) = q(x)$] if and only if all the corresponding coefficients are equal—that is, $a_0 = b_0$, $a_1 = b_1$, $a_2 = b_2$, and so on. In particular, $a_0 + a_1x + a_2x^2 + \cdots = 0$ means that $a_0 = 0$, $a_1 = 0$, $a_2 = 0$, ..., and this is the reason for calling x an **indeterminate**. The set **P** has an addition and scalar multiplication defined on it as follows: if $p(x)$ and $q(x)$ are as before and a is a real number,

$$p(x) + q(x) = (a_0 + b_0) + (a_1 + b_1)x + (a_2 + b_2)x^2 + \cdots$$
$$ap(x) = aa_0 + (aa_1)x + (aa_2)x^2 + \cdots$$

Evidently, these are again polynomials, so **P** is closed under these operations, called **pointwise** addition and scalar multiplication. The other vector space axioms are easily verified, and we have

EXAMPLE 5

The set **P** of all polynomials is a vector space with the foregoing addition and scalar multiplication. The zero vector is the zero polynomial, and the negative of a polynomial $p(x) = a_0 + a_1x + a_2x^2 + \cdots$ is the polynomial $-p(x) = -a_0 - a_1x - a_2x^2 - \cdots$ obtained by negating all the coefficients.

There is another vector space of polynomials that will be referred to later.

EXAMPLE 6

Given $n \geq 1$, let $\mathbf{P}_n$ denote the set of all polynomials of degree at most n, together with the zero polynomial. That is

$$\mathbf{P}_n = \{a_0 + a_1x + a_2x^2 + \cdots + a_nx^n \mid a_0, a_1, a_2, ..., a_n \text{ in } \mathbb{R}\}.$$

Then $\mathbf{P}_n$ is a vector space. Indeed, sums and scalar multiples of polynomials in $\mathbf{P}_n$ are again in $\mathbf{P}_n$, and the other vector space axioms are inherited from **P**. In particular, the zero vector and the negative of a polynomial in $\mathbf{P}_n$ are the same as those in **P**.

If a and b are real numbers and $a < b$, the **interval** $[a, b]$ is defined to be the set of all real numbers x such that $a \leq x \leq b$. A (real-valued) **function** f on $[a, b]$ is a rule that associates to every number x in $[a, b]$ a real number denoted $f(x)$. The rule is frequently specified by giving a formula for $f(x)$ in terms of x. For example, $f(x) = 2^x$, $f(x) = \sin x$, and $f(x) = x^2 + 1$ are familiar functions. In fact, every polynomial $p(x)$ can be regarded as the formula for a function p.

The set of all functions on $[a, b]$ is denoted $\mathbf{F}[a, b]$. Two functions f and g in $\mathbf{F}[a, b]$ are **equal** if $f(x) = g(x)$ for every x in $[a, b]$, and we describe this by saying that f and g have the **same action**. Note that two polynomials are equal in **P** (defined prior to Example 5) if and only if they are equal as functions.

If f and g are two functions in $\mathbf{F}[a, b]$, and if r is a real number, define the sum $f + g$ and the scalar product rf by

$$(f + g)(x) = f(x) + g(x) \quad \text{for each } x \text{ in } [a, b]$$
$$(rf)(x) = rf(x) \quad \text{for each } x \text{ in } [a, b]$$

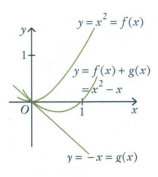

In other words, the action of $f + g$ upon x is to associate x with the number $f(x) + g(x)$, and rf associates x with $rf(x)$. The sum of $f(x) = x^2$ and $g(x) = -x$ is shown in the diagram. These operations on $\mathbf{F}[a, b]$ are called **pointwise addition and scalar multiplication** of functions and they are the usual operations familiar from elementary algebra and calculus.

EXAMPLE 7

The set $\mathbf{F}[a, b]$ of all functions on the interval $[a, b]$ is a vector space using pointwise addition and scalar multiplication. The zero function (in axiom A4), denoted 0, is the constant function defined by

$$0(x) = 0 \quad \text{for each } x \text{ in } [a, b]$$

The negative of a function f is denoted $-f$ and has action defined by

$$(-f)(x) = -f(x) \quad \text{for each } x \text{ in } [a, b]$$

Axioms A1 and S1 are clearly satisfied because, if f and g are functions on $[a, b]$, then $f + g$ and rf are again such functions. The verification of the remaining axioms is left as Exercise 14.

Other examples of vector spaces will appear later, but these are sufficiently varied to indicate the scope of the concept and to illustrate the properties of vector spaces to be discussed. With such a variety of examples, it may come as a surprise that a well-developed *theory* of vector spaces exists. That is, many properties can be shown to hold for *all* vector spaces and hence hold in every example. Such properties are called *theorems* and can be deduced from the axioms. Here is an important example.

Theorem 1

Cancellation
Let $\mathbf{u}$, $\mathbf{v}$, and $\mathbf{w}$ be vectors in a vector space V. If $\mathbf{v} + \mathbf{u} = \mathbf{v} + \mathbf{w}$, then $\mathbf{u} = \mathbf{w}$.

PROOF

We are given $\mathbf{v} + \mathbf{u} = \mathbf{v} + \mathbf{w}$. If these were numbers instead of vectors, we would simply subtract $\mathbf{v}$ from both sides of the equation to obtain $\mathbf{u} = \mathbf{w}$. This can be accomplished with vectors by adding $-\mathbf{v}$ to both sides of the equation. The steps (using only the axioms) are as follows:

$$\begin{aligned} \mathbf{v} + \mathbf{u} &= \mathbf{v} + \mathbf{w} & \\ -\mathbf{v} + (\mathbf{v} + \mathbf{u}) &= -\mathbf{v} + (\mathbf{v} + \mathbf{w}) & \text{(axiom A5)} \\ (-\mathbf{v} + \mathbf{v}) + \mathbf{u} &= (-\mathbf{v} + \mathbf{v}) + \mathbf{w} & \text{(axiom A3)} \\ \mathbf{0} + \mathbf{u} &= \mathbf{0} + \mathbf{w} & \text{(axiom A5)} \\ \mathbf{u} &= \mathbf{w} & \text{(axiom A4)} \end{aligned}$$

This is the desired conclusion.[3]

---

[3] Observe that none of the scalar multiplication axioms are needed here.

SECTION 6.1 Examples and Basic Properties

As with many good mathematical theorems, the technique of the proof of Theorem 1 is at least as important as the theorem itself. The idea was to mimic the well-known process of numerical subtraction in a vector space V as follows: To subtract a vector $\mathbf{v}$ from both sides of a vector equation, we added $-\mathbf{v}$ to both sides. With this in mind, we define **difference** $\mathbf{u} - \mathbf{v}$ of two vectors in V as

$$\mathbf{u} - \mathbf{v} = \mathbf{u} + (-\mathbf{v}).$$

We shall say that this vector is the result of having **subtracted** $\mathbf{v}$ from $\mathbf{u}$ and, as in arithmetic, this operation has the property given in Theorem 2.

Theorem 2

If $\mathbf{u}$ and $\mathbf{v}$ are vectors in a vector space V, the equation

$$\mathbf{x} + \mathbf{v} = \mathbf{u}$$

has one and only one solution $\mathbf{x}$ in V given by

$$\mathbf{x} = \mathbf{u} - \mathbf{v}.$$

PROOF

The difference $\mathbf{x} = \mathbf{u} - \mathbf{v}$ is indeed a solution to the equation because (using several axioms)

$$\mathbf{x} + \mathbf{v} = (\mathbf{u} - \mathbf{v}) + \mathbf{v} = [\mathbf{u} + (-\mathbf{v})] + \mathbf{v} = \mathbf{u} + (-\mathbf{v} + \mathbf{v}) = \mathbf{u} + \mathbf{0} = \mathbf{u}.$$

To see that this is the only solution, suppose $\mathbf{x}_1$ is another solution so that $\mathbf{x}_1 + \mathbf{v} = \mathbf{u}$. Then $\mathbf{x} + \mathbf{v} = \mathbf{x}_1 + \mathbf{v}$ (they both equal $\mathbf{u}$), so $\mathbf{x} = \mathbf{x}_1$ by cancellation.

Similarly, cancellation shows that there is only one zero vector in any vector space and only one negative of each vector (Exercises 10 and 11). Hence we speak of *the* zero vector and *the* negative of a vector.

The next theorem derives some basic properties of scalar multiplication that hold in every vector space, and will be used extensively.

Theorem 3

Let $\mathbf{v}$ denote a vector in a vector space V and let a denote a real number.
1. $0\mathbf{v} = \mathbf{0}$.
2. $a\mathbf{0} = \mathbf{0}$.
3. If $a\mathbf{v} = \mathbf{0}$, then either $a = 0$ or $\mathbf{v} = \mathbf{0}$.
4. $(-1)\mathbf{v} = -\mathbf{v}$.
5. $(-a)\mathbf{v} = -(a\mathbf{v}) = a(-\mathbf{v})$.

PROOF

1. Observe that $0\mathbf{v} + 0\mathbf{v} = (0 + 0)\mathbf{v} = 0\mathbf{v} = 0\mathbf{v} + \mathbf{0}$ where the first equality is by axiom S3. It follows that $0\mathbf{v} = \mathbf{0}$ by cancellation.

2. The proof is similar to that of (1), and is left as Exercise 12(a).

3. Assume that $a\mathbf{v} = \mathbf{0}$. If $a = 0$, there is nothing to prove; if $a \neq 0$, we must show that $\mathbf{v} = \mathbf{0}$. But $a \neq 0$ means we can scalar-multiply the equation $a\mathbf{v} = \mathbf{0}$ by the scalar $\frac{1}{a}$. The result (using Axioms S5, S4, and (2)) is
$$\mathbf{v} = 1\mathbf{v} = (\tfrac{1}{a}a)\mathbf{v} = \tfrac{1}{a}(a\mathbf{v}) = \tfrac{1}{a}\mathbf{0} = \mathbf{0}.$$

4. We have $-\mathbf{v} + \mathbf{v} = \mathbf{0}$ by axiom A5. On the other hand,
$$(-1)\mathbf{v} + \mathbf{v} = (-1)\mathbf{v} + 1\mathbf{v} = (-1 + 1)\mathbf{v} = 0\mathbf{v} = \mathbf{0}$$
using (1) and axioms S5 and S3. Hence $(-1)\mathbf{v} + \mathbf{v} = -\mathbf{v} + \mathbf{v}$ (because both are equal to $\mathbf{0}$), so $(-1)\mathbf{v} = -\mathbf{v}$ by cancellation.

5. The proof is left as Exercise 12.

The properties in Theorem 3 are familiar for matrices; the point here is that they hold in *every* vector space.

Axiom A3 ensures that the sum $\mathbf{u} + (\mathbf{v} + \mathbf{w}) = (\mathbf{u} + \mathbf{v}) + \mathbf{w}$ is the same however it is formed, and we write it simply as $\mathbf{u} + \mathbf{v} + \mathbf{w}$. Similarly, there are different ways to form any sum $\mathbf{v}_1 + \mathbf{v}_2 + \cdots + \mathbf{v}_n$, and Axiom A3 guarantees that they are all equal. Moreover, Axiom A2 shows that the order in which the vectors are written does not matter (for example: $\mathbf{u} + \mathbf{v} + \mathbf{w} + \mathbf{z} = \mathbf{z} + \mathbf{u} + \mathbf{w} + \mathbf{v}$).

Similarly, Axioms S2 and S3 extend. For example $a(\mathbf{u} + \mathbf{v} + \mathbf{w}) = a\mathbf{u} + a\mathbf{v} + a\mathbf{w}$ and $(a + b + c)\mathbf{v} = a\mathbf{v} + b\mathbf{v} + c\mathbf{v}$ hold for all values of the scalars and vectors involved (verify). More generally,
$$a(\mathbf{v}_1 + \mathbf{v}_2 + \cdots + \mathbf{v}_n) = a\mathbf{v}_1 + a\mathbf{v}_2 + \cdots + a\mathbf{v}_n$$
$$(a_1 + a_2 + \cdots + a_n)\mathbf{v} = a_1\mathbf{v} + a_2\mathbf{v} + \cdots + a_n\mathbf{v}$$
hold for all $n \geq 1$, all numbers $a, a_1, \ldots, a_n$, and all vectors, $\mathbf{v}, \mathbf{v}_1, \ldots, \mathbf{v}_n$. The verifications are by induction and are left to the reader (Exercise 13). These facts—together with the axioms, Theorem 3, and the definition of subtraction—enable us to simplify expressions involving sums of scalar multiples of vectors by collecting like terms, expanding, and taking out common factors. This has been discussed for the vector space of matrices in Section 2.1 (and for geometric vectors in Section 4.1); the manipulations in an arbitrary vector space are carried out in the same way. Here is an illustration.

EXAMPLE 8

If $\mathbf{u}$, $\mathbf{v}$, and $\mathbf{w}$ are vectors in a vector space V, simplify the expression
$$2(\mathbf{u} + 3\mathbf{w}) - 3(2\mathbf{w} - \mathbf{v}) - 3[2(2\mathbf{u} + \mathbf{v} - 4\mathbf{w}) - 4(\mathbf{u} - 2\mathbf{w})].$$

Solution ▶ The reduction proceeds as though $\mathbf{u}$, $\mathbf{v}$, and $\mathbf{w}$ were matrices or variables.
$$2(\mathbf{u} + 3\mathbf{w}) - 3(2\mathbf{w} - \mathbf{v}) - 3[2(2\mathbf{u} + \mathbf{v} - 4\mathbf{w}) - 4(\mathbf{u} - 2\mathbf{w})]$$
$$= 2\mathbf{u} + 6\mathbf{w} - 6\mathbf{w} + 3\mathbf{v} - 3[4\mathbf{u} + 2\mathbf{v} - 8\mathbf{w} - 4\mathbf{u} + 8\mathbf{w}]$$
$$= 2\mathbf{u} + 3\mathbf{v} - 3[2\mathbf{v}]$$
$$= 2\mathbf{u} + 3\mathbf{v} - 6\mathbf{v}$$
$$= 2\mathbf{u} - 3\mathbf{v}.$$

Condition (2) in Theorem 3 points to another example of a vector space.

SECTION 6.1 Examples and Basic Properties

EXAMPLE 9

A set $\{0\}$ with one element becomes a vector space if we define
$$0 + 0 = 0 \quad \text{and} \quad a0 = 0 \text{ for all scalars } a.$$
The resulting space is called the **zero vector space** and is denoted $\{0\}$.

The vector space axioms are easily verified for $\{0\}$. In any vector space V, Theorem 3(2) shows that the zero subspace (consisting of the zero vector of V alone) is a copy of the zero vector space.

EXERCISES 6.1

1. Let V denote the set of ordered triples (x, y, z) and define addition in V as in $\mathbb{R}^3$. For each of the following definitions of scalar multiplication, decide whether V is a vector space.

   (a) $a(x, y, z) = (ax, y, az)$

   ◆(b) $a(x, y, z) = (ax, 0, az)$

   (c) $a(x, y, z) = (0, 0, 0)$

   ◆(d) $a(x, y, z) = (2ax, 2ay, 2az)$

2. Are the following sets vector spaces with the indicated operations? If not, why not?

   (a) The set V of nonnegative real numbers; ordinary addition and scalar multiplication.

   ◆(b) The set V of all polynomials of degree ≥ 3, together with 0; operations of **P**.

   (c) The set of all polynomials of degree ≤ 3; operations of **P**.

   ◆(d) The set $\{1, x, x^2, \ldots\}$; operations of **P**.

   (e) The set V of all 2×2 matrices of the form $\begin{bmatrix} a & b \\ 0 & c \end{bmatrix}$; operations of $\mathbf{M}_{22}$.

   ◆(f) The set V of 2×2 matrices with equal column sums; operations of $\mathbf{M}_{22}$.

   (g) The set V of 2×2 matrices with zero determinant; usual matrix operations.

   ◆(h) The set V of real numbers; usual operations.

   (i) The set V of complex numbers; usual addition and multiplication by a real number.

   ◆(j) The set V of all ordered pairs (x, y) with the addition of $\mathbb{R}^2$, but scalar multiplication $a(x, y) = (ax, -ay)$.

   (k) The set V of all ordered pairs (x, y) with the addition of $\mathbb{R}^2$, but scalar multiplication $a(x, y) = (x, y)$ for all a in $\mathbb{R}$.

   ◆(l) The set V of all functions $f: \mathbb{R} \to \mathbb{R}$ with pointwise addition, but scalar multiplication defined by $(af)(x) = f(ax)$.

   (m) The set V of all 2×2 matrices whose entries sum to 0; operations of $\mathbf{M}_{22}$.

   ◆(n) The set V of all 2×2 matrices with the addition of $\mathbf{M}_{22}$ but scalar multiplication $*$ defined by $a * X = aX^T$.

3. Let V be the set of positive real numbers with vector addition being ordinary multiplication, and scalar multiplication being $a \cdot v = v^a$. Show that V is a vector space.

◆4. If V is the set of ordered pairs (x, y) of real numbers, show that it is a vector space if $(x, y) + (x_1, y_1) = (x + x_1, y + y_1 + 1)$ and $a(x, y) = (ax, ay + a - 1)$. What is the zero vector in V?

5. Find **x** and **y** (in terms of **u** and **v**) such that:

   (a) $2\mathbf{x} + \mathbf{y} = \mathbf{u}$
       $5\mathbf{x} + 3\mathbf{y} = \mathbf{v}$

   ◆(b) $3\mathbf{x} - 2\mathbf{y} = \mathbf{u}$
       $4\mathbf{x} - 5\mathbf{y} = \mathbf{v}$

6. In each case show that the condition $a\mathbf{u} + b\mathbf{v} + c\mathbf{w} = \mathbf{0}$ in V implies that $a = b = c = 0$.

   (a) $V = \mathbb{R}^4$; $\mathbf{u} = (2, 1, 0, 2)$, $\mathbf{v} = (1, 1, -1, 0)$, $\mathbf{w} = (0, 1, 2, 1)$

   ◆(b) $V = \mathbf{M}_{22}$; $\mathbf{u} = \begin{bmatrix} 1 & 0 \\ 0 & 1 \end{bmatrix}$, $\mathbf{v} = \begin{bmatrix} 0 & 1 \\ 1 & 0 \end{bmatrix}$, $\mathbf{w} = \begin{bmatrix} 1 & 1 \\ 1 & -1 \end{bmatrix}$

   (c) $V = \mathbf{P}$; $\mathbf{u} = x^3 + x$, $\mathbf{v} = x^2 + 1$, $\mathbf{w} = x^3 - x^2 + x + 1$

◆(d) $V = \mathbf{F}[0, \pi]$; $\mathbf{u} = \sin x$, $\mathbf{v} = \cos x$, $\mathbf{w} = 1-$ the constant funciton

7. Simplify each of the following.
   (a) $3[2(\mathbf{u} - 2\mathbf{v} - \mathbf{w}) + 3(\mathbf{w} - \mathbf{v})] - 7(\mathbf{u} - 3\mathbf{v} - \mathbf{w})$
   ◆(b) $4(3\mathbf{u} - \mathbf{v} + \mathbf{w}) - 2[(3\mathbf{u} - 2\mathbf{v}) - 3(\mathbf{v} - \mathbf{w})] + 6(\mathbf{w} - \mathbf{u} - \mathbf{v})$

8. Show that $\mathbf{x} = \mathbf{v}$ is the only solution to the equation $\mathbf{x} + \mathbf{x} = 2\mathbf{v}$ in a vector space V. Cite all axioms used.

9. Show that $-\mathbf{0} = \mathbf{0}$ in any vector space. Cite all axioms used.

◆10. Show that the zero vector $\mathbf{0}$ is uniquely determined by the property in axiom A4.

11. Given a vector $\mathbf{v}$, show that its negative $-\mathbf{v}$ is uniquely determined by the property in axiom A5.

12. (a) Prove (2) of Theorem 3. [*Hint*: Axiom S2.]
    ◆(b) Prove that $(-a)\mathbf{v} = -(a\mathbf{v})$ in Theorem 3 by first computing $(-a)\mathbf{v} + a\mathbf{v}$. Then do it using (4) of Theorem 3 and axiom S4.
    (c) Prove that $a(-\mathbf{v}) = -(a\mathbf{v})$ in Theorem 3 in two ways, as in part (b).

13. Let $\mathbf{v}, \mathbf{v}_1, \ldots, \mathbf{v}_n$ denote vectors in a vector space V and let $a, a_1, \ldots, a_n$ denote numbers. Use induction on n to prove each of the following.
    (a) $a(\mathbf{v}_1 + \mathbf{v}_2 + \cdots + \mathbf{v}_n) = a\mathbf{v}_1 + a\mathbf{v}_2 + \cdots + a\mathbf{v}_n$
    ◆(b) $(a_1 + a_2 + \cdots + a_n)\mathbf{v} = a_1\mathbf{v} + a_2\mathbf{v} + \cdots + a_n\mathbf{v}$

14. Verify axioms A2—A5 and S2—S5 for the space $\mathbf{F}[a, b]$ of functions on $[a, b]$ (Example 7).

15. Prove each of the following for vectors $\mathbf{u}$ and $\mathbf{v}$ and scalars a and b.
    (a) If $a\mathbf{v} = \mathbf{0}$, then $a = 0$ or $\mathbf{v} = \mathbf{0}$.
    (b) If $a\mathbf{v} = b\mathbf{v}$ and $\mathbf{v} \neq \mathbf{0}$, then $a = b$.
    ◆(c) If $a\mathbf{v} = a\mathbf{w}$ and $a \neq 0$, then $\mathbf{v} = \mathbf{w}$.

16. By calculating $(1 + 1)(\mathbf{v} + \mathbf{w})$ in two ways (using axioms S2 and S3), show that axiom A2 follows from the other axioms.

17. Let V be a vector space, and define V^n to be the set of all n-tuples $(\mathbf{v}_1, \mathbf{v}_2, \ldots, \mathbf{v}_n)$ of n vectors $\mathbf{v}_i$, each belonging to V. Define addition and scalar multiplication in V^n as follows:
$$(\mathbf{u}_1, \mathbf{u}_2, \ldots, \mathbf{u}_n) + (\mathbf{v}_1, \mathbf{v}_2, \ldots, \mathbf{v}_n)$$
$$= (\mathbf{u}_1 + \mathbf{v}_1, \mathbf{u}_2 + \mathbf{v}_2, \ldots, \mathbf{u}_n + \mathbf{v}_n)$$
$$a(\mathbf{v}_1, \mathbf{v}_2, \ldots, \mathbf{v}_n) = (a\mathbf{v}_1, a\mathbf{v}_2, \ldots, a\mathbf{v}_n)$$
Show that V^n is a vector space.

18. Let V^n be the vector space of n-tuples from the preceding exercise, written as columns. If A is an $m \times n$ matrix, and X is in V^n, define AX in V^m by matrix multiplication. More precisely, if

$$A = [a_{ij}] \text{ and } X = \begin{bmatrix} \mathbf{v}_1 \\ \vdots \\ \mathbf{v}_n \end{bmatrix}, \text{ let } AX = \begin{bmatrix} \mathbf{u}_1 \\ \vdots \\ \mathbf{u}_n \end{bmatrix}, \text{ where}$$

$\mathbf{u}_i = a_{i1}\mathbf{v}_1 + a_{i2}\mathbf{v}_2 + \cdots + a_{in}\mathbf{v}_n$ for each i. Prove that:
(a) $B(AX) = (BA)X$
(b) $(A + A_1)X = AX + A_1X$
(c) $A(X + X_1) = AX + AX_1$
(d) $(kA)X = k(AX) = A(kX)$ if k is any number
(e) $IX = X$ if I is the $n \times n$ identity matrix
(f) Let E be an elementary matrix obtained by performing a row operation on the rows of I_n (see Section 2.5). Show that EX is the column resulting from performing that same row operation on the vectors (call them rows) of X. [*Hint*: Lemma 1 Section 2.5.]

SECTION 6.2 Subspaces and Spanning Sets

Definition 6.2 *If V is a vector space, a nonempty subset $U \subseteq V$ is called a **subspace** of V if U is itself a vector space using the addition and scalar multiplication of V.*

Subspaces of $\mathbb{R}^n$ (as defined in Section 5.1) are subspaces in the present sense by Example 3 Section 6.1. Moreover, the defining properties for a subspace of $\mathbb{R}^n$ actually *characterize* subspaces in general.

SECTION 6.2 *Subspaces and Spanning Sets*

> ### Theorem 1
>
> **Subspace Test**
> A subset U of a vector space is a subspace of V if and only if it satisfies the following three conditions:
>
> 1. $\mathbf{0}$ lies in U where $\mathbf{0}$ is the zero vector of V.
> 2. If $\mathbf{u}_1$ and $\mathbf{u}_2$ are in U, then $\mathbf{u}_1 + \mathbf{u}_2$ is also in U.
> 3. If $\mathbf{u}$ is in U, then $a\mathbf{u}$ is also in U for each scalar a.

> **PROOF**
>
> If U is a subspace of V, then (2) and (3) hold by axioms A1 and S1 respectively, applied to the vector space U. Since U is nonempty (it is a vector space), choose $\mathbf{u}$ in U. Then (1) holds because $\mathbf{0} = 0\mathbf{u}$ is in U by (3) and Theorem 3 Section 6.1.
>
> Conversely, if (1), (2), and (3) hold, then axioms A1 and S1 hold because of (2) and (3), and axioms A2, A3, S2, S3, S4, and S5 hold in U because they hold in V. Axiom A4 holds because the zero vector $\mathbf{0}$ of V is actually in U by (1), and so serves as the zero of U. Finally, given $\mathbf{u}$ in U, then its negative $-\mathbf{u}$ in V is again in U by (3) because $-\mathbf{u} = (-1)\mathbf{u}$ (again using Theorem 3 Section 6.1). Hence $-\mathbf{u}$ serves as the negative of $\mathbf{u}$ in U.

Note that the proof of Theorem 1 shows that if U is a subspace of V, then U and V share the same zero vector, and that the negative of a vector in the space U is the same as its negative in V.

EXAMPLE 1

If V is any vector space, show that $\{\mathbf{0}\}$ and V are subspaces of V.

Solution ▶ $U = V$ clearly satisfies the conditions of the test. As to $U = \{\mathbf{0}\}$, it satisfies the conditions because $\mathbf{0} + \mathbf{0} = \mathbf{0}$ and $a\mathbf{0} = \mathbf{0}$ for all a in $\mathbb{R}$.

The vector space $\{\mathbf{0}\}$ is called the **zero subspace** of V.

EXAMPLE 2

Let $\mathbf{v}$ be a vector in a vector space V. Show that the set
$$\mathbb{R}\mathbf{v} = \{a\mathbf{v} \mid a \text{ in } \mathbb{R}\}$$
of all scalar multiples of $\mathbf{v}$ is a subspace of V.

Solution ▶ Because $\mathbf{0} = 0\mathbf{v}$, it is clear that $\mathbf{0}$ lies in $\mathbb{R}\mathbf{v}$. Given two vectors $a\mathbf{v}$ and $a_1\mathbf{v}$ in $\mathbb{R}\mathbf{v}$, their sum $a\mathbf{v} + a_1\mathbf{v} = (a + a_1)\mathbf{v}$ is also a scalar multiple of $\mathbf{v}$ and so lies in $\mathbb{R}\mathbf{v}$. Hence $\mathbb{R}\mathbf{v}$ is closed under addition. Finally, given $a\mathbf{v}$, $r(a\mathbf{v}) = (ra)\mathbf{v}$ lies in $\mathbb{R}\mathbf{v}$ for all $r \in \mathbb{R}$, so $\mathbb{R}\mathbf{v}$ is closed under scalar multiplication. Hence the subspace test applies.

In particular, given $\mathbf{d} \neq \mathbf{0}$ in $\mathbb{R}^3$, $\mathbb{R}\mathbf{d}$ is the line through the origin with direction vector $\mathbf{d}$.

The space $\mathbb{R}\mathbf{v}$ in Example 2 is described by giving the *form* of each vector in $\mathbb{R}\mathbf{v}$. The next example describes a subset U of the space $\mathbf{M}_{nn}$ by giving a *condition* that each matrix of U must satisfy.

EXAMPLE 3

Let A be a fixed matrix in $\mathbf{M}_{nn}$. Show that $U = \{X \text{ in } \mathbf{M}_{nn} \mid AX = XA\}$ is a subspace of $\mathbf{M}_{nn}$.

Solution ▶ If 0 is the $n \times n$ zero matrix, then $A0 = 0A$, so 0 satisfies the condition for membership in U. Next suppose that X and X_1 lie in U so that $AX = XA$ and $AX_1 = X_1 A$. Then

$$A(X + X_1) = AX + AX_1 = XA + X_1 A = (X + X_1)A$$
$$A(aX) = a(AX) = a(XA) = (aX)A$$

for all a in $\mathbb{R}$, so both $X + X_1$ and aX lie in U. Hence U is a subspace of $\mathbf{M}_{nn}$.

Suppose $p(x)$ is a polynomial and a is a number. Then the number $p(a)$ obtained by replacing x by a in the expression for $p(x)$ is called the **evaluation** of $p(x)$ at a. For example, if $p(x) = 5 - 6x + 2x^2$, then the evaluation of $p(x)$ at $a = 2$ is $p(2) = 5 - 12 + 8 = 1$. If $p(a) = 0$, the number a is called a **root** of $p(x)$.

EXAMPLE 4

Consider the set U of all polynomials in $\mathbf{P}$ that have 3 as a root:

$$U = \{p(x) \text{ in } \mathbf{P} \mid p(3) = 0\}.$$

Show that U is a subspace of $\mathbf{P}$.

Solution ▶ Clearly, the zero polynomial lies in U. Now let $p(x)$ and $q(x)$ lie in U so $p(3) = 0$ and $q(3) = 0$. We have $(p + q)(x) = p(x) + q(x)$ for all x, so $(p + q)(3) = p(3) + q(3) = 0 + 0 = 0$, and U is closed under addition. The verification that U is closed under scalar multiplication is similar.

Recall that the space $\mathbf{P}_n$ consists of all polynomials of the form

$$a_0 + a_1 x + a_2 x^2 + \cdots + a_n x^n$$

where $a_0, a_1, a_2, \ldots, a_n$ are real numbers, and so is closed under the addition and scalar multiplication in $\mathbf{P}$. Moreover, the zero polynomial is included in $\mathbf{P}_n$. Thus the subspace test gives Example 5.

EXAMPLE 5

$\mathbf{P}_n$ is a subspace of $\mathbf{P}$ for each $n \geq 0$.

SECTION 6.2 Subspaces and Spanning Sets

The next example involves the notion of the derivative f' of a function f. (If the reader is not familiar with calculus, this example may be omitted.) A function f defined on the interval $[a, b]$ is called **differentiable** if the derivative $f'(r)$ exists at every r in $[a, b]$.

EXAMPLE 6

Show that the subset $\mathbf{D}[a, b]$ of all **differentiable functions** on $[a, b]$ is a subspace of the vector space $\mathbf{F}[a, b]$ of all functions on $[a, b]$.

Solution ▶ The derivative of any constant function is the constant function 0; in particular, 0 itself is differentiable and so lies in $\mathbf{D}[a, b]$. If f and g both lie in $\mathbf{D}[a, b]$ (so that f' and g' exist), then it is a theorem of calculus that $f + g$ and af are both differentiable [in fact, $(f + g)' = f' + g'$ and $(af)' = af'$], so both lie in $\mathbf{D}[a, b]$. This shows that $\mathbf{D}[a, b]$ is a subspace of $\mathbf{F}[a, b]$.

Linear Combinations and Spanning Sets

Definition 6.3 *Let $\{\mathbf{v}_1, \mathbf{v}_2, ..., \mathbf{v}_n\}$ be a set of vectors in a vector space V. As in $\mathbb{R}^n$, a vector $\mathbf{v}$ is called a* **linear combination** *of the vectors $\mathbf{v}_1, \mathbf{v}_2, ..., \mathbf{v}_n$ if it can be expressed in the form*

$$\mathbf{v} = a_1\mathbf{v}_1 + a_2\mathbf{v}_2 + \cdots + a_n\mathbf{v}_n$$

where $a_1, a_2, ..., a_n$ are scalars, called the **coefficients** *of $\mathbf{v}_1, \mathbf{v}_2, ..., \mathbf{v}_n$. The set of all linear combinations of these vectors is called their* **span**, *and is denoted by*

$$\text{span}\{\mathbf{v}_1, \mathbf{v}_2, ..., \mathbf{v}_n\} = \{a_1\mathbf{v}_1 + a_2\mathbf{v}_2 + \cdots + a_n\mathbf{v}_n \mid a_i \text{ in } \mathbb{R}\}.$$

If it happens that $V = \text{span}\{\mathbf{v}_1, \mathbf{v}_2, ..., \mathbf{v}_n\}$, these vectors are called a **spanning set** for V. For example, the span of two vectors $\mathbf{v}$ and $\mathbf{w}$ is the set

$$\text{span}\{\mathbf{v}, \mathbf{w}\} = \{s\mathbf{v} + t\mathbf{w} \mid s \text{ and } t \text{ in } \mathbb{R}\}$$

of all sums of scalar multiples of these vectors.

EXAMPLE 7

Consider the vectors $p_1 = 1 + x + 4x^2$ and $p_2 = 1 + 5x + x^2$ in $\mathbf{P}_2$. Determine whether p_1 and p_2 lie in $\text{span}\{1 + 2x - x^2, 3 + 5x + 2x^2\}$.

Solution ▶ For p_1, we want to determine if s and t exist such that

$$p_1 = s(1 + 2x - x^2) + t(3 + 5x + 2x^2).$$

Equating coefficients of powers of x (where $x^0 = 1$) gives

$$1 = s + 3t, \quad 1 = 2s + 5t, \quad \text{and} \quad 4 = -s + 2t.$$

These equations have the solution $s = -2$ and $t = 1$, so p_1 is indeed in $\text{span}\{1 + 2x - x^2, 3 + 5x + 2x^2\}$.

Turning to $p_2 = 1 + 5x + x^2$, we are looking for s and t such that $p_2 = s(1 + 2x - x^2) + t(3 + 5x + 2x^2)$. Again equating coefficients of powers of x gives equations $1 = s + 3t$, $5 = 2s + 5t$, and $1 = -s + 2t$. But in this case there is no solution, so p_2 is *not* in $\text{span}\{1 + 2x - x^2, 3 + 5x + 2x^2\}$.

We saw in Example 6 Section 5.1 that $\mathbb{R}^m = \text{span}\{\mathbf{e}_1, \mathbf{e}_2, \ldots, \mathbf{e}_m\}$ where the vectors $\mathbf{e}_1, \mathbf{e}_2, \ldots, \mathbf{e}_m$ are the columns of the $m \times m$ identity matrix. Of course $\mathbb{R}^m = \mathbf{M}_{m1}$ is the set of all $m \times 1$ matrices, and there is an analogous spanning set for each space $\mathbf{M}_{mn}$. For example, each 2×2 matrix has the form

$$\begin{bmatrix} a & b \\ c & d \end{bmatrix} = a\begin{bmatrix} 1 & 0 \\ 0 & 0 \end{bmatrix} + b\begin{bmatrix} 0 & 1 \\ 0 & 0 \end{bmatrix} + c\begin{bmatrix} 0 & 0 \\ 1 & 0 \end{bmatrix} + d\begin{bmatrix} 0 & 0 \\ 0 & 1 \end{bmatrix}$$

so

$$\mathbf{M}_{22} = \text{span}\left\{\begin{bmatrix} 1 & 0 \\ 0 & 0 \end{bmatrix}, \begin{bmatrix} 0 & 1 \\ 0 & 0 \end{bmatrix}, \begin{bmatrix} 0 & 0 \\ 1 & 0 \end{bmatrix}, \begin{bmatrix} 0 & 0 \\ 0 & 1 \end{bmatrix}\right\}.$$

Similarly, we obtain

EXAMPLE 8

$\mathbf{M}_{mn}$ is the span of the set of all $m \times n$ matrices with exactly one entry equal to 1, and all other entries zero.

The fact that every polynomial in $\mathbf{P}_n$ has the form $a_0 + a_1 x + a_2 x^2 + \cdots + a_n x^n$ where each a_i is in $\mathbb{R}$ shows that

EXAMPLE 9

$\mathbf{P}_n = \text{span}\{1, x, x^2, \ldots, x^n\}$.

In Example 2 we saw that $\text{span}\{\mathbf{v}\} = \{a\mathbf{v} \mid a \text{ in } \mathbb{R}\} = \mathbb{R}\mathbf{v}$ is a subspace for any vector $\mathbf{v}$ in a vector space V. More generally, the span of *any* set of vectors is a subspace. In fact, the proof of Theorem 1 Section 5.1 goes through to prove:

Theorem 2

Let $U = \text{span}\{\mathbf{v}_1, \mathbf{v}_2, \ldots, \mathbf{v}_n\}$ in a vector space V. Then:

1. U is a subspace of V containing each of $\mathbf{v}_1, \mathbf{v}_2, \ldots, \mathbf{v}_n$.
2. U is the "smallest" subspace containing these vectors in the sense that any subspace that contains each of $\mathbf{v}_1, \mathbf{v}_2, \ldots, \mathbf{v}_n$ must contain U.

Theorem 2 is used frequently to determine spanning sets, as the following examples show.

EXAMPLE 10

Show that $\mathbf{P}_3 = \text{span}\{x^2 + x^3, x, 2x^2 + 1, 3\}$.

Solution ▶ Write $U = \text{span}\{x^2 + x^3, x, 2x^2 + 1, 3\}$. Then $U \subseteq \mathbf{P}_3$, and we use the fact that $\mathbf{P}_3 = \text{span}\{1, x, x^2, x^3\}$ to show that $\mathbf{P}_3 \subseteq U$. In fact, x and $1 = \frac{1}{3} \cdot 3$ clearly lie in U. But then successively,

$$x^2 = \tfrac{1}{2}[(2x^2 + 1) - 1] \quad \text{and} \quad x^3 = (x^2 + x^3) - x^2$$

also lie in U. Hence $\mathbf{P}_3 \subseteq U$ by Theorem 2.

EXAMPLE 11

Let **u** and **v** be two vectors in a vector space V. Show that
$$\text{span}\{\mathbf{u}, \mathbf{v}\} = \text{span}\{\mathbf{u} + 2\mathbf{v}, \mathbf{u} - \mathbf{v}\}.$$

Solution ▶ We have $\text{span}\{\mathbf{u} + 2\mathbf{v}, \mathbf{u} - \mathbf{v}\} \subseteq \text{span}\{\mathbf{u}, \mathbf{v}\}$ by Theorem 2 because both $\mathbf{u} + 2\mathbf{v}$ and $\mathbf{u} - \mathbf{v}$ lie in $\text{span}\{\mathbf{u}, \mathbf{v}\}$. On the other hand,
$$\mathbf{u} = \tfrac{1}{3}(\mathbf{u} + 2\mathbf{v}) + \tfrac{2}{3}(\mathbf{u} - \mathbf{v}) \quad \text{and} \quad \mathbf{v} = \tfrac{1}{3}(\mathbf{u} + 2\mathbf{v}) - \tfrac{1}{3}(\mathbf{u} - \mathbf{v})$$
so $\text{span}\{\mathbf{u}, \mathbf{v}\} \subseteq \text{span}\{\mathbf{u} + 2\mathbf{v}, \mathbf{u} - \mathbf{v}\}$, again by Theorem 2.

EXERCISES 6.2

1. Which of the following are subspaces of $\mathbf{P}_3$? Support your answer.
   - (a) $U = \{f(x) \mid f(x) \text{ in } \mathbf{P}_3, f(2) = 1\}$
   - ◆(b) $U = \{xg(x) \mid g(x) \text{ in } \mathbf{P}_2\}$
   - (c) $U = \{xg(x) \mid g(x) \text{ in } \mathbf{P}_3\}$
   - ◆(d) $U = \{xg(x) + (1 - x)h(x) \mid g(x) \text{ and } h(x) \text{ in } \mathbf{P}_2\}$
   - (e) $U =$ The set of all polynomials in $\mathbf{P}_3$ with constant term 0
   - ◆(f) $U = \{f(x) \mid f(x) \text{ in } \mathbf{P}_3, \deg f(x) = 3\}$

2. Which of the following are subspaces of $\mathbf{M}_{22}$? Support your answer.
   - (a) $U = \left\{ \begin{bmatrix} a & b \\ 0 & c \end{bmatrix} \middle| a, b, \text{ and } c \text{ in } \mathbb{R} \right\}$
   - ◆(b) $U = \left\{ \begin{bmatrix} a & b \\ c & d \end{bmatrix} \middle| a + b = c + d; a, b, c, \text{ and } d \text{ in } \mathbb{R} \right\}$
   - (c) $U = \{A \mid A \text{ in } \mathbf{M}_{22}, A = A^T\}$
   - ◆(d) $U = \{A \mid A \text{ in } \mathbf{M}_{22}, AB = 0\}$, B a fixed 2×2 matrix
   - (e) $U = \{A \mid A \text{ in } \mathbf{M}_{22}, A^2 = A\}$
   - ◆(f) $U = \{A \mid A \text{ in } \mathbf{M}_{22}, A \text{ is not invertible}\}$
   - (g) $U = \{A \mid A \text{ in } \mathbf{M}_{22}, BAC = CAB\}$, B and C fixed 2×2 matrices

3. Which of the following are subspaces of $\mathbf{F}[0, 1]$? Support your answer.
   - (a) $U = \{f \mid f(0) = 0\}$
   - ◆(b) $U = \{f \mid f(0) = 1\}$
   - (c) $U = \{f \mid f(0) = f(1)\}$
   - ◆(d) $U = \{f \mid f(x) \geq 0 \text{ for all } x \text{ in } [0, 1]\}$
   - (e) $U = \{f \mid f(x) = f(y) \text{ for all } x \text{ and } y \text{ in } [0, 1]\}$
   - ◆(f) $U = \{f \mid f(x + y) = f(x) + f(y) \text{ for all } x \text{ and } y \text{ in } [0, 1]\}$
   - (g) $U = \{f \mid f \text{ is integrable and } \int_0^1 f(x)dx = 0\}$

4. Let A be an $m \times n$ matrix. For which columns **b** in $\mathbb{R}^m$ is $U = \{\mathbf{x} \mid \mathbf{x} \text{ in } \mathbb{R}^n, A\mathbf{x} = \mathbf{b}\}$ a subspace of $\mathbb{R}^n$? Support your answer.

5. Let **x** be a vector in $\mathbb{R}^n$ (written as a column), and define $U = \{A\mathbf{x} \mid A \text{ in } \mathbf{M}_{mn}\}$.
   - (a) Show that U is a subspace of $\mathbb{R}^m$.
   - ◆(b) Show that $U = \mathbb{R}^m$ if $\mathbf{x} \neq \mathbf{0}$.

6. Write each of the following as a linear combination of $x + 1$, $x^2 + x$, and $x^2 + 2$.
   - (a) $x^2 + 3x + 2$
   - ◆(b) $2x^2 - 3x + 1$
   - (c) $x^2 + 1$
   - ◆(d) x

7. Determine whether **v** lies in span$\{\mathbf{u}, \mathbf{w}\}$ in each case.
   - (a) $\mathbf{v} = 3x^2 - 2x - 1$; $\mathbf{u} = x^2 + 1$, $\mathbf{w} = x + 2$
   - ◆(b) $\mathbf{v} = x$; $\mathbf{u} = x^2 + 1$, $\mathbf{w} = x + 2$
   - (c) $\mathbf{v} = \begin{bmatrix} 1 & 3 \\ -1 & 1 \end{bmatrix}$; $\mathbf{u} = \begin{bmatrix} 1 & -1 \\ 2 & 1 \end{bmatrix}$, $\mathbf{w} = \begin{bmatrix} 2 & 1 \\ 1 & 0 \end{bmatrix}$
   - ◆(d) $\mathbf{v} = \begin{bmatrix} 1 & -4 \\ 5 & 3 \end{bmatrix}$; $\mathbf{u} = \begin{bmatrix} 1 & -1 \\ 2 & 1 \end{bmatrix}$, $\mathbf{w} = \begin{bmatrix} 2 & 1 \\ 1 & 0 \end{bmatrix}$

8. Which of the following functions lie in span$\{\cos^2 x, \sin^2 x\}$? (Work in $\mathbf{F}[0, \pi]$.)

   (a) $\cos 2x$
   ◆(b) 1
   (c) x^2
   ◆(d) $1 + x^2$

9. (a) Show that $\mathbb{R}^3$ is spanned by $\{(1, 0, 1), (1, 1, 0), (0, 1, 1)\}$.

   ◆(b) Show that $\mathbf{P}_2$ is spanned by $\{1 + 2x^2, 3x, 1 + x\}$.

   (c) Show that $\mathbf{M}_{22}$ is spanned by
   $$\left\{\begin{bmatrix} 1 & 0 \\ 0 & 0 \end{bmatrix}, \begin{bmatrix} 1 & 0 \\ 0 & 1 \end{bmatrix}, \begin{bmatrix} 0 & 1 \\ 1 & 0 \end{bmatrix}, \begin{bmatrix} 1 & 1 \\ 0 & 1 \end{bmatrix}\right\}.$$

10. If X and Y are two sets of vectors in a vector space V, and if $X \subseteq Y$, show that span $X \subseteq$ span Y.

11. Let $\mathbf{u}$, $\mathbf{v}$, and $\mathbf{w}$ denote vectors in a vector space V. Show that:

    (a) span$\{\mathbf{u}, \mathbf{v}, \mathbf{w}\}$ = span$\{\mathbf{u} + \mathbf{v}, \mathbf{u} + \mathbf{w}, \mathbf{v} + \mathbf{w}\}$

    ◆(b) span$\{\mathbf{u}, \mathbf{v}, \mathbf{w}\}$ = span$\{\mathbf{u} - \mathbf{v}, \mathbf{u} + \mathbf{w}, \mathbf{w}\}$

12. Show that
    span$\{\mathbf{v}_1, \mathbf{v}_2, \ldots, \mathbf{v}_n, \mathbf{0}\}$ = span$\{\mathbf{v}_1, \mathbf{v}_2, \ldots, \mathbf{v}_n\}$
    holds for any set of vectors $\{\mathbf{v}_1, \mathbf{v}_2, \ldots, \mathbf{v}_n\}$.

13. If X and Y are nonempty subsets of a vector space V such that span X = span $Y = V$, must there be a vector common to both X and Y? Justify your answer.

◆14. Is it possible that $\{(1, 2, 0), (1, 1, 1)\}$ can span the subspace $U = \{(a, b, 0) \mid a$ and b in $\mathbb{R}\}$?

15. Describe span$\{\mathbf{0}\}$.

16. Let $\mathbf{v}$ denote any vector in a vector space V. Show that span$\{\mathbf{v}\}$ = span$\{a\mathbf{v}\}$ for any $a \neq 0$.

17. Determine all subspaces of $\mathbb{R}\mathbf{v}$ where $\mathbf{v} \neq \mathbf{0}$ in some vector space V.

◆18. Suppose V = span$\{\mathbf{v}_1, \mathbf{v}_2, \ldots, \mathbf{v}_n\}$. If $\mathbf{u} = a_1\mathbf{v}_1 + a_2\mathbf{v}_2 + \cdots + a_n\mathbf{v}_n$ where the a_i are in $\mathbb{R}$ and $a_1 \neq 0$, show that V = span$\{\mathbf{u}, \mathbf{v}_2, \ldots, \mathbf{v}_n\}$.

19. If $\mathbf{M}_{nn}$ = span$\{A_1, A_2, \ldots, A_k\}$, show that $\mathbf{M}_{nn}$ = span$\{A_1^T, A_2^T, \ldots, A_k^T\}$.

20. If $\mathbf{P}_n$ = span$\{p_1(x), p_2(x), \ldots, p_k(x)\}$ and a is in $\mathbb{R}$, show that $p_i(a) \neq 0$ for some i.

21. Let U be a subspace of a vector space V.

    (a) If $a\mathbf{u}$ is in U where $a \neq 0$, show that $\mathbf{u}$ is in U.

    ◆(b) If $\mathbf{u}$ and $\mathbf{u} + \mathbf{v}$ are in U, show that $\mathbf{v}$ is in U.

◆22. Let U be a nonempty subset of a vector space V. Show that U is a subspace of V if and only if $\mathbf{u}_1 + a\mathbf{u}_2$ lies in U for all $\mathbf{u}_1$ and $\mathbf{u}_2$ in U and all a in $\mathbb{R}$.

23. Let $U = \{p(x)$ in $\mathbf{P} \mid p(3) = 0\}$ be the set in Example 4. Use the factor theorem (see Section 6.5) to show that U consists of multiples of $x - 3$; that is, show that $U = \{(x - 3)q(x) \mid q(x)$ in $\mathbf{P}\}$. Use this to show that U is a subspace of $\mathbf{P}$.

24. Let $A_1, A_2, \ldots, A_m$ denote $n \times n$ matrices. If $\mathbf{y}$ is a nonzero column in $\mathbb{R}^n$ and $A_1\mathbf{y} = A_2\mathbf{y} = \cdots = A_m\mathbf{y} = \mathbf{0}$, show that $\{A_1, A_2, \ldots, A_m\}$ cannot span $\mathbf{M}_{nn}$.

25. Let $\{\mathbf{v}_1, \mathbf{v}_2, \ldots, \mathbf{v}_n\}$ and $\{\mathbf{u}_1, \mathbf{u}_2, \ldots, \mathbf{u}_n\}$ be sets of vectors in a vector space, and let

    $$X = \begin{bmatrix} \mathbf{v}_1 \\ \vdots \\ \mathbf{v}_n \end{bmatrix} \quad Y = \begin{bmatrix} \mathbf{u}_1 \\ \vdots \\ \mathbf{u}_n \end{bmatrix}$$

    as in Exercise 18 Section 6.1.

    (a) Show that span$\{\mathbf{v}_1, \ldots, \mathbf{v}_n\} \subseteq$ span$\{\mathbf{u}_1, \ldots, \mathbf{u}_n\}$ if and only if $AY = X$ for some $n \times n$ matrix A.

    (b) If $X = AY$ where A is invertible, show that span$\{\mathbf{v}_1, \ldots, \mathbf{v}_n\}$ = span$\{\mathbf{u}_1, \ldots, \mathbf{u}_n\}$.

26. If U and W are subspaces of a vector space V, let $U \cup W = \{\mathbf{v} \mid \mathbf{v}$ is in U or $\mathbf{v}$ is in $W\}$. Show that $U \cup W$ is a subspace if and only if $U \subseteq W$ or $W \subseteq U$.

27. Show that $\mathbf{P}$ cannot be spanned by a finite set of polynomials.

SECTION 6.3 Linear Independence and Dimension

Definition 6.4

As in $\mathbb{R}^n$, a set of vectors $\{\mathbf{v}_1, \mathbf{v}_2, \ldots, \mathbf{v}_n\}$ in a vector space V is called **linearly independent** (or simply **independent**) if it satisfies the following condition:

If $s_1\mathbf{v}_1 + s_2\mathbf{v}_2 + \cdots + s_n\mathbf{v}_n = \mathbf{0}$, then $s_1 = s_2 = \cdots = s_n = 0$.

A set of vectors that is not linearly independent is said to be **linearly dependent** (or simply **dependent**).

The **trivial linear combination** of the vectors $\mathbf{v}_1, \mathbf{v}_2, \ldots, \mathbf{v}_n$ is the one with every coefficient zero:

$$0\mathbf{v}_1 + 0\mathbf{v}_2 + \cdots + 0\mathbf{v}_n.$$

This is obviously one way of expressing $\mathbf{0}$ as a linear combination of the vectors $\mathbf{v}_1, \mathbf{v}_2, \ldots, \mathbf{v}_n$, and they are linearly independent when it is the *only* way.

EXAMPLE 1

Show that $\{1 + x, 3x + x^2, 2 + x - x^2\}$ is independent in $\mathbf{P}_2$.

Solution ▶ Suppose a linear combination of these polynomials vanishes.

$$s_1(1 + x) + s_2(3x + x^2) + s_3(2 + x - x^2) = 0.$$

Equating the coefficients of 1, x, and x^2 gives a set of linear equations.

$$\begin{aligned} s_1 \phantom{{}+3s_2} + 2s_3 &= 0 \\ s_1 + 3s_2 + s_3 &= 0 \\ s_2 - s_3 &= 0 \end{aligned}$$

The only solution is $s_1 = s_2 = s_3 = 0$.

EXAMPLE 2

Show that $\{\sin x, \cos x\}$ is independent in the vector space $\mathbf{F}[0, 2\pi]$ of functions defined on the interval $[0, 2\pi]$.

Solution ▶ Suppose that a linear combination of these functions vanishes.

$$s_1(\sin x) + s_2(\cos x) = 0.$$

This must hold for *all* values of x in $[0, 2\pi]$ (by the definition of equality in $\mathbf{F}[0, 2\pi]$). Taking $x = 0$ yields $s_2 = 0$ (because $\sin 0 = 0$ and $\cos 0 = 1$). Similarly, $s_1 = 0$ follows from taking $x = \frac{\pi}{2}$ (because $\sin \frac{\pi}{2} = 1$ and $\cos \frac{\pi}{2} = 0$).

EXAMPLE 3

Suppose that $\{\mathbf{u}, \mathbf{v}\}$ is an independent set in a vector space V. Show that $\{\mathbf{u} + 2\mathbf{v}, \mathbf{u} - 3\mathbf{v}\}$ is also independent.

Solution ▶ Suppose a linear combination of $\mathbf{u} + 2\mathbf{v}$ and $\mathbf{u} - 3\mathbf{v}$ vanishes:
$$s(\mathbf{u} + 2\mathbf{v}) + t(\mathbf{u} - 3\mathbf{v}) = \mathbf{0}.$$
We must deduce that $s = t = 0$. Collecting terms involving $\mathbf{u}$ and $\mathbf{v}$ gives
$$(s + t)\mathbf{u} + (2s - 3t)\mathbf{v} = \mathbf{0}.$$
Because $\{\mathbf{u}, \mathbf{v}\}$ is independent, this yields linear equations $s + t = 0$ and $2s - 3t = 0$. The only solution is $s = t = 0$.

EXAMPLE 4

Show that any set of polynomials of distinct degrees is independent.

Solution ▶ Let $p_1, p_2, \ldots, p_m$ be polynomials where $\deg(p_i) = d_i$. By relabelling if necessary, we may assume that $d_1 > d_2 > \cdots > d_m$. Suppose that a linear combination vanishes:
$$t_1 p_1 + t_2 p_2 + \cdots + t_m p_m = 0$$
where each t_i is in $\mathbb{R}$. As $\deg(p_1) = d_1$, let ax^{d_1} be the term in p_1 of highest degree, where $a \neq 0$. Since $d_1 > d_2 > \cdots > d_m$, it follows that $t_1 a x^{d_1}$ is the only term of degree d_1 in the linear combination $t_1 p_1 + t_2 p_2 + \cdots + t_m p_m = 0$. This means that $t_1 a x^{d_1} = 0$, whence $t_1 a = 0$, hence $t_1 = 0$ (because $a \neq 0$). But then $t_2 p_2 + \cdots + t_m p_m = 0$ so we can repeat the argument to show that $t_2 = 0$. Continuing, we obtain $t_i = 0$ for each i, as desired.

EXAMPLE 5

Suppose that A is an $n \times n$ matrix such that $A^k = 0$ but $A^{k-1} \neq 0$. Show that $B = \{I, A, A^2, \ldots, A^{k-1}\}$ is independent in $\mathbf{M}_{nn}$.

Solution ▶ Suppose $r_0 I + r_1 A^1 + r_2 A^2 + \cdots + r_{k-1} A^{k-1} = 0$. Multiply by A^{k-1}:
$$r_0 A^{k-1} + r_1 A^k + r_2 A^{k+1} + \cdots + r_{k-1} A^{2k-2} = 0$$
Since $A^k = 0$, all the higher powers are zero, so this becomes $r_0 A^{k-1} = 0$. But $A^{k-1} \neq 0$, so $r_0 = 0$, and we have $r_1 A^1 + r_2 A^2 + \cdots + r_{k-1} A^{k-1} = 0$. Now multiply by A^{k-2} to conclude that $r_1 = 0$. Continuing, we obtain $r_i = 0$ for each i, so B is independent.

The next example collects several useful properties of independence for reference.

SECTION 6.3 Linear Independence and Dimension

EXAMPLE 6

Let V denote a vector space.
1. If $\mathbf{v} \neq \mathbf{0}$ in V, then $\{\mathbf{v}\}$ is an independent set.
2. No independent set of vectors in V can contain the zero vector.

Solution ▶

1. Let $t\mathbf{v} = \mathbf{0}$, t in $\mathbb{R}$. If $t \neq 0$, then $\mathbf{v} = 1\mathbf{v} = \frac{1}{t}(t\mathbf{v}) = \frac{1}{t}\mathbf{0} = \mathbf{0}$, contrary to assumption. So $t = 0$.
2. If $\{\mathbf{v}_1, \mathbf{v}_2, \ldots, \mathbf{v}_k\}$ is independent and (say) $\mathbf{v}_2 = \mathbf{0}$, then $0\mathbf{v}_1 + 1\mathbf{v}_2 + \cdots + 0\mathbf{v}_k = \mathbf{0}$ is a nontrivial linear combination that vanishes, contrary to the independence of $\{\mathbf{v}_1, \mathbf{v}_2, \ldots, \mathbf{v}_k\}$.

A set of vectors is independent if $\mathbf{0}$ is a linear combination in a unique way. The following theorem shows that *every* linear combination of these vectors has uniquely determined coefficients, and so extends Theorem 1 Section 5.2.

Theorem 1

Let $\{\mathbf{v}_1, \mathbf{v}_2, \ldots, \mathbf{v}_n\}$ *be a linearly independent set of vectors in a vector space V. If a vector $\mathbf{v}$ has two (ostensibly different) representations*

$$\mathbf{v} = s_1\mathbf{v}_1 + s_2\mathbf{v}_2 + \cdots + s_n\mathbf{v}_n$$
$$\mathbf{v} = t_1\mathbf{v}_1 + t_2\mathbf{v}_2 + \cdots + t_n\mathbf{v}_n$$

as linear combinations of these vectors, then $s_1 = t_1$, $s_2 = t_2$, ..., $s_n = t_n$. In other words, every vector in V can be written in a unique *way as a linear comination of the $\mathbf{v}_i$.*

PROOF

Subtracting the equations given in the theorem gives

$$(s_1 - t_1)\mathbf{v}_1 + (s_2 - t_2)\mathbf{v}_2 + \cdots + (s_n - t_n)\mathbf{v}_n = \mathbf{0}$$

The independence of $\{\mathbf{v}_1, \mathbf{v}_2, \ldots, \mathbf{v}_n\}$ gives $s_i - t_i = 0$ for each i, as required.

The following theorem extends (and proves) Theorem 4 Section 5.2, and is one of the most useful results in linear algebra.

Theorem 2

Fundamental Theorem
Suppose a vector space V can be spanned by n vectors. If any set of m vectors in V is linearly independent, then $m \leq n$.

PROOF

Let $V = \text{span}\{\mathbf{v}_1, \mathbf{v}_2, ..., \mathbf{v}_n\}$, and suppose that $\{\mathbf{u}_1, \mathbf{u}_2, ..., \mathbf{u}_m\}$ is an independent set in V. Then $\mathbf{u}_1 = a_1\mathbf{v}_1 + a_2\mathbf{v}_2 + \cdots + a_n\mathbf{v}_n$ where each a_i is in $\mathbb{R}$. As $\mathbf{u}_1 \neq \mathbf{0}$ (Example 6), not all of the a_i are zero, say $a_1 \neq 0$ (after relabelling the $\mathbf{v}_i$). Then $V = \text{span}\{\mathbf{u}_1, \mathbf{v}_2, \mathbf{v}_3, ..., \mathbf{v}_n\}$ as the reader can verify. Hence, write $\mathbf{u}_2 = b_1\mathbf{u}_1 + c_2\mathbf{v}_2 + c_3\mathbf{v}_3 + \cdots + c_n\mathbf{v}_n$. Then some $c_i \neq 0$ because $\{\mathbf{u}_1, \mathbf{u}_2\}$ is independent; so, as before, $V = \text{span}\{\mathbf{u}_1, \mathbf{u}_2, \mathbf{v}_3, ..., \mathbf{v}_n\}$, again after possible relabelling of the $\mathbf{v}_i$. If $m > n$, this procedure continues until all the vectors $\mathbf{v}_i$ are replaced by the vectors $\mathbf{u}_1, \mathbf{u}_2, ..., \mathbf{u}_n$. In particular, $V = \text{span}\{\mathbf{u}_1, \mathbf{u}_2, ..., \mathbf{u}_n\}$. But then $\mathbf{u}_{n+1}$ is a linear combination of $\mathbf{u}_1, \mathbf{u}_2, ..., \mathbf{u}_n$ contrary to the independence of the $\mathbf{u}_i$. Hence, the assumption $m > n$ cannot be valid, so $m \leq n$ and the theorem is proved.

If $V = \text{span}\{\mathbf{v}_1, \mathbf{v}_2, ..., \mathbf{v}_n\}$, and if $\{\mathbf{u}_1, \mathbf{u}_2, ..., \mathbf{u}_m\}$ is an independent set in V, the above proof shows not only that $m \leq n$ but also that m of the (spanning) vectors $\mathbf{v}_1, \mathbf{v}_2, ..., \mathbf{v}_n$ can be replaced by the (independent) vectors $\mathbf{u}_1, \mathbf{u}_2, ..., \mathbf{u}_m$ and the resulting set will still span V. In this form the result is called the **Steinitz Exchange Lemma**.

Definition 6.5 As in $\mathbb{R}^n$, a set $\{\mathbf{e}_1, \mathbf{e}_2, ..., \mathbf{e}_n\}$ of vectors in a vector space V is called a **basis** of V if it satisfies the following two conditions:

1. $\{\mathbf{e}_1, \mathbf{e}_2, ..., \mathbf{e}_n\}$ is linearly independent
2. $V = \text{span}\{\mathbf{e}_1, \mathbf{e}_2, ..., \mathbf{e}_n\}$

Thus if a set of vectors $\{\mathbf{e}_1, \mathbf{e}_2, ..., \mathbf{e}_n\}$ is a basis, then *every* vector in V can be written as a linear combination of these vectors in a *unique* way (Theorem 1). But even more is true: Any two (finite) bases of V contain the same number of vectors.

Theorem 3

Invariance Theorem
Let $\{\mathbf{e}_1, \mathbf{e}_2, ..., \mathbf{e}_n\}$ and $\{\mathbf{f}_1, \mathbf{f}_2, ..., \mathbf{f}_m\}$ be two bases of a vector space V. Then $n = m$.

PROOF

Because $V = \text{span}\{\mathbf{e}_1, \mathbf{e}_2, ..., \mathbf{e}_n\}$ and $\{\mathbf{f}_1, \mathbf{f}_2, ..., \mathbf{f}_m\}$ is independent, it follows from Theorem 2 that $m \leq n$. Similarly $n \leq m$, so $n = m$, as asserted.

Theorem 3 guarantees that no matter which basis of V is chosen it contains the same number of vectors as any other basis. Hence there is no ambiguity about the following definition.

SECTION 6.3 *Linear Independence and Dimension*

Definition 6.6 If $\{\mathbf{e}_1, \mathbf{e}_2, ..., \mathbf{e}_n\}$ is a basis of the nonzero vector space V, the number n of vectors in the basis is called the **dimension** of V, and we write

$$\dim V = n.$$

The zero vector space $\{\mathbf{0}\}$ is defined to have dimension 0:

$$\dim \{\mathbf{0}\} = 0.$$

In our discussion to this point we have always assumed that a basis is nonempty and hence that the dimension of the space is at least 1. However, the zero space $\{\mathbf{0}\}$ has *no* basis (by Example 6) so our insistence that $\dim\{\mathbf{0}\} = 0$ amounts to saying that the *empty* set of vectors is a basis of $\{\mathbf{0}\}$. Thus the statement that "the dimension of a vector space is the number of vectors in any basis" holds even for the zero space.

We saw in Example 9 Section 5.2 that $\dim(\mathbb{R}^n) = n$ and, if $\mathbf{e}_j$ denotes column j of I_n, that $\{\mathbf{e}_1, \mathbf{e}_2, ..., \mathbf{e}_n\}$ is a basis (called the standard basis). In Example 7 below, similar considerations apply to the space $\mathbf{M}_{mn}$ of all $m \times n$ matrices; the verifications are left to the reader.

EXAMPLE 7

The space $\mathbf{M}_{mn}$ has dimension mn, and one basis consists of all $m \times n$ matrices with exactly one entry equal to 1 and all other entries equal to 0. We call this the **standard basis** of $\mathbf{M}_{mn}$.

EXAMPLE 8

Show that $\dim \mathbf{P}_n = n + 1$ and that $\{1, x, x^2, ..., x^n\}$ is a basis, called the **standard basis** of $\mathbf{P}_n$.

Solution ▶ Each polynomial $p(x) = a_0 + a_1 x + \cdots + a_n x^n$ in $\mathbf{P}_n$ is clearly a linear combination of $1, x, ..., x^n$, so $\mathbf{P}_n = \text{span}\{1, x, ..., x^n\}$. However, if a linear combination of these vectors vanishes, $a_0 1 + a_1 x + \cdots + a_n x^n = 0$, then $a_0 = a_1 = \cdots = a_n = 0$ because x is an indeterminate. So $\{1, x, ..., x^n\}$ is linearly independent and hence is a basis containing $n + 1$ vectors. Thus, $\dim(\mathbf{P}_n) = n + 1$.

EXAMPLE 9

If $\mathbf{v} \neq \mathbf{0}$ is any nonzero vector in a vector space V, show that $\text{span}\{\mathbf{v}\} = \mathbb{R}\mathbf{v}$ has dimension 1.

Solution ▶ $\{\mathbf{v}\}$ clearly spans $\mathbb{R}\mathbf{v}$, and it is linearly independent by Example 6. Hence $\{\mathbf{v}\}$ is a basis of $\mathbb{R}\mathbf{v}$, and so $\dim \mathbb{R}\mathbf{v} = 1$.

EXAMPLE 10

Let $A = \begin{bmatrix} 1 & 1 \\ 0 & 0 \end{bmatrix}$ and consider the subspace

$$U = \{X \text{ in } \mathbf{M}_{22} \mid AX = XA\}$$

of $\mathbf{M}_{22}$. Show that dim $U = 2$ and find a basis of U.

Solution ▶ It was shown in Example 3 Section 6.2 that U is a subspace for any choice of the matrix A. In the present case, if $X = \begin{bmatrix} x & y \\ z & w \end{bmatrix}$ is in U, the condition $AX = XA$ gives $z = 0$ and $x = y + w$. Hence each matrix X in U can be written

$$X = \begin{bmatrix} y + w & y \\ 0 & w \end{bmatrix} = y\begin{bmatrix} 1 & 1 \\ 0 & 0 \end{bmatrix} + w\begin{bmatrix} 1 & 0 \\ 0 & 1 \end{bmatrix}$$

so $U = \text{span } B$ where $B = \left\{ \begin{bmatrix} 1 & 1 \\ 0 & 0 \end{bmatrix}, \begin{bmatrix} 1 & 0 \\ 0 & 1 \end{bmatrix} \right\}$. Moreover, the set B is linearly independent (verify this), so it is a basis of U and dim $U = 2$.

EXAMPLE 11

Show that the set V of all symmetric 2×2 matrices is a vector space, and find the dimension of V.

Solution ▶ A matrix A is symmetric if $A^T = A$. If A and B lie in V, then

$$(A + B)^T = A^T + B^T = A + B \quad \text{and} \quad (kA)^T = kA^T = kA$$

using Theorem 2 Section 2.1. Hence $A + B$ and kA are also symmetric. As the 2×2 zero matrix is also in V, this shows that V is a vector space (being a subspace of $\mathbf{M}_{22}$). Now a matrix A is symmetric when entries directly across the main diagonal are equal, so each 2×2 symmetric matrix has the form

$$\begin{bmatrix} a & c \\ c & b \end{bmatrix} = a\begin{bmatrix} 1 & 0 \\ 0 & 0 \end{bmatrix} + b\begin{bmatrix} 0 & 0 \\ 0 & 1 \end{bmatrix} + c\begin{bmatrix} 0 & 1 \\ 1 & 0 \end{bmatrix}$$

Hence the set $B = \left\{ \begin{bmatrix} 1 & 0 \\ 0 & 0 \end{bmatrix}, \begin{bmatrix} 0 & 0 \\ 0 & 1 \end{bmatrix}, \begin{bmatrix} 0 & 1 \\ 1 & 0 \end{bmatrix} \right\}$ spans V, and the reader can verify that B is linearly independent. Thus B is a basis of V, so dim $V = 3$.

It is frequently convenient to alter a basis by multiplying each basis vector by a nonzero scalar. The next example shows that this always produces another basis. The proof is left as Exercise 22.

EXAMPLE 12

Let $B = \{\mathbf{v}_1, \mathbf{v}_2, \ldots, \mathbf{v}_n\}$ be vectors in a vector space V. Given nonzero scalars $a_1, a_2, \ldots, a_n$, write $D = \{a_1\mathbf{v}_1, a_2\mathbf{v}_2, \ldots, a_n\mathbf{v}_n\}$. If B is independent or spans V, the same is true of D. In particular, if B is a basis of V, so also is D.

SECTION 6.3 Linear Independence and Dimension

EXERCISES 6.3

1. Show that each of the following sets of vectors is independent.

   (a) $\{1 + x, 1 - x, x + x^2\}$ in $\mathbf{P}_2$

   ◆(b) $\{x^2, x + 1, 1 - x - x^2\}$ in $\mathbf{P}_2$

   (c) $\left\{\begin{bmatrix} 1 & 1 \\ 0 & 0 \end{bmatrix}, \begin{bmatrix} 1 & 0 \\ 1 & 0 \end{bmatrix}, \begin{bmatrix} 0 & 0 \\ 1 & -1 \end{bmatrix}, \begin{bmatrix} 0 & 1 \\ 0 & 1 \end{bmatrix}\right\}$ in $\mathbf{M}_{22}$

   ◆(d) $\left\{\begin{bmatrix} 1 & 1 \\ 1 & 0 \end{bmatrix}, \begin{bmatrix} 0 & 1 \\ 1 & 1 \end{bmatrix}, \begin{bmatrix} 1 & 0 \\ 1 & 1 \end{bmatrix}, \begin{bmatrix} 1 & 1 \\ 0 & 1 \end{bmatrix}\right\}$ in $\mathbf{M}_{22}$

2. Which of the following subsets of V are independent?

   (a) $V = \mathbf{P}_2$; $\{x^2 + 1, x + 1, x\}$

   ◆(b) $V = \mathbf{P}_2$; $\{x^2 - x + 3, 2x^2 + x + 5, x^2 + 5x + 1\}$

   (c) $V = \mathbf{M}_{22}$; $\left\{\begin{bmatrix} 1 & 1 \\ 0 & 1 \end{bmatrix}, \begin{bmatrix} 1 & 0 \\ 1 & 1 \end{bmatrix}, \begin{bmatrix} 1 & 0 \\ 0 & 1 \end{bmatrix}\right\}$

   ◆(d) $V = \mathbf{M}_{22}$;
   $\left\{\begin{bmatrix} -1 & 0 \\ 0 & -1 \end{bmatrix}, \begin{bmatrix} 1 & -1 \\ -1 & 1 \end{bmatrix}, \begin{bmatrix} 1 & 1 \\ 1 & 1 \end{bmatrix}, \begin{bmatrix} 0 & -1 \\ -1 & 0 \end{bmatrix}\right\}$

   (e) $V = \mathbf{F}[1, 2]$; $\left\{\frac{1}{x}, \frac{1}{x^2}, \frac{1}{x^3}\right\}$

   ◆(f) $V = \mathbf{F}[0, 1]$;
   $\left\{\frac{1}{x^2 + x - 6}, \frac{1}{x^2 - 5x + 6}, \frac{1}{x^2 - 9}\right\}$

3. Which of the following are independent in $\mathbf{F}[0, 2\pi]$?

   (a) $\{\sin^2 x, \cos^2 x\}$

   ◆(b) $\{1, \sin^2 x, \cos^2 x\}$

   (c) $\{x, \sin^2 x, \cos^2 x\}$

4. Find all values of x such that the following are independent in $\mathbb{R}^3$.

   (a) $\{(1, -1, 0), (x, 1, 0), (0, 2, 3)\}$

   ◆(b) $\{(2, x, 1), (1, 0, 1), (0, 1, 3)\}$

5. Show that the following are bases of the space V indicated.

   (a) $\{(1, 1, 0), (1, 0, 1), (0, 1, 1)\}$; $V = \mathbb{R}^3$

   ◆(b) $\{(-1, 1, 1), (1, -1, 1), (1, 1, -1)\}$; $V = \mathbb{R}^3$

   (c) $\left\{\begin{bmatrix} 1 & 0 \\ 0 & 1 \end{bmatrix}, \begin{bmatrix} 0 & 1 \\ 1 & 0 \end{bmatrix}, \begin{bmatrix} 1 & 1 \\ 0 & 1 \end{bmatrix}, \begin{bmatrix} 1 & 0 \\ 0 & 0 \end{bmatrix}\right\}$; $V = \mathbf{M}_{22}$

   ◆(d) $\{1 + x, x + x^2, x^2 + x^3, x^3\}$; $V = \mathbf{P}_3$

6. Exhibit a basis and calculate the dimension of each of the following subspaces of $\mathbf{P}_2$.

   (a) $\{a(1 + x) + b(x + x^2) \mid a \text{ and } b \text{ in } \mathbb{R}\}$

   ◆(b) $\{a + b(x + x^2) \mid a \text{ and } b \text{ in } \mathbb{R}\}$

   (c) $\{p(x) \mid p(1) = 0\}$

   ◆(d) $\{p(x) \mid p(x) = p(-x)\}$

7. Exhibit a basis and calculate the dimension of each of the following subspaces of $\mathbf{M}_{22}$.

   (a) $\{A \mid A^T = -A\}$

   ◆(b) $\left\{A \mid A\begin{bmatrix} 1 & 1 \\ -1 & 0 \end{bmatrix} = \begin{bmatrix} 1 & 1 \\ -1 & 0 \end{bmatrix}A\right\}$

   (c) $\left\{A \mid A\begin{bmatrix} 1 & 0 \\ -1 & 0 \end{bmatrix} = \begin{bmatrix} 0 & 0 \\ 0 & 0 \end{bmatrix}\right\}$

   ◆(d) $\left\{A \mid A\begin{bmatrix} 1 & 1 \\ -1 & 0 \end{bmatrix} = \begin{bmatrix} 0 & 1 \\ -1 & 1 \end{bmatrix}A\right\}$

8. Let $A = \begin{bmatrix} 1 & 1 \\ 0 & 0 \end{bmatrix}$ and define
   $U = \{X \mid X \text{ is in } \mathbf{M}_{22} \text{ and } AX = X\}$.

   (a) Find a basis of U containing A.

   ◆(b) Find a basis of U not containing A.

9. Show that the set $\mathbb{C}$ of all complex numbers is a vector space with the usual operations, and find its dimension.

10. (a) Let V denote the set of all 2×2 matrices with equal column sums. Show that V is a subspace of $\mathbf{M}_{22}$, and compute dim V.

    ◆(b) Repeat part (a) for 3×3 matrices.

    (c) Repeat part (a) for $n \times n$ matrices.

11. (a) Let $V = \{(x^2 + x + 1)p(x) \mid p(x) \text{ in } \mathbf{P}_2\}$. Show that V is a subspace of $\mathbf{P}_4$ and find dim V. [Hint: If $f(x)g(x) = 0$ in $\mathbf{P}$, then $f(x) = 0$ or $g(x) = 0$.]

    ◆(b) Repeat with $V = \{(x^2 - x)p(x) \mid p(x) \text{ in } \mathbf{P}_3\}$, a subset of $\mathbf{P}_5$.

    (c) Generalize.

12. In each case, either prove the assertion or give an example showing that it is false.

   (a) Every set of four nonzero polynomials in $\mathbf{P}_3$ is a basis.

   ◆(b) $\mathbf{P}_2$ has a basis of polynomials $f(x)$ such that $f(0) = 0$.

   (c) $\mathbf{P}_2$ has a basis of polynomials $f(x)$ such that $f(0) = 1$.

   ◆(d) Every basis of $\mathbf{M}_{22}$ contains a noninvertible matrix.

   (e) No independent subset of $\mathbf{M}_{22}$ contains a matrix A with $A^2 = 0$.

   ◆(f) If $\{\mathbf{u}, \mathbf{v}, \mathbf{w}\}$ is independent then, $a\mathbf{u} + b\mathbf{v} + c\mathbf{w} = \mathbf{0}$ for some a, b, c.

   (g) $\{\mathbf{u}, \mathbf{v}, \mathbf{w}\}$ is independent if $a\mathbf{u} + b\mathbf{v} + c\mathbf{w} = \mathbf{0}$ for some a, b, c.

   ◆(h) If $\{\mathbf{u}, \mathbf{v}\}$ is independent, so is $\{\mathbf{u}, \mathbf{u} + \mathbf{v}\}$.

   (i) If $\{\mathbf{u}, \mathbf{v}\}$ is independent, so is $\{\mathbf{u}, \mathbf{v}, \mathbf{u} + \mathbf{v}\}$.

   ◆(j) If $\{\mathbf{u}, \mathbf{v}, \mathbf{w}\}$ is independent, so is $\{\mathbf{u}, \mathbf{v}\}$.

   (k) If $\{\mathbf{u}, \mathbf{v}, \mathbf{w}\}$ is independent, so is $\{\mathbf{u} + \mathbf{w}, \mathbf{v} + \mathbf{w}\}$.

   ◆(l) If $\{\mathbf{u}, \mathbf{v}, \mathbf{w}\}$ is independent, so is $\{\mathbf{u} + \mathbf{v} + \mathbf{w}\}$.

   (m) If $\mathbf{u} \neq \mathbf{0}$ and $\mathbf{v} \neq \mathbf{0}$ then $\{\mathbf{u}, \mathbf{v}\}$ is dependent if and only if one is a scalar multiple of the other.

   ◆(n) If $\dim V = n$, then no set of more than n vectors can be independent.

   (o) If $\dim V = n$, then no set of fewer than n vectors can span V.

13. Let $A \neq 0$ and $B \neq 0$ be $n \times n$ matrices, and assume that A is symmetric and B is skew-symmetric (that is, $B^T = -B$). Show that $\{A, B\}$ is independent.

14. Show that every set of vectors containing a dependent set is again dependent.

◆15. Show that every nonempty subset of an independent set of vectors is again independent.

16. Let f and g be functions on $[a, b]$, and assume that $f(a) = 1 = g(b)$ and $f(b) = 0 = g(a)$. Show that $\{f, g\}$ is independent in $\mathbf{F}[a, b]$.

17. Let $\{A_1, A_2, \ldots, A_k\}$ be independent in $\mathbf{M}_{mn}$, and suppose that U and V are invertible matrices of size $m \times m$ and $n \times n$, respectively. Show that $\{UA_1V, UA_2V, \ldots, UA_kV\}$ is independent.

18. Show that $\{\mathbf{v}, \mathbf{w}\}$ is independent if and only if neither $\mathbf{v}$ nor $\mathbf{w}$ is a scalar multiple of the other.

◆19. Assume that $\{\mathbf{u}, \mathbf{v}\}$ is independent in a vector space V. Write $\mathbf{u}' = a\mathbf{u} + b\mathbf{v}$ and $\mathbf{v}' = c\mathbf{u} + d\mathbf{v}$, where $a, b, c,$ and d are numbers. Show that $\{\mathbf{u}', \mathbf{v}'\}$ is independent if and only if the matrix $\begin{bmatrix} a & c \\ b & d \end{bmatrix}$ is invertible. [Hint: Theorem 5 Section 2.4.]

20. If $\{\mathbf{v}_1, \mathbf{v}_2, \ldots, \mathbf{v}_k\}$ is independent and $\mathbf{w}$ is not in span$\{\mathbf{v}_1, \mathbf{v}_2, \ldots, \mathbf{v}_k\}$, show that:

   (a) $\{\mathbf{w}, \mathbf{v}_1, \mathbf{v}_2, \ldots, \mathbf{v}_k\}$ is independent.

   (b) $\{\mathbf{v}_1 + \mathbf{w}, \mathbf{v}_2 + \mathbf{w}, \ldots, \mathbf{v}_k + \mathbf{w}\}$ is independent.

21. If $\{\mathbf{v}_1, \mathbf{v}_2, \ldots, \mathbf{v}_k\}$ is independent, show that $\{\mathbf{v}_1, \mathbf{v}_1 + \mathbf{v}_2, \ldots, \mathbf{v}_1 + \mathbf{v}_2 + \cdots + \mathbf{v}_k\}$ is also independent.

22. Prove Example 12.

23. Let $\{\mathbf{u}, \mathbf{v}, \mathbf{w}, \mathbf{z}\}$ be independent. Which of the following are dependent?

   (a) $\{\mathbf{u} - \mathbf{v}, \mathbf{v} - \mathbf{w}, \mathbf{w} - \mathbf{u}\}$

   ◆(b) $\{\mathbf{u} + \mathbf{v}, \mathbf{v} + \mathbf{w}, \mathbf{w} + \mathbf{u}\}$

   (c) $\{\mathbf{u} - \mathbf{v}, \mathbf{v} - \mathbf{w}, \mathbf{w} - \mathbf{z}, \mathbf{z} - \mathbf{u}\}$

   ◆(d) $\{\mathbf{u} + \mathbf{v}, \mathbf{v} + \mathbf{w}, \mathbf{w} + \mathbf{z}, \mathbf{z} + \mathbf{u}\}$

24. Let U and W be subspaces of V with bases $\{\mathbf{u}_1, \mathbf{u}_2, \mathbf{u}_3\}$ and $\{\mathbf{w}_1, \mathbf{w}_2\}$ respectively. If U and W have only the zero vector in common, show that $\{\mathbf{u}_1, \mathbf{u}_2, \mathbf{u}_3, \mathbf{w}_1, \mathbf{w}_2\}$ is independent.

25. Let $\{p, q\}$ be independent polynomials. Show that $\{p, q, pq\}$ is independent if and only if $\deg p \geq 1$ and $\deg q \geq 1$.

◆26. If z is a complex number, show that $\{z, z^2\}$ is independent if and only if z is not real.

27. Let $B = \{A_1, A_2, \ldots, A_n\} \subseteq \mathbf{M}_{mn}$, and write $B' = \{A_1^T, A_2^T, \ldots, A_n^T\} \subseteq \mathbf{M}_{nm}$. Show that:

   (a) B is independent if and only if B' is independent.

   (b) B spans $\mathbf{M}_{mn}$ if and only if B' spans $\mathbf{M}_{nm}$.

28. If $V = \mathbf{F}[a, b]$ as in Example 7 Section 6.1, show that the set of constant functions is a subspace of dimension 1 (f is **constant** if there is a number c such that $f(x) = c$ for all x).

29. (a) If U is an invertible $n \times n$ matrix and $\{A_1, A_2, \ldots, A_{mn}\}$ is a basis of $\mathbf{M}_{mn}$, show that $\{A_1U, A_2U, \ldots, A_{mn}U\}$ is also a basis.

    ◆(b) Show that part (a) fails if U is not invertible. [*Hint*: Theorem 5 Section 2.4.]

30. Show that $\{(a, b), (a_1, b_1)\}$ is a basis of $\mathbb{R}^2$ if and only if $\{a + bx, a_1 + b_1x\}$ is a basis of $\mathbf{P}_1$.

31. Find the dimension of the subspace span$\{1, \sin^2 \theta, \cos 2\theta\}$ of $\mathbf{F}[0, 2\pi]$.

32. Show that $\mathbf{F}[0, 1]$ is not finite dimensional.

33. If U and W are subspaces of V, define their intersection $U \cap W$ as follows:
    $$U \cap W = \{\mathbf{v} \mid \mathbf{v} \text{ is in both } U \text{ and } W\}$$
    (a) Show that $U \cap W$ is a subspace contained in U and W.

    ◆(b) Show that $U \cap W = \{\mathbf{0}\}$ if and only if $\{\mathbf{u}, \mathbf{w}\}$ is independent for any nonzero vectors $\mathbf{u}$ in U and $\mathbf{w}$ in W.

    (c) If B and D are bases of U and W, and if $U \cap W = \{\mathbf{0}\}$, show that $B \cup D = \{\mathbf{v} \mid \mathbf{v} \text{ is in } B \text{ or } D\}$ is independent.

34. If U and W are vector spaces, let $V = \{(\mathbf{u}, \mathbf{w}) \mid \mathbf{u} \text{ in } U \text{ and } \mathbf{w} \text{ in } W\}$.

    (a) Show that V is a vector space if $(\mathbf{u}, \mathbf{w}) + (\mathbf{u}_1, \mathbf{w}_1) = (\mathbf{u} + \mathbf{u}_1, \mathbf{w} + \mathbf{w}_1)$ and $a(\mathbf{u}, \mathbf{w}) = (a\mathbf{u}, a\mathbf{w})$.

    (b) If dim $U = m$ and dim $W = n$, show that dim $V = m + n$.

    (c) If $V_1, \ldots, V_m$ are vector spaces, let
    $V = V_1 \times \cdots \times V_m = \{(\mathbf{v}_1, \ldots, \mathbf{v}_m) \mid \mathbf{v}_i \text{ in } V_i \text{ for each } i\}$ denote the space of n-tuples from the V_i with componentwise operations (see Exercise 17 Section 6.1). If dim $V_i = n_i$ for each i, show that dim $V = n_1 + \cdots + n_m$.

35. Let $\mathbf{D}_n$ denote the set of all functions f from the set $\{1, 2, \ldots, n\}$ to $\mathbb{R}$.

    (a) Show that $\mathbf{D}_n$ is a vector space with pointwise addition and scalar multiplication.

    (b) Show that $\{S_1, S_2, \ldots, S_n\}$ is a basis of $\mathbf{D}_n$ where, for each $k = 1, 2, \ldots, n$, the function S_k is defined by $S_k(k) = 1$, whereas $S_k(j) = 0$ if $j \neq k$.

36. A polynomial $p(x)$ is **even** if $p(-x) = p(x)$ and **odd** if $p(-x) = -p(x)$. Let E_n and O_n denote the sets of even and odd polynomials in $\mathbf{P}_n$.

    (a) Show that E_n is a subspace of $\mathbf{P}_n$ and find dim E_n.

    ◆(b) Show that O_n is a subspace of $\mathbf{P}_n$ and find dim O_n.

37. Let $\{\mathbf{v}_1, \ldots, \mathbf{v}_n\}$ be independent in a vector space V, and let A be an $n \times n$ matrix. Define $\mathbf{u}_1, \ldots, \mathbf{u}_n$ by
    $$\begin{bmatrix} \mathbf{u}_1 \\ \vdots \\ \mathbf{u}_n \end{bmatrix} = A \begin{bmatrix} \mathbf{v}_1 \\ \vdots \\ \mathbf{v}_n \end{bmatrix}$$
    (See Exercise 18 Section 6.1.) Show that $\{\mathbf{u}_1, \ldots, \mathbf{u}_n\}$ is independent if and only if A is invertible.

SECTION 6.4 Finite Dimensional Spaces

Up to this point, we have had no guarantee that an arbitrary vector space *has* a basis—and hence no guarantee that one can speak *at all* of the dimension of V. However, Theorem 1 will show that any space that is spanned by a finite set of vectors has a (finite) basis: The proof requires the following basic lemma, of interest in itself, that gives a way to enlarge a given independent set of vectors.

Lemma 1

Independent Lemma

Let $\{\mathbf{v}_1, \mathbf{v}_2, \ldots, \mathbf{v}_k\}$ be an independent set of vectors in a vector space V. If $\mathbf{u} \in V$ but[4] $\mathbf{u} \notin \text{span}\{\mathbf{v}_1, \mathbf{v}_2, \ldots, \mathbf{v}_k\}$, then $\{\mathbf{u}, \mathbf{v}_1, \mathbf{v}_2, \ldots, \mathbf{v}_k\}$ is also independent.

PROOF

Let $t\mathbf{u} + t_1\mathbf{v}_1 + t_2\mathbf{v}_2 + \cdots + t_k\mathbf{v}_k = \mathbf{0}$; we must show that all the coefficients are zero. First, $t = 0$ because, otherwise, $\mathbf{u} = -\frac{t_1}{t}\mathbf{v}_1 - \frac{t_2}{t}\mathbf{v}_2 - \cdots - \frac{t_k}{t}\mathbf{v}_k$ is in $\text{span}\{\mathbf{v}_1, \mathbf{v}_2, \ldots, \mathbf{v}_k\}$, contrary to our assumption. Hence $t = 0$. But then $t_1\mathbf{v}_1 + t_2\mathbf{v}_2 + \cdots + t_k\mathbf{v}_k = \mathbf{0}$ so the rest of the t_i are zero by the independence of $\{\mathbf{v}_1, \mathbf{v}_2, \ldots, \mathbf{v}_k\}$. This is what we wanted.

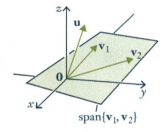

Note that the converse of Lemma 1 is also true: if $\{\mathbf{u}, \mathbf{v}_1, \mathbf{v}_2, \ldots, \mathbf{v}_k\}$ is independent, then $\mathbf{u}$ is not in $\text{span}\{\mathbf{v}_1, \mathbf{v}_2, \ldots, \mathbf{v}_k\}$.

As an illustration, suppose that $\{\mathbf{v}_1, \mathbf{v}_2\}$ is independent in $\mathbb{R}^3$. Then $\mathbf{v}_1$ and $\mathbf{v}_2$ are not parallel, so $\text{span}\{\mathbf{v}_1, \mathbf{v}_2\}$ is a plane through the origin (shaded in the diagram). By Lemma 1, $\mathbf{u}$ is not in this plane if and only if $\{\mathbf{u}, \mathbf{v}_1, \mathbf{v}_2\}$ is independent.

Definition 6.7   *A vector space V is called **finite dimensional** if it is spanned by a finite set of vectors. Otherwise, V is called **infinite dimensional**.*

Thus the zero vector space $\{\mathbf{0}\}$ is finite dimensional because $\{\mathbf{0}\}$ is a spanning set.

Lemma 2

Let V be a finite dimensional vector space. If U is any subspace of V, then any independent subset of U can be enlarged to a finite basis of U.

PROOF

Suppose that I is an independent subset of U. If $\text{span } I = U$ then I is already a basis of U. If $\text{span } I \neq U$, choose $\mathbf{u}_1 \in U$ such that $\mathbf{u}_1 \notin \text{span } I$. Hence the set $I \cup \{\mathbf{u}_1\}$ is independent by Lemma 1. If $\text{span}\{I \cup \{\mathbf{u}_1\}\} = U$ we are done; otherwise choose $\mathbf{u}_2 \in U$ such that $\mathbf{u}_2 \notin \text{span}\{I \cup \{\mathbf{u}_1\}\}$. Hence $\{I \cup \{\mathbf{u}_1, \mathbf{u}_2\}\}$ is independent, and the process continues. We claim that a basis of U will be reached eventually. Indeed, if no basis of U is ever reached, the process creates arbitrarily large independent sets in V. But this is impossible by the fundamental theorem because V is finite dimensional and so is spanned by a finite set of vectors.

---

[4] If X is a set, we write $a \in X$ to indicate that a is an element of the set X. If a is not an element of X, we write $a \notin X$.

SECTION 6.4 Finite Dimensional Spaces

Theorem 1

Let V be a finite dimensional vector space spanned by m vectors.
(1) V has a finite basis, and $\dim V \leq m$.
(2) Every independent set of vectors in V can be enlarged to a basis of V by adding vectors from any fixed basis of V.
(3) If U is a subspace of V, then
  (a) U is finite dimensional and $\dim U \leq \dim V$.
  (b) Every basis of U is part of a basis of V.

PROOF

(1) If $V = \{\mathbf{0}\}$, then V has an empty basis and $\dim V = 0 \leq m$. Otherwise, let $\mathbf{v} \neq \mathbf{0}$ be a vector in V. Then $\{\mathbf{v}\}$ is independent, so (1) follows from Lemma 2 with $U = V$.

(2) We refine the proof of Lemma 2. Fix a basis B of V and let I be an independent subset of V. If span $I = V$ then I is already a basis of V. If span $I \neq V$, then B is not contained in I (because B spans V). Hence choose $\mathbf{b}_1 \in B$ such that $\mathbf{b}_1 \notin$ span I. Hence the set $I \cup \{\mathbf{b}_1\}$ is independent by Lemma 1. If span$\{I \cup \{\mathbf{b}_1\}\} = V$ we are done; otherwise a similar argument shows that $\{I \cup \{\mathbf{b}_1, \mathbf{b}_2\}\}$ is independent for some $\mathbf{b}_2 \in B$. Continue this process. As in the proof of Lemma 2, a basis of V will be reached eventually.

(3a) This is clear if $U = \{\mathbf{0}\}$. Otherwise, let $\mathbf{u} \neq \mathbf{0}$ in U. Then $\{\mathbf{u}\}$ can be enlarged to a finite basis B of U by Lemma 2, proving that U is finite dimensional. But B is independent in V, so $\dim U \leq \dim V$ by the fundamental theorem.

(3b) This is clear if $U = \{\mathbf{0}\}$ because V *has* a basis; otherwise, it follows from (2).

Theorem 1 shows that a vector space V is finite dimensional if and only if it has a finite basis (possibly empty), and that every subspace of a finite dimensional space is again finite dimensional.

EXAMPLE 1

Enlarge the independent set $D = \left\{ \begin{bmatrix} 1 & 1 \\ 1 & 0 \end{bmatrix}, \begin{bmatrix} 0 & 1 \\ 1 & 1 \end{bmatrix}, \begin{bmatrix} 1 & 0 \\ 1 & 1 \end{bmatrix} \right\}$ to a basis of $\mathbf{M}_{22}$.

Solution ▶ The standard basis of $\mathbf{M}_{22}$ is $\left\{ \begin{bmatrix} 1 & 0 \\ 0 & 0 \end{bmatrix}, \begin{bmatrix} 0 & 1 \\ 0 & 0 \end{bmatrix}, \begin{bmatrix} 0 & 0 \\ 1 & 0 \end{bmatrix}, \begin{bmatrix} 0 & 0 \\ 0 & 1 \end{bmatrix} \right\}$, so including one of these in D will produce a basis by Theorem 1. In fact including *any* of these matrices in D produces an independent set (verify), and hence a basis by Theorem 4. Of course these vectors are not the only possibilities, for example, including $\begin{bmatrix} 1 & 1 \\ 0 & 1 \end{bmatrix}$ works as well.

EXAMPLE 2

Find a basis of $\mathbf{P}_3$ containing the independent set $\{1 + x, 1 + x^2\}$.

Solution ▶ The standard basis of $\mathbf{P}_3$ is $\{1, x, x^2, x^3\}$, so including two of these vectors will do. If we use 1 and x^3, the result is $\{1, 1 + x, 1 + x^2, x^3\}$. This is independent because the polynomials have distinct degrees (Example 4 Section 6.3), and so is a basis by Theorem 1. Of course, including $\{1, x\}$ or $\{1, x^2\}$ would *not* work!

EXAMPLE 3

Show that the space $\mathbf{P}$ of all polynomials is infinite dimensional.

Solution ▶ For each $n \geq 1$, $\mathbf{P}$ has a subspace $\mathbf{P}_n$ of dimension $n + 1$. Suppose $\mathbf{P}$ is finite dimensional, say dim $\mathbf{P} = m$. Then dim $\mathbf{P}_n \leq$ dim $\mathbf{P}$ by Theorem 1, that is $n + 1 \leq m$. This is impossible since n is arbitrary, so $\mathbf{P}$ must be infinite dimensional.

The next example illustrates how (2) of Theorem 1 can be used.

EXAMPLE 4

If $\mathbf{c}_1, \mathbf{c}_2, \ldots, \mathbf{c}_k$ are independent columns in $\mathbb{R}^n$, show that they are the first k columns in some invertible $n \times n$ matrix.

Solution ▶ By Theorem 1, expand $\{\mathbf{c}_1, \mathbf{c}_2, \ldots, \mathbf{c}_k\}$ to a basis $\{\mathbf{c}_1, \mathbf{c}_2, \ldots, \mathbf{c}_k, \mathbf{c}_{k+1}, \ldots, \mathbf{c}_n\}$ of $\mathbb{R}^n$. Then the matrix $A = [\mathbf{c}_1 \ \mathbf{c}_2 \ \cdots \ \mathbf{c}_k \ \mathbf{c}_{k+1} \ \cdots \ \mathbf{c}_n]$ with this basis as its columns is an $n \times n$ matrix and it is invertible by Theorem 3 Section 5.2.

Theorem 2

Let U and W be subspaces of the finite dimensional space V.
1. If $U \subseteq W$, then dim $U \leq$ dim W.
2. If $U \subseteq W$ and dim $U =$ dim W, then $U = W$.

PROOF

Since W is finite dimensional, (1) follows by taking $V = W$ in part (3) of Theorem 1. Now assume dim $U =$ dim $W = n$, and let B be a basis of U. Then B is an independent set in W. If $U \neq W$, then span $B \neq W$, so B can be extended to an independent set of $n + 1$ vectors in W by Lemma 1. This contradicts the fundamental theorem (Theorem 2 Section 6.3) because W is spanned by dim $W = n$ vectors. Hence $U = W$, proving (2).

SECTION 6.4 Finite Dimensional Spaces

Theorem 2 is very useful. This was illustrated in Example 13 Section 5.2 for $\mathbb{R}^2$ and $\mathbb{R}^3$; here is another example.

EXAMPLE 5

If a is a number, let W denote the subspace of all polynomials in $\mathbf{P}_n$ that have a as a root:
$$W = \{p(x) \mid p(x) \text{ is in } \mathbf{P}_n \text{ and } p(a) = 0\}.$$
Show that $\{(x - a), (x - a)^2, \ldots, (x - a)^n\}$ is a basis of W.

Solution ▶ Observe first that $(x - a), (x - a)^2, \ldots, (x - a)^n$ are members of W, and that they are independent because they have distinct degrees (Example 4 Section 6.3). Write
$$U = \text{span}\{(x - a), (x - a)^2, \ldots, (x - a)^n\}$$
Then we have $U \subseteq W \subseteq \mathbf{P}_n$, dim $U = n$, and dim $\mathbf{P}_n = n + 1$. Hence $n \leq \dim W \leq n + 1$ by Theorem 2. Since dim W is an integer, we must have dim $W = n$ or dim $W = n + 1$. But then $W = U$ or $W = \mathbf{P}_n$, again by Theorem 2. Because $W \neq \mathbf{P}_n$, it follows that $W = U$, as required.

A set of vectors is called **dependent** if it is *not* independent, that is if some nontrivial linear combination vanishes. The next result is a convenient test for dependence.

Lemma 3

Dependent Lemma
A set $D = \{\mathbf{v}_1, \mathbf{v}_2, \ldots, \mathbf{v}_k\}$ of vectors in a vector space V is dependent if and only if some vector in D is a linear combination of the others.

PROOF

Let $\mathbf{v}_2$ (say) be a linear combination of the rest: $\mathbf{v}_2 = s_1\mathbf{v}_1 + s_3\mathbf{v}_3 + \cdots + s_k\mathbf{v}_k$. Then $s_1\mathbf{v}_1 + (-1)\mathbf{v}_2 + s_3\mathbf{v}_3 + \cdots + s_k\mathbf{v}_k = \mathbf{0}$ is a nontrivial linear combination that vanishes, so D is dependent. Conversely, if D is dependent, let $t_1\mathbf{v}_1 + t_2\mathbf{v}_2 + \cdots + t_k\mathbf{v}_k = \mathbf{0}$ where some coefficient is nonzero. If (say) $t_2 \neq 0$, then $\mathbf{v}_2 = -\frac{t_1}{t_2}\mathbf{v}_1 - \frac{t_3}{t_2}\mathbf{v}_3 - \cdots - \frac{t_k}{t_2}\mathbf{v}_k$ is a linear combination of the others.

Lemma 1 gives a way to enlarge independent sets to a basis; by contrast, Lemma 3 shows that spanning sets can be cut down to a basis.

Theorem 3

Let V be a finite dimensional vector space. Any spanning set for V can be cut down (by deleting vectors) to a basis of V.

PROOF

Since V is finite dimensional, it has a finite spanning set S. Among all spanning sets contained in S, choose S_0 containing the smallest number of vectors. It suffices to show that S_0 is independent (then S_0 is a basis, proving the theorem). Suppose, on the contrary, that S_0 is not independent. Then, by Lemma 3, some vector $\mathbf{u} \in S_0$ is a linear combination of the set $S_1 = S_0 \setminus \{\mathbf{u}\}$ of vectors in S_0 other than $\mathbf{u}$. It follows that span $S_0 = $ span S_1, that is, $V = $ span S_1. But S_1 has fewer elements than S_0 so this contradicts the choice of S_0. Hence S_0 is independent after all.

Note that, with Theorem 1, Theorem 3 completes the promised proof of Theorem 6 Section 5.2 for the case $V = \mathbb{R}^n$.

EXAMPLE 6

Find a basis of $\mathbf{P}_3$ in the spanning set $S = \{1, x + x^2, 2x - 3x^2, 1 + 3x - 2x^2, x^3\}$.

Solution ▶ Since dim $\mathbf{P}_3 = 4$, we must eliminate one polynomial from S. It cannot be x^3 because the span of the rest of S is contained in $\mathbf{P}_2$. But eliminating $1 + 3x - 2x^2$ does leave a basis (verify). Note that $1 + 3x - 2x^2$ is the sum of the first three polynomials in S.

Theorems 1 and 3 have other useful consequences.

Theorem 4

Let V be a vector space with dim $V = n$, and suppose S is a set of exactly n vectors in V. Then S is independent if and only if S spans V.

PROOF

Assume first that S is independent. By Theorem 1, S is contained in a basis B of V. Hence $|S| = n = |B|$ so, since $S \subseteq B$, it follows that $S = B$. In particular S spans V.

Conversely, assume that S spans V, so S contains a basis B by Theorem 3. Again $|S| = n = |B|$ so, since $S \supseteq B$, it follows that $S = B$. Hence S is independent.

One of independence or spanning is often easier to establish than the other when showing that a set of vectors is a basis. For example if $V = \mathbb{R}^n$ it is easy to check whether a subset S of $\mathbb{R}^n$ is orthogonal (hence independent) but checking spanning can be tedious. Here are three more examples.

SECTION 6.4 Finite Dimensional Spaces

EXAMPLE 7

Consider the set $S = \{p_0(x), p_1(x), \ldots, p_n(x)\}$ of polynomials in $\mathbf{P}_n$. If $\deg p_k(x) = k$ for each k, show that S is a basis of $\mathbf{P}_n$.

Solution ▶ The set S is independent—the degrees are distinct—see Example 4 Section 6.3. Hence S is a basis of $\mathbf{P}_n$ by Theorem 4 because $\dim \mathbf{P}_n = n + 1$.

EXAMPLE 8

Let V denote the space of all symmetric 2×2 matrices. Find a basis of V consisting of invertible matrices.

Solution ▶ We know that $\dim V = 3$ (Example 11 Section 6.3), so what is needed is a set of three invertible, symmetric matrices that (using Theorem 4) is either independent or spans V. The set $\left\{ \begin{bmatrix} 1 & 0 \\ 0 & 1 \end{bmatrix}, \begin{bmatrix} 1 & 0 \\ 0 & -1 \end{bmatrix}, \begin{bmatrix} 0 & 1 \\ 1 & 0 \end{bmatrix} \right\}$ is independent (verify) and so is a basis of the required type.

EXAMPLE 9

Let A be any $n \times n$ matrix. Show that there exist $n^2 + 1$ scalars $a_0, a_1, a_2, \ldots, a_{n^2}$ not all zero, such that

$$a_0 I + a_1 A + a_2 A^2 + \cdots + a_{n^2} A^{n^2} = 0$$

where I denotes the $n \times n$ identity matrix.

Solution ▶ The space $\mathbf{M}_{nn}$ of all $n \times n$ matrices has dimension n^2 by Example 7 Section 6.3. Hence the $n^2 + 1$ matrices $I, A, A^2, \ldots, A^{n^2}$ cannot be independent by Theorem 4, so a nontrivial linear combination vanishes. This is the desired conclusion.

Note that the result in Example 9 can be written as $f(A) = 0$ where $f(x) = a_0 + a_1 x + a_2 x^2 + \cdots + a_{n^2} x^{n^2}$. In other words, A satisfies a nonzero polynomial $f(x)$ of degree at most n^2. In fact we know that A satisfies a nonzero polynomial of degree n (this is the Cayley-Hamilton theorem—see Theorem 10 Section 8.6 or Theorem 2 Section 9.4), but the brevity of the solution in Example 6 is an indication of the power of these methods.

If U and W are subspaces of a vector space V, there are two related subspaces that are of interest, their **sum** $U + W$ and their **intersection** $U \cap W$, defined by

$$U + W = \{\mathbf{u} + \mathbf{w} \mid \mathbf{u} \text{ in } U, \text{ and } \mathbf{w} \text{ in } W\}$$
$$U \cap W = \{\mathbf{v} \text{ in } V \mid \mathbf{v} \text{ in both } U \text{ and } W\}$$

It is routine to verify that these are indeed subspaces of V, that $U \cap W$ is contained in both U and W, and that $U + W$ contains both U and W. We conclude this section with a useful fact about the dimensions of these spaces. The proof is a good illustration of how the theorems in this section are used.

Theorem 5

Suppose that U and W are finite dimensional subspaces of a vector space V. Then $U + W$ is finite dimensional and

$$\dim(U + W) = \dim U + \dim W - \dim(U \cap W).$$

PROOF

Since $U \cap W \subseteq U$, it has a finite basis, say $\{\mathbf{x}_1, \ldots, \mathbf{x}_d\}$. Extend it to a basis $\{\mathbf{x}_1, \ldots, \mathbf{x}_d, \mathbf{u}_1, \ldots, \mathbf{u}_m\}$ of U by Theorem 1. Similarly extend $\{\mathbf{x}_1, \ldots, \mathbf{x}_d\}$ to a basis $\{\mathbf{x}_1, \ldots, \mathbf{x}_d, \mathbf{w}_1, \ldots, \mathbf{w}_p\}$ of W. Then

$$U + W = \text{span}\{\mathbf{x}_1, \ldots, \mathbf{x}_d, \mathbf{u}_1, \ldots, \mathbf{u}_m, \mathbf{w}_1, \ldots, \mathbf{w}_p\}$$

as the reader can verify, so $U + W$ is finite dimensional. For the rest, it suffices to show that $\{\mathbf{x}_1, \ldots, \mathbf{x}_d, \mathbf{u}_1, \ldots, \mathbf{u}_m, \mathbf{w}_1, \ldots, \mathbf{w}_p\}$ is independent (verify). Suppose that

$$r_1\mathbf{x}_1 + \cdots + r_d\mathbf{x}_d + s_1\mathbf{u}_1 + \cdots + s_m\mathbf{u}_m + t_1\mathbf{w}_1 + \cdots + t_p\mathbf{w}_p = \mathbf{0} \quad (*)$$

where the r_i, s_j, and t_k are scalars. Then

$$r_1\mathbf{x}_1 + \cdots + r_d\mathbf{x}_d + s_1\mathbf{u}_1 + \cdots + s_m\mathbf{u}_m = -(t_1\mathbf{w}_1 + \cdots + t_p\mathbf{w}_p)$$

is in U (left side) and also in W (right side), and so is in $U \cap W$. Hence $(t_1\mathbf{w}_1 + \cdots + t_p\mathbf{w}_p)$ is a linear combination of $\{\mathbf{x}_1, \ldots, \mathbf{x}_d\}$, so $t_1 = \cdots = t_p = 0$, because $\{\mathbf{x}_1, \ldots, \mathbf{x}_d, \mathbf{w}_1, \ldots, \mathbf{w}_p\}$ is independent. Similarly, $s_1 = \cdots = s_m = 0$, so $(*)$ becomes $r_1\mathbf{x}_1 + \cdots + r_d\mathbf{x}_d = \mathbf{0}$. It follows that $r_1 = \cdots = r_d = 0$, as required.

Theorem 5 is particularly interesting if $U \cap W = \{\mathbf{0}\}$. Then there are *no* vectors $\mathbf{x}_i$ in the above proof, and the argument shows that if $\{\mathbf{u}_1, \ldots, \mathbf{u}_m\}$ and $\{\mathbf{w}_1, \ldots, \mathbf{w}_p\}$ are bases of U and W respectively, then $\{\mathbf{u}_1, \ldots, \mathbf{u}_m, \mathbf{w}_1, \ldots, \mathbf{w}_p\}$ is a basis of $U + W$. In this case $U + W$ is said to be a **direct sum** (written $U \oplus W$); we return to this in Chapter 9.

EXERCISES 6.4

1. In each case, find a basis for V that includes the vector $\mathbf{v}$.
   - (a) $V = \mathbb{R}^3$, $\mathbf{v} = (1, -1, 1)$
   - ◆(b) $V = \mathbb{R}^3$, $\mathbf{v} = (0, 1, 1)$
   - (c) $V = \mathbf{M}_{22}$, $\mathbf{v} = \begin{bmatrix} 1 & 1 \\ 1 & 1 \end{bmatrix}$
   - ◆(d) $V = \mathbf{P}_2$, $\mathbf{v} = x^2 - x + 1$

2. In each case, find a basis for V among the given vectors.
   - (a) $V = \mathbb{R}^3$, $\{(1, 1, -1), (2, 0, 1), (-1, 1, -2), (1, 2, 1)\}$
   - ◆(b) $V = \mathbf{P}_2$, $\{x^2 + 3, x + 2, x^2 - 2x - 1, x^2 + x\}$

3. In each case, find a basis of V containing $\mathbf{v}$ and $\mathbf{w}$.
   - (a) $V = \mathbb{R}^4$, $\mathbf{v} = (1, -1, 1, -1)$, $\mathbf{w} = (0, 1, 0, 1)$
   - ◆(b) $V = \mathbb{R}^4$, $\mathbf{v} = (0, 0, 1, 1)$, $\mathbf{w} = (1, 1, 1, 1)$
   - (c) $V = \mathbf{M}_{22}$, $\mathbf{v} = \begin{bmatrix} 1 & 0 \\ 0 & 1 \end{bmatrix}$, $\mathbf{w} = \begin{bmatrix} 0 & 1 \\ 1 & 0 \end{bmatrix}$
   - ◆(d) $V = \mathbf{P}_3$, $\mathbf{v} = x^2 + 1$, $\mathbf{w} = x^2 + x$

4. (a) If z is not a real number, show that $\{z, z^2\}$ is a basis of the real vector space $\mathbb{C}$ of all complex numbers.

◆(b) If z is neither real nor pure imaginary, show that $\{z, \bar{z}\}$ is a basis of $\mathbb{C}$.

5. In each case use Theorem 4 to decide if S is a basis of V.

   (a) $V = \mathbf{M}_{22}$;
   $$S = \left\{ \begin{bmatrix} 1 & 1 \\ 1 & 1 \end{bmatrix}, \begin{bmatrix} 0 & 1 \\ 1 & 1 \end{bmatrix}, \begin{bmatrix} 0 & 0 \\ 1 & 1 \end{bmatrix}, \begin{bmatrix} 0 & 0 \\ 0 & 1 \end{bmatrix} \right\}.$$

   ◆(b) $V = \mathbf{P}_3$; $S = \{2x^2, 1 + x, 3, 1 + x + x^2 + x^3\}$.

6. (a) Find a basis of $\mathbf{M}_{22}$ consisting of matrices with the property that $A^2 = A$.

   ◆(b) Find a basis of $\mathbf{P}_3$ consisting of polynomials whose coefficients sum to 4. What if they sum to 0?

7. If $\{\mathbf{u}, \mathbf{v}, \mathbf{w}\}$ is a basis of V, determine which of the following are bases.

   (a) $\{\mathbf{u} + \mathbf{v}, \mathbf{u} + \mathbf{w}, \mathbf{v} + \mathbf{w}\}$

   ◆(b) $\{2\mathbf{u} + \mathbf{v} + 3\mathbf{w}, 3\mathbf{u} + \mathbf{v} - \mathbf{w}, \mathbf{u} - 4\mathbf{w}\}$

   (c) $\{\mathbf{u}, \mathbf{u} + \mathbf{v} + \mathbf{w}\}$

   ◆(d) $\{\mathbf{u}, \mathbf{u} + \mathbf{w}, \mathbf{u} - \mathbf{w}, \mathbf{v} + \mathbf{w}\}$

8. (a) Can two vectors span $\mathbb{R}^3$? Can they be linearly independent? Explain.

   ◆(b) Can four vectors span $\mathbb{R}^3$? Can they be linearly independent? Explain.

9. Show that any nonzero vector in a finite dimensional vector space is part of a basis.

◆10. If A is a square matrix, show that $\det A = 0$ if and only if some row is a linear combination of the others.

11. Let D, I, and X denote finite, nonempty sets of vectors in a vector space V. Assume that D is dependent and I is independent. In each case answer yes or no, and defend your answer.

    (a) If $X \supseteq D$, must X be dependent?

    ◆(b) If $X \subseteq D$, must X be dependent?

    (c) If $X \supseteq I$, must X be independent?

    ◆(d) If $X \subseteq I$, must X be independent?

12. If U and W are subspaces of V and $\dim U = 2$, show that either $U \subseteq W$ or $\dim(U \cap W) \leq 1$.

13. Let A be a nonzero 2×2 matrix and write $U = \{X \text{ in } \mathbf{M}_{22} \mid XA = AX\}$. Show that $\dim U \geq 2$. [*Hint*: I and A are in U.]

14. If $U \subseteq \mathbb{R}^2$ is a subspace, show that $U = \{\mathbf{0}\}$, $U = \mathbb{R}^2$, or U is a line through the origin.

◆15. Given $\mathbf{v}_1, \mathbf{v}_2, \mathbf{v}_3, \ldots, \mathbf{v}_k$, and $\mathbf{v}$, let $U = \text{span}\{\mathbf{v}_1, \mathbf{v}_2, \ldots, \mathbf{v}_k\}$ and $W = \text{span}\{\mathbf{v}_1, \mathbf{v}_2, \ldots, \mathbf{v}_k, \mathbf{v}\}$. Show that either $\dim W = \dim U$ or $\dim W = 1 + \dim U$.

16. Suppose U is a subspace of $\mathbf{P}_1$, $U \neq \{\mathbf{0}\}$, and $U \neq \mathbf{P}_1$. Show that either $U = \mathbb{R}$ or $U = \mathbb{R}(a + x)$ for some a in $\mathbb{R}$.

17. Let U be a subspace of V and assume $\dim V = 4$ and $\dim U = 2$. Does every basis of V result from adding (two) vectors to some basis of U? Defend your answer.

18. Let U and W be subspaces of a vector space V.

    (a) If $\dim V = 3$, $\dim U = \dim W = 2$, and $U \neq W$, show that $\dim(U \cap W) = 1$.

    ◆(b) Interpret (a) geometrically if $V = \mathbb{R}^3$.

19. Let $U \subseteq W$ be subspaces of V with $\dim U = k$ and $\dim W = m$, where $k < m$. If $k < l < m$, show that a subspace X exists where $U \subseteq X \subseteq W$ and $\dim X = l$.

20. Let $B = \{\mathbf{v}_1, \ldots, \mathbf{v}_n\}$ be a *maximal* independent set in a vector space V. That is, no set of more than n vectors S is independent. Show that B is a basis of V.

21. Let $B = \{\mathbf{v}_1, \ldots, \mathbf{v}_n\}$ be a *minimal* spanning set for a vector space V. That is, V cannot be spanned by fewer than n vectors. Show that B is a basis of V.

22. (a) Let $p(x)$ and $q(x)$ lie in $\mathbf{P}_1$ and suppose that $p(1) \neq 0$, $q(2) \neq 0$, and $p(2) = 0 = q(1)$. Show that $\{p(x), q(x)\}$ is a basis of $\mathbf{P}_1$. [*Hint*: If $rp(x) + sq(x) = 0$, evaluate at $x = 1$, $x = 2$.]

    (b) Let $B = \{p_0(x), p_1(x), \ldots, p_n(x)\}$ be a set of polynomials in $\mathbf{P}_n$. Assume that there exist numbers $a_0, a_1, \ldots, a_n$ such that $p_i(a_i) \neq 0$ for each i but $p_i(a_j) = 0$ if i is different from j. Show that B is a basis of $\mathbf{P}_n$.

23. Let V be the set of all infinite sequences $(a_0, a_1, a_2, \ldots)$ of real numbers. Define addition and scalar multiplication by
    $(a_0, a_1, \ldots) + (b_0, b_1, \ldots) = (a_0 + b_0, a_1 + b_1, \ldots)$
    and $r(a_0, a_1, \ldots) = (ra_0, ra_1, \ldots)$.

    (a) Show that V is a vector space.

    ◆(b) Show that V is not finite dimensional.

    (c) [For those with some calculus.] Show that the set of convergent sequences (that is, $\lim_{n \to \infty} a_n$ exists) is a subspace, also of infinite dimension.

24. Let A be an $n \times n$ matrix of rank r. If $U = \{X \text{ in } \mathbf{M}_{nn} \mid AX = 0\}$, show that $\dim U = n(n - r)$. [Hint: Exercise 34 Section 6.3.]

25. Let U and W be subspaces of V.

    (a) Show that $U + W$ is a subspace of V containing U and W.

    ◆(b) Show that $\text{span}\{\mathbf{u}, \mathbf{w}\} = \mathbb{R}\mathbf{u} + \mathbb{R}\mathbf{w}$ for any vectors $\mathbf{u}$ and $\mathbf{w}$.

    (c) Show that $\text{span}\{\mathbf{u}_1, \ldots, \mathbf{u}_m, \mathbf{w}_1, \ldots, \mathbf{w}_n\}$ $= \text{span}\{\mathbf{u}_1, \ldots, \mathbf{u}_m\} + \text{span}\{\mathbf{w}_1, \ldots, \mathbf{w}_n\}$ for any vectors $\mathbf{u}_i$ in U and $\mathbf{w}_j$ in W.

26. If A and B are $m \times n$ matrices, show that $\text{rank}(A + B) \leq \text{rank } A + \text{rank } B$. [Hint: If U and V are the column spaces of A and B, respectively, show that the column space of $A + B$ is contained in $U + V$ and that $\dim(U + V) \leq \dim U + \dim V$. (See Theorem 5.)]

SECTION 6.5 An Application to Polynomials

The vector space of all polynomials of degree at most n is denoted $\mathbf{P}_n$, and it was established in Section 6.3 that $\mathbf{P}_n$ has dimension $n + 1$; in fact, $\{1, x, x^2, \ldots, x^n\}$ is a basis. More generally, *any $n + 1$ polynomials of distinct degrees form a basis*, by Theorem 4 Section 6.4 (they are independent by Example 4 Section 6.3). This proves

Theorem 1

Let $p_0(x), p_1(x), p_2(x), \ldots, p_n(x)$ be polynomials in $\mathbf{P}_n$ of degrees $0, 1, 2, \ldots, n$, respectively. Then $\{p_0(x), \ldots, p_n(x)\}$ is a basis of $\mathbf{P}_n$.

An immediate consequence is that $\{1, (x - a), (x - a)^2, \ldots, (x - a)^n\}$ is a basis of $\mathbf{P}_n$ for any number a. Hence we have the following:

Corollary 1

If a is any number, every polynomial $f(x)$ of degree at most n has an expansion in powers of $(x - a)$:

$$f(x) = a_0 + a_1(x - a) + a_2(x - a)^2 + \cdots + a_n(x - a)^n. \qquad (*)$$

If $f(x)$ is evaluated at $x = a$, then equation $(*)$ becomes

$$f(a) = a_0 + a_1(a - a) + \cdots + a_n(a - a)^n = a_0.$$

Hence $a_0 = f(a)$, and equation $(*)$ can be written $f(x) = f(a) + (x - a)g(x)$, where $g(x)$ is a polynomial of degree $n - 1$ (this assumes that $n \geq 1$). If it happens that

SECTION 6.5 An Application to Polynomials

$f(a) = 0$, then it is clear that $f(x)$ has the form $f(x) = (x - a)g(x)$. Conversely, every such polynomial certainly satisfies $f(a) = 0$, and we obtain:

Corollary 2

Let $f(x)$ be a polynomial of degree $n \geq 1$ and let a be any number. Then:
Remainder Theorem
1. $f(x) = f(a) + (x - a)g(x)$ for some polynomial $g(x)$ of degree $n - 1$.

Factor Theorem
2. $f(a) = 0$ if and only if $f(x) = (x - a)g(x)$ for some polynomial $g(x)$.

The polynomial $g(x)$ can be computed easily by using "long division" to divide $f(x)$ by $(x - a)$—see Appendix D.

All the coefficients in the expansion (∗) of $f(x)$ in powers of $(x - a)$ can be determined in terms of the derivatives of $f(x)$.[5] These will be familiar to students of calculus. Let $f^{(n)}(x)$ denote the nth derivative of the polynomial $f(x)$, and write $f^{(0)}(x) = f(x)$. Then, if

$$f(x) = a_0 + a_1(x - a) + a_2(x - a)^2 + \cdots + a_n(x - a)^n,$$

it is clear that $a_0 = f(a) = f^{(0)}(a)$. Differentiation gives

$$f^{(1)}(x) = a_1 + 2a_2(x - a) + 3a_3(x - a)^2 + \cdots + na_n(x - a)^{n-1}$$

and substituting $x = a$ yields $a_1 = f^{(1)}(a)$. This process continues to give

$$a_2 = \frac{f^{(2)}(a)}{2!}, \quad a_3 = \frac{f^{(3)}(a)}{3!}, \quad \ldots, \quad = \frac{f^{(k)}(a)}{k!}, \text{ where } k! \text{ is defined as } k! = k(k-1) \cdots 2 \cdot 1.$$

Hence we obtain the following:

Corollary 3

Taylor's Theorem
If $f(x)$ is a polynomial of degree n, then

$$f(x) = f(a) + \frac{f^{(1)}(a)}{1!}(x - a) + \frac{f^{(2)}(a)}{2!}(x - a)^2 + \cdots + \frac{f^{(n)}(a)}{n!}(x - a)^n.$$

EXAMPLE 1

Expand $f(x) = 5x^3 + 10x + 2$ as a polynomial in powers of $x - 1$.

Solution ▶ The derivatives are $f^{(1)}(x) = 15x^2 + 10$, $f^{(2)}(x) = 30x$, and $f^{(3)}(x) = 30$. Hence the Taylor expansion is

$$f(x) = f(1) + \frac{f^{(1)}(1)}{1!}(x - 1) + \frac{f^{(2)}(1)}{2!}(x - 1)^2 + \frac{f^{(3)}(1)}{3!}(x - 1)^3$$
$$= 17 + 25(x - 1) + 15(x - 1)^2 + 5(x - 1)^3.$$

---

5 The discussion of Taylor's theorem can be omitted with no loss of continuity.

Taylor's theorem is useful in that it provides a formula for the coefficients in the expansion. It is dealt with in calculus texts and will not be pursued here.

Theorem 1 produces bases of $\mathbf{P}_n$ consisting of polynomials of distinct degrees. A different criterion is involved in the next theorem.

Theorem 2

Let $f_0(x), f_1(x), \ldots, f_n(x)$ be nonzero polynomials in $\mathbf{P}_n$. Assume that numbers $a_0, a_1, \ldots, a_n$ exist such that

$$f_i(a_i) \neq 0 \quad \text{for each } i$$
$$f_i(a_j) = 0 \quad \text{if } i \neq j$$

Then

1. $\{f_0(x), \ldots, f_n(x)\}$ is a basis of $\mathbf{P}_n$.
2. If $f(x)$ is any polynomial in $\mathbf{P}_n$, its expansion as a linear combination of these basis vectors is

$$f(x) = \frac{f(a_0)}{f_0(a_0)} f_0(x) + \frac{f(a_1)}{f_1(a_1)} f_1(x) + \cdots + \frac{f(a_n)}{f_n(a_n)} f_n(x).$$

PROOF

1. It suffices (by Theorem 4 Section 6.4) to show that $\{f_0(x), \ldots, f_n(x)\}$ is linearly independent (because $\dim \mathbf{P}_n = n + 1$). Suppose that

$$r_0 f_0(x) + r_1 f_1(x) + \cdots + r_n f_n(x) = 0, \quad r_i \in \mathbb{R}.$$

Because $f_i(a_0) = 0$ for all $i > 0$, taking $x = a_0$ gives $r_0 f_0(a_0) = 0$. But then $r_0 = 0$ because $f_0(a_0) \neq 0$. The proof that $r_i = 0$ for $i > 0$ is analogous.

2. By (1), $f(x) = r_0 f_0(x) + \cdots + r_n f_n(x)$ for *some* numbers r_i. Again, evaluating at a_0 gives $f(a_0) = r_0 f_0(a_0)$, so $r_0 = f(a_0)/f_0(a_0)$. Similarly, $r_i = f(a_i)/f_i(a_i)$ for each i.

EXAMPLE 2

Show that $\{x^2 - x, x^2 - 2x, x^2 - 3x + 2\}$ is a basis of $\mathbf{P}_2$.

Solution ▶ Write $f_0(x) = x^2 - x = x(x - 1)$, $f_1(x) = x^2 - 2x = x(x - 2)$, and $f_2(x) = x^2 - 3x + 2 = (x - 1)(x - 2)$. Then the conditions of Theorem 2 are satisfied with $a_0 = 2$, $a_1 = 1$, and $a_2 = 0$.

We investigate one natural choice of the polynomials $f_i(x)$ in Theorem 2. To illustrate, let a_0, a_1, and a_2 be distinct numbers and write

$$f_0(x) = \frac{(x - a_1)(x - a_2)}{(a_0 - a_1)(a_0 - a_2)} \quad f_1(x) = \frac{(x - a_0)(x - a_2)}{(a_1 - a_0)(a_1 - a_2)} \quad f_2(x) = \frac{(x - a_0)(x - a_1)}{(a_2 - a_0)(a_2 - a_1)}$$

Then $f_0(a_0) = f_1(a_1) = f_2(a_2) = 1$, and $f_i(a_j) = 0$ for $i \neq j$. Hence Theorem 2 applies, and because $f_i(a_i) = 1$ for each i, the formula for expanding any polynomial is simplified.

SECTION 6.5 An Application to Polynomials

In fact, this can be generalized with no extra effort. If $a_0, a_1, \ldots, a_n$ are distinct numbers, define the **Lagrange polynomials** $\delta_0(x), \delta_1(x), \ldots, \delta_n(x)$ relative to these numbers as follows:

$$\delta_k(x) = \frac{\prod_{i \neq k}(x - a_i)}{\prod_{i \neq k}(a_k - a_i)} \quad k = 0, 1, 2, \ldots, n$$

Here the numerator is the product of all the terms $(x - a_0), (x - a_1), \ldots, (x - a_n)$ with $(x - a_k)$ omitted, and a similar remark applies to the denominator. If $n = 2$, these are just the polynomials in the preceding paragraph. For another example, if $n = 3$, the polynomial $\delta_1(x)$ takes the form

$$\delta_1(x) = \frac{(x - a_0)(x - a_2)(x - a_3)}{(a_1 - a_0)(a_1 - a_2)(a_1 - a_3)}$$

In the general case, it is clear that $\delta_i(a_i) = 1$ for each i and that $\delta_i(a_j) = 0$ if $i \neq j$. Hence Theorem 2 specializes as Theorem 3.

Theorem 3

Lagrange Interpolation Expansion
Let $a_0, a_1, \ldots, a_n$ be distinct numbers. The corresponding set

$$\{\delta_0(x), \delta_1(x), \ldots, \delta_n(x)\}$$

of Lagrange polynomials is a basis of $\mathbf{P}_n$, and any polynomial $f(x)$ in $\mathbf{P}_n$ has the following unique expansion as a linear combination of these polynomials.

$$f(x) = f(a_0)\delta_0(x) + f(a_1)\delta_1(x) + \cdots + f(a_n)\delta_n(x)$$

EXAMPLE 3

Find the Lagrange interpolation expansion for $f(x) = x^2 - 2x + 1$ relative to $a_0 = -1$, $a_1 = 0$, and $a_2 = 1$.

Solution ▶ The Lagrange polynomials are

$$\delta_0(x) = \frac{(x - 0)(x - 1)}{(-1 - 0)(-1 - 1)} = \tfrac{1}{2}(x^2 - x)$$

$$\delta_1(x) = \frac{(x + 1)(x - 1)}{(0 + 1)(0 - 1)} = -(x^2 - 1)$$

$$\delta_2(x) = \frac{(x + 1)(x - 0)}{(1 + 1)(1 - 0)} = \tfrac{1}{2}(x^2 + x)$$

Because $f(-1) = 4$, $f(0) = 1$, and $f(1) = 0$, the expansion is
$$f(x) = 2(x^2 - x) - (x^2 - 1).$$

The Lagrange interpolation expansion gives an easy proof of the following important fact.

Theorem 4

Let $f(x)$ be a polynomial in $\mathbf{P}_n$, and let $a_0, a_1, \ldots, a_n$ denote distinct numbers. If $f(a_i) = 0$ for all i, then $f(x)$ is the zero polynomial (that is, all coefficients are zero).

PROOF

All the coefficients in the Lagrange expansion of $f(x)$ are zero.

EXERCISES 6.5

1. If polynomials $f(x)$ and $g(x)$ satisfy $f(a) = g(a)$, show that $f(x) - g(x) = (x - a)h(x)$ for some polynomial $h(x)$.

Exercises 2, 3, 4, and 5 require polynomial differentiation.

2. Expand each of the following as a polynomial in powers of $x - 1$.

   (a) $f(x) = x^3 - 2x^2 + x - 1$

   ◆(b) $f(x) = x^3 + x + 1$

   (c) $f(x) = x^4$

   ◆(d) $f(x) = x^3 - 3x^2 + 3x$

3. Prove Taylor's theorem for polynomials.

4. Use Taylor's theorem to derive the **binomial theorem**:
$$(1+x)^n = \binom{n}{0} + \binom{n}{1}x + \binom{n}{2}x^2 + \cdots + \binom{n}{n}x^n$$
   Here the **binomial coefficients** $\binom{n}{r}$ are defined by $\binom{n}{r} = \frac{n!}{r!(n-r)!}$ where $n! = n(n-1) \cdots 2 \cdot 1$ if $n \geq 1$ and $0! = 1$.

5. Let $f(x)$ be a polynomial of degree n. Show that, given any polynomial $g(x)$ in $\mathbf{P}_n$, there exist numbers $b_0, b_1, \ldots, b_n$ such that
$$g(x) = b_0 f(x) + b_1 f^{(1)}(x) + \cdots + b_n f^{(n)}(x)$$
   where $f^{(k)}(x)$ denotes the kth derivative of $f(x)$.

6. Use Theorem 2 to show that the following are bases of $\mathbf{P}_2$.

   (a) $\{x^2 - 2x, x^2 + 2x, x^2 - 4\}$

   ◆(b) $\{x^2 - 3x + 2, x^2 - 4x + 3, x^2 - 5x + 6\}$

7. Find the Lagrange interpolation expansion of $f(x)$ relative to $a_0 = 1$, $a_1 = 2$, and $a_2 = 3$ if:

   (a) $f(x) = x^2 + 1$ ◆(b) $f(x) = x^2 + x + 1$

8. Let $a_0, a_1, \ldots, a_n$ be distinct numbers. If $f(x)$ and $g(x)$ in $\mathbf{P}_n$ satisfy $f(a_i) = g(a_i)$ for all i, show that $f(x) = g(x)$. [Hint: See Theorem 4.]

9. Let $a_0, a_1, \ldots, a_n$ be distinct numbers. If $f(x)$ in $\mathbf{P}_{n+1}$ satisfies $f(a_i) = 0$ for each $i = 0, 1, \ldots, n$, show that $f(x) = r(x - a_0)(x - a_1) \cdots (x - a_n)$ for some r in $\mathbb{R}$. [Hint: r is the coefficient of x^{n+1} in $f(x)$. Consider $f(x) - r(x - a_0) \cdots (x - a_n)$ and use Theorem 4.]

10. Let a and b denote distinct numbers.

    (a) Show that $\{(x - a), (x - b)\}$ is a basis of $\mathbf{P}_1$.

    ◆(b) Show that $\{(x - a)^2, (x - a)(x - b), (x - b)^2\}$ is a basis of $\mathbf{P}_2$.

    (c) Show that $\{(x - a)^n, (x - a)^{n-1}(x - b), \ldots, (x - a)(x - b)^{n-1}, (x - b)^n\}$ is a basis of $\mathbf{P}_n$. [Hint: If a linear combination vanishes, evaluate at $x = a$ and $x = b$. Then reduce to the case $n - 2$ by using the fact that if $p(x)q(x) = 0$ in $\mathbf{P}$, then either $p(x) = 0$ or $q(x) = 0$.]

11. Let a and b be two distinct numbers. Assume that $n \geq 2$ and let
$$U_n = \{f(x) \text{ in } \mathbf{P}_n \mid f(a) = 0 = f(b)\}.$$

    (a) Show that $U_n = \{(x - a)(x - b)p(x) \mid p(x) \text{ in } \mathbf{P}_{n-2}\}$.

    ◆(b) Show that $\dim U_n = n - 1$. [Hint: If $p(x)q(x) = 0$ in $\mathbf{P}$, then either $p(x) = 0$, or $q(x) = 0$.]

    (c) Show that $\{(x - a)^{n-1}(x - b), (x - a)^{n-2}(x - b)^2, \ldots, (x - a)^2(x - b)^{n-2}, (x - a)(x - b)^{n-1}\}$ is a basis of U_n. [Hint: Exercise 10.]

SECTION 6.6 An Application to Differential Equations

Call a function $f: \mathbb{R} \to \mathbb{R}$ **differentiable** if it can be differentiated as many times as we want. If f is a differentiable function, the nth derivative $f^{(n)}$ of f is the result of differentiating n times. Thus $f^{(0)} = f, f^{(1)} = f', f^{(2)} = f^{(1)\prime}, \ldots$ and, in general, $f^{(n+1)} = f^{(n)\prime}$ for each $n \geq 0$. For small values of n these are often written as $f, f', f'', f''', \ldots$.

If a, b, and c are numbers, the differential equations

$$f'' + af' + bf = 0 \quad \text{or} \quad f''' + af'' + bf' + cf = 0,$$

are said to be of **second-order** and **third-order**, respectively. In general, an equation

$$f^{(n)} + a_{n-1}f^{(n-1)} + a_{n-2}f^{(n-2)} + \cdots + a_2f^{(2)} + a_1f^{(1)} + a_0f^{(0)} = 0, \, a_i \text{ in } \mathbb{R}, \quad (*)$$

is called a **differential equation of order n**. In this section we investigate the set of solutions to $(*)$ and, if n is 1 or 2, find explicit solutions. Of course an acquaintance with calculus is required.

Let f and g be solutions to $(*)$. Then $f + g$ is also a solution because $(f + g)^{(k)} = f^{(k)} + g^{(k)}$ for all k, and af is a solution for any a in $\mathbb{R}$ because $(af)^{(k)} = af^{(k)}$. It follows that the set of solutions to $(*)$ is a vector space, and we ask for the dimension of this space.

We have already dealt with the simplest case (see Theorem 1 Section 3.5):

Theorem 1

The set of solutions of the first-order differential equation $f' + af = 0$ is a one-dimensional vector space and $\{e^{-ax}\}$ is a basis.

There is a far-reaching generalization of Theorem 1 that will be proved in Section 7.4 (Theorem 1).

Theorem 2

The set of solutions to the nth order equation $(*)$ has dimension n.

Remark Every differential equation of order n can be converted into a system of n linear first-order equations (see Exercises 6 and 7 in Section 3.5). In the case that the matrix of this system is diagonalizable, this approach provides a proof of Theorem 2. But if the matrix is not diagonalizable, Theorem 1 Section 7.4 is required.

Theorem 1 suggests that we look for solutions to $(*)$ of the form $e^{\lambda x}$ for some number λ. This is a good idea. If we write $f(x) = e^{\lambda x}$, it is easy to verify that $f^{(k)}(x) = \lambda^k e^{\lambda x}$ for each $k \geq 0$, so substituting f in $(*)$ gives

$$(\lambda^n + a_{n-1}\lambda^{n-1} + a_{n-2}\lambda^{n-2} + \cdots + a_2\lambda^2 + a_1\lambda^1 + a_0)e^{\lambda x} = 0.$$

Since $e^{\lambda x} \neq 0$ for all x, this shows that $e^{\lambda x}$ is a solution of $(*)$ if and only if λ is a root of the **characteristic polynomial** $c(x)$, defined to be

$$c(x) = x^n + a_{n-1}x^{n-1} + a_{n-2}x^{n-2} + \cdots + a_2x^2 + a_1x + a_0.$$

This proves Theorem 3.

Theorem 3

If λ is real, the function $e^{\lambda x}$ is a solution of $(*)$ if and only if λ is a root of the characteristic polynomial $c(x)$.

EXAMPLE 1

Find a basis of the space U of solutions of $f''' - 2f'' - f' - 2f = 0$.

Solution ▶ The characteristic polynomial is $x^3 - 2x^2 - x - 1 = (x-1)(x+1)(x-2)$, with roots $\lambda_1 = 1$, $\lambda_2 = -1$, and $\lambda_3 = 2$. Hence e^x, e^{-x}, and e^{2x} are all in U. Moreover they are independent (by Lemma 1 below) so, since $\dim(U) = 3$ by Theorem 2, $\{e^x, e^{-x}, e^{2x}\}$ is a basis of U.

Lemma 1

If $\lambda_1, \lambda_2, \ldots, \lambda_k$ are distinct, then $\{e^{\lambda_1 x}, e^{\lambda_2 x}, \ldots, e^{\lambda_k x}\}$ is linearly independent.

PROOF

If $r_1 e^{\lambda_1 x} + r_2 e^{\lambda_2 x} + \cdots + r_k e^{\lambda_k x} = 0$ for all x, then $r_1 + r_2 e^{(\lambda_2 - \lambda_1)x} + \cdots + r_k e^{(\lambda_k - \lambda_1)x} = 0$; that is, $r_2 e^{(\lambda_2 - \lambda_1)x} + \cdots + r_k e^{(\lambda_k - \lambda_1)x}$ is a constant. Since the λ_i are distinct, this forces $r_2 = \cdots = r_k = 0$, whence $r_1 = 0$ also. This is what we wanted.

Theorem 4

Let U denote the space of solutions to the second-order equation
$$f'' + af' + bf = 0$$
where a and b are real constants. Assume that the characteristic polynomial $x^2 + ax + b$ has two real roots λ and μ. Then

(1) If $\lambda \neq \mu$, then $\{e^{\lambda x}, e^{\mu x}\}$ is a basis of U.
(2) If $\lambda = \mu$, then $\{e^{\lambda x}, xe^{\lambda x}\}$ is a basis of U.

PROOF

Since $\dim(U) = 2$ by Theorem 2, (1) follows by Lemma 1, and (2) follows because the set $\{e^{\lambda x}, xe^{\lambda x}\}$ is independent (Exercise 3).

EXAMPLE 2

Find the solution of $f'' + 4f' + 4f = 0$ that satisfies the **boundary conditions** $f(0) = 1$, $f(1) = -1$.

SECTION 6.6 An Application to Differential Equations

Solution ▶ The characteristic polynomial is $x^2 + 4x + 4 = (x + 2)^2$, so -2 is a double root. Hence $\{e^{-2x}, xe^{-2x}\}$ is a basis for the space of solutions, and the general solution takes the form $f(x) = ce^{-2x} + dxe^{-2x}$. Applying the boundary conditions gives $1 = f(0) = c$ and $-1 = f(1) = (c + d)e^{-2}$. Hence $c = 1$ and $d = -(1 + e^2)$, so the required solution is
$$f(x) = e^{-2x} - (1 + e^2)xe^{-2x}.$$

One other question remains: What happens if the roots of the characteristic polynomial are not real? To answer this, we must first state precisely what $e^{\lambda x}$ means when λ is not real. If q is a real number, define
$$e^{iq} = \cos q + i \sin q$$
where $i^2 = -1$. Then the relationship $e^{iq}e^{iq_1} = e^{i(q+q_1)}$ holds for all real q and q_1, as is easily verified. If $\lambda = p + iq$, where p and q are real numbers, we define
$$e^{\lambda} = e^p e^{iq} = e^p(\cos q + i \sin q).$$
Then it is a routine exercise to show that

1. $e^{\lambda}e^{\mu} = e^{\lambda+\mu}$
2. $e^{\lambda} = 1$ if and only if $\lambda = 0$
3. $(e^{\lambda x})' = \lambda e^{\lambda x}$

These easily imply that $f(x) = e^{\lambda x}$ is a solution to $f'' + af' + bf = 0$ if λ is a (possibly complex) root of the characteristic polynomial $x^2 + ax + b$. Now write $\lambda = p + iq$ so that
$$f(x) = e^{\lambda x} = e^{px}\cos(qx) + ie^{px}\sin(qx).$$

For convenience, denote the real and imaginary parts of $f(x)$ as $u(x) = e^{px}\cos(qx)$ and $v(x) = e^{px}\sin(qx)$. Then the fact that $f(x)$ satisfies the differential equation gives
$$0 = f'' + af' + bf = (u'' + au' + bu) + i(v'' + av' + bv).$$

Equating real and imaginary parts shows that $u(x)$ and $v(x)$ are both solutions to the differential equation. This proves part of Theorem 5.

Theorem 5

Let U denote the space of solutions of the second-order differential equation
$$f'' + af' + bf = 0$$
where a and b are real. Suppose λ is a nonreal root of the characteristic polynomial $x^2 + ax + b$. If $\lambda = p + iq$, where p and q are real, then
$$\{e^{px}\cos(qx), e^{px}\sin(qx)\}$$
is a basis of U.

PROOF

The foregoing discussion shows that these functions lie in U. Because $\dim U = 2$ by Theorem 2, it suffices to show that they are linearly independent. But if

$$re^{px}\cos(qx) + se^{px}\sin(qx) = 0$$

for all x, then $r\cos(qx) + s\sin(qx) = 0$ for all x (because $e^{px} \neq 0$). Taking $x = 0$ gives $r = 0$, and taking $x = \frac{\pi}{2q}$ gives $s = 0$ ($q \neq 0$ because λ is not real). This is what we wanted.

EXAMPLE 3

Find the solution $f(x)$ to $f'' - 2f' + 2f = 0$ that satisfies $f(0) = 2$ and $f(\frac{\pi}{2}) = 0$.

Solution ▶ The characteristic polynomial $x^2 - 2x + 2$ has roots $1 + i$ and $1 - i$. Taking $\lambda = 1 + i$ (quite arbitrarily) gives $p = q = 1$ in the notation of Theorem 5, so $\{e^x \cos x, e^x \sin x\}$ is a basis for the space of solutions. The general solution is thus $f(x) = e^x(r \cos x + s \sin x)$. The boundary conditions yield $2 = f(0) = r$ and $0 = f(\frac{\pi}{2}) = e^{\pi/2} s$. Thus $r = 2$ and $s = 0$, and the required solution is $f(x) = 2e^x \cos x$.

The following theorem is an important special case of Theorem 5.

Theorem 6

If $q \neq 0$ is a real number, the space of solutions to the differential equation $f'' + q^2 f = 0$ has basis $\{\cos(qx), \sin(qx)\}$.

PROOF

The characteristic polynomial $x^2 + q^2$ has roots qi and $-qi$, so Theorem 5 applies with $p = 0$.

In many situations, the displacement $s(t)$ of some object at time t turns out to have an oscillating form $s(t) = c \sin(at) + d \cos(at)$. These are called **simple harmonic motions**. An example follows.

EXAMPLE 4

A weight is attached to an extension spring (see diagram). If it is pulled from the equilibrium position and released, it is observed to oscillate up and down. Let $d(t)$ denote the distance of the weight below the equilibrium position t seconds later. It is known (**Hooke's law**) that the acceleration $d''(t)$ of the weight is proportional to the displacement $d(t)$ and in the opposite direction. That is,

$$d''(t) = -kd(t)$$

where $k > 0$ is called the **spring constant**. Find $d(t)$ if the maximum extension is 10 cm below the equilibrium position and find the **period** of the oscillation (time taken for the weight to make a full oscillation).

Solution ▶ It follows from Theorem 6 (with $q^2 = k$) that

$$d(t) = r\sin(\sqrt{k}\,t) + s\cos(\sqrt{k}\,t)$$

where r and s are constants. The condition $d(0) = 0$ gives $s = 0$, so $d(t) = r\sin(\sqrt{k}\,t)$. Now the maximum value of the function $\sin x$ is 1 (when $x = \frac{\pi}{2}$), so $r = 10$ (when $t = \frac{\pi}{2\sqrt{k}}$). Hence

$$d(t) = 10\sin(\sqrt{k}\,t).$$

Finally, the weight goes through a full oscillation as $\sqrt{k}\,t$ increases from 0 to 2π. The time taken is $t = \frac{2\pi}{\sqrt{k}}$, the period of the oscillation.

EXERCISES 6.6

1. Find a solution f to each of the following differential equations satisfying the given boundary conditions.

   (a) $f' - 3f = 0; f(1) = 2$

   ◆(b) $f' + f = 0; f(1) = 1$

   (c) $f'' + 2f' - 15f = 0; f(1) = f(0) = 0$

   ◆(d) $f'' + f' - 6f = 0; f(0) = 0, f(1) = 1$

   (e) $f'' - 2f' + f = 0; f(1) = f(0) = 1$

   ◆(f) $f'' - 4f' + 4f = 0; f(0) = 2, f(-1) = 0$

   (g) $f'' - 3af' + 2a^2 f = 0; a \neq 0; f(0) = 0, f(1) = 1 - e^a$

   ◆(h) $f'' - a^2 f = 0, a \neq 0; f(0) = 1, f(1) = 0$

   (i) $f'' - 2f' + 5f = 0; f(0) = 1, f(\frac{\pi}{4}) = 0$

   ◆(j) $f'' + 4f' + 5f = 0; f(0) = 0, f(\frac{\pi}{2}) = 1$

2. If the characteristic polynomial of $f'' + af' + bf = 0$ has real roots, show that $f = 0$ is the only solution satisfying $f(0) = 0 = f(1)$.

3. Complete the proof of Theorem 2. [*Hint*: If λ is a double root of $x^2 + ax + b$, show that $a = -2\lambda$ and $b = \lambda^2$. Hence $xe^{\lambda x}$ is a solution.]

4. (a) Given the equation $f' + af = b, (a \neq 0)$, make the substitution $f(x) = g(x) + b/a$ and obtain a differential equation for g. Then derive the general solution for $f' + af = b$.

   ◆(b) Find the general solution to $f' + f = 2$.

5. Consider the differential equation $f'' + af' + bf = g$, where g is some fixed function. Assume that f_0 is one solution of this equation.

   (a) Show that the general solution is $cf_1 + df_2 + f_0$, where c and d are constants and $\{f_1, f_2\}$ is any basis for the solutions to $f'' + af' + bf = 0$.

   ◆(b) Find a solution to $f'' + f' - 6f = 2x^3 - x^2 - 2x$. [*Hint*: Try $f(x) = \frac{-1}{3}x^3$.]

6. A radioactive element decays at a rate proportional to the amount present. Suppose an initial mass of 10 grams decays to 8 grams in 3 hours.

   (a) Find the mass t hours later.

   ◆(b) Find the *half-life* of the element—the time it takes to decay to half its mass.

7. The population $N(t)$ of a region at time t increases at a rate proportional to the population. If the population doubles in 5 years and is 3 million initially, find $N(t)$.

◆8. Consider a spring, as in Example 4. If the period of the oscillation is 30 seconds, find the spring constant k.

9. As a pendulum swings (see the diagram), let t measure the time since it was vertical. The angle $\theta = \theta(t)$ from the vertical can be shown to satisfy the equation $\theta'' + k\theta = 0$, provided that θ is small. If the maximal angle is $\theta = 0.05$ radians, find $\theta(t)$ in terms of k. If the period is 0.5 seconds, find k. [Assume that $\theta = 0$ when $t = 0$.]

SUPPLEMENTARY EXERCISES FOR CHAPTER 6

1. (Requires calculus) Let V denote the space of all functions $f: \mathbb{R} \to \mathbb{R}$ for which the derivatives f' and f'' exist. Show that $f_1, f_2,$ and f_3 in V are linearly independent provided that their **wronskian** $w(x)$ is nonzero for some x, where

$$w(x) = \det \begin{bmatrix} f_1(x) & f_2(x) & f_3(x) \\ f_1'(x) & f_2'(x) & f_3'(x) \\ f_1''(x) & f_2''(x) & f_3''(x) \end{bmatrix}$$

2. Let $\{\mathbf{v}_1, \mathbf{v}_2, \ldots, \mathbf{v}_n\}$ be a basis of $\mathbb{R}^n$ (written as columns), and let A be an $n \times n$ matrix.

   (a) If A is invertible, show that $\{A\mathbf{v}_1, A\mathbf{v}_2, \ldots, A\mathbf{v}_n\}$ is a basis of $\mathbb{R}^n$.

   ◆(b) If $\{A\mathbf{v}_1, A\mathbf{v}_2, \ldots, A\mathbf{v}_n\}$ is a basis of $\mathbb{R}^n$, show that A is invertible.

3. If A is an $m \times n$ matrix, show that A has rank m if and only if col A contains every column of I_m.

◆4. Show that null $A = \text{null}(A^TA)$ for any real matrix A.

5. Let A be an $m \times n$ matrix of rank r. Show that $\dim(\text{null } A) = n - r$ (Theorem 3 Section 5.4) as follows. Choose a basis $\{\mathbf{x}_1, \ldots, \mathbf{x}_k\}$ of null A and extend it to a basis $\{\mathbf{x}_1, \ldots, \mathbf{x}_k, \mathbf{z}_1, \ldots, \mathbf{z}_m\}$ of $\mathbb{R}^n$. Show that $\{A\mathbf{z}_1, \ldots, A\mathbf{z}_m\}$ is a basis of col A.

7 Linear Transformations

If V and W are vector spaces, a function $T: V \to W$ is a rule that assigns to each vector $\mathbf{v}$ in V a uniquely determined vector $T(\mathbf{v})$ in W. As mentioned in Section 2.2, two functions $S: V \to W$ and $T: V \to W$ are equal if $S(\mathbf{v}) = T(\mathbf{v})$ for every $\mathbf{v}$ in V. A function $T: V \to W$ is called a *linear transformation* if $T(\mathbf{v} + \mathbf{v}_1) = T(\mathbf{v}) + T(\mathbf{v}_1)$ for all $\mathbf{v}, \mathbf{v}_1$ in V and $T(r\mathbf{v}) = rT(\mathbf{v})$ for all $\mathbf{v}$ in V and all scalars r. $T(\mathbf{v})$ is called the *image* of $\mathbf{v}$ under T. We have already studied linear transformations $T: \mathbb{R}^n \to \mathbb{R}^m$ and shown (in Section 2.6) that they all are given by multiplication by a uniquely determined $m \times n$ matrix A; that is, $T(\mathbf{x}) = A\mathbf{x}$ for all $\mathbf{x}$ in $\mathbb{R}^n$. In the case of linear operators $\mathbb{R}^2 \to \mathbb{R}^2$, this yields an important way to describe geometric functions such as rotations about the origin and reflections in a line through the origin.

In the present chapter we will describe linear transformations in general, introduce the *kernel* and *image* of a linear transformation, and prove a useful result (called the *dimension theorem*) that relates the dimensions of the kernel and image, and unifies and extends several earlier results. Finally we study the notion of *isomorphic* vector spaces, that is, spaces that are identical except for notation, and relate this to composition of transformations that was introduced in Section 2.3.

SECTION 7.1 Examples and Elementary Properties

Definition 7.1

If V and W are two vector spaces, a function $T: V \to W$ is called a **linear transformation** if it satisfies the following axioms.

T1. $T(\mathbf{v} + \mathbf{v}_1) = T(\mathbf{v}) + T(\mathbf{v}_1)$ for all $\mathbf{v}$ and $\mathbf{v}_1$ in V.

T2. $T(r\mathbf{v}) = rT(\mathbf{v})$ for all $\mathbf{v}$ in V and r in $\mathbb{R}$.

A linear transformation $T: V \to V$ is called a **linear operator** on V. The situation can be visualized as in the diagram.

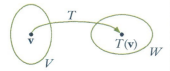

Axiom T1 is just the requirement that T *preserves* vector addition. It asserts that the result $T(\mathbf{v} + \mathbf{v}_1)$ of adding $\mathbf{v}$ and $\mathbf{v}_1$ first and then applying T is the same as applying T first to get $T(\mathbf{v})$ and $T(\mathbf{v}_1)$ and then adding. Similarly, axiom T2 means that T *preserves* scalar multiplication. Note that, even though the additions in axiom T1 are both denoted by the same symbol $+$, the addition on the left forming $\mathbf{v} + \mathbf{v}_1$ is carried out in V, whereas the addition $T(\mathbf{v}) + T(\mathbf{v}_1)$ is done in W. Similarly, the scalar multiplications $r\mathbf{v}$ and $rT(\mathbf{v})$ in axiom T2 refer to the spaces V and W, respectively.

Chapter 7   Linear Transformations

We have already seen many examples of linear transformations $T : \mathbb{R}^n \to \mathbb{R}^m$. In fact, writing vectors in $\mathbb{R}^n$ as columns, Theorem 2 Section 2.6 shows that, for each such T, there is an $m \times n$ matrix A such that $T(\mathbf{x}) = A\mathbf{x}$ for every $\mathbf{x}$ in $\mathbb{R}^n$. Moreover, the matrix A is given by $A = [T(\mathbf{e}_1) \; T(\mathbf{e}_2) \; \cdots \; T(\mathbf{e}_n)]$ where $\{\mathbf{e}_1, \mathbf{e}_2, \ldots, \mathbf{e}_n\}$ is the standard basis of $\mathbb{R}^n$. We denote this transformation by $T_A : \mathbb{R}^n \to \mathbb{R}^m$, defined by

$$T_A(\mathbf{x}) = A\mathbf{x} \quad \text{for all } \mathbf{x} \text{ in } \mathbb{R}^n.$$

Example 1 lists three important linear transformations that will be referred to later. The verification of axioms T1 and T2 is left to the reader.

EXAMPLE 1

If V and W are vector spaces, the following are linear transformations:

Identity operator $V \to V$ $1_V : V \to V$ where $1_V(\mathbf{v}) = \mathbf{v}$ for all $\mathbf{v}$ in V
Zero transformation $V \to W$ $0 : V \to W$ where $0(\mathbf{v}) = \mathbf{0}$ for all $\mathbf{v}$ in V
Scalar operator $V \to V$ $a : V \to V$ where $a(\mathbf{v}) = a\mathbf{v}$ for all $\mathbf{v}$ in V
                                                                       (Here a is any real number.)

The symbol 0 will be used to denote the zero transformation from V to W for *any* spaces V and W. It was also used earlier to denote the zero function $[a, b] \to \mathbb{R}$.

The next example gives two important transformations of matrices. Recall that the trace tr A of an $n \times n$ matrix A is the sum of the entries on the main diagonal.

EXAMPLE 2

Show that the transposition and trace are linear transformations. More precisely,

$$R : \mathbf{M}_{mn} \to \mathbf{M}_{nm} \quad \text{where } R(A) = A^T \text{ for all } A \text{ in } \mathbf{M}_{mn}$$

$$S : \mathbf{M}_{mn} \to \mathbb{R} \quad \text{where } S(A) = \text{tr } A \text{ for all } A \text{ in } \mathbf{M}_{nn}$$

are both linear transformations.

Solution ▶ Axioms T1 and T2 for transposition are $(A + B)^T = A^T + B^T$ and $(rA)^T = r(A^T)$, respectively (using Theorem 2 Section 2.1). The verifications for the trace are left to the reader.

EXAMPLE 3

If a is a scalar, define $E_a : \mathbf{P}_n \to \mathbb{R}$ by $E_a(p) = p(a)$ for each polynomial p in $\mathbf{P}_n$. Show that E_a is a linear transformation (called **evaluation** at a).

Solution ▶ If p and q are polynomials and r is in $\mathbb{R}$, we use the fact that the sum $p + q$ and scalar product rp are defined as for functions:

$$(p + q)(x) = p(x) + q(x) \quad \text{and} \quad (rp)(x) = rp(x)$$

for all x. Hence, for all p and q in $\mathbf{P}_n$ and all r in $\mathbb{R}$:

SECTION 7.1 Examples and Elementary Properties

$$E_a(p+q) = (p+q)(a) = p(a) + q(a) = E_a(p) + E_a(q), \text{ and}$$
$$E_a(rp) = (rp)(a) = rp(a) = rE_a(p).$$

Hence E_a is a linear transformation.

The next example involves some calculus.

EXAMPLE 4

Show that the differentiation and integration operations on $\mathbf{P}_n$ are linear transformations. More precisely,

$$D : \mathbf{P}_n \to \mathbf{P}_{n-1} \quad \text{where } D[p(x)] = p'(x) \text{ for all } p(x) \text{ in } \mathbf{P}_n$$
$$I : \mathbf{P}_n \to \mathbf{P}_{n+1} \quad \text{where } I[p(x)] = \int_0^x p(t)dt \text{ for all } p(x) \text{ in } \mathbf{P}_n$$

are linear transformations.

Solution ▶ These restate the following fundamental properties of differentiation and integration.

$$[p(x) + q(x)]' = p'(x) + q'(x) \quad \text{and} \quad [rp(x)]' = (rp)'(x)$$
$$\int_0^x [p(t) + q(t)]dt = \int_0^x p(t)dt + \int_0^x q(t)dt \quad \text{and} \quad \int_0^x rp(t)dt = r\int_0^x p(t)dt$$

The next theorem collects three useful properties of *all* linear transformations. They can be described by saying that, in addition to preserving addition and scalar multiplication (these are the axioms), linear transformations preserve the zero vector, negatives, and linear combinations.

Theorem 1

Let $T : V \to W$ be a linear transformation.
1. $T(\mathbf{0}) = \mathbf{0}$.
2. $T(-\mathbf{v}) = -T(\mathbf{v})$ for all $\mathbf{v}$ in V.
3. $T(r_1\mathbf{v}_1 + r_2\mathbf{v}_2 + \cdots + r_k\mathbf{v}_k) = r_1 T(\mathbf{v}_1) + r_2 T(\mathbf{v}_2) + \cdots + r_k T(\mathbf{v}_k)$ for all $\mathbf{v}_i$ in V and all r_i in $\mathbb{R}$.

PROOF

1. $T(\mathbf{0}) = T(0\mathbf{v}) = 0T(\mathbf{v}) = \mathbf{0}$ for any $\mathbf{v}$ in V.
2. $T(-\mathbf{v}) = T[(-1)\mathbf{v}] = (-1)T(\mathbf{v}) = -T(\mathbf{v})$ for any $\mathbf{v}$ in V.
3. The proof of Theorem 1 Section 2.6 goes through.

The ability to use the last part of Theorem 1 effectively is vital to obtaining the benefits of linear transformations. Example 5 and Theorem 2 provide illustrations.

EXAMPLE 5

Let $T: V \to W$ be a linear transformation. If $T(\mathbf{v} - 3\mathbf{v}_1) = \mathbf{w}$ and $T(2\mathbf{v} - \mathbf{v}_1) = \mathbf{w}_1$, find $T(\mathbf{v})$ and $T(\mathbf{v}_1)$ in terms of $\mathbf{w}$ and $\mathbf{w}_1$.

Solution ▶ The given relations imply that

$$T(\mathbf{v}) - 3T(\mathbf{v}_1) = \mathbf{w}$$
$$2T(\mathbf{v}) - T(\mathbf{v}_1) = \mathbf{w}_1$$

by Theorem 1. Subtracting twice the first from the second gives $T(\mathbf{v}_1) = \frac{1}{5}(\mathbf{w}_1 - 2\mathbf{w})$. Then substitution gives $T(\mathbf{v}) = \frac{1}{5}(3\mathbf{w}_1 - \mathbf{w})$.

The full effect of property (3) in Theorem 1 is this: If $T: V \to W$ is a linear transformation and $T(\mathbf{v}_1), T(\mathbf{v}_2), \ldots, T(\mathbf{v}_n)$ are known, then $T(\mathbf{v})$ can be computed for *every* vector $\mathbf{v}$ in span$\{\mathbf{v}_1, \mathbf{v}_2, \ldots, \mathbf{v}_n\}$. In particular, if $\{\mathbf{v}_1, \mathbf{v}_2, \ldots, \mathbf{v}_n\}$ spans V, then $T(\mathbf{v})$ is determined for all $\mathbf{v}$ in V by the choice of $T(\mathbf{v}_1), T(\mathbf{v}_2), \ldots, T(\mathbf{v}_n)$. The next theorem states this somewhat differently. As for functions in general, two linear transformations $T: V \to W$ and $S: V \to W$ are called **equal** (written $T = S$) if they have the same **action**; that is, if $T(\mathbf{v}) = S(\mathbf{v})$ for all $\mathbf{v}$ in V.

Theorem 2

Let $T: V \to W$ and $S: V \to W$ be two linear transformations. Suppose that $V = \text{span}\{\mathbf{v}_1, \mathbf{v}_2, \ldots, \mathbf{v}_n\}$. If $T(\mathbf{v}_i) = S(\mathbf{v}_i)$ for each i, then $T = S$.

PROOF

If $\mathbf{v}$ is any vector in $V = \text{span}\{\mathbf{v}_1, \mathbf{v}_2, \ldots, \mathbf{v}_n\}$, write $\mathbf{v} = a_1\mathbf{v}_1 + a_2\mathbf{v}_2 + \cdots + a_n\mathbf{v}_n$ where each a_i is in $\mathbb{R}$. Since $T(\mathbf{v}_i) = S(\mathbf{v}_i)$ for each i, Theorem 1 gives

$$\begin{aligned} T(\mathbf{v}) &= T(a_1\mathbf{v}_1 + a_2\mathbf{v}_2 + \cdots + a_n\mathbf{v}_n) \\ &= a_1 T(\mathbf{v}_1) + a_2 T(\mathbf{v}_2) + \cdots + a_n T(\mathbf{v}_n) \\ &= a_1 S(\mathbf{v}_1) + a_2 S(\mathbf{v}_2) + \cdots + a_n S(\mathbf{v}_n) \\ &= S(a_1\mathbf{v}_1 + a_2\mathbf{v}_2 + \cdots + a_n\mathbf{v}_n) \\ &= S(\mathbf{v}). \end{aligned}$$

Since $\mathbf{v}$ was arbitrary in V, this shows that $T = S$.

EXAMPLE 6

Let $V = \text{span}\{\mathbf{v}_1, \ldots, \mathbf{v}_n\}$. Let $T: V \to W$ be a linear transformation. If $T(\mathbf{v}_1) = \cdots = T(\mathbf{v}_n) = \mathbf{0}$, show that $T = 0$, the zero transformation from V to W.

Solution ▶ The zero transformation $0: V \to W$ is defined by $0(\mathbf{v}) = \mathbf{0}$ for all $\mathbf{v}$ in V (Example 1), so $T(\mathbf{v}_i) = 0(\mathbf{v}_i)$ holds for each i. Hence $T = 0$ by Theorem 2.

Theorem 2 can be expressed as follows: If we know what a linear transformation $T: V \to W$ does to each vector in a spanning set for V, then we know what T does

SECTION 7.1 Examples and Elementary Properties

to *every* vector in V. If the spanning set is a basis, we can say much more.

Theorem 3

Let V and W be vector spaces and let $\{\mathbf{b}_1, \mathbf{b}_2, \ldots, \mathbf{b}_n\}$ be a basis of V. Given any vectors $\mathbf{w}_1, \mathbf{w}_2, \ldots, \mathbf{w}_n$ in W (they need not be distinct), there exists a unique linear transformation $T : V \to W$ satisfying $T(\mathbf{b}_i) = \mathbf{w}_i$ for each $i = 1, 2, \ldots, n$. In fact, the action of T is as follows:
Given $\mathbf{v} = v_1\mathbf{b}_1 + v_2\mathbf{b}_2 + \cdots + v_n\mathbf{b}_n$ in V, v_i in $\mathbb{R}$, then

$$T(\mathbf{v}) = T(v_1\mathbf{b}_1 + v_2\mathbf{b}_2 + \cdots + v_n\mathbf{b}_n) = v_1\mathbf{w}_1 + v_2\mathbf{w}_2 + \cdots + v_n\mathbf{w}_n.$$

PROOF

If a transformation T *does* exist with $T(\mathbf{b}_i) = \mathbf{w}_i$ for each i, and if S is any other such transformation, then $T(\mathbf{b}_i) = \mathbf{w}_i = S(\mathbf{b}_i)$ holds for each i, so $S = T$ by Theorem 2. Hence T is unique if it exists, and it remains to show that there really is such a linear transformation. Given $\mathbf{v}$ in V, we must specify $T(\mathbf{v})$ in W. Because $\{\mathbf{b}_1, \ldots, \mathbf{b}_n\}$ is a basis of V, we have $\mathbf{v} = v_1\mathbf{b}_1 + \cdots + v_n\mathbf{b}_n$, where $v_1, \ldots, v_n$ are *uniquely* determined by $\mathbf{v}$ (this is Theorem 1 Section 6.3). Hence we may define $T : V \to W$ by

$$T(\mathbf{v}) = T(v_1\mathbf{b}_1 + v_2\mathbf{b}_2 + \cdots + v_n\mathbf{b}_n) = v_1\mathbf{w}_1 + v_2\mathbf{w}_2 + \cdots + v_n\mathbf{w}_n$$

for all $\mathbf{v} = v_1\mathbf{b}_1 + \cdots + v_n\mathbf{b}_n$ in V. This satisfies $T(\mathbf{b}_i) = \mathbf{w}_i$ for each i; the verification that T is linear is left to the reader.

This theorem shows that linear transformations can be defined almost at will: Simply specify where the basis vectors go, and the rest of the action is dictated by the linearity. Moreover, Theorem 2 shows that deciding whether two linear transformations are equal comes down to determining whether they have the same effect on the basis vectors. So, given a basis $\{\mathbf{b}_1, \ldots, \mathbf{b}_n\}$ of a vector space V, there is a different linear transformation $V \to W$ for every ordered selection $\mathbf{w}_1, \mathbf{w}_2, \ldots, \mathbf{w}_n$ of vectors in W (not necessarily distinct).

EXAMPLE 7

Find a linear transformation $T : \mathbf{P}_2 \to \mathbf{M}_{22}$ such that

$$T(1 + x) = \begin{bmatrix} 1 & 0 \\ 0 & 0 \end{bmatrix}, \quad T(x + x^2) = \begin{bmatrix} 0 & 1 \\ 1 & 0 \end{bmatrix}, \quad \text{and} \quad T(1 + x^2) = \begin{bmatrix} 0 & 0 \\ 0 & 1 \end{bmatrix}.$$

Solution ▶ The set $\{1 + x, x + x^2, 1 + x^2\}$ is a basis of $\mathbf{P}_2$, so every vector $p = a + bx + cx^2$ in $\mathbf{P}_2$ is a linear combination of these vectors. In fact

$$p(x) = \tfrac{1}{2}(a + b - c)(1 + x) + \tfrac{1}{2}(-a + b + c)(x + x^2) + \tfrac{1}{2}(a - b + c)(1 + x^2)$$

Hence Theorem 3 gives

$$T[p(x)] = \tfrac{1}{2}(a + b - c)\begin{bmatrix} 1 & 0 \\ 0 & 0 \end{bmatrix} + \tfrac{1}{2}(-a + b + c)\begin{bmatrix} 0 & 1 \\ 1 & 0 \end{bmatrix} + \tfrac{1}{2}(a - b + c)\begin{bmatrix} 0 & 0 \\ 0 & 1 \end{bmatrix}$$

$$= \tfrac{1}{2}\begin{bmatrix} a + b - c & -a + b + c \\ -a + b + c & a - b + c \end{bmatrix}$$

EXERCISES 7.1

1. Show that each of the following functions is a linear transformation.

   (a) $T: \mathbb{R}^2 \to \mathbb{R}^2$; $T(x, y) = (x, -y)$ (reflection in the x axis)

   ◆(b) $T: \mathbb{R}^3 \to \mathbb{R}^3$; $T(x, y, z) = (x, y, -z)$ (reflection in the x-y plane)

   (c) $T: \mathbb{C} \to \mathbb{C}$; $T(z) = \bar{z}$ (conjugation)

   ◆(d) $T: \mathbf{M}_{mn} \to \mathbf{M}_{kl}$; $T(A) = PAQ$, P a $k \times m$ matrix, Q an $n \times l$ matrix, both fixed

   (e) $T: \mathbf{M}_{nm} \to \mathbf{M}_{mn}$; $T(A) = A^T + A$

   ◆(f) $T: \mathbf{P}_n \to \mathbb{R}$; $T[p(x)] = p(0)$

   (g) $T: \mathbf{P}_n \to \mathbb{R}$; $T(r_0 + r_1 x + \cdots + r_n x^n) = r_n$

   ◆(h) $T: \mathbb{R}^n \to \mathbb{R}$; $T(\mathbf{x}) = \mathbf{x} \cdot \mathbf{z}$, $\mathbf{z}$ a fixed vector in $\mathbb{R}^n$

   (i) $T: \mathbf{P}_n \to \mathbf{P}_n$; $T[p(x)] = p(x + 1)$

   ◆(j) $T: \mathbb{R}^n \to V$; $T(r_1, \ldots, r_n) = r_1 \mathbf{e}_1 + \cdots + r_n \mathbf{e}_n$ where $\{\mathbf{e}_1, \ldots, \mathbf{e}_n\}$ is a fixed basis of V.

   (k) $T: V \to \mathbb{R}$; $T(r_1 \mathbf{e}_1 + \cdots + r_n \mathbf{e}_n) = r_1$, where $\{\mathbf{e}_1, \ldots, \mathbf{e}_n\}$ is a fixed basis of V

2. In each case, show that T is *not* a linear transformation.

   (a) $T: \mathbf{M}_{nn} \to \mathbb{R}$; $T(A) = \det A$

   ◆(b) $T: \mathbf{M}_{mn} \to \mathbb{R}$; $T(A) = \operatorname{rank} A$

   (c) $T: \mathbb{R} \to \mathbb{R}$; $T(x) = x^2$

   ◆(d) $T: V \to V$; $T(\mathbf{v}) = \mathbf{v} + \mathbf{u}$ where $\mathbf{u} \neq \mathbf{0}$ is a fixed vector in V (T is called the **translation** by $\mathbf{u}$)

3. In each case, assume that T is a linear transformation.

   (a) If $T: V \to \mathbb{R}$ and $T(\mathbf{v}_1) = 1$, $T(\mathbf{v}_2) = -1$, find $T(3\mathbf{v}_1 - 5\mathbf{v}_2)$.

   ◆(b) If $T: V \to \mathbb{R}$ and $T(\mathbf{v}_1) = 2$, $T(\mathbf{v}_2) = -3$, find $T(3\mathbf{v}_1 + 2\mathbf{v}_2)$.

   (c) If $T: \mathbb{R}^2 \to \mathbb{R}^2$ and $T\begin{bmatrix} 1 \\ 3 \end{bmatrix} = \begin{bmatrix} 1 \\ 1 \end{bmatrix}$, $T\begin{bmatrix} 1 \\ 1 \end{bmatrix} = \begin{bmatrix} 0 \\ 1 \end{bmatrix}$, find $T\begin{bmatrix} -1 \\ 3 \end{bmatrix}$.

   ◆(d) If $T: \mathbb{R}^2 \to \mathbb{R}^2$ and $T\begin{bmatrix} 1 \\ -1 \end{bmatrix} = \begin{bmatrix} 0 \\ 1 \end{bmatrix}$, $T\begin{bmatrix} 1 \\ 1 \end{bmatrix} = \begin{bmatrix} 1 \\ 0 \end{bmatrix}$, find $T\begin{bmatrix} 1 \\ -7 \end{bmatrix}$.

   (e) If $T: \mathbf{P}_2 \to \mathbf{P}_2$ and $T(x + 1) = x$, $T(x - 1) = 1$, $T(x^2) = 0$, find $T(2 + 3x - x^2)$.

   ◆(f) If $T: \mathbf{P}_2 \to \mathbb{R}$ and $T(x + 2) = 1$, $T(1) = 5$, $T(x^2 + x) = 0$, find $T(2 - x + 3x^2)$.

4. In each case, find a linear transformation with the given properties and compute $T(\mathbf{v})$.

   (a) $T: \mathbb{R}^2 \to \mathbb{R}^3$; $T(1, 2) = (1, 0, 1)$, $T(-1, 0) = (0, 1, 1)$; $\mathbf{v} = (2, 1)$

   ◆(b) $T: \mathbb{R}^2 \to \mathbb{R}^3$; $T(2, -1) = (1, -1, 1)$, $T(1, 1) = (0, 1, 0)$; $\mathbf{v} = (-1, 2)$

   (c) $T: \mathbf{P}_2 \to \mathbf{P}_3$; $T(x^2) = x^3$, $T(x + 1) = 0$, $T(x - 1) = x$; $\mathbf{v} = x^2 + x + 1$

   ◆(d) $T: \mathbf{M}_{22} \to \mathbb{R}$; $T\begin{bmatrix} 1 & 0 \\ 0 & 0 \end{bmatrix} = 3$, $T\begin{bmatrix} 0 & 1 \\ 1 & 0 \end{bmatrix} = -1$, $T\begin{bmatrix} 1 & 0 \\ 1 & 0 \end{bmatrix} = 0 = T\begin{bmatrix} 0 & 0 \\ 0 & 1 \end{bmatrix}$; $\mathbf{v} = \begin{bmatrix} a & b \\ c & d \end{bmatrix}$

5. If $T: V \to V$ is a linear transformation, find $T(\mathbf{v})$ and $T(\mathbf{w})$ if:

   (a) $T(\mathbf{v} + \mathbf{w}) = \mathbf{v} - 2\mathbf{w}$ and $T(2\mathbf{v} - \mathbf{w}) = 2\mathbf{v}$

   ◆(b) $T(\mathbf{v} + 2\mathbf{w}) = 3\mathbf{v} - \mathbf{w}$ and $T(\mathbf{v} - \mathbf{w}) = 2\mathbf{v} - 4\mathbf{w}$

6. If $T: V \to W$ is a linear transformation, show that $T(\mathbf{v} - \mathbf{v}_1) = T(\mathbf{v}) - T(\mathbf{v}_1)$ for all $\mathbf{v}$ and $\mathbf{v}_1$ in V.

7. Let $\{\mathbf{e}_1, \mathbf{e}_2\}$ be the standard basis of $\mathbb{R}^2$. Is it possible to have a linear transformation T such that $T(\mathbf{e}_1)$ lies in $\mathbb{R}$ while $T(\mathbf{e}_2)$ lies in $\mathbb{R}^2$? Explain your answer.

8. Let $\{\mathbf{v}_1, \ldots, \mathbf{v}_n\}$ be a basis of V and let $T: V \to V$ be a linear transformation.

   (a) If $T(\mathbf{v}_i) = \mathbf{v}_i$ for each i, show that $T = 1_V$.

   ◆(b) If $T(\mathbf{v}_i) = -\mathbf{v}_i$ for each i, show that $T = -1$ is the scalar operator (see Example 1).

9. If A is an $m \times n$ matrix, let $C_k(A)$ denote column k of A. Show that $C_k : \mathbf{M}_{mn} \to \mathbb{R}^m$ is a linear transformation for each $k = 1, \ldots, n$.

10. Let $\{\mathbf{e}_1, \ldots, \mathbf{e}_n\}$ be a basis of $\mathbb{R}^n$. Given k, $1 \leq k \leq n$, define $P_k: \mathbb{R}^n \to \mathbb{R}^n$ by $P_k(r_1 \mathbf{e}_1 + \cdots + r_n \mathbf{e}_n) = r_k \mathbf{e}_k$. Show that P_k a linear transformation for each k.

11. Let $S: V \to W$ and $T: V \to W$ be linear transformations. Given a in $\mathbb{R}$, define functions $(S + T): V \to W$ and $(aT): V \to W$ by $(S + T)(\mathbf{v}) = S(\mathbf{v}) + T(\mathbf{v})$ and $(aT)(\mathbf{v}) = aT(\mathbf{v})$ for all $\mathbf{v}$ in V. Show that $S + T$ and aT are linear transformations.

◆12. Describe all linear transformations $T: \mathbb{R} \to V$.

13. Let V and W be vector spaces, let V be finite dimensional, and let $\mathbf{v} \neq \mathbf{0}$ in V. Given any $\mathbf{w}$ in W, show that there exists a linear transformation $T: V \to W$ with $T(\mathbf{v}) = \mathbf{w}$. [Hint: Theorem 1(2) Section 6.4 and Theorem 3.]

14. Given $\mathbf{y}$ in $\mathbb{R}^n$, define $S_\mathbf{y}: \mathbb{R}^n \to \mathbb{R}$ by $S_\mathbf{y}(\mathbf{x}) = \mathbf{x} \cdot \mathbf{y}$ for all $\mathbf{x}$ in $\mathbb{R}^n$ (where • is the dot product introduced in Section 5.3).

    (a) Show that $S_\mathbf{y}: \mathbb{R}^n \to \mathbb{R}$ is a linear transformation for any $\mathbf{y}$ in $\mathbb{R}^n$.

    (b) Show that every linear transformation $T: \mathbb{R}^n \to \mathbb{R}$ arises in this way; that is, $T = S_\mathbf{y}$ for some $\mathbf{y}$ in $\mathbb{R}^n$. [Hint: If $\{\mathbf{e}_1, \ldots, \mathbf{e}_n\}$ is the standard basis of $\mathbb{R}^n$, write $S_\mathbf{y}(\mathbf{e}_i) = y_i$ for each i. Use Theorem 1.]

15. Let $T: V \to W$ be a linear transformation.

    (a) If U is a subspace of V, show that $T(U) = \{T(\mathbf{u}) \mid \mathbf{u} \text{ in } U\}$ is a subspace of W (called the **image** of U under T).

    ◆(b) If P is a subspace of W, show that $\{\mathbf{v} \text{ in } V \mid T(\mathbf{v}) \text{ in } P\}$ is a subspace of V (called the **preimage** of P under T).

16. Show that differentiation is the only linear transformation $\mathbf{P}_n \to \mathbf{P}_n$ that satisfies $T(x^k) = kx^{k-1}$ for each $k = 0, 1, 2, \ldots, n$.

17. Let $T: V \to W$ be a linear transformation and let $\mathbf{v}_1, \ldots, \mathbf{v}_n$ denote vectors in V.

    (a) If $\{T(\mathbf{v}_1), \ldots, T(\mathbf{v}_n)\}$ is linearly independent, show that $\{\mathbf{v}_1, \ldots, \mathbf{v}_n\}$ is also independent.

    (b) Find $T: \mathbb{R}^2 \to \mathbb{R}^2$ for which the converse of part (a) is false.

◆18. Suppose $T: V \to V$ is a linear operator with the property that $T[T(\mathbf{v})] = \mathbf{v}$ for all $\mathbf{v}$ in V. (For example, transposition in $\mathbf{M}_{nn}$ or conjugation in $\mathbb{C}$.) If $\mathbf{v} \neq \mathbf{0}$ in V, show that $\{\mathbf{v}, T(\mathbf{v})\}$ is linearly independent if and only if $T(\mathbf{v}) \neq \mathbf{v}$ and $T(\mathbf{v}) \neq -\mathbf{v}$.

19. If a and b are real numbers, define $T_{a,b}: \mathbb{C} \to \mathbb{C}$ by $T_{a,b}(r + si) = ra + sbi$ for all $r + si$ in $\mathbb{C}$.

    (a) Show that $T_{a,b}$ is linear and $T_{a,b}(\overline{z}) = \overline{T_{a,b}(z)}$ for all z in $\mathbb{C}$. (Here $\overline{z}$ denotes the conjugate of z.)

    (b) If $T: \mathbb{C} \to \mathbb{C}$ is linear and $T(\overline{z}) = \overline{T(z)}$ for all z in $\mathbb{C}$, show that $T = T_{a,b}$ for some real a and b.

20. Show that the following conditions are equivalent for a linear transformation $T: \mathbf{M}_{22} \to \mathbf{M}_{22}$.

    (1) $\text{tr}[T(A)] = \text{tr } A$ for all A in $\mathbf{M}_{22}$.

    (2) $T\begin{bmatrix} r_{11} & r_{12} \\ r_{21} & r_{22} \end{bmatrix} = r_{11}B_{11} + r_{12}B_{12} + r_{21}B_{21} + r_{22}B_{22}$ for matrices B_{ij} such that $\text{tr } B_{11} = 1 = \text{tr } B_{22}$ and $\text{tr } B_{12} = 0 = \text{tr } B_{21}$.

21. Given a in $\mathbb{R}$, consider the **evaluation** map $E_a: \mathbf{P}_n \to \mathbb{R}$ defined in Example 3.

    (a) Show that E_a is a linear transformation satisfying the additional condition that $E_a(x^k) = [E_a(x)]^k$ holds for all $k = 0, 1, 2, \ldots$. [Note: $x^0 = 1$.]

    ◆(b) If $T: \mathbf{P}_n \to \mathbb{R}$ is a linear transformation satisfying $T(x^k) = [T(x)]^k$ for all $k = 0, 1, 2, \ldots$, show that $T = E_a$ for some a in R.

22. If $T: \mathbf{M}_{nn} \to \mathbb{R}$ is any linear transformation satisfying $T(AB) = T(BA)$ for all A and B in $\mathbf{M}_{nn}$, show that there exists a number k such that $T(A) = k \text{ tr } A$ for all A. (See Lemma 1 Section 5.5.) [Hint: Let E_{ij} denote the $n \times n$ matrix with 1 in the (i, j) position and zeros elsewhere. Show that $E_{ik}E_{lj} = \begin{cases} 0 & \text{if } k \neq l \\ E_{ij} & \text{if } k = l \end{cases}$. Use this to show that $T(E_{ij}) = 0$ if $i \neq j$ and $T(E_{11}) = T(E_{22}) = \cdots = T(E_{nn})$. Put $k = T(E_{11})$ and use the fact that $\{E_{ij} \mid 1 \leq i, j \leq n\}$ is a basis of $\mathbf{M}_{nn}$.]

23. Let $T: \mathbb{C} \to \mathbb{C}$ be a linear transformation of the real vector space $\mathbb{C}$, and assume that $T(a) = a$ for every real number a. Show that the following are equivalent:

    (a) $T(zw) = T(z)T(w)$ for all z and w in $\mathbb{C}$.

    (b) Either $T = 1_\mathbb{C}$ or $T(z) = \overline{z}$ for each z in $\mathbb{C}$ (where $\overline{z}$ denotes the conjugate).

SECTION 7.2 Kernel and Image of a Linear Transformation

This section is devoted to two important subspaces associated with a linear transformation $T : V \to W$.

Definition 7.2 *The **kernel** of T (denoted $\ker T$) and the **image** of T (denoted $\operatorname{im} T$ or $T(V)$) are defined by*

$$\ker T = \{\mathbf{v} \text{ in } V \mid T(\mathbf{v}) = \mathbf{0}\}$$
$$\operatorname{im} T = \{T(\mathbf{v}) \mid \mathbf{v} \text{ in } V\} = T(V)$$

The kernel of T is often called the **nullspace** of T. It consists of all vectors $\mathbf{v}$ in V satisfying the *condition* that $T(\mathbf{v}) = \mathbf{0}$. The image of T is often called the **range** of T and consists of all vectors $\mathbf{w}$ in W of the *form* $\mathbf{w} = T(\mathbf{v})$ for some $\mathbf{v}$ in V. These subspaces are depicted in the diagrams.

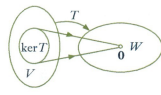
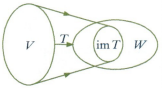

EXAMPLE 1

Let $T_A : \mathbb{R}^n \to \mathbb{R}^m$ be the linear transformation induced by the $m \times n$ matrix A, that is $T_A(\mathbf{x}) = A\mathbf{x}$ for all columns $\mathbf{x}$ in $\mathbb{R}^n$. Then

$$\ker T_A = \{\mathbf{x} \mid A\mathbf{x} = \mathbf{0}\} = \operatorname{null} A \quad \text{and} \quad \operatorname{im} T_A = \{A\mathbf{x} \mid \mathbf{x} \text{ in } \mathbb{R}^n\} = \operatorname{im} A$$

Hence the following theorem extends Example 2 Section 5.1.

Theorem 1

Let $T : V \to W$ be a linear transformation.
1. $\ker T$ is a subspace of V.
2. $\operatorname{im} T$ is a subspace of W.

PROOF

The fact that $T(\mathbf{0}) = \mathbf{0}$ shows that $\ker T$ and $\operatorname{im} T$ contain the zero vector of V and W respectively.

1. If $\mathbf{v}$ and $\mathbf{v}_1$ lie in $\ker T$, then $T(\mathbf{v}) = \mathbf{0} = T(\mathbf{v}_1)$, so

$$T(\mathbf{v} + \mathbf{v}_1) = T(\mathbf{v}) + T(\mathbf{v}_1) = \mathbf{0} + \mathbf{0} = \mathbf{0}$$
$$T(r\mathbf{v}) = rT(\mathbf{v}) = r\mathbf{0} = \mathbf{0} \quad \text{for all } r \text{ in } \mathbb{R}$$

Hence $\mathbf{v} + \mathbf{v}_1$ and $r\mathbf{v}$ lie in $\ker T$ (they satisfy the required condition), so $\ker T$ is a subspace of V by the subspace test (Theorem 1 Section 6.2).

2. If $\mathbf{w}$ and $\mathbf{w}_1$ lie in $\operatorname{im} T$, write $\mathbf{w} = T(\mathbf{v})$ and $\mathbf{w}_1 = T(\mathbf{v}_1)$ where $\mathbf{v}, \mathbf{v}_1 \in V$. Then

$$\mathbf{w} + \mathbf{w}_1 = T(\mathbf{v}) + T(\mathbf{v}_1) = T(\mathbf{v} + \mathbf{v}_1)$$
$$r\mathbf{w} = rT(\mathbf{v}) = T(r\mathbf{v}) \quad \text{for all } r \text{ in } \mathbb{R}$$

Hence $\mathbf{w} + \mathbf{w}_1$ and $r\mathbf{w}$ both lie in $\operatorname{im} T$ (they have the required form), so $\operatorname{im} T$ is a subspace of W.

SECTION 7.2 Kernel and Image of a Linear Transformation

Given a linear transformation $T : V \to W$:

dim(ker T) is called the **nullity** of T and denoted as nullity(T)

dim(im T) is called the **rank** of T and denoted as rank(T)

The rank of a matrix A was defined earlier to be the dimension of col A, the column space of A. The two usages of the word *rank* are consistent in the following sense. Recall the definition of T_A in Example 1.

EXAMPLE 2

Given an $m \times n$ matrix A, show that im T_A = col A, so rank T_A = rank A.

Solution ▶ Write $A = [\mathbf{c}_1 \ \cdots \ \mathbf{c}_n]$ in terms of its columns. Then

$$\text{im } T_A = \{A\mathbf{x} \mid \mathbf{x} \text{ in } \mathbb{R}^n\} = \{x_1 \mathbf{c}_1 + \cdots + x_n \mathbf{c}_n \mid x_i \text{ in } \mathbb{R}\}$$

using Definition 2.5. Hence im T_A is the column space of A; the rest follows.

Often, a useful way to study a subspace of a vector space is to exhibit it as the kernel or image of a linear transformation. Here is an example.

EXAMPLE 3

Define a transformation $P : \mathbf{M}_{mm} \to \mathbf{M}_{mm}$ by $P(A) = A - A^T$ for all A in $\mathbf{M}_{mm}$. Show that P is linear and that:

(a) ker P consists of all symmetric matrices.

(b) im P consists of all skew-symmetric matrices.

Solution ▶ The verification that P is linear is left to the reader. To prove part (a), note that a matrix A lies in ker P just when $0 = P(A) = A - A^T$, and this occurs if and only if $A = A^T$—that is, A is symmetric. Turning to part (b), the space im P consists of all matrices $P(A)$, A in $\mathbf{M}_{mm}$. Every such matrix is skew-symmetric because

$$P(A)^T = (A - A^T)^T = A^T - A = -P(A)$$

On the other hand, if S is skew-symmetric (that is, $S^T = -S$), then S lies in im P. In fact,

$$P[\tfrac{1}{2}S] = \tfrac{1}{2}S - [\tfrac{1}{2}S]^T = \tfrac{1}{2}(S - S^T) = \tfrac{1}{2}(S + S) = S.$$

One-to-One and Onto Transformations

Definition 7.3 Let $T : V \to W$ be a linear transformation.

1. T is said to be **onto** if im $T = W$.
2. T is said to be **one-to-one** if $T(\mathbf{v}) = T(\mathbf{v}_1)$ implies $\mathbf{v} = \mathbf{v}_1$.

A vector **w** in W is said to be **hit** by T if $\mathbf{w} = T(\mathbf{v})$ for some **v** in V. Then T is onto if every vector in W is hit at least once, and T is one-to-one if no element of W gets hit twice. Clearly the onto transformations T are those for which im $T = W$ is as large a subspace of W as possible. By contrast, Theorem 2 shows that the one-to-one transformations T are the ones with ker T as *small* a subspace of V as possible.

Theorem 2

If $T: V \to W$ is a linear transformation, then T is one-to-one if and only if ker $T = \{\mathbf{0}\}$.

PROOF

If T is one-to-one, let **v** be any vector in ker T. Then $T(\mathbf{v}) = \mathbf{0}$, so $T(\mathbf{v}) = T(\mathbf{0})$. Hence $\mathbf{v} = \mathbf{0}$ because T is one-to-one. Hence ker $T = \{\mathbf{0}\}$.

Conversely, assume that ker $T = \{\mathbf{0}\}$ and let $T(\mathbf{v}) = T(\mathbf{v}_1)$ with **v** and $\mathbf{v}_1$ in V. Then $T(\mathbf{v} - \mathbf{v}_1) = T(\mathbf{v}) - T(\mathbf{v}_1) = \mathbf{0}$, so $\mathbf{v} - \mathbf{v}_1$ lies in ker $T = \{\mathbf{0}\}$. This means that $\mathbf{v} - \mathbf{v}_1 = \mathbf{0}$, so $\mathbf{v} = \mathbf{v}_1$, proving that T is one-to-one.

EXAMPLE 4

The identity transformation $1_V: V \to V$ is both one-to-one and onto for any vector space V.

EXAMPLE 5

Consider the linear transformations

$$S: \mathbb{R}^3 \to \mathbb{R}^2 \quad \text{given by} \quad S(x, y, z) = (x + y, x - y)$$
$$T: \mathbb{R}^2 \to \mathbb{R}^3 \quad \text{given by} \quad T(x, y) = (x + y, x - y, x)$$

Show that T is one-to-one but not onto, whereas S is onto but not one-to-one.

Solution ▶ The verification that they are linear is omitted. T is one-to-one because

$$\ker T = \{(x, y) \mid x + y = x - y = x = 0\} = \{(0, 0)\}$$

However, it is not onto. For example $(0, 0, 1)$ does not lie in im T because if $(0, 0, 1) = (x + y, x - y, x)$ for some x and y, then $x + y = 0 = x - y$ and $x = 1$, an impossibility. Turning to S, it is not one-to-one by Theorem 2 because $(0, 0, 1)$ lies in ker S. But every element (s, t) in $\mathbb{R}^2$ lies in im S because $(s, t) = (x + y, x - y) = S(x, y, z)$ for some x, y, and z (in fact, $x = \frac{1}{2}(s + t), y = \frac{1}{2}(s - t)$, and $z = 0$. Hence S is onto.

SECTION 7.2 Kernel and Image of a Linear Transformation

EXAMPLE 6

Let U be an invertible $m \times m$ matrix and define
$$T: \mathbf{M}_{mn} \to \mathbf{M}_{mn} \quad \text{by} \quad T(X) = UX \text{ for all } X \text{ in } \mathbf{M}_{mn}$$
Show that T is a linear transformation that is both one-to-one and onto.

Solution ▶ The verification that T is linear is left to the reader. To see that T is one-to-one, let $T(X) = 0$. Then $UX = 0$, so left-multiplication by U^{-1} gives $X = 0$. Hence ker $T = \{0\}$, so T is one-to-one. Finally, if Y is any member of $\mathbf{M}_{mn}$, then $U^{-1}Y$ lies in $\mathbf{M}_{mn}$ too, and $T(U^{-1}Y) = U(U^{-1}Y) = Y$. This shows that T is onto.

The linear transformations $\mathbb{R}^n \to \mathbb{R}^m$ all have the form T_A for some $m \times n$ matrix A (Theorem 2 Section 2.6). The next theorem gives conditions under which they are onto or one-to-one. Note the connection with Theorems 3 and 4 in Section 5.4.

Theorem 3

Let A be an $m \times n$ matrix, and let $T_A: \mathbb{R}^n \to \mathbb{R}^m$ be the linear transformation induced by A, that is $T_A(\mathbf{x}) = A\mathbf{x}$ for all columns $\mathbf{x}$ in $\mathbb{R}^n$.
1. T_A is onto if and only if rank $A = m$.
2. T_A is one-to-one if and only if rank $A = n$.

PROOF

1. We have that im T_A is the column space of A (see Example 2), so T_A is onto if and only if the column space of A is $\mathbb{R}^m$. Because the rank of A is the dimension of the column space, this holds if and only if rank $A = m$.
2. ker $T_A = \{\mathbf{x} \text{ in } \mathbb{R}^n \mid A\mathbf{x} = \mathbf{0}\}$, so (using Theorem 2) T_A is one-to-one if and only if $A\mathbf{x} = \mathbf{0}$ implies $\mathbf{x} = \mathbf{0}$. This is equivalent to rank $A = n$ by Theorem 3 Section 5.4.

The Dimension Theorem

Let A denote an $m \times n$ matrix of rank r and let $T_A: \mathbb{R}^n \to \mathbb{R}^m$ denote the corresponding matrix transformation given by $T_A(\mathbf{x}) = A\mathbf{x}$ for all columns $\mathbf{x}$ in $\mathbb{R}^n$. It follows from Examples 1 and 2 that im $T_A = \text{col } A$, so $\dim(\text{im } T_A) = \dim(\text{col } A) = r$. But Theorem 2 Section 5.4 shows that $\dim(\ker T_A) = \dim(\text{null } A) = n - r$. Combining these we see that
$$\dim(\text{im } T_A) + \dim(\ker T_A) = n \quad \text{for every } m \times n \text{ matrix } A.$$
The main result of this section is a deep generalization of this observation.

> **Theorem 4**
>
> **Dimension Theorem**
> Let $T: V \to W$ be any linear transformation and assume that ker T and im T are both finite dimensional. Then V is also finite dimensional and
> $$\dim V = \dim(\ker T) + \dim(\operatorname{im} T)$$
> In other words, $\dim V = \operatorname{nullity}(T) + \operatorname{rank}(T)$.

> **PROOF**
>
> Every vector in im $T = T(V)$ has the form $T(\mathbf{v})$ for some $\mathbf{v}$ in V. Hence let $\{T(\mathbf{e}_1), T(\mathbf{e}_2), \ldots, T(\mathbf{e}_r)\}$ be a basis of im T, where the $\mathbf{e}_i$ lie in V. Let $\{\mathbf{f}_1, \mathbf{f}_2, \ldots, \mathbf{f}_k\}$ be any basis of ker T. Then $\dim(\operatorname{im} T) = r$ and $\dim(\ker T) = k$, so it suffices to show that $B = \{\mathbf{e}_1, \ldots, \mathbf{e}_r, \mathbf{f}_1, \ldots, \mathbf{f}_k\}$ is a basis of V.
>
> 1. *B spans V.* If $\mathbf{v}$ lies in V, then $T(\mathbf{v})$ lies in im V, so
> $$T(\mathbf{v}) = t_1 T(\mathbf{e}_1) + t_2 T(\mathbf{e}_2) + \cdots + t_r T(\mathbf{e}_r) \quad t_i \text{ in } \mathbb{R}$$
> This implies that $\mathbf{v} - t_1 \mathbf{e}_1 - t_2 \mathbf{e}_2 - \cdots - t_r \mathbf{e}_r$ lies in ker T and so is a linear combination of $\mathbf{f}_1, \ldots, \mathbf{f}_k$. Hence $\mathbf{v}$ is a linear combination of the vectors in B.
>
> 2. *B is linearly independent.* Suppose that t_i and s_j in $\mathbb{R}$ satisfy
> $$t_1 \mathbf{e}_1 + \cdots + t_r \mathbf{e}_r + s_1 \mathbf{f}_1 + \cdots + s_k \mathbf{f}_k = \mathbf{0} \quad (*)$$
> Applying T gives $t_1 T(\mathbf{e}_1) + \cdots + t_r T(\mathbf{e}_r) = \mathbf{0}$ (because $T(\mathbf{f}_i) = \mathbf{0}$ for each i). Hence the independence of $\{T(\mathbf{e}_1), \ldots, T(\mathbf{e}_r)\}$ yields $t_1 = \cdots = t_r = 0$. But then $(*)$ becomes
> $$s_1 \mathbf{f}_1 + \cdots + s_k \mathbf{f}_k = \mathbf{0}$$
> so $s_1 = \cdots = s_k = 0$ by the independence of $\{\mathbf{f}_1, \ldots, \mathbf{f}_k\}$. This proves that B is linearly independent.

Note that the vector space V is not assumed to be finite dimensional in Theorem 4. In fact, verifying that ker T and im T are both finite dimensional is often an important way to *prove* that V is finite dimensional.

Note further that $r + k = n$ in the proof so, after relabelling, we end up with a basis
$$B = \{\mathbf{e}_1, \mathbf{e}_2, \ldots, \mathbf{e}_r, \mathbf{e}_{r+1}, \ldots, \mathbf{e}_n\}$$
of V with the property that $\{\mathbf{e}_{r+1}, \ldots, \mathbf{e}_n\}$ is a basis of ker T and $\{T(\mathbf{e}_1), \ldots, T(\mathbf{e}_r)\}$ is a basis of im T. In fact, if V is known in advance to be finite dimensional, then *any* basis $\{\mathbf{e}_{r+1}, \ldots, \mathbf{e}_n\}$ of ker T can be extended to a basis $\{\mathbf{e}_1, \mathbf{e}_2, \ldots, \mathbf{e}_r, \mathbf{e}_{r+1}, \ldots, \mathbf{e}_n\}$ of V by Theorem 1 Section 6.4. Moreover, it turns out that, no matter how this is done, the vectors $\{T(\mathbf{e}_1), \ldots, T(\mathbf{e}_r)\}$ will be a basis of im T. This result is useful, and we record it for reference. The proof is much like that of Theorem 4 and is left as Exercise 26.

SECTION 7.2 *Kernel and Image of a Linear Transformation*

> **Theorem 5**
>
> Let $T : V \to W$ be a linear transformation, and let $\{e_1, \ldots, e_r, e_{r+1}, \ldots, e_n\}$ be a basis of V such that $\{e_{r+1}, \ldots, e_n\}$ is a basis of ker T. Then $\{T(e_1), \ldots, T(e_r)\}$ is a basis of im T, and hence r = rank T.

The dimension theorem is one of the most useful results in all of linear algebra. It shows that if either dim(ker T) or dim(im T) can be found, then the other is automatically known. In many cases it is easier to compute one than the other, so the theorem is a real asset. The rest of this section is devoted to illustrations of this fact. The next example uses the dimension theorem to give a different proof of the first part of Theorem 2 Section 5.4.

EXAMPLE 7

Let A be an $m \times n$ matrix of rank r. Show that the space null A of all solutions of the system $A\mathbf{x} = \mathbf{0}$ of m homogeneous equations in n variables has dimension $n - r$.

Solution ▶ The space in question is just ker T_A, where $T_A : \mathbb{R}^n \to \mathbb{R}^m$ is defined by $T_A(\mathbf{x}) = A\mathbf{x}$ for all columns $\mathbf{x}$ in $\mathbb{R}^n$. But dim(im T_A) = rank T_A = rank $A = r$ by Example 2, so dim(ker T_A) = $n - r$ by the dimension theorem.

EXAMPLE 8

If $T : V \to W$ is a linear transformation where V is finite dimensional, then
$$\dim(\ker T) \le \dim V \quad \text{and} \quad \dim(\operatorname{im} T) \le \dim V$$
Indeed, dim V = dim(ker T) + dim(im T) by Theorem 4. Of course, the first inequality also follows because ker T is a subspace of V.

EXAMPLE 9

Let $D : \mathbf{P}_n \to \mathbf{P}_{n-1}$ be the differentiation map defined by $D[p(x)] = p'(x)$. Compute ker D and hence conclude that D is onto.

Solution ▶ Because $p'(x) = 0$ means $p(x)$ is constant, we have dim(ker D) = 1. Since dim $\mathbf{P}_n = n + 1$, the dimension theorem gives
$$\dim(\operatorname{im} D) = (n + 1) - \dim(\ker D) = n = \dim(\mathbf{P}_{n-1})$$
This implies that im $D = \mathbf{P}_{n-1}$, so D is onto.

Of course it is not difficult to verify directly that each polynomial $q(x)$ in $\mathbf{P}_{n-1}$ is the derivative of some polynomial in $\mathbf{P}_n$ (simply integrate $q(x)$!), so the dimension theorem is not needed in this case. However, in some situations it is difficult to see directly that a linear transformation is onto, and the method used in Example 9 may be by far the easiest way to prove it. Here is another illustration.

EXAMPLE 10

Given a in $\mathbb{R}$, the evaluation map $E_a : \mathbf{P}_n \to \mathbb{R}$ is given by $E_a[p(x)] = p(a)$. Show that E_a is linear and onto, and hence conclude that $\{(x - a), (x - a)^2, \ldots, (x - a)^n\}$ is a basis of ker E_a, the subspace of all polynomials $p(x)$ for which $p(a) = 0$.

Solution ▶ E_a is linear by Example 3 Section 7.1; the verification that it is onto is left to the reader. Hence dim(im E_a) = dim($\mathbb{R}$) = 1, so dim(ker E_a) = $(n + 1) - 1 = n$ by the dimension theorem. Now each of the n polynomials $(x - a), (x - a)^2, \ldots, (x - a)^n$ clearly lies in ker E_a, and they are linearly independent (they have distinct degrees). Hence they are a basis because dim(ker E_a) = n.

We conclude by applying the dimension theorem to the rank of a matrix.

EXAMPLE 11

If A is any $m \times n$ matrix, show that rank A = rank $A^T A$ = rank AA^T.

Solution ▶ It suffices to show that rank A = rank $A^T A$ (the rest follows by replacing A with A^T). Write $B = A^T A$, and consider the associated matrix transformations

$$T_A : \mathbb{R}^n \to \mathbb{R}^m \quad \text{and} \quad T_B : \mathbb{R}^n \to \mathbb{R}^n$$

The dimension theorem and Example 2 give

$$\text{rank } A = \text{rank } T_A = \dim(\text{im } T_A) = n - \dim(\text{ker } T_A)$$
$$\text{rank } B = \text{rank } T_B = \dim(\text{im } T_B) = n - \dim(\text{ker } T_B)$$

so it suffices to show that ker T_A = ker T_B. Now $A\mathbf{x} = \mathbf{0}$ implies that $B\mathbf{x} = A^T A\mathbf{x} = \mathbf{0}$, so ker T_A is contained in ker T_B. On the other hand, if $B\mathbf{x} = \mathbf{0}$, then $A^T A\mathbf{x} = \mathbf{0}$, so

$$\|A\mathbf{x}\|^2 = (A\mathbf{x})^T (A\mathbf{x}) = \mathbf{x}^T A^T A\mathbf{x} = \mathbf{x}^T \mathbf{0} = 0$$

This implies that $A\mathbf{x} = \mathbf{0}$, so ker T_B is contained in ker T_A.

EXERCISES 7.2

1. For each matrix A, find a basis for the kernel and image of T_A, and find the rank and nullity of T_A.

   (a) $\begin{bmatrix} 1 & 2 & -1 & 1 \\ 3 & 1 & 0 & 2 \\ 1 & -3 & 2 & 0 \end{bmatrix}$
   ◆(b) $\begin{bmatrix} 2 & 1 & -1 & 3 \\ 1 & 0 & 3 & 1 \\ 1 & 1 & -4 & 2 \end{bmatrix}$

   (c) $\begin{bmatrix} 1 & 2 & -1 \\ 3 & 1 & 2 \\ 4 & -1 & 5 \\ 0 & 2 & -2 \end{bmatrix}$
   ◆(d) $\begin{bmatrix} 2 & 1 & 0 \\ 1 & -1 & 3 \\ 1 & 2 & -3 \\ 0 & 3 & -6 \end{bmatrix}$

2. In each case, (i) find a basis of ker T, and (ii) find a basis of im T. You may assume that T is linear.

   (a) $T : \mathbf{P}_2 \to \mathbb{R}^2$; $T(a + bx + cx^2) = (a, b)$

   ◆(b) $T : \mathbf{P}_2 \to \mathbb{R}^2$; $T(p(x)) = (p(0), p(1))$

   (c) $T : \mathbb{R}^3 \to \mathbb{R}^3$; $T(x, y, z) = (x + y, x + y, 0)$

   ◆(d) $T : \mathbb{R}^3 \to \mathbb{R}^4$; $T(x, y, z) = (x, x, y, y)$

   (e) $T : \mathbf{M}_{22} \to \mathbf{M}_{22}$; $T \begin{bmatrix} a & b \\ c & d \end{bmatrix} = \begin{bmatrix} a+b & b+c \\ c+d & d+a \end{bmatrix}$

- (f) $T: \mathbf{M}_{22} \to \mathbb{R}$; $T\begin{bmatrix} a & b \\ c & d \end{bmatrix} = a + d$
- (g) $T: \mathbf{P}_n \to \mathbb{R}$; $T(r_0 + r_1 x + \cdots + r_n x^n) = r_n$
- ◆(h) $T: \mathbb{R}^n \to \mathbb{R}$;
  $T(r_1, r_2, \ldots, r_n) = r_1 + r_2 + \cdots + r_n$
- (i) $T: \mathbf{M}_{22} \to \mathbf{M}_{22}$; $T(X) = XA - AX$, where $A = \begin{bmatrix} 0 & 1 \\ 1 & 0 \end{bmatrix}$
- ◆(j) $T: \mathbf{M}_{22} \to \mathbf{M}_{22}$; $T(X) = XA$, where $A = \begin{bmatrix} 1 & 1 \\ 0 & 0 \end{bmatrix}$

3. Let $P: V \to \mathbb{R}$ and $Q: V \to \mathbb{R}$ be linear transformations, where V is a vector space. Define $T: V \to \mathbb{R}^2$ by $T(\mathbf{v}) = (P(\mathbf{v}), Q(\mathbf{v}))$.
   (a) Show that T is a linear transformation.
   ◆(b) Show that $\ker T = \ker P \cap \ker Q$, the set of vectors in both $\ker P$ and $\ker Q$.

4. In each case, find a basis $B = \{\mathbf{e}_1, \ldots, \mathbf{e}_r, \mathbf{e}_{r+1}, \ldots, \mathbf{e}_n\}$ of V such that $\{\mathbf{e}_{r+1}, \ldots, \mathbf{e}_n\}$ is a basis of $\ker T$, and verify Theorem 5.
   (a) $T: \mathbb{R}^3 \to \mathbb{R}^4$; $T(x, y, z) = (x - y + 2z, x + y - z, 2x + z, 2y - 3z)$
   ◆(b) $T: \mathbb{R}^3 \to \mathbb{R}^4$; $T(x, y, z) = (x + y + z, 2x - y + 3z, z - 3y, 3x + 4z)$

5. Show that every matrix X in $\mathbf{M}_{nn}$ has the form $X = A^T - 2A$ for some matrix A in $\mathbf{M}_{nn}$. [Hint: The dimension theorem.]

6. In each case either prove the statement or give an example in which it is false. Throughout, let $T: V \to W$ be a linear transformation where V and W are finite dimensional.
   (a) If $V = W$, then $\ker T \subseteq \text{im } T$.
   ◆(b) If $\dim V = 5$, $\dim W = 3$, and $\dim(\ker T) = 2$, then T is onto.
   (c) If $\dim V = 5$ and $\dim W = 4$, then $\ker T \neq \{\mathbf{0}\}$.
   ◆(d) If $\ker T = V$, then $W = \{\mathbf{0}\}$.
   (e) If $W = \{\mathbf{0}\}$, then $\ker T = V$.
   ◆(f) If $W = V$, and $\text{im } T \subseteq \ker T$, then $T = 0$.
   (g) If $\{\mathbf{e}_1, \mathbf{e}_2, \mathbf{e}_3\}$ is a basis of V and $T(\mathbf{e}_1) = \mathbf{0} = T(\mathbf{e}_2)$, then $\dim(\text{im } T) \leq 1$.
   ◆(h) If $\dim(\ker T) \leq \dim W$, then $\dim W \geq \frac{1}{2} \dim V$.
   (i) If T is one-to-one, then $\dim V \leq \dim W$.
   ◆(j) If $\dim V \leq \dim W$, then T is one-to-one.
   (k) If T is onto, then $\dim V \geq \dim W$.
   ◆(l) If $\dim V \geq \dim W$, then T is onto.
   (m) If $\{T(\mathbf{v}_1), \ldots, T(\mathbf{v}_k)\}$ is independent, then $\{\mathbf{v}_1, \ldots, \mathbf{v}_k\}$ is independent.
   ◆(n) If $\{\mathbf{v}_1, \ldots, \mathbf{v}_k\}$ spans V, then $\{T(\mathbf{v}_1), \ldots, T(\mathbf{v}_k)\}$ spans W.

7. Show that linear independence is preserved by one-to-one transformations and that spanning sets are preserved by onto transformations. More precisely, if $T: V \to W$ is a linear transformation, show that:
   (a) If T is one-to-one and $\{\mathbf{v}_1, \ldots, \mathbf{v}_n\}$ is independent in V, then $\{T(\mathbf{v}_1), \ldots, T(\mathbf{v}_n)\}$ is independent in W.
   ◆(b) If T is onto and $V = \text{span}\{\mathbf{v}_1, \ldots, \mathbf{v}_n\}$, then $W = \text{span}\{T(\mathbf{v}_1), \ldots, T(\mathbf{v}_n)\}$.

8. Given $\{\mathbf{v}_1, \ldots, \mathbf{v}_n\}$ in a vector space V, define $T: \mathbb{R}^n \to V$ by $T(r_1, \ldots, r_n) = r_1 \mathbf{v}_1 + \cdots + r_n \mathbf{v}_n$. Show that T is linear, and that:
   (a) T is one-to-one if and only if $\{\mathbf{v}_1, \ldots, \mathbf{v}_n\}$ is independent.
   ◆(b) T is onto if and only if $V = \text{span}\{\mathbf{v}_1, \ldots, \mathbf{v}_n\}$.

9. Let $T: V \to V$ be a linear transformation where V is finite dimensional. Show that exactly one of (i) and (ii) holds: (i) $T(\mathbf{v}) = \mathbf{0}$ for some $\mathbf{v} \neq \mathbf{0}$ in V; (ii) $T(\mathbf{x}) = \mathbf{v}$ has a solution $\mathbf{x}$ in V for every $\mathbf{v}$ in V.

◆10. Let $T: \mathbf{M}_{nn} \to \mathbb{R}$ denote the trace map: $T(A) = \text{tr } A$ for all A in $\mathbf{M}_{nn}$. Show that $\dim(\ker T) = n^2 - 1$.

11. Show that the following are equivalent for a linear transformation $T: V \to W$.
    (a) $\ker T = V$ (b) $\text{im } T = \{\mathbf{0}\}$
    (c) $T = 0$

◆12. Let A and B be $m \times n$ and $k \times n$ matrices, respectively. Assume that $A\mathbf{x} = \mathbf{0}$ implies $B\mathbf{x} = \mathbf{0}$ for every n-column $\mathbf{x}$. Show that $\text{rank } A \geq \text{rank } B$. [Hint: Theorem 4.]

13. Let A be an $m \times n$ matrix of rank r. Thinking of $\mathbb{R}^n$ as rows, define $V = \{\mathbf{x} \text{ in } \mathbb{R}^m \mid \mathbf{x}A = \mathbf{0}\}$. Show that $\dim V = m - r$.

14. Consider $V = \left\{ \begin{bmatrix} a & b \\ c & d \end{bmatrix} \middle| a + c = b + d \right\}$.

    (a) Consider $S : \mathbf{M}_{22} \to \mathbb{R}$ with
    $S\begin{bmatrix} a & b \\ c & d \end{bmatrix} = a + c - b - d$. Show that
    S is linear and onto and that V is a subspace of $\mathbf{M}_{22}$. Compute $\dim V$.

    (b) Consider $T : V \to \mathbb{R}$ with $T\begin{bmatrix} a & b \\ c & d \end{bmatrix} = a + c$.
    Show that T is linear and onto, and use this information to compute $\dim(\ker T)$.

15. Define $T : \mathbf{P}_n \to \mathbb{R}$ by $T[p(x)] = $ the sum of all the coefficients of $p(x)$.

    (a) Use the dimension theorem to show that $\dim(\ker T) = n$.

    ◆(b) Conclude that $\{x - 1, x^2 - 1, \ldots, x^n - 1\}$ is a basis of $\ker T$.

16. Use the dimension theorem to prove Theorem 1 Section 1.3: If A is an $m \times n$ matrix with $m < n$, the system $A\mathbf{x} = \mathbf{0}$ of m homogeneous equations in n variables always has a nontrivial solution.

17. Let B be an $n \times n$ matrix, and consider the subspaces $U = \{A \mid A \text{ in } \mathbf{M}_{mn}, AB = 0\}$ and $V = \{AB \mid A \text{ in } \mathbf{M}_{mn}\}$. Show that $\dim U + \dim V = mn$.

18. Let U and V denote, respectively, the spaces of even and odd polynomials in $\mathbf{P}_n$. Show that $\dim U + \dim V = n + 1$. [Hint: Consider $T : \mathbf{P}_n \to \mathbf{P}_n$ where $T[p(x)] = p(x) - p(-x)$.]

19. Show that every polynomial $f(x)$ in $\mathbf{P}_{n-1}$ can be written as $f(x) = p(x + 1) - p(x)$ for some polynomial $p(x)$ in $\mathbf{P}_n$. [Hint: Define $T : \mathbf{P}_n \to \mathbf{P}_{n-1}$ by $T[p(x)] = p(x + 1) - p(x)$.]

◆20. Let U and V denote the spaces of symmetric and skew-symmetric $n \times n$ matrices. Show that $\dim U + \dim V = n^2$.

21. Assume that B in $\mathbf{M}_{nn}$ satisfies $B^k = 0$ for some $k \geq 1$. Show that every matrix in $\mathbf{M}_{nn}$ has the form $BA - A$ for some A in $\mathbf{M}_{nn}$. [Hint: Show that $T : \mathbf{M}_{nn} \to \mathbf{M}_{nn}$ is linear and one-to-one where $T(A) = BA - A$ for each A.]

◆22. Fix a column $\mathbf{y} \neq \mathbf{0}$ in $\mathbb{R}^n$ and let $U = \{A \text{ in } \mathbf{M}_{nn} \mid A\mathbf{y} = \mathbf{0}\}$. Show that $\dim U = n(n - 1)$.

23. If B in $\mathbf{M}_{mn}$ has rank r, let $U = \{A \text{ in } \mathbf{M}_{nm} \mid BA = 0\}$ and $W = \{BA \mid A \text{ in } \mathbf{M}_{nm}\}$. Show that $\dim U = n(n - r)$ and $\dim W = nr$. [Hint: Show that U consists of all matrices A whose columns are in the null space of B. Use Example 7.]

24. Let $T : V \to V$ be a linear transformation where $\dim V = n$. If $\ker T \cap \operatorname{im} T = \{\mathbf{0}\}$, show that every vector $\mathbf{v}$ in V can be written $\mathbf{v} = \mathbf{u} + \mathbf{w}$ for some $\mathbf{u}$ in $\ker T$ and $\mathbf{w}$ in $\operatorname{im} T$. [Hint: Choose bases $B \subseteq \ker T$ and $D \subseteq \operatorname{im} T$, and use Exercise 33 Section 6.3.]

25. Let $T : \mathbb{R}^n \to \mathbb{R}^n$ be a linear operator of rank 1, where $\mathbb{R}^n$ is written as rows. Show that there exist numbers $a_1, a_2, \ldots, a_n$ and $b_1, b_2, \ldots, b_n$ such that $T(X) = XA$ for all rows X in $\mathbb{R}^n$, where

$$A = \begin{bmatrix} a_1 b_1 & a_1 b_2 & \cdots & a_1 b_n \\ a_2 b_1 & a_2 b_2 & \cdots & a_2 b_n \\ \vdots & \vdots & & \vdots \\ a_n b_1 & a_n b_2 & \cdots & a_n b_n \end{bmatrix}$$

[Hint: $\operatorname{im} T = \mathbb{R}\mathbf{w}$ for $\mathbf{w} = (b_1, \ldots, b_n)$ in $\mathbb{R}^n$.]

26. Prove Theorem 5.

27. Let $T : V \to \mathbb{R}$ be a nonzero linear transformation, where $\dim V = n$. Show that there is a basis $\{\mathbf{e}_1, \ldots, \mathbf{e}_n\}$ of V such that $T(r_1\mathbf{e}_1 + r_2\mathbf{e}_2 + \cdots + r_n\mathbf{e}_n) = r_1$.

28. Let $f \neq 0$ be a fixed polynomial of degree $m \geq 1$. If p is any polynomial, recall that $(p \circ f)(x) = p[f(x)]$. Define $T_f : \mathbf{P}_n \to \mathbf{P}_{n+m}$ by $T_f(p) = p \circ f$.

    (a) Show that T_f is linear.

    (b) Show that T_f is one-to-one.

29. Let U be a subspace of a finite dimensional vector space V.

    (a) Show that $U = \ker T$ for some linear operator $T : V \to V$.

    ◆(b) Show that $U = \operatorname{im} S$ for some linear operator $S : V \to V$. [Hint: Theorems 1 Section 6.4 and 3 Section 7.1.]

30. Let V and W be finite dimensional vector spaces.

   (a) Show that dim $W \leq$ dim V if and only if there exists an onto linear transformation $T: V \to W$. [*Hint*: Theorems 1 Section 6.4 and 3 Section 7.1.]

   (b) Show that dim $W \geq$ dim V if and only if there exists a one-to-one linear transformation $T: V \to W$. [*Hint*: Theorems 1 Section 6.4 and 3 Section 7.1.]

SECTION 7.3 Isomorphisms and Composition

Often two vector spaces can consist of quite different types of vectors but, on closer examination, turn out to be the same underlying space displayed in different symbols. For example, consider the spaces

$$\mathbb{R}^2 = \{(a, b) \mid a, b \in \mathbb{R}\} \quad \text{and} \quad \mathbf{P}_1 = \{a + bx \mid a, b \in \mathbb{R}\}.$$

Compare the addition and scalar multiplication in these spaces:

$$(a, b) + (a_1, b_1) = (a + a_1, b + b_1) \qquad (a + bx) + (a_1 + b_1 x) = (a + a_1) + (b_1 + b_1)x$$
$$r(a, b) = (ra, rb) \qquad r(a + bx) = (ra) + (rb)x$$

Clearly these are the *same* vector space expressed in different notation: if we change each (a, b) in $\mathbb{R}^2$ to $a + bx$, then $\mathbb{R}^2$ *becomes* $\mathbf{P}_1$, complete with addition and scalar multiplication. This can be expressed by noting that the map $(a, b) \mapsto a + bx$ is a linear transformation $\mathbb{R}^2 \to \mathbf{P}_1$ that is both one-to-one and onto. In this form, we can describe the general situation.

Definition 7.4 *A linear transformation $T: V \to W$ is called an* **isomorphism** *if it is both onto and one-to-one. The vector spaces V and W are said to be* **isomorphic** *if there exists an isomorphism $T: V \to W$, and we write $V \cong W$ when this is the case.*

EXAMPLE 1

The identity transformation $1_V : V \to V$ is an isomorphism for any vector space V.

EXAMPLE 2

If $T: \mathbf{M}_{mn} \to \mathbf{M}_{nm}$ is defined by $T(A) = A^T$ for all A in $\mathbf{M}_{mn}$, then T is an isomorphism (verfiy). Hence $\mathbf{M}_{mn} \cong \mathbf{M}_{nm}$.

EXAMPLE 3

Isomorphic spaces can "look" quite different. For example, $\mathbf{M}_{22} \cong \mathbf{P}_3$ because the map $T: \mathbf{M}_{22} \to \mathbf{P}_3$ given by $T\begin{bmatrix} a & b \\ c & d \end{bmatrix} = a + bx + cx^2 + dx^3$ is an isomorphism (verify).

The word *isomorphism* comes from two Greek roots: *iso*, meaning "same," and *morphos*, meaning "form." An isomorphism $T : V \to W$ induces a pairing

$$\mathbf{v} \leftrightarrow T(\mathbf{v})$$

between vectors $\mathbf{v}$ in V and vectors $T(\mathbf{v})$ in W that preserves vector addition and scalar multiplication. Hence, *as far as their vector space properties are concerned*, the spaces V and W are identical except for notation. Because addition and scalar multiplication in either space are completely determined by the same operations in the other space, all *vector space* properties of either space are completely determined by those of the other.

One of the most important examples of isomorphic spaces was considered in Chapter 4. Let A denote the set of all "arrows" with tail at the origin in space, and make A into a vector space using the parallelogram law and the scalar multiple law (see Section 4.1). Then define a transformation $T : \mathbb{R}^3 \to A$ by taking

$$T \begin{bmatrix} x \\ y \\ z \end{bmatrix} \text{ to be the arrow } \mathbf{v} \text{ from the origin to the point } P(x, y, z).$$

In Section 4.1 matrix addition and scalar multiplication were shown to correspond to the parallelogram law and the scalar multiplication law for these arrows, so the map T is a linear transformation. Moreover T is an isomorphism: it is one-to-one by Theorem 2 Section 4.1, and it is onto because, given an arrow $\mathbf{v}$ in A with tip $P(x, y, z)$, we have $T \begin{bmatrix} x \\ y \\ z \end{bmatrix} = \mathbf{v}$. This justifies the identification $\mathbf{v} = \begin{bmatrix} x \\ y \\ z \end{bmatrix}$ in Chapter 4 of the geometric arrows with the algebraic matrices. This identification is very useful. The arrows give a "picture" of the matrices and so bring geometric intuition into $\mathbb{R}^3$; the matrices are useful for detailed calculations and so bring analytic precision into geometry. This is one of the best examples of the power of an isomorphism to shed light on both spaces being considered.

The following theorem gives a very useful characterization of isomorphisms: They are the linear transformations that preserve bases.

Theorem 1

If V and W are finite dimensional spaces, the following conditions are equivalent for a linear transformation $T : V \to W$.

1. T is an isomorphism.
2. If $\{\mathbf{e}_1, \mathbf{e}_2, ..., \mathbf{e}_n\}$ is any basis of V, then $\{T(\mathbf{e}_1), T(\mathbf{e}_2), ..., T(\mathbf{e}_n)\}$ is a basis of W.
3. There exists a basis $\{\mathbf{e}_1, \mathbf{e}_2, ..., \mathbf{e}_n\}$ of V such that $\{T(\mathbf{e}_1), T(\mathbf{e}_2), ..., T(\mathbf{e}_n)\}$ is a basis of W.

PROOF

(1) $\Rightarrow$ (2). Let $\{\mathbf{e}_1, ..., \mathbf{e}_n\}$ be a basis of V. If $t_1 T(\mathbf{e}_1) + \cdots + t_n T(\mathbf{e}_n) = \mathbf{0}$ with t_i in $\mathbb{R}$, then $T(t_1 \mathbf{e}_1 + \cdots + t_n \mathbf{e}_n) = \mathbf{0}$, so $t_1 \mathbf{e}_1 + \cdots + t_n \mathbf{e}_n = \mathbf{0}$ (because ker $T = \{\mathbf{0}\}$). But then each $t_i = 0$ by the independence of the $\mathbf{e}_i$, so $\{T(\mathbf{e}_1), ..., T(\mathbf{e}_n)\}$ is independent. To show that it spans W, choose $\mathbf{w}$ in W. Because T is onto, $\mathbf{w} = T(\mathbf{v})$ for some $\mathbf{v}$ in V, so write $\mathbf{v} = t_1 \mathbf{e}_1 + \cdots + t_n \mathbf{e}_n$. Then $\mathbf{w} = T(\mathbf{v}) = t_1 T(\mathbf{e}_1) + \cdots + t_n T(\mathbf{e}_n)$, proving that $\{T(\mathbf{e}_1), ..., T(\mathbf{e}_n)\}$ spans W.

SECTION 7.3 *Isomorphisms and Composition*

(2) $\Rightarrow$ (3). This is because V *has* a basis.

(3) $\Rightarrow$ (1). If $T(\mathbf{v}) = \mathbf{0}$, write $\mathbf{v} = v_1\mathbf{e}_1 + \cdots + v_n\mathbf{e}_n$ where each v_i is in $\mathbb{R}$. Then $\mathbf{0} = T(\mathbf{v}) = v_1 T(\mathbf{e}_1) + \cdots + v_n T(\mathbf{e}_n)$, so $v_1 = \cdots = v_n = 0$ by (3). Hence $\mathbf{v} = \mathbf{0}$, so ker $T = \{\mathbf{0}\}$ and T is one-to-one. To show that T is onto, let $\mathbf{w}$ be any vector in W. By (3) there exist $w_1, \ldots, w_n$ in $\mathbb{R}$ such that

$$\mathbf{w} = w_1 T(\mathbf{e}_1) + \cdots + w_n T(\mathbf{e}_n) = T(w_1\mathbf{e}_1 + \cdots + w_n\mathbf{e}_n).$$

Thus T is onto.

Theorem 1 dovetails nicely with Theorem 3 Section 7.1 as follows. Let V and W be vector spaces of dimension n, and suppose that $\{\mathbf{e}_1, \mathbf{e}_2, \ldots, \mathbf{e}_n\}$ and $\{\mathbf{f}_1, \mathbf{f}_2, \ldots, \mathbf{f}_n\}$ are bases of V and W, respectively. Theorem 3 Section 7.1 asserts that there exists a linear transformation $T : V \to W$ such that

$$T(\mathbf{e}_i) = \mathbf{f}_i \quad \text{for each } i = 1, 2, \ldots, n$$

Then $\{T(\mathbf{e}_1), \ldots, T(\mathbf{e}_n)\}$ is evidently a basis of W, so T is an isomorphism by Theorem 1. Furthermore, the action of T is prescribed by

$$T(r_1\mathbf{e}_1 + \cdots + r_n\mathbf{e}_n) = r_1\mathbf{f}_1 + \cdots + r_n\mathbf{f}_n$$

so isomorphisms between spaces of equal dimension can be easily defined as soon as bases are known. In particular, this shows that if two vector spaces V and W have the same dimension then they are isomorphic, that is $V \cong W$. This is half of the following theorem.

Theorem 2

If V and W are finite dimensional vector spaces, then $V \cong W$ if and only if $\dim V = \dim W$.

PROOF

It remains to show that if $V \cong W$ then $\dim V = \dim W$. But if $V \cong W$, then there exists an isomorphism $T : V \to W$. Since V is finite dimensional, let $\{\mathbf{e}_1, \ldots, \mathbf{e}_n\}$ be a basis of V. Then $\{T(\mathbf{e}_1), \ldots, T(\mathbf{e}_n)\}$ is a basis of W by Theorem 1, so $\dim W = n = \dim V$.

Corollary 1

Let U, V, and W denote vector spaces. Then:

1. $V \cong V$ for every vector space V.
2. If $V \cong W$ then $W \cong V$.
3. If $U \cong V$ and $V \cong W$, then $U \cong W$.

The proof is left to the reader. By virtue of these properties, the relation $\cong$ is called an *equivalence relation* on the class of finite dimensional vector spaces. Since $\dim(\mathbb{R}^n) = n$ it follows that

Corollary 2

If V is a vector space and $\dim V = n$, then V is isomorphic to $\mathbb{R}^n$.

If V is a vector space of dimension n, note that there are important explicit isomorphisms $V \to \mathbb{R}^n$. Fix a basis $B = \{\mathbf{b}_1, \mathbf{b}_2, \ldots, \mathbf{b}_n\}$ of V and write $\{\mathbf{e}_1, \mathbf{e}_2, \ldots, \mathbf{e}_n\}$ for the standard basis of $\mathbb{R}^n$. By Theorem 3 Section 7.1 there is a unique linear transformation $C_B : V \to \mathbb{R}^n$ given by

$$C_B(v_1 \mathbf{b}_1 + v_2 \mathbf{b}_2 + \cdots + v_n \mathbf{b}_n) = v_1 \mathbf{e}_1 + v_2 \mathbf{e}_2 + \cdots + v_n \mathbf{e}_n = \begin{bmatrix} v_1 \\ v_2 \\ \vdots \\ v_n \end{bmatrix}$$

where each v_i is in $\mathbb{R}$. Moreover, $C_B(\mathbf{b}_i) = \mathbf{e}_i$ for each i so C_B is an isomorphism by Theorem 1, called the **coordinate isomorphism** corresponding to the basis B. These isomorphisms will play a central role in Chapter 9.

The conclusion in the above corollary can be phrased as follows: As far as vector space properties are concerned, every n-dimensional vector space V is essentially the same as $\mathbb{R}^n$; they are the "same" vector space except for a change of symbols. This appears to make the process of abstraction seem less important—just study $\mathbb{R}^n$ and be done with it! But consider the different "feel" of the spaces $\mathbf{P}_8$ and $\mathbf{M}_{33}$ even though they are both the "same" as $\mathbb{R}^9$: For example, vectors in $\mathbf{P}_8$ can have roots, while vectors in $\mathbf{M}_{33}$ can be multiplied. So the merit in the abstraction process lies in identifying *common* properties of the vector spaces in the various examples. This is important even for finite dimensional spaces. However, the payoff from abstraction is much greater in the infinite dimensional case, particularly for spaces of functions.

EXAMPLE 4

Let V denote the space of all 2×2 symmetric matrices. Find an isomorphism $T : \mathbf{P}_2 \to V$ such that $T(1) = I$, where I is the 2×2 identity matrix

Solution ▶ $\{1, x, x^2\}$ is a basis of $\mathbf{P}_2$, and we want a basis of V containing I. The set $\left\{ \begin{bmatrix} 1 & 0 \\ 0 & 1 \end{bmatrix}, \begin{bmatrix} 0 & 1 \\ 1 & 0 \end{bmatrix}, \begin{bmatrix} 0 & 0 \\ 0 & 1 \end{bmatrix} \right\}$ is independent in V, so it is a basis because $\dim V = 3$ (by Example 11 Section 6.3). Hence define $T : \mathbf{P}_2 \to V$ by taking $T(1) = \begin{bmatrix} 1 & 0 \\ 0 & 1 \end{bmatrix}$, $T(x) = \begin{bmatrix} 0 & 1 \\ 1 & 0 \end{bmatrix}$, $T(x^2) = \begin{bmatrix} 0 & 0 \\ 0 & 1 \end{bmatrix}$, and extending linearly as in Theorem 3 Section 7.1. Then T is an isomorphism by Theorem 1, and its action is given by $T(a + bx + cx^2) = aT(1) + bT(x) + cT(x^2) = \begin{bmatrix} a & b \\ b & a+c \end{bmatrix}$.

The dimension theorem (Theorem 4 Section 7.2) gives the following useful fact about isomorphisms.

Theorem 3

If V and W have the same dimension n, a linear transformation $T : V \to W$ is an isomorphism if it is either one-to-one or onto.

SECTION 7.3 Isomorphisms and Composition

> **PROOF**
>
> The dimension theorem asserts that $\dim(\ker T) + \dim(\operatorname{im} T) = n$, so $\dim(\ker T) = 0$ if and only if $\dim(\operatorname{im} T) = n$. Thus T is one-to-one if and only if T is onto, and the result follows.

Composition

Suppose that $T : V \to W$ and $S : W \to U$ are linear transformations. They link together as in the diagram so, as in Section 2.3, it is possible to define a new function $V \to U$ by first applying T and then S.

Definition 7.5

> Given linear transformations $V \xrightarrow{T} W \xrightarrow{S} U$, the **composite** $ST : V \to U$ of T and S is defined by
>
> $$ST(\mathbf{v}) = S[T(\mathbf{v})] \quad \text{for all } \mathbf{v} \text{ in } V.$$
>
> The operation of forming the new function ST is called **composition**.[1]

The action of ST can be described compactly as follows: ST means first T then S.

Not all pairs of linear transformations can be composed. For example, if $T : V \to W$ and $S : W \to U$ are linear transformations then $ST : V \to U$ is defined, but TS *cannot* be formed unless $U = V$ because $S : W \to U$ and $T : V \to W$ do not "link" in that order.[2]

Moreover, even if ST and TS can both be formed, they may not be equal. In fact, if $S : \mathbb{R}^m \to \mathbb{R}^n$ and $T : \mathbb{R}^n \to \mathbb{R}^m$ are induced by matrices A and B respectively, then ST and TS can both be formed (they are induced by AB and BA respectively), but the matrix products AB and BA may not be equal (they may not even be the same size). Here is another example.

EXAMPLE 5

Define: $S : \mathbf{M}_{22} \to \mathbf{M}_{22}$ and $T : \mathbf{M}_{22} \to \mathbf{M}_{22}$ by $S\begin{bmatrix} a & b \\ c & d \end{bmatrix} = \begin{bmatrix} c & d \\ a & b \end{bmatrix}$ and $T(A) = A^T$ for $A \in \mathbf{M}_{22}$. Describe the action of ST and TS, and show that $ST \neq TS$.

Solution ▸ $ST\begin{bmatrix} a & b \\ c & d \end{bmatrix} = S\begin{bmatrix} a & c \\ b & d \end{bmatrix} = \begin{bmatrix} b & d \\ a & c \end{bmatrix}$, whereas $TS\begin{bmatrix} a & b \\ c & d \end{bmatrix} = T\begin{bmatrix} c & d \\ a & b \end{bmatrix} = \begin{bmatrix} c & a \\ d & b \end{bmatrix}$.

It is clear that $TS\begin{bmatrix} a & b \\ c & d \end{bmatrix}$ need not equal $ST\begin{bmatrix} a & b \\ c & d \end{bmatrix}$, so $TS \neq ST$.

The next theorem collects some basic properties[3] of the composition operation.

---

[1] In Section 2.3 we denoted the composite as $S \circ T$. However, it is more convenient to use the simpler notation ST.

[2] Actually, all that is required is $U \subseteq V$.

[3] Theorem 4 can be expressed by saying that vector spaces and linear transformations are an example of a category. In general a category consists of certain objects and, for any two objects X and Y, a set mor(X, Y). The elements α of mor(X, Y) are called morphisms from X to Y and are written $\alpha : X \to Y$. It is assumed that identity morphisms and composition are defined in such a way that Theorem 4 holds. Hence, in the category of vector spaces the objects are the vector spaces themselves and the morphisms are the linear transformations. Another example is the category of metric spaces, in which the objects are sets equipped with a distance function (called a metric), and the morphisms are continuous functions (with respect to the metric). The category of sets and functions is a very basic example.

Theorem 4

Let $V \xrightarrow{T} W \xrightarrow{S} U \xrightarrow{R} Z$ be linear transformations.
1. The composite ST is again a linear transformation.
2. $T1_V = T$ and $1_W T = T$.
3. $(RS)T = R(ST)$.

PROOF

The proofs of (1) and (2) are left as Exercise 25. To prove (3), observe that, for all $\mathbf{v}$ in V:

$$\{(RS)T\}(\mathbf{v}) = (RS)[T(\mathbf{v})] = R\{S[T(\mathbf{v})]\} = R\{(ST)(\mathbf{v})\} = \{R(ST)\}(\mathbf{v})$$

Up to this point, composition seems to have no connection with isomorphisms. In fact, the two notions are closely related.

Theorem 5

Let V and W be finite dimensional vector spaces. The following conditions are equivalent for a linear transformation $T : V \to W$.
1. T is an isomorphism.
2. There exists a linear transformation $S : W \to V$ such that $ST = 1_V$ and $TS = 1_W$.

Moreover, in this case S is also an isomorphism and is uniquely determined by T:

If $\mathbf{w}$ in W is written as $\mathbf{w} = T(\mathbf{v})$, then $S(\mathbf{w}) = \mathbf{v}$.

PROOF

(1) $\Rightarrow$ (2). If $B = \{\mathbf{e}_1, \ldots, \mathbf{e}_n\}$ is a basis of V, then $D = \{T(\mathbf{e}_1), \ldots, T(\mathbf{e}_n)\}$ is a basis of W by Theorem 1. Hence (using Theorem 3 Section 7.1), define a linear transformation $S : W \to V$ by

$$S[T(\mathbf{e}_i)] = \mathbf{e}_i \quad \text{for each } i \tag{$*$}$$

Since $\mathbf{e}_i = 1_V(\mathbf{e}_i)$, this gives $ST = 1_V$ by Theorem 2 Section 7.1. But applying T gives $T[S[T(\mathbf{e}_i)]] = T(\mathbf{e}_i)$ for each i, so $TS = 1_W$ (again by Theorem 2 Section 7.1, using the basis D of W).

(2) $\Rightarrow$ (1). If $T(\mathbf{v}) = T(\mathbf{v}_1)$, then $S[T(\mathbf{v})] = S[T(\mathbf{v}_1)]$. Because $ST = 1_V$ by (2), this reads $\mathbf{v} = \mathbf{v}_1$; that is, T is one-to-one. Given $\mathbf{w}$ in W, the fact that $TS = 1_W$ means that $\mathbf{w} = T[S(\mathbf{w})]$, so T is onto.

Finally, S is uniquely determined by the condition $ST = 1_V$ because this condition implies $(*)$. S is an isomorphism because it carries the basis D to B. As to the last assertion, given $\mathbf{w}$ in W, write $\mathbf{w} = r_1 T(\mathbf{e}_1) + \cdots + r_n T(\mathbf{e}_n)$. Then $\mathbf{w} = T(\mathbf{v})$, where $\mathbf{v} = r_1 \mathbf{e}_1 + \cdots + r_n \mathbf{e}_n$. Then $S(\mathbf{w}) = \mathbf{v}$ by $(*)$.

SECTION 7.3 Isomorphisms and Composition

Given an isomorphism $T : V \to W$, the unique isomorphism $S : W \to V$ satisfying condition (2) of Theorem 5 is called the **inverse** of T and is denoted by T^{-1}. Hence $T : V \to W$ and $T^{-1} : W \to V$ are related by the **fundamental identities**:

$$T^{-1}[T(\mathbf{v})] = \mathbf{v} \text{ for all } \mathbf{v} \text{ in } V \quad \text{and} \quad T[T^{-1}(\mathbf{w})] = \mathbf{w} \text{ for all } \mathbf{w} \text{ in } W$$

In other words, each of T and T^{-1} reverses the action of the other. In particular, equation (∗) in the proof of Theorem 5 shows how to define T^{-1} using the image of a basis under the isomorphism T. Here is an example.

EXAMPLE 6

Define $T : \mathbf{P}_1 \to \mathbf{P}_1$ by $T(a + bx) = (a - b) + ax$. Show that T has an inverse, and find the action of T^{-1}.

Solution ▶ The transformation T is linear (verify). Because $T(1) = 1 + x$ and $T(x) = -1$, T carries the basis $B = \{1, x\}$ to the basis $D = \{1 + x, -1\}$. Hence T is an isomorphism, and T^{-1} carries D back to B, that is,

$$T^{-1}(1 + x) = 1 \quad \text{and} \quad T^{-1}(-1) = x.$$

Because $a + bx = b(1 + x) + (b - a)(-1)$, we obtain

$$T^{-1}(a + bx) = bT^{-1}(1 + x) + (b - a)T^{-1}(-1) = b + (b - a)x.$$

Sometimes the action of the inverse of a transformation is apparent.

EXAMPLE 7

If $B = \{\mathbf{b}_1, \mathbf{b}_2, \ldots, \mathbf{b}_n\}$ is a basis of a vector space V, the coordinate transformation $C_B : V \to \mathbb{R}^n$ is an isomorphism defined by

$$C_B(v_1\mathbf{b}_1 + v_2\mathbf{b}_2 + \cdots + v_n\mathbf{b}_n) = (v_1, v_2, \ldots, v_n)^T.$$

The way to reverse the action of C_B is clear: $C_B^{-1} : \mathbb{R}^n \to V$ is given by

$$C_B^{-1}(v_1, v_2, \ldots, v_n) = v_1\mathbf{b}_1 + v_2\mathbf{b}_2 + \cdots + v_n\mathbf{b}_n \quad \text{for all } v_i \text{ in } V.$$

Condition (2) in Theorem 5 characterizes the inverse of a linear transformation $T : V \to W$ as the (unique) transformation $S : W \to V$ that satisfies $ST = 1_V$ and $TS = 1_W$. This often determines the inverse.

EXAMPLE 8

Define $T : \mathbb{R}^3 \to \mathbb{R}^3$ by $T(x, y, z) = (z, x, y)$. Show that $T^3 = 1_{\mathbb{R}^3}$, and hence find T^{-1}.

Solution ▶ $T^2(x, y, z) = T[T(x, y, z)] = T(z, x, y) = (y, z, x)$. Hence

$$T^3(x, y, z) = T[T^2(x, y, z)] = T(y, z, x) = (x, y, z)$$

Since this holds for all (x, y, z), it shows that $T^3 = 1_{\mathbb{R}^3}$, so $T(T^2) = 1_{\mathbb{R}^3} = (T^2)T$. Thus $T^{-1} = T^2$ by (2) of Theorem 5.

EXAMPLE 9

Define $T: \mathbf{P}_n \to \mathbb{R}^{n+1}$ by $T(p) = (p(0), p(1), \ldots, p(n))$ for all p in $\mathbf{P}_n$. Show that T^{-1} exists.

Solution ▶ The verification that T is linear is left to the reader. If $T(p) = 0$, then $p(k) = 0$ for $k = 0, 1, \ldots, n$, so p has $n + 1$ distinct roots. Because p has degree at most n, this implies that $p = 0$ is the zero polynomial (Theorem 4 Section 6.5) and hence that T is one-to-one. But $\dim \mathbf{P}_n = n + 1 = \dim \mathbb{R}^{n+1}$, so this means that T is also onto and hence is an isomorphism. Thus T^{-1} exists by Theorem 5. Note that we have not given a description of the action of T^{-1}, we have merely shown that such a description exists. To give it explicitly requires some ingenuity; one method involves the Lagrange interpolation expansion (Theorem 3 Section 6.5).

EXERCISES 7.3

1. Verify that each of the following is an isomorphism (Theorem 3 is useful).

   (a) $T: \mathbb{R}^3 \to \mathbb{R}^3$;
   $T(x, y, z) = (x + y, y + z, z + x)$

   ◆(b) $T: \mathbb{R}^3 \to \mathbb{R}^3$;
   $T(x, y, z) = (x, x + y, x + y + z)$

   (c) $T: \mathbb{C} \to \mathbb{C}$; $T(z) = \bar{z}$

   ◆(d) $T: \mathbf{M}_{mn} \to \mathbf{M}_{mn}$; $T(X) = UXV$, U and V invertible

   (e) $T: \mathbf{P}_1 \to \mathbb{R}^2$; $T[p(x)] = [p(0), p(1)]$

   ◆(f) $T: V \to V$; $T(\mathbf{v}) = k\mathbf{v}$, $k \neq 0$ a fixed number, V any vector space

   (g) $T: \mathbf{M}_{22} \to \mathbb{R}^4$; $T\begin{bmatrix} a & b \\ c & d \end{bmatrix} = (a + b, d, c, a - b)$

   ◆(h) $T: \mathbf{M}_{mn} \to \mathbf{M}_{nm}$; $T(A) = A^T$

2. Show that $\{a + bx + cx^2, a_1 + b_1 x + c_1 x^2, a_2 + b_2 x + c_2 x^2\}$ is a basis of $\mathbf{P}_2$ if and only if $\{(a, b, c), (a_1, b_1, c_1), (a_2, b_2, c_2)\}$ is a basis of $\mathbb{R}^3$.

3. If V is any vector space, let V^n denote the space of all n-tuples $(\mathbf{v}_1, \mathbf{v}_2, \ldots, \mathbf{v}_n)$, where each $\mathbf{v}_i$ lies in V. (This is a vector space with component-wise operations; see Exercise 17 Section 6.1.) If $C_j(A)$ denotes the jth column of the $m \times n$ matrix A, show that $T: \mathbf{M}_{mn} \to (\mathbb{R}^m)^n$ is an isomorphism if $T(A) = [C_1(A) \ C_2(A) \ \cdots \ C_n(A)]$. (Here $\mathbb{R}^m$ consists of columns.)

4. In each case, compute the action of ST and TS, and show that $ST \neq TS$.

   (a) $S: \mathbb{R}^2 \to \mathbb{R}^2$ with $S(x, y) = (y, x)$;
   $T: \mathbb{R}^2 \to \mathbb{R}^2$ with $T(x, y) = (x, 0)$

   ◆(b) $S: \mathbb{R}^3 \to \mathbb{R}^3$ with $S(x, y, z) = (x, 0, z)$;
   $T: \mathbb{R}^3 \to \mathbb{R}^3$ with $T(x, y, z) = (x + y, 0, y + z)$

   (c) $S: \mathbf{P}_2 \to \mathbf{P}_2$ with $S(p) = p(0) + p(1)x + p(2)x^2$;
   $T: \mathbf{P}_2 \to \mathbf{P}_2$ with
   $T(a + bx + cx^2) = b + cx + ax^2$

   ◆(d) $S: \mathbf{M}_{22} \to \mathbf{M}_{22}$ with $S\begin{bmatrix} a & b \\ c & d \end{bmatrix} = \begin{bmatrix} a & 0 \\ 0 & d \end{bmatrix}$;
   $T: \mathbf{M}_{22} \to \mathbf{M}_{22}$ with $T\begin{bmatrix} a & b \\ c & d \end{bmatrix} = \begin{bmatrix} c & a \\ d & b \end{bmatrix}$

5. In each case, show that the linear transformation T satisfies $T^2 = T$.

   (a) $T: \mathbb{R}^4 \to \mathbb{R}^4$; $T(x, y, z, w) = (x, 0, z, 0)$

   ◆(b) $T: \mathbb{R}^2 \to \mathbb{R}^2$; $T(x, y) = (x + y, 0)$

   (c) $T: \mathbf{P}_2 \to \mathbf{P}_2$;
   $T(a + bx + cx^2) = (a + b - c) + cx + cx^2$

   ◆(d) $T: \mathbf{M}_{22} \to \mathbf{M}_{22}$; $T\begin{bmatrix} a & b \\ c & d \end{bmatrix} = \frac{1}{2}\begin{bmatrix} a + c & b + d \\ a + c & b + d \end{bmatrix}$

6. Determine whether each of the following transformations T has an inverse and, if so, determine the action of T^{-1}.

   (a) $T: \mathbb{R}^3 \to \mathbb{R}^3$;
   $T(x, y, z) = (x + y, y + z, z + x)$

◆(b) $T: \mathbb{R}^4 \to \mathbb{R}^4$;
$T(x, y, z, t) = (x + y, y + z, z + t, t + x)$

(c) $T: \mathbf{M}_{22} \to \mathbf{M}_{22}$; $T\begin{bmatrix} a & b \\ c & d \end{bmatrix} = \begin{bmatrix} a - c & b - d \\ 2a - c & 2b - d \end{bmatrix}$

◆(d) $T: \mathbf{M}_{22} \to \mathbf{M}_{22}$; $T\begin{bmatrix} a & b \\ c & d \end{bmatrix} = \begin{bmatrix} a + 2c & b + 2d \\ 3c - a & 3d - b \end{bmatrix}$

(e) $T: \mathbf{P}_2 \to \mathbb{R}^3$;
$T(a + bx + cx^2) = (a - c, 2b, a + c)$

◆(f) $T: \mathbf{P}_2 \to \mathbb{R}^3$; $T(p) = [p(0), p(1), p(-1)]$

7. In each case, show that T is self-inverse: $T^{-1} = T$.

   (a) $T: \mathbb{R}^4 \to \mathbb{R}^4$; $T(x, y, z, w) = (x, -y, -z, w)$

   ◆(b) $T: \mathbb{R}^2 \to \mathbb{R}^2$; $T(x, y) = (ky - x, y)$, k any fixed number

   (c) $T: \mathbf{P}_n \to \mathbf{P}_n$; $T(p(x)) = p(3 - x)$

   ◆(d) $T: \mathbf{M}_{22} \to \mathbf{M}_{22}$; $T(X) = AX$ where
   $A = \frac{1}{4}\begin{bmatrix} 5 & -3 \\ 3 & -5 \end{bmatrix}$

8. In each case, show that $T^6 = 1_{\mathbb{R}^4}$ and so determine T^{-1}.

   (a) $T: \mathbb{R}^4 \to \mathbb{R}^4$; $T(x, y, z, w) = (-x, z, w, y)$

   ◆(b) $T: \mathbb{R}^4 \to \mathbb{R}^4$;
   $T(x, y, z, w) = (-y, x - y, z, -w)$

9. In each case, show that T is an isomorphism by defining T^{-1} explicitly.

   (a) $T: \mathbf{P}_n \to \mathbf{P}_n$ is given by $T[p(x)] = p(x + 1)$.

   ◆(b) $T: \mathbf{M}_{nn} \to \mathbf{M}_{nn}$ is given by $T(A) = UA$ where U is invertible in $\mathbf{M}_{nn}$.

10. Given linear transformations $V \xrightarrow{T} W \xrightarrow{S} U$:

    (a) If S and T are both one-to-one, show that ST is one-to-one.

    ◆(b) If S and T are both onto, show that ST is onto.

11. Let $T: V \to W$ be a linear transformation.

    (a) If T is one-to-one and $TR = TR_1$ for transformations R and $R_1: U \to V$, show that $R = R_1$.

    (b) If T is onto and $ST = S_1 T$ for transformations S and $S_1: W \to U$, show that $S = S_1$.

12. Consider the linear transformations
    $V \xrightarrow{T} W \xrightarrow{R} U$.

    (a) Show that ker $T \subseteq$ ker RT.

    ◆(b) Show that im $RT \subseteq$ im R.

13. Let $V \xrightarrow{T} U \xrightarrow{S} W$ be linear transformations.

    (a) If ST is one-to-one, show that T is one-to-one and that dim $V \leq$ dim U.

    ◆(b) If ST is onto, show that S is onto and that dim $W \leq$ dim U.

◆14. Let $T: V \to V$ be a linear transformation. Show that $T^2 = 1_V$ if and only if T is invertible and $T = T^{-1}$.

15. Let N be a nilpotent $n \times n$ matrix (that is, $N^k = 0$ for some k). Show that $T: \mathbf{M}_{nm} \to \mathbf{M}_{nm}$ is an isomorphism if $T(X) = X - NX$. [Hint: If X is in ker T, show that $X = NX = N^2 X = \cdots$. Then use Theorem 3.]

◆16. Let $T: V \to W$ be a linear transformation, and let $\{e_1, \ldots, e_r, e_{r+1}, \ldots, e_n\}$ be any basis of V such that $\{e_{r+1}, \ldots, e_n\}$ is a basis of ker T. Show that im $T \cong$ span$\{e_1, \ldots, e_r\}$. [Hint: See Theorem 5 Section 7.2.]

17. Is every isomorphism $T: \mathbf{M}_{22} \to \mathbf{M}_{22}$ given by an invertible matrix U such that $T(X) = UX$ for all X in $\mathbf{M}_{22}$? Prove your answer.

18. Let $\mathbf{D}_n$ denote the space of all functions f from $\{1, 2, \ldots, n\}$ to $\mathbb{R}$ (see Exercise 35 Section 6.3). If $T: \mathbf{D}_n \to \mathbb{R}^n$ is defined by
    $$T(f) = (f(1), f(2), \ldots, f(n)),$$
    show that T is an isomorphism.

19. (a) Let V be the vector space of Exercise 3 Section 6.1. Find an isomorphism $T: V \to \mathbb{R}^1$.

    ◆(b) Let V be the vector space of Exercise 4 Section 6.1. Find an isomorphism $T: V \to \mathbb{R}^2$.

20. Let $V \xrightarrow{T} W \xrightarrow{S} V$ be linear transformations such that $ST = 1_V$. If dim $V =$ dim $W = n$, show that $S = T^{-1}$ and $T = S^{-1}$. [Hint: Exercise 13 and Theorems 3, 4, and 5.]

21. Let $V \xrightarrow{T} W \xrightarrow{S} V$ be functions such that $TS = 1_W$ and $ST = 1_V$. If T is linear, show that S is also linear.

22. Let A and B be matrices of size $p \times m$ and $n \times q$. Assume that $mn = pq$. Define $R : \mathbf{M}_{mn} \to \mathbf{M}_{pq}$ by $R(X) = AXB$.

   (a) Show that $\mathbf{M}_{mn} \cong \mathbf{M}_{pq}$ by comparing dimensions.

   (b) Show that R is a linear transformation.

   (c) Show that if R is an isomorphism, then $m = p$ and $n = q$. [Hint: Show that $T : \mathbf{M}_{mn} \to \mathbf{M}_{pn}$ given by $T(X) = AX$ and $S : \mathbf{M}_{mn} \to \mathbf{M}_{mq}$ given by $S(X) = XB$ are both one-to-one, and use the dimension theorem.]

23. Let $T : V \to V$ be a linear transformation such that $T^2 = 0$ is the zero transformation.

   (a) If $V \neq \{0\}$, show that T cannot be invertible.

   (b) If $R : V \to V$ is defined by $R(\mathbf{v}) = \mathbf{v} + T(\mathbf{v})$ for all $\mathbf{v}$ in V, show that R is linear and invertible.

24. Let V consist of all sequences $[x_0, x_1, x_2, \ldots)$ of numbers, and define vector operations

   $$[x_0, x_1, \ldots) + [y_0, y_1, \ldots) = [x_0 + y_0, x_1 + y_1, \ldots)$$
   $$r[x_0, x_1, \ldots) = [rx_0, rx_1, \ldots)$$

   (a) Show that V is a vector space of infinite dimension.

   ◆(b) Define $T : V \to V$ and $S : V \to V$ by $T[x_0, x_1, \ldots) = [x_1, x_2, \ldots)$ and $S[x_0, x_1, \ldots) = [0, x_0, x_1, \ldots)$. Show that $TS = 1_V$, so TS is one-to-one and onto, but that T is not one-to-one and S is not onto.

25. Prove (1) and (2) of Theorem 4.

26. Define $T : \mathbf{P}_n \to \mathbf{P}_n$ by $T(p) = p(x) + xp'(x)$ for all p in $\mathbf{P}_n$.

   (a) Show that T is linear.

   ◆(b) Show that $\ker T = \{0\}$ and conclude that T is an isomorphism. [Hint: Write $p(x) = a_0 + a_1 x + \cdots + a_n x^n$ and compare coefficients if $p(x) = -xp'(x)$.]

   (c) Conclude that each $q(x)$ in $\mathbf{P}_n$ has the form $q(x) = p(x) + xp'(x)$ for some unique polynomial $p(x)$.

   (d) Does this remain valid if T is defined by $T[p(x)] = p(x) - xp'(x)$? Explain.

27. Let $T : V \to W$ be a linear transformation, where V and W are finite dimensional.

   (a) Show that T is one-to-one if and only if there exists a linear transformation $S : W \to V$ with $ST = 1_V$. [Hint: If $\{\mathbf{e}_1, \ldots, \mathbf{e}_n\}$ is a basis of V and T is one-to-one, show that W has a basis $\{T(\mathbf{e}_1), \ldots, T(\mathbf{e}_n), \mathbf{f}_{n+1}, \ldots, \mathbf{f}_{n+k}\}$ and use Theorems 2 and 3, Section 7.1.]

   ◆(b) Show that T is onto if and only if there exists a linear transformation $S : W \to V$ with $TS = 1_W$. [Hint: Let $\{\mathbf{e}_1, \ldots, \mathbf{e}_r, \ldots, \mathbf{e}_n\}$ be a basis of V such that $\{\mathbf{e}_{r+1}, \ldots, \mathbf{e}_n\}$ is a basis of $\ker T$. Use Theorem 5 Section 7.2 and Theorems 2 and 3, Section 7.1.]

28. Let S and T be linear transformations $V \to W$, where $\dim V = n$ and $\dim W = m$.

   (a) Show that $\ker S = \ker T$ if and only if $T = RS$ for some isomorphism $R : W \to W$. [Hint: Let $\{\mathbf{e}_1, \ldots, \mathbf{e}_r, \ldots, \mathbf{e}_n\}$ be a basis of V such that $\{\mathbf{e}_{r+1}, \ldots, \mathbf{e}_n\}$ is a basis of $\ker S = \ker T$. Use Theorem 5 Section 7.2 to extend $\{S(\mathbf{e}_1), \ldots, S(\mathbf{e}_r)\}$ and $\{T(\mathbf{e}_1), \ldots, T(\mathbf{e}_r)\}$ to bases of W.]

   ◆(b) Show that $\operatorname{im} S = \operatorname{im} T$ if and only if $T = SR$ for some isomorphism $R : V \to V$. [Hint: Show that $\dim(\ker S) = \dim(\ker T)$ and choose bases $\{\mathbf{e}_1, \ldots, \mathbf{e}_r, \ldots, \mathbf{e}_n\}$ and $\{\mathbf{f}_1, \ldots, \mathbf{f}_r, \ldots, \mathbf{f}_n\}$ of V where $\{\mathbf{e}_{r+1}, \ldots, \mathbf{e}_n\}$ and $\{\mathbf{f}_{r+1}, \ldots, \mathbf{f}_n\}$ are bases of $\ker S$ and $\ker T$, respectively. If $1 \leq i \leq r$, show that $S(\mathbf{e}_i) = T(\mathbf{g}_i)$ for some $\mathbf{g}_i$ in V, and prove that $\{\mathbf{g}_1, \ldots, \mathbf{g}_r, \mathbf{f}_{r+1}, \ldots, \mathbf{f}_n\}$ is a basis of V.]

◆29. If $T : V \to V$ is a linear transformation where $\dim V = n$, show that $TST = T$ for some isomorphism $S : V \to V$. [Hint: Let $\{\mathbf{e}_1, \ldots, \mathbf{e}_r, \mathbf{e}_{r+1}, \ldots, \mathbf{e}_n\}$ be as in Theorem 5 Section 7.2. Extend $\{T(\mathbf{e}_1), \ldots, T(\mathbf{e}_r)\}$ to a basis of V, and use Theorem 1 and Theorems 2 and 3, Section 7.1.]

30. Let A and B denote $m \times n$ matrices. In each case show that (1) and (2) are equivalent.

   (a) (1) A and B have the same null space.
   (2) $B = PA$ for some invertible $m \times m$ matrix P.

   (b) (1) A and B have the same range.
   (2) $B = AQ$ for some invertible $n \times n$ matrix Q.

   [Hint: Use Exercise 28.]

SECTION 7.4 A Theorem about Differential Equations

Differential equations are instrumental in solving a variety of problems throughout science, social science, and engineering. In this brief section, we will see that the set of solutions of a linear differential equation (with constant coefficients) is a vector space and we will calculate its dimension. The proof is pure linear algebra, although the applications are primarily in analysis. However, a key result (Lemma 3 below) can be applied much more widely.

We denote the derivative of a function $f: \mathbb{R} \to \mathbb{R}$ by f', and f will be called **differentiable** if it can be differentiated any number of times. If f is a differentiable function, the nth derivative $f^{(n)}$ of f is the result of differentiating n times. Thus $f^{(0)} = f, f^{(1)} = f', f^{(2)} = f^{(1)'}, \ldots$, and in general $f^{(n+1)} = f^{(n)'}$ for each $n \geq 0$. For small values of n these are often written as $f, f', f'', f''', \ldots$.

If a, b, and c are numbers, the differential equations

$$f'' - af' - bf = 0 \quad \text{or} \quad f''' - af'' - bf' - cf = 0$$

are said to be of **second order** and **third order**, respectively. In general, an equation

$$f^{(n)} - a_{n-1}f^{(n-1)} - a_{n-2}f^{(n-2)} - \cdots - a_2 f^{(2)} - a_1 f^{(1)} - a_0 f^{(0)} = 0, \, a_i \text{ in } \mathbb{R}, \quad (*)$$

is called a **differential equation of order** n. We want to describe all solutions of this equation. Of course a knowledge of calculus is required.

The set $\mathbf{F}$ of all functions $\mathbb{R} \to \mathbb{R}$ is a vector space with operations as described in Example 7 Section 6.1. If f and g are differentiable, we have $(f + g)' = f' + g'$ and $(af)' = af'$ for all a in $\mathbb{R}$. With this it is a routine matter to verify that the following set is a subspace of $\mathbf{F}$:

$$\mathbf{D}_n = \{f: \mathbb{R} \to \mathbb{R} \mid f \text{ is differentiable and is a solution to } (*)\}$$

Our sole objective in this section is to prove

Theorem 1

The space $\mathbf{D}_n$ has dimension n.

We have already used this theorem in Section 3.5.

As will be clear later, the proof of Theorem 1 requires that we enlarge $\mathbf{D}_n$ somewhat and allow our differentiable functions to take values in the set $\mathbb{C}$ of complex numbers. To do this, we must clarify what it means for a function $f: \mathbb{R} \to \mathbb{C}$ to be differentiable. For each real number x write $f(x)$ in terms of its real and imaginary parts $f_r(x)$ and $f_i(x)$:

$$f(x) = f_r(x) + i f_i(x).$$

This produces new functions $f_r: \mathbb{R} \to \mathbb{R}$ and $f_i: \mathbb{R} \to \mathbb{R}$, called the **real** and **imaginary parts** of f, respectively. We say that f is **differentiable** if both f_r and f_i are differentiable (as real functions), and we define the **derivative** f' of f by

$$f' = f_r' + i f_i'. \quad (**)$$

We refer to this frequently in what follows.[4]

With this, write $\mathbf{D}_\infty$ for the set of all differentiable complex valued functions $f: \mathbb{R} \to \mathbb{C}$. This is a *complex* vector space using pointwise addition (see Example 7 Section 6.1), and the following scalar multiplication: For any w in $\mathbb{C}$ and f in $\mathbf{D}_\infty$, we define $wf: \mathbb{R} \to \mathbb{C}$ by $(wf)(x) = wf(x)$ for all x in $\mathbb{R}$. We will be working in $\mathbf{D}_\infty$ for the rest of this section. In particular, consider the following complex subspace of $\mathbf{D}_\infty$:

$$\mathbf{D}_n^* = \{f: \mathbb{R} \to \mathbb{C} \mid f \text{ is a solution to } (*)\}$$

Clearly, $\mathbf{D}_n \subseteq \mathbf{D}_n^*$, and our interest in $\mathbf{D}_n^*$ comes from

Lemma 1

If $\dim_\mathbb{C}(\mathbf{D}_n^*) = n$, then $\dim_\mathbb{R}(\mathbf{D}_n) = n$.

PROOF

Observe first that if $\dim_\mathbb{C}(\mathbf{D}_n^*) = n$, then $\dim_\mathbb{R}(\mathbf{D}_n^*) = 2n$. [In fact, if $\{g_1, \ldots, g_n\}$ is a $\mathbb{C}$-basis of $\mathbf{D}_n^*$ then $\{g_1, \ldots, g_n, ig_1, \ldots, ig_n\}$ is a $\mathbb{R}$-basis of $\mathbf{D}_n^*$]. Now observe that the set $\mathbf{D}_n \times \mathbf{D}_n$ of all ordered pairs (f, g) with f and g in $\mathbf{D}_n$ is a real vector space with componentwise operations. Define

$$\theta : \mathbf{D}_n^* \to \mathbf{D}_n \times \mathbf{D}_n \quad \text{given by} \quad \theta(f) = (f_r, f_i) \text{ for } f \text{ in } \mathbf{D}_n^*.$$

One verifies that θ is onto and one-to-one, and it is $\mathbb{R}$-linear because $f \to f_r$ and $f \to f_i$ are both $\mathbb{R}$-linear. Hence $\mathbf{D}_n^* \cong \mathbf{D}_n \times \mathbf{D}_n$ as $\mathbb{R}$-spaces. Since $\dim_\mathbb{R}(\mathbf{D}_n^*)$ is finite, it follows that $\dim_\mathbb{R}(\mathbf{D}_n)$ is finite, and we have

$$2 \dim_\mathbb{R}(\mathbf{D}_n) = \dim_\mathbb{R}(\mathbf{D}_n \times \mathbf{D}_n) = \dim_\mathbb{R}(\mathbf{D}_n^*) = 2n.$$

Hence $\dim_\mathbb{R}(\mathbf{D}_n) = n$, as required.

It follows that to prove Theorem 1 it suffices to show that $\dim_\mathbb{C}(\mathbf{D}_n^*) = n$.

There is one function that arises frequently in any discussion of differential equations. Given a complex number $w = a + ib$ (where a and b are real), we have $e^w = e^a(\cos b + i \sin b)$. The law of exponents, $e^w e^v = e^{w+v}$ for all w, v in $\mathbb{C}$ is easily verified using the formulas for $\sin(b + b_1)$ and $\cos(b + b_1)$. If x is a variable and $w = a + ib$ is a complex number, define the **exponential function** e^{wx} by

$$e^{wx} = e^{ax}(\cos bx + i \sin bx).$$

Hence e^{wx} is differentiable because its real and imaginary parts are differentiable for all x. Moreover, the following can be proved using $(**)$:

$$(e^{wx})' = we^{wx}$$

In addition, $(**)$ gives the **product rule** for differentiation:

If f and g are in $\mathbf{D}_\infty$, then $(fg)' = f'g + fg'$.

---

[4] Write $|w|$ for the absolute value of any complex number w. As for functions $\mathbb{R} \to \mathbb{R}$, we say that $\lim_{t \to 0} f(t) = w$ if, for all $\varepsilon > 0$ there exists $\delta > 0$ such that $|f(t) - w| < \varepsilon$ whenever $|t| < \delta$. (Note that t represents a real number here.) In particular, given a real number x, we define the *derivative* f' of a function $f: \mathbb{R} \to \mathbb{C}$ by $f'(x) = \lim_{t \to 0} \{\frac{1}{t}[f(x + t) - f(x)]\}$, and we say that f is *differentiable* if $f'(x)$ exists for all x in $\mathbb{R}$. Then we can *prove* that f is differentiable if and only if both f_r and f_i are differentiable, and that $f' = f'_r + if'_i$ in this case.

SECTION 7.4 A Theorem about Differential Equations

We omit the verifications.

To prove that $\dim_{\mathbb{C}}(\mathbf{D}_n^*) = n$, two preliminary results are required. Here is the first.

Lemma 2

Given f in $\mathbf{D}_\infty$ and w in $\mathbb{C}$, there exists g in $\mathbf{D}_\infty$ such that $g' - wg = f$.

PROOF

Define $p(x) = f(x)e^{-wx}$. Then p is differentiable, whence p_r and p_i are both differentiable, hence continuous, and so both have antiderivatives, say $p_r = q_r'$ and $p_i = q_i'$. Then the function $q = q_r + iq_i$ is in $\mathbf{D}_\infty$, and $q' = p$ by (**). Finally define $g(x) = q(x)e^{wx}$. Then $g' = q'e^{wx} + qwe^{wx} = pe^{wx} + w(qe^{wx}) = f + wg$ by the product rule, as required.

The second preliminary result is important in its own right.

Lemma 3

Kernel Lemma

Let V be a vector space, and let S and T be linear operators $V \to V$. If S is onto and both $\ker(S)$ and $\ker(T)$ are finite dimensional, then $\ker(TS)$ is also finite dimensional and $\dim[\ker(TS)] = \dim[\ker(T)] + \dim[\ker(S)]$.

PROOF

Let $\{\mathbf{u}_1, \mathbf{u}_2, \ldots, \mathbf{u}_m\}$ be a basis of $\ker(T)$ and let $\{\mathbf{v}_1, \mathbf{v}_2, \ldots, \mathbf{v}_n\}$ be a basis of $\ker(S)$. Since S is onto, let $\mathbf{u}_i = S(\mathbf{w}_i)$ for some $\mathbf{w}_i$ in V. It suffices to show that

$$B = \{\mathbf{w}_1, \mathbf{w}_2, \ldots, \mathbf{w}_m, \mathbf{v}_1, \mathbf{v}_2, \ldots, \mathbf{v}_n\}$$

is a basis of $\ker(TS)$. Note that $B \subseteq \ker(TS)$ because $TS(\mathbf{w}_i) = T(\mathbf{u}_i) = \mathbf{0}$ for each i and $TS(\mathbf{v}_j) = T(\mathbf{0}) = \mathbf{0}$ for each j.

Spanning. If $\mathbf{v}$ is in $\ker(TS)$, then $S(\mathbf{v})$ is in $\ker(T)$, say $S(\mathbf{v}) = \sum r_i \mathbf{u}_i = \sum r_i S(\mathbf{w}_i) = S(\sum r_i \mathbf{w}_i)$. It follows that $\mathbf{v} - \sum r_i \mathbf{w}_i$ is in $\ker(S) = \text{span}\{\mathbf{v}_1, \mathbf{v}_2, \ldots, \mathbf{v}_n\}$, proving that $\mathbf{v}$ is in span(B).

Independence. Let $\sum r_i \mathbf{w}_i + \sum t_j \mathbf{v}_j = \mathbf{0}$. Applying S, and noting that $S(\mathbf{v}_j) = \mathbf{0}$ for each j, yields $\mathbf{0} = \sum r_i S(\mathbf{w}_i) = \sum r_i \mathbf{u}_i$. Hence $r_i = 0$ for each i, and so $\sum t_j \mathbf{v}_j = \mathbf{0}$. This implies that each $t_j = 0$, and so proves the independence of B.

PROOF OF THEOREM 1

By Lemma 1, it suffices to prove that $\dim_{\mathbb{C}}(\mathbf{D}_n^*) = n$. This holds for $n = 1$ because the proof of Theorem 1 Section 3.5 goes through to show that $\mathbf{D}_1^* = \mathbb{C}e^{a_0 x}$. Hence we proceed by induction on n. With an eye on (*), consider the polynomial

$$p(t) = t^n - a_{n-1}t^{n-1} - a_{n-2}t^{n-2} - \cdots - a_2 t^2 - a_1 t - 0$$

(called the *characteristic polynomial* of equation (*)). Now define a map $D : \mathbf{D}_\infty \to \mathbf{D}_\infty$ by $D(f) = f'$ for all f in $\mathbf{D}_\infty$. Then D is a linear operator, whence $p(D): \mathbf{D}_\infty \to \mathbf{D}_\infty$ is also a linear operator. Moreover, since $D^k(f) = f^{(k)}$ for each $k \geq 0$, equation (*) takes the form $p(D)(f) = 0$. In other words,

$$\mathbf{D}_n^* = \ker[p(D)].$$

By the fundamental theorem of algebra,[5] let w be a complex root of $p(t)$, so that $p(t) = q(t)(t - w)$ for some complex polynomial $q(t)$ of degree $n - 1$. It follows that $p(D) = q(D)(D - w1_{\mathbf{D}_\infty})$. Moreover $D - w1_{\mathbf{D}_\infty}$ is onto by Lemma 2, $\dim_{\mathbb{C}}[\ker(D - w1_{\mathbf{D}_\infty})] = 1$ by the case $n = 1$ above, and $\dim_{\mathbb{C}}(\ker[q(D)]) = n - 1$ by induction. Hence Lemma 3 shows that $\ker[P(D)]$ is also finite dimensional and

$$\dim_{\mathbb{C}}(\ker[p(D)]) = \dim_{\mathbb{C}}(\ker[q(D)]) + \dim_{\mathbb{C}}(\ker[D - w1_{\mathbf{D}_\infty}]) = (n - 1) + 1 = n.$$

Since $\mathbf{D}_n^* = \ker[p(D)]$, this completes the induction, and so proves Theorem 1.

SECTION 7.5 More on Linear Recurrences[6]

In Section 3.4 we used diagonalization to study linear recurrences, and gave several examples. We now apply the theory of vector spaces and linear transformations to study the problem in more generality.

Consider the linear recurrence

$$x_{n+2} = 6x_n - x_{n+1} \quad \text{for } n \geq 0$$

If the initial values x_0 and x_1 are prescribed, this gives a sequence of numbers. For example, if $x_0 = 1$ and $x_1 = 1$ the sequence continues

$$x_2 = 5, \, x_3 = 1, \, x_4 = 29, \, x_5 = -23, \, x_6 = 197, \, \ldots$$

as the reader can verify. Clearly, the entire sequence is uniquely determined by the recurrence and the two initial values. In this section we define a vector space structure on the set of *all* sequences, and study the subspace of those sequences that satisfy a particular recurrence.

Sequences will be considered entities in their own right, so it is useful to have a special notation for them. Let

$$[x_n] \quad \text{denote the sequence } x_0, x_1, x_2, \ldots, x_n, \ldots$$

---

5   This is the reason for allowing our solutions to (*) to be *complex* valued.
6   This section requires only Sections 7.1–7.3.

SECTION 7.5 More on Linear Recurrences

EXAMPLE 1

$[n)$	is the sequence 0, 1, 2, 3, ...
$[n + 1)$	is the sequence 1, 2, 3, 4, ...
$[2^n)$	is the sequence 1, 2, 2^2, 2^3, ...
$[(-1)^n)$	is the sequence 1, −1, 1, −1, ...
$[5)$	is the sequence 5, 5, 5, 5, ...

Sequences of the form $[c)$ for a fixed number c will be referred to as **constant sequences**, and those of the form $[\lambda^n)$, λ some number, are **power sequences**.

Two sequences are regarded as **equal** when they are identical:

$$[x_n) = [y_n) \quad \text{means} \quad x_n = y_n \quad \text{for all } n = 0, 1, 2, \ldots$$

Addition and scalar multiplication of sequences are defined by

$$[x_n) + [y_n) = [x_n + y_n)$$
$$r[x_n) = [rx_n)$$

These operations are analogous to the addition and scalar multiplication in $\mathbb{R}^n$, and it is easy to check that the vector-space axioms are satisfied. The zero vector is the constant sequence $[0)$, and the negative of a sequence $[x_n)$ is given by $-[x_n) = [-x_n)$.

Now suppose k real numbers $r_0, r_1, \ldots, r_{k-1}$ are given, and consider the **linear recurrence relation** determined by these numbers.

$$x_{n+k} = r_0 x_n + r_1 x_{n+1} + \cdots + r_{k-1} x_{n+k-1} \tag{*}$$

When $r_0 \neq 0$, we say this recurrence has **length** k.[7] For example, the relation $x_{n+2} = 2x_n + x_{n+1}$ is of length 2.

A sequence $[x_n)$ is said to **satisfy** the relation (*) if (*) holds for all $n \geq 0$. Let V denote the set of all sequences that satisfy the relation. In symbols,

$$V = \{[x_n) \mid x_{n+k} = r_0 x_n + r_1 x_{n+1} + \cdots + r_{k-1} x_{n+k-1} \text{ hold for all } n \geq 0\}$$

It is easy to see that the constant sequence $[0)$ lies in V and that V is closed under addition and scalar multiplication of sequences. Hence V is vector space (being a subspace of the space of all sequences). The following important observation about V is needed (it was used implicitly earlier): If the first k terms of two sequences agree, then the sequences are identical. More formally,

Lemma 1

Let $[x_n)$ and $[y_n)$ denote two sequences in V. Then

$$[x_n) = [y_n) \quad \text{if and only if} \quad x_0 = y_0, x_1 = y_1, \ldots, x_{k-1} = y_{k-1}$$

PROOF

If $[x_n) = [y_n)$ then $x_n = y_n$ for *all* $n = 0, 1, 2, \ldots$. Conversely, if $x_i = y_i$ for all $i = 0, 1, \ldots, k - 1$, use the recurrence (*) for $n = 0$.

$$x_k = r_0 x_0 + r_1 x_1 + \cdots + r_{k-1} x_{k-1} = r_0 y_0 + r_1 y_1 + \cdots + r_{k-1} y_{k-1} = y_k$$

Next the recurrence for $n = 1$ establishes $x_{k+1} = y_{k+1}$. The process continues to show that $x_{n+k} = y_{n+k}$ holds for *all* $n \geq 0$ by induction on n. Hence $[x_n) = [y_n)$.

---

[7] We shall usually assume that $r_0 \neq 0$; otherwise, we are essentially dealing with a recurrence of shorter length than k.

This shows that a sequence in V is completely determined by its first k terms. In particular, given a k-tuple $\mathbf{v} = (v_0, v_1, \ldots, v_{k-1})$ in $\mathbb{R}^k$, define

$T(\mathbf{v})$ to be the sequence in V whose first k terms are $v_0, v_1, \ldots, v_{k-1}$.

The rest of the sequence $T(\mathbf{v})$ is determined by the recurrence, so $T: \mathbb{R}^k \to V$ is a function. In fact, it is an isomorphism.

Theorem 1

Given real numbers $r_0, r_1, \ldots, r_{k-1}$, let

$$V = \{[x_n] \mid x_{n+k} = r_0 x_n + r_1 x_{n+1} + \cdots + r_{k-1} x_{n+k-1}, \text{ for all } n \geq 0\}$$

denote the vector space of all sequences satisfying the linear recurrence relation (∗) determined by $r_0, r_1, \ldots, r_{k-1}$. Then the function

$$T: \mathbb{R}^k \to V$$

defined above is an isomorphism. In particular:

1. $\dim V = k$.
2. If $\{\mathbf{v}_1, \ldots, \mathbf{v}_k\}$ is any basis of $\mathbb{R}^k$, then $\{T(\mathbf{v}_1), \ldots, T(\mathbf{v}_k)\}$ is a basis of V.

PROOF

(1) and (2) will follow from Theorems 1 and 2, Section 7.3 as soon as we show that T is an isomorphism. Given $\mathbf{v}$ and $\mathbf{w}$ in $\mathbb{R}^k$, write $\mathbf{v} = (v_0, v_1, \ldots, v_{k-1})$ and $\mathbf{w} = (w_0, w_1, \ldots, w_{k-1})$. The first k terms of $T(\mathbf{v})$ and $T(\mathbf{w})$ are $v_0, v_1, \ldots, v_{k-1}$ and $w_0, w_1, \ldots, w_{k-1}$, respectively, so the first k terms of $T(\mathbf{v}) + T(\mathbf{w})$ are $v_0 + w_0$, $v_1 + w_1, \ldots, v_{k-1} + w_{k-1}$. Because these terms agree with the first k terms of $T(\mathbf{v} + \mathbf{w})$, Lemma 1 implies that $T(\mathbf{v} + \mathbf{w}) = T(\mathbf{v}) + T(\mathbf{w})$. The proof that $T(r\mathbf{v}) + rT(\mathbf{v})$ is similar, so T is linear.

Now let $[x_n]$ be any sequence in V, and let $\mathbf{v} = (x_0, x_1, \ldots, x_{k-1})$. Then the first k terms of $[x_n]$ and $T(\mathbf{v})$ agree, so $T(\mathbf{v}) = [x_n]$. Hence T is onto. Finally, if $T(\mathbf{v}) = [0]$ is the zero sequence, then the first k terms of $T(\mathbf{v})$ are all zero (*all* terms of $T(\mathbf{v})$ are zero!) so $\mathbf{v} = \mathbf{0}$. This means that $\ker T = \{\mathbf{0}\}$, so T is one-to-one.

EXAMPLE 2

Show that the sequences $[1]$, $[n]$, and $[(-1)^n]$ are a basis of the space V of all solutions of the recurrence

$$x_{n+3} = -x_n + x_{n+1} + x_{n+2}.$$

Then find the solution satisfying $x_0 = 1, x_1 = 2, x_2 = 5$.

Solution ▶ The verifications that these sequences satisfy the recurrence (and hence lie in V) are left to the reader. They are a basis because $[1] = T(1, 1, 1)$, $[n] = T(0, 1, 2)$, and $[(-1)^n] = T(1, -1, 1)$; and $\{(1, 1, 1), (0, 1, 2), (1, -1, 1)\}$ is a basis of $\mathbb{R}^3$. Hence the sequence $[x_n]$ in V satisfying $x_0 = 1, x_1 = 2, x_2 = 5$ is a linear combination of this basis:

SECTION 7.5 More on Linear Recurrences

$$[x_n) = t_1[1) + t_2[n) + t_3[(-1)^n)$$

The nth term is $x_n = t_1 + nt_2 + (-1)^n t_3$, so taking $n = 0, 1, 2$ gives

$$1 = x_0 = t_1 + 0 + t_3$$
$$2 = x_1 = t_1 + t_2 - t_3$$
$$5 = x_2 = t_1 + 2t_2 + t_3$$

This has the solution $t_1 = t_3 = \frac{1}{2}, t_2 = 2$, so $x_n = \frac{1}{2} + 2n + \frac{1}{2}(-1)^n$.

This technique clearly works for any linear recurrence of length k: Simply take your favourite basis $\{\mathbf{v}_1, \ldots, \mathbf{v}_k\}$ of $\mathbb{R}^k$—perhaps the standard basis—and compute $T(\mathbf{v}_1), \ldots, T(\mathbf{v}_k)$. This is a basis of V all right, but the nth term of $T(\mathbf{v}_i)$ is not usually given as an explicit function of n. (The basis in Example 2 was carefully chosen so that the nth terms of the three sequences were $1, n$, and $(-1)^n$, respectively, each a simple function of n.)

However, it turns out that an explicit basis of V can be given in the general situation. Given the recurrence (∗) again:

$$x_{n+k} = r_0 x_n + r_1 x_{n+1} + \cdots + r_{k-1} x_{n+k-1}$$

the idea is to look for numbers λ such that the power sequence $[\lambda^n)$ satisfies (∗). This happens if and only if

$$\lambda^{n+k} = r_0 \lambda^n + r_1 \lambda^{n+1} + \cdots + r_{k-1} \lambda^{n+k-1}$$

holds for all $n \geq 0$. This is true just when the case $n = 0$ holds; that is,

$$\lambda^k = r_0 + r_1 \lambda + \cdots + r_{k-1} \lambda^{k-1}$$

The polynomial

$$p(x) = x^k - r_{k-1} x^{k-1} - \cdots - r_1 x - r_0$$

is called the polynomial **associated** with the linear recurrence (∗). Thus every root λ of $p(x)$ provides a sequence $[\lambda^n)$ satisfying (∗). If there are k distinct roots, the power sequences provide a basis. Incidentally, if $\lambda = 0$, the sequence $[\lambda^n)$ is $1, 0, 0, \ldots$; that is, we accept the convention that $0^0 = 1$.

Theorem 2

Let $r_0, r_1, \ldots, r_{k-1}$ be real numbers; let

$$V = \{[x_n) \mid x_{n+k} = r_0 x_n + r_1 x_{n+1} + \cdots + r_{k-1} x_{n+k-1} \text{ for all } n \geq 0\}$$

denote the vector space of all sequences satisfying the linear recurrence relation determined by $r_0, r_1, \ldots, r_{k-1}$; and let

$$p(x) = x^k - r_{k-1} x^{k-1} - \cdots - r_1 x - r_0$$

denote the polynomial associated with the recurrence relation. Then

1. $[\lambda^n)$ lies in V if and only if λ is a root of $p(x)$.
2. If $\lambda_1, \lambda_2, \ldots, \lambda_k$ are distinct real roots of $p(x)$, then $\{[\lambda_1^n), [\lambda_2^n), \ldots, [\lambda_k^n)\}$ is a basis of V.

PROOF

It remains to prove (2). But $[\lambda_i^n] = T(\mathbf{v}_i)$ where $\mathbf{v}_i = (1, \lambda_i, \lambda_i^2, ..., \lambda_i^{k-1})$, so (2) follows by Theorem 1, provided that $(\mathbf{v}_1, \mathbf{v}_2, ..., \mathbf{v}_n)$ is a basis of $\mathbb{R}^k$. This is true provided that the matrix with the $\mathbf{v}_i$ as its rows

$$\begin{bmatrix} 1 & \lambda_1 & \lambda_1^2 & \cdots & \lambda_1^{k-1} \\ 1 & \lambda_2 & \lambda_2^2 & \cdots & \lambda_2^{k-1} \\ \vdots & \vdots & \vdots & \ddots & \vdots \\ 1 & \lambda_k & \lambda_k^2 & \cdots & \lambda_k^{k-1} \end{bmatrix}$$

is invertible. But this is a Vandermonde matrix and so is invertible if the λ_i are distinct (Theorem 7 Section 3.2). This proves (2).

EXAMPLE 3

Find the solution of $x_{n+2} = 2x_n + x_{n+1}$ that satisfies $x_0 = a$, $x_1 = b$.

Solution ▶ The associated polynomial is $p(x) = x^2 - x - 2 = (x - 2)(x + 1)$. The roots are $\lambda_1 = 2$ and $\lambda_2 = -1$, so the sequences $[2^n)$ and $[(-1)^n)$ are a basis for the space of solutions by Theorem 2. Hence every solution $[x_n]$ is a linear combination

$$[x_n) = t_1[2^n) + t_2[(-1)^n)$$

This means that $x_n = t_1 2^n + t_2(-1)^n$ holds for $n = 0, 1, 2, ...$, so (taking $n = 0, 1$) $x_0 = a$ and $x_1 = b$ give

$$t_1 + t_2 = a$$
$$2t_1 - t_2 = b$$

These are easily solved: $t_1 = \frac{1}{3}(a + b)$ and $t_2 = \frac{1}{3}(2a - b)$, so

$$t_n = \tfrac{1}{3}[(a + b)2^n + (2a - b)(-1)^n]$$

The Shift Operator

If $p(x)$ is the polynomial associated with a linear recurrence relation of length k, and if $p(x)$ has k distinct roots $\lambda_1, \lambda_2, ..., \lambda_k$, then $p(x)$ factors completely:

$$p(x) = (x - \lambda_1)(x - \lambda_2)\cdots(x - \lambda_k)$$

Each root λ_i provides a sequence $[\lambda_i^n)$ satisfying the recurrence, and they are a basis of V by Theorem 2. In this case, each λ_i has multiplicity 1 as a root of $p(x)$. In general, a root λ has **multiplicity** m if $p(x) = (x - \lambda)^m q(x)$, where $q(\lambda) \neq 0$. In this case, there are fewer than k distinct roots and so fewer than k sequences $[\lambda^n)$ satisfying the recurrence. However, we can still obtain a basis because, if λ has multiplicity m (and $\lambda \neq 0$), it provides m linearly independent sequences that satisfy the recurrence. To prove this, it is convenient to give another way to describe the space V of all sequences satisfying a given linear recurrence relation.

Let **S** denote the vector space of *all* sequences and define a function

$$S : \mathbf{S} \to \mathbf{S} \quad \text{by} \quad S[x_n) = [x_{n+1}) = [x_1, x_2, x_3, ...)$$

S is clearly a linear transformation and is called the **shift operator** on **S**. Note that powers of S shift the sequence further: $S^2[x_n) = S[x_{n+1}) = [x_{n+2})$. In general,

SECTION 7.5 More on Linear Recurrences

$$S^k[x_n] = [x_{n+k}] = [x_k, x_{k+1}, \ldots) \quad \text{for all } k = 0, 1, 2, \ldots$$

But then a linear recurrence relation

$$x_{n+k} = r_0 x_n + r_1 x_{n+1} + \cdots + r_{k-1} x_{n+k-1} \quad \text{for all } n = 0, 1, \ldots$$

can be written

$$S^k[x_n] = r_0[x_n] + r_1 S[x_n] + \cdots + r_{k-1} S^{k-1}[x_n] \qquad (**)$$

Now let $p(x) = x^k - r_{k-1} x^{k-1} - \cdots - r_1 x - r_0$ denote the polynomial associated with the recurrence relation. The set $\mathbf{L}[\mathbf{S}, \mathbf{S}]$ of all linear transformations from $\mathbf{S}$ to itself is a vector space (verify[8]) that is closed under composition. In particular,

$$p(S) = S^k - r_{k-1} S^{k-1} - \cdots - r_1 S - r_0$$

is a linear transformation called the **evaluation** of p at S. The point is that condition $(**)$ can be written as

$$p(S)[x_n] = 0$$

In other words, the space V of all sequences satisfying the recurrence relation is just $\ker[p(S)]$. This is the first assertion in the following theorem.

Theorem 3

Let $r_0, r_1, \ldots, r_{k-1}$ be real numbers, and let

$$V = \{[x_n] \mid x_{n+k} = r_0 x_n + r_1 x_{n+1} + \cdots + r_{k-1} x_{n+k-1} \text{ for all } n \geq 0\}$$

denote the space of all sequences satisfying the linear recurrence relation determined by $r_0, r_1, \ldots, r_{k-1}$. Let

$$p(x) = x^k - r_{k-1} x^{k-1} - \cdots - r_1 x - r_0$$

denote the corresponding polynomial. Then:

1. $V = \ker[p(S)]$, where S is the shift operator.
2. If $p(x) = (x - \lambda)^m q(x)$, where $\lambda \neq 0$ and $m > 1$, then the sequences

$$\{[\lambda^n], [n\lambda^n], [n^2 \lambda^n], \ldots, [n^{m-1} \lambda^n]\}$$

all lie in V and are linearly independent.

PROOF

(Sketch) It remains to prove (2). If $\binom{n}{k} = \dfrac{n(n-1)\cdots(n-k+1)}{k!}$ denotes the binomial coefficient, the idea is to use (1) to show that the sequence $s_k = \left[\binom{n}{k}\lambda^n\right]$ is a solution for each $k = 0, 1, \ldots, m-1$. Then (2) of Theorem 1 can be applied to show that $\{s_0, s_1, \ldots, s_{m-1}\}$ is linearly independent. Finally, the sequences $t_k = [n^k \lambda^n]$, $k = 0, 1, \ldots, m-1$, in the present theorem can be given by $t_k = \sum_{j=0}^{m-1} a_{kj} s_j$, where $A = [a_{ij}]$ is an invertible matrix. Then (2) follows. We omit the details.

---

[8] See Exercises 19 and 20, Section 9.1.

This theorem combines with Theorem 2 to give a basis for V when $p(x)$ has k real roots (not necessarily distinct) none of which is zero. This last requirement means $r_0 \neq 0$, a condition that is unimportant in practice (see Remark 1 below).

Theorem 4

Let $r_0, r_1, \ldots, r_{k-1}$ be real numbers with $r_0 \neq 0$; let
$$V = \{[x_n] \mid x_{n+k} = r_0 x_n + r_1 x_{n+1} + \cdots + r_{k-1} x_{n+k-1} \text{ for all } n \geq 0\}$$
denote the space of all sequences satisfying the linear recurrence relation of length k determined by $r_0, \ldots, r_{k-1}$; and assume that the polynomial
$$p(x) = x^k - r_{k-1} x^{k-1} - \cdots - r_1 x - r_0$$
factors completely as
$$p(x) = (x - \lambda_1)^{m_1} (x - \lambda_2)^{m_2} \cdots (x - \lambda_p)^{m_p}$$
where $\lambda_1, \lambda_2, \ldots, \lambda_p$ are distinct real numbers and each $m_i \geq 1$. Then $\lambda_i \neq 0$ for each i, and
$$[\lambda_1^n], [n\lambda_1^n], \ldots, [n^{m_1-1} \lambda_1^n]$$
$$[\lambda_2^n], [n\lambda_2^n], \ldots, [n^{m_2-1} \lambda_2^n]$$
$$\vdots$$
$$[\lambda_p^n], [n\lambda_p^n], \ldots, [n^{m_p-1} \lambda_p^n]$$
is a basis of V.

PROOF

There are $m_1 + m_2 + \cdots + m_p = k$ sequences in all so, because dim $V = k$, it suffices to show that they are linearly independent. The assumption that $r_0 \neq 0$, implies that 0 is not a root of $p(x)$. Hence each $\lambda_i \neq 0$, so $\{[\lambda_i^n], [n\lambda_i^n], \ldots [n^{m_i-1}\lambda_i^n]\}$ is linearly independent by Theorem 3.
The proof that the whole set of sequences is linearly independent is omitted.

EXAMPLE 4

Find a basis for the space V of all sequences $[x_n]$ satisfying
$$x_{n+3} = -9x_n - 3x_{n+1} + 5x_{n+2}.$$

Solution ▶ The associated polynomial is
$$p(x) = x^3 - 5x^2 + 3x + 9 = (x - 3)^2 (x + 1).$$
Hence 3 is a double root, so $[3^n]$ and $[n3^n]$ both lie in V by Theorem 3 (the reader should verify this). Similarly, $\lambda = -1$ is a root of multiplicity 1, so $[(-1)^n]$ lies in V. Hence $\{[3^n], [n3^n], [(-1)^n]\}$ is a basis by Theorem 4.

Remark 1 If $r_0 = 0$ [so $p(x)$ has 0 as a root], the recurrence reduces to one of shorter length. For example, consider
$$x_{n+4} = 0x_n + 0x_{n+1} + 3x_{n+2} + 2x_{n+3} \tag{***}$$

SECTION 7.5 More on Linear Recurrences

If we set $y_n = x_{n+2}$, this recurrence becomes $y_{n+2} = 3y_n + 2y_{n+1}$, which has solutions $[3^n]$ and $[(-1)^n]$. These give the following solution to (∗):

$$[0, 0, 1, 3, 3^2, \ldots]$$
$$[0, 0, 1, -1, (-1)^2, \ldots]$$

In addition, it is easy to verify that

$$[1, 0, 0, 0, 0, \ldots]$$
$$[0, 1, 0, 0, 0, \ldots]$$

are also solutions to (∗∗∗). The space of all solutions of (∗) has dimension 4 (Theorem 1), so these sequences are a basis. This technique works whenever $r_0 = 0$.

Remark 2 Theorem 4 completely describes the space V of sequences that satisfy a linear recurrence relation for which the associated polynomial $p(x)$ has all real roots. However, in many cases of interest, $p(x)$ has complex roots that are not real. If $p(\mu) = 0$, μ complex, then $p(\overline{\mu}) = 0$ too ($\overline{\mu}$ the conjugate), and the main observation is that $[\mu^n + \overline{\mu}^n]$ and $[i(\mu^n + \overline{\mu}^n)]$ are *real* solutions. Analogs of the preceding theorems can then be proved.

EXERCISES 7.5

1. Find a basis for the space V of sequences $[x_n]$ satisfying the following recurrences, and use it to find the sequence satisfying $x_0 = 1$, $x_1 = 2$, $x_2 = 1$.

    (a) $x_{n+3} = -2x_n + x_{n+1} + 2x_{n+2}$
    ◆(b) $x_{n+3} = -6x_n + 7x_{n+1}$
    (c) $x_{n+3} = -36x_n + 7x_{n+2}$

2. In each case, find a basis for the space V of all sequences $[x_n]$ satisfying the recurrence, and use it to find x_n if $x_0 = 1$, $x_1 = -1$, and $x_2 = 1$.

    (a) $x_{n+3} = x_n + x_{n+1} - x_{n+2}$
    ◆(b) $x_{n+3} = -2x_n + 3x_{n+1}$
    (c) $x_{n+3} = -4x_n + 3x_{n+2}$
    ◆(d) $x_{n+3} = x_n - 3x_{n+1} + 3x_{n+2}$
    (e) $x_{n+3} = 8x_n - 12x_{n+1} + 6x_{n+2}$

3. Find a basis for the space V of sequences $[x_n]$ satisfying each of the following recurrences.

    (a) $x_{n+2} = -a^2 x_n + 2ax_{n+1}$, $a \neq 0$
    ◆(b) $x_{n+2} = -abx_n + (a+b)x_{n+1}$, $(a \neq b)$

4. In each case, find a basis of V.

    (a) $V = \{[x_n] \mid x_{n+4} = 2x_{n+2} - x_{n+3},\ \text{for}\ n \geq 0\}$
    ◆(b) $V = \{[x_n] \mid x_{n+4} = -x_{n+2} + 2x_{n+3},\ \text{for}\ n \geq 0\}$

5. Suppose that $[x_n]$ satisfies a linear recurrence relation of length k. If $\{e_0 = (1, 0, \ldots, 0), e_1 = (0, 1, \ldots, 0), e_{k-1} = (0, 0, \ldots, 1)\}$ is the standard basis of $\mathbb{R}^k$, show that
$$x_n = x_0 T(e_0) + x_1 T(e_1) + \cdots + x_{k-1} T(e_{k-1})$$
holds for all $n \geq k$. (Here T is as in Theorem 1.)

6. Show that the shift operator S is onto but not one-to-one. Find ker S.

◆7. Find a basis for the space V of all sequences $[x_n]$ satisfying $x_{n+2} = -x_n$.

8 Orthogonality

In Section 5.3 we introduced the dot product in $\mathbb{R}^n$ and extended the basic geometric notions of length and distance. A set $\{\mathbf{f}_1, \mathbf{f}_2, \ldots, \mathbf{f}_m\}$ of nonzero vectors in $\mathbb{R}^n$ was called an **orthogonal set** if $\mathbf{f}_i \cdot \mathbf{f}_j = 0$ for all $i \neq j$, and it was proved that every orthogonal set is independent. In particular, it was observed that the expansion of a vector as a linear combination of orthogonal basis vectors is easy to obtain because formulas exist for the coefficients. Hence the orthogonal bases are the "nice" bases, and much of this chapter is devoted to extending results about bases to orthogonal bases. This leads to some very powerful methods and theorems. Our first task is to show that every subspace of $\mathbb{R}^n$ *has* an orthogonal basis.

SECTION 8.1 Orthogonal Complements and Projections

If $\{\mathbf{v}_1, \ldots, \mathbf{v}_m\}$ is linearly independent in a general vector space, and if $\mathbf{v}_{m+1}$ is not in span$\{\mathbf{v}_1, \ldots, \mathbf{v}_m\}$, then $\{\mathbf{v}_1, \ldots, \mathbf{v}_m, \mathbf{v}_{m+1}\}$ is independent (Lemma 1 Section 6.4). Here is the analog for *orthogonal* sets in $\mathbb{R}^n$.

Lemma 1

Orthogonal Lemma

Let $\{\mathbf{f}_1, \mathbf{f}_2, \ldots, \mathbf{f}_m\}$ be an orthogonal set in $\mathbb{R}^n$. Given $\mathbf{x}$ in $\mathbb{R}^n$, write

$$\mathbf{f}_{m+1} = \mathbf{x} - \frac{\mathbf{x} \cdot \mathbf{f}_1}{\|\mathbf{f}_1\|^2}\mathbf{f}_1 - \frac{\mathbf{x} \cdot \mathbf{f}_2}{\|\mathbf{f}_2\|^2}\mathbf{f}_2 - \cdots - \frac{\mathbf{x} \cdot \mathbf{f}_m}{\|\mathbf{f}_m\|^2}\mathbf{f}_m$$

Then:

1. $\mathbf{f}_{m+1} \cdot \mathbf{f}_k = 0$ for $k = 1, 2, \ldots, m$.
2. If $\mathbf{x}$ is not in span$\{\mathbf{f}_1, \ldots, \mathbf{f}_m\}$, then $\mathbf{f}_{m+1} \neq \mathbf{0}$ and $\{\mathbf{f}_1, \ldots, \mathbf{f}_m, \mathbf{f}_{m+1}\}$ is an orthogonal set.

PROOF

For convenience, write $t_i = (\mathbf{x} \cdot \mathbf{f}_i)/\|\mathbf{f}_i\|^2$ for each i. Given $1 \leq k \leq m$:

$$\begin{aligned}\mathbf{f}_{m+1} \cdot \mathbf{f}_k &= (\mathbf{x} - t_1\mathbf{f}_1 - \cdots - t_k\mathbf{f}_k - \cdots - t_m\mathbf{f}_m) \cdot \mathbf{f}_k \\ &= \mathbf{x} \cdot \mathbf{f}_k - t_1(\mathbf{f}_1 \cdot \mathbf{f}_k) - \cdots - t_k(\mathbf{f}_k \cdot \mathbf{f}_k) - \cdots - t_m(\mathbf{f}_m \cdot \mathbf{f}_k) \\ &= \mathbf{x} \cdot \mathbf{f}_k - t_k\|\mathbf{f}_k\|^2 \\ &= 0\end{aligned}$$

This proves (1), and (2) follows because $\mathbf{f}_{m+1} \neq \mathbf{0}$ if $\mathbf{x}$ is not in span$\{\mathbf{f}_1, \ldots, \mathbf{f}_m\}$.

SECTION 8.1 Orthogonal Complements and Projections

The orthogonal lemma has three important consequences for $\mathbb{R}^n$. The first is an extension for orthogonal sets of the fundamental fact that any independent set is part of a basis (Theorem 1 Section 6.4).

Theorem 1

Let U be a subspace of $\mathbb{R}^n$.

1. Every orthogonal subset $\{\mathbf{f}_1, \ldots, \mathbf{f}_m\}$ in U is a subset of an orthogonal basis of U.
2. U has an orthogonal basis.

PROOF

1. If span$\{\mathbf{f}_1, \ldots, \mathbf{f}_m\} = U$, it is *already* a basis. Otherwise, there exists $\mathbf{x}$ in U outside span$\{\mathbf{f}_1, \ldots, \mathbf{f}_m\}$. If $\mathbf{f}_{m+1}$ is as given in the orthogonal lemma, then $\mathbf{f}_{m+1}$ is in U and $\{\mathbf{f}_1, \ldots, \mathbf{f}_m, \mathbf{f}_{m+1}\}$ is orthogonal. If span$\{\mathbf{f}_1, \ldots, \mathbf{f}_m, \mathbf{f}_{m+1}\} = U$, we are done. Otherwise, the process continues to create larger and larger orthogonal subsets of U. They are all independent by Theorem 5 Section 5.3, so we have a basis when we reach a subset containing dim U vectors.

2. If $U = \{\mathbf{0}\}$, the empty basis is orthogonal. Otherwise, if $\mathbf{f} \neq \mathbf{0}$ is in U, then $\{\mathbf{f}\}$ is orthogonal, so (2) follows from (1).

We can improve upon (2) of Theorem 1. In fact, the second consequence of the orthogonal lemma is a procedure by which *any* basis $\{\mathbf{x}_1, \ldots, \mathbf{x}_m\}$ of a subspace U of $\mathbb{R}^n$ can be systematically modified to yield an orthogonal basis $\{\mathbf{f}_1, \ldots, \mathbf{f}_m\}$ of U. The $\mathbf{f}_i$ are constructed one at a time from the $\mathbf{x}_i$.

To start the process, take $\mathbf{f}_1 = \mathbf{x}_1$. Then $\mathbf{x}_2$ is not in span$\{\mathbf{f}_1\}$ because $\{\mathbf{x}_1, \mathbf{x}_2\}$ is independent, so take

$$\mathbf{f}_2 = \mathbf{x}_2 - \frac{\mathbf{x}_2 \cdot \mathbf{f}_1}{\|\mathbf{f}_1\|^2} \mathbf{f}_1$$

Thus $\{\mathbf{f}_1, \mathbf{f}_2\}$ is orthogonal by Lemma 1. Moreover, span$\{\mathbf{f}_1, \mathbf{f}_2\}$ = span$\{\mathbf{x}_1, \mathbf{x}_2\}$ (verify), so $\mathbf{x}_3$ is not in span$\{\mathbf{f}_1, \mathbf{f}_2\}$. Hence $\{\mathbf{f}_1, \mathbf{f}_2, \mathbf{f}_3\}$ is orthogonal where

$$\mathbf{f}_3 = \mathbf{x}_3 - \frac{\mathbf{x}_3 \cdot \mathbf{f}_1}{\|\mathbf{f}_1\|^2} \mathbf{f}_1 - \frac{\mathbf{x}_3 \cdot \mathbf{f}_2}{\|\mathbf{f}_2\|^2} \mathbf{f}_2$$

Again, span$\{\mathbf{f}_1, \mathbf{f}_2, \mathbf{f}_3\}$ = span$\{\mathbf{x}_1, \mathbf{x}_2, \mathbf{x}_3\}$, so $\mathbf{x}_4$ is not in span$\{\mathbf{f}_1, \mathbf{f}_2, \mathbf{f}_3\}$ and the process continues. At the mth iteration we construct an orthogonal set $\{\mathbf{f}_1, \ldots, \mathbf{f}_m\}$ such that

$$\text{span}\{\mathbf{f}_1, \mathbf{f}_2, \ldots, \mathbf{f}_m\} = \text{span}\{\mathbf{x}_1, \mathbf{x}_2, \ldots, \mathbf{x}_m\} = U$$

Hence $\{\mathbf{f}_1, \mathbf{f}_2, \ldots, \mathbf{f}_m\}$ is the desired orthogonal basis of U. The procedure can be summarized as follows.

Theorem 2

Gram-Schmidt Orthogonalization Algorithm[1]

If $\{\mathbf{x}_1, \mathbf{x}_2, \ldots, \mathbf{x}_m\}$ is any basis of a subspace U of $\mathbb{R}^n$, construct $\mathbf{f}_1, \mathbf{f}_2, \ldots, \mathbf{f}_m$ in U successively as follows:

---

1   Erhardt Schmidt (1876–1959) was a German mathematician who studied under the great David Hilbert and later developed the theory of Hilbert spaces. He first described the present algorithm in 1907. Jörgen Pederson Gram (1850–1916) was a Danish actuary.

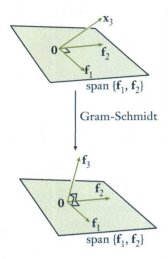

Gram-Schmidt

$$\mathbf{f}_1 = \mathbf{x}_1$$
$$\mathbf{f}_2 = \mathbf{x}_2 - \frac{\mathbf{x}_2 \cdot \mathbf{f}_1}{\|\mathbf{f}_1\|^2}\mathbf{f}_1$$
$$\mathbf{f}_3 = \mathbf{x}_3 - \frac{\mathbf{x}_3 \cdot \mathbf{f}_1}{\|\mathbf{f}_1\|^2}\mathbf{f}_1 - \frac{\mathbf{x}_3 \cdot \mathbf{f}_2}{\|\mathbf{f}_2\|^2}\mathbf{f}_2$$
$$\vdots$$
$$\mathbf{f}_k = \mathbf{x}_k - \frac{\mathbf{x}_k \cdot \mathbf{f}_1}{\|\mathbf{f}_1\|^2}\mathbf{f}_1 - \frac{\mathbf{x}_k \cdot \mathbf{f}_2}{\|\mathbf{f}_2\|^2}\mathbf{f}_2 - \cdots - \frac{\mathbf{x}_k \cdot \mathbf{f}_{k-1}}{\|\mathbf{f}_{k-1}\|^2}\mathbf{f}_{k-1}$$

for each $k = 2, 3, \ldots, m$. Then

1. $\{\mathbf{f}_1, \mathbf{f}_2, \ldots, \mathbf{f}_m\}$ *is an orthogonal basis of U.*
2. $span\{\mathbf{f}_1, \mathbf{f}_2, \ldots, \mathbf{f}_k\} = span\{\mathbf{x}_1, \mathbf{x}_2, \ldots, \mathbf{x}_k\}$ *for each $k = 1, 2, \ldots, m$.*

The process (for $k = 3$) is depicted in the diagrams. Of course, the algorithm converts any basis of $\mathbb{R}^n$ itself into an orthogonal basis.

EXAMPLE 1

Find an orthogonal basis of the row space of $A = \begin{bmatrix} 1 & 1 & -1 & -1 \\ 3 & 2 & 0 & 1 \\ 1 & 0 & 1 & 0 \end{bmatrix}$.

Solution ▶ Let $\mathbf{x}_1, \mathbf{x}_2, \mathbf{x}_3$ denote the rows of A and observe that $\{\mathbf{x}_1, \mathbf{x}_2, \mathbf{x}_3\}$ is linearly independent. Take $\mathbf{f}_1 = \mathbf{x}_1$. The algorithm gives

$$\mathbf{f}_2 = \mathbf{x}_2 - \frac{\mathbf{x}_2 \cdot \mathbf{f}_1}{\|\mathbf{f}_1\|^2}\mathbf{f}_1 = (3, 2, 0, 1) - \tfrac{4}{4}(1, 1, -1, -1) = (2, 1, 1, 2)$$

$$\mathbf{f}_3 = \mathbf{x}_3 - \frac{\mathbf{x}_3 \cdot \mathbf{f}_1}{\|\mathbf{f}_1\|^2}\mathbf{f}_1 - \frac{\mathbf{x}_3 \cdot \mathbf{f}_2}{\|\mathbf{f}_2\|^2}\mathbf{f}_2 = \mathbf{x}_3 - \tfrac{0}{4}\mathbf{f}_1 - \tfrac{3}{10}\mathbf{f}_2 = \tfrac{1}{10}(4, -3, 7, -6)$$

Hence $\{(1, 1, -1, -1), (2, 1, 1, 2), \tfrac{1}{10}(4, -3, 7, -6)\}$ is the orthogonal basis provided by the algorithm. In hand calculations it may be convenient to eliminate fractions, so $\{(1, 1, -1, -1), (2, 1, 1, 2), (4, -3, 7, -6)\}$ is also an orthogonal basis for row A.

Remark Observe that the vector $\frac{\mathbf{x} \cdot \mathbf{f}_i}{\|\mathbf{f}_i\|^2}\mathbf{f}_i$ is unchanged if a nonzero scalar multiple of $\mathbf{f}_i$ is used in place of $\mathbf{f}_i$. Hence, if a newly constructed $\mathbf{f}_i$ is multiplied by a nonzero scalar at some stage of the Gram-Schmidt algorithm, the subsequent $\mathbf{f}$s will be unchanged. This is useful in actual calculations.

Projections

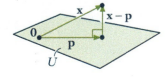

Suppose a point $\mathbf{x}$ and a plane U through the origin in $\mathbb{R}^3$ are given, and we want to find the point $\mathbf{p}$ in the plane that is closest to $\mathbf{x}$. Our geometric intuition assures us that such a point $\mathbf{p}$ exists. In fact (see the diagram), $\mathbf{p}$ must be chosen in such a way that $\mathbf{x} - \mathbf{p}$ is *perpendicular* to the plane.

SECTION 8.1 Orthogonal Complements and Projections

Now we make two observations: first, the plane U is a *subspace of* $\mathbb{R}^3$ (because U contains the origin); and second, that the condition that $\mathbf{x} - \mathbf{p}$ is perpendicular to the plane U means that $\mathbf{x} - \mathbf{p}$ is *orthogonal* to every vector in U. In these terms the whole discussion makes sense in $\mathbb{R}^n$. Furthermore, the orthogonal lemma provides exactly what is needed to find $\mathbf{p}$ in this more general setting.

Definition 8.1   *If U is a subspace of $\mathbb{R}^n$, define the **orthogonal complement** $U^\perp$ of U (pronounced "U-perp") by*

$$U^\perp = \{\mathbf{x} \text{ in } \mathbb{R}^n \mid \mathbf{x} \cdot \mathbf{y} = 0 \text{ for all } \mathbf{y} \text{ in } U\}.$$

The following lemma collects some useful properties of the orthogonal complement; the proof of (1) and (2) is left as Exercise 6.

Lemma 2

Let U be a subspace of $\mathbb{R}^n$.
1. $U^\perp$ is a subspace of $\mathbb{R}^n$.
2. $\{\mathbf{0}\}^\perp = \mathbb{R}^n$ and $(\mathbb{R}^n)^\perp = \{\mathbf{0}\}$.
3. If $U = \text{span}\{\mathbf{x}_1, \mathbf{x}_2, ..., \mathbf{x}_k\}$, then $U^\perp = \{\mathbf{x} \text{ in } \mathbb{R}^n \mid \mathbf{x} \cdot \mathbf{x}_i = 0 \text{ for } i = 1, 2, ..., k\}$.

PROOF

3. Let $U = \text{span}\{\mathbf{x}_1, \mathbf{x}_2, ..., \mathbf{x}_k\}$; we must show that $U^\perp = \{\mathbf{x} \mid \mathbf{x} \cdot \mathbf{x}_i = 0 \text{ for each } i\}$. If $\mathbf{x}$ is in $U^\perp$ then $\mathbf{x} \cdot \mathbf{x}_i = 0$ for all i because each $\mathbf{x}_i$ is in U. Conversely, suppose that $\mathbf{x} \cdot \mathbf{x}_i = 0$ for all i; we must show that $\mathbf{x}$ is in $U^\perp$, that is, $\mathbf{x} \cdot \mathbf{y} = 0$ for each $\mathbf{y}$ in U. Write $\mathbf{y} = r_1\mathbf{x}_1 + r_2\mathbf{x}_2 + \cdots + r_k\mathbf{x}_k$, where each r_i is in $\mathbb{R}$. Then, using Theorem 1 Section 5.3,

$$\mathbf{x} \cdot \mathbf{y} = r_1(\mathbf{x} \cdot \mathbf{x}_1) + r_2(\mathbf{x} \cdot \mathbf{x}_2) + \cdots + r_k(\mathbf{x} \cdot \mathbf{x}_k) = r_1 0 + r_2 0 + \cdots + r_k 0 = 0,$$

as required.

EXAMPLE 2

Find $U^\perp$ if $U = \text{span}\{(1, -1, 2, 0), (1, 0, -2, 3)\}$ in $\mathbb{R}^4$.

Solution ▶ By Lemma 2, $\mathbf{x} = (x, y, z, w)$ is in $U^\perp$ if and only if it is orthogonal to both $(1, -1, 2, 0)$ and $(1, 0, -2, 3)$; that is,

$$\begin{aligned} x - y + 2z &= 0 \\ x \quad\quad - 2z + 3w &= 0 \end{aligned}$$

Gaussian elimination gives $U^\perp = \text{span}\{(2, 4, 1, 0), (3, 3, 0, -1)\}$.

Now consider vectors $\mathbf{x}$ and $\mathbf{d} \neq \mathbf{0}$ in $\mathbb{R}^3$. The projection $\mathbf{p} = \text{proj}_\mathbf{d}(\mathbf{x})$ of $\mathbf{x}$ on $\mathbf{d}$ was defined in Section 4.2 as in the diagram. The following formula for $\mathbf{p}$ was derived in Theorem 4 Section 4.2

$$\mathbf{p} = \text{proj}_\mathbf{d}(\mathbf{x}) = \left(\frac{\mathbf{x} \cdot \mathbf{d}}{\|\mathbf{d}\|^2}\right)\mathbf{d},$$

where it is shown that $\mathbf{x} - \mathbf{p}$ is orthogonal to $\mathbf{d}$. Now observe that the line $U = \mathbb{R}\mathbf{d} = \{t\mathbf{d} \mid t \in \mathbb{R}\}$ a subspace of $\mathbb{R}^3$, that $\{\mathbf{d}\}$ is an orthogonal basis of U, and that $\mathbf{p} \in U$ and $\mathbf{x} - \mathbf{p} \in U^\perp$ (by Theorem 4 Section 4.2).

In this form, this makes sense for any vector $\mathbf{x}$ in $\mathbb{R}^n$ and any subspace U of $\mathbb{R}^n$, so we generalize it as follows. If $\{\mathbf{f}_1, \mathbf{f}_2, \ldots, \mathbf{f}_m\}$ is an orthogonal basis of U, we define the projection $\mathbf{p}$ of $\mathbf{x}$ on U by the formula

$$\mathbf{p} = \left(\frac{\mathbf{x} \cdot \mathbf{f}_1}{\|\mathbf{f}_1\|^2}\right)\mathbf{f}_1 + \left(\frac{\mathbf{x} \cdot \mathbf{f}_2}{\|\mathbf{f}_2\|^2}\right)\mathbf{f}_2 + \cdots + \left(\frac{\mathbf{x} \cdot \mathbf{f}_m}{\|\mathbf{f}_m\|^2}\right)\mathbf{f}_m. \quad (*)$$

Then $\mathbf{p} \in U$ and (by the orthogonal lemma) $\mathbf{x} - \mathbf{p} \in U^\perp$, so it looks like we have a generalization of Theorem 4 Section 4.2.

However there is a potential problem: the formula $(*)$ for $\mathbf{p}$ must be shown to be independent of the choice of the orthogonal basis $\{\mathbf{f}_1, \mathbf{f}_2, \ldots, \mathbf{f}_m\}$. To verify this, suppose that $\{\mathbf{f}'_1, \mathbf{f}'_2, \ldots, \mathbf{f}'_m\}$ is another orthogonal basis of U, and write

$$\mathbf{p}' = \left(\frac{\mathbf{x} \cdot \mathbf{f}'_1}{\|\mathbf{f}'_1\|^2}\right)\mathbf{f}'_1 + \left(\frac{\mathbf{x} \cdot \mathbf{f}'_2}{\|\mathbf{f}'_2\|^2}\right)\mathbf{f}'_2 + \cdots + \left(\frac{\mathbf{x} \cdot \mathbf{f}'_m}{\|\mathbf{f}'_m\|^2}\right)\mathbf{f}'_m.$$

As before, $\mathbf{p}' \in U$ and $\mathbf{x} - \mathbf{p}' \in U^\perp$, and we must show that $\mathbf{p}' = \mathbf{p}$. To see this, write the vector $\mathbf{p} - \mathbf{p}'$ as follows:

$$\mathbf{p} - \mathbf{p}' = (\mathbf{x} - \mathbf{p}') - (\mathbf{x} - \mathbf{p}).$$

This vector is in U (because $\mathbf{p}$ and $\mathbf{p}'$ are in U) and it is in $U^\perp$ (because $\mathbf{x} - \mathbf{p}'$ and $\mathbf{x} - \mathbf{p}$ are in $U^\perp$), and so it must be zero (it is orthogonal to itself!). This means $\mathbf{p}' = \mathbf{p}$ as desired.

Hence, the vector $\mathbf{p}$ in equation $(*)$ depends only on $\mathbf{x}$ and the subspace U, and *not* on the choice of orthogonal basis $\{\mathbf{f}_1, \ldots, \mathbf{f}_m\}$ of U used to compute it. Thus, we are entitled to make the following definition:

Definition 8.2 Let U be a subspace of $\mathbb{R}^n$ with orthogonal basis $\{\mathbf{f}_1, \mathbf{f}_2, \ldots, \mathbf{f}_m\}$. If $\mathbf{x}$ is in $\mathbb{R}^n$, the vector

$$\text{proj}_U(\mathbf{x}) = \frac{\mathbf{x} \cdot \mathbf{f}_1}{\|\mathbf{f}_1\|^2}\mathbf{f}_1 + \frac{\mathbf{x} \cdot \mathbf{f}_2}{\|\mathbf{f}_2\|^2}\mathbf{f}_2 + \cdots + \frac{\mathbf{x} \cdot \mathbf{f}_m}{\|\mathbf{f}_m\|^2}\mathbf{f}_m$$

is called the **orthogonal projection** *of* $\mathbf{x}$ *on* U. For the zero subspace $U = \{\mathbf{0}\}$, we define

$$\text{proj}_{\{\mathbf{0}\}}(\mathbf{x}) = \mathbf{0}.$$

The preceding discussion proves (1) of the following theorem.

Theorem 3

Projection Theorem
If U is a subspace of $\mathbb{R}^n$ and $\mathbf{x}$ is in $\mathbb{R}^n$, write $\mathbf{p} = \text{proj}_U(\mathbf{x})$. Then
1. $\mathbf{p}$ is in U and $\mathbf{x} - \mathbf{p}$ is in $U^\perp$.
2. $\mathbf{p}$ is the vector in U closest to $\mathbf{x}$ in the sense that

$$\|\mathbf{x} - \mathbf{p}\| < \|\mathbf{x} - \mathbf{y}\| \quad \text{for all } \mathbf{y} \in U, \mathbf{y} \neq \mathbf{p}$$

SECTION 8.1 Orthogonal Complements and Projections

PROOF

1. This is proved in the preceding discussion (it is clear if $U = \{\mathbf{0}\}$).
2. Write $\mathbf{x} - \mathbf{y} = (\mathbf{x} - \mathbf{p}) + (\mathbf{p} - \mathbf{y})$. Then $\mathbf{p} - \mathbf{y}$ is in U and so is orthogonal to $\mathbf{x} - \mathbf{p}$ by (1). Hence, the pythagorean theorem gives
$$\|\mathbf{x} - \mathbf{y}\|^2 = \|\mathbf{x} - \mathbf{p}\|^2 + \|\mathbf{p} - \mathbf{y}\|^2 > \|\mathbf{x} - \mathbf{p}\|^2$$
because $\mathbf{p} - \mathbf{y} \neq \mathbf{0}$. This gives (2).

EXAMPLE 3

Let $U = \text{span}\{\mathbf{x}_1, \mathbf{x}_2\}$ in $\mathbb{R}^4$ where $\mathbf{x}_1 = (1, 1, 0, 1)$ and $\mathbf{x}_2 = (0, 1, 1, 2)$. If $\mathbf{x} = (3, -1, 0, 2)$, find the vector in U closest to $\mathbf{x}$ and express $\mathbf{x}$ as the sum of a vector in U and a vector orthogonal to U.

Solution ▶ $\{\mathbf{x}_1, \mathbf{x}_2\}$ is independent but not orthogonal. The Gram-Schmidt process gives an orthogonal basis $\{\mathbf{f}_1, \mathbf{f}_2\}$ of U where $\mathbf{f}_1 = \mathbf{x}_1 = (1, 1, 0, 1)$ and
$$\mathbf{f}_2 = \mathbf{x}_2 - \frac{\mathbf{x}_2 \cdot \mathbf{f}_1}{\|\mathbf{f}_1\|^2} \mathbf{f}_1 = \mathbf{x}_2 - \tfrac{3}{3}\mathbf{f}_1 = (-1, 0, 1, 1)$$
Hence, we can compute the projection using $\{\mathbf{f}_1, \mathbf{f}_2\}$:
$$\mathbf{p} = \text{proj}_U(\mathbf{x}) = \frac{\mathbf{x} \cdot \mathbf{f}_1}{\|\mathbf{f}_1\|^2}\mathbf{f}_1 + \frac{\mathbf{x} \cdot \mathbf{f}_2}{\|\mathbf{f}_2\|^2}\mathbf{f}_2 = \tfrac{4}{3}\mathbf{f}_1 + \tfrac{-1}{3}\mathbf{f}_2 = \tfrac{1}{3}[5 \ 4 \ -1 \ 3]$$
Thus, $\mathbf{p}$ is the vector in U closest to $\mathbf{x}$, and $\mathbf{x} - \mathbf{p} = \tfrac{1}{3}(4, -7, 1, 3)$ is orthogonal to every vector in U. (This can be verified by checking that it is orthogonal to the generators $\mathbf{x}_1$ and $\mathbf{x}_2$ of U.) The required decomposition of $\mathbf{x}$ is thus
$$\mathbf{x} = \mathbf{p} + (\mathbf{x} - \mathbf{p}) = \tfrac{1}{3}(5, 4, -1, 3) + \tfrac{1}{3}(4, -7, 1, 3).$$

EXAMPLE 4

Find the point in the plane with equation $2x + y - z = 0$ that is closest to the point $(2, -1, -3)$.

Solution ▶ We write $\mathbb{R}^3$ as rows. The plane is the subspace U whose points (x, y, z) satisfy $z = 2x + y$. Hence
$$U = \{(s, t, 2s + t) \mid s, t \text{ in } \mathbb{R}\} = \text{span}\{(0, 1, 1), (1, 0, 2)\}$$
The Gram-Schmidt process produces an orthogonal basis $\{\mathbf{f}_1, \mathbf{f}_2\}$ of U where $\mathbf{f}_1 = (0, 1, 1)$ and $\mathbf{f}_2 = (1, -1, 1)$. Hence, the vector in U closest to $\mathbf{x} = (2, -1, -3)$ is
$$\text{proj}_U(\mathbf{x}) = \frac{\mathbf{x} \cdot \mathbf{f}_1}{\|\mathbf{f}_1\|^2}\mathbf{f}_1 + \frac{\mathbf{x} \cdot \mathbf{f}_2}{\|\mathbf{f}_2\|^2}\mathbf{f}_2 = -2\mathbf{f}_1 + 0\mathbf{f}_2 = (0, -2, -2)$$
Thus, the point in U closest to $(2, -1, -3)$ is $(0, -2, -2)$.

The next theorem shows that projection on a subspace of $\mathbb{R}^n$ is actually a linear operator $\mathbb{R}^n \to \mathbb{R}^n$.

Theorem 4

Let U be a fixed subspace of $\mathbb{R}^n$. If we define $T: \mathbb{R}^n \to \mathbb{R}^n$ by

$$T(\mathbf{x}) = \text{proj}_U(\mathbf{x}) \quad \text{for all } \mathbf{x} \text{ in } \mathbb{R}^n.$$

1. T is a linear operator.
2. $\text{im } T = U$ and $\ker T = U^\perp$.
3. $\dim U + \dim U^\perp = n$.

PROOF

If $U = \{\mathbf{0}\}$, then $U^\perp = \mathbb{R}^n$, and so $T(\mathbf{x}) = \text{proj}_{\{\mathbf{0}\}}(\mathbf{x}) = \mathbf{0}$ for all $\mathbf{x}$. Thus $T = 0$ is the zero (linear) operator, so (1), (2), and (3) hold. Hence assume that $U \neq \{\mathbf{0}\}$.

1. If $\{\mathbf{f}_1, \mathbf{f}_2, \ldots, \mathbf{f}_m\}$ is an orthonormal basis of U, then

$$T(\mathbf{x}) = (\mathbf{x} \cdot \mathbf{f}_1)\mathbf{f}_1 + (\mathbf{x} \cdot \mathbf{f}_2)\mathbf{f}_2 + \cdots + (\mathbf{x} \cdot \mathbf{f}_m)\mathbf{f}_m \quad \text{for all } \mathbf{x} \text{ in } \mathbb{R}^n \quad (*)$$

by the definition of the projection. Thus T is linear because

$$(\mathbf{x} + \mathbf{y}) \cdot \mathbf{f}_i = \mathbf{x} \cdot \mathbf{f}_i + \mathbf{y} \cdot \mathbf{f}_i \quad \text{and} \quad (r\mathbf{x}) \cdot \mathbf{f}_i = r(\mathbf{x} \cdot \mathbf{f}_i) \quad \text{for each } i.$$

2. We have $\text{im } T \subseteq U$ by $(*)$ because each $\mathbf{f}_i$ is in U. But if $\mathbf{x}$ is in U, then $\mathbf{x} = T(\mathbf{x})$ by $(*)$ and the expansion theorem applied to the space U. This shows that $U \subseteq \text{im } T$, so $\text{im } T = U$.

Now suppose that $\mathbf{x}$ is in $U^\perp$. Then $\mathbf{x} \cdot \mathbf{f}_i = 0$ for each i (again because each $\mathbf{f}_i$ is in U) so $\mathbf{x}$ is in $\ker T$ by $(*)$. Hence $U^\perp \subseteq \ker T$. On the other hand, Theorem 3 shows that $\mathbf{x} - T(\mathbf{x})$ is in $U^\perp$ for all $\mathbf{x}$ in $\mathbb{R}^n$, and it follows that $\ker T \subseteq U^\perp$. Hence $\ker T = U^\perp$, proving (2).

3. This follows from (1), (2), and the dimension theorem (Theorem 4 Section 7.2).

EXERCISES 8.1

1. In each case, use the Gram-Schmidt algorithm to convert the given basis B of V into an orthogonal basis.

    (a) $V = \mathbb{R}^2$, $B = \{(1, -1), (2, 1)\}$

    ◆(b) $V = \mathbb{R}^2$, $B = \{(2, 1), (1, 2)\}$

    (c) $V = \mathbb{R}^3$, $B = \{(1, -1, 1), (1, 0, 1), (1, 1, 2)\}$

    ◆(d) $V = \mathbb{R}^3$, $B = \{(0, 1, 1), (1, 1, 1), (1, -2, 2)\}$

2. In each case, write $\mathbf{x}$ as the sum of a vector in U and a vector in $U^\perp$.

    (a) $\mathbf{x} = (1, 5, 7)$, $U = \text{span}\{(1, -2, 3), (-1, 1, 1)\}$

    ◆(b) $\mathbf{x} = (2, 1, 6)$, $U = \text{span}\{(3, -1, 2), (2, 0, -3)\}$

    (c) $\mathbf{x} = (3, 1, 5, 9)$, $U = \text{span}\{(1, 0, 1, 1), (0, 1, -1, 1), (-2, 0, 1, 1)\}$

    ◆(d) $\mathbf{x} = (2, 0, 1, 6)$, $U = \text{span}\{(1, 1, 1, 1), (1, 1, -1, -1), (1, -1, 1, -1)\}$

    (e) $\mathbf{x} = (a, b, c, d)$, $U = \text{span}\{(1, 0, 0, 0), (0, 1, 0, 0), (0, 0, 1, 0)\}$

    ◆(f) $\mathbf{x} = (a, b, c, d)$, $U = \text{span}\{(1, -1, 2, 0), (-1, 1, 1, 1)\}$

3. Let $\mathbf{x} = (1, -2, 1, 6)$ in $\mathbb{R}^4$, and let $U = \text{span}\{(2, 1, 3, -4), (1, 2, 0, 1)\}$.

    ◆(a) Compute $\text{proj}_U(\mathbf{x})$.

    (b) Show that $\{(1, 0, 2, -3), (4, 7, 1, 2)\}$ is another orthogonal basis of U.

    ◆(c) Use the basis in part (b) to compute $\text{proj}_U(\mathbf{x})$.

SECTION 8.1 Orthogonal Complements and Projections

4. In each case, use the Gram-Schmidt algorithm to find an orthogonal basis of the subspace U, and find the vector in U closest to $\mathbf{x}$.

   (a) $U = \text{span}\{(1, 1, 1), (0, 1, 1)\}$, $\mathbf{x} = (-1, 2, 1)$

   ◆(b) $U = \text{span}\{(1, -1, 0), (-1, 0, 1)\}$, $\mathbf{x} = (2, 1, 0)$

   (c) $U = \text{span}\{(1, 0, 1, 0), (1, 1, 1, 0), (1, 1, 0, 0)\}$, $\mathbf{x} = (2, 0, -1, 3)$

   ◆(d) $U = \text{span}\{(1, -1, 0, 1), (1, 1, 0, 0), (1, 1, 0, 1)\}$, $\mathbf{x} = (2, 0, 3, 1)$

5. Let $U = \text{span}\{\mathbf{v}_1, \mathbf{v}_2, \ldots, \mathbf{v}_k\}$, $\mathbf{v}_i$ in $\mathbb{R}^n$, and let A be the $k \times n$ matrix with the $\mathbf{v}_i$ as rows.

   (a) Show that $U^\perp = \{\mathbf{x} \mid \mathbf{x} \text{ in } \mathbb{R}^n, A\mathbf{x}^T = \mathbf{0}\}$.

   ◆(b) Use part (a) to find $U^\perp$ if $U = \text{span}\{(1, -1, 2, 1), (1, 0, -1, 1)\}$.

6. (a) Prove part 1 of Lemma 2.

   (b) Prove part 2 of Lemma 2.

7. Let U be a subspace of $\mathbb{R}^n$. If $\mathbf{x}$ in $\mathbb{R}^n$ can be written in any way at all as $\mathbf{x} = \mathbf{p} + \mathbf{q}$ with $\mathbf{p}$ in U and $\mathbf{q}$ in $U^\perp$, show that necessarily $\mathbf{p} = \text{proj}_U(\mathbf{x})$.

◆8. Let U be a subspace of $\mathbb{R}^n$ and let $\mathbf{x}$ be a vector in $\mathbb{R}^n$. Using Exercise 7, or otherwise, show that $\mathbf{x}$ is in U if and only if $\mathbf{x} = \text{proj}_U(\mathbf{x})$.

9. Let U be a subspace of $\mathbb{R}^n$.

   (a) Show that $U^\perp = \mathbb{R}^n$ if and only if $U = \{\mathbf{0}\}$.

   (b) Show that $U^\perp = \{\mathbf{0}\}$ if and only if $U = \mathbb{R}^n$.

◆10. If U is a subspace of $\mathbb{R}^n$, show that $\text{proj}_U(\mathbf{x}) = \mathbf{x}$ for all $\mathbf{x}$ in U.

11. If U is a subspace of $\mathbb{R}^n$, show that $\mathbf{x} = \text{proj}_U(\mathbf{x}) + \text{proj}_{U^\perp}(\mathbf{x})$ for all $\mathbf{x}$ in $\mathbb{R}^n$.

12. If $\{\mathbf{f}_1, \ldots, \mathbf{f}_n\}$ is an orthogonal basis of $\mathbb{R}^n$ and $U = \text{span}\{\mathbf{f}_1, \ldots, \mathbf{f}_m\}$, show that $U^\perp = \text{span}\{\mathbf{f}_{m+1}, \ldots, \mathbf{f}_n\}$.

13. If U is a subspace of $\mathbb{R}^n$, show that $U^{\perp\perp} = U$. [*Hint*: Show that $U \subseteq U^{\perp\perp}$, then use Theorem 4(3) twice.]

◆14. If U is a subspace of $\mathbb{R}^n$, show how to find an $n \times n$ matrix A such that $U = \{\mathbf{x} \mid A\mathbf{x} = \mathbf{0}\}$. [*Hint*: Exercise 13.]

15. Write $\mathbb{R}^n$ as rows. If A is an $n \times n$ matrix, write its null space as null $A = \{\mathbf{x} \text{ in } \mathbb{R}^n \mid A\mathbf{x}^T = \mathbf{0}\}$. Show that:

    (a) null $A = (\text{row } A)^\perp$;

    (b) null $A^T = (\text{col } A)^\perp$.

16. If U and W are subspaces, show that $(U + W)^\perp = U^\perp \cap W^\perp$. [See Exercise 22 Section 5.1.]

17. Think of $\mathbb{R}^n$ as consisting of rows.

    (a) Let E be an $n \times n$ matrix, and let $U = \{\mathbf{x}E \mid \mathbf{x} \text{ in } \mathbb{R}^n\}$. Show that the following are equivalent.

    (i) $E^2 = E = E^T$ (E is a **projection matrix**).

    (ii) $(\mathbf{x} - \mathbf{x}E) \cdot (\mathbf{y}E) = 0$ for all $\mathbf{x}$ and $\mathbf{y}$ in $\mathbb{R}^n$.

    (iii) $\text{proj}_U(\mathbf{x}) = \mathbf{x}E$ for all $\mathbf{x}$ in $\mathbb{R}^n$.

    [*Hint*: For (ii) implies (iii): Write $\mathbf{x} = \mathbf{x}E + (\mathbf{x} - \mathbf{x}E)$ and use the uniqueness argument preceding the definition of $\text{proj}_U(\mathbf{x})$. For (iii) implies (ii): $\mathbf{x} - \mathbf{x}E$ is in $U^\perp$ for all $\mathbf{x}$ in $\mathbb{R}^n$.]

    (b) If E is a projection matrix, show that $I - E$ is also a projection matrix.

    (c) If $EF = 0 = FE$ and E and F are projection matrices, show that $E + F$ is also a projection matrix.

    ◆(d) If A is $m \times n$ and AA^T is invertible, show that $E = A^T(AA^T)^{-1}A$ is a projection matrix.

18. Let A be an $n \times n$ matrix of rank r. Show that there is an invertible $n \times n$ matrix U such that UA is a row-echelon matrix with the property that the first r rows are orthogonal. [*Hint*: Let R be the row-echelon form of A, and use the Gram-Schmidt process on the nonzero rows of R from the bottom up. Use Lemma 1 Section 2.4.]

19. Let A be an $(n - 1) \times n$ matrix with rows $\mathbf{x}_1, \mathbf{x}_2, \ldots, \mathbf{x}_{n-1}$ and let A_i denote the $(n - 1) \times (n - 1)$ matrix obtained from A by deleting column i. Define the vector $\mathbf{y}$ in $\mathbb{R}^n$ by
$$\mathbf{y} = [\det A_1 \ -\det A_2 \ \det A_3 \ \cdots \ (-1)^{n+1} \det A_n]$$
Show that:

    (a) $\mathbf{x}_i \cdot \mathbf{y} = 0$ for all $i = 1, 2, \ldots, n - 1$. [*Hint*: Write $B_i = \begin{bmatrix} \mathbf{x}_i \\ A \end{bmatrix}$ and show that $\det B_i = 0$.]

(b) $\mathbf{y} \neq \mathbf{0}$ if and only if $\{\mathbf{x}_1, \mathbf{x}_2, ..., \mathbf{x}_{n-1}\}$ is linearly independent. [*Hint*: If some $\det A_i \neq 0$, the rows of A_i are linearly independent. Conversely, if the $\mathbf{x}_i$ are independent, consider $A = UR$ where R is in reduced row-echelon form.]

(c) If $\{\mathbf{x}_1, \mathbf{x}_2, ..., \mathbf{x}_{n-1}\}$ is linearly independent, use Theorem 3(3) to show that all solutions to the system of $n - 1$ homogeneous equations
$$A\mathbf{x}^T = \mathbf{0}$$
are given by $t\mathbf{y}$, t a parameter.

SECTION 8.2 Orthogonal Diagonalization

Recall (Theorem 3 Section 5.5) that an $n \times n$ matrix A is diagonalizable if and only if it has n linearly independent eigenvectors. Moreover, the matrix P with these eigenvectors as columns is a diagonalizing matrix for A, that is

$$P^{-1}AP \text{ is diagonal.}$$

As we have seen, the really nice bases of $\mathbb{R}^n$ are the orthogonal ones, so a natural question is: which $n \times n$ matrices have an *orthogonal* basis of eigenvectors? These turn out to be precisely the symmetric matrices, and this is the main result of this section.

Before proceeding, recall that an orthogonal set of vectors is called *orthonormal* if $\|\mathbf{v}\| = 1$ for each vector $\mathbf{v}$ in the set, and that any orthogonal set $\{\mathbf{v}_1, \mathbf{v}_2, ..., \mathbf{v}_k\}$ can be *"normalized"*, that is converted into an orthonormal set $\left\{\frac{1}{\|\mathbf{v}_1\|}\mathbf{v}_1, \frac{1}{\|\mathbf{v}_2\|}\mathbf{v}_2, ..., \frac{1}{\|\mathbf{v}_k\|}\mathbf{v}_k\right\}$. In particular, if a matrix A has n orthogonal eigenvectors, they can (by normalizing) be taken to be orthonormal. The corresponding diagonalizing matrix P has orthonormal columns, and such matrices are very easy to invert.

Theorem 1

The following conditions are equivalent for an $n \times n$ matrix P.

1. *P is invertible and $P^{-1} = P^T$.*
2. *The rows of P are orthonormal.*
3. *The columns of P are orthonormal.*

PROOF

First recall that condition (1) is equivalent to $PP^T = I$ by Corollary 1 of Theorem 5 Section 2.4. Let $\mathbf{x}_1, \mathbf{x}_2, ..., \mathbf{x}_n$ denote the rows of P. Then $\mathbf{x}_j^T$ is the jth column of P^T, so the (i, j)-entry of PP^T is $\mathbf{x}_i \cdot \mathbf{x}_j$. Thus $PP^T = I$ means that $\mathbf{x}_i \cdot \mathbf{x}_j = 0$ if $i \neq j$ and $\mathbf{x}_i \cdot \mathbf{x}_j = 1$ if $i = j$. Hence condition (1) is equivalent to (2). The proof of the equivalence of (1) and (3) is similar.

Definition 8.3 *An $n \times n$ matrix P is called an* **orthogonal matrix**[2] *if it satisfies one (and hence all) of the conditions in Theorem 1.*

---

2 In view of (2) and (3) of Theorem 1, *orthonormal matrix* might be a better name. But *orthogonal matrix* is standard.

SECTION 8.2 Orthogonal Diagonalization

EXAMPLE 1

The rotation matrix $\begin{bmatrix} \cos\theta & -\sin\theta \\ \sin\theta & \cos\theta \end{bmatrix}$ is orthogonal for any angle θ.

These orthogonal matrices have the virtue that they are easy to invert—simply take the transpose. But they have many other important properties as well. If $T : \mathbb{R}^n \to \mathbb{R}^n$ is a linear operator, we will prove (Theorem 3 Section 10.4) that T is distance preserving if and only if its matrix is orthogonal. In particular, the matrices of rotations and reflections about the origin in $\mathbb{R}^2$ and $\mathbb{R}^3$ are all orthogonal (see Example 1).

It is not enough that the rows of a matrix A are merely orthogonal for A to be an orthogonal matrix. Here is an example.

EXAMPLE 2

The matrix $\begin{bmatrix} 2 & 1 & 1 \\ -1 & 1 & 1 \\ 0 & -1 & 1 \end{bmatrix}$ has orthogonal rows but the columns are not orthogonal.

However, if the rows are normalized, the resulting matrix $\begin{bmatrix} \frac{2}{\sqrt{6}} & \frac{1}{\sqrt{6}} & \frac{1}{\sqrt{6}} \\ \frac{-1}{\sqrt{3}} & \frac{1}{\sqrt{3}} & \frac{1}{\sqrt{3}} \\ 0 & \frac{-1}{\sqrt{2}} & \frac{1}{\sqrt{2}} \end{bmatrix}$ is orthogonal (so the columns are now orthonormal as the reader can verify).

EXAMPLE 3

If P and Q are orthogonal matrices, then PQ is also orthogonal, as is $P^{-1} = P^T$.

Solution ▶ P and Q are invertible, so PQ is also invertible and $(PQ)^{-1} = Q^{-1}P^{-1} = Q^T P^T = (PQ)^T$. Hence PQ is orthogonal. Similarly, $(P^{-1})^{-1} = P = (P^T)^T = (P^{-1})^T$ shows that P^{-1} is orthogonal.

Definition 8.4 An $n \times n$ matrix A is said to be **orthogonally diagonalizable** when an orthogonal matrix P can be found such that $P^{-1}AP = P^T AP$ is diagonal.

This condition turns out to characterize the symmetric matrices.

Theorem 2

Principal Axis Theorem
The following conditions are equivalent for an $n \times n$ matrix A.
1. A has an orthonormal set of n eigenvectors.
2. A is orthogonally diagonalizable.
3. A is symmetric.

PROOF

(1) ⇔ (2). Given (1), let $\mathbf{x}_1, \mathbf{x}_2, \ldots, \mathbf{x}_n$ be orthonormal eigenvectors of A. Then $P = [\mathbf{x}_1 \ \mathbf{x}_2 \ \cdots \ \mathbf{x}_n]$ is orthogonal, and $P^{-1}AP$ is diagonal by Theorem 4 Section 3.3. This proves (2). Conversely, given (2) let $P^{-1}AP$ be diagonal where P is orthogonal. If $\mathbf{x}_1, \mathbf{x}_2, \ldots, \mathbf{x}_n$ are the columns of P then $\{\mathbf{x}_1, \mathbf{x}_2, \ldots, \mathbf{x}_n\}$ is an orthonormal basis of $\mathbb{R}^n$ that consists of eigenvectors of A by Theorem 4 Section 3.3. This proves (1).

(2) ⇒ (3). If $P^T AP = D$ is diagonal, where $P^{-1} = P^T$, then $A = PDP^T$. But $D^T = D$, so this gives $A^T = P^{TT}D^TP^T = PDP^T = A$.

(3) ⇒ (2). If A is an $n \times n$ symmetric matrix, we proceed by induction on n. If $n = 1$, A is already diagonal. If $n > 1$, assume that (3) ⇒ (2) for $(n - 1) \times (n - 1)$ symmetric matrices. By Theorem 7 Section 5.5 let λ_1 be a (real) eigenvalue of A, and let $A\mathbf{x}_1 = \lambda_1 \mathbf{x}_1$, where $\|\mathbf{x}_1\| = 1$. Use the Gram-Schmidt algorithm to find an orthonormal basis $\{\mathbf{x}_1, \mathbf{x}_2, \ldots, \mathbf{x}_n\}$ for $\mathbb{R}^n$.

Let $P_1 = [\mathbf{x}_1 \ \mathbf{x}_2 \ \cdots \ \mathbf{x}_n]$, so P_1 is an orthogonal matrix and $P_1^T AP_1 = \begin{bmatrix} \lambda_1 & B \\ 0 & A_1 \end{bmatrix}$

in block form by Lemma 2 Section 5.5. But $P_1^T AP_1$ is symmetric (A is), so it follows that $B = 0$ and A_1 is symmetric. Then, by induction, there exists an $(n - 1) \times (n - 1)$ orthogonal matrix Q such that $Q^T A_1 Q = D_1$ is diagonal. Observe that $P_2 = \begin{bmatrix} 1 & 0 \\ 0 & Q \end{bmatrix}$ is orthogonal, and compute:

$$(P_1P_2)^T A(P_1P_2) = P_2^T(P_1^T AP_1)P_2$$
$$= \begin{bmatrix} 1 & 0 \\ 0 & Q^T \end{bmatrix}\begin{bmatrix} \lambda_1 & 0 \\ 0 & A_1 \end{bmatrix}\begin{bmatrix} 1 & 0 \\ 0 & Q \end{bmatrix}$$
$$= \begin{bmatrix} \lambda_1 & 0 \\ 0 & D_1 \end{bmatrix}$$

is diagonal. Because P_1P_2 is orthogonal, this proves (2).

A set of orthonormal eigenvectors of a symmetric matrix A is called a set of **principal axes** for A. The name comes from geometry, and this is discussed in Section 8.8. Because the eigenvalues of a (real) symmetric matrix are real, Theorem 2 is also called the **real spectral theorem**, and the set of distinct eigenvalues is called the **spectrum** of the matrix. In full generality, the spectral theorem is a similar result for matrices with complex entries (Theorem 8 Section 8.6).

EXAMPLE 4

Find an orthogonal matrix P such that $P^{-1}AP$ is diagonal, where $A = \begin{bmatrix} 1 & 0 & -1 \\ 0 & 1 & 2 \\ -1 & 2 & 5 \end{bmatrix}$.

SECTION 8.2 Orthogonal Diagonalization

Solution ▶ The characteristic polynomial of A is (adding twice row 1 to row 2):

$$c_A(x) = \det \begin{bmatrix} x-1 & 0 & 1 \\ 0 & x-1 & -2 \\ 1 & -2 & x-5 \end{bmatrix} = x(x-1)(x-6)$$

Thus the eigenvalues are $\lambda = 0, 1$, and 6, and corresponding eigenvectors are

$$\mathbf{x}_1 = \begin{bmatrix} 1 \\ -2 \\ 1 \end{bmatrix} \quad \mathbf{x}_2 = \begin{bmatrix} 2 \\ 1 \\ 0 \end{bmatrix} \quad \mathbf{x}_3 = \begin{bmatrix} -1 \\ 2 \\ 5 \end{bmatrix}$$

respectively. Moreover, by what appears to be remarkably good luck, these eigenvectors are *orthogonal*. We have $\|\mathbf{x}_1\|^2 = 6$, $\|\mathbf{x}_2\|^2 = 5$, and $\|\mathbf{x}_3\|^2 = 30$, so

$$P = \begin{bmatrix} \tfrac{1}{\sqrt{6}}\mathbf{x}_1 & \tfrac{1}{\sqrt{5}}\mathbf{x}_2 & \tfrac{1}{\sqrt{30}}\mathbf{x}_3 \end{bmatrix} = \tfrac{1}{\sqrt{30}} \begin{bmatrix} \sqrt{5} & 2\sqrt{6} & -1 \\ -2\sqrt{5} & \sqrt{6} & 2 \\ \sqrt{5} & 0 & 5 \end{bmatrix}$$

is an orthogonal matrix. Thus $P^{-1} = P^T$ and

$$P^T A P = \begin{bmatrix} 0 & 0 & 0 \\ 0 & 1 & 0 \\ 0 & 0 & 6 \end{bmatrix}$$

by the diagonalization algorithm.

Actually, the fact that the eigenvectors in Example 4 are orthogonal is no coincidence. Theorem 4 Section 5.5 guarantees they are linearly independent (they correspond to distinct eigenvalues); the fact that the matrix is *symmetric* implies that they are orthogonal. To prove this we need the following useful fact about symmetric matrices.

Theorem 3

If A is an $n \times n$ symmetric matrix, then

$$(A\mathbf{x}) \cdot \mathbf{y} = \mathbf{x} \cdot (A\mathbf{y})$$

for all columns $\mathbf{x}$ and $\mathbf{y}$ in $\mathbb{R}^n$.[3]

PROOF

Recall that $\mathbf{x} \cdot \mathbf{y} = \mathbf{x}^T \mathbf{y}$ for all columns $\mathbf{x}$ and $\mathbf{y}$. Because $A^T = A$, we get

$$(A\mathbf{x}) \cdot \mathbf{y} = (A\mathbf{x})^T \mathbf{y} = \mathbf{x}^T A^T \mathbf{y} = \mathbf{x}^T A \mathbf{y} = \mathbf{x} \cdot (A\mathbf{y}).$$

Theorem 4

If A is a symmetric matrix, then eigenvectors of A corresponding to distinct eigenvalues are orthogonal.

---

[3] The converse also holds (Exercise 15).

PROOF

Let $A\mathbf{x} = \lambda \mathbf{x}$ and $A\mathbf{y} = \mu \mathbf{y}$, where $\lambda \neq \mu$. Using Theorem 3, we compute

$$\lambda(\mathbf{x} \cdot \mathbf{y}) = (\lambda \mathbf{x}) \cdot \mathbf{y} = (A\mathbf{x}) \cdot \mathbf{y} = \mathbf{x} \cdot (A\mathbf{y}) = \mathbf{x} \cdot (\mu \mathbf{y}) = \mu(\mathbf{x} \cdot \mathbf{y})$$

Hence $(\lambda - \mu)(\mathbf{x} \cdot \mathbf{y}) = 0$, and so $\mathbf{x} \cdot \mathbf{y} = 0$ because $\lambda \neq \mu$.

Now the procedure for diagonalizing a symmetric $n \times n$ matrix is clear. Find the distinct eigenvalues (all real by Theorem 7 Section 5.5) and find orthonormal bases for each eigenspace (the Gram-Schmidt algorithm may be needed). Then the set of all these basis vectors is orthonormal (by Theorem 4) and contains n vectors. Here is an example.

EXAMPLE 5

Orthogonally diagonalize the symmetric matrix $A = \begin{bmatrix} 8 & -2 & 2 \\ -2 & 5 & 4 \\ 2 & 4 & 5 \end{bmatrix}$.

Solution ▶ The characteristic polynomial is

$$c_A(x) = \det \begin{bmatrix} x-8 & 2 & -2 \\ 2 & x-5 & -4 \\ -2 & -4 & x-5 \end{bmatrix} = x(x-9)^2.$$

Hence the distinct eigenvalues are 0 and 9 of multiplicities 1 and 2, respectively, so $\dim(E_0) = 1$ and $\dim(E_9) = 2$ by Theorem 6 Section 5.5 (A is diagonalizable, being symmetric). Gaussian elimination gives

$$E_0(A) = \text{span}\{\mathbf{x}_1\}, \mathbf{x}_1 = \begin{bmatrix} 1 \\ 2 \\ -2 \end{bmatrix}, \quad \text{and} \quad E_9(A) = \text{span}\left\{ \begin{bmatrix} -2 \\ 1 \\ 0 \end{bmatrix}, \begin{bmatrix} 2 \\ 0 \\ 1 \end{bmatrix} \right\}.$$

The eigenvectors in E_9 are both orthogonal to $\mathbf{x}_1$ as Theorem 4 guarantees, but not to each other. However, the Gram-Schmidt process yields and orthogonal basis

$$\{\mathbf{x}_2, \mathbf{x}_3\} \text{ of } E_9(A) \quad \text{where} \quad \mathbf{x}_2 = \begin{bmatrix} -2 \\ 1 \\ 0 \end{bmatrix} \text{ and } \mathbf{x}_3 = \begin{bmatrix} 2 \\ 4 \\ 5 \end{bmatrix}.$$

Normalizing gives orthonormal vectors $\{\frac{1}{3}\mathbf{x}_1, \frac{1}{\sqrt{5}}\mathbf{x}_2, \frac{1}{3\sqrt{5}}\mathbf{x}_3\}$, so

$$P = \begin{bmatrix} \frac{1}{3}\mathbf{x}_1 & \frac{1}{\sqrt{5}}\mathbf{x}_2 & \frac{1}{3\sqrt{5}}\mathbf{x}_3 \end{bmatrix} = \frac{1}{3\sqrt{5}} \begin{bmatrix} \sqrt{5} & -6 & 2 \\ 2\sqrt{5} & 3 & 4 \\ -2\sqrt{5} & 0 & 5 \end{bmatrix}$$

is an orthogonal matrix such that $P^{-1}AP$ is diagonal.

It is worth noting that other, more convenient, diagonalizing matrices P exist. For example, $\mathbf{y}_2 = \begin{bmatrix} 2 \\ 1 \\ 2 \end{bmatrix}$ and $\mathbf{y}_3 = \begin{bmatrix} -2 \\ 2 \\ 1 \end{bmatrix}$ lie in $E_9(A)$ and they are orthogonal. Moreover, they both have norm 3 (as does $\mathbf{x}_1$), so

$$Q = \begin{bmatrix} \frac{1}{3}\mathbf{x}_1 & \frac{1}{3}\mathbf{y}_2 & \frac{1}{3}\mathbf{y}_3 \end{bmatrix} = \frac{1}{3}\begin{bmatrix} 1 & 2 & -2 \\ 2 & 1 & 2 \\ -2 & 2 & 1 \end{bmatrix}$$

is a nicer orthogonal matrix with the property that $Q^{-1}AQ$ is diagonal.

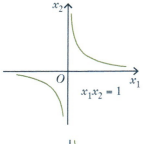

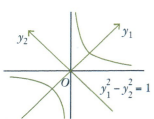

If A is symmetric and a set of orthogonal eigenvectors of A is given, the eigenvectors are called **principal axes** of A. The name comes from geometry. An expression $q = ax_1^2 + bx_1x_2 + cx_2^2$ is called a **quadratic form** in the variables x_1 and x_2, and the graph of the equation $q = 1$ is called a conic in these variables. For example, if $q = x_1x_2$, the graph of $q = 1$ is given in the first diagram.

But if we introduce new variables y_1 and y_2 by setting $x_1 = y_1 + y_2$ and $x_2 = y_1 - y_2$, then q becomes $q = y_1^2 - y_2^2$, a diagonal form with no cross term y_1y_2 (see the second diagram). Because of this, the y_1 and y_2 axes are called the principal axes for the conic (hence the name). Orthogonal diagonalization provides a systematic method for finding principal axes. Here is an illustration.

EXAMPLE 6

Find principal axes for the quadratic form $q = x_1^2 - 4x_1x_2 + x_2^2$.

Solution ▶ In order to utilize diagonalization, we first express q in matrix form. Observe that

$$q = \begin{bmatrix} x_1 & x_2 \end{bmatrix} \begin{bmatrix} 1 & -4 \\ 0 & 1 \end{bmatrix} \begin{bmatrix} x_1 \\ x_2 \end{bmatrix}.$$

The matrix here is not symmetric, but we can remedy that by writing

$$q = x_1^2 - 2x_1x_2 - 2x_2x_1 + x_2^2.$$

Then we have

$$q = \begin{bmatrix} x_1 & x_2 \end{bmatrix} \begin{bmatrix} 1 & -2 \\ -2 & 1 \end{bmatrix} \begin{bmatrix} x_1 \\ x_2 \end{bmatrix} = \mathbf{x}^T A \mathbf{x}$$

where $\mathbf{x} = \begin{bmatrix} x_1 \\ x_2 \end{bmatrix}$ and $A = \begin{bmatrix} 1 & -2 \\ -2 & 1 \end{bmatrix}$ is symmetric. The eigenvalues of A are $\lambda_1 = 3$ and $\lambda_2 = -1$, with corresponding (orthogonal) eigenvectors $\mathbf{x}_1 = \begin{bmatrix} 1 \\ -1 \end{bmatrix}$ and $\mathbf{x}_2 = \begin{bmatrix} 1 \\ 1 \end{bmatrix}$. Since $\|\mathbf{x}_1\| = \|\mathbf{x}_2\| = \sqrt{2}$, so

$$P = \frac{1}{\sqrt{2}}\begin{bmatrix} 1 & 1 \\ -1 & 1 \end{bmatrix} \text{ is orthogonal and } P^T A P = D = \begin{bmatrix} 3 & 0 \\ 0 & -1 \end{bmatrix}.$$

Now define new variables $\begin{bmatrix} y_1 \\ y_2 \end{bmatrix} = \mathbf{y}$ by $\mathbf{y} = P^T \mathbf{x}$, equivalently $\mathbf{x} = P\mathbf{y}$ (since $P^{-1} = P^T$). Hence

$$y_1 = \tfrac{1}{\sqrt{2}}(x_1 - x_2) \quad \text{and} \quad y_2 = \tfrac{1}{\sqrt{2}}(x_1 + x_2).$$

In terms of y_1 and y_2, q takes the form

$$q = \mathbf{x}^T A \mathbf{x} = (P\mathbf{y})^T A (P\mathbf{y}) = \mathbf{y}^T (P^T A P) \mathbf{y} = \mathbf{y}^T D \mathbf{y} = 3y_1^2 - y_2^2.$$

Note that $\mathbf{y} = P^T \mathbf{x}$ is obtained from $\mathbf{x}$ by a counterclockwise rotation of $\frac{\pi}{4}$ (see Theorem 6 Section 2.4).

Observe that the quadratic form q in Example 6 can be diagonalized in other ways. For example
$$q = x_1^2 - 4x_1x_2 + x_2^2 = z_1^2 - \tfrac{1}{3}z_2^2$$
where $z_1 = x_1 - 2x_2$ and $z_2 = 3x_2$. We examine this more carefully in Section 8.8.

If we are willing to replace "diagonal" by "upper triangular" in the principal axis theorem, we can weaken the requirement that A is symmetric to insisting only that A has real eigenvalues.

Theorem 5

Triangulation Theorem
If A is an $n \times n$ matrix with n real eigenvalues, an orthogonal matrix P exists such that $P^T AP$ is upper triangular.[4]

PROOF

We modify the proof of Theorem 2. If $A\mathbf{x}_1 = \lambda_1 \mathbf{x}_1$ where $\|\mathbf{x}_1\| = 1$, let $\{\mathbf{x}_1, \mathbf{x}_2, \ldots, \mathbf{x}_n\}$ be an orthonormal basis of $\mathbb{R}^n$, and let $P_1 = [\mathbf{x}_1 \ \mathbf{x}_2 \ \cdots \ \mathbf{x}_n]$. Then P_1 is orthogonal and $P_1^T AP_1 = \begin{bmatrix} \lambda_1 & B \\ 0 & A_1 \end{bmatrix}$ in block form. By induction, let $Q^T A_1 Q = T_1$ be upper triangular where Q is orthogonal of size $(n-1) \times (n-1)$. Then $P_2 = \begin{bmatrix} 1 & 0 \\ 0 & Q \end{bmatrix}$ is orthogonal, so $P = P_1 P_2$ is also orthogonal and $P^T AP = \begin{bmatrix} \lambda_1 & BQ \\ 0 & T_1 \end{bmatrix}$ is upper triangular.

The proof of Theorem 5 gives no way to construct the matrix P. However, an algorithm will be given in Section 11.1 where an improved version of Theorem 5 is presented. In a different direction, a version of Theorem 5 holds for an arbitrary matrix with complex entries (Schur's Theorem in Section 8.6).

As for a diagonal matrix, the eigenvalues of an upper triangular matrix are displayed along the main diagonal. Because A and $P^T AP$ have the same determinant and trace whenever P is orthogonal, Theorem 5 gives:

Corollary 1

If A is an $n \times n$ matrix with real eigenvalues $\lambda_1, \lambda_2, \ldots, \lambda_n$ (possibly not all distinct), then $\det A = \lambda_1 \lambda_2 \cdots \lambda_n$ *and* $\operatorname{tr} A = \lambda_1 + \lambda_2 + \cdots + \lambda_n$.

This corollary remains true even if the eigenvalues are not real (using Schur's theorem).

---

[4] There is also a lower triangular version.

EXERCISES 8.2

1. Normalize the rows to make each of the following matrices orthogonal.

   (a) $A = \begin{bmatrix} 1 & 1 \\ -1 & 1 \end{bmatrix}$ (b) $A = \begin{bmatrix} 3 & -4 \\ 4 & 3 \end{bmatrix}$

   (c) $A = \begin{bmatrix} 1 & 2 \\ -4 & 2 \end{bmatrix}$ (d) $A = \begin{bmatrix} a & b \\ -b & a \end{bmatrix}$, $(a, b) \neq (0, 0)$

   (e) $A = \begin{bmatrix} \cos\theta & -\sin\theta & 0 \\ \sin\theta & \cos\theta & 0 \\ 0 & 0 & 2 \end{bmatrix}$ (f) $A = \begin{bmatrix} 2 & 1 & -1 \\ 1 & -1 & 1 \\ 0 & 1 & 1 \end{bmatrix}$

   (g) $A = \begin{bmatrix} -1 & 2 & 2 \\ 2 & -1 & 2 \\ 2 & 2 & -1 \end{bmatrix}$ (h) $A = \begin{bmatrix} 2 & 6 & -3 \\ 3 & 2 & 6 \\ -6 & 3 & 2 \end{bmatrix}$

2. If P is a triangular orthogonal matrix, show that P is diagonal and that all diagonal entries are 1 or -1.

3. If P is orthogonal, show that kP is orthogonal if and only if $k = 1$ or $k = -1$.

4. If the first two rows of an orthogonal matrix are $(\frac{1}{3}, \frac{2}{3}, \frac{2}{3})$ and $(\frac{2}{3}, \frac{1}{3}, \frac{-2}{3})$, find all possible third rows.

5. For each matrix A, find an orthogonal matrix P such that $P^{-1}AP$ is diagonal.

   (a) $A = \begin{bmatrix} 0 & 1 \\ 1 & 0 \end{bmatrix}$ (b) $A = \begin{bmatrix} 1 & -1 \\ -1 & 1 \end{bmatrix}$

   (c) $A = \begin{bmatrix} 3 & 0 & 0 \\ 0 & 2 & 2 \\ 0 & 2 & 5 \end{bmatrix}$ (d) $A = \begin{bmatrix} 3 & 0 & 7 \\ 0 & 5 & 0 \\ 7 & 0 & 3 \end{bmatrix}$

   (e) $A = \begin{bmatrix} 1 & 1 & 0 \\ 1 & 1 & 0 \\ 0 & 0 & 2 \end{bmatrix}$ (f) $A = \begin{bmatrix} 5 & -2 & -4 \\ -2 & 8 & -2 \\ -4 & -2 & 5 \end{bmatrix}$

   (g) $A = \begin{bmatrix} 5 & 3 & 0 & 0 \\ 3 & 5 & 0 & 0 \\ 0 & 0 & 7 & 1 \\ 0 & 0 & 1 & 7 \end{bmatrix}$ (h) $A = \begin{bmatrix} 3 & 5 & -1 & 1 \\ 5 & 3 & 1 & -1 \\ -1 & 1 & 3 & 5 \\ 1 & -1 & 5 & 3 \end{bmatrix}$

6. Consider $A = \begin{bmatrix} 0 & a & 0 \\ a & 0 & c \\ 0 & c & 0 \end{bmatrix}$ where one of $a, c \neq 0$.
   Show that $c_A(x) = x(x - k)(x + k)$, where $k = \sqrt{a^2 + c^2}$ and find an orthogonal matrix P such that $P^{-1}AP$ is diagonal.

7. Consider $A = \begin{bmatrix} 0 & 0 & a \\ 0 & b & 0 \\ a & 0 & 0 \end{bmatrix}$. Show that $c_A(x) = (x - b)(x - a)(x + a)$ and find an orthogonal matrix P such that $P^{-1}AP$ is diagonal.

8. Given $A = \begin{bmatrix} b & a \\ a & b \end{bmatrix}$, show that $c_A(x) = (x - a - b)(x + a - b)$ and find an orthogonal matrix P such that $P^{-1}AP$ is diagonal.

9. Consider $A = \begin{bmatrix} b & 0 & a \\ 0 & b & 0 \\ a & 0 & b \end{bmatrix}$. Show that $c_A(x) = (x - b)(x - b - a)(x - b + a)$ and find an orthogonal matrix P such that $P^{-1}AP$ is diagonal.

10. In each case find new variables y_1 and y_2 that diagonalize the quadratic form q.

    (a) $q = x_1^2 + 6x_1x_2 + x_2^2$

    (b) $q = x_1^2 + 4x_1x_2 - 2x_2^2$

11. Show that the following are equivalent for a symmetric matrix A.

    (a) A is orthogonal.   (b) $A^2 = I$.

    (c) All eigenvalues of A are ± 1.

    [Hint: For (b) if and only if (c), use Theorem 2.]

12. We call matrices A and B **orthogonally similar** (and write $A \overset{\circ}{\sim} B$) if $B = P^T A P$ for an orthogonal matrix P.

    (a) Show that $A \overset{\circ}{\sim} A$ for all A; $A \overset{\circ}{\sim} B \Rightarrow B \overset{\circ}{\sim} A$; and $A \overset{\circ}{\sim} B$ and $B \overset{\circ}{\sim} C \Rightarrow A \overset{\circ}{\sim} C$.

    (b) Show that the following are equivalent for two symmetric matrices A and B.

    (i) A and B are similar.

    (ii) A and B are orthogonally similar.

    (iii) A and B have the same eigenvalues.

13. Assume that A and B are orthogonally similar (Exercise 12).

    (a) If A and B are invertible, show that A^{-1} and B^{-1} are orthogonally similar.

    (b) Show that A^2 and B^2 are orthogonally similar.

    (c) Show that, if A is symmetric, so is B.

14. If A is symmetric, show that every eigenvalue of A is nonnegative if and only if $A = B^2$ for some symmetric matrix B.

◆15. Prove the converse of Theorem 3:
If $(A\mathbf{x}) \cdot \mathbf{y} = \mathbf{x} \cdot (A\mathbf{y})$ for all n-columns $\mathbf{x}$ and $\mathbf{y}$, then A is symmetric.

16. Show that every eigenvalue of A is zero if and only if A is nilpotent ($A^k = 0$ for some $k \geq 1$).

17. If A has real eigenvalues, show that $A = B + C$ where B is symmetric and C is nilpotent. [Hint: Theorem 5.]

18. Let P be an orthogonal matrix.
    (a) Show that $\det P = 1$ or $\det P = -1$.
    ◆(b) Give 2×2 examples of P such that $\det P = 1$ and $\det P = -1$.
    (c) If $\det P = -1$, show that $I + P$ has no inverse. [Hint: $P^T(I + P) = (I + P)^T$.]
    ◆(d) If P is $n \times n$ and $\det P \neq (-1)^n$, show that $I - P$ has no inverse.
    [Hint: $P^T(I - P) = -(I - P)^T$.]

19. We call a square matrix E a **projection matrix** if $E^2 = E = E^T$.
    (a) If E is a projection matrix, show that $P = I - 2E$ is orthogonal and symmetric.
    (b) If P is orthogonal and symmetric, show that $E = \frac{1}{2}(I - P)$ is a projection matrix.
    (c) If U is $m \times n$ and $U^T U = I$ (for example, a unit column in $\mathbb{R}^n$), show that $E = UU^T$ is a projection matrix.

20. A matrix that we obtain from the identity matrix by writing its rows in a different order is called a **permutation matrix**. Show that every permutation matrix is orthogonal.

◆21. If the rows $\mathbf{r}_1, \ldots, \mathbf{r}_n$ of the $n \times n$ matrix $A = [a_{ij}]$ are orthogonal, show that the (i, j)-entry of A^{-1} is $\dfrac{a_{ji}}{\|\mathbf{r}_j\|^2}$.

22. (a) Let A be an $m \times n$ matrix. Show that the following are equivalent.
    (i) A has orthogonal rows.
    (ii) A can be factored as $A = DP$, where D is invertible and diagonal and P has orthonormal rows.
    (iii) AA^T is an invertible, diagonal matrix.
    (b) Show that an $n \times n$ matrix A has orthogonal rows if and only if A can be factored as $A = DP$, where P is orthogonal and D is diagonal and invertible.

23. Let A be a skew-symmetric matrix; that is, $A^T = -A$. Assume that A is an $n \times n$ matrix.
    (a) Show that $I + A$ is invertible. [Hint: By Theorem 5 Section 2.4, it suffices to show that $(I + A)\mathbf{x} = \mathbf{0}$, $\mathbf{x}$ in $\mathbb{R}^n$, implies $\mathbf{x} = \mathbf{0}$. Compute $\mathbf{x} \cdot \mathbf{x} = \mathbf{x}^T\mathbf{x}$, and use the fact that $A\mathbf{x} = -\mathbf{x}$ and $A^2\mathbf{x} = \mathbf{x}$.]
    ◆(b) Show that $P = (I - A)(I + A)^{-1}$ is orthogonal.
    (c) Show that every orthogonal matrix P such that $I + P$ is invertible arises as in part (b) from some skew-symmetric matrix A.
    [Hint: Solve $P = (I - A)(I + A)^{-1}$ for A.]

24. Show that the following are equivalent for an $n \times n$ matrix P.
    (a) P is orthogonal.
    (b) $\|P\mathbf{x}\| = \|\mathbf{x}\|$ for all columns $\mathbf{x}$ in $\mathbb{R}^n$.
    (c) $\|P\mathbf{x} - P\mathbf{y}\| = \|\mathbf{x} - \mathbf{y}\|$ for all columns $\mathbf{x}$ and $\mathbf{y}$ in $\mathbb{R}^n$.
    (d) $(P\mathbf{x}) \cdot (P\mathbf{y}) = \mathbf{x} \cdot \mathbf{y}$ for all columns $\mathbf{x}$ and $\mathbf{y}$ in $\mathbb{R}^n$.

    [Hints: For (c) $\Rightarrow$ (d), see Exercise 14(a) Section 5.3. For (d) $\Rightarrow$ (a), show that column i of P equals $P\mathbf{e}_i$, where $\mathbf{e}_i$ is column i of the identity matrix.]

25. Show that every 2×2 orthogonal matrix has the form $\begin{bmatrix} \cos\theta & -\sin\theta \\ \sin\theta & \cos\theta \end{bmatrix}$ or $\begin{bmatrix} \cos\theta & \sin\theta \\ \sin\theta & -\cos\theta \end{bmatrix}$ for some angle θ. [Hint: If $a^2 + b^2 = 1$, then $a = \cos\theta$ and $b = \sin\theta$ for some angle θ.]

26. Use Theorem 5 to show that every symmetric matrix is orthogonally diagonalizable.

SECTION 8.3 Positive Definite Matrices

All the eigenvalues of any symmetric matrix are real; this section is about the case in which the eigenvalues are positive. These matrices, which arise whenever optimization (maximum and minimum) problems are encountered, have countless applications throughout science and engineering. They also arise in statistics (for example, in factor analysis used in the social sciences) and in geometry (see Section 8.8). We will encounter them again in Chapter 10 when describing all inner products in $\mathbb{R}^n$.

Definition 8.5 *A square matrix is called **positive definite** if it is symmetric and all its eigenvalues λ are positive, that is $\lambda > 0$.*

Because these matrices are symmetric, the principal axis theorem plays a central role in the theory.

Theorem 1

If A is positive definite, then it is invertible and $\det A > 0$.

PROOF

If A is $n \times n$ and the eigenvalues are $\lambda_1, \lambda_2, \ldots, \lambda_n$, then $\det A = \lambda_1 \lambda_2 \cdots \lambda_n > 0$ by the principal axis theorem (or the corollary to Theorem 5 Section 8.2).

If $\mathbf{x}$ is a column in $\mathbb{R}^n$ and A is any real $n \times n$ matrix, we view the 1×1 matrix $\mathbf{x}^T A \mathbf{x}$ as a real number. With this convention, we have the following characterization of positive definite matrices.

Theorem 2

A symmetric matrix A is positive definite if and only if $\mathbf{x}^T A \mathbf{x} > 0$ for every column $\mathbf{x} \neq \mathbf{0}$ in $\mathbb{R}^n$.

PROOF

A is symmetric so, by the principal axis theorem, let $P^T A P = D = \text{diag}(\lambda_1, \lambda_2, \ldots, \lambda_n)$ where $P^{-1} = P^T$ and the λ_i are the eigenvalues of A. Given a column $\mathbf{x}$ in $\mathbb{R}^n$, write $\mathbf{y} = P^T \mathbf{x} = [y_1 \ y_2 \ \cdots \ y_n]^T$. Then

$$\mathbf{x}^T A \mathbf{x} > \mathbf{x}^T (PDP^T) \mathbf{x} = \mathbf{y}^T D \mathbf{y} = \lambda_1 y_1^2 + \lambda_2 y_2^2 + \cdots + \lambda_n y_n^2 \qquad (*)$$

If A is positive definite and $\mathbf{x} \neq \mathbf{0}$, then $\mathbf{x}^T A \mathbf{x} > 0$ by $(*)$ because some $y_j \neq 0$ and every $\lambda_i > 0$. Conversely, if $\mathbf{x}^T A \mathbf{x} > 0$ whenever $\mathbf{x} \neq \mathbf{0}$, let $\mathbf{x} = P \mathbf{e}_j \neq \mathbf{0}$ where $\mathbf{e}_j$ is column j of I_n. Then $\mathbf{y} = \mathbf{e}_j$, so $(*)$ reads $\lambda_j = \mathbf{x}^T A \mathbf{x} > 0$.

Note that Theorem 2 shows that the positive definite matrices are exactly the symmetric matrices A for which the quadratic form $q = \mathbf{x}^T A \mathbf{x}$ takes only positive values.

EXAMPLE 1

If U is any invertible $n \times n$ matrix, show that $A = U^T U$ is positive definite.

Solution ▶ If $\mathbf{x}$ is in $\mathbb{R}^n$ and $\mathbf{x} \neq \mathbf{0}$, then
$$\mathbf{x}^T A \mathbf{x} = \mathbf{x}^T (U^T U) \mathbf{x} = (U\mathbf{x})^T (U\mathbf{x}) = \|U\mathbf{x}\|^2 > 0$$
because $U\mathbf{x} \neq \mathbf{0}$ (U is invertible). Hence Theorem 2 applies.

It is remarkable that the converse to Example 1 is also true. In fact every positive definite matrix A can be factored as $A = U^T U$ where U is an upper triangular matrix with positive elements on the main diagonal. However, before verifying this, we introduce another concept that is central to any discussion of positive definite matrices.

If A is any $n \times n$ matrix, let $^{(r)}A$ denote the $r \times r$ submatrix in the upper left corner of A; that is, $^{(r)}A$ is the matrix obtained from A by deleting the last $n - r$ rows and columns. The matrices $^{(1)}A$, $^{(2)}A$, $^{(3)}A$, ..., $^{(n)}A = A$ are called the **principal submatrices** of A.

EXAMPLE 2

If $A = \begin{bmatrix} 10 & 5 & 2 \\ 5 & 3 & 2 \\ 2 & 2 & 3 \end{bmatrix}$ then $^{(1)}A = [10]$, $^{(2)}A = \begin{bmatrix} 10 & 5 \\ 5 & 3 \end{bmatrix}$ and $^{(3)}A = A$.

Lemma 1

If A is positive definite, so is each principal submatrix $^{(r)}A$ for $r = 1, 2, \ldots, n$.

PROOF

Write $A = \begin{bmatrix} ^{(r)}A & P \\ Q & R \end{bmatrix}$ in block form. If $\mathbf{y} \neq \mathbf{0}$ in $\mathbb{R}^r$, write $\mathbf{x} = \begin{bmatrix} \mathbf{y} \\ \mathbf{0} \end{bmatrix}$ in $\mathbb{R}^n$.

Then $\mathbf{x} \neq \mathbf{0}$, so the fact that A is positive definite gives
$$0 < \mathbf{x}^T A \mathbf{x} = [\mathbf{y}^T \ \mathbf{0}] \begin{bmatrix} ^{(r)}A & P \\ Q & R \end{bmatrix} \begin{bmatrix} \mathbf{y} \\ \mathbf{0} \end{bmatrix} = \mathbf{y}^T (^{(r)}A) \mathbf{y}$$

This shows that $^{(r)}A$ is positive definite by Theorem 2.[5]

If A is positive definite, Lemma 1 and Theorem 1 show that $\det(^{(r)}A) > 0$ for every r. This proves part of the following theorem which contains the converse to Example 1, and characterizes the positive definite matrices among the symmetric ones.

---

[5] A similar argument shows that, if B is any matrix obtained from a positive definite matrix A by deleting certain rows and deleting the *same* columns, then B is also positive definite.

SECTION 8.3 Positive Definite Matrices

Theorem 3

The following conditions are equivalent for a symmetric $n \times n$ matrix A:

1. *A is positive definite.*
2. *$\det(^{(r)}A) > 0$ for each $r = 1, 2, \ldots, n$.*
3. *$A = U^T U$ where U is an upper triangular matrix with positive entries on the main diagonal.*

*Furthermore, the factorization in (3) is unique (called the **Cholesky factorization**[6] of A).*

PROOF

First, (3) $\Rightarrow$ (1) by Example 1, and (1) $\Rightarrow$ (2) by Lemma 1 and Theorem 1.

(2) $\Rightarrow$ (3). Assume (2) and proceed by induction on n. If $n = 1$, then $A = [a]$ where $a > 0$ by (2), so take $U = [\sqrt{a}]$. If $n > 1$, write $B = {}^{(n-1)}A$. Then B is symmetric and satisfies (2) so, by induction, we have $B = U^T U$ as in (3) where U is of size $(n-1) \times (n-1)$. Then, as A is symmetric, it has block form
$$A = \begin{bmatrix} B & \mathbf{p} \\ \mathbf{p}^T & b \end{bmatrix}.$$
where $\mathbf{p}$ is a column in $\mathbb{R}^{n-1}$ and b is in $\mathbb{R}$. If we write $\mathbf{x} = (U^T)^{-1}\mathbf{p}$ and $c = b - \mathbf{x}^T\mathbf{x}$, block multiplication gives
$$A = \begin{bmatrix} U^T U & \mathbf{p} \\ \mathbf{p}^T & b \end{bmatrix} = \begin{bmatrix} U^T & 0 \\ \mathbf{x}^T & 1 \end{bmatrix} \begin{bmatrix} U & \mathbf{x} \\ 0 & c \end{bmatrix}$$
as the reader can verify. Taking determinants and applying Theorem 5 Section 3.1 gives $\det A = \det(U^T) \det U \cdot c = c(\det U)^2$. Hence $c > 0$ because $\det A > 0$ by (2), so the above factorization can be written $A = \begin{bmatrix} U^T & 0 \\ \mathbf{x}^T & \sqrt{c} \end{bmatrix} \begin{bmatrix} U & \mathbf{x} \\ 0 & \sqrt{c} \end{bmatrix}$. Since U has positive diagonal entries, this proves (3).

As to the uniqueness, suppose that $A = U^T U = U_1^T U_1$ are two Cholesky factorizations. Write $D = UU_1^{-1} = (U^T)^{-1}U_1^T$. Then D is upper triangular, because $D = UU_1^{-1}$, and lower triangular, because $D = (U^T)^{-1}U_1^T$, and so it is a diagonal matrix. Thus $U = DU_1$ and $U_1 = DU$, so it suffices to show that $D = I$. But eliminating U_1 gives $U = D^2 U$, so $D^2 = I$ because U is invertible. Since the diagonal entries of D are positive (this is true of U and U_1), it follows that $D = I$.

The remarkable thing is that the matrix U in the Cholesky factorization is easy to obtain from A using row operations. The key is that Step 1 of the following algorithm is *possible* for any positive definite matrix A. A proof of the algorithm is given following Example 3.

---

6   Andre-Louis Cholesky (1875–1918), was a French mathematician who died in World War I. His factorization was published in 1924 by a fellow officer.

Algorithm for the Cholesky Factorization

If A is a positive definite matrix, the Cholesky factorization $A = U^T U$ can be obtained as follows:

Step 1. Carry A to an upper triangular matrix U_1 with positive diagonal entries using row operations each of which adds a multiple of a row to a lower row.

Step 2. Obtain U from U_1 by dividing each row of U_1 by the square root of the diagonal entry in that row.

EXAMPLE 3

Find the Cholesky factorization of $A = \begin{bmatrix} 10 & 5 & 2 \\ 5 & 3 & 2 \\ 2 & 2 & 3 \end{bmatrix}$.

Solution ▶ The matrix A is positive definite by Theorem 3 because $\det{}^{(1)}A = 10 > 0$, $\det{}^{(2)}A = 5 > 0$, and $\det{}^{(3)}A = \det A = 3 > 0$. Hence Step 1 of the algorithm is carried out as follows:

$$A = \begin{bmatrix} 10 & 5 & 2 \\ 5 & 3 & 2 \\ 2 & 2 & 3 \end{bmatrix} \rightarrow \begin{bmatrix} 10 & 5 & 2 \\ 0 & \frac{1}{2} & 1 \\ 0 & 1 & \frac{13}{5} \end{bmatrix} \rightarrow \begin{bmatrix} 10 & 5 & 2 \\ 0 & \frac{1}{2} & 1 \\ 0 & 0 & \frac{3}{5} \end{bmatrix} = U_1$$

Now carry out Step 2 on U_1 to obtain $U = \begin{bmatrix} \sqrt{10} & \frac{5}{\sqrt{10}} & \frac{2}{\sqrt{10}} \\ 0 & \frac{1}{\sqrt{2}} & \sqrt{2} \\ 0 & 0 & \frac{\sqrt{3}}{\sqrt{5}} \end{bmatrix}$.

The reader can verify that $U^T U = A$.

PROOF OF THE CHOLESKY ALGORITHM

If A is positive definite, let $A = U^T U$ be the Cholesky factorization, and let $D = \text{diag}(d_1, \ldots, d_n)$ be the common diagonal of U and U^T. Then $U^T D^{-1}$ is lower triangular with ones on the diagonal (call such matrices LT-1). Hence $L = (U^T D^{-1})^{-1}$ is also LT-1, and so $I_n \rightarrow L$ by a sequence of row operations each of which adds a multiple of a row to a lower row (verify; modify columns right to left). But then $A \rightarrow LA$ by the same sequence of row operations (see the discussion preceding Theorem 1 Section 2.5). Since $LA = [D(U^T)^{-1}][U^T U] = DU$ is upper triangular with positive entries on the diagonal, this shows that Step 1 of the algorithm is possible.

Turning to Step 2, let $A \rightarrow U_1$ as in Step 1 so that $U_1 = L_1 A$ where L_1 is LT-1. Since A is symmetric, we get

$$L_1 U_1^T = L_1(L_1 A)^T = L_1 A^T L_1^T = L_1 A L_1^T = U_1 L_1^T \qquad (*)$$

Let $D_1 = \text{diag}(e_1, ..., e_n)$ denote the diagonal of U_1. Then $(*)$ gives $L_1(U_1^T D_1^{-1}) = U_1 L_1^T D_1^{-1}$. This is both upper triangular (right side) and LT-1 (left side), and so must equal I_n. In particular, $U_1^T D_1^{-1} = L_1^{-1}$. Now let $D_2 = \text{diag}(\sqrt{e_1}, ..., \sqrt{e_n})$, so that $D_2^2 = D_1$. If we write $U = D_2^{-1} U_1$ we have

$$U^T U = (U_1^T D_2^{-1})(D_2^{-1} U_1) = U_1^T (D_2^2)^{-1} U_1 = (U_1^T D_1^{-1}) U_1 = (L_1^{-1}) U_1 = A$$

This proves Step 2 because $U = D_2^{-1} U_1$ is formed by dividing each row of U_1 by the square root of its diagonal entry (verify).

EXERCISES 8.3

1. Find the Cholesky decomposition of each of the following matrices.

   (a) $\begin{bmatrix} 4 & 3 \\ 3 & 5 \end{bmatrix}$

   ◆(b) $\begin{bmatrix} 2 & -1 \\ -1 & 1 \end{bmatrix}$

   (c) $\begin{bmatrix} 12 & 4 & 3 \\ 4 & 2 & -1 \\ 3 & -1 & 7 \end{bmatrix}$

   ◆(d) $\begin{bmatrix} 20 & 4 & 5 \\ 4 & 2 & 3 \\ 5 & 3 & 5 \end{bmatrix}$

2. (a) If A is positive definite, show that A^k is positive definite for all $k \geq 1$.

   ◆(b) Prove the converse to (a) when k is odd.

   (c) Find a symmetric matrix A such that A^2 is positive definite but A is not.

3. Let $A = \begin{bmatrix} 1 & a \\ a & b \end{bmatrix}$. If $a^2 < b$, show that A is positive definite and find the Cholesky factorization.

◆4. If A and B are positive definite and $r > 0$, show that $A + B$ and rA are both positive definite.

5. If A and B are positive definite, show that $\begin{bmatrix} A & 0 \\ 0 & B \end{bmatrix}$ is positive definite.

◆6. If A is an $n \times n$ positive definite matrix and U is an $n \times m$ matrix of rank m, show that $U^T A U$ is positive definite.

7. If A is positive definite, show that each diagonal entry is positive.

8. Let A_0 be formed from A by deleting rows 2 and 4 and deleting columns 2 and 4. If A is positive definite, show that A_0 is positive definite.

9. If A is positive definite, show that $A = CC^T$ where C has orthogonal columns.

◆10. If A is positive definite, show that $A = C^2$ where C is positive definite.

11. Let A be a positive definite matrix. If a is a real number, show that aA is positive definite if and only if $a > 0$.

12. (a) Suppose an invertible matrix A can be factored in $\mathbf{M}_{nn}$ as $A = LDU$ where L is lower triangular with 1s on the diagonal, U is upper triangular with 1s on the diagonal, and D is diagonal with positive diagonal entries. Show that the factorization is unique: If $A = L_1 D_1 U_1$ is another such factorization, show that $L_1 = L$, $D_1 = D$, and $U_1 = U$.

    ◆(b) Show that a matrix A is positive definite if and only if A is symmetric and admits a factorization $A = LDU$ as in (a).

13. Let A be positive definite and write $d_r = \det {}^{(r)}A$ for each $r = 1, 2, ..., n$. If U is the upper triangular matrix obtained in step 1 of the algorithm, show that the diagonal elements $u_{11}, u_{22}, ..., u_{nn}$ of U are given by $u_{11} = d_1$, $u_{jj} = d_j/d_{j-1}$ if $j > 1$. [Hint: If $LA = U$ where L is lower triangular with 1s on the diagonal, use block multiplication to show that $\det {}^{(r)}A = \det {}^{(r)}U$ for each r.]

SECTION 8.4 QR-Factorization[7]

One of the main virtues of orthogonal matrices is that they can be easily inverted—the transpose is the inverse. This fact, combined with the factorization theorem in this section, provides a useful way to simplify many matrix calculations (for example, in least squares approximation).

Definition 8.6

*Let A be an $m \times n$ matrix with independent columns. A **QR-factorization** of A expresses it as $A = QR$ where Q is $m \times n$ with orthonormal columns and R is an invertible and upper triangular matrix with positive diagonal entries.*

The importance of the factorization lies in the fact that there are computer algorithms that accomplish it with good control over round-off error, making it particularly useful in matrix calculations. The factorization is a matrix version of the Gram-Schmidt process.

Suppose $A = [\mathbf{c}_1 \ \mathbf{c}_2 \ \cdots \ \mathbf{c}_n]$ is an $m \times n$ matrix with linearly independent columns $\mathbf{c}_1, \mathbf{c}_2, \ldots, \mathbf{c}_n$. The Gram-Schmidt algorithm can be applied to these columns to provide orthogonal columns $\mathbf{f}_1, \mathbf{f}_2, \ldots, \mathbf{f}_n$ where $\mathbf{f}_1 = \mathbf{c}_1$ and

$$\mathbf{f}_k = \mathbf{c}_k - \frac{\mathbf{c}_k \cdot \mathbf{f}_1}{\|\mathbf{f}_1\|^2}\mathbf{f}_1 + \frac{\mathbf{c}_k \cdot \mathbf{f}_2}{\|\mathbf{f}_2\|^2}\mathbf{f}_2 - \cdots - \frac{\mathbf{c}_k \cdot \mathbf{f}_{k-1}}{\|\mathbf{f}_{k-1}\|^2}\mathbf{f}_{k-1}$$

for each $k = 2, 3, \ldots, n$. Now write $\mathbf{q}_k = \frac{1}{\|\mathbf{f}_k\|}\mathbf{f}_k$ for each k. Then $\mathbf{q}_1, \mathbf{q}_2, \ldots, \mathbf{q}_n$ are orthonormal columns, and the above equation becomes

$$\|\mathbf{f}_k\|\mathbf{q}_k = \mathbf{c}_k - (\mathbf{c}_k \cdot \mathbf{q}_1)\mathbf{q}_1 - (\mathbf{c}_k \cdot \mathbf{q}_2)\mathbf{q}_2 - \cdots - (\mathbf{c}_k \cdot \mathbf{q}_{k-1})\mathbf{q}_{k-1}$$

Using these equations, express each $\mathbf{c}_k$ as a linear combination of the $\mathbf{q}_i$:

$$\mathbf{c}_1 = \|\mathbf{f}_1\|\mathbf{q}_1$$
$$\mathbf{c}_2 = (\mathbf{c}_2 \cdot \mathbf{q}_1)\mathbf{q}_1 + \|\mathbf{f}_2\|\mathbf{q}_2$$
$$\mathbf{c}_3 = (\mathbf{c}_3 \cdot \mathbf{q}_1)\mathbf{q}_1 + (\mathbf{c}_3 \cdot \mathbf{q}_2)\mathbf{q}_2 + \|\mathbf{f}_3\|\mathbf{q}_3$$
$$\vdots$$
$$\mathbf{c}_n = (\mathbf{c}_n \cdot \mathbf{q}_1)\mathbf{q}_1 + (\mathbf{c}_n \cdot \mathbf{q}_2)\mathbf{q}_2 + (\mathbf{c}_n \cdot \mathbf{q}_3)\mathbf{q}_3 + \cdots + \|\mathbf{f}_n\|\mathbf{q}_n$$

These equations have a matrix form that gives the required factorization:

$$A = [\mathbf{c}_1 \ \mathbf{c}_2 \ \mathbf{c}_3 \ \cdots \ \mathbf{c}_n]$$

$$= [\mathbf{q}_1 \ \mathbf{q}_2 \ \mathbf{q}_3 \ \cdots \ \mathbf{q}_n] \begin{bmatrix} \|\mathbf{f}_1\| & \mathbf{c}_2 \cdot \mathbf{q}_1 & \mathbf{c}_3 \cdot \mathbf{q}_1 & \cdots & \mathbf{c}_n \cdot \mathbf{q}_1 \\ 0 & \|\mathbf{f}_2\| & \mathbf{c}_3 \cdot \mathbf{q}_2 & \cdots & \mathbf{c}_n \cdot \mathbf{q}_2 \\ 0 & 0 & \|\mathbf{f}_3\| & \cdots & \mathbf{c}_n \cdot \mathbf{q}_3 \\ \vdots & \vdots & \vdots & \ddots & \vdots \\ 0 & 0 & 0 & \cdots & \|\mathbf{f}_n\| \end{bmatrix} \quad (*)$$

Here the first factor $Q = [\mathbf{q}_1 \ \mathbf{q}_2 \ \mathbf{q}_3 \ \cdots \ \mathbf{q}_n]$ has orthonormal columns, and the second factor is an $n \times n$ upper triangular matrix R with positive diagonal entries (and so is invertible). We record this in the following theorem.

Theorem 1

QR-Factorization

Every $m \times n$ matrix A with linearly independent columns has a QR-factorization $A = QR$ where Q has orthonormal columns and R is upper triangular with positive diagonal entries.

---

[7] This section is not used elsewhere in this book.

SECTION 8.4 QR-Factorization

The matrices Q and R in Theorem 1 are uniquely determined by A; we return to this below.

EXAMPLE 1

Find the QR-factorization of $A = \begin{bmatrix} 1 & 1 & 0 \\ -1 & 0 & 1 \\ 0 & 1 & 1 \\ 0 & 0 & 1 \end{bmatrix}$.

Solution ▶ Denote the columns of A as c_1, c_2, and c_3, and observe that $\{c_1, c_2, c_3\}$ is independent. If we apply the Gram-Schmidt algorithm to these columns, the result is:

$$f_1 = c_1 = \begin{bmatrix} 1 \\ -1 \\ 0 \\ 0 \end{bmatrix}, \quad f_2 = c_2 - \tfrac{1}{2}f_1 = \begin{bmatrix} \tfrac{1}{2} \\ \tfrac{1}{2} \\ 1 \\ 0 \end{bmatrix}, \quad \text{and} \quad f_3 = c_3 + \tfrac{1}{2}f_1 - f_2 = \begin{bmatrix} 0 \\ 0 \\ 0 \\ 1 \end{bmatrix}.$$

Write $q_j = \dfrac{1}{\|f_j\|^2} f_j$ for each j, so $\{q_1, q_2, q_3\}$ is orthonormal. Then equation (*) preceding Theorem 1 gives $A = QR$ where

$$Q = [q_1 \ q_2 \ q_3] = \begin{bmatrix} \tfrac{1}{\sqrt{2}} & \tfrac{1}{\sqrt{6}} & 0 \\ \tfrac{-1}{\sqrt{2}} & \tfrac{1}{\sqrt{6}} & 0 \\ 0 & \tfrac{2}{\sqrt{6}} & 0 \\ 0 & 0 & 1 \end{bmatrix} = \tfrac{1}{\sqrt{6}} \begin{bmatrix} \sqrt{3} & 1 & 0 \\ -\sqrt{3} & 1 & 0 \\ 0 & 2 & 0 \\ 0 & 0 & \sqrt{6} \end{bmatrix}$$

$$R = \begin{bmatrix} \|f_1\| & c_2 \cdot q_1 & c_3 \cdot q_1 \\ 0 & \|f_2\| & c_3 \cdot q_2 \\ 0 & 0 & \|f_3\| \end{bmatrix} = \begin{bmatrix} \sqrt{2} & \tfrac{1}{\sqrt{2}} & \tfrac{-1}{\sqrt{2}} \\ 0 & \tfrac{\sqrt{3}}{\sqrt{2}} & \tfrac{\sqrt{3}}{\sqrt{2}} \\ 0 & 0 & 1 \end{bmatrix} = \tfrac{1}{\sqrt{2}} \begin{bmatrix} 2 & 1 & -1 \\ 0 & \sqrt{3} & \sqrt{3} \\ 0 & 0 & \sqrt{2} \end{bmatrix}$$

The reader can verify that indeed $A = QR$.

If a matrix A has independent rows and we apply QR-factorization to A^T, the result is:

Corollary 1

If A has independent rows, then A factors uniquely as $A = LP$ where P has orthonormal rows and L is an invertible lower triangular matrix with positive main diagonal entries.

Since a square matrix with orthonormal columns is orthogonal, we have

Theorem 2

Every square, invertible matrix A has factorizations $A = QR$ and $A = LP$ where Q and P are orthogonal, R is upper triangular with positive diagonal entries, and L is lower triangular with positive diagonal entries.

Remark

In Section 5.6 we found how to find a best approximation $\mathbf{z}$ to a solution of a (possibly inconsistent) system $A\mathbf{x} = \mathbf{b}$ of linear equations: take $\mathbf{z}$ to be any solution of the "normal" equations $(A^TA)\mathbf{z} = A^T\mathbf{b}$. If A has independent columns this $\mathbf{z}$ is unique (A^TA is invertible by Theorem 3 Section 5.4), so it is often desirable to compute $(A^TA)^{-1}$. This is particularly useful in least squares approximation (Section 5.6). This is simplified if we have a QR-factorization of A (and is one of the main reasons for the importance of Theorem 1). For if $A = QR$ is such a factorization, then $Q^TQ = I_n$ because Q has orthonormal columns (verify), so we obtain

$$A^TA = R^TQ^TQR = R^TR.$$

Hence computing $(A^TA)^{-1}$ amounts to finding R^{-1}, and this is a routine matter because R is upper triangular. Thus the difficulty in computing $(A^TA)^{-1}$ lies in obtaining the QR-factorization of A.

We conclude by proving the uniqueness of the QR-factorization.

Theorem 3

Let A be an $m \times n$ matrix with independent columns. If $A = QR$ and $A = Q_1R_1$ are QR-factorizations of A, then $Q_1 = Q$ and $R_1 = R$.

PROOF

Write $Q = [\mathbf{c}_1 \ \mathbf{c}_2 \ \cdots \ \mathbf{c}_n]$ and $Q_1 = [\mathbf{d}_1 \ \mathbf{d}_2 \ \cdots \ \mathbf{d}_n]$ in terms of their columns, and observe first that $Q^TQ = I_n = Q_1^TQ_1$ because Q and Q_1 have orthonormal columns. Hence it suffices to show that $Q_1 = Q$ (then $R_1 = Q_1^TA = Q^TA = R$). Since $Q_1^TQ_1 = I_n$, the equation $QR = Q_1R_1$ gives $Q_1^TQ = R_1R^{-1}$; for convenience we write this matrix as

$$Q_1^TQ = R_1R^{-1} = [t_{ij}].$$

This matrix is upper triangular with positive diagonal elements (since this is true for R and R_1), so $t_{ii} > 0$ for each i and $t_{ij} = 0$ if $i > j$. On the other hand, the (i, j)-entry of Q_1^TQ is $\mathbf{d}_i^T\mathbf{c}_j = \mathbf{d}_i \cdot \mathbf{c}_j$, so we have $\mathbf{d}_i \cdot \mathbf{c}_j = t_{ij}$ for all i and j. But each $\mathbf{c}_j$ is in span$\{\mathbf{d}_1, \mathbf{d}_2, \ldots, \mathbf{d}_n\}$ because $Q = Q_1(R_1R^{-1})$. Hence the expansion theorem gives

$$\mathbf{c}_j = (\mathbf{d}_1 \cdot \mathbf{c}_j)\mathbf{d}_1 + (\mathbf{d}_2 \cdot \mathbf{c}_j)\mathbf{d}_2 + \cdots + (\mathbf{d}_n \cdot \mathbf{c}_j)\mathbf{d}_n = t_{1j}\mathbf{d}_1 + t_{2j}\mathbf{d}_2 + \cdots + t_{jj}\mathbf{d}_j$$

because $\mathbf{d}_i \cdot \mathbf{c}_j = t_{ij} = 0$ if $i > j$. The first few equations here are

$$\mathbf{c}_1 = t_{11}\mathbf{d}_1$$
$$\mathbf{c}_2 = t_{12}\mathbf{d}_1 + t_{22}\mathbf{d}_2$$
$$\mathbf{c}_3 = t_{13}\mathbf{d}_1 + t_{23}\mathbf{d}_2 + t_{33}\mathbf{d}_3$$
$$\mathbf{c}_4 = t_{14}\mathbf{d}_1 + t_{24}\mathbf{d}_2 + t_{34}\mathbf{d}_3 + t_{44}\mathbf{d}_4$$
$$\vdots \qquad \vdots$$

The first of these equations gives $1 = \|\mathbf{c}_1\| = \|t_{11}\mathbf{d}_1\| = |t_{11}|\|\mathbf{d}_1\| = t_{11}$, whence $\mathbf{c}_1 = \mathbf{d}_1$. But then $t_{12} = \mathbf{d}_1 \cdot \mathbf{c}_2 = \mathbf{c}_1 \cdot \mathbf{c}_2 = 0$, so the second equation becomes $\mathbf{c}_2 = t_{22}\mathbf{d}_2$. Now a similar argument gives $\mathbf{c}_2 = \mathbf{d}_2$, and then $t_{13} = 0$ and $t_{23} = 0$ follows in the same way. Hence $\mathbf{c}_3 = t_{33}\mathbf{d}_3$ and $\mathbf{c}_3 = \mathbf{d}_3$. Continue in this way to get $\mathbf{c}_i = \mathbf{d}_i$ for all i. This means that $Q_1 = Q$, which is what we wanted.

EXERCISES 8.4

1. In each case find the QR-factorization of A.

   (a) $A = \begin{bmatrix} 1 & -1 \\ -1 & 0 \end{bmatrix}$ ◆(b) $A = \begin{bmatrix} 2 & 1 \\ 1 & 1 \end{bmatrix}$

   (c) $A = \begin{bmatrix} 1 & 1 & 1 \\ 1 & 1 & 0 \\ 1 & 0 & 0 \\ 0 & 0 & 0 \end{bmatrix}$ ◆(d) $A = \begin{bmatrix} 1 & 1 & 0 \\ -1 & 0 & 1 \\ 0 & 1 & 1 \\ 1 & -1 & 0 \end{bmatrix}$

2. Let A and B denote matrices.

   (a) If A and B have independent columns, show that AB has independent columns. [*Hint*: Theorem 3 Section 5.4.]

   ◆(b) Show that A has a QR-factorization if and only if A has independent columns.

   (c) If AB has a QR-factorization, show that the same is true of B but not necessarily A.
   [*Hint*: Consider AA^T where $A = \begin{bmatrix} 1 & 0 & 0 \\ 1 & 1 & 1 \end{bmatrix}$.]

3. If R is upper triangular and invertible, show that there exists a diagonal matrix D with diagonal entries ± 1 such that $R_1 = DR$ is invertible, upper triangular, and has positive diagonal entries.

4. If A has independent columns, let $A = QR$ where Q has orthonormal columns and R is invertible and upper triangular. [Some authors call *this* a QR-factorization of A.] Show that there is a diagonal matrix D with diagonal entries ± 1 such that $A = (QD)(DR)$ is the QR-factorization of A. [*Hint*: Preceding exercise.]

SECTION 8.5 Computing Eigenvalues

In practice, the problem of finding eigenvalues of a matrix is virtually never solved by finding the roots of the characteristic polynomial. This is difficult for large matrices and iterative methods are much better. Two such methods are described briefly in this section.

The Power Method

In Chapter 3 our initial rationale for diagonalizing matrices was to be able to compute the powers of a square matrix, and the eigenvalues were needed to do this. In this section, we are interested in efficiently computing eigenvalues, and it may come as no surprise that the first method we discuss uses the powers of a matrix.

Recall that an eigenvalue λ of an $n \times n$ matrix A is called a **dominant eigenvalue** if λ has multiplicity 1, and

$$|\lambda| > |\mu| \quad \text{for all eigenvalues } \mu \neq \lambda.$$

Any corresponding eigenvector is called a **dominant eigenvector** of A. When such an eigenvalue exists, one technique for finding it is as follows: Let $\mathbf{x}_0$ in $\mathbb{R}^n$ be a first approximation to a dominant eigenvector λ, and compute successive approximations $\mathbf{x}_1, \mathbf{x}_2, \ldots$ as follows:

$$\mathbf{x}_1 = A\mathbf{x}_0 \quad \mathbf{x}_2 = A\mathbf{x}_1 \quad \mathbf{x}_3 = A\mathbf{x}_2 \quad \cdots$$

In general, we define

$$\mathbf{x}_{k+1} = A\mathbf{x}_k \quad \text{for each } k \geq 0.$$

If the first estimate $\mathbf{x}_0$ is good enough, these vectors $\mathbf{x}_n$ will approximate the dominant eigenvector λ (see below). This technique is called the **power method** (because $\mathbf{x}_k = A^k \mathbf{x}_0$ for each $k \geq 1$). Observe that if $\mathbf{z}$ is any eigenvector corresponding to λ, then

$$\frac{\mathbf{z} \cdot (A\mathbf{z})}{\|\mathbf{z}\|^2} = \frac{\mathbf{z} \cdot (\lambda \mathbf{z})}{\|\mathbf{z}\|^2} = \lambda.$$

Because the vectors $x_1, x_2, \ldots, x_n, \ldots$ approximate dominant eigenvectors, this suggests that we define the **Rayleigh quotients** as follows:

$$r_k = \frac{x_k \cdot x_{k+1}}{\|x_k\|^2} \quad \text{for } k \geq 1.$$

Then the numbers r_k approximate the dominant eigenvalue λ.

EXAMPLE 1

Use the power method to approximate a dominant eigenvector and eigenvalue of $A = \begin{bmatrix} 1 & 1 \\ 2 & 0 \end{bmatrix}$.

Solution ▶ The eigenvalues of A are 2 and -1, with eigenvectors $\begin{bmatrix} 1 \\ 1 \end{bmatrix}$ and $\begin{bmatrix} 1 \\ -2 \end{bmatrix}$. Take $x_0 = \begin{bmatrix} 1 \\ 0 \end{bmatrix}$ as the first approximation and compute $x_1, x_2, \ldots$, successively, from $x_1 = Ax_0$, $x_2 = Ax_1$, ... The result is

$$x_1 = \begin{bmatrix} 1 \\ 2 \end{bmatrix}, \quad x_2 = \begin{bmatrix} 3 \\ 2 \end{bmatrix}, \quad x_3 = \begin{bmatrix} 5 \\ 6 \end{bmatrix}, \quad x_4 = \begin{bmatrix} 11 \\ 10 \end{bmatrix}, \quad x_5 = \begin{bmatrix} 21 \\ 22 \end{bmatrix}, \ldots$$

These vectors are approaching scalar multiples of the dominant eigenvector $\begin{bmatrix} 1 \\ 1 \end{bmatrix}$. Moreover, the Rayleigh quotients are

$$r_1 = \tfrac{7}{5}, \; r_2 = \tfrac{27}{13}, \; r_3 = \tfrac{115}{61}, \; r_4 = \tfrac{451}{221}, \ldots$$

and these are approaching the dominant eigenvalue 2.

To see why the power method works, let $\lambda_1, \lambda_2, \ldots, \lambda_m$ be eigenvalues of A with λ_1 dominant and let $y_1, y_2, \ldots, y_m$ be corresponding eigenvectors. What is required is that the first approximation x_0 be a linear combination of these eigenvectors:

$$x_0 = a_1 y_1 + a_2 y_2 + \cdots + a_m y_m \quad \text{with } a_1 \neq 0$$

If $k \geq 1$, the fact that $x_k = A^k x_0$ and $A^k y_i = \lambda_i^k y_i$ for each i gives

$$x_k = a_1 \lambda_1^k y_1 + a_2 \lambda_2^k y_2 + \cdots + a_m \lambda_m^k y_m \quad \text{for } k \geq 1$$

Hence

$$\frac{1}{\lambda_1^k} x_k = a_1 y_1 + a_2 \left(\frac{\lambda_2}{\lambda_1}\right)^k y_2 + \cdots + a_m \left(\frac{\lambda_m}{\lambda_1}\right)^k y_m$$

The right side approaches $a_1 y_1$ as k increases because λ_1 is dominant $\left(\left|\frac{\lambda_i}{\lambda_1}\right| < 1 \text{ for each } i > 1\right)$. Because $a_1 \neq 0$, this means that x_k approximates the dominant eigenvector $a_1 \lambda_1^k y_1$.

The power method requires that the first approximation x_0 be a linear combination of eigenvectors. (In Example 1 the eigenvectors form a basis of $\mathbb{R}^2$.) But even in this case the method fails if $a_1 = 0$, where a_1 is the coefficient of the dominant eigenvector (try $x_0 = \begin{bmatrix} -1 \\ 2 \end{bmatrix}$ in Example 1). In general, the rate of convergence is quite slow if any of the ratios $\left|\frac{\lambda_i}{\lambda_1}\right|$ is near 1. Also, because the method requires repeated multiplications by A, it is not recommended unless these multiplications are easy to carry out (for example, if most of the entries of A are zero).

QR-Algorithm

A much better method for approximating the eigenvalues of an invertible matrix A depends on the factorization (using the Gram-Schmidt algorithm) of A in the form

$$A = QR$$

where Q is orthogonal and R is invertible and upper triangular (see Theorem 2 Section 8.4). The **QR-algorithm** uses this repeatedly to create a sequence of matrices $A_1 = A, A_2, A_3, \ldots$, as follows:

1. Define $A_1 = A$ and factor it as $A_1 = Q_1 R_1$.
2. Define $A_2 = R_1 Q_1$ and factor it as $A_2 = Q_2 R_2$.
3. Define $A_3 = R_2 Q_2$ and factor it as $A_3 = Q_3 R_3$.

   $\vdots$

In general, A_k is factored as $A_k = Q_k R_k$ and we define $A_{k+1} = R_k Q_k$. Then A_{k+1} is similar to A_k [in fact, $A_{k+1} = R_k Q_k = (Q_k^{-1} A_k) Q_k$], and hence each A_k has the same eigenvalues as A. If the eigenvalues of A are real and have distinct absolute values, the remarkable thing is that the sequence of matrices $A_1, A_2, A_3, \ldots$ converges to an upper triangular matrix with these eigenvalues on the main diagonal. [See below for the case of complex eigenvalues.]

EXAMPLE 2

If $A = \begin{bmatrix} 1 & 1 \\ 2 & 0 \end{bmatrix}$ as in Example 1, use the QR-algorithm to approximate the eigenvalues.

Solution ▶ The matrices A_1, A_2, and A_3 are as follows:

$$A_1 = \begin{bmatrix} 1 & 1 \\ 2 & 0 \end{bmatrix} = Q_1 R_1 \quad \text{where } Q_1 = \tfrac{1}{\sqrt{5}} \begin{bmatrix} 1 & 2 \\ 2 & -1 \end{bmatrix} \text{ and } R_1 = \tfrac{1}{\sqrt{5}} \begin{bmatrix} 5 & 1 \\ 0 & 2 \end{bmatrix}$$

$$A_2 = \tfrac{1}{5} \begin{bmatrix} 7 & 9 \\ 4 & -2 \end{bmatrix} = \begin{bmatrix} 1.4 & -1.8 \\ -0.8 & -0.4 \end{bmatrix} = Q_2 R_2$$

$$\text{where } Q_2 = \tfrac{1}{\sqrt{65}} \begin{bmatrix} 7 & 4 \\ 4 & -7 \end{bmatrix} \text{ and } R_2 = \tfrac{1}{\sqrt{65}} \begin{bmatrix} 13 & 11 \\ 0 & 10 \end{bmatrix}$$

$$A_3 = \tfrac{1}{13} \begin{bmatrix} 27 & -5 \\ 8 & -14 \end{bmatrix} = \begin{bmatrix} 2.08 & -0.38 \\ 0.62 & -1.08 \end{bmatrix}$$

This is converging to $\begin{bmatrix} 2 & * \\ 0 & -1 \end{bmatrix}$ and so is approximating the eigenvalues 2 and -1 on the main diagonal.

It is beyond the scope of this book to pursue a detailed discussion of these methods. The reader is referred to J. M. Wilkinson, *The Algebraic Eigenvalue Problem* (Oxford, England: Oxford University Press, 1965) or G. W. Stewart, *Introduction to Matrix Computations* (New York: Academic Press, 1973). We conclude with some remarks on the QR-algorithm.

Shifting. Convergence is accelerated if, at stage k of the algorithm, a number s_k is chosen and $A_k - s_k I$ is factored in the form $Q_k R_k$ rather than A_k itself. Then

$$Q_k^{-1} A_k Q_k = Q_k^{-1}(Q_k R_k + s_k I) Q_k = R_k Q_k + s_k I$$

so we take $A_{k+1} = R_k Q_k + s_k I$. If the shifts s_k are carefully chosen, convergence can be greatly improved.

Preliminary Preparation. A matrix such as

$$\begin{bmatrix} * & * & * & * & * \\ * & * & * & * & * \\ 0 & * & * & * & * \\ 0 & 0 & * & * & * \\ 0 & 0 & 0 & * & * \end{bmatrix}$$

is said to be in **upper Hessenberg** form, and the QR-factorizations of such matrices are greatly simplified. Given an $n \times n$ matrix A, a series of orthogonal matrices $H_1, H_2, \ldots, H_m$ (called **Householder matrices**) can be easily constructed such that

$$B = H_m^T \cdots H_1^T A H_1 \cdots H_m$$

is in upper Hessenberg form. Then the QR-algorithm can be efficiently applied to B and, because B is similar to A, it produces the eigenvalues of A.

Complex Eigenvalues. If some of the eigenvalues of a real matrix A are not real, the QR-algorithm converges to a block upper triangular matrix where the diagonal blocks are either 1×1 (the real eigenvalues) or 2×2 (each providing a pair of conjugate complex eigenvalues of A).

EXERCISES 8.5

1. In each case, find the exact eigenvalues and determine corresponding eigenvectors. Then start with $\mathbf{x}_0 = \begin{bmatrix} 1 \\ 1 \end{bmatrix}$ and compute $\mathbf{x}_4$ and r_3 using the power method.

   (a) $A = \begin{bmatrix} 2 & -4 \\ -3 & 3 \end{bmatrix}$ ◆(b) $A = \begin{bmatrix} 5 & 2 \\ -3 & -2 \end{bmatrix}$

   (c) $A = \begin{bmatrix} 1 & 2 \\ 2 & 1 \end{bmatrix}$ ◆(d) $A = \begin{bmatrix} 3 & 1 \\ 1 & 0 \end{bmatrix}$

2. In each case, find the exact eigenvalues and then approximate them using the QR-algorithm.

   (a) $A = \begin{bmatrix} 1 & 1 \\ 1 & 0 \end{bmatrix}$ ◆(b) $A = \begin{bmatrix} 3 & 1 \\ 1 & 0 \end{bmatrix}$

3. Apply the power method to $A = \begin{bmatrix} 0 & 1 \\ -1 & 0 \end{bmatrix}$, starting at $\mathbf{x}_0 = \begin{bmatrix} 1 \\ 1 \end{bmatrix}$. Does it converge? Explain.

◆4. If A is symmetric, show that each matrix A_k in the QR-algorithm is also symmetric. Deduce that they converge to a diagonal matrix.

5. Apply the QR-algorithm to $A = \begin{bmatrix} 2 & -3 \\ 1 & -2 \end{bmatrix}$. Explain.

6. Given a matrix A, let A_k, Q_k, and R_k, $k \geq 1$, be the matrices constructed in the QR-algorithm. Show that $A^k = (Q_1 Q_2 \cdots Q_k)(R_k \cdots R_2 R_1)$ for each $k \geq 1$ and hence that this is a QR-factorization of A^k. [*Hint*: Show that $Q_k R_k = R_{k-1} Q_{k-1}$ for each $k \geq 2$, and use this equality to compute $(Q_1 Q_2 \cdots Q_k)(R_k \cdots R_2 R_1)$ "from the centre out." Use the fact that $(AB)^{n+1} = A(BA)^n B$ for any square matrices A and B.]

SECTION 8.6 Complex Matrices

If A is an $n \times n$ matrix, the characteristic polynomial $c_A(x)$ is a polynomial of degree n and the eigenvalues of A are just the roots of $c_A(x)$. In most of our examples these roots have been *real* numbers (in fact, the examples have been carefully chosen so this will be the case!); but it need not happen, even when the characteristic polynomial has real coefficients. For example, if $A = \begin{bmatrix} 0 & 1 \\ -1 & 0 \end{bmatrix}$ then $c_A(x) = x^2 + 1$ has roots i and $-i$, where i is a complex number satisfying $i^2 = -1$. Therefore, we have to deal with the possibility that the eigenvalues of a (real) square matrix might be complex numbers.

In fact, nearly everything in this book would remain true if the phrase *real number* were replaced by *complex number* wherever it occurs. Then we would deal with matrices with complex entries, systems of linear equations with complex coefficients (and complex solutions), determinants of complex matrices, and vector spaces with scalar multiplication by any complex number allowed. Moreover, the proofs of most theorems about (the real version of) these concepts extend easily to the complex case. It is not our intention here to give a full treatment of complex linear algebra. However, we will carry the theory far enough to give another proof that the eigenvalues of a real symmetric matrix A are real (Theorem 7 Section 5.5) and to prove the spectral theorem, an extension of the principal axis theorem (Theorem 2 Section 8.2).

The set of complex numbers is denoted $\mathbb{C}$. We will use only the most basic properties of these numbers (mainly conjugation and absolute values), and the reader can find this material in Appendix A.

If $n \geq 1$, we denote the set of all n-tuples of complex numbers by $\mathbb{C}^n$. As with $\mathbb{R}^n$, these n-tuples will be written either as row or column matrices and will be referred to as **vectors**. We define vector operations on $\mathbb{C}^n$ as follows:

$$(v_1, v_2, ..., v_n) + (w_1, w_2, ..., w_n) = (v_1 + w_1, v_2 + w_2, ..., v_n + w_n)$$
$$u(v_1, v_2, ..., v_n) = (uv_1, uv_2, ..., uv_n) \quad \text{for } u \text{ in } \mathbb{C}$$

With these definitions, $\mathbb{C}^n$ satisfies the axioms for a vector space (with complex scalars) given in Chapter 6. Thus we can speak of spanning sets for $\mathbb{C}^n$, of linearly independent subsets, and of bases. In all cases, the definitions are identical to the real case, except that the scalars are allowed to be complex numbers. In particular, the standard basis of $\mathbb{R}^n$ remains a basis of $\mathbb{C}^n$, called the **standard basis** of $\mathbb{C}^n$.

The Standard Inner Product

There is a natural generalization to $\mathbb{C}^n$ of the dot product in $\mathbb{R}^n$.

Definition 8.7 Given $\mathbf{z} = (z_1, z_2, ..., z_n)$ and $\mathbf{w} = (w_1, w_2, ..., w_n)$ in $\mathbb{C}^n$, define their **standard inner product** $\langle \mathbf{z}, \mathbf{w} \rangle$ by

$$\langle \mathbf{z}, \mathbf{w} \rangle = z_1 \overline{w}_1 + z_2 \overline{w}_2 + \cdots + z_n \overline{w}_n$$

where $\overline{w}$ is the conjugate of the complex number w.

Clearly, if $\mathbf{z}$ and $\mathbf{w}$ actually lie in $\mathbb{R}^n$, then $\langle \mathbf{z}, \mathbf{w} \rangle = \mathbf{z} \cdot \mathbf{w}$ is the usual dot product.

EXAMPLE 1

If $\mathbf{z} = (2, 1 - i, 2i, 3 - i)$ and $\mathbf{w} = (1 - i, -1, -i, 3 + 2i)$, then

$$\langle \mathbf{z}, \mathbf{w} \rangle = 2(1 + i) + (1 - i)(-1) + (2i)(i) + (3 - i)(3 - 2i) = 6 - 6i$$
$$\langle \mathbf{z}, \mathbf{z} \rangle = 2 \cdot 2 + (1 - i)(1 + i) + (2i)(-2i) + (3 - i)(3 + i) = 20$$

Note that $\langle \mathbf{z}, \mathbf{w} \rangle$ is a complex number in general. However, if $\mathbf{w} = \mathbf{z} = (z_1, z_2, \ldots, z_n)$, the definition gives $\langle \mathbf{z}, \mathbf{z} \rangle = |z_1|^2 + \cdots + |z_n|^2$ which is a nonnegative real number, equal to 0 if and only if $\mathbf{z} = \mathbf{0}$. This explains the conjugation in the definition of $\langle \mathbf{z}, \mathbf{w} \rangle$, and it gives (4) of the following theorem.

Theorem 1

Let $\mathbf{z}, \mathbf{z}_1, \mathbf{w},$ and $\mathbf{w}_1$ denote vectors in $\mathbb{C}^n$, and let λ denote a complex number.

1. $\langle \mathbf{z} + \mathbf{z}_1, \mathbf{w} \rangle = \langle \mathbf{z}, \mathbf{w} \rangle + \langle \mathbf{z}_1, \mathbf{w} \rangle$ and $\langle \mathbf{z}, \mathbf{w} + \mathbf{w}_1 \rangle = \langle \mathbf{z}, \mathbf{w} \rangle + \langle \mathbf{z}, \mathbf{w}_1 \rangle$.
2. $\langle \lambda \mathbf{z}, \mathbf{w} \rangle = \lambda \langle \mathbf{z}, \mathbf{w} \rangle$ and $\langle \mathbf{z}, \lambda \mathbf{w} \rangle = \overline{\lambda} \langle \mathbf{z}, \mathbf{w} \rangle$.
3. $\langle \mathbf{z}, \mathbf{w} \rangle = \overline{\langle \mathbf{w}, \mathbf{z} \rangle}$.
4. $\langle \mathbf{z}, \mathbf{z} \rangle \geq 0$, and $\langle \mathbf{z}, \mathbf{z} \rangle = 0$ if and only if $\mathbf{z} = \mathbf{0}$.

PROOF

We leave (1) and (2) to the reader (Exercise 10), and (4) has already been proved. To prove (3), write $\mathbf{z} = (z_1, z_2, \ldots, z_n)$ and $\mathbf{w} = (w_1, w_2, \ldots, w_n)$. Then

$$\overline{\langle \mathbf{w}, \mathbf{z} \rangle} = \overline{(w_1 \overline{z}_1 + \cdots + w_n \overline{z}_n)} = \overline{w}_1 \overline{\overline{z}_1} + \cdots + \overline{w}_n \overline{\overline{z}_n}$$
$$= z_1 \overline{w}_1 + \cdots + z_n \overline{w}_n = \langle \mathbf{z}, \mathbf{w} \rangle$$

Definition 8.8 As for the dot product on $\mathbb{R}^n$, property (4) enables us to define the **norm** or **length** $\|\mathbf{z}\|$ of a vector $\mathbf{z} = (z_1, z_2, \ldots, z_n)$ in $\mathbb{C}^n$:

$$\|\mathbf{z}\| = \sqrt{\langle \mathbf{z}, \mathbf{z} \rangle} = \sqrt{|z_1|^2 + |z_2|^2 + \cdots + |z_n|^2}$$

The only properties of the norm function we will need are the following (the proofs are left to the reader):

Theorem 2

If $\mathbf{z}$ is any vector in $\mathbb{C}^n$, then

1. $\|\mathbf{z}\| \geq 0$, and $\|\mathbf{z}\| = 0$ if and only if $\mathbf{z} = \mathbf{0}$.
2. $\|\lambda \mathbf{z}\| = |\lambda| \|\mathbf{z}\|$ for all complex numbers λ.

A vector $\mathbf{u}$ in $\mathbb{C}^n$ is called a **unit vector** if $\|\mathbf{u}\| = 1$. Property (2) in Theorem 2 then shows that if $\mathbf{z} \neq \mathbf{0}$ is any nonzero vector in $\mathbb{C}^n$, then $\mathbf{u} = \frac{1}{\|\mathbf{z}\|} \mathbf{z}$ is a unit vector.

SECTION 8.6 Complex Matrices

EXAMPLE 2

In $\mathbb{C}^4$, find a unit vector **u** that is a positive real multiple of
$\mathbf{z} = (1 - i, i, 2, 3 + 4i)$.

Solution ▶ $\|\mathbf{z}\| = \sqrt{2 + 1 + 4 + 25} = \sqrt{32} = 4\sqrt{2}$, so take $\mathbf{u} = \frac{1}{4\sqrt{2}}\mathbf{z}$.

A matrix $A = [a_{ij}]$ is called a **complex matrix** if every entry a_{ij} is a complex number. The notion of conjugation for complex numbers extends to matrices as follows: Define the **conjugate** of $A = [a_{ij}]$ to be the matrix
$$\overline{A} = [\overline{a}_{ij}]$$
obtained from A by conjugating every entry. Then (using Appendix A)
$$\overline{A + B} = \overline{A} + \overline{B} \quad \text{and} \quad \overline{AB} = \overline{A}\,\overline{B}$$
holds for all (complex) matrices of appropriate size.

Transposition of complex matrices is defined just as in the real case, and the following notion is fundamental.

Definition 8.9 The **conjugate transpose** A^H of a complex matrix A is defined by
$$A^H = (\overline{A})^T = \overline{(A^T)}$$

Observe that $A^H = A^T$ when A is real.[8]

EXAMPLE 3

$$\begin{bmatrix} 3 & 1-i & 2+i \\ 2i & 5+2i & -i \end{bmatrix}^H = \begin{bmatrix} 3 & -2i \\ 1+i & 5-2i \\ 2-i & i \end{bmatrix}$$

The following properties of A^H follow easily from the rules for transposition of real matrices and extend these rules to complex matrices. Note the conjugate in property (3).

Theorem 3

Let A and B denote complex matrices, and let λ be a complex number.
1. $(A^H)^H = A$.
2. $(A + B)^H = A^H + B^H$.
3. $(\lambda A)^H = \overline{\lambda} A^H$.
4. $(AB)^H = B^H A^H$.

---

[8] Other notations for A^H are A^* and $A^\dagger$.

Hermitian and Unitary Matrices

If A is a real symmetric matrix, it is clear that $A^H = A$. The complex matrices that satisfy this condition turn out to be the most natural generalization of the real symmetric matrices:

Definition 8.10 *A square complex matrix A is called* **hermitian**[9] *if $A^H = A$, equivalently $\overline{A} = A^T$.*

Hermitian matrices are easy to recognize because the entries on the main diagonal must be real, and the "reflection" of each nondiagonal entry in the main diagonal must be the conjugate of that entry.

EXAMPLE 4

$\begin{bmatrix} 3 & i & 2+i \\ -i & -2 & -7 \\ 2-i & -7 & 1 \end{bmatrix}$ is hermitian, whereas $\begin{bmatrix} 1 & i \\ i & -2 \end{bmatrix}$ and $\begin{bmatrix} 1 & i \\ -i & i \end{bmatrix}$ are not.

The following Theorem extends Theorem 3 Section 8.2, and gives a very useful characterization of hermitian matrices in terms of the standard inner product in $\mathbb{C}^n$.

Theorem 4

An $n \times n$ complex matrix A is hermitian if and only if
$$\langle A\mathbf{z}, \mathbf{w} \rangle = \langle \mathbf{z}, A\mathbf{w} \rangle$$
for all n-tuples $\mathbf{z}$ and $\mathbf{w}$ in $\mathbb{C}^n$.

PROOF

If A is hermitian, we have $A^T = \overline{A}$. If $\mathbf{z}$ and $\mathbf{w}$ are columns in $\mathbb{C}^n$, then $\langle \mathbf{z}, \mathbf{w} \rangle = \mathbf{z}^T \overline{\mathbf{w}}$, so
$$\langle A\mathbf{z}, \mathbf{w} \rangle = (A\mathbf{z})^T \overline{\mathbf{w}} = \mathbf{z}^T A^T \overline{\mathbf{w}} = \mathbf{z}^T \overline{A} \, \overline{\mathbf{w}} = \mathbf{z}^T (\overline{A\mathbf{w}}) = \langle \mathbf{z}, A\mathbf{w} \rangle.$$

To prove the converse, let $\mathbf{e}_j$ denote column j of the identity matrix. If $A = [a_{ij}]$, the condition gives
$$\overline{a}_{ij} = \langle \mathbf{e}_i, A\mathbf{e}_j \rangle = \langle A\mathbf{e}_i, \mathbf{e}_j \rangle = a_{ij}.$$
Hence $\overline{A} = A^T$, so A is hermitian.

Let A be an $n \times n$ complex matrix. As in the real case, a complex number λ is called an **eigenvalue** of A if $A\mathbf{x} = \lambda \mathbf{x}$ holds for some column $\mathbf{x} \neq \mathbf{0}$ in $\mathbb{C}^n$. In this case $\mathbf{x}$ is called an **eigenvector** of A corresponding to λ. The **characteristic polynomial** $c_A(x)$ is defined by
$$c_A(x) = \det(xI - A).$$

---

[9] The name hermitian honours Charles Hermite (1822–1901), a French mathematician who worked primarily in analysis and is remembered as the first to show that the number *e* from calculus is transcendental—that is, *e* is not a root of any polynomial with integer coefficients.

SECTION 8.6 Complex Matrices

This polynomial has complex coefficients (possibly nonreal). However, the proof of Theorem 2 Section 3.3 goes through to show that the eigenvalues of A are the roots (possibly complex) of $c_A(x)$.

It is at this point that the advantage of working with complex numbers becomes apparent. The real numbers are incomplete in the sense that the characteristic polynomial of a real matrix may fail to have all its roots real. However, this difficulty does not occur for the complex numbers. The so-called fundamental theorem of algebra ensures that *every* polynomial of positive degree with complex coefficients has a complex root. Hence every square complex matrix A has a (complex) eigenvalue. Indeed (Appendix A), $c_A(x)$ factors completely as follows:

$$c_A(x) = (x - \lambda_1)(x - \lambda_2)\cdots(x - \lambda_n)$$

where $\lambda_1, \lambda_2, \ldots, \lambda_n$ are the eigenvalues of A (with possible repetitions due to multiple roots).

The next result shows that, for hermitian matrices, the eigenvalues are actually real. Because symmetric real matrices are hermitian, this re-proves Theorem 7 Section 5.5. It also extends Theorem 4 Section 8.2, which asserts that eigenvectors of a symmetric real matrix corresponding to distinct eigenvalues are actually orthogonal. In the complex context, two n-tuples $\mathbf{z}$ and $\mathbf{w}$ in $\mathbb{C}^n$ are said to be **orthogonal** if $\langle \mathbf{z}, \mathbf{w} \rangle = 0$.

Theorem 5

Let A denote a hermitian matrix.
1. The eigenvalues of A are real.
2. Eigenvectors of A corresponding to distinct eigenvalues are orthogonal.

PROOF

Let λ and μ be eigenvalues of A with (nonzero) eigenvectors $\mathbf{z}$ and $\mathbf{w}$. Then $A\mathbf{z} = \lambda\mathbf{z}$ and $A\mathbf{w} = \mu\mathbf{w}$, so Theorem 4 gives

$$\lambda\langle \mathbf{z}, \mathbf{w}\rangle = \langle \lambda\mathbf{z}, \mathbf{w}\rangle = \langle A\mathbf{z}, \mathbf{w}\rangle = \langle \mathbf{z}, A\mathbf{w}\rangle = \langle \mathbf{z}, \mu\mathbf{w}\rangle = \overline{\mu}\langle \mathbf{z}, \mathbf{w}\rangle \quad (*)$$

If $\mu = \lambda$ and $\mathbf{w} = \mathbf{z}$, this becomes $\lambda\langle \mathbf{z}, \mathbf{z}\rangle = \overline{\lambda}\langle \mathbf{z}, \mathbf{z}\rangle$. Because $\langle \mathbf{z}, \mathbf{z}\rangle = \|\mathbf{z}\|^2 \neq 0$, this implies $\lambda = \overline{\lambda}$. Thus λ is real, proving (1). Similarly, μ is real, so equation $(*)$ gives $\lambda\langle \mathbf{z}, \mathbf{w}\rangle = \mu\langle \mathbf{z}, \mathbf{w}\rangle$. If $\lambda \neq \mu$, this implies $\langle \mathbf{z}, \mathbf{w}\rangle = 0$, proving (2).

The principal axis theorem (Theorem 2 Section 8.2) asserts that every real symmetric matrix A is orthogonally diagonalizable—that is $P^T A P$ is diagonal where P is an orthogonal matrix ($P^{-1} = P^T$). The next theorem identifies the complex analogs of these orthogonal real matrices.

Definition 8.11 *As in the real case, a set of nonzero vectors $\{\mathbf{z}_1, \mathbf{z}_2, \ldots, \mathbf{z}_m\}$ in $\mathbb{C}^n$ is called **orthogonal** if $\langle \mathbf{z}_i, \mathbf{z}_j\rangle = 0$ whenever $i \neq j$, and it is **orthonormal** if, in addition, $\|\mathbf{z}_i\| = 1$ for each i.*

Chapter 8 Orthogonality

Theorem 6

The following are equivalent for an $n \times n$ complex matrix A.

1. A is invertible and $A^{-1} = A^H$.
2. The rows of A are an orthonormal set in $\mathbb{C}^n$.
3. The columns of A are an orthonormal set in $\mathbb{C}^n$.

PROOF

If $A = [\mathbf{c}_1 \; \mathbf{c}_2 \; \cdots \; \mathbf{c}_n]$ is a complex matrix with jth column $\mathbf{c}_j$, then $A^T A = [\langle \mathbf{c}_i, \mathbf{c}_j \rangle]$, as in Theorem 1 Section 8.2. Now (1) $\Leftrightarrow$ (2) follows, and (1) $\Leftrightarrow$ (3) is proved in the same way.

Definition 8.12 A square complex matrix U is called **unitary** if $U^{-1} = U^H$.

Thus a real matrix is unitary if and only if it is orthogonal.

EXAMPLE 5

The matrix $A = \begin{bmatrix} 1+i & 1 \\ 1-i & i \end{bmatrix}$ has orthogonal columns, but the rows are not orthogonal. Normalizing the columns gives the unitary matrix $\frac{1}{2}\begin{bmatrix} 1+i & \sqrt{2} \\ 1-i & \sqrt{2}i \end{bmatrix}$.

Given a real symmetric matrix A, the diagonalization algorithm in Section 3.3 leads to a procedure for finding an orthogonal matrix P such that $P^T A P$ is diagonal (see Example 4, Section 8.2). The following example illustrates Theorem 5 and shows that the technique works for complex matrices.

EXAMPLE 6

Consider the hermitian matrix $A = \begin{bmatrix} 3 & 2+i \\ 2-i & 7 \end{bmatrix}$. Find the eigenvalues of A, find two orthonormal eigenvectors, and so find a unitary matrix U such that $U^H A U$ is diagonal.

Solution ▶ The characteristic polynomial of A is

$$c_A(x) = \det(xI - A) = \det\begin{bmatrix} x-3 & -2-i \\ -2+i & x-7 \end{bmatrix} = (x-2)(x-8)$$

Hence the eigenvalues are 2 and 8 (both real as expected), and corresponding eigenvectors are $\begin{bmatrix} 2+i \\ -1 \end{bmatrix}$ and $\begin{bmatrix} 1 \\ 2-i \end{bmatrix}$ (orthogonal as expected). Each has length $\sqrt{6}$ so, as in the (real) diagonalization algorithm, let $U = \frac{1}{\sqrt{6}}\begin{bmatrix} 2+i & 1 \\ -1 & 2-i \end{bmatrix}$ be the unitary matrix with the normalized eigenvectors as columns. Then $U^H A U = \begin{bmatrix} 2 & 0 \\ 0 & 8 \end{bmatrix}$ is diagonal.

Unitary Diagonalization

An $n \times n$ complex matrix A is called **unitarily diagonalizable** if $U^H AU$ is diagonal for some unitary matrix U. As Example 6 suggests, we are going to prove that every hermitian matrix is unitarily diagonalizable. However, with only a little extra effort, we can get a very important theorem that has this result as an easy consequence.

A complex matrix is called **upper triangular** if every entry below the main diagonal is zero. We owe the following theorem to Issai Schur.[10]

Theorem 7

Schur's Theorem

If A is any $n \times n$ complex matrix, there exists a unitary matrix U such that
$$U^H AU = T$$
is upper triangular. Moreover, the entries on the main diagonal of T are the eigenvalues $\lambda_1, \lambda_2, \ldots, \lambda_n$ of A (including multiplicities).

PROOF

We use induction on n. If $n = 1$, A is already upper triangular. If $n > 1$, assume the theorem is valid for $(n-1) \times (n-1)$ complex matrices. Let λ_1 be an eigenvalue of A, and let $\mathbf{y}_1$ be an eigenvector with $\|\mathbf{y}_1\| = 1$. Then $\mathbf{y}_1$ is part of a basis of $\mathbb{C}^n$ (by the analog of Theorem 1 Section 6.4), so the (complex analog of the) Gram-Schmidt process provides $\mathbf{y}_2, \ldots, \mathbf{y}_n$ such that $\{\mathbf{y}_1, \mathbf{y}_2, \ldots, \mathbf{y}_n\}$ is an orthonormal basis of $\mathbb{C}^n$. If $U_1 = [\mathbf{y}_1 \ \mathbf{y}_2 \ \cdots \ \mathbf{y}_n]$ is the matrix with these vectors as its columns, then (see Lemma 3)
$$U_1^H A U_1 = \begin{bmatrix} \lambda_1 & X_1 \\ 0 & A_1 \end{bmatrix}$$
in block form. Now apply induction to find a unitary $(n-1) \times (n-1)$ matrix W_1 such that $W_1^H A_1 W_1 = T_1$ is upper triangular. Then $U_2 = \begin{bmatrix} 1 & 0 \\ 0 & W_1 \end{bmatrix}$ is a unitary $n \times n$ matrix. Hence $U = U_1 U_2$ is unitary (using Theorem 6), and
$$U^H AU = U_2^H (U_1^H A U_1) U_2$$
$$= \begin{bmatrix} 1 & 0 \\ 0 & W_1^H \end{bmatrix} \begin{bmatrix} \lambda_1 & X_1 \\ 0 & A_1 \end{bmatrix} \begin{bmatrix} 1 & 0 \\ 0 & W_1 \end{bmatrix} = \begin{bmatrix} \lambda_1 & X_1 W_1 \\ 0 & T_1 \end{bmatrix}$$
is upper triangular. Finally, A and $U^H AU = T$ have the same eigenvalues by (the complex version of) Theorem 1 Section 5.5, and they are the diagonal entries of T because T is upper triangular.

The fact that similar matrices have the same traces and determinants gives the following consequence of Schur's theorem.

---

10 Issai Schur (1875–1941) was a German mathematician who did fundamental work in the theory of representations of groups as matrices.

Corollary 1

Let A be an $n \times n$ complex matrix, and let $\lambda_1, \lambda_2, \ldots, \lambda_n$ denote the eigenvalues of A, including multiplicities. Then
$$\det A = \lambda_1 \lambda_2 \cdots \lambda_n \quad \text{and} \quad \operatorname{tr} A = \lambda_1 + \lambda_2 + \cdots + \lambda_n$$

Schur's theorem asserts that every complex matrix can be "unitarily triangularized." However, we cannot substitute "unitarily diagonalized" here. In fact, if $A = \begin{bmatrix} 1 & 1 \\ 0 & 1 \end{bmatrix}$, there is no invertible complex matrix U at all such that $U^{-1}AU$ is diagonal. However, the situation is much better for hermitian matrices.

Theorem 8

Spectral Theorem
If A is hermitian, there is a unitary matrix U such that $U^H A U$ is diagonal.

PROOF

By Schur's theorem, let $U^H A U = T$ be upper triangular where U is unitary. Since A is hermitian, this gives
$$T^H = (U^H A U)^H = U^H A^H U^{HH} = U^H A U = T$$
This means that T is both upper and lower triangular. Hence T is actually diagonal.

The principal axis theorem asserts that a real matrix A is symmetric if and only if it is orthogonally diagonalizable (that is, $P^T A P$ is diagonal for some real orthogonal matrix P). Theorem 8 is the complex analog of half of this result. However, the converse is false for complex matrices: There exist unitarily diagonalizable matrices that are not hermitian.

EXAMPLE 7

Show that the non-hermitian matrix $A = \begin{bmatrix} 0 & 1 \\ -1 & 0 \end{bmatrix}$ is unitarily diagonalizable.

Solution ▶ The characteristic polynomial is $c_A(x) = x^2 + 1$. Hence the eigenvalues are i and $-i$, and it is easy to verify that $\begin{bmatrix} i \\ -1 \end{bmatrix}$ and $\begin{bmatrix} -1 \\ i \end{bmatrix}$ are corresponding eigenvectors. Moreover, these eigenvectors are orthogonal and both have length $\sqrt{2}$, so $U = \frac{1}{\sqrt{2}} \begin{bmatrix} i & -1 \\ -1 & i \end{bmatrix}$ is a unitary matrix such that $U^H A U = \begin{bmatrix} i & 0 \\ 0 & -i \end{bmatrix}$ is diagonal.

SECTION 8.6 Complex Matrices

There is a very simple way to characterize those complex matrices that are unitarily diagonalizable. To this end, an $n \times n$ complex matrix N is called **normal** if $NN^H = N^H N$. It is clear that every hermitian or unitary matrix is normal, as is the matrix $\begin{bmatrix} 0 & 1 \\ -1 & 0 \end{bmatrix}$ in Example 7. In fact we have the following result.

Theorem 9

An $n \times n$ complex matrix A is unitarily diagonalizable if and only if A is normal.

PROOF

Assume first that $U^H A U = D$, where U is unitary and D is diagonal. Then $DD^H = D^H D$ as is easily verified. Because $DD^H = U^H(AA^H)U$ and $D^H D = U^H(A^H A)U$, it follows by cancellation that $AA^H = A^H A$.

Conversely, assume A is normal—that is, $AA^H = A^H A$. By Schur's theorem, let $U^H A U = T$, where T is upper triangular and U is unitary. Then T is normal too:
$$TT^H = U^H(AA^H)U = U^H(A^H A)U = T^H T$$

Hence it suffices to show that a normal $n \times n$ upper triangular matrix T must be diagonal. We induct on n; it is clear if $n = 1$. If $n > 1$ and $T = [t_{ij}]$, then equating $(1, 1)$-entries in TT^H and $T^H T$ gives
$$|t_{11}|^2 + |t_{12}|^2 + \cdots + |t_{1n}|^2 = |t_{11}|^2$$

This implies $t_{12} = t_{13} = \cdots = t_{1n} = 0$, so $T = \begin{bmatrix} t_{11} & 0 \\ 0 & T_1 \end{bmatrix}$ in block form. Hence $T = \begin{bmatrix} \bar{t}_{11} & 0 \\ 0 & T_1^H \end{bmatrix}$ so $TT^H = T^H T$ implies $T_1 T_1^H = T_1 T_1^H$. Thus T_1 is diagonal by induction, and the proof is complete.

We conclude this section by using Schur's theorem (Theorem 7) to prove a famous theorem about matrices. Recall that the characteristic polynomial of a square matrix A is defined by $c_A(x) = \det(xI - A)$, and that the eigenvalues of A are just the roots of $c_A(x)$.

Theorem 10

Cayley-Hamilton Theorem[11]

If A is an $n \times n$ complex matrix, then $c_A(A) = 0$; that is, A is a root of its characteristic polynomial.

---

11 Named after the English mathematician Arthur Cayley (1821–1895), see page 32, and William Rowan Hamilton (1805–1865), an Irish mathematician famous for his work on physical dynamics.

> **PROOF**
>
> If $p(x)$ is any polynomial with complex coefficients, then $p(P^{-1}AP) = P^{-1}p(A)P$ for any invertible complex matrix P. Hence, by Schur's theorem, we may assume that A is upper triangular. Then the eigenvalues $\lambda_1, \lambda_2, \ldots, \lambda_n$ of A appear along the main diagonal, so $c_A(x) = (x - \lambda_1)(x - \lambda_2)(x - \lambda_3)\cdots(x - \lambda_n)$. Thus
>
> $$c_A(A) = (A - \lambda_1 I)(A - \lambda_2 I)(A - \lambda_3 I)\cdots(A - \lambda_n I)$$
>
> Note that each matrix $A - \lambda_i I$ is upper triangular. Now observe:
>
> 1. $A - \lambda_1 I$ has zero first column because column 1 of A is $(\lambda_1, 0, 0, \ldots, 0)^T$.
>
> 2. Then $(A - \lambda_1 I)(A - \lambda_2 I)$ has the first two columns zero because column 2 of $(A - \lambda_2 I)$ is $(b, 0, 0, \ldots, 0)^T$ for some constant b.
>
> 3. Next $(A - \lambda_1 I)(A - \lambda_2 I)(A - \lambda_3 I)$ has the first three columns zero because column 3 of $(A - \lambda_3 I)$ is $(c, d, 0, \ldots, 0)^T$ for some constants c and d.
>
> Continuing in this way we see that $(A - \lambda_1 I)(A - \lambda_2 I)(A - \lambda_3 I) \cdots (A - \lambda_n I)$ has all n columns zero; that is, $c_A(A) = 0$.

EXERCISES 8.6

1. In each case, compute the norm of the complex vector.

   (a) $(1, 1 - i, -2, i)$

   ◆(b) $(1 - i, 1 + i, 1, -1)$

   (c) $(2 + i, 1 - i, 2, 0, -i)$

   ◆(d) $(-2, -i, 1 + i, 1 - i, 2i)$

2. In each case, determine whether the two vectors are orthogonal.

   (a) $(4, -3i, 2 + i), (i, 2, 2 - 4i)$

   ◆(b) $(i, -i, 2 + i), (i, i, 2 - i)$

   (c) $(1, 1, i, i), (1, i, -i, 1)$

   ◆(d) $(4 + 4i, 2 + i, 2i), (-1 + i, 2, 3 - 2i)$

3. A subset U of $\mathbb{C}^n$ is called a **complex subspace** of $\mathbb{C}^n$ if it contains 0 and if, given $\mathbf{v}$ and $\mathbf{w}$ in U, both $\mathbf{v} + \mathbf{w}$ and $z\mathbf{v}$ lie in U (z any complex number). In each case, determine whether U is a complex subspace of $\mathbb{C}^3$.

   (a) $U = \{(w, \overline{w}, 0) \mid w \text{ in } \mathbb{C}\}$

   ◆(b) $U = \{(w, 2w, a) \mid w \text{ in } \mathbb{C}, a \text{ in } \mathbb{R}\}$

   (c) $U = \mathbb{R}^3$

   ◆(d) $U = \{(v + w, v - 2w, v) \mid v, w \text{ in } \mathbb{C}\}$

4. In each case, find a basis over $\mathbb{C}$, and determine the dimension of the complex subspace U of $\mathbb{C}^3$ (see the previous exercise).

   (a) $U = \{(w, v + w, v - iw) \mid v, w \text{ in } \mathbb{C}\}$

   ◆(b) $U = \{(iv + w, 0, 2v - w) \mid v, w \text{ in } \mathbb{C}\}$

   (c) $U = \{(u, v, w) \mid iu - 3v + (1 - i)w = 0; u, v, w \text{ in } \mathbb{C}\}$

   ◆(d) $U = \{(u, v, w) \mid 2u + (1 + i)v - iw = 0; u, v, w \text{ in } \mathbb{C}\}$

5. In each case, determine whether the given matrix is hermitian, unitary, or normal.

   (a) $\begin{bmatrix} 1 & -i \\ i & i \end{bmatrix}$

   ◆(b) $\begin{bmatrix} 2 & 3 \\ -3 & 2 \end{bmatrix}$

   (c) $\begin{bmatrix} 1 & i \\ -i & 2 \end{bmatrix}$

   ◆(d) $\begin{bmatrix} 1 & -i \\ i & -1 \end{bmatrix}$

   (e) $\frac{1}{\sqrt{2}} \cdot \begin{bmatrix} 1 & -1 \\ 1 & 1 \end{bmatrix}$

   ◆(f) $\begin{bmatrix} 1 & 1 + i \\ 1 + i & i \end{bmatrix}$

   (g) $\begin{bmatrix} 1 + i & 1 \\ -i & -1 + i \end{bmatrix}$

   ◆(h) $\frac{1}{\sqrt{2}|z|}\begin{bmatrix} z & z \\ \overline{z} & -\overline{z} \end{bmatrix}, z \neq 0$

6. Show that a matrix N is normal if and only if $\overline{N}N^T = N^T\overline{N}$.

SECTION 8.6 Complex Matrices

7. Let $A = \begin{bmatrix} z & \overline{v} \\ v & w \end{bmatrix}$ where v, w, and z are complex numbers. Characterize in terms of v, w, and z when A is

   (a) hermitian   (b) unitary

   (c) normal.

8. In each case, find a unitary matrix U such that $U^H A U$ is diagonal.

   (a) $A = \begin{bmatrix} 1 & i \\ -i & 1 \end{bmatrix}$
   ◆(b) $A = \begin{bmatrix} 4 & 3-i \\ 3+i & 1 \end{bmatrix}$

   (c) $A = \begin{bmatrix} a & b \\ -b & a \end{bmatrix}$
   ◆(d) $A = \begin{bmatrix} 2 & 1+i \\ 1-i & 3 \end{bmatrix}$

   a, b, real

   (e) $A = \begin{bmatrix} 1 & 0 & 1+i \\ 0 & 2 & 0 \\ 1-i & 0 & 0 \end{bmatrix}$
   ◆(f) $A = \begin{bmatrix} 1 & 0 & 0 \\ 0 & 1 & 1+i \\ 0 & 1-i & 2 \end{bmatrix}$

9. Show that $\langle A\mathbf{x}, \mathbf{y}\rangle = \langle \mathbf{x}, A^H \mathbf{y}\rangle$ holds for all $n \times n$ matrices A and for all n-tuples $\mathbf{x}$ and $\mathbf{y}$ in $\mathbb{C}^n$.

10. (a) Prove (1) and (2) of Theorem 1.
    ◆(b) Prove Theorem 2.
    (c) Prove Theorem 3.

11. (a) Show that A is hermitian if and only if $\overline{A} = A^T$.
    ◆(b) Show that the diagonal entries of any hermitian matrix are real.

12. (a) Show that every complex matrix Z can be written uniquely in the form $Z = A + iB$, where A and B are real matrices.
    (b) If $Z = A + iB$ as in (a), show that Z is hermitian if and only if A is symmetric, and B is skew-symmetric (that is, $B^T = -B$).

13. If Z is any complex $n \times n$ matrix, show that ZZ^H and $Z + Z^H$ are hermitian.

14. A complex matrix B is called **skew-hermitian** if $B^H = -B$.
    (a) Show that $Z - Z^H$ is skew-hermitian for any square complex matrix Z.
    ◆(b) If B is skew-hermitian, show that B^2 and iB are hermitian.
    (c) If B is skew-hermitian, show that the eigenvalues of B are pure imaginary ($i\lambda$ for real λ).

◆(d) Show that every $n \times n$ complex matrix Z can be written uniquely as $Z = A + B$, where A is hermitian and B is skew-hermitian.

15. Let U be a unitary matrix. Show that:
    (a) $\|U\mathbf{x}\| = \|\mathbf{x}\|$ for all columns $\mathbf{x}$ in $\mathbb{C}^n$.
    (b) $|\lambda| = 1$ for every eigenvalue λ of U.

16. (a) If Z is an invertible complex matrix, show that Z^H is invertible and that $(Z^H)^{-1} = (Z^{-1})^H$.
    ◆(b) Show that the inverse of a unitary matrix is again unitary.
    (c) If U is unitary, show that U^H is unitary.

17. Let Z be an $m \times n$ matrix such that $Z^H Z = I_n$ (for example, Z is a unit column in $\mathbb{C}^n$).
    (a) Show that $V = ZZ^H$ is hermitian and satisfies $V^2 = V$.
    (b) Show that $U = I - 2ZZ^H$ is both unitary and hermitian (so $U^{-1} = U^H = U$).

18. (a) If N is normal, show that zN is also normal for all complex numbers z.
    ◆(b) Show that (a) fails if *normal* is replaced by *hermitian*.

19. Show that a real 2×2 normal matrix is either symmetric or has the form $\begin{bmatrix} a & b \\ -b & a \end{bmatrix}$.

20. If A is hermitian, show that all the coefficients of $c_A(x)$ are real numbers.

21. (a) If $A = \begin{bmatrix} 1 & 1 \\ 0 & 1 \end{bmatrix}$, show that $U^{-1}AU$ is not diagonal for any invertible complex matrix U.
    ◆(b) If $A = \begin{bmatrix} 0 & 1 \\ -1 & 0 \end{bmatrix}$, show that $U^{-1}AU$ is not upper triangular for any *real* invertible matrix U.

22. If A is any $n \times n$ matrix, show that $U^H A U$ is lower triangular for some unitary matrix U.

23. If A is a 3×3 matrix, show that $A^2 = 0$ if and only if there exists a unitary matrix U such that $U^H A U$ has the form $\begin{bmatrix} 0 & 0 & u \\ 0 & 0 & v \\ 0 & 0 & 0 \end{bmatrix}$ or the form $\begin{bmatrix} 0 & u & v \\ 0 & 0 & 0 \\ 0 & 0 & 0 \end{bmatrix}$.

24. If $A^2 = A$, show that rank $A = \text{tr } A$. [*Hint*: Use Schur's theorem.]

SECTION 8.7 An Application to Linear Codes over Finite Fields

For centuries mankind has been using codes to transmit messages. In many cases, for example transmitting financial, medical, or military information, the message is disguised in such a way that it cannot be understood by an intruder who intercepts it, but can be easily "decoded" by the intended receiver. This subject is called *cryptography* and, while intriguing, is not our focus here. Instead, we investigate methods for detecting and correcting errors in the transmission of the message.

The stunning photos of the planet Saturn sent by the space probe are a very good example of how successful these methods can be. These messages are subject to "noise" such as solar interference which causes errors in the message. The signal is received on Earth with errors that must be detected and corrected before the high-quality pictures can be printed. This is done using error-correcting codes. To see how, we first discuss a system of adding and multiplying integers while ignoring multiples of a fixed integer.

Modular Arithmetic

We work in the set $\mathbb{Z} = \{0, \pm 1, \pm 2, \pm 3, ...\}$ of **integers**, that is the set of whole numbers. Everyone is familiar with the process of "long division" from arithmetic. For example, we can divide an integer a by 5 and leave a remainder "modulo 5" in the set $\{0, 1, 2, 3, 4\}$. As an illustration

$$19 = 3 \cdot 5 + 4,$$

so the remainder of 19 modulo 5 is 4. Similarly, the remainder of 137 modulo 5 is 2 because $137 = 27 \cdot 5 + 2$. This works even for negative integers: For example,

$$-17 = (-4) \cdot 5 + 3,$$

so the remainder of -17 modulo 5 is 3.

This process is called the **division algorithm**. More formally, let $n \geq 2$ denote an integer. Then every integer a can be written uniquely in the form

$$a = qn + r \quad \text{where } q \text{ and } r \text{ are integers and } 0 \leq r \leq n - 1.$$

Here q is called the **quotient** of a **modulo** n, and r is called the **remainder** of a **modulo** n. We refer to n as the **modulus**. Thus, if $n = 6$, the fact that $134 = 22 \cdot 6 + 2$ means that 134 has quotient 22 and remainder 2 modulo 6.

Our interest here is in the set of *all* possible remainders modulo n. This set is denoted

$$\mathbb{Z}_n = \{0, 1, 2, 3, ..., n - 1\}$$

and is called the set of **integers modulo** n. Thus every integer is uniquely represented in $\mathbb{Z}_n$ by its remainder modulo n.

We are going to show how to do arithmetic in $\mathbb{Z}_n$ by adding and multiplying modulo n. That is, we add or multiply two numbers in $\mathbb{Z}_n$ by calculating the usual sum or product in $\mathbb{Z}$ and taking the remainder modulo n. It is proved in books on abstract algebra that the usual laws of arithmetic hold in $\mathbb{Z}_n$ for any modulus $n \geq 2$. This seems remarkable until we remember that these laws are true for ordinary addition and multiplication and all we are doing is reducing modulo n.

To illustrate, consider the case $n = 6$, so that $\mathbb{Z}_6 = \{0, 1, 2, 3, 4, 5\}$. Then $2 + 5 = 1$ in $\mathbb{Z}_6$ because 7 leaves a remainder of 1 when divided by 6. Similarly, $2 \cdot 5 = 4$ in $\mathbb{Z}_6$, while $3 + 5 = 2$, and $3 + 3 = 0$. In this way we can fill in the addition and multiplication tables for $\mathbb{Z}_6$; the result is:

SECTION 8.7 An Application to Linear Codes over Finite Fields

Tables for $\mathbb{Z}_6$

+	0	1	2	3	4	5
0	0	1	2	3	4	5
1	1	2	3	4	5	0
2	2	3	4	5	0	1
3	3	4	5	0	1	2
4	4	5	0	1	2	3
5	5	0	1	2	3	4

×	0	1	2	3	4	5
0	0	0	0	0	0	0
1	0	1	2	3	4	5
2	0	2	4	0	2	4
3	0	3	0	3	0	3
4	0	4	2	0	4	2
5	0	5	4	3	2	1

Calculations in $\mathbb{Z}_6$ are carried out much as in $\mathbb{Z}$. As an illustration, consider the "distributive law" $a(b + c) = ab + ac$ familiar from ordinary arithmetic. This holds for all a, b, and c in $\mathbb{Z}_6$; we verify a particular case:

$$3(5 + 4) = 3 \cdot 5 + 3 \cdot 4 \quad \text{in } \mathbb{Z}_6$$

In fact, the left side is $3(5 + 4) = 3 \cdot 3 = 3$, and the right side is $(3 \cdot 5) + (3 \cdot 4) = 3 + 0 = 3$ too. Hence doing arithmetic in $\mathbb{Z}_6$ is familiar. However, there are differences. For example, $3 \cdot 4 = 0$ in $\mathbb{Z}_6$, in contrast to the fact that $a \cdot b = 0$ in $\mathbb{Z}$ can only happen when either $a = 0$ or $b = 0$. Similarly, $3^2 = 3$ in $\mathbb{Z}_6$, unlike $\mathbb{Z}$.

Note that we will make statements like $-30 = 19$ in $\mathbb{Z}_7$; it means that -30 and 19 leave the same remainder 5 when divided by 7, and so are equal in $\mathbb{Z}_7$ because they both equal 5. In general, if $n \geq 2$ is any modulus, the operative fact is that

$$a = b \text{ in } \mathbb{Z}_n \quad \text{if and only if} \quad a - b \text{ is a nultiple of } n.$$

In this case we say that a and b are **equal modulo** n, and write $a = b \pmod{n}$.

Arithmetic in $\mathbb{Z}_n$ is, in a sense, simpler than that for the integers. For example, consider negatives. Given the element 8 in $\mathbb{Z}_{17}$, what is -8? The answer lies in the observation that $8 + 9 = 0$ in $\mathbb{Z}_{17}$, so $-8 = 9$ (and $-9 = 8$). In the same way, finding negatives is not difficult in $\mathbb{Z}_n$ for any modulus n.

Finite Fields

In our study of linear algebra so far the scalars have been real (possibly complex) numbers. The set $\mathbb{R}$ of real numbers has the property that it is closed under addition and multiplication, that the usual laws of arithmetic hold, and that every nonzero real number has an inverse in $\mathbb{R}$. Such a system is called a **field**. Hence the real numbers $\mathbb{R}$ form a field, as does the set $\mathbb{C}$ of complex numbers. Another example is the set $\mathbb{Q}$ of all rational numbers (fractions); however the set $\mathbb{Z}$ of integers is *not* a field—for example, 2 has no inverse *in* the set $\mathbb{Z}$ because $2 \cdot x = 1$ has no solution x in $\mathbb{Z}$.

Our motivation for isolating the concept of a field is that nearly everything we have done remains valid if the scalars are restricted to some field: The gaussian algorithm can be used to solve systems of linear equations with coefficients in the field; a square matrix with entries from the field is invertible if and only if its determinant is nonzero; the matrix inversion algorithm works in the same way; and so on. The reason is that the field has all the properties used in the proofs of these results for the field $\mathbb{R}$, so all the theorems remain valid.

It turns out that there are *finite* fields—that is, finite sets that satisfy the usual laws of arithmetic and in which every nonzero element a has an **inverse**, that is an element b in the field such that $ab = 1$. If $n \geq 2$ is an integer, the modular system $\mathbb{Z}_n$ certainly satisfies the basic laws of arithmetic, but it need not be a field. For example we have $2 \cdot 3 = 0$ in $\mathbb{Z}_6$ so 3 has no inverse in $\mathbb{Z}_6$ (if $3a = 1$ then $2 = 2 \cdot 1 = 2(3a) = 0a = 0$ in $\mathbb{Z}_6$, a contradiction). The problem is that $6 = 2 \cdot 3$ can be properly factored in $\mathbb{Z}$.

An integer $p \geq 2$ is called a **prime** if p *cannot* be factored as $p = ab$ where a and b are positive integers and neither a nor b equals 1. Thus the first few primes are 2, 3, 5, 7, 11, 13, 17, ... If $n \geq 2$ is not a prime and $n = ab$ where $2 \leq a, b \leq n - 1$, then $ab = 0$ in $\mathbb{Z}_n$ and it follows (as above in the case $n = 6$) that b cannot have an inverse in $\mathbb{Z}_n$, and hence that $\mathbb{Z}_n$ is not a field. In other words, if $\mathbb{Z}_n$ is a field, then n must be a prime. Surprisingly, the converse is true:

Theorem 1

If p is a prime, then $\mathbb{Z}_p$ is a field using addition and multiplication modulo p.

The proof can be found in books on abstract algebra.[12] If p is a prime, the field $\mathbb{Z}_p$ is called the **field of integers modulo** p.

For example, consider the case $n = 5$. Then $\mathbb{Z}_5 = \{0, 1, 2, 3, 4\}$ and the addition and multiplication tables are:

+	0	1	2	3	4		×	0	1	2	3	4
0	0	1	2	3	4		0	0	0	0	0	0
1	1	2	3	4	0		1	0	1	2	3	4
2	2	3	4	0	1		2	0	2	4	1	3
3	3	4	0	1	2		3	0	3	1	4	2
4	4	0	1	2	3		4	0	4	3	2	1

Hence 1 and 4 are self-inverse in $\mathbb{Z}_5$, and 2 and 3 are inverses of each other, so $\mathbb{Z}_5$ is indeed a field. Here is another important example.

EXAMPLE 1

If $p = 2$, then $\mathbb{Z}_2 = \{0, 1\}$ is a field with addition and multiplication modulo 2 given by the tables

+	0	1		×	0	1
0	0	1	and	0	0	0
1	1	0		1	0	1

This is binary arithmetic, the basic algebra of computers.

While it is routine to find negatives of elements of $\mathbb{Z}_p$, it is a bit more difficult to find inverses in $\mathbb{Z}_p$. For example, how does one find 14^{-1} in $\mathbb{Z}_{17}$? Since we want $14^{-1} \cdot 14 = 1$ in $\mathbb{Z}_{17}$, we are looking for an integer a with the property that $a \cdot 14 = 1$ modulo 17. Of course we can try all possibilities in $\mathbb{Z}_{17}$ (there are only 17 of them!), and the result is $a = 11$ (verify). However this method is of little use for large primes p, and it is a comfort to know that there is a systematic procedure (called the **euclidean algorithm**) for finding inverses in $\mathbb{Z}_p$ for any prime p. Furthermore, this algorithm is easy to program for a computer. To illustrate the method, let us once again find the inverse of 14 in $\mathbb{Z}_{17}$.

---

12 See, for example, W. K. Nicholson, *Introduction to Abstract Algebra*, 4th ed., (New York: Wiley, 2012).

EXAMPLE 2

Find the inverse of 14 in $\mathbb{Z}_{17}$.

Solution ▶ The idea is to first divide $p = 17$ by 14:
$$17 = 1 \cdot 14 + 3.$$
Now divide (the previous divisor) 14 by the new remainder 3 to get
$$14 = 4 \cdot 3 + 2,$$
and then divide (the previous divisor) 3 by the new remainder 2 to get
$$3 = 1 \cdot 2 + 1.$$
It is a theorem of number theory that, because 17 is a prime, this procedure will *always* lead to a remainder of 1. At this point we eliminate remainders in these equations from the bottom up:

$$\begin{aligned} 1 &= 3 - 1 \cdot 2 & &\text{since } 3 = 1 \cdot 2 + 1 \\ &= 3 - 1 \cdot (14 - 4 \cdot 3) = 5 \cdot 3 - 1 \cdot 14 & &\text{since } 2 = 14 - 4 \cdot 3 \\ &= 5 \cdot (17 - 1 \cdot 14) - 1 \cdot 14 = 5 \cdot 17 - 6 \cdot 14 & &\text{since } 3 = 17 - 1 \cdot 14 \end{aligned}$$

Hence $(-6) \cdot 14 = 1$ in $\mathbb{Z}_{17}$, that is, $11 \cdot 14 = 1$. So $14^{-1} = 11$ in $\mathbb{Z}_{17}$.

As mentioned above, nearly everything we have done with matrices over the field of real numbers can be done in the same way for matrices with entries from $\mathbb{Z}_p$. We illustrate this with one example. Again the reader is referred to books on abstract algebra.

EXAMPLE 3

Determine if the matrix $A = \begin{bmatrix} 1 & 4 \\ 6 & 5 \end{bmatrix}$ from $\mathbb{Z}_7$ is invertible and, if so, find its inverse.

Solution ▶ Working in $\mathbb{Z}_7$ we have $\det A = 1 \cdot 5 - 6 \cdot 4 = 5 - 3 = 2 \neq 0$ in $\mathbb{Z}_7$, so A is invertible. Hence Example 4 Section 2.4 gives $A^{-1} = 2^{-1} \begin{bmatrix} 5 & -4 \\ -6 & 1 \end{bmatrix}$. Note that $2^{-1} = 4$ in $\mathbb{Z}_7$ (because $2 \cdot 4 = 1$ in $\mathbb{Z}_7$). Note also that $-4 = 3$ and $-6 = 1$ in $\mathbb{Z}_7$, so finally $A^{-1} = 4 \begin{bmatrix} 5 & 3 \\ 1 & 1 \end{bmatrix} = \begin{bmatrix} 6 & 5 \\ 4 & 4 \end{bmatrix}$. The reader can verify that indeed $\begin{bmatrix} 1 & 4 \\ 6 & 5 \end{bmatrix} \begin{bmatrix} 6 & 5 \\ 4 & 4 \end{bmatrix} = \begin{bmatrix} 1 & 0 \\ 0 & 1 \end{bmatrix}$ in $\mathbb{Z}_7$.

While we shall not use them, there are finite fields other than $\mathbb{Z}_p$ for the various primes p. Surprisingly, for every prime p and every integer $n \geq 1$, there *exists* a field with exactly p^n elements, and this field is *unique*.[13] It is called the **Galois field** of order p^n, and is denoted $GF(p^n)$.

---

13 See, for example, W. K. Nicholson, *Introduction to Abstract Algebra*, 4th ed., (New York: Wiley, 2012).

Error Correcting Codes

Coding theory is concerned with the transmission of information over a *channel* that is affected by *noise*. The noise causes errors, so the aim of the theory is to find ways to detect such errors and correct at least some of them. General coding theory originated with the work of Claude Shannon (1916–2001) who showed that information can be transmitted at near optimal rates with arbitrarily small chance of error.

Let F denote a finite field and, if $n \geq 1$, let

$$F^n \text{ denote the } F\text{-vector space of } 1 \times n \text{ row matrices over } F$$

with the usual componentwise addition and scalar multiplication. In this context, the rows in F^n are called **words** (or *n*-**words**) and, as the name implies, will be written as $[a\ b\ c\ d] = abcd$. The individual components of a word are called its **digits**. A nonempty subset C of F^n is called a **code** (or an *n*-**code**), and the elements in C are called **code words**. If $F = \mathbb{Z}_2$, these are called **binary** codes.

If a code word $\mathbf{w}$ is transmitted and an error occurs, the resulting word $\mathbf{v}$ is decoded as the code word "closest" to $\mathbf{v}$ in F^n. To make sense of what "closest" means, we need a distance function on F^n analogous to that in $\mathbb{R}^n$ (see Theorem 3 Section 5.3). The usual definition in $\mathbb{R}^n$ does not work in this situation. For example, if $\mathbf{w} = 1111$ in $(\mathbb{Z}_2)^4$ then the square of the distance of $\mathbf{w}$ from $\mathbf{0}$ is $(1-0)^2 + (1-0)^2 + (1-0)^2 + (1-0)^2 = 0$, even though $\mathbf{w} \neq \mathbf{0}$.

However there is a satisfactory notion of distance in F^n due to Richard Hamming (1915–1998). Given a word $\mathbf{w} = a_1 a_2 \cdots a_n$ in F^n, we first define the **Hamming weight** $wt(\mathbf{w})$ to be the number of nonzero digits in $\mathbf{w}$:

$$wt(\mathbf{w}) = wt(a_1 a_2 \cdots a_n) = |\{i \mid a_i \neq 0\}|$$

Clearly, $0 \leq wt(\mathbf{w}) \leq n$ for every word $\mathbf{w}$ in F^n. Given another word $\mathbf{v} = b_1 b_2 \cdots b_n$ in F^n, the **Hamming distance** $d(\mathbf{v}, \mathbf{w})$ between $\mathbf{v}$ and $\mathbf{w}$ is defined by

$$d(\mathbf{v}, \mathbf{w}) = wt(\mathbf{v} - \mathbf{w}) = |\{i \mid b_i \neq a_i\}|.$$

In other words, $d(\mathbf{v}, \mathbf{w})$ is the number of places at which the digits of $\mathbf{v}$ and $\mathbf{w}$ differ. The next result justifies using the term *distance* for this function d.

Theorem 2

Let $\mathbf{u}, \mathbf{v},$ and $\mathbf{w}$ denote words in F^n. Then:

1. $d(\mathbf{v}, \mathbf{w}) \geq 0$.
2. $d(\mathbf{v}, \mathbf{w}) = 0$ if and only if $\mathbf{v} = \mathbf{w}$.
3. $d(\mathbf{v}, \mathbf{w}) = d(\mathbf{w}, \mathbf{v})$.
4. $d(\mathbf{v}, \mathbf{w}) \leq d(\mathbf{v}, \mathbf{u}) + d(\mathbf{u}, \mathbf{w})$

PROOF

(1) and (3) are clear, and (2) follows because $wt(\mathbf{v}) = 0$ if and only if $\mathbf{v} = \mathbf{0}$. To prove (4), write $\mathbf{x} = \mathbf{v} - \mathbf{u}$ and $\mathbf{y} = \mathbf{u} - \mathbf{w}$. Then (4) reads $wt(\mathbf{x} + \mathbf{y}) \leq wt(\mathbf{x}) + wt(\mathbf{y})$. If $\mathbf{x} = a_1 a_2 \cdots a_n$ and $\mathbf{y} = b_1 b_2 \cdots b_n$, this follows because $a_i + b_i \neq 0$ implies that either $a_i \neq 0$ or $b_i \neq 0$.

SECTION 8.7 An Application to Linear Codes over Finite Fields

Given a word $\mathbf{w}$ in F^n and a real number $r > 0$, define the **ball** $B_r(\mathbf{w})$ of radius r (or simply the r-**ball**) about $\mathbf{w}$ as follows:

$$B_r(\mathbf{w}) = \{\mathbf{x} \in F^n \mid d(\mathbf{w}, \mathbf{x}) \leq r\}.$$

Using this we can describe one of the most useful decoding methods.

Nearest Neighbour Decoding

Let C be an n-code, and suppose a word $\mathbf{v}$ is transmitted and $\mathbf{w}$ is received. Then $\mathbf{w}$ is decoded as the code word in C closest to it. (If there is a tie, choose arbitrarily.)

Using this method, we can describe how to construct a code C that can detect (or correct) t errors. Suppose a code word $\mathbf{c}$ is transmitted and a word $\mathbf{w}$ is received with s errors where $1 \leq s \leq t$. Then s is the number of places at which the $\mathbf{c}$- and $\mathbf{w}$-digits differ, that is, $s = d(\mathbf{c}, \mathbf{w})$. Hence $B_t(\mathbf{c})$ consists of all possible received words where at most t errors have occurred.

Assume first that C has the property that no code word lies in the t-ball of another code word. Because $\mathbf{w}$ is in $B_t(\mathbf{c})$ and $\mathbf{w} \neq \mathbf{c}$, this means that $\mathbf{w}$ is not a code word and the error has been detected. If we strengthen the assumption on C to require that the t-balls about code words are pairwise disjoint, then $\mathbf{w}$ belongs to a unique ball (the one about $\mathbf{c}$), and so $\mathbf{w}$ will be correctly decoded as $\mathbf{c}$.

To describe when this happens, let C be an n-code. The **minimum distance** d of C is defined to be the smallest distance between two distinct code words in C; that is,

$$d = \min\{d(\mathbf{v}, \mathbf{w}) \mid \mathbf{v} \text{ and } \mathbf{w} \text{ in } C; \mathbf{v} \neq \mathbf{w}\}.$$

Theorem 3

Let C be an n-code with minimum distance d. Assume that nearest neighbour decoding is used. Then:

1. *If $t < d$, then C can detect t errors.*[14]
2. *If $2t < d$, then C can correct t errors.*

PROOF

1. Let $\mathbf{c}$ be a code word in C. If $\mathbf{w} \in B_t(\mathbf{c})$, then $d(\mathbf{w}, \mathbf{c}) \leq t < d$ by hypothesis. Thus the t-ball $B_t(\mathbf{c})$ contains no other code word, so C can detect t errors by the preceding discussion.

2. If $2t < d$, it suffices (again by the preceding discussion) to show that the t-balls about distinct code words are pairwise disjoint. But if $\mathbf{c} \neq \mathbf{c}'$ are code words in C and $\mathbf{w}$ is in $B_t(\mathbf{c}') \cap B_t(\mathbf{c})$, then Theorem 2 gives

   $$d(\mathbf{c}, \mathbf{c}') \leq d(\mathbf{c}, \mathbf{w}) + d(\mathbf{w}, \mathbf{c}') \leq t + t = 2t < d$$

   by hypothesis, contradicting the minimality of d.

---

14 We say that C detects (corrects) t errors if C can detect (or correct) t or *fewer* errors.

EXAMPLE 4

If $F = \mathbb{Z}_3 = \{0, 1, 2\}$, the 6-code $\{111111, 111222, 222111\}$ has minimum distance 3 and so can detect 2 errors and correct 1 error.

Let $\mathbf{c}$ be any word in F^n. A word $\mathbf{w}$ satisfies $d(\mathbf{w}, \mathbf{c}) = r$ if and only if $\mathbf{w}$ and $\mathbf{c}$ differ in exactly r digits. If $|F| = q$, there are exactly $\binom{n}{r}(q-1)^r$ such words where $\binom{n}{r}$ is the binomial coefficient. Indeed, choose the r places where they differ in $\binom{n}{r}$ ways, and then fill those places in $\mathbf{w}$ in $(q-1)^r$ ways. It follows that the number of words in the t-ball about $\mathbf{c}$ is

$$|B_t(\mathbf{c})| = \binom{n}{0} + \binom{n}{1}(q-1) + \binom{n}{2}(q-1)^2 + \cdots + \binom{n}{t}(q-1)^t = \sum_{i=0}^{t}\binom{n}{i}(q-1)^i.$$

This leads to a useful bound on the size of error-correcting codes.

Theorem 4

Hamming Bound
Let C be an n-code over a field F that can correct t errors using nearest neighbour decoding. If $|F| = q$, then

$$|C| \leq \frac{q^n}{\sum_{i=0}^{t}\binom{n}{i}(q-1)^i}.$$

PROOF

Write $k = \sum_{i=0}^{t}\binom{n}{i}(q-1)^i$. The t-balls centred at distinct code words each contain k words, and there are $|C|$ of them. Moreover they are pairwise disjoint because the code corrects t errors (see the discussion preceding Theorem 3). Hence they contain $k \cdot |C|$ distinct words, and so $k \cdot |C| \leq |F^n| = q^n$, proving the theorem.

A code is called **perfect** if there is equality in the Hamming bound; equivalently, if every word in F^n lies in exactly one t-ball about a code word. For example, if $F = \mathbb{Z}_2$, $n = 3$, and $t = 1$, then $q = 2$ and $\binom{3}{0} + \binom{3}{1} = 4$, so the Hamming bound is $\frac{2^3}{4} = 2$. The 3-code $C = \{000, 111\}$ has minimum distance 3 and so can correct 1 error by Theorem 3. Hence C is perfect.

Linear Codes

Up to this point we have been regarding *any* nonempty subset of the F-vector space F^n as a code. However many important codes are actually subspaces. A subspace $C \subseteq F^n$ of dimension $k \geq 1$ over F is called an (n, k)-**linear code**, or simply an (n, k)-**code**. We do not regard the zero subspace (that is, $k = 0$) as a code.

SECTION 8.7 An Application to Linear Codes over Finite Fields

EXAMPLE 5

If $F = \mathbb{Z}_2$ and $n \geq 2$, the n-**parity-check code** is constructed as follows: An extra digit is added to each word in F^{n-1} to make the number of 1s in the resulting word even (we say such words have **even parity**). The resulting $(n, n-1)$-code is linear because the sum of two words of even parity again has even parity.

Many of the properties of general codes take a simpler form for linear codes. The following result gives a much easier way to find the minimal distance of a linear code, and sharpens the results in Theorem 3.

Theorem 5

Let C be an (n, k)-code with minimum distance d over a finite field F, and use nearest neighbour decoding.

1. $d = \min\{wt(\mathbf{w}) \mid \mathbf{0} \neq \mathbf{w} \text{ in } C\}$.
2. C can detect $t \geq 1$ errors if and only if $t < d$.
3. C can correct $t \geq 1$ errors if and only if $2t < d$.
4. If C can correct $t \geq 1$ errors and $|F| = q$, then

$$\binom{n}{0} + \binom{n}{1}(q-1) + \binom{n}{2}(q-1)^2 + \cdots + \binom{n}{t}(q-1)^t \leq q^{n-k}.$$

PROOF

1. Write $d' = \min\{wt(\mathbf{w}) \mid \mathbf{0} \neq \mathbf{w} \text{ in } C\}$. If $\mathbf{v} \neq \mathbf{w}$ are words in C, then $d(\mathbf{v}, \mathbf{w}) = wt(\mathbf{v} - \mathbf{w}) \geq d'$ because $\mathbf{v} - \mathbf{w}$ is in the subspace C. Hence $d \geq d'$. Conversely, given $\mathbf{w} \neq \mathbf{0}$ in C then, since $\mathbf{0}$ is in C, we have $wt(\mathbf{w}) = d(\mathbf{w}, \mathbf{0}) \geq d$ by the definition of d. Hence $d' \geq d$ and (1) is proved.

2. Assume that C can detect t errors. Given $\mathbf{w} \neq \mathbf{0}$ in C, the t-ball $B_t(\mathbf{w})$ about $\mathbf{w}$ contains no other code word (see the discussion preceding Theorem 3). In particular, it does not contain the code word $\mathbf{0}$, so $t < d(\mathbf{w}, \mathbf{0}) = wt(\mathbf{w})$. Hence $t < d$ by (1). The converse is part of Theorem 3.

3. We require a result of interest in itself.

   Claim. Suppose $\mathbf{c}$ in C has $wt(\mathbf{c}) \leq 2t$. Then $B_t(\mathbf{0}) \cap B_t(\mathbf{c})$ is nonempty.

   Proof. If $wt(\mathbf{c}) \leq t$, then $\mathbf{c}$ itself is in $B_t(\mathbf{0}) \cap B_t(\mathbf{c})$. So assume $t < wt(\mathbf{c}) \leq 2t$. Then $\mathbf{c}$ has more than t nonzero digits, so we can form a new word $\mathbf{w}$ by changing exactly t of these nonzero digits to zero. Then $d(\mathbf{w}, \mathbf{c}) = t$, so $\mathbf{w}$ is in $B_t(\mathbf{c})$. But $wt(\mathbf{w}) = wt(\mathbf{c}) - t \leq t$, so $\mathbf{w}$ is also in $B_t(\mathbf{0})$. Hence $\mathbf{w}$ is in $B_t(\mathbf{0}) \cap B_t(\mathbf{c})$, proving the Claim.

   If C corrects t errors, the t-balls about code words are pairwise disjoint (see the discussion preceding Theorem 3). Hence the claim shows that $wt(\mathbf{c}) > 2t$ for all $\mathbf{c} \neq \mathbf{0}$ in C, from which $d > 2t$ by (1). The other inequality comes from Theorem 3.

4. We have $|C| = q^k$ because $\dim_F C = k$, so this assertion restates Theorem 4.

EXAMPLE 6

If $F = \mathbb{Z}_2$, then

$$C = \{0000000, 0101010, 1010101, 1110000,$$
$$1011010, 0100101, 0001111, 1111111\}$$

is a (7, 3)-code; in fact $C = \text{span}\{0101010, 1010101, 1110000\}$. The minimum distance for C is 3, the minimum weight of a nonzero word in C.

Matrix Generators

Given a linear n-code C over a finite field F, the way encoding works in practice is as follows. A message stream is blocked off into segments of length $k \leq n$ called **messages**. Each message **u** in F^k is encoded as a code word, the code word is transmitted, the receiver decodes the received word as the nearest code word, and then re-creates the original message. A fast and convenient method is needed to encode the incoming messages, to decode the received word after transmission (with or without error), and finally to retrieve messages from code words. All this can be achieved for any linear code using matrix multiplication.

Let G denote a $k \times n$ matrix over a finite field F, and encode each message **u** in F^k as the word **u**G in F^n using matrix multiplication (thinking of words as rows). This amounts to saying that the set of code words is the subspace $C = \{\mathbf{u}G \mid \mathbf{u} \text{ in } F^k\}$ of F^n. This subspace need not have dimension k for every $k \times n$ matrix G. But, if $\{\mathbf{e}_1, \mathbf{e}_2, \ldots, \mathbf{e}_k\}$ is the standard basis of F^k, then $\mathbf{e}_i G$ is row i of G for each I and $\{\mathbf{e}_1 G, \mathbf{e}_2 G, \ldots, \mathbf{e}_k G\}$ spans C. Hence $\dim C = k$ if and only if the rows of G are independent in F^n, and these matrices turn out to be exactly the ones we need. For reference, we state their main properties in Lemma 1 below (see Theorem 4 Section 5.4).

Lemma 1

The following are equivalent for a $k \times n$ matrix G over a finite field F:

1. rank $G = k$.
2. The columns of G span F^k.
3. The rows of G are independent in F^n.
4. The system $GX = B$ is consistent for every column B in $\mathbb{R}^k$.
5. $GK = I_k$ for some $n \times k$ matrix K.

PROOF

(1) $\Rightarrow$ (2). This is because $\dim(\text{col } G) = k$ by (1).

(2) $\Rightarrow$ (4). $G[x_1 \cdots x_n]^T = x_1 \mathbf{c}_1 + \cdots + x_n \mathbf{c}_n$ where $\mathbf{c}_j$ is column j of G.

(4) $\Rightarrow$ (5). $G[\mathbf{k}_1 \cdots \mathbf{k}_k] = [G\mathbf{k}_1 \cdots G\mathbf{k}_k]$ for columns $\mathbf{k}_j$.

(5) $\Rightarrow$ (3). If $a_1 R_1 + \cdots + a_k R_k = 0$ where R_i is row i of G, then $[a_1 \cdots a_k]G = 0$, so $[a_1 \cdots a_k] = 0$, by (5). Hence each $a_i = 0$, proving (3).

(3) $\Rightarrow$ (1). rank $G = \dim(\text{row } G) = k$ by (3).

SECTION 8.7 An Application to Linear Codes over Finite Fields

Note that Theorem 4 Section 5.4 asserts that, over the real field $\mathbb{R}$, the properties in Lemma 1 hold if and only if GG^T is invertible. But this need not be true in general. For example, if $F = \mathbb{Z}_2$ and $G = \begin{bmatrix} 1 & 0 & 1 & 0 \\ 0 & 1 & 0 & 1 \end{bmatrix}$, then $GG^T = 0$. The reason is that the dot product $\mathbf{w} \cdot \mathbf{w}$ can be zero for $\mathbf{w}$ in F^n even if $\mathbf{w} \neq \mathbf{0}$. However, even though GG^T is not invertible, we do have $GK = I_2$ for some 4×2 matrix K over F as Lemma 1 asserts (in fact, $K = \begin{bmatrix} 1 & 0 & 0 & 0 \\ 0 & 1 & 0 & 0 \end{bmatrix}^T$ is one such matrix).

Let $C \subseteq F^n$ be an (n, k)-code over a finite field F. If $\{\mathbf{w}_1, \ldots, \mathbf{w}_k\}$ is a basis of C, let $G = \begin{bmatrix} \mathbf{w}_1 \\ \vdots \\ \mathbf{w}_k \end{bmatrix}$ be the $k \times n$ matrix with the $\mathbf{w}_i$ as its rows. Let $\{\mathbf{e}_1, \ldots, \mathbf{e}_k\}$ is the standard basis of F^k regarded as rows. Then $\mathbf{w}_i = \mathbf{e}_i G$ for each i, so $C = \text{span}\{\mathbf{w}_1, \ldots, \mathbf{w}_k\} = \text{span}\{\mathbf{e}_1 G, \ldots, \mathbf{e}_k G\}$. It follows (verify) that

$$C = \{\mathbf{u}G \mid \mathbf{u} \text{ in } F^k\}.$$

Because of this, the $k \times n$ matrix G is called a **generator** of the code C, and G has rank k by Lemma 1 because its rows $\mathbf{w}_i$ are independent.

In fact, every linear code C in F^n has a generator of a simple, convenient form. If G is a generator matrix for C, let R be the reduced row-echelon form of G. We claim that C is also generated by R. Since $G \to R$ by row operations, Theorem 1 Section 2.5 shows that these same row operations $[G\ I_k] \to [R\ W]$, performed on $[G\ I_k]$, produce an invertible $k \times k$ matrix W such that $R = WG$. This shows that $C = \{\mathbf{u}R \mid \mathbf{u} \text{ in } F^k\}$. [In fact, if $\mathbf{u}$ is in F^k, then $\mathbf{u}G = \mathbf{u}_1 R$ where $\mathbf{u}_1 = \mathbf{u}W^{-1}$ is in F^k, and $\mathbf{u}R = \mathbf{u}_2 G$ where $\mathbf{u}_2 = \mathbf{u}W$ is in F^k]. Thus R is a generator of C, so we may assume that G is in reduced row-echelon form.

In that case, G has no row of zeros (since rank $G = k$) and so contains all the columns of I_k. Hence a series of *column* interchanges will carry G to the block form $G'' = [I_k\ A]$ for some $k \times (n - k)$ matrix A. Hence the code $C'' = \{\mathbf{u}G'' \mid \mathbf{u} \text{ in } F^k\}$ is essentially the same as C; the code words in C'' are obtained from those in C by a series of column interchanges. Hence if C is a linear (n, k)-code, we may (and shall) assume that the generator matrix G has the form

$$G = [I_k\ A] \quad \text{for some} \quad k \times (n - k) \text{ matrix } A.$$

Such a matrix is called a **standard generator**, or a **systematic generator**, for the code C. In this case, if $\mathbf{u}$ is a message word in F^k, the first k digits of the encoded word $\mathbf{u}G$ are just the first k digits of $\mathbf{u}$, so retrieval of $\mathbf{u}$ from $\mathbf{u}G$ is very simple indeed. The last $n - k$ digits of $\mathbf{u}G$ are called **parity digits**.

Parity-Check Matrices

We begin with an important theorem about matrices over a finite field.

Theorem 6

Let F be a finite field, let G be a $k \times n$ matrix of rank k, let H be an $(n - k) \times n$ matrix of rank $n - k$, and let $C = \{\mathbf{u}G \mid \mathbf{u} \text{ in } F^k\}$ and $D = \{\mathbf{v}H \mid \mathbf{V} \text{ in } F^{n-k}\}$ be the codes they generate. Then the following conditions are equivalent:

1. $GH^T = 0$.

2. $HG^T = 0$.
3. $C = \{\mathbf{w} \text{ in } F^n \mid \mathbf{w}H^T = \mathbf{0}\}$.
4. $D = \{\mathbf{w} \text{ in } F^n \mid \mathbf{w}G^T = \mathbf{0}\}$.

PROOF

First, (1) $\Leftrightarrow$ (2) holds because HG^T and GH^T are transposes of each other.

(1) $\Rightarrow$ (3). Consider the linear transformation $T : F^n \to F^{n-k}$ defined by $T(\mathbf{w}) = \mathbf{w}H^T$ for all $\mathbf{w}$ in F^n. To prove (3) we must show that $C = \ker T$. We have $C \subseteq \ker T$ by (1) because $T(\mathbf{u}G) = \mathbf{u}GH^T = \mathbf{0}$ for all $\mathbf{u}$ in F^k. Since $\dim C = \operatorname{rank} G = k$, it is enough (by Theorem 2 Section 6.4) to show that $\dim(\ker T) = k$. However the dimension theorem (Theorem 4 Section 7.2) shows that $\dim(\ker T) = n - \dim(\operatorname{im} T)$, so it is enough to show that $\dim(\operatorname{im} T) = n - k$. But if $R_1, \ldots, R_n$ are the rows of H^T, then block multiplication gives

$$\operatorname{im} T = \{\mathbf{w}H^T \mid \mathbf{w} \text{ in } \mathbb{R}^n\} = \operatorname{span}\{R_1, \ldots, R_n\} = \operatorname{row}(H^T).$$

Hence $\dim(\operatorname{im} T) = \operatorname{rank}(H^T) = \operatorname{rank} H = n - k$, as required. This proves (3).

(3) $\Rightarrow$ (1). If $\mathbf{u}$ is in F^k, then $\mathbf{u}G$ is in C so, by (3), $\mathbf{u}(GH^T) = (\mathbf{u}G)H^T = \mathbf{0}$. Since $\mathbf{u}$ is arbitrary in F^k, it follows that $GH^T = 0$.

(2) $\Leftrightarrow$ (4). The proof is analogous to (1) $\Leftrightarrow$ (3).

The relationship between the codes C and D in Theorem 6 will be characterized in another way in the next subsection.

If C is an (n, k)-code, an $(n - k) \times n$ matrix H is called a **parity-check matrix** for C if $C = \{\mathbf{w} \mid \mathbf{w}H^T = \mathbf{0}\}$ as in Theorem 6. Such matrices are easy to find for a given code C. If $G = [I_k \ A]$ is a standard generator for C where A is $k \times (n - k)$, the $(n - k) \times n$ matrix

$$H = [-A^T \ I_{n-k}]$$

is a parity-check matrix for C. Indeed, rank $H = n - k$ because the rows of H are independent (due to the presence of I_{n-k}), and

$$GH^T = [I_k \ A]\begin{bmatrix} -A \\ I_{n-k} \end{bmatrix} = -A + A = 0$$

by block multiplication. Hence H is a parity-check matrix for C and we have $C = \{\mathbf{w} \text{ in } F^n \mid \mathbf{w}H^T = \mathbf{0}\}$. Since $\mathbf{w}H^T$ and $H\mathbf{w}^T$ are transposes of each other, this shows that C can be characterized as follows:

$$C = \{\mathbf{w} \text{ in } F^n \mid H\mathbf{w}^T = \mathbf{0}\}$$

by Theorem 6.

This is useful in decoding. The reason is that decoding is done as follows: If a code word $\mathbf{c}$ is transmitted and $\mathbf{v}$ is received, then $\mathbf{z} = \mathbf{v} - \mathbf{c}$ is called the **error**. Since $H\mathbf{c}^T = \mathbf{0}$, we have $H\mathbf{z}^T = H\mathbf{v}^T$ and this word

$$\mathbf{s} = H\mathbf{z}^T = H\mathbf{v}^T$$

is called the **syndrome**. The receiver knows $\mathbf{v}$ and $\mathbf{s} = H\mathbf{v}^T$, and wants to recover $\mathbf{c}$. Since $\mathbf{c} = \mathbf{v} - \mathbf{z}$, it is enough to find $\mathbf{z}$. But the possibilities for $\mathbf{z}$ are the solutions of the linear system

$$H\mathbf{z}^T = \mathbf{s}$$

where $\mathbf{s}$ is known. Now recall that Theorem 3 Section 2.2 shows that these solutions have the form $\mathbf{z} = \mathbf{x} + \mathbf{s}$ where $\mathbf{x}$ is any solution of the homogeneous system $H\mathbf{x}^T = \mathbf{0}$, that is, $\mathbf{x}$ is any word in C (by Lemma 1). In other words, the errors $\mathbf{z}$ are the elements of the set

$$C + \mathbf{s} = \{\mathbf{c} + \mathbf{s} \mid \mathbf{c} \text{ in } C\}.$$

The set $C + \mathbf{s}$ is called a **coset** of C. Let $|F| = q$. Since $|C + \mathbf{s}| = |C| = q^{n-k}$ the search for $\mathbf{z}$ is reduced from q^n possibilities in F^n to q^{n-k} possibilities in $C + \mathbf{s}$. This is called **syndrome decoding**, and various methods for improving efficiency and accuracy have been devised. The reader is referred to books on coding for more details.[15]

Orthogonal Codes

Let F be a finite field. Given two words $\mathbf{v} = a_1 a_2 \cdots a_n$ and $\mathbf{w} = b_1 b_2 \cdots b_n$ in F^n, the dot product $\mathbf{v} \cdot \mathbf{w}$ is defined (as in $\mathbb{R}^n$) by

$$\mathbf{v} \cdot \mathbf{w} = a_1 b_1 + a_2 b_2 + \cdots + a_n b_n.$$

Note that $\mathbf{v} \cdot \mathbf{w}$ is an element of F, and it can be computed as a matrix product: $\mathbf{v} \cdot \mathbf{w} = \mathbf{v}\mathbf{w}^T$.

If $C \subseteq F^n$ is an (n, k)-code, the **orthogonal complement** $C^\perp$ is defined as in $\mathbb{R}^n$:

$$C^\perp = \{\mathbf{v} \text{ in } F^n \mid \mathbf{v} \cdot \mathbf{c} = 0 \text{ for all } \mathbf{c} \text{ in } C\}.$$

This is easily seen to be a subspace of F^n, and it turns out to be an $(n, n-k)$-code. This follows when $F = \mathbb{R}$ because we showed (in the projection theorem) that $n = \dim U^\perp + \dim U$ for any subspace U of $\mathbb{R}^n$. However the proofs break down for a finite field F because the dot product in F^n has the property that $\mathbf{w} \cdot \mathbf{w} = 0$ can happen even if $\mathbf{w} \neq \mathbf{0}$. Nonetheless, the result remains valid.

Theorem 7

Let C be an (n, k)-code over a finite field F, let $G = [I_k \ A]$ be a standard generator for C where A is $k \times (n - k)$, and write $H = [-A^T \ I_{n-k}]$ for the parity-check matrix. Then:

1. H is a generator of $C^\perp$.
2. $\dim(C^\perp) = n - k = \text{rank } H$.
3. $C^{\perp\perp} = C$ and $\dim(C^\perp) + \dim C = n$.

PROOF

As in Theorem 6, let $D = \{\mathbf{v}H \mid \mathbf{v} \text{ in } F^{n-k}\}$ denote the code generated by H. Observe first that, for all $\mathbf{w}$ in F^n and all $\mathbf{u}$ in F^k, we have

$$\mathbf{w} \cdot (\mathbf{u}G) = \mathbf{w}(\mathbf{u}G)^T = \mathbf{w}(G^T \mathbf{u}^T) = (\mathbf{w}G^T) \cdot \mathbf{u}.$$

---

[15] For an elementary introduction, see V. Pless, *Introduction to the Theory of Error-Correcting Codes*, 3rd ed., (New York: Wiley, 1998). For a more detailed treatment, see A. A. Bruen and M. A. Forcinito, *Cryptography, Information Theory, and Error-Correction*, (New York: Wiley, 2005).

> Since $C = \{\mathbf{u}G \mid \mathbf{u} \text{ in } F^k\}$, this shows that $\mathbf{w}$ is in $C^\perp$ if and only if $(\mathbf{w}G^T) \cdot \mathbf{u} = 0$ for all $\mathbf{u}$ in F^k; if and only if[16] $\mathbf{w}G^T = \mathbf{0}$; if and only if $\mathbf{w}$ is in D (by Theorem 6). Thus $C^\perp = D$ and a similar argument shows that $D^\perp = C$.
>
> 1. H generates $C^\perp$ because $C^\perp = D = \{\mathbf{v}H \mid \mathbf{v} \text{ in } F^{n-k}\}$.
>
> 2. This follows from (1) because, as we observed above, rank $H = n - k$.
>
> 3. Since $C^\perp = D$ and $D^\perp = C$, we have $C^{\perp\perp} = (C^\perp)^\perp = D^\perp = C$. Finally the second equation in (3) restates (2) because dim $C = k$.

We note in passing that, if C is a subspace of $\mathbb{R}^k$, we have $C + C^\perp = \mathbb{R}^k$ by the projection theorem (Theorem 3 Section 8.1), and $C \cap C^\perp = \{\mathbf{0}\}$ because any vector $\mathbf{x}$ in $C \cap C^\perp$ satisfies $\|\mathbf{x}\|^2 = \mathbf{x} \cdot \mathbf{x} = 0$. However, this fails in general. For example, if $F = \mathbb{Z}_2$ and $C = \text{span}\{1010, 0101\}$ in F^4 then $C^\perp = C$, so $C + C^\perp = C = C \cap C^\perp$.

We conclude with one more example. If $F = \mathbb{Z}_2$, consider the standard matrix G below, and the corresponding parity-check matrix H:

$$G = \begin{bmatrix} 1 & 0 & 0 & 0 & 1 & 1 & 1 \\ 0 & 1 & 0 & 0 & 1 & 1 & 0 \\ 0 & 0 & 1 & 0 & 1 & 0 & 1 \\ 0 & 0 & 0 & 1 & 0 & 1 & 1 \end{bmatrix} \quad \text{and} \quad H = \begin{bmatrix} 1 & 1 & 1 & 0 & 1 & 0 & 0 \\ 1 & 1 & 0 & 1 & 0 & 1 & 0 \\ 1 & 0 & 1 & 1 & 0 & 0 & 1 \end{bmatrix}$$

The code $C = \{\mathbf{u}G \mid \mathbf{u} \text{ in } F^4\}$ generated by G has dimension $k = 4$, and is called the **Hamming (7, 4)-code**. The vectors in C are listed in the first table below. The dual code generated by H has dimension $n - k = 3$ and is listed in the second table.

C:

u	uG
0000	0000000
0001	0001011
0010	0010101
0011	0011110
0100	0100110
0101	0101101
0110	0110011
0111	0111000
1000	1000111
1001	1001100
1010	1010010
1011	1011001
1100	1100001
1101	1101010
1110	1110100
1111	1111111

$C^\perp$:

v	vH
000	0000000
001	1011001
010	1101010
011	0110011
100	1110100
101	0101101
110	0011110
111	1000111

Clearly each nonzero code word in C has weight at least 3, so C has minimum distance $d = 3$. Hence C can detect two errors and correct one error by Theorem 5. The dual code has minimum distance 4 and so can detect 3 errors and correct 1 error.

---

16 If $\mathbf{v} \cdot \mathbf{u} = 0$ for every $\mathbf{u}$ in F^k, then $\mathbf{v} = \mathbf{0}$—let $\mathbf{u}$ range over the standard basis of F^k.

EXERCISES 8.7

1. Find all a in $\mathbb{Z}_{10}$ such that:
   (a) $a^2 = a$
   ◆(b) a has an inverse (and find the inverse)
   (c) $a^k = 0$ for some $k \geq 1$
   ◆(d) $a = 2^k$ for some $k \geq 1$
   (e) $a = b^2$ for some b in $\mathbb{Z}_{10}$

2. (a) Show that if $3a = 0$ in $\mathbb{Z}_{10}$, then necessarily $a = 0$ in $\mathbb{Z}_{10}$.
   ◆(b) Show that $2a = 0$ in $\mathbb{Z}_{10}$ holds in $\mathbb{Z}_{10}$ if and only if $a = 0$ or $a = 5$.

3. Find the inverse of:
   (a) 8 in $\mathbb{Z}_{13}$;   ◆(b) 11 in $\mathbb{Z}_{19}$.

4. If $ab = 0$ in a field F, show that either $a = 0$ or $b = 0$.

5. Show that the entries of the last column of the multiplication table of $\mathbb{Z}_n$ are $0, n-1, n-2, \ldots, 2, 1$ in that order.

6. In each case show that the matrix A is invertible over the given field, and find A^{-1}.
   (a) $A = \begin{bmatrix} 1 & 4 \\ 2 & 1 \end{bmatrix}$ over $\mathbb{Z}_5$.
   ◆(b) $A = \begin{bmatrix} 5 & 6 \\ 4 & 3 \end{bmatrix}$ over $\mathbb{Z}_7$.

7. Consider the linear system $\begin{matrix} 3x + y + 4z = 3 \\ 4x + 3y + z = 1 \end{matrix}$.
   In each case solve the system by reducing the augmented matrix to reduced row-echelon form over the given field:
   (a) $\mathbb{Z}_5$.   ◆(b) $\mathbb{Z}_7$.

8. Let K be a vector space over $\mathbb{Z}_2$ with basis $\{1, t\}$, so $K = \{a + bt \mid a, b, \text{ in } \mathbb{Z}_2\}$. It is known that K becomes a field of four elements if we define $t^2 = 1 + t$. Write down the multiplication table of K.

9. Let K be a vector space over $\mathbb{Z}_3$ with basis $\{1, t\}$, so $K = \{a + bt \mid a, b, \text{ in } \mathbb{Z}_3\}$. It is known that K becomes a field of nine elements if we define $t^2 = -1$ in $\mathbb{Z}_3$. In each case find the inverse of the element x of K:
   (a) $x = 1 + 2t$.   ◆(b) $x = 1 + t$.

10. How many errors can be detected or corrected by each of the following binary linear codes?
    (a) $C = \{0000000, 0011110, 0100111, 0111001, 1001011, 1010101, 1101100, 1110010\}$
    ◆(b) $C = \{0000000000, 0010011111, 0101100111, 0111111000, 1001110001, 1011101110, 1100010110, 1110001001\}$

11. (a) If a binary linear $(n, 2)$-code corrects one error, show that $n \geq 5$. [*Hint*: Hamming bound.]
    ◆(b) Find a $(5, 2)$-code that corrects one error.

12. (a) If a binary linear $(n, 3)$-code corrects two errors, show that $n \geq 9$. [*Hint*: Hamming bound.]
    ◆(b) If $G = \begin{bmatrix} 1 & 0 & 0 & 1 & 1 & 1 & 1 & 0 & 0 & 0 \\ 0 & 1 & 0 & 1 & 1 & 0 & 0 & 1 & 1 & 0 \\ 0 & 0 & 1 & 1 & 0 & 1 & 0 & 1 & 1 & 1 \end{bmatrix}$, show that the binary $(10, 3)$-code generated by G corrects two errors. [It can be shown that no binary $(9, 3)$-code corrects two errors.]

13. (a) Show that no binary linear $(4, 2)$-code can correct single errors.
    ◆(b) Find a binary linear $(5, 2)$-code that can correct one error.

14. Find the standard generator matrix G and the parity-check matrix H for each of the following systematic codes:
    (a) $\{00000, 11111\}$ over $\mathbb{Z}_2$.
    ◆(b) Any systematic $(n, 1)$-code where $n \geq 2$.
    (c) The code in Exercise 10(a).
    (d) The code in Exercise 10(b).

15. Let $\mathbf{c}$ be a word in F^n. Show that $B_t(\mathbf{c}) = \mathbf{c} + B_t(\mathbf{0})$, where we write $\mathbf{c} + B_t(\mathbf{0}) = \{\mathbf{c} + \mathbf{v} \mid \mathbf{v} \text{ in } B_t(\mathbf{0})\}$.

16. If a (n, k)-code has two standard generator matrices G and G_1, show that $G = G_1$.

17. Let C be a binary linear n-code (over $\mathbb{Z}_2$). Show that either each word in C has even weight, or half the words in C have even weight and half have odd weight. [*Hint*: The dimension theorem.]

SECTION 8.8 An Application to Quadratic Forms[17]

An expression like $x_1^2 + x_2^2 + x_3^2 - 2x_1x_3 + x_2x_3$ is called a quadratic form in the variables x_1, x_2, and x_3. In this section we show that new variables y_1, y_2, and y_3 can always be found so that the quadratic form, when expressed in terms of the new variables, has no cross terms y_1y_2, y_1y_3, or y_2y_3. Moreover, we do this for forms involving any finite number of variables using orthogonal diagonalization. This has far-reaching applications; quadratic forms arise in such diverse areas as statistics, physics, the theory of functions of several variables, number theory, and geometry.

Definition 8.13 *A* **quadratic form** *q in the n variables x_1, x_2, ..., x_n is a linear combination of terms x_1^2, x_2^2, ..., x_n^2, and cross terms x_1x_2, x_1x_3, x_2x_3, ...*

If $n = 3$, q has the form

$$q = a_{11}x_1^2 + a_{22}x_2^2 + a_{33}x_3^2 + a_{12}x_1x_2 + a_{21}x_2x_1 \\ + a_{13}x_1x_3 + a_{31}x_3x_1 + a_{23}x_2x_3 + a_{32}x_3x_2$$

In general

$$q = a_{11}x_1^2 + a_{22}x_2^2 + \cdots + a_{nn}x_n^2 + a_{12}x_1x_2 + a_{13}x_1x_3 + \cdots$$

This sum can be written compactly as a matrix product

$$q = q(\mathbf{x}) = \mathbf{x}^T A \mathbf{x}$$

where $\mathbf{x} = (x_1, x_2, ..., x_n)$ is thought of as a column, and $A = [a_{ij}]$ is a real $n \times n$ matrix. Note that if $i \neq j$, two separate terms $a_{ij}x_ix_j$ and $a_{ji}x_jx_i$ are listed, each of which involves x_ix_j, and they can (rather cleverly) be replaced by

$$\tfrac{1}{2}(a_{ij} + a_{ji})x_ix_j \quad \text{and} \quad \tfrac{1}{2}(a_{ij} + a_{ji})x_jx_i$$

respectively, *without altering the quadratic form*. Hence there is no loss of generality in assuming that x_ix_j and x_jx_i have the same coefficient in the sum for q. In other words, **we may assume that A is symmetric**.

EXAMPLE 1

Write $q = x_1^2 + 3x_3^2 + 2x_1x_2 - x_1x_3$ in the form $q(\mathbf{x}) = \mathbf{x}^T A \mathbf{x}$, where A is a symmetric 3×3 matrix.

Solution ▸ The cross terms are $2x_1x_2 = x_1x_2 + x_2x_1$ and $-x_1x_3 = -\tfrac{1}{2}x_1x_3 - \tfrac{1}{2}x_3x_1$. Of course, x_2x_3 and x_3x_2 both have coefficient zero, as does x_2^2. Hence

$$q(\mathbf{x}) = [x_1 \ x_2 \ x_3] \begin{bmatrix} 1 & 1 & -\tfrac{1}{2} \\ 1 & 0 & 0 \\ -\tfrac{1}{2} & 0 & 3 \end{bmatrix} \begin{bmatrix} x_1 \\ x_2 \\ x_3 \end{bmatrix}$$

is the required form (verify).

We shall assume from now on that all quadratic forms are given by

$$q(\mathbf{x}) = \mathbf{x}^T A \mathbf{x}$$

---

[17] This section requires only Section 8.2.

SECTION 8.8 An Application to Quadratic Forms

where A is symmetric. Given such a form, the problem is to find new variables $y_1, y_2, \ldots, y_n$, related to $x_1, x_2, \ldots, x_n$, with the property that when q is expressed in terms of $y_1, y_2, \ldots, y_n$, there are no cross terms. If we write

$$\mathbf{y} = (y_1, y_2, \ldots, y_n)^T$$

this amounts to asking that $q = \mathbf{y}^T D \mathbf{y}$ where D is diagonal. It turns out that this can always be accomplished and, not surprisingly, that D is the matrix obtained when the symmetric matrix A is othogonally diagonalized. In fact, as Theorem 2 Section 8.2 shows, a matrix P can be found that is orthogonal (that is, $P^{-1} = P^T$) and diagonalizes A:

$$P^T A P = D = \begin{bmatrix} \lambda_1 & 0 & \cdots & 0 \\ 0 & \lambda_2 & \cdots & 0 \\ \vdots & \vdots & & \vdots \\ 0 & 0 & \cdots & \lambda_n \end{bmatrix}$$

The diagonal entries $\lambda_1, \lambda_2, \ldots, \lambda_n$ are the (not necessarily distinct) eigenvalues of A, repeated according to their multiplicities in $c_A(x)$, and the columns of P are corresponding (orthonormal) eigenvectors of A. As A is symmetric, the λ_i are real by Theorem 7 Section 5.5.

Now define new variables $\mathbf{y}$ by the equations

$$\mathbf{x} = P\mathbf{y} \quad \text{equivalently} \quad \mathbf{y} = P^T \mathbf{x}$$

Then substitution in $q(\mathbf{x}) = \mathbf{x}^T A \mathbf{x}$ gives

$$q = (P\mathbf{y})^T A (P\mathbf{y}) = \mathbf{y}^T (P^T A P) \mathbf{y} = \mathbf{y}^T D \mathbf{y} = \lambda_1 y_1^2 + \lambda_2 y_2^2 + \cdots + \lambda_n y_n^2$$

Hence this change of variables produces the desired simplification in q.

Theorem 1

Diagonalization Theorem
Let $q = \mathbf{x}^T A \mathbf{x}$ be a quadratic form in the variables $x_1, x_2, \ldots, x_n$, where $\mathbf{x} = (x_1, x_2, \ldots, x_n)^T$ and A is a symmetric $n \times n$ matrix. Let P be an orthogonal matrix such that $P^T A P$ is diagonal, and define new variables $\mathbf{y} = (y_1, y_2, \ldots, y_n)^T$ by

$$\mathbf{x} = P\mathbf{y} \quad \text{equivalently} \quad \mathbf{y} = P^T \mathbf{x}$$

If q is expressed in terms of these new variables $y_1, y_2, \ldots, y_n$, the result is

$$q = \lambda_1 y_1^2 + \lambda_2 y_2^2 + \cdots + \lambda_n y_n^2$$

where $\lambda_1, \lambda_2, \ldots, \lambda_n$ are the eigenvalues of A repeated according to their multiplicities.

Let $q = \mathbf{x}^T A \mathbf{x}$ be a quadratic form where A is a symmetric matrix and let $\lambda_1, \ldots, \lambda_n$ be the (real) eigenvalues of A repeated according to their multiplicities. A corresponding set $\{\mathbf{f}_1, \ldots, \mathbf{f}_n\}$ of orthonormal eigenvectors for A is called a set of **principal axes** for the quadratic form q. (The reason for the name will become clear later.) The orthogonal matrix P in Theorem 1 is given as $P = [\mathbf{f}_1 \cdots \mathbf{f}_n]$, so the variables X and Y are related by

$$\mathbf{x} = P\mathbf{y} = [\mathbf{f}_1 \ \mathbf{f}_2 \ \cdots \ \mathbf{f}_n] \begin{bmatrix} y_1 \\ y_2 \\ \vdots \\ y_n \end{bmatrix} = y_1 \mathbf{f}_1 + y_2 \mathbf{f}_2 + \cdots + y_n \mathbf{f}_n.$$

Thus the new variables y_i are the coefficients when $\mathbf{x}$ is expanded in terms of the orthonormal basis $\{\mathbf{f}_1, \ldots, \mathbf{f}_n\}$ of $\mathbb{R}^n$. In particular, the coefficients y_i are given by $y_i = \mathbf{x} \cdot \mathbf{f}_i$ by the expansion theorem (Theorem 6 Section 5.3). Hence q itself is easily computed from the eigenvalues λ_i and the principal axes $\mathbf{f}_i$:

$$q = q(\mathbf{x}) = \lambda_1(\mathbf{x} \cdot \mathbf{f}_1)^2 + \cdots + \lambda_n(\mathbf{x} \cdot \mathbf{f}_n)^2.$$

EXAMPLE 2

Find new variables y_1, y_2, y_3, and y_4 such that
$$q = 3(x_1^2 + x_2^2 + x_3^2 + x_4^2)$$
$$+ 2x_1x_2 - 10x_1x_3 + 10x_1x_4 + 10x_2x_3 - 10x_2x_4 + 2x_3x_4$$
has diagonal form, and find the corresponding principal axes.

Solution ▶ The form can be written as $q = \mathbf{x}^T A\mathbf{x}$, where

$$\mathbf{x} = \begin{bmatrix} x_1 \\ x_2 \\ x_3 \\ x_4 \end{bmatrix} \quad \text{and} \quad A = \begin{bmatrix} 3 & 1 & -5 & 5 \\ 1 & 3 & 5 & -5 \\ -5 & 5 & 3 & 1 \\ 5 & -5 & 1 & 3 \end{bmatrix}$$

A routine calculation yields
$$c_A(x) = \det(xI - A) = (x - 12)(x + 8)(x - 4)^2$$
so the eigenvalues are $\lambda_1 = 12$, $\lambda_2 = -8$, and $\lambda_3 = \lambda_4 = 4$. Corresponding orthonormal eigenvectors are the principal axes:

$$\mathbf{f}_1 = \tfrac{1}{2}\begin{bmatrix} 1 \\ -1 \\ -1 \\ 1 \end{bmatrix} \quad \mathbf{f}_2 = \tfrac{1}{2}\begin{bmatrix} 1 \\ -1 \\ 1 \\ -1 \end{bmatrix} \quad \mathbf{f}_3 = \tfrac{1}{2}\begin{bmatrix} 1 \\ 1 \\ 1 \\ 1 \end{bmatrix} \quad \mathbf{f}_4 = \tfrac{1}{2}\begin{bmatrix} 1 \\ 1 \\ -1 \\ -1 \end{bmatrix}$$

The matrix

$$P = [\mathbf{f}_1 \; \mathbf{f}_2 \; \mathbf{f}_3 \; \mathbf{f}_4] = \tfrac{1}{2}\begin{bmatrix} 1 & 1 & 1 & 1 \\ -1 & -1 & 1 & 1 \\ -1 & 1 & 1 & -1 \\ 1 & -1 & 1 & -1 \end{bmatrix}$$

is thus orthogonal, and $P^{-1}AP = P^T AP$ is diagonal. Hence the new variables $\mathbf{y}$ and the old variables $\mathbf{x}$ are related by $\mathbf{y} = P^T\mathbf{x}$ and $\mathbf{x} = P\mathbf{y}$. Explicitly,

$$y_1 = \tfrac{1}{2}(x_1 - x_2 - x_3 + x_4) \qquad x_1 = \tfrac{1}{2}(y_1 + y_2 + y_3 + y_4)$$
$$y_2 = \tfrac{1}{2}(x_1 - x_2 + x_3 - x_4) \qquad x_2 = \tfrac{1}{2}(-y_1 - y_2 + y_3 + y_4)$$
$$y_3 = \tfrac{1}{2}(x_1 + x_2 + x_3 + x_4) \qquad x_3 = \tfrac{1}{2}(-y_1 + y_2 + y_3 - y_4)$$
$$y_3 = \tfrac{1}{2}(x_1 + x_2 - x_3 - x_4) \qquad x_4 = \tfrac{1}{2}(y_1 - y_2 + y_3 - y_4)$$

If these x_i are substituted in the original expression for q, the result is
$$q = 12y_1^2 - 8y_2^2 + 4y_3^2 + 4y_4^2$$
This is the required diagonal form.

SECTION 8.8 An Application to Quadratic Forms

It is instructive to look at the case of quadratic forms in two variables x_1 and x_2. Then the principal axes can always be found by rotating the x_1 and x_2 axes counterclockwise about the origin through an angle θ. This rotation is a linear transformation $R_\theta : \mathbb{R}^2 \to \mathbb{R}^2$, and it is shown in Theorem 4 Section 2.6 that R_θ has matrix $P = \begin{bmatrix} \cos\theta & -\sin\theta \\ \sin\theta & \cos\theta \end{bmatrix}$. If $\{\mathbf{e}_1, \mathbf{e}_2\}$ denotes the standard basis of $\mathbb{R}^2$, the rotation produces a new basis $\{\mathbf{f}_1, \mathbf{f}_2\}$ given by

$$\mathbf{f}_1 = R_\theta(\mathbf{e}_1) = \begin{bmatrix} \cos\theta \\ \sin\theta \end{bmatrix} \quad \text{and} \quad \mathbf{f}_2 = R_\theta(\mathbf{e}_2) = \begin{bmatrix} -\sin\theta \\ \cos\theta \end{bmatrix} \quad (*)$$

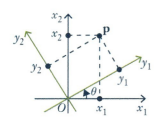

Given a point $\mathbf{p} = \begin{bmatrix} x_1 \\ x_2 \end{bmatrix} = x_1\mathbf{e}_1 + x_2\mathbf{e}_2$ in the original system, let y_1 and y_2 be the coordinates of $\mathbf{p}$ in the new system (see the diagram). That is,

$$\begin{bmatrix} x_1 \\ x_2 \end{bmatrix} = \mathbf{p} = y_1\mathbf{f}_1 + y_2\mathbf{f}_2 = \begin{bmatrix} \cos\theta & -\sin\theta \\ \sin\theta & \cos\theta \end{bmatrix}\begin{bmatrix} y_1 \\ y_2 \end{bmatrix} \quad (**)$$

Writing $\mathbf{x} = \begin{bmatrix} x_1 \\ x_2 \end{bmatrix}$ and $\mathbf{y} = \begin{bmatrix} y_1 \\ y_2 \end{bmatrix}$, this reads $\mathbf{x} = P\mathbf{y}$ so, since P is orthogonal, this is the change of variables formula for the rotation as in Theorem 1.

If $r \neq 0 \neq s$, the graph of the equation $rx_1^2 + sx_2^2 = 1$ is called an **ellipse** if $rs > 0$ and a **hyperbola** if $rs < 0$. More generally, given a quadratic form

$$q = ax_1^2 + bx_1x_2 + cx_2^2 \quad \text{where not all of } a, b, \text{ and } c \text{ are zero,}$$

the graph of the equation $q = 1$ is called a **conic**. We can now completely describe this graph. There are two special cases which we leave to the reader.

1. If exactly one of a and c is zero, then the graph of $q = 1$ is a **parabola**.

So we assume that $a \neq 0$ and $c \neq 0$. In this case, the description depends on the quantity $b^2 - 4ac$, called the **discriminant** of the quadratic form q.

2. If $b^2 - 4ac = 0$, then either both $a \geq 0$ and $c \geq 0$, or both $a \leq 0$ and $c \leq 0$. Hence $q = (\sqrt{a}x_1 + \sqrt{c}x_2)^2$ or $q = (\sqrt{-a}x_1 + \sqrt{-c}x_2)^2$, so the graph of $q = 1$ is a **pair of straight lines** in either case.

So we also assume that $b^2 - 4ac \neq 0$. But then the next theorem asserts that there exists a rotation of the plane about the origin which transforms the equation $ax_1^2 + bx_1x_2 + cx_2^2 = 1$ into either an ellipse or a hyperbola, and the theorem also provides a simple way to decide which conic it is.

Theorem 2

Consider the quadratic form $q = ax_1^2 + bx_1x_2 + cx_2^2$ where a, c, and $b^2 - 4ac$ are all nonzero.

1. There is a counterclockwise rotation of the coordinate axes about the origin such that, in the new coordinate system, q has no cross term.
2. The graph of the equation

$$ax_1^2 + bx_1x_2 + cx_2^2 = 1$$

is an ellipse if $b^2 - 4ac < 0$ and an hyperbola if $b^2 - 4ac > 0$.

> **PROOF**
>
> If $b = 0$, q already has no cross term and (1) and (2) are clear. So assume $b \neq 0$. The matrix $A = \begin{bmatrix} a & \frac{1}{2}b \\ \frac{1}{2}b & c \end{bmatrix}$ of q has characteristic polynomial
>
> $c_A(x) = x^2 - (a + c)x - \frac{1}{4}(b^2 - 4ac)$. If we write $d = \sqrt{b^2 + (a - c)^2}$ for convenience; then the quadratic formula gives the eigenvalues
>
> $$\lambda_1 = \tfrac{1}{2}[a + c - d] \quad \text{and} \quad \lambda_2 = \tfrac{1}{2}[a + c + d]$$
>
> with corresponding principal axes
>
> $$\mathbf{f}_1 = \frac{1}{\sqrt{b^2 + (a - c - d)^2}} \begin{bmatrix} a - c - d \\ b \end{bmatrix} \quad \text{and}$$
>
> $$\mathbf{f}_2 = \frac{1}{\sqrt{b^2 + (a - c - d)^2}} \begin{bmatrix} -b \\ a - c - d \end{bmatrix}$$
>
> as the reader can verify. These agree with equation (∗) above if θ is an angle such that
>
> $$\cos \theta = \frac{a - c - d}{\sqrt{b^2 + (a - c - d)^2}} \quad \text{and} \quad \sin \theta = \frac{b}{\sqrt{b^2 + (a - c - d)^2}}$$
>
> Then $P = [\mathbf{f}_1 \ \mathbf{f}_2] = \begin{bmatrix} \cos \theta & -\sin \theta \\ \sin \theta & \cos \theta \end{bmatrix}$ diagonalizes A and equation (∗∗) becomes the formula $\mathbf{x} = P\mathbf{y}$ in Theorem 1. This proves (1).
>
> Finally, A is similar to $\begin{bmatrix} \lambda_1 & 0 \\ 0 & \lambda_2 \end{bmatrix}$ so $\lambda_1 \lambda_2 = \det A = \tfrac{1}{4}(4ac - b^2)$. Hence the graph of $\lambda_1 y_1^2 + \lambda_2 y_2^2 = 1$ is an ellipse if $b^2 < 4ac$ and an hyperbola if $b^2 > 4ac$. This proves (2).

EXAMPLE 3

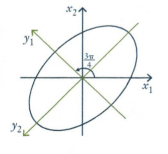

Consider the equation $x^2 + xy + y^2 = 1$. Find a rotation so that the equation has no cross term.

Solution ▶ Here $a = b = c = 1$ in the notation of Theorem 2, so $\cos \theta = \frac{-1}{\sqrt{2}}$ and $\sin \theta = \frac{1}{\sqrt{2}}$. Hence $\theta = \frac{3\pi}{4}$ will do it. The new variables are $y_1 = \frac{1}{\sqrt{2}}(x_2 - x_1)$ and $y_2 = \frac{-1}{\sqrt{2}}(x_2 + x_1)$ by (∗∗), and the equation becomes $y_1^2 + 3y_2^2 = 2$. The angle θ has been chosen such that the new y_1 and y_2 axes are the axes of symmetry of the ellipse (see the diagram). The eigenvectors $\mathbf{f}_1 = \frac{1}{\sqrt{2}}\begin{bmatrix} -1 \\ 1 \end{bmatrix}$ and $\mathbf{f}_2 = \frac{1}{\sqrt{2}}\begin{bmatrix} -1 \\ -1 \end{bmatrix}$ point along these axes of symmetry, and this is the reason for the name *principal axes*.

The determinant of any orthogonal matrix P is either 1 or -1 (because $PP^T = I$). The orthogonal matrices $\begin{bmatrix} \cos \theta & -\sin \theta \\ \sin \theta & \cos \theta \end{bmatrix}$ arising from rotations all have determinant 1. More generally, given any quadratic form $q = \mathbf{x}^T A\mathbf{x}$, the orthogonal matrix P such that $P^T A P$ is diagonal can always be chosen so that $\det P = 1$ by

SECTION 8.8 An Application to Quadratic Forms

interchanging two eigenvalues (and hence the corresponding columns of P). It is shown in Theorem 4 Section 10.4 that orthogonal 2×2 matrices with determinant 1 correspond to rotations. Similarly, it can be shown that orthogonal 3×3 matrices with determinant 1 correspond to rotations about a line through the origin. This extends Theorem 2: Every quadratic form in two or three variables can be diagonalized by a rotation of the coordinate system.

Congruence

We return to the study of quadratic forms in general.

Theorem 3

If $q(\mathbf{x}) = \mathbf{x}^T A \mathbf{x}$ is a quadratic form given by a symmetric matrix A, then A is uniquely determined by q.

PROOF

Let $q(\mathbf{x}) = \mathbf{x}^T B \mathbf{x}$ for all $\mathbf{x}$ where $B^T = B$. If $C = A - B$, then $C^T = C$ and $\mathbf{x}^T C \mathbf{x} = 0$ for all $\mathbf{x}$. We must show that $C = 0$. Given $\mathbf{y}$ in $\mathbb{R}^n$,

$$0 = (\mathbf{x} + \mathbf{y})^T C (\mathbf{x} + \mathbf{y}) = \mathbf{x}^T C \mathbf{x} + \mathbf{x}^T C \mathbf{y} + \mathbf{y}^T C \mathbf{x} + \mathbf{y}^T C \mathbf{y}$$
$$= \mathbf{x}^T C \mathbf{y} + \mathbf{y}^T C \mathbf{x}$$

But $\mathbf{y}^T C \mathbf{x} = (\mathbf{x}^T C \mathbf{y})^T = \mathbf{x}^T C \mathbf{y}$ (it is 1×1). Hence $\mathbf{x}^T C \mathbf{y} = 0$ for all $\mathbf{x}$ and $\mathbf{y}$ in $\mathbb{R}^n$. If $\mathbf{e}_j$ is column j of I_n, then the (i, j)-entry of C is $\mathbf{e}_i^T C \mathbf{e}_j = 0$. Thus $C = 0$.

Hence we can speak of *the* symmetric matrix of a quadratic form.

On the other hand, a quadratic form q in variables x_i can be written in several ways as a linear combination of squares of new variables, even if the new variables are required to be linear combinations of the x_i. For example, if $q = 2x_1^2 - 4x_1 x_2 + x_2^2$ then

$$q = 2(x_1 - x_2)^2 - x_2^2 \quad \text{and} \quad q = -2x_1^2 + (2x_1 - x_2)^2$$

The question arises: How are these changes of variables related, and what properties do they share? To investigate this, we need a new concept.

Let a quadratic form $q = q(\mathbf{x}) = \mathbf{x}^T A \mathbf{x}$ be given in terms of variables $\mathbf{x} = (x_1, x_2, \ldots, x_n)^T$. If the new variables $\mathbf{y} = (y_1, y_2, \ldots, y_n)^T$ are to be linear combinations of the x_i, then $\mathbf{y} = A\mathbf{x}$ for some $n \times n$ matrix A. Moreover, since we want to be able to solve for the x_i in terms of the y_i, we ask that the matrix A be invertible. Hence suppose U is an invertible matrix and that the new variables $\mathbf{y}$ are given by

$$\mathbf{y} = U^{-1} \mathbf{x}, \quad \text{equivalently} \quad \mathbf{x} = U \mathbf{y}.$$

In terms of these new variables, q takes the form

$$q = q(\mathbf{x}) = (U\mathbf{y})^T A (U\mathbf{y}) = \mathbf{y}^T (U^T A U) \mathbf{y}.$$

That is, q has matrix $U^T A U$ with respect to the new variables $\mathbf{y}$. Hence, to study changes of variables in quadratic forms, we study the following relationship on matrices: Two $n \times n$ matrices A and B are called **congruent**, written $A \stackrel{c}{\sim} B$, if $B = U^T A U$ for some invertible matrix U. Here are some properties of congruence:

1. $A \stackrel{c}{\sim} A$ for all A.
2. If $A \stackrel{c}{\sim} B$, then $B \stackrel{c}{\sim} A$.
3. If $A \stackrel{c}{\sim} B$ and $B \stackrel{c}{\sim} C$, then $A \stackrel{c}{\sim} C$.
4. If $A \stackrel{c}{\sim} B$, then A is symmetric if and only if B is symmetric.
5. If $A \stackrel{c}{\sim} B$, then rank A = rank B.

The converse to (5) can fail even for symmetric matrices.

EXAMPLE 4

The symmetric matrices $A = \begin{bmatrix} 1 & 0 \\ 0 & 1 \end{bmatrix}$ and $B = \begin{bmatrix} 1 & 0 \\ 0 & -1 \end{bmatrix}$ have the same rank but are not congruent. Indeed, if $A \stackrel{c}{\sim} B$, an invertible matrix U exists such that $B = U^T A U = U^T U$. But then $-1 = \det B = (\det U)^2$, a contradiction.

The key distinction between A and B in Example 4 is that A has two positive eigenvalues (counting multiplicities) whereas B has only one.

Theorem 4

Sylvester's Law of Inertia
If $A \stackrel{c}{\sim} B$, then A and B have the same number of positive eigenvalues, counting multiplicities.

The proof is given at the end of this section.

The **index** of a symmetric matrix A is the number of positive eigenvalues of A. If $q = q(\mathbf{x}) = \mathbf{x}^T A \mathbf{x}$ is a quadratic form, the **index** and **rank** of q are defined to be, respectively, the index and rank of the matrix A. As we saw before, if the variables expressing a quadratic form q are changed, the new matrix is congruent to the old one. Hence the index and rank depend only on q and not on the way it is expressed.

Now let $q = q(\mathbf{x}) = \mathbf{x}^T A \mathbf{x}$ be any quadratic form in n variables, of index k and rank r, where A is symmetric. We claim that new variables $\mathbf{z}$ can be found so that q is **completely diagonalized**—that is,

$$q(\mathbf{z}) = z_1^2 + \cdots + z_k^2 - z_{k+1}^2 - \cdots - z_r^2$$

If $k \leq r \leq n$, let $D_n(k, r)$ denote the $n \times n$ diagonal matrix whose main diagonal consists of k ones, followed by $r - k$ minus ones, followed by $n - r$ zeros. Then we seek new variables $\mathbf{z}$ such that

$$q(\mathbf{z}) = \mathbf{z}^T D_n(k, r) \mathbf{z}$$

To determine $\mathbf{z}$, first diagonalize A as follows: Find an orthogonal matrix P_0 such that

$$P_0^T A P_0 = D = \operatorname{diag}(\lambda_1, \lambda_2, \ldots, \lambda_r, 0, \ldots, 0)$$

is diagonal with the nonzero eigenvalues $\lambda_1, \lambda_2, \ldots, \lambda_r$ of A on the main diagonal (followed by $n - r$ zeros). By reordering the columns of P_0, if necessary, we may assume that $\lambda_1, \ldots, \lambda_k$ are positive and $\lambda_{k+1}, \ldots, \lambda_r$ are negative. This being the case, let D_0 be the $n \times n$ diagonal matrix

$$D_0 = \operatorname{diag}\left(\frac{1}{\sqrt{\lambda_1}}, \ldots, \frac{1}{\sqrt{\lambda_k}}, \frac{1}{\sqrt{-\lambda_{k+1}}}, \ldots, \frac{1}{\sqrt{-\lambda_r}}, 1, \ldots, 1\right)$$

SECTION 8.8 An Application to Quadratic Forms

Then $D_0^T D D_0 = D_n(k, r)$, so if new variables $\mathbf{z}$ are given by $\mathbf{x} = (P_0 D_0)\mathbf{z}$, we obtain

$$q(\mathbf{z}) = \mathbf{z}^T D_n(k, r)\mathbf{z} = z_1^2 + \cdots + z_k^2 - z_{k+1}^2 - \cdots - z_r^2$$

as required. Note that the change-of-variables matrix $P_0 D_0$ from $\mathbf{z}$ to $\mathbf{x}$ has orthogonal columns (in fact, scalar multiples of the columns of P_0).

EXAMPLE 5

Completely diagonalize the quadratic form q in Example 2 and find the index and rank.

Solution ▶ In the notation of Example 2, the eigenvalues of the matrix A of q are $12, -8, 4, 4$; so the index is 3 and the rank is 4. Moreover, the corresponding orthogonal eigenvectors are $\mathbf{f}_1, \mathbf{f}_2, \mathbf{f}_3$ (see Example 2), and $\mathbf{f}_4$. Hence $P_0 = [\mathbf{f}_1 \; \mathbf{f}_3 \; \mathbf{f}_4 \; \mathbf{f}_2]$ is orthogonal and

$$P_0^T A P_0 = \text{diag}(12, 4, 4, -8)$$

As before, take $D_0 = \text{diag}(\frac{1}{\sqrt{12}}, \frac{1}{2}, \frac{1}{2}, \frac{1}{\sqrt{8}})$ and define the new variables $\mathbf{z}$ by $\mathbf{x} = (P_0 D_0)\mathbf{z}$. Hence the new variables are given by $\mathbf{z} = D_0^{-1} P_0^T \mathbf{x}$. The result is

$$z_1 = \sqrt{3}(x_1 - x_2 - x_3 + x_4)$$
$$z_2 = x_1 + x_2 + x_3 + x_4$$
$$z_3 = x_1 + x_2 - x_3 - x_4$$
$$z_4 = \sqrt{2}(x_1 - x_2 + x_3 - x_4)$$

This discussion gives the following information about symmetric matrices.

Theorem 5

Let A and B be symmetric $n \times n$ matrices, and let $0 \leq k \leq r \leq n$.
1. A has index k and rank r if and only if $A \stackrel{c}{\sim} D_n(k, r)$.
2. $A \stackrel{c}{\sim} B$ if and only if they have the same rank and index.

PROOF

1. If A has index k and rank r, take $U = P_0 D_0$ where P_0 and D_0 are as described prior to Example 5. Then $U^T A U = D_n(k, r)$. The converse is true because $D_n(k, r)$ has index k and rank r (using Theorem 4).
2. If A and B both have index k and rank r, then $A \stackrel{c}{\sim} D_n(k, r) \stackrel{c}{\sim} B$ by (1). The converse was given earlier.

PROOF OF THEOREM 4

By Theorem 1, $A \stackrel{c}{\sim} D_1$ and $B \stackrel{c}{\sim} D_2$ where D_1 and D_2 are diagonal and have the same eigenvalues as A and B, respectively. We have $D_1 \stackrel{c}{\sim} D_2$, (because $A \stackrel{c}{\sim} B$), so we may assume that A and B are both diagonal. Consider the quadratic form $q(\mathbf{x}) = \mathbf{x}^T A \mathbf{x}$. If A has k positive eigenvalues, q has the form

$$q(\mathbf{x}) = a_1 x_1^2 + \cdots + a_k x_k^2 - a_{k+1} x_{k+1}^2 - \cdots - a_r x_r^2, \quad a_i > 0$$

where $r = \operatorname{rank} A = \operatorname{rank} B$. The subspace $W_1 = \{\mathbf{x} \mid x_{k+1} = \cdots = x_r = 0\}$ of $\mathbb{R}^n$ has dimension $n - r + k$ and satisfies $q(\mathbf{x}) > 0$ for all $\mathbf{x} \neq \mathbf{0}$ in W_1.

On the other hand, if $B = U^T A U$, define new variables $\mathbf{y}$ by $\mathbf{x} = U\mathbf{y}$. If B has k' positive eigenvalues, q has the form

$$q(\mathbf{x}) = b_1 y_1^2 + \cdots + b_{k'} y_{k'}^2 - b_{k'+1} y_{k'+1}^2 - \cdots - b_r y_r^2, \quad b_i > 0$$

Let $\mathbf{f}_1, \ldots, \mathbf{f}_n$ denote the columns of U. They are a basis of $\mathbb{R}^n$ and

$$\mathbf{x} = U\mathbf{y} = [\mathbf{f}_1 \cdots \mathbf{f}_n] \begin{bmatrix} y_1 \\ \vdots \\ y_n \end{bmatrix} = y_1 \mathbf{f}_1 + \cdots + y_n \mathbf{f}_n$$

Hence the subspace $W_2 = \operatorname{span}\{\mathbf{f}_{k'+1}, \ldots, \mathbf{f}_r\}$ satisfies $q(\mathbf{x}) < 0$ for all $\mathbf{x} \neq \mathbf{0}$ in W_2. Note that $\dim W_2 = r - k'$. It follows that W_1 and W_2 have only the zero vector in common. Hence, if B_1 and B_2 are bases of W_1 and W_2, respectively, then (Exercise 33 Section 6.3) $B_1 \cup B_2$ is an independent set of $(n - r + k) + (r - k') = n + k - k'$ vectors in $\mathbb{R}^n$. This implies that $k \leq k'$, and a similar argument shows $k' \leq k$.

EXERCISES 8.8

1. In each case, find a symmetric matrix A such that $q = \mathbf{x}^T B \mathbf{x}$ takes the form $q = \mathbf{x}^T A \mathbf{x}$.

   (a) $\begin{bmatrix} 1 & 1 \\ 0 & 1 \end{bmatrix}$ ◆(b) $\begin{bmatrix} 1 & 1 \\ -1 & 2 \end{bmatrix}$

   (c) $\begin{bmatrix} 1 & 0 & 1 \\ 1 & 1 & 0 \\ 0 & 1 & 1 \end{bmatrix}$ ◆(d) $\begin{bmatrix} 1 & 2 & -1 \\ 4 & 1 & 0 \\ 5 & -2 & 3 \end{bmatrix}$

2. In each case, find a change of variables that will diagonalize the quadratic form q. Determine the index and rank of q.

   (a) $q = x_1^2 + 2x_1 x_2 + x_2^2$

   ◆(b) $q = x_1^2 + 4x_1 x_2 + x_2^2$

   (c) $q = x_1^2 + x_2^2 + x_3^2 - 4(x_1 x_2 + x_1 x_3 + x_2 x_3)$

   ◆(d) $q = 7x_1^2 + x_2^2 + x_3^2 + 8x_1 x_2 + 8x_1 x_3 - 16x_2 x_3$

   (e) $q = 2(x_1^2 + x_2^2 + x_3^2 - x_1 x_2 + x_1 x_3 - x_2 x_3)$

   ◆(f) $q = 5x_1^2 + 8x_2^2 + 5x_3^2 - 4(x_1 x_2 + 2x_1 x_3 + x_2 x_3)$

   (g) $q = x_1^2 - x_3^2 - 4x_1 x_2 + 4x_2 x_3$

   ◆(h) $q = x_1^2 + x_3^2 - 2x_1 x_2 + 2x_2 x_3$

3. For each of the following, write the equation in terms of new variables so that it is in standard position, and identify the curve.

   (a) $xy = 1$

   ◆(b) $3x^2 - 4xy = 2$

   (c) $6x^2 + 6xy - 2y^2 = 5$

   ◆(d) $2x^2 + 4xy + 5y^2 = 1$

4. Consider the equation $ax^2 + bxy + cy^2 = d$, where $b \neq 0$. Introduce new variables x_1 and y_1 by rotating the axes counterclockwise through an angle θ. Show that the resulting equation has no $x_1 y_1$-term if θ is given by

   $$\cos 2\theta = \frac{a - c}{\sqrt{b^2 + (a - c)^2}},$$

   $$\sin 2\theta = \frac{b}{\sqrt{b^2 + (a - c)^2}}$$

   [Hint: Use equation (∗∗) preceding Theorem 2 to get x and y in terms of x_1 and y_1, and substitute.]

5. Prove properties (1)–(5) preceding Example 4.

6. If $A \stackrel{c}{\sim} B$ show that A is invertible if and only if B is invertible.

7. If $\mathbf{x} = (x_1, \ldots, x_n)^T$ is a column of variables, $A = A^T$ is $n \times n$, B is $1 \times n$, and c is a constant, $\mathbf{x}^T A \mathbf{x} + B \mathbf{x} = c$ is called a **quadratic equation** in the variables x_i.

   (a) Show that new variables $y_1, \ldots, y_n$ can be found such that the equation takes the form $\lambda_1 y_1^2 + \cdots + \lambda_r y_r^2 + k_1 y_1 + \cdots + k_n y_n = c$.

   ◆(b) Write $x_1^2 + 3x_2^2 + 3x_3^2 + 4x_1 x_2 - 4x_1 x_3 + 5x_1 - 6x_3 = 7$ in this form and find variables y_1, y_2, y_3 as in (a).

8. Given a symmetric matrix A, define $q_A(\mathbf{x}) = \mathbf{x}^T A \mathbf{x}$. Show that $B \stackrel{c}{\sim} A$ if and only if B is symmetric and there is an invertible matrix U such that $q_B(\mathbf{x}) = q_A(U\mathbf{x})$ for all $\mathbf{x}$. [*Hint*: Theorem 3.]

9. Let $q(\mathbf{x}) = \mathbf{x}^T A \mathbf{x}$ be a quadratic form, $A = A^T$.

   (a) Show that $q(\mathbf{x}) > 0$ for all $\mathbf{x} \neq \mathbf{0}$, if and only if A is positive definite (all eigenvalues are positive). In this case, q is called **positive definite**.

   ◆(b) Show that new variables $\mathbf{y}$ can be found such that $q = \|\mathbf{y}\|^2$ and $\mathbf{y} = U\mathbf{x}$ where U is upper triangular with positive diagonal entries. [*Hint*: Theorem 3 Section 8.3.]

10. A **bilinear form** β on $\mathbb{R}^n$ is a function that assigns to every pair $\mathbf{x}$, $\mathbf{y}$ of columns in $\mathbb{R}^n$ a number $\beta(\mathbf{x}, \mathbf{y})$ in such a way that

$$\beta(r\mathbf{x} + s\mathbf{y}, \mathbf{z}) = r\beta(\mathbf{x}, \mathbf{z}) + s\beta(\mathbf{y}, \mathbf{z})$$
$$\beta(\mathbf{x}, r\mathbf{y} + s\mathbf{z}) = r\beta(\mathbf{x}, \mathbf{y}) + s\beta(\mathbf{x}, \mathbf{z})$$

for all $\mathbf{x}$, $\mathbf{y}$, $\mathbf{z}$ in $\mathbb{R}^n$ and r, s in $\mathbb{R}$. If $\beta(\mathbf{x}, \mathbf{y}) = \beta(\mathbf{y}, \mathbf{x})$ for all $\mathbf{x}$, $\mathbf{y}$, β is called **symmetric**.

   (a) If β is a bilinear form, show that an $n \times n$ matrix A exists such that $\beta(\mathbf{x}, \mathbf{y}) = \mathbf{x}^T A \mathbf{y}$ for all $\mathbf{x}$, $\mathbf{y}$.

   (b) Show that A is uniquely determined by β.

   (c) Show that β is symmetric if and only if $A = A^T$.

SECTION 8.9 An Application to Constrained Optimization

It is a frequent occurrence in applications that a function $q = q(x_1, x_2, \ldots, x_n)$ of n variables, called an **objective function**, is to be made as large or as small as possible among all vectors $\mathbf{x} = (x_1, x_2, \ldots, x_n)$ lying in a certain region of $\mathbb{R}^n$ called the **feasible region**. A wide variety of objective functions q arise in practice; our primary concern here is to examine one important situation where q is a quadratic form. The next example gives some indication of how such problems arise.

EXAMPLE 1

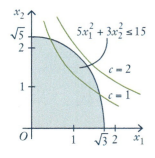

A politician proposes to spend x_1 dollars annually on health care and x_2 dollars annually on education. She is constrained in her spending by various budget pressures, and one model of this is that the expenditures x_1 and x_2 should satisfy a constraint like

$$5x_1^2 + 3x_2^2 \leq 15.$$

Since $x_i \geq 0$ for each i, the feasible region is the shaded area shown in the diagram. Any choice of feasible point (x_1, x_2) in this region will satisfy the budget constraints. However, these choices have different effects on voters, and the politician wants to choose $\mathbf{x} = (x_1, x_2)$ to maximize some measure $q = q(x_1, x_2)$ of voter satisfaction. Thus the assumption is that, for any value of c, all points on the graph of $q(x_1, x_2) = c$ have the same appeal to voters.

Hence the goal is to find the largest value of c for which the graph of $q(x_1, x_2) = c$ contains a feasible point.

The choice of the function q depends upon many factors; we will show how to solve the problem for any quadratic form q (even with more than two variables). In the diagram the function q is given by

$$q(x_1, x_2) = x_1 x_2,$$

and the graphs of $q(x_1, x_2) = c$ are shown for $c = 1$ and $c = 2$. As c increases the graph of $q(x_1, x_2) = c$ moves up and to the right. From this it is clear that there will be a solution for some value of c between 1 and 2 (in fact the largest value is $c = \frac{1}{2}\sqrt{15} = 1.94$ to two decimal places).

The constraint $5x_1^2 + 3x_2^2 \leq 15$ in Example 9 can be put in a standard form. If we divide through by 15, it becomes $\left(\frac{x_1}{\sqrt{3}}\right)^2 + \left(\frac{x_2}{\sqrt{5}}\right)^2 \leq 1$. This suggests that we introduce new variables $\mathbf{y} = (y_1, y_2)$ where $y_1 = \frac{x_1}{\sqrt{3}}$ and $y_2 = \frac{x_2}{\sqrt{5}}$. Then the constraint becomes $\|\mathbf{y}\|^2 \leq 1$, equivalently $\|\mathbf{y}\| \leq 1$. In terms of these new variables, the objective function is $q = \sqrt{15} y_1 y_2$, and we want to maximize this subject to $\|\mathbf{y}\| \leq 1$. When this is done, the maximizing values of x_1 and x_2 are obtained from $x_1 = \sqrt{3} y_1$ and $x_2 = \sqrt{5} y_2$.

Hence, for constraints like that in Example 1, there is no real loss in generality in assuming that the constraint takes the form $\|\mathbf{x}\| \leq 1$. In this case the principal axis theorem solves the problem. Recall that a vector in $\mathbb{R}^n$ of length 1 is called a *unit vector*.

Theorem 1

Consider the quadratic form $q = q(\mathbf{x}) = \mathbf{x}^T A \mathbf{x}$ where A is an $n \times n$ symmetric matrix, and let λ_1 and λ_n denote the largest and smallest eigenvalues of A, respectively. Then:

(1) $\max\{q(\mathbf{x}) \mid \|\mathbf{x}\| \leq 1\} = \lambda_1$, and $q(\mathbf{f}_1) = \lambda_1$ where $\mathbf{f}_1$ is any unit eigenvector corresponding to λ_1.

(2) $\min\{q(\mathbf{x}) \mid \|\mathbf{x}\| \leq 1\} = \lambda_n$, and $q(\mathbf{f}_n) = \lambda_n$ where $\mathbf{f}_n$ is any unit eigenvector corresponding to λ_n.

PROOF

Since A is symmetric, let the (real) eigenvalues λ_i of A be ordered as to size as follows: $\lambda_1 \geq \lambda_2 \geq \cdots \geq \lambda_n$. By the principal axis theorem, let P be an orthogonal matrix such that $P^T A P = D = \text{diag}(\lambda_1, \lambda_2, \ldots, \lambda_n)$. Define $\mathbf{y} = P^T \mathbf{x}$, equivalently $\mathbf{x} = P\mathbf{y}$, and note that $\|\mathbf{y}\| = \|\mathbf{x}\|$ because $\|\mathbf{y}\|^2 = \mathbf{y}^T \mathbf{y} = \mathbf{x}^T(PP^T)\mathbf{x} = \mathbf{x}^T \mathbf{x} = \|\mathbf{x}\|^2$. If we write $\mathbf{y} = (y_1, y_2, \ldots, y_n)^T$, then

$$\begin{aligned} q(\mathbf{x}) &= q(P\mathbf{y}) = (P\mathbf{y})^T A (P\mathbf{y}) \\ &= \mathbf{y}^T (P^T A P) \mathbf{y} = \mathbf{y}^T D \mathbf{y} \\ &= \lambda_1 y_1^2 + \lambda_2 y_2^2 + \cdots + \lambda_n y_n^2. \end{aligned} \quad (*)$$

Now assume that $\|\mathbf{x}\| \leq 1$. Since $\lambda_i \leq \lambda_1$ for each i, (*) gives
$$q(\mathbf{x}) = \lambda_1 y_1^2 + \lambda_2 y_2^2 + \cdots + \lambda_n y_n^2 \leq \lambda_1 y_1^2 + \lambda_1 y_2^2 + \cdots + \lambda_1 y_n^2 = \lambda_1 \|\mathbf{y}\|^2 \leq \lambda_1$$
because $\|\mathbf{y}\| = \|\mathbf{x}\| \leq 1$. This shows that $q(\mathbf{x})$ cannot exceed λ_1 when $\|\mathbf{x}\| \leq 1$. To see that this maximum is actually achieved, let $\mathbf{f}_1$ be a unit eigenvector corresponding to λ_1. Then
$$q(\mathbf{f}_1) = \mathbf{f}_1^T A \mathbf{f}_1 = \mathbf{f}_1^T(\lambda_1 \mathbf{f}_1) = \lambda_1(\mathbf{f}_1^T \mathbf{f}_1) = \lambda_1 \|\mathbf{f}_1\|^2 = \lambda_1.$$
Hence λ_1 is the maximum value of $q(\mathbf{x})$ when $\|\mathbf{x}\| \leq 1$, proving (1). The proof of (2) is analogous.

The set of all vectors $\mathbf{x}$ in $\mathbb{R}^n$ such that $\|\mathbf{x}\| \leq 1$ is called the **unit ball**. If $n = 2$, it is often called the unit disk and consists of the unit circle and its interior; if $n = 3$, it is the unit sphere and its interior. It is worth noting that the maximum value of a quadratic form $q(\mathbf{x})$ as $\mathbf{x}$ ranges *throughout* the unit ball is (by Theorem 1) actually attained for a unit vector $\mathbf{x}$ on the *boundary* of the unit ball.

Theorem 1 is important for applications involving vibrations in areas as diverse as aerodynamics and particle physics, and the maximum and minimum values in the theorem are often found using advanced calculus to minimize the quadratic form on the unit ball. The algebraic approach using the principal axis theorem gives a geometrical interpretation of the optimal values because they are eigenvalues.

EXAMPLE 2

Maximize and minimize the form $q(\mathbf{x}) = 3x_1^2 + 14x_1x_2 + 3x_2^2$ subject to $\|\mathbf{x}\| \leq 1$.

Solution ▶ The matrix of q is $A = \begin{bmatrix} 3 & 7 \\ 7 & 3 \end{bmatrix}$, with eigenvalues $\lambda_1 = 10$ and $\lambda_2 = -4$, and corresponding unit eigenvectors $\mathbf{f}_1 = \frac{1}{\sqrt{2}}(1, 1)$ and $\mathbf{f}_2 = \frac{1}{\sqrt{2}}(1, -1)$. Hence, among all unit vectors $\mathbf{x}$ in $\mathbb{R}^2$, $q(\mathbf{x})$ takes its maximal value 10 at $\mathbf{x} = \mathbf{f}_1$, and the minimum value of $q(\mathbf{x})$ is -4 when $\mathbf{x} = \mathbf{f}_2$.

As noted above, the objective function in a constrained optimization problem need not be a quadratic form. We conclude with an example where the objective function is linear, and the feasible region is determined by linear constraints.

EXAMPLE 3

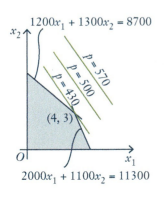

A manufacturer makes x_1 units of product 1, and x_2 units of product 2, at a profit of \$70 and \$50 per unit respectively, and wants to choose x_1 and x_2 to maximize the total profit $p(x_1, x_2) = 70x_1 + 50x_2$. However x_1 and x_2 are not arbitrary; for example, $x_1 \geq 0$ and $x_2 \geq 0$. Other conditions also come into play. Each unit of product 1 costs \$1200 to produce and requires 2000 square feet of warehouse space; each unit of product 2 costs \$1300 to produce and requires 1100 square feet of space. If the total warehouse space is 11 300 square feet, and if the total production budget is \$8700, x_1 and x_2 must also satisfy the conditions

$$2000x_1 + 1100x_2 \leq 11\ 300,$$
$$1200x_1 + 1300x_2 \leq 8700.$$

The feasible region in the plane satisfying these constraints (and $x_1 \geq 0$, $x_2 \geq 0$) is shaded in the diagram. If the profit equation $70x_1 + 50x_2 = p$ is plotted for various values of p, the resulting lines are parallel, with p increasing with distance from the origin. Hence the best choice occurs for the line $70x_1 + 50x_2 = 430$ that touches the shaded region at the point $(4, 3)$. So the profit p has a maximum of $p = 430$ for $x_1 = 4$ units and $x_2 = 3$ units.

Example 3 is a simple case of the general **linear programming** problem[18] which arises in economic, management, network, and scheduling applications. Here the objective function is a linear combination $q = a_1x_1 + a_2x_2 + \cdots + a_nx_n$ of the variables, and the feasible region consists of the vectors $\mathbf{x} = (x_1, x_2, \ldots, x_n)^T$ in $\mathbb{R}^n$ which satisfy a set of linear inequalities of the form $b_1x_1 + b_2x_2 + \cdots + b_nx_n \leq b$. There is a good method (an extension of the gaussian algorithm) called the **simplex algorithm** for finding the maximum and minimum values of q when $\mathbf{x}$ ranges over such a feasible set. As Example 3 suggests, the optimal values turn out to be vertices of the feasible set. In particular, they are on the boundary of the feasible region, as is the case in Theorem 1.

More detailed discussion of linear programming is available on the Online Learning Centre (www.mcgrawhill.ca/olc/nicholson).

SECTION 8.10 An Application to Statistical Principal Component Analysis

Linear algebra is important in multivariate analysis in statistics, and we conclude with a very short look at one application of diagonalization in this area. A main feature of probability and statistics is the idea of a *random variable X*, that is a real-valued function which takes its values according to a probability law (called its *distribution*). Random variables occur in a wide variety of contexts; examples include the number of meteors falling per square kilometre in a given region, the price of a share of a stock, or the duration of a long distance telephone call from a certain city.

The values of a random variable X are distributed about a central number μ, called the *mean* of X. The mean can be calculated from the distribution as the *expectation* $E(X) = \mu$ of the random variable X. Functions of a random variable are again random variables. In particular, $(X - \mu)^2$ is a random variable, and the *variance* of the random variable X, denoted $var(X)$, is defined to be the number

$$var(X) = E\{(X - \mu)^2\} \quad \text{where } \mu = E(X).$$

It is not difficult to see that $var(X) \geq 0$ for every random variable X. The number $\sigma = \sqrt{var(X)}$ is called the *standard deviation* of X, and is a measure of how much the values of X are spread about the mean μ of X. A main goal of statistical inference is finding reliable methods for estimating the mean and the standard deviation of a random variable X by sampling the values of X.

If two random variables X and Y are given, and their joint distribution is known, then functions of X and Y are also random variables. In particular, $X + Y$ and aX are random variables for any real number a, and we have

---

[18] A good introduction can be found at http://www.mcgrawhill.ca/olc/nicholson, and more information is available in "Linear Programming and Extensions" by N. Wu and R. Coppins, McGraw-Hill, 1981.

SECTION 8.10 An Application to Statistical Principal Component Analysis

$$E(X + Y) = E(X) + E(Y) \quad \text{and} \quad E(aX) = aE(X).\text{[19]}$$

An important question is how much the random variables X and Y depend on each other. One measure of this is the *covariance* of X and Y, denoted $cov(X, Y)$, defined by

$$cov(X, Y) = E\{(X - \mu)(Y - \upsilon)\} \quad \text{where } \mu = E(X) \text{ and } \upsilon = E(Y).$$

Clearly, $cov(X, X) = var(X)$. If $cov(X, Y) = 0$ then X and Y have little relationship to each other and are said to be *uncorrelated*.[20]

Multivariate statistical analysis deals with a family $X_1, X_2, \ldots, X_n$ of random variables with means $\mu_i = E(X_i)$ and variances $\sigma_i^2 = var(X_i)$ for each i. We denote the covariance of X_i and X_j by $\sigma_{ij} = cov(X_i, X_j)$. Then the *covariance matrix* of the random variables $X_1, X_2, \ldots, X_n$ is defined to be the $n \times n$ matrix

$$\Sigma = [\sigma_{ij}]$$

whose (i, j)-entry is σ_{ij}. The matrix Σ is clearly symmetric; in fact it can be shown that Σ is **positive semidefinite** in the sense that $\lambda \geq 0$ for every eigenvalue λ of Σ. (In reality, Σ is positive definite in most cases of interest.) So suppose that the eigenvalues of Σ are $\lambda_1 \geq \lambda_2 \geq \cdots \geq \lambda_n \geq 0$. The principal axis theorem (Theorem 2 Section 8.2) shows that an orthogonal matrix P exists such that

$$P^T \Sigma P = \text{diag}(\lambda_1, \lambda_2, \ldots, \lambda_n).$$

If we write $\overline{X} = (X_1, X_2, \ldots, X_n)$, the procedure for diagonalizing a quadratic form gives new variables $\overline{Y} = (Y_1, Y_2, \ldots, Y_n)$ defined by

$$\overline{Y} = P^T \overline{X}.$$

These new random variables $Y_1, Y_2, \ldots, Y_n$ are called the **principal components** of the original random variables X_i, and are linear combinations of the X_i. Furthermore, it can be shown that

$$cov(Y_i, Y_j) = 0 \text{ if } i \neq j \quad \text{and} \quad var(Y_i) = \lambda_i \quad \text{for each } i.$$

Of course the principal components Y_i point along the principal axes of the quadratic form $q = \overline{X}^T \Sigma \overline{X}$.

The sum of the variances of a set of random variables is called the **total variance** of the variables, and determining the source of this total variance is one of the benefits of principal component analysis. The fact that the matrices Σ and $\text{diag}(\lambda_1, \lambda_2, \ldots, \lambda_n)$ are similar means that they have the same trace, that is,

$$\sigma_{11} + \sigma_{22} + \cdots + \sigma_{nn} = \lambda_1 + \lambda_2 + \cdots + \lambda_n$$

This means that the principal components Y_i have the same total variance as the original random variables X_i. Moreover, the fact that $\lambda_1 \geq \lambda_2 \geq \cdots \geq \lambda_n \geq 0$ means that most of this variance resides in the first few Y_i. In practice, statisticians find that studying these first few Y_i (and ignoring the rest) gives an accurate analysis of the total system variability. This results in substantial data reduction since often only a few Y_i suffice for all practical purposes. Furthermore, these Y_i are easily obtained as linear combinations of the X_i. Finally, the analysis of the principal components often reveals relationships among the X_i that were not previously suspected, and so results in interpretations that would not otherwise have been made.

---

[19] Hence $E(\ )$ is a linear transformation from the vector space of all random variables to the space of real numbers.

[20] If X and Y are independent in the sense of probability theory, then they are uncorrelated; however, the converse is not true in general.

9 Change of Basis

If A is an $m \times n$ matrix, the corresponding **matrix transformation** $T_A : \mathbb{R}^n \to \mathbb{R}^m$ is defined by

$$T_A(\mathbf{x}) = A\mathbf{x} \quad \text{for all columns } \mathbf{x} \text{ in } \mathbb{R}^n.$$

It was shown in Theorem 2 Section 2.6 that every linear transformation $T : \mathbb{R}^n \to \mathbb{R}^m$ is a matrix transformation; that is, $T = T_A$ for some $m \times n$ matrix A. Furthermore, the matrix A is uniquely determined by T. In fact A is given in terms of its columns by

$$A = [T(\mathbf{e}_1) \ T(\mathbf{e}_2) \ \cdots \ T(\mathbf{e}_n)]$$

where $\{\mathbf{e}_1, \mathbf{e}_2, \ldots, \mathbf{e}_n\}$ is the standard basis of $\mathbb{R}^n$.

In this chapter we show how to associate a matrix with *any* linear transformation $T : V \to W$ where V and W are finite-dimensional vector spaces, and we describe how the matrix can be used to compute $T(\mathbf{v})$ for any $\mathbf{v}$ in V. The matrix depends on the choice of a basis B in V and a basis D in W, and is denoted $M_{DB}(T)$. The case when $W = V$ is particularly important. If B and D are two bases of V, we show that the matrices $M_{BB}(T)$ and $M_{DD}(T)$ are similar, that is $M_{DD}(T) = P^{-1}M_{BB}(T)P$ for some invertible matrix P. Moreover, we give an explicit method for constructing P depending only on the bases B and D. This leads to some of the most important theorems in linear algebra, as we shall see in Chapter 11.

SECTION 9.1 The Matrix of a Linear Transformation

Let $T : V \to W$ be a linear transformation where $\dim V = n$ and $\dim W = m$. The aim in this section is to describe the action of T as multiplication by an $m \times n$ matrix A. The idea is to convert a vector $\mathbf{v}$ in V into a column in $\mathbb{R}^n$, multiply that column by A to get a column in $\mathbb{R}^m$, and convert this column back to get $T(\mathbf{v})$ in W.

Converting vectors to columns is a simple matter, but one small change is needed. Up to now the *order* of the vectors in a basis has been of no importance. However, in this section, we shall speak of an **ordered basis** $\{\mathbf{b}_1, \mathbf{b}_2, \ldots, \mathbf{b}_n\}$, which is just a basis where the order in which the vectors are listed is taken into account. Hence $\{\mathbf{b}_2, \mathbf{b}_1, \mathbf{b}_3\}$ is a different *ordered* basis from $\{\mathbf{b}_1, \mathbf{b}_2, \mathbf{b}_3\}$.

If $B = \{\mathbf{b}_1, \mathbf{b}_2, \ldots, \mathbf{b}_n\}$ is an ordered basis in a vector space V, and if

$$\mathbf{v} = v_1\mathbf{b}_1 + v_2\mathbf{b}_2 + \cdots + v_n\mathbf{b}_n, \quad v_i \in \mathbb{R}$$

is a vector in V, then the (uniquely determined) numbers $v_1, v_2, \ldots, v_n$ are called the **coordinates** of $\mathbf{v}$ with respect to the basis B.

SECTION 9.1 *The Matrix of a Linear Transformation*

Definition 9.1 *The **coordinate vector** of $\mathbf{v}$ with respect to B is defined to be*

$$C_B(\mathbf{v}) = (v_1\mathbf{b}_1 + v_2\mathbf{b}_2 + \cdots + v_n\mathbf{b}_n) = \begin{bmatrix} v_1 \\ v_2 \\ \vdots \\ v_n \end{bmatrix}$$

The reason for writing $C_B(\mathbf{v})$ as a column instead of a row will become clear later. Note that $C_B(\mathbf{b}_i) = \mathbf{e}_i$ is column i of I_n.

EXAMPLE 1

The coordinate vector for $\mathbf{v} = (2, 1, 3)$ with respect to the ordered basis $B = \{(1, 1, 0), (1, 0, 1), (0, 1, 1)\}$ of $\mathbb{R}^3$ is $C_B(\mathbf{v}) = \begin{bmatrix} 0 \\ 2 \\ 1 \end{bmatrix}$ because

$$\mathbf{v} = (2, 1, 3) = 0(1, 1, 0) + 2(1, 0, 1) + 1(0, 1, 1).$$

Theorem 1

If V has dimension n and $B = \{\mathbf{b}_1, \mathbf{b}_2, \ldots, \mathbf{b}_n\}$ is any ordered basis of V, the coordinate transformation $C_B : V \to \mathbb{R}^n$ is an isomorphism. In fact, $C_B^{-1} : \mathbb{R}^n \to V$ is given by

$$C_B^{-1}\begin{bmatrix} v_1 \\ v_2 \\ \vdots \\ v_n \end{bmatrix} = v_1\mathbf{b}_1 + v_2\mathbf{b}_2 + \cdots + v_n\mathbf{b}_n \quad \text{for all} \quad \begin{bmatrix} v_1 \\ v_2 \\ \vdots \\ v_n \end{bmatrix} \text{ in } \mathbb{R}^n.$$

PROOF

The verification that C_B is linear is Exercise 13. If $T : \mathbb{R}^n \to V$ is the map denoted C_B^{-1} in the theorem, one verifies (Exercise 13) that $TC_B = 1_V$ and $C_B T = 1_{\mathbb{R}^n}$. Note that $C_B(\mathbf{b}_j)$ is column j of the identity matrix, so C_B carries the basis B to the standard basis of $\mathbb{R}^n$, proving again that it is an isomorphism (Theorem 1 Section 7.3).

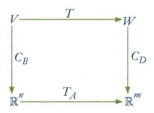

Now let $T : V \to W$ be any linear transformation where $\dim V = n$ and $\dim W = m$, and let $B = \{\mathbf{b}_1, \mathbf{b}_2, \ldots, \mathbf{b}_n\}$ and D be ordered bases of V and W, respectively. Then $C_B : V \to \mathbb{R}^n$ and $C_D : W \to \mathbb{R}^m$ are isomorphisms and we have the situation shown in the diagram where A is an $m \times n$ matrix (to be determined). In fact, the composite

$$C_D T C_B^{-1} : \mathbb{R}^n \to \mathbb{R}^m \text{ is a linear transformation}$$

so Theorem 2 Section 2.6 shows that a unique $m \times n$ matrix A exists such that

$$C_D T C_B^{-1} = T_A, \quad \text{equivalently } C_D T = T_A C_B$$

T_A acts by left multiplication by A, so this latter condition is

$$C_D[T(\mathbf{v})] = A C_B(\mathbf{v}) \text{ for all } \mathbf{v} \text{ in } V$$

This requirement completely determines A. Indeed, the fact that $C_B(\mathbf{b}_j)$ is column j of the identity matrix gives

$$\text{column } j \text{ of } A = A C_B(\mathbf{b}_j) = C_D[T(\mathbf{b}_j)]$$

for all j. Hence, in terms of its columns,
$$A = [\, C_D[T(\mathbf{b}_1)] \quad C_D[T(\mathbf{b}_2)] \quad \cdots \quad C_D[T(\mathbf{b}_n)] \,].$$

Definition 9.2 *This is called the **matrix of T corresponding to the ordered bases B and D**, and we use the following notation:*
$$M_{DB}(T) = [\, C_D[T(\mathbf{b}_1)] \quad C_D[T(\mathbf{b}_2)] \quad \cdots \quad C_D[T(\mathbf{b}_n)] \,]$$

This discussion is summarized in the following important theorem.

Theorem 2

Let $T: V \to W$ be a linear transformation where $\dim V = n$ and $\dim W = m$, and let $B = \{\mathbf{b}_1, \ldots, \mathbf{b}_n\}$ and D be ordered bases of V and W, respectively. Then the matrix $M_{DB}(T)$ just given is the unique $m \times n$ matrix A that satisfies
$$C_D T = T_A C_B.$$
Hence the defining property of $M_{DB}(T)$ is
$$C_D[T(\mathbf{v})] = M_{DB}(T) C_B(\mathbf{v}) \quad \text{for all } \mathbf{v} \text{ in } V.$$
The matrix $M_{DB}[T]$ is given in terms of its columns by
$$M_{DB}(T) = [\, C_D[T(\mathbf{b}_1)] \quad C_D[T(\mathbf{b}_2)] \quad \cdots \quad C_D[T(\mathbf{b}_n)] \,]$$

The fact that $T = C_D^{-1} T_A C_B$ means that the action of T on a vector $\mathbf{v}$ in V can be performed by first taking coordinates (that is, applying C_B to $\mathbf{v}$), then multiplying by A (applying T_A), and finally converting the resulting m-tuple back to a vector in W (applying C_D^{-1}).

EXAMPLE 2

Define $T: \mathbf{P}_2 \to \mathbb{R}^2$ by $T(a + bx + cx^2) = (a + c, b - a - c)$ for all polynomials $a + bx + cx^2$. If $B = \{\mathbf{b}_1, \mathbf{b}_2, \mathbf{b}_3\}$ and $D = \{\mathbf{d}_1, \mathbf{d}_2\}$ where
$$\mathbf{b}_1 = 1, \ \mathbf{b}_2 = x, \ \mathbf{b}_3 = x^2 \quad \text{and} \quad \mathbf{d}_1 = (1, 0), \ \mathbf{d}_2 = (0, 1)$$
compute $M_{DB}(T)$ and verify Theorem 2.

Solution ▶ We have $T(\mathbf{b}_1) = \mathbf{d}_1 - \mathbf{d}_2$, $T(\mathbf{b}_2) = \mathbf{d}_2$, and $T(\mathbf{b}_3) = \mathbf{d}_1 - \mathbf{d}_2$. Hence
$$M_{DB}(T) = [\, C_D[T(\mathbf{b}_1)] \quad C_D[T(\mathbf{b}_2)] \quad C_D[T(\mathbf{b}_3)] \,] = \begin{bmatrix} 1 & 0 & 1 \\ -1 & 1 & -1 \end{bmatrix}.$$
If $\mathbf{v} = a + bx + cx^2 = a\mathbf{b}_1 + b\mathbf{b}_2 + c\mathbf{b}_3$, then $T(\mathbf{v}) = (a + c)\mathbf{d}_1 + (b - a - c)\mathbf{d}_2$, so
$$C_D[T(\mathbf{v})] = \begin{bmatrix} a + c \\ b - a - c \end{bmatrix} = \begin{bmatrix} 1 & 0 & 1 \\ -1 & 1 & -1 \end{bmatrix} \begin{bmatrix} a \\ b \\ c \end{bmatrix} = M_{DB}(T) C_B(\mathbf{v})$$
as Theorem 2 asserts.

SECTION 9.1 The Matrix of a Linear Transformation

The next example shows how to determine the action of a transformation from its matrix.

EXAMPLE 3

Suppose $T: \mathbf{M}_{22}(\mathbb{R}) \to \mathbb{R}^3$ is linear with matrix $M_{DB}(T) = \begin{bmatrix} 1 & -1 & 0 & 0 \\ 0 & 1 & -1 & 0 \\ 0 & 0 & 1 & -1 \end{bmatrix}$ where

$B = \left\{ \begin{bmatrix} 1 & 0 \\ 0 & 0 \end{bmatrix}, \begin{bmatrix} 0 & 1 \\ 0 & 0 \end{bmatrix}, \begin{bmatrix} 0 & 0 \\ 1 & 0 \end{bmatrix}, \begin{bmatrix} 0 & 0 \\ 0 & 1 \end{bmatrix} \right\}$ and $D = \{(1, 0, 0), (0, 1, 0), (0, 0, 1)\}$.

Compute $T(\mathbf{v})$ where $\mathbf{v} = \begin{bmatrix} a & b \\ c & d \end{bmatrix}$.

Solution ▶ The idea is to compute $C_D[T(\mathbf{v})]$ first, and then obtain $T(\mathbf{v})$. We have

$$C_D[T(\mathbf{v})] = M_{DB}(T) C_B(\mathbf{v}) = \begin{bmatrix} 1 & -1 & 0 & 0 \\ 0 & 1 & -1 & 0 \\ 0 & 0 & 1 & -1 \end{bmatrix} \begin{bmatrix} a \\ b \\ c \\ d \end{bmatrix} = \begin{bmatrix} a - b \\ b - c \\ c - d \end{bmatrix}$$

Hence $T(\mathbf{v}) = (a - b)(1, 0, 0) + (b - c)(0, 1, 0) + (c - d)(0, 0, 1)$
$= (a - b, b - c, c - d)$

The next two examples will be referred to later.

EXAMPLE 4

Let A be an $m \times n$ matrix, and let $T_A : \mathbb{R}^n \to \mathbb{R}^m$ be the matrix transformation induced by A : $T_A(\mathbf{x}) = A\mathbf{x}$ for all columns $\mathbf{x}$ in $\mathbb{R}^n$. If B and D are the standard bases of $\mathbb{R}^n$ and $\mathbb{R}^m$, respectively (ordered as usual), then

$$M_{DB}(T_A) = A$$

In other words, the matrix of T_A corresponding to the standard bases is A itself.

Solution ▶ Write $B = \{\mathbf{e}_1, \dots, \mathbf{e}_n\}$. Because D is the standard basis of $\mathbb{R}^m$, it is easy to verify that $C_D(\mathbf{y}) = \mathbf{y}$ for all columns $\mathbf{y}$ in $\mathbb{R}^m$. Hence

$$M_{DB}(T_A) = [T_A(\mathbf{e}_1) \; T_A(\mathbf{e}_2) \; \cdots \; T_A(\mathbf{e}_n)] = [A\mathbf{e}_1 \; A\mathbf{e}_2 \; \cdots \; A\mathbf{e}_n] = A$$

because $A\mathbf{e}_j$ is the jth column of A.

EXAMPLE 5

Let V and W have ordered bases B and D, respectively. Let $\dim V = n$.
1. The identity transformation $1_V : V \to V$ has matrix $M_{BB}(1_V) = I_n$.
2. The zero transformation $0 : V \to W$ has matrix $M_{DB}(0) = 0$.

The first result in Example 5 is false if the two bases of V are not equal. In fact, if B is the standard basis of $\mathbb{R}^n$, then the basis D of $\mathbb{R}^n$ can be chosen so that $M_{BD}(1_{\mathbb{R}^n})$ turns out to be any invertible matrix we wish (Exercise 14).

The next two theorems show that composition of linear transformations is compatible with multiplication of the corresponding matrices.

Theorem 3

Let $V \xrightarrow{T} W \xrightarrow{S} U$, be linear transformations and let B, D, and E be finite ordered bases of V, W, and U, respectively. Then
$$M_{EB}(ST) = M_{ED}(S) \cdot M_{DB}(T)$$

PROOF

We use the property in Theorem 2 three times. If $\mathbf{v}$ is in V,
$$M_{ED}(S)M_{DB}(T)C_B(\mathbf{v}) = M_{ED}(S)C_D[T(\mathbf{v})] = C_E[ST(\mathbf{v})] = M_{EB}(ST)C_B(\mathbf{v})$$
If $B = \{\mathbf{e}_1, \ldots, \mathbf{e}_n\}$, then $C_B(\mathbf{e}_j)$ is column j of I_n. Hence taking $\mathbf{v} = \mathbf{e}_j$ shows that $M_{ED}(S)M_{DB}(T)$ and $M_{EB}(ST)$ have equal jth columns. The theorem follows.

Theorem 4

Let $T: V \to W$ be a linear transformation, where $\dim V = \dim W = n$. The following are equivalent.

1. T is an isomorphism.
2. $M_{DB}(T)$ is invertible for all ordered bases B and D of V and W.
3. $M_{DB}(T)$ is invertible for some pair of ordered bases B and D of V and W.

When this is the case, $[M_{DB}(T)]^{-1} = M_{BD}(T^{-1})$.

PROOF

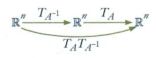

(1) $\Rightarrow$ (2). We have $V \xrightarrow{T} W \xrightarrow{T^{-1}} V$, so Theorem 3 and Example 5 give
$$M_{BD}(T^{-1})M_{DB}(T) = M_{BB}(T^{-1}T) = M_{BB}(1_V) = I_n$$
Similarly, $M_{DB}(T)M_{BD}(T^{-1}) = I_n$, proving (2) (and the last statement in the theorem).

(2) $\Rightarrow$ (3). This is clear.

(3) $\Rightarrow$ (1). Suppose that $T_{DB}(T)$ is invertible for some bases B and D and, for convenience, write $A = M_{DB}(T)$. Then we have $C_D T = T_A C_B$ by Theorem 2, so
$$T = (C_D)^{-1}T_A C_B$$
by Theorem 1 where $(C_D)^{-1}$ and C_B are isomorphisms. Hence (1) follows if we can show that $T_A: \mathbb{R}^n \to \mathbb{R}^n$ is also an isomorphism. But A is invertible by (3) and one verifies that $T_A T_{A^{-1}} = 1_{\mathbb{R}^n} = T_{A^{-1}} T_A$. So T_A is indeed invertible (and $(T_A)^{-1} = T_{A^{-1}}$).

SECTION 9.1 The Matrix of a Linear Transformation

In Section 7.2 we defined the rank of a linear transformation $T: V \to W$ by rank $T = \dim(\text{im } T)$. Moreover, if A is any $m \times n$ matrix and $T_A: \mathbb{R}^n \to \mathbb{R}^m$ is the matrix transformation, we showed that $\text{rank}(T_A) = \text{rank } A$. So it may not be surprising that rank T equals the rank of any matrix of T.

Theorem 5

Let $T: V \to W$ be a linear transformation where $\dim V = n$ and $\dim W = m$. If B and D are any ordered bases of V and W, then $\text{rank } T = \text{rank}[M_{DB}(T)]$.

PROOF

Write $A = M_{DB}(T)$ for convenience. The column space of A is $U = \{A\mathbf{x} \mid \mathbf{x} \text{ in } \mathbb{R}^n\}$. Hence $\text{rank } A = \dim U$ and so, because rank $T = \dim(\text{im } T)$, it suffices to find an isomorphism $S: \text{im } T \to U$. Now every vector in im T has the form $T(\mathbf{v})$, $\mathbf{v}$ in V. By Theorem 2, $C_D[T(\mathbf{v})] = AC_B(\mathbf{v})$ lies in U. So define $S: \text{im } T \to U$ by

$$S[T(\mathbf{v})] = C_D[T(\mathbf{v})] \quad \text{for all vectors } T(\mathbf{v}) \text{ in im } T$$

The fact that C_D is linear and one-to-one implies immediately that S is linear and one-to-one. To see that S is onto, let $A\mathbf{x}$ be any member of U, $\mathbf{x}$ in $\mathbb{R}^n$. Then $\mathbf{x} = C_B(\mathbf{v})$ for some $\mathbf{v}$ in V because C_B is onto. Hence $A\mathbf{x} = AC_B(\mathbf{v}) = C_D[T(\mathbf{v})] = S[T(\mathbf{v})]$, so S is onto. This means that S is an isomorphism.

EXAMPLE 6

Define $T: \mathbf{P}_2 \to \mathbb{R}^3$ by $T(a + bx + cx^2) = (a - 2b, 3c - 2a, 3c - 4b)$ for $a, b, c \in \mathbb{R}$. Compute rank T.

Solution ▶ Since rank $T = \text{rank } [M_{DB}(T)]$ for *any* bases $B \subseteq \mathbf{P}_2$ and $D \subseteq \mathbb{R}^3$, we choose the most convenient ones: $B = \{1, x, x^2\}$ and $D = \{(1, 0, 0), (0, 1, 0), (0, 0, 1)\}$. Then $M_{DB}(T) = [C_D[T(1)] \quad C_D[T(x)] \quad C_D[T(x^2)]] = A$ where

$$A = \begin{bmatrix} 1 & -2 & 0 \\ -2 & 0 & 3 \\ 0 & -4 & 3 \end{bmatrix}. \text{ Since } A \to \begin{bmatrix} 1 & -2 & 0 \\ 0 & -4 & 3 \\ 0 & -4 & 3 \end{bmatrix} \to \begin{bmatrix} 1 & -2 & 0 \\ 0 & 1 & -\frac{3}{4} \\ 0 & 0 & 0 \end{bmatrix}$$

we have rank $A = 2$. Hence rank $T = 2$ as well.

We conclude with an example showing that the matrix of a linear transformation can be made very simple by a careful choice of the two bases.

EXAMPLE 7

Let $T: V \to W$ be a linear transformation where $\dim V = n$ and $\dim W = m$. Choose an ordered basis $B = \{\mathbf{b}_1, \ldots, \mathbf{b}_r, \mathbf{b}_{r+1}, \ldots, \mathbf{b}_n\}$ of V in which $\{\mathbf{b}_{r+1}, \ldots, \mathbf{b}_n\}$ is a basis of ker T, possibly empty. Then $\{T(\mathbf{b}_1), \ldots, T(\mathbf{b}_r)\}$ is a basis of im T by Theorem 5 Section 7.2, so extend it to an ordered basis $D = \{T(\mathbf{b}_1), \ldots, T(\mathbf{b}_r), \mathbf{f}_{r+1}, \ldots, \mathbf{f}_m\}$ of W. Because $T(\mathbf{b}_{r+1}) = \cdots = T(\mathbf{b}_n) = \mathbf{0}$, we have

$$M_{DB}(T) = [C_D[T(\mathbf{b}_1)] \cdots C_D[T(\mathbf{b}_r)] \ C_D[T(\mathbf{b}_{r+1})] \cdots C_D[T(\mathbf{b}_n)]] = \begin{bmatrix} I_r & 0 \\ 0 & 0 \end{bmatrix}.$$

Incidentally, this shows that rank $T = r$ by Theorem 5.

EXERCISES 9.1

1. In each case, find the coordinates of **v** with respect to the basis B of the vector space V.

   (a) $V = \mathbf{P}_2$, $\mathbf{v} = 2x^2 + x - 1$, $B = \{x + 1, x^2, 3\}$

   ◆(b) $V = \mathbf{P}_2$, $\mathbf{v} = ax^2 + bx + c$,
   $B = \{x^2, x + 1, x + 2\}$

   (c) $V = \mathbb{R}^3$, $\mathbf{v} = (1, -1, 2)$,
   $B = \{(1, -1, 0), (1, 1, 1), (0, 1, 1)\}$

   ◆(d) $V = \mathbb{R}^3$, $\mathbf{v} = (a, b, c)$,
   $B = \{(1, -1, 2), (1, 1, -1), (0, 0, 1)\}$

   (e) $V = \mathbf{M}_{22}$, $\mathbf{v} = \begin{bmatrix} 1 & 2 \\ -1 & 0 \end{bmatrix}$,
   $B = \left\{ \begin{bmatrix} 1 & 1 \\ 0 & 0 \end{bmatrix}, \begin{bmatrix} 1 & 0 \\ 1 & 0 \end{bmatrix}, \begin{bmatrix} 0 & 0 \\ 1 & 1 \end{bmatrix}, \begin{bmatrix} 1 & 0 \\ 0 & 1 \end{bmatrix} \right\}$

2. Suppose $T: \mathbf{P}_2 \to \mathbb{R}^2$ is a linear transformation. If $B = \{1, x, x^2\}$ and $D = \{(1, 1), (0, 1)\}$, find the action of T given:

   (a) $M_{DB}(T) = \begin{bmatrix} 1 & 2 & -1 \\ -1 & 0 & 1 \end{bmatrix}$

   ◆(b) $M_{DB}(T) = \begin{bmatrix} 2 & 1 & 3 \\ -1 & 0 & -2 \end{bmatrix}$

3. In each case, find the matrix of $T: V \to W$ corresponding to the bases B and D of V and W, respectively.

   (a) $T: \mathbf{M}_{22} \to \mathbb{R}$, $T(A) = \text{tr } A$;
   $B = \left\{ \begin{bmatrix} 1 & 0 \\ 0 & 0 \end{bmatrix}, \begin{bmatrix} 0 & 1 \\ 0 & 0 \end{bmatrix}, \begin{bmatrix} 0 & 0 \\ 1 & 0 \end{bmatrix}, \begin{bmatrix} 0 & 0 \\ 0 & 1 \end{bmatrix} \right\}$,
   $D = \{1\}$

   ◆(b) $T: \mathbf{M}_{22} \to \mathbf{M}_{22}$, $T(A) = A^T$;
   $B = D = \left\{ \begin{bmatrix} 1 & 0 \\ 0 & 0 \end{bmatrix}, \begin{bmatrix} 0 & 1 \\ 0 & 0 \end{bmatrix}, \begin{bmatrix} 0 & 0 \\ 1 & 0 \end{bmatrix}, \begin{bmatrix} 0 & 0 \\ 0 & 1 \end{bmatrix} \right\}$

   (c) $T: \mathbf{P}_2 \to \mathbf{P}_3$, $T[p(x)] = xp(x)$; $B = \{1, x, x^2\}$ and $D = \{1, x, x^2, x^3\}$

   ◆(d) $T: \mathbf{P}_2 \to \mathbf{P}_2$, $T[p(x)] = p(x + 1)$;
   $B = D = \{1, x, x^2\}$

4. In each case, find the matrix of $T: V \to W$ corresponding to the bases B and D, respectively, and use it to compute $C_D[T(\mathbf{v})]$, and hence $T(\mathbf{v})$.

   (a) $T: \mathbb{R}^3 \to \mathbb{R}^4$,
   $T(x, y, z) = (x + z, 2z, y - z, x + 2y)$; B and D standard; $\mathbf{v} = (1, -1, 3)$

   ◆(b) $T: \mathbb{R}^2 \to \mathbb{R}^4$, $T(x, y) = (2x - y, 3x + 2y, 4y, x)$;
   $B = \{(1, 1), (1, 0)\}$, D standard; $\mathbf{v} = (a, b)$

   (c) $T: \mathbf{P}_2 \to \mathbb{R}^2$, $T(a + bx + cx^2) = (a + c, 2b)$;
   $B = \{1, x, x^2\}$, $D = \{(1, 0), (1, -1)\}$;
   $\mathbf{v} = a + bx + cx^2$

   ◆(d) $T: \mathbf{P}_2 \to \mathbb{R}^2$, $T(a + bx + cx^2) = (a + b, c)$;
   $B = \{1, x, x^2\}$, $D = \{(1, -1), (1, 1)\}$;
   $\mathbf{v} = a + bx + cx^2$

   (e) $T: \mathbf{M}_{22} \to \mathbb{R}$, $T\begin{bmatrix} a & b \\ c & d \end{bmatrix} = a + b + c + d$;
   $B = \left\{ \begin{bmatrix} 1 & 0 \\ 0 & 0 \end{bmatrix}, \begin{bmatrix} 0 & 1 \\ 0 & 0 \end{bmatrix}, \begin{bmatrix} 0 & 0 \\ 1 & 0 \end{bmatrix}, \begin{bmatrix} 0 & 0 \\ 0 & 1 \end{bmatrix} \right\}$,
   $D = \{1\}$; $\mathbf{v} = \begin{bmatrix} a & b \\ c & d \end{bmatrix}$

◆(f) $T: \mathbf{M}_{22} \to \mathbf{M}_{22}$, $T\begin{bmatrix} a & b \\ c & d \end{bmatrix} = \begin{bmatrix} a & b+c \\ b+c & d \end{bmatrix}$;

$B = D = \left\{ \begin{bmatrix} 1 & 0 \\ 0 & 0 \end{bmatrix}, \begin{bmatrix} 0 & 1 \\ 0 & 0 \end{bmatrix}, \begin{bmatrix} 0 & 0 \\ 1 & 0 \end{bmatrix}, \begin{bmatrix} 0 & 0 \\ 0 & 1 \end{bmatrix} \right\}$;

$\mathbf{v} = \begin{bmatrix} a & b \\ c & d \end{bmatrix}$

5. In each case, verify Theorem 3. Use the standard basis in $\mathbb{R}^n$ and $\{1, x, x^2\}$ in $\mathbf{P}_2$.

   (a) $\mathbb{R}^3 \xrightarrow{T} \mathbb{R}^2 \xrightarrow{S} \mathbb{R}^4$; $T(a, b, c) = (a + b, b - c)$, $S(a, b) = (a, b - 2a, 3b, a + b)$

   ◆(b) $\mathbb{R}^3 \xrightarrow{T} \mathbb{R}^4 \xrightarrow{S} \mathbb{R}^2$; $T(a, b, c) = (a + b, c + b, a + c, b - a)$, $S(a, b, c, d) = (a + b, c - d)$

   (c) $\mathbf{P}_2 \xrightarrow{T} \mathbb{R}^3 \xrightarrow{S} \mathbf{P}_2$; $T(a + bx + cx^2) = (a, b - c, c - a)$, $S(a, b, c) = b + cx + (a - c)x^2$

   ◆(d) $\mathbb{R}^3 \xrightarrow{T} \mathbf{P}_2 \xrightarrow{S} \mathbb{R}^2$; $T(a, b, c) = (a - b) + (c - a)x + bx^2$, $S(a + bx + cx^2) = (a - b, c)$

6. Verify Theorem 3 for $\mathbf{M}_{22} \xrightarrow{T} \mathbf{M}_{22} \xrightarrow{S} \mathbf{P}_2$ where $T(A) = A^T$ and $S\begin{bmatrix} a & b \\ c & d \end{bmatrix} = b + (a + d)x + cx^2$.

   Use the bases $B = D = \left\{ \begin{bmatrix} 1 & 0 \\ 0 & 0 \end{bmatrix}, \begin{bmatrix} 0 & 1 \\ 0 & 0 \end{bmatrix}, \begin{bmatrix} 0 & 0 \\ 1 & 0 \end{bmatrix}, \begin{bmatrix} 0 & 0 \\ 0 & 1 \end{bmatrix} \right\}$ and $E = \{1, x, x^2\}$.

7. In each case, find T^{-1} and verify that $[M_{DB}(T)]^{-1} = M_{BD}(T^{-1})$.

   (a) $T: \mathbb{R}^2 \to \mathbb{R}^2$, $T(a, b) = (a + 2b, 2a + 5b)$; $B = D =$ standard

   ◆(b) $T: \mathbb{R}^3 \to \mathbb{R}^3$, $T(a, b, c) = (b + c, a + c, a + b)$; $B = D =$ standard

   (c) $T: \mathbf{P}_2 \to \mathbb{R}^3$, $T(a + bx + cx^2) = (a - c, b, 2a - c)$; $B = \{1, x, x^2\}$, $D =$ standard

   ◆(d) $T: \mathbf{P}_2 \to \mathbb{R}^3$, $T(a + bx + cx^2) = (a + b + c, b + c, c)$; $B = \{1, x, x^2\}$, $D =$ standard

8. In each case, show that $M_{DB}(T)$ is invertible and use the fact that $M_{BD}(T^{-1}) = [M_{BD}(T)]^{-1}$ to determine the action of T^{-1}.

   (a) $T: \mathbf{P}_2 \to \mathbb{R}^3$, $T(a + bx + cx^2) = (a + c, c, b - c)$; $B = \{1, x, x^2\}$, $D =$ standard

   ◆(b) $T: \mathbf{M}_{22} \to \mathbb{R}^4$, $T\begin{bmatrix} a & b \\ c & d \end{bmatrix} = (a + b + c, b + c, c, d)$; $B = \left\{ \begin{bmatrix} 1 & 0 \\ 0 & 0 \end{bmatrix}, \begin{bmatrix} 0 & 1 \\ 0 & 0 \end{bmatrix}, \begin{bmatrix} 0 & 0 \\ 1 & 0 \end{bmatrix}, \begin{bmatrix} 0 & 0 \\ 0 & 1 \end{bmatrix} \right\}$, $D =$ standard

9. Let $D: \mathbf{P}_3 \to \mathbf{P}_2$ be the differentiation map given by $D[p(x)] = p'(x)$. Find the matrix of D corresponding to the bases $B = \{1, x, x^2, x^3\}$ and $E = \{1, x, x^2\}$, and use it to compute $D(a + bx + cx^2 + dx^3)$.

10. Use Theorem 4 to show that $T: V \to V$ is not an isomorphism if $\ker T \neq 0$ (assume $\dim V = n$). [*Hint*: Choose any ordered basis B containing a vector in $\ker T$.]

11. Let $T: V \to \mathbb{R}$ be a linear transformation, and let $D = \{1\}$ be the basis of $\mathbb{R}$. Given any ordered basis $B = \{\mathbf{e}_1, \ldots, \mathbf{e}_n\}$ of V, show that $M_{DB}(T) = [T(\mathbf{e}_1) \cdots T(\mathbf{e}_n)]$.

◆12. Let $T: V \to W$ be an isomorphism, let $B = \{\mathbf{e}_1, \ldots, \mathbf{e}_n\}$ be an ordered basis of V, and let $D = \{T(\mathbf{e}_1), \ldots, T(\mathbf{e}_n)\}$. Show that $M_{DB}(T) = I_n$—the $n \times n$ identity matrix.

13. Complete the proof of Theorem 1.

14. Let U be any invertible $n \times n$ matrix, and let $D = \{\mathbf{f}_1, \mathbf{f}_2, \ldots, \mathbf{f}_n\}$ where $\mathbf{f}_j$ is column j of U. Show that $M_{BD}(1_{\mathbb{R}^n}) = U$ when B is the standard basis of $\mathbb{R}^n$.

15. Let B be an ordered basis of the n-dimensional space V and let $C_B: V \to \mathbb{R}^n$ be the coordinate transformation. If D is the standard basis of $\mathbb{R}^n$, show that $M_{DB}(C_B) = I_n$.

16. Let $T: \mathbf{P}_2 \to \mathbb{R}^3$ be defined by $T(p) = (p(0), p(1), p(2))$ for all p in $\mathbf{P}_2$. Let $B = \{1, x, x^2\}$ and $D = \{(1, 0, 0), (0, 1, 0), (0, 0, 1)\}$.

    (a) Show that $M_{DB}(T) = \begin{bmatrix} 1 & 0 & 0 \\ 1 & 1 & 1 \\ 1 & 2 & 4 \end{bmatrix}$ and conclude that T is an isomorphism.

◆(b) Generalize to $T: \mathbf{P}_n \to \mathbb{R}^{n+1}$ where $T(p) = (p(a_0), p(a_1), \ldots, p(a_n))$ and $a_0, a_1, \ldots, a_n$ are distinct real numbers. [*Hint*: Theorem 7 Section 3.2.]

17. Let $T: \mathbf{P}_n \to \mathbf{P}_n$ be defined by $T[p(x)] = p(x) + xp'(x)$, where $p'(x)$ denotes the derivative. Show that T is an isomorphism by finding $M_{BB}(T)$ when $B = \{1, x, x^2, \ldots, x^n\}$.

18. If k is any number, define $T_k : \mathbf{M}_{22} \to \mathbf{M}_{22}$ by $T_k(A) = A + kA^T$.

    (a) If $B = \left\{\begin{bmatrix}1 & 0\\0 & 0\end{bmatrix}, \begin{bmatrix}0 & 0\\0 & 1\end{bmatrix}, \begin{bmatrix}0 & 1\\1 & 0\end{bmatrix}, \begin{bmatrix}0 & 1\\-1 & 0\end{bmatrix}\right\}$

    find $M_{BB}(T_k)$, and conclude that T_k is invertible if $k \neq 1$ and $k \neq -1$.

    (b) Repeat for $T_k: \mathbf{M}_{33} \to \mathbf{M}_{33}$. Can you generalize?

The remaining exercises require the following definitions. If V and W are vector spaces, the set of all linear transformations from V to W will be denoted by

$$\mathbf{L}(V, W) = \{T \mid T: V \to W \text{ is a linear transformation}\}$$

Given S and T in $\mathbf{L}(V, W)$ and a in $\mathbb{R}$, define $S + T: V \to W$ and $aT: V \to W$ by

$$(S + T)(\mathbf{v}) = S(\mathbf{v}) + T(\mathbf{v}) \quad \text{for all } \mathbf{v} \text{ in } V$$
$$(aT)(\mathbf{v}) = aT(\mathbf{v}) \quad \text{for all } \mathbf{v} \text{ in } V$$

19. Show that $\mathbf{L}(V, W)$ is a vector space.

20. Show that the following properties hold provided that the transformations link together in such a way that all the operations are defined.

    (a) $R(ST) = (RS)T$

    (b) $1_W T = T = T 1_V$

    (c) $R(S + T) = RS + RT$

    ◆(d) $(S + T)R = SR + TR$

    (e) $(aS)T = a(ST) = S(aT)$

21. Given S and T in $\mathbf{L}(V, W)$, show that:

    (a) $\ker S \cap \ker T \subseteq \ker(S + T)$

    ◆(b) $\operatorname{im}(S + T) \subseteq \operatorname{im} S + \operatorname{im} T$

22. Let V and W be vector spaces. If X is a subset of V, define
    $$X^0 = \{T \text{ in } \mathbf{L}(V, W) \mid T(\mathbf{v}) = 0 \text{ for all } \mathbf{v} \text{ in } X\}$$

    (a) Show that X^0 is a subspace of $\mathbf{L}(V, W)$.

    ◆(b) If $X \subseteq X_1$, show that $X_1^0 \subseteq X^0$.

    (c) If U and U_1 are subspaces of V, show that $(U + U_1)^0 = U^0 \cap U_1^0$.

23. Define $R : \mathbf{M}_{mn} \to \mathbf{L}(\mathbb{R}^n, \mathbb{R}^m)$ by $R(A) = T_A$ for each $m \times n$ matrix A, where $T_A : \mathbb{R}^n \to \mathbb{R}^m$ is given by $T_A(\mathbf{x}) = A\mathbf{x}$ for all $\mathbf{x}$ in $\mathbb{R}^n$. Show that R is an isomorphism.

24. Let V be any vector space (we do not assume it is finite dimensional). Given $\mathbf{v}$ in V, define $S_\mathbf{v}: \mathbb{R} \to V$ by $S_\mathbf{v}(r) = r\mathbf{v}$ for all r in $\mathbb{R}$.

    (a) Show that $S_\mathbf{v}$ lies in $\mathbf{L}(\mathbb{R}, V)$ for each $\mathbf{v}$ in V.

    ◆(b) Show that the map $R : V \to \mathbf{L}(\mathbb{R}, V)$ given by $R(\mathbf{v}) = S_\mathbf{v}$ is an isomorphism. [*Hint*: To show that R is onto, if T lies in $\mathbf{L}(\mathbb{R}, V)$, show that $T = S_\mathbf{v}$ where $\mathbf{v} = T(1)$.]

25. Let V be a vector space with ordered basis $B = \{\mathbf{b}_1, \mathbf{b}_2, \ldots, \mathbf{b}_n\}$. For each $i = 1, 2, \ldots, n$, define $S_i: \mathbb{R} \to V$ by $S_i(r) = r\mathbf{b}_i$ for all r in $\mathbb{R}$.

    (a) Show that each S_i lies in $\mathbf{L}(\mathbb{R}, V)$ and $S_i(1) = \mathbf{b}_i$.

    ◆(b) Given T in $\mathbf{L}(\mathbb{R}, V)$, let $T(1) = a_1\mathbf{b}_1 + a_2\mathbf{b}_2 + \cdots + a_n\mathbf{b}_n$, a_i in $\mathbb{R}$. Show that $T = a_1 S_1 + a_2 S_2 + \cdots + a_n S_n$.

    (c) Show that $\{S_1, S_2, \ldots, S_n\}$ is a basis of $\mathbf{L}(\mathbb{R}, V)$.

26. Let $\dim V = n$, $\dim W = m$, and let B and D be ordered bases of V and W, respectively. Show that $M_{DB}: \mathbf{L}(V, W) \to \mathbf{M}_{mn}$ is an isomorphism of vector spaces. [*Hint*: Let $B = \{\mathbf{b}_1, \ldots, \mathbf{b}_n\}$ and $D = \{\mathbf{d}_1, \ldots, \mathbf{d}_m\}$. Given $A = [a_{ij}]$ in $\mathbf{M}_{mn}$, show that $A = M_{DB}(T)$ where $T: V \to W$ is defined by $T(\mathbf{b}_j) = a_{1j}\mathbf{d}_1 + a_{2j}\mathbf{d}_2 + \cdots + a_{mj}\mathbf{d}_m$ for each j.]

27. If V is a vector space, the space $V^* = \mathbf{L}(V, \mathbb{R})$ is called the **dual** of V. Given a basis $B = \{\mathbf{b}_1, \mathbf{b}_2, \ldots, \mathbf{b}_n\}$ of V, let $E_i : V \to \mathbb{R}$ for each $i = 1, 2, \ldots, n$ be the linear transformation satisfying
    $$E_i(\mathbf{b}_j) = \begin{cases} 0 & \text{if } i \neq j \\ 1 & \text{if } i = j \end{cases}$$
    (each E_i exists by Theorem 3 Section 7.1). Prove the following:

    (a) $E_i(r_1\mathbf{b}_1 + \cdots + r_n\mathbf{b}_n) = r_i$ for each $i = 1, 2, \ldots, n$

    ◆(b) $\mathbf{v} = E_1(\mathbf{v})\mathbf{b}_1 + E_2(\mathbf{v})\mathbf{b}_2 + \cdots + E_n(\mathbf{v})\mathbf{b}_n$ for all $\mathbf{v}$ in V

(c) $T = T(\mathbf{b}_1)E_1 + T(\mathbf{b}_2)E_2 + \cdots + T(\mathbf{b}_n)E_n$ for all T in V^*

(d) $\{E_1, E_2, ..., E_n\}$ is a basis of V^* (called the **dual basis** of B).

Given $\mathbf{v}$ in V, define $\mathbf{v}^* : V \to \mathbb{R}$ by
$\mathbf{v}^*(\mathbf{w}) = E_1(\mathbf{v})E_1(\mathbf{w}) + E_2(\mathbf{v})E_2(\mathbf{w}) + \cdots + E_n(\mathbf{v})E_n(\mathbf{w})$
for all $\mathbf{w}$ in V. Show that:

(e) $\mathbf{v}^* : V \to \mathbb{R}$ is linear, so $\mathbf{v}^*$ lies in V^*.

(f) $\mathbf{b}_i^* = E_i$ for each $i = 1, 2, ..., n$.

(g) The map $R : V \to V^*$ with $R(\mathbf{v}) = \mathbf{v}^*$ is an isomorphism. [*Hint*: Show that R is linear and one-to-one and use Theorem 3 Section 7.3. Alternatively, show that $R^{-1}(T) = T(\mathbf{b}_1)\mathbf{b}_1 + \cdots + T(\mathbf{b}_n)\mathbf{b}_n$.]

SECTION 9.2 Operators and Similarity

While the study of linear transformations from one vector space to another is important, the central problem of linear algebra is to understand the structure of a linear transformation $T : V \to V$ from a space V to itself. Such transformations are called **linear operators**. If $T : V \to V$ is a linear operator where $\dim(V) = n$, it is possible to choose bases B and D of V such that the matrix $M_{DB}(T)$ has a very simple form: $M_{DB}(T) = \begin{bmatrix} I_r & 0 \\ 0 & 0 \end{bmatrix}$ where $r = \text{rank } T$ (see Example 7 Section 9.1).

Consequently, only the rank of T is revealed by determining the simplest matrices $M_{DB}(T)$ of T where the bases B and D can be chosen arbitrarily. But if we insist that $B = D$ and look for bases B such that $M_{BB}(T)$ is as simple as possible, we learn a great deal about the operator T. We begin this task in this section.

The B-matrix of an Operator

Definition 9.3 If $T : V \to V$ is an operator on a vector space V, and if B is an ordered basis of V, define $M_B(T) = M_{BB}(T)$ and call this the **B-matrix** of T.

Recall that if $T : \mathbb{R}^n \to \mathbb{R}^n$ is a linear operator and $E = \{\mathbf{e}_1, \mathbf{e}_2, ..., \mathbf{e}_n\}$ is the standard basis of $\mathbb{R}^n$, then $C_E(\mathbf{x}) = \mathbf{x}$ for every $\mathbf{x} \in \mathbb{R}^n$, so $M_E(T) = [T(\mathbf{e}_1), T(\mathbf{e}_2), ..., T(\mathbf{e}_n)]$ is the matrix obtained in Theorem 2 Section 2.6. Hence $M_E(T)$ will be called the **standard matrix** of the operator T.

For reference the following theorem collects some results from Theorems 2, 3, and 4 in Section 9.1, specialized for operators. As before, $C_B(\mathbf{v})$ denoted the coordinate vector of $\mathbf{v}$ with respect to the basis B.

Theorem 1

Let $T : V \to V$ be an operator where $\dim V = n$, and let B be an ordered basis of V.

1. $C_B(T(\mathbf{v})) = M_B(T)C_B(\mathbf{v})$ for all $\mathbf{v}$ in V.
2. If $S : V \to V$ is another operator on V, then $M_B(ST) = M_B(S)M_B(T)$.
3. T is an isomorphism if and only if $M_B(T)$ is invertible. In this case $M_D[T]$ is invertible for every ordered basis D of V.
4. If T is an isomorphism, then $M_B(T^{-1}) = [M_B(T)]^{-1}$.
5. If $B = \{\mathbf{b}_1, \mathbf{b}_2, ..., \mathbf{b}_n\}$, then $M_B(T) = [\,C_B[T(\mathbf{b}_1)] \ \ C_B[T(\mathbf{b}_2)] \ \cdots \ C_B[T(\mathbf{b}_n)]\,]$.

Chapter 9   Change of Basis

For a fixed operator T on a vector space V, we are going to study how the matrix $M_B(T)$ changes when the basis B changes. This turns out to be closely related to how the coordinates $C_B(\mathbf{v})$ change for a vector $\mathbf{v}$ in V. If B and D are two ordered bases of V, and if we take $T = 1_V$ in Theorem 2 Section 9.1, we obtain

$$C_D(\mathbf{v}) = M_{DB}(1_V)C_B(\mathbf{v}) \quad \text{for all } \mathbf{v} \text{ in } V.$$

Definition 9.4

With this in mind, define the **change matrix** $P_{D \leftarrow B}$ by

$$P_{D \leftarrow B} = M_{DB}(1_V) \quad \text{for any ordered bases } B \text{ and } D \text{ of } V.$$

This proves (∗∗) in the following theorem:

Theorem 2

Let $B = \{\mathbf{b}_1, \mathbf{b}_2, \ldots, \mathbf{b}_n\}$ and D denote ordered bases of a vector space V. Then the change matrix $P_{D \leftarrow B}$ is given in terms of its columns by

$$P_{D \leftarrow B} = [C_D(\mathbf{b}_1) \; C_D(\mathbf{b}_2) \; \cdots \; C_D(\mathbf{b}_n)] \tag{*}$$

and has the property that

$$C_D(\mathbf{v}) = P_{D \leftarrow B} C_B(\mathbf{v}) \quad \text{for all } \mathbf{v} \text{ in } V. \tag{**}$$

Moreover, if E is another ordered basis of V, we have

1. $P_{B \leftarrow B} = I_n$.
2. $P_{D \leftarrow B}$ is invertible and $(P_{D \leftarrow B})^{-1} = P_{B \leftarrow D}$.
3. $P_{E \leftarrow D} P_{D \leftarrow B} = P_{E \leftarrow B}$.

PROOF

The formula (∗∗) is derived above, and (∗) is immediate from the definition of $P_{D \leftarrow B}$ and the formula for $M_{DB}(T)$ in Theorem 2 Section 9.1.

1. $P_{B \leftarrow B} = M_{BB}(1_V) = I_n$ as is easily verified.
2. This follows from (1) and (3).
3. Let $V \xrightarrow{T} W \xrightarrow{S} U$ be operators, and let B, D, and E be ordered bases of V, W, and U respectively. We have $M_{EB}(ST) = M_{ED}(S)M_{DB}(T)$ by Theorem 3 Section 9.1. Now (3) is the result of specializing $V = W = U$ and $T = S = 1_V$.

Property (3) in Theorem 2 explains the notation $\mathbf{P}_{D \leftarrow B}$.

EXAMPLE 1

In $\mathbf{P}_2$ find $P_{D \leftarrow B}$ if $B = \{1, x, x^2\}$ and $D = \{1, (1-x), (1-x)^2\}$. Then use this to express $p = p(x) = a + bx + cx^2$ as a polynomial in powers of $(1-x)$.

SECTION 9.2 Operators and Similarity

Solution ▶ To compute the change matrix $P_{D\leftarrow B}$, express $1, x, x^2$ in the basis D:
$$1 = 1 + 0(1-x) + 0(1-x)^2$$
$$x = 1 - 1(1-x) + 0(1-x)^2$$
$$x^2 = 1 - 2(1-x) + 1(1-x)^2$$

Hence $P_{D\leftarrow B} = [C_D(1), C_D(x), C_D(x^2)] = \begin{bmatrix} 1 & 1 & 1 \\ 0 & -1 & -2 \\ 0 & 0 & 1 \end{bmatrix}$. We have $C_B(p) = \begin{bmatrix} a \\ b \\ c \end{bmatrix}$, so

$$C_D(p) = P_{D\leftarrow B} C_D(p) = \begin{bmatrix} 1 & 1 & 1 \\ 0 & -1 & -2 \\ 0 & 0 & 1 \end{bmatrix} \begin{bmatrix} a \\ b \\ c \end{bmatrix} = \begin{bmatrix} a+b+c \\ -b-2c \\ c \end{bmatrix}$$

Hence $p(x) = (a+b+c) - (b+2c)(1-x) + c(1-x)^2$ by Definition 9.1.[1]

Now let $B = \{\mathbf{b}_1, \mathbf{b}_2, \ldots, \mathbf{b}_n\}$ and B_0 be two ordered bases of a vector space V. An operator $T: V \to V$ has different matrices $M_B[T]$ and $M_{B_0}[T]$ with respect to B and B_0. We can now determine how these matrices are related. Theorem 2 asserts that

$$C_{B_0}(\mathbf{v}) = P_{B_0\leftarrow B} C_B(\mathbf{v}) \quad \text{for all } \mathbf{v} \text{ in } V.$$

On the other hand, Theorem 1 gives

$$C_B[T(\mathbf{v})] = M_B(T) C_B(\mathbf{v}) \quad \text{for all } \mathbf{v} \text{ in } V.$$

Combining these (and writing $P = P_{B_0\leftarrow B}$ for convenience) gives

$$PM_B(T)C_B(\mathbf{v}) = PC_B[T(\mathbf{v})]$$
$$= C_{B_0}[T(\mathbf{v})]$$
$$= M_{B_0}(T)C_{B_0}(\mathbf{v})$$
$$= M_{B_0}(T)PC_B(\mathbf{v})$$

This holds for all $\mathbf{v}$ in V. Because $C_B(\mathbf{b}_j)$ is the jth column of the identity matrix, it follows that

$$PM_B(T) = M_{B_0}(T)P$$

Moreover P is invertible (in fact, $P^{-1} = P_{B\leftarrow B_0}$ by Theorem 2), so this gives

$$M_B(T) = P^{-1}M_{B_0}(T)P$$

This asserts that $M_{B_0}(T)$ and $M_B(T)$ are similar matrices, and proves Theorem 3.

Theorem 3

Let B_0 and B be two ordered bases of a finite dimensional vector space V. If $T: V \to V$ is any linear operator, the matrices $M_B(T)$ and $M_{B_0}(T)$ of T with respect to these bases are similar. More precisely,

$$M_B(T) = P^{-1}M_{B_0}(T)P$$

where $P = P_{B_0\leftarrow B}$ is the change matrix from B to B_0.

---

1 This also follows from Taylor's Theorem (Corollary 3 of Theorem 1 Section 6.5 with $a = 1$).

EXAMPLE 2

Let $T : \mathbb{R}^3 \to \mathbb{R}^3$ be defined by $T(a, b, c) = (2a - b, b + c, c - 3a)$. If B_0 denotes the standard basis of $\mathbb{R}^3$ and $B = \{(1, 1, 0), (1, 0, 1), (0, 1, 0)\}$, find an invertible matrix P such that $P^{-1}M_{B_0}(T)P = M_B(T)$.

Solution ▶ We have

$$M_{B_0}(T) = [C_{B_0}(2, 0, -3) \ \ C_{B_0}(-1, 1, 0) \ \ C_{B_0}(0, 1, 1)] = \begin{bmatrix} 2 & -1 & 0 \\ 0 & 1 & 1 \\ -3 & 0 & 1 \end{bmatrix}$$

$$M_B(T) = [C_B(1, 1, -3) \ \ C_B(2, 1, -2) \ \ C_B(-1, 1, 0)] = \begin{bmatrix} 4 & 4 & -1 \\ -3 & -2 & 0 \\ -3 & -3 & 2 \end{bmatrix}$$

$$P = P_{B_0 \leftarrow B} = [C_{B_0}(1, 1, 0) \ \ C_{B_0}(1, 0, 1) \ \ C_{B_0}(0, 1, 0)] = \begin{bmatrix} 1 & 1 & 0 \\ 1 & 0 & 1 \\ 0 & 1 & 0 \end{bmatrix}$$

The reader can verify that $P^{-1}M_{B_0}(T)P = M_B(T)$; equivalently that $M_{B_0}(T)P = PM_B(T)$.

A square matrix is diagonalizable if and only if it is similar to a diagonal matrix. Theorem 3 comes into this as follows: Suppose an $n \times n$ matrix $A = M_{B_0}(T)$ is the matrix of some operator $T : V \to V$ with respect to an ordered basis B_0. If another ordered basis B of V can be found such that $M_B(T) = D$ is diagonal, then Theorem 3 shows how to find an invertible P such that $P^{-1}AP = D$. In other words, the "algebraic" problem of finding P such that $P^{-1}AP$ is diagonal comes down to the "geometric" problem of finding a basis B such that $M_B(T)$ is diagonal. This shift of emphasis is one of the most important techniques in linear algebra.

Each $n \times n$ matrix A can be easily realized as the matrix of an operator. In fact, (Example 4 Section 9.1),

$$M_E(T_A) = A$$

where $T_A : \mathbb{R}^n \to \mathbb{R}^n$ is the matrix operator given by $T_A(\mathbf{x}) = A\mathbf{x}$, and E is the standard basis of $\mathbb{R}^n$. The first part of the next theorem gives the converse of Theorem 3: Any pair of similar matrices can be realized as the matrices of the same linear operator with respect to different bases.

Theorem 4

Let A be an $n \times n$ matrix and let E be the standard basis of $\mathbb{R}^n$.

1. Let A' be similar to A, say $A' = P^{-1}AP$, and let B be the ordered basis of $\mathbb{R}^n$ consisting of the columns of P in order. Then $T_A : \mathbb{R}^n \to \mathbb{R}^n$ is linear and

$$M_E(T_A) = A \quad \text{and} \quad M_B(T_A) = A'$$

2. If B is any ordered basis of $\mathbb{R}^n$, let P be the (invertible) matrix whose columns are the vectors in B in order. Then

$$M_B(T_A) = P^{-1}AP$$

SECTION 9.2 Operators and Similarity

PROOF

1. We have $M_E(T_A) = A$ by Example 4 Section 9.1. Write $P = [\mathbf{b}_1 \cdots \mathbf{b}_n]$ in terms of its columns so $B = \{\mathbf{b}_1, \ldots, \mathbf{b}_n\}$ is a basis of $\mathbb{R}^n$. Since E is the standard basis,

$$P_{E \leftarrow B} = [C_E(\mathbf{b}_1) \cdots C_E(\mathbf{b}_n)] = [\mathbf{b}_1 \cdots \mathbf{b}_n] = P.$$

Hence Theorem 3 (with $B_0 = E$) gives $M_B(T_A) = P^{-1}M_E(T_A)P = P^{-1}AP = A'$.

2. Here P and B are as above, so again $P_{E \leftarrow B} = P$ and $M_B(T_A) = P^{-1}AP$.

EXAMPLE 3

Given $A = \begin{bmatrix} 10 & 6 \\ -18 & -11 \end{bmatrix}$, $P = \begin{bmatrix} 2 & -1 \\ -3 & 2 \end{bmatrix}$, and $D = \begin{bmatrix} 1 & 0 \\ 0 & -2 \end{bmatrix}$, verify that $P^{-1}AP = D$ and use this fact to find a basis B of $\mathbb{R}^2$ such that $M_B(T_A) = D$.

Solution ▶ $P^{-1}AP = D$ holds if $AP = PD$; this verification is left to the reader. Let B consist of the columns of P in order, that is $P : B = \left\{ \begin{bmatrix} 2 \\ -3 \end{bmatrix}, \begin{bmatrix} -1 \\ 2 \end{bmatrix} \right\}$.

Then Theorem 4 gives $M_B(T_A) = P^{-1}AP = D$. More explicitly,

$$M_B(T_A) = \left[C_B\left(T_A\begin{bmatrix} 2 \\ -3 \end{bmatrix}\right) \ C_B\left(T_A\begin{bmatrix} -1 \\ 2 \end{bmatrix}\right) \right] = \left[C_B\begin{bmatrix} 2 \\ -3 \end{bmatrix} \ C_B\begin{bmatrix} 2 \\ -4 \end{bmatrix} \right] = \begin{bmatrix} 1 & 0 \\ 0 & -2 \end{bmatrix} = D.$$

Let A be an $n \times n$ matrix. As in Example 3, Theorem 4 provides a new way to find an invertible matrix P such that $P^{-1}AP$ is diagonal. The idea is to find a basis $B = \{\mathbf{b}_1, \mathbf{b}_2, \ldots, \mathbf{b}_n\}$ of $\mathbb{R}^n$ such that $M_B(T_A) = D$ is diagonal and take $P = [\mathbf{b}_1 \ \mathbf{b}_2 \ \cdots \ \mathbf{b}_n]$ to be the matrix with the $\mathbf{b}_j$ as columns. Then, by Theorem 4,

$$P^{-1}AP = M_B(T_A) = D.$$

As mentioned above, this converts the algebraic problem of diagonalizing A into the geometric problem of finding the basis B. This new point of view is very powerful and will be explored in the next two sections.

Theorem 4 enables facts about matrices to be deduced from the corresponding properties of operators. Here is an example.

EXAMPLE 4

1. If $T : V \to V$ is an operator where V is finite dimensional, show that $TST = T$ for some invertible operator $S : V \to V$.
2. If A is an $n \times n$ matrix, show that $AUA = A$ for some invertible matrix U.

Solution ▶

1. Let $B = \{\mathbf{b}_1, \ldots, \mathbf{b}_r, \mathbf{b}_{r+1}, \ldots, \mathbf{b}_n\}$ be a basis of V chosen so that $\ker T = \text{span}\{\mathbf{b}_{r+1}, \ldots, \mathbf{b}_n\}$. Then $\{T(\mathbf{b}_1), \ldots, T(\mathbf{b}_r)\}$ is independent (Theorem 5 Section 7.2), so complete it to a basis $\{T(\mathbf{b}_1), \ldots, T(\mathbf{b}_r), \mathbf{f}_{r+1}, \ldots, \mathbf{f}_n\}$ of V.

By Theorem 3 Section 7.1, define $S : V \to V$ by
$$S[T(\mathbf{b}_i)] = \mathbf{b}_i \quad \text{for } 1 \leq i \leq r$$
$$S(\mathbf{f}_j) = \mathbf{b}_j \quad \text{for } r < j \leq n$$

Then S is an isomorphism by Theorem 1 Section 7.3, and $TST = T$ because these operators agree on the basis B. In fact,
$$(TST)(\mathbf{b}_i) = T[ST(\mathbf{b}_i)] = T(\mathbf{b}_i) \text{ if } 1 \leq i \leq r, \text{ and}$$
$$(TST)(\mathbf{b}_j) = TS[T(\mathbf{b}_j)] = TS(\mathbf{0}) = \mathbf{0} = T(\mathbf{b}_j) \text{ for } r < j \leq n.$$

2. Given A, let $T = T_A : \mathbb{R}^n \to \mathbb{R}^n$. By (1) let $TST = T$ where $S : \mathbb{R}^n \to \mathbb{R}^n$ is an isomorphism. If E is the standard basis of $\mathbb{R}^n$, then $A = M_E(T)$ by Theorem 4. If $U = M_E(S)$ then, by Theorem 1, U is invertible and
$$AUA = M_E(T)M_E(S)M_E(T) = M_E(TST) = M_E(T) = A$$

as required.

The reader will appreciate the power of these methods if he/she tries to find U directly in part 2 of Example 4, even if A is 2×2.

A property of $n \times n$ matrices is called a **similarity invariant** if, whenever a given $n \times n$ matrix A has the property, every matrix similar to A also has the property. Theorem 1 Section 5.5 shows that rank, determinant, trace, and characteristic polynomial are all similarity invariants.

To illustrate how such similarity invariants are related to linear operators, consider the case of rank. If $T : V \to V$ is a linear operator, the matrices of T with respect to various bases of V all have the same rank (being similar), so it is natural to regard the common rank of all these matrices as a property of T itself and not of the particular matrix used to describe T. Hence the rank of T could be *defined* to be the rank of A, where A is *any* matrix of T. This would be unambiguous because rank is a similarity invariant. Of course, this is unnecessary in the case of rank because rank T was defined earlier to be the dimension of im T, and this was *proved* to equal the rank of every matrix representing T (Theorem 5 Section 9.1). This definition of rank T is said to be *intrinsic* because it makes no reference to the matrices representing T. However, the technique serves to identify an intrinsic property of T with *every* similarity invariant, and some of these properties are not so easily defined directly.

In particular, if $T : V \to V$ is a linear operator on a finite dimensional space V, define the **determinant** of T (denoted det T) by
$$\det T = \det M_B(T), \quad B \text{ any basis of } V$$

This is independent of the choice of basis B because, if D is any other basis of V, the matrices $M_B(T)$ and $M_D(T)$ are similar and so have the same determinant. In the same way, the **trace** of T (denoted tr T) can be defined by
$$\operatorname{tr} T = \operatorname{tr} M_B(T), \quad B \text{ any basis of } V$$

This is unambiguous for the same reason.

Theorems about matrices can often be translated to theorems about linear operators. Here is an example.

EXAMPLE 5

Let S and T denote linear operators on the finite dimensional space V. Show that
$$\det(ST) = \det S \det T$$

SECTION 9.2 Operators and Similarity

Solution ▶ Choose a basis B of V and use Theorem 1.
$$\det(ST) = \det M_B(ST) = \det[M_B(S)M_B(T)]$$
$$= \det[M_B(S)]\det[M_B(T)] = \det S \det T$$

Recall next that the characteristic polynomial of a matrix is another similarity invariant: If A and A' are similar matrices, then $c_A(x) = c_{A'}(x)$ (Theorem 1 Section 5.5). As discussed above, the discovery of a similarity invariant means the discovery of a property of linear operators. In this case, if $T : V \to V$ is a linear operator on the finite dimensional space V, define the **characteristic polynomial** of T by
$$c_T(x) = c_A(x) \quad \text{where } A = M_B(T), B \text{ any basis of } V$$
In other words, the characteristic polynomial of an operator T is the characteristic polynomial of *any* matrix representing T. This is unambiguous because any two such matrices are similar by Theorem 3.

EXAMPLE 6

Compute the characteristic polynomial $c_T(x)$ of the operator $T : \mathbf{P}_2 \to \mathbf{P}_2$ given by $T(a + bx + cx^2) = (b + c) + (a + c)x + (a + b)x^2$.

Solution ▶ If $B = \{1, x, x^2\}$, the corresponding matrix of T is
$$M_B(T) = [C_B[T(1)] \quad C_B[T(x)] \quad C_B[T(x^2)]] = \begin{bmatrix} 0 & 1 & 1 \\ 1 & 0 & 1 \\ 1 & 1 & 0 \end{bmatrix}$$
Hence $c_T(x) = \det[xI - M_B(T)] = x^3 - 3x - 2 = (x + 1)^2(x - 2)$

In Section 4.4 we computed the matrix of various projections, reflections, and rotations in $\mathbb{R}^3$. However, the methods available then were not adequate to find the matrix of a rotation about a line through the origin. We conclude this section with an example of how Theorem 3 can be used to compute such a matrix.

EXAMPLE 7

Let L be the line in $\mathbb{R}^3$ through the origin with (unit) direction vector $\mathbf{d} = \frac{1}{3}[2 \ 1 \ 2]^T$. Compute the matrix of the rotation about L through an angle θ measured counterclockwise when viewed in the direction of $\mathbf{d}$.

Solution ▶ Let $R : \mathbb{R}^3 \to \mathbb{R}^3$ be the rotation. The idea is to first find a basis B_0 for which the matrix of $M_{B_0}(R)$ of R is easy to compute, and then use Theorem 3 to compute the "standard" matrix $M_E(R)$ with respect to the standard basis $E = \{\mathbf{e}_1, \mathbf{e}_2, \mathbf{e}_3\}$ of $\mathbb{R}^3$.

To construct the basis B_0, let K denote the plane through the origin with $\mathbf{d}$ as normal, shaded in the diagram. Then the vectors $\mathbf{f} = \frac{1}{3}[-2 \ 2 \ 1]^T$ and $\mathbf{g} = \frac{1}{3}[1 \ 2 \ -2]^T$ are both in K (they are orthogonal to $\mathbf{d}$) and are independent (they are orthogonal to each other).

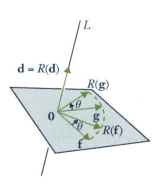

Hence $B_0 = \{\mathbf{d}, \mathbf{f}, \mathbf{g}\}$ is an orthonormal basis of $\mathbb{R}^3$, and the effect of R on B_0 is easy to determine. In fact $R(\mathbf{d}) = \mathbf{d}$ and (as in Theorem 4 Section 2.6) the second diagram gives

$$R(\mathbf{f}) = \cos\theta\, \mathbf{f} + \sin\theta\, \mathbf{g} \quad \text{and} \quad R(\mathbf{g}) = -\sin\theta\, \mathbf{f} + \cos\theta\, \mathbf{g}$$

because $\|\mathbf{f}\| = 1 = \|\mathbf{g}\|$. Hence

$$M_{B_0}(R) = [C_{B_0}(\mathbf{d}) \ \ C_{B_0}(\mathbf{f}) \ \ C_{B_0}(\mathbf{g})] = \begin{bmatrix} 1 & 0 & 0 \\ 0 & \cos\theta & -\sin\theta \\ 0 & \sin\theta & \cos\theta \end{bmatrix}.$$

Now Theorem 3 (with $B = E$) asserts that $M_E(R) = P^{-1}M_{B_0}(R)P$ where

$$P = P_{B_0 \leftarrow E} = [C_{B_0}(\mathbf{e}_1) \ \ C_{B_0}(\mathbf{e}_2) \ \ C_{B_0}(\mathbf{e}_3)] = \tfrac{1}{3}\begin{bmatrix} 2 & 1 & 2 \\ -2 & 2 & 1 \\ 1 & 2 & -2 \end{bmatrix}$$

using the expansion theorem (Theorem 6 Section 5.3). Since $P^{-1} = P^T$ (P is orthogonal), the matrix of R with respect to E is

$$M_E(R) = P^T M_{B_0}(R) P$$
$$= \tfrac{1}{9}\begin{bmatrix} 5\cos\theta+4 & 6\sin\theta-2\cos\theta+2 & 4-3\sin\theta-4\cos\theta \\ 2-6\sin\theta-2\cos\theta & 8\cos\theta+1 & 6\sin\theta-2\cos\theta+2 \\ 3\sin\theta-4\cos\theta+4 & 2-6\sin\theta-2\cos\theta & 5\cos\theta+4 \end{bmatrix}$$

As a check one verifies that this is the identity matrix when $\theta = 0$, as it should.

Note that in Example 7 not much motivation was given to the choices of the (orthonormal) vectors $\mathbf{f}$ and $\mathbf{g}$ in the basis B_0, which is the key to the solution. However, if we begin with *any* basis containing $\mathbf{d}$ the Gram-Schmidt algorithm will produce an orthogonal basis containing $\mathbf{d}$, and the other two vectors will automatically be in $L^\perp = K$.

EXERCISES 9.2

1. In each case find $P_{D \leftarrow B}$, where B and D are ordered bases of V. Then verify that $C_D(\mathbf{v}) = P_{D \leftarrow B} C_B(\mathbf{v})$.

   (a) $V = \mathbb{R}^2$, $B = \{(0, -1), (2, 1)\}$,
   $D = \{(0, 1), (1, 1)\}$, $\mathbf{v} = (3, -5)$

   ◆(b) $V = \mathbf{P}_2$, $B = \{x, 1 + x, x^2\}$,
   $D = \{2, x + 3, x^2 - 1\}$, $\mathbf{v} = 1 + x + x^2$

   (c) $V = \mathbf{M}_{22}$,

   $B = \left\{ \begin{bmatrix} 1 & 0 \\ 0 & 0 \end{bmatrix}, \begin{bmatrix} 0 & 1 \\ 0 & 0 \end{bmatrix}, \begin{bmatrix} 0 & 0 \\ 0 & 1 \end{bmatrix}, \begin{bmatrix} 0 & 0 \\ 1 & 0 \end{bmatrix} \right\}$,

   $D = \left\{ \begin{bmatrix} 1 & 1 \\ 0 & 0 \end{bmatrix}, \begin{bmatrix} 1 & 0 \\ 1 & 0 \end{bmatrix}, \begin{bmatrix} 1 & 0 \\ 0 & 1 \end{bmatrix}, \begin{bmatrix} 0 & 1 \\ 1 & 0 \end{bmatrix} \right\}$,

   $\mathbf{v} = \begin{bmatrix} 3 & -1 \\ 1 & 4 \end{bmatrix}$

2. In $\mathbb{R}^3$ find $P_{D \leftarrow B}$, where
   $B = \{(1, 0, 0), (1, 1, 0), (1, 1, 1)\}$ and
   $D = \{(1, 0, 1), (1, 0, -1), (0, 1, 0)\}$.
   If $\mathbf{v} = (a, b, c)$, show that

   $C_D(\mathbf{v}) = \tfrac{1}{2}\begin{bmatrix} a+c \\ a-c \\ 2b \end{bmatrix}$ and $C_B(\mathbf{v}) = \begin{bmatrix} a-b \\ b-c \\ c \end{bmatrix}$, and

   verify that $C_D(\mathbf{v}) = P_{D \leftarrow B} C_B(\mathbf{v})$.

3. In $\mathbf{P}_3$ find $P_{D \leftarrow B}$ if $B = \{1, x, x^2, x^3\}$ and
   $D = \{1, (1-x), (1-x)^2, (1-x)^3\}$.
   Then express $p = a + bx + cx^2 + dx^3$ as a polynomial in powers of $(1 - x)$.

4. In each case verify that $P_{D \leftarrow B}$ is the inverse of $P_{B \leftarrow D}$ and that $P_{E \leftarrow D} P_{D \leftarrow B} = P_{E \leftarrow B}$, where B, D, and E are ordered bases of V.

(a) $V = \mathbb{R}^3$, $B = \{(1, 1, 1), (1, -2, 1), (1, 0, -1)\}$,
D = standard basis,
$E = \{(1, 1, 1), (1, -1, 0), (-1, 0, 1)\}$

◆(b) $V = \mathbf{P}_2$, $B = \{1, x, x^2\}$,
$D = \{1 + x + x^2, 1 - x, -1 + x^2\}$. $E = \{x^2, x, 1\}$

5. Use property (2) of Theorem 2, with D the standard basis of $\mathbb{R}^n$, to find the inverse of:

(a) $A = \begin{bmatrix} 1 & 1 & 0 \\ 1 & 0 & 1 \\ 0 & 1 & 1 \end{bmatrix}$ ◆(b) $A = \begin{bmatrix} 1 & 2 & 1 \\ 2 & 3 & 0 \\ -1 & 0 & 2 \end{bmatrix}$

6. Find $P_{D \leftarrow B}$ if $B = \{\mathbf{b}_1, \mathbf{b}_2, \mathbf{b}_3, \mathbf{b}_4\}$ and $D = \{\mathbf{b}_2, \mathbf{b}_3, \mathbf{b}_1, \mathbf{b}_4\}$. Change matrices arising when the bases differ only in the *order* of the vectors are called **permutation matrices**.

7. In each case, find $P = P_{B_0 \leftarrow B}$ and verify that $P^{-1}M_{B_0}(T)P = M_B(T)$ for the given operator T.

(a) $T: \mathbb{R}^3 \to \mathbb{R}^3$, $T(a, b, c) = (2a - b, b + c, c - 3a)$;
$B_0 = \{(1, 1, 0), (1, 0, 1), (0, 1, 0)\}$ and B is the standard basis.

◆(b) $T: \mathbf{P}_2 \to \mathbf{P}_2$,
$T(a + bx + cx^2) = (a + b) + (b + c)x + (c + a)x^2$;
$B_0 = \{1, x, x^2\}$ and $B = \{1 - x^2, 1 + x, 2x + x^2\}$.

(c) $T: \mathbf{M}_{22} \to \mathbf{M}_{22}$, $T\begin{bmatrix} a & b \\ c & d \end{bmatrix} = \begin{bmatrix} a + d & b + c \\ a + c & b + d \end{bmatrix}$;
$B_0 = \left\{ \begin{bmatrix} 1 & 0 \\ 0 & 0 \end{bmatrix}, \begin{bmatrix} 0 & 1 \\ 0 & 0 \end{bmatrix}, \begin{bmatrix} 0 & 0 \\ 1 & 0 \end{bmatrix}, \begin{bmatrix} 0 & 0 \\ 0 & 1 \end{bmatrix} \right\}$,
and $B = \left\{ \begin{bmatrix} 1 & 1 \\ 0 & 0 \end{bmatrix}, \begin{bmatrix} 0 & 0 \\ 1 & 1 \end{bmatrix}, \begin{bmatrix} 1 & 0 \\ 0 & 1 \end{bmatrix}, \begin{bmatrix} 0 & 1 \\ 1 & 1 \end{bmatrix} \right\}$

8. In each case, verify that $P^{-1}AP = D$ and find a basis B of $\mathbb{R}^2$ such that $M_B(T_A) = D$.

(a) $A = \begin{bmatrix} 11 & -6 \\ 12 & -6 \end{bmatrix}$ $P = \begin{bmatrix} 2 & 3 \\ 3 & 4 \end{bmatrix}$ $D = \begin{bmatrix} 2 & 0 \\ 0 & 3 \end{bmatrix}$

◆(b) $A = \begin{bmatrix} 29 & -12 \\ 70 & -29 \end{bmatrix}$ $P = \begin{bmatrix} 3 & 2 \\ 7 & 5 \end{bmatrix}$ $D = \begin{bmatrix} 1 & 0 \\ 0 & -1 \end{bmatrix}$

9. In each case, compute the characteristic polynomial $c_T(x)$.

(a) $T: \mathbb{R}^2 \to \mathbb{R}^2$, $T(a, b) = (a - b, 2b - a)$

◆(b) $T: \mathbb{R}^2 \to \mathbb{R}^2$, $T(a, b) = (3a + 5b, 2a + 3b)$

(c) $T: \mathbf{P}_2 \to \mathbf{P}_2$, $T(a + bx + cx^2)$
$= (a - 2c) + (2a + b + c)x + (c - a)x^2$

◆(d) $T: \mathbf{P}_2 \to \mathbf{P}_2$, $T(a + bx + cx^2)$
$= (a + b - 2c) + (a - 2b + c)x + (b - 2a)x^2$

(e) $T: \mathbb{R}^3 \to \mathbb{R}^3$, $T(a, b, c) = (b, c, a)$

◆(f) $T: \mathbf{M}_{22} \to \mathbf{M}_{22}$, $T\begin{bmatrix} a & b \\ c & d \end{bmatrix} = \begin{bmatrix} a - c & b - d \\ a - c & b - d \end{bmatrix}$;

10. If V is finite dimensional, show that a linear operator T on V has an inverse if and only if $\det T \neq 0$.

11. Let S and T be linear operators on V where V is finite dimensional.

(a) Show that $\text{tr}(ST) = \text{tr}(TS)$.
[*Hint*: Lemma 1 Section 5.5.]

(b) [See Exercise 19 Section 9.1.] For a in $\mathbb{R}$, show that $\text{tr}(S + T) = \text{tr } S + \text{tr } T$, and $\text{tr}(aT) = a \, \text{tr}(T)$.

◆12. If A and B are $n \times n$ matrices, show that they have the same null space if and only if $A = UB$ for some invertible matrix U. [*Hint*: Exercise 28 Section 7.3.]

13. If A and B are $n \times n$ matrices, show that they have the same column space if and only if $A = BU$ for some invertible matrix U. [*Hint*: Exercise 28 Section 7.3.]

14. Let $E = \{\mathbf{e}_1, \ldots, \mathbf{e}_n\}$ be the standard ordered basis of $\mathbb{R}^n$, written as columns. If $D = \{\mathbf{d}_1, \ldots, \mathbf{d}_n\}$ is any ordered basis, show that $P_{E \leftarrow D} = [\mathbf{d}_1 \; \cdots \; \mathbf{d}_n]$.

15. Let $B = \{\mathbf{b}_1, \mathbf{b}_2, \ldots, \mathbf{b}_n\}$ be any ordered basis of $\mathbb{R}^n$, written as columns. If $Q = [\mathbf{b}_1 \; \mathbf{b}_2 \; \cdots \; \mathbf{b}_n]$ is the matrix with the $\mathbf{b}_i$ as columns, show that $QC_B(\mathbf{v}) = \mathbf{v}$ for all $\mathbf{v}$ in $\mathbb{R}^n$.

16. Given a complex number w, define $T_w: \mathbb{C} \to \mathbb{C}$ by $T_w(z) = wz$ for all z in $\mathbb{C}$.

(a) Show that T_w is a linear operator for each w in $\mathbb{C}$, viewing $\mathbb{C}$ as a real vector space.

◆(b) If B is any ordered basis of $\mathbb{C}$, define $S: \mathbb{C} \to \mathbf{M}_{22}$ by $S(w) = M_B(T_w)$ for all w in $\mathbb{C}$. Show that S is a one-to-one linear transformation with the additional property that $S(wv) = S(w)S(v)$ holds for all w and v in $\mathbb{C}$.

(c) Taking $B = \{1, i\}$ show that
$S(a + bi) = \begin{bmatrix} a & -b \\ b & a \end{bmatrix}$ for all complex numbers $a + bi$. This is called the **regular representation** of the complex numbers as 2×2 matrices. If θ is any angle,

describe $S(e^{i\theta})$ geometrically. Show that $S(\overline{w}) = S(w)^T$ for all w in $\mathbb{C}$; that is, that conjugation corresponds to transposition.

17. Let $B = \{\mathbf{b}_1, \mathbf{b}_2, \ldots, \mathbf{b}_n\}$ and $D = \{\mathbf{d}_1, \mathbf{d}_2, \ldots, \mathbf{d}_n\}$ be two ordered bases of a vector space V. Prove that $C_D(\mathbf{v}) = P_{D \leftarrow B} C_B(\mathbf{v})$ holds for all $\mathbf{v}$ in V as follows: Express each $\mathbf{b}_j$ in the form $\mathbf{b}_j = p_{1j}\mathbf{d}_1 + p_{2j}\mathbf{d}_2 + \cdots + p_{nj}\mathbf{d}_n$

and write $P = [p_{ij}]$. Show that $P = [C_D(\mathbf{b}_1) \; C_D(\mathbf{b}_1) \; \cdots \; C_D(\mathbf{b}_1)]$ and that $C_D(\mathbf{v}) = PC_B(\mathbf{v})$ for all $\mathbf{v}$ in B.

18. Find the standard matrix of the rotation R about the line through the origin with direction vector $\mathbf{d} = [2 \; 3 \; 6]^T$. [*Hint*: Consider $\mathbf{f} = [6 \; 2 \; -3]^T$ and $\mathbf{g} = [3 \; -6 \; 2]^T$.]

SECTION 9.3 Invariant Subspaces and Direct Sums

A fundamental question in linear algebra is the following: If $T : V \to V$ is a linear operator, how can a basis B of V be chosen so the matrix $M_B(T)$ is as simple as possible? A basic technique for answering such questions will be explained in this section. If U is a subspace of V, write its image under T as

$$T(U) = \{T(\mathbf{u}) \mid \mathbf{u} \text{ in } U\}.$$

Definition 9.5 Let $T : V \to V$ be an operator. A subspace $U \subseteq V$ is called **T-invariant** if $T(U) \subseteq U$, that is, $T(\mathbf{u}) \in U$ for every vector $\mathbf{u} \in U$. Hence T is a linear operator on the vector space U.

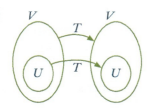

This is illustrated in the diagram, and the fact that $T : U \to U$ is an operator on U is the primary reason for our interest in T-invariant subspaces.

EXAMPLE 1

Let $T : V \to V$ be any linear operator. Then:

1. $\{\mathbf{0}\}$ and V are T-invariant subspaces.
2. Both ker T and im $T = T(V)$ are T-invariant subspaces.
3. If U and W are T-invariant subspaces, so are $T(U)$, $U \cap W$, and $U + W$.

Solution ▶ Item 1 is clear, and the rest is left as Exercises 1 and 2.

EXAMPLE 2

Define $T : \mathbb{R}^3 \to \mathbb{R}^3$ by $T(a, b, c) = (3a + 2b, b - c, 4a + 2b - c)$. Then $U = \{(a, b, a) \mid a, b \text{ in } \mathbb{R}\}$ is T-invariant because

$$T(a, b, a) = (3a + 2b, b - a, 3a + 2b)$$

is in U for all a and b (the first and last entries are equal).

If a spanning set for a subspace U is known, it is easy to check whether U is T-invariant.

SECTION 9.3 Invariant Subspaces and Direct Sums

EXAMPLE 3

Let $T: V \to V$ be a linear operator, and suppose that $U = \text{span}\{\mathbf{u}_1, \mathbf{u}_2, ..., \mathbf{u}_k\}$ is a subspace of V. Show that U is T-invariant if and only if $T(\mathbf{u}_i)$ lies in U for each $i = 1, 2, ..., k$.

Solution ▶ Given $\mathbf{u}$ in U, write it as $\mathbf{u} = r_1\mathbf{u}_1 + \cdots + r_k\mathbf{u}_k$, r_i in $\mathbb{R}$. Then
$$T(\mathbf{u}) = r_1 T(\mathbf{u}_1) + \cdots + r_k T(\mathbf{u}_k)$$
and this lies in U if each $T(\mathbf{u}_i)$ lies in U. This shows that U is T-invariant if each $T(\mathbf{u}_i)$ lies in U; the converse is clear.

EXAMPLE 4

Define $T: \mathbb{R}^2 \to \mathbb{R}^2$ by $T(a, b) = (b, -a)$. Show that $\mathbb{R}^2$ contains no T-invariant subspace except 0 and $\mathbb{R}^2$.

Solution ▶ Suppose, if possible, that U is T-invariant, but $U \neq 0$, $U \neq \mathbb{R}^2$. Then U has dimension 1 so $U = \mathbb{R}\mathbf{x}$ where $\mathbf{x} \neq \mathbf{0}$. Now $T(\mathbf{x})$ lies in U—say $T(\mathbf{x}) = r\mathbf{x}$, r in $\mathbb{R}$. If we write $\mathbf{x} = (a, b)$, this is $(b, -a) = r(a, b)$, which gives $b = ra$ and $-a = rb$. Eliminating b gives $r^2 a = rb = -a$, so $(r^2 + 1)a = 0$. Hence $a = 0$. Then $b = ra = 0$ too, contrary to the assumption that $\mathbf{x} \neq \mathbf{0}$. Hence no one-dimensional T-invariant subspace exists.

Definition 9.6   Let $T: V \to V$ be a linear operator. If U is any T-invariant subspace of V, then
$$T: U \to U$$
is a linear operator on the subspace U, called the **restriction** of T to U.

This is the reason for the importance of T-invariant subspaces and is the first step toward finding a basis that simplifies the matrix of T.

Theorem 1

Let $T: V \to V$ be a linear operator where V has dimension n and suppose that U is any T-invariant subspace of V. Let $B_1 = \{\mathbf{b}_1, ..., \mathbf{b}_k\}$ be any basis of U and extend it to a basis $B = \{\mathbf{b}_1, ..., \mathbf{b}_k, \mathbf{b}_{k+1}, ..., \mathbf{b}_n\}$ of V in any way. Then $M_B(T)$ has the block triangular form
$$M_B(T) = \begin{bmatrix} M_{B_1}(T) & Y \\ 0 & Z \end{bmatrix}$$
where Z is $(n-k) \times (n-k)$ and $M_{B_1}(T)$ is the matrix of the restriction of T to U.

PROOF

The matrix of (the restriction) $T : U \to U$ with respect to the basis B_1 is the $k \times k$ matrix

$$M_{B_1}(T) = [C_{B_1}[T(\mathbf{b}_1)] \quad C_{B_1}[T(\mathbf{b}_2)] \quad \cdots \quad C_{B_1}[T(\mathbf{b}_k)]]$$

Now compare the first column $C_{B_1}[T(\mathbf{b}_1)]$ here with the first column $C_B[T(\mathbf{b}_1)]$ of $M_B(T)$. The fact that $T(\mathbf{b}_1)$ lies in U (because U is T-invariant) means that $T(\mathbf{b}_1)$ has the form

$$T(\mathbf{b}_1) = t_1 \mathbf{b}_1 + t_2 \mathbf{b}_2 + \cdots + t_k \mathbf{b}_k + 0 \mathbf{b}_{k+1} + \cdots + 0 \mathbf{b}_n$$

Consequently,

$$C_{B_1}[T(\mathbf{b}_1)] = \begin{bmatrix} t_1 \\ t_2 \\ \vdots \\ t_k \end{bmatrix} \text{ in } \mathbb{R}^k \quad \text{whereas} \quad C_B[T(\mathbf{b}_1)] = \begin{bmatrix} t_1 \\ t_2 \\ \vdots \\ t_k \\ 0 \\ \vdots \\ 0 \end{bmatrix} \text{ in } \mathbb{R}^n$$

This shows that the matrices $M_B(T)$ and $\begin{bmatrix} M_{B_1}(T) & Y \\ 0 & Z \end{bmatrix}$ have identical first columns.

Similar statements apply to columns 2, 3, ..., k, and this proves the theorem.

The block upper triangular form for the matrix $M_B(T)$ in Theorem 1 is very useful because the determinant of such a matrix equals the product of the determinants of each of the diagonal blocks. This is recorded in Theorem 2 for reference, together with an important application to characteristic polynomials.

Theorem 2

Let A be a block upper triangular matrix, say

$$A = \begin{bmatrix} A_{11} & A_{12} & A_{13} & \cdots & A_{1n} \\ 0 & A_{22} & A_{23} & \cdots & A_{2n} \\ 0 & 0 & A_{33} & \cdots & A_{3n} \\ \vdots & \vdots & \vdots & & \vdots \\ 0 & 0 & 0 & \cdots & A_{nn} \end{bmatrix}$$

where the diagonal blocks are square. Then:

1. $\det A = (\det A_{11})(\det A_{22})(\det A_{33}) \cdots (\det A_{nn})$.
2. $c_A(x) = c_{A_{11}}(x) c_{A_{22}}(x) c_{A_{33}}(x) \cdots c_{A_{nn}}(x)$.

SECTION 9.3 Invariant Subspaces and Direct Sums

PROOF

If $n = 2$, (1) is Theorem 5 Section 3.1; the general case (by induction on n) is left to the reader. Then (2) follows from (1) because

$$xI - A = \begin{bmatrix} xI - A_{11} & -A_{12} & -A_{13} & \cdots & -A_{1n} \\ 0 & xI - A_{22} & -A_{23} & \cdots & -A_{2n} \\ 0 & 0 & xI - A_{33} & \cdots & -A_{3n} \\ \vdots & \vdots & \vdots & & \vdots \\ 0 & 0 & 0 & \cdots & xI - A_{nn} \end{bmatrix}$$

where, in each diagonal block, the symbol I stands for the identity matrix of the appropriate size.

EXAMPLE 5

Consider the linear operator $T : \mathbf{P}_2 \to \mathbf{P}_2$ given by

$$T(a + bx + cx^2) = (-2a - b + 2c) + (a + b)x + (-6a - 2b + 5c)x^2$$

Show that $U = \text{span}\{x, 1 + 2x^2\}$ is T-invariant, use it to find a block upper triangular matrix for T, and use that to compute $c_T(x)$.

Solution ▶ U is T-invariant by Example 3 because $U = \text{span}\{x, 1 + 2x^2\}$ and both $T(x)$ and $T(1 + 2x^2)$ lie in U:

$$T(x) = -1 + x - 2x^2 = x - (1 + 2x^2)$$
$$T(1 + 2x^2) = 2 + x + 4x^2 = x + 2(1 + 2x^2)$$

Extend the basis $B_1 = \{x, 1 + 2x^2\}$ of U to a basis B of $\mathbf{P}_2$ in any way at all—say, $B = \{x, 1 + 2x^2, x^2\}$. Then

$$M_B(T) = [C_B[T(x)] \quad C_B[T(1 + 2x^2)] \quad C_B[T(x^2)]]$$
$$= [C_B(-1 + x - 2x^2) \quad C_B(2 + x + 4x^2) \quad C_B(2 + 5x^2)]$$
$$= \begin{bmatrix} 1 & 1 & 0 \\ -1 & 2 & 2 \\ 0 & 0 & 1 \end{bmatrix}$$

is in block upper triangular form as expected. Finally,

$$c_T(x) = \det \begin{bmatrix} x - 1 & -1 & 0 \\ 1 & x - 2 & -2 \\ 0 & 0 & x - 1 \end{bmatrix} = (x^2 - 3x + 3)(x - 1)$$

Eigenvalues

Let $T : V \to V$ be a linear operator. A one-dimensional subspace $\mathbb{R}\mathbf{v}$, $\mathbf{v} \neq \mathbf{0}$, is T-invariant if and only if $T(r\mathbf{v}) = rT(\mathbf{v})$ lies in $\mathbb{R}\mathbf{v}$ for all r in $\mathbb{R}$. This holds if and only if $T(\mathbf{v})$ lies in $\mathbb{R}\mathbf{v}$; that is, $T(\mathbf{v}) = \lambda \mathbf{v}$ for some λ in $\mathbb{R}$. A real number λ is called an **eigenvalue** of an operator $T : V \to V$ if

$$T(\mathbf{v}) = \lambda \mathbf{v}$$

holds for some nonzero vector $\mathbf{v}$ in V. In this case, $\mathbf{v}$ is called an **eigenvector** of T corresponding to λ. The subspace

$$E_\lambda(T) = \{\mathbf{v} \text{ in } V \mid T(\mathbf{v}) = \lambda \mathbf{v}\}$$

is called the **eigenspace** of T corresponding to λ. These terms are consistent with those used in Section 5.5 for matrices. If A is an $n \times n$ matrix, a real number λ is an eigenvalue of the matrix operator $T_A : \mathbb{R}^n \to \mathbb{R}^n$ if and only if λ is an eigenvalue of the matrix A. Moreover, the eigenspaces agree:

$$E_\lambda(T_A) = \{\mathbf{x} \text{ in } \mathbb{R}^n \mid A\mathbf{x} = \lambda\mathbf{x}\} = E_\lambda(A)$$

The following theorem reveals the connection between the eigenspaces of an operator T and those of the matrices representing T.

Theorem 3

Let $T : V \to V$ be a linear operator where $\dim V = n$, let B denote any ordered basis of V, and let $C_B : V \to \mathbb{R}^n$ denote the coordinate isomorphism. Then:

1. The eigenvalues λ of T are precisely the eigenvalues of the matrix $M_B(T)$ and thus are the roots of the characteristic polynomial $c_T(x)$.
2. In this case the eigenspaces $E_\lambda(T)$ and $E_\lambda[M_B(T)]$ are isomorphic via the restriciton $C_B : E_\lambda(T) \to E_\lambda[M_B(T)]$.

PROOF

Write $A = M_B(T)$ for convenience. If $T(\mathbf{v}) = \lambda \mathbf{v}$, then applying C_B gives $\lambda C_B(\mathbf{v}) = C_B[T(\mathbf{v})] = AC_B(\mathbf{v})$ because C_B is linear. Hence $C_B(\mathbf{v})$ lies in $E_\lambda(A)$, so we do indeed have a function $C_B : E_\lambda(T) \to E_\lambda(A)$. It is clearly linear and one-to-one; we claim it is onto. If $\mathbf{x}$ is in $E_\lambda(A)$, write $\mathbf{x} = C_B(\mathbf{v})$ for some $\mathbf{v}$ in V (C_B is onto). This $\mathbf{v}$ actually lies in $E_\lambda(T)$. To see why, observe that

$$C_B[T(\mathbf{v})] = AC_B(\mathbf{v}) = A\mathbf{x} = \lambda\mathbf{x} = \lambda C_B(\mathbf{v}) = C_B(\lambda\mathbf{v})$$

Hence $T(\mathbf{v}) = \lambda \mathbf{v}$ because C_B is one-to-one, and this proves (2). As to (1), we have already shown that eigenvalues of T are eigenvalues of A. The converse follows, as in the foregoing proof that C_B is onto.

Theorem 3 shows how to pass back and forth between the eigenvectors of an operator T and the eigenvectors of any matrix $M_B(T)$ of T:

$$\mathbf{v} \text{ lies in } E_\lambda(T) \quad \text{if and only if} \quad C_B(\mathbf{v}) \text{ lies in } E_\lambda[M_B(T)]$$

EXAMPLE 6

Find the eigenvalues and eigenspaces for $T : \mathbf{P}_2 \to \mathbf{P}_2$ given by

$$T(a + bx + cx^2) = (2a + b + c) + (2a + b - 2c)x - (a + 2c)x^2$$

Solution ▶ If $B = \{1, x, x^2\}$, then

$$M_B(T) = \begin{bmatrix} C_B[T(1)] & C_B[T(x)] & C_B[T(x^2)] \end{bmatrix} = \begin{bmatrix} 2 & 1 & 1 \\ 2 & 1 & -2 \\ -1 & 0 & -2 \end{bmatrix}$$

SECTION 9.3 Invariant Subspaces and Direct Sums

Hence $c_T(x) = \det[xI - M_B(T)] = (x + 1)^2(x - 3)$ as the reader can verify.
Moreover, $E_{-1}[M_B(T)] = \mathbb{R}\begin{bmatrix} -1 \\ 2 \\ 1 \end{bmatrix}$ and $E_3[M_B(T)] = \mathbb{R}\begin{bmatrix} 5 \\ 6 \\ -1 \end{bmatrix}$, so Theorem 3 gives
$E_{-1}(T) = \mathbb{R}(-1 + 2x + x^2)$ and $E_3(T) = \mathbb{R}(5 + 6x - x^2)$.

Theorem 4

Each eigenspace of a linear operator $T: V \to V$ is a T-invariant subspace of V.

PROOF

If $\mathbf{v}$ lies in the eigenspace $E_\lambda(T)$, then $T(\mathbf{v}) = \lambda\mathbf{v}$, so $T[T(\mathbf{v})] = T(\lambda\mathbf{v}) = \lambda T(\mathbf{v})$. This shows that $T(\mathbf{v})$ lies in $E_\lambda(T)$ too.

Direct Sums

Sometimes vectors in a space V can be written naturally as a sum of vectors in two subspaces. For example, in the space $\mathbf{M}_{nn}$ of all $n \times n$ matrices, we have subspaces

$$U = \{P \text{ in } \mathbf{M}_{nn} \mid P \text{ is symmetric}\} \text{ and } W = \{Q \text{ in } \mathbf{M}_{nn} \mid Q \text{ is skew symmetric}\}$$

where a matrix Q is called **skew-symmetric** if $Q^T = -Q$. Then every matrix A in $\mathbf{M}_{nn}$ can be written as the sum of a matrix in U and a matrix in W; indeed,

$$A = \tfrac{1}{2}(A + A^T) + \tfrac{1}{2}(A - A^T)$$

where $\tfrac{1}{2}(A + A^T)$ is symmetric and $\tfrac{1}{2}(A - A^T)$ is skew symmetric. Remarkably, this representation is unique: If $A = P + Q$ where $P^T = P$ and $Q^T = -Q$, then $A^T = P^T + Q^T = P - Q$; adding this to $A = P + Q$ gives $P = \tfrac{1}{2}(A + A^T)$, and subtracting gives $Q = \tfrac{1}{2}(A - A^T)$. In addition, this uniqueness turns out to be closely related to the fact that the only matrix in both U and W is 0. This is a useful way to view matrices, and the idea generalizes to the important notion of a direct sum of subspaces.

If U and W are subspaces of V, their *sum* $U + W$ and their *intersection* $U \cap W$ were defined in Section 6.4 as follows:

$$U + W = \{\mathbf{u} + \mathbf{w} \mid \mathbf{u} \text{ in } U \text{ and } \mathbf{w} \text{ in } W\}$$
$$U \cap W = \{\mathbf{v} \mid \mathbf{v} \text{ lies in both } U \text{ and } W\}$$

These are subspaces of V, the sum containing both U and W and the intersection contained in both U and W. It turns out that the most interesting pairs U and W are those for which $U \cap W$ is as small as possible and $U + W$ is as large as possible.

Definition 9.7 A vector space V is said to be the **direct sum** of subspaces U and W if

$$U \cap W = \{\mathbf{0}\} \quad \text{and} \quad U + W = V$$

In this case we write $V = U \oplus W$. Given a subspace U, any subspace W such that $V = U \oplus W$ is called a **complement** of U in V.

EXAMPLE 7

In the space $\mathbb{R}^5$, consider the subspaces $U = \{(a, b, c, 0, 0) \mid a, b, \text{ and } c \text{ in } \mathbb{R}\}$ and $W = \{(0, 0, 0, d, e) \mid d \text{ and } e \text{ in } \mathbb{R}\}$. Show that $\mathbb{R}^5 = U \oplus W$.

Solution ▶ If $\mathbf{x} = (a, b, c, d, e)$ is any vector in $\mathbb{R}^5$, then $\mathbf{x} = (a, b, c, 0, 0) + (0, 0, 0, d, e)$, so $\mathbf{x}$ lies in $U + W$. Hence $\mathbb{R}^5 = U + W$. To show that $U \cap W = \{\mathbf{0}\}$, let $\mathbf{x} = (a, b, c, d, e)$ lie in $U \cap W$. Then $d = e = 0$ because $\mathbf{x}$ lies in U, and $a = b = c = 0$ because $\mathbf{x}$ lies in W. Thus $\mathbf{x} = (0, 0, 0, 0, 0) = \mathbf{0}$, so $\mathbf{0}$ is the only vector in $U \cap W$. Hence $U \cap W = \{\mathbf{0}\}$.

EXAMPLE 8

If U is a subspace of $\mathbb{R}^n$, show that $\mathbb{R}^n = U \oplus U^\perp$.

Solution ▶ The equation $\mathbb{R}^n = U + U^\perp$ holds because, given $\mathbf{x}$ in $\mathbb{R}^n$, the vector $\text{proj}_U(\mathbf{x})$ lies in U and $\mathbf{x} - \text{proj}_U(\mathbf{x})$ lies in $U^\perp$. To see that $U \cap U^\perp = \{\mathbf{0}\}$, observe that any vector in $U \cap U^\perp$ is orthogonal to itself and hence must be zero.

EXAMPLE 9

Let $\{\mathbf{e}_1, \mathbf{e}_2, \dots, \mathbf{e}_n\}$ be a basis of a vector space V, and partition it into two parts: $\{\mathbf{e}_1, \dots, \mathbf{e}_k\}$ and $\{\mathbf{e}_{k+1}, \dots, \mathbf{e}_n\}$. If $U = \text{span}\{\mathbf{e}_1, \dots, \mathbf{e}_k\}$ and $W = \text{span}\{\mathbf{e}_{k+1}, \dots, \mathbf{e}_n\}$, show that $V = U \oplus W$.

Solution ▶ If $\mathbf{v}$ lies in $U \cap W$, then $\mathbf{v} = a_1\mathbf{e}_1 + \cdots + a_k\mathbf{e}_k$ and $\mathbf{v} = b_{k+1}\mathbf{e}_{k+1} + \cdots + b_n\mathbf{e}_n$ hold for some a_i and b_j in $\mathbb{R}$. The fact that the $\mathbf{e}_i$ are linearly independent forces all $a_i = b_j = 0$, so $\mathbf{v} = \mathbf{0}$. Hence $U \cap W = \{\mathbf{0}\}$. Now, given $\mathbf{v}$ in V, write $\mathbf{v} = v_1\mathbf{e}_1 + \cdots + v_n\mathbf{e}_n$ where the v_i are in $\mathbb{R}$. Then $\mathbf{v} = \mathbf{u} + \mathbf{w}$, where $\mathbf{u} = v_1\mathbf{e}_1 + \cdots + v_k\mathbf{e}_k$ lies in U and $\mathbf{w} = v_{k+1}\mathbf{e}_{k+1} + \cdots + v_n\mathbf{e}_n$ lies in W. This proves that $V = U + W$.

Example 9 is typical of all direct sum decompositions.

Theorem 5

Let U and W be subspaces of a finite dimensional vector space V. The following three conditions are equivalent:

1. $V = U \oplus W$.
2. Each vector $\mathbf{v}$ in V can be written uniquely in the form
$$\mathbf{v} = \mathbf{u} + \mathbf{w} \quad \mathbf{u} \text{ in } U, \mathbf{w} \text{ in } W$$
3. If $\{\mathbf{u}_1, \dots, \mathbf{u}_k\}$ and $\{\mathbf{w}_1, \dots, \mathbf{w}_m\}$ are bases of U and W, respectively, then $B = \{\mathbf{u}_1, \dots, \mathbf{u}_k, \mathbf{w}_1, \dots, \mathbf{w}_m\}$ is a basis of V.

SECTION 9.3 Invariant Subspaces and Direct Sums

(The uniqueness in 2 means that if $\mathbf{v} = \mathbf{u}_1 + \mathbf{w}_1$ is another such representation, then $\mathbf{u}_1 = \mathbf{u}$ and $\mathbf{w}_1 = \mathbf{w}$.)

PROOF

Example 9 shows that (3) $\Rightarrow$ (1).

(1) $\Rightarrow$ (2). Given $\mathbf{v}$ in V, we have $\mathbf{v} = \mathbf{u} + \mathbf{w}$, $\mathbf{u}$ in U, $\mathbf{w}$ in W, because $V = U + W$. If also $\mathbf{v} = \mathbf{u}_1 + \mathbf{w}_1$, then $\mathbf{u} - \mathbf{u}_1 = \mathbf{w}_1 - \mathbf{w}$ lies in $U \cap W = \{\mathbf{0}\}$, so $\mathbf{u} = \mathbf{u}_1$ and $\mathbf{w} = \mathbf{w}_1$.

(2) $\Rightarrow$ (3). Given $\mathbf{v}$ in V, we have $\mathbf{v} = \mathbf{u} + \mathbf{w}$, $\mathbf{u}$ in U, $\mathbf{w}$ in W. Hence $\mathbf{v}$ lies in span B; that is, $V = $ span B. To see that B is independent, let $a_1 \mathbf{u}_1 + \cdots + a_k \mathbf{u}_k + b_1 \mathbf{w}_1 + \cdots + b_m \mathbf{w}_m = \mathbf{0}$. Write $\mathbf{u} = a_1 \mathbf{u}_1 + \cdots + a_k \mathbf{u}_k$ and $\mathbf{w} = b_1 \mathbf{w}_1 + \cdots + b_m \mathbf{w}_m$. Then $\mathbf{u} + \mathbf{w} = \mathbf{0}$, and so $\mathbf{u} = \mathbf{0}$ and $\mathbf{w} = \mathbf{0}$ by the uniqueness in (2). Hence $a_i = 0$ for all i and $b_j = 0$ for all j.

Condition (3) in Theorem 5 gives the following useful result.

Theorem 6

If a finite dimensional vector space V is the direct sum $V = U \oplus W$ of subspaces U and W, then

$$\dim V = \dim U + \dim W$$

These direct sum decompositions of V play an important role in any discussion of invariant subspaces. If $T : V \to V$ is a linear operator and if U_1 is a T-invariant subspace, the block upper triangular matrix

$$M_B(T) = \begin{bmatrix} M_{B_1}(T) & Y \\ 0 & Z \end{bmatrix} \qquad (*)$$

in Theorem 1 is achieved by choosing any basis $B_1 = \{\mathbf{b}_1, \ldots, \mathbf{b}_k\}$ of U_1 and completing it to a basis $B = \{\mathbf{b}_1, \ldots, \mathbf{b}_k, \mathbf{b}_{k+1}, \ldots, \mathbf{b}_n\}$ of V in any way at all. The fact that U_1 is T-invariant ensures that the first k columns of $M_B(T)$ have the form in $(*)$ (that is, the last $n - k$ entries are zero), and the question arises whether the additional basis vectors $\mathbf{b}_{k+1}, \ldots, \mathbf{b}_n$ can be chosen such that

$$U_2 = \text{span}\{\mathbf{b}_{k+1}, \ldots, \mathbf{b}_n\}$$

is *also* T-invariant. In other words, does each T-invariant subspace of V have a T-invariant complement? Unfortunately the answer in general is no (see Example 11 below); but when it is possible, the matrix $M_B(T)$ simplifies further. The assumption that the complement $U_2 = \text{span}\{\mathbf{b}_{k+1}, \ldots, \mathbf{b}_n\}$ is T-invariant too means that $Y = 0$ in equation $(*)$ above, and that $Z = M_{B_2}(T)$ is the matrix of the restriction of T to U_2 (where $B_2 = \{\mathbf{b}_{k+1}, \ldots, \mathbf{b}_n\}$). The verification is the same as in the proof of Theorem 1.

Theorem 7

Let $T: V \to V$ be a linear operator where V has dimension n. Suppose $V = U_1 \oplus U_2$ where both U_1 and U_2 are T-invariant. If $B_1 = \{\mathbf{b}_1, \dots, \mathbf{b}_k\}$ and $B_2 = \{\mathbf{b}_{k+1}, \dots, \mathbf{b}_n\}$ are bases of U_1 and U_2 respectively, then

$$B = \{\mathbf{b}_1, \dots, \mathbf{b}_k, \mathbf{b}_{k+1}, \dots, \mathbf{b}_n\}$$

is a basis of V, and $M_B(T)$ has the block diagonal form

$$M_B(T) = \begin{bmatrix} M_{B_1}(T) & 0 \\ 0 & M_{B_1}(T) \end{bmatrix}$$

where $M_{B_1}(T)$ and $M_{B_2}(T)$ are the matrices of the restrictions of T to U_1 and to U_2 respectively.

Definition 9.8 *The linear operator $T: V \to V$ is said to be **reducible** if nonzero T-invariant subspaces U_1 and U_2 can be found such that $V = U_1 \oplus U_2$.*

Then T has a matrix in block diagonal form as in Theorem 7, and the study of T is reduced to studying its restrictions to the lower-dimensional spaces U_1 and U_2. If these can be determined, so can T. Here is an example in which the action of T on the invariant subspaces U_1 and U_2 is very simple indeed. The result for operators is used to derive the corresponding similarity theorem for matrices.

EXAMPLE 10

Let $T: V \to V$ be a linear operator satisfying $T^2 = 1_V$ (such operators are called **involutions**). Define

$$U_1 = \{\mathbf{v} \mid T(\mathbf{v}) = \mathbf{v}\} \quad \text{and} \quad U_2 = \{\mathbf{v} \mid T(\mathbf{v}) = -\mathbf{v}\}$$

(a) Show that $V = U_1 \oplus U_2$.

(b) If $\dim V = n$, find a basis B of V such that $M_B(T) = \begin{bmatrix} I_k & 0 \\ 0 & -I_{n-k} \end{bmatrix}$ for some k.

(c) Conclude that, if A is an $n \times n$ matrix such that $A^2 = I$, then A is similar to $\begin{bmatrix} I_k & 0 \\ 0 & -I_{n-k} \end{bmatrix}$ for some k.

Solution ▶

(a) The verification that U_1 and U_2 are subspaces of V is left to the reader. If $\mathbf{v}$ lies in $U_1 \cap U_2$, then $\mathbf{v} = T(\mathbf{v}) = -\mathbf{v}$, and it follows that $\mathbf{v} = \mathbf{0}$. Hence $U_1 \cap U_2 = \{\mathbf{0}\}$. Given $\mathbf{v}$ in V, write

$$\mathbf{v} = \tfrac{1}{2}\{[\mathbf{v} + T(\mathbf{v})] + [\mathbf{v} - T(\mathbf{v})]\}$$

Then $\mathbf{v} + T(\mathbf{v})$ lies in U_1, because $T[\mathbf{v} + T(\mathbf{v})] = T(\mathbf{v}) + T^2(\mathbf{v}) = \mathbf{v} + T(\mathbf{v})$. Similarly, $\mathbf{v} - T(\mathbf{v})$ lies in U_2, and it follows that $V = U_1 + U_2$. This proves part (a).

(b) U_1 and U_2 are easily shown to be T-invariant, so the result follows from Theorem 7 if bases $B_1 = \{\mathbf{b}_1, \ldots, \mathbf{b}_k\}$ and $B_2 = \{\mathbf{b}_{k+1}, \ldots, \mathbf{b}_n\}$ of U_1 and U_2 can be found such that $M_{B_1}(T) = I_k$ and $M_{B_2}(T) = -I_{n-k}$. But this is true for *any* choice of B_1 and B_2:

$$\begin{aligned} M_{B_1}(T) &= [C_{B_1}[T(\mathbf{b}_1)] \quad C_{B_1}[T(\mathbf{b}_2)] \quad \cdots \quad C_{B_1}[T(\mathbf{b}_k)]] \\ &= [C_{B_1}(\mathbf{b}_1) \quad C_{B_1}(\mathbf{b}_2) \quad \cdots \quad C_{B_1}(\mathbf{b}_k)] \\ &= I_k \end{aligned}$$

A similar argument shows that $M_{B_2}(T) = -I_{n-k}$, so part (b) follows with $B = \{\mathbf{b}_1, \mathbf{b}_2, \ldots, \mathbf{b}_n\}$.

(c) Given A such that $A^2 = I$, consider $T_A : \mathbb{R}^n \to \mathbb{R}^n$. Then $(T_A)^2(\mathbf{x}) = A^2\mathbf{x} = \mathbf{x}$ for all $\mathbf{x}$ in $\mathbb{R}^n$, so $(T_A)^2 = 1_V$. Hence, by part (b), there exists a basis B of $\mathbb{R}^n$ such that

$$M_B(T_A) = \begin{bmatrix} I_r & 0 \\ 0 & -I_{n-r} \end{bmatrix}$$

But Theorem 4 Section 9.2 shows that $M_B(T_A) = P^{-1}AP$ for some invertible matrix P, and this proves part (c).

Note that the passage from the result for operators to the analogous result for matrices is routine and can be carried out in any situation, as in the verification of part (c) of Example 10. The key is the analysis of the operators. In this case, the involutions are just the operators satisfying $T^2 = 1_V$, and the simplicity of this condition means that the invariant subspaces U_1 and U_2 are easy to find.

Unfortunately, not every linear operator $T : V \to V$ is reducible. In fact, the linear operator in Example 4 has *no* invariant subspaces except 0 and V. On the other hand, one might expect that this is the only type of nonreducible operator; that is, if the operator *has* an invariant subspace that is not 0 or V, then *some* invariant complement must exist. The next example shows that even this is not valid.

EXAMPLE 11

Consider the operator $T : \mathbb{R}^2 \to \mathbb{R}^2$ given by $T\begin{bmatrix} a \\ b \end{bmatrix} = \begin{bmatrix} a+b \\ b \end{bmatrix}$. Show that $U_1 = \mathbb{R}\begin{bmatrix} 1 \\ 0 \end{bmatrix}$ is T-invariant but that U_1 has no T-invariant complement in $\mathbb{R}^2$.

Solution ▶ Because $U_1 = \text{span}\left\{\begin{bmatrix} 1 \\ 0 \end{bmatrix}\right\}$ and $T\begin{bmatrix} 1 \\ 0 \end{bmatrix} = \begin{bmatrix} 1 \\ 0 \end{bmatrix}$, it follows (by Example 3) that U_1 is T-invariant. Now assume, if possible, that U_1 has a T-invariant complement U_2 in $\mathbb{R}^2$. Then $U_1 \oplus U_2 = \mathbb{R}^2$ and $T(U_2) \subseteq U_2$. Theorem 6 gives

$$2 = \dim \mathbb{R}^2 = \dim U_1 + \dim U_2 = 1 + \dim U_2$$

so $\dim U_2 = 1$. Let $U_2 = \mathbb{R}\mathbf{u}_2$, and write $\mathbf{u}_2 = \begin{bmatrix} p \\ q \end{bmatrix}$. We claim that $\mathbf{u}_2$ is not U_1. For if $\mathbf{u}_2 \in U_1$, then $\mathbf{u}_2 \in U_1 \cap U_2 = \{\mathbf{0}\}$, so $\mathbf{u}_2 = \mathbf{0}$. But then $U_2 = \mathbb{R}\mathbf{u}_2 = \{\mathbf{0}\}$, a contradiction, as $\dim U_2 = 1$. So $\mathbf{u}_2 \notin U_1$, from which $q \neq 0$. On the other hand, $T(\mathbf{u}_2) \in U_2 = \mathbb{R}\mathbf{u}_2$ (because U_2 is T-invariant), say $T(\mathbf{u}_2) = \lambda \mathbf{u}_2 = \lambda\begin{bmatrix} p \\ q \end{bmatrix}$.

Thus
$$\begin{bmatrix} p+q \\ q \end{bmatrix} = T\begin{bmatrix} p \\ q \end{bmatrix} = \lambda \begin{bmatrix} p \\ q \end{bmatrix} \quad \text{where } \lambda \in \mathbb{R}.$$

Hence $p + q = \lambda p$ and $q = \lambda q$. Because $q \neq 0$, the second of these equations implies that $\lambda = 1$, so the first equation implies $q = 0$, a contradiction. So a T-invariant complement of U_1 does not exist.

This is as far as we take the theory here, but in Chapter 11 the techniques introduced in this section will be refined to show that every matrix is similar to a very nice matrix indeed—its Jordan canonical form.

EXERCISES 9.3

1. If $T : V \to V$ is any linear operator, show that ker T and im T are T-invariant subspaces.

2. Let T be a linear operator on V. If U and W are T-invariant, show that
   (a) $U \cap W$ and $U + W$ are also T-invariant.
   ◆(b) $T(U)$ is T-invariant.

3. Let S and T be linear operators on V and assume that $ST = TS$.
   (a) Show that im S and ker S are T-invariant.
   ◆(b) If U is T-invariant, show that $S(U)$ is T-invariant.

4. Let $T : V \to V$ be a linear operator. Given $\mathbf{v}$ in V, let U denote the set of vectors in V that lie in every T-invariant subspace that contains $\mathbf{v}$.
   (a) Show that U is a T-invariant subspace of V containing $\mathbf{v}$.
   (b) Show that U is contained in every T-invariant subspace of V that contains $\mathbf{v}$.

5. (a) If T is a scalar operator (see Example 1 Section 7.1) show that every subspace is T-invariant.
   (b) Conversely, if every subspace is T-invariant, show that T is scalar.

◆6. Show that the only subspaces of V that are T-invariant for every operator $T : V \to V$ are 0 and V. Assume that V is finite dimensional. [*Hint*: Theorem 3 Section 7.1.]

7. Suppose that $T : V \to V$ is a linear operator and that U is a T-invariant subspace of V. If S is an invertible operator, put $T' = STS^{-1}$. Show that $S(U)$ is a T'-invariant subspace.

8. In each case, show that U is T-invariant, use it to find a block upper triangular matrix for T, and use that to compute $c_T(x)$.
   (a) $T : \mathbf{P}_2 \to \mathbf{P}_2$, $T(a + bx + cx^2)$
   $= (-a + 2b + c) + (a + 3b + c)x + (a + 4b)x^2$,
   $U = \text{span}\{1, x + x^2\}$
   ◆(b) $T : \mathbf{P}_2 \to \mathbf{P}_2$, $T(a + bx + cx^2)$
   $= (5a - 2b + c) + (5a - b + c)x + (a + 2c)x^2$,
   $U = \text{span}\{1 - 2x^2, x + x^2\}$

9. In each case, show that $T_A : \mathbb{R}^2 \to \mathbb{R}^2$ has no invariant subspaces except 0 and $\mathbb{R}^2$.
   (a) $A = \begin{bmatrix} 1 & 2 \\ -1 & -1 \end{bmatrix}$
   ◆(b) $A = \begin{bmatrix} \cos \theta & -\sin \theta \\ \sin \theta & \cos \theta \end{bmatrix}$, $0 < \theta < \pi$

10. In each case, show that $V = U \oplus W$.
    (a) $V = \mathbb{R}^4$, $U = \text{span}\{(1, 1, 0, 0), (0, 1, 1, 0)\}$,
    $W = \text{span}\{(0, 1, 0, 1), (0, 0, 1, 1)\}$
    ◆(b) $V = \mathbb{R}^4$, $U = \{(a, a, b, b) \mid a, b \text{ in } \mathbb{R}\}$,
    $W = \{(c, d, c, -d) \mid c, d \text{ in } \mathbb{R}\}$
    (c) $V = \mathbf{P}_3$, $U = \{a + bx \mid a, b \text{ in } \mathbb{R}\}$,
    $W = \{ax^2 + bx^3 \mid a, b \text{ in } \mathbb{R}\}$

◆(d) $V = \mathbf{M}_{22}$, $U = \left\{\begin{bmatrix} a & a \\ b & b \end{bmatrix} \middle| a, b \text{ in } \mathbb{R}\right\}$,
$W = \left\{\begin{bmatrix} a & b \\ -a & b \end{bmatrix} \middle| a, b \text{ in } \mathbb{R}\right\}$

11. Let $U = \text{span}\{(1, 0, 0, 0), (0, 1, 0, 0)\}$ in $\mathbb{R}^4$. Show that $\mathbb{R}^4 = U \oplus W_1$ and $\mathbb{R}^4 = U \oplus W_2$, where $W_1 = \text{span}\{(0, 0, 1, 0), (0, 0, 0, 1)\}$ and $W_2 = \text{span}\{(1, 1, 1, 1), (1, 1, 1, -1)\}$.

12. Let U be a subspace of V, and suppose that $V = U \oplus W_1$ and $V = U \oplus W_2$ hold for subspaces W_1 and W_2. Show that $\dim W_1 = \dim W_2$.

13. If U and W denote the subspaces of even and odd polynomials in $\mathbf{P}_m$, respectively, show that $\mathbf{P}_n = U \oplus W$. (See Exercise 36 Section 6.3.) [Hint: $f(x) + f(-x)$ is even.]

◆14. Let E be a 2×2 matrix such that $E^2 = E$. Show that $\mathbf{M}_{22} = U \oplus W$, where $U = \{A \mid AE = A\}$ and $W = \{B \mid BE = 0\}$. [Hint: XE lies in U for every matrix X.]

15. Let U and W be subspaces of V. Show that $U \cap W = \{\mathbf{0}\}$ if and only if $\{\mathbf{u}, \mathbf{w}\}$ is independent for all $\mathbf{u} \neq \mathbf{0}$ in U and all $\mathbf{w} \neq \mathbf{0}$ in W.

16. Let $V \xrightarrow{T} W \xrightarrow{S} V$ be linear transformations, and assume that $\dim V$ and $\dim W$ are finite.

   (a) If $ST = 1_V$, show that $W = \text{im } T \oplus \ker S$. [Hint: Given $\mathbf{w}$ in W, show that $\mathbf{w} - TS(\mathbf{w})$ lies in $\ker S$.]

   (b) Illustrate with $\mathbb{R}^2 \xrightarrow{T} \mathbb{R}^3 \xrightarrow{S} \mathbb{R}^2$ where $T(x, y) = (x, y, 0)$ and $S(x, y, z) = (x, y)$.

17. Let U and W be subspaces of V, let $\dim V = n$, and assume that $\dim U + \dim W = n$.

   (a) If $U \cap W = \{\mathbf{0}\}$, show that $V = U \oplus W$.

   ◆(b) If $U + W = V$, show that $V = U \oplus W$. [Hint: Theorem 5 Section 6.4.]

18. Let $A = \begin{bmatrix} 0 & 1 \\ 0 & 0 \end{bmatrix}$ and consider $T_A: \mathbb{R}^2 \to \mathbb{R}^2$.

   (a) Show that the only eigenvalue of T_A is $\lambda = 0$.

   ◆(b) Show that $\ker(T_A) = \mathbb{R}\begin{bmatrix} 1 \\ 0 \end{bmatrix}$ is the unique T_A-invariant subspace of $\mathbb{R}^2$ (except for 0 and $\mathbb{R}^2$).

19. If $A = \begin{bmatrix} 2 & -5 & 0 & 0 \\ 1 & -2 & 0 & 0 \\ 0 & 0 & -1 & -2 \\ 0 & 0 & 1 & 1 \end{bmatrix}$, show that $T_A: \mathbb{R}^4 \to \mathbb{R}^4$ has two-dimensional T-invariant subspaces U and W such that $\mathbb{R}^4 = U \oplus W$, but A has no real eigenvalue.

◆20. Let $T: V \to V$ be a linear operator where $\dim V = n$. If U is a T-invariant subspace of V, let $T_1: U \to U$ denote the restriction of T to U (so $T_1(\mathbf{u}) = T(\mathbf{u})$ for all $\mathbf{u}$ in U). Show that $c_T(x) = c_{T_1}(x) \cdot q(x)$ for some polynomial $q(x)$. [Hint: Theorem 1.]

21. Let $T: V \to V$ be a linear operator where $\dim V = n$. Show that V has a basis of eigenvectors if and only if V has a basis B such that $M_B(T)$ is diagonal.

22. In each case, show that $T^2 = 1$ and find (as in Example 10) an ordered basis B such that $M_B(T)$ has the given block form.

   (a) $T: \mathbf{M}_{22} \to \mathbf{M}_{22}$ where $T(A) = A^T$, $M_B(T) = \begin{bmatrix} I_3 & 0 \\ 0 & -1 \end{bmatrix}$

   ◆(b) $T: \mathbf{P}_3 \to \mathbf{P}_3$ where $T[p(x)] = p(-x)$, $M_B(T) = \begin{bmatrix} I_2 & 0 \\ 0 & -I_2 \end{bmatrix}$

   (c) $T: \mathbb{C} \to \mathbb{C}$ where $T(a + bi) = a - bi$, $M_B(T) = \begin{bmatrix} 1 & 0 \\ 0 & -1 \end{bmatrix}$

   ◆(d) $T: \mathbb{R}^3 \to \mathbb{R}^3$ where $T(a, b, c) = (-a + 2b + c, b + c, -c)$, $M_B(T) = \begin{bmatrix} 1 & 0 \\ 0 & -I_2 \end{bmatrix}$

   (e) $T: V \to V$ where $T(\mathbf{v}) = -\mathbf{v}$, $\dim V = n$, $M_B(T) = -I_n$

23. Let U and W denote subspaces of a vector space V.

   (a) If $V = U \oplus W$, define $T: V \to V$ by $T(\mathbf{v}) = \mathbf{w}$ where $\mathbf{v}$ is written (uniquely) as $\mathbf{v} = \mathbf{u} + \mathbf{w}$ with $\mathbf{u}$ in U and $\mathbf{w}$ in W. Show that T is a linear transformation, $U = \ker T$, $W = \text{im } T$, and $T^2 = T$.

◆(b) Conversely, if $T: V \to V$ is a linear transformation such that $T^2 = T$, show that $V = \ker T \oplus \operatorname{im} T$. [*Hint*: $\mathbf{v} - T(\mathbf{v})$ lies in $\ker T$ for all $\mathbf{v}$ in V.]

24. Let $T: V \to V$ be a linear operator satisfying $T^2 = T$ (such operators are called **idempotents**). Define $U_1 = \{\mathbf{v} \mid T(\mathbf{v}) = \mathbf{v}\}$ and $U_2 = \ker T = \{\mathbf{v} \mid T(\mathbf{v}) = \mathbf{0}\}$.

    (a) Show that $V = U_1 \oplus U_2$.

    (b) If $\dim V = n$, find a basis B of V such that $M_B(T) = \begin{bmatrix} I_r & 0 \\ 0 & 0 \end{bmatrix}$, where $r = \operatorname{rank} T$.

    (c) If A is an $n \times n$ matrix such that $A^2 = A$, show that A is similar to $\begin{bmatrix} I_r & 0 \\ 0 & 0 \end{bmatrix}$, where $r = \operatorname{rank} A$. [*Hint*: Example 10.]

25. In each case, show that $T^2 = T$ and find (as in the preceding exercise) an ordered basis B such that $M_B(T)$ has the form given (0_k is the $k \times k$ zero matrix).

    (a) $T: \mathbf{P}_2 \to \mathbf{P}_2$ where
    $T(a + bx + cx^2) = (a - b + c)(1 + x + x^2)$,
    $M_B(T) = \begin{bmatrix} 1 & 0 \\ 0 & 0_2 \end{bmatrix}$

    ◆(b) $T: \mathbb{R}^3 \to \mathbb{R}^3$ where
    $T(a, b, c) = (a + 2b, 0, 4b + c)$,
    $M_B(T) = \begin{bmatrix} I_2 & 0 \\ 0 & 0 \end{bmatrix}$

    (c) $T: \mathbf{M}_{22} \to \mathbf{M}_{22}$ where
    $T\begin{bmatrix} a & b \\ c & d \end{bmatrix} = \begin{bmatrix} -5 & -15 \\ 2 & 6 \end{bmatrix}\begin{bmatrix} a & b \\ c & d \end{bmatrix}$,
    $M_B(T) = \begin{bmatrix} I_2 & 0 \\ 0 & 0_2 \end{bmatrix}$

26. Let $T: V \to V$ be an operator satisfying $T^2 = cT$, $c \neq 0$.

    (a) Show that $V = U \oplus \ker T$, where $U = \{\mathbf{u} \mid T(\mathbf{u}) = c\mathbf{u}\}$. [*Hint*: Compute $T(\mathbf{v} - \frac{1}{c}T(\mathbf{v}))$.]

    (b) If $\dim V = n$, show that V has a basis B such that $M_B(T) = \begin{bmatrix} cI_r & 0 \\ 0 & 0 \end{bmatrix}$, where $r = \operatorname{rank} T$.

    (c) If A is any $n \times n$ matrix of rank r such that $A^2 = cA$, $c \neq 0$, show that A is similar to $\begin{bmatrix} cI_r & 0 \\ 0 & 0 \end{bmatrix}$.

27. Let $T: V \to V$ be an operator such that $T^2 = c^2$, $c \neq 0$.

    (a) Show that $V = U_1 \oplus U_2$, where
    $U_1 = \{\mathbf{v} \mid T(\mathbf{v}) = c\mathbf{v}\}$ and
    $U_2 = \{\mathbf{v} \mid T(\mathbf{v}) = -c\mathbf{v}\}$.
    [*Hint*: $\mathbf{v} = \frac{1}{2c}\{[T(\mathbf{v}) + c\mathbf{v}] - [T(\mathbf{v}) - c\mathbf{v}]\}$.]

    (b) If $\dim V = n$, show that V has a basis B such that $M_B(T) = \begin{bmatrix} cI_k & 0 \\ 0 & -cI_{n-k} \end{bmatrix}$ for some k.

    (c) If A is an $n \times n$ matrix such that $A^2 = c^2I$, $c \neq 0$, show that A is similar to $\begin{bmatrix} cI_k & 0 \\ 0 & -cI_{n-k} \end{bmatrix}$ or some k.

28. If P is a fixed $n \times n$ matrix, define $T: \mathbf{M}_{nn} \to \mathbf{M}_{nn}$ by $T(A) = PA$. Let U_j denote the subspace of $\mathbf{M}_{nn}$ consisting of all matrices with all columns zero except possibly column j.

    (a) Show that each U_j is T-invariant.

    (b) Show that $\mathbf{M}_{nn}$ has a basis B such that $M_B(T)$ is block diagonal with each block on the diagonal equal to P.

29. Let V be a vector space. If $f: V \to \mathbb{R}$ is a linear transformation and $\mathbf{z}$ is a vector in V, define $T_{f,\mathbf{z}}: V \to V$ by $T_{f,\mathbf{z}}(\mathbf{v}) = f(\mathbf{v})\mathbf{z}$ for all $\mathbf{v}$ in V. Assume that $f \neq 0$ and $\mathbf{z} \neq \mathbf{0}$.

    (a) Show that $T_{f,\mathbf{z}}$ is a linear operator of rank 1.

    ◆(b) If $f \neq 0$, show that $T_{f,\mathbf{z}}$ is an idempotent if and only if $f(\mathbf{z}) = 1$. (Recall that $T: V \to V$ is called an idempotent if $T^2 = T$.)

    (c) Show that every idempotent $T: V \to V$ of rank 1 has the form $T = T_{f,\mathbf{z}}$ for some $f: V \to \mathbb{R}$ and some $\mathbf{z}$ in V with $f(\mathbf{z}) = 1$. [*Hint*: Write $\operatorname{im} T = \mathbb{R}\mathbf{z}$ and show that $T(\mathbf{z}) = \mathbf{z}$. Then use Exercise 23.]

30. Let U be a fixed $n \times n$ matrix, and consider the operator $T: \mathbf{M}_{nn} \to \mathbf{M}_{nn}$ given by $T(A) = UA$.

    (a) Show that λ is an eigenvalue of T if and only if it is an eigenvalue of U.

    ◆(b) If λ is an eigenvalue of T, show that $E_\lambda(T)$ consists of all matrices whose columns lie in $E_\lambda(U)$:

    $E_\lambda(T) = \{[P_1 \; P_2 \; \cdots \; P_n] \mid P_i \text{ in } E_\lambda(U) \text{ for each } i\}$

(c) Show that if $\dim[E_\lambda(U)] = d$, then $\dim[E_\lambda(T)] = nd$. [*Hint*: If $B = \{\mathbf{x}_1, ..., \mathbf{x}_d\}$ is a basis of $E_\lambda(U)$, consider the set of all matrices with one column from B and the other columns zero.]

31. Let $T: V \to V$ be a linear operator where V is finite dimensional. If $U \subseteq V$ is a subspace, let $\overline{U} = \{\mathbf{u}_0 + T(\mathbf{u}_1) + T^2(\mathbf{u}_2) + \cdots + T^k(\mathbf{u}_k) \mid \mathbf{u}_i$ in $U, k \geq 0\}$. Show that $\overline{U}$ is the smallest T-invariant subspace containing U (that is, it is T-invariant, contains U, and is contained in every such subspace).

32. Let $U_1, ..., U_m$ be subspaces of V and assume that $V = U_1 + \cdots + U_m$; that is, every $\mathbf{v}$ in V can be written (in at least one way) in the form $\mathbf{v} = \mathbf{u}_1 + \cdots + \mathbf{u}_m$, $\mathbf{u}_i$ in U_i. Show that the following conditions are equivalent.

    (i) If $\mathbf{u}_1 + \cdots + \mathbf{u}_m = \mathbf{0}$, $\mathbf{u}_i$ in U_i, then $\mathbf{u}_i = \mathbf{0}$ for each i.

    (ii) If $\mathbf{u}_1 + \cdots + \mathbf{u}_m = \mathbf{u}'_1 + \cdots + \mathbf{u}'_m$, $\mathbf{u}_i$ and $\mathbf{u}'_i$ in U_i, then $\mathbf{u}_i = \mathbf{u}'_i$ for each i.

    (iii) $U_i \cap (U_1 + \cdots + U_{i-1} + U_{i+1} + \cdots + U_m) = \{\mathbf{0}\}$ for each $i = 1, 2, ..., m$.

    (iv) $U_i \cap (U_{i+1} + \cdots + U_m) = \{\mathbf{0}\}$ for each $i = 1, 2, ..., m - 1$.

    When these conditions are satisfied, we say that V is the **direct sum** of the subspaces U_i, and write $V = U_1 \oplus U_2 \oplus \cdots \oplus U_m$.

33. (a) Let B be a basis of V and let $B = B_1 \cup B_2 \cup \cdots \cup B_m$ where the B_i are pairwise disjoint, nonempty subsets of B. If $U_i = \text{span } B_i$ for each i, show that $V = U_1 \oplus U_2 \oplus \cdots \oplus U_m$ (preceding exercise).

    (b) Conversely if $V = U_1 \oplus \cdots \oplus U_m$ and B_i is a basis of U_i for each i, show that $B = B_1 \cup \cdots \cup B_m$ is a basis of V as in (a).

10 Inner Product Spaces

SECTION 10.1 Inner Products and Norms

The dot product was introduced in $\mathbb{R}^n$ to provide a natural generalization of the geometrical notions of length and orthogonality that were so important in Chapter 4. The plan in this chapter is to define an *inner product* on an arbitrary real vector space V (of which the dot product is an example in $\mathbb{R}^n$) and use it to introduce these concepts in V.

Definition 10.1

*An **inner product** on a real vector space V is a function that assigns a real number $\langle \mathbf{v}, \mathbf{w} \rangle$ to every pair $\mathbf{v}, \mathbf{w}$ of vectors in V in such a way that the following axioms are satisfied.*

P1. $\langle \mathbf{v}, \mathbf{w} \rangle$ is a real number for all $\mathbf{v}$ and $\mathbf{w}$ in V.

P2. $\langle \mathbf{v}, \mathbf{w} \rangle = \langle \mathbf{w}, \mathbf{v} \rangle$ for all $\mathbf{v}$ and $\mathbf{w}$ in V.

P3. $\langle \mathbf{v} + \mathbf{w}, \mathbf{u} \rangle = \langle \mathbf{v}, \mathbf{u} \rangle + \langle \mathbf{w}, \mathbf{u} \rangle$ for all $\mathbf{u}, \mathbf{v}$, and $\mathbf{w}$ in V.

P4. $\langle r\mathbf{v}, \mathbf{w} \rangle = r\langle \mathbf{v}, \mathbf{w} \rangle$ for all $\mathbf{v}$ and $\mathbf{w}$ in V and all r in $\mathbb{R}$.

P5. $\langle \mathbf{v}, \mathbf{v} \rangle > 0$ for all $\mathbf{v} \neq \mathbf{0}$ in V.

A real vector space V with an inner product $\langle , \rangle$ will be called an **inner product space**. Note that every subspace of an inner product space is again an inner product space using the same inner product.[1]

EXAMPLE 1

$\mathbb{R}^n$ is an inner product space with the dot product as inner product:

$$\langle \mathbf{x}, \mathbf{y} \rangle = \mathbf{x} \cdot \mathbf{y} \quad \text{for all } \mathbf{v}, \mathbf{w} \in \mathbb{R}^n$$

See Theorem 1 Section 5.3. This is also called the **euclidean** inner product, and $\mathbb{R}^n$, equipped with the dot product, is called **euclidean** n-**space**.

---

[1] If we regard $\mathbb{C}^n$ as a vector space over the field $\mathbb{C}$ of complex numbers, then the "standard inner product" on $\mathbb{C}^n$ defined in Section 8.6 does not satisfy Axiom P4 (see Theorem 1(3) Section 8.6).

SECTION 10.1 Inner Products and Norms

EXAMPLE 2

If A and B are $m \times n$ matrices, define $\langle A, B \rangle = \text{tr}(AB^T)$ where $\text{tr}(X)$ is the trace of the square matrix X. Show that $\langle , \rangle$ is an inner product in $\mathbf{M}_{mn}$.

Solution ▶ P1 is clear. Since $\text{tr}(P) = \text{tr}(P^T)$ for every $m \times n$ matrix P, we have P2:
$$\langle A, B \rangle = \text{tr}(AB^T) = \text{tr}[(AB^T)^T] = \text{tr}(BA^T) = \langle B, A \rangle.$$
Next, P3 and P4 follow because trace is a linear transformation $\mathbf{M}_{mn} \to \mathbb{R}$ (Exercise 19). Turning to P5, let $\mathbf{r}_1, \mathbf{r}_2, \ldots, \mathbf{r}_m$ denote the rows of the matrix A. Then the (i, j)-entry of AA^T is $\mathbf{r}_i \cdot \mathbf{r}_j$, so
$$\langle A, A \rangle = \text{tr}(AA^T) = \mathbf{r}_1 \cdot \mathbf{r}_1 + \mathbf{r}_2 \cdot \mathbf{r}_2 + \cdots + \mathbf{r}_m \cdot \mathbf{r}_m$$
But $\mathbf{r}_j \cdot \mathbf{r}_j$ is the sum of the squares of the entries of $\mathbf{r}_j$, so this shows that $\langle A, A \rangle$ is the sum of the squares of all nm entries of A. Axiom P5 follows.

The next example is important in analysis.

EXAMPLE 3[2]

Let $\mathbf{C}[a, b]$ denote the vector space of **continuous functions** from $[a, b]$ to $\mathbb{R}$, a subspace of $\mathbf{F}[a, b]$. Show that
$$\langle f, g \rangle = \int_a^b f(x) g(x) \, dx$$
defines an inner product on $\mathbf{C}[a, b]$.

Solution ▶ Axioms P1 and P2 are clear. As to axiom P4,
$$\langle rf, g \rangle = \int_a^b rf(x)g(x) \, dx = r \int_a^b f(x)g(x) \, dx = r\langle f, g \rangle$$
Axiom P3 is similar. Finally, theorems of calculus show that $\langle f, f \rangle = \int_a^b f(x)^2 \, dx \geq 0$ and, if f is continuous, that this is zero if and only if f is the zero function. This gives axiom P5.

If $\mathbf{v}$ is any vector, then, using axiom P3, we get
$$\langle \mathbf{0}, \mathbf{v} \rangle = \langle \mathbf{0} + \mathbf{0}, \mathbf{v} \rangle = \langle \mathbf{0}, \mathbf{v} \rangle + \langle \mathbf{0}, \mathbf{v} \rangle$$
and it follows that the number $\langle \mathbf{0}, \mathbf{v} \rangle$ must be zero. This observation is recorded for reference in the following theorem, along with several other properties of inner products. The other proofs are left as Exercise 20.

Theorem 1

Let $\langle , \rangle$ be an inner product on a space V; let $\mathbf{v}$, $\mathbf{u}$, and $\mathbf{w}$ denote vectors in V; and let r denote a real number.

1. $\langle \mathbf{u}, \mathbf{v} + \mathbf{w} \rangle = \langle \mathbf{u}, \mathbf{v} \rangle + \langle \mathbf{u}, \mathbf{w} \rangle$.
2. $\langle \mathbf{v}, r\mathbf{w} \rangle = r \langle \mathbf{v}, \mathbf{w} \rangle = \langle r\mathbf{v}, \mathbf{w} \rangle$.

---

[2] This example (and others later that refer to it) can be omitted with no loss of continuity by students with no calculus background.

3. $\langle \mathbf{v}, \mathbf{0} \rangle = 0 = \langle \mathbf{0}, \mathbf{v} \rangle$.
4. $\langle \mathbf{v}, \mathbf{v} \rangle = 0$ if and only if $\mathbf{v} = \mathbf{0}$.

If $\langle , \rangle$ is an inner product on a space V, then, given $\mathbf{u}$, $\mathbf{v}$, and $\mathbf{w}$ in V,

$$\langle r\mathbf{u} + s\mathbf{v}, \mathbf{w} \rangle = \langle r\mathbf{u}, \mathbf{w} \rangle + \langle s\mathbf{v}, \mathbf{w} \rangle = r\langle \mathbf{u}, \mathbf{w} \rangle + s\langle \mathbf{v}, \mathbf{w} \rangle$$

for all r and s in $\mathbb{R}$ by axioms P3 and P4. Moreover, there is nothing special about the fact that there are two terms in the linear combination or that it is in the first component:

$$\langle r_1\mathbf{v}_1 + r_2\mathbf{v}_2 + \cdots + r_n\mathbf{v}_n, \mathbf{w} \rangle = r_1\langle \mathbf{v}_1, \mathbf{w} \rangle + r_2\langle \mathbf{v}_2, \mathbf{w} \rangle + \cdots + r_n\langle \mathbf{v}_n, \mathbf{w} \rangle$$

and

$$\langle \mathbf{v}, s_1\mathbf{w}_1 + s_2\mathbf{w}_2 + \cdots + s_m\mathbf{w}_m \rangle = s_1\langle \mathbf{v}, \mathbf{w}_1 \rangle + s_2\langle \mathbf{v}, \mathbf{w}_2 \rangle + \cdots + s_m\langle \mathbf{v}, \mathbf{w}_m \rangle$$

hold for all r_i and s_i in $\mathbb{R}$ and all $\mathbf{v}$, $\mathbf{w}$, $\mathbf{v}_i$, and $\mathbf{w}_j$ in V. These results are described by saying that inner products "preserve" linear combinations. For example,

$$\langle 2\mathbf{u} - \mathbf{v}, 3\mathbf{u} + 2\mathbf{v} \rangle = \langle 2\mathbf{u}, 3\mathbf{u} \rangle + \langle 2\mathbf{u}, 2\mathbf{v} \rangle + \langle -\mathbf{v}, 3\mathbf{u} \rangle + \langle -\mathbf{v}, 2\mathbf{v} \rangle$$
$$= 6\langle \mathbf{u}, \mathbf{u} \rangle + 4\langle \mathbf{u}, \mathbf{v} \rangle - 3\langle \mathbf{v}, \mathbf{u} \rangle - 2\langle \mathbf{v}, \mathbf{v} \rangle$$
$$= 6\langle \mathbf{u}, \mathbf{u} \rangle + \langle \mathbf{u}, \mathbf{v} \rangle - 2\langle \mathbf{v}, \mathbf{v} \rangle$$

If A is a symmetric $n \times n$ matrix and $\mathbf{x}$ and $\mathbf{y}$ are columns in $\mathbb{R}^n$, we regard the 1×1 matrix $\mathbf{x}^T A \mathbf{y}$ as a number. If we write

$$\langle \mathbf{x}, \mathbf{y} \rangle = \mathbf{x}^T A \mathbf{y} \quad \text{for all columns } \mathbf{x}, \mathbf{y} \text{ in } \mathbb{R}^n$$

then axioms P1–P4 follow from matrix arithmetic (only P2 requires that A is symmetric). Axiom P5 reads

$$\mathbf{x}^T A \mathbf{x} > 0 \quad \text{for all columns } \mathbf{x} \neq \mathbf{0} \text{ in } \mathbb{R}^n$$

and this condition characterizes the positive definite matrices (Theorem 2 Section 8.3). This proves the first assertion in the next theorem.

Theorem 2

If A is any $n \times n$ positive definite matrix, then

$$\langle \mathbf{x}, \mathbf{y} \rangle = \mathbf{x}^T A \mathbf{y} \quad \text{for all columns } \mathbf{x}, \mathbf{y} \text{ in } \mathbb{R}^n$$

defines an inner product on $\mathbb{R}^n$, and every inner product on $\mathbb{R}^n$ arises in this way.

PROOF

Given an inner product $\langle , \rangle$ on $\mathbb{R}^n$, let $\{\mathbf{e}_1, \mathbf{e}_2, \ldots, \mathbf{e}_n\}$ be the standard basis of $\mathbb{R}^n$. If $\mathbf{x} = \sum_{i=1}^n x_i \mathbf{e}_i$ and $\mathbf{y} = \sum_{j=1}^n y_j \mathbf{e}_j$ are two vectors in $\mathbb{R}^n$, compute $\langle \mathbf{x}, \mathbf{y} \rangle$ by adding the inner product of each term $x_i \mathbf{e}_i$ to each term $y_j \mathbf{e}_j$. The result is a double sum.

$$\langle \mathbf{x}, \mathbf{y} \rangle = \sum_{i=1}^n \sum_{j=1}^n \langle x_i \mathbf{e}_i, y_j \mathbf{e}_j \rangle = \sum_{i=1}^n \sum_{j=1}^n x_i \langle \mathbf{e}_i, \mathbf{e}_j \rangle y_j$$

SECTION 10.1 Inner Products and Norms

As the reader can verify, this is a matrix product:

$$\langle \mathbf{x}, \mathbf{y} \rangle = [x_1 \ x_2 \ \cdots \ x_n] \begin{bmatrix} \langle \mathbf{e}_1, \mathbf{e}_1 \rangle & \langle \mathbf{e}_1, \mathbf{e}_2 \rangle & \cdots & \langle \mathbf{e}_1, \mathbf{e}_n \rangle \\ \langle \mathbf{e}_2, \mathbf{e}_1 \rangle & \langle \mathbf{e}_2, \mathbf{e}_2 \rangle & \cdots & \langle \mathbf{e}_2, \mathbf{e}_n \rangle \\ \vdots & \vdots & & \vdots \\ \langle \mathbf{e}_n, \mathbf{e}_1 \rangle & \langle \mathbf{e}_n, \mathbf{e}_2 \rangle & \cdots & \langle \mathbf{e}_n, \mathbf{e}_n \rangle \end{bmatrix} \begin{bmatrix} y_1 \\ y_2 \\ \vdots \\ y_n \end{bmatrix}$$

Hence $\langle \mathbf{x}, \mathbf{y} \rangle = \mathbf{x}^T A \mathbf{y}$, where A is the $n \times n$ matrix whose (i, j)-entry is $\langle \mathbf{e}_i, \mathbf{e}_j \rangle$. The fact that $\langle \mathbf{e}_i, \mathbf{e}_j \rangle = \langle \mathbf{e}_j, \mathbf{e}_i \rangle$ shows that A is symmetric. Finally, A is positive definite by Theorem 2 Section 8.3.

Thus, just as every linear operator $\mathbb{R}^n \to \mathbb{R}^n$ corresponds to an $n \times n$ matrix, every inner product on $\mathbb{R}^n$ corresponds to a positive definite $n \times n$ matrix. In particular, the dot product corresponds to the identity matrix I_n.

Remark If we refer to the inner product space $\mathbb{R}^n$ without specifying the inner product, we mean that the dot product is to be used.

EXAMPLE 4

Let the inner product $\langle , \rangle$ be defined on $\mathbb{R}^2$ by

$$\left\langle \begin{bmatrix} v_1 \\ v_2 \end{bmatrix}, \begin{bmatrix} w_1 \\ w_2 \end{bmatrix} \right\rangle = 2v_1 w_1 - v_1 w_2 - v_2 w_1 - v_2 w_2$$

Find a symmetric 2×2 matrix A such that $\langle \mathbf{x}, \mathbf{y} \rangle = \mathbf{x}^T A \mathbf{y}$ for all $\mathbf{x}, \mathbf{y}$ in $\mathbb{R}^2$.

Solution ▶ The (i, j)-entry of the matrix A is the coefficient of $v_i w_j$ in the expression, so $A = \begin{bmatrix} 2 & -1 \\ -1 & 1 \end{bmatrix}$. Incidentally, if $\mathbf{x} = \begin{bmatrix} x \\ y \end{bmatrix}$, then

$$\langle \mathbf{x}, \mathbf{x} \rangle = 2x^2 - 2xy + y^2 = x^2 + (x - y)^2 \geq 0$$

for all $\mathbf{x}$, so $\langle \mathbf{x}, \mathbf{x} \rangle = 0$ implies $\mathbf{x} = \mathbf{0}$. Hence $\langle , \rangle$ is indeed an inner product, so A is positive definite.

Let $\langle , \rangle$ be an inner product on $\mathbb{R}^n$ given as in Theorem 2 by a positive definite matrix A. If $\mathbf{x} = [x_1 \ x_2 \ \cdots \ x_n]^T$, then $\langle \mathbf{x}, \mathbf{x} \rangle = \mathbf{x}^T A \mathbf{x}$ is an expression in the variables $x_1, x_2, \ldots, x_n$ called a **quadratic form**. These are studied in detail in Section 8.8.

Norm and Distance

Definition 10.2 As in $\mathbb{R}^n$, if $\langle , \rangle$ is an inner product on a space V, the **norm**[3] $\|\mathbf{v}\|$ of a vector $\mathbf{v}$ in V is defined by

$$\|\mathbf{v}\| = \sqrt{\langle \mathbf{v}, \mathbf{v} \rangle}$$

We define the **distance** between vectors $\mathbf{v}$ and $\mathbf{w}$ in an inner product space V to be

$$d(\mathbf{v}, \mathbf{w}) = \|\mathbf{v} - \mathbf{w}\|$$

Note that axiom P5 guarantees that $\langle \mathbf{v}, \mathbf{v} \rangle \geq 0$, so $\|\mathbf{v}\|$ is a real number.

---

3 If the dot product is used in $\mathbb{R}^n$, the norm $\|\mathbf{x}\|$ of a vector $\mathbf{x}$ is usually called the **length** of $\mathbf{x}$.

EXAMPLE 5

The norm of a continuous function $f = f(x)$ in $\mathbf{C}[a, b]$ (with the inner product from Example 3) is given by

$$\|f\| = \sqrt{\int_a^b f(x)^2 dx}$$

Hence $\|f\|^2$ is the area beneath the graph of $y = f(x)^2$ between $x = a$ and $x = b$ (see the diagram).

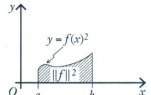

EXAMPLE 6

Show that $\langle \mathbf{u} + \mathbf{v}, \mathbf{u} - \mathbf{v} \rangle = \|\mathbf{u}\|^2 - \|\mathbf{v}\|^2$ in any inner product space.

Solution ▶
$$\langle \mathbf{u} + \mathbf{v}, \mathbf{u} - \mathbf{v} \rangle = \langle \mathbf{u}, \mathbf{u} \rangle - \langle \mathbf{u}, \mathbf{v} \rangle + \langle \mathbf{v}, \mathbf{u} \rangle - \langle \mathbf{v}, \mathbf{v} \rangle$$
$$= \|\mathbf{u}\|^2 - \langle \mathbf{u}, \mathbf{v} \rangle + \langle \mathbf{u}, \mathbf{v} \rangle - \|\mathbf{v}\|^2$$
$$= \|\mathbf{u}\|^2 - \|\mathbf{v}\|^2$$

A vector $\mathbf{v}$ in an inner product space V is called a **unit vector** if $\|\mathbf{v}\| = 1$. The set of all unit vectors in V is called the **unit ball** in V. For example, if $V = \mathbb{R}^2$ (with the dot product) and $\mathbf{v} = (x, y)$, then

$$\|\mathbf{v}\| = 1 \quad \text{if and only if} \quad x^2 + y^2 = 1$$

Hence the unit ball in $\mathbb{R}^2$ is the **unit circle** $x^2 + y^2 = 1$ with centre at the origin and radius 1. However, the shape of the unit ball varies with the choice of inner product.

EXAMPLE 7

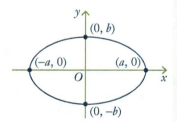

Let $a > 0$ and $b > 0$. If $\mathbf{v} = (x, y)$ and $\mathbf{w} = (x_1, y_1)$, define an inner product on $\mathbb{R}^2$ by

$$\langle \mathbf{v}, \mathbf{w} \rangle = \frac{xx_1}{a^2} + \frac{yy_1}{b^2}$$

The reader can verify (Exercise 5) that this is indeed an inner product. In this case

$$\|\mathbf{v}\| = 1 \quad \text{if and only if} \quad \frac{x^2}{a^2} + \frac{y^2}{b^2} = 1$$

so the unit ball is the ellipse shown in the diagram.

Example 7 graphically illustrates the fact that norms and distances in an inner product space V vary with the choice of inner product in V.

Theorem 3

If $\mathbf{v} \neq \mathbf{0}$ is any vector in an inner product space V, then $\dfrac{1}{\|\mathbf{v}\|}\mathbf{v}$ is the unique unit vector that is a positive multiple of $\mathbf{v}$.

SECTION 10.1 Inner Products and Norms

The next theorem reveals an important and useful fact about the relationship between norms and inner products, extending the Cauchy inequality for $\mathbb{R}^n$ (Theorem 2 Section 5.3).

Theorem 4

Cauchy-Schwarz Inequality[4]

If $\mathbf{v}$ and $\mathbf{w}$ are two vectors in an inner product space V, then
$$\langle \mathbf{v}, \mathbf{w} \rangle^2 \leq \|\mathbf{v}\|^2 \|\mathbf{w}\|^2$$
Moreover, equality occurs if and only if one of $\mathbf{v}$ and $\mathbf{w}$ is a scalar multiple of the other.

PROOF

Write $\|\mathbf{v}\| = a$ and $\|\mathbf{w}\| = b$. Using Theorem 1 we compute:
$$\|b\mathbf{v} - a\mathbf{w}\|^2 = b^2\|\mathbf{v}\|^2 - 2ab\langle \mathbf{v}, \mathbf{w} \rangle + a^2\|\mathbf{w}\|^2 = 2ab(ab - \langle \mathbf{v}, \mathbf{w} \rangle)$$
$$\|b\mathbf{v} + a\mathbf{w}\|^2 = b^2\|\mathbf{v}\|^2 + 2ab\langle \mathbf{v}, \mathbf{w} \rangle + a^2\|\mathbf{w}\|^2 = 2ab(ab + \langle \mathbf{v}, \mathbf{w} \rangle)$$
(∗)

It follows that $ab - \langle \mathbf{v}, \mathbf{w} \rangle \geq 0$ and $ab + \langle \mathbf{v}, \mathbf{w} \rangle \geq 0$, and hence that $-ab \leq \langle \mathbf{v}, \mathbf{w} \rangle \leq ab$. But then $|\langle \mathbf{v}, \mathbf{w} \rangle| \leq ab = \|\mathbf{v}\|\|\mathbf{w}\|$, as desired.

Conversely, if $|\langle \mathbf{v}, \mathbf{w} \rangle| = \|\mathbf{v}\|\|\mathbf{w}\| = ab$ then $\langle \mathbf{v}, \mathbf{w} \rangle = \pm ab$. Hence (∗) shows that $b\mathbf{v} - a\mathbf{w} = \mathbf{0}$ or $b\mathbf{v} + a\mathbf{w} = \mathbf{0}$. It follows that one of $\mathbf{v}$ and $\mathbf{w}$ is a scalar multiple of the other, even if $a = 0$ or $b = 0$.

EXAMPLE 8

If f and g are continuous functions on the interval $[a, b]$, then (see Example 3)
$$\left\{ \int_a^b f(x)g(x)dx \right\}^2 \leq \int_a^b f(x)^2 dx \int_a^b g(x)^2 dx$$

Another famous inequality, the so-called *triangle inequality*, also comes from the Cauchy-Schwarz inequality. It is included in the following list of basic properties of the norm of a vector.

Theorem 5

If V is an inner product space, the norm $\|\cdot\|$ has the following properties.
1. $\|\mathbf{v}\| \geq 0$ for every vector $\mathbf{v}$ in V.
2. $\|\mathbf{v}\| = 0$ if and only if $\mathbf{v} = \mathbf{0}$.
3. $\|r\mathbf{v}\| = |r|\|\mathbf{v}\|$ for every $\mathbf{v}$ in V and every r in $\mathbb{R}$.
4. $\|\mathbf{v} + \mathbf{w}\| \leq \|\mathbf{v}\| + \|\mathbf{w}\|$ for all $\mathbf{v}$ and $\mathbf{w}$ in V (**triangle inequality**).

---

[4] Hermann Amandus Schwarz (1843–1921) was a German mathematician at the University of Berlin. He had strong geometric intuition, which he applied with great ingenuity to particular problems. A version of the inequality appeared in 1885.

PROOF

Because $\|\mathbf{v}\| = \sqrt{\langle \mathbf{v}, \mathbf{v} \rangle}$, properties (1) and (2) follow immediately from (3) and (4) of Theorem 1. As to (3), compute

$$\|r\mathbf{v}\|^2 = \langle r\mathbf{v}, r\mathbf{v} \rangle = r^2 \langle \mathbf{v}, \mathbf{v} \rangle = r^2 \|\mathbf{v}\|^2$$

Hence (3) follows by taking positive square roots. Finally, the fact that $\langle \mathbf{v}, \mathbf{w} \rangle \leq \|\mathbf{v}\|\|\mathbf{w}\|$ by the Cauchy-Schwarz inequality gives

$$\|\mathbf{v} + \mathbf{w}\|^2 = \langle \mathbf{v} + \mathbf{w}, \mathbf{v} + \mathbf{w} \rangle = \|\mathbf{v}\|^2 + 2\langle \mathbf{v}, \mathbf{w} \rangle + \|\mathbf{w}\|^2$$
$$\leq \|\mathbf{v}\|^2 + 2\|\mathbf{v}\|\|\mathbf{w}\| + \|\mathbf{w}\|^2$$
$$= (\|\mathbf{v}\| + \|\mathbf{w}\|)^2$$

Hence (4) follows by taking positive square roots.

It is worth noting that the usual triangle inequality for absolute values,

$$|r + s| \leq |r| + |s| \text{ for all real numbers } r \text{ and } s,$$

is a special case of (4) where $V = \mathbb{R} = \mathbb{R}^1$ and the dot product $\langle r, s \rangle = rs$ is used.

In many calculations in an inner product space, it is required to show that some vector $\mathbf{v}$ is zero. This is often accomplished most easily by showing that its norm $\|\mathbf{v}\|$ is zero. Here is an example.

EXAMPLE 9

Let $\{\mathbf{v}_1, \ldots, \mathbf{v}_n\}$ be a spanning set for an inner product space V. If $\mathbf{v}$ in V satisfies $\langle \mathbf{v}, \mathbf{v}_i \rangle = 0$ for each $i = 1, 2, \ldots, n$, show that $\mathbf{v} = \mathbf{0}$.

Solution ▸ Write $\mathbf{v} = r_1 \mathbf{v}_1 + \cdots + r_n \mathbf{v}_n$, r_i in $\mathbb{R}$. To show that $\mathbf{v} = \mathbf{0}$, we show that $\|\mathbf{v}\|^2 = \langle \mathbf{v}, \mathbf{v} \rangle = 0$. Compute:

$$\langle \mathbf{v}, \mathbf{v} \rangle = \langle \mathbf{v}, r_1 \mathbf{v}_1 + \cdots + r_n \mathbf{v}_n \rangle = r_1 \langle \mathbf{v}, \mathbf{v}_1 \rangle + \cdots + r_n \langle \mathbf{v}, \mathbf{v}_n \rangle = 0$$

by hypothesis, and the result follows.

The norm properties in Theorem 5 translate to the following properties of distance familiar from geometry. The proof is Exercise 21.

Theorem 6

Let V be an inner product space.
1. $d(\mathbf{v}, \mathbf{w}) \geq 0$ for all $\mathbf{v}, \mathbf{w}$ in V.
2. $d(\mathbf{v}, \mathbf{w}) = 0$ if and only if $\mathbf{v} = \mathbf{w}$.
3. $d(\mathbf{v}, \mathbf{w}) = d(\mathbf{w}, \mathbf{v})$ for all $\mathbf{v}$ and $\mathbf{w}$ in V.
4. $d(\mathbf{v}, \mathbf{w}) \leq d(\mathbf{v}, \mathbf{u}) + d(\mathbf{u}, \mathbf{w})$ for all $\mathbf{v}, \mathbf{u}$, and $\mathbf{w}$ in V.

SECTION 10.1 Inner Products and Norms

EXERCISES 10.1

1. In each case, determine which of axioms P1–P5 fail to hold.
   (a) $V = \mathbb{R}^2$, $\langle (x_1, y_1), (x_2, y_2) \rangle = x_1 y_1 x_2 y_2$
   ◆(b) $V = \mathbb{R}^3$, $\langle (x_1, x_2, x_3), (y_1, y_2, y_3) \rangle$
   $\quad = x_1 y_1 - x_2 y_2 + x_3 y_3$
   (c) $V = \mathbb{C}$, $\langle z, w \rangle = z\bar{w}$, where $\bar{w}$ is complex conjugation
   ◆(d) $V = \mathbf{P}_3$, $\langle p(x), q(x) \rangle = p(1)q(1)$
   (e) $V = \mathbf{M}_{22}$, $\langle A, B \rangle = \det(AB)$
   ◆(f) $V = \mathbf{F}[0, 1]$, $\langle f, g \rangle = f(1)g(0) + f(0)g(1)$

◆2. Let V be an inner product space. If $U \subseteq V$ is a subspace, show that U is an inner product space using the same inner product.

3. In each case, find a scalar multiple of $\mathbf{v}$ that is a unit vector.
   (a) $\mathbf{v} = f$ in $\mathbf{C}[0, 1]$ where $f(x) = x^2$
   $\langle \mathbf{f}, \mathbf{g} \rangle = \int_0^1 f(x)g(x)dx$
   ◆(b) $\mathbf{v} = f$ in $\mathbf{C}[-\pi, \pi]$ where $f(x) = \cos x$
   $\langle \mathbf{f}, \mathbf{g} \rangle = \int_{-\pi}^{\pi} f(x)g(x)dx$
   (c) $\mathbf{v} = \begin{bmatrix} 1 \\ 3 \end{bmatrix}$ in $\mathbb{R}^2$ where $\langle \mathbf{v}, \mathbf{w} \rangle = \mathbf{v}^T \begin{bmatrix} 1 & 1 \\ 1 & 2 \end{bmatrix} \mathbf{w}$
   ◆(d) $\mathbf{v} = \begin{bmatrix} 3 \\ -1 \end{bmatrix}$ in $\mathbb{R}^2$, $\langle \mathbf{v}, \mathbf{w} \rangle = \mathbf{v}^T \begin{bmatrix} 1 & -1 \\ -1 & 2 \end{bmatrix} \mathbf{w}$

4. In each case, find the distance between $\mathbf{u}$ and $\mathbf{v}$.
   (a) $\mathbf{u} = (3, -1, 2, 0)$, $\mathbf{v} = (1, 1, 1, 3)$;
   $\langle \mathbf{u}, \mathbf{v} \rangle = \mathbf{u} \cdot \mathbf{v}$
   ◆(b) $\mathbf{u} = (1, 2, -1, 2)$, $\mathbf{v} = (2, 1, -1, 3)$;
   $\langle \mathbf{u}, \mathbf{v} \rangle = \mathbf{u} \cdot \mathbf{v}$
   (c) $\mathbf{u} = f$, $\mathbf{v} = g$ in $\mathbf{C}[0, 1]$ where $f(x) = x^2$ and $g(x) = 1 - x$; $\langle \mathbf{f}, \mathbf{g} \rangle = \int_0^1 f(x)g(x)dx$
   ◆(d) $\mathbf{u} = f$, $\mathbf{v} = g$ in $\mathbf{C}[-\pi, \pi]$ where $f(x) = 1$ and $g(x) = \cos x$; $\langle \mathbf{f}, \mathbf{g} \rangle = \int_{-\pi}^{\pi} f(x)g(x)dx$

5. Let $a_1, a_2, \ldots, a_n$ be positive numbers. Given $\mathbf{v} = (v_1, v_2, \ldots, v_n)$ and $\mathbf{w} = (w_1, w_2, \ldots, w_n)$, define $\langle \mathbf{v}, \mathbf{w} \rangle = a_1 v_1 w_1 + \cdots + a_n v_n w_n$. Show that this is an inner product on $\mathbb{R}^n$.

6. If $\{\mathbf{b}_1, \ldots, \mathbf{b}_n\}$ is a basis of V and if
   $\mathbf{v} = v_1 \mathbf{b}_1 + \cdots + v_n \mathbf{b}_n$ and
   $\mathbf{w} = w_1 \mathbf{b}_1 + \cdots + w_n \mathbf{b}_n$ are vectors in V, define

   $\langle \mathbf{v}, \mathbf{w} \rangle = v_1 w_1 + \cdots + v_n w_n$.

   Show that this is an inner product on V.

7. If $p = p(x)$ and $q = q(x)$ are polynomials in $\mathbf{P}_n$, define

   $\langle p, q \rangle = p(0)q(0) + p(1)q(1) + \cdots + p(n)q(n)$

   Show that this is an inner product on $\mathbf{P}_n$. [Hint for P5: Theorem 4 Section 6.5 or Appendix D.]

◆8. Let $\mathbf{D}_n$ denote the space of all functions from the set $\{1, 2, 3, \ldots, n\}$ to $\mathbb{R}$ with pointwise addition and scalar multiplication (see Exercise 35 Section 6.3). Show that $\langle , \rangle$ is an inner product on $\mathbf{D}_n$ if $\langle f, g \rangle = f(1)g(1) + f(2)g(2) + \cdots + f(n)g(n)$.

9. Let $\text{re}(z)$ denote the real part of the complex number z. Show that $\langle , \rangle$ is an inner product on $\mathbb{C}$ if $\langle z, w \rangle = \text{re}(z\bar{w})$.

10. If $T: V \to V$ is an isomorphism of the inner product space V, show that

    $\langle \mathbf{v}, \mathbf{w} \rangle_1 = \langle T(\mathbf{v}), T(\mathbf{w}) \rangle$

    defines a new inner product $\langle , \rangle_1$ on V.

11. Show that every inner product $\langle , \rangle$ on $\mathbb{R}^n$ has the form $\langle \mathbf{x}, \mathbf{y} \rangle = (U\mathbf{x}) \cdot (U\mathbf{y})$ for some upper triangular matrix U with positive diagonal entries. [Hint: Theorem 3 Section 8.3.]

12. In each case, show that $\langle \mathbf{v}, \mathbf{w} \rangle = \mathbf{v}^T A \mathbf{w}$ defines an inner product on $\mathbb{R}^2$ and hence show that A is positive definite.
    (a) $A = \begin{bmatrix} 2 & 1 \\ 1 & 1 \end{bmatrix}$ ◆(b) $A = \begin{bmatrix} 5 & -3 \\ -3 & 2 \end{bmatrix}$
    (c) $A = \begin{bmatrix} 3 & 2 \\ 2 & 3 \end{bmatrix}$ ◆(d) $A = \begin{bmatrix} 3 & 4 \\ 4 & 6 \end{bmatrix}$

13. In each case, find a symmetric matrix A such that $\langle \mathbf{v}, \mathbf{w} \rangle = \mathbf{v}^T A \mathbf{w}$.
    (a) $\left\langle \begin{bmatrix} v_1 \\ v_2 \end{bmatrix}, \begin{bmatrix} w_1 \\ w_2 \end{bmatrix} \right\rangle = v_1 w_1 + 2 v_1 w_2 + 2 v_2 w_1 + 5 v_2 w_2$
    ◆(b) $\left\langle \begin{bmatrix} v_1 \\ v_2 \end{bmatrix}, \begin{bmatrix} w_1 \\ w_2 \end{bmatrix} \right\rangle = v_1 w_1 - v_1 w_2 - v_2 w_1 + 2 v_2 w_2$
    (c) $\left\langle \begin{bmatrix} v_1 \\ v_2 \\ v_3 \end{bmatrix}, \begin{bmatrix} w_1 \\ w_2 \\ w_3 \end{bmatrix} \right\rangle = 2v_1 w_1 + v_2 w_2 + v_3 w_3 - v_1 w_2$
    $\quad - v_2 w_1 + v_2 w_3 + v_3 w_2$

◆(d) $\left\langle \begin{bmatrix} v_1 \\ v_2 \\ v_3 \end{bmatrix}, \begin{bmatrix} w_1 \\ w_2 \\ w_3 \end{bmatrix} \right\rangle = v_1w_1 + 2v_2w_2 + 5v_3w_3 - 2v_1w_3 - 2v_3w_1$

◆14. If A is symmetric and $\mathbf{x}^T A \mathbf{x} = 0$ for all columns $\mathbf{x}$ in $\mathbb{R}^n$, show that $A = 0$. [Hint: Consider $\langle \mathbf{x} + \mathbf{y}, \mathbf{x} + \mathbf{y} \rangle$ where $\langle \mathbf{x}, \mathbf{y} \rangle = \mathbf{x}^T A \mathbf{y}$.]

15. Show that the sum of two inner products on V is again an inner product.

16. Let $\|\mathbf{u}\| = 1$, $\|\mathbf{v}\| = 2$, $\|\mathbf{w}\| = \sqrt{3}$, $\langle \mathbf{u}, \mathbf{v} \rangle = -1$, $\langle \mathbf{u}, \mathbf{w} \rangle = 0$ and $\langle \mathbf{v}, \mathbf{w} \rangle = 3$. Compute:

(a) $\langle \mathbf{v} + \mathbf{w}, 2\mathbf{u} - \mathbf{v} \rangle$

◆(b) $\langle \mathbf{u} - 2\mathbf{v} - \mathbf{w}, 3\mathbf{w} - \mathbf{v} \rangle$

17. Given the data in Exercise 16, show that $\mathbf{u} + \mathbf{v} = \mathbf{w}$.

18. Show that no vectors exist such that $\|\mathbf{u}\| = 1$, $\|\mathbf{v}\| = 2$, and $\langle \mathbf{u}, \mathbf{v} \rangle = -3$.

19. Complete Example 2.

◆20. Prove Theorem 1.

21. Prove Theorem 6.

22. Let $\mathbf{u}$ and $\mathbf{v}$ be vectors in an inner product space V.

(a) Expand $\langle 2\mathbf{u} - 7\mathbf{v}, 3\mathbf{u} + 5\mathbf{v} \rangle$.

◆(b) Expand $\langle 3\mathbf{u} - 4\mathbf{v}, 5\mathbf{u} + \mathbf{v} \rangle$.

(c) Show that $\|\mathbf{u} + \mathbf{v}\|^2 = \|\mathbf{u}\|^2 + 2\langle \mathbf{u}, \mathbf{v} \rangle + \|\mathbf{v}\|^2$.

◆(d) Show that $\|\mathbf{u} - \mathbf{v}\|^2 = \|\mathbf{u}\|^2 - 2\langle \mathbf{u}, \mathbf{v} \rangle + \|\mathbf{v}\|^2$.

23. Show that $\|\mathbf{v}\|^2 + \|\mathbf{w}\|^2 = \frac{1}{2}\{\|\mathbf{v} + \mathbf{w}\|^2 + \|\mathbf{v} - \mathbf{w}\|^2\}$ for any $\mathbf{v}$ and $\mathbf{w}$ in an inner product space.

24. Let $\langle , \rangle$ be an inner product on a vector space V. Show that the corresponding distance function is translation invariant. That is, show that $d(\mathbf{v}, \mathbf{w}) = d(\mathbf{v} + \mathbf{u}, \mathbf{w} + \mathbf{u})$ for all $\mathbf{v}, \mathbf{w}$, and $\mathbf{u}$ in V.

25. (a) Show that $\langle \mathbf{u}, \mathbf{v} \rangle = \frac{1}{4}[\|\mathbf{u} + \mathbf{v}\|^2 - \|\mathbf{u} - \mathbf{v}\|^2]$ for all $\mathbf{u}, \mathbf{v}$ in an inner product space V.

(b) If $\langle , \rangle$ and $\langle , \rangle'$ are two inner products on V that have equal associated norm functions, show that $\langle \mathbf{u}, \mathbf{v} \rangle = \langle \mathbf{u}, \mathbf{v} \rangle'$ holds for all $\mathbf{u}$ and $\mathbf{v}$.

26. Let $\mathbf{v}$ denote a vector in an inner product space V.

(a) Show that $W = \{\mathbf{w} \mid \mathbf{w} \text{ in } V, \langle \mathbf{v}, \mathbf{w} \rangle = 0\}$ is a subspace of V.

◆(b) If $V = \mathbb{R}^3$ with the dot product, and if $\mathbf{v} = (1, -1, 2)$, find a basis for W (W as in (a)).

27. Given vectors $\mathbf{w}_1, \mathbf{w}_2, \ldots, \mathbf{w}_n$ and $\mathbf{v}$, assume that $\langle \mathbf{v}, \mathbf{w}_i \rangle = 0$ for each i. Show that $\langle \mathbf{v}, \mathbf{w} \rangle = 0$ for all $\mathbf{w}$ in $\text{span}\{\mathbf{w}_1, \mathbf{w}_2, \ldots, \mathbf{w}_n\}$.

◆28. If $V = \text{span}\{\mathbf{v}_1, \mathbf{v}_2, \ldots, \mathbf{v}_n\}$ and $\langle \mathbf{v}, \mathbf{v}_i \rangle = \langle \mathbf{w}, \mathbf{v}_i \rangle$ holds for each i. Show that $\mathbf{v} = \mathbf{w}$.

29. Use the Cauchy-Schwarz inequality in an inner product space to show that:

(a) If $\|\mathbf{u}\| \leq 1$, then $\langle \mathbf{u}, \mathbf{v} \rangle^2 \leq \|\mathbf{v}\|^2$ for all $\mathbf{v}$ in V.

◆(b) $(x \cos \theta + y \sin \theta)^2 \leq x^2 + y^2$ for all real x, y, and θ.

(c) $\|r_1\mathbf{v}_1 + \cdots + r_n\mathbf{v}_n\|^2 \leq [r_1\|\mathbf{v}_1\| + \cdots + r_n\|\mathbf{v}_n\|]^2$ for all vectors $\mathbf{v}_i$, and all $r_i > 0$ in $\mathbb{R}$.

30. If A is a $2 \times n$ matrix, let $\mathbf{u}$ and $\mathbf{v}$ denote the rows of A.

(a) Show that $AA^T = \begin{bmatrix} \|\mathbf{u}\|^2 & \mathbf{u} \cdot \mathbf{v} \\ \mathbf{u} \cdot \mathbf{v} & \|\mathbf{v}\|^2 \end{bmatrix}$.

(b) Show that $\det(AA^T) \geq 0$.

31. (a) If $\mathbf{v}$ and $\mathbf{w}$ are nonzero vectors in an inner product space V, show that

$$-1 \leq \frac{\langle \mathbf{v}, \mathbf{w} \rangle}{\|\mathbf{v}\|\|\mathbf{w}\|} \leq 1,$$ and hence that a unique angle θ exists such that $\frac{\langle \mathbf{v}, \mathbf{w} \rangle}{\|\mathbf{v}\|\|\mathbf{w}\|} = \cos \theta$ and $0 \leq \theta \leq \pi$. This angle θ is called the **angle between** $\mathbf{v}$ **and** $\mathbf{w}$.

(b) Find the angle between $\mathbf{v} = (1, 2, -1, 1, 3)$ and $\mathbf{w} = (2, 1, 0, 2, 0)$ in $\mathbb{R}^5$ with the dot product.

(c) If θ is the angle between $\mathbf{v}$ and $\mathbf{w}$, show that the **law of cosines** is valid: $\|\mathbf{v} - \mathbf{w}\|^2 = \|\mathbf{v}\|^2 + \|\mathbf{w}\|^2 - 2\|\mathbf{v}\|\|\mathbf{w}\|\cos \theta$.

32. If $V = \mathbb{R}^2$, define $\|(x, y)\| = |x| + |y|$.

(a) Show that $\|\cdot\|$ satisfies the conditions in Theorem 5.

(b) Show that $\|\cdot\|$ does not arise from an inner product on $\mathbb{R}^2$ given by a matrix A. [Hint: If it did, use Theorem 2 to find numbers a, b, and c such that $\|(x, y)\|^2 = ax^2 + bxy + cy^2$ for all x and y.]

SECTION 10.2 Orthogonal Sets of Vectors

The idea that two lines can be perpendicular is fundamental in geometry, and this section is devoted to introducing this notion into a general inner product space V. To motivate the definition, recall that two nonzero geometric vectors $\mathbf{x}$ and $\mathbf{y}$ in $\mathbb{R}^n$ are perpendicular (or orthogonal) if and only if $\mathbf{x} \cdot \mathbf{y} = 0$. In general, two vectors $\mathbf{v}$ and $\mathbf{w}$ in an inner product space V are said to be **orthogonal** if

$$\langle \mathbf{v}, \mathbf{w} \rangle = 0$$

A set $\{\mathbf{f}_1, \mathbf{f}_2, ..., \mathbf{f}_n\}$ of vectors is called an **orthogonal set of vectors** if

1. Each $\mathbf{f}_i \neq \mathbf{0}$.
2. $\langle \mathbf{f}_i, \mathbf{f}_j \rangle = 0$ for all $i \neq j$.

If, in addition, $\|\mathbf{f}_i\| = 1$ for each i, the set $\{\mathbf{f}_1, \mathbf{f}_2, ..., \mathbf{f}_n\}$ is called an **orthonormal** set.

EXAMPLE 1

$\{\sin x, \cos x\}$ is orthogonal in $\mathbf{C}[-\pi, \pi]$ because

$$\int_{-\pi}^{\pi} \sin x \cos x \, dx = [-\tfrac{1}{4} \cos 2x]_{-\pi}^{\pi} = 0$$

The first result about orthogonal sets extends Pythagoras' theorem in $\mathbb{R}^n$ (Theorem 4 Section 5.3) and the same proof works.

Theorem 1

Pythagoras' Theorem
If $\{\mathbf{f}_1, \mathbf{f}_2, ..., \mathbf{f}_n\}$ is an orthogonal set of vectors, then

$$\|\mathbf{f}_1 + \mathbf{f}_2 + \cdots + \mathbf{f}_n\|^2 = \|\mathbf{f}_1\|^2 + \|\mathbf{f}_2\|^2 + \cdots + \|\mathbf{f}_n\|^2$$

The proof of the next result is left to the reader.

Theorem 2

Let $\{\mathbf{f}_1, \mathbf{f}_2, ..., \mathbf{f}_n\}$ be an orthogonal set of vectors.

1. $\{r_1 \mathbf{f}_1, r_2 \mathbf{f}_2, ..., r_n \mathbf{f}_n\}$ is also orthogonal for any $r_i \neq 0$ in $\mathbb{R}$.
2. $\left\{ \dfrac{1}{\|\mathbf{f}_1\|} \mathbf{f}_1, \dfrac{1}{\|\mathbf{f}_2\|} \mathbf{f}_2, ..., \dfrac{1}{\|\mathbf{f}_n\|} \mathbf{f}_n \right\}$ is an orthonormal set.

As before, the process of passing from an orthogonal set to an orthonormal one is called **normalizing** the orthogonal set. The proof of Theorem 5 Section 5.3 goes through to give

Theorem 3

Every orthogonal set of vectors is linearly independent.

EXAMPLE 2

Show that $\left\{ \begin{bmatrix} 2 \\ -1 \\ 0 \end{bmatrix}, \begin{bmatrix} 0 \\ 1 \\ 1 \end{bmatrix}, \begin{bmatrix} 0 \\ -1 \\ 2 \end{bmatrix} \right\}$ is an orthogonal basis of $\mathbb{R}^3$ with inner product

$$\langle \mathbf{v}, \mathbf{w} \rangle = \mathbf{v}^T A \mathbf{w}, \text{ where } A = \begin{bmatrix} 1 & 1 & 0 \\ 1 & 2 & 0 \\ 0 & 0 & 1 \end{bmatrix}.$$

Solution ▶ We have

$$\left\langle \begin{bmatrix} 2 \\ -1 \\ 0 \end{bmatrix}, \begin{bmatrix} 0 \\ 1 \\ 1 \end{bmatrix} \right\rangle = [2 \ -1 \ 0] \begin{bmatrix} 1 & 1 & 0 \\ 1 & 2 & 0 \\ 0 & 0 & 1 \end{bmatrix} \begin{bmatrix} 0 \\ 1 \\ 1 \end{bmatrix} = [1 \ 0 \ 0] \begin{bmatrix} 0 \\ 1 \\ 1 \end{bmatrix} = 0$$

and the reader can verify that the other pairs are orthogonal too. Hence the set is orthogonal, so it is linearly independent by Theorem 3. Because dim $\mathbb{R}^3 = 3$, it is a basis.

The proof of Theorem 6 Section 5.3 generalizes to give the following:

Theorem 4

Expansion Theorem
Let $\{\mathbf{f}_1, \mathbf{f}_2, ..., \mathbf{f}_n\}$ be an orthogonal basis of an inner product space V. If $\mathbf{v}$ is any vector in V, then

$$\mathbf{v} = \frac{\langle \mathbf{v}, \mathbf{f}_1 \rangle}{\|\mathbf{f}_1\|^2} \mathbf{f}_1 + \frac{\langle \mathbf{v}, \mathbf{f}_2 \rangle}{\|\mathbf{f}_2\|^2} \mathbf{f}_2 + \cdots + \frac{\langle \mathbf{v}, \mathbf{f}_n \rangle}{\|\mathbf{f}_n\|^2} \mathbf{f}_n$$

is the expansion of $\mathbf{v}$ as a linear combination of the basis vectors.

The coefficients $\frac{\langle \mathbf{v}, \mathbf{f}_1 \rangle}{\|\mathbf{f}_1\|^2}, \frac{\langle \mathbf{v}, \mathbf{f}_2 \rangle}{\|\mathbf{f}_2\|^2}, ..., \frac{\langle \mathbf{v}, \mathbf{f}_n \rangle}{\|\mathbf{f}_n\|^2}$ in the expansion theorem are sometimes called the **Fourier coefficients** of $\mathbf{v}$ with respect to the orthogonal basis $\{\mathbf{f}_1, \mathbf{f}_2, ..., \mathbf{f}_n\}$. This is in honour of the French mathematician J.B.J. Fourier (1768–1830). His original work was with a particular orthogonal set in the space $\mathbf{C}[a, b]$, about which there will be more to say in Section 10.5.

EXAMPLE 3

If $a_0, a_1, ..., a_n$ are distinct numbers and $p(x)$ and $q(x)$ are in $\mathbf{P}_n$, define

$$\langle p(x), q(x) \rangle = p(a_0)q(a_0) + p(a_1)q(a_1) + \cdots + p(a_n)q(a_n)$$

SECTION 10.2 Orthogonal Sets of Vectors

This is an inner product on $\mathbf{P}_n$. (Axioms P1–P4 are routinely verified, and P5 holds because 0 is the only polynomial of degree n with $n + 1$ distinct roots. See Theorem 4 Section 6.5 or Appendix D.)

Recall that the **Lagrange polynomials** $\delta_0(x)$, $\delta_1(x)$, ..., $\delta_n(x)$ relative to the numbers $a_0, a_1, \ldots, a_n$ are defined as follows (see Section 6.5):

$$\delta_k(x) = \frac{\prod_{i \neq k}(x - a_i)}{\prod_{i \neq k}(a_k - a_i)} \quad k = 0, 1, 2, \ldots, n$$

where $\prod_{i \neq k}(x - a_i)$ means the product of all the terms

$$(x - a_0), (x - a_1), (x - a_2), \ldots, (x - a_n)$$

except that the kth term is omitted. Then $\{\delta_0(x), \delta_1(x), \ldots, \delta_n(x)\}$ is orthonormal with respect to $\langle , \rangle$ because $\delta_k(a_i) = 0$ if $i \neq k$ and $\delta_k(a_k) = 1$. These facts also show that $\langle p(x), \delta_k(x) \rangle = p(a_k)$ so the expansion theorem gives

$$p(x) = p(a_0)\delta_0(x) + p(a_1)\delta_1(x) + \cdots + p(a_n)\delta_n(x)$$

for each $p(x)$ in $\mathbf{P}_n$. This is the **Lagrange interpolation expansion** of $p(x)$, Theorem 3 Section 6.5, which is important in numerical integration.

Lemma 1

Orthogonal Lemma
Let $\{\mathbf{f}_1, \mathbf{f}_2, \ldots, \mathbf{f}_m\}$ be an orthogonal set of vectors in an inner product space V, and let $\mathbf{v}$ be any vector not in span$\{\mathbf{f}_1, \mathbf{f}_2, \ldots, \mathbf{f}_m\}$. Define

$$\mathbf{f}_{m+1} = \mathbf{v} - \frac{\langle \mathbf{v}, \mathbf{f}_1 \rangle}{\|\mathbf{f}_1\|^2}\mathbf{f}_1 - \frac{\langle \mathbf{v}, \mathbf{f}_2 \rangle}{\|\mathbf{f}_2\|^2}\mathbf{f}_2 - \cdots - \frac{\langle \mathbf{v}, \mathbf{f}_m \rangle}{\|\mathbf{f}_m\|^2}\mathbf{f}_m$$

Then $\{\mathbf{f}_1, \mathbf{f}_2, \ldots, \mathbf{f}_m, \mathbf{f}_{m+1}\}$ is an orthogonal set of vectors.

The proof of this result (and the next) is the same as for the dot product in $\mathbb{R}^n$ (Lemma 1 and Theorem 2 in Section 8.1).

Theorem 5

Gram-Schmidt Orthogonalization Algorithm
Let V be an inner product space and let $\{\mathbf{v}_1, \mathbf{v}_2, \ldots, \mathbf{v}_n\}$ be any basis of V. Define vectors $\mathbf{f}_1, \mathbf{f}_2, \ldots, \mathbf{f}_n$ in V successively as follows:

$$\mathbf{f}_1 = \mathbf{v}_1$$
$$\mathbf{f}_2 = \mathbf{v}_2 - \frac{\langle \mathbf{v}_2, \mathbf{f}_1 \rangle}{\|\mathbf{f}_1\|^2}\mathbf{f}_1$$
$$\mathbf{f}_3 = \mathbf{v}_3 - \frac{\langle \mathbf{v}_3, \mathbf{f}_1 \rangle}{\|\mathbf{f}_1\|^2}\mathbf{f}_1 - \frac{\langle \mathbf{v}_3, \mathbf{f}_2 \rangle}{\|\mathbf{f}_2\|^2}\mathbf{f}_2$$
$$\vdots \qquad \vdots$$
$$\mathbf{f}_k = \mathbf{v}_k - \frac{\langle \mathbf{v}_k, \mathbf{f}_1 \rangle}{\|\mathbf{f}_1\|^2}\mathbf{f}_1 - \frac{\langle \mathbf{v}_k, \mathbf{f}_2 \rangle}{\|\mathbf{f}_2\|^2}\mathbf{f}_2 - \cdots - \frac{\langle \mathbf{v}_k, \mathbf{f}_{k-1} \rangle}{\|\mathbf{f}_{k-1}\|^2}\mathbf{f}_{k-1}$$

for each $k = 2, 3, \ldots, n$. Then

1. $\{\mathbf{f}_1, \mathbf{f}_2, \ldots, \mathbf{f}_n\}$ is an orthogonal basis of V.
2. span$\{\mathbf{f}_1, \mathbf{f}_2, \ldots, \mathbf{f}_k\}$ = span$\{\mathbf{v}_1, \mathbf{v}_2, \ldots, \mathbf{v}_k\}$ holds for each $k = 1, 2, \ldots, n$.

The purpose of the Gram-Schmidt algorithm is to convert a basis of an inner product space into an *orthogonal* basis. In particular, it shows that every finite dimensional inner product space *has* an orthogonal basis.

EXAMPLE 4

Consider $V = \mathbf{P}_3$ with the inner product $\langle p, q \rangle = \int_{-1}^{1} p(x)q(x)dx$. If the Gram-Schmidt algorithm is applied to the basis $\{1, x, x^2, x^3\}$, show that the result is the orthogonal basis

$$\{1, x, \tfrac{1}{3}(3x^2 - 1), \tfrac{1}{5}(5x^3 - 3x)\}.$$

Solution ▶ Take $\mathbf{f}_1 = 1$. Then the algorithm gives

$$\mathbf{f}_2 = x - \frac{\langle x, \mathbf{f}_1 \rangle}{\|\mathbf{f}_1\|^2}\mathbf{f}_1 = x - \tfrac{0}{2}\mathbf{f}_1 = x$$

$$\mathbf{f}_3 = x^2 - \frac{\langle x^2, \mathbf{f}_1 \rangle}{\|\mathbf{f}_1\|^2}\mathbf{f}_1 - \frac{\langle x^2, \mathbf{f}_2 \rangle}{\|\mathbf{f}_2\|^2}\mathbf{f}_2$$

$$= x^2 - \frac{\tfrac{2}{3}}{2}1 - \frac{0}{\tfrac{2}{3}}x$$

$$= \tfrac{1}{3}(3x^2 - 1)$$

The verification that $\mathbf{f}_4 = \tfrac{1}{5}(5x^3 - 3x)$ is omitted.

The polynomials in Example 4 are such that the leading coefficient is 1 in each case. In other contexts (the study of differential equations, for example) it is customary to take multiples $p(x)$ of these polynomials such that $p(1) = 1$. The resulting orthogonal basis of $\mathbf{P}_3$ is

$$\{1, x, \tfrac{1}{2}(3x^2 - 1), \tfrac{1}{2}(5x^3 - 3x)\}$$

and these are the first four **Legendre polynomials**, so called to honour the French mathematician A. M. Legendre (1752–1833). They are important in the study of differential equations.

If V is an inner product space of dimension n, let $E = \{\mathbf{f}_1, \mathbf{f}_2, ..., \mathbf{f}_n\}$ be an orthonormal basis of V (by Theorem 5). If $\mathbf{v} = v_1\mathbf{f}_1 + v_2\mathbf{f}_2 + \cdots + v_n\mathbf{f}_n$ and $\mathbf{w} = w_1\mathbf{f}_1 + w_2\mathbf{f}_2 + \cdots + w_n\mathbf{f}_n$ are two vectors in V, we have $C_E(\mathbf{v}) = [v_1 \; v_2 \; \cdots \; v_n]^T$ and $C_E(\mathbf{w}) = [w_1 \; w_2 \; \cdots \; w_n]^T$. Hence

$$\langle \mathbf{v}, \mathbf{w} \rangle = \langle \textstyle\sum_i v_i\mathbf{f}_i, \sum_j w_j\mathbf{f}_j \rangle = \sum_{i,j} v_i w_j \langle \mathbf{f}_i, \mathbf{f}_j \rangle = \sum_i v_i w_i = C_E(\mathbf{v}) \cdot C_E(\mathbf{w}).$$

This shows that the coordinate isomorphism $C_E : V \to \mathbb{R}^n$ preserves inner products, and so proves

Corollary 1

If V is any n-dimensional inner product space, then V is isomorphic to $\mathbb{R}^n$ as inner product spaces. More precisely, if E is any orthonormal basis of V, the coordinate isomorphism

$$C_E : V \to \mathbb{R}^n \text{ satisfies } \langle \mathbf{v}, \mathbf{w} \rangle = C_E(\mathbf{v}) \cdot C_E(\mathbf{w})$$

for all $\mathbf{v}$ and $\mathbf{w}$ in V.

SECTION 10.2 *Orthogonal Sets of Vectors*

The orthogonal complement of a subspace U of $\mathbb{R}^n$ was defined (in Chapter 8) to be the set of all vectors in $\mathbb{R}^n$ that are orthogonal to every vector in U. This notion has a natural extension in an arbitrary inner product space. Let U be a subspace of an inner product space V. As in $\mathbb{R}^n$, the **orthogonal complement** $U^{\perp}$ of U in V is defined by

$$U^{\perp} = \{\mathbf{v} \mid \mathbf{v} \text{ in } V, \langle \mathbf{v}, \mathbf{u} \rangle = 0 \text{ for all } \mathbf{u} \text{ in } U\}.$$

Theorem 6

Let U be a finite dimensional subspace of an inner product space V.
1. $U^{\perp}$ is a subspace of V and $V = U \oplus U^{\perp}$.
2. If $\dim V = n$, then $\dim U + \dim U^{\perp} = n$.
3. If $\dim V = n$, then $U^{\perp\perp} = U$.

PROOF

1. $U^{\perp}$ is a subspace by Theorem 1 Section 10.1. If $\mathbf{v}$ is in $U \cap U^{\perp}$, then $\langle \mathbf{v}, \mathbf{v} \rangle = 0$, so $\mathbf{v} = \mathbf{0}$ again by Theorem 1 Section 10.1. Hence $U \cap U^{\perp} = \{\mathbf{0}\}$, and it remains to show that $U + U^{\perp} = V$. Given $\mathbf{v}$ in V, we must show that $\mathbf{v}$ is in $U + U^{\perp}$, and this is clear if $\mathbf{v}$ is in U. If $\mathbf{v}$ is not in U, let $\{\mathbf{f}_1, \mathbf{f}_2, \ldots, \mathbf{f}_m\}$ be an orthogonal basis of U. Then the orthogonal lemma shows that $\mathbf{v} - \left(\dfrac{\langle \mathbf{v}, \mathbf{f}_1 \rangle}{\|\mathbf{f}_1\|^2} \mathbf{f}_1 + \dfrac{\langle \mathbf{v}, \mathbf{f}_2 \rangle}{\|\mathbf{f}_2\|^2} \mathbf{f}_2 + \cdots + \dfrac{\langle \mathbf{v}, \mathbf{f}_m \rangle}{\|\mathbf{f}_m\|^2} \mathbf{f}_m \right)$ is in $U^{\perp}$, so $\mathbf{v}$ is in $U + U^{\perp}$ as required.

2. This follows from Theorem 6 Section 9.3.

3. We have $\dim U^{\perp\perp} = n - \dim U^{\perp} = n - (n - \dim U) = \dim U$, using (2) twice. As $U \subseteq U^{\perp\perp}$ always holds (verify), (3) follows by Theorem 2 Section 6.4.

We digress briefly and consider a subspace U of an arbitrary vector space V. As in Section 9.3, if W is any complement of U in V, that is, $V = U \oplus W$, then each vector $\mathbf{v}$ in V has a *unique* representation as a sum $\mathbf{v} = \mathbf{u} + \mathbf{w}$ where $\mathbf{u}$ is in U and $\mathbf{w}$ is in W. Hence we may define a function $T : V \to V$ as follows:

$$T(\mathbf{v}) = \mathbf{u} \quad \text{where } \mathbf{v} = \mathbf{u} + \mathbf{w}, \mathbf{u} \text{ in } U, \mathbf{w} \text{ in } W$$

Thus, to compute $T(\mathbf{v})$, express $\mathbf{v}$ in any way at all as the sum of a vector $\mathbf{u}$ in U and a vector in W; then $T(\mathbf{v}) = \mathbf{u}$.

This function T is a linear operator on V. Indeed, if $\mathbf{v}_1 = \mathbf{u}_1 + \mathbf{w}_1$ where $\mathbf{u}_1$ is in U and $\mathbf{w}_1$ is in W, then $\mathbf{v} + \mathbf{v}_1 = (\mathbf{u} + \mathbf{u}_1) + (\mathbf{w} + \mathbf{w}_1)$ where $\mathbf{u} + \mathbf{u}_1$ is in U and $\mathbf{w} + \mathbf{w}_1$ is in W, so

$$T(\mathbf{v} + \mathbf{v}_1) = \mathbf{u} + \mathbf{u}_1 = T(\mathbf{v}) + T(\mathbf{v}_1)$$

Similarly, $T(a\mathbf{v}) = aT(\mathbf{v})$ for all a in $\mathbb{R}$, so T is a linear operator. Furthermore, $\text{im } T = U$ and $\ker T = W$ as the reader can verify, and T is called the **projection on U with kernel W**.

If U is a subspace of V, there are many projections on U, one for each complementary subspace W with $V = U \oplus W$. If V is an *inner product space*, we single out one for special attention. Let U be a finite dimensional subspace of an inner product space V.

Definition 10.3 *The projection on U with kernel $U^\perp$ is called the **orthogonal projection** on U (or simply the **projection** on U) and is denoted $\text{proj}_U : V \to V$.*

Theorem 7

Projection Theorem
Let U be a finite dimensional subspace of an inner product space V and let $\mathbf{v}$ be a vector in V.

1. $\text{proj}_U : V \to V$ is a linear operator with image U and kernel $U^\perp$.
2. $\text{proj}_U(\mathbf{v})$ is in U and $\mathbf{v} - \text{proj}_U(\mathbf{v})$ is in $U^\perp$.
3. If $\{\mathbf{f}_1, \mathbf{f}_2, \ldots, \mathbf{f}_m\}$ is any orthogonal basis of U, then

$$\text{proj}_U(\mathbf{v}) = \frac{\langle \mathbf{v}, \mathbf{f}_1 \rangle}{\|\mathbf{f}_1\|^2} \mathbf{f}_1 + \frac{\langle \mathbf{v}, \mathbf{f}_2 \rangle}{\|\mathbf{f}_2\|^2} \mathbf{f}_2 + \cdots + \frac{\langle \mathbf{v}, \mathbf{f}_m \rangle}{\|\mathbf{f}_m\|^2} \mathbf{f}_m.$$

PROOF

Only (3) remains to be proved. But since $\{\mathbf{f}_1, \mathbf{f}_2, \ldots, \mathbf{f}_n\}$ is an orthogonal basis of U and since $\text{proj}_U(\mathbf{v})$ is in U, the result follows from the expansion theorem (Theorem 4) applied to the finite dimensional space U.

Note that there is no requirement in Theorem 7 that V is finite dimensional.

EXAMPLE 5

Let U be a subspace of the finite dimensional inner product space V. Show that $\text{proj}_{U^\perp}(\mathbf{v}) = \mathbf{v} - \text{proj}_U(\mathbf{v})$ for all $\mathbf{v}$ in V.

Solution ▶ We have $V = U^\perp \oplus U^{\perp\perp}$ by Theorem 6. If we write $\mathbf{p} = \text{proj}_U(\mathbf{v})$, then $\mathbf{v} = (\mathbf{v} - \mathbf{p}) + \mathbf{p}$ where $\mathbf{v} - \mathbf{p}$ is in $U^\perp$ and $\mathbf{p}$ is in $U = U^{\perp\perp}$ by Theorem 7. Hence $\text{proj}_{U^\perp}(\mathbf{v}) = \mathbf{v} - \mathbf{p}$. See Exercise 7 Section 8.1.

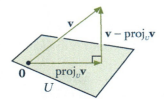

The vectors $\mathbf{v}$, $\text{proj}_U(\mathbf{v})$, and $\mathbf{v} - \text{proj}_U(\mathbf{v})$ in Theorem 7 can be visualized geometrically as in the diagram (where U is shaded and $\dim U = 2$). This suggests that $\text{proj}_U(\mathbf{v})$ is the vector in U closest to $\mathbf{v}$. This is, in fact, the case.

Theorem 8

Approximation Theorem
Let U be a finite dimensional subspace of an inner product space V. If $\mathbf{v}$ is any vector in V, then $\text{proj}_U(\mathbf{v})$ is the vector in U that is closest to $\mathbf{v}$. Here *closest* means that

$$\|\mathbf{v} - \text{proj}_U(\mathbf{v})\| < \|\mathbf{v} - \mathbf{u}\|$$

for all $\mathbf{u}$ in U, $\mathbf{u} \neq \text{proj}_U(\mathbf{v})$.

PROOF

Write $\mathbf{p} = \operatorname{proj}_U(\mathbf{v})$, and consider $\mathbf{v} - \mathbf{u} = (\mathbf{v} - \mathbf{p}) + (\mathbf{p} - \mathbf{u})$. Because $\mathbf{v} - \mathbf{p}$ is in $U^\perp$ and $\mathbf{p} - \mathbf{u}$ is in U, Pythagoras' theorem gives

$$\|\mathbf{v} - \mathbf{u}\|^2 = \|\mathbf{v} - \mathbf{p}\|^2 + \|\mathbf{p} - \mathbf{u}\|^2 > \|\mathbf{v} - \mathbf{p}\|^2$$

because $\mathbf{p} - \mathbf{u} \neq 0$. The result follows.

EXAMPLE 6

Consider the space $\mathbf{C}[-1, 1]$ of real-valued continuous functions on the interval $[-1, 1]$ with inner product $\langle f, g \rangle = \int_{-1}^{1} f(x)g(x)\,dx$. Find the polynomial $p = p(x)$ of degree at most 2 that best approximates the absolute-value function f given by $f(x) = |x|$.

Solution ▶ Here we want the vector p in the subspace $U = \mathbf{P}_2$ of $\mathbf{C}[-1, 1]$ that is closest to f. In Example 4 the Gram-Schmidt algorithm was applied to give an orthogonal basis $\{\mathbf{f}_1 = 1, \mathbf{f}_2 = x, \mathbf{f}_3 = 3x^2 - 1\}$ of $\mathbf{P}_2$ (where, for convenience, we have changed $\mathbf{f}_3$ by a numerical factor). Hence the required polynomial is

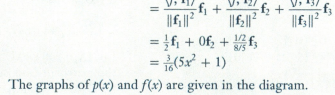

$$\begin{aligned} p &= \operatorname{proj}_{\mathbf{P}_2}(f) \\ &= \frac{\langle f, \mathbf{f}_1 \rangle}{\|\mathbf{f}_1\|^2}\mathbf{f}_1 + \frac{\langle f, \mathbf{f}_2 \rangle}{\|\mathbf{f}_2\|^2}\mathbf{f}_2 + \frac{\langle f, \mathbf{f}_3 \rangle}{\|\mathbf{f}_3\|^2}\mathbf{f}_3 \\ &= \tfrac{1}{2}\mathbf{f}_1 + 0\mathbf{f}_2 + \tfrac{1/2}{8/5}\mathbf{f}_3 \\ &= \tfrac{3}{16}(5x^2 + 1) \end{aligned}$$

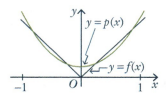

The graphs of $p(x)$ and $f(x)$ are given in the diagram.

If polynomials of degree at most n are allowed in Example 6, the polynomial in $\mathbf{P}_n$ is $\operatorname{proj}_{\mathbf{P}_n}(f)$, and it is calculated in the same way. Because the subspaces $\mathbf{P}_n$ get larger as n increases, it turns out that the approximating polynomials $\operatorname{proj}_{\mathbf{P}_n}(f)$ get closer and closer to f. In fact, solving many practical problems comes down to approximating some interesting vector $\mathbf{v}$ (often a function) in an infinite dimensional inner product space V by vectors in finite dimensional subspaces (which can be computed). If $U_1 \subseteq U_2$ are finite dimensional subspaces of V, then

$$\|\mathbf{v} - \operatorname{proj}_{U_2}(\mathbf{v})\| < \|\mathbf{v} - \operatorname{proj}_{U_1}(\mathbf{v})\|$$

by Theorem 8 (because $\operatorname{proj}_{U_1}(\mathbf{v})$ lies in U_1 and hence in U_2). Thus $\operatorname{proj}_{U_2}(\mathbf{v})$ is a better approximation to $\mathbf{v}$ than $\operatorname{proj}_{U_1}(\mathbf{v})$. Hence a general method in approximation theory might be described as follows: Given $\mathbf{v}$, use it to construct a sequence of finite dimensional subspaces

$$U_1 \subseteq U_2 \subseteq U_3 \subseteq \cdots$$

of V in such a way that $\|\mathbf{v} - \operatorname{proj}_{U_k}(\mathbf{v})\|$ approaches zero as k increases. Then $\operatorname{proj}_{U_k}(\mathbf{v})$ is a suitable approximation to $\mathbf{v}$ if k is large enough. For more information, the interested reader may wish to consult *Interpolation and Approximation* by Philip J. Davis (New York: Blaisdell, 1963).

EXERCISES 10.2

Use the dot product in $\mathbb{R}^n$ unless otherwise instructed.

1. In each case, verify that B is an orthogonal basis of V with the given inner product and use the expansion theorem to express $\mathbf{v}$ as a linear combination of the basis vectors.

   (a) $\mathbf{v} = \begin{bmatrix} a \\ b \end{bmatrix}$, $B = \left\{ \begin{bmatrix} 1 \\ -1 \end{bmatrix}, \begin{bmatrix} 1 \\ 0 \end{bmatrix} \right\}$, $V = \mathbb{R}^2$,

   $\langle \mathbf{v}, \mathbf{w} \rangle = \mathbf{v}^T A \mathbf{w}$ where $A = \begin{bmatrix} 2 & 2 \\ 2 & 5 \end{bmatrix}$

   ◆(b) $\mathbf{v} = \begin{bmatrix} a \\ b \\ c \end{bmatrix}$, $B = \left\{ \begin{bmatrix} 1 \\ 1 \\ 1 \end{bmatrix}, \begin{bmatrix} -1 \\ 0 \\ 1 \end{bmatrix}, \begin{bmatrix} 1 \\ -6 \\ 1 \end{bmatrix} \right\}$, $V = \mathbb{R}^3$,

   $\langle \mathbf{v}, \mathbf{w} \rangle = \mathbf{v}^T A \mathbf{w}$ where $A = \begin{bmatrix} 2 & 0 & 1 \\ 0 & 1 & 0 \\ 1 & 0 & 2 \end{bmatrix}$

   (c) $\mathbf{v} = a + bx + cx^2$, $B = \{1, x, 2 - 3x^2\}$, $V = \mathbf{P}_2$, $\langle p, q \rangle = p(0)q(0) + p(1)q(1) + p(-1)q(-1)$

   ◆(d) $\mathbf{v} = \begin{bmatrix} a & b \\ c & d \end{bmatrix}$,

   $B = \left\{ \begin{bmatrix} 1 & 0 \\ 0 & 1 \end{bmatrix}, \begin{bmatrix} 1 & 0 \\ 0 & -1 \end{bmatrix}, \begin{bmatrix} 0 & 1 \\ 1 & 0 \end{bmatrix}, \begin{bmatrix} 0 & 1 \\ -1 & 0 \end{bmatrix} \right\}$,

   $V = \mathbf{M}_{22}$, $\langle X, Y \rangle = \text{tr}(XY^T)$

2. Let $\mathbb{R}^3$ have the inner product
   $\langle (x, y, z), (x', y', z') \rangle = 2xx' + yy' + 3zz'$.
   In each case, use the Gram-Schmidt algorithm to transform B into an orthogonal basis.

   (a) $B = \{(1, 1, 0), (1, 0, 1), (0, 1, 1)\}$

   ◆(b) $B = \{(1, 1, 1), (1, -1, 1), (1, 1, 0)\}$

3. Let $\mathbf{M}_{22}$ have the inner product $\langle X, Y \rangle = \text{tr}(XY^T)$. In each case, use the Gram-Schmidt algorithm to transform B into an orthogonal basis.

   (a) $B = \left\{ \begin{bmatrix} 1 & 1 \\ 0 & 0 \end{bmatrix}, \begin{bmatrix} 1 & 0 \\ 1 & 0 \end{bmatrix}, \begin{bmatrix} 0 & 1 \\ 0 & 1 \end{bmatrix}, \begin{bmatrix} 1 & 0 \\ 0 & 1 \end{bmatrix} \right\}$

   ◆(b) $B = \left\{ \begin{bmatrix} 1 & 1 \\ 0 & 1 \end{bmatrix}, \begin{bmatrix} 1 & 0 \\ 1 & 1 \end{bmatrix}, \begin{bmatrix} 1 & 0 \\ 0 & 1 \end{bmatrix}, \begin{bmatrix} 1 & 0 \\ 0 & 0 \end{bmatrix} \right\}$

4. In each case, use the Gram-Schmidt process to convert the basis $B = \{1, x, x^2\}$ into an orthogonal basis of $\mathbf{P}_2$.

   (a) $\langle p, q \rangle = p(0)q(0) + p(1)q(1) + p(2)q(2)$

   ◆(b) $\langle p, q \rangle = \int_0^2 p(x)q(x)\,dx$

5. Show that $\{1, x - \frac{1}{2}, x^2 - x + \frac{1}{6}\}$, is an orthogonal basis of $\mathbf{P}_2$ with the inner product $\langle p, q \rangle = \int_0^1 p(x)q(x)\,dx$, and find the corresponding orthonormal basis.

6. In each case find $U^\perp$ and compute $\dim U$ and $\dim U^\perp$.

   (a) $U = \text{span}\{(1, 1, 2, 0), (3, -1, 2, 1), (1, -3, -2, 1)\}$ in $\mathbb{R}^4$

   ◆(b) $U = \text{span}\{(1, 1, 0, 0)\}$ in $\mathbb{R}^4$

   (c) $U = \text{span}\{1, x\}$ in $\mathbf{P}_2$ with $\langle p, q \rangle = p(0)q(0) + p(1)q(1) + p(2)q(2)$

   ◆(d) $U = \text{span}\{x\}$ in $\mathbf{P}_2$ with $\langle p, q \rangle = \int_0^1 p(x)q(x)\,dx$

   (e) $U = \text{span}\left\{ \begin{bmatrix} 1 & 0 \\ 0 & 1 \end{bmatrix}, \begin{bmatrix} 1 & 1 \\ 0 & 0 \end{bmatrix} \right\}$ in $\mathbf{M}_{22}$ with $\langle X, Y \rangle = \text{tr}(XY^T)$

   ◆(f) $U = \text{span}\left\{ \begin{bmatrix} 1 & 1 \\ 0 & 0 \end{bmatrix}, \begin{bmatrix} 1 & 0 \\ 1 & 0 \end{bmatrix}, \begin{bmatrix} 1 & 0 \\ 1 & 1 \end{bmatrix} \right\}$ in $\mathbf{M}_{22}$ with $\langle X, Y \rangle = \text{tr}(XY^T)$

7. Let $\langle X, Y \rangle = \text{tr}(XY^T)$ in $\mathbf{M}_{22}$. In each case find the matrix in U closest to A.

   (a) $U = \text{span}\left\{ \begin{bmatrix} 1 & 0 \\ 0 & 1 \end{bmatrix}, \begin{bmatrix} 1 & 1 \\ 1 & 1 \end{bmatrix} \right\}$, $A = \begin{bmatrix} 1 & -1 \\ 2 & 3 \end{bmatrix}$

   ◆(b) $U = \text{span}\left\{ \begin{bmatrix} 1 & 0 \\ 0 & 1 \end{bmatrix}, \begin{bmatrix} 1 & 1 \\ 1 & -1 \end{bmatrix}, \begin{bmatrix} 1 & 1 \\ 0 & 0 \end{bmatrix} \right\}$,

   $A = \begin{bmatrix} 2 & 1 \\ 3 & 2 \end{bmatrix}$

8. Let $\langle p(x), q(x) \rangle = p(0)q(0) + p(1)q(1) + p(2)q(2)$ in $\mathbf{P}_2$. In each case find the polynomial in U closest to $f(x)$.

   (a) $U = \text{span}\{1 + x, x^2\}$, $f(x) = 1 + x^2$

   ◆(b) $U = \text{span}\{1, 1 + x^2\}$; $f(x) = x$

9. Using the inner product $\langle p, q \rangle = \int_0^1 p(x)q(x)\,dx$ on $\mathbf{P}_2$, write $\mathbf{v}$ as the sum of a vector in U and a vector in $U^\perp$.

   (a) $\mathbf{v} = x^2$, $U = \text{span}\{x + 1, 9x - 5\}$

   ◆(b) $\mathbf{v} = x^2 + 1$, $U = \text{span}\{1, 2x - 1\}$

10. (a) Show that $\{\mathbf{u}, \mathbf{v}\}$ is orthogonal if and only if $\|\mathbf{u} + \mathbf{v}\|^2 = \|\mathbf{u}\|^2 + \|\mathbf{v}\|^2$.

(b) If $\mathbf{u} = \mathbf{v} = (1, 1)$ and $\mathbf{w} = (-1, 0)$, show that $\|\mathbf{u} + \mathbf{v} + \mathbf{w}\|^2 = \|\mathbf{u}\|^2 + \|\mathbf{v}\|^2 + \|\mathbf{w}\|^2$ but $\{\mathbf{u}, \mathbf{v}, \mathbf{w}\}$ is *not* orthogonal. Hence the converse to Pythagoras' theorem need not hold for more than two vectors.

11. Let $\mathbf{v}$ and $\mathbf{w}$ be vectors in an inner product space V. Show that:

(a) $\mathbf{v}$ is orthogonal to $\mathbf{w}$ if and only if $\|\mathbf{v} + \mathbf{w}\| = \|\mathbf{v} - \mathbf{w}\|$.

◆(b) $\mathbf{v} + \mathbf{w}$ and $\mathbf{v} - \mathbf{w}$ are orthogonal if and only if $\|\mathbf{v}\| = \|\mathbf{w}\|$.

12. Let U and W be subspaces of an n-dimensional inner product space V. If $\dim U + \dim W = n$ and $\langle \mathbf{u}, \mathbf{v} \rangle = 0$ for all $\mathbf{u}$ in U and $\mathbf{w}$ in W, show that $U^\perp = W$.

13. If U and W are subspaces of an inner product space, show that $(U + W)^\perp = U^\perp \cap W^\perp$.

14. If X is any set of vectors in an inner product space V, define
$X^\perp = \{\mathbf{v} \mid \mathbf{v} \text{ in } V, \langle \mathbf{v}, \mathbf{x} \rangle = 0 \text{ for all } \mathbf{x} \text{ in } X\}$.

(a) Show that $X^\perp$ is a subspace of V.

◆(b) If $U = \text{span}\{\mathbf{u}_1, \mathbf{u}_2, \ldots, \mathbf{u}_m\}$, show that $U^\perp = \{\mathbf{u}_1, \ldots, \mathbf{u}_m\}^\perp$.

(c) If $X \subseteq Y$, show that $Y^\perp \subseteq X^\perp$.

(d) Show that $X^\perp \cap Y^\perp = (X \cup Y)^\perp$.

15. If $\dim V = n$ and $\mathbf{w} \neq \mathbf{0}$ in V, show that $\dim\{\mathbf{v} \mid \mathbf{v} \text{ in } V, \langle \mathbf{v}, \mathbf{w} \rangle = 0\} = n - 1$.

16. If the Gram-Schmidt process is used on an orthogonal basis $\{\mathbf{v}_1, \ldots, \mathbf{v}_n\}$ of V, show that $\mathbf{f}_k = \mathbf{v}_k$ holds for each $k = 1, 2, \ldots, n$. That is, show that the algorithm reproduces the same basis.

17. If $\{\mathbf{f}_1, \mathbf{f}_2, \ldots, \mathbf{f}_{n-1}\}$ is orthonormal in an inner product space of dimension n, prove that there are exactly two vectors $\mathbf{f}_n$ such that $\{\mathbf{f}_1, \mathbf{f}_2, \ldots, \mathbf{f}_{n-1}, \mathbf{f}_n\}$ is an orthonormal basis.

18. Let U be a finite dimensional subspace of an inner product space V, and let $\mathbf{v}$ be a vector in V.

(a) Show that $\mathbf{v}$ lies in U if and only if $\mathbf{v} = \text{proj}_U(\mathbf{v})$.

◆(b) If $V = \mathbb{R}^3$, show that $(-5, 4, -3)$ lies in $\text{span}\{(3, -2, 5), (-1, 1, 1)\}$ but that $(-1, 0, 2)$ does not.

19. Let $\mathbf{n} \neq \mathbf{0}$ and $\mathbf{w} \neq \mathbf{0}$ be nonparallel vectors in $\mathbb{R}^3$ (as in Chapter 4).

(a) Show that $\left\{\mathbf{n}, \mathbf{n} \times \mathbf{w}, \mathbf{w} - \frac{\mathbf{n} \cdot \mathbf{w}}{\|\mathbf{n}\|^2}\mathbf{n}\right\}$ is an orthogonal basis of $\mathbb{R}^3$.

◆(b) Show that $\text{span}\left\{\mathbf{n} \times \mathbf{w}, \mathbf{w} - \frac{\mathbf{n} \cdot \mathbf{w}}{\|\mathbf{n}\|^2}\mathbf{n}\right\}$ is the plane through the origin with normal $\mathbf{n}$.

20. Let $E = \{\mathbf{f}_1, \mathbf{f}_2, \ldots, \mathbf{f}_n\}$ be an orthonormal basis of V.

(a) Show that $\langle \mathbf{v}, \mathbf{w} \rangle = C_E(\mathbf{v}) \cdot C_E(\mathbf{w})$ for all $\mathbf{v}, \mathbf{w}$ in V.

◆(b) If $P = [p_{ij}]$ is an $n \times n$ matrix, define $\mathbf{b}_i = p_{i1}\mathbf{f}_1 + \cdots + p_{in}\mathbf{f}_n$ for each i. Show that $B = \{\mathbf{b}_1, \mathbf{b}_2, \ldots, \mathbf{b}_n\}$ is an orthonormal basis if and only if P is an orthogonal matrix.

21. Let $\{\mathbf{f}_1, \ldots, \mathbf{f}_n\}$ be an orthogonal basis of V. If $\mathbf{v}$ and $\mathbf{w}$ are in V, show that
$$\langle \mathbf{v}, \mathbf{w} \rangle = \frac{\langle \mathbf{v}, \mathbf{f}_1 \rangle \langle \mathbf{w}, \mathbf{f}_1 \rangle}{\|\mathbf{f}_1\|^2} + \cdots + \frac{\langle \mathbf{v}, \mathbf{f}_n \rangle \langle \mathbf{w}, \mathbf{f}_n \rangle}{\|\mathbf{f}_n\|^2}.$$

22. Let $\{\mathbf{f}_1, \ldots, \mathbf{f}_n\}$ be an orthonormal basis of V, and let $\mathbf{v} = v_1\mathbf{f}_1 + \cdots + v_n\mathbf{f}_n$ and $\mathbf{w} = w_1\mathbf{f}_1 + \cdots + w_n\mathbf{f}_n$. Show that $\langle \mathbf{v}, \mathbf{w} \rangle = v_1w_1 + \cdots + v_nw_n$ and $\|\mathbf{v}\|^2 = v_1^2 + \cdots + v_n^2$ (**Parseval's formula**).

23. Let $\mathbf{v}$ be a vector in an inner product space V.

(a) Show that $\|\mathbf{v}\| \geq \|\text{proj}_U(\mathbf{v})\|$ holds for all finite dimensional subspaces U. [*Hint*: Pythagoras' theorem.]

◆(b) If $\{\mathbf{f}_1, \mathbf{f}_2, \ldots, \mathbf{f}_m\}$ is any orthogonal set in V, prove **Bessel's inequality**:
$$\frac{\langle \mathbf{v}, \mathbf{f}_1 \rangle^2}{\|\mathbf{f}_1\|^2} + \cdots + \frac{\langle \mathbf{v}, \mathbf{f}_m \rangle^2}{\|\mathbf{f}_m\|^2} \leq \|\mathbf{v}\|^2$$

24. Let $B = \{\mathbf{f}_1, \mathbf{f}_2, \ldots, \mathbf{f}_n\}$ be an orthogonal basis of an inner product space V. Given $\mathbf{v} \in V$, let θ_i be the angle between $\mathbf{v}$ and $\mathbf{f}_i$ for each i (see Exercise 31 Section 10.1). Show that $\cos^2 \theta_1 + \cos^2 \theta_2 + \cdots + \cos^2 \theta_n = 1$. [The $\cos \theta_i$ are called **direction cosines** for $\mathbf{v}$ corresponding to B.]

25. (a) Let S denote a set of vectors in a finite dimensional inner product space V, and suppose that $\langle \mathbf{u}, \mathbf{v} \rangle = 0$ for all $\mathbf{u}$ in S implies $\mathbf{v} = \mathbf{0}$. Show that $V = \operatorname{span} S$. [*Hint*: Write $U = \operatorname{span} S$ and use Theorem 6.]

(b) Let $A_1, A_2, \ldots, A_k$ be $n \times n$ matrices. Show that the following are equivalent.

   (i) If $A_i \mathbf{b} = \mathbf{0}$ for all i (where $\mathbf{b}$ is a column in $\mathbb{R}^n$), then $\mathbf{b} = \mathbf{0}$.

   (ii) The set of all rows of the matrices A_i spans $\mathbb{R}^n$.

26. Let $[x_i] = (x_1, x_2, \ldots)$ denote a sequence of real numbers x_i, and let $V = \{[x_i] \mid \text{only finitely many } x_i \neq 0\}$. Define componentwise addition and scalar multiplication on V as follows: $[x_i] + [y_i] = [x_i + y_i]$, and $a[x_i] = [ax_i]$ for a in $\mathbb{R}$.

Given $[x_i]$ and $[y_i]$ in V, define $\langle [x_i], [y_i] \rangle = \sum_{i=0}^{\infty} x_i y_i$.

(Note that this makes sense since only finitely many x_i and y_i are nonzero.) Finally define

$$U = \{[x_i] \text{ in } V \mid \sum_{i=0}^{\infty} x_i = 0\}.$$

(a) Show that V is a vector space and that U is a subspace.

(b) Show that $\langle , \rangle$ is an inner product on V.

(c) Show that $U^{\perp} = \{\mathbf{0}\}$.

(d) Hence show that $U \oplus U^{\perp} \neq V$ and $U \neq U^{\perp\perp}$.

SECTION 10.3 Orthogonal Diagonalization

There is a natural way to define a symmetric linear operator T on a finite dimensional inner product space V. If T is such an operator, it is shown in this section that V has an orthogonal basis consisting of eigenvectors of T. This yields another proof of the principal axis theorem in the context of inner product spaces.

Theorem 1

Let $T: V \to V$ be a linear operator on a finite dimensional space V. Then the following conditions are equivalent.

1. V has a basis consisting of eigenvectors of T.
2. There exists a basis B of V such that $M_B(T)$ is diagonal.

PROOF

We have $M_B(T) = [C_B[T(\mathbf{b}_1)] \ C_B[T(\mathbf{b}_2)] \ \cdots \ C_B[T(\mathbf{b}_n)]]$ where $B = \{\mathbf{b}_1, \mathbf{b}_2, \ldots, \mathbf{b}_n\}$ is any basis of V. By comparing columns:

$$M_B(T) = \begin{bmatrix} \lambda_1 & 0 & \cdots & 0 \\ 0 & \lambda_2 & \cdots & 0 \\ \vdots & \vdots & & \vdots \\ 0 & 0 & \cdots & \lambda_n \end{bmatrix} \quad \text{if and only if} \quad T(\mathbf{b}_i) = \lambda_i \mathbf{b}_i \text{ for each } i$$

Theorem 1 follows.

Definition 10.4

A linear operator T on a finite dimensional space V is called **diagonalizable** if V has a basis consisting of eigenvectors of T.

SECTION 10.3 Orthogonal Diagonalization

EXAMPLE 1

Let $T: \mathbf{P}_2 \to \mathbf{P}_2$ be given by
$$T(a + bx + cx^2) = (a + 4c) - 2bx + (3a + 2c)x^2$$
Find the eigenspaces of T and hence find a basis of eigenvectors.

Solution ▶ If $B_0 = \{1, x, x^2\}$, then
$$M_{B_0}(T) = \begin{bmatrix} 1 & 0 & 4 \\ 0 & -2 & 0 \\ 3 & 0 & 2 \end{bmatrix}$$
so $c_T(x) = (x + 2)^2(x - 5)$, and the eigenvalues of T are $\lambda = -2$ and $\lambda = 5$.

One sees that $\left\{ \begin{bmatrix} 0 \\ 1 \\ 0 \end{bmatrix}, \begin{bmatrix} 4 \\ 0 \\ -3 \end{bmatrix}, \begin{bmatrix} 1 \\ 0 \\ 1 \end{bmatrix} \right\}$ is a basis of eigenvectors of $M_{B_0}(T)$, so $B = \{x, 4 - 3x^2, 1 + x^2\}$ is a basis of $\mathbf{P}_2$ consisting of eigenvectors of T.

If V is an inner product space, the expansion theorem gives a simple formula for the matrix of a linear operator with respect to an orthogonal basis.

Theorem 2

Let $T: V \to V$ be a linear operator on an inner product space V. If $B = \{\mathbf{b}_1, \mathbf{b}_2, \ldots, \mathbf{b}_n\}$ is an orthogonal basis of V, then
$$M_B(T) = \left[\frac{\langle \mathbf{b}_i, T(\mathbf{b}_j) \rangle}{\|\mathbf{b}_i\|^2} \right]$$

PROOF

Write $M_B(T) = [a_{ij}]$. The jth column of $M_B(T)$ is $C_B[T(\mathbf{e}_j)]$, so
$$T(\mathbf{b}_j) = a_{1j}\mathbf{b}_1 + \cdots + a_{ij}\mathbf{b}_i + \cdots + a_{nj}\mathbf{b}_n$$
On the other hand, the expansion theorem (Theorem 4 Section 10.2) gives
$$\mathbf{v} = \frac{\langle \mathbf{b}_1, \mathbf{v} \rangle}{\|\mathbf{b}_1\|^2}\mathbf{b}_1 + \cdots + \frac{\langle \mathbf{b}_i, \mathbf{v} \rangle}{\|\mathbf{b}_i\|^2}\mathbf{b}_i + \cdots + \frac{\langle \mathbf{b}_n, \mathbf{v} \rangle}{\|\mathbf{b}_n\|^2}\mathbf{b}_n$$
for any $\mathbf{v}$ in V. The result follows by taking $\mathbf{v} = T(\mathbf{b}_j)$.

EXAMPLE 2

Let $T: \mathbb{R}^3 \to \mathbb{R}^3$ be given by
$$T(a, b, c) = (a + 2b - c, 2a + 3c, -a + 3b + 2c)$$
If the dot product in $\mathbb{R}^3$ is used, find the matrix of T with respect to the standard basis $B = \{\mathbf{e}_1, \mathbf{e}_2, \mathbf{e}_3\}$ where $\mathbf{e}_1 = (1, 0, 0)$, $\mathbf{e}_2 = (0, 1, 0)$, $\mathbf{e}_3 = (0, 0, 1)$.

Solution ▶ The basis B is orthonormal, so Theorem 2 gives

$$M_B(T) = \begin{bmatrix} \mathbf{e}_1 \cdot T(\mathbf{e}_1) & \mathbf{e}_1 \cdot T(\mathbf{e}_2) & \mathbf{e}_1 \cdot T(\mathbf{e}_3) \\ \mathbf{e}_2 \cdot T(\mathbf{e}_1) & \mathbf{e}_2 \cdot T(\mathbf{e}_2) & \mathbf{e}_2 \cdot T(\mathbf{e}_3) \\ \mathbf{e}_3 \cdot T(\mathbf{e}_1) & \mathbf{e}_3 \cdot T(\mathbf{e}_2) & \mathbf{e}_3 \cdot T(\mathbf{e}_3) \end{bmatrix} = \begin{bmatrix} 1 & 2 & -1 \\ 2 & 0 & 3 \\ -1 & 3 & 2 \end{bmatrix}$$

Of course, this can also be found in the usual way.

It is not difficult to verify that an $n \times n$ matrix A is symmetric if and only if $\mathbf{x} \cdot (A\mathbf{y}) = (A\mathbf{x}) \cdot \mathbf{y}$ holds for all columns $\mathbf{x}$ and $\mathbf{y}$ in $\mathbb{R}^n$. The analog for operators is as follows:

Theorem 3

Let V be a finite dimensional inner product space. The following conditions are equivalent for a linear operator $T : V \to V$.

1. $\langle \mathbf{v}, T(\mathbf{w}) \rangle = \langle T(\mathbf{v}), \mathbf{w} \rangle$ for all $\mathbf{v}$ and $\mathbf{w}$ in V.
2. The matrix of T is symmetric with respect to every orthonormal basis of V.
3. The matrix of T is symmetric with respect to some orthonormal basis of V.
4. There is an orthonormal basis $B = \{\mathbf{f}_1, \mathbf{f}_2, \ldots, \mathbf{f}_n\}$ of V such that $\langle \mathbf{f}_i, T(\mathbf{f}_j) \rangle = \langle T(\mathbf{f}_i), \mathbf{f}_j \rangle$ holds for all i and j.

PROOF

(1) $\Rightarrow$ (2). Let $B = \{\mathbf{f}_1, \ldots, \mathbf{f}_n\}$ be an orthonormal basis of V, and write $M_B(T) = [a_{ij}]$. Then $a_{ij} = \langle \mathbf{f}_i, T(\mathbf{f}_j) \rangle$ by Theorem 2. Hence (1) and axiom P2 give

$$a_{ij} = \langle \mathbf{f}_i, T(\mathbf{f}_j) \rangle = \langle T(\mathbf{f}_i), \mathbf{f}_j \rangle = \langle \mathbf{f}_j, T(\mathbf{f}_i) \rangle = a_{ji}$$

for all i and j. This shows that $M_B(T)$ is symmetric.

(2) $\Rightarrow$ (3). This is clear.

(3) $\Rightarrow$ (4). Let $B = \{\mathbf{f}_1, \ldots, \mathbf{f}_n\}$ be an orthonormal basis of V such that $M_B(T)$ is symmetric. By (3) and Theorem 2, $\langle \mathbf{f}_i, T(\mathbf{f}_j) \rangle = \langle \mathbf{f}_j, T(\mathbf{f}_i) \rangle$ for all i and j, so (4) follows from axiom P2.

(4) $\Rightarrow$ (1). Let $\mathbf{v}$ and $\mathbf{w}$ be vectors in V and write them as $\mathbf{v} = \sum_{i=1}^{n} v_i \mathbf{f}_i$ and $\mathbf{w} = \sum_{j=1}^{n} w_j \mathbf{f}_j$. Then

$$\langle \mathbf{v}, T(\mathbf{w}) \rangle = \left\langle \sum_i v_i \mathbf{f}_i, \sum_j w_j T(\mathbf{f}_j) \right\rangle = \sum_i \sum_j v_i w_j \langle \mathbf{f}_i, T(\mathbf{f}_j) \rangle$$

$$= \sum_i \sum_j v_i w_j \langle T(\mathbf{f}_i), \mathbf{f}_j \rangle$$

$$= \left\langle \sum_i v_i T(\mathbf{f}_i), \sum_j w_j \mathbf{f}_j \right\rangle$$

$$= \langle T(\mathbf{v}), \mathbf{w} \rangle$$

where we used (4) at the third stage. This proves (1).

SECTION 10.3 Orthogonal Diagonalization

A linear operator T on an inner product space V is called **symmetric** if $\langle \mathbf{v}, T(\mathbf{w}) \rangle = \langle T(\mathbf{v}), \mathbf{w} \rangle$ holds for all $\mathbf{v}$ and $\mathbf{w}$ in V.

EXAMPLE 3

If A is an $n \times n$ matrix, let $T_A : \mathbb{R}^n \to \mathbb{R}^n$ be the matrix operator given by $T_A(\mathbf{v}) = A\mathbf{v}$ for all columns $\mathbf{v}$. If the dot product is used in $\mathbb{R}^n$, then T_A is a symmetric operator if and only if A is a symmetric matrix.

Solution ▶ If E is the standard basis of $\mathbb{R}^n$, then E is orthonormal when the dot product is used. We have $M_E(T_A) = A$ (by Example 4 Section 9.1), so the result follows immediately from part (3) of Theorem 3.

It is important to note that whether an operator is symmetric depends on which inner product is being used (see Exercise 2).

If V is a finite dimensional inner product space, the eigenvalues of an operator $T : V \to V$ are the same as those of $M_B(T)$ for any orthonormal basis B (see Theorem 3 Section 9.3). If T is symmetric, $M_B(T)$ is a symmetric matrix and so has real eigenvalues by Theorem 7 Section 5.5. Hence we have the following:

Theorem 4

A symmetric linear operator on a finite dimensional inner product space has real eigenvalues.

If U is a subspace of an inner product space V, recall that its orthogonal complement is the subspace $U^\perp$ of V defined by

$$U^\perp = \{\mathbf{v} \text{ in } V \mid \langle \mathbf{v}, \mathbf{u} \rangle = 0 \text{ for all } \mathbf{u} \text{ in } U\}.$$

Theorem 5

Let $T : V \to V$ be a symmetric linear operator on an inner product space V, and let U be a T-invariant subspace of V. Then:

1. *The restriction of T to U is a symmetric linear operator on U.*
2. *$U^\perp$ is also T-invariant.*

PROOF

1. U is itself an inner product space using the same inner product, and condition 1 in Theorem 3 that T is symmetric is clearly preserved.

2. If $\mathbf{v}$ is in $U^\perp$, our task is to show that $T(\mathbf{v})$ is also in $U^\perp$; that is, $\langle T(\mathbf{v}), \mathbf{u} \rangle = 0$ for all $\mathbf{u}$ in U. But if $\mathbf{u}$ is in U, then $T(\mathbf{u})$ also lies in U because U is T-invariant, so

$$\langle T(\mathbf{v}), \mathbf{u} \rangle = \langle \mathbf{v}, T(\mathbf{u}) \rangle = 0$$

using the symmetry of T and the definition of $U^\perp$.

The principal axis theorem (Theorem 2 Section 8.2) asserts that an $n \times n$ matrix A is symmetric if and only if $\mathbb{R}^n$ has an orthogonal basis of eigenvectors of A. The following result not only extends this theorem to an arbitrary n-dimensional inner product space, but the proof is much more intuitive.

Theorem 6

Principal Axis Theorem
The following conditions are equivalent for a linear operator T on a finite dimensional inner product space V.

1. *T is symmetric.*
2. *V has an orthogonal basis consisting of eigenvectors of T.*

PROOF

(1) $\Rightarrow$ (2). Assume that T is symmetric and proceed by induction on $n = \dim V$. If $n = 1$, *every* nonzero vector in V is an eigenvector of T, so there is nothing to prove. If $n \geq 2$, assume inductively that the theorem holds for spaces of dimension less than n. Let λ_1 be a real eigenvalue of T (by Theorem 4) and choose an eigenvector $\mathbf{f}_1$ corresponding to λ_1. Then $U = \mathbb{R}\mathbf{f}_1$ is T-invariant, so $U^\perp$ is also T-invariant by Theorem 5 (T is symmetric). Because $\dim U^\perp = n - 1$ (Theorem 6 Section 10.2), and because the restriction of T to $U^\perp$ is a symmetric operator (Theorem 5), it follows by induction that $U^\perp$ has an orthogonal basis $\{\mathbf{f}_2, \ldots, \mathbf{f}_n\}$ of eigenvectors of T. Hence $B = \{\mathbf{f}_1, \mathbf{f}_2, \ldots, \mathbf{f}_n\}$ is an orthogonal basis of V, which proves (2).

(2) $\Rightarrow$ (1). If $B = \{\mathbf{f}_1, \ldots, \mathbf{f}_n\}$ is a basis as in (2), then $M_B(T)$ is symmetric (indeed diagonal), so T is symmetric by Theorem 3.

The matrix version of the principal axis theorem is an immediate consequence of Theorem 6. If A is an $n \times n$ symmetric matrix, then $T_A : \mathbb{R}^n \to \mathbb{R}^n$ is a symmetric operator, so let B be an orthonormal basis of $\mathbb{R}^n$ consisting of eigenvectors of T_A (and hence of A). Then $P^T A P$ is diagonal where P is the orthogonal matrix whose columns are the vectors in B (see Theorem 4 Section 9.2).

Similarly, let $T : V \to V$ be a symmetric linear operator on the n-dimensional inner product space V and let B_0 be any convenient orthonormal basis of V. Then an orthonormal basis of eigenvectors of T can be computed from $M_{B_0}(T)$. In fact, if $P^T M_{B_0}(T) P$ is diagonal where P is orthogonal, let $B = \{\mathbf{f}_1, \ldots, \mathbf{f}_n\}$ be the vectors in V such that $C_{B_0}(\mathbf{f}_j)$ is column j of P for each j. Then B consists of eigenvectors of T by Theorem 3 Section 9.3, and they are orthonormal because B_0 is orthonormal. Indeed

$$\langle \mathbf{f}_i, \mathbf{f}_j \rangle = C_{B_0}(\mathbf{f}_i) \cdot C_{B_0}(\mathbf{f}_j)$$

holds for all i and j, as the reader can verify. Here is an example.

SECTION 10.3 Orthogonal Diagonalization

EXAMPLE 4

Let $T: \mathbf{P}_2 \to \mathbf{P}_2$ be given by

$$T(a + bx + cx^2) = (8a - 2b + 2c) + (-2a + 5b + 4c)x + (2a + 4b + 5c)x^2$$

Using the inner product $\langle a + bx + cx^2, a' + b'x + c'x^2 \rangle = aa' + bb' + cc'$, show that T is symmetric and find an orthonormal basis of $\mathbf{P}_2$ consisting of eigenvectors.

Solution ▸ If $B_0 = \{1, x, x^2\}$, then $M_{B_0}(T) = \begin{bmatrix} 8 & -2 & 2 \\ -2 & 5 & 4 \\ 2 & 4 & 5 \end{bmatrix}$ is symmetric, so

T is symmetric. This matrix was analyzed in Example 5 Section 8.2, where it was found that an *orthonormal* basis of eigenvectors is $\{\frac{1}{3}[1 \ 2 \ -2]^T, \frac{1}{3}[2 \ 1 \ 2]^T, \frac{1}{3}[-2 \ 2 \ 1]^T\}$. Because B_0 is orthonormal, the corresponding orthonormal basis of $\mathbf{P}_2$ is

$$B = \{\tfrac{1}{3}(1 + 2x - 2x^2), \tfrac{1}{3}(2 + x + 2x^2), \tfrac{1}{3}(-2 + 2x + x^2)\}.$$

EXERCISES 10.3

1. In each case, show that T is symmetric by calculating $M_B(T)$ for some orthonormal basis B.

   (a) $T: \mathbb{R}^3 \to \mathbb{R}^3$;
   $T(a, b, c) = (a - 2b, -2a + 2b + 2c, 2b - c)$;
   dot product

   ◆(b) $T: \mathbf{M}_{22} \to \mathbf{M}_{22}$; $T\begin{bmatrix} a & b \\ c & d \end{bmatrix} = \begin{bmatrix} c - a & d - b \\ a + 2c & b + 2d \end{bmatrix}$;
   inner product
   $\left\langle \begin{bmatrix} x & y \\ z & w \end{bmatrix}, \begin{bmatrix} x' & y' \\ z' & w' \end{bmatrix} \right\rangle = xx' + yy' + zz' + ww'$

   (c) $T: \mathbf{P}_2 \to \mathbf{P}_2$; $T(a + bx + cx^2)$
   $= (b + c) + (a + c)x + (a + b)x^2$; inner product
   $\langle a + bx + cx^2, a' + b'x + c'x^2 \rangle = aa' + bb' + cc'$

2. Let $T: \mathbb{R}^2 \to \mathbb{R}^2$ be given by
   $$T(a, b) = (2a + b, a - b).$$
   (a) Show that T is symmetric if the dot product is used.
   (b) Show that T is *not* symmetric if
   $\langle \mathbf{x}, \mathbf{y} \rangle = \mathbf{x} A \mathbf{y}^T$, where $A = \begin{bmatrix} 1 & 1 \\ 1 & 2 \end{bmatrix}$.
   [*Hint*: Check that $B = \{(1, 0), (1, -1)\}$ is an orthonormal basis.]

3. Let $T: \mathbb{R}^2 \to \mathbb{R}^2$ be given by
   $T(a, b) = (a - b, b - a)$.
   Use the dot product in $\mathbb{R}^2$.

   (a) Show that T is symmetric.
   (b) Show that $M_B(T)$ is *not* symmetric if the orthogonal basis $B = \{(1, 0), (0, 2)\}$ is used. Why does this not contradict Theorem 3?

4. Let V be an n-dimensional inner product space, and let T and S denote symmetric linear operators on V. Show that:

   (a) The identity operator is symmetric.
   ◆(b) rT is symmetric for all r in $\mathbb{R}$.
   (c) $S + T$ is symmetric.
   ◆(d) If T is invertible, then T^{-1} is symmetric.
   (e) If $ST = TS$, then ST is symmetric.

5. In each case, show that T is symmetric and find an orthonormal basis of eigenvectors of T.

   (a) $T: \mathbb{R}^3 \to \mathbb{R}^3$; $T(a, b, c) = (2a + 2c, 3b, 2a + 5c)$;
   use the dot product
   ◆(b) $T: \mathbb{R}^3 \to \mathbb{R}^3$; $T(a, b, c) = (7a - b, -a + 7b, 2c)$;
   use the dot product
   (c) $T: \mathbf{P}_2 \to \mathbf{P}_2$; $T(a + bx + cx^2)$
   $= 3b + (3a + 4c)x + 4bx^2$; inner product
   $\langle a + bx + cx^2, a' + b'x + c'x^2 \rangle = aa' + bb' + cc'$

◆(d) $T: \mathbf{P}_2 \to \mathbf{P}_2$;
$T(a + bx + cx^2) = (c - a) + 3bx + (a - c)x^2$;
inner product as in part (c)

6. If A is any $n \times n$ matrix, let $T_A: \mathbb{R}^n \to \mathbb{R}^n$ be given by $T_A(\mathbf{x}) = A\mathbf{x}$. Suppose an inner product on $\mathbb{R}^n$ is given by $\langle \mathbf{x}, \mathbf{y} \rangle = \mathbf{x}^T P \mathbf{y}$, where P is a positive definite matrix.

   (a) Show that T_A is symmetric if and only if $PA = A^T P$.

   (b) Use part (a) to deduce Example 3.

7. Let $T: \mathbf{M}_{22} \to \mathbf{M}_{22}$ be given by $T(X) = AX$, where A is a fixed 2×2 matrix.

   (a) Compute $M_B(T)$, where
   $$B = \left\{ \begin{bmatrix} 1 & 0 \\ 0 & 0 \end{bmatrix}, \begin{bmatrix} 0 & 0 \\ 1 & 0 \end{bmatrix}, \begin{bmatrix} 0 & 1 \\ 0 & 0 \end{bmatrix}, \begin{bmatrix} 0 & 0 \\ 0 & 1 \end{bmatrix} \right\}.$$
   Note the order!

   ◆(b) Show that $c_T(x) = [c_A(x)]^2$.

   (c) If the inner product on $\mathbf{M}_{22}$ is $\langle X, Y \rangle = \operatorname{tr}(XY^T)$, show that T is symmetric if and only if A is a symmetric matrix.

8. Let $T: \mathbb{R}^2 \to \mathbb{R}^2$ be given by $T(a, b) = (b - a, a + 2b)$. Show that T is symmetric if the dot product is used in $\mathbb{R}^2$ but that it is not symmetric if the following inner product is used:
   $$\langle \mathbf{x}, \mathbf{y} \rangle = \mathbf{x} A \mathbf{y}^T, A = \begin{bmatrix} 1 & -1 \\ -1 & 2 \end{bmatrix}.$$

9. If $T: V \to V$ is symmetric, write $T^{-1}(W) = \{\mathbf{v} \mid T(\mathbf{v}) \text{ is in } W\}$. Show that $T(U)^\perp = T^{-1}(U^\perp)$ holds for every subspace U of V.

10. Let $T: \mathbf{M}_{22} \to \mathbf{M}_{22}$ be defined by $T(X) = PXQ$, where P and Q are nonzero 2×2 matrices. Use the inner product $\langle X, Y \rangle = \operatorname{tr}(XY^T)$. Show that T is symmetric if and only if either P and Q are both symmetric or both are scalar multiples of $\begin{bmatrix} 0 & 1 \\ -1 & 0 \end{bmatrix}$. [Hint: If B is as in part (a) of Exercise 7, then $M_B(T) = \begin{bmatrix} aP & cP \\ bP & dP \end{bmatrix}$ in block form, where $Q = \begin{bmatrix} a & b \\ c & d \end{bmatrix}$.

    If $B_0 = \left\{ \begin{bmatrix} 1 & 0 \\ 0 & 0 \end{bmatrix}, \begin{bmatrix} 0 & 1 \\ 0 & 0 \end{bmatrix}, \begin{bmatrix} 0 & 0 \\ 1 & 0 \end{bmatrix}, \begin{bmatrix} 0 & 0 \\ 0 & 1 \end{bmatrix} \right\},$

    then $M_B(T) = \begin{bmatrix} pQ^T & qQ^T \\ rQ^T & sQ^T \end{bmatrix}$, where $P = \begin{bmatrix} p & q \\ r & s \end{bmatrix}$.
    Use the fact that $cP = bP^T \Rightarrow (c^2 - b^2)P = 0$.]

11. Let $T: V \to W$ be any linear transformation and let $B = \{\mathbf{b}_1, \ldots, \mathbf{b}_n\}$ and $D = \{\mathbf{d}_1, \ldots, \mathbf{d}_m\}$ be bases of V and W, respectively. If W is an inner product space and D is orthogonal, show that
    $$M_{DB}(T) = \left[\frac{\langle \mathbf{d}_i, T(\mathbf{b}_j) \rangle}{\|\mathbf{d}_i\|^2} \right]$$
    This is a generalization of Theorem 2.

12. Let $T: V \to V$ be a linear operator on an inner product space V of finite dimension. Show that the following are equivalent.

    (1) $\langle \mathbf{v}, T(\mathbf{w}) \rangle = -\langle T(\mathbf{v}), \mathbf{w} \rangle$ for all $\mathbf{v}$ and $\mathbf{w}$ in V.

    ◆(2) $M_B(T)$ is skew-symmetric for every orthonormal basis B.

    (3) $M_B(T)$ is skew-symmetric for some orthonormal basis B.

    Such operators T are called **skew-symmetric** operators.

13. Let $T: V \to V$ be a linear operator on an n-dimensional inner product space V.

    (a) Show that T is symmetric if and only if it satisfies the following two conditions.

       (i) $c_T(x)$ factors completely over $\mathbb{R}$.

       (ii) If U is a T-invariant subspace of V, then $U^\perp$ is also T-invariant.

    (b) Using the standard inner product on $\mathbb{R}^2$, show that $T: \mathbb{R}^2 \to \mathbb{R}^2$ with $T(a, b) = (a, a + b)$ satisfies condition (i) and that $S: \mathbb{R}^2 \to \mathbb{R}^2$ with $S(a, b) = (b, -a)$ satisfies condition (ii), but that neither is symmetric. (Example 4 Section 9.3 is useful for S.)

[Hint for part (a): If conditions (i) and (ii) hold, proceed by induction on n. By condition (i), let $\mathbf{e}_1$ be an eigenvector of T. If $U = \mathbb{R}\mathbf{e}_1$, then $U^\perp$ is T-invariant by condition (ii), so show that the restriction of T to $U^\perp$ satisfies conditions (i) and (ii). (Theorem 1 Section 9.3 is helpful for part (i)). Then apply induction to show that V has an orthogonal basis of eigenvectors (as in Theorem 6)].

14. Let $B = \{f_1, f_2, \ldots, f_n\}$ be an orthonormal basis of an inner product space V. Given $T : V \to V$, define $T' : V \to V$ by

$$T'(v) = \langle v, T(f_1)\rangle f_1 + \langle v, T(f_2)\rangle f_2$$
$$+ \cdots + \langle v, T(f_n)\rangle f_n = \sum_{i=1}^{n} \langle v, T(f_i)\rangle f_i$$

(a) Show that $(aT)' = aT'$.

(b) Show that $(S + T)' = S' + T'$.

♦(c) Show that $M_B(T')$ is the transpose of $M_B(T)$.

(d) Show that $(T')' = T$, using part (c). [*Hint*: $M_B(S) = M_B(T)$ implies that $S = T$.]

(e) Show that $(ST)' = T'S'$, using part (c).

(f) Show that T is symmetric if and only if $T = T'$. [*Hint*: Use the expansion theorem and Theorem 3.]

(g) Show that $T + T'$ and TT' are symmetric, using parts (b) through (e).

(h) Show that $T'(v)$ is independent of the choice of orthonormal basis B. [*Hint*: If $D = \{g_1, \ldots, g_n\}$ is also orthonormal, use the fact that $f_i = \sum_{j=1}^{n} \langle f_i, g_j \rangle g_j$ for each i.]

15. Let V be a finite dimensional inner product space. Show that the following conditions are equivalent for a linear operator $T : V \to V$.

(1) T is symmetric and $T^2 = T$.

(2) $M_B(T) = \begin{bmatrix} I_r & 0 \\ 0 & 0 \end{bmatrix}$ for some orthonormal basis B of V.

An operator is called a **projection** if it satisfies these conditions. [*Hint*: If $T^2 = T$ and $T(v) = \lambda v$, apply T to get $\lambda v = \lambda^2 v$. Hence show that 0, 1 are the only eigenvalues of T.]

16. Let V denote a finite dimensional inner product space. Given a subspace U, define $\text{proj}_U : V \to V$ as in Theorem 7 Section 10.2.

(a) Show that proj_U is a projection in the sense of Exercise 15.

(b) If T is any projection, show that $T = \text{proj}_U$, where $U = \text{im } T$. [*Hint*: Use $T^2 = T$ to show that $V = \text{im } T \oplus \ker T$ and $T(u) = u$ for all u in $\text{im } T$. Use the fact that T is symmetric to show that $\ker T \subseteq (\text{im } T)^\perp$ and hence that these are equal because they have the same dimension.]

SECTION 10.4 Isometries

We saw in Section 2.6 that rotations about the origin and reflections in a line through the origin are linear operators on $\mathbb{R}^2$. Similar geometric arguments (in Section 4.4) establish that, in $\mathbb{R}^3$, rotations about a line through the origin and reflections in a plane through the origin are linear. We are going to give an algebraic proof of these results that is valid in any inner product space. The key observation is that reflections and rotations are distance preserving in the following sense. If V is an inner product space, a transformation $S : V \to V$ (not necessarily linear) is said to be **distance preserving** if the distance between $S(v)$ and $S(w)$ is the same as the distance between v and w for all vectors v and w; more formally, if

$$\|S(v) - S(w)\| = \|v - w\| \quad \text{for all } v \text{ and } w \text{ in } V. \qquad (*)$$

Distance-preserving maps need not be linear. For example, if u is any vector in V, the transformation $S_u : V \to V$ defined by $S_u(v) = v + u$ for all v in V is called **translation** by u, and it is routine to verify that S_u is distance preserving for any u. However, S_u is linear only if $u = 0$ (since then $S_u(0) = 0$). Remarkably, distance-preserving operators that do fix the origin are necessarily linear.

Lemma 1

Let V be an inner product space of dimension n, and consider a distance-preserving transformation $S : V \to V$. If $S(0) = 0$, then S is linear.

PROOF

We have $\|S(\mathbf{v}) - S(\mathbf{w})\|^2 = \|\mathbf{v} - \mathbf{w}\|^2$ for all $\mathbf{v}$ and $\mathbf{w}$ in V by (*), which gives

$$\langle S(\mathbf{v}), S(\mathbf{w}) \rangle = \langle \mathbf{v}, \mathbf{w} \rangle \quad \text{for all } \mathbf{v} \text{ and } \mathbf{w} \text{ in } V. \quad (**)$$

Now let $\{\mathbf{f}_1, \mathbf{f}_2, \ldots, \mathbf{f}_n\}$ be an orthonormal basis of V. Then $\{S(\mathbf{f}_1), S(\mathbf{f}_2), \ldots, S(\mathbf{f}_n)\}$ is orthonormal by (**) and so is a basis because $\dim V = n$. Now compute:

$$\langle S(\mathbf{v} + \mathbf{w}) - S(\mathbf{v}) - S(\mathbf{w}), S(\mathbf{f}_i) \rangle = \langle S(\mathbf{v} + \mathbf{w}), S(\mathbf{f}_i) \rangle - \langle S(\mathbf{v}), S(\mathbf{f}_i) \rangle - \langle S(\mathbf{w}), S(\mathbf{f}_i) \rangle$$
$$= \langle \mathbf{v} + \mathbf{w}, \mathbf{f}_i \rangle - \langle \mathbf{v}, \mathbf{f}_i \rangle - \langle \mathbf{w}, \mathbf{f}_i \rangle$$
$$= 0$$

for each i. It follows from the expansion theorem (Theorem 4 Section 10.2) that $S(\mathbf{v} + \mathbf{w}) - S(\mathbf{v}) - S(\mathbf{w}) = 0$; that is, $S(\mathbf{v} + \mathbf{w}) = S(\mathbf{v}) + S(\mathbf{w})$. A similar argument shows that $S(a\mathbf{v}) = aS(\mathbf{v})$ holds for all a in $\mathbb{R}$ and $\mathbf{v}$ in V, so S is linear after all.

Definition 10.5 *Distance-preserving linear operators are called* **isometries**.

It is routine to verify that the composite of two distance-preserving transformations is again distance preserving. In particular the composite of a translation and an isometry is distance preserving. Surprisingly, the converse is true.

Theorem 1

If V is a finite dimensional inner product space, then every distance-preserving transformation $S : V \to V$ is the composite of a translation and an isometry.

PROOF

If $S : V \to V$ is distance preserving, write $S(\mathbf{0}) = \mathbf{u}$ and define $T : V \to V$ by $T(\mathbf{v}) = S(\mathbf{v}) - \mathbf{u}$ for all $\mathbf{v}$ in V. Then $\|T(\mathbf{v}) - T(\mathbf{w})\| = \|\mathbf{v} - \mathbf{w}\|$ for all vectors $\mathbf{v}$ and $\mathbf{w}$ in V as the reader can verify; that is, T is distance preserving. Clearly, $T(\mathbf{0}) = \mathbf{0}$, so it is an isometry by Lemma 1. Since $S(\mathbf{v}) = \mathbf{u} + T(\mathbf{v}) = (S_{\mathbf{u}} \circ T)(\mathbf{v})$ for all $\mathbf{v}$ in V, we have $S = S_{\mathbf{u}} \circ T$, and the theorem is proved.

In Theorem 1, $S = S_{\mathbf{u}} \circ T$ factors as the composite of an isometry T followed by a translation $S_{\mathbf{u}}$. More is true: this factorization is unique in that $\mathbf{u}$ and T are uniquely determined by S; and $\mathbf{w} \in V$ exists such that $S = T \circ S_{\mathbf{w}}$ is uniquely the composite of translation by $\mathbf{w}$ followed by the same isometry T (Exercise 12).

Theorem 1 focuses our attention on the isometries, and the next theorem shows that, while they preserve distance, they are characterized as those operators that preserve other properties.

Theorem 2

Let $T : V \to V$ be a linear operator on a finite dimensional inner product space V. The following conditions are equivalent:

1. *T is an isometry.*   *(T preserves distance)*

2. $\|T(\mathbf{v})\| = \|\mathbf{v}\|$ for all $\mathbf{v}$ in V.   (*T* preserves norms)
3. $\langle T(\mathbf{v}), T(\mathbf{w})\rangle = \langle \mathbf{v}, \mathbf{w}\rangle$ for all $\mathbf{v}$ and $\mathbf{w}$ in V.   (*T* preserves inner products)
4. If $\{\mathbf{f}_1, \mathbf{f}_2, \ldots, \mathbf{f}_n\}$ is an orthonormal basis of V, then $\{T(\mathbf{f}_1), T(\mathbf{f}_2), \ldots, T(\mathbf{f}_n)\}$ is also an orthonormal basis.   (*T* preserves orthonormal bases)
5. *T* carries some orthonormal basis to an orthonormal basis.

PROOF

(1) $\Rightarrow$ (2). Take $\mathbf{w} = \mathbf{0}$ in (∗).

(2) $\Rightarrow$ (3). Since *T* is linear, (2) gives $\|T(\mathbf{v}) - T(\mathbf{w})\|^2 = \|T(\mathbf{v} - \mathbf{w})\|^2 = \|\mathbf{v} - \mathbf{w}\|^2$. Now (3) follows.

(3) $\Rightarrow$ (4). By (3), $\{T(\mathbf{f}_1), T(\mathbf{f}_2), \ldots, T(\mathbf{f}_n)\}$ is orthogonal and $\|T(\mathbf{f}_i)\|^2 = \|\mathbf{f}_i\|^2 = 1$. Hence it is a basis because $\dim V = n$.

(4) $\Rightarrow$ (5). This needs no proof.

(5) $\Rightarrow$ (1). By (5), let $\{\mathbf{f}_1, \ldots, \mathbf{f}_n\}$ be an orthonormal basis of V such that $\{T(\mathbf{f}_1), \ldots, T(\mathbf{f}_n)\}$ is also orthonormal. Given $\mathbf{v} = v_1\mathbf{f}_1 + \cdots + v_n\mathbf{f}_n$ in V, we have $T(\mathbf{v}) = v_1 T(\mathbf{f}_1) + \cdots + v_n T(\mathbf{f}_n)$ so Pythagoras' theorem gives

$$\|T(\mathbf{v})\|^2 = v_1^2 + \cdots + v_n^2 = \|\mathbf{v}\|^2.$$

Hence $\|T(\mathbf{v})\| = \|\mathbf{v}\|$ for all $\mathbf{v}$, and (1) follows by replacing $\mathbf{v}$ by $\mathbf{v} - \mathbf{w}$.

Before giving examples, we note some consequences of Theorem 2.

Corollary 1

Let V be a finite dimensional inner product space.
1. Every isometry of V is an isomorphism.[5]
2. (a) $1_V : V \to V$ is an isometry.
   (b) The composite of two isometries of V is an isometry.
   (c) The inverse of an isometry of V is an isometry.

PROOF

(1) is by (4) of Theorem 2 and Theorem 1 Section 7.3. (2a) is clear, and (2b) is left to the reader. If $T : V \to V$ is an isometry and $\{\mathbf{f}_1, \ldots, \mathbf{f}_n\}$ is an orthonormal basis of V, then (2c) follows because T^{-1} carries the orthonormal basis $\{T(\mathbf{f}_1), \ldots, T(\mathbf{f}_n)\}$ back to $\{\mathbf{f}_1, \ldots, \mathbf{f}_n\}$.

The conditions in part (2) of the corollary assert that the set of isometries of a finite dimensional inner product space forms an algebraic system called a **group**. The theory of groups is well developed, and groups of operators are important in geometry. In fact, geometry itself can be fruitfully viewed as the

---

[5] *V* must be finite dimensional—see Exercise 13.

study of those properties of a vector space that are preserved by a group of invertible linear operators.

EXAMPLE 1

Rotations of $\mathbb{R}^2$ about the origin are isometries, as are reflections in lines through the origin: They clearly preserve distance and so are linear by Lemma 1. Similarly, rotations about lines through the origin and reflections in planes through the origin are isometries of $\mathbb{R}^3$.

EXAMPLE 2

Let $T: \mathbf{M}_{nn} \to \mathbf{M}_{nn}$ be the transposition operator: $T(A) = A^T$. Then T is an isometry if the inner product is $\langle A, B \rangle = \text{tr}(AB^T) = \sum_{i,j} a_{ij} b_{ij}$. In fact, T permutes the basis consisting of all matrices with one entry 1 and the other entries 0.

The proof of the next result requires the fact (see Theorem 2) that, if B is an orthonormal basis, then $\langle \mathbf{v}, \mathbf{w} \rangle = C_B(\mathbf{v}) \cdot C_B(\mathbf{w})$ for all vectors $\mathbf{v}$ and $\mathbf{w}$.

Theorem 3

Let $T: V \to V$ be an operator where V is a finite dimensional inner product space. The following conditions are equivalent.
1. T is an isometry.
2. $M_B(T)$ is an orthogonal matrix for every orthonormal basis B.
3. $M_B(T)$ is an orthogonal matrix for some orthonormal basis B.

PROOF

(1) $\Rightarrow$ (2). Let $B = \{\mathbf{e}_1, \ldots, \mathbf{e}_n\}$ be an orthonormal basis. Then the jth column of $M_B(T)$ is $C_B[T(\mathbf{e}_j)]$, and we have
$$C_B[T(\mathbf{e}_j)] \cdot C_B[T(\mathbf{e}_k)] = \langle T(\mathbf{e}_j), T(\mathbf{e}_k) \rangle = \langle \mathbf{e}_j, \mathbf{e}_k \rangle$$
using (1). Hence the columns of $M_B(T)$ are orthonormal in $\mathbb{R}^n$, which proves (2).

(2) $\Rightarrow$ (3). This is clear.

(3) $\Rightarrow$ (1). Let $B = \{\mathbf{e}_1, \ldots, \mathbf{e}_n\}$ be as in (3). Then, as before,
$$\langle T(\mathbf{e}_j), T(\mathbf{e}_k) \rangle = C_B[T(\mathbf{e}_j)] \cdot C_B[T(\mathbf{e}_k)]$$
so $\{T(\mathbf{e}_1), \ldots, T(\mathbf{e}_n)\}$ is orthonormal by (3). Hence Theorem 2 gives (1).

It is important that B is *orthonormal* in Theorem 3. For example, $T: V \to V$ given by $T(\mathbf{v}) = 2\mathbf{v}$ preserves *orthogonal* sets but is not an isometry, as is easily checked.

SECTION 10.4 *Isometries*

If P is an orthogonal square matrix, then $P^{-1} = P^T$. Taking determinants yields $(\det P)^2 = 1$, so $\det P = \pm 1$. Hence:

Corollary 2

If $T: V \to V$ is an isometry where V is a finite dimensional inner product space, then $\det T = \pm 1$.

EXAMPLE 3

If A is any $n \times n$ matrix, the matrix operator $T_A: \mathbb{R}^n \to \mathbb{R}^n$ is an isometry if and only if A is orthogonal using the dot product in $\mathbb{R}^n$. Indeed, if E is the standard basis of $\mathbb{R}^n$, then $M_E(T_A) = A$ by Theorem 4, Section 9.2.

Rotations and reflections that fix the origin are isometries in $\mathbb{R}^2$ and $\mathbb{R}^3$ (Example 1); we are going to show that these isometries (and compositions of them in $\mathbb{R}^3$) are the only possibilities. In fact, this will follow from a general structure theorem for isometries. Surprisingly enough, much of the work involves the two-dimensional case.

Theorem 4

Let $T: V \to V$ be an isometry on the two-dimensional inner product space V. Then there are two possibilities.

Either (1) *There is an orthonormal basis B of V such that*

$$M_B(T) = \begin{bmatrix} \cos\theta & -\sin\theta \\ \sin\theta & \cos\theta \end{bmatrix}, \quad 0 \le \theta < 2\pi$$

or (2) *There is an orthonormal basis B of V such that*

$$M_B(T) = \begin{bmatrix} 1 & 0 \\ 0 & -1 \end{bmatrix}$$

Furthermore, type (1) occurs if and only if $\det T = 1$, and type (2) occurs if and only if $\det T = -1$.

PROOF

The final statement follows from the rest because $\det T = \det[M_B(T)]$ for any basis B. Let $B_0 = \{\mathbf{e}_1, \mathbf{e}_2\}$ be any ordered orthonormal basis of V and write

$$A = M_{B_0}(T) = \begin{bmatrix} a & b \\ c & d \end{bmatrix}; \quad \text{that is,} \quad \begin{aligned} T(\mathbf{e}_1) &= a\mathbf{e}_1 + c\mathbf{e}_2 \\ T(\mathbf{e}_2) &= b\mathbf{e}_1 + d\mathbf{e}_2 \end{aligned}$$

Then A is orthogonal by Theorem 3, so its columns (and rows) are orthonormal. Hence $a^2 + c^2 = 1 = b^2 + d^2$, so (a, c) and (d, b) lie on the unit circle. Thus angles θ and φ exist such that

$$a = \cos\theta, \quad c = \sin\theta \quad 0 \le \theta < 2\pi$$
$$d = \cos\varphi, \quad b = \sin\varphi \quad 0 \le \varphi < 2\pi$$

Then $\sin(\theta + \varphi) = cd + ab = 0$ because the columns of A are orthogonal, so $\theta + \varphi = k\pi$ for some integer k. This gives $d = \cos(k\pi - \theta) = (-1)^k \cos \theta$ and $b = \sin(k\pi - \theta) = (-1)^{k+1} \sin \theta$. Finally

$$A = \begin{bmatrix} \cos \theta & (-1)^{k+1} \sin \theta \\ \sin \theta & (-1)^k \cos \theta \end{bmatrix}$$

If k is even we are in type (1) with $B = B_0$, so assume k is odd. Then $A = \begin{bmatrix} a & c \\ c & -a \end{bmatrix}$. If $a = -1$ and $c = 0$, we are in type (1) with $B = \{\mathbf{e}_2, \mathbf{e}_2\}$. Otherwise A has eigenvalues $\lambda_1 = 1$ and $\lambda_2 = -1$ with corresponding eigenvectors $\mathbf{x}_1 = \begin{bmatrix} 1 + a \\ c \end{bmatrix}$ and $\mathbf{x}_2 = \begin{bmatrix} -c \\ 1 + a \end{bmatrix}$ as the reader can verify. Write

$$\mathbf{f}_1 = (1 + a)\mathbf{e}_1 + c\mathbf{e}_2 \quad \text{and} \quad \mathbf{f}_2 = -c\mathbf{e}_2 + (1 + a)\mathbf{e}_2.$$

Then $\mathbf{f}_1$ and $\mathbf{f}_2$ are orthogonal (verify) and $C_{B_0}(\mathbf{f}_i) = C_{B_0}(\lambda_i \mathbf{f}_i) = \mathbf{x}_i$ for each i. Moreover

$$C_{B_0}[T(\mathbf{f}_i)] = AC_{B_0}(\mathbf{f}_i) = A\mathbf{x}_i = \lambda_i \mathbf{x}_i = \lambda_i C_{B_0}(\mathbf{f}_i) = C_{B_0}(\lambda_i \mathbf{f}_i),$$

so $T(\mathbf{f}_i) = \lambda_i \mathbf{f}_i$ for each i. Hence $M_B(T) = \begin{bmatrix} \lambda_1 & 0 \\ 0 & \lambda_2 \end{bmatrix} = \begin{bmatrix} 1 & 0 \\ 0 & -1 \end{bmatrix}$ and we are in type (2) with $B = \left\{ \frac{1}{\|\mathbf{f}_1\|}\mathbf{f}_1, \frac{1}{\|\mathbf{f}_2\|}\mathbf{f}_2 \right\}$.

Corollary 3

An operator $T: \mathbb{R}^2 \to \mathbb{R}^2$ is an isometry if and only if T is a rotation or a reflection.

In fact, if E is the standard basis of $\mathbb{R}^2$, then the clockwise rotation R_θ about the origin through an angle θ has matrix

$$M_E(R_\theta) = \begin{bmatrix} \cos \theta & -\sin \theta \\ \sin \theta & \cos \theta \end{bmatrix}$$

(see Theorem 4 Section 2.6). On the other hand, if $S: \mathbb{R}^2 \to \mathbb{R}^2$ is the reflection in a line through the origin (called the **fixed line** of the reflection), let $\mathbf{f}_1$ be a unit vector pointing along the fixed line and let $\mathbf{f}_2$ be a unit vector perpendicular to the fixed line. Then $B = \{\mathbf{f}_1, \mathbf{f}_2\}$ is an orthonormal basis, $S(\mathbf{f}_1) = \mathbf{f}_1$ and $S(\mathbf{f}_2) = -\mathbf{f}_2$, so

$$M_B(S) = \begin{bmatrix} 1 & 0 \\ 0 & -1 \end{bmatrix}$$

Thus S is of type 2. Note that, in this case, 1 is an eigenvalue of S, and any eigenvector corresponding to 1 is a direction vector for the fixed line.

EXAMPLE 4

In each case, determine whether $T_A : \mathbb{R}^2 \to \mathbb{R}^2$ is a rotation or a reflection, and then find the angle or fixed line:

$$\text{(a) } A = \tfrac{1}{2}\begin{bmatrix} 1 & \sqrt{3} \\ -\sqrt{3} & 1 \end{bmatrix} \quad \text{(b) } A = \tfrac{1}{5}\begin{bmatrix} -3 & 4 \\ 4 & 3 \end{bmatrix}$$

Solution ▶ Both matrices are orthogonal, so (because $M_E(T_A) = A$, where E is the standard basis) T_A is an isometry in both cases. In the first case, $\det A = 1$, so T_A is a counterclockwise rotation through θ, where $\cos \theta = \tfrac{1}{2}$ and $\sin \theta = -\tfrac{\sqrt{3}}{2}$. Thus $\theta = -\tfrac{\pi}{3}$. In (b), $\det A = -1$, so T_A is a reflection in this case. We verify that $\mathbf{d} = \begin{bmatrix} 1 \\ 2 \end{bmatrix}$ is an eigenvector corresponding to the eigenvalue 1. Hence the fixed line $\mathbb{R}\mathbf{d}$ has equation $y = 2x$.

We now give a structure theorem for isometries. The proof requires three preliminary results, each of interest in its own right.

Lemma 2

Let $T : V \to V$ be an isometry of a finite dimensional inner product space V. If U is a T-invariant subspace of V, then $U^\perp$ is also T-invariant.

PROOF

Let $\mathbf{w}$ lie in $U^\perp$. We are to prove that $T(\mathbf{w})$ is also in $U^\perp$; that is, $\langle T(\mathbf{w}), \mathbf{u} \rangle = 0$ for all $\mathbf{u}$ in U. At this point, observe that the restriction of T to U is an isometry $U \to U$ and so is an isomorphism by the corollary to Theorem 2. In particular, each $\mathbf{u}$ in U can be written in the form $\mathbf{u} = T(\mathbf{u}_1)$ for some $\mathbf{u}_1$ in U, so

$$\langle T(\mathbf{w}), \mathbf{u} \rangle = \langle T(\mathbf{w}), T(\mathbf{u}_1) \rangle = \langle \mathbf{w}, \mathbf{u}_1 \rangle = 0$$

because $\mathbf{w}$ is in $U^\perp$. This is what we wanted.

To employ Lemma 2 above to analyze an isometry $T : V \to V$ when $\dim V = n$, it is necessary to show that a T-invariant subspace U exists such that $U \neq 0$ and $U \neq V$. We will show, in fact, that such a subspace U can always be found of dimension 1 or 2. If T has a real eigenvalue λ then $\mathbb{R}\mathbf{u}$ is T-invariant where $\mathbf{u}$ is any λ-eigenvector. But, in case (1) of Theorem 4, the eigenvalues of T are $e^{i\theta}$ and $e^{-i\theta}$ (the reader should check this), and these are nonreal if $\theta \neq 0$ and $\theta \neq \pi$. It turns out that every complex eigenvalue λ of T has absolute value 1 (Lemma 3 below); and that U has a T-invariant subspace of dimension 2 if λ is not real (Lemma 4).

Lemma 3

Let $T : V \to V$ be an isometry of the finite dimensional inner product space V. If λ is a complex eigenvalue of T, then $|\lambda| = 1$.

PROOF

Choose an orthonormal basis B of V, and let $A = M_B(T)$. Then A is a real orthogonal matrix so, using the standard inner product $\langle \mathbf{x}, \mathbf{y} \rangle = \mathbf{x}^T \overline{\mathbf{y}}$ in $\mathbb{C}^n$, we get

$$\|A\mathbf{x}\|^2 = (A\mathbf{x})^T(\overline{A\mathbf{x}}) = \mathbf{x}^T A^T \overline{A}\overline{\mathbf{x}} = \mathbf{x}^T I \mathbf{x} = \|\mathbf{x}\|^2$$

for all $\mathbf{x}$ in $\mathbb{C}^n$. But $A\mathbf{x} = \lambda \mathbf{x}$ for some $\mathbf{x} \neq \mathbf{0}$, whence $\|\mathbf{x}\|^2 = \|\lambda \mathbf{x}\|^2 = |\lambda|^2 \|\mathbf{x}\|^2$. This gives $|\lambda| = 1$, as required.

Lemma 4

Let $T: V \to V$ be an isometry of the n-dimensional inner product space V. If T has a nonreal eigenvalue, then V has a two-dimensional T-invariant subspace.

PROOF

Let B be an orthonormal basis of V, let $A = M_B(T)$, and (using Lemma 3) let $\lambda = e^{i\alpha}$ be a nonreal eigenvalue of A, say $A\mathbf{x} = \lambda \mathbf{x}$ where $\mathbf{x} \neq \mathbf{0}$ in $\mathbb{C}^n$. Because A is real, complex conjugation gives $A\overline{\mathbf{x}} = \overline{\lambda}\,\overline{\mathbf{x}}$, so $\overline{\lambda}$ is also an eigenvalue. Moreover $\lambda \neq \overline{\lambda}$ (λ is nonreal), so $\{\mathbf{x}, \overline{\mathbf{x}}\}$ is linearly independent in $\mathbb{C}^n$ (the argument in the proof of Theorem 4 Section 5.5 works). Now define

$$\mathbf{z}_1 = \mathbf{x} + \overline{\mathbf{x}} \quad \text{and} \quad \mathbf{z}_2 = i(\mathbf{x} - \overline{\mathbf{x}})$$

Then $\mathbf{z}_1$ and $\mathbf{z}_2$ lie in $\mathbb{R}^n$, and $\{\mathbf{z}_1, \mathbf{z}_2\}$ is linearly independent over $\mathbb{R}$ because $\{\mathbf{x}, \overline{\mathbf{x}}\}$ is linearly independent over $\mathbb{C}$. Moreover

$$\mathbf{x} = \tfrac{1}{2}(\mathbf{z}_1 - i\mathbf{z}_2) \quad \text{and} \quad \overline{\mathbf{x}} = \tfrac{1}{2}(\mathbf{z}_1 + i\mathbf{z}_2)$$

Now $\lambda + \overline{\lambda} = 2\cos\alpha$ and $\lambda - \overline{\lambda} = 2i\sin\alpha$, and a routine computation gives

$$A\mathbf{z}_1 = \mathbf{z}_1 \cos\alpha + \mathbf{z}_2 \sin\alpha$$
$$A\mathbf{z}_2 = -\mathbf{z}_1 \sin\alpha + \mathbf{z}_2 \cos\alpha$$

Finally, let $\mathbf{e}_1$ and $\mathbf{e}_2$ in V be such that $\mathbf{z}_1 = C_B(\mathbf{e}_1)$ and $\mathbf{z}_2 = C_B(\mathbf{e}_2)$. Then

$$C_B[T(\mathbf{e}_1)] = AC_B(\mathbf{e}_1) = A\mathbf{z}_1 = C_B(\mathbf{e}_1 \cos\alpha + \mathbf{e}_2 \sin\alpha)$$

using Theorem 2 Section 9.1. Because C_B is one-to-one, this gives the first of the following equations (the other is similar):

$$T(\mathbf{e}_1) = \mathbf{e}_1 \cos\alpha + \mathbf{e}_2 \sin\alpha$$
$$T(\mathbf{e}_2) = -\mathbf{e}_1 \sin\alpha + \mathbf{e}_2 \cos\alpha$$

Thus $U = \text{span}\{\mathbf{e}_1, \mathbf{e}_2\}$ is T-invariant and two-dimensional.

We can now prove the structure theorem for isometries.

Theorem 5

Let $T: V \to V$ be an isometry of the n-dimensional inner product space V. Given an angle θ, write $R(\theta) = \begin{bmatrix} \cos\theta & -\sin\theta \\ \sin\theta & \cos\theta \end{bmatrix}$. Then there exists an orthonormal basis B of V such that $M_B(T)$ has one of the following block diagonal forms, classified for convenience by whether n is odd or even:

$$n = 2k+1 \quad \begin{bmatrix} 1 & 0 & \cdots & 0 \\ 0 & R(\theta_1) & \cdots & 0 \\ \vdots & \vdots & \ddots & \vdots \\ 0 & 0 & \cdots & R(\theta_k) \end{bmatrix} \text{ or } \begin{bmatrix} -1 & 0 & \cdots & 0 \\ 0 & R(\theta_1) & \cdots & 0 \\ \vdots & \vdots & \ddots & \vdots \\ 0 & 0 & \cdots & R(\theta_k) \end{bmatrix}$$

$$n = 2k \quad \begin{bmatrix} R(\theta_1) & 0 & \cdots & 0 \\ 0 & R(\theta_2) & \cdots & 0 \\ \vdots & \vdots & \ddots & \vdots \\ 0 & 0 & \cdots & R(\theta_k) \end{bmatrix} \text{ or } \begin{bmatrix} -1 & 0 & 0 & \cdots & 0 \\ 0 & 1 & 0 & \cdots & 0 \\ 0 & 0 & R(\theta_1) & \cdots & 0 \\ \vdots & \vdots & \vdots & \ddots & \vdots \\ 0 & 0 & 0 & \cdots & R(\theta_{k-1}) \end{bmatrix}$$

PROOF

We show first, by induction on n, that an orthonormal basis B of V can be found such that $M_B(T)$ is a block diagonal matrix of the following form:

$$M_B(T) = \begin{bmatrix} I_r & 0 & 0 & \cdots & 0 \\ 0 & -I_s & 0 & \cdots & 0 \\ 0 & 0 & R(\theta_1) & \cdots & 0 \\ \vdots & \vdots & \vdots & \ddots & \vdots \\ 0 & 0 & 0 & \cdots & R(\theta_t) \end{bmatrix}$$

where the identity matrix I_r, the matrix $-I_s$, or the matrices $R(\theta_i)$ may be missing. If $n=1$ and $V = \mathbb{R}\mathbf{v}$, this holds because $T(\mathbf{v}) = \lambda\mathbf{v}$ and $\lambda = \pm 1$ by Lemma 3. If $n=2$, this follows from Theorem 4. If $n \geq 3$, either T has a real eigenvalue and therefore has a one-dimensional T-invariant subspace $U = \mathbb{R}\mathbf{u}$ for any eigenvector $\mathbf{u}$, or T has no real eigenvalue and therefore has a two-dimensional T-invariant subspace U by Lemma 4. In either case $U^\perp$ is T-invariant (Lemma 2) and $\dim U^\perp = n - \dim U < n$. Hence, by induction, let B_1 and B_2 be orthonormal bases of U and $U^\perp$ such that $M_{B_1}(T)$ and $M_{B_2}(T)$ have the form given. Then $B = B_1 \cup B_2$ is an orthonormal basis of V, and $M_B(T)$ has the desired form with a suitable ordering of the vectors in B.

Now observe that $R(0) = \begin{bmatrix} 1 & 0 \\ 0 & 1 \end{bmatrix}$ and $R(\pi) = \begin{bmatrix} -1 & 0 \\ 0 & -1 \end{bmatrix}$. It follows that an even number of 1s or -1s can be written as $R(\theta_1)$-blocks. Hence, with a suitable reordering of the basis B, the theorem follows.

As in the dimension 2 situation, these possibilities can be given a geometric interpretation when $V = \mathbb{R}^3$ is taken as euclidean space. As before, this entails looking carefully at reflections and rotations in $\mathbb{R}^3$. If $Q: \mathbb{R}^3 \to \mathbb{R}^3$ is any reflection in a plane through the origin (called the **fixed plane** of the reflection), take

$\{f_2, f_3\}$ to be any orthonormal basis of the fixed plane and take f_1 to be a unit vector perpendicular to the fixed plane. Then $Q(f_1) = -f_1$, whereas $Q(f_2) = f_2$ and $Q(f_3) = f_3$. Hence $B = \{f_1, f_2, f_3\}$ is an orthonormal basis such that

$$M_B(Q) = \begin{bmatrix} -1 & 0 & 0 \\ 0 & 1 & 0 \\ 0 & 0 & 1 \end{bmatrix}$$

Similarly, suppose that $R : \mathbb{R}^3 \to \mathbb{R}^3$ is any rotation about a line through the origin (called the **axis** of the rotation), and let f_1 be a unit vector pointing along the axis, so $R(f_1) = f_1$. Now the plane through the origin perpendicular to the axis is an R-invariant subspace of $\mathbb{R}^2$ of dimension 2, and the restriction of R to this plane is a rotation. Hence, by Theorem 4, there is an orthonormal basis $B_1 = \{f_2, f_3\}$ of this plane such that $M_{B_1}(R) = \begin{bmatrix} \cos\theta & -\sin\theta \\ \sin\theta & \cos\theta \end{bmatrix}$. But then $B = \{f_1, f_2, f_3\}$ is an orthonormal basis of $\mathbb{R}^3$ such that the matrix of R is

$$M_B(R) = \begin{bmatrix} 1 & 0 & 0 \\ 0 & \cos\theta & -\sin\theta \\ 0 & \sin\theta & \cos\theta \end{bmatrix}$$

However, Theorem 5 shows that there are isometries T in $\mathbb{R}^3$ of a third type: those with a matrix of the form

$$M_B(T) = \begin{bmatrix} -1 & 0 & 0 \\ 0 & \cos\theta & -\sin\theta \\ 0 & \sin\theta & \cos\theta \end{bmatrix}$$

If $B = \{f_1, f_2, f_3\}$, let Q be the reflection in the plane spanned by f_2 and f_3, and let R be the rotation corresponding to θ about the line spanned by f_1. Then $M_B(Q)$ and $M_B(R)$ are as above, and $M_B(Q) M_B(R) = M_B(T)$ as the reader can verify. This means that $M_B(QR) = M_B(T)$ by Theorem 1 Section 9.2, and this in turn implies that $QR = T$ because M_B is one-to-one (see Exercise 26 Section 9.1). A similar argument shows that $RQ = T$, and we have Theorem 6.

Theorem 6

If $T : \mathbb{R}^3 \to \mathbb{R}^3$ is an isometry, there are three possibilities.

(a) T is a rotation, and $M_B(T) = \begin{bmatrix} 1 & 0 & 0 \\ 0 & \cos\theta & -\sin\theta \\ 0 & \sin\theta & \cos\theta \end{bmatrix}$ for some orthonormal basis B.

(b) T is a reflection, and $M_B(T) = \begin{bmatrix} -1 & 0 & 0 \\ 0 & 1 & 0 \\ 0 & 0 & 1 \end{bmatrix}$ for some orthonormal basis B.

(c) $T = QR = RQ$ where Q is a reflection, R is a rotation about an axis perpendicular to the fixed plane of Q and $M_B(T) = \begin{bmatrix} -1 & 0 & 0 \\ 0 & \cos\theta & -\sin\theta \\ 0 & \sin\theta & \cos\theta \end{bmatrix}$ for some orthonormal basis B.

Hence T is a rotation if and only if $\det T = 1$.

SECTION 10.4 Isometries

PROOF

It remains only to verify the final observation that T is a rotation if and only if $\det T = 1$. But clearly $\det T = -1$ in parts (b) and (c).

A useful way of analyzing a given isometry $T : \mathbb{R}^3 \to \mathbb{R}^3$ comes from computing the eigenvalues of T. Because the characteristic polynomial of T has degree 3, it must have a real root. Hence, there must be at least one real eigenvalue, and the only possible real eigenvalues are ± 1 by Lemma 3. Thus Table 1 includes all possibilities.

TABLE 1

Eigenvalues of T	Action of T
(1) 1, no other real eigenvalues	Rotation about the line $\mathbb{R}\mathbf{f}$ where $\mathbf{f}$ is an eigenvector corresponding to 1. [Case (a) of Theorem 6.]
(2) -1, no other real eigenvalues	Rotation about the line $\mathbb{R}\mathbf{f}$ followed by reflection in the plane $(\mathbb{R}\mathbf{f})^\perp$ where $\mathbf{f}$ is an eigenvector corresponding to -1. [Case (c) of Theorem 6.]
(3) $-1, 1, 1$	Reflection in the plane $(\mathbb{R}\mathbf{f})^\perp$ where $\mathbf{f}$ is an eigenvector corresponding to -1. [Case (b) of Theorem 6.]
(4) $1, -1, -1$	This is as in (1) with a rotation of π.
(5) $-1, -1, -1$	Here $T(\mathbf{x}) = -\mathbf{x}$ for all $\mathbf{x}$. This is (2) with a rotation of π.
(6) $1, 1, 1$	Here T is the identity isometry.

EXAMPLE 5

Analyze the isometry $T : \mathbb{R}^3 \to \mathbb{R}^3$ given by $T \begin{bmatrix} x \\ y \\ z \end{bmatrix} = \begin{bmatrix} y \\ z \\ -x \end{bmatrix}$.

Solution ▶ If B_0 is the standard basis of $\mathbb{R}^3$, then $M_{B_0}(T) = \begin{bmatrix} 0 & 1 & 0 \\ 0 & 0 & 1 \\ -1 & 0 & 0 \end{bmatrix}$, so $c_T(x) = x^3 + 1 = (x+1)(x^2 - x + 1)$. This is (2) in Table 1. Write:

$$\mathbf{f}_1 = \tfrac{1}{\sqrt{3}} \begin{bmatrix} 1 \\ -1 \\ 1 \end{bmatrix} \quad \mathbf{f}_2 = \tfrac{1}{\sqrt{6}} \begin{bmatrix} 1 \\ 2 \\ 1 \end{bmatrix} \quad \mathbf{f}_3 = \tfrac{1}{\sqrt{2}} \begin{bmatrix} 1 \\ 0 \\ -1 \end{bmatrix}$$

Here $\mathbf{f}_1$ is a unit eigenvector corresponding to $\lambda_1 = -1$, so T is a rotation (through an angle θ) about the line $L = \mathbb{R}\mathbf{f}_1$, followed by reflection in the plane U through the origin perpendicular to $\mathbf{f}_1$ (with equation $x - y + z = 0$). Then, $\{\mathbf{f}_1, \mathbf{f}_2\}$ is chosen as an orthonormal basis of U, so $B = \{\mathbf{f}_1, \mathbf{f}_2, \mathbf{f}_3\}$ is an orthonormal basis of $\mathbb{R}^3$ and

$$M_B(T) = \begin{bmatrix} -1 & 0 & 0 \\ 0 & \tfrac{1}{2} & -\tfrac{\sqrt{3}}{2} \\ 0 & \tfrac{\sqrt{3}}{2} & \tfrac{1}{2} \end{bmatrix}$$

Hence θ is given by $\cos\theta = \tfrac{1}{2}$, $\sin\theta = \tfrac{\sqrt{3}}{2}$, so $\theta = \tfrac{\pi}{3}$.

Let V be an n-dimensional inner product space. A subspace of V of dimension $n - 1$ is called a **hyperplane** in V. Thus the hyperplanes in $\mathbb{R}^3$ and $\mathbb{R}^2$ are, respectively, the planes and lines through the origin. Let $Q : V \to V$ be an isometry with matrix

$$M_B(Q) = \begin{bmatrix} -1 & 0 \\ 0 & I_{n-1} \end{bmatrix}$$

for some orthonormal basis $B = \{\mathbf{f}_1, \mathbf{f}_2, \ldots, \mathbf{f}_n\}$. Then $Q(\mathbf{f}_1) = -\mathbf{f}_1$ whereas $Q(\mathbf{u}) = \mathbf{u}$ for each $\mathbf{u}$ in $U = \text{span}\{\mathbf{f}_2, \ldots, \mathbf{f}_n\}$. Hence U is called the **fixed hyperplane** of Q, and Q is called **reflection** in U. Note that each hyperplane in V is the fixed hyperplane of a (unique) reflection of V. Clearly, reflections in $\mathbb{R}^2$ and $\mathbb{R}^3$ are reflections in this more general sense.

Continuing the analogy with $\mathbb{R}^2$ and $\mathbb{R}^3$, an isometry $T: V \to V$ is called a **rotation** if there exists an orthonormal basis $\{\mathbf{f}_1, \ldots, \mathbf{f}_n\}$ such that

$$M_B(T) = \begin{bmatrix} I_r & 0 & 0 \\ 0 & R(\theta) & 0 \\ 0 & 0 & I_s \end{bmatrix}$$

in block form, where $R(\theta) = \begin{bmatrix} \cos \theta & -\sin \theta \\ \sin \theta & \cos \theta \end{bmatrix}$, and where either I_r or I_s (or both) may be missing. If $R(\theta)$ occupies columns i and $i + 1$ of $M_B(T)$, and if $W = \text{span}\{\mathbf{f}_i, \mathbf{f}_{i+1}\}$, then W is T-invariant and the matrix of $T : W \to W$ with respect to $\{\mathbf{f}_i, \mathbf{f}_{i+1}\}$ is $R(\theta)$. Clearly, if W is viewed as a copy of $\mathbb{R}^2$, then T is a rotation in W. Moreover, $T(\mathbf{u}) = \mathbf{u}$ holds for all vectors $\mathbf{u}$ in the $(n - 2)$-dimensional subspace $U = \text{span}\{\mathbf{f}_1, \ldots, \mathbf{f}_{i-1}, \mathbf{f}_{i+1}, \ldots, \mathbf{f}_n\}$, and U is called the **fixed axis** of the rotation T. In $\mathbb{R}^3$, the axis of any rotation is a line (one-dimensional), whereas in $\mathbb{R}^2$ the axis is $U = \{\mathbf{0}\}$.

With these definitions, the following theorem is an immediate consequence of Theorem 5 (the details are left to the reader).

Theorem 7

Let $T : V \to V$ be an isometry of a finite dimensional inner product space V. Then there exist isometries $T_1, \ldots, T_k$ such that

$$T = T_k T_{k-1} \cdots T_2 T_1$$

where each T_i is either a rotation or a reflection, at most one is a reflection, and $T_i T_j = T_j T_i$ holds for all i and j. Furthermore, T is a composite of rotations if and only if $\det T = 1$.

EXERCISES 10.4

Throughout these exercises, V denotes a finite dimensional inner product space.

1. Show that the following linear operators are isometries.

    (a) $T : \mathbb{C} \to \mathbb{C}$; $T(z) = \bar{z}$; $\langle z, w \rangle = \text{re}(z\bar{w})$

    (b) $T : \mathbb{R}^n \to \mathbb{R}^n$; $T(a_1, a_2, \ldots, a_n) = (a_n, a_{n-1}, \ldots, a_2, a_1)$; dot product

    (c) $T : \mathbf{M}_{22} \to \mathbf{M}_{22}$; $T \begin{bmatrix} a & b \\ c & d \end{bmatrix} = \begin{bmatrix} c & d \\ b & a \end{bmatrix}$;
    $\langle A, B \rangle = \text{tr}(AB^T)$

    (d) $T : \mathbb{R}^3 \to \mathbb{R}^3$; $T(a, b, c) = \frac{1}{9}(2a + 2b - c, 2a + 2c - b, 2b + 2c - a)$; dot product

2. In each case, show that T is an isometry of $\mathbb{R}^2$, determine whether it is a rotation or a reflection, and find the angle or the fixed line. Use the dot product.

   (a) $T\begin{bmatrix} a \\ b \end{bmatrix} = \begin{bmatrix} -a \\ b \end{bmatrix}$ (b) $T\begin{bmatrix} a \\ b \end{bmatrix} = \begin{bmatrix} -a \\ -b \end{bmatrix}$

   (c) $T\begin{bmatrix} a \\ b \end{bmatrix} = \begin{bmatrix} b \\ -a \end{bmatrix}$ (d) $T\begin{bmatrix} a \\ b \end{bmatrix} = \begin{bmatrix} -b \\ -a \end{bmatrix}$

   (e) $T\begin{bmatrix} a \\ b \end{bmatrix} = \frac{1}{\sqrt{2}}\begin{bmatrix} a+b \\ b-a \end{bmatrix}$ (f) $T\begin{bmatrix} a \\ b \end{bmatrix} = \frac{1}{\sqrt{2}}\begin{bmatrix} a-b \\ a+b \end{bmatrix}$

3. In each case, show that T is an isometry of $\mathbb{R}^3$, determine the type (Theorem 6), and find the axis of any rotations and the fixed plane of any reflections involved.

   (a) $T\begin{bmatrix} a \\ b \\ c \end{bmatrix} = \begin{bmatrix} a \\ -b \\ c \end{bmatrix}$ (b) $T\begin{bmatrix} a \\ b \\ c \end{bmatrix} = \frac{1}{2}\begin{bmatrix} \sqrt{3}c - a \\ \sqrt{3}a + c \\ 2b \end{bmatrix}$

   (c) $T\begin{bmatrix} a \\ b \\ c \end{bmatrix} = \begin{bmatrix} b \\ c \\ a \end{bmatrix}$ (d) $T\begin{bmatrix} a \\ b \\ c \end{bmatrix} = \begin{bmatrix} a \\ -b \\ -c \end{bmatrix}$

   (e) $T\begin{bmatrix} a \\ b \\ c \end{bmatrix} = \frac{1}{2}\begin{bmatrix} a + \sqrt{3}b \\ b - \sqrt{3}a \\ 2c \end{bmatrix}$

   (f) $T\begin{bmatrix} a \\ b \\ c \end{bmatrix} = \frac{1}{\sqrt{2}}\begin{bmatrix} a+c \\ -\sqrt{2}b \\ c-a \end{bmatrix}$

4. Let $T: \mathbb{R}^2 \to \mathbb{R}^2$ be an isometry. A vector $\mathbf{x}$ in $\mathbb{R}^2$ is said to be **fixed** by T if $T(\mathbf{x}) = \mathbf{x}$. Let E_1 denote the set of all vectors in $\mathbb{R}^2$ fixed by T. Show that:

   (a) E_1 is a subspace of $\mathbb{R}^2$.

   (b) $E_1 = \mathbb{R}^2$ if and only if $T = 1$ is the identity map.

   (c) $\dim E_1 = 1$ if and only if T is a reflection (about the line E_1).

   (d) $E_1 = \{0\}$ if and only if T is a rotation ($T \neq 1$).

5. Let $T: \mathbb{R}^3 \to \mathbb{R}^3$ be an isometry, and let E_1 be the subspace of all fixed vectors in $\mathbb{R}^3$ (see Exercise 4). Show that:

   (a) $E_1 = \mathbb{R}^3$ if and only if $T = 1$.

   (b) $\dim E_1 = 2$ if and only if T is a reflection (about the plane E_1).

   (c) $\dim E_1 = 1$ if and only if T is a rotation ($T \neq 1$) (about the line E_1).

   (d) $\dim E_1 = 0$ if and only if T is a reflection followed by a (nonidentity) rotation.

6. If T is an isometry, show that aT is an isometry if and only if $a = \pm 1$.

7. Show that every isometry preserves the angle between any pair of nonzero vectors (see Exercise 31 Section 10.1). Must an angle-preserving isomorphism be an isometry? Support your answer.

8. If $T: V \to V$ is an isometry, show that $T^2 = 1_V$ if and only if the only complex eigenvalues of T are 1 and -1.

9. Let $T: V \to V$ be a linear operator. Show that any two of the following conditions implies the third:

   (1) T is symmetric.

   (2) T is an involution ($T^2 = 1_V$).

   (3) T is an isometry.

   [*Hint*: In all cases, use the definition $\langle \mathbf{v}, T(\mathbf{w}) \rangle = \langle T(\mathbf{v}), \mathbf{w} \rangle$ of a symmetric operator. For (1) and (3) $\Rightarrow$ (2), use the fact that, if $\langle T^2(\mathbf{v}) - \mathbf{v}, \mathbf{w} \rangle = 0$ for all $\mathbf{w}$, then $T^2(\mathbf{v}) = \mathbf{v}$.]

10. If B and D are any orthonormal bases of V, show that there is an isometry $T: V \to V$ that carries B to D.

11. Show that the following are equivalent for a linear transformation $S: V \to V$ where V is finite dimensional and $S \neq 0$:

    (1) $\langle S(\mathbf{v}), S(\mathbf{w}) \rangle = 0$ whenever $\langle \mathbf{v}, \mathbf{w} \rangle = 0$;

    (2) $S = aT$ for some isometry $T: V \to V$ and some $a \neq 0$ in $\mathbb{R}$.

    (3) S is an isomorphism and preserves angles between nonzero vectors.

    [*Hint*: Given (1), show that $\|S(\mathbf{e})\| = \|S(\mathbf{f})\|$ for all unit vectors $\mathbf{e}$ and $\mathbf{f}$ in V.]

12. Let $S: V \to V$ be a distance preserving transformation where V is finite dimensional.

    (a) Show that the factorization in the proof of Theorem 1 is unique. That is, if $S = S_\mathbf{u} \circ T$ and $S = S_{\mathbf{u}'} \circ T'$ where $\mathbf{u}, \mathbf{u}' \in V$ and T, $T': V \to V$ are isometries, show that $\mathbf{u} = \mathbf{u}'$ and $T = T'$.

    (b) If $S = S_\mathbf{u} \circ T$, $\mathbf{u} \in V$, T an isometry, show that $\mathbf{w} \in V$ exists such that $S = T \circ S_\mathbf{w}$.

13. Define $T: \mathbf{P} \to \mathbf{P}$ by $T(f) = xf(x)$ for all $f \in \mathbf{P}$, and define an inner product on $\mathbf{P}$ as follows: If
$f = a_0 + a_1 x + a_2 x^2 + \cdots$ and
$g = b_0 + b_1 x + b_2 x^2 + \cdots$ are in $\mathbf{P}$, define
$\langle f, g \rangle = a_0 b_0 + a_1 b_1 + a_2 b_2 + \cdots$.

(a) Show that $\langle , \rangle$ is an inner product on $\mathbf{P}$.

(b) Show that T is an isometry of $\mathbf{P}$.

(c) Show that T is one-to-one but not onto.

SECTION 10.5 An Application to Fourier Approximation[6]

In this section we shall investigate an important orthogonal set in the space $\mathbf{C}[-\pi, \pi]$ of continuous functions on the interval $[-\pi, \pi]$, using the inner product.

$$\langle f, g \rangle = \int_{-\pi}^{\pi} f(x)g(x)\, dx$$

Of course, calculus will be needed. The orthogonal set in question is

$$\{1, \sin x, \cos x, \sin(2x), \cos(2x), \sin(3x), \cos(3x), \ldots\}$$

Standard techniques of integration give

$$\|1\|^2 = \int_{-\pi}^{\pi} 1^2\, dx = 2\pi$$
$$\|\sin kx\|^2 = \int_{-\pi}^{\pi} \sin^2(kx)\, dx = \pi \quad \text{for any } k = 1, 2, 3, \ldots$$
$$\|\cos kx\|^2 = \int_{-\pi}^{\pi} \cos^2(kx)\, dx = \pi \quad \text{for any } k = 1, 2, 3, \ldots$$

We leave the verifications to the reader, together with the task of showing that these functions are orthogonal:

$$\langle \sin(kx), \sin(mx) \rangle = 0 = \langle \cos(kx), \cos(mx) \rangle \quad \text{if } k \neq m$$

and

$$\langle \sin(kx), \cos(mx) \rangle = 0 \quad \text{for all } k \geq 0 \text{ and } m \geq 0$$

(Note that $1 = \cos(0x)$, so the constant function 1 is included.)

Now define the following subspace of $\mathbf{C}[-\pi, \pi]$:

$$F_n = \text{span}\{1, \sin x, \cos x, \sin(2x), \cos(2x), \ldots, \sin(nx), \cos(nx)\}$$

The aim is to use the approximation theorem (Theorem 8 Section 10.2); so, given a function f in $\mathbf{C}[-\pi, \pi]$, define the **Fourier coefficients** of f by

$$a_0 = \frac{\langle f(x), 1 \rangle}{\|1\|^2} = \frac{1}{2\pi}\int_{-\pi}^{\pi} f(x)\, dx$$

$$a_k = \frac{\langle f(x), \cos(kx) \rangle}{\|\cos(kx)\|^2} = \frac{1}{\pi}\int_{-\pi}^{\pi} f(x)\cos(kx)\, dx \quad k = 1, 2, \ldots$$

$$b_k = \frac{\langle f(x), \sin(kx) \rangle}{\|\sin(kx)\|^2} = \frac{1}{\pi}\int_{-\pi}^{\pi} f(x)\sin(kx)\, dx \quad k = 1, 2, \ldots$$

Then the approximation theorem (Theorem 8 Section 10.2) gives Theorem 1.

---

[6] The name honours the French mathematician J.B.J. Fourier (1768–1830) who used these techniques in 1822 to investigate heat conduction in solids.

SECTION 10.5 An Application to Fourier Approximation

> **Theorem 1**
>
> Let f be any continuous real-valued function defined on the interval $[-\pi, \pi]$. If $a_0, a_1, \ldots,$ and $b_0, b_1, \ldots$ are the Fourier coefficients of f, then given $n \geq 0$,
>
> $$f_n(x) = a_0 + a_1 \cos x + b_1 \sin x + a_2 \cos(2x) + b_2 \sin(2x) + \cdots \\ + a_n \cos(nx) + b_n \sin(nx)$$
>
> is a function in F_n that is closest to f in the sense that
>
> $$\|f - f_n\| \leq \|f - g\|$$
>
> holds for all functions g in F_n.

The function f_n is called the nth **Fourier approximation** to the function f.

EXAMPLE 1

Find the fifth Fourier approximation to the function $f(x)$ defined on $[-\pi, \pi]$ as follows:

$$f(x) = \begin{cases} \pi + x & \text{if } -\pi \leq x < 0 \\ \pi - x & \text{if } 0 \leq x \leq \pi \end{cases}$$

Solution ▶ The graph of $y = f(x)$ appears in the top diagram. The Fourier coefficients are computed as follows. The details of the integrations (usually by parts) are omitted.

$$a_0 = \tfrac{1}{2\pi} \int_{-\pi}^{\pi} f(x)\, dx = \tfrac{\pi}{2}$$

$$a_k = \tfrac{1}{\pi} \int_{-\pi}^{\pi} f(x)\cos(kx)\, dx = \tfrac{2}{\pi k^2}[1 - \cos(k\pi)] = \begin{cases} 0 & \text{if } k \text{ is even} \\ \tfrac{4}{\pi k^2} & \text{if } k \text{ is odd} \end{cases}$$

$$b_k = \tfrac{1}{\pi} \int_{-\pi}^{\pi} f(x)\sin(kx)\, dx = 0 \quad \text{for all } k = 1, 2, \ldots$$

Hence the fifth Fourier approximation is

$$f_5(x) = \tfrac{\pi}{2} + \tfrac{4}{\pi}\left\{\cos x + \tfrac{1}{3^2} \cos(3x) + \tfrac{1}{5^2} \cos(5x)\right\}$$

This is plotted in the middle diagram and is already a reasonable approximation to $f(x)$. By comparison, $f_{13}(x)$ is also plotted in the bottom diagram.

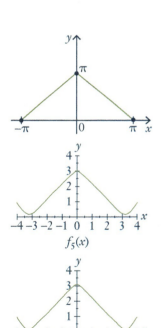

We say that a function f is an **even function** if $f(x) = f(-x)$ holds for all x; f is called an **odd function** if $f(-x) = -f(x)$ holds for all x. Examples of even functions are constant functions, the even powers $x^2, x^4, \ldots$, and $\cos(kx)$; these functions are characterized by the fact that the graph of $y = f(x)$ is symmetric about the y axis. Examples of odd functions are the odd powers $x, x^3, \ldots$, and $\sin(kx)$ where $k > 0$, and the graph of $y = f(x)$ is symmetric about the origin if f is odd. The usefulness of these functions stems from the fact that

$$\int_{-\pi}^{\pi} f(x)\, dx = 0 \qquad \text{if } f \text{ is odd}$$

$$\int_{-\pi}^{\pi} f(x)\, dx = 2\int_{0}^{\pi} f(x)\, dx \qquad \text{if } f \text{ is even}$$

These facts often simplify the computations of the Fourier coefficients. For example:

1. The Fourier sine coefficients b_k all vanish if f is even.
2. The Fourier cosine coefficients a_k all vanish if f is odd.

This is because $f(x) \sin(kx)$ is odd in the first case and $f(x) \cos(kx)$ is odd in the second case.

The functions 1, $\cos(kx)$, and $\sin(kx)$ that occur in the Fourier approximation for $f(x)$ are all easy to generate as an electrical voltage (when x is time). By summing these signals (with the amplitudes given by the Fourier coefficients), it is possible to produce an electrical signal with (the approximation to) $f(x)$ as the voltage. Hence these Fourier approximations play a fundamental role in electronics.

Finally, the Fourier approximations $f_1, f_2, \ldots$ of a function f get better and better as n increases. The reason is that the subspaces F_n increase:

$$F_1 \subseteq F_2 \subseteq F_3 \subseteq \cdots \subseteq F_n \subseteq \cdots$$

So, because $f_n = \text{proj}_{F_n}(f)$, we get (see the discussion following Example 6 Section 10.2)

$$\|f - f_1\| \geq \|f - f_2\| \geq \cdots \geq \|f - f_n\| \geq \cdots$$

These numbers $\|f - f_n\|$ approach zero; in fact, we have the following fundamental theorem.[7]

Theorem 2

Let f be any continuous function in $\mathbf{C}[-\pi, \pi]$. Then

$$f_n(x) \text{ approaches } f(x) \text{ for all } x \text{ such that } -\pi < x < \pi.\text{[8]}$$

It shows that f has a representation as an infinite series, called the **Fourier series** of f:

$$f(x) = a_0 + a_1 \cos x + b_1 \sin x + a_2 \cos(2x) + b_2 \sin(2x) + \cdots$$

whenever $-\pi < x < \pi$. A full discussion of Theorem 2 is beyond the scope of this book. This subject had great historical impact on the development of mathematics, and has become one of the standard tools in science and engineering.

Thus the Fourier series for the function f in Example 1 is

$$f(x) = \frac{\pi}{2} + \frac{4}{\pi}\left\{\cos x + \frac{1}{3^2}\cos(3x) + \frac{1}{5^2}\cos(5x) + \frac{1}{7^2}\cos(7x) + \cdots\right\}$$

Since $f(0) = \pi$ and $\cos(0) = 1$, taking $x = 0$ leads to the series

$$\frac{\pi^2}{8} = 1 + \frac{1}{3^2} + \frac{1}{5^2} + \frac{1}{7^2} + \cdots$$

---

[7] See, for example, J. W. Brown and R. V. Churchill, *Fourier Series and Boundary Value Problems*, 7th ed., (New York: McGraw-Hill, 2008).

[8] We have to be careful at the end points $x = \pi$ or $x = -\pi$ because $\sin(k\pi) = \sin(-k\pi)$ and $\cos(k\pi) = \cos(-k\pi)$.

EXAMPLE 2

Expand $f(x) = x$ on the interval $[-\pi, \pi]$ in a Fourier series, and so obtain a series expansion of $\frac{\pi}{4}$.

Solution ▶ Here f is an odd function so all the Fourier cosine coefficients a_k are zero. As to the sine coefficients:

$$b_k = \tfrac{1}{\pi} \int_{-\pi}^{\pi} x \sin(kx)\, dx = \tfrac{2}{k}(-1)^{k+1} \quad \text{for } k \geq 1$$

where we omit the details of the integration by parts. Hence the Fourier series for x is

$$x = 2\left[\sin x - \tfrac{1}{2}\sin(2x) + \tfrac{1}{3}\sin(3x) - \tfrac{1}{4}\sin(4x) + \ldots\right]$$

for $-\pi < x < \pi$. In particular, taking $x = \frac{\pi}{2}$ gives an infinite series for $\frac{\pi}{4}$.

$$\tfrac{\pi}{4} = 1 - \tfrac{1}{3} + \tfrac{1}{5} - \tfrac{1}{7} + \tfrac{1}{9} - \cdots$$

Many other such formulas can be proved using Theorem 2.

EXERCISES 10.5

1. In each case, find the Fourier approximation f_5 of the given function in $\mathbf{C}[-\pi, \pi]$.

    (a) $f(x) = \pi - x$

    ◆(b) $f(x) = |x| = \begin{cases} x & \text{if } 0 \leq x \leq \pi \\ -x & \text{if } -\pi \leq x < 0 \end{cases}$

    (c) $f(x) = x^2$

    ◆(d) $f(x) = \begin{cases} 0 & \text{if } -\pi \leq x < 0 \\ x & \text{if } 0 \leq x \leq \pi \end{cases}$

2. (a) Find f_5 for the even function f on $[-\pi, \pi]$ satisfying $f(x) = x$ for $0 \leq x \leq \pi$.

    ◆(b) Find f_6 for the even function f on $[-\pi, \pi]$ satisfying $f(x) = \sin x$ for $0 \leq x \leq \pi$.

    [*Hint*: If $k > 1$, $\int \sin x \cos(kx)$
    $= \tfrac{1}{2}\left[\dfrac{\cos[(k-1)x]}{k-1} - \dfrac{\cos[(k+1)x]}{k+1}\right]$.]

3. (a) Prove that $\int_{-\pi}^{\pi} f(x)dx = 0$ if f is odd and that $\int_{-\pi}^{\pi} f(x)dx = 2\int_{0}^{\pi} f(x)dx$ if f is even.

    (b) Prove that $\tfrac{1}{2}[f(x) + f(-x)]$ is even and that $\tfrac{1}{2}[f(x) - f(-x)]$ is odd for any function f. Note that they sum to $f(x)$.

◆4. Show that $\{1, \cos x, \cos(2x), \cos(3x), \ldots\}$ is an orthogonal set in $\mathbf{C}[0, \pi]$ with respect to the inner product $\langle f, g \rangle = \int_{0}^{\pi} f(x)g(x)dx$.

5. (a) Show that $\frac{\pi^2}{8} = 1 + \frac{1}{3^2} + \frac{1}{5^2} + \cdots$ using Exercise 1(b).

    (b) Show that $\frac{\pi^2}{12} = 1 - \frac{1}{2^2} + \frac{1}{3^2} - \frac{1}{4^2} + \cdots$ using Exercise 1(c).

Canonical Forms

11

Given a matrix A, the effect of a sequence of row-operations on A is to produce UA where U is invertible. Under this "row-equivalence" operation the best that can be achieved is the reduced row-echelon form for A. If column operations are also allowed, the result is UAV where both U and V are invertible, and the best outcome under this "equivalence" operation is called the Smith canonical form of A (Theorem 3 Section 2.5). There are other kinds of operations on a matrix and, in many cases, there is a "canonical" best possible result.

If A is square, the most important operation of this sort is arguably "similarity" wherein A is carried to $U^{-1}AU$ where U is invertible. In this case we say that matrices A and B are *similar*, and write $A \sim B$, when $B = U^{-1}AU$ for some invertible matrix U. Under similarity the canonical matrices, called *Jordan canonical matrices*, are block triangular with upper triangular "Jordan" blocks on the main diagonal. In this short chapter we are going to define these Jordan blocks and prove that every matrix is similar to a Jordan canonical matrix.

Here is the key to the method. Let $T : V \rightarrow V$ be an operator on an n-dimensional vector space V, and suppose that we can find an ordered basis B of V so that the matrix $M_B(T)$ is as simple as possible. Then, if B_0 is *any* ordered basis of V, the matrices $M_B(T)$ and $M_{B_0}(T)$ are similar; that is,

$$M_B(T) = P^{-1}M_{B_0}(T)P \quad \text{for some invertible matrix } P.$$

Moreover, $P = P_{B_0 \leftarrow B}$ is easily computed from the bases B and D (Theorem 3 Section 9.2). This, combined with the invariant subspaces and direct sums studied in Section 9.3, enables us to calculate the Jordan canonical form of any square matrix A. Along the way we derive an explicit construction of an invertible matrix P such that $P^{-1}AP$ is block triangular.

This technique is important in many ways. For example, if we want to diagonalize an $n \times n$ matrix A, let $T_A : \mathbb{R}^n \rightarrow \mathbb{R}^n$ be the operator given by $T_A(\mathbf{x}) = A\mathbf{x}$ for all $\mathbf{x}$ in $\mathbb{R}^n$, and look for a basis B of $\mathbb{R}^n$ such that $M_B(T_A)$ is diagonal. If $B_0 = E$ is the standard basis of $\mathbb{R}^n$, then $M_E(T_A) = A$, so

$$P^{-1}AP = P^{-1}M_E(T_A)P = M_B(T_A),$$

and we have diagonalized A. Thus the "algebraic" problem of finding an invertible matrix P such that $P^{-1}AP$ is diagonal is converted into the "geometric" problem of finding a basis B such that $M_B(T_A)$ is diagonal. This change of perspective is one of the most important techniques in linear algebra.

SECTION 11.1 Block Triangular Form

We have shown (Theorem 5 Section 8.2) that any $n \times n$ matrix A with every eigenvalue real is orthogonally similar to an upper triangular matrix U. The following theorem shows that U can be chosen in a special way.

Theorem 1

Block Triangulation Theorem
Let A be an $n \times n$ matrix with every eigenvalue real and let
$$c_A(x) = (x - \lambda_1)^{m_1}(x - \lambda_2)^{m_2}\cdots(x - \lambda_k)^{m_k}$$
where $\lambda_1, \lambda_2, \ldots, \lambda_k$ are the distinct eigenvalues of A. Then an invertible matrix P exists such that
$$P^{-1}AP = \begin{bmatrix} U_1 & 0 & 0 & \cdots & 0 \\ 0 & U_2 & 0 & \cdots & 0 \\ 0 & 0 & U_3 & \cdots & 0 \\ \vdots & \vdots & \vdots & & \vdots \\ 0 & 0 & 0 & \cdots & U_k \end{bmatrix}$$
where, for each i, U_i is an $m_i \times m_i$ upper triangular matrix with every entry on the main diagonal equal to λ_i.

The proof is given at the end of this section. For now, we focus on a method for *finding* the matrix P. The key concept is as follows.

Definition 11.1 If A is as in Theorem 1, the **generalized eigenspace** $G_{\lambda_i}(A)$ is defined by
$$G_{\lambda_i}(A) = \text{null}[(\lambda_i I - A)^{m_i}]$$
where m_i is the multiplicity of λ_i.

Observe that the eigenspace $E_{\lambda_i}(A) = \text{null}(\lambda_i I - A)$ is a subspace of $G_{\lambda_i}(A)$. We need three technical results.

Lemma 1

Using the notation of Theorem 1, we have $\dim[G_{\lambda_i}(A)] = m_i$.

PROOF

Write $A_i = (\lambda_i I - A)^{m_i}$ for convenience and let P be as in Theorem 1. The spaces $G_{\lambda_i}(A) = \text{null}(A_i)$ and $\text{null}(P^{-1}A_iP)$ are isomorphic via $\mathbf{x} \leftrightarrow P^{-1}\mathbf{x}$, so we show that $\dim[\text{null}(P^{-1}A_iP)] = m_i$. Now $P^{-1}A_iP = (\lambda_i I - P^{-1}AP)^{m_i}$. If we use the block form in Theorem 1, this becomes

$$P^{-1}A_iP = \begin{bmatrix} \lambda_i I - U_1 & 0 & \cdots & 0 \\ 0 & \lambda_i I - U_2 & \cdots & 0 \\ \vdots & \vdots & & \vdots \\ 0 & 0 & \cdots & \lambda_i I - U_k \end{bmatrix}^{m_i}$$

$$= \begin{bmatrix} (\lambda_i I - U_1)^{m_i} & 0 & \cdots & 0 \\ 0 & (\lambda_i I - U_2)^{m_i} & \cdots & 0 \\ \vdots & \vdots & & \vdots \\ 0 & 0 & \cdots & (\lambda_i I - U_k)^{m_i} \end{bmatrix}$$

The matrix $(\lambda_i I - U_j)^{m_i}$ is invertible if $j \neq i$ and zero if $j = i$ (because then U_i is an $m_i \times m_i$ upper triangular matrix with each entry on the main diagonal equal to λ_i). It follows that $m_i = \dim[\text{null}(P^{-1}A_iP)]$, as required.

Lemma 2

If P is as in Theorem 1, denote the columns of P as follows:

$$\mathbf{p}_{11}, \mathbf{p}_{12}, \ldots, \mathbf{p}_{1m_1}; \quad \mathbf{p}_{21}, \mathbf{p}_{22}, \ldots, \mathbf{p}_{2m_2}; \quad \ldots; \quad \mathbf{p}_{k1}, \mathbf{p}_{k2}, \ldots, \mathbf{p}_{km_k}$$

Then $\{\mathbf{p}_{i1}, \mathbf{p}_{i2}, \ldots, \mathbf{p}_{im_i}\}$ is a basis of $G_{\lambda_i}(A)$.

PROOF

It suffices by Lemma 1 to show that each $\mathbf{p}_{ij}$ is in $G_{\lambda_i}(A)$. Write the matrix in Theorem 1 as $P^{-1}AP = \text{diag}(U_1, U_2, \ldots, U_k)$. Then

$$AP = P\,\text{diag}(U_1, U_2, \ldots, U_k)$$

Comparing columns gives, successively:

$A\mathbf{p}_{11} = \lambda_1 \mathbf{p}_{11}$, $\quad$ so $(\lambda_1 I - A)\mathbf{p}_{11} = \mathbf{0}$
$A\mathbf{p}_{12} = u\mathbf{p}_{11} + \lambda_1 \mathbf{p}_{12}$, $\quad$ so $(\lambda_1 I - A)^2 \mathbf{p}_{12} = \mathbf{0}$
$A\mathbf{p}_{13} = w\mathbf{p}_{11} + v\mathbf{p}_{12} + \lambda_1 \mathbf{p}_{13}$ $\quad$ so $(\lambda_1 I - A)^3 \mathbf{p}_{13} = \mathbf{0}$
$\vdots$ $\quad\quad\quad\quad\quad\quad$ $\vdots$

where u, v, w are in $\mathbb{R}$. In general, $(\lambda_1 I - A)^j \mathbf{p}_{1j} = \mathbf{0}$ for $j = 1, 2, \ldots, m_1$, so $\mathbf{p}_{1j}$ is in $G_{\lambda_1}(A)$. Similarly, $\mathbf{p}_{ij}$ is in $G_{\lambda_i}(A)$ for each i and j.

Lemma 3

If B_i is any basis of $G_{\lambda_i}(A)$, then $B = B_1 \cup B_2 \cup \cdots \cup B_k$ is a basis of $\mathbb{R}^n$.

SECTION 11.1 Block Triangular Form

PROOF

It suffices by Lemma 1 to show that B is independent. If a linear combination from B vanishes, let $\mathbf{x}_i$ be the sum of the terms from B_i. Then $\mathbf{x}_1 + \cdots + \mathbf{x}_k = \mathbf{0}$. But $\mathbf{x}_i = \sum_j r_{ij} \mathbf{p}_{ij}$ by Lemma 2, so $\sum_{i,j} r_{ij} \mathbf{p}_{ij} = \mathbf{0}$. Hence each $\mathbf{x}_i = \mathbf{0}$, so each coefficient in $\mathbf{x}_i$ is zero.

Lemma 2 suggests an algorithm for finding the matrix P in Theorem 1. Observe that there is an ascending chain of subspaces leading from $E_{\lambda_i}(A)$ to $G_{\lambda_i}(A)$:

$$E_{\lambda_i}(A) = \text{null}[(\lambda_i I - A)] \subseteq \text{null}[(\lambda_i I - A)^2] \subseteq \cdots \subseteq \text{null}[(\lambda_i I - A)^{m_i}] = G_{\lambda_i}(A)$$

We construct a basis for $G_{\lambda_i}(A)$ by climbing up this chain.

Triangulation Algorithm

Suppose A has characteristic polynomial

$$c_A(x) = (x - \lambda_1)^{m_1}(x - \lambda_2)^{m_2}\cdots(x - \lambda_k)^{m_k}$$

1. *Choose a basis of $\text{null}[(\lambda_1 I - A)]$; enlarge it by adding vectors (possibly none) to a basis of $\text{null}[(\lambda_1 I - A)^2]$; enlarge that to a basis of $\text{null}[(\lambda_1 I - A)^3]$, and so on. Continue to obtain an ordered basis $\{\mathbf{p}_{11}, \mathbf{p}_{12}, \ldots, \mathbf{p}_{1m_1}\}$ of $G_{\lambda_1}(A)$.*
2. *As in (1) choose a basis $\{\mathbf{p}_{i1}, \mathbf{p}_{i2}, \ldots, \mathbf{p}_{im_i}\}$ of $G_{\lambda_i}(A)$ for each i.*
3. *Let $P = [\mathbf{p}_{11}\ \mathbf{p}_{12}\cdots\mathbf{p}_{1m_1};\ \mathbf{p}_{21}\ \mathbf{p}_{22}\cdots\mathbf{p}_{2m_2};\ \ldots;\ \mathbf{p}_{k1}\ \mathbf{p}_{k2}\cdots\mathbf{p}_{km_k}]$ be the matrix with these basis vectors (in order) as columns.*

Then $P^{-1}AP = \text{diag}(U_1, U_2, \ldots, U_k)$ as in Theorem 1.

PROOF

Lemma 3 guarantees that $B = \{\mathbf{p}_{11}, \ldots, \mathbf{p}_{km_k}\}$ is a basis of $\mathbb{R}^n$, and Theorem 4 Section 9.2 shows that $P^{-1}AP = M_B(T_A)$. Now $G_{\lambda_i}(A)$ is T_A-invariant for each i because

$$(\lambda_i I - A)^{m_i}\mathbf{x} = \mathbf{0} \quad \text{implies} \quad (\lambda_i I - A)^{m_i}(A\mathbf{x}) = A(\lambda_i I - A)^{m_i}\mathbf{x} = \mathbf{0}$$

By Theorem 7 Section 9.3 (and induction), we have

$$P^{-1}AP = M_B(T_A) = \text{diag}(U_1, U_2, \ldots, U_k)$$

where U_i is the matrix of the restriction of T_A to $G_{\lambda_i}(A)$, and it remains to show that U_i has the desired upper triangular form. Given s, let $\mathbf{p}_{ij}$ be a basis vector in $\text{null}[(\lambda_i I - A)^{s+1}]$. Then $(\lambda_i I - A)\mathbf{p}_{ij}$ is in $\text{null}[(\lambda_i I - A)^s]$, and therefore is a linear combination of the basis vectors $\mathbf{p}_{it}$ coming *before* $\mathbf{p}_{ij}$. Hence

$$T_A(\mathbf{p}_{ij}) = A\mathbf{p}_{ij} = \lambda_i \mathbf{p}_{ij} - (\lambda_i I - A)\mathbf{p}_{ij}$$

shows that the column of U_i corresponding to $\mathbf{p}_{ij}$ has λ_i on the main diagonal and zeros below the main diagonal. This is what we wanted.

EXAMPLE 1

If $A = \begin{bmatrix} 2 & 0 & 0 & 1 \\ 0 & 2 & 0 & -1 \\ -1 & 1 & 2 & 0 \\ 0 & 0 & 0 & 2 \end{bmatrix}$, find P such that $P^{-1}AP$ is block triangular.

Solution ▶ $c_A(x) = \det[xI - A] = (x - 2)^4$, so $\lambda_1 = 2$ is the only eigenvalue and we are in the case $k = 1$ of Theorem 1. Compute:

$$(2I - A) = \begin{bmatrix} 0 & 0 & 0 & -1 \\ 0 & 0 & 0 & 1 \\ 1 & -1 & 0 & 0 \\ 0 & 0 & 0 & 0 \end{bmatrix} \quad (2I - A)^2 = \begin{bmatrix} 0 & 0 & 0 & 0 \\ 0 & 0 & 0 & 0 \\ 0 & 0 & 0 & -2 \\ 0 & 0 & 0 & 0 \end{bmatrix} \quad (2I - A)^3 = 0$$

By gaussian elimination find a basis $\{\mathbf{p}_{11}, \mathbf{p}_{12}\}$ of null$(2I - A)$; then extend in any way to a basis $\{\mathbf{p}_{11}, \mathbf{p}_{12}, \mathbf{p}_{13}\}$ of null$[(2I - A)^2]$; and finally get a basis $\{\mathbf{p}_{11}, \mathbf{p}_{12}, \mathbf{p}_{13}, \mathbf{p}_{14}\}$ of null$[(2I - A)^3] = \mathbb{R}^4$. One choice is

$$\mathbf{p}_{11} = \begin{bmatrix} 1 \\ 1 \\ 0 \\ 0 \end{bmatrix} \quad \mathbf{p}_{12} = \begin{bmatrix} 0 \\ 0 \\ 1 \\ 0 \end{bmatrix} \quad \mathbf{p}_{13} = \begin{bmatrix} 0 \\ 1 \\ 0 \\ 0 \end{bmatrix} \quad \mathbf{p}_{14} = \begin{bmatrix} 0 \\ 0 \\ 0 \\ 1 \end{bmatrix}$$

Hence $P = [\mathbf{p}_{11} \; \mathbf{p}_{12} \; \mathbf{p}_{13} \; \mathbf{p}_{14}] = \begin{bmatrix} 1 & 0 & 0 & 0 \\ 1 & 0 & 1 & 0 \\ 0 & 1 & 0 & 0 \\ 0 & 0 & 0 & 1 \end{bmatrix}$ gives $P^{-1}AP = \begin{bmatrix} 2 & 0 & 0 & 1 \\ 0 & 2 & 1 & 0 \\ 0 & 0 & 2 & -2 \\ 0 & 0 & 0 & 2 \end{bmatrix}$

EXAMPLE 2

If $A = \begin{bmatrix} 2 & 0 & 1 & 1 \\ 3 & 5 & 4 & 1 \\ -4 & -3 & -3 & -1 \\ 1 & 0 & 1 & 2 \end{bmatrix}$, find P such that $P^{-1}AP$ is block triangular.

Solution ▶ The eigenvalues are $\lambda_1 = 1$ and $\lambda_2 = 2$ because

$$c_A(x) = \begin{vmatrix} x-2 & 0 & -1 & -1 \\ -3 & x-5 & -4 & -1 \\ 4 & 3 & x+3 & 1 \\ -1 & 0 & -1 & x-2 \end{vmatrix} = \begin{vmatrix} x-1 & 0 & 0 & -x+1 \\ -3 & x-5 & -4 & -1 \\ 4 & 3 & x+3 & 1 \\ -1 & 0 & -1 & x-2 \end{vmatrix}$$

$$= \begin{vmatrix} x-1 & 0 & 0 & 0 \\ -3 & x-5 & -4 & -4 \\ 4 & 3 & x+3 & 5 \\ -1 & 0 & -1 & x-3 \end{vmatrix} = (x-1)\begin{vmatrix} x-5 & -4 & -4 \\ 3 & x+3 & 5 \\ 0 & -1 & x-3 \end{vmatrix}$$

$$= (x-1)\begin{vmatrix} x-5 & -4 & 0 \\ 3 & x+3 & -x+2 \\ 0 & -1 & x-2 \end{vmatrix} = (x-1)\begin{vmatrix} x-5 & -4 & 0 \\ 3 & x+2 & 0 \\ 0 & -1 & x-2 \end{vmatrix}$$

SECTION 11.1 Block Triangular Form

$$= (x-1)(x-2)\begin{vmatrix} x-5 & -4 \\ 3 & x+2 \end{vmatrix} = (x-1)^2(x-2)^2$$

By solving equations, we find $\text{null}(I - A) = \text{span}\{\mathbf{p}_{11}\}$ and $\text{null}(I - A)^2 = \text{span}\{\mathbf{p}_{11}, \mathbf{p}_{12}\}$ where

$$\mathbf{p}_{11} = \begin{bmatrix} 1 \\ 1 \\ -2 \\ 1 \end{bmatrix} \quad \mathbf{p}_{12} = \begin{bmatrix} 0 \\ 3 \\ -4 \\ 1 \end{bmatrix}$$

Since $\lambda_1 = 1$ has multiplicity 2 as a root of $c_A(x)$, $\dim G_{\lambda_1}(A) = 2$ by Lemma 1. Since $\mathbf{p}_{11}$ and $\mathbf{p}_{12}$ both lie in $G_{\lambda_1}(A)$, we have $G_{\lambda_1}(A) = \text{span}\{\mathbf{p}_{11}, \mathbf{p}_{12}\}$. Turning to $\lambda_2 = 2$, we find that $\text{null}(2I - A) = \text{span}\{\mathbf{p}_{21}\}$ and $\text{null}[(2I - A)^2] = \text{span}\{\mathbf{p}_{21}, \mathbf{p}_{22}\}$ where

$$\mathbf{p}_{21} = \begin{bmatrix} 1 \\ 0 \\ -1 \\ 1 \end{bmatrix} \quad \text{and} \quad \mathbf{p}_{22} = \begin{bmatrix} 0 \\ -4 \\ 3 \\ 0 \end{bmatrix}$$

Again, $\dim G_{\lambda_2}(A) = 2$ as λ_2 has multiplicity 2, so $G_{\lambda_2}(A) = \text{span}\{\mathbf{p}_{21}, \mathbf{p}_{22}\}$.

Hence $P = \begin{bmatrix} 1 & 0 & 1 & 0 \\ 1 & 3 & 0 & -4 \\ -2 & -4 & -1 & 3 \\ 1 & 1 & 1 & 0 \end{bmatrix}$ gives $P^{-1}AP = \begin{bmatrix} 1 & -3 & 0 & 0 \\ 0 & 1 & 0 & 0 \\ 0 & 0 & 2 & 3 \\ 0 & 0 & 0 & 2 \end{bmatrix}$.

If $p(x)$ is a polynomial and A is an $n \times n$ matrix, then $p(A)$ is also an $n \times n$ matrix if we interpret $A^0 = I_n$. For example, if $p(x) = x^2 - 2x + 3$, then $p(A) = A^2 - 2A + 3I$. Theorem 1 provides another proof of the Cayley-Hamilton theorem (see also Theorem 10 Section 8.6). As before, let $c_A(x)$ denote the characteristic polynomial of A.

Theorem 2

Cayley-Hamilton Theorem
If A is a square matrix with every eigenvalue real, then $c_A(A) = 0$.

PROOF

As in Theorem 1, write $c_A(x) = (x - \lambda_1)^{m_1} \cdots (x - \lambda_k)^{m_k} = \prod_{i=1}^{k}(x - \lambda_i)^{m_i}$, and write $P^{-1}AP = D = \text{diag}(U_1, \ldots, U_k)$. Hence

$$c_A(U_i) = \prod_{i=1}^{k}(U_i - \lambda_i I_{m_i})^{m_i} = 0 \text{ for each } i$$

because the factor $(U_i - \lambda_i I_{m_i})^{m_i} = 0$. In fact $U_i - \lambda_i I_{m_i}$ is $m_i \times m_i$ and has zeros on the main diagonal. But then

$$P^{-1}c_A(A)P = c_A(D) = c_A[\text{diag}(U_1, \ldots, U_k)]$$
$$= \text{diag}[c_A(U_1), \ldots, c_A(U_k)]$$
$$= 0$$

It follows that $c_A(A) = 0$.

EXAMPLE 3

If $A = \begin{bmatrix} 1 & 3 \\ -1 & 2 \end{bmatrix}$, then $c_A(x) = \det\begin{bmatrix} x-1 & -3 \\ 1 & x-2 \end{bmatrix} = x^2 - 3x + 5$. Then
$$c_A(A) = A^2 - 3A + 5I_2 = \begin{bmatrix} -2 & 9 \\ -3 & 1 \end{bmatrix} - \begin{bmatrix} 3 & 9 \\ -3 & 6 \end{bmatrix} + \begin{bmatrix} 5 & 0 \\ 0 & 5 \end{bmatrix} = \begin{bmatrix} 0 & 0 \\ 0 & 0 \end{bmatrix}.$$

Theorem 1 will be refined even further in the next section.

Proof of Theorem 1

The proof of Theorem 1 requires the following simple fact about bases, the proof of which we leave to the reader.

Lemma 4

If $\{\mathbf{v}_1, \mathbf{v}_2, \ldots, \mathbf{v}_n\}$ is a basis of a vector space V, so also is $\{\mathbf{v}_1 + s\mathbf{v}_2, \mathbf{v}_2, \ldots, \mathbf{v}_n\}$ for any scalar s.

PROOF OF THEOREM 1

Let A be as in Theorem 1, and let $T = T_A : \mathbb{R}^n \to \mathbb{R}^n$ be the matrix transformation induced by A. For convenience, call a matrix a λ-m-ut matrix if it is an $m \times m$ upper triangular matrix and every diagonal entry equals λ. Then we must find a basis B of $\mathbb{R}^n$ such that $M_B(T) = \text{diag}(U_1, U_2, \ldots, U_k)$ where U_i is a λ_i-m_i-ut matrix for each i. We proceed by induction on n. If $n = 1$, take $B = \{\mathbf{v}\}$ where $\mathbf{v}$ is any eigenvector of T.

If $n > 1$, let $\mathbf{v}_1$ be a λ_1-eigenvector of T, and let $B_0 = \{\mathbf{v}_1, \mathbf{w}_1, \ldots, \mathbf{w}_{n-1}\}$ be any basis of $\mathbb{R}^n$ containing $\mathbf{v}_1$. Then (see Lemma 2 Section 5.5)
$$M_{B_0}(T) = \begin{bmatrix} \lambda_1 & X \\ 0 & A_1 \end{bmatrix}$$
in block form where A_1 is $(n-1) \times (n-1)$. Moreover, A and $M_{B_0}(T)$ are similar, so
$$c_A(x) = c_{M_{B_0}(T)}(x) = (x - \lambda_1)c_{A_1}(x)$$
Hence $c_{A_1}(x) = (x - \lambda_1)^{m_1-1}(x - \lambda_2)^{m_2}\cdots(x - \lambda_k)^{m_k}$, so (by induction) let
$$Q^{-1}A_1Q = \text{diag}(Z_1, U_2, \ldots, U_k)$$
where Z_1 is a λ_1-$(m_1 - 1)$-ut matrix and U_i is a λ_i-m_i-ut matrix for each $i > 1$. If $P = \begin{bmatrix} 1 & 0 \\ 0 & Q \end{bmatrix}$, then $P^{-1}M_{B_0}(T) = \begin{bmatrix} \lambda_1 & XQ \\ 0 & Q^{-1}A_1Q \end{bmatrix} = A'$, say. Hence $A' \sim M_{B_0}(T) \sim A$ so by Theorem 4(2) Section 9.2 there is a basis B of $\mathbb{R}^n$ such that $M_{B_1}(T_A) = A'$, that is $M_{B_1}(T) = A'$. Hence $M_{B_1}(T)$ takes the block form

SECTION 11.1 Block Triangular Form

$$M_{B_1}(T) = \begin{bmatrix} \lambda_1 & XQ \\ 0 & \text{diag}(Z_1, U_2, \ldots, U_k) \end{bmatrix} = \left[\begin{array}{cc|cccc} \lambda_1 & X_1 & \multicolumn{4}{c}{Y} \\ 0 & Z_1 & 0 & 0 & 0 \\ \hline & & U_2 & \cdots & 0 \\ & 0 & & \vdots & \vdots \\ & & 0 & \cdots & U_k \end{array}\right] \quad (*)$$

If we write $U_1 = \begin{bmatrix} \lambda_1 & X_1 \\ 0 & Z_1 \end{bmatrix}$, the basis B_1 fulfills our needs except that the row matrix Y may not be zero.

We remedy this defect as follows. Observe that the first vector in the basis B_1 is a λ_1 eigenvector of T, which we continue to denote as $\mathbf{v}_1$. The idea is to add suitable scalar multiples of $\mathbf{v}_1$ to the other vectors in B_1. This results in a new basis by Lemma 4, and the multiples can be chosen so that the new matrix of T is the same as $(*)$ except that $Y = 0$. Let $\{\mathbf{w}_1, \ldots, \mathbf{w}_{m_2}\}$ be the vectors in B_1 corresponding to λ_2 (giving rise to U_2 in $(*)$). Write

$$U_2 = \begin{bmatrix} \lambda_2 & u_{12} & u_{13} & \cdots & u_{1m_2} \\ 0 & \lambda_2 & u_{23} & \cdots & u_{2m_2} \\ 0 & 0 & \lambda_2 & \cdots & u_{3m_2} \\ \vdots & \vdots & \vdots & & \vdots \\ 0 & 0 & 0 & \cdots & \lambda_2 \end{bmatrix} \quad \text{and} \quad Y = [y_1 \; y_2 \; \cdots \; y_{m_2}]$$

We first replace $\mathbf{w}_1$ by $\mathbf{w}'_1 = \mathbf{w}_1 + s\mathbf{v}_1$ where s is to be determined. Then $(*)$ gives

$$\begin{aligned} T(\mathbf{w}'_1) &= T(\mathbf{w}_1) + sT(\mathbf{v}_1) \\ &= (y_1 \mathbf{v}_1 + \lambda_2 \mathbf{w}_1) + s\lambda_1 \mathbf{v}_1 \\ &= y_1 \mathbf{v}_1 + \lambda_2(\mathbf{w}'_1 - s\mathbf{v}_1) + s\lambda_1 \mathbf{v}_1 \\ &= \lambda_2 \mathbf{w}'_1 + [(y_1 - s(\lambda_2 - \lambda_1)] \mathbf{v}_1 \end{aligned}$$

Because $\lambda_2 \neq \lambda_1$ we can choose s such that $T(\mathbf{w}'_1) = \lambda_2 \mathbf{w}'_1$. Similarly, let $\mathbf{w}'_2 = \mathbf{w}_2 + t\mathbf{v}_1$ where t is to be chosen. Then, as before,

$$\begin{aligned} T(\mathbf{w}'_2) &= T(\mathbf{w}_2) + tT(\mathbf{v}_1) \\ &= (y_2 \mathbf{v}_1 + u_{12} \mathbf{w}_1 + \lambda_2 \mathbf{w}_2) + t\lambda_1 \mathbf{v}_1 \\ &= u_{12} \mathbf{w}'_1 + \lambda_2 \mathbf{w}'_2 + [(y_2 - u_{12}s) - t(\lambda_2 - \lambda_1)] \mathbf{v}_1 \end{aligned}$$

Again, t can be chosen so that $T(\mathbf{w}'_2) = u_{12} \mathbf{w}'_1 + \lambda_2 \mathbf{w}'_2$. Continue in this way to eliminate $y_1, \ldots, y_{m_2}$. This procedure also works for $\lambda_3, \lambda_4, \ldots$ and so produces a new basis B such that $M_B(T)$ is as in $(*)$ but with $Y = 0$.

EXERCISES 11.1

1. In each case, find a matrix P such that $P^{-1}AP$ is in block triangular form as in Theorem 1.

(a) $A = \begin{bmatrix} 2 & 3 & 2 \\ -1 & -1 & -1 \\ 1 & 2 & 2 \end{bmatrix}$ ◆(b) $A = \begin{bmatrix} -5 & 3 & 1 \\ -4 & 2 & 1 \\ -4 & 3 & 0 \end{bmatrix}$

(c) $A = \begin{bmatrix} 0 & 1 & 1 \\ 2 & 3 & 6 \\ -1 & -1 & -2 \end{bmatrix}$ ◆(d) $A = \begin{bmatrix} -3 & -1 & 0 \\ 4 & -1 & 3 \\ 4 & -2 & 4 \end{bmatrix}$

(e) $A = \begin{bmatrix} -1 & -1 & -1 & 0 \\ 3 & 2 & 3 & -1 \\ 2 & 1 & 3 & -1 \\ 2 & 1 & 4 & -2 \end{bmatrix}$ ◆(f) $A = \begin{bmatrix} -3 & 6 & 3 & 2 \\ -2 & 3 & 2 & 2 \\ -1 & 3 & 0 & 1 \\ -1 & 1 & 2 & 0 \end{bmatrix}$

2. Show that the following conditions are equivalent for a linear operator T on a finite dimensional space V.

   (1) $M_B(T)$ is upper triangular for some ordered basis B of E.

   (2) A basis $\{\mathbf{b}_1, \ldots, \mathbf{b}_n\}$ of V exists such that, for each i, $T(\mathbf{b}_i)$ is a linear combination of $\mathbf{b}_1, \ldots, \mathbf{b}_i$.

   (3) There exist T-invariant subspaces $V_1 \subseteq V_2 \subseteq \cdots \subseteq V_n = V$ such that $\dim V_i = i$ for each i.

3. If A is an $n \times n$ invertible matrix, show that $A^{-1} = r_0 I + r_1 A + \cdots + r_{n-1} A^{n-1}$ for some scalars $r_0, r_1, \ldots, r_{n-1}$. [*Hint*: Cayley-Hamilton theorem.]

•4. If $T: V \to V$ is a linear operator where V is finite dimensional, show that $c_T(T) = 0$. [*Hint*: Exercise 26 Section 9.1.]

5. Define $T: \mathbf{P} \to \mathbf{P}$ by $T[p(x)] = xp(x)$. Show that:

   (a) T is linear and $f(T)[p(x)] = f(x)p(x)$ for all polynomials $f(x)$.

   (b) Conclude that $f(T) \neq 0$ for all nonzero polynomials $f(x)$. [See Exercise 4.]

SECTION 11.2 The Jordan Canonical Form

Two $m \times n$ matrices A and B are called row-equivalent if A can be carried to B using row operations and, equivalently, if $B = UA$ for some invertible matrix U. We know (Theorem 4 Section 2.6) that each $m \times n$ matrix is row-equivalent to a unique matrix in reduced row-echelon form, and we say that these reduced row-echelon matrices are *canonical forms* for $m \times n$ matrices using row operations. If we allow column operations as well, then $A \to UAV = \begin{bmatrix} I_r & 0 \\ 0 & 0 \end{bmatrix}$ for invertible U and V, and the canonical forms are the matrices $\begin{bmatrix} I_r & 0 \\ 0 & 0 \end{bmatrix}$ where r is the rank (this is the Smith normal form and is discussed in Theorem 3 Section 2.6). In this section, we discover the canonical forms for square matrices under similarity: $A \to P^{-1}AP$.

If A is an $n \times n$ matrix with distinct real eigenvalues $\lambda_1, \lambda_2, \ldots, \lambda_k$, we saw in Theorem 1 Section 11.1 that A is similar to a block triangular matrix; more precisely, an invertible matrix P exists such that

$$P^{-1}AP = \begin{bmatrix} U_1 & 0 & \cdots & 0 \\ 0 & U_2 & \cdots & 0 \\ \vdots & \vdots & \ddots & \vdots \\ 0 & 0 & \cdots & U_k \end{bmatrix} = \text{diag}(U_1, U_2, \ldots, U_k) \qquad (*)$$

where, for each i, U_i is upper triangular with λ_i repeated on the main diagonal. The Jordan canonical form is a refinement of this theorem. The proof we gave of $(*)$ is matrix theoretic because we wanted to give an algorithm for actually finding the matrix P. However, we are going to employ abstract methods here. Consequently, we reformulate Theorem 1 Section 11.1 as follows:

Theorem 1

Let $T: V \to V$ be a linear operator where $\dim V = n$. Assume that $\lambda_1, \lambda_2, \ldots, \lambda_k$ are the distinct eigenvalues of T, and that the λ_i are all real. Then there exists a basis F of V such that $M_F(T) = \text{diag}(U_1, U_2, \ldots, U_k)$ where, for each i, U_i is square, upper triangular, with λ_i repeated on the main diagonal.

SECTION 11.2 The Jordan Canonical Form

PROOF

Choose any basis $B = \{\mathbf{b}_1, \mathbf{b}_2, \ldots, \mathbf{b}_n\}$ of V and write $A = M_B(T)$. Since A has the same eigenvalues as T, Theorem 1 Section 11.1 shows that an invertible matrix P exists such that $P^{-1}AP = \text{diag}(U_1, U_2, \ldots, U_k)$ where the U_i are as in the statement of the Theorem. If $\mathbf{p}_j$ denotes column j of P and $C_B : V \to \mathbb{R}^n$ is the coordinate isomorphism, let $\mathbf{f}_j = C_B^{-1}(\mathbf{p}_j)$ for each j. Then $F = \{\mathbf{f}_1, \mathbf{f}_2, \ldots, \mathbf{f}_n\}$ is a basis of V and $C_B(\mathbf{f}_j) = \mathbf{p}_j$ for each j. This means that $P_{B \leftarrow F} = [C_B(\mathbf{f}_j)] = [\mathbf{p}_j] = P$, and hence (by Theorem 2 Section 9.2) that $P_{F \leftarrow B} = P^{-1}$. With this, column j of $M_F(T)$ is

$$C_F(T(\mathbf{f}_j)) = P_{F \leftarrow B} C_B(T(\mathbf{f}_j)) = P^{-1} M_B(T) C_B(\mathbf{f}_j) = P^{-1} A \mathbf{p}_j$$

for all j. Hence

$$M_F(T) = [C_F(T(\mathbf{f}_j))] = [P^{-1} A \mathbf{p}_j] = P^{-1} A [\mathbf{p}_j] = P^{-1} A P = \text{diag}(U_1, U_2, \ldots, U_k)$$

as required.

Definition 11.2 If $n \geq 1$, define the **Jordan block** $J_n(\lambda)$ to be the $n \times n$ matrix with λs on the main diagonal, 1s on the diagonal above, and 0s elsewhere. We take $J_1(\lambda) = [\lambda]$.

Hence

$$J_1(\lambda) = [\lambda], \quad J_2(\lambda) = \begin{bmatrix} \lambda & 1 \\ 0 & \lambda \end{bmatrix}, \quad J_3(\lambda) = \begin{bmatrix} \lambda & 1 & 0 \\ 0 & \lambda & 1 \\ 0 & 0 & \lambda \end{bmatrix}, \quad J_4(\lambda) = \begin{bmatrix} \lambda & 1 & 0 & 0 \\ 0 & \lambda & 1 & 0 \\ 0 & 0 & \lambda & 1 \\ 0 & 0 & 0 & \lambda \end{bmatrix}, \ldots$$

We are going to show that Theorem 1 holds with each block U_i replaced by Jordan blocks corresponding to eigenvalues. It turns out that the whole thing hinges on the case $\lambda = 0$. An operator T is called **nilpotent** if $T^m = 0$ for some $m \geq 1$, and in this case $\lambda = 0$ for every eigenvalue λ of T. Moreover, the converse holds by Theorem 1 Section 11.1. Hence the following lemma is crucial.

Lemma 1

Let $T : V \to V$ be a linear operator where $\dim V = n$, and assume that T is nilpotent; that is, $T^m = 0$ for some $m \geq 1$. Then V has a basis B such that

$$M_B(T) = \text{diag}(J_1, J_2, \ldots, J_k)$$

where each J_i is a Jordan block corresponding to $\lambda = 0$.[1]

A proof is given at the end of this section.

---

[1] The converse is true too: If $M_B(T)$ has this form for some basis B of V, then T is nilpotent.

Theorem 2

Real Jordan Canonical Form
Let $T: V \to V$ be a linear operator where $\dim V = n$, and assume that $\lambda_1, \lambda_2, \ldots, \lambda_m$ are the distinct eigenvalues of T and that the λ_i are all real. Then there exists a basis E of V such that

$$M_E(T) = \mathrm{diag}(U_1, U_2, \ldots, U_k)$$

in block form. Moreover, each U_j is itself block diagonal:

$$U_j = \mathrm{diag}(J_1, J_2, \ldots, J_k)$$

where each J_i is a Jordan block corresponding to some λ_j.

PROOF

Let $E = \{\mathbf{e}_1, \mathbf{e}_2, \ldots, \mathbf{e}_n\}$ be a basis of V as in Theorem 1, and assume that U_i is an $n_i \times n_i$ matrix for each i. Let

$$E_1 = \{\mathbf{e}_1, \ldots, \mathbf{e}_{n_1}\}, \quad E_2 = \{\mathbf{e}_{n_1+1}, \ldots, \mathbf{e}_{n_2}\}, \quad \ldots, \quad E_k = \{\mathbf{e}_{n_{k-1}+1}, \ldots, \mathbf{e}_{n_k}\}$$

where $n_k = n$, and define $V_i = \mathrm{span}\{E_i\}$ for each i. Because the matrix $M_E(T) = \mathrm{diag}(U_1, U_2, \ldots, U_m)$ is block diagonal, it follows that each V_i is T-invariant and $M_{E_i}(T) = U_i$ for each i. Let U_i have λ_i repeated along the main diagonal, and consider the restriction $T: V_i \to V_i$. Then $M_{E_i}(T - \lambda_i I_{n_i})$ is a nilpotent matrix, and hence $(T - \lambda_i I_{n_i})$ is a nilpotent operator on V_i. But then Lemma 1 shows that V_i has a basis B_i such that $M_{B_i}(T - \lambda_i I_{n_i}) = \mathrm{diag}(K_1, K_2, \ldots, K_t)$ where each K_i is a Jordan block corresponding to $\lambda = 0$. Hence

$$\begin{aligned} M_{B_i}(T) &= M_{B_i}(\lambda_i I_{n_i}) + M_{B_i}(T - \lambda_i I_{n_i}) \\ &= \lambda_i I_{n_i} + \mathrm{diag}(K_1, K_2, \ldots, K_t) = \mathrm{diag}(J_1, J_2, \ldots, J_k) \end{aligned}$$

where $J_i = \lambda_i I_{f_i} + K_i$ is a Jordan block corresponding to λ_i (where K_i is $f_i \times f_i$). Finally, $B = B_1 \cup B_2 \cup \cdots \cup B_k$ is a basis of V with respect to which T has the desired matrix.

Corollary 1

If A is an $n \times n$ matrix with real eigenvalues, an invertible matrix P exists such that $P^{-1}AP = \mathrm{diag}(J_1, J_2, \ldots, J_k)$ where each J_i is a Jordan block corresponding to an eigenvalue λ_i.

PROOF

Apply Theorem 2 to the matrix transformation $T_A: \mathbb{R}^n \to \mathbb{R}^n$ to find a basis B of $\mathbb{R}^n$ such that $M_B(T_A)$ has the desired form. If P is the (invertible) $n \times n$ matrix with the vectors of B as its columns, then $P^{-1}AP = M_B(T_A)$ by Theorem 4 Section 9.2.

Of course if we work over the field $\mathbb{C}$ of complex numbers rather than $\mathbb{R}$, the characteristic polynomial of a (complex) matrix A splits completely as a product of linear factors. The proof of Theorem 2 goes through to give

Theorem 3

Jordan Canonical Form[2]
Let $T: V \to V$ be a linear operator where $\dim V = n$, and assume that $\lambda_1, \lambda_2, \dots, \lambda_m$ are the distinct eigenvalues of T. Then there exists a basis F of V such that
$$M_F(T) = \mathrm{diag}(U_1, U_2, \dots, U_k)$$
in block form. Moreover, each U_j is itself block diagonal:
$$U_j = \mathrm{diag}(J_1, J_2, \dots, J_{t_j})$$
where each J_i is a Jordan block corresponding to some λ_i.

Camille Jordan. Photo © Corbis.

Except for the order of the Jordan blocks J_i, the Jordan canonical form is uniquely determined by the operator T. That is, for each eigenvalue λ the number and size of the Jordan blocks corresponding to λ is uniquely determined. Thus, for example, two matrices (or two operators) are similar if and only if they have the same Jordan canonical form. We omit the proof of uniqueness; it is best presented using modules in a course on abstract algebra.

Proof of Lemma 1

Lemma 1

Let $T: V \to V$ be a linear operator where $\dim V = n$, and assume that T is nilpotent; that is, $T^m = 0$ for some $m \geq 1$. Then V has a basis B such that
$$M_B(T) = \mathrm{diag}(J_1, J_2, \dots, J_k)$$
where each $J_i = J_{n_i}(0)$ is a Jordan block corresponding to $\lambda = 0$.

PROOF

The proof proceeds by induction on n. If $n = 1$, then T is a scalar operator, and so $T = 0$ and the lemma holds. If $n \geq 1$, we may assume that $T \neq 0$, so $m \geq 1$ and we may assume that m is chosen such that $T^m = 0$, but $T^{m-1} \neq 0$. Suppose $T^{m-1}\mathbf{u} \neq \mathbf{0}$ for some $\mathbf{u}$ in V.[3]

Claim. $\{\mathbf{u}, T\mathbf{u}, T^2\mathbf{u}, \dots, T^{m-1}\mathbf{u}\}$ is independent.

Proof. Suppose $a_0\mathbf{u} + a_1 T\mathbf{u} + a_2 T^2\mathbf{u} + \cdots + a_{m-1}T^{m-1}\mathbf{u} = \mathbf{0}$ where each a_i is in $\mathbb{R}$. Since $T^m = 0$, applying T^{m-1} gives $\mathbf{0} = T^{m-1}\mathbf{0} = a_0\, T^{m-1}\mathbf{u}$, whence $a_0 = 0$.

---

[2] This was first proved in 1870 by the French mathematician Camille Jordan (1838–1922) in his monumental *Traité des substitutions et des équations algébriques.*

[3] If $S: V \to V$ is an operator, we abbreviate $S(\mathbf{u})$ by $S\mathbf{u}$ for simplicity.

Hence $a_1 T\mathbf{u} + a_2 T^2 \mathbf{u} + \cdots + a_{m-1} T^{m-1} \mathbf{u} = \mathbf{0}$ and applying T^{m-2} gives $a_1 = 0$ in the same way. Continue in this fashion to obtain $a_i = 0$ for each i. This proves the Claim.

Now define $P = \text{span}\{\mathbf{u}, T\mathbf{u}, T^2\mathbf{u}, \ldots, T^{m-1}\mathbf{u}\}$. Then P is a T-invariant subspace (because $T^m = 0$), and $T : P \to P$ is nilpotent with matrix $M_B(T) = J_m(0)$ where $B = \{\mathbf{u}, T\mathbf{u}, T^2\mathbf{u}, \ldots, T^{m-1}\mathbf{u}\}$. Hence we are done, by induction, if $V = P \oplus Q$ where Q is T-invariant (then $\dim Q = n - \dim P < n$ because $P \neq 0$, and $T : Q \to Q$ is nilpotent). With this in mind, choose a T-invariant subspace Q of maximal dimension such that $P \cap Q = \{\mathbf{0}\}$.[4] We assume that $V \neq P \oplus Q$ and look for a contradiction.

Choose $\mathbf{x} \in V$ such that $\mathbf{x} \notin P \oplus Q$. Then $T^m \mathbf{x} = \mathbf{0} \in P \oplus Q$ while $T^0 \mathbf{x} = \mathbf{x} \notin P \oplus Q$. Hence there exists k, $1 \leq k \leq m$, such that $T^k \mathbf{x} \in P \oplus Q$ but $T^{k-1}\mathbf{x} \notin P \oplus Q$. Write $\mathbf{v} = T^{k-1}\mathbf{x}$, so that

$$\mathbf{v} \notin P \oplus Q \quad \text{and} \quad T\mathbf{v} \in P \oplus Q$$

Let $T\mathbf{v} = \mathbf{p} + \mathbf{q}$ with $\mathbf{p}$ in P and $\mathbf{q}$ in Q. Then $\mathbf{0} = T^{m-1}(T\mathbf{v}) = T^{m-1}\mathbf{p} + T^{m-1}\mathbf{q}$ so, since P and Q are T-invariant, $T^{m-1}\mathbf{p} = -T^{m-1}\mathbf{q} \in P \cap Q = \{\mathbf{0}\}$. Hence

$$T^{m-1}\mathbf{p} = \mathbf{0}.$$

Since $\mathbf{p} \in P$ we have $\mathbf{p} = a_0 \mathbf{u} + a_1 T\mathbf{u} + a_2 T^2 \mathbf{u} + \cdots + a_{m-1} T^{m-1} \mathbf{u}$ for $a_i \in \mathbb{R}$. Since $T^m = 0$, applying T^{m-1} gives $\mathbf{0} = T^{m-1}\mathbf{p} = a_0 T^{m-1}\mathbf{u}$, whence $a_0 = 0$. Thus $\mathbf{p} = T(\mathbf{p}_1)$ where $\mathbf{p}_1 = a_1 \mathbf{u} + a_2 T\mathbf{u} + \cdots + a_{m-1} T^{m-2} \mathbf{u} \in P$. If we write $\mathbf{v}_1 = \mathbf{v} - \mathbf{p}_1$ we have

$$T(\mathbf{v}_1) = T(\mathbf{v} - \mathbf{p}_1) = T\mathbf{v} - \mathbf{p} = \mathbf{q} \in Q.$$

Since $T(Q) \subseteq Q$, it follows that $T(Q + \mathbb{R}\mathbf{v}_1) \subseteq Q \subseteq Q + \mathbb{R}\mathbf{v}_1$. Moreover $\mathbf{v}_1 \notin Q$ (otherwise $\mathbf{v} = \mathbf{v}_1 + \mathbf{p}_1 \in P \oplus Q$, a contradiction). Hence $Q \subset Q + \mathbb{R}\mathbf{v}_1$ so, by the maximality of Q, we must have $(Q + \mathbb{R}\mathbf{v}_1) \cap P \neq \{\mathbf{0}\}$, say

$$\mathbf{0} \neq \mathbf{p}_2 = \mathbf{q}_1 + a\mathbf{v}_1 \quad \text{where} \quad \mathbf{p}_2 \in P, \quad \mathbf{q}_1 \in Q, \quad \text{and} \quad a \in \mathbb{R}.$$

Thus $a\mathbf{v}_1 = \mathbf{p}_2 - \mathbf{q}_1 \in P \oplus Q$. But since $\mathbf{v}_1 = \mathbf{v} - \mathbf{p}_1$ we have

$$a\mathbf{v} = a\mathbf{v}_1 + a\mathbf{p}_1 \in (P \oplus Q) + P = P \oplus Q$$

Since $\mathbf{v} \notin P \oplus Q$, this implies that $a = 0$. But then $\mathbf{p}_2 = \mathbf{q}_1 \in P \cap Q = \{\mathbf{0}\}$, a contradiction. This completes the proof.

EXERCISES 11.2

1. By direct computation, show that there is no invertible complex matrix C such that
$$C^{-1} \begin{bmatrix} 1 & 1 & 0 \\ 0 & 1 & 1 \\ 0 & 0 & 1 \end{bmatrix} C = \begin{bmatrix} 1 & 1 & 0 \\ 0 & 1 & 0 \\ 0 & 0 & 1 \end{bmatrix}.$$

♦2. Show that $\begin{bmatrix} a & 1 & 0 \\ 0 & a & 0 \\ 0 & 0 & b \end{bmatrix}$ is similar to $\begin{bmatrix} b & 0 & 0 \\ 0 & a & 1 \\ 0 & 0 & a \end{bmatrix}$.

3. (a) Show that every complex matrix is similar to its transpose.

   (b) Show every real matrix is similar to its transpose. [*Hint*: Show that $J_k(0)Q = Q[J_k(0)]^T$ where Q is the $k \times k$ matrix with 1s down the "counter diagonal", that is from the $(1, k)$-position to the $(k, 1)$-position.]

---

[4] Observe that there *is* at least one such subspace: $Q = \{\mathbf{0}\}$.

Appendix A: Complex Numbers

The fact that the square of every real number is nonnegative shows that the equation $x^2 + 1 = 0$ has no real root; in other words, there is no real number u such that $u^2 = -1$. So the set of real numbers is inadequate for finding all roots of all polynomials. This kind of problem arises with other number systems as well. The set of integers contains no solution of the equation $3x + 2 = 0$, and the rational numbers had to be invented to solve such equations. But the set of rational numbers is also incomplete because, for example, it contains no root of the polynomial $x^2 - 2$. Hence the real numbers were invented. In the same way, the set of complex numbers was invented, which contains all real numbers together with a root of the equation $x^2 + 1 = 0$. However, the process ends here: the complex numbers have the property that *every* polynomial with complex coefficients has a (complex) root. This fact is known as the fundamental theorem of algebra.

One pleasant aspect of the complex numbers is that, whereas describing the real numbers in terms of the rationals is a rather complicated business, the complex numbers are quite easy to describe in terms of real numbers. Every **complex number** has the form

$$a + bi$$

where a and b are real numbers, and i is a root of the polynomial $x^2 + 1$. Here a and b are called the **real part** and the **imaginary part** of the complex number, respectively. The real numbers are now regarded as special complex numbers of the form $a + 0i = a$, with zero imaginary part. The complex numbers of the form $0 + bi = bi$ with zero real part are called **pure imaginary** numbers. The complex number i itself is called the **imaginary unit** and is distinguished by the fact that

$$i^2 = -1$$

As the terms *complex* and *imaginary* suggest, these numbers met with some resistance when they were first used. This has changed; now they are essential in science and engineering as well as mathematics, and they are used extensively. The names persist, however, and continue to be a bit misleading: These numbers are no more "*complex*" than the real numbers, and the number i is no more "*imaginary*" than -1.

Much as for polynomials, two complex numbers are declared to be **equal** if and only if they have the same real parts and the same imaginary parts. In symbols,

$$a + bi = a' + b'i \quad \text{if and only if} \quad a = a' \text{ and } b = b'$$

The addition and subtraction of complex numbers is accomplished by adding and subtracting real and imaginary parts:

$$(a + bi) + (a' + b'i) = (a + a') + (b + b')i$$
$$(a + bi) - (a' + b'i) = (a - a') + (b - b')i$$

This is analogous to these operations for linear polynomials $a + bx$ and $a' + b'x$, and the multiplication of complex numbers is also analogous with one difference: $i^2 = -1$. The definition is

$$(a + bi)(a' + b'i) = (aa' - bb') + (ab' + ba')i$$

With these definitions of equality, addition, and multiplication, the complex numbers *satisfy all the basic arithmetical axioms adhered to by the real numbers* (the verifications are omitted). One consequence of this is that they can be manipulated in the obvious fashion, except that i^2 is replaced by -1 wherever it occurs, and the rule for equality must be observed.

EXAMPLE 1

If $z = 2 - 3i$ and $w = -1 + i$, write each of the following in the form $a + bi$:
$z + w$, $z - w$, zw, $\frac{1}{3}z$, and z^2.

Solution ▶
$$z + w = (2 - 3i) + (-1 + i) = (2 - 1) + (-3 + 1)i = 1 - 2i$$
$$z - w = (2 - 3i) - (-1 + i) = (2 + 1) + (-3 - 1)i = 3 - 4i$$
$$zw = (2 - 3i)(-1 + i) = (-2 - 3i^2) + (2 + 3)i = 1 + 5i$$
$$\tfrac{1}{3}z = \tfrac{1}{3}(2 - 3i) = \tfrac{2}{3} - i$$
$$z^2 = (2 - 3i)(2 - 3i) = (4 + 9i^2) + (-6 - 6)i = -5 - 12i$$

EXAMPLE 2

Find all complex numbers z such as that $z^2 = i$.

Solution ▶ Write $z = a + bi$; we must determine a and b. Now $z^2 = (a^2 - b^2) + (2ab)i$, so the condition $z^2 = i$ becomes

$$(a^2 - b^2) + (2ab)i = 0 + i$$

Equating real and imaginary parts, we find that $a^2 = b^2$ and $2ab = 1$. The solution is $a = b = \pm\frac{1}{\sqrt{2}}$, so the complex numbers required are $z = \frac{1}{\sqrt{2}} + \frac{1}{\sqrt{2}}i$ and $z = -\frac{1}{\sqrt{2}} - \frac{1}{\sqrt{2}}i$.

As for real numbers, it is possible to divide by every nonzero complex number z. That is, there exists a complex number w such that $wz = 1$. As in the real case, this number w is called the **inverse** of z and is denoted by z^{-1} or $\frac{1}{z}$. Moreover, if $z = a + bi$, the fact that $z \neq 0$ means that $a \neq 0$ or $b \neq 0$. Hence $a^2 + b^2 \neq 0$, and an explicit formula for the inverse is

$$\frac{1}{z} = \frac{a}{a^2 + b^2} - \frac{b}{a^2 + b^2}i$$

In actual calculations, the work is facilitated by two useful notions: the conjugate and the absolute value of a complex number. The next example illustrates the technique.

Appendix A: Complex Numbers 525

EXAMPLE 3

Write $\dfrac{3+2i}{2+5i}$ in the form $a+bi$.

Solution ▶ Multiply top and bottom by the complex number $2-5i$ (obtained from the denominator by negating the imaginary part). The result is

$$\frac{3+2i}{2+5i} = \frac{(2-5i)(3+2i)}{(2-5i)(2+5i)} = \frac{(6+10)+(4-15)i}{2^2-(5i)^2} = \tfrac{16}{29} - \tfrac{11}{29}i$$

Hence the simplified form is $\tfrac{16}{29} - \tfrac{11}{29}i$, as required.

The key to this technique is that the product $(2-5i)(2+5i) = 29$ in the denominator turned out to be a *real* number. The situation in general leads to the following notation: If $z = a + bi$ is a complex number, the **conjugate** of z is the complex number, denoted $\bar{z}$, given by

$$\bar{z} = a - bi \quad \text{where } z = a + bi$$

Hence $\bar{z}$ is obtained from z by negating the imaginary part. For example, $\overline{(2+3i)} = 2 - 3i$ and $\overline{(1-i)} = 1 + i$. If we multiply $z = a + bi$ by $\bar{z}$, we obtain

$$z\bar{z} = a^2 + b^2 \quad \text{where } z = a + bi$$

The real number $a^2 + b^2$ is always nonnegative, so we can state the following definition: The **absolute value** or **modulus** of a complex number $z = a + bi$, denoted by $|z|$, is the positive square root $\sqrt{a^2 + b^2}$; that is,

$$|z| = \sqrt{a^2 + b^2} \quad \text{where } z = a + bi$$

For example, $|2 - 3i| = \sqrt{2^2 + (-3)^2} = \sqrt{13}$ and $|1 + i| = \sqrt{1^2 + 1^2} = \sqrt{2}$.

Note that if a real number a is viewed as the complex number $a + 0i$, its absolute value (as a complex number) is $|a| = \sqrt{a^2}$, which agrees with its absolute value as a *real* number.

With these notions in hand, we can describe the technique applied in Example 3 as follows: When converting a quotient $\tfrac{z}{w}$ of complex numbers to the form $a + bi$, multiply top and bottom by the conjugate $\bar{w}$ of the denominator.

The following list contains the most important properties of conjugates and absolute values. Throughout, z and w denote complex numbers.

C1. $\overline{z \pm w} = \bar{z} \pm \bar{w}$

C2. $\overline{zw} = \bar{z}\,\bar{w}$

C3. $\overline{\left(\dfrac{z}{w}\right)} = \dfrac{\bar{z}}{\bar{w}}$

C4. $\overline{(\bar{z})} = z$

C5. z is real if and only if $\bar{z} = z$

C6. $z\bar{z} = |z|^2$

C7. $\dfrac{1}{z} = \dfrac{1}{|z|^2}\bar{z}$

C8. $|z| \geq 0$ *for all complex numbers z*

C9. $|z| = 0$ *if and only if $z = 0$*

C10. $|zw| = |z||w|$

C11. $\left|\dfrac{z}{w}\right| = \dfrac{|z|}{|w|}$

C12. $|z + w| = |z| + |w|$ **(triangle inequality)**

All these properties (except property C12) can (and should) be verified by the reader for arbitrary complex numbers $z = a + bi$ and $w = c + di$. They are not independent; for example, property C10 follows from properties C2 and C6.

The triangle inequality, as its name suggests, comes from a geometric representation of the complex numbers analogous to identification of the real numbers with the points of a line. The representation is achieved as follows: Introduce a rectangular coordinate system in the plane (Figure A.1), and identify the complex number $a + bi$ with the point (a, b). When this is done, the plane is called the **complex plane**. Note that the point $(a, 0)$ on the x axis now represents the *real* number $a = a + 0i$, and for this reason, the x axis is called the **real axis**. Similarly, the y axis is called the **imaginary axis**. The identification $(a, b) = a + bi$ of the geometric point (a, b) and the complex number $a + bi$ will be used in what follows without comment. For example, the origin will be referred to as 0.

This representation of the complex numbers in the complex plane gives a useful way of describing the absolute value and conjugate of a complex number $z = a + bi$. The absolute value $|z| = \sqrt{a^2 + b^2}$ is just the distance from z to the origin. This makes properties C8 and C9 quite obvious. The conjugate $\bar{z} = a - bi$ of z is just the reflection of z in the real axis (x axis), a fact that makes properties C4 and C5 clear.

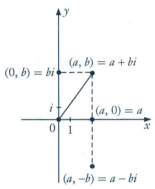
FIGURE A.1

Given two complex numbers $z_1 = a_1 + b_1 i = (a_1, b_1)$ and $z_2 = a_2 + b_2 i = (a_2, b_2)$, the absolute value of their difference

$$|z_1 - z_2| = \sqrt{(a_1 - a_2)^2 + (b_1 - b_2)^2}$$

is just the distance between them. This gives the **complex distance formula**:

$$|z_1 - z_2| \text{ is the distance between } z_1 \text{ and } z_2$$

This useful fact yields a simple verification of the triangle inequality, property C12. Suppose z and w are given complex numbers. Consider the triangle in Figure A.2 whose vertices are 0, w, and $z + w$. The three sides have lengths $|z|$, $|w|$, and $|z + w|$ by the complex distance formula, so the inequality

$$|z + w| \leq |z| + |w|$$

expresses the obvious geometric fact that the sum of the lengths of two sides of a triangle is at least as great as the length of the third side.

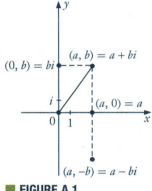
FIGURE A.2

The representation of complex numbers as points in the complex plane has another very useful property: It enables us to give a geometric description of the sum and product of two complex numbers. To obtain the description for the sum, let

$$z = a + bi = (a, b)$$
$$w = c + di = (c, d)$$

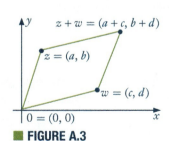
FIGURE A.3

denote two complex numbers. We claim that the four points 0, z, w, and $z + w$ form the vertices of a parallelogram. In fact, in Figure A.3 the lines from 0 to z and from w to $z + w$ have slopes

$$\frac{b - 0}{a - 0} = \frac{b}{a} \quad \text{and} \quad \frac{(b + d) - d}{(a + c) - c} = \frac{b}{a}$$

respectively, so these lines are parallel. (If it happens that $a = 0$, then both these lines are vertical.) Similarly, the lines from z to $z + w$ and from 0 to w are also parallel, so the figure with vertices 0, z, w, and $z + w$ is indeed a parallelogram. Hence, the complex number $z + w$ can be obtained geometrically from z and w by *completing* the parallelogram. This is sometimes called the **parallelogram law** of complex addition. Readers who have studied mechanics will recall that velocities and accelerations add in the same way; in fact, these are all special cases of *vector* addition.

Polar Form

The geometric description of what happens when two complex numbers are multiplied is at least as elegant as the parallelogram law of addition, but it requires that the complex numbers be represented in polar form. Before discussing this, we pause to recall the general definition of the trigonometric functions sine and cosine. An angle θ in the complex plane is in **standard position** if it is measured counterclockwise from the positive real axis as indicated in Figure A.4. Rather than using degrees to measure angles, it is more natural to use radian measure. This is defined as follows: The circle with its centre at the origin and radius 1 (called the **unit circle**) is drawn in Figure A.4. It has circumference 2π, and the **radian measure** of θ is the length of the arc on the unit circle counterclockwise from 1 to the point P on the unit circle determined by θ. Hence $90° = \frac{\pi}{2}$, $45° = \frac{\pi}{4}$, $180° = \pi$, and a full circle has the angle $360° = 2\pi$. Angles measured clockwise from 1 are negative; for example, $-i$ corresponds to $-\frac{\pi}{2}$ (or to $\frac{2\pi}{3}$).

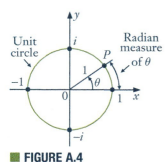

FIGURE A.4

Consider an angle θ in the range $0 \leq \theta \leq \frac{\pi}{2}$. If θ is plotted in standard position as in Figure A.4, it determines a unique point P on the unit circle, and P has coordinates $(\cos\theta, \sin\theta)$ by elementary trigonometry. However, *any* angle θ (acute or not) determines a unique point on the unit circle, so we *define* the **cosine** and **sine** of θ (written $\cos\theta$ and $\sin\theta$) to be the x and y coordinates of this point. For example, the points

$$1 = (1, 0) \qquad i = (0, 1) \qquad -1 = (-1, 0) \qquad -i = (0, -1)$$

plotted in Figure A.4 are determined by the angles $0, \frac{\pi}{2}, \pi, \frac{3\pi}{2}$, respectively. Hence

$$\cos 0 = 1 \qquad \cos\tfrac{\pi}{2} = 0 \qquad \cos\pi = -1 \qquad \cos\tfrac{3\pi}{2} = 0$$
$$\sin 0 = 0 \qquad \sin\tfrac{\pi}{2} = 1 \qquad \sin\pi = 0 \qquad \sin\tfrac{3\pi}{2} = -1$$

Now we can describe the polar form of a complex number. Let $z = a + bi$ be a complex number, and write the absolute value of z as

$$r = |z| = \sqrt{a^2 + b^2}$$

If $z \neq 0$, the angle θ shown in Figure A.5 is called an **argument** of z and is denoted

$$\theta = \arg z$$

This angle is not unique ($\theta + 2\pi k$ would do as well for any $k = 0, \pm 1, \pm 2, \ldots$). However, there is only one argument θ in the range $-\pi \leq \theta \leq \pi$, and this is sometimes called the **principal argument** of z.

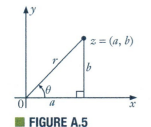

FIGURE A.5

Returning to Figure A.5, we find that the real and imaginary parts a and b of z are related to r and θ by

$$a = r\cos\theta$$
$$b = r\sin\theta$$

Hence the complex number $z = a + bi$ has the form

$$z = r(\cos\theta + i\sin\theta) \quad r = |z|, \theta = \arg(z)$$

The combination $\cos\theta + i\sin\theta$ is so important that a special notation is used:

$$e^{i\theta} = \cos\theta + i\sin\theta$$

is called **Euler's formula** after the great Swiss mathematician Leonhard Euler (1707–1783). With this notation, z is written

$$z = re^{i\theta} \quad r = |z|, \theta = \arg(z)$$

This is a **polar form** of the complex number z. Of course it is not unique, because the argument can be changed by adding a multiple of 2π.

Appendix A: Complex Numbers

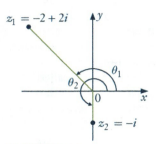

FIGURE A.6

EXAMPLE 4

Write $z_1 = -2 + 2i$ and $z_2 = -i$ in polar form.

Solution ▶ The two numbers are plotted in the complex plane in Figure A.6. The absolute values are

$$r_1 = |-2 + 2i| = \sqrt{(-2)^2 + 2^2} = 2\sqrt{2}$$

$$r_2 = |-i| = \sqrt{0^2 + (-1)^2} = 1$$

By inspection of Figure A.6, arguments of z_1 and z_2 are

$$\theta_1 = \arg(-2 + 2i) = \tfrac{3\pi}{4}$$

$$\theta_2 = \arg(-i) = \tfrac{3\pi}{2}$$

The corresponding polar forms are $z_1 = -2 + 2i = 2\sqrt{2}e^{3\pi i/4}$ and $z_2 = -i = e^{3\pi i/2}$. Of course, we could have taken the argument $-\tfrac{\pi}{2}$ for z_2 and obtained the polar form $z_2 = e^{-\pi i/2}$.

In Euler's formula $e^{i\theta} = \cos\theta + i\sin\theta$, the number e is the familiar constant $e = 2.71828\ldots$ from calculus. The reason for using e will not be given here; the reason why $\cos\theta + i\sin\theta$ is written as an *exponential* function of θ is that the **law of exponents** holds:

$$e^{i\theta} \cdot e^{i\phi} = e^{i(\theta+\phi)}$$

where θ and ϕ are any two angles. In fact, this is an immediate consequence of the addition identities for $\sin(\theta + \phi)$ and $\cos(\theta + \phi)$:

$$\begin{aligned} e^{i\theta}e^{i\phi} &= (\cos\theta + i\sin\theta)(\cos\phi + i\sin\phi) \\ &= (\cos\theta\cos\phi - \sin\theta\sin\phi) + i(\cos\theta\sin\phi + \sin\theta\cos\phi) \\ &= \cos(\theta + \phi) + i\sin(\theta + \phi) \\ &= e^{i(\theta+\phi)} \end{aligned}$$

This is analogous to the rule $e^a e^b = e^{a+b}$, which holds for real numbers a and b, so it is not unnatural to use the exponential notation $e^{i\theta}$ for the expression $\cos\theta + i\sin\theta$. In fact, a whole theory exists wherein functions such as e^z, $\sin z$, and $\cos z$ are studied, where z is a *complex* variable. Many deep and beautiful theorems can be proved in this theory, one of which is the so-called fundamental theorem of algebra mentioned later (Theorem 4). We shall not pursue this here.

The geometric description of the multiplication of two complex numbers follows from the law of exponents.

Theorem 1

Multiplication Rule

If $z_1 = r_1 e^{i\theta_1}$ and $z_2 = r_2 e^{i\theta_2}$ are complex numbers in polar form, then

$$z_1 z_2 = r_1 r_2 e^{i(\theta_2 + \theta_2)}$$

In other words, to multiply two complex numbers, simply multiply the absolute values and add the arguments. This simplifies calculations considerably, particularly when we observe that it is valid for *any* arguments θ_1 and θ_2.

EXAMPLE 5

Multiply $(1 - i)(1 + \sqrt{3}i)$ in two ways.

Solution ▶ We have $|1 - i| = \sqrt{2}$ and $|1 + \sqrt{3}i| = 2$ so, from Figure A.7,
$$1 - i = \sqrt{2}e^{-i\pi/4}$$
$$1 + \sqrt{3}i = 2e^{i\pi/3}$$

Hence, by the multiplication rule,
$$(1 - i)(1 + \sqrt{3}i) = (\sqrt{2}e^{-i\pi/4})(2e^{i\pi/3})$$
$$= 2\sqrt{2}e^{i(-\pi/4+\pi/3)}$$
$$= 2\sqrt{2}e^{i\pi/12}$$

This gives the required product in polar form. Of course, direct multiplication gives $(1 - i)(1 + \sqrt{3}i) = (\sqrt{3} + 1) + (\sqrt{3} - 1)i$. Hence, equating real and imaginary parts gives the formulas $\cos(\frac{\pi}{12}) = \frac{\sqrt{3}+1}{2\sqrt{2}}$ and $\sin(\frac{\pi}{12}) = \frac{\sqrt{3}-1}{2\sqrt{2}}$.

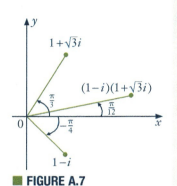

FIGURE A.7

Roots of Unity

If a complex number $z = re^{i\theta}$ is given in polar form, the powers assume a particularly simple form. In fact, $z^2 = (re^{i\theta})(re^{i\theta}) = r^2e^{2i\theta}$, $z^3 = z^2 \cdot z = (r^2e^{2i\theta})(re^{i\theta}) = r^3e^{3i\theta}$, and so on. Continuing in this way, it follows by induction that the following theorem holds for any positive integer n. The name honours Abraham De Moivre (1667–1754).

Theorem 2

De Moivre's Theorem
If θ is any angle, then $(e^{i\theta})^n = e^{in\theta}$ holds for all integers n.

PROOF

The case $n > 0$ has been discussed, and the reader can verify the result for $n = 0$. To derive it for $n < 0$, first observe that
$$\text{if} \quad z = re^{i\theta} \neq 0 \quad \text{then} \quad z^{-1} = \tfrac{1}{r}e^{-i\theta}$$

In fact, $(re^{i\theta})(\tfrac{1}{r}e^{-i\theta}) = 1e^{i0} = 1$ by the multiplication rule. Now assume that n is negative and write it as $n = -m$, $m > 0$. Then
$$(re^{i\theta})^n = [(re^{i\theta})^{-1}]^m = (\tfrac{1}{r}e^{-i\theta})^m = r^{-m}e^{i(-m\theta)} = r^ne^{in\theta}$$

If $r = 1$, this is De Moivre's theorem for negative n.

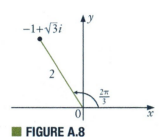

FIGURE A.8

EXAMPLE 6

Verify that $(-1 + \sqrt{3}i)^3 = 8$.

Solution ▶ We have $|-1 + \sqrt{3}i| = 8$, so $-1 + \sqrt{3}i = 2e^{2\pi i/3}$ (see Figure A.8). Hence De Moivre's theorem gives

$$(-1 + \sqrt{3}i)^3 = (2e^{2\pi i/3})^3 = 8e^{3(2\pi i/3)} = 8e^{2\pi i} = 8$$

De Moivre's theorem can be used to find nth roots of complex numbers where n is positive. The next example illustrates this technique.

EXAMPLE 7

Find the cube roots of unity; that is, find all complex numbers z such that $z^3 = 1$.

Solution ▶ First write $z = re^{i\theta}$ and $1 = 1e^{i0}$ in polar form. We must use the condition $z^3 = 1$ to determine r and θ. Because $z^3 = r^3 e^{3i\theta}$ by De Moivre's theorem, this requirement becomes

$$r^3 e^{3i\theta} = 1e^{0i}$$

These two complex numbers are equal, so their absolute values must be equal and the arguments must either be equal or differ by an integral multiple of 2π:

$$r^3 = 1$$

$$3\theta = 0 + 2k\pi, \quad k \text{ some integer}$$

Because r is real and positive, the condition $r^3 = 1$ implies that $r = 1$. However,

$$\theta = \frac{2k\pi}{3}, \quad k \text{ some integer}$$

seems at first glance to yield infinitely many different angles for z. However, choosing $k = 0, 1, 2$ gives three possible arguments θ (where $0 \leq \theta < 2\pi$), and the corresponding roots are

$$1e^{0i} = 1$$
$$1e^{2\pi i/3} = -\tfrac{1}{2} + \tfrac{\sqrt{3}}{2}i$$
$$1e^{4\pi i/3} = -\tfrac{1}{2} - \tfrac{\sqrt{3}}{2}i$$

These are displayed in Figure A.9. All other values of k yield values of θ that differ from one of these by a multiple of 2π—and so do not give new roots. Hence we have found all the roots.

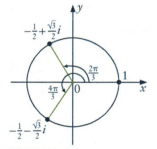

FIGURE A.9

The same type of calculation gives all complex n**th roots of unity**; that is, all complex numbers z such that $z^n = 1$. As before, write $1 = 1e^{0i}$ and

$$z = re^{i\theta}$$

in polar form. Then $z^n = 1$ takes the form

$$r^n e^{n\theta i} = 1e^{0i}$$

using De Moivre's theorem. Comparing absolute values and arguments yields

Appendix A: Complex Numbers

$$r^n = 1$$
$$n\theta = 0 + 2k\pi, \quad k \text{ some integer}$$

Hence $r = 1$, and the n values

$$\theta = \frac{2k\pi}{n}, \quad k = 0, 1, 2, \ldots, n-1$$

of θ all lie in the range $0 \leq \theta < 2\pi$. As in Example 7, *every* choice of k yields a value of θ that differs from one of these by a multiple of 2π, so these give the arguments of *all* the possible roots.

Theorem 3

nth Roots of Unity
If $n \geq 1$ is an integer, the nth roots of unity (that is, the solutions to $z^n = 1$) are given by

$$z = e^{2\pi k i/n}, \quad k = 0, 1, 2, \ldots, n-1$$

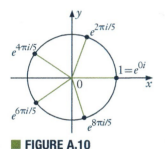

FIGURE A.10

The nth roots of unity can be found geometrically as the points on the unit circle that cut the circle into n equal sectors, starting at 1. The case $n = 5$ is shown in Figure A.10, where the five fifth roots of unity are plotted.

The method just used to find the nth roots of unity works equally well to find the nth roots of any complex number in polar form. We give one example.

EXAMPLE 8

Find the fourth roots of $\sqrt{2} + \sqrt{2}i$.

Solution ▶ First write $\sqrt{2} + \sqrt{2}i = 2e^{\pi i/4}$ in polar form. If $z = re^{i\theta}$ satisfies $z^4 = \sqrt{2} + \sqrt{2}i$, then De Moivre's theorem gives

$$r^4 e^{i(4\theta)} = 2e^{\pi i/4}$$

Hence $r^4 = 2$ and $4\theta = \frac{\pi}{4} + 2k\pi$, k an integer. We obtain four distinct roots (and hence all) by

$$r = \sqrt[4]{2}, \quad \theta = \frac{\pi}{16} + \frac{2k\pi}{16}, k = 0, 1, 2, 3$$

Thus the four roots are

$$\sqrt[4]{2}e^{\pi i/16} \quad \sqrt[4]{2}e^{9\pi i/16} \quad \sqrt[4]{2}e^{17\pi i/16} \quad \sqrt[4]{2}e^{25\pi i/16}$$

Of course, reducing these roots to the form $a + bi$ would require the computation of $\sqrt[4]{2}$ and the sine and cosine of the various angles.

An expression of the form $ax^2 + bx + c$, where the coefficients $a \neq 0$, b, and c are real numbers, is called a **real quadratic**. A complex number u is called a **root** of the quadratic if $au^2 + bu + c = 0$. The roots are given by the famous **quadratic formula**:

$$u = \frac{-b \pm \sqrt{b^2 - 4ac}}{2a}$$

The quantity $d = b^2 - 4ac$ is called the **discriminant** of the quadratic $ax^2 + bx + c$, and there is no real root if and only if $d < 0$. In this case the quadratic is said to be **irreducible**. Moreover, the fact that $d < 0$ means that $\sqrt{d} = i\sqrt{|d|}$, so the two (complex) roots are conjugates of each other:

Appendix A: Complex Numbers

$$u = \tfrac{1}{2a}(-b + i\sqrt{|d|}) \quad \text{and} \quad \bar{u} = \tfrac{1}{2a}(-b - i\sqrt{|d|})$$

The converse of this is true too: Given any nonreal complex number u, then u and $\bar{u}$ are the roots of some real irreducible quadratic. Indeed, the quadratic

$$x^2 - (u + \bar{u})x + u\bar{u} = (x - u)(x - \bar{u})$$

has real coefficients ($u\bar{u} = |u|^2$ and $u + \bar{u}$ is twice the real part of u) and so is irreducible because its roots u and $\bar{u}$ are not real.

EXAMPLE 9

Find a real irreducible quadratic with $u = 3 - 4i$ as a root.

Solution ▶ We have $u + \bar{u} = 6$ and $|u|^2 = 25$, so $x^2 - 6x + 25$ is irreducible with u and $\bar{u} = 3 + 4i$ as roots.

Fundamental Theorem of Algebra

As we mentioned earlier, the complex numbers are the culmination of a long search by mathematicians to find a set of numbers large enough to contain a root of every polynomial. The fact that the complex numbers have this property was first proved by Gauss in 1797 when he was 20 years old. The proof is omitted.

Theorem 4

Fundamental Theorem of Algebra
Every polynomial of positive degree with complex coefficients has a complex root.

If $f(x)$ is a polynomial with complex coefficients, and if u_1 is a root, then the factor theorem (Section 6.5) asserts that

$$f(x) = (x - u_1)g(x)$$

where $g(x)$ is a polynomial with complex coefficients and with degree one less than the degree of $f(x)$. Suppose that u_2 is a root of $g(x)$, again by the fundamental theorem. Then $g(x) = (x - u_2)h(x)$, so

$$f(x) = (x - u_1)(x - u_2)h(x)$$

This process continues until the last polynomial to appear is linear. Thus $f(x)$ has been expressed as a product of linear factors. The last of these factors can be written in the form $u(x - u_n)$, where u and u_n are complex (verify this), so the fundamental theorem takes the following form.

Theorem 5

Every complex polynomial $f(x)$ of degree $n \geq 1$ has the form

$$f(x) = u(x - u_1)(x - u_2)\cdots(x - u_n)$$

where, $u, u_1, \ldots, u_n$ are complex numbers and $u \neq 0$. The numbers $u_1, u_2, \ldots, u_n$ are the roots of $f(x)$ (and need not all be distinct), and u is the coefficient of x^n.

Appendix A: Complex Numbers

This form of the fundamental theorem, when applied to a polynomial $f(x)$ with *real* coefficients, can be used to deduce the following result.

Theorem 6

Every polynomial $f(x)$ of positive degree with real coefficients can be factored as a product of linear and irreducible quadratic factors.

In fact, suppose $f(x)$ has the form
$$f(x) = a_n x^n + a_{n-1} x^{n-1} + \cdots + a_1 x + a_0$$
where the coefficients a_i are real. If u is a complex root of $f(x)$, then we claim first that $\bar{u}$ is also a root. In fact, we have $f(u) = 0$, so
$$0 = \bar{0} = \overline{f(u)} = \overline{a_n u^n + a_{n-1} u^{n-1} + \cdots + a_1 u + a_0}$$
$$= \overline{a_n u^n} + \overline{a_{n-1} u^{n-1}} + \cdots + \overline{a_1 u} + \overline{a_0}$$
$$= \bar{a}_n \bar{u}^n + \bar{a}_{n-1} \bar{u}^{n-1} + \cdots + \bar{a}_1 \bar{u} + \bar{a}_0$$
$$= a_n \bar{u}^n + a_{n-1} \bar{u}^{n-1} + \cdots + a_1 \bar{u} + a_0$$
$$= f(\bar{u})$$

where $\bar{a}_i = a_i$ for each i because the coefficients a_i are real. Thus if u is a root of $f(x)$, so is its conjugate $\bar{u}$. Of course some of the roots of $f(x)$ may be real (and so equal their conjugates), but the nonreal roots come in pairs, u and $\bar{u}$. By Theorem 6, we can thus write $f(x)$ as a product:
$$f(x) = a_n(x - r_1) \cdots (x - r_k)(x - u_1)(x - \bar{u}_1) \cdots (x - u_m)(x - \bar{u}_m) \quad (*)$$
where a_n is the coefficient of x^n in $f(x)$; $r_1, r_2, \ldots, r_k$ are the real roots; and $u_1, \bar{u}_1, u_2, \bar{u}_2, \ldots, u_m, \bar{u}_m$ are the nonreal roots. But the product
$$(x - u_j)(x - \bar{u}_j) = x^2 - (u_j + \bar{u}_j)x + u_j \bar{u}_j$$
is a real irreducible quadratic for each j (see the discussion preceding Example 9). Hence $(*)$ shows that $f(x)$ is a product of linear and irreducible quadratic factors, each with real coefficients. This is the conclusion in Theorem 6.

EXERCISES A

1. Solve each of the following for the real number x.
   (a) $x - 4i = (2 - i)^2$
   ◆(b) $(2 + xi)(3 - 2i) = 12 + 5i$
   (c) $(2 + xi)^2 = 4$
   ◆(d) $(2 + xi)(2 - xi) = 5$

2. Convert each of the following to the form $a + bi$.
   (a) $(2 - 3i) - 2(2 - 3i) + 9$
   ◆(b) $(3 - 2i)(1 + i) + |3 + 4i|$
   (c) $\dfrac{1+i}{2-3i} + \dfrac{1-i}{-2+3i}$
   ◆(d) $\dfrac{3-2i}{1-i} + \dfrac{3-7i}{2-3i}$
   (e) i^{131}
   ◆(f) $(2 - i)^3$
   (g) $(1 + i)^4$
   ◆(h) $(1 - i)^2(2 + i)^2$
   (i) $\dfrac{3\sqrt{3} - i}{\sqrt{3} + i} + \dfrac{\sqrt{3} + 7i}{\sqrt{3} - i}$

3. In each case, find the complex number z.
   (a) $iz - (1 + i)^2 = 3 - i$
   ◆(b) $(i + z) - 3i(2 - z) = iz + 1$
   (c) $z^2 = -i$
   ◆(d) $z^2 = 3 - 4i$
   (e) $z(1 + i) = \bar{z} + (3 + 2i)$
   ◆(f) $z(2 - i) = (\bar{z} + 1)(1 + i)$

4. In each case, find the roots of the real quadratic equation.
   (a) $x^2 - 2x + 3 = 0$
   ◆(b) $x^2 - x + 1 = 0$
   (c) $3x^2 - 4x + 2 = 0$
   ◆(d) $2x^2 - 5x + 2 = 0$

5. Find all numbers x in each case.
   (a) $x^3 = 8$
   ◆(b) $x^3 = -8$
   (c) $x^4 = 16$
   ◆(d) $x^4 = 64$

6. In each case, find a real quadratic with u as a root, and find the other root.
   (a) $u = 1 + i$
   ◆(b) $u = 2 - 3i$
   (c) $u = -i$
   ◆(d) $u = 3 - 4i$

7. Find the roots of $x^2 - 2\cos\theta\, x + 1 = 0$, θ any angle.

◆8. Find a real polynomial of degree 4 with $2 - i$ and $3 - 2i$ as roots.

9. Let re z and im z denote, respectively, the real and imaginary parts of z. Show that:
   (a) im$(iz) = $ re z
   (b) re$(iz) = -$im z
   (c) $z + \bar{z} = 2$ re z
   (d) $z - \bar{z} = 2i$ im z
   (e) re$(z + w) = $ re $z + $ re w, and re$(tz) = t \cdot $ re z if t is real
   (f) im$(z + w) = $ im $z + $ im w, and im$(tz) = t \cdot $ im z if t is real

10. In each case, show that u is a root of the quadratic equation, and find the other root.
    (a) $x^2 - 3ix + (-3 + i) = 0$; $u = 1 + i$
    ◆(b) $x^2 + ix - (4 - 2i) = 0$; $u = -2$
    (c) $x^2 - (3 - 2i)x + (5 - i) = 0$; $u = 2 - 3i$
    ◆(d) $x^2 + 3(1 - i)x - 5i = 0$; $u = -2 + i$

11. Find the roots of each of the following complex quadratic equations.
    (a) $x^2 + 2x + (1 + i) = 0$
    ◆(b) $x^2 - x + (1 - i) = 0$
    (c) $x^2 - (2 - i)x + (3 - i) = 0$
    ◆(d) $x^2 - 3(1 - i)x - 5i = 0$

12. In each case, describe the graph of the equation (where z denotes a complex number).
    (a) $|z| = 1$
    ◆(b) $|z - 1| = 2$
    (c) $z = i\bar{z}$
    ◆(d) $z = -\bar{z}$
    (e) $z = |z|$
    ◆(f) im $z = m \cdot $ re z, m a real number

13. (a) Verify $|zw| = |z||w|$ directly for $z = a + bi$ and $w = c + di$.
    (b) Deduce (a) from properties C2 and C6.

14. Prove that $|z + w|^2 = |z|^2 + |w|^2 + w\bar{z} + \bar{w}z$ for all complex numbers w and z.

15. If zw is real and $z \neq 0$, show that $w = a\bar{z}$ for some real number a.

16. If $z\bar{w} = \bar{z}v$ and $z \neq 0$, show that $w = uv$ for some u in $\mathbb{C}$ with $|u| = 1$.

17. Use property C5 to show that $(1 + i)^n + (1 - i)^n$ is real for all n.

18. Express each of the following in polar form (use the principal argument).
    (a) $3 - 3i$
    ◆(b) $-4i$
    (c) $-\sqrt{3} + i$
    ◆(d) $-4 + 4\sqrt{3}i$
    (e) $-7i$
    ◆(f) $-6 + 6i$

19. Express each of the following in the form $a + bi$.
    (a) $3e^{\pi i}$
    ◆(b) $e^{7\pi i/3}$
    (c) $2e^{3\pi i/4}$
    ◆(d) $\sqrt{2}e^{-\pi i/4}$
    (e) $e^{5\pi i/4}$
    ◆(f) $2\sqrt{3}e^{-2\pi i/6}$

20. Express each of the following in the form $a + bi$.
    (a) $(-1 + \sqrt{3}i)^2$
    ◆(b) $(1 + \sqrt{3}i)^{-4}$
    (c) $(1 + i)^8$
    ◆(d) $(1 - i)^{10}$
    (e) $(1 - i)^6(\sqrt{3} + i)^3$
    ◆(f) $(\sqrt{3} - i)^9(2 - 2i)^5$

21. Use De Moivre's theorem to show that:
    (a) $\cos 2\theta = \cos^2\theta - \sin^2\theta$;
        $\sin 2\theta = 2\cos\theta\sin\theta$
    (b) $\cos 3\theta = \cos^3\theta - 3\cos\theta\sin^2\theta$;
        $\sin 3\theta = 3\cos^2\theta\sin\theta - \sin^3\theta$

22. (a) Find the fourth roots of unity.
    (b) Find the sixth roots of unity.

23. Find all complex numbers z such that:
    (a) $z^4 = -1$
    ◆(b) $z^4 = 2(\sqrt{3}i - 1)$
    (c) $z^3 = -27i$
    ◆(d) $z^6 = -64$

24. If $z = re^{i\theta}$ in polar form, show that:
    (a) $\bar{z} = re^{-i\theta}$
    ◆(b) $z^{-1} = \frac{1}{r}e^{-i\theta}$ if $z \neq 0$

25. Show that the sum of the nth roots of unity is zero. [*Hint:* $1 - z^n = (1 - z)(1 + z + z^2 + \cdots + z^{n-1})$ for any complex number z.]

26. (a) Suppose $z_1, z_2, z_3, z_4,$ and z_5 are equally spaced around the unit circle. Show that $z_1 + z_2 + z_3 + z_4 + z_5 = 0$. [*Hint:* $(1 - z)(1 + z + z^2 + z^3 + z^4) = 1 - z^5$ for any complex number z.]
    ◆(b) Repeat (a) for any $n \geq 2$ points equally spaced around the unit circle.

(c) If $|w| = 1$, show that the sum of the roots of $z^n = w$ is zero.

27. If z^n is real, $n \geq 1$, show that $(\bar{z})^n$ is real.

28. If $\bar{z}^2 = z^2$, show that z is real or pure imaginary.

29. If a and b are *rational* numbers, let p and q denote numbers of the form $a + b\sqrt{2}$. If $p = a + b\sqrt{2}$, define $\tilde{p} = a - b\sqrt{2}$ and $[p] = a^2 - 2b^2$. Show that each of the following holds.
    (a) $a + b\sqrt{2} = a_1 + b_1\sqrt{2}$ only if $a = a_1$ and $b = b_1$
    (b) $\widetilde{p \pm q} = \tilde{p} \pm \tilde{q}$
    (c) $\widetilde{pq} = \tilde{p}\tilde{q}$
    (d) $[p] = p\tilde{p}$
    (e) $[pq] = [p][q]$
    (f) If $f(x)$ is a polynomial with rational coefficients and $p = a + b\sqrt{2}$ is a root of $f(x)$, then $\tilde{p}$ is also a root of $f(x)$.

Appendix B: Proofs

Logic plays a basic role in human affairs. Scientists use logic to draw conclusions from experiments, judges use it to deduce consequences of the law, and mathematicians use it to prove theorems. Logic arises in ordinary speech with assertions such as "If John studies hard, he will pass the course," or "If an integer n is divisible by 6, then n is divisible by 3."[1] In each case, the aim is to assert that if a certain statement is true, then another statement must also be true. In fact, if p and q denote statements, most theorems take the form of an **implication**: "If p is true, then q is true." We write this in symbols as

$$p \Rightarrow q$$

and read it as "p implies q." Here p is the **hypothesis** and q the **conclusion** of the implication. The verification that $p \Rightarrow q$ is valid is called the **proof** of the implication. In this section we examine the most common methods of proof[2] and illustrate each technique with some examples.

Method of Direct Proof

To prove that $p \Rightarrow q$, demonstrate directly that q is true whenever p is true.

EXAMPLE 1

If n is an odd integer, show that n^2 is odd.

Solution ▶ If n is odd, it has the form $n = 2k + 1$ for some integer k. Then $n^2 = 4k^2 + 4k + 1 = 2(2k^2 + 2k) + 1$ also is odd because $2k^2 + 2k$ is an integer.

Note that the computation $n^2 = 4k^2 + 4k + 1$ in Example 1 involves some simple properties of arithmetic that we did not prove. These properties, in turn, can be proved from certain more basic properties of numbers (called axioms)—more about that later. Actually, a whole body of mathematical information lies behind nearly every proof of any complexity, although this fact usually is not stated explicitly. Here is a geometrical example.

---

1. By an *integer* we mean a "whole number"; that is, a number in the set 0, ±1, ±2, ±3, ...
2. For a more detailed look at proof techniques see D. Solow, How to Read and Do Proofs, 2nd ed. (New York: Wiley, 1990); or J. F. Lucas. *Introduction to Abstract Mathematics*, Chapter 2 (Belmont, CA: Wadsworth, 1986).

Appendix B: Proofs

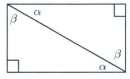

EXAMPLE 2

In a right triangle, show that the sum of the two acute angles is 90 degrees.

Solution ▶ The right triangle is shown in the diagram. Construct a rectangle with sides of the same length as the short sides of the original triangle, and draw a diagonal as shown. The original triangle appears on the bottom of the rectangle, and the top triangle is identical to the original (but rotated). Now it is clear that $\alpha + \beta$ is a right angle.

Geometry was one of the first subjects in which formal proofs were used—Euclid's *Elements* was published about 300 B.C. The *Elements* is the most successful textbook ever written, and contains many of the basic geometrical theorems that are taught in school today. In particular, Euclid included a proof of an earlier theorem (about 500 B.C.) due to Pythagoras. Recall that, in a right triangle, the side opposite the right angle is called the *hypotenuse* of the triangle.

EXAMPLE 3

Pythagoras' Theorem
In a right-angled triangle, show that the square of the length of the hypotenuse equals the sum of the squares of the lengths of the other two sides.

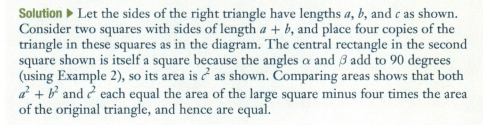

Solution ▶ Let the sides of the right triangle have lengths a, b, and c as shown. Consider two squares with sides of length $a + b$, and place four copies of the triangle in these squares as in the diagram. The central rectangle in the second square shown is itself a square because the angles α and β add to 90 degrees (using Example 2), so its area is c^2 as shown. Comparing areas shows that both $a^2 + b^2$ and c^2 each equal the area of the large square minus four times the area of the original triangle, and hence are equal.

Sometimes it is convenient (or even necessary) to break a proof into parts, and deal with each case separately. We formulate the general method as follows:

Method of Reduction to Cases

To prove that $p \Rightarrow q$, show that p implies at least one of a list $p_1, p_2, \ldots, p_n$ of statements (the cases) and then show that $p_i \Rightarrow q$ for each i.

EXAMPLE 4

Show that $n^2 \geq 0$ for every integer n.

Solution ▶ This statement can be expressed as an implication: If n is an integer, then $n^2 \geq 0$. To prove it, consider the following three cases:

(1) $n > 0$;   (2) $n = 0$;   (3) $n < 0$.

Then $n^2 > 0$ in Cases (1) and (3) because the product of two positive (or two negative) integers is positive. In Case (2) $n^2 = 0^2 = 0$, so $n^2 \geq 0$ in every case.

EXAMPLE 5

If n is an integer, show that $n^2 - n$ is even.

Solution ▶ We consider two cases:

(1) n is even; (2) n is odd.

We have $n^2 - n = n(n - 1)$, so this is even in Case (1) because any multiple of an even number is again even. Similarly, $n - 1$ is even in Case (2) so $n(n - 1)$ is again even for the same reason. Hence $n^2 - n$ is even in any case.

The statements used in mathematics are required to be either true or false. This leads to a proof technique which causes consternation in many beginning students. The method is a formal version of a debating strategy whereby the debater assumes the truth of an opponent's position and shows that it leads to an absurd conclusion.

Method of Proof by Contradiction

To prove that $p \Rightarrow q$, show that the assumption that both p is true and q is false leads to a contradiction. In other words, if p is true, then q must be true; that is, $p \Rightarrow q$.

EXAMPLE 6

If r is a rational number (fraction), show that $r^2 \neq 2$.

Solution ▶ To argue by contradiction, we assume that r is a rational number and that $r^2 = 2$, and show that this assumption leads to a contradiction. Let m and n be integers such that $r = m/n$ is in lowest terms (so, in particular, m and n are not both even). Then $r^2 = 2$ gives $m^2 = 2n^2$, so m^2 is even. This means m is even (Example 1), say $m = 2k$. But then $2n^2 = m^2 = 4k^2$, so $n^2 = 2k^2$ is even, and hence n is even. This shows that n and m are both even, contrary to the choice of these numbers.

EXAMPLE 7

Pigeonhole Principle
If $n + 1$ pigeons are placed in n holes, then some hole contains at least 2 pigeons.

Solution ▶ Assume the conclusion is false. Then each hole contains at most one pigeon and so, since there are n holes, there must be at most n pigeons, contrary to assumption.

The next example involves the notion of a *prime* number, that is an integer that is greater than 1 which cannot be factored as the product of two smaller positive integers both greater than 1. The first few primes are 2, 3, 5, 7, 11, ...

EXAMPLE 8

If $2^n - 1$ is a prime number, show that n is a prime number.

Solution ▶ We must show that $p \Rightarrow q$ where p is the statement "$2^n - 1$ *is a* *prime*", and q is the statement "*n is a prime.*" Suppose that p is true but q is false so that n is not a prime, say $n = ab$ where $a \geq 2$ and $b \geq 2$ are integers. If we write $2^a = x$, then $2^n = 2^{ab} = (2^a)^b = x^b$. Hence $2^n - 1$ factors:

$$2^n - 1 = x^b - 1 = (x - 1)(x^{b-1} + x^{b-2} + \cdots + x^2 + x + 1)$$

As $x \geq 4$, this expression is a factorization of $2^n - 1$ into smaller positive integers, contradicting the assumption that $2^n - 1$ is prime.

The next example exhibits one way to show that an implication is *not* valid.

EXAMPLE 9

Show that the implication "*n is a prime* $\Rightarrow$ $2^n - 1$ *is a prime*" is false.

Solution ▶ The first four primes are 2, 3, 5, and 7, and the corresponding values for $2^n - 1$ are 3, 7, 31, 127 (when $n = 2, 3, 5, 7$). These are all prime as the reader can verify. This result seems to be evidence that the implication is true. However, the next prime is 11 and $2^{11} - 1 = 2047 = 23 \cdot 89$, which is clearly *not* a prime.

We say that $n = 11$ is a **counterexample** to the (proposed) implication in Example 9. Note that, if you can find even one example for which an implication is not valid, the implication is false. Thus disproving implications is in a sense easier than proving them.

The implications in Examples 8 and 9 are closely related: They have the form $p \Rightarrow q$ and $q \Rightarrow p$, where p and q are statements. Each is called the **converse** of the other and, as these examples show, an implication can be valid even though its converse is not valid. If *both* $p \Rightarrow q$ and $q \Rightarrow p$ are valid, the statements p and q are called **logically equivalent**. This is written in symbols as

$$p \Leftrightarrow q$$

and is read "*p if and only if q*". Many of the most satisfying theorems make the assertion that two statements, ostensibly quite different, are in fact logically equivalent.

EXAMPLE 10

If n is an integer, show that "n is odd $\Leftrightarrow n^2$ is odd."

Solution ▶ In Example 1 we proved the implication "n is odd $\Rightarrow n^2$ is odd." Here we prove the converse by contradiction. If n^2 is odd, we assume that n is not odd. Then n is even, say $n = 2k$, so $n^2 = 4k^2$, which is also even, a contradiction.

Many more examples of proofs can be found in this book and, although they are often more complex, most are based on one of these methods. In fact, linear algebra is one of the best topics on which the reader can sharpen his or her skill at constructing proofs. Part of the reason for this is that much of linear algebra is developed using the **axiomatic method**. That is, in the course of studying various examples it is observed that they all have certain properties in common. Then a general, abstract system is studied in which these basic properties are *assumed* to hold (and are called **axioms**). In this system, statements (called **theorems**) are deduced from the axioms using the methods presented in this appendix. These theorems will then be true in *all* the concrete examples, because the axioms hold in each case. But this procedure is more than just an efficient method for finding theorems in the examples. By reducing the proof to its essentials, we gain a better understanding of why the theorem is true and how it relates to analogous theorems in other abstract systems.

The axiomatic method is not new. Euclid first used it in about 300 B.C. to derive all the propositions of (euclidean) geometry from a list of 10 axioms. The method lends itself well to linear algebra. The axioms are simple and easy to understand, and there are only a few of them. For example, the theory of vector spaces contains a large number of theorems derived from only ten simple axioms.

EXERCISES B

1. In each case prove the result and either prove the converse or give a counterexample.

   (a) If n is an even integer, then n^2 is a multiple of 4.

   ◆(b) If m is an even integer and n is an odd integer, then $m + n$ is odd.

   (c) If $x = 2$ or $x = 3$, then $x^3 - 6x^2 + 11x - 6 = 0$.

   ◆(d) If $x^2 - 5x + 6 = 0$, then $x = 2$ or $x = 3$.

2. In each case either prove the result by splitting into cases, or give a counterexample.

   (a) If n is any integer, then $n^2 = 4k + 1$ for some integer k.

   ◆(b) If n is any odd integer, then $n^2 = 8k + 1$ for some integer k.

   (c) If n is any integer, $n^3 - n = 3k$ for some integer k. [*Hint*: Use the fact that each integer has one of the forms $3k$, $3k + 1$, or $3k + 2$, where k is an integer.]

3. In each case prove the result by contradiction and either prove the converse or give a counterexample.

   (a) If $n > 2$ is a prime integer, then n is odd.

   ◆(b) If $n + m = 25$ where n and m are integers, then one of n and m is greater than 12.

   (c) If a and b are positive numbers and $a \leq b$, then $\sqrt{a} \leq \sqrt{b}$.

   (d) If m and n are integers and mn is even, then m is even or n is even.

4. Prove each implication by contradiction.

   (a) If x and y are positive numbers, then $\sqrt{x + y} \neq \sqrt{x} + \sqrt{y}$.

   ◆(b) If x is irrational and y is rational, then $x + y$ is irrational.

   (c) If 13 people are selected, at least 2 have birthdays in the same month.

5. Disprove each statement by giving a counterexample.

   (a) $n^2 + n + 11$ is a prime for all positive integers n.

   ◆(b) $n^3 \geq 2^n$ for all integers $n \geq 2$.

   (c) If $n \geq 2$ points are arranged on a circle in such a way that no three of the lines joining them have a common point, then these lines divide the circle into 2^{n-1} regions. [The cases $n = 2, 3,$ and 4 are shown in the diagram.]

$n = 2 \qquad\qquad n = 3 \qquad\qquad n = 4$

6. The number e from calculus has a series expansion

$$e = 1 + \frac{1}{1!} + \frac{1}{2!} + \frac{1}{3!} + \cdots$$

where $n! = n(n-1) \cdots 3 \cdot 2 \cdot 1$ for each integer $n \geq 1$. Prove that e is irrational by contradiction. [*Hint*: If $e = m/n$, consider

$$k = n!\left(e - 1 - \frac{1}{1!} - \frac{1}{2!} - \frac{1}{3!} - \cdots - \frac{1}{n!}\right).$$

Show that k is a positive integer and that

$$k = \frac{1}{n+1} + \frac{1}{(n+1)(n+2)} + \cdots < \frac{1}{n}.]$$

Appendix C: Mathematical Induction

Suppose one is presented with the following sequence of equations:

$$1 = 1$$
$$1 + 3 = 4$$
$$1 + 3 + 5 = 9$$
$$1 + 3 + 5 + 7 = 16$$
$$1 + 3 + 5 + 7 + 9 = 25$$

It is clear that there is a pattern. The numbers on the right side of the equations are the squares 1^2, 2^2, 3^2, 4^2, and 5^2 and, in the equation with n^2 on the right side, the left side is the sum of the first n odd numbers. The odd numbers are

$$1 = 2 \cdot 1 - 1$$
$$3 = 2 \cdot 2 - 1$$
$$5 = 2 \cdot 3 - 1$$
$$7 = 2 \cdot 4 - 1$$
$$9 = 2 \cdot 5 - 1$$

and from this it is clear that the nth odd number is $2n - 1$. Hence, at least for $n = 1, 2, 3, 4,$ or 5, the following is true:

$$1 + 3 + \cdots + (2n - 1) = n^2 \qquad (S_n)$$

The question arises whether the statement S_n is true for *every* n. There is no hope of separately verifying all these statements because there are infinitely many of them. A more subtle approach is required.

The idea is as follows: Suppose it is verified that the statement S_{n+1} will be true whenever S_n is true. That is, suppose we prove that, *if* S_n is true, then it necessarily follows that S_{n+1} is also true. Then, if we can show that S_1 is true, it follows that S_2 is true, and from this that S_3 is true, hence that S_4 is true, and so on and on. This is the principle of induction. To express it more compactly, it is useful to have a short way to express the assertion "If S_n is true, then S_{n+1} is true." As in Appendix B, we write this assertion as

$$S_n \Rightarrow S_{n+1}$$

and read it as "S_n implies S_{n+1}." We can now state the principle of mathematical induction.

Appendix C: Mathematical Induction

The Principle of Mathematical Induction

Suppose S_n is a statement about the natural number n for each $n = 1, 2, 3, \ldots$. Suppose further that:

1. S_1 is true.
2. $S_n \Rightarrow S_{n+1}$ for every $n \geq 1$.

Then S_n is true for every $n \geq 1$.

This is one of the most useful techniques in all of mathematics. It applies in a wide variety of situations, as the following examples illustrate.

EXAMPLE 1

Show that $1 + 2 + \cdots + n = \frac{1}{2}n(n + 1)$ for $n \geq 1$.

Solution ▶ Let S_n be the statement: $1 + 2 + \cdots + n = \frac{1}{2}n(n + 1)$ for $n \geq 1$. We apply induction.

1. S_1 is true. The statement S_1 is $1 = \frac{1}{2}1(1 + 1)$, which is true.
2. $S_n \Rightarrow S_{n+1}$. We *assume* that S_n is true for some $n \geq 1$—that is, that
$$1 + 2 + \cdots + n = \tfrac{1}{2}n(n + 1)$$

We must prove that the statement
$$S_{n+1}: 1 + 2 + \cdots + (n + 1) = \tfrac{1}{2}(n + 1)(n + 2)$$

is also true, and we are entitled to use S_n to do so. Now the left side of S_{n+1} is the sum of the first $n + 1$ positive integers. Hence the second-to-last term is n, so we can write

$$\begin{aligned}
1 + 2 + \cdots + (n + 1) &= (1 + 2 + \cdots + n) + (n + 1) \\
&= \tfrac{1}{2}n(n + 1) + (n + 1) \quad \text{using } S_n \\
&= \tfrac{1}{2}(n + 1)(n + 2)
\end{aligned}$$

This shows that S_{n+1} is true and so completes the induction.

In the verification that $S_n \Rightarrow S_{n+1}$, we *assume* that S_n is true and use it to deduce that S_{n+1} is true. The assumption that S_n is true is sometimes called the **induction hypothesis**.

EXAMPLE 2

If x is any number such that $x \neq 1$, show that $1 + x + x^2 + \cdots + x^n = \dfrac{x^{n+1} - 1}{x - 1}$ for $n \geq 1$.

Solution ▶ Let S_n be the statement: $1 + x + x^2 + \cdots + x^n = \dfrac{x^{n+1} - 1}{x - 1}$.

1. S_1 is true. S_1 reads $1 + x = \dfrac{x^2 - 1}{x - 1}$, which is true because $x^2 - 1 = (x - 1)(x + 1)$.

Appendix C: Mathematical Induction

2. $S_n \Rightarrow S_{n+1}$. Assume the truth of S_n: $1 + x + x^2 + \cdots + x^n = \dfrac{x^{n+1} - 1}{x - 1}$.

We must *deduce* from this the truth of S_{n+1}: $1 + x + x^2 + \cdots + x^{n+1} = \dfrac{x^{n+2} - 1}{x - 1}$.

Starting with the left side of S_{n+1} and using the induction hypothesis, we find
$$1 + x + x^2 + \cdots + x^{n+1} = (1 + x + x^2 + \cdots + x^n) + x^{n+1}$$
$$= \frac{x^{n+1} - 1}{x - 1} + x^{n+1}$$
$$= \frac{x^{n+1} - 1 + x^{n+1}(x - 1)}{x - 1}$$
$$= \frac{x^{n+2} - 1}{x - 1}$$

This shows that S_{n+1} is true and so completes the induction.

Both of these examples involve formulas for a certain sum, and it is often convenient to use summation notation. For example, $\sum_{k=1}^{n}(2k - 1)$ means that in the expression $(2k - 1)$, k is to be given the values $k = 1$, $k = 2$, $k = 3$, ..., $k = n$, and then the resulting n numbers are to be added. The same thing applies to other expressions involving k. For example,
$$\sum_{k=1}^{n} k^3 = 1^3 + 2^3 + \cdots + n^3$$
$$\sum_{k=1}^{5}(3k - 1) = (3 \cdot 1 - 1) + (3 \cdot 2 - 1) + (3 \cdot 3 - 1) + (3 \cdot 4 - 1) + (3 \cdot 5 - 1)$$

The next example involves this notation.

EXAMPLE 3

Show that $\sum_{k=1}^{n}(3k^2 - k) = n^2(n + 1)$ for each $n \geq 1$.

Solution ▶ Let S_n be the statement: $\sum_{k=1}^{n}(3k^2 - k) = n^2(n + 1)$.

1. S_1 is true. S_1 reads $(3 \cdot 1^2 - 1) = 1^2(1 + 1)$, which is true.
2. $S_n \Rightarrow S_{n+1}$. Assume that S_n is true. We must prove S_{n+1}:
$$\sum_{k=1}^{n+1}(3k^2 - k) = \sum_{k=1}^{n}(3k^2 - k) + [3(n + 1)^2 - (n + 1)]$$
$$= n^2(n + 1) + (n + 1)[3(n + 1) - 1] \quad \text{using } S_n$$
$$= (n + 1)[n^2 + 3n + 2]$$
$$= (n + 1)[(n + 1)(n + 2)]$$
$$= (n + 1)^2(n + 2)$$

This proves that S_{n+1} is true.

We now turn to examples wherein induction is used to prove propositions that do not involve sums.

EXAMPLE 4

Show that $7^n + 2$ is a multiple of 3 for all $n \geq 1$.

Solution ▶

1. S_1 is true: $7^1 + 2 = 9$ is a multiple of 3.
2. $S_n \Rightarrow S_{n+1}$. Assume that $7^n + 2$ is a multiple of 3 for some $n \geq 1$; say, $7^n + 2 = 3m$ for some integer m. Then
$$7^{n+1} + 2 = 7(7^n) + 2 = 7(3m - 2) + 2 = 21m - 12 = 3(7m - 4)$$
so $7^{n+1} + 2$ is also a multiple of 3. This proves that S_{n+1} is true.

In all the foregoing examples, we have used the principle of induction starting at 1; that is, we have verified that S_1 is true and that $S_n \Rightarrow S_{n+1}$ for each $n \geq 1$, and then we have concluded that S_n is true for every $n \geq 1$. But there is nothing special about 1 here. If m is some fixed integer and we verify that

1. S_m is true.
2. $S_n \Rightarrow S_{n+1}$ for every $n \geq m$.

then it follows that S_n is true for every $n \geq m$. This "extended" induction principle is just as plausible as the induction principle and can, in fact, be proved by induction. The next example will illustrate it. Recall that if n is a positive integer, the number $n!$ (which is read "n-factorial") is the product
$$n! = n(n - 1)(n - 2)\cdots 3 \cdot 2 \cdot 1$$
of all the numbers from n to 1. Thus $2! = 2$, $3! = 6$, and so on.

EXAMPLE 5

Show that $2^n < n!$ for all $n \geq 4$.

Solution ▶ Observe that $2^n < n!$ is actually false if $n = 1, 2, 3$.

1. S_4 is true. $2^4 = 16 < 24 = 4!$.
2. $S_n \Rightarrow S_{n+1}$ if $n \geq 4$. Assume that S_n is true; that is, $2^n < n!$. Then
$$\begin{aligned} 2^{n+1} &= 2 \cdot 2^n \\ &< 2 \cdot n! &&\text{because } 2^n < n! \\ &< (n + 1)n! &&\text{because } 2 < n + 1 \\ &= (n + 1)! \end{aligned}$$

Hence S_{n+1} is true.

EXERCISES C

In Exercises 1–19, prove the given statement by induction for all $n \geq 1$.

1. $1 + 3 + 5 + 7 + \cdots + (2n - 1) = n^2$
2. $1^2 + 2^2 + \cdots + n^2 = \frac{1}{6}n(n + 1)(2n + 1)$
3. $1^3 + 2^3 + \cdots + n^3 = (1 + 2 + \cdots + n)^2$
4. $1 \cdot 2 + 2 \cdot 3 + \cdots + n(n + 1) = \frac{1}{3}n(n + 1)(n + 2)$
5. $1 \cdot 2^2 + 2 \cdot 3^2 + \cdots + n(n + 1)^2$
   $= \frac{1}{12}n(n + 1)(n + 2)(3n + 5)$
◆6. $\dfrac{1}{1 \cdot 2} + \dfrac{1}{2 \cdot 3} + \cdots + \dfrac{1}{n(n + 1)} = \dfrac{n}{n + 1}$
7. $1^2 + 3^2 + \cdots + (2n - 1)^2 = \frac{n}{3}(4n^2 - 1)$
8. $\dfrac{1}{1 \cdot 2 \cdot 3} + \dfrac{1}{2 \cdot 3 \cdot 4} + \cdots + \dfrac{1}{n(n + 1)(n + 2)}$
   $= \dfrac{n(n + 3)}{4(n + 1)(n + 2)}$
9. $1 + 2 + 2^2 + \cdots + 2^{n-1} = 2^n - 1$
10. $3 + 3^3 + 3^5 + \cdots + 3^{2n-1} = \frac{3}{8}(9^n - 1)$
11. $\dfrac{1}{1^2} + \dfrac{1}{2^2} + \cdots + \dfrac{1}{n^2} \leq 2 - \dfrac{1}{n}$
12. $n < 2^n$
13. For any integer $m > 0$, $m!n! < (m + n)!$
◆14. $\dfrac{1}{\sqrt{1}} + \dfrac{1}{\sqrt{2}} + \cdots + \dfrac{1}{\sqrt{n}} \leq 2\sqrt{n} - 1$
15. $\dfrac{1}{\sqrt{1}} + \dfrac{1}{\sqrt{2}} + \cdots + \dfrac{1}{\sqrt{n}} \geq \sqrt{n}$
16. $n^3 + (n + 1)^3 + (n + 2)^3$ is a multiple of 9.
17. $5n + 3$ is a multiple of 4.
◆18. $n^3 - n$ is a multiple of 3.
19. $3^{2n+1} + 2^{n+2}$ is a multiple of 7.

◆20. Let $B_n = 1 \cdot 1! + 2 \cdot 2! + 3 \cdot 3! + \cdots + n \cdot n!$. Find a formula for B_n and prove it.

21. Let $A_n = (1 - \frac{1}{2})(1 - \frac{1}{3})(1 - \frac{1}{4}) \cdots (1 - \frac{1}{n})$. Find a formula for A_n and prove it.

22. Suppose S_n is a statement about n for each $n \geq 1$. Explain what must be done to prove that S_n is true for all $n \geq 1$ if it is known that:
    (a) $S_n \Rightarrow S_{n+2}$ for each $n \geq 1$.
    ◆(b) $S_n \Rightarrow S_{n+8}$ for each $n \geq 1$.
    (c) $S_n \Rightarrow S_{n+1}$ for each $n \geq 10$.
    (d) Both S_n and $S_{n+1} \Rightarrow S_{n+2}$ for each $n \geq 1$.

23. If S_n is a statement for each $n \geq 1$, argue that S_n is true for all $n \geq 1$ if it is known that the following two conditions hold:
    (1) $S_n \Rightarrow S_{n-1}$ for each $n \geq 2$.
    (2) S_n is true for infinitely many values of n.

24. Suppose a sequence $a_1, a_2, \ldots$ of numbers is given that satisfies:
    (1) $a_1 = 2$.
    (2) $a_{n+1} = 2a_n$ for each $n \geq 1$.
    Formulate a theorem giving a_n in terms of n, and prove your result by induction.

25. Suppose a sequence $a_1, a_2, \ldots$ of numbers is given that satisfies:
    (1) $a_1 = b$.
    (2) $a_{n+1} = ca_n + b$ for $n = 1, 2, 3, \ldots$.
    Formulate a theorem giving a_n in terms of n, and prove your result by induction.

26. (a) Show that $n^2 \leq 2^n$ for all $n \geq 4$.
    (b) Show that $n^3 \leq 2^n$ for all $n \geq 10$.

Appendix D: Polynomials

Expressions like $3 - 5x$ and $1 + 3x - 2x^2$ are examples of polynomials. In general, a **polynomial** is an expression of the form
$$f(x) = a_0 + a_1x + a_2x^2 + \cdots + a_nx^n$$
where the a_i are numbers, called the **coefficients** of the polynomial, and x is a variable called an **indeterminate**. The number a_0 is called the **constant** coefficient of the polynomial. The polynomial with every coefficient zero is called the **zero polynomial**, and is denoted simply as 0.

If $f(x) \neq 0$, the coefficient of the highest power of x appearing in $f(x)$ is called the **leading** coefficient of $f(x)$, and the highest power itself is called the **degree** of the polynomial and is denoted $\deg(f(x))$. Hence

$-1 + 5x + 3x^2$ has constant coefficient -1, leading coefficient 3, and degree 2,
7 has constant coefficient 7, leading coefficient 7, and degree 0,
$6x - 3x^3 + x^4 - x^5$ has constant coefficient 0, leading coefficient -1, and degree 5.

We do not define the degree of the zero polynomial.

Two polynomials $f(x)$ and $g(x)$ are called **equal** if every coefficient of $f(x)$ is the same as the corresponding coefficient of $g(x)$. More precisely, if
$$f(x) = a_0 + a_1x + a_2x^2 + \cdots \quad \text{and} \quad g(x) = b_0 + b_1x + b_2x^2 + \cdots$$
are polynomials, then
$$f(x) = g(x) \quad \text{if and only if} \quad a_0 = b_0, a_1 = b_1, a_2 = b_2, \ldots.$$
In particular, this means that
$$f(x) = 0 \text{ is the zero polynomial if and only if } a_0 = 0, a_1 = 0, a_2 = 0, \ldots.$$
This is the reason for calling x an indeterminate.

Let $f(x)$ and $g(x)$ denote nonzero polynomials of degrees n and m respectively, say
$$f(x) = a_0 + a_1x + a_2x^2 + \cdots + a_nx^n \quad \text{and} \quad g(x) = b_0 + b_1x + b_2x^2 + \cdots + b_mx^m$$
where $a_n \neq 0$ and $b_m \neq 0$. If these expressions are multiplied, the result is
$$f(x)\,g(x) = a_0b_0 + (a_0b_1 + a_1b_0)x + (a_0b_2 + a_1b_1 + a_2b_0)x^2 + \cdots + a_nb_mx^{n+m}.$$
Since a_n and b_m are nonzero numbers, their product $a_nb_m \neq 0$ and we have

Appendix D: Polynomials

Theorem 1

If $f(x)$ and $g(x)$ are nonzero polynomials of degrees n and m respectively, their product $f(x) g(x)$ is also nonzero and

$$\deg[f(x)g(x)] = n + m.$$

EXAMPLE 1

$(2 - x + 3x^2)(3 + x^2 - 5x^3) = 6 - 3x + 11x^2 - 11x^3 + 8x^4 - 15x^5.$

If $f(x)$ is any polynomial, the next theorem shows that $f(x) - f(a)$ is a multiple of the polynomial $x - a$. In fact we have

Theorem 2

Remainder Theorem

If $f(x)$ is a polynomial of degree $n \geq 1$ and a is any number, then there exists a polynomial $q(x)$ such that

$$f(x) = (x - a)q(x) + f(a)$$

where $\deg(q(x)) = n - 1$.

PROOF

Write $f(x) = a_0 + a_1 x + a_2 x^2 + \cdots + a_n x^n$ where the a_i are numbers, so that $f(a) = a_0 + a_1 a + a_2 a^2 + \cdots + a_n a^n$. If these expressions are subtracted, the constant terms cancel and we obtain

$$f(x) - f(a) = a_1(x - a) + a_2(x^2 - a^2) + \cdots + a_n(x^n - a^n).$$

Hence it suffices to show that, for each $k \geq 1$, $x^k - a^k = (x - a)p(x)$ for some polynomial $p(x)$ of degree $k - 1$. This is clear if $k = 1$. If it holds for some value k, the fact that

$$x^{k+1} - a^{k+1} = (x - a)x^k + a(x^k - a^k)$$

shows that it holds for $k + 1$. Hence the proof is complete by induction.

There is a systematic procedure for finding the polynomial $q(x)$ in the remainder theorem. It is illustrated below for $f(x) = x^3 - 3x^2 + x - 1$ and $a = 2$. The polynomial $q(x)$ is generated on the top line one term at a time as follows: First x^2 is chosen because $x^2(x - 2)$ has the same x^3-term as $f(x)$, and this is substracted from $f(x)$ to leave a "remainder" of $-x^2 + x - 1$. Next, the second term on top is $-x$ because $-x(x - 2)$ has the same x^2-term, and this is subtracted to leave $-x - 1$. Finally, the third term on top is -1, and the process ends with a "remainder" of -3.

$$\begin{array}{r}
x^2 - x - 1 \\
x-2{\overline{\smash{\big)}\,x^3 - 3x^2 + x - 1}} \\
\underline{x^3 - 2x^2} \\
-x^2 + x - 1 \\
\underline{-x^2 + 2x} \\
-x - 1 \\
\underline{-x + 2} \\
-3
\end{array}$$

Hence $x^3 - 3x^2 + x - 1 = (x - 2)(x^2 - x - 1) + (-3)$. The final remainder is $-3 = f(2)$ as is easily verified. This procedure is called the **division algorithm**.[1]

A real number a is called a **root** of the polynomial $f(x)$ if $f(a) = 0$. Hence for example, 1 is a root of $f(x) = 2 - x + 3x^2 - 4x^3$, but -1 is not a root because $f(-1) = 10 \neq 0$. If $f(x)$ is a multiple of $x - a$, we say that $x - a$ is a **factor** of $f(x)$. Hence the remainder theorem shows immediately that if a is root of $f(x)$, then $x - a$ is factor of $f(x)$. But the converse is also true: If $x - a$ is a factor of $f(x)$, say $f(x) = (x - a) q(x)$, then $f(a) = (a - a)q(a) = 0$. This proves the

Theorem 3

Factor Theorem
If $f(x)$ is a polynomial and a is a number, then $x - a$ is a factor of $f(x)$ if and only if a is a root of $f(x)$.

EXAMPLE 2

If $f(x) = x^3 - 2x^2 - 6x + 4$, then $f(-2) = 0$, so $x - (-2) = x + 2$ is a factor of $f(x)$. In fact, the division algorithm gives $f(x) = (x + 2)(x^2 - 4x + 2)$.

Consider the polynomial $f(x) = x^3 - 3x + 2$. Then 1 is clearly a root of $f(x)$, and the division algorithm gives $f(x) = (x - 1)(x^2 + x - 2)$. But 1 is also a root of $x^2 + x - 2$; in fact, $x^2 + x - 2 = (x - 1)(x + 2)$. Hence

$$f(x) = (x - 1)^2(x + 2)$$

and we say that the root 1 has **multiplicity** 2.

Note that non-zero constant polynomials $f(x) = b \neq 0$ have *no* roots. However, there do exist nonconstant polynomials with no roots. For example, if $g(x) = x^2 + 1$, then $g(a) = a^2 + 1 \geq 1$ for every real number a, so a is not a root. However the *complex* number i is a root of $g(x)$; we return to this below.

Now suppose that $f(x)$ is any nonzero polynomial. We claim that it can be factored in the following form:

$$f(x) = (x - a_1)(x - a_2) \cdots (x - a_m)g(x)$$

where $a_1, a_2, \ldots, a_m$ are the roots of $f(x)$ and $g(x)$ has no root (where the a_i may have repetitions, and my not appear at all if $f(x)$ has no real root).

---

[1] This procedure can be used to divide $f(x)$ by any nonzero polynomial $d(x)$ in place of $x - a$; the remainder then is a polynomial that is either zero or of degree less than the degree of $d(x)$.

Appendix D: Polynomials

By the above calculation $f(x) = x^3 - 3x + 2 = (x-1)^2(x+2)$ has roots 1 and -2, with 1 of multiplicity two (and $g(x) = 1$). Counting the root -2 once, we say that $f(x)$ has three roots counting multiplicities. The next theorem shows that no polynomial can have more roots than its degree even if multiplicities are counted.

Theorem 4

If $f(x)$ is a nonzero polynomial of degree n, then $f(x)$ has at most n roots counting multiplicities.

PROOF

If $n = 0$, then $f(x)$ is a constant and has no roots. So the theorem is true if $n = 0$. (It also holds for $n = 1$ because, if $f(x) = a + bx$ where $b \neq 0$, then the only root is $-\frac{a}{b}$.) In general, suppose inductively that the theorem holds for some value of $n \geq 0$, and let $f(x)$ have degree $n + 1$. We must show that $f(x)$ has at most $n + 1$ roots counting multiplicities. This is certainly true if $f(x)$ has no root. On the other hand, if a is a root of $f(x)$, the factor theorem shows that $f(x) = (x - a)\, q(x)$ for some polynomial $q(x)$, and $q(x)$ has degree n by Theorem 1. By induction, $q(x)$ has at most n roots. But if b is any root of $f(x)$, then

$$(b - a)q(b) = f(b) = 0$$

so either $b = a$ or b is a root of $q(x)$. It follows that $f(x)$ has at most n roots. This completes the induction and so proves Theorem 4.

As we have seen, a polynomial may have *no* root, for example $f(x) = x^2 + 1$. Of course $f(x)$ has complex roots i and $-i$, where i is the complex number such that $i^2 = -1$. But Theorem 4 even holds for complex roots: the number of complex roots (counting multiplicities) cannot exceed the degree of the polynomial. Moreover, the fundamental theorem of algebra asserts that the only nonzero polynomials with no complex root are the non-zero constant polynomials. This is discussed more in Appendix A, Theorems 4 and 5.

Selected Answers

Exercises 1.1 Solutions and Elementary Operations (Page 7)

1. **(b)** $2(2s + 12t + 13) + 5s + 9(-s - 3t - 3) + 3t = -1$;
   $(2s + 12t + 13) + 2s + 4(-s - 3t - 3) = 1$
2. **(b)** $x = t, y = \frac{1}{3}(1 - 2t)$ or $x = \frac{1}{2}(1 - 3s), y = s$
   (d) $x = 1 + 2s - 5t, y = s, z = t$ or $x = s, y = t,$
   $z = \frac{1}{5}(1 - s + 2t)$
4. $x = \frac{1}{4}(3 + 2s), y = s, z = t$
5. **(a)** No solution if $b \neq 0$. If $b = 0$, *any* x is a solution.
   (b) $x = \frac{b}{a}$
7. **(b)** $\begin{bmatrix} 1 & 2 & 0 \\ 0 & 1 & 1 \end{bmatrix}$ **(d)** $\begin{bmatrix} 1 & 1 & 0 & 1 \\ 0 & 1 & 1 & 0 \\ -1 & 0 & 1 & 2 \end{bmatrix}$
8. **(b)** $\begin{aligned} 2x - y &= -1 \\ -3x + 2y + z &= 0 \\ y + z &= 3 \end{aligned}$ or $\begin{aligned} 2x_1 - x_2 &= -1 \\ -3x_1 + 2x_2 + x_3 &= 0 \\ x_2 + x_3 &= 3 \end{aligned}$
9. **(b)** $x = -3, y = 2$ **(d)** $x = -17, y = 13$
10. **(b)** $x = \frac{1}{9}, y = \frac{10}{9}, z = \frac{-7}{3}$
11. **(b)** No solution
14. **(b)** F. $x + y = 0, x - y = 0$ has a unique solution.
    (d) T. Theorem 1.
16. $x' = 5, y' = 1$, so $x = 23, y = -32$
17. $a = -\frac{1}{9}, b = -\frac{5}{9}, c = \frac{11}{9}$
19. $4.50, $5.20

Exercises 1.2 Gaussian Elimination (Page 16)

1. **(b)** No, no **(d)** No, yes **(f)** No, no
2. **(b)** $\begin{bmatrix} 0 & 1 & -3 & 0 & 0 & 0 & 0 \\ 0 & 0 & 0 & 1 & 0 & 0 & -1 \\ 0 & 0 & 0 & 0 & 1 & 0 & 0 \\ 0 & 0 & 0 & 0 & 0 & 1 & 1 \end{bmatrix}$
3. **(b)** $x_1 = 2r - 2s - t + 1, x_2 = r, x_3 = -5s + 3t - 1,$
   $x_4 = s, x_5 = -6t + 1, x_6 = t$ **(d)** $x_1 = -4s - 5t - 4,$
   $x_2 = -2s + t - 2, x_3 = s, x_4 = 1, x_5 = t$
4. **(b)** $x = -\frac{1}{7}, y = -\frac{3}{7}$ **(d)** $x = \frac{1}{3}(t + 2), y = t$
   (f) No solution
5. **(b)** $x = -15t - 21, y = -11t - 17, z = t$
   (d) No solution **(f)** $x = -7, y = -9, z = 1$
   (h) $x = 4, y = 3 + 2t, z = t$
6. **(b)** Denote the equations as $E_1, E_2,$ and E_3. Apply gaussian elimination to column 1 of the augmented matrix, and observe that $E_3 - E_1 = -4(E_2 - E_1)$. Hence $E_3 = 5E_1 - 4E_2$.
7. **(b)** $x_1 = 0, x_2 = -t, x_3 = 0, x_4 = t$
   (d) $x_1 = 1, x_2 = 1 - t, x_3 = 1 + t, x_4 = t$
8. **(b)** If $ab \neq 2$, unique solution $x = \frac{-2 - 5b}{2 - ab},$
   $y = \frac{a + 5}{2 - ab}$. If $ab = 2$: no solution if $a \neq -5$; if $a = -5$, the solutions are $x = -1 + \frac{2}{5}t, y = t$.
   (d) If $a \neq 2$, unique solution $x = \frac{1 - b}{a - 2}, y = \frac{ab - 2}{a - 2}$. If $a = 2$, no solution if $b \neq 1$; if $b = 1$, the solutions are $x = \frac{1}{2}(1 - t), y = t$.
9. **(b)** Unique solution $x = -2a + b + 5c, y = 3a - b - 6c,$
   $z = -2a + b + 4c$, for any a, b, c.
   (d) If $abc \neq -1$, unique solution $x = y = z = 0$; if $abc = -1$ the solutions are $x = abt, y = -bt, z = t$.
   (f) If $a = 1$, solutions $x = -t, y = t, z = -1$. If $a = 0$, there is no solution. If $a \neq 1$ and $a \neq 0$, unique solution $x = \frac{a - 1}{a}, y = 0, z = \frac{-1}{a}$.
10. **(b)** 1 **(d)** 3 **(f)** 1
11. **(b)** 2 **(d)** 3
    (f) 2 if $a = 0$ or $a = 2$; 3, otherwise.
12. **(b)** False. $A = \begin{bmatrix} 1 & 0 & 1 \\ 0 & 1 & 1 \\ 0 & 0 & 0 \end{bmatrix}$ **(d)** False. $A = \begin{bmatrix} 1 & 0 & 1 \\ 0 & 1 & 0 \\ 0 & 0 & 0 \end{bmatrix}$
    (f) False. $\begin{aligned} 2x - y &= 0 \\ -4x + 2y &= 0 \end{aligned}$ is consistent
    but $\begin{aligned} 2x - y &= 1 \\ -4x + 2y &= 1 \end{aligned}$ is not.
    (h) True, A has 3 rows, so there are at most 3 leading 1's.

Selected Answers

14. **(b)** Since one of $b - a$ and $c - a$ is nonzero, then
$$\begin{bmatrix} 1 & a & b+c \\ 1 & b & c+a \\ 1 & b & c+a \end{bmatrix} \to \begin{bmatrix} 1 & a & b+c \\ 0 & b-a & a-b \\ 0 & c-a & a-c \end{bmatrix} \to$$
$$\begin{bmatrix} 1 & a & b+c \\ 0 & 1 & -1 \\ 0 & 0 & 0 \end{bmatrix} \to \begin{bmatrix} 1 & 0 & b+c+a \\ 0 & 1 & -1 \\ 0 & 0 & 0 \end{bmatrix}$$

16. **(b)** $x^2 + y^2 - 2x + 6y - 6 = 0$
18. $\frac{5}{20}$ in A, $\frac{7}{20}$ in B, $\frac{8}{20}$ in C

Exercises 1.3 Homogeneous Equations (Page 24)

1. **(b)** False. $A = \begin{bmatrix} 1 & 0 & 1 & 0 \\ 0 & 1 & 1 & 0 \end{bmatrix}$ **(d)** False. $A = \begin{bmatrix} 1 & 0 & 1 & 1 \\ 0 & 1 & 1 & 0 \end{bmatrix}$

   (f) False. $A = \begin{bmatrix} 1 & 0 & 0 \\ 0 & 1 & 0 \end{bmatrix}$ **(h)** False. $A = \begin{bmatrix} 1 & 0 & 0 \\ 0 & 1 & 0 \\ 0 & 0 & 0 \end{bmatrix}$

2. **(b)** $a = -3$, $x = 9t$, $y = -5t$, $z = t$ **(d)** $a = 1$, $x = -t$, $y = t$, $z = 0$; or $a = -1$, $x = t$, $y = 0$, $z = t$

3. **(b)** Not a linear combination. **(d)** $\mathbf{v} = \mathbf{x} + 2\mathbf{y} - \mathbf{z}$

4. **(b)** $\mathbf{y} = 2\mathbf{a}_1 - \mathbf{a}_2 + 4\mathbf{a}_3$.

5. **(b)** $r \begin{bmatrix} -2 \\ 1 \\ 0 \\ 0 \\ 0 \end{bmatrix} + s \begin{bmatrix} -2 \\ 0 \\ -1 \\ 1 \\ 0 \end{bmatrix} + t \begin{bmatrix} -3 \\ 0 \\ -2 \\ 0 \\ 1 \end{bmatrix}$ **(d)** $s \begin{bmatrix} 0 \\ 2 \\ 1 \\ 0 \\ 0 \end{bmatrix} + t \begin{bmatrix} -1 \\ 3 \\ 0 \\ 1 \\ 0 \end{bmatrix}$

6. **(b)** The system in (a) has nontrivial solutions.
7. **(b)** By Theorem 2 Section 1.2, there are $n - r = 6 - 1 = 5$ parameters and thus infinitely many solutions.

   (d) If R is the row-echelon form of A, then R has a row of zeros and 4 rows in all. Hence R has $r = \operatorname{rank} A = 1, 2$, or 3. Thus there are $n - r = 6 - r = 5, 4$, or 3 parameters and thus infinitely many solutions.

9. **(b)** That the graph of $ax + by + cz = d$ contains three points leads to 3 linear equations homogeneous in variables $a, b, c,$ and d. Apply Theorem 1.
11. There are $n - r$ parameters (Theorem 2 Section 1.2), so there are nontrivial solutions if and only if $n - r > 0$.

Exercises 1.4 An Application to Network Flow (Page 27)

1. **(b)** $f_1 = 85 - f_4 - f_7$
   $f_2 = 60 - f_4 - f_7$
   $f_3 = -75 + f_4 + f_6$
   $f_5 = 40 - f_6 - f_7$ f_4, f_6, f_7 parameters

2. **(b)** $f_5 = 15$
   $25 \leq f_4 \leq 30$

3. **(b)** CD

Exercises 1.5 An Application to Electrical Networks (Page 29)

2. $I_1 = -\frac{1}{5}$, $I_2 = \frac{3}{5}$, $I_3 = \frac{4}{5}$
4. $I_1 = 2$, $I_2 = 1$, $I_3 = \frac{1}{2}$, $I_4 = \frac{3}{2}$, $I_5 = \frac{3}{2}$, $I_6 = \frac{1}{2}$

Exercises 1.6 An Application to Chemical Reactions (Page 30)

2. $2NH_3 + 3CuO \to N_2 + 3Cu + 3H_2O$
4. $15Pb(N_3)_2 + 44Cr(MnO_4)_2 \to 22Cr_2O_3 + 88MnO_2 + 5Pb_3O_4 + 90NO$

Supplementary Exercises for Chapter 1 (Page 30)

1. **(b)** No. If the corresponding planes are parallel and distinct, there is no solution. Otherwise they either coincide or have a whole common line of solutions, that is, at least one parameter.

2. **(b)** $x_1 = \frac{1}{10}(-6s - 6t + 16)$, $x_2 = \frac{1}{10}(4s - t + 1)$, $x_3 = s$, $x_4 = t$

3. **(b)** If $a = 1$, no solution. If $a = 2$, $x = 2 - 2t$, $y = -t$, $z = t$. If $a \neq 1$ and $a \neq 2$, the unique solution is
$$x = \frac{8 - 5a}{3(a-1)}, y = \frac{-2-a}{3(a-1)}, z = \frac{a+2}{3}$$

4. $\begin{bmatrix} R_1 \\ R_2 \end{bmatrix} \to \begin{bmatrix} R_1 + R_2 \\ R_2 \end{bmatrix} \to \begin{bmatrix} R_1 + R_2 \\ -R_1 \end{bmatrix} \to \begin{bmatrix} R_2 \\ -R_1 \end{bmatrix} \to \begin{bmatrix} R_2 \\ R_1 \end{bmatrix}$

6. $a = 1, b = 2, c = -1$
9. **(b)** 5 of brand 1, 0 of brand 2, 3 of brand 3
8. The (real) solution is $x = 2$, $y = 3 - t$, $z = t$ where t is a parameter. The given complex solution occurs when $t = 3 - i$ is complex. If the real system has a unique solution, that solution is real because the coefficients and constants are all real.

Exercises 2.1 Matrix Addition, Scalar Multiplication, and Transposition (Page 40)

1. **(b)** $(a\ b\ c\ d) = (-2, -4, -6, 0) + t(1, 1, 1, 1)$, t arbitrary **(d)** $a = b = c = d = t$, t arbitrary

2. **(b)** $\begin{bmatrix} -14 \\ -20 \end{bmatrix}$ **(d)** $(-12, 4, -12)$

   (f) $\begin{bmatrix} 0 & 1 & -2 \\ -1 & 0 & 4 \\ 2 & -4 & 0 \end{bmatrix}$ **(h)** $\begin{bmatrix} 4 & -1 \\ -1 & -6 \end{bmatrix}$

3. **(b)** $\begin{bmatrix} 15 & -5 \\ 10 & 0 \end{bmatrix}$ **(d)** Impossible **(f)** $\begin{bmatrix} 5 & 2 \\ 0 & -1 \end{bmatrix}$

   (h) Impossible

4. **(b)** $\begin{bmatrix} 4 \\ \frac{1}{2} \end{bmatrix}$

5. **(b)** $A = -\frac{11}{3}B$
6. **(b)** $X = 4A - 3B$, $Y = 4B - 5A$
7. **(b)** $Y = (s, t)$, $X = \frac{1}{2}(1 + 5s, 2 + 5t)$; s and t arbitrary
8. **(b)** $20A - 7B + 2C$
9. **(b)** If $A = \begin{bmatrix} a & b \\ c & d \end{bmatrix}$, then $(p, q, r, s) = \frac{1}{2}(2d, a + b - c - d, a - b + c - d, -a + b + c + d)$.
11. **(b)** If $A + A' = 0$ then $-A = -A + 0 = -A + (A + A') = (-A + A) + A' = 0 + A' = A'$

13. **(b)** Write $A = \text{diag}(a_1, \ldots, a_n)$, where $a_1, \ldots, a_n$ are the main diagonal entries. If $B = \text{diag}(b_1, \ldots, b_n)$ then $kA = \text{diag}(ka_1, \ldots, ka_n)$.

14. **(b)** $s = 1$ or $t = 0$ **(d)** $s = 0$, and $t = 3$

15. **(b)** $\begin{bmatrix} 2 & 0 \\ 1 & -1 \end{bmatrix}$ **(d)** $\begin{bmatrix} 2 & 7 \\ -\frac{9}{2} & -5 \end{bmatrix}$

16. **(b)** $A = A^T$, so using Theorem 2 Section 2.1, $(kA)^T = kA^T = kA$.

19. **(b)** False. Take $B = -A$ for any $A \neq 0$.
    (d) True. Transposing fixes the main diagonal.
    (f) True. $(kA + mB)^T = (kA)^T + (mB)^T = kA^T + mB^T = kA + mB$

20. **(c)** Suppose $A = S + W$, where $S = S^T$ and $W = -W^T$. Then $A^T = S^T + W^T = S - W$, so $A + A^T = 2S$ and $A - A^T = 2W$. Hence $S = \frac{1}{2}(A + A^T)$ and $W = \frac{1}{2}(A - A^T)$ are uniquely determined by A.

22. **(b)** If $A = [a_{ij}]$ then $(kp)A = [(kp)a_{ij}] = [k(pa_{ij})] = k[pa_{ij}] = k(pA)$.

Exercises 2.2 Equations, Matrices, and Transformations (Page 54)

1. **(b)** $\begin{aligned} x_1 - 3x_2 - 3x_3 + 3x_4 &= 5 \\ 8x_2 \phantom{{}+{}} + 2x_4 &= 1 \\ x_1 + 2x_2 + 2x_3 \phantom{{}+{}} &= 2 \\ x_2 + 2x_3 - 5x_4 &= 0 \end{aligned}$

2. **(b)** $x_1 \begin{bmatrix} 1 \\ -1 \\ 2 \\ 3 \end{bmatrix} + x_2 \begin{bmatrix} -2 \\ 0 \\ -2 \\ -4 \end{bmatrix} + x_3 \begin{bmatrix} -1 \\ 1 \\ 7 \\ 9 \end{bmatrix} + x_4 \begin{bmatrix} 1 \\ -2 \\ 0 \\ -2 \end{bmatrix} = \begin{bmatrix} 5 \\ -3 \\ 8 \\ 12 \end{bmatrix}$

3. **(b)** $Ax = \begin{bmatrix} 1 & 2 & 3 \\ 0 & -4 & 5 \end{bmatrix} \begin{bmatrix} x_1 \\ x_2 \\ x_3 \end{bmatrix} = x_1 \begin{bmatrix} 1 \\ 0 \end{bmatrix} + x_2 \begin{bmatrix} 2 \\ -4 \end{bmatrix} + x_3 \begin{bmatrix} 3 \\ 5 \end{bmatrix}$

$= \begin{bmatrix} x_1 + 2x_2 + 3x_3 \\ -4x_2 + 5x_3 \end{bmatrix}$

(d) $Ax = \begin{bmatrix} 3 & -4 & 1 & 6 \\ 0 & 2 & 1 & 5 \\ -8 & 7 & -3 & 0 \end{bmatrix} \begin{bmatrix} x_1 \\ x_2 \\ x_3 \\ x_4 \end{bmatrix} = x_1 \begin{bmatrix} 3 \\ 0 \\ -8 \end{bmatrix} + x_2 \begin{bmatrix} -4 \\ 2 \\ 7 \end{bmatrix} +$

$x_3 \begin{bmatrix} 1 \\ 1 \\ -3 \end{bmatrix} + x_4 \begin{bmatrix} 6 \\ 5 \\ 0 \end{bmatrix} = \begin{bmatrix} 3x_1 - 4x_2 + x_3 + 6x_4 \\ 2x_2 + x_3 + 5x_4 \\ -8x_1 + 7x_2 - 3x_3 \end{bmatrix}$

4. **(b)** To solve $Ax = b$ the reduction is

$\begin{bmatrix} 1 & 3 & 2 & 0 & | & 4 \\ 1 & 0 & -1 & -3 & | & 1 \\ -1 & 2 & 3 & 5 & | & 1 \end{bmatrix} \rightarrow \begin{bmatrix} 1 & 0 & -1 & -3 & | & 1 \\ 0 & 1 & 1 & 1 & | & 1 \\ 0 & 0 & 0 & 0 & | & 0 \end{bmatrix}$, so the general solution is $\begin{bmatrix} 1 + s + 3t \\ 1 - s - t \\ s \\ t \end{bmatrix}$.

Hence $(1 + s + 3t)a_1 + (1 - s - t)a_2 + sa_3 + ta_4 = b$ for any choice of s and t. If $s = t = 0$, we get $a_1 + a_2 = b$; if $s = 1$ and $t = 0$, we have $2a_1 + a_3 = b$.

5. **(b)** $\begin{bmatrix} -2 \\ 2 \\ 0 \end{bmatrix} + t \begin{bmatrix} 1 \\ -3 \\ 1 \end{bmatrix}$ **(d)** $\begin{bmatrix} 3 \\ -9 \\ -2 \\ 0 \end{bmatrix} + t \begin{bmatrix} -1 \\ 4 \\ 1 \\ 1 \end{bmatrix}$

6. We have $Ax_0 = 0$ and $Ax_1 = 0$ and so $A(sx_0 + tx_1) = s(Ax_0) + t(Ax_1) = s \cdot 0 + t \cdot 0 = 0$.

8. **(b)** $x = \begin{bmatrix} -3 \\ 0 \\ -1 \\ 0 \\ 0 \end{bmatrix} + s \begin{bmatrix} 2 \\ 1 \\ 0 \\ 0 \\ 0 \end{bmatrix} + t \begin{bmatrix} -5 \\ 0 \\ 2 \\ 0 \\ 1 \end{bmatrix}$

10. **(b)** False. $\begin{bmatrix} 1 & 2 \\ 2 & 4 \end{bmatrix} \begin{bmatrix} 2 \\ -1 \end{bmatrix} = \begin{bmatrix} 0 \\ 0 \end{bmatrix}$.

(d) True. The linear combination $x_1a_1 + \cdots + x_na_n$ equals Ax where $A = [a_1 \cdots a_n]$ by Theorem 1.

(f) False. If $A = \begin{bmatrix} 1 & 1 & -1 \\ 2 & 2 & 0 \end{bmatrix}$ and $x = \begin{bmatrix} 2 \\ 0 \\ 1 \end{bmatrix}$, then

$Ax = \begin{bmatrix} 1 \\ 4 \end{bmatrix} \neq s \begin{bmatrix} 1 \\ 2 \end{bmatrix} + t \begin{bmatrix} 1 \\ 2 \end{bmatrix}$ for any s and t.

(h) False. If $A = \begin{bmatrix} 1 & -1 & 1 \\ -1 & 1 & -1 \end{bmatrix}$, there is a solution for $b = \begin{bmatrix} 0 \\ 0 \end{bmatrix}$ but not for $b = \begin{bmatrix} 1 \\ 0 \end{bmatrix}$.

11. **(b)** Here $T\begin{bmatrix} x \\ y \end{bmatrix} = \begin{bmatrix} y \\ x \end{bmatrix} = \begin{bmatrix} 0 & 1 \\ 1 & 0 \end{bmatrix} \begin{bmatrix} x \\ y \end{bmatrix}$

(d) Here $T\begin{bmatrix} x \\ y \end{bmatrix} = \begin{bmatrix} y \\ -x \end{bmatrix} = \begin{bmatrix} 0 & 1 \\ -1 & 0 \end{bmatrix} \begin{bmatrix} x \\ y \end{bmatrix}$.

13. **(b)** Here $T\begin{bmatrix} x \\ y \\ z \end{bmatrix} = \begin{bmatrix} -x \\ y \\ z \end{bmatrix} = \begin{bmatrix} -1 & 0 & 0 \\ 0 & 1 & 0 \\ 0 & 0 & 1 \end{bmatrix} \begin{bmatrix} x \\ y \\ z \end{bmatrix}$, so the matrix is $\begin{bmatrix} -1 & 0 & 0 \\ 0 & 1 & 0 \\ 0 & 0 & 1 \end{bmatrix}$.

16. Write $A = [a_1 \; a_2 \; \cdots \; a_n]$ in terms of its columns. If $b = x_1a_1 + x_2a_2 + \cdots + x_na_n$ where the x_i are scalars, then $Ax = b$ by Theorem 1 where $x = [x_1 \; x_2 \; \cdots \; x_n]^T$. That is, x is a solution to the system $Ax = b$.

18. **(b)** By Theorem 3, $A(tx_1) = t(Ax_1) = t \cdot 0 = 0$; that is, tx_1 is a solution to $Ax = 0$.

22. If A is $m \times n$ and x and y are n-vectors, we must show that $A(x + y) = Ax + Ay$. Denote the columns of A by $a_1, a_2, \ldots, a_n$, and write $x = [x_1 \; x_2 \; \cdots \; x_n]^T$ and $y = [y_1 \; y_2 \; \cdots \; y_n]^T$. Then $x + y = [x_1 + y_1 \; x_2 + y_2 \; \cdots \; x_n + y_n]^T$, so Definition 1 and Theorem 1 §2.1 give
$A(x + y) = (x_1 + y_1)a_1 + (x_2 + y_2)a_2 + \cdots + (x_n + y_n)a_n$
$= (x_1a_1 + x_2a_2 + \cdots + x_na_n) +$
$(y_1a_1 + y_2a_2 + \cdots + y_na_n) = Ax + Ay$.

Exercises 2.3 Matrix Multiplication (Page 67)

1. (b) $\begin{bmatrix} -1 & -6 & -2 \\ 0 & 6 & 10 \end{bmatrix}$ (d) $[-3\ -15]$ (f) $[-23]$

   (h) $\begin{bmatrix} 1 & 0 \\ 0 & 1 \end{bmatrix}$ (j) $\begin{bmatrix} aa' & 0 & 0 \\ 0 & bb' & 0 \\ 0 & 0 & cc' \end{bmatrix}$

2. (b) $BA = \begin{bmatrix} -1 & 4 & -10 \\ 1 & 2 & 4 \end{bmatrix}$ $B^2 = \begin{bmatrix} 7 & -6 \\ -1 & 6 \end{bmatrix}$ $CB = \begin{bmatrix} -2 & 12 \\ 2 & -6 \\ 1 & 6 \end{bmatrix}$

   $AC = \begin{bmatrix} 4 & 10 \\ -2 & -1 \end{bmatrix}$ $CA = \begin{bmatrix} 2 & 4 & 8 \\ -1 & -1 & -5 \\ 1 & 4 & 2 \end{bmatrix}$

3. (b) $(a, b, a_1, b_1) = (3, 0, 1, 2)$

4. (b) $A^2 - A - 6I = \begin{bmatrix} 8 & 2 \\ 2 & 5 \end{bmatrix} - \begin{bmatrix} 2 & 2 \\ 2 & -1 \end{bmatrix} - \begin{bmatrix} 6 & 0 \\ 0 & 6 \end{bmatrix} = \begin{bmatrix} 0 & 0 \\ 0 & 0 \end{bmatrix}$

5. (b) $A(BC) = \begin{bmatrix} 1 & -1 \\ 0 & 1 \end{bmatrix}\begin{bmatrix} -9 & -16 \\ 5 & 1 \end{bmatrix} = \begin{bmatrix} -14 & -17 \\ 5 & 1 \end{bmatrix} =$

   $\begin{bmatrix} -2 & -1 & -2 \\ 3 & 1 & 0 \end{bmatrix}\begin{bmatrix} 1 & 0 \\ 2 & 1 \\ 5 & 8 \end{bmatrix} = (AB)C$

6. (b) If $A = \begin{bmatrix} a & b \\ c & d \end{bmatrix}$ and $E = \begin{bmatrix} 0 & 0 \\ 1 & 0 \end{bmatrix}$, compare entries an AE and EA.

7. (b) $m \times n$ and $n \times m$ for some m and n

8. (b) (i) $\begin{bmatrix} 1 & 0 \\ 0 & 1 \end{bmatrix}, \begin{bmatrix} 1 & 0 \\ 0 & -1 \end{bmatrix}, \begin{bmatrix} 1 & 1 \\ 0 & -1 \end{bmatrix}$

   (ii) $\begin{bmatrix} 1 & 0 \\ 0 & 0 \end{bmatrix}, \begin{bmatrix} 1 & 0 \\ 0 & 1 \end{bmatrix}, \begin{bmatrix} 1 & 1 \\ 0 & 0 \end{bmatrix}$

12. (b) $A^{2k} = \begin{bmatrix} 1 & -2k & 0 & 0 \\ 0 & 1 & 0 & 0 \\ 0 & 0 & 1 & 0 \\ 0 & 0 & 0 & 1 \end{bmatrix}$ for $k = 0, 1, 2, \ldots, A^{2k+1} =$

    $A^{2k}A = \begin{bmatrix} 1 & -(2k+1) & 2 & -1 \\ 0 & 1 & 0 & 0 \\ 0 & 0 & -1 & 1 \\ 0 & 0 & 0 & 1 \end{bmatrix}$ for $k = 0, 1, 2, \ldots$

13. (b) $\begin{bmatrix} I & 0 \\ 0 & I \end{bmatrix} = I_{2k}$ (d) 0_k (f) $\begin{bmatrix} X^m & 0 \\ 0 & X^m \end{bmatrix}$ if $n = 2m$;

    $\begin{bmatrix} 0 & X^{m+1} \\ X^m & 0 \end{bmatrix}$ if $n = 2m + 1$.

14. (b) If Y is row i of the identity matrix I, then YA is row i of $IA = A$.

16. (b) $AB - BA$ (d) 0

18. (b) $(kA)C = k(AC) = k(CA) = C(kA)$

20. We have $A^T = A$ and $B^T = B$, so $(AB)^T = B^T A^T = BA$. Hence AB is symmetric if and only if $AB = BA$.

22. (b) $A = 0$

24. If $BC = I$, then $AB = 0$ gives $0 = 0C = (AB)C = A(BC) = AI = A$, contrary to the assumption that $A \neq 0$.

26. 3 paths $v_1 \to v_4$, 0 paths $v_2 \to v_3$

27. (b) False. If $A = \begin{bmatrix} 1 & 0 \\ 0 & 0 \end{bmatrix} = J$, then $AJ = A$ but $J \neq I$.

(d) True. Since $A^T = A$, we have $(I + A)^T = I^T + A^T = I + A$. (f) False. If $A = \begin{bmatrix} 0 & 1 \\ 0 & 0 \end{bmatrix}$, then $A \neq 0$ but $A^2 = 0$.

(h) True. We have $A(A + B) = (A + B)A$; that is, $A^2 + AB = A^2 + BA$. Subtracting A^2 gives $AB = BA$.

(j) False. $A = \begin{bmatrix} 1 & -2 \\ 2 & 4 \end{bmatrix}, B = \begin{bmatrix} 2 & 4 \\ 1 & 2 \end{bmatrix}$ (l) False. See (j).

28. (b) If $A = [a_{ij}]$ and $B = [b_{ij}]$ and $\sum_j a_{ij} = 1 = \sum_j b_{ij}$, then the (i,j)-entry of AB is $c_{ij} = \sum_k a_{ik} b_{kj}$, whence $\sum_j c_{ij} = \sum_j \sum_k a_{ik} b_{kj} = \sum_k a_{ik}(\sum_j b_{kj}) = \sum_k a_{ik} = 1$. Alternatively: If $\mathbf{e} = (1, 1, \ldots, 1)$, then the rows of A sum to 1 if and only if $A\mathbf{e} = \mathbf{e}$. If also $B\mathbf{e} = \mathbf{e}$ then $(AB)\mathbf{e} = A(B\mathbf{e}) = A\mathbf{e} = \mathbf{e}$.

30. (b) If $A = [a_{ij}]$, then $\text{tr}(kA) = \text{tr}[ka_{ij}] = \sum_{i=1}^{n} ka_{ii} = k\sum_{i=1}^{n} a_{ii} = k\,\text{tr}(A)$.

    (e) Write $A^T = [a'_{ij}]$, where $a'_{ij} = a_{ji}$. Then $AA^T = (\sum_k a_{ik} a'_{kj})$, so $\text{tr}(AA^T) = \sum_{i=1}^{n}[\sum_{k=1}^{n} a_{ik} a'_{ki}] = \sum_{i=1}^{n}\sum_{k=1}^{n} a_{ik}^2$.

32. (e) Observe that $PQ = P^2 + PAP - P^2AP = P$, so $Q^2 = PQ + APQ - PAPQ = P + AP - PAP = Q$.

34. (b) $(A + B)(A - B) = A^2 - AB + BA - B^2$, and $(A - B)(A + B) = A^2 + AB - BA - B^2$. These are equal if and only if $-AB + BA = AB - BA$; that is, $2BA = 2AB$; that is, $BA = AB$.

35. (b) $(A + B)(A - B) = A^2 - AB + BA - B^2$ and $(A - B)(A + B) = A^2 - BA + AB - B^2$. These are equal if and only if $-AB + BA = -BA + AB$, that is $2AB = 2BA$, that is $AB = BA$.

36. See V. Camillo, Communications in Algebra 25(6), (1997), 1767–1782; Theorem 2.

Exercises 2.4 Matrix Inverses (Page 80)

2. (b) $\frac{1}{5}\begin{bmatrix} 2 & -1 \\ -3 & 4 \end{bmatrix}$ (d) $\begin{bmatrix} 2 & -1 & 3 \\ 3 & 1 & -1 \\ 1 & 1 & -2 \end{bmatrix}$ (f) $\frac{1}{10}\begin{bmatrix} 1 & 4 & -1 \\ -2 & 2 & 2 \\ -9 & 14 & -1 \end{bmatrix}$

   (h) $\frac{1}{4}\begin{bmatrix} 2 & 0 & -2 \\ -5 & 2 & 5 \\ -3 & 2 & -1 \end{bmatrix}$ (j) $\begin{bmatrix} 0 & 0 & 1 & -2 \\ -1 & -2 & -1 & -3 \\ 1 & 2 & 1 & 2 \\ 0 & -1 & 0 & 0 \end{bmatrix}$

   (l) $\begin{bmatrix} 1 & -2 & 6 & -30 & 210 \\ 0 & 1 & -3 & 15 & -105 \\ 0 & 0 & 1 & -5 & 35 \\ 0 & 0 & 0 & 1 & -7 \\ 0 & 0 & 0 & 0 & 1 \end{bmatrix}$

3. (b) $\begin{bmatrix} x \\ y \end{bmatrix} = \frac{1}{5}\begin{bmatrix} 4 & -3 \\ 1 & -2 \end{bmatrix}\begin{bmatrix} 0 \\ 1 \end{bmatrix} = \frac{1}{5}\begin{bmatrix} -3 \\ -2 \end{bmatrix}$

   (d) $\begin{bmatrix} x \\ y \\ z \end{bmatrix} = \frac{1}{5}\begin{bmatrix} 9 & -14 & 6 \\ 4 & -4 & 1 \\ -10 & 15 & -5 \end{bmatrix}\begin{bmatrix} 1 \\ -1 \\ 0 \end{bmatrix} = \frac{1}{5}\begin{bmatrix} 23 \\ 8 \\ -25 \end{bmatrix}$

4. (b) $B = A^{-1}AB = \begin{bmatrix} 4 & -2 & 1 \\ 7 & -2 & 4 \\ -1 & 2 & -1 \end{bmatrix}$

5. (b) $\frac{1}{10}\begin{bmatrix} 3 & -2 \\ 1 & 1 \end{bmatrix}$ (d) $\frac{1}{2}\begin{bmatrix} 0 & 1 \\ 1 & -1 \end{bmatrix}$

(f) $\frac{1}{2}\begin{bmatrix} 2 & 0 \\ -6 & 1 \end{bmatrix}$ (h) $-\frac{1}{2}\begin{bmatrix} 1 & 1 \\ 1 & 0 \end{bmatrix}$

6. (b) $A = \frac{1}{2}\begin{bmatrix} 2 & -1 & 3 \\ 0 & 1 & -1 \\ -2 & 1 & -1 \end{bmatrix}$

8. (b) A and B are inverses.

9. (b) False. $\begin{bmatrix} 1 & 0 \\ 0 & 1 \end{bmatrix} + \begin{bmatrix} 1 & 0 \\ 0 & -1 \end{bmatrix}$ (d) True. $A^{-1} = \frac{1}{3}A^3$

(f) False. $A = B = \begin{bmatrix} 1 & 0 \\ 0 & 0 \end{bmatrix}$

(h) True. If $(A^2)B = I$, then $A(AB) = I$; use Theorem 5.

10. (b) $(C^T)^{-1} = (C^{-1})^T = A^T$ because $C^{-1} = (A^{-1})^{-1} = A$.

11. (b) (i) Inconsistent. (ii) $\begin{bmatrix} x_1 \\ x_2 \end{bmatrix} = \begin{bmatrix} 2 \\ -1 \end{bmatrix}$

15. (b) $B^4 = I$, so $B^{-1} = B^3 = \begin{bmatrix} 0 & 1 \\ -1 & 0 \end{bmatrix}$

16. $\begin{bmatrix} c^2 - 2 & -c & 1 \\ -c & 1 & 0 \\ 3 - c^2 & c & -1 \end{bmatrix}$

18. (b) If column j of A is zero, $A\mathbf{y} = \mathbf{0}$ where $\mathbf{y}$ is column j of the identity matrix. Use Theorem 5.
(d) If each column of A sums to 0, $XA = 0$ where X is the row of 1s. Hence $A^T X^T = 0$ so A has no inverse by Theorem 5 ($X^T \neq 0$).

19. (b) (ii) $(-1, 1, 1)A = 0$

20. (b) Each power A^k is invertible by Theorem 4 (because A is invertible). Hence A^k cannot be 0.

21. (b) By (a), if one has an inverse the other is zero and so has no inverse.

22. If $A = \begin{bmatrix} a & 0 \\ 0 & 1 \end{bmatrix}$, $a > 1$, then $A^{-1} = \begin{bmatrix} \frac{1}{a} & 0 \\ 0 & 1 \end{bmatrix}$ is an x-compression because $\frac{1}{a} < 1$.

24. (b) $A^{-1} = \frac{1}{4}(A^3 + 2A^2 - I)$

25. (b) If $B\mathbf{x} = \mathbf{0}$, then $(AB)\mathbf{x} = (A)B\mathbf{x} = \mathbf{0}$, so $\mathbf{x} = \mathbf{0}$ because AB is invertible. Hence B is invertible by Theorem 5. But then $A = (AB)B^{-1}$ is invertible by Theorem 4.

26. (b) $\begin{bmatrix} 2 & -1 & 0 \\ -5 & 3 & 0 \\ -13 & 8 & -1 \end{bmatrix}$ (d) $\begin{bmatrix} 1 & -1 & -14 & 8 \\ -1 & 2 & 16 & -9 \\ 0 & 0 & 2 & -1 \\ 0 & 0 & 1 & -1 \end{bmatrix}$

28. (d) If $A^n = 0$, $(I - A)^{-1} = I + A + \cdots + A^{n-1}$.

30. (b) $A[B(AB)^{-1}] = I = [(BA)^{-1}B]A$, so A is invertible by Exercise 10.

32. (a) Have $AC = CA$. Left-multiply by A^{-1} to get $C = A^{-1}CA$. Then right-multiply by A^{-1} to get $CA^{-1} = A^{-1}C$.

33. (b) Given $ABAB = AABB$. Left multiply by A^{-1}, then right multiply by B^{-1}.

34. If $B\mathbf{x} = \mathbf{0}$ where $\mathbf{x}$ is $n \times 1$, then $AB\mathbf{x} = \mathbf{0}$ so $\mathbf{x} = \mathbf{0}$ as AB is invertible. Hence B is invertible by Theorem 5, so $A = (AB)B^{-1}$ is invertible.

35. (b) $B\begin{bmatrix} -1 \\ 3 \\ -1 \end{bmatrix} = 0$ so B is not invertible by Theorem 5.

38. (b) Write $U = I_n - 2XX^T$. Then $U^T = I_n^T - 2X^{TT}X^T = U$, and $U^2 = I_n^2 - (2XX^T)I_n - I_n(2XX^T) + 4(XX^T)(XX^T) = I_n - 4XX^T + 4XX^T = I_n$.

39. (b) $(I - 2P)^2 = I - 4P + 4P^2$, and this equals I if and only if $P^2 = P$.

41. (b) $(A^{-1} + B^{-1})^{-1} = B(A + B)^{-1}A$

Exercises 2.5 Elementary Matrices (Page 89)

1. (b) Interchange rows 1 and 3 of I. $E^{-1} = E$.

(d) Add (-2) times row 1 of I to row 2. $E^{-1} = \begin{bmatrix} 1 & 0 & 0 \\ 2 & 1 & 0 \\ 0 & 0 & 1 \end{bmatrix}$

(f) Multiply row 3 of I by 5. $E^{-1} = \begin{bmatrix} 1 & 0 & 0 \\ 0 & 1 & 0 \\ 0 & 0 & \frac{1}{5} \end{bmatrix}$

2. (b) $\begin{bmatrix} -1 & 0 \\ 0 & 1 \end{bmatrix}$ (d) $\begin{bmatrix} 1 & -1 \\ 0 & 1 \end{bmatrix}$ (f) $\begin{bmatrix} 0 & 1 \\ 1 & 0 \end{bmatrix}$

3. (b) The only possibilities for E are $\begin{bmatrix} 0 & 1 \\ 1 & 0 \end{bmatrix}$, $\begin{bmatrix} k & 0 \\ 0 & 1 \end{bmatrix}$, $\begin{bmatrix} 1 & 0 \\ 0 & k \end{bmatrix}$, $\begin{bmatrix} 1 & k \\ 0 & 1 \end{bmatrix}$, and $\begin{bmatrix} 1 & 0 \\ k & 1 \end{bmatrix}$. In each case, EA has a row different from C.

5. (b) No, 0 is not invertible.

6. (b) $\begin{bmatrix} 1 & -2 \\ 0 & 1 \end{bmatrix}\begin{bmatrix} 1 & 0 \\ 0 & \frac{1}{2} \end{bmatrix}\begin{bmatrix} 1 & 0 \\ -5 & 1 \end{bmatrix}A = \begin{bmatrix} 1 & 0 & 7 \\ 0 & 1 & -3 \end{bmatrix}$.

Alternatively, $\begin{bmatrix} 1 & 0 \\ 0 & \frac{1}{2} \end{bmatrix}\begin{bmatrix} 1 & -1 \\ 0 & 1 \end{bmatrix}\begin{bmatrix} 1 & 0 \\ -5 & 1 \end{bmatrix}A = \begin{bmatrix} 1 & 0 & 7 \\ 0 & 1 & -3 \end{bmatrix}$.

(d) $\begin{bmatrix} 1 & 2 & 0 \\ 0 & 1 & 0 \\ 0 & 0 & 1 \end{bmatrix}\begin{bmatrix} 1 & 0 & 0 \\ 0 & \frac{1}{5} & 0 \\ 0 & 0 & 1 \end{bmatrix}\begin{bmatrix} 1 & 0 & 0 \\ 0 & 1 & 0 \\ 0 & -1 & 1 \end{bmatrix}\begin{bmatrix} 1 & 0 & 0 \\ 0 & 1 & 0 \\ -2 & 0 & 1 \end{bmatrix}\begin{bmatrix} 1 & 0 & 0 \\ -3 & 1 & 0 \\ 0 & 0 & 1 \end{bmatrix}$

$\begin{bmatrix} 0 & 0 & 1 \\ 0 & 1 & 0 \\ 1 & 0 & 0 \end{bmatrix}A = \begin{bmatrix} 1 & 0 & \frac{1}{5} & \frac{1}{5} \\ 0 & 1 & -\frac{7}{5} & -\frac{2}{5} \\ 0 & 0 & 0 & 0 \end{bmatrix}$

7. (b) $U = \begin{bmatrix} 1 & 1 \\ 1 & 0 \end{bmatrix} = \begin{bmatrix} 1 & 1 \\ 0 & 1 \end{bmatrix}\begin{bmatrix} 0 & 1 \\ 1 & 0 \end{bmatrix}$

8. (b) $A = \begin{bmatrix} 0 & 1 \\ 1 & 0 \end{bmatrix}\begin{bmatrix} 1 & 0 \\ 2 & 1 \end{bmatrix}\begin{bmatrix} 1 & 0 \\ 0 & -1 \end{bmatrix}\begin{bmatrix} 1 & 2 \\ 0 & 1 \end{bmatrix}$

(d) $A = \begin{bmatrix} 1 & 0 & 0 \\ 0 & 1 & 0 \\ -2 & 0 & 1 \end{bmatrix}\begin{bmatrix} 1 & 0 & 0 \\ 0 & 1 & 0 \\ 0 & 2 & 1 \end{bmatrix}\begin{bmatrix} 1 & 0 & -3 \\ 0 & 1 & 0 \\ 0 & 0 & 1 \end{bmatrix}\begin{bmatrix} 1 & 0 & 0 \\ 0 & 1 & 4 \\ 0 & 0 & 1 \end{bmatrix}$

10. $UA = R$ by Theorem 1, so $A = U^{-1}R$.

12. (b) $U = A^{-1}$, $V = I$; rank $A = 2$

(d) $U = \begin{bmatrix} -2 & 1 & 0 \\ 3 & -1 & 0 \\ 2 & -1 & 1 \end{bmatrix}$, $V = \begin{bmatrix} 1 & 0 & -1 & -3 \\ 0 & 1 & 1 & 4 \\ 0 & 0 & 1 & 0 \\ 0 & 0 & 0 & 1 \end{bmatrix}$; rank $A = 2$

16. Write $U^{-1} = E_k E_{k-1} \cdots E_2 E_1$, E_i elementary. Then $[I \ U^{-1}A] = [U^{-1}U \ U^{-1}A] = U^{-1}[U \ A] = E_k E_{k-1} \cdots E_2 E_1 [U \ A]$. So $[U \ A] \to [I \ U^{-1}A]$ by row operations (Lemma 1).

17. (b) (i) $A \stackrel{r}{\sim} A$ because $A = IA$ **(ii)** If $A \stackrel{r}{\sim} B$, then $A = UB$, U invertible, so $B = U^{-1}A$. Thus $B \stackrel{r}{\sim} A$. **(iii)** If $A \stackrel{r}{\sim} B$ and $B \stackrel{r}{\sim} C$, then $A = UB$ and $B = VC$, U and V invertible. Hence $A = U(VC) = (UV)C$, so $A \stackrel{r}{\sim} C$.

19. (b) If $B \stackrel{r}{\sim} A$, let $B = UA$, U invertible. If $U = \begin{bmatrix} d & b \\ -b & d \end{bmatrix}$,

$B = UA = \begin{bmatrix} 0 & 0 & b \\ 0 & 0 & d \end{bmatrix}$ where b and d are not both zero (as U is invertible). Every such matrix B arises in this way: Use $U = \begin{bmatrix} a & b \\ -b & a \end{bmatrix}$—it is invertible by Example 4 Section 2.3.

22. (b) Multiply column i by $1/k$.

Exercises 2.6 Linear Transformations (Page 101)

1. (b) $\begin{bmatrix} 5 \\ 6 \\ -13 \end{bmatrix} = 3\begin{bmatrix} 3 \\ 2 \\ -1 \end{bmatrix} - 2\begin{bmatrix} 2 \\ 0 \\ 5 \end{bmatrix}$, so

$T\begin{bmatrix} 5 \\ 6 \\ -13 \end{bmatrix} = 3T\begin{bmatrix} 3 \\ 2 \\ -1 \end{bmatrix} - 2T\begin{bmatrix} 2 \\ 0 \\ 5 \end{bmatrix} = 3\begin{bmatrix} 3 \\ 5 \end{bmatrix} - 2\begin{bmatrix} -1 \\ 2 \end{bmatrix} = \begin{bmatrix} 11 \\ 11 \end{bmatrix}$

2. (b) As in 1(b), $T\begin{bmatrix} 5 \\ -1 \\ 2 \\ -4 \end{bmatrix} = \begin{bmatrix} 4 \\ 2 \\ -9 \end{bmatrix}$.

3. (b) $T(\mathbf{e}_1) = -\mathbf{e}_2$ and $T(\mathbf{e}_2) = -\mathbf{e}_1$.
So $A[T(\mathbf{e}_1) \; T(\mathbf{e}_2)] = [-\mathbf{e}_2 \; -\mathbf{e}_1] = \begin{bmatrix} -1 & 0 \\ 0 & -1 \end{bmatrix}$.

(d) $T(\mathbf{e}_1) = \begin{bmatrix} \frac{\sqrt{2}}{2} \\ \frac{\sqrt{2}}{2} \end{bmatrix}$ and $T(\mathbf{e}_2) = \begin{bmatrix} -\frac{\sqrt{2}}{2} \\ \frac{\sqrt{2}}{2} \end{bmatrix}$.

So $A = [T(\mathbf{e}_1) \; T(\mathbf{e}_2)] = \frac{\sqrt{2}}{2}\begin{bmatrix} 1 & -1 \\ 1 & 1 \end{bmatrix}$.

4. (b) $T(\mathbf{e}_1) = -\mathbf{e}_1$, $T(\mathbf{e}_2) = \mathbf{e}_2$ and $T(\mathbf{e}_3) = \mathbf{e}_3$. Hence Theorem 2 gives $A[T(\mathbf{e}_1) \; T(\mathbf{e}_2) \; T(\mathbf{e}_3)] = [-\mathbf{e}_1 \; \mathbf{e}_2 \; \mathbf{e}_3] = \begin{bmatrix} -1 & 0 & 0 \\ 0 & 1 & 0 \\ 0 & 0 & 1 \end{bmatrix}$.

5. (b) We have $\mathbf{y}_1 = T(\mathbf{x}_1)$ for some $\mathbf{x}_1$ in $\mathbb{R}^n$, and $\mathbf{y}_2 = T(\mathbf{x}_2)$ for some $\mathbf{x}_2$ in $\mathbb{R}^n$. So $a\mathbf{y}_1 + b\mathbf{y}_2 = aT(\mathbf{x}_1) + bT(\mathbf{x}_2) = T(a\mathbf{x}_1 + b\mathbf{x}_2)$. Hence $a\mathbf{y}_1 + b\mathbf{y}_2$ is also in the image of T.

7. (b) $T\left(2\begin{bmatrix} 0 \\ 1 \end{bmatrix}\right) \neq 2\begin{bmatrix} 0 \\ -1 \end{bmatrix}$.

8. (b) $A = \frac{1}{\sqrt{2}}\begin{bmatrix} 1 & 1 \\ -1 & 1 \end{bmatrix}$, rotation through $\theta = -\frac{\pi}{4}$.

(d) $A = \frac{1}{10}\begin{bmatrix} -8 & -6 \\ -6 & 8 \end{bmatrix}$, reflection in the line $y = -3x$.

10. (b) $\begin{bmatrix} \cos\theta & 0 & -\sin\theta \\ 0 & 1 & 0 \\ \sin\theta & 0 & \cos\theta \end{bmatrix}$

12. (b) Reflection in the y axis **(d)** Reflection in $y = x$ **(f)** Rotation through $\frac{\pi}{2}$

13. (b) $T(\mathbf{x}) = aR(\mathbf{x}) = a(A\mathbf{x}) = (aA)\mathbf{x}$ for all $\mathbf{x}$ in $\mathbb{R}$. Hence T is induced by aA.

14. (b) If $\mathbf{x}$ is in $\mathbb{R}^n$, then
$T(-\mathbf{x}) = T[(-1)\mathbf{x}] = (-1)T(\mathbf{x}) = -T(\mathbf{x})$.

17. (b) If $B^2 = I$ then $T^2(\mathbf{x}) = T[T(\mathbf{x})] = B(B\mathbf{x}) = B^2\mathbf{x} = I\mathbf{x} = \mathbf{x} = 1_{\mathbb{R}^n}(\mathbf{x})$ for all $\mathbf{x}$ in $\mathbb{R}^n$. Hence $T^2 = 1_{\mathbb{R}^2}$. If $T^2 = 1_{\mathbb{R}^2}$, then $B^2\mathbf{x} = T^2(\mathbf{x}) = 1_{\mathbb{R}^2}(\mathbf{x}) = \mathbf{x} = I\mathbf{x}$ for all $\mathbf{x}$, so $B^2 = I$ by Theorem 5 Section 2.2.

18. (b) The matrix of $Q_1 \circ Q_0$ is $\begin{bmatrix} 0 & 1 \\ 1 & 0 \end{bmatrix}\begin{bmatrix} 1 & 0 \\ 0 & -1 \end{bmatrix} = \begin{bmatrix} 0 & -1 \\ 1 & 0 \end{bmatrix}$, which is the matrix of $R_{\frac{\pi}{2}}$. **(d)** The matrix of $Q_0 \circ R_{\frac{\pi}{2}}$ is $\begin{bmatrix} 1 & 0 \\ 0 & -1 \end{bmatrix}\begin{bmatrix} 0 & -1 \\ 1 & 0 \end{bmatrix} = \begin{bmatrix} 0 & -1 \\ -1 & 0 \end{bmatrix}$, which is the matrix of Q_{-1}.

20. We have $T(\mathbf{x}) = x_1 + x_2 + \cdots + x_n = [1 \; 1 \; \cdots \; 1]\begin{bmatrix} x_1 \\ x_2 \\ \vdots \\ x_n \end{bmatrix}$, so T is the matrix transformation induced by the matrix $A = [1 \; 1 \; \cdots \; 1]$. In particular, T is linear. On the other hand, we can use Theorem 2 to get A, but to do this we must first show directly that T is linear.

If we write $\mathbf{x} = \begin{bmatrix} x_1 \\ x_2 \\ \vdots \\ x_n \end{bmatrix}$ and $\mathbf{y} = \begin{bmatrix} y_1 \\ y_2 \\ \vdots \\ y_n \end{bmatrix}$. Then

$T(\mathbf{x} + \mathbf{y}) = T\begin{bmatrix} x_1 + y_1 \\ x_2 + y_2 \\ \vdots \\ x_n + y_n \end{bmatrix}$
$= (x_1 + y_1) + (x_2 + y_2) + \cdots + (x_n + y_n)$
$= (x_1 + x_2 + \cdots + x_n) + (y_1 + y_2 + \cdots + y_n)$
$= T(\mathbf{x}) + T(\mathbf{y})$

Similarly, $T(a\mathbf{x}) = aT(\mathbf{x})$ for any scalar a, so T is linear. By Theorem 2, T has matrix $A = [T(\mathbf{e}_1) \; T(\mathbf{e}_2) \; \cdots \; T(\mathbf{e}_n)] = [1 \; 1 \; \cdots \; 1]$, as before.

22. (b) If $T : \mathbb{R}^n \to \mathbb{R}$ is linear, write $T(\mathbf{e}_j) = w_j$ for each $j = 1, 2, \ldots, n$ where $\{\mathbf{e}_1, \mathbf{e}_2, \ldots, \mathbf{e}_n\}$ is the standard basis of $\mathbb{R}^n$. Since $\mathbf{x} = x_1\mathbf{e}_1 + x_2\mathbf{e}_2 + \cdots + x_n\mathbf{e}_n$, Theorem 1 gives
$T(\mathbf{x}) = T(x_1\mathbf{e}_1 + x_2\mathbf{e}_2 + \cdots + x_n\mathbf{e}_n)$
$= x_1T(\mathbf{e}_1) + x_2T(\mathbf{e}_2) + \cdots + x_nT(\mathbf{e}_n)$
$= x_1w_1 + x_2w_2 + \cdots + x_nw_n$
$= \mathbf{w} \cdot \mathbf{x} = T_{\mathbf{w}}(\mathbf{x})$

where $\mathbf{w} = \begin{bmatrix} w_1 \\ w_2 \\ \vdots \\ w_n \end{bmatrix}$. Since this holds for all $\mathbf{x}$ in $\mathbb{R}^n$, it shows that $T = T_{\mathbf{W}}$. This also follows from Theorem 2, but we have first to verify that T is linear. (This comes to showing that
$\mathbf{w} \cdot (\mathbf{x} + \mathbf{y}) = \mathbf{w} \cdot \mathbf{s} + \mathbf{w} \cdot \mathbf{y}$ and $\mathbf{w} \cdot (a\mathbf{x}) = a(\mathbf{w} \cdot \mathbf{x})$
for all $\mathbf{x}$ and $\mathbf{y}$ in $\mathbb{R}^n$ and all a in $\mathbb{R}$.) Then T has matrix $A = [T(\mathbf{e}_1) \; T(\mathbf{e}_2) \; \cdots \; T(\mathbf{e}_n)] = [w_1 \; w_2 \; \cdots \; w_n]$ by Theorem 2. Hence if $\mathbf{x} = \begin{bmatrix} x_1 \\ x_2 \\ \vdots \\ x_n \end{bmatrix}$ in $\mathbb{R}$, then
$T(\mathbf{x}) = A\mathbf{x} = \mathbf{w} \cdot \mathbf{x}$, as required.

23. (b) Given $\mathbf{x}$ in $\mathbb{R}$ and a in $\mathbb{R}$, we have
$$\begin{aligned}(S \circ T)(a\mathbf{x}) &= S[T(a\mathbf{x})] & \text{Definition of } S \circ T \\ &= S[aT(\mathbf{x})] & \text{Because } T \text{ is linear.} \\ &= a[S[T(\mathbf{x})]] & \text{Because } S \text{ is linear.} \\ &= a[(S \circ T)(\mathbf{x})] & \text{Definition of } S \circ T\end{aligned}$$

Exercises 2.7 LU-Factorization (Page 111)

1. (b) $\begin{bmatrix} 2 & 0 & 0 \\ 1 & -3 & 0 \\ -1 & 9 & 1 \end{bmatrix}\begin{bmatrix} 1 & 2 & 1 \\ 0 & 1 & -\frac{2}{3} \\ 0 & 0 & 0 \end{bmatrix}$

(d) $\begin{bmatrix} -1 & 0 & 0 & 0 \\ 1 & 1 & 0 & 0 \\ 1 & -1 & 1 & 0 \\ 0 & -2 & 0 & 1 \end{bmatrix}\begin{bmatrix} 1 & 3 & -1 & 0 & 1 \\ 0 & 1 & 2 & 1 & 0 \\ 0 & 0 & 0 & 0 & 0 \\ 0 & 0 & 0 & 0 & 0 \end{bmatrix}$

(f) $\begin{bmatrix} 2 & 0 & 0 & 0 \\ 1 & -2 & 0 & 0 \\ 3 & -2 & 1 & 0 \\ 0 & 2 & 0 & 1 \end{bmatrix}\begin{bmatrix} 1 & 1 & -1 & 2 & 1 \\ 0 & 1 & -\frac{1}{2} & 0 & 0 \\ 0 & 0 & 0 & 0 & 0 \\ 0 & 0 & 0 & 0 & 0 \end{bmatrix}$

2. (b) $P = \begin{bmatrix} 0 & 0 & 1 \\ 1 & 0 & 0 \\ 0 & 1 & 0 \end{bmatrix}$

$PA = \begin{bmatrix} -1 & 2 & 1 \\ 0 & -1 & 2 \\ 0 & 0 & 4 \end{bmatrix} = \begin{bmatrix} -1 & 0 & 0 \\ 0 & -1 & 0 \\ 0 & 0 & 4 \end{bmatrix}\begin{bmatrix} 1 & -2 & -1 \\ 0 & 1 & 2 \\ 0 & 0 & 1 \end{bmatrix}$

(d) $P = \begin{bmatrix} 1 & 0 & 0 & 0 \\ 0 & 0 & 1 & 0 \\ 0 & 0 & 0 & 1 \\ 0 & 1 & 0 & 0 \end{bmatrix}$ $PA = \begin{bmatrix} -1 & -2 & 3 & 0 \\ 1 & 1 & -1 & 3 \\ 2 & 5 & -10 & 1 \\ 2 & 4 & -6 & 5 \end{bmatrix} =$

$\begin{bmatrix} -1 & 0 & 0 & 0 \\ 1 & -1 & 0 & 0 \\ 2 & 1 & -2 & 0 \\ 2 & 0 & 0 & 5 \end{bmatrix}\begin{bmatrix} 1 & 2 & -3 & 0 \\ 0 & 1 & -2 & -3 \\ 0 & 0 & 1 & -2 \\ 0 & 0 & 0 & 1 \end{bmatrix}$

3. (b) $\mathbf{y} = \begin{bmatrix} -1 \\ 0 \\ 0 \end{bmatrix}$ $\mathbf{x} = \begin{bmatrix} -1+2t \\ -t \\ s \\ t \end{bmatrix}$ s and t arbitray

(d) $\mathbf{y} = \begin{bmatrix} 2 \\ 8 \\ -1 \\ 0 \end{bmatrix}$ $\mathbf{x} = \begin{bmatrix} 8-2t \\ 6-t \\ -1-t \\ t \end{bmatrix}$ t arbitrary

5. $\begin{bmatrix} R_1 \\ R_2 \end{bmatrix} \to \begin{bmatrix} R_1 + R_2 \\ R_2 \end{bmatrix} \to \begin{bmatrix} R_1 + R_2 \\ -R_1 \end{bmatrix} \to \begin{bmatrix} R_2 \\ -R_1 \end{bmatrix} \to \begin{bmatrix} R_2 \\ R_1 \end{bmatrix}$

6. (b) Let $A = LU = L_1U_1$ be LU-factorizations of the invertible matrix A. Then U and U_1 have no row of zeros and so (being row-echelon) are upper triangular with 1's on the main diagonal. Thus, using **(a)**, the diagonal matrix $D = UU_1^{-1}$ has 1's on the main diagonal. Thus $D = I$, $U = U_1$, and $L = L_1$.

7. If $A = \begin{bmatrix} a & 0 \\ X & A_1 \end{bmatrix}$ and $B = \begin{bmatrix} b & 0 \\ Y & B_1 \end{bmatrix}$ in block form, then $AB = \begin{bmatrix} ab & 0 \\ Xb + A_1Y & A_1B_1 \end{bmatrix}$, and A_1B_1 is lower triangular by induction.

9. (b) Let $A = LU = L_1U_1$ be two such factorizations. Then $UU_1^{-1} = L^{-1}L_1$; write this matrix as $D = UU_1^{-1} = L^{-1}L_1$. Then D is lower triangular (apply Lemma 1 to $D = L^{-1}L_1$); and D is also upper triangular (consider UU_1^{-1}). Hence D is diagonal, and so $D = I$ because L^{-1} and L_1 are unit triangular. Since $A = LU$; this completes the proof.

Exercises 2.8 An Application to Input-Output Economic Models (Page 116)

1. (b) $\begin{bmatrix} t \\ 3t \\ t \end{bmatrix}$ **(d)** $\begin{bmatrix} 14t \\ 17t \\ 47t \\ 23t \end{bmatrix}$ **2.** $\begin{bmatrix} t \\ t \\ t \end{bmatrix}$

4. $P = \begin{bmatrix} bt \\ (1-a)t \end{bmatrix}$ is nonzero (for some t) unless $b = 0$ and $a = 1$. In that case, $\begin{bmatrix} 1 \\ 1 \end{bmatrix}$ is a solution. If the entries of E are positive, then $P = \begin{bmatrix} b \\ 1-a \end{bmatrix}$ has positive entries.

7. (b) $\begin{bmatrix} 0.4 & 0.8 \\ 0.7 & 0.2 \end{bmatrix}$

8. If $E = \begin{bmatrix} a & b \\ c & d \end{bmatrix}$, then $I - E = \begin{bmatrix} 1-a & -b \\ -c & 1-d \end{bmatrix}$, so $\det(I - E) = (1-a)(1-d) - bc = 1 - \operatorname{tr} E + \det E$. If $\det(I - E) \neq 0$, then
$(I - E)^{-1} = \frac{1}{\det(I-E)}\begin{bmatrix} 1-d & b \\ c & 1-a \end{bmatrix}$, so $(I - E)^{-1} \geq 0$ if $\det(I - E) > 0$, that is, $\operatorname{tr} E < 1 + \det E$. The converse is now clear.

9. (b) Use $\mathbf{p} = \begin{bmatrix} 3 \\ 2 \\ 1 \end{bmatrix}$ in Theorem 2.

(d) $\mathbf{p} = \begin{bmatrix} 3 \\ 2 \\ 2 \end{bmatrix}$ in Theorem 2.

Exercises 2.9 An Application to Markov Chains (Page 123)

1. (b) Not regular

2. (b) $\frac{1}{3}\begin{bmatrix} 2 \\ 1 \end{bmatrix}, \frac{3}{8}$ **(d)** $\frac{1}{3}\begin{bmatrix} 1 \\ 1 \\ 1 \end{bmatrix}, 0.312$ **(f)** $\frac{1}{20}\begin{bmatrix} 5 \\ 7 \\ 8 \end{bmatrix}, 0.306$

4. (b) 50% middle, 25% upper, 25% lower

6. $\frac{7}{16}, \frac{9}{16}$

8. (a) $\frac{7}{75}$ **(b)** He spends most of his time in compartment 3; steady state $\frac{1}{16}\begin{bmatrix} 3 \\ 2 \\ 5 \\ 4 \\ 2 \end{bmatrix}$.

12. (a) Direct verification.
(b) Since $0 < p < 1$ and $0 < q < 1$ we get $0 < p + q < 2$ whence $-1 < p + q - 1 < 1$. Finally, $-1 < 1 - p - q < 1$, so $(1 - p - q)^m$ converges to zero as m increases.

Supplementary Exercises for Chapter 2 (Page 124)

2. **(b)** $U^{-1} = \frac{1}{4}(U^2 - 5U + 11I)$.
4. **(b)** If $x_k = x_m$, then $y + k(y - z) = y + m(y - z)$. So $(k - m)(y - z) = 0$. But $y - z$ is not zero (because y and z are distinct), so $k - m = 0$ by Example 7 Section 2.1.
6. **(d)** Using parts **(c)** and **(b)** gives $I_{pq}AI_{rs} = \sum_{i=1}^{n}\sum_{j=1}^{n} a_{ij}I_{pq}I_{ij}I_{rs}$. The only nonzero term occurs when $i = q$ and $j = r$, so $I_{pq}AI_{rs} = a_{qr}I_{ps}$.
7. **(b)** If $A = [a_{ij}] = \sum_{i,j} a_{ij}I_{ij}$, then $I_{pq}AI_{rs} = a_{qr}I_{ps}$ by 6(d). But then $a_{qr}I_{ps} = AI_{pq}I_{rs} = 0$ if $q \neq r$, so $a_{qr} = 0$ if $q \neq r$. If $q = r$, then $a_{qq}I_{ps} = AI_{pq}I_{rs} = AI_{ps}$ is independent of q. Thus $a_{qq} = a_{11}$ for all q.

Exercises 3.1 The Cofactor Expansion (Page 135)

1. **(b)** 0 **(d)** -1 **(f)** -39 **(h)** 0 **(j)** $2abc$ **(l)** 0 **(n)** -56 **(p)** $abcd$
5. **(b)** -17 **(d)** 106 6. **(b)** 0 7. **(b)** 12
8. **(b)** $\det\begin{bmatrix} 2a+p & 2b+q & 2c+r \\ 2p+x & 2q+y & 2r+z \\ 2x+a & 2y+b & 2z+c \end{bmatrix} =$
$3\det\begin{bmatrix} a+p+x & b+q+y & c+r+z \\ 2p+x & 2q+y & 2r+z \\ 2x+a & 2y+b & 2z+c \end{bmatrix} =$
$3\det\begin{bmatrix} a+p+x & b+q+y & c+r+z \\ p-a & q-b & r-c \\ x-p & y-q & z-r \end{bmatrix} =$
$3\det\begin{bmatrix} 3x & 3y & 3z \\ p-a & q-b & r-c \\ x-p & y-q & z-r \end{bmatrix} \cdots$

9. **(b)** F, $A = \begin{bmatrix} 1 & 1 \\ 2 & 2 \end{bmatrix}$ **(d)** F, $A = \begin{bmatrix} 2 & 0 \\ 0 & 1 \end{bmatrix} \to R = \begin{bmatrix} 1 & 0 \\ 0 & 1 \end{bmatrix}$
(f) F, $A = \begin{bmatrix} 1 & 1 \\ 0 & 1 \end{bmatrix}$ **(h)** F, $A = \begin{bmatrix} 1 & 1 \\ 0 & 1 \end{bmatrix}$ and $B = \begin{bmatrix} 1 & 0 \\ 1 & 1 \end{bmatrix}$
10. **(b)** 35
11. **(b)** -6 **(d)** -6
14. **(b)** $-(x - 2)(x^2 + 2x - 12)$
15. **(b)** -7
16. **(b)** $\pm\frac{\sqrt{6}}{2}$ **(d)** $x = \pm y$
21. Let $\mathbf{x} = \begin{bmatrix} x_1 \\ x_2 \\ \vdots \\ x_n \end{bmatrix}$, $\mathbf{y} = \begin{bmatrix} y_1 \\ y_2 \\ \vdots \\ y_n \end{bmatrix}$, and $A = [\mathbf{c}_1 \cdots \mathbf{x} + \mathbf{y} \cdots \mathbf{c}_n]$ where $\mathbf{x} + \mathbf{y}$ is in column j. Expanding $\det A$ along column j (the one containing $\mathbf{x} + \mathbf{y}$):

$T(\mathbf{x} + \mathbf{y}) = \det A = \sum_{i=1}^{n}(x_i + y_i)c_{ij}(A)$
$= \sum_{i=1}^{n} x_i c_{ij}(A) + \sum_{i=1}^{n} y_i c_{ij}(A)$
$= T(\mathbf{x}) + T(\mathbf{y})$

Similarly for $T(a\mathbf{x}) = aT(\mathbf{x})$.

24. If A is $n \times n$, then $\det B = (-1)^k \det A$ where $n = 2k$ or $n = 2k + 1$.

Exercises 3.2 Determinants and Matrix Inverses (Page 148)

1. **(b)** $\begin{bmatrix} 1 & -1 & -2 \\ -3 & 1 & 6 \\ -3 & 1 & 4 \end{bmatrix}$ **(d)** $\frac{1}{3}\begin{bmatrix} -1 & 2 & 2 \\ 2 & -1 & 2 \\ 2 & 2 & -1 \end{bmatrix} = A$
2. **(b)** $c \neq 0$ **(d)** any c **(f)** $c \neq -1$
3. **(b)** -2 4. **(b)** 1 6. **(b)** $\frac{4}{9}$ 7. **(b)** 16
8. **(b)** $\frac{1}{11}\begin{bmatrix} 5 \\ 21 \end{bmatrix}$ **(d)** $\frac{1}{79}\begin{bmatrix} 12 \\ -37 \\ -2 \end{bmatrix}$ 9. **(b)** $\frac{4}{51}$
10. **(b)** $\det A = 1, -1$ **(d)** $\det A = 1$
(f) $\det A = 0$ if n is odd; nothing can be said if n is even
15. dA where $d = \det A$

19. **(b)** $\frac{1}{c}\begin{bmatrix} 1 & 0 & 1 \\ 0 & c & 1 \\ -1 & c & 1 \end{bmatrix}, c \neq 0$ **(d)** $\frac{1}{2}\begin{bmatrix} 8-c^2 & -c & c^2-6 \\ c & 1 & -c \\ c^2-10 & c & 8-c^2 \end{bmatrix}$

(f) $\frac{1}{c^3+1}\begin{bmatrix} 1-c & c^2+1 & -c-1 \\ c^2 & -c & c+1 \\ -c & 1 & c^2-1 \end{bmatrix}, c \neq -1$

20. **(b)** T. $\det AB = \det A \det B = \det B \det A = \det BA$.
(d) T. $\det A \neq 0$ means A^{-1} exists, so $AB = AC$ implies that $B = C$. **(f)** F. If $A = \begin{bmatrix} 1 & 1 & 1 \\ 1 & 1 & 1 \\ 1 & 1 & 1 \end{bmatrix}$, then adj $A = 0$. **(h)** F. If $A = \begin{bmatrix} 1 & 1 \\ 0 & 0 \end{bmatrix}$, then adj $A = \begin{bmatrix} 0 & -1 \\ 0 & 1 \end{bmatrix}$.
(j) F. If $A = \begin{bmatrix} -1 & 1 \\ 1 & -1 \end{bmatrix}$, then $\det(I + A) = -1$ but $1 + \det A = 1$. **(l)** F. If $A = \begin{bmatrix} 1 & 1 \\ 0 & 1 \end{bmatrix}$, then $\det A = 1$ but adj $A = \begin{bmatrix} 1 & -1 \\ 0 & 1 \end{bmatrix} \neq A$.

22. **(b)** $5 - 4x + 2x^2$.
23. **(b)** $1 - \frac{5}{3}x + \frac{1}{2}x^2 + \frac{7}{6}x^3$.
24. **(b)** $1 - 0.51x + 2.1x^2 - 1.1x^3$; 1.25, so $y = 1.25$
26. **(b)** Use induction on n where A is $n \times n$. It is clear if $n = 1$. If $n > 1$, write $A = \begin{bmatrix} a & X \\ 0 & B \end{bmatrix}$ in block form where B is $(n-1) \times (n-1)$. Then $A^{-1} = \begin{bmatrix} a^{-1} & -a^{-1}XB^{-1} \\ 0 & B^{-1} \end{bmatrix}$, and this is upper triangular because B is upper triangular by induction.

28. $-\frac{1}{21}\begin{bmatrix} 3 & 0 & 1 \\ 0 & 2 & 3 \\ 3 & 1 & -1 \end{bmatrix}$

34. **(b)** Have $(\text{adj } A)A = (\det A)I$; so taking inverses, $A^{-1} \cdot (\text{adj } A)^{-1} = \frac{1}{\det A}I$. On the other hand, $A^{-1}\text{adj}(A^{-1}) = \det(A^{-1})I = \frac{1}{\det A}I$. Comparison yields $A^{-1}(\text{adj } A)^{-1} = A^{-1}\text{adj}(A^{-1})$, and part **(b)** follows.
(d) Write $\det A = d$, $\det B = e$. By the adjugate formula $AB \text{ adj}(AB) = deI$, and $AB \text{ adj } B \text{ adj } A = A[eI] \text{ adj } A = (eI)(dI) = deI$. Done as AB is invertible.

Exercises 3.3 Diagonalization and Eigenvalues (Page 166)

1. **(b)** $(x-3)(x+2)$; $3, -2$; $\begin{bmatrix} 4 \\ -1 \end{bmatrix}, \begin{bmatrix} 1 \\ 1 \end{bmatrix}$;

$P = \begin{bmatrix} 4 & 1 \\ -1 & 1 \end{bmatrix}$; $P^{-1}AP = \begin{bmatrix} 3 & 0 \\ 0 & -2 \end{bmatrix}$.

(d) $(x-2)^3$; 2; $\begin{bmatrix} 1 \\ 1 \\ 0 \end{bmatrix}, \begin{bmatrix} -3 \\ 0 \\ 1 \end{bmatrix}$; No such P; Not diagonalizable.

(f) $(x+1)^2(x-2)$; $-1, 2$; $\begin{bmatrix} -1 \\ 1 \\ 2 \end{bmatrix}, \begin{bmatrix} 1 \\ 2 \\ 1 \end{bmatrix}$; No such P;

Not diagonalizable. Note that this matrix and the matrix in Example 9 have the same characteristic polynomial, but that matrix is diagonalizable.

(h) $(x-1)^2(x-3)$; $1, 3$; $\begin{bmatrix} -1 \\ 0 \\ 1 \end{bmatrix}, \begin{bmatrix} 1 \\ 0 \\ 1 \end{bmatrix}$; No such P; Not diagonalizable.

2. **(b)** $V_k = \frac{7}{3}2^k\begin{bmatrix} 2 \\ 1 \end{bmatrix}$. **(d)** $V_k = \frac{3}{2}3^k\begin{bmatrix} 1 \\ 0 \\ 1 \end{bmatrix}$.

4. $A\mathbf{x} = \lambda\mathbf{x}$ if and only if $(A - \alpha I)\mathbf{x} = (\lambda - \alpha)\mathbf{x}$. Same eigenvectors.

8. **(b)** $P^{-1}AP = \begin{bmatrix} 1 & 0 \\ 0 & 2 \end{bmatrix}$, so $A^n = P\begin{bmatrix} 1 & 0 \\ 0 & 2^n \end{bmatrix}P^{-1} = \begin{bmatrix} 9 - 8 \cdot 2^n & 12(1 - 2^n) \\ 6(2^n - 1) & 9 \cdot 2^n - 8 \end{bmatrix}$.

9. **(b)** $A = \begin{bmatrix} 0 & 1 \\ 0 & 2 \end{bmatrix}$

11. **(b)** and **(d)** If $PAP^{-1} = D$ is diagonal, then
(b) $P^{-1}(kA)P = kD$ is diagonal, and
(d) $Q(U^{-1}AU)Q = D$ where $Q = PU$.

12. $\begin{bmatrix} 1 & 1 \\ 0 & 1 \end{bmatrix}$ is not diagonalizable by Example 8. But

$\begin{bmatrix} 1 & 1 \\ 0 & 1 \end{bmatrix} = \begin{bmatrix} 2 & 1 \\ 0 & -1 \end{bmatrix} + \begin{bmatrix} -1 & 0 \\ 0 & 2 \end{bmatrix}$ where $\begin{bmatrix} 2 & 1 \\ 0 & -1 \end{bmatrix}$ has

diagonalizing matrix $P = \begin{bmatrix} 1 & -1 \\ 0 & 3 \end{bmatrix}$, and $\begin{bmatrix} -1 & 0 \\ 0 & 2 \end{bmatrix}$
is already diagonal.

14. We have $\lambda^2 = \lambda$ for every eigenvalue λ (as $\lambda = 0, 1$) so $D^2 = D$, and so $A^2 = A$ as in Example 9.

18. **(b)** $c_{rA}(x) = \det[xI - rA] = r^n\det[\frac{x}{r}I - A] = r^n c_A[\frac{x}{r}]$.

20. **(b)** If $\lambda \neq 0$, $A\mathbf{x} = \lambda\mathbf{x}$ if and only if $A^{-1}\mathbf{x} = \frac{1}{\lambda}\mathbf{x}$. The result follows.

21. **(b)** $(A^3 - 2A - 3I)\mathbf{x} = A^3\mathbf{x} - 2A\mathbf{x} + 3\mathbf{x} = \lambda^3\mathbf{x} - 2\lambda\mathbf{x} + 3\mathbf{x} = (\lambda^3 - 2\lambda - 3)\mathbf{x}$.

23. **(b)** If $A^m = 0$ and $A\mathbf{x} = \lambda\mathbf{x}$, $\mathbf{x} \neq 0$, then $A^2\mathbf{x} = A(\lambda\mathbf{x}) = \lambda A\mathbf{x} = \lambda^2\mathbf{x}$. In general, $A^k\mathbf{x} = \lambda^k\mathbf{x}$ for all $k \geq 1$. Hence, $\lambda^m\mathbf{x} = A^m\mathbf{x} = 0\mathbf{x} = 0$, so $\lambda = 0$ (because $\mathbf{x} \neq 0$).

24. **(a)** If $A\mathbf{x} = \lambda\mathbf{x}$, then $A^k\mathbf{x} = \lambda^k\mathbf{x}$ for each k. Hence $\lambda^m\mathbf{x} = A^m\mathbf{x} = \mathbf{x}$, so $\lambda^m = 1$. As λ is real, $\lambda = \pm 1$ by the Hint. So if $P^{-1}AP = D$ is diagonal, then $D^2 = I$ by Theorem 4. Hence $A^2 = PD^2P = I$.

27. **(a)** We have $P^{-1}AP = \lambda I$ by the diagonalization algorithm, so $A = P(\lambda I)P^{-1} = \lambda PP^{-1} = \lambda I$.
(b) No. $\lambda = 1$ is the only eigenvalue.

31. **(b)** $\lambda_1 = 1$, stabilizes. **(d)** $\lambda_1 = \frac{1}{24}(3 + \sqrt{69}) = 1.13$, diverges.

34. Extinct if $\alpha < \frac{1}{5}$, stable if $\alpha = \frac{1}{5}$, diverges if $\alpha > \frac{1}{5}$.

Exercises 3.4 An Application to Linear Recurrences (Page 172)

1. **(b)** $x_k = \frac{1}{3}[4 - (-2)^k]$ **(d)** $x_k = \frac{1}{5}[2^{k+2} + (-3)^k]$.
2. **(b)** $x_k = \frac{1}{2}[(-1)^k + 1]$
3. **(b)** $x_{k+4} = x_k + x_{k+2} + x_{k+3}$; $x_{10} = 169$
5. $\frac{1}{2\sqrt{5}}[3 + \sqrt{5}]\lambda_1^k + (-3 + \sqrt{5})\lambda_2^k$ where $\lambda_1 = \frac{1}{2}(1 + \sqrt{5})$ and $\lambda_2 = \frac{1}{2}(1 - \sqrt{5})$.
7. $\frac{1}{2\sqrt{3}}[2 + \sqrt{3}]\lambda_1^k + (-2 + \sqrt{3})\lambda_2^k$ where $\lambda_1 = 1 + \sqrt{3}$ and $\lambda_2 = 1 - \sqrt{3}$.
9. $\frac{34}{3} - \frac{4}{3}(-\frac{1}{2})^k$. Long term $11\frac{1}{3}$ million tons.

11. **(b)** $A\begin{bmatrix} 1 \\ \lambda \\ \lambda^2 \end{bmatrix} = \begin{bmatrix} \lambda \\ \lambda^2 \\ a + b\lambda + c\lambda^2 \end{bmatrix} = \begin{bmatrix} \lambda \\ \lambda^2 \\ \lambda^3 \end{bmatrix} = \lambda\begin{bmatrix} 1 \\ \lambda \\ \lambda^2 \end{bmatrix}$

12. **(b)** $x_k = \frac{11}{10}3^k + \frac{11}{15}(-2)^k - \frac{5}{6}$

13. **(a)** $p_{k+2} + q_{k+2} = [ap_{k+1} + bp_k + c(k)] + [aq_{k+1} + bq_k] = a(p_{k+1} + q_{k+1}) + b(p_k + q_k) + c(k)$

Section 3.5 An Application to Systems of Differential Equations (Page 178)

1. **(b)** $c_1\begin{bmatrix} 1 \\ 1 \end{bmatrix}e^{4x} + c_2\begin{bmatrix} 5 \\ -1 \end{bmatrix}e^{-2x}$; $c_1 = -\frac{2}{3}$, $c_2 = \frac{1}{3}$

(d) $c_1\begin{bmatrix} -8 \\ 10 \\ 7 \end{bmatrix}e^{-x} + c_2\begin{bmatrix} 1 \\ -2 \\ 1 \end{bmatrix}e^{2x} + c_3\begin{bmatrix} 1 \\ 0 \\ 1 \end{bmatrix}e^{4x}$; $c_1 = 0$, $c_2 = -\frac{1}{2}$, $c_3 = \frac{3}{2}$

3. **(b)** The solution to (a) is $m(t) = 10(\frac{4}{5})^{t/3}$. Hence we want t such that $10(\frac{4}{5})^{t/3} = 5$. We solve for t by taking natural logarithms: $t = \frac{3\ln(\frac{1}{2})}{\ln(\frac{4}{5})} = 9.32$ hours.

5. **(a)** If $\mathbf{g}' = A\mathbf{g}$, put $\mathbf{f} = \mathbf{g} - A^{-1}\mathbf{b}$. Then $\mathbf{f}' = \mathbf{g}'$ and $A\mathbf{f} = A\mathbf{g} - \mathbf{b}$, so $\mathbf{f}' = \mathbf{g}' = A\mathbf{g} = A\mathbf{f} + \mathbf{b}$, as required.

6. **(b)** Assume that $f_1' = a_1f_1 + f_2$ and $f_2' = a_2f_1$. Differentiating gives $f_1'' = a_1f_1' + f_2' = a_1f_1' + a_2f_1$, proving that f_1 satisfies (∗).

Exercises 3.6 Proof of the Cofactor Expansion Theorem (Page 182)

2. Consider the rows $R_p, R_{p+1}, ..., R_{q-1}, R_q$. In $q - p$ adjacent interchanges they can be put in the order $R_{p+1}, ..., R_{q-1}, R_q, R_p$. Then in $q - p - 1$ adjacent interchanges we can obtain the order $R_q, R_{p+1}, ..., R_{q-1}, R_p$. This uses $2(q - p) - 1$ adjacent interchanges in all.

Supplementary Exercises for Chapter 3 (Page 182)

2. **(b)** If A is 1×1, then $A^T = A$. In general, $\det[A_{ij}] = \det[(A_{ij})^T] = \det[(A^T)_{ji}]$ by (a) and induction. Write $A^T = [a'_{ij}]$ where $a'_{ij} = a_{ji}$, and expand $\det A^T$ along column 1.

$$\det A^T = \sum_{j=1}^{n} a'_{j1}(-1)^{j+1}\det[(A^T)_{j1}]$$
$$= \sum_{j=1}^{n} a_{1j}(-1)^{1+j}\det[A_{1j}] = \det A$$

where the last equality is the expansion of $\det A$ along row 1.

Exercises 4.1 Vectors and Lines (Page 195)

1. **(b)** $\sqrt{6}$ **(d)** $\sqrt{5}$ **(f)** $3\sqrt{6}$.
2. **(b)** $\frac{1}{3}\begin{bmatrix}-2\\-1\\2\end{bmatrix}$.
4. **(b)** $\sqrt{2}$ **(d)** 3.
6. **(b)** $\vec{FE} = \vec{FC} + \vec{CE} = \frac{1}{2}\vec{AC} + \frac{1}{2}\vec{CB} = \frac{1}{2}(\vec{AC} + \vec{CB}) = \frac{1}{2}\vec{AB}$.
7. **(b)** Yes **(d)** Yes
8. **(b)** $\mathbf{p}$ **(d)** $-(\mathbf{p} + \mathbf{q})$.
9. **(b)** $\begin{bmatrix}-1\\-1\\5\end{bmatrix}, \sqrt{27}$ **(d)** $\begin{bmatrix}0\\0\\0\end{bmatrix}, 0$ **(f)** $\begin{bmatrix}-2\\2\\2\end{bmatrix}, \sqrt{12}$
10. **(b)** (i) $Q(5, -1, 2)$ (ii) $Q(1, 1, -4)$.
11. **(b)** $\mathbf{x} = \mathbf{u} - 6\mathbf{v} + 5\mathbf{w} = \begin{bmatrix}-26\\4\\19\end{bmatrix}$.
12. **(b)** $\begin{bmatrix}a\\b\\c\end{bmatrix} = \begin{bmatrix}-5\\8\\6\end{bmatrix}$
13. **(b)** If it holds then $\begin{bmatrix}3a + 4b + c\\-a + c\\b + c\end{bmatrix} = \begin{bmatrix}x_1\\x_2\\x_3\end{bmatrix}$.

$\begin{bmatrix}3 & 4 & 1 & x_1\\-1 & 0 & 1 & x_2\\0 & 1 & 1 & x_3\end{bmatrix} \rightarrow \begin{bmatrix}0 & 4 & 4 & x_1 + 3x_2\\-1 & 0 & 1 & x_2\\0 & 1 & 1 & x_3\end{bmatrix}$

If there is to be a solution then $x_1 + 3x_2 = 4x_3$ must hold. This is not satisfied.

14. **(b)** $\frac{1}{4}\begin{bmatrix}5\\-5\\-2\end{bmatrix}$. 17. **(b)** $Q(0, 7, 3)$. 18. **(b)** $\mathbf{x} = \frac{1}{40}\begin{bmatrix}-20\\-13\\14\end{bmatrix}$.
20. **(b)** $S(-1, 3, 2)$.
21. **(b)** T. $\|\mathbf{v} - \mathbf{w}\| = 0$ implies that $\mathbf{v} - \mathbf{w} = \mathbf{0}$. **(d)** F. $\|\mathbf{v}\| = \|-\mathbf{v}\|$ for all $\mathbf{v}$ but $\mathbf{v} = -\mathbf{v}$ only holds if $\mathbf{v} = \mathbf{0}$.
   (f) F. If $t < 0$ they have the *opposite* direction.
   (h) F. $\|-5\mathbf{v}\| = 5\|\mathbf{v}\|$ for all $\mathbf{v}$, so it fails if $\mathbf{v} \neq \mathbf{0}$.
   (j) F. Take $\mathbf{w} = -\mathbf{v}$ where $\mathbf{v} \neq \mathbf{0}$.
22. **(b)** $\begin{bmatrix}3\\-1\\4\end{bmatrix} + t\begin{bmatrix}2\\-1\\5\end{bmatrix}$; $x = 3 + 2t, y = -1 - t, z = 4 + 5t$
   (d) $\begin{bmatrix}1\\1\\1\end{bmatrix} + t\begin{bmatrix}1\\1\\1\end{bmatrix}$; $x = y = z = 1 + t$
   (f) $\begin{bmatrix}2\\-1\\1\end{bmatrix} + t\begin{bmatrix}-1\\0\\1\end{bmatrix}$; $x = 2 - t, y = -1, z = 1 + t$
23. **(b)** P corresponds to $t = 2$; Q corresponds to $t = 5$.
24. **(b)** No intersection
   (d) $P(2, -1, 3)$; $t = -2, s = -3$
29. $P(3, 1, 0)$ or $P(\frac{5}{3}, \frac{-1}{3}, \frac{4}{3})$
31. **(b)** $\vec{CP_k} = -\vec{CP_{n+k}}$ if $1 \leq k \leq n$, where there are $2n$ points.
33. $\vec{DA} = 2\vec{EA}$ and $2\vec{AF} = \vec{FC}$, so $2\vec{EF} = 2(\vec{EF} + \vec{AF})$
   $= \vec{DA} + \vec{FC} = \vec{CB} + \vec{FC} = \vec{FC} + \vec{CB} = \vec{FB}$.
   Hence $\vec{EF} = \frac{1}{2}\vec{FB}$. So F is the trisection point of both AC and EB.

Exercises 4.2 Projections and Planes (Page 209)

1. **(b)** 6 **(d)** 0 **(f)** 0
2. **(b)** π or $180°$ **(d)** $\frac{\pi}{3}$ or $60°$ **(f)** $\frac{2\pi}{3}$ or $120°$
3. **(b)** 1 or -17
4. **(b)** $t\begin{bmatrix}-1\\1\\2\end{bmatrix}$ **(d)** $s\begin{bmatrix}1\\2\\0\end{bmatrix} + t\begin{bmatrix}0\\3\\1\end{bmatrix}$
6. **(b)** $29 + 57 = 86$
8. **(b)** $A = B = C = \frac{\pi}{3}$ or $60°$
10. **(b)** $\frac{11}{18}\mathbf{v}$ **(d)** $-\frac{1}{2}\mathbf{v}$
11. **(b)** $\frac{5}{21}\begin{bmatrix}2\\-1\\-4\end{bmatrix} + \frac{1}{21}\begin{bmatrix}53\\26\\20\end{bmatrix}$ **(d)** $\frac{27}{53}\begin{bmatrix}6\\-4\\1\end{bmatrix} + \frac{1}{53}\begin{bmatrix}-3\\2\\26\end{bmatrix}$
12. **(b)** $\frac{1}{26}\sqrt{5642}$, $Q(\frac{71}{26}, \frac{15}{26}, \frac{34}{26})$
13. **(b)** $\begin{bmatrix}0\\0\\0\end{bmatrix}$ **(b)** $\begin{bmatrix}4\\-15\\8\end{bmatrix}$
14. **(b)** $-23x + 32y + 11z = 11$ **(d)** $2x - y + z = 5$
   (f) $2x + 3y + 2z = 7$ **(h)** $2x - 7y - 3z = -1$
   (j) $x - y - z = 3$

15. **(b)** $\begin{bmatrix} x \\ y \\ z \end{bmatrix} = \begin{bmatrix} 2 \\ -1 \\ 3 \end{bmatrix} + t\begin{bmatrix} 2 \\ 1 \\ 0 \end{bmatrix}$ **(d)** $\begin{bmatrix} x \\ y \\ z \end{bmatrix} = \begin{bmatrix} 1 \\ 1 \\ -1 \end{bmatrix} + t\begin{bmatrix} 1 \\ 1 \\ 1 \end{bmatrix}$

(f) $\begin{bmatrix} x \\ y \\ z \end{bmatrix} = \begin{bmatrix} 1 \\ 1 \\ 2 \end{bmatrix} + t\begin{bmatrix} 4 \\ 1 \\ -5 \end{bmatrix}$

16. **(b)** $\frac{\sqrt{6}}{3}$, $Q(\frac{7}{3}, \frac{2}{3}, \frac{-2}{3})$
17. **(b)** Yes. The equation is $5x - 3y - 4z = 0$.
19. **(b)** $(-2, 7, 0) + t(3, -5, 2)$
20. **(b)** None **(d)** $P(\frac{13}{19}, \frac{-78}{19}, \frac{65}{19})$
21. **(b)** $3x + 2z = d$, d arbitrary **(d)** $a(x - 3) + b(y - 2) + c(z + 4) = 0$; a, b, and c not all zero
 (f) $ax + by + (b - a)z = a$; a and b not both zero
 (h) $ax + by + (a - 2b)z = 5a - 4b$; a and b not both zero
23. **(b)** $\sqrt{10}$
24. **(b)** $\frac{\sqrt{14}}{2}$, $A(3, 1, 2)$, $B(\frac{7}{2}, -\frac{1}{2}, 3)$
 (d) $\frac{\sqrt{6}}{6}$, $A(\frac{19}{3}, 2, \frac{1}{3})$, $B(\frac{37}{6}, \frac{13}{6}, 0)$
26. **(b)** Consider the diagonal $\mathbf{d} = \begin{bmatrix} a \\ a \\ a \end{bmatrix}$. The six face diagonals in question are $\pm\begin{bmatrix} a \\ 0 \\ -a \end{bmatrix}, \pm\begin{bmatrix} 0 \\ a \\ -a \end{bmatrix}, \pm\begin{bmatrix} a \\ -a \\ 0 \end{bmatrix}$. All of these are orthogonal to $\mathbf{d}$. The result works for the other diagonals by symmetry.

28. The four diagonals are (a, b, c), $(-a, b, c)$, $(a, -b, c)$ and $(a, b, -c)$ or their negatives. The dot products are $\pm(-a^2 + b^2 + c^2)$, $\pm(a^2 - b^2 + c^2)$, and $\pm(a^2 + b^2 - c^2)$.
34. **(b)** The sum of the squares of the lengths of the diagonals equals the sum of the squares of the lengths of the four sides.
38. **(b)** The angle θ between $\mathbf{u}$ and $(\mathbf{u} + \mathbf{v} + \mathbf{w})$ is given by
$$\cos\theta = \frac{\mathbf{u} \cdot (\mathbf{u} + \mathbf{v} + \mathbf{w})}{\|\mathbf{u}\| \|\mathbf{u} + \mathbf{v} + \mathbf{w}\|} = \frac{\|\mathbf{u}\|}{\sqrt{\|\mathbf{u}\|^2 + \|\mathbf{v}\|^2 + \|\mathbf{w}\|^2}}$$
$= \frac{1}{\sqrt{3}}$, because $\|\mathbf{u}\| = \|\mathbf{v}\| = \|\mathbf{w}\|$. Similar remarks apply to the other angles.
39. **(b)** Let $\mathbf{p}_0$, $\mathbf{p}_1$ be the vectors of P_0, P_1, so $\mathbf{u} = \mathbf{p}_0 - \mathbf{p}_1$. Then $\mathbf{u} \cdot \mathbf{n} = \mathbf{p}_0 \cdot \mathbf{n} - \mathbf{p}_1 \cdot \mathbf{n} = (ax_0 + by_0) - (ax_1 + by_1) = ax_0 + by_0 + c$. Hence the distance is
$$\left\| \left(\frac{\mathbf{u} \cdot \mathbf{n}}{\|\mathbf{n}\|^2}\right)\mathbf{n} \right\| = \frac{|\mathbf{u} \cdot \mathbf{n}|}{\|\mathbf{n}\|}, \text{ as required.}$$
41. **(b)** This follows from **(a)** because $\|\mathbf{v}\|^2 = a^2 + b^2 + c^2$.
44. **(d)** Take $\begin{bmatrix} x_1 \\ y_1 \\ z_1 \end{bmatrix} = \begin{bmatrix} x \\ y \\ z \end{bmatrix}$ and $\begin{bmatrix} x_2 \\ y_2 \\ z_2 \end{bmatrix} = \begin{bmatrix} y \\ z \\ x \end{bmatrix}$ in **(c)**.

Exercises 4.3 More on the Cross Product (Page 217)

3. **(b)** $\pm\frac{\sqrt{3}}{3}\begin{bmatrix} 1 \\ -1 \\ -1 \end{bmatrix}$. 4. **(b)** 0 **(d)** $\sqrt{5}$ 5. **(b)** 7

6. **(b)** The distance is $\|\mathbf{p} - \mathbf{p}_0\|$; use **(a)**.
10. $\|\overrightarrow{AB} \times \overrightarrow{AC}\|$ is the area of the parallelogram determined by A, B, and C.

12. Because $\mathbf{u}$ and $\mathbf{v} \times \mathbf{w}$ are parallel, the angle θ between them is 0 or π. Hence $\cos(\theta) = \pm 1$, so the volume is $|\mathbf{u} \cdot (\mathbf{v} \times \mathbf{w})| = \|\mathbf{u}\| \|\mathbf{v} \times \mathbf{w}\| \cos(\theta) = \|\mathbf{u}\| \|(\mathbf{v} \times \mathbf{w})\|$. But the angle between $\mathbf{v}$ and $\mathbf{w}$ is $\frac{\pi}{2}$ so $\|\mathbf{v} \times \mathbf{w}\| = \|\mathbf{v}\| \|\mathbf{w}\| \cos(\frac{\pi}{2}) = \|\mathbf{v}\| \|\mathbf{w}\|$. The result follows.

15. **(b)** If $\mathbf{u} = \begin{bmatrix} u_1 \\ u_2 \\ u_3 \end{bmatrix}$, $\mathbf{v} = \begin{bmatrix} v_1 \\ v_2 \\ v_3 \end{bmatrix}$, and $\mathbf{w} = \begin{bmatrix} w_1 \\ w_2 \\ w_3 \end{bmatrix}$, then

$\mathbf{u} \times (\mathbf{v} + \mathbf{w}) = \det\begin{bmatrix} \mathbf{i} & u_1 & v_1 + w_1 \\ \mathbf{j} & u_2 & v_2 + w_2 \\ \mathbf{k} & u_3 & v_3 + w_3 \end{bmatrix} =$

$\det\begin{bmatrix} \mathbf{i} & u_1 & v_1 \\ \mathbf{j} & u_2 & v_2 \\ \mathbf{k} & u_3 & v_3 \end{bmatrix} + \det\begin{bmatrix} \mathbf{i} & u_1 & w_1 \\ \mathbf{j} & u_2 & w_2 \\ \mathbf{k} & u_3 & w_3 \end{bmatrix} = (\mathbf{u} \times \mathbf{v}) + (\mathbf{u} \times \mathbf{w})$

where we used Exercise 21 Section 3.1.
16. **(b)** $(\mathbf{v} - \mathbf{w}) \cdot [(\mathbf{u} \times \mathbf{v}) + (\mathbf{v} \times \mathbf{w}) + (\mathbf{w} \times \mathbf{u})]$
$= (\mathbf{v} - \mathbf{w}) \cdot (\mathbf{u} \times \mathbf{v}) + (\mathbf{v} - \mathbf{w}) \cdot (\mathbf{v} \times \mathbf{w}) + (\mathbf{v} - \mathbf{w}) \cdot (\mathbf{w} \times \mathbf{u})$
$= -\mathbf{w} \cdot (\mathbf{u} \times \mathbf{v}) + 0 + \mathbf{v} \cdot (\mathbf{w} \times \mathbf{u}) = 0$.

22. Let $\mathbf{p}_1$ and $\mathbf{p}_2$ be vectors of points in the planes, so $\mathbf{p}_1 \cdot \mathbf{n} = d_1$ and $\mathbf{p}_2 \cdot \mathbf{n} = d_2$. The distance is the length of the projection of $\mathbf{p}_2 - \mathbf{p}_1$ along $\mathbf{n}$; that is
$\frac{|(\mathbf{p}_2 - \mathbf{p}_1) \cdot \mathbf{n}|}{\|\mathbf{n}\|} = \frac{|d_1 - d_2|}{\|\mathbf{n}\|}$.

Exercises 4.4 Linear Operators on $\mathbb{R}^3$ (Page 224)

1. **(b)** $A = \begin{bmatrix} 1 & -1 \\ -1 & 1 \end{bmatrix}$, projection on $y = -x$.
 (d) $A = \frac{1}{5}\begin{bmatrix} -3 & 4 \\ 4 & 3 \end{bmatrix}$, reflection in $y = 2x$.
 (f) $A = \frac{1}{2}\begin{bmatrix} 1 & -\sqrt{3} \\ \sqrt{3} & 1 \end{bmatrix}$, rotation through $\frac{\pi}{3}$.

2. **(b)** The zero transformation.

3. **(b)** $\frac{1}{21}\begin{bmatrix} 17 & 2 & -8 \\ 2 & 20 & 4 \\ -8 & 4 & 5 \end{bmatrix}\begin{bmatrix} 0 \\ 1 \\ -3 \end{bmatrix}$

 (d) $\frac{1}{30}\begin{bmatrix} 22 & -4 & 20 \\ -4 & 28 & 10 \\ 20 & 10 & -20 \end{bmatrix}\begin{bmatrix} 0 \\ 1 \\ -3 \end{bmatrix}$

 (f) $\frac{1}{25}\begin{bmatrix} 9 & 0 & 12 \\ 0 & 0 & 0 \\ 12 & 0 & 16 \end{bmatrix}\begin{bmatrix} 1 \\ -1 \\ 7 \end{bmatrix}$

 (h) $\frac{1}{11}\begin{bmatrix} -9 & 2 & -6 \\ 2 & -9 & -6 \\ -6 & -6 & 7 \end{bmatrix}\begin{bmatrix} 2 \\ -5 \\ 0 \end{bmatrix}$

4. **(b)** $\frac{1}{2}\begin{bmatrix} \sqrt{3} & -1 & 0 \\ 1 & \sqrt{3} & 0 \\ 0 & 0 & 1 \end{bmatrix}\begin{bmatrix} 1 \\ 0 \\ 3 \end{bmatrix}$

6. $\begin{bmatrix} \cos\theta & 0 & -\sin\theta \\ 0 & 1 & 0 \\ \sin\theta & 0 & \cos\theta \end{bmatrix}$

9. (a) Write $\mathbf{v} = \begin{bmatrix} x \\ y \end{bmatrix}$. $P_L(\mathbf{v}) = \left(\dfrac{\mathbf{v} \cdot \mathbf{d}}{\|\mathbf{d}\|^2}\right)\mathbf{d} = \dfrac{ax+by}{a^2+b^2}\begin{bmatrix} a \\ b \end{bmatrix}$
$= \dfrac{1}{a^2+b^2}\begin{bmatrix} a^2x + aby \\ abx + b^2y \end{bmatrix} = \dfrac{1}{a^2+b^2}\begin{bmatrix} a^2 & ab \\ ab & b^2 \end{bmatrix}\begin{bmatrix} x \\ y \end{bmatrix}$

Exercises 4.5 An Application to Computer Graphics (page 228)

1. (b) $\dfrac{1}{2}\begin{bmatrix} \sqrt{2}+2 & 7\sqrt{2}+2 & 3\sqrt{2}+2 & -\sqrt{2}+2 & -5\sqrt{2}+2 \\ -3\sqrt{2}+4 & 3\sqrt{2}+4 & 5\sqrt{2}+4 & \sqrt{2}+4 & 9\sqrt{2}+4 \\ 2 & 2 & 2 & 2 & 2 \end{bmatrix}$

5. (b) $P(\tfrac{9}{5}, \tfrac{18}{5})$

Supplementary Exercises for Chapter 4 (Page 228)

4. 125 knots in a direction θ degrees east of north, where $\cos\theta = 0.6$ ($\theta = 53°$ or 0.93 radians).
6. $(12, 5)$. Actual speed 12 knots.

Exercises 5.1 Subspaces and Spanning (Page 234)

1. (b) Yes (d) No (f) No.
2. (b) No (d) Yes, $\mathbf{x} = 3\mathbf{y} + 4\mathbf{z}$.
3. (b) No
10. $\text{span}\{a_1\mathbf{x}_1, a_2\mathbf{x}_2, \ldots, a_k\mathbf{x}_k\} \subseteq \text{span}\{\mathbf{x}_1, \mathbf{x}_2, \ldots, \mathbf{x}_k\}$ by Theorem 1 because, for each i, $a_i\mathbf{x}_i$ is in $\text{span}\{\mathbf{x}_1, \mathbf{x}_2, \ldots, \mathbf{x}_k\}$. Similarly, the fact that $\mathbf{x}_i = a_i^{-1}(a_i\mathbf{x}_i)$ is in $\text{span}\{a_1\mathbf{x}_1, a_2\mathbf{x}_2, \ldots, a_k\mathbf{x}_k\}$ for each i shows that $\text{span}\{\mathbf{x}_1, \mathbf{x}_2, \ldots, \mathbf{x}_k\} \subseteq \text{span}\{a_1\mathbf{x}_1, a_2\mathbf{x}_2, \ldots, a_k\mathbf{x}_k\}$, again by Theorem 1.
12. If $\mathbf{y} = r_1\mathbf{x}_1 + \cdots + r_k\mathbf{x}_k$ then $A\mathbf{y} = r_1(A\mathbf{x}_1) + \cdots + r_k(A\mathbf{x}_k) = \mathbf{0}$.
15. (b) $\mathbf{x} = (\mathbf{x} + \mathbf{y}) - \mathbf{y} = (\mathbf{x} + \mathbf{y}) + (-\mathbf{y})$ is in U because U is a subspace and both $\mathbf{x} + \mathbf{y}$ and $-\mathbf{y} = (-1)\mathbf{y}$ are in U.
16. (b) True. $\mathbf{x} = 1\mathbf{x}$ is in U. (d) True. Always $\text{span}\{\mathbf{y}, \mathbf{z}\} \subseteq \text{span}\{\mathbf{x}, \mathbf{y}, \mathbf{z}\}$ by Theorem 1. Since $\mathbf{x}$ is in $\text{span}\{\mathbf{x}, \mathbf{y}\}$ we have $\text{span}\{\mathbf{x}, \mathbf{y}, \mathbf{z}\} \subseteq \text{span}\{\mathbf{y}, \mathbf{z}\}$, again by Theorem 1.
(f) False. $a\begin{bmatrix} 1 \\ 0 \end{bmatrix} + b\begin{bmatrix} 2 \\ 0 \end{bmatrix} = \begin{bmatrix} a + 2b \\ 0 \end{bmatrix}$ cannot equal $\begin{bmatrix} 0 \\ 1 \end{bmatrix}$.
20. If U is a subspace, then S2 and S3 certainly hold. Conversely, assume that S2 and S3 hold for U. Since U is nonempty, choose $\mathbf{x}$ in U. Then $\mathbf{0} = 0\mathbf{x}$ is in U by S3, so S1 also holds. This means that U is a subspace.
22. (b) The zero vector $\mathbf{0}$ is in $U + W$ because $\mathbf{0} = \mathbf{0} + \mathbf{0}$. Let $\mathbf{p}$ and $\mathbf{q}$ be vectors in $U + W$, say $\mathbf{p} = \mathbf{x}_1 + \mathbf{y}_1$ and $\mathbf{q} = \mathbf{x}_2 + \mathbf{y}_2$ where $\mathbf{x}_1$ and $\mathbf{x}_2$ are in U, and $\mathbf{y}_1$ and $\mathbf{y}_2$ are in W. Then $\mathbf{p} + \mathbf{q} = (\mathbf{x}_1 + \mathbf{x}_2) + (\mathbf{y}_1 + \mathbf{y}_2)$ is in $U + W$ because $\mathbf{x}_1 + \mathbf{x}_2$ is in U and $\mathbf{y}_1 + \mathbf{y}_2$ is in W. Similarly, $a(\mathbf{p} + \mathbf{q}) = a\mathbf{p} + a\mathbf{q}$ is in $U + W$ for any scalar a because $a\mathbf{p}$ is in U and $a\mathbf{q}$ is in W. Hence $U + W$ is indeed a subspace of $\mathbb{R}^n$.

Exercises 5.2 Independence and Dimension (Page 244)

1. (b) Yes. If $r\begin{bmatrix} 1 \\ 1 \\ 1 \end{bmatrix} + s\begin{bmatrix} 1 \\ 1 \\ 1 \end{bmatrix} + t\begin{bmatrix} 0 \\ 0 \\ 1 \end{bmatrix} = \begin{bmatrix} 0 \\ 0 \\ 0 \end{bmatrix}$, then $r + s = 0$, $r - s = 0$, and $r + s + t = 0$. These equations give $r = s = t = 0$.
(d) No. Indeed: $\begin{bmatrix} 1 \\ 1 \\ 0 \\ 0 \end{bmatrix} - \begin{bmatrix} 1 \\ 0 \\ 1 \\ 0 \end{bmatrix} + \begin{bmatrix} 0 \\ 0 \\ 1 \\ 1 \end{bmatrix} - \begin{bmatrix} 0 \\ 1 \\ 0 \\ 1 \end{bmatrix} = \begin{bmatrix} 0 \\ 0 \\ 0 \\ 0 \end{bmatrix}$.
2. (b) Yes. If $r(\mathbf{x} + \mathbf{y}) + s(\mathbf{y} + \mathbf{z}) + t(\mathbf{z} + \mathbf{x}) = \mathbf{0}$, then $(r + t)\mathbf{x} + (r + s)\mathbf{y} + (s + t)\mathbf{z} = \mathbf{0}$. Since $\{\mathbf{x}, \mathbf{y}, \mathbf{z}\}$ is independent, this implies that $r + t = 0$, $r + s = 0$, and $s + t = 0$. The only solution is $r = s = t = 0$.
(d) No. In fact, $(\mathbf{x} + \mathbf{y}) - (\mathbf{y} + \mathbf{z}) + (\mathbf{z} + \mathbf{w}) - (\mathbf{w} + \mathbf{x}) = \mathbf{0}$.
3. (b) $\left\{\begin{bmatrix} 2 \\ 1 \\ 0 \\ -1 \end{bmatrix}, \begin{bmatrix} -1 \\ 1 \\ 1 \\ 1 \end{bmatrix}\right\}$; dimension 2.

(d) $\left\{\begin{bmatrix} -2 \\ 0 \\ 3 \\ 1 \end{bmatrix}, \begin{bmatrix} 1 \\ 2 \\ -1 \\ 0 \end{bmatrix}\right\}$; dimension 2.

4. (b) $\left\{\begin{bmatrix} 1 \\ 1 \\ 0 \\ 1 \end{bmatrix}, \begin{bmatrix} 1 \\ -1 \\ 1 \\ 0 \end{bmatrix}\right\}$; dimension 2. (d) $\left\{\begin{bmatrix} 1 \\ 0 \\ 1 \\ 0 \end{bmatrix}, \begin{bmatrix} -1 \\ 1 \\ 0 \\ 1 \end{bmatrix}, \begin{bmatrix} 0 \\ 1 \\ 0 \\ 1 \end{bmatrix}\right\}$; dimension 3. (f) $\left\{\begin{bmatrix} -1 \\ 1 \\ 0 \\ 0 \end{bmatrix}, \begin{bmatrix} 1 \\ 0 \\ 1 \\ 0 \end{bmatrix}, \begin{bmatrix} 1 \\ 0 \\ 0 \\ 1 \end{bmatrix}\right\}$; dimension 3.

5. (b) If $r(\mathbf{x} + \mathbf{w}) + s(\mathbf{y} + \mathbf{w}) + t(\mathbf{z} + \mathbf{w}) + u(\mathbf{w}) = \mathbf{0}$, then $r\mathbf{x} + s\mathbf{y} + t\mathbf{z} + (r + s + t + u)\mathbf{w} = \mathbf{0}$, so $r = 0$, $s = 0$, $t = 0$, and $r + s + t + u = 0$. The only solution is $r = s = t = u = 0$, so the set is independent. Since $\dim \mathbb{R}^4 = 4$, the set is a basis by Theorem 7.
6. (b) Yes (d) Yes (f) No.
7. (b) T. If $r\mathbf{y} + s\mathbf{z} = \mathbf{0}$, then $0\mathbf{x} + r\mathbf{y} + s\mathbf{z} = \mathbf{0}$ so $r = s = 0$ because $\{\mathbf{x}, \mathbf{y}, \mathbf{z}\}$ is independent.
(d) F. If $\mathbf{x} \neq \mathbf{0}$, take $k = 2$, $\mathbf{x}_1 = \mathbf{x}$ and $\mathbf{x}_2 = -\mathbf{x}$.
(f) F. If $\mathbf{y} = -\mathbf{x}$ and $\mathbf{z} = \mathbf{0}$, then $1\mathbf{x} + 1\mathbf{y} + 1\mathbf{z} = \mathbf{0}$.
(h) T. This is a nontrivial, vanishing linear combination, so the $\mathbf{x}_i$ cannot be independent.
10. If $r\mathbf{x}_2 + s\mathbf{x}_3 + t\mathbf{x}_5 = \mathbf{0}$ then $0\mathbf{x}_1 + r\mathbf{x}_2 + s\mathbf{x}_3 + 0\mathbf{x}_4 + t\mathbf{x}_5 + 0\mathbf{x}_6 = \mathbf{0}$ so $r = s = t = 0$.
12. If $t_1\mathbf{x}_1 + t_2(\mathbf{x}_1 + \mathbf{x}_2) + \cdots + t_k(\mathbf{x}_1 + \mathbf{x}_2 + \cdots + \mathbf{x}_k) = \mathbf{0}$, then $(t_1 + t_2 + \cdots + t_k)\mathbf{x}_1 + (t_2 + \cdots + t_k)\mathbf{x}_2 + \cdots + (t_{k-1} + t_k)\mathbf{x}_{k-1} + (t_k)\mathbf{x}_k = \mathbf{0}$. Hence all these coefficients are zero, so we obtain successively $t_k = 0$, $t_{k-1} = 0$, ..., $t_2 = 0$, $t_1 = 0$.
16. (b) We show A^T is invertible (then A is invertible). Let $A^T\mathbf{x} = \mathbf{0}$ where $\mathbf{x} = [s\ t]^T$. This means $as + ct = 0$ and $bs + dt = 0$, so $s(a\mathbf{x} + b\mathbf{y}) + t(c\mathbf{x} + d\mathbf{y}) = (sa + tc)\mathbf{x} + (sb + td)\mathbf{y} = \mathbf{0}$. Hence $s = t = 0$ by hypothesis.

17. **(b)** Each $V^{-1}\mathbf{x}_i$ is in null(AV) because $AV(V^{-1}\mathbf{x}_i) = A\mathbf{x}_i = \mathbf{0}$. The set $\{V^{-1}\mathbf{x}_1, \ldots, V^{-1}\mathbf{x}_k\}$ is independent as V^{-1} is invertible. If $\mathbf{y}$ is in null(AV), then $V\mathbf{y}$ is in null(A) so let $V\mathbf{y} = t_1\mathbf{x}_1 + \cdots + t_k\mathbf{x}_k$ where each t_k is in $\mathbb{R}$. Thus $\mathbf{y} = t_1V^{-1}\mathbf{x}_1 + \cdots + t_kV^{-1}\mathbf{x}_k$ is in span$\{V^{-1}\mathbf{x}_1, \ldots, V^{-1}\mathbf{x}_k\}$.

20. We have $\{\mathbf{0}\} \subseteq U \subseteq W$ where dim$\{\mathbf{0}\} = 0$ and dim $W = 1$. Hence dim $U = 0$ or dim $U = 1$ by Theorem 8, that is $U = 0$ or $U = W$, again by Theorem 8.

Exercises 5.3 Orthogonality (Page 252)

1. **(b)** $\left\{ \frac{1}{\sqrt{3}}\begin{bmatrix}1\\1\\1\end{bmatrix}, \frac{1}{\sqrt{42}}\begin{bmatrix}4\\1\\-5\end{bmatrix}, \frac{1}{\sqrt{14}}\begin{bmatrix}2\\-3\\1\end{bmatrix} \right\}$.

3. **(b)** $\begin{bmatrix}a\\b\\c\end{bmatrix} = \frac{1}{2}(a-c)\begin{bmatrix}1\\0\\-1\end{bmatrix} + \frac{1}{18}(a+4b+c)\begin{bmatrix}1\\4\\1\end{bmatrix} + \frac{1}{9}(2a-b+2c)\begin{bmatrix}2\\-1\\2\end{bmatrix}$.

 (d) $\begin{bmatrix}a\\b\\c\end{bmatrix} = \frac{1}{3}(a+b+c)\begin{bmatrix}1\\1\\1\end{bmatrix} + \frac{1}{2}(a-b)\begin{bmatrix}1\\-1\\0\end{bmatrix} + \frac{1}{6}(a+b-2c)\begin{bmatrix}1\\1\\-2\end{bmatrix}$.

4. **(b)** $\begin{bmatrix}14\\1\\-8\\5\end{bmatrix} = 3\begin{bmatrix}2\\-1\\0\\3\end{bmatrix} + 4\begin{bmatrix}2\\1\\-2\\-1\end{bmatrix}$.

5. **(b)** $t\begin{bmatrix}-1\\3\\10\\11\end{bmatrix}$, t in $\mathbb{R}$.

6. **(b)** $\sqrt{29}$ **(d)** 19

7. **(b)** F. $\mathbf{x} = \begin{bmatrix}1\\0\end{bmatrix}$ and $\mathbf{y} = \begin{bmatrix}0\\1\end{bmatrix}$.

 (d) T. Every $\mathbf{x}_i \cdot \mathbf{y}_j = 0$ by assumption, every $\mathbf{x}_i \cdot \mathbf{x}_j = 0$ if $i \neq j$ because the $\mathbf{x}_i$ are orthogonal, and every $\mathbf{y}_i \cdot \mathbf{y}_j = 0$ if $i \neq j$ because the $\mathbf{y}_i$ are orthogonal. As all the vectors are nonzero, this does it.

 (f) T. Every pair of *distinct* vectors in the set $\{\mathbf{x}\}$ has dot product zero (there are no such pairs).

9. Let $\mathbf{c}_1, \ldots, \mathbf{c}_n$ be the columns of A. Then row i of A^T is $\mathbf{c}_i^T$, so the (i,j)-entry of A^TA is $\mathbf{c}_i^T\mathbf{c}_j = \mathbf{c}_i \cdot \mathbf{c}_j = 0, 1$ according as $i \neq j$, $i = j$. So $A^TA = I$.

11. **(b)** Take $n = 3$ in **(a)**, expand, and simplify.

12. **(b)** We have $(\mathbf{x} + \mathbf{y}) \cdot (\mathbf{x} - \mathbf{y}) = \|\mathbf{x}\|^2 - \|\mathbf{y}\|^2$. Hence $(\mathbf{x} + \mathbf{y}) \cdot (\mathbf{x} - \mathbf{y}) = 0$ if and only if $\|\mathbf{x}\|^2 = \|\mathbf{y}\|^2$; if and only if $\|\mathbf{x}\| = \|\mathbf{y}\|$—where we used the fact that $\|\mathbf{x}\| \geq 0$ and $\|\mathbf{y}\| \geq 0$.

15. If $A^TA\mathbf{x} = \lambda\mathbf{x}$, then $\|A\mathbf{x}\|^2 = (A\mathbf{x}) \cdot (A\mathbf{x}) = \mathbf{x}^TA^TA\mathbf{x} = \mathbf{x}^T(\lambda\mathbf{x}) = \lambda\|\mathbf{x}\|^2$.

Exercises 5.4 Rank of a Matrix (Page 260)

1. **(b)** $\left\{\begin{bmatrix}2\\-1\\1\end{bmatrix}, \begin{bmatrix}0\\0\\1\end{bmatrix}\right\}; \left\{\begin{bmatrix}2\\-2\\4\\-6\end{bmatrix}, \begin{bmatrix}1\\1\\3\\0\end{bmatrix}\right\}; 2$

 (d) $\left\{\begin{bmatrix}1\\2\\-1\\3\end{bmatrix}, \begin{bmatrix}0\\0\\0\\1\end{bmatrix}\right\}; \left\{\begin{bmatrix}1\\-3\end{bmatrix}, \begin{bmatrix}3\\-2\end{bmatrix}\right\}; 2$

2. **(b)** $\left\{\begin{bmatrix}1\\1\\0\\0\end{bmatrix}, \begin{bmatrix}0\\-2\\2\\5\\1\end{bmatrix}, \begin{bmatrix}0\\0\\2\\-3\\6\end{bmatrix}\right\}$ **(d)** $\left\{\begin{bmatrix}1\\5\\-6\end{bmatrix}, \begin{bmatrix}0\\1\\-1\end{bmatrix}, \begin{bmatrix}0\\0\\1\end{bmatrix}\right\}$

3. **(b)** No; no **(d)** No

 (f) Otherwise, if A is $m \times n$, we have $m = \dim(\text{row } A) = \text{rank } A = \dim(\text{col } A) = n$.

4. Let $A = [\mathbf{c}_1 \cdots \mathbf{c}_n]$. Then col $A = \text{span}\{\mathbf{c}_1, \ldots, \mathbf{c}_n\} = \{x_1\mathbf{c}_1 + \cdots + x_n\mathbf{c}_n \mid x_i \text{ in } \mathbb{R}\} = \{A\mathbf{x} \mid \mathbf{x} \text{ in } \mathbb{R}^n\}$.

7. **(b)** The basis is $\left\{\begin{bmatrix}6\\0\\-4\\1\\0\end{bmatrix}, \begin{bmatrix}5\\0\\-3\\0\\1\end{bmatrix}\right\}$, so the dimension is 2.

 Have rank $A = 3$ and $n - 3 = 2$.

8. **(b)** $n - 1$

9. **(b)** If $r_1\mathbf{c}_1 + \cdots + r_n\mathbf{c}_n = \mathbf{0}$, let $\mathbf{x} = [r_1, \ldots, r_n]^T$. Then $C\mathbf{x} = r_1\mathbf{c}_1 + \cdots + r_n\mathbf{c}_n = \mathbf{0}$, so $\mathbf{x}$ is in null $A = 0$. Hence each $r_i = 0$.

10. **(b)** Write $r = \text{rank } A$. Then **(a)** gives $r = \dim(\text{col } A) \leq \dim(\text{null } A) = n - r$.

12. We have rank$(A) = \dim[\text{col}(A)]$ and rank $(A^T) = \dim[\text{row}(A^T)]$. Let $\{\mathbf{c}_1, \mathbf{c}_2, \ldots, \mathbf{c}_k\}$ be a basis of col(A); it suffices to show that $\{\mathbf{c}_1^T, \mathbf{c}_2^T, \ldots, \mathbf{c}_k^T\}$ is a basis of row(A^T). But if $t_1\mathbf{c}_1^T + t_2\mathbf{c}_2^T + \cdots + t_k\mathbf{c}_k^T = \mathbf{0}$, t_j in $\mathbb{R}$, then (taking transposes) $t_1\mathbf{c}_1 + t_2\mathbf{c}_2 + \cdots + t_k\mathbf{c}_k = \mathbf{0}$ so each $t_j = 0$. Hence $\{\mathbf{c}_1^T, \mathbf{c}_2^T, \ldots, \mathbf{c}_k^T\}$ is independent. Given $\mathbf{v}$ in row(A^T) then $\mathbf{v}^T$ is in col(A); say $\mathbf{v}^T = s_1\mathbf{c}_1 + s_2\mathbf{c}_2 + \cdots + s_k\mathbf{c}_k$, s_j in $\mathbb{R}$: Hence $\mathbf{v} = s_1\mathbf{c}_1^T + s_2\mathbf{c}_2^T + \cdots + s_k\mathbf{c}_k^T$, so $\{\mathbf{c}_1^T, \mathbf{c}_2^T, \ldots, \mathbf{c}_k^T\}$ spans row(A^T), as required.

15. **(b)** Let $\{\mathbf{u}_1, \ldots, \mathbf{u}_r\}$ be a basis of col(A). Then $\mathbf{b}$ is *not* in col(A), so $\{\mathbf{u}_1, \ldots, \mathbf{u}_r, \mathbf{b}\}$ is linearly independent. Show that col$[A \ \mathbf{b}] = \text{span}\{\mathbf{u}_1, \ldots, \mathbf{u}_r, \mathbf{b}\}$.

Exercises 5.5 Similarity and Diagonalization (Page 271)

1. **(b)** traces = 2, ranks = 2, but det $A = -5$, det $B = -1$
 (d) ranks = 2, determinants = 7, but tr $A = 5$, tr $B = 4$
 (f) traces = -5, determinants = 0, but rank $A = 2$, rank $B = 1$

Selected Answers

3. **(b)** If $B = P^{-1}AP$, then $B^{-1} = P^{-1}A^{-1}(P^{-1})^{-1} = P^{-1}A^{-1}P$.

4. **(b)** Yes, $P = \begin{bmatrix} -1 & 0 & 6 \\ 0 & 1 & 0 \\ 1 & 0 & 5 \end{bmatrix}$, $P^{-1}AP = \begin{bmatrix} -3 & 0 & 0 \\ 0 & -3 & 0 \\ 0 & 0 & 8 \end{bmatrix}$.
   (d) No, $c_A(x) = (x+1)(x-4)^2$ so $\lambda = 4$ has multiplicity 2. But $\dim(E_4) = 1$ so Theorem 6 applies.

8. **(b)** If $B = P^{-1}AP$ and $A^k = 0$, then $B^k = (P^{-1}AP)^k = P^{-1}A^kP = P^{-1}0P = 0$.

9. **(b)** The eigenvalues of A are all equal (they are the diagonal elements), so if $P^{-1}AP = D$ is diagonal, then $D = \lambda I$. Hence $A = P^{-1}(\lambda I)P = \lambda I$.

10. **(b)** A is similar to $D = \text{diag}(\lambda_1, \lambda_2, \ldots, \lambda_n)$ so (Theorem 1) $\text{tr } A = \text{tr } D = \lambda_1 + \lambda_2 + \cdots + \lambda_n$.

12. **(b)** $T_P(A)T_P(B) = (P^{-1}AP)(P^{-1}BP) = P^{-1}(AB)P = T_P(AB)$.

13. **(b)** If A is diagonalizable, so is A^T, and they have the same eigenvalues. Use **(a)**.

17. **(b)** $c_B(x) = [x - (a+b+c)][x^2 - k]$ where $k = a^2 + b^2 + c^2 - [ab + ac + bc]$. Use Theorem 7.

Exercises 5.6 Best Approximation and Least Squares (Page 281)

1. **(b)** $\frac{1}{12}\begin{bmatrix} -20 \\ 46 \\ 95 \end{bmatrix}$, $(A^TA)^{-1} = \frac{1}{12}\begin{bmatrix} 8 & -10 & -18 \\ -10 & 14 & 24 \\ -18 & 24 & 43 \end{bmatrix}$

2. **(b)** $\frac{64}{13} - \frac{6}{13}x$ **(d)** $-\frac{4}{10} - \frac{17}{10}x$

3. **(b)** $y = 0.127 - 0.024x + 0.194x^2$,
   $(M^TM)^{-1} = \frac{1}{4248}\begin{bmatrix} 3348 & 642 & -426 \\ 642 & 571 & -187 \\ -426 & -187 & 91 \end{bmatrix}$

4. **(b)** $\frac{1}{92}(-46x + 66x^2 + 60 \cdot 2^x)$,
   $(M^TM)^{-1} = \frac{1}{46}\begin{bmatrix} 115 & 0 & -46 \\ 0 & 17 & -18 \\ -46 & -18 & 38 \end{bmatrix}$

5. **(b)** $\frac{1}{20}\left[18 + 21x^2 + 28\sin(\frac{\pi x}{2})\right]$,
   $(M^TM)^{-1} = \frac{1}{40}\begin{bmatrix} 24 & -2 & 14 \\ -2 & 1 & 3 \\ 14 & 3 & 49 \end{bmatrix}$

7. $s = 99.71 - 4.87x$; the estimate of g is 9.74. [The true value of g is 9.81]. If a quadratic in s is fit, the result is $s = 101 - \frac{3}{2}t - \frac{9}{2}t^2$ giving $g = 9$;
   $(M^TM)^{-1} = \frac{1}{2}\begin{bmatrix} 38 & -42 & 10 \\ -42 & 49 & -12 \\ 10 & -12 & 3 \end{bmatrix}$.

9. $y = -5.19 + 0.34x_1 + 0.51x_2 + 0.71x_3$,
   $(A^TA)^{-1} = \frac{1}{25080}\begin{bmatrix} 517860 & -8016 & 5040 & -22650 \\ -8016 & 208 & -316 & 400 \\ 5040 & -316 & 1300 & -1090 \\ -22650 & 400 & -1090 & 1975 \end{bmatrix}$

10. **(b)** $f(x) = a_0$ here, so the sum of squares is $S = \sum(y_i - a_0)^2 = na_0^2 - 2a_0\sum y_i + \sum y_i^2$. Completing the square gives $S = n\left[a_0 - \frac{1}{n}\sum y_i\right]^2 + \left[\sum y_i^2 - \frac{1}{n}(\sum y_i)^2\right]$. This is minimal when $a_0 = \frac{1}{n}\sum y_i$.

13. **(b)** Here $f(x) = r_0 + r_1 e^x$. If $f(x_1) = 0 = f(x_2)$ where $x_1 \neq x_2$, then $r_0 + r_1 \cdot e^{x_1} = 0 = r_0 + r_1 \cdot e^{x_2}$ so $r_1(e^{x_1} - e^{x_2}) = 0$. Hence $r_1 = 0 = r_0$.

Exercises 5.7 An Application to Correlation and Variance (Page 287)

2. Let X denote the number of years of education, and let Y denote the yearly income (in 1000's). Then $\bar{x} = 15.3$, $s_x^2 = 9.12$ and $s_x = 3.02$, while $\bar{y} = 40.3$, $s_y^2 = 114.23$ and $s_y = 10.69$. The correlation is $r(X, Y) = 0.599$.

4. **(b)** Given the sample vector $\mathbf{x} = \begin{bmatrix} x_1 \\ x_2 \\ \vdots \\ x_n \end{bmatrix}$, let $\mathbf{z} = \begin{bmatrix} z_1 \\ z_2 \\ \vdots \\ z_n \end{bmatrix}$
   where $z_i = a + bx_i$ for each i. By **(a)** we have $\bar{z} = a + b\bar{x}$, so $s_z^2 = \frac{1}{n-1}\sum_i(z_i - \bar{z})^2$
   $= \frac{1}{n-1}\sum_i[(a + bx_i) - (a + b\bar{x})]^2 = \frac{1}{n-1}\sum_i b^2(x_i - \bar{x})^2$
   $= b^2 s_x^2$. Now **(b)** follows because $\sqrt{b^2} = |b|$.

Supplementary Exercises for Chapter 5 (Page 287)

(b) F **(d)** T **(f)** T **(h)** F **(j)** F **(l)** T **(n)** F **(p)** F **(r)** F

Exercises 6.1 Examples and Basic Properties (Page 295)

1. **(b)** No; S5 fails. **(d)** No; S4 and S5 fail.
2. **(b)** No; only A1 fails. **(d)** No
   (f) Yes **(h)** Yes **(j)** No
   (l) No; only S3 fails.
   (n) No; only S4 and S5 fail.
4. The zero vector is $(0, -1)$; the negative of (x, y) is $(-x, -2 - y)$.
5. **(b)** $\mathbf{x} = \frac{1}{7}(5\mathbf{u} - 2\mathbf{v})$, $\mathbf{y} = \frac{1}{7}(4\mathbf{u} - 3\mathbf{v})$
6. **(b)** Equating entries gives $a + c = 0$, $b + c = 0$, $b + c = 0$, $a - c = 0$. The solution is $a = b = c = 0$.
   (d) If $a\sin x + b\cos y + c = 0$ in $\mathbf{F}[0, \pi]$, then this must hold for *every* x in $[0, \pi]$. Taking $x = 0$, $\frac{\pi}{2}$, and π, respectively, gives $b + c = 0$, $a + c = 0$, $-b + c = 0$ whence, $a = b = c = 0$.
7. **(b)** $4\mathbf{w}$
10. If $\mathbf{z} + \mathbf{v} = \mathbf{v}$ for all $\mathbf{v}$, then $\mathbf{z} + \mathbf{v} = \mathbf{0} + \mathbf{v}$, so $\mathbf{z} = \mathbf{0}$ by cancellation.
12. **(b)** $(-a)\mathbf{v} + a\mathbf{v} = (-a + a)\mathbf{v} = 0\mathbf{v} = \mathbf{0}$ by Theorem 3. Because also $-(a\mathbf{v}) + a\mathbf{v} = \mathbf{0}$ (by the definition of $-(a\mathbf{v})$ in axiom A5), this means that $(-a)\mathbf{v} = -(a\mathbf{v})$ by cancellation. Alternatively, use Theorem 3(4) to give $(-a)\mathbf{v} = [(-1)a]\mathbf{v} = (-1)(a\mathbf{v}) = -(a\mathbf{v})$.

13. **(b)** The case $n = 1$ is clear, and $n = 2$ is axiom S3. If $n > 2$, then $(a_1 + a_2 + \cdots + a_n)\mathbf{v} = [a_1 + (a_2 + \cdots + a_n)]\mathbf{v} = a_1\mathbf{v} + (a_2 + \cdots + a_n)\mathbf{v} = a_1\mathbf{v} + (a_2\mathbf{v} + \cdots + a_n\mathbf{v})$ using the induction hypothesis; so it holds for all n.

15. **(c)** If $a\mathbf{v} = a\mathbf{w}$, then $\mathbf{v} = 1\mathbf{v} = (a^{-1}a)\mathbf{v} = a^{-1}(a\mathbf{v}) = a^{-1}(a\mathbf{w}) = (a^{-1}a)\mathbf{w} = 1\mathbf{w} = \mathbf{w}$.

Exercises 6.2 Subspaces and Spanning Sets (Page 301)

1. **(b)** Yes **(d)** Yes **(f)** No; not closed under addition or scalar multiplication, and 0 is not in the set.
2. **(b)** Yes **(d)** Yes **(f)** No; not closed under addition.
3. **(b)** No; not closed under addition. **(d)** No; not closed under scalar multiplication. **(f)** Yes
5. **(b)** If entry k of $\mathbf{x}$ is $x_k \neq 0$, and if $\mathbf{y}$ is in $\mathbb{R}^n$, then $\mathbf{y} = A\mathbf{x}$ where the column of A is $x_k^{-1}\mathbf{y}$, and the other columns are zero.
6. **(b)** $-3(x + 1) + 0(x^2 + x) + 2(x^2 + 2)$
   (d) $\frac{2}{3}(x + 1) + \frac{1}{3}(x^2 + x) - \frac{1}{3}(x^2 + 2)$
7. **(b)** No **(d)** Yes; $\mathbf{v} = 3\mathbf{u} - \mathbf{w}$.
8. **(b)** Yes; $1 = \cos^2 x + \sin^2 x$
   (d) No. If $1 + x^2 = a\cos^2 x + b\sin^2 x$, then taking $x = 0$ and $x = \pi$ gives $a = 1$ and $a = 1 + \pi^2$.
9. **(b)** Because $\mathbf{P}_2 = \text{span}\{1, x, x^2\}$, it suffices to show that $\{1, x, x^2\} \subseteq \text{span}\{1 + 2x^2, 3x, 1 + x\}$. But $x = \frac{1}{3}(3x)$; $1 = (1 + x) - x$ and $x^2 = \frac{1}{2}[(1 + 2x^2) - 1]$.
11. **(b)** $\mathbf{u} = (\mathbf{u} + \mathbf{w}) - \mathbf{w}$, $\mathbf{v} = -(\mathbf{u} - \mathbf{v}) + (\mathbf{u} + \mathbf{w}) - \mathbf{w}$, and $\mathbf{w} = \mathbf{w}$
14. No
17. **(b)** Yes.
18. $\mathbf{v}_1 = \frac{1}{a_1}\mathbf{u} - \frac{a_2}{a_1}\mathbf{v}_2 - \cdots - \frac{a_n}{a_1}\mathbf{v}_n$, so $V \subseteq \text{span}\{\mathbf{u}, \mathbf{v}_2, \ldots, \mathbf{v}_n\}$.
21. **(b)** $\mathbf{v} = (\mathbf{u} + \mathbf{v}) - \mathbf{u}$ is in U.
22. Given the condition and $\mathbf{u} \in U$, $\mathbf{0} = \mathbf{u} + (-1)\mathbf{u} \in U$. The converse holds by the subspace test.

Exercises 6.3 Linear Independence and Dimension (Page 309)

1. **(b)** If $ax^2 + b(x + 1) + c(1 - x - x^2) = 0$, then $a + c = 0, b - c = 0, b + c = 0$, so $a = b = c = 0$.
   (d) If $a\begin{bmatrix} 1 & 1 \\ 1 & 0 \end{bmatrix} + b\begin{bmatrix} 0 & 1 \\ 1 & 1 \end{bmatrix} + c\begin{bmatrix} 1 & 0 \\ 1 & 1 \end{bmatrix} + d\begin{bmatrix} 1 & 1 \\ 0 & 1 \end{bmatrix} = \begin{bmatrix} 0 & 0 \\ 0 & 0 \end{bmatrix}$, then $a + c + d = 0, a + b + d = 0, a + b + c = 0$, and $b + c + d = 0$, so $a = b = c = d = 0$.
2. **(b)** $3(x^2 - x + 3) - 2(2x^2 + x + 5) + (x^2 + 5x + 1) = 0$
   (d) $2\begin{bmatrix} -1 & 0 \\ 0 & -1 \end{bmatrix} + \begin{bmatrix} 1 & -1 \\ -1 & 1 \end{bmatrix} + \begin{bmatrix} 1 & 1 \\ 1 & 1 \end{bmatrix} = \begin{bmatrix} 0 & 0 \\ 0 & 0 \end{bmatrix}$
   (f) $\dfrac{5}{x^2 + x - 6} + \dfrac{1}{x^2 - 5x + 6} - \dfrac{6}{x^2 - 9} = 0$
3. **(b)** Dependent: $1 - \sin^2 x - \cos^2 x = 0$
4. **(b)** $x \neq -\frac{1}{3}$

5. **(b)** If $r(-1, 1, 1) + s(1, -1, 1) + t(1, 1, -1) = (0, 0, 0)$, then $-r + s + t = 0, r - s + t = 0$, and $r - s - t = 0$, and this implies that $r = s = t = 0$. This proves independence. To prove that they span $\mathbb{R}^3$, observe that $(0, 0, 1) = \frac{1}{2}[(-1, 1, 1) + (1, -1, 1)]$ so $(0, 0, 1)$ lies in $\text{span}\{(-1, 1, 1), (1, -1, 1), (1, 1, -1)\}$. The proof is similar for $(0, 1, 0)$ and $(1, 0, 0)$.
   (d) If $r(1 + x) + s(x + x^2) + t(x^2 + x^3) + ux^3 = 0$, then $r = 0, r + s = 0, s + t = 0$, and $t + u = 0$, so $r = s = t = u = 0$. This proves independence. To show that they span $\mathbf{P}_3$, observe that $x^2 = (x^2 + x^3) - x^3$, $x = (x + x^2) - x^2$, and $1 = (1 + x) - x$, so $\{1, x, x^2, x^3\} \subseteq \text{span}\{1 + x, x + x^2, x^2 + x^3, x^3\}$.
6. **(b)** $\{1, x + x^2\}$; dimension $= 2$
   (d) $\{1, x^2\}$; dimension $= 2$
7. **(b)** $\left\{\begin{bmatrix} 1 & 1 \\ -1 & 0 \end{bmatrix}, \begin{bmatrix} 1 & 0 \\ 0 & 1 \end{bmatrix}\right\}$; dimension $= 2$
   (d) $\left\{\begin{bmatrix} 1 & 0 \\ 1 & 1 \end{bmatrix}, \begin{bmatrix} 0 & 1 \\ -1 & 0 \end{bmatrix}\right\}$; dimension $= 2$
8. **(b)** $\left\{\begin{bmatrix} 1 & 0 \\ 0 & 0 \end{bmatrix}, \begin{bmatrix} 0 & 1 \\ 0 & 0 \end{bmatrix}\right\}$
10. **(b)** $\dim V = 7$
11. **(b)** $\{x^2 - x, x(x^2 - x), x^2(x^2 - x), x^3(x^2 - x)\}$; $\dim V = 4$
12. **(b)** No. Any linear combination f of such polynomials has $f(0) = 0$.
    (d) No. $\left\{\begin{bmatrix} 1 & 0 \\ 0 & 1 \end{bmatrix}, \begin{bmatrix} 1 & 1 \\ 0 & 1 \end{bmatrix}, \begin{bmatrix} 1 & 0 \\ 1 & 1 \end{bmatrix}, \begin{bmatrix} 0 & 1 \\ 1 & 1 \end{bmatrix}\right\}$; consists of invertible matrices.
    (f) Yes. $0\mathbf{u} + 0\mathbf{v} + 0\mathbf{w} = \mathbf{0}$ for every set $\{\mathbf{u}, \mathbf{v}, \mathbf{w}\}$.
    (h) Yes. $s\mathbf{u} + t(\mathbf{u} + \mathbf{v}) = \mathbf{0}$ gives $(s + t)\mathbf{u} + t\mathbf{v} = \mathbf{0}$, whence $s + t = 0 = t$. **(j)** Yes. If $r\mathbf{u} + s\mathbf{v} = \mathbf{0}$, then $r\mathbf{u} + s\mathbf{v} + 0\mathbf{w} = \mathbf{0}$, so $r = 0 = s$.
    (l) Yes. $\mathbf{u} + \mathbf{v} + \mathbf{w} \neq \mathbf{0}$ because $\{\mathbf{u}, \mathbf{v}, \mathbf{w}\}$ is independent.
    (n) Yes. If I is independent, then $|I| \leq n$ by the fundamental theorem because any basis spans V.
15. If a linear combination of the subset vanishes, it is a linear combination of the vectors in the larger set (coefficients outside the subset are zero) so it is trivial.
19. Because $\{\mathbf{u}, \mathbf{v}\}$ is linearly independent, $s\mathbf{u}' + t\mathbf{v}' = \mathbf{0}$ is equivalent to $\begin{bmatrix} a & c \\ b & d \end{bmatrix}\begin{bmatrix} s \\ t \end{bmatrix} = \begin{bmatrix} 0 \\ 0 \end{bmatrix}$. Now apply Theorem 5 Section 2.4.
23. **(b)** Independent **(d)** Dependent. For example, $(\mathbf{u} + \mathbf{v}) - (\mathbf{v} + \mathbf{w}) + (\mathbf{w} + \mathbf{z}) - (\mathbf{z} + \mathbf{u}) = \mathbf{0}$.
26. If z is not real and $az + bz^2 = 0$, then $a + bz = 0$ $(z \neq 0)$. Hence if $b \neq 0$, then $z = -ab^{-1}$ is real. So $b = 0$, and so $a = 0$. Conversely, if z is real, say $z = a$, then $(-a)z + 1z^2 = 0$, contrary to the independence of $\{z, z^2\}$.
29. **(b)** If $U\mathbf{x} = \mathbf{0}, \mathbf{x} \neq \mathbf{0}$ in $\mathbb{R}^n$, then $R\mathbf{x} = \mathbf{0}$ where $R \neq 0$ is row 1 of U. If $B \in \mathbf{M}_{mn}$ has each row equal to R, then $B\mathbf{x} \neq \mathbf{0}$. But if $B = \sum r_i A_i U$, then $B\mathbf{x} = \sum r_i A_i U\mathbf{x} = \mathbf{0}$. So $\{A_i U\}$ cannot span $\mathbf{M}_{mn}$.

Selected Answers

33. **(b)** If $U \cap W = 0$ and $r\mathbf{u} + s\mathbf{w} = \mathbf{0}$, then $r\mathbf{u} = -s\mathbf{w}$ is in $U \cap W$, so $r\mathbf{u} = \mathbf{0} = s\mathbf{w}$. Hence $r = 0 = s$ because $\mathbf{u} \neq \mathbf{0} \neq \mathbf{w}$. Conversely, if $\mathbf{v} \neq \mathbf{0}$ lies in $U \cap W$, then $1\mathbf{v} + (-1)\mathbf{v} = \mathbf{0}$, contrary to hypothesis.

36. **(b)** $\dim O_n = \frac{n}{2}$ if n is even and $\dim O_n = \frac{n+1}{2}$ if n is odd.

Exercises 6.4 Finite Dimensional Spaces (Page 318)

1. **(b)** $\{(0, 1, 1), (1, 0, 0), (0, 1, 0)\}$ **(d)** $\{x^2 - x + 1, 1, x\}$
2. **(b)** Any three except $\{x^2 + 3, x + 2, x^2 - 2x - 1\}$
3. **(b)** Add $(0, 1, 0, 0)$ and $(0, 0, 1, 0)$.   **(d)** Add 1 and x^3.
4. **(b)** If $z = a + bi$, then $a \neq 0$ and $b \neq 0$. If $rz + s\bar{z} = 0$, then $(r + s)a = 0$ and $(r - s)b = 0$. This means that $r + s = 0 = r - s$, so $r = s = 0$. Thus $\{z, \bar{z}\}$ is independent; it is a basis because $\dim \mathbb{C} = 2$.
5. **(b)** The polynomials in S have distinct degrees.
6. **(b)** $\{4, 4x, 4x^2, 4x^3\}$ is one such basis of $\mathbf{P}_3$. However, there is *no* basis of $\mathbf{P}_3$ consisting of polynomials that have the property that their coefficients sum to zero. For if such a basis exists, then every polynomial in $\mathbf{P}_3$ would have this property (because sums and scalar multiples of such polynomials have the same property).
7. **(b)** Not a basis   **(d)** Not a basis
8. **(b)** Yes; no
10. $\det A = 0$ if and only if A is not invertible; if and only if the rows of A are dependent (Theorem 3 Section 5.2); if and only if some row is a linear combination of the others (Lemma 2).
11. **(b)** No. $\{(0, 1), (1, 0)\} \subseteq \{(0, 1), (1, 0), (1, 1)\}$.
    (d) Yes. See Exercise 15 Section 6.3.
15. If $\mathbf{v} \in U$ then $W = U$; if $\mathbf{v} \notin U$ then $\{\mathbf{v}_1, \mathbf{v}_2, \ldots, \mathbf{v}_k, \mathbf{v}\}$ is a basis of W by the independent lemma.
18. **(b)** Two distinct planes through the origin (U and W) meet in a line through the origin ($U \cap W$).
23. **(b)** The set $\{(1, 0, 0, 0, \ldots), (0, 1, 0, 0, 0, \ldots), (0, 0, 1, 0, 0, \ldots), \ldots\}$ contains independent subsets of arbitrary size.
25. **(b)** $\mathbb{R}\mathbf{u} + \mathbb{R}\mathbf{w} = \{r\mathbf{u} + s\mathbf{w} \mid r, s \text{ in } \mathbb{R}\} = \text{span}\{\mathbf{u}, \mathbf{w}\}$

Exercises 6.5 An Application to Polynomials (Page 324)

2. **(b)** $3 + 4(x - 1) + 3(x - 1)^2 + (x - 1)^3$ **(d)** $1 + (x - 1)^3$
6. **(b)** The polynomials are $(x - 1)(x - 2)$, $(x - 1)(x - 3)$, $(x - 2)(x - 3)$. Use $a_0 = 3$, $a_1 = 2$, and $a_2 = 1$.
7. **(b)** $f(x) = \frac{3}{2}(x - 2)(x - 3) - 7(x - 1)(x - 3) + \frac{13}{2}(x - 1)(x - 2)$.
10. **(b)** If $r(x - a)^2 + s(x - a)(x - b) + t(x - b)^2 = 0$, then evaluation at $x = a$ ($x = b$) gives $t = 0$ ($r = 0$). Thus $s(x - a)(x - b) = 0$, so $s = 0$. Use Theorem 4 Section 6.4.

11. **(b)** Suppose $\{p_0(x), p_1(x), \ldots, p_{n-2}(x)\}$ is a basis of $\mathbf{P}_{n-2}$. We show that $\{(x - a)(x - b)p_0(x), (x - a)(x - b)p_1(x), \ldots, (x - a)(x - b)p_{n-2}(x)\}$ is a basis of U_n. It is a spanning set by part **(a)**, so assume that a linear combination vanishes with coefficients $r_0, r_1, \ldots, r_{n-2}$. Then $(x - a)(x - b)[r_0 p_0(x) + \cdots + r_{n-2} p_{n-2}(x)] = 0$, so $r_0 p_0(x) + \cdots + r_{n-2} p_{n-2}(x) = 0$ by the Hint. This implies that $r_0 = \cdots = r_{n-2} = 0$.

Exercises 6.6 An Application to Differential Equations (Page 329)

1. **(b)** e^{1-x} **(d)** $\dfrac{e^{2x} - e^{-3x}}{e^2 - e^{-3}}$ **(f)** $2e^{2x}(1 + x)$
   (h) $\dfrac{e^{ax} - e^{a(2-x)}}{1 - e^{2a}}$ **(j)** $e^{\pi - 2x} \sin x$
4. **(b)** $ce^{-x} + 2$, c a constant
5. **(b)** $ce^{-3x} + de^{2x} - \dfrac{x^3}{3}$
6. **(b)** $t = \dfrac{3 \ln(\frac{1}{2})}{\ln(\frac{4}{5})} = 9.32$ hours   **8.** $k = \left(\dfrac{\pi}{15}\right)^2 = 0.044$

Supplementary Exercises for Chapter 6 (Page 330)

2. **(b)** If $YA = 0$, Y a row, we show that $Y = 0$; thus A^T (and hence A) is invertible. Given a column $\mathbf{c}$ in $\mathbb{R}^n$ write $\mathbf{c} = \sum_i r_i(A\mathbf{v}_i)$ where each r_i is in $\mathbb{R}$. Then $Y\mathbf{c} = \sum_i r_i Y A \mathbf{v}_i$, so $Y = YI_n = Y[\mathbf{e}_1 \ \mathbf{e}_2 \ \cdots \ \mathbf{e}_n] = [Y\mathbf{e}_1 \ Y\mathbf{e}_2 \ \cdots \ Y\mathbf{e}_n] = [0 \ 0 \ \cdots \ 0] = 0$, as required.
4. We have $\text{null } A \subseteq \text{null}(A^T A)$ because $A\mathbf{x} = \mathbf{0}$ implies $(A^T A)\mathbf{x} = \mathbf{0}$. Conversely, if $(A^T A)\mathbf{x} = \mathbf{0}$, then $\|A\mathbf{x}\|^2 = (A\mathbf{x})^T(A\mathbf{x}) = \mathbf{x}^T A^T A \mathbf{x} = 0$. Thus $A\mathbf{x} = \mathbf{0}$.

Exercises 7.1 Examples and Elementary Properties (Page 336)

1. **(b)** $T(\mathbf{v}) = \mathbf{v}A$ where $A = \begin{bmatrix} 1 & 0 & 0 \\ 0 & 1 & 0 \\ 0 & 0 & -1 \end{bmatrix}$
   (d) $T(A + B) = P(A + B)Q = PAQ + PBQ = T(A) + T(B)$; $T(rA) = P(rA)Q = rPAQ = rT(A)$
   (f) $T[(p + q)(x)] = (p + q)(0) = p(0) + q(0) = T[p(x)] + T[q(x)]$; $T[(rp)(x)] = (rp)(0) = r(p(0)) = rT[p(x)]$
   (h) $T(X + Y) = (X + Y) \cdot Z = X \cdot Z + Y \cdot Z = T(X) + T(Y)$, and $T(rX) = (rX) \cdot Z = r(X \cdot Z) = rT(X)$
   (j) If $\mathbf{v} = (v_1, \ldots, v_n)$ and $\mathbf{w} = (w_1, \ldots, w_n)$, then $T(\mathbf{v} + \mathbf{w}) = (v_1 + w_1)\mathbf{e}_1 + \cdots + (v_n + w_n)\mathbf{e}_n = (v_1 \mathbf{e}_1 + \cdots + v_n \mathbf{e}_n) + (w_1 \mathbf{e}_1 + \cdots + w_n \mathbf{e}_n) = T(\mathbf{v}) + T(\mathbf{w})$
   $T(a\mathbf{v}) = (av_1)\mathbf{e} + \cdots + (av_n)\mathbf{e}_n = a(v\mathbf{e} + \cdots + v_n \mathbf{e}_n) = aT(\mathbf{v})$

2. **(b)** rank($A + B$) ≠ rank A + rank B in general. For example, $A = \begin{bmatrix} 1 & 0 \\ 0 & 1 \end{bmatrix}$ and $B = \begin{bmatrix} 1 & 0 \\ 0 & -1 \end{bmatrix}$.
   (d) $T(\mathbf{0}) = \mathbf{0} + \mathbf{u} = \mathbf{u} \neq \mathbf{0}$, so T is not linear by Theorem 1.

3. **(b)** $T(3\mathbf{v}_1 + 2\mathbf{v}_2) = 0$ **(d)** $T\begin{bmatrix} 1 \\ -7 \end{bmatrix} = \begin{bmatrix} -3 \\ 4 \end{bmatrix}$
   (f) $T(2 - x + 3x^2) = 46$

4. **(b)** $T(x, y) = \frac{1}{3}(x - y, 3y, x - y)$; $T(-1, 2) = (-1, 2, -1)$
   (d) $T\begin{bmatrix} a & b \\ c & d \end{bmatrix} = 3a - 3c + 2b$

5. **(b)** $T(\mathbf{v}) = \frac{1}{3}(7\mathbf{v} - 9\mathbf{w})$, $T(\mathbf{w}) = \frac{1}{3}(\mathbf{v} + 3\mathbf{w})$

8. **(b)** $T(\mathbf{v}) = (-1)\mathbf{v}$ for all $\mathbf{v}$ in V, so T is the scalar operator -1

12. If $T(1) = \mathbf{v}$, then $T(r) = T(r \cdot 1) = rT(1) = r\mathbf{v}$ for all r in $\mathbb{R}$.

15. **(b)** $\mathbf{0}$ is in $U = \{\mathbf{v} \in V | T(\mathbf{v}) \in P\}$ because $T(\mathbf{0}) = \mathbf{0}$ is in P. If $\mathbf{v}$ and $\mathbf{w}$ are in U, then $T(\mathbf{v})$ and $T(\mathbf{w})$ are in P. Hence $T(\mathbf{v} + \mathbf{w}) = T(\mathbf{v}) + T(\mathbf{w})$ is in P and $T(r\mathbf{v}) = rT(\mathbf{v})$ is in P, so $\mathbf{v} + \mathbf{w}$ and $r\mathbf{v}$ are in U.

18. Suppose $r\mathbf{v} + sT(\mathbf{v}) = \mathbf{0}$. If $s = 0$, then $r = 0$ (because $\mathbf{v} \neq \mathbf{0}$). If $s \neq 0$, then $T(\mathbf{v}) = a\mathbf{v}$ where $a = -s^{-1}r$. Thus $\mathbf{v} = T^2(\mathbf{v}) = T(a\mathbf{v}) = a^2\mathbf{v}$, so $a^2 = 1$, again because $\mathbf{v} \neq \mathbf{0}$. Hence $a = \pm 1$. Conversely, if $T(\mathbf{v}) = \pm\mathbf{v}$, then $\{\mathbf{v}, T(\mathbf{v})\}$ is certainly not independent.

21. **(b)** Given such a T, write $T(x) = a$. If $p = p(x) = \sum_{i=0}^{n} a_i x^i$, then $T(p) = \sum a_i T(x^i) = \sum a_i [T(x)]^i = \sum a_i a^i = p(a) = E_a(p)$. Hence $T = E_a$.

Exercises 7.2 Kernel and Image of a Linear Transformation (Page 344)

1. **(b)** $\left\{\begin{bmatrix} -3 \\ 7 \\ 1 \\ 0 \end{bmatrix}, \begin{bmatrix} 1 \\ 1 \\ 0 \\ -1 \end{bmatrix}\right\}; \left\{\begin{bmatrix} 1 \\ 0 \\ 1 \end{bmatrix}, \begin{bmatrix} 0 \\ 1 \\ -1 \end{bmatrix}\right\}$; 2, 2
   (d) $\left\{\begin{bmatrix} -1 \\ 2 \\ 1 \end{bmatrix}\right\}; \left\{\begin{bmatrix} 1 \\ 0 \\ 1 \\ 1 \end{bmatrix}, \begin{bmatrix} 0 \\ 1 \\ -1 \\ -2 \end{bmatrix}\right\}$; 2, 1

2. **(b)** $\{x^2 - x\}$; $\{(1, 0), (0, 1)\}$
   (d) $\{(0, 0, 1)\}$; $\{(1, 1, 0, 0), (0, 0, 1, 1)\}$
   (f) $\left\{\begin{bmatrix} 1 & 0 \\ 0 & -1 \end{bmatrix}, \begin{bmatrix} 0 & 1 \\ 0 & 0 \end{bmatrix}, \begin{bmatrix} 0 & 0 \\ 1 & 0 \end{bmatrix}\right\}$; $\{1\}$
   (h) $\{(1, 0, 0, \ldots, 0, -1), (0, 1, 0, \ldots, 0, -1), \ldots, (0, 0, 0, \ldots, 1, -1)\}$; $\{1\}$
   (j) $\left\{\begin{bmatrix} 0 & 1 \\ 0 & 0 \end{bmatrix}, \begin{bmatrix} 0 & 0 \\ 0 & 1 \end{bmatrix}\right\}; \left\{\begin{bmatrix} 1 & 1 \\ 0 & 0 \end{bmatrix}, \begin{bmatrix} 0 & 0 \\ 1 & 1 \end{bmatrix}\right\}$

3. **(b)** $T(\mathbf{v}) = \mathbf{0} = (0, 0)$ if and only if $P(\mathbf{v}) = 0$ and $Q(\mathbf{v}) = 0$; that is, if and only if $\mathbf{v}$ is in ker $P \cap$ ker Q.

4. **(b)** ker $T = \text{span}\{(-4, 1, 3)\}$; $B = \{(1, 0, 0), (0, 1, 0), (-4, 1, 3)\}$, im $T = \text{span}\{(1, 2, 0, 3), (1, -1, -3, 0)\}$

6. **(b)** Yes. dim(im T) = $5 - $ dim(ker T) = 3, so im $T = W$ as dim $W = 3$.   **(d)** No. $T = 0 : \mathbb{R}^2 \to \mathbb{R}^2$
   (f) No. $T : \mathbb{R}^2 \to \mathbb{R}^2$, $T(x, y) = (y, 0)$. Then ker $T = $ im T
   (h) Yes. dim $V = $ dim(ker T) + dim(im T) ≤ dim W + dim $W = 2$ dim W
   (j) No. Consider $T : \mathbb{R}^2 \to \mathbb{R}^2$ with $T(x, y) = (y, 0)$.
   (l) No. Same example as **(j)**.
   (n) No. Define $T : \mathbb{R}^2 \to \mathbb{R}^2$ by $T(x, y) = (x, 0)$. If $\mathbf{v}_1 = (1, 0)$ and $\mathbf{v}_2 = (0, 1)$, then $\mathbb{R}^2 = \text{span}\{\mathbf{v}_1, \mathbf{v}_2\}$ but $\mathbb{R}^2 \neq \text{span}\{T(\mathbf{v}_1), T(\mathbf{v}_2)\}$.

7. **(b)** Given $\mathbf{w}$ in W, let $\mathbf{w} = T(\mathbf{v})$, $\mathbf{v}$ in V, and write $\mathbf{v} = r_1\mathbf{v}_1 + \cdots + r_n\mathbf{v}_n$. Then $\mathbf{w} = T(\mathbf{v}) = r_1T(\mathbf{v}_1) + \cdots + r_nT(\mathbf{v}_n)$.

8. **(b)** im $T = \{\sum_i r_i\mathbf{v}_i | r_i \text{ in } \mathbb{R}\} = \text{span}\{\mathbf{v}_i\}$.

10. T is linear and onto. Hence $1 = $ dim $\mathbb{R} = $ dim(im T) = dim($\mathbf{M}_{nn}$) $-$ dim(ker T) = $n^2 - $ dim(ker T).

12. The condition means ker $(T_A) \subseteq$ ker(T_B), so dim[ker(T_A)] ≤ dim[ker(T_B)]. Then Theorem 4 gives dim[im(T_A)] ≥ dim[im(T_B)]; that is, rank A ≥ rank B.

15. **(b)** $B = \{x - 1, \ldots, x^n - 1\}$ is independent (distinct degrees) and contained in ker T. Hence B is a basis of ker T by **(a)**.

20. Define $T : \mathbf{M}_{nn} \to \mathbf{M}_{nn}$ by $T(A) = A - A^T$ for all A in $\mathbf{M}_{nn}$. Then ker $T = U$ and im $T = V$ by Example 3, so the dimension theorem gives $n^2 = $ dim $\mathbf{M}_{nn} = $ dim(U) + dim(V).

22. Define $T : \mathbf{M}_{nn} \to \mathbb{R}^n$ by $T(A) = A\mathbf{y}$ for all A in $\mathbf{M}_{nn}$. Then T is linear with ker $T = U$, so it is enough to show that T is onto (then dim $U = n^2 - $ dim(im T) = $n^2 - n$). We have $T(0) = \mathbf{0}$. Let $\mathbf{y} = [y_1 \ y_2 \ \cdots \ y_n]^T \neq \mathbf{0}$ in $\mathbb{R}^n$. If $y_k \neq 0$ let $\mathbf{c}_k = y_k^{-1}\mathbf{y}$, and let $\mathbf{c}_j = \mathbf{0}$ if $j \neq k$. If $A = [\mathbf{c}_1 \ \mathbf{c}_2 \ \cdots \ \mathbf{c}_n]$, then $T(A) = A\mathbf{y} = y_1\mathbf{c}_1 + \cdots + y_k\mathbf{c}_k + \cdots + y_n\mathbf{c}_n = \mathbf{y}$. This shows that T is onto, as required.

29. **(b)** By Lemma 2 Section 6.4, let $\{\mathbf{u}_1, \ldots, \mathbf{u}_m, \ldots, \mathbf{u}_n\}$ be a basis of V where $\{\mathbf{u}_1, \ldots, \mathbf{u}_m\}$ is a basis of U. By Theorem 3 Section 7.1 there is a linear transformation $S : V \to V$ such that $S(\mathbf{u}_i) = \mathbf{u}_i$ for $1 \leq i \leq m$, and $S(\mathbf{u}_i) = \mathbf{0}$ if $i > m$. Because each $\mathbf{u}_i$ is in im S, $U \subseteq$ im S. But if $S(\mathbf{v})$ is in im S, write $\mathbf{v} = r_1\mathbf{u}_1 + \cdots + r_m\mathbf{u}_m + \cdots + r_n\mathbf{u}_n$. Then $S(\mathbf{v}) = r_1S(\mathbf{u}_1) + \cdots + r_mS(\mathbf{u}_m) = r_1\mathbf{u}_1 + \cdots + r_m\mathbf{u}_m$ is in U. So im $S \subseteq U$.

Exercises 7.3 Isomorphisms and Composition (Page 354)

1. **(b)** T is onto because $T(1, -1, 0) = (1, 0, 0)$, $T(0, 1, -1) = (0, 1, 0)$, and $T(0, 0, 1) = (0, 0, 1)$. Use Theorem 3.
   (d) T is one-to-one because $0 = T(X) = UXV$ implies that $X = 0$ (U and V are invertible). Use Theorem 3.

(f) T is one-to-one because $\mathbf{0} = T(\mathbf{v}) = k\mathbf{v}$ implies that $\mathbf{v} = \mathbf{0}$ (because $k \neq 0$). T is onto because $T(\frac{1}{k}\mathbf{v}) = \mathbf{v}$ for all $\mathbf{v}$. [Here Theorem 3 does not apply if dim V is not finite.]
(h) T is one-to-one because $T(A) = 0$ implies $A^T = 0$, whence $A = 0$. Use Theorem 3.

4. **(b)** $ST(x, y, z) = (x + y, 0, y + z)$, $TS(x, y, z) = (x, 0, z)$
   (d) $ST\begin{bmatrix} a & b \\ c & d \end{bmatrix} = \begin{bmatrix} c & 0 \\ 0 & d \end{bmatrix}$, $TS\begin{bmatrix} a & b \\ c & d \end{bmatrix} = \begin{bmatrix} 0 & a \\ d & 0 \end{bmatrix}$

5. **(b)** $T^2(x, y) = T(x + y, 0) = (x + y, 0) = T(x, y)$. Hence $T^2 = T$.
   (d) $T^2\begin{bmatrix} a & b \\ c & d \end{bmatrix} = \frac{1}{2}T\begin{bmatrix} a+c & b+d \\ a+c & b+d \end{bmatrix} = \frac{1}{2}\begin{bmatrix} a+c & b+d \\ a+c & b+d \end{bmatrix}$

6. **(b)** No inverse; $(1, -1, 1, -1)$ is in ker T.
   (d) $T^{-1}\begin{bmatrix} a & b \\ c & d \end{bmatrix} = \frac{1}{5}\begin{bmatrix} 3a - 2c & 3b - 2d \\ a + c & b + d \end{bmatrix}$
   (f) $T^{-1}(a, b, c) = \frac{1}{2}[2a + (b-c)x - (2a - b - c)x^2]$

7. **(b)** $T^2(x, y) = T(ky - x, y) = (ky - (ky - x), y) = (x, y)$
   (d) $T^2(X) = A^2X = IX = X$

8. **(b)** $T^3(x, y, z, w) = (x, y, z, -w)$ so $T^6(x, y, z, w) = T^3[T^3(x, y, z, w)] = (x, y, z, w)$. Hence $T^{-1} = T^5$. So $T^{-1}(x, y, z, w) = (y - x, -x, z, -w)$.

9. **(b)** $T^{-1}(A) = U^{-1}A$.

10. **(b)** Given $\mathbf{u}$ in U, write $\mathbf{u} = S(\mathbf{w})$, $\mathbf{w}$ in W (because S is onto). Then write $\mathbf{w} = T(\mathbf{v})$, $\mathbf{v}$ in V (T is onto). Hence $\mathbf{u} = ST(\mathbf{v})$, so ST is onto.

12. **(b)** For all $\mathbf{v}$ in V, $(RT)(\mathbf{v}) = R[T(\mathbf{v})]$ is in im(R).

13. **(b)** Given $\mathbf{w}$ in W, write $\mathbf{w} = ST(\mathbf{v})$, $\mathbf{v}$ in V (ST is onto). Then $\mathbf{w} = S[T(\mathbf{v})]$, $T(\mathbf{v})$ in U, so S is onto. But then im $S = W$, so dim $U = $ dim(ker S) + dim(im S) $\geq$ dim(im S) = dim W.

16. $\{T(\mathbf{e}_1), T(\mathbf{e}_2), \ldots, T(\mathbf{e}_r)\}$ is a basis of im T by Theorem 5 Section 7.2. So $T: \text{span}\{\mathbf{e}_1, \ldots, \mathbf{e}_r\} \to \text{im } T$ is an isomorphism by Theorem 1.

19. **(b)** $T(x, y) = (x, y + 1)$

24. **(b)** $TS[x_0, x_1, \ldots] = T[0, x_0, x_1, \ldots] = [x_0, x_1, \ldots]$, so $TS = 1_V$. Hence TS is both onto and one-to-one, so T is onto and S is one-to-one by Exercise 13. But $[1, 0, 0, \ldots]$ is in ker T while $[1, 0, 0, \ldots]$ is not in im S.

26. **(b)** If $T(p) = 0$, then $p(x) = -xp'(x)$. We write $p(x) = a_0 + a_1x + a_2x^2 + \cdots + a_nx^n$, and this becomes
$a_0 + a_1x + a_2x^2 + \cdots + a_nx^n =$
$-a_1x - 2a_2x^2 - \cdots - na_nx^n$.
Equating coefficients yields $a_0 = 0$, $2a_1 = 0$, $3a_2 = 0$, $\ldots$, $(n + 1)a_n = 0$, whence $p(x) = 0$. This means that ker $T = 0$, so T is one-to-one. But then T is an isomorphism by Theorem 3.

27. **(b)** If $ST = 1_V$ for some S, then T is onto by Exercise 13. If T is onto, let $\{\mathbf{e}_1, \ldots, \mathbf{e}_r, \ldots, \mathbf{e}_n\}$ be a basis of V such that $\{\mathbf{e}_{r+1}, \ldots, \mathbf{e}_n\}$ is a basis of ker T. Since T is onto, $\{T(\mathbf{e}_1), \ldots, T(\mathbf{e}_r)\}$ is a basis of im $T = W$ by Theorem 5 Section 7.2. Thus $S: W \to V$ is an isomorphism where by $S\{T(\mathbf{e}_i)\} = \mathbf{e}_i$ for $i = 1, 2, \ldots, r$. Hence $TS[T(\mathbf{e}_i)] = T(\mathbf{e}_i)$ for each i, that is $TS[T(\mathbf{e}_i)] = 1_W[T(\mathbf{e}_i)]$. This means that $TS = 1_W$ because they agree on the basis $\{T(\mathbf{e}_1), \ldots, T(\mathbf{e}_r)\}$ of W.

28. **(b)** If $T = SR$, then every vector $T(\mathbf{v})$ in im T has the form $T(\mathbf{v}) = S[R(\mathbf{v})]$, whence im $T \subseteq$ im S. Since R is invertible, $S = TR^{-1}$ implies im $S \subseteq$ im T.
Conversely, assume that im $S = $ im T. Then dim(ker S) = dim(ker T) by the dimension theorem. Let $\{\mathbf{e}_1, \ldots, \mathbf{e}_r, \mathbf{e}_{r+1}, \ldots, \mathbf{e}_n\}$ and $\{\mathbf{f}_1, \ldots, \mathbf{f}_r, \mathbf{f}_{r+1}, \ldots, \mathbf{f}_n\}$ be bases of V such that $\{\mathbf{e}_{r+1}, \ldots, \mathbf{e}_n\}$ and $\{\mathbf{f}_{r+1}, \ldots, \mathbf{f}_n\}$ are bases of ker S and ker T, respectively. By Theorem 5, Section 7.2, $\{S(\mathbf{e}_1), \ldots, S(\mathbf{e}_r)\}$ and $\{T(\mathbf{f}_1), \ldots, T(\mathbf{f}_r)\}$ are both bases of im $S = $ im T. So let $\mathbf{g}_1, \ldots, \mathbf{g}_r$ in V be such that $S(\mathbf{e}_i) = T(\mathbf{g}_i)$ for each $i = 1, 2, \ldots, r$. Show that
$$B = \{\mathbf{g}_1, \ldots, \mathbf{g}_r, \mathbf{f}_{r+1}, \ldots, \mathbf{f}_n\} \text{ is a basis of } V.$$
Then define $R: V \to V$ by $R(\mathbf{g}_i) = \mathbf{e}_i$ for $i = 1, 2, \ldots, r$, and $R(\mathbf{f}_j) = \mathbf{e}_j$ for $j = r + 1, \ldots, n$. Then R is an isomorphism by Theorem 1, Section 7.3. Finally $SR = T$ since they have the same effect on the basis B.

29. Let $B = \{\mathbf{e}_1, \ldots, \mathbf{e}_r, \mathbf{e}_{r+1}, \ldots, \mathbf{e}_n\}$ be a basis of V with $\{\mathbf{e}_{r+1}, \ldots, \mathbf{e}_n\}$ a basis of ker T. If $\{T(\mathbf{e}_1), \ldots, T(\mathbf{e}_r), \mathbf{w}_{r+1}, \ldots, \mathbf{w}_n\}$ is a basis of V, define S by $S[T(\mathbf{e}_i)] = \mathbf{e}_i$ for $1 \leq i \leq r$, and $S(\mathbf{w}_j) = \mathbf{e}_j$ for $r + 1 \leq j \leq n$. Then S is an isomorphism by Theorem 1, and $TST(\mathbf{e}_i) = T(\mathbf{e}_i)$ clearly holds for $1 \leq i \leq r$. But if $i \geq r + 1$, then $T(\mathbf{e}_i) = \mathbf{0} = TST(\mathbf{e}_i)$, so $T = TST$ by Theorem 2 Section 7.1.

Exercises 7.5 More on Linear Recurrences (Page 367)

1. **(b)** $\{(1), (2^n), ((-3)^n)\}$; $x_n = \frac{1}{20}(15 + 2^{n+3} + (-3)^{n+1})$
2. **(b)** $\{(1), (n), ((-2)^n)\}$; $x_n = \frac{1}{9}(5 - 6n + (-2)^{n+2})$
   (d) $\{(1), (n), (n^2)\}$; $x_n = 2(n-1)^2 - 1$
3. **(b)** $\{(a^n), (b^n)\}$
4. **(b)** $[1, 0, 0, 0, \ldots)$, $[0, 1, 0, 0, 0, \ldots)$, $[0, 0, 1, 1, 1, \ldots)$, $[0, 0, 1, 2, 3, \ldots)$
7. By Remark 2,
$[i^n + (-i)^n] = [2, 0, -2, 0, 2, 0, -2, 0, \ldots)$
$[i(i^n - (-i)^n)] = [0, -2, 0, 2, 0, -2, 0, 2, \ldots)$
are solutions. They are linearly independent and so are a basis.

Exercises 8.1 Orthogonal Complements and Projections (Page 374)

1. **(b)** $\{(2, 1), \frac{3}{5}(-1, 2)\}$ **(d)** $\{(0, 1, 1), (1, 0, 0), (0, -2, 2)\}$
2. **(b)** $\mathbf{x} = \frac{1}{182}(271, -221, 1030) + \frac{1}{182}(93, 403, 62)$
   (d) $\mathbf{x} = \frac{1}{4}(1, 7, 11, 17) + \frac{1}{4}(7, -7, -7, 7)$
   (f) $\mathbf{x} = \frac{1}{12}(5a - 5b + c - 3d, -5a + 5b - c + 3d,$
   $a - b + 11c + 3d, -3a + 3b + 3c + 3d)$
   $+ \frac{1}{12}(7a + 5b - c + 3d, 5a + 7b + c - 3d,$
   $-a + b + c - 3d, 3a - 3b - 3c + 9d)$
3. **(a)** $\frac{1}{10}(-9, 3, -21, 33) = \frac{3}{10}(-3, 1, -7, 11)$
   (c) $\frac{1}{70}(-63, 21, -147, 231) = \frac{3}{10}(-3, 1, -7, 11)$
4. **(b)** $\{(1, -1, 0), \frac{1}{2}(-1, -1, 2)\}$; $\text{proj}_U(\mathbf{x}) = (1, 0, -1)$
   (d) $\{(1, -1, 0, 1), (1, 1, 0, 0), \frac{1}{3}(-1, 1, 0, 2)\}$;
   $\text{proj}_U(\mathbf{x}) = (2, 0, 0, 1)$
5. **(b)** $U^\perp = \text{span}\{(1, 3, 1, 0), (-1, 0, 0, 1)\}$
8. Write $\mathbf{p} = \text{proj}_U(\mathbf{x})$. Then $\mathbf{p}$ is in U by definition. If $\mathbf{x}$ is U, then $\mathbf{x} - \mathbf{p}$ is in U. But $\mathbf{x} - \mathbf{p}$ is also in $U^\perp$ by Theorem 3, so $\mathbf{x} - \mathbf{p}$ is in $U \cap U^\perp = \{\mathbf{0}\}$. Thus $\mathbf{x} = \mathbf{p}$.
10. Let $\{\mathbf{f}_1, \mathbf{f}_2, ..., \mathbf{f}_m\}$ be an orthonormal basis of U. If $\mathbf{x}$ is in U the expansion theorem gives
    $\mathbf{x} = (\mathbf{x} \cdot \mathbf{f}_1)\mathbf{f}_1 + (\mathbf{x} \cdot \mathbf{f}_2)\mathbf{f}_2 + \cdots + (\mathbf{x} \cdot \mathbf{f}_m)\mathbf{f}_m = \text{proj}_U(\mathbf{x})$.
14. Let $\{\mathbf{y}_1, \mathbf{y}_2, ..., \mathbf{y}_m\}$ be a basis of $U^\perp$, and let A be the $n \times n$ matrix with rows $\mathbf{y}_1^T, \mathbf{y}_2^T, ..., \mathbf{y}_m^T, \mathbf{0}, ..., \mathbf{0}$. Then $A\mathbf{x} = \mathbf{0}$ if and only if $\mathbf{y}_i \cdot \mathbf{x} = 0$ for each $i = 1, 2, ..., m$; if and only if $\mathbf{x}$ is in $U^{\perp\perp} = U$.
17. **(d)** $E^T = A^T[(AA^T)^{-1}]^T(A^T)^T = A^T[(AA^T)^T]^{-1}A = A^T[AA^T]^{-1}A = E$
    $E^2 = A^T(AA^T)^{-1}AA^T(AA^T)^{-1}A = A^T(AA^T)^{-1}A = E$

Exercises 8.2 Orthogonal Diagonalization (Page 383)

1. **(b)** $\frac{1}{5}\begin{bmatrix} 3 & -4 \\ 4 & 3 \end{bmatrix}$ **(d)** $\frac{1}{\sqrt{a^2 + b^2}}\begin{bmatrix} a & b \\ -b & a \end{bmatrix}$
   (f) $\begin{bmatrix} \frac{2}{\sqrt{6}} & \frac{1}{\sqrt{6}} & -\frac{1}{\sqrt{6}} \\ \frac{1}{\sqrt{3}} & -\frac{1}{\sqrt{3}} & \frac{1}{\sqrt{3}} \\ 0 & \frac{1}{\sqrt{2}} & \frac{1}{\sqrt{2}} \end{bmatrix}$ **(h)** $\frac{1}{7}\begin{bmatrix} 2 & 6 & -3 \\ 3 & 2 & 6 \\ -6 & 3 & 2 \end{bmatrix}$

2. We have $P^T = P^{-1}$; this matrix is lower triangular (left side) and also upper triangular (right side–see Lemma 1 Section 2.7), and so is diagonal. But then $P = P^T = P^{-1}$, so $P^2 = I$. This implies that the diagonal entries of P are all ± 1.

5. **(b)** $\frac{1}{\sqrt{2}}\begin{bmatrix} 1 & -1 \\ 1 & 1 \end{bmatrix}$ **(d)** $\frac{1}{\sqrt{2}}\begin{bmatrix} 0 & 1 & 1 \\ \sqrt{2} & 0 & 0 \\ 0 & 1 & -1 \end{bmatrix}$
   (f) $\frac{1}{3\sqrt{2}}\begin{bmatrix} 2\sqrt{2} & 3 & 1 \\ \sqrt{2} & 0 & -4 \\ 2\sqrt{2} & -3 & 1 \end{bmatrix}$ or $\frac{1}{3}\begin{bmatrix} 2 & -2 & 1 \\ 1 & 2 & 2 \\ 2 & 1 & -2 \end{bmatrix}$

(h) $\frac{1}{2}\begin{bmatrix} 1 & -1 & \sqrt{2} & 0 \\ -1 & 1 & \sqrt{2} & 0 \\ -1 & -1 & 0 & \sqrt{2} \\ 1 & 1 & 0 & \sqrt{2} \end{bmatrix}$

6. $P = \frac{1}{\sqrt{2k}}\begin{bmatrix} c\sqrt{2} & a & a \\ 0 & k & -k \\ -a\sqrt{2} & c & c \end{bmatrix}$

10. **(b)** $y_1 = \frac{1}{\sqrt{5}}(-x_1 + 2x_2)$ and $y_2 = \frac{1}{\sqrt{5}}(2x_1 + x_2)$;
    $q = -3y_1^2 + 2y_2^2$.
11. **(c)** $\Rightarrow$ **(a)** By Theorem 1 let $P^{-1}AP = D = \text{diag}(\lambda_1, ..., \lambda_n)$ where the λ_i are the eigenvalues of A. By **(c)** we have $\lambda_i = \pm 1$ for each i, whence $D^2 = I$. But then $A^2 = (PDP^{-1})^2 = PD^2P^{-1} = I$. Since A is symmetric this is $AA^T = I$, proving **(a)**.
13. **(b)** If $B = P^TAP = P^{-1}$, then $B^2 = P^TAPP^TAP = P^TA^2P$.
15. If $\mathbf{x}$ and $\mathbf{y}$ are respectively columns i and j of I_n, then $\mathbf{x}^TA^T\mathbf{y} = \mathbf{x}^TA\mathbf{y}$ shows that the (i, j)-entries of A^T and A are equal.
18. **(b)** $\det\begin{bmatrix} \cos\theta & -\sin\theta \\ \sin\theta & \cos\theta \end{bmatrix} = 1$ and $\det\begin{bmatrix} \cos\theta & \sin\theta \\ \sin\theta & -\cos\theta \end{bmatrix} = -1$
    [*Remark*: These are the *only* 2×2 examples.]
    (d) Use the fact that $P^{-1} = P^T$ to show that $P^T(I - P) = -(I - P)^T$. Now take determinants and use the hypothesis that $\det P \neq (-1)^n$.
21. We have $AA^T = D$, where D is diagonal with main diagonal entries $\|R_1\|^2, ..., \|R_n\|^2$. Hence $A^{-1} = A^TD^{-1}$, and the result follows because D^{-1} has diagonal entries $1/\|R_1\|^2, ..., 1/\|R_n\|^2$.
23. **(b)** Because $I - A$ and $I + A$ commute,
    $PP^T = (I - A)(I + A)^{-1}[(I + A)^{-1}]^T(I - A)^T = (I - A)(I + A)^{-1}(I - A)^{-1}(I + A) = I$.

Exercises 8.3 Positive Definite Matrices (Page 389)

1. **(b)** $U = \frac{\sqrt{2}}{2}\begin{bmatrix} 2 & -1 \\ 0 & 1 \end{bmatrix}$ **(d)** $U = \frac{1}{30}\begin{bmatrix} 60\sqrt{5} & 12\sqrt{5} & 15\sqrt{5} \\ 0 & 6\sqrt{30} & 10\sqrt{30} \\ 0 & 0 & 5\sqrt{15} \end{bmatrix}$

2. **(b)** If $\lambda^k > 0$, k odd, then $\lambda > 0$.
4. If $\mathbf{x} \neq \mathbf{0}$, then $\mathbf{x}^TA\mathbf{x} > 0$ and $\mathbf{x}^TB\mathbf{x} > 0$. Hence $\mathbf{x}^T(A + B)\mathbf{x} = \mathbf{x}^TA\mathbf{x} + \mathbf{x}^TB\mathbf{x} > 0$ and $\mathbf{x}^T(rA)\mathbf{x} = r(\mathbf{x}^TA\mathbf{x}) > 0$, as $r > 0$.
6. Let $\mathbf{x} \neq \mathbf{0}$ in $\mathbb{R}^n$. Then $\mathbf{x}^T(U^TAU)\mathbf{x} = (U\mathbf{x})^TA(U\mathbf{x}) > 0$ provided $U\mathbf{x} \neq \mathbf{0}$. But if $U = [\mathbf{c}_1 \ \mathbf{c}_2 \ \cdots \ \mathbf{c}_n]$ and $\mathbf{x} = (x_1, x_2, ..., x_n)$, then $U\mathbf{x} = x_1\mathbf{c}_1 + x_2\mathbf{c}_2 + \cdots + x_n\mathbf{c}_n \neq \mathbf{0}$ because $\mathbf{x} \neq \mathbf{0}$ and the $\mathbf{c}_i$ are independent.
10. Let $P^TAP = D = \text{diag}(\lambda_1, ..., \lambda_n)$ where $P^T = P$. Since A is positive definite, each eigenvalue $\lambda_i > 0$. If $B = \text{diag}(\sqrt{\lambda_1}, ..., \sqrt{\lambda_n})$ then $B^2 = D$, so $A = PB^2P^T = (PBP^T)^2$. Take $C = PBP^T$. Since C has eigenvalues $\sqrt{\lambda_i} > 0$, it is positive definite.

Selected Answers

12. (b) If A is positive definite, use Theorem 1 to write $A = U^TU$ where U is upper triangular with positive diagonal D. Then $A = (D^{-1}U)^TD^2(D^{-1}U)$ so $A = L_1D_1U_1$ is such a factorization if $U_1 = D^{-1}U$, $D_1 = D^2$, and $L_1 = U_1^T$. Conversely, let $A^T = A = LDU$ be such a factorization. Then $U^TD^TL^T = A^T = A = LDU$, so $L = U^T$ by **(a)**. Hence $A = LDL^T = V^TV$ where $V = LD_0$ and D_0 is diagonal with $D_0^2 = D$ (the matrix D_0 exists because D has positive diagonal entries). Hence A is symmetric, and it is positive definite by Example 1.

Exercises 8.4 QR-Factorization (Page 393)

1. (b) $Q = \frac{1}{\sqrt{5}}\begin{bmatrix} 2 & -1 \\ 1 & 2 \end{bmatrix}$, $R = \frac{1}{\sqrt{5}}\begin{bmatrix} 5 & 3 \\ 0 & 1 \end{bmatrix}$

(d) $Q = \frac{1}{\sqrt{3}}\begin{bmatrix} 1 & 1 & 0 \\ -1 & 0 & 1 \\ 0 & 1 & 1 \\ 1 & -1 & 1 \end{bmatrix}$, $R = \frac{1}{\sqrt{3}}\begin{bmatrix} 3 & 0 & -1 \\ 0 & 3 & 1 \\ 0 & 0 & 2 \end{bmatrix}$

2. If A has a QR-factorization, use (a). For the converse use Theorem 1.

Exercises 8.5 Computing Eigenvalues (Page 396)

1. (b) Eigenvalues $4, -1$; eigenvectors $\begin{bmatrix} 2 \\ -1 \end{bmatrix}, \begin{bmatrix} 1 \\ -3 \end{bmatrix}$;

$\mathbf{x}_4 = \begin{bmatrix} 409 \\ -203 \end{bmatrix}$; $r_3 = 3.94$

(d) Eigenvalues $\lambda_1 = \frac{1}{2}(3 + \sqrt{13})$, $\lambda_2 = \frac{1}{2}(3 - \sqrt{13})$;

eigenvectors $\begin{bmatrix} \lambda_1 \\ 1 \end{bmatrix}, \begin{bmatrix} \lambda_2 \\ 1 \end{bmatrix}$; $\mathbf{x}_4 = \begin{bmatrix} 142 \\ 43 \end{bmatrix}$; $r_3 = 3.3027750$

(The true value is $\lambda_1 = 3.3027756$, to seven decimal places.)

2. (b) Eigenvalues $\lambda_1 = \frac{1}{2}(3 + \sqrt{13}) = 3.302776$, $\lambda_2 = \frac{1}{2}(3 - \sqrt{13}) = -0.302776$

$A_1 = \begin{bmatrix} 3 & 1 \\ 1 & 0 \end{bmatrix}$, $Q_1 = \frac{1}{\sqrt{10}}\begin{bmatrix} 3 & -1 \\ 1 & 3 \end{bmatrix}$, $R_1 = \frac{1}{\sqrt{10}}\begin{bmatrix} 10 & 3 \\ 0 & -1 \end{bmatrix}$

$A_2 = \frac{1}{10}\begin{bmatrix} 33 & -1 \\ -1 & -3 \end{bmatrix}$, $Q_2 = \frac{1}{\sqrt{1090}}\begin{bmatrix} 33 & 1 \\ -1 & 33 \end{bmatrix}$,

$R_2 = \frac{1}{\sqrt{1090}}\begin{bmatrix} 109 & -3 \\ 0 & -10 \end{bmatrix}$

$A_3 = \frac{1}{109}\begin{bmatrix} 360 & 1 \\ 1 & -33 \end{bmatrix} = \begin{bmatrix} 3.302775 & 0.009174 \\ 0.009174 & -0.302775 \end{bmatrix}$

4. Use induction on k. If $k = 1$, $A_1 = A$. In general $A_{k+1} = Q_k^{-1}A_kQ_k = Q_k^TA_kQ_k$, so the fact that $A_k^T = A_k$ implies $A_{k+1}^T = A_{k+1}$. The eigenvalues of A are all real (Theorem 7 Section 5.5), so the A_k converge to an upper triangular matrix T. But T must also be symmetric (it is the limit of symmetric matrices), so it is diagonal.

Exercises 8.6 Complex Matrices (Page 406)

1. (b) $\sqrt{6}$ **(d)** $\sqrt{13}$

2. (b) Not orthogonal   **(d)** Orthogonal

3. (b) Not a subspace. For example, $i(0, 0, 1) = (0, 0, i)$ is not in U.   **(d)** This is a subspace.

4. (b) Basis $\{(i, 0, 2), (1, 0, -1)\}$; dimension 2   **(d)** Basis $\{(1, 0, -2i), (0, 1, 1 - i)\}$; dimension 2

5. (b) Normal only   **(d)** Hermitian (and normal), not unitary   **(f)** None   **(h)** Unitary (and normal); hermitian if and only if z is real

8. (b) $U = \frac{1}{\sqrt{14}}\begin{bmatrix} -2 & 3-i \\ 3+i & 2 \end{bmatrix}$, $U^HAU = \begin{bmatrix} -1 & 0 \\ 0 & 6 \end{bmatrix}$

(d) $U = \frac{1}{\sqrt{3}}\begin{bmatrix} 1+i & 1 \\ -1 & 1-i \end{bmatrix}$, $U^HAU = \begin{bmatrix} 1 & 0 \\ 0 & 4 \end{bmatrix}$

(f) $U = \frac{1}{\sqrt{3}}\begin{bmatrix} \sqrt{3} & 0 & 0 \\ 0 & 1+i & 1 \\ 0 & -1 & 1-i \end{bmatrix}$, $U^HAU = \begin{bmatrix} 1 & 0 & 0 \\ 0 & 0 & 0 \\ 0 & 0 & 3 \end{bmatrix}$

10. (b) $\|\lambda Z\|^2 = \langle \lambda Z, \lambda Z \rangle = \lambda \overline{\lambda} \langle Z, Z \rangle = |\lambda|^2 \|Z\|^2$

11. (b) If the (k, k)-entry of A is a_{kk}, then the (k, k)-entry of $\overline{A}$ is $\overline{a}_{kk}$, so the (k, k)-entry of $(\overline{A})^T = A^H$ is $\overline{a}_{kk}$. This equals a, so a_{kk} is real.

14. (b) Show that $(B^2)^H = B^HB^H = (-B)(-B) = B^2$; $(iB)^H = \overline{i}B^H = (-i)(-B) = iB$.   **(d)** If $Z = A + B$, as given, first show that $Z^H = A - B$, and hence that $A = \frac{1}{2}(Z + Z^H)$ and $B = \frac{1}{2}(Z - Z^H)$.

16. (b) If U is unitary, $(U^{-1})^{-1} = (U^H)^{-1} = (U^{-1})^H$, so U^{-1} is unitary.

18. (b) $H = \begin{bmatrix} 1 & i \\ -i & 0 \end{bmatrix}$ is hermitian but $iH = \begin{bmatrix} i & -1 \\ 1 & 0 \end{bmatrix}$ is not.

21. (b) Let $U = \begin{bmatrix} a & b \\ c & d \end{bmatrix}$ be real and invertible, and assume that $U^{-1}AU = \begin{bmatrix} \lambda & \mu \\ 0 & \nu \end{bmatrix}$. Then $AU = U\begin{bmatrix} \lambda & \mu \\ 0 & \nu \end{bmatrix}$, and first column entries are $c = a\lambda$ and $-a = c\lambda$. Hence λ is real (c and a are both real and are not both 0), and $(1 + \lambda^2)a = 0$. Thus $a = 0$, $c = a\lambda = 0$, a contradiction.

Section 8.7 An Application to Linear Codes over Finite Fields (Page 421)

1. (b) $1^{-1} = 1$, $9^{-1} = 9$, $3^{-1} = 7$, $7^{-1} = 3$.

(d) $2^1 = 2$, $2^2 = 4$, $2^3 = 8$, $2^4 = 16 = 6$, $2^5 = 12 = 2$, $2^6 = 2^2$... so $a = 2^k$ if and only if $a = 2, 4, 6, 8$.

2. (b) If $2a = 0$ in $\mathbb{Z}_{10}$, then $2a = 10k$ for some integer k. Thus $a = 5k$.

3. (b) $11^{-1} = 7$ in $\mathbb{Z}_{19}$.

6. (b) $\det A = 15 - 24 = 1 + 4 = 5 \neq 0$ in $\mathbb{Z}_7$, so A^{-1} exists. Since $5^{-1} = 3$ in $\mathbb{Z}_7$, we have $A^{-1} = 3\begin{bmatrix} 3 & -6 \\ 3 & 5 \end{bmatrix} = 3\begin{bmatrix} 3 & 1 \\ 3 & 5 \end{bmatrix} = \begin{bmatrix} 2 & 3 \\ 2 & 1 \end{bmatrix}$.

7. (b) We have $5 \cdot 3 = 1$ in $\mathbb{Z}_7$ so the reduction of the augmented matrix is:

$$\begin{bmatrix} 3 & 1 & 4 & 3 \\ 4 & 3 & 1 & 1 \end{bmatrix} \to \begin{bmatrix} 1 & 5 & 6 & 1 \\ 4 & 3 & 1 & 1 \end{bmatrix} \to \begin{bmatrix} 1 & 5 & 6 & 1 \\ 0 & 4 & 5 & 4 \end{bmatrix} \to$$
$$\begin{bmatrix} 1 & 5 & 6 & 1 \\ 0 & 1 & 3 & 1 \end{bmatrix} \to \begin{bmatrix} 1 & 0 & 5 & 3 \\ 0 & 1 & 3 & 1 \end{bmatrix}.$$

Hence $x = 3 + 2t$, $y = 1 + 4t$, $z = t$; t in $\mathbb{Z}_7$.

9. (b) $(1 + t)^{-1} = 2 + t$.

10. (b) The minimum weight of C is 5, so it detects 4 errors and corrects 2 errors.

11. (b) $\{00000, 01110, 10011, 11101\}$.

12. (b) The code is $\{0000000000, 1001111000,$ $0101100110, 0011010111, 1100011110, 1010101111,$ $0110110001, 1111001001\}$. This has minimum distance 5 and so corrects 2 errors.

13. (b) $\{00000, 10110, 01101, 11011\}$ is a (5,2)-code of minimal weight 3, so it corrects single errors.

14. (b) $G = [1 \ \mathbf{u}]$ where $\mathbf{u}$ is any nonzero vector in the code. $H = \begin{bmatrix} \mathbf{u} \\ I_{n-1} \end{bmatrix}$.

Exercises 8.8 An Application to Quadratic Forms (Page 430)

1. (b) $A = \begin{bmatrix} 1 & 0 \\ 0 & 2 \end{bmatrix}$ **(d)** $A = \begin{bmatrix} 1 & 3 & 2 \\ 3 & 1 & -1 \\ 2 & -1 & 3 \end{bmatrix}$

2. (b) $P = \frac{1}{\sqrt{2}}\begin{bmatrix} 1 & 1 \\ 1 & -1 \end{bmatrix}$; $\mathbf{y} = \frac{1}{\sqrt{2}}\begin{bmatrix} x_1 + x_2 \\ x_1 - x_2 \end{bmatrix}$; $q = 3y_1^2 - y_2^2$; 1, 2

(d) $P = \frac{1}{3}\begin{bmatrix} 2 & 2 & -1 \\ 2 & -1 & 2 \\ -1 & 2 & 2 \end{bmatrix}$; $\mathbf{y} = \frac{1}{3}\begin{bmatrix} 2x_1 + 2x_2 - x_3 \\ 2x_1 - x_2 + 2x_3 \\ -x_1 + 2x_2 + 2x_3 \end{bmatrix}$;
$q = 9y_1^2 + 9y_2^2 - 9y_3^2$; 2, 3

(f) $P = \frac{1}{3}\begin{bmatrix} -2 & 1 & 2 \\ 2 & 2 & 1 \\ 1 & -2 & 2 \end{bmatrix}$; $\mathbf{y} = \frac{1}{3}\begin{bmatrix} -2x_1 + 2x_2 + x_3 \\ x_1 + 2x_2 - 2x_3 \\ 2x_1 + x_2 + 2x_3 \end{bmatrix}$;
$q = 9y_1^2 + 9y_2^2$; 2, 2

(h) $P = \frac{1}{\sqrt{6}}\begin{bmatrix} -\sqrt{2} & \sqrt{3} & 1 \\ \sqrt{2} & 0 & 2 \\ \sqrt{2} & \sqrt{3} & -1 \end{bmatrix}$;
$\mathbf{y} = \frac{1}{\sqrt{6}}\begin{bmatrix} -\sqrt{2}x_1 + \sqrt{2}x_2 + \sqrt{2}x_3 \\ \sqrt{3}x_1 + \sqrt{3}x_3 \\ x_1 + 2x_2 - x_3 \end{bmatrix}$;
$q = 2y_1^2 + y_2^2 - y_3^2$; 2, 3

3. (b) $x_1 = \frac{1}{\sqrt{5}}(2x - y)$, $y_1 = \frac{1}{\sqrt{5}}(x + 2y)$; $4x_1^2 - y_1^2 = 2$; hyperbola **(d)** $x_1 = \frac{1}{\sqrt{5}}(x + 2y)$, $y_1 = \frac{1}{\sqrt{5}}(2x - y)$; $6x_1^2 + y_1^2 = 1$; ellipse

4. (b) Basis $\{(i, 0, i), (1, 0, -1)\}$, dimension 2
(d) Basis $\{(1, 0, -2i), (0, 1, 1 - i)\}$, dimension 2

7. (b) $3y_1^2 + 5y_2^2 - y_3^2 - 3\sqrt{2}y_1 + \frac{11}{3}\sqrt{3}y_2 + \frac{2}{3}\sqrt{6}y_3 = 7$
$y_1 = \frac{1}{\sqrt{2}}(x_2 + x_3)$, $y_2 = \frac{1}{\sqrt{3}}(x_1 + x_2 - x_3)$,
$y_3 = \frac{1}{\sqrt{6}}(2x_1 - x_2 + x_3)$

9. (b) By Theorem 3 Section 8.3 let $A = U^T U$ where U is upper triangular with positive diagonal entries. Then $q = \mathbf{x}^T(U^T U)\mathbf{x} = (U\mathbf{x})^T U\mathbf{x} = \|U\mathbf{x}\|^2$.

Exercises 9.1 The Matrix of a Linear Transformation (Page 442)

1. (b) $\begin{bmatrix} a \\ 2b - c \\ c - b \end{bmatrix}$ **(d)** $\frac{1}{2}\begin{bmatrix} a - b \\ a + b \\ -a + 3b + 2c \end{bmatrix}$

2. (b) Let $\mathbf{v} = a + bx + cx^2$. Then $C_D[T(\mathbf{v})] =$
$M_{DB}(T)C_B(\mathbf{v}) = \begin{bmatrix} 2 & 1 & 3 \\ -1 & 0 & -2 \end{bmatrix}\begin{bmatrix} a \\ b \\ c \end{bmatrix} = \begin{bmatrix} 2a + b + 3c \\ -a - 2c \end{bmatrix}$
Hence $T(\mathbf{v}) = (2a + b + 3c)(1, 1) + (-a - 2c)(0, 1)$
$= (2a + b + 3c, a + b + c)$.

3. (b) $\begin{bmatrix} 1 & 0 & 0 & 0 \\ 0 & 0 & 1 & 0 \\ 0 & 1 & 0 & 0 \\ 0 & 0 & 0 & 1 \end{bmatrix}$ **(d)** $\begin{bmatrix} 1 & 1 & 1 \\ 0 & 1 & 2 \\ 0 & 0 & 1 \end{bmatrix}$

4. (b) $\begin{bmatrix} 1 & 2 \\ 5 & 3 \\ 4 & 0 \\ 1 & 1 \end{bmatrix}$; $C_D[T(a, b)] = \begin{bmatrix} 1 & 2 \\ 5 & 3 \\ 4 & 0 \\ 1 & 1 \end{bmatrix}\begin{bmatrix} b \\ a - b \end{bmatrix} = \begin{bmatrix} 2a - b \\ 3a + 2b \\ 4b \\ a \end{bmatrix}$

(d) $\frac{1}{2}\begin{bmatrix} 1 & 1 & -1 \\ 1 & 1 & 1 \end{bmatrix}$; $C_D[T(a + bx + cx^2)] =$
$\frac{1}{2}\begin{bmatrix} 1 & 1 & -1 \\ 1 & 1 & 1 \end{bmatrix}\begin{bmatrix} a \\ b \\ c \end{bmatrix} = \frac{1}{2}\begin{bmatrix} a + b - c \\ a + b + c \end{bmatrix}$

(f) $\begin{bmatrix} 1 & 0 & 0 & 0 \\ 0 & 1 & 1 & 0 \\ 0 & 1 & 1 & 0 \\ 0 & 0 & 0 & 1 \end{bmatrix}$; $C_D\left(T\begin{bmatrix} a & b \\ c & d \end{bmatrix}\right) = \begin{bmatrix} 1 & 0 & 0 & 0 \\ 0 & 1 & 1 & 0 \\ 0 & 1 & 1 & 0 \\ 0 & 0 & 0 & 1 \end{bmatrix}\begin{bmatrix} a \\ b \\ c \\ d \end{bmatrix} = \begin{bmatrix} a \\ b + c \\ b + c \\ d \end{bmatrix}$

5. (b) $M_{ED}(S)M_{DB}(T) =$
$\begin{bmatrix} 1 & 1 & 0 & 0 \\ 0 & 0 & 1 & -1 \end{bmatrix}\begin{bmatrix} 1 & 1 & 0 \\ 0 & 1 & 1 \\ 1 & 0 & 1 \\ -1 & 1 & 0 \end{bmatrix} = \begin{bmatrix} 1 & 2 & 1 \\ 2 & -1 & 1 \end{bmatrix} = M_{EB}(ST)$

(d) $M_{ED}(S)M_{DB}(T) =$
$\begin{bmatrix} 1 & -1 & 0 \\ 0 & 0 & 1 \end{bmatrix}\begin{bmatrix} 1 & -1 & 0 \\ -1 & 0 & 1 \\ 0 & 1 & 0 \end{bmatrix} = \begin{bmatrix} 2 & -1 & -1 \\ 0 & 1 & 0 \end{bmatrix} = M_{EB}(ST)$

7. (b) $T^{-1}(a, b, c) = \frac{1}{2}(b + c - a, a + c - b, a + b - c)$;
$M_{DB}(T) = \begin{bmatrix} 0 & 1 & 1 \\ 1 & 0 & 1 \\ 1 & 1 & 0 \end{bmatrix}$; $M_{BD}(T^{-1}) = \frac{1}{2}\begin{bmatrix} -1 & 1 & 1 \\ 1 & -1 & 1 \\ 1 & 1 & -1 \end{bmatrix}$

(d) $T^{-1}(a, b, c) = (a - b) + (b - c)x + cx^2$;
$M_{DB}(T) = \begin{bmatrix} 1 & 1 & 1 \\ 0 & 1 & 1 \\ 0 & 0 & 1 \end{bmatrix}$; $M_{BD}(T^{-1}) = \begin{bmatrix} 1 & -1 & 0 \\ 0 & 1 & -1 \\ 0 & 0 & 1 \end{bmatrix}$

Selected Answers

8. (b) $M_{DB}(T^{-1}) = [M_{BD}[(T)]^{-1} =$
$\begin{bmatrix} 1 & 1 & 1 & 0 \\ 0 & 1 & 1 & 0 \\ 0 & 0 & 1 & 0 \\ 0 & 0 & 0 & 1 \end{bmatrix}^{-1} = \begin{bmatrix} 1 & -1 & 0 & 0 \\ 0 & 1 & -1 & 0 \\ 0 & 0 & 1 & 0 \\ 0 & 0 & 0 & 1 \end{bmatrix}$.

Hence $C_B[T^{-1}(a, b, c, d)] = M_{BD}(T^{-1})C_D(a, b, c, d) =$
$\begin{bmatrix} 1 & -1 & 0 & 0 \\ 0 & 1 & -1 & 0 \\ 0 & 0 & 1 & 0 \\ 0 & 0 & 0 & 1 \end{bmatrix} \begin{bmatrix} a \\ b \\ c \\ d \end{bmatrix} = \begin{bmatrix} a-b \\ b-c \\ c \\ d \end{bmatrix}$, so

$T^{-1}(a, b, c, d) = \begin{bmatrix} a-b & b-c \\ c & d \end{bmatrix}$.

12. Have $C_D[T(\mathbf{e}_j)] = $ column j of I_n. Hence $M_{DB}(T) = [C_D[T(\mathbf{e}_1)] \; C_D[T(\mathbf{e}_2)] \; \cdots \; C_D[T(\mathbf{e}_n)]] = I_n$.

16. (b) If D is the standard basis of $\mathbb{R}^{n+1}$ and $B = \{1, x, x^2, \ldots, x^n\}$, then
$M_{DB}(T) = [C_D[T(1)] \; C_D[T(x)] \; \cdots \; C_D[T(x^n)]] =$
$\begin{bmatrix} 1 & a_0 & a_0^2 & \cdots & a_0^n \\ 1 & a_1 & a_1^2 & \cdots & a_1^n \\ 1 & a_2 & a_2^2 & \cdots & a_2^n \\ \vdots & \vdots & \vdots & & \vdots \\ 1 & a_n & a_n^2 & \cdots & a_n^n \end{bmatrix}$.

This matrix has nonzero determinant by Theorem 7 Section 3.2 (since the a_i are distinct), so T is an isomorphism.

20. (d) $[(S + T)R](\mathbf{v}) = (S + T)(R(\mathbf{v})) = S[(R(\mathbf{v}))] + T[(R(\mathbf{v}))] = SR(\mathbf{v}) + TR(\mathbf{v}) = [SR + TR](\mathbf{v})$ holds for all $\mathbf{v}$ in V. Hence $(S + T)R = SR + TR$.

21. (b) If $\mathbf{w}$ lies in $\text{im}(S + T)$, then $\mathbf{w} = (S + T)(\mathbf{v})$ for some $\mathbf{v}$ in V. But then $\mathbf{w} = S(\mathbf{v}) + T(\mathbf{v})$, so $\mathbf{w}$ lies in $\text{im } S + \text{im } T$.

22. (b) If $X \subseteq X_1$, let T lie in X_1^0. Then $T(\mathbf{v}) = \mathbf{0}$ for all $\mathbf{v}$ in X_1, whence $T(\mathbf{v}) = \mathbf{0}$ for all $\mathbf{v}$ in X. Thus T is in X^0 and we have shown that $X_1^0 \subseteq X^0$.

24. (b) R is linear means $S_{\mathbf{v}+\mathbf{w}} = S_\mathbf{v} + S_\mathbf{w}$ and $S_{a\mathbf{v}} = aS_\mathbf{v}$. These are proved as follows: $S_{\mathbf{v}+\mathbf{w}}(r) = r(\mathbf{v} + \mathbf{w}) = r\mathbf{v} + r\mathbf{w} = S_\mathbf{v}(r) + S_\mathbf{w}(r) = (S_\mathbf{v} + S_\mathbf{w})(r)$, and $S_{a\mathbf{v}}(r) = r(a\mathbf{v}) = a(r\mathbf{v}) = (aS_\mathbf{v})(r)$ for all r in $\mathbb{R}$. To show R is one-to-one, let $R(\mathbf{v}) = \mathbf{0}$. This means $S_\mathbf{v} = 0$ so $0 = S_\mathbf{v}(r) = r\mathbf{v}$ for all r. Hence $\mathbf{v} = \mathbf{0}$ (take $r = 1$). Finally, to show R is onto, let T lie in $\mathbf{L}(\mathbb{R}, V)$. We must find $\mathbf{v}$ such that $R(\mathbf{v}) = T$, that is $S_\mathbf{v} = T$. In fact, $\mathbf{v} = T(1)$ works since then $T(r) = T(r \cdot 1) = rT(1) = r\mathbf{v} = S_\mathbf{v}(r)$ holds for all r, so $T = S_\mathbf{v}$.

25. (b) Given $T : \mathbb{R} \to V$, let $T(1) = a_1\mathbf{b}_1 + \cdots + a_n\mathbf{b}_n$, a_i in $\mathbb{R}$. For all r in $\mathbb{R}$, we have $(a_1S_1 + \cdots + a_nS_n)(r) = a_1S_1(r) + \cdots + a_nS_n(r) = (a_1r\mathbf{b}_1 + \cdots + a_nr\mathbf{b}_n) = rT(1) = T(r)$. This shows that $a_1S_1 + \cdots + a_nS_n = T$.

27. (b) Write $\mathbf{v} = v_1\mathbf{b}_1 + \cdots + v_n\mathbf{b}_n$, v_j in $\mathbb{R}$. Apply E_i to get $E_i(\mathbf{v}) = v_1E_i(\mathbf{b}_1) + \cdots + v_nE_i(\mathbf{b}_n) = v_i$ by the definition of the E_i.

Exercises 9.2 Operators and Similarity (Page 452)

1. (b) $\frac{1}{2}\begin{bmatrix} -3 & -2 & 1 \\ 2 & 2 & 0 \\ 0 & 0 & 2 \end{bmatrix}$

4. (b) $P_{B \leftarrow D} = \begin{bmatrix} 1 & 1 & -1 \\ 1 & -1 & 0 \\ 1 & 0 & 1 \end{bmatrix}$, $P_{D \leftarrow B} = \frac{1}{3}\begin{bmatrix} 1 & 1 & 1 \\ 1 & -2 & 1 \\ -1 & -1 & 2 \end{bmatrix}$,

$P_{E \leftarrow D} = \begin{bmatrix} 1 & 0 & 1 \\ 1 & -1 & 0 \\ 1 & 1 & -1 \end{bmatrix}$, $P_{E \leftarrow B} = \begin{bmatrix} 0 & 0 & 1 \\ 0 & 1 & 0 \\ 1 & 0 & 0 \end{bmatrix}$.

5. (b) $A = P_{D \leftarrow B}$, where $B = \{(1, 2, -1), (2, 3, 0), (1, 0, 2)\}$.

Hence $A^{-1} = P_{B \leftarrow D} = \begin{bmatrix} 6 & -4 & -3 \\ -4 & 3 & 2 \\ 3 & -2 & -1 \end{bmatrix}$

7. (b) $P = \begin{bmatrix} 1 & 1 & 0 \\ 0 & 1 & 2 \\ -1 & 0 & 1 \end{bmatrix}$ **8. (b)** $B = \left\{ \begin{bmatrix} 3 \\ 7 \end{bmatrix}, \begin{bmatrix} 2 \\ 5 \end{bmatrix} \right\}$

9. (b) $c_T(x) = x^2 - 6x - 1$
   (d) $c_T(x) = x^3 + x^2 - 8x - 3$ **(f)** $c_T(x) = x^4$

12. Define $T_A : \mathbb{R}^n \to \mathbb{R}^n$ by $T_A(\mathbf{x}) = A\mathbf{x}$ for all $\mathbf{x}$ in $\mathbb{R}^n$. If null $A = $ null B, then $\ker(T_A) = $ null $A = $ null $B = \ker(T_B)$ so, by Exercise 28 Section 7.3, $T_A = ST_B$ for some isomorphism $S : \mathbb{R}^n \to \mathbb{R}^n$. If B_0 is the standard basis of $\mathbb{R}^n$, we have $A = M_{B_0}(T_A) = M_{B_0}(ST_B) = M_{B_0}(S)M_{B_0}(T_B) = UB$ where $U = M_{B_0}(S)$ is invertible by Theorem 1. Conversely, if $A = UB$ with U invertible, then $A\mathbf{x} = \mathbf{0}$ if and only if $B\mathbf{x} = \mathbf{0}$, so null $A = $ null B.

16. (b) Showing $S(w + v) = S(w) + S(v)$ means $M_B(T_{w+v}) = M_B(T_w) + M_B(T_v)$. If $B = \{b_1, b_2\}$, then column j of $M_B(T_{w+v})$ is $C_B[(w + v)b_j] = C_B(wb_j + vb_j) = C_B(wb_j) + C_B(vb_j)$ because C_B is linear. This is column j of $M_B(T_w) + M_B(T_v)$. Similarly $M_B(T_{aw}) = aM_B(T_w)$; so $S(aw) = aS(w)$. Finally $T_wT_v = T_{wv}$ so $S(wv) = M_B(T_wT_v) = M_B(T_w)M_B(T_v) = S(w)S(v)$ by Theorem 1.

Exercises 9.3 Invariant Subspaces and Direct Sums (Page 464)

2. (b) $T(U) \subseteq U$, so $T[T(U)] \subseteq T(U)$.

3. (b) If $\mathbf{v}$ is in $S(U)$, write $\mathbf{v} = S(\mathbf{u})$, $\mathbf{u}$ in U. Then $T(\mathbf{v}) = T[S(\mathbf{u})] = (TS)(\mathbf{u}) = (ST)(\mathbf{u}) = S[T(\mathbf{u})]$ and this lies in $S(U)$ because $T(\mathbf{u})$ lies in U (U is T-invariant).

6. Suppose U is T-invariant for every T. If $U \neq 0$, choose $\mathbf{u} \neq \mathbf{0}$ in U. Choose a basis $B = \{\mathbf{u}, \mathbf{u}_2, \ldots, \mathbf{u}_n\}$ of V containing $\mathbf{u}$. Given any $\mathbf{v}$ in V, there is (by Theorem 3 Section 7.1) a linear transformation $T : V \to V$ such that $T(\mathbf{u}) = \mathbf{v}$, $T(\mathbf{u}_2) = \cdots = T(\mathbf{u}_n) = \mathbf{0}$. Then $\mathbf{v} = T(\mathbf{u})$ lies in U because U is T-invariant. This shows that $V = U$.

8. (b) $T(1 - 2x^2) = 3 + 3x - 3x^2 = 3(1 - 2x^2) + 3(x + x^2)$ and $T(x + x^2) = -(1 - 2x^2)$, so both are in U. Hence U is T-invariant by Example 3. If
$B = \{1 - 2x^2, x + x^2, x^2\}$ then $M_B(T) = \begin{bmatrix} 3 & -1 & 1 \\ 3 & 0 & 1 \\ 0 & 0 & 3 \end{bmatrix}$, so
$c_T(x) = \det \begin{bmatrix} x-3 & 1 & -1 \\ -3 & x & -1 \\ 0 & 0 & x-3 \end{bmatrix} = (x - 3)\det \begin{bmatrix} x-3 & 1 \\ -3 & x \end{bmatrix}$
$= (x - 3)(x^2 - 3x + 3)$

9. (b) Suppose $\mathbb{R}\mathbf{u}$ is T_A-invariant where $\mathbf{u} \neq \mathbf{0}$. Then $T_A(\mathbf{u}) = r\mathbf{u}$ for some r in $\mathbb{R}$, so $(rI - A)\mathbf{u} = \mathbf{0}$. But $\det(rI - A) = (r - \cos \theta)^2 + \sin^2 \theta \neq 0$ because $0 < \theta < \pi$. Hence $\mathbf{u} = \mathbf{0}$, a contradiction.

10. (b) $U = \text{span}\{(1, 1, 0, 0), (0, 0, 1, 1)\}$ and $W = \text{span}\{(1, 0, 1, 0), (0, 1, 0, -1)\}$, and these four vectors form a basis of $\mathbb{R}^4$. Use Example 9.
(d) $U = \text{span}\left\{\begin{bmatrix} 1 & 1 \\ 0 & 0 \end{bmatrix}, \begin{bmatrix} 0 & 0 \\ 1 & 1 \end{bmatrix}\right\}$ and
$W = \text{span}\left\{\begin{bmatrix} 1 & 0 \\ -1 & 0 \end{bmatrix}, \begin{bmatrix} 0 & 1 \\ 0 & 1 \end{bmatrix}\right\}$, and these vectors are a basis of $\mathbf{M}_{22}$. Use Example 9.

14. The fact that U and W are subspaces is easily verified using the subspace test. If A lies in $U \cap V$, then $A = AE = 0$; that is, $U \cap V = 0$. To show that $\mathbf{M}_{22} = U + V$, choose any A in $\mathbf{M}_{22}$. Then $A = AE + (A - AE)$, and AE lies in U [because $(AE)E = AE^2 = AE$], and $A - AE$ lies in W [because $(A - AE)E = AE - AE^2 = 0$].

17. (b) By (a) it remains to show $U + W = V$; we show that $\dim(U + W) = n$ and invoke Theorem 2 Section 6.4. But $U + W = U \oplus W$ because $U \cap W = 0$, so $\dim(U + W) = \dim U + \dim W = n$.

18. (b) First, $\ker(T_A)$ is T_A-invariant. Let $U = \mathbb{R}\mathbf{p}$ be T_A-invariant. Then $T_A(\mathbf{p})$ is in U, say $T_A(\mathbf{p}) = \lambda \mathbf{p}$. Hence $A\mathbf{p} = \lambda \mathbf{p}$ so λ is an eigenvalue of A. This means that $\lambda = 0$ by (a), so $\mathbf{p}$ is in $\ker(T_A)$. Thus $U \subseteq \ker(T_A)$. But $\dim[\ker(T_A)] \neq 2$ because $T_A \neq 0$, so $\dim[\ker(T_A)] = 1 = \dim(U)$. Hence $U = \ker(T_A)$.

20. Let B_1 be a basis of U and extend it to a basis B of V. Then $M_B(T) = \begin{bmatrix} M_{B_1}(T) & Y \\ 0 & Z \end{bmatrix}$, so
$c_T(x) = \det[xI - M_B(T)] = \det[xI - M_{B_1}(T)]\det[xI - Z] = c_{T_1}(x)q(x)$.

22. (b) $T^2[p(x)] = p[-(-x)] = p(x)$, so $T^2 = 1$; $B = \{1, x^2; x, x^3\}$
(d) $T^2(a, b, c) = T(-a + 2b + c, b + c, -c) = (a, b, c)$, so $T^2 = 1$; $B = \{(1, 1, 0); (1, 0, 0), (0, -1, 2)\}$

23. (b) Use the Hint and Exercise 2.

25. (b) $T^2(a, b, c) = T(a + 2b, 0, 4b + c) = (a + 2b, 0, 4b + c) = T(a, b, c)$, so $T^2 = T$; $B = \{(1, 0, 0), (0, 0, 1); (2, -1, 4)\}$

29. (b) $T_{f;z}[T_{f;z}(\mathbf{v})] = T_{f;z}[f(\mathbf{v})\mathbf{z}] = f[f(\mathbf{v})\mathbf{z}]\mathbf{z} = f(\mathbf{v})\{f[\mathbf{z}]\mathbf{z}\}$ $= f(\mathbf{v})f(\mathbf{z})\mathbf{z}$. This equals $T_{f;z}(\mathbf{v}) = f(\mathbf{v})\mathbf{z}$ for all $\mathbf{v}$ if and only if $f(\mathbf{v})f(\mathbf{z}) = f(\mathbf{v})$ for all $\mathbf{v}$. Since $f \neq 0$, this holds if and only if $f(\mathbf{z}) = 1$.

30. (b) If $A = [\mathbf{p}_1 \ \mathbf{p}_2 \ \cdots \ \mathbf{p}_n]$ where $U\mathbf{p}_i = \lambda \mathbf{p}_i$ for each i, then $UA = \lambda A$. Conversely, $UA = \lambda A$ means that $U\mathbf{p} = \lambda \mathbf{p}$ for every column $\mathbf{p}$ of A.

Exercises 10.1 Inner Products and Norms (Page 475)

1. (b) P5 fails. (d) P5 fails. (f) P5 fails.
2. Axioms P1–P5 hold in U because they hold in V.
3. (b) $\frac{1}{\sqrt{\pi}}f$ (d) $\frac{1}{\sqrt{17}}\begin{bmatrix} 3 \\ -1 \end{bmatrix}$
4. (b) $\sqrt{3}$ (d) $\sqrt{3\pi}$
8. P1 and P2 are clear since $f(i)$ and $g(i)$ are real numbers.
P3: $\langle f + g, h \rangle = \sum_i (f + g)(i) \cdot h(i)$
$= \sum_i (f(i) + g(i)) \cdot h(i)$
$= \sum_i [f(i)h(i) + g(i)h(i)]$
$= \sum_i f(i)h(i) + \sum_i g(i)h(i) = \langle f, h \rangle + \langle g, h \rangle$.
P4: $\langle rf, g \rangle = \sum_i (rf)(i) \cdot g(i) = \sum_i rf(i) \cdot g(i)$
$= r\sum_i f(i) \cdot g(i) = r\langle f, g \rangle$.
P5: If $f \neq 0$, then $\langle f, f \rangle = \sum_i f(i)^2 > 0$ because some $f(i) \neq 0$.

12. (b) $\langle \mathbf{v}, \mathbf{v} \rangle = 5v_1^2 - 6v_1v_2 + 2v_2^2 = \frac{1}{5}[(5v_1 - 3v_2)^2 + v_2^2]$
(d) $\langle \mathbf{v}, \mathbf{v} \rangle = 3v_1^2 + 8v_1v_2 + 6v_2^2 = \frac{1}{3}[(3v_1 + 4v_2)^2 + 2v_2^2]$

13. (b) $\begin{bmatrix} 1 & -2 \\ -2 & 1 \end{bmatrix}$
(d) $\begin{bmatrix} 1 & 0 & -2 \\ 0 & 2 & 0 \\ -2 & 0 & 5 \end{bmatrix}$

14. By the condition, $\langle \mathbf{x}, \mathbf{y} \rangle = \frac{1}{2}\langle \mathbf{x} + \mathbf{y}, \mathbf{x} + \mathbf{y} \rangle = 0$ for all $\mathbf{x}, \mathbf{y}$. Let $\mathbf{e}_i$ denote column i of I. If $A = [a_{ij}]$, then $a_{ij} = \mathbf{e}_i^T A \mathbf{e}_j = \langle \mathbf{e}_i, \mathbf{e}_j \rangle = 0$ for all i and j.

16. (b) -15

20. 1. Using P2: $\langle \mathbf{u}, \mathbf{v} + \mathbf{w} \rangle = \langle \mathbf{v} + \mathbf{w}, \mathbf{u} \rangle = \langle \mathbf{v}, \mathbf{u} \rangle + \langle \mathbf{w}, \mathbf{u} \rangle$
$= \langle \mathbf{u}, \mathbf{v} \rangle + \langle \mathbf{u}, \mathbf{w} \rangle$.
2. Using P2 and P4:
$\langle \mathbf{v}, r\mathbf{w} \rangle = \langle r\mathbf{w}, \mathbf{v} \rangle = r\langle \mathbf{w}, \mathbf{v} \rangle = r\langle \mathbf{v}, \mathbf{w} \rangle$.
3. Using P3: $\langle \mathbf{0}, \mathbf{v} \rangle = \langle \mathbf{0} + \mathbf{0}, \mathbf{v} \rangle = \langle \mathbf{0}, \mathbf{v} \rangle + \langle \mathbf{0}, \mathbf{v} \rangle$, so $\langle \mathbf{0}, \mathbf{v} \rangle = 0$. The rest is P2.
4. Assume that $\langle \mathbf{v}, \mathbf{v} \rangle = 0$. If $\mathbf{v} \neq \mathbf{0}$ this contradicts P5, so $\mathbf{v} = \mathbf{0}$. Conversely, if $\mathbf{v} = \mathbf{0}$, then $\langle \mathbf{v}, \mathbf{v} \rangle = 0$ by Part 3 of this theorem.

22. (b) $15\|\mathbf{u}\|^2 - 17\langle \mathbf{u}, \mathbf{v} \rangle - 4\|\mathbf{v}\|^2$
(d) $\|\mathbf{u} + \mathbf{v}\|^2 = \langle \mathbf{u} + \mathbf{v}, \mathbf{u} + \mathbf{v} \rangle = \|\mathbf{u}\|^2 + 2\langle \mathbf{u}, \mathbf{v} \rangle + \|\mathbf{v}\|^2$

26. (b) $\{(1, 1, 0), (0, 2, 1)\}$

28. $\langle \mathbf{v} - \mathbf{w}, \mathbf{v}_i \rangle = \langle \mathbf{v}, \mathbf{v}_i \rangle - \langle \mathbf{w}, \mathbf{v}_i \rangle = 0$ for each i, so $\mathbf{v} = \mathbf{w}$ by Exercise 27.

29. **(b)** If $\mathbf{u} = (\cos\theta, \sin\theta)$ in $\mathbb{R}^2$ (with the dot product) then $\|\mathbf{u}\| = 1$. Use **(a)** with $\mathbf{v} = (x, y)$.

Exercises 10.2 Orthogonal Sets of Vectors (Page 484)

1. **(b)** $\frac{1}{14}\left\{(6a + 2b + 6c)\begin{bmatrix}1\\1\\1\end{bmatrix} + (7c - 7a)\begin{bmatrix}-1\\0\\1\end{bmatrix} + (a - 2b + c)\begin{bmatrix}1\\-6\\1\end{bmatrix}\right\}$

    (d) $\left(\frac{a+d}{2}\right)\begin{bmatrix}1 & 0\\0 & 1\end{bmatrix} + \left(\frac{a-d}{2}\right)\begin{bmatrix}1 & 0\\0 & -1\end{bmatrix} + \left(\frac{b+c}{2}\right)\begin{bmatrix}0 & 1\\1 & 0\end{bmatrix} + \left(\frac{b-c}{2}\right)\begin{bmatrix}0 & 1\\-1 & 0\end{bmatrix}$

2. **(b)** $\{(1, 1, 1), (1, -5, 1), (3, 0, -2)\}$

3. **(b)** $\left\{\begin{bmatrix}1 & 1\\0 & 1\end{bmatrix}, \begin{bmatrix}1 & -2\\3 & 1\end{bmatrix}, \begin{bmatrix}1 & -2\\-2 & 1\end{bmatrix}, \begin{bmatrix}1 & 0\\0 & -1\end{bmatrix}\right\}$

4. **(b)** $\{1, x - 1, x^2 - 2x + \frac{2}{3}\}$

6. **(b)** $U^\perp = \text{span}\{[1\ -1\ 0\ 0], [0\ 0\ 1\ 0], [0\ 0\ 0\ 1]\}$, $\dim U^\perp = 3$, $\dim U = 1$

    (d) $U^\perp = \text{span}\{2 - 3x, 1 - 2x^2\}$, $\dim U^\perp = 2$, $\dim U = 1$

    (f) $U^\perp = \text{span}\left\{\begin{bmatrix}1 & -1\\-1 & 0\end{bmatrix}\right\}$, $\dim U^\perp = 1$, $\dim U = 3$

7. **(b)** $U = \text{span}\left\{\begin{bmatrix}1 & 0\\0 & 1\end{bmatrix}, \begin{bmatrix}1 & 1\\1 & -1\end{bmatrix}, \begin{bmatrix}0 & 1\\-1 & 0\end{bmatrix}\right\}$; $\text{proj}_U(A) = \begin{bmatrix}3 & 0\\2 & 1\end{bmatrix}$

8. **(b)** $U = \text{span}\{1, 5 - 3x^2\}$; $\text{proj}_U(x) = \frac{3}{13}(1 + 2x^2)$

9. **(b)** $B = \{1, 2x - 1\}$ is an orthogonal basis of U because $\int_0^1(2x - 1)\,dx = 0$. Using it, we get $\text{proj}_U(x^2 + 1) = x + \frac{5}{6}$, so $x^2 + 1 = (x + \frac{5}{6}) + (x^2 - x + \frac{1}{6})$.

11. **(b)** This follows from $\langle \mathbf{v} + \mathbf{w}, \mathbf{v} - \mathbf{w} \rangle = \|\mathbf{v}\|^2 - \|\mathbf{w}\|^2$.

14. **(b)** $U^\perp \subseteq \{\mathbf{u}_1, \ldots, \mathbf{u}_m\}^\perp$ because each $\mathbf{u}_i$ is in U. Conversely, if $\langle \mathbf{v}, \mathbf{u}_i \rangle = 0$ for each i, and $\mathbf{u} = r_1\mathbf{u}_1 + \cdots + r_m\mathbf{u}_m$ is any vector in U, then $\langle \mathbf{v}, \mathbf{u} \rangle = r_1\langle \mathbf{v}, \mathbf{u}_1\rangle + \cdots + r_m\langle \mathbf{v}, \mathbf{u}_m\rangle = 0$.

18. **(b)** $\text{proj}_U(-5, 4, -3) = (-5, 4, -3)$; $\text{proj}_U(-1, 0, 2) = \frac{1}{38}(-17, 24, 73)$

19. **(b)** The plane is $U = \{\mathbf{x} \mid \mathbf{x} \cdot \mathbf{n} = 0\}$ so $\text{span}\left\{\mathbf{n} \times \mathbf{w}, \mathbf{w} - \frac{\mathbf{n} \cdot \mathbf{w}}{\|\mathbf{n}\|^2}\mathbf{n}\right\} \subseteq U$. This is equality because both spaces have dimension 2 (using **(a)**).

20. **(b)** $C_E(\mathbf{b}_i)$ is column i of P. Since $C_E(\mathbf{b}_i) \cdot C_E(\mathbf{b}_j) = \langle \mathbf{b}_i, \mathbf{b}_j \rangle$ by **(a)**, the result follows.

23. **(b)** If $U = \text{span}\{\mathbf{f}_1, \mathbf{f}_2, \ldots, \mathbf{f}_m\}$, then $\text{proj}_U(\mathbf{v}) = \sum_{i=1}^m \frac{\langle \mathbf{v}_1, \mathbf{f}_i \rangle}{\|\mathbf{f}_i\|^2} \mathbf{f}_i$ by Theorem 7. Hence $\|\text{proj}_U(\mathbf{v})\|^2 = \sum_{i=1}^m \frac{\langle \mathbf{v}_1, \mathbf{f}_i \rangle^2}{\|\mathbf{f}_i\|^2}$ by Pythagoras' theorem. Now use **(a)**.

Exercises 10.3 Orthogonal Diagonalization (Page 491)

1. **(b)** $B = \left\{\begin{bmatrix}1 & 0\\0 & 0\end{bmatrix}, \begin{bmatrix}0 & 1\\0 & 0\end{bmatrix}, \begin{bmatrix}0 & 0\\1 & 0\end{bmatrix}, \begin{bmatrix}0 & 0\\0 & 1\end{bmatrix}\right\}$; $M_B(T) = \begin{bmatrix}-1 & 0 & 1 & 0\\0 & -1 & 0 & 1\\1 & 0 & 2 & 0\\0 & 1 & 0 & 2\end{bmatrix}$

4. **(b)** $\langle \mathbf{v}, (rT)(\mathbf{w})\rangle = \langle \mathbf{v}, rT(\mathbf{w})\rangle = r\langle \mathbf{v}, T(\mathbf{w})\rangle = r\langle T(\mathbf{v}), \mathbf{w}\rangle = \langle rT(\mathbf{v}), \mathbf{w}\rangle = \langle (rT)(\mathbf{v}), \mathbf{w}\rangle$

    (d) Given $\mathbf{v}$ and $\mathbf{w}$, write $T^{-1}(\mathbf{v}) = \mathbf{v}_1$ and $T^{-1}(\mathbf{w}) = \mathbf{w}_1$. Then $\langle T^{-1}(\mathbf{v}), \mathbf{w}\rangle = \langle \mathbf{v}_1, T(\mathbf{w}_1)\rangle = \langle T(\mathbf{v}_1), \mathbf{w}_1\rangle = \langle \mathbf{v}, T^{-1}(\mathbf{w})\rangle$.

5. **(b)** If $B_0 = \{(1, 0, 0), (0, 1, 0), (0, 0, 1)\}$, then $M_{B_0}(T) = \begin{bmatrix}7 & -1 & 0\\-1 & 7 & 0\\0 & 0 & 2\end{bmatrix}$ has an orthonormal basis of eigenvectors $\left\{\frac{1}{\sqrt{2}}\begin{bmatrix}1\\1\\0\end{bmatrix}, \frac{1}{\sqrt{2}}\begin{bmatrix}1\\-1\\0\end{bmatrix}, \begin{bmatrix}0\\0\\1\end{bmatrix}\right\}$. Hence an orthonormal basis of eigenvectors of T is $\{\frac{1}{\sqrt{2}}(1, 1, 0), \frac{1}{\sqrt{2}}(1, -1, 0), (0, 0, 1)\}$.

    (d) If $B_0 = \{1, x, x^2\}$, then $M_{B_0}(T) = \begin{bmatrix}-1 & 0 & 1\\0 & 3 & 0\\1 & 0 & -1\end{bmatrix}$ has an orthonormal basis of eigenvectors $\left\{\begin{bmatrix}0\\1\\0\end{bmatrix}, \frac{1}{\sqrt{2}}\begin{bmatrix}1\\0\\1\end{bmatrix}, \frac{1}{\sqrt{2}}\begin{bmatrix}1\\0\\-1\end{bmatrix}\right\}$. Hence an orthonormal basis of eigenvectors of T is $\{x, \frac{1}{\sqrt{2}}(1 + x^2), \frac{1}{\sqrt{2}}(1 - x^2)\}$.

7. **(b)** $M_B(T) = \begin{bmatrix}A & 0\\0 & A\end{bmatrix}$, so $c_T(x) = \det\begin{bmatrix}xI_2 - A & 0\\0 & xI_2 - A\end{bmatrix} = [c_A(x)]^2$.

12. (1)$\Rightarrow$(2). If $B = \{\mathbf{f}_1, \ldots, \mathbf{f}_n\}$ is an orthonormal basis of V, then $M_B(T) = [a_{ij}]$ where $a_{ij} = \langle \mathbf{f}_i, T(\mathbf{f}_j)\rangle$ by Theorem 2. If (1) holds, then $a_{ji} = \langle \mathbf{f}_j, T(\mathbf{f}_i)\rangle = -\langle T(\mathbf{f}_j), \mathbf{f}_i\rangle = -\langle \mathbf{f}_i, T(\mathbf{f}_j)\rangle = -a_{ij}$. Hence $[M_V(T)]^T = -M_V(T)$, proving (2).

14. **(c)** The coefficients in the definition of $T'(\mathbf{f}_j) = \sum_{i=1}^n \langle \mathbf{f}_j, T(\mathbf{f}_i)\rangle \mathbf{f}_i$ are the entries in the jth column $C_B[T'(\mathbf{f}_j)]$ of $M_B(T')$. Hence $M_B(T') = [\langle \mathbf{f}_j, T(\mathbf{f}_i)\rangle]$, and this is the transpose of $M_B(T)$ by Theorem 2.

Exercises 10.4 Isometries (Page 504)

2. **(b)** Rotation through π **(d)** Reflection in the line $y = -x$ **(f)** Rotation through $\frac{\pi}{4}$

3. **(b)** $c_T(x) = (x - 1)(x^2 + \frac{3}{2}x + 1)$. If $\mathbf{e} = [1\ \sqrt{3}\ \sqrt{3}]^T$, then T is a rotation about $\mathbb{R}\mathbf{e}$.

    (d) $c_T(x) = (x + 1)(x + 1)^2$. Rotation (of π) about the x axis.

(f) $c_T(x) = (x+1)(x^2 - \sqrt{2}x + 1)$. Rotation (of $-\frac{\pi}{4}$) about the y axis followed by a reflection in the x-z plane.

6. If $\|\mathbf{v}\| = \|(aT)(\mathbf{v})\| = |a|\|T(\mathbf{v})\| = |a|\|\mathbf{v}\|$ for some $\mathbf{v} \neq \mathbf{0}$, then $|a| = 1$ so $a = \pm 1$.

12. (b) Assume that $S = S_\mathbf{u} \circ T$, $\mathbf{u} \in V$, T an isometry of V. Since T is onto (by Theorem 2), let $\mathbf{u} = T(\mathbf{w})$ where $\mathbf{w} \in V$. Then for any $\mathbf{v} \in V$, we have $(T \circ S_\mathbf{w})(\mathbf{v}) = T(\mathbf{w} + \mathbf{v}) = T(\mathbf{w}) + T(\mathbf{v}) = S_{T(\mathbf{w})}(T(\mathbf{v})) = (S_{T(\mathbf{w})} \circ T)(\mathbf{v})$, and it follows that $T \circ S_\mathbf{w} = S_{T(\mathbf{w})} \circ T$.

Exercises 10.5 An Application to Fourier Approximation (Page 509)

1. (b) $\frac{\pi}{2} - \frac{4}{\pi}\left[\cos x + \frac{\cos 3x}{3^2} + \frac{\cos 5x}{5^2}\right]$

   (d) $\frac{\pi}{4} + \left[\sin x - \frac{\sin 2x}{2} + \frac{\sin 3x}{3} - \frac{\sin 4x}{4} + \frac{\sin 5x}{5}\right] - \frac{2}{\pi}\left[\cos x + \frac{\cos 3x}{3^2} + \frac{\cos 5x}{5^2}\right]$

2. (b) $\frac{2}{\pi} - \frac{8}{\pi}\left[\frac{\cos 2x}{2^2 - 1} + \frac{\cos 4x}{4^2 - 1} + \frac{\cos 6x}{6^2 - 1}\right]$

4. $\int \cos kx \cos lx\, dx = \frac{1}{2}\left[\frac{\sin[(k+l)x]}{k+l} - \frac{\sin[(k-l)x]}{k-l}\right]_0^\pi = 0$ provided that $k \neq l$.

Exercises 11.1 Block Triangular Form (Page 517)

1. (b) $c_A(x) = (x+1)^3$; $P = \begin{bmatrix} 1 & 0 & 0 \\ 1 & 1 & 0 \\ 1 & -3 & 1 \end{bmatrix}$;

   $P^{-1}AP = \begin{bmatrix} -1 & 0 & 1 \\ 0 & -1 & 0 \\ 0 & 0 & -1 \end{bmatrix}$

   (d) $c_A(x) = (x-1)^2(x+2)$; $P = \begin{bmatrix} -1 & 0 & -1 \\ 4 & 1 & 1 \\ 4 & 2 & 1 \end{bmatrix}$;

   $P^{-1}AP = \begin{bmatrix} 1 & 1 & 0 \\ 0 & 1 & 0 \\ 0 & 0 & -2 \end{bmatrix}$

   (f) $c_A(x) = (x+1)^2(x-1)^2$; $P = \begin{bmatrix} 1 & 1 & 5 & 1 \\ 0 & 0 & 2 & -1 \\ 0 & 1 & 2 & 0 \\ 1 & 0 & 1 & 1 \end{bmatrix}$;

   $P^{-1}AP = \begin{bmatrix} -1 & 1 & 0 & 0 \\ 0 & -1 & 1 & 0 \\ 0 & 0 & 1 & -2 \\ 0 & 0 & 0 & 1 \end{bmatrix}$

4. If B is any ordered basis of V, write $A = M_B(T)$. Then $c_T(x) = c_A(x) = a_0 + a_1 x + \cdots + a_n x^n$ for scalars a_i in $\mathbb{R}$. Since M_B is linear and $M_B(T^k) = M_B(T)^k$, we have $M_B[c_T(T)] = M_V[a_0 + a_1 T + \cdots + a_n T^n] = a_0 I + a_1 A + \cdots + a_n A^n = c_A(A) = 0$ by the Cayley-Hamilton theorem. Hence $c_T(T) = 0$ because M_B is one-to-one.

Exercises 11.2 The Jordan Canonical Form (Page 522)

2. $\begin{bmatrix} a & 1 & 0 \\ 0 & a & 0 \\ 0 & 0 & b \end{bmatrix}\begin{bmatrix} 0 & 1 & 0 \\ 0 & 0 & 1 \\ 1 & 0 & 0 \end{bmatrix} = \begin{bmatrix} 0 & 1 & 0 \\ 0 & 0 & 1 \\ 1 & 0 & 0 \end{bmatrix}\begin{bmatrix} a & 0 & 0 \\ 0 & a & 1 \\ 0 & 0 & a \end{bmatrix}$

Appendix A Complex Numbers (Page 533)

1. (b) $x = 3$ (d) $x = \pm 1$
2. (b) $10 + i$ (d) $\frac{11}{26} + \frac{23}{26}i$
   (f) $2 - 11i$ (h) $8 - 6i$
3. (b) $\frac{11}{5} + \frac{3}{5}i$ (d) $\pm(2 - i)$ (f) $1 + i$
4. (b) $\frac{1}{2} \pm \frac{\sqrt{3}}{2}i$ (d) $2, \frac{1}{2}$
5. (b) $-2, 1 \pm \sqrt{3}i$
   (d) $\pm 2\sqrt{2}, \pm 2\sqrt{2}i$
6. (b) $x^2 - 4x + 13$; $2 + 3i$
   (d) $x^2 - 6x + 25$; $3 + 4i$
8. $x^4 - 10x^3 + 42x^2 - 82x + 65$
10. (b) $(-2)^2 + 2i - (4 - 2i) = 0$; $2 - i$
    (d) $(-2+i)^2 + 3(1-i)(-1+2i) - 5i = 0$; $-1 + 2i$
11. (b) $-i, 1 + i$
    (d) $2 - i, 1 - 2i$
12. (b) Circle, centre at 1, radius 2
    (d) Imaginary axis
    (f) Line $y = mx$
18. (b) $4e^{-\pi i/2}$ (d) $8e^{2\pi i/3}$ (f) $6\sqrt{2}e^{3\pi i/4}$
19. (b) $\frac{1}{2} + \frac{\sqrt{3}}{2}i$ (d) $1 - i$ (f) $\sqrt{3} - 3i$
20. (b) $-\frac{1}{32} + \frac{\sqrt{3}}{32}i$ (d) $-32i$ (f) $-2^{16}(1+i)$
23. (b) $\pm\frac{\sqrt{2}}{2}(\sqrt{3}+i), \pm\frac{\sqrt{2}}{2}(-1+\sqrt{3}i)$
    (d) $\pm 2i, \pm(\sqrt{3}+i), \pm(\sqrt{3}-i)$
26. (b) The argument in (a) applies using $\beta = \frac{2\pi}{n}$. Then $1 + z + \cdots + z^{n-1} = \frac{1 - z^n}{1 - z} = 0$.

Appendix B Proofs (Page 540)

1. (b) If $m = 2p$ and $n = 2q + 1$ where p and q are integers, then $m + n = 2(p + q) + 1$ is odd. The converse is false: $m = 1$ and $n = 2$ is a counterexample.
   (d) $x^2 - 5x + 6 = (x-2)(x-3)$ so, if this is zero, then $x = 2$ or $x = 3$. The converse is true: each of 2 and 3 satisfies $x^2 - 5x + 6 = 0$.

2. (b) This implication is true. If $n = 2t + 1$ where t is an integer, then $n^2 = 4t^2 + 4t + 1 = 4t(t+1) + 1$. Now t is either even or odd, say $t = 2m$ or $t = 2m + 1$. If $t = 2m$, then $n^2 = 8m(2m+1) + 1$; if $t = 2m + 1$, then $n^2 = 8(2m+1)(m+1) + 1$. Either way, n^2 has the form $n^2 = 8k + 1$ for some integer k.

Selected Answers

3. **(b)** Assume that the statement "one of m and n is greater than 12" is false. Then both $n \leq 12$ and $m \leq 12$, so $n + m \leq 24$, contradicting the hypothesis that $n + m = 25$. This proves the implication. The converse is false: $n = 13$ and $m = 13$ is a counterexample.

   (d) Assume that the statement "m is even or n is even" is false. Then both m and n are odd, so mn is odd, contradicting the hypothesis. The converse is true: If m or n is even, then mn is even.

4. **(b)** If x is irrational and y is rational, assume that $x + y$ is rational. Then $x = (x + y) - y$ is the difference of two rationals, and so is rational, contrary to the hypothesis.

5. **(b)** $n = 10$ is a counterexample because $10^3 = 1000$ while $2^{10} = 1024$, so the statement $n^3 \geq 2^n$ is false if $n = 10$. Note that $n^3 \geq 2^n$ *does* hold for $2 \leq n \leq 9$.

Appendix C Mathematical Induction (Page 545)

6. $\dfrac{n}{n+1} + \dfrac{1}{(n+1)(n+2)} = \dfrac{n(n+2)+1}{(n+1)(n+2)}$
   $= \dfrac{(n+1)^2}{(n+1)(n+2)} = \dfrac{n+1}{n+2}$

14. $2\sqrt{n} - 1 + \dfrac{1}{\sqrt{n+1}} = \dfrac{2\sqrt{n^2+n}+1}{\sqrt{n+1}} - 1$
    $< \dfrac{2(n+1)}{\sqrt{n+1}} - 1 = 2\sqrt{n+1} - 1$

18. If $n^3 - n = 3k$, then
    $(n+1)^3 - (n+1) = 3k + 3n^2 + 3n = 3(k + n^2 + n)$

20. $B_n = (n+1)! - 1$

22. **(b)** Verify each of $S_1, S_2, \ldots, S_8$.

Index

A

A-invariance, 154–156
absolute value
  complex number, 358n, 525, 526
  notation, 97n
  real number, 185n
  symmetric matrices, 270n
  triangle inequality, 473–474
abstract vector space, 288
action
  same action, 52, 291, 334
  transformations, 52, 436, 438–439
addition
  see also sum
  closed under, 43, 289
  closed under addition, 229
  complex number, 523–524
  matrix addition, 34–36
  pointwise addition, 291, 292
  transformations, preserving addition, 91
  vector addition, 289, 526
adjacency matrix, 66
adjugate, 71, 140–142
adjugate formula, 142
adult survival rate, 150
aerodynamics, 433
The Algebraic Eigenvalue Problem (Wilkinson), 395
algebraic method, 4, 9
algebraic multiplicity, 267n
algebraic sum, 28
altitude, 228
analytic geometry, 42
angles
  angle between two vectors, 199, 476
  radian measure, 53n, 97, 527
  standard position, 97, 527
  unit circle, 97, 527
approximation theorem, 482–483, 506–507
Archimedes, 10n
area
  linear transformations of, 223–224
  parallelogram, equal to zero, 215–216
argument, 527
arrows, 184
associated homogeneous system, 47
associative law, 34, 62
attractor, 164
augmented matrix, 3, 4–5, 12, 13
auxiliary theorems, 84n
axiomatic method, 539–540

axioms, 540
axis, 184, 502

B

B-matrix, 445–452
back substitution, 13, 103
balanced reaction, 29
ball, 413
Banach, Stephan, 288
bases, 241n
  see also basis
basic eigenvectors, 153
basic solutions, 22, 23
basis
  choice of basis, 436, 441–442, 450
  dual basis, 445
  enlarging subset to, 242
  geometric problem of finding, 448, 449, 510
  independent set, 369
  isomorphisms, 348
  linear operators, and choice of basis, 450
  matrix of T corresponding to the ordered bases B and D, 438
  ordered basis, 436, 438
  orthogonal basis, 249n, 251–252, 369, 479–480
  orthonormal basis, 480, 488
  standard basis, 93, 233, 241, 242, 250, 307, 397
  of subspace, 241
  vector spaces, 306
Bessel's inequality, 485
best approximation, 273–275
best approximation theorem, 273–274
bilinear form, 431
binary codes, 412
Binet formula, 171–172
binomial coefficients, 324
binomial theorem, 324
block matrix, 134–135
block multiplication, 64–65
block triangular form, 511–517
block triangular matrix, 455–456
block triangulation theorem, 511
blocks, 64
boundary condition, 174, 326–327
Brown, J.W., 508n
Bruen, A.A., 419n

C

Calcolo Geometrico (Peano), 288

cancellation, 292, 293–294
cancellation laws, 74
canonical forms
  block triangular form, 511–517
  Jordan canonical form, 518–522
  $m \times n$ matrix, 518
Cartesian coordinates, 184
cartesian geometry, 184
category, 351n
Cauchy, Augustin Louis, 137, 248f, 248n, 270n
Cauchy inequality, 248, 285
Cauchy-Schwarz inequality, 213, 473
Cayley, Arthur, 32, 32f, 32n, 126, 405n
Cayley-Hamilton theorem, 160, 405–406, 515
centred, 283
centroid, 198
change matrix, 446
channel, 412
characteristic polynomial
  block triangular matrix, 455–456
  complex matrix, 400
  diagonalizable matrix, 264
  eigenvalues, 152
  root of, 152, 325, 326
  similarity invariant, 451
  square matrix, 152, 405
chemical reactions, 29–30
choice of basis, 436, 441–442, 450
Cholesky, Andre-Louis, 387n
Cholesky algorithm, 388
Cholesky factorization, 387
Churchill, R.V., 508n
circuit rule, 28
classical adjoint, 140n
closed economy, 113
closed under addition, 43, 229, 289
closed under scalar multiplication, 43, 229, 289
code
  binary codes, 412
  decoding, 418
  defined, 412
  error correcting codes, 412–414
  Hamming (7,4)-code, 420
  linear codes, 414–416
  matrix generators, 416–417
  minimum distance, 413
  n-code, 412
  (n, k)-code, 414
  nearest neighbour decoding, 413
  orthogonal codes, 419–420
  parity-check code, 415

parity-check matrices, 417–419
perfect, 414
syndrome decoding, 419
use of, 408
code words, 412, 413, 416, 420
coding theory, 412
coefficient matrix, 3, 143
coefficients
  binomial coefficients, 324
  constant coefficient, 546
  correlation coefficients. See correlation coefficients
  Fourier coefficients, 251–252, 478, 506, 507
  leading coefficient, 145n, 290, 546
  linear combination, 232, 236, 299
  in linear equation, 1
  of the polynomial, 279–280, 290, 546
  sample correlation coefficient, 284–285
  of vectors, 299
cofactor, 127
cofactor expansion, 126–135, 129n, 179–183
cofactor expansion theorem, 129, 179–183
cofactor matrix, 140
column matrix, 33, 42
coiumn space, 253
column vectors, 151
columns
  convention, 33
  elementary column operations, 130
  equal, 20
  (i, j)-entry, 33
  leading column, 105
  as notations for ordered n-tuples, 239n
  shape of matrix, 32
  Smith normal form, 87–88
  transpose, 38
commutative law, 34
commute, 60, 63
companion matrix, 137
compatibility rule, 59
compatible
  blocks, 64
  for multiplication, 59
complement, 459
completely diagonalized, 428
complex conjugation, 270n
complex distance formula, 526
complex eigenvalues, 269, 396

Index

complex matrix
  Cayley-Hamilton theorem, 405–406
  characteristic polynomial, 400
  conjugate, 399
  conjugate transpose, 399
  defined, 399
  eigenvalue, 400
  eigenvector, 400
  hermitan matrix, 400–401
  normal, 405
  Schur's theorem, 403, 405–406
  skew-hermitian, 407
  spectral theorem, 404
  standard inner product, 397–399
  unitarily diagonalizable, 403
  unitary diagonalization, 403–406
  unitary matrix, 402
  upper triangular matrix, 403
complex number
  absolute value, 358n, 525, 526
  addition, 523–524
  advantage of working with, 401
  in complex plane, 525
  conjugate, 525
  equal, 523
  extension of concepts to, 397
  form, 523
  fundamental theorem of algebra, 532–533
  imaginary axis, 526
  imaginary part, 523
  imaginary unit, 523
  inverse, 524
  modulus, 525
  multiplication, 524
  parallelogram law, 526
  polar form, 527–529
  product, 526
  pure imaginary numbers, 523
  real axis, 526
  real part, 523
  regular representation, 526
  root of the quadratic, 531
  roots of unity, 529–532
  scalars, 409
  subtraction, 523–524
  sum, 526
  triangle inequality, 525
complex plane, 526
complex subspace, 406
composite, 57–59, 95, 351, 351n
composition, 57, 351–354, 441–442
computer graphics, 225–227
conclusion, 536
congruence, 427–430
congruent matrices, 427–428
conic graph, 20, 425
conjugate, 399, 525, 526
conjugate matrix, 270
conjugate transpose, 399
consistent system, 2, 15
constant, 311, 546
constant matrix, 3
constant sequences, 361
constant term, 1
constrained optimization, 431–434
continuous functions, 469
contraction, 56
contradiction, proof by, 538–540
convergence, 394
converges, 121
converse, 539
coordinate isomorphism, 350
coordinate transformation, 437
coordinate vectors, 189, 207, 224, 233, 437
coordinates, 184, 436
Coppins, R., 434n

correlation, 282–286
correlation coefficients
  computation, with dot product, 285–286
  Pearson correlation coefficient, 284n
  sample correlation coefficient, 284–285
correlation formula, 286
coset, 419
cosine, 97, 199–200, 485, 527
counterexample, 8n, 539
covariance, 435
covariance matrix, 435
Cramer, Gabriel, 143n
Cramer's rule, 137, 143–144
cross product
  coordinate-free description, 217
  coordinate vectors, 207
  defined, 207, 213
  determinant form, 207, 213–214
  and dot product, 206–209, 215
  Lagrange identity, 215
  properties of, 214
  right-hand rule, 217
  shortest distance between nonparallel lines, 209
cryptography, 408
Cryptography, Information Theory, and Error-Correction (Bruen and Forcinito), 419n

D

data scaling, 286
Davis, Philip J., 483
De Moivre, Abraham, 529
De Moivre's Theorem, 529
decoding, 418
defined, 59
defining transformation, 52
degree of the polynomial, 145n, 290, 546
demand matrix, 115
dependent, 238, 303, 315
dependent lemma, 315
derivative, 357, 358n
Descartes, René, 184n
determinants
  adjugate, 71, 140
  block matrix, 134–135
  coefficient matrix, 145
  cofactor expansion, 126–135, 179–183
  Cramer's rule, 137, 143–144
  cross product, 207, 213
  defined, 71, 126, 132, 179, 450
  and eigenvalues, 126
  inductive method of determination, 127
  initial development of, 224
  and inverses, 71, 126, 137–147
  notation, 126n
  polynomial interpolation, 144–147
  product of matrices (product theorem), 138, 147
  similarity invariant, 450
  square matrices, 51, 126, 139
  theory of determinants, 32n
  triangular matrix, 134
  Vandermonde determinant, 146
  Vandermonde matrix, 133
deviation, 283
diagonal matrices, 41, 69, 151, 156–160, 264, 448
diagonalizable linear operator, 486–487
diagonalizable matrix, 156, 264, 448
diagonalization
  completely diagonalized, 428

described, 156–160, 264–265
eigenvalues, 126, 151–154, 264–265
example, 150
general differential systems, 174–175
linear dynamical system, 160–164
matrix, 151–152
multivariate analysis, 434
orthogonal diagonalization, 376–382, 486–491
quadratic form, 422
test, 265–266
unitary diagonalization, 403–406
diagonalization algorithm, 159
diagonalization theorem, 423
diagonalizing matrix, 156
difference
  $m \times n$ matrices, 35
  of two vectors, 188, 293
differentiable function, 173, 299, 325, 357
differential equation of order n, 325, 357
differential equations, 173–178, 325–329, 357–360
differential system
  defined, 173
  exponential function, 173–174
  general differential systems, 174–178
  general solution, 177–178
  simplest differential system, 173–174
differentiation, 333
digits, 412
dilation, 56
dimension, 240–244, 307, 357
dimension theorem, 331, 341–344, 350–351
direct proof, 536–537
direct sum, 318, 459–464, 467
directed graphs, 66
direction, 185
direction cosines, 213, 485
direction vector, 191–194
discriminant, 425, 531
Disquisitiones Arithmeticae (Gauss), 10n, 126
distance, 249, 471–474
distance function, 351n
distance preserving, 219–220, 493
distance preserving isometries, 493–504
distribution, 434
distributive laws, 37, 62
division algorithm, 408, 548
dominant eigenvalue, 162, 393
dominant eigenvector, 393
dot product
  basic properties, 198–199
  correlation coefficients, computation of, 285–286
  and cross product, 206–209, 215
  defined, 198
  dot product rule, 49, 59
  as inner product, 468
  inner product space, 471
  length, 471n
  and matrix multiplication, 57–59
  in set of all ordered n-tuples ($\mathbb{R}^n$), 246–249, 397
  of two ordered n-tuples, 48
  of two vectors, 198–202
  variances, computation of, 285–286
doubly stochastic matrix, 124
dual, 445
dual basis, 445

E

economic models, input-output, 112–116
economic system, 112
edges, 66
eigenspace, 231, 267, 458, 511
eigenvalues
  complex eigenvalues, 155–156, 269, 396
  complex matrix, 400
  computation of, 393–396
  defined, 151, 264
  and determinants, 126
  and diagonalizable matrices, 157–158, 264
  dominant eigenvalue, 162, 393
  and eigenspace, 231, 267
  and Google PageRank, 165–166
  iterative methods, 393–396
  linear operator, 457–459
  multiple eigenvalues, 266
  multiplicity, 158, 267
  power method, 393–394
  real eigenvalues, 270–271
  root of the characteristic polynomial, 152
  solving for, 151–154
  spectrum of the matrix, 378
  symmetric linear operator on finite dimensional inner product space, 486
eigenvector
  basic eigenvectors, 153
  complex matrix, 400
  defined, 151, 264
  dominant eigenvector, 393
  fractions, 153n
  linear combination, 394
  linear operator, 457
  nonzero linear combination, 153n
  nonzero multiple, 153, 153n
  nonzero vectors, 231, 264
  orthogonal basis, 376
  orthogonal eigenvectors, 377–381
  orthonormal basis, 488
  principal axes, 378
electrical networks, 27–28
elementary matrix
  defined, 84
  and inverses, 85–86
  LU-factorization, 103–111
  operating corresponding to, 84
  permutation matrix, 107–109
  self-inverse, 85
  Smith normal form, 87–88
  uniqueness of reduced row-echelon form, 88–89
elementary operations, 4–7
elementary row operations
  corresponding, 84
  inverses, 6–7, 85
  matrices, 5–7
  reversed, 6–7
  scalar product, 20
  sum, 20
Elements (Euclid), 537
elements of the set, 229n
ellipse, 425
entries of the matrix, 32
equal
  columns, 20
  complex number, 523
  fractions, 186n
  functions, 291
  linear transformations, 334
  matrices, 33
  polynomials, 291, 546
  sequences, 361

sets, 229n
transformation, 52, 334
equal modulo, 409
equilibrium, 113
equilibrium condition, 113
equilibrium price structures, 113, 114
equivalence relation, 262
equivalent
  matrices, 90–91
  statements, 77n
  systems of linear equations, 4
error, 275, 418
error correcting codes, 412–414
Euclid, 537, 540
euclidean algorithm, 410
euclidean geometry, 246
euclidean inner product, 468
euclidean n-space, 468
Euler, Leonhard, 527
Euler's formula, 527
evaluation, 160, 271, 298, 332, 337, 365
even function, 507
even parity, 415
even polynomial, 311
exact formula, 162
expansion theorem, 251, 478, 487
expectation, 434
exponential function, 173–174, 358

F

factor, 267, 548
factor theorem, 321, 548
feasible region, 431
Fibonacci sequence, 171, 171n
field, 289n, 409
field of integers modulo, 410
finite dimensional spaces, 311–318
finite fields, 409–411, 416
finite sets, 409
fixed axis, 504
fixed hyperplane, 504
fixed line, 498
fixed plane, 501
fixed vectors, 505
Forcinito, M.A., 419n
formal proofs, 537
forward substitution, 103
Fourier, J.B.J., 478, 506n
Fourier approximation, 506–509
Fourier coefficients, 251–252, 478, 506, 507
Fourier expansion, 251–252
Fourier series, 508
Fourier Series and Boundary Value Problems (Brown and Churchill), 508n
fractions
  eigenvectors, 153n
  equal fractions, 186n
  field, 409
  probabilities, 117
free variables, 12
function
  of a complex variable, 248n
  composition, 351
  continuous functions, 469
  defined, 291
  derivative, 357, 358n
  differentiable function, 173, 299, 325, 357
  equal, 291
  even function, 507
  exponential function, 173–174, 358
  inner product. See inner product
  objective function, 431, 433

odd function, 507
pointwise addition, 292
real-valued function, 291
scalar multiplication, 292
fundamental identities, 80, 353
fundamental theorem, 240, 243, 305–306
fundamental theorem of algebra, 155n, 269, 401, 532–533

G

Galois field, 411
Galton, Francis, 284n
Gauss, Carl Friedrich, 10f, 10n, 126, 155n, 269n, 532
gaussian algorithm, 10, 22–23, 267, 409
gaussian elimination
  defined, 13
  example, 13–14
  LU-factorization, 103
  normal equations, 273
  scalar multiple, 36
  systems of linear equations and, 9–16
general differential systems, 174–178
general solution, 2, 13, 177
general theory of relativity, 288
generalized eigenspace, 511
generalized inverse, 274, 282
generator, 417
geometric vectors
  see also vector geometry
  defined, 186
  described, 186
  difference, 188
  intrinsic descriptions, 187
  midpoint, 190
  parallelogram law, 187–191
  Pythagoras' theorem, 194–195
  scalar multiple law, 189, 191
  scalar multiplication, 189–191
  scalar product, 189
  standard position, 188
  sum, 188
  tip-to-tail rule, 187
  unit vector, 189
  vector subtraction, 188
geometry, 32
Google PageRank, 165–166, 166n
Gram, Jörgen Pederson, 369n
Gram-Schmidt orthogonalization algorithm, 252, 369–370, 378, 380, 390, 452, 479
graphs
  attractor, 164
  conic, 20
  directed graphs, 66
  ellipse, 425
  hyperbola, 425
  linear dynamical systems, 164–165
  saddle point, 165
  trajectory, 164
Grassman, Hermann, 288
groups, 495
group theory, 32

H

Hamilton, William Rowan, 405n
Hamming, Richard, 412
Hamming (7,4)-code, 420
Hamming bound, 414
Hamming distance, 412
Hamming, Richard, 412
Hamming weight, 412
heat conduction in solids, 506n
Hermite, Charles, 400n

hermitian matrix, 400–402
higher-dimensional geometry, 32n
Hilbert, David, 369n
Hilbert spaces, 369n
hit, 340
homogenous coordinates, 227
homogenous equations, 18–24
  associated homogenous system, 47
  basic solutions, 22–23
  defined, 18
  general solution, 21–22
  linear combinations, 20–24
  nontrivial solutions, 18, 152
  trivial solution, 18
homogenous system, 23–24
Hooke's law, 328
Householder matrices, 396
How to Read and Do Proofs (Solow), 536n
hyperbola, 425
hyperplanes, 3, 504
hypotenuse, 537
hypothesis, 536

I

(i, j)-entry, 33
idempotents, 69, 466
identity matrix, 46, 50, 107
identity operator, 332
identity transformation, 53, 101
"if and only if," 33n
image
  of linear transformations, 52, 337, 338–344
  of the parallelogram, 223
image space, 230
imaginary axis, 526
imaginary parts, 357, 523
imaginary unit, 523
implication, 536
implies, 77n
inconsistent system, 2
independence, 236–240, 303–308
independence test, 236–237
independent, 236, 303, 312
independent lemma, 312
indeterminate, 291, 546
index, 428, 429
induction
  cofactor expansion theorem, 180–181
  determinant, determination of, 127
  mathematical induction, 92, 541–544
  on path of length r, 66
induction hypothesis, 542
infinite dimensional, 312
initial condition, 174
initial state vector, 119
inner product
  coordinate isomorphism, 350
  defined, 468
  euclidean inner product, 468
  and norms, 471
  positive definite $n \times n$ matrix, 470–471
  properties of, 469–470
inner product space
  defined, 468
  distance, 471
  dot product, use of, 471
  Fourier approximation, 506–509
  isometries, 493–504
  norms, 471–474
  orthogonal diagonalization, 486–491
  orthogonal sets of vectors, 477–483
  unit vector, 472

input-output economic models, 112–116
input-output matrix, 113
integers, 408, 410, 536n
integers modulo, 408
integration, 333
interpolating polynomial, 145
Interpolation and Approximation (Davis), 483
intersection, 235, 317–318
interval, 291
intrinsic descriptions, 186
Introduction to Abstract Algebra (Nicholson), 410n, 411n
Introduction to Abstract Mathematics (Lucas), 536n
Introduction to Matrix Computations (Stewart), 395
Introduction to the Theory of Error-Correcting Codes (Pless), 419n
invariance theorem, 241
invariant subspaces, 454–464
invariants, 52n, 154–155
inverse theorem, 77
inverses
  see also invertible matrix
  adjugate, 140–142
  cancellation laws, 74
  complex numbers, 524
  Cramer's rule, 137, 143–144
  defined, 70
  determinants, 126, 137–147
  and elementary matrices, 85–86
  elementary row operations, 6, 83–85
  finite fields, 409
  generalized inverse, 274, 282
  inverse theorem, 77
  inversion algorithm, 73–74, 86
  and linear systems, 72
  linear transformation, 79–80, 353
  matrix transformations, 79–80
  Moore-Penrose inverse, 282
  nonzero matrices, 70
  properties of inverses, 74–79
  square matrices, application to, 70n, 72–74, 138
  and zero matrices, 70
inversion algorithm, 73–74
invertibility condition, 138
invertible matrix
  see also inverses
  defined, 70
  determinants, 126
  left cancelled, 74
  LU-factorization, 108
  "mixed" cancellation, 74
  orthogonal matrices, 391–392
  product of elementary matrix, 85–86
  right cancelled, 74
involutions, 462
irreducible, 531
isometries, 220, 493–504
isomorphic, 347
isomorphism, 347–351, 437, 440–441, 495

J

Jacobi identity, 218
Jordan, Camille, 521f, 521n
Jordan blocks, 519
Jordan canonical form, 518–522
Jordan canonical matrices, 510
junction rule, 25, 28
juvenile survival rate, 150

K

Kemeny, J., 122n
kernel, 338–344
kernel lemma, 359
Kirchhoff's Laws, 28

L

Lagrange, Joseph Louis, 215f, 215n
Lagrange identity, 215
Lagrange interpolation expansion, 323, 479
Lagrange polynomials, 323, 479
Lancaster, P., 114n
Laplace, Pierre Simon de, 129n
law of cosines, 199–200, 476
law of exponents, 528
law of sines, 219
leading 1, 9
leading coefficient, 145n, 290, 546
leading column, 105
leading variables, 12
least squares approximating line, 276
least squares approximating polynomials, 277–279
least squares approximation, 275–277
least squares best approximation, 280
left cancelled invertible matrix, 74
Legendre, A.M., 480
Legendre polynomials, 480
Leibnitz, 126
lemma, 84n
length
  geometric vector, 186
  linear recurrence relation, 169
  linear recurrences, 361
  norm, where dot product used, 398, 471n
  path of length, 66
  recurrence, 361
  vector, 184–185, 246, 398
Leontief, Wassily, 112, 112n
line
  fixed line, 498
  least squares approximating line, 276
  parametric operations of a line, 192
  perpendicular lines, 198–202
  point-slope formula, 194
  shortest distance between nonparallel lines, 208–209
  in space, 191–194
  straight, pair of, 425
  through the origin, 230
  vector equation of a line, 192
linear codes, 414–416
linear combinations
  of columns of coefficient matrix, 44, 45
  defined, 20, 91
  eigenvectors, 394
  homogeneous equations, 20–23, 55
  linear transformations and, 91
  of orthogonal basis, 251–252
  random variables, 435n
  of solutions to homogeneous system, 21, 43n
  spanning sets, 232, 299–301
  trivial, 236, 303
  unique, 236
  vanishes, 236, 239
  vectors, 299
linear discrete dynamical system, 151n
linear dynamical system, 151, 160–164
linear equation
  conic graph, 20
  constant term, 1
  Cramer's rule, 143–144
  defined, 1
  vs. linear inequalities, 16
  system. See system of linear equations
linear independence
  dependent, 238, 303, 315
  geometric description, 238
  independent, 236, 303, 312
  orthogonal sets, 251, 478
  properties of, 303
  set of vectors, 236, 303
  vector spaces, 303–306
linear inequalities, 16
linear operator
  B-matrix, 445–452
  change matrix, 446
  choice of basis, 450
  defined, 219, 331, 445
  diagonalizable, 486–487
  distance preserving, 219
  distance preserving isometries, 493–504
  eigenvalues, 457–459
  eigenvector, 457
  on finite dimensional inner product space, 486–491
  idempotents, 466
  involutions, 462
  isometries, 220, 494
  projection, 220–222, 373–374, 493
  properties of matrices, 448–449
  reducible, 462
  reflections, 220–222
  restriction, 455
  rotations, 222–223
  standard matrix, 445
  symmetric, 489
  transformations of areas and volumes, 223–224
linear programming, 434
"Linear Programming and Extensions" (Wu and Coppins), 434n
linear recurrence relation, 169, 361
linear recurrences
  diagonalization, 169–172
  length, 169, 361
  linear transformations, 360–367
  polynomials associated with the linear recurrence, 363
  shift operator, 364–367
  vector spaces, 360–367
linear system of differential equations, 175
  see also differential systems
linear systems. See system of linear equations
linear transformations
  see also transformations
  action of a transformation, 438–439
  of areas, 223–224
  association with matrix, 436
  as category, 351n
  composite, 57–64, 95, 351
  composition, 351–354, 440–441
  in computer graphics, 226
  coordinate transformation, 437
  defined, 52, 91, 331, 335
  described, 32, 91–96
  differentiation, 333
  dimension theorem, 341–344
  distance preserving, 219–220
  equal, 52, 334
  evaluation, 332, 365
  examples, 331–335
  fundamental identities, 353
  hit, 340
  identity operator, 332
  image, 52, 337, 338–344
  integration, 333
  inverses, 79–80, 353
  isomorphism, 347–351
  kernel, 338–344
  linear operators. See linear operator
  linear recurrences, 360–367
  $m \times n$ matrix, 436
  matrix of a linear transformation, 332–333, 436–441
  matrix of T corresponding to the ordered bases B and D, 438
  as matrix transformation, 436
  matrix transformation induced, 53, 219
  matrix transformations, another perspective on, 91–96
  nullity, 339
  nullspace, 338
  one-to-one transformations, 339–341
  onto transformations, 339–341
  projections, 100, 220–222
  properties, 333–334
  range, 338
  rank, 339, 441
  reflections, 99–100, 220–222
  rotations, 97–98
  scalar multiple law, 97
  scalar operator, 332
  of volume, 223–224
  zero transformation, 53, 332
linearly dependent, 238, 303, 315
linearly independent, 236, 303, 312
logically equivalent, 539
lower reduced, 104
lower triangular matrix, 103, 134
LU-algorithm, 105–106, 109–110
LU-factorization, 103–111
Lucas, J.F., 536n

M

$m \times n$ matrix
  canonical forms, 518
  defined, 32
  difference, 35
  elementary row operation, 84
  main diagonal, 39
  matrix transformation, 436
  negative, 35
  subspaces, 230–231
  transpose, 38
  zero matrix, 34
magnitude, 186n
main diagonal, 39, 103
Markov, Andrei Andreyevich, 117n
Markov chains, 117–123
mathematical induction, 92, 541–544
mathematical statistics, 282
matrices, 32
  see also matrix
matrix
  adjacency matrix, 66
  algebra. See matrix algebra
  augmented matrix, 3, 4–5, 12, 13
  block matrix, 64–65
  change matrix, 446
  coefficient matrix, 3, 143
  column matrix, 33, 42
  companion matrix, 137
  complex matrices. See complex matrix
  congruent matrices, 427–428
  conjugate matrix, 270
  constant matrix, 3
  covariance matrix, 435
  defined, 3n, 32
  demand matrix, 115
  diagonal matrices, 41, 69, 151, 156–160, 264, 448
  diagonalizable matrix, 156
  diagonalizing matrix, 156
  doubly stochastic matrix, 124
  elementary matrix, 83–89
  elementary row operations, 5–7
  entries of a matrix, 33
  entries of the matrix, 32
  equal matrices, 33
  equivalent matrices, 90–91
  hermitian matrix, 400–402
  Householder matrices, 396
  (i, j)-entry, 33
  identity matrix, 46, 50, 107
  input-output matrix, 113
  invertible matrix, 70, 74
  linear transformation, association with, 436
  lower triangular matrix, 103, 134
  $m \times n$ matrix. See $m \times n$ matrix
  nilpotent matrix, 167
  orthogonal matrix, 140, 376–377, 376n, 383
  orthonormal matrix, 140, 376n
  over finite field, 417–418
  parity-check matrices, 417–419
  partitioned into blocks, 64
  permutation matrix, 107–110, 384, 453
  projection matrix, 56, 384
  rank, 14–16, 253–260, 339, 428
  reduced row-echelon matrix, 9, 13, 88–89
  row-echelon matrix, 9, 10, 13, 254
  row-equivalent matrices, 90
  row matrix, 32
  scalar matrix, 125
  shapes, 32
  similar matrices, 262–264
  spectrum, 378
  square matrix. See square matrix ($n \times n$ matrix)
  standard generator, 417
  standard matrix, 445
  stochastic matrices, 113, 119, 122
  submatrix, 261
  subtracting, 35
  symmetric matrix. See symmetric matrix
  systematic generator, 417
  transition matrix, 118, 119
  transpose, 38–40
  triangular matrices, 103–104, 111, 134
  unitary matrix, 402
  upper triangular matrix, 103, 134, 403
  Vandermonde matrix, 133
  zero matrix, 34
  zeroes, creating in matrix, 130
matrix addition, 34–36, 37
matrix algebra
  dot product, 48–50
  elementary matrix, 83–89
  input-output economic models, application to, 112–116
  inverses, 72
  LU-factorization, 103–111
  Markov chains, application to, 117–123
  matrices as entities, 32
  matrix addition, 34–36, 37

matrix multiplication, 56–66
matrix subtraction, 35
matrix transformations. See linear transformations
matrix-vector multiplication, 43–48
   numerical division, 69–80
   scalar multiplication, 36–38
   size of matrices, 36
   transformations, 51–54
   transpose of a matrix, 38–40
   usefulness of, 32
matrix form
   defined, 45
   reduced row-echelon form, 9, 10, 12, 14, 88–89
   row-echelon form, 9
   upper Hessenberg form, 396
matrix generators, 416–417
matrix inverses. See inverses
matrix inversion algorithm, 73–74, 86
matrix multiplication
   associative law, 62
   block, 64–65
   commute, 60, 63
   compatibility rule, 59
   and composition of transformations, 57–64
   definition, 57, 59
   directed graphs, 66
   distributive laws, 62–64
   dot product rule, 49, 59
   left-multiplication, 76
   matrix of composite of two linear transformations, 95
   matrix products, 58–62
   non-commutative, 76
   order of the factors, 62
   results of, 61–62
   right-multiplication, 76
matrix of a linear transformation, 436–441
matrix of T corresponding to the ordered bases B and D, 438
matrix recurrence, 151
matrix theory, 32n
matrix transformation induced, 53, 91, 219
matrix transformations. See linear transformations
matrix-vector products, 44
mean
   "average" of the sample values, 283n
   calculation, 434
   sample mean, 283
Mécanique Analytique (Legrange), 215n
median
   tetrahedron, 198
   triangle, 198
messages, 416
methods of proof. See proof
metric, 351n
midpoint, 190
minimum distance, 413
Mirkil, H., 122n
"mixed" cancellation, 74
modular arithmetic, 408–409
modulo, 408
modulus, 408, 525
Moler, Cleve, 166n
Moore-Penrose inverse, 282
morphisms, 351n
multiplication
   block multiplication, 64–65
   compatible, 59
   matrix multiplication, 56–66
   matrix-vector multiplication, 43–48

matrix-vector products, 44
scalar multiplication, 36–38, 289, 292
multiplication rule, 528–529
multiplicity, 158, 267, 364, 548
multivariate analysis, 434

N

n-code. See code
(n, k)-code. See code
n-parity-check code, 415
n-tuples, 229, 253, 289n
   see also set of all ordered ($\mathbb{R}^n$)
n-vectors. See vectors
n-words, 412
$n \times 1$ matrix. See zero n-vector
$n \times n$ matrix. See square matrix ($n \times n$ matrix)
nearest neighbour decoding, 413
negative
   correlation, 284, 285
   of $m \times n$ matrix, 35
   vector, 43, 289
negative x, 43
negative x-shear, 54
network flow, 25–26
Newton, Sir Isaac, 10n
Nicholson, W.K., 410n, 411n
nilpotent, 167, 519
noise, 408
nonleading variable, 19
nonlinear recurrences, 172
nontrivial solution, 18, 152
nonzero scalar multiple of a basic solution, 23
nonzero vectors, 191, 231n, 248, 249n
norm, 398, 471–474
normal, 204, 405
normal equations, 273
normalizing the orthogonal set, 250, 477
nth roots of unity, 530–531
null space, 231
nullity, 339
nullspace, 338

O

objective function, 431, 433
objects, 351n
odd function, 507
odd polynomial, 311
Ohm's law, 27
one-to-one transformations, 339–341
onto transformations, 339–341
open model of the economy, 114–116
open sector, 114
ordered basis, 436, 437–438
ordered n-tuple, 42, 239n
   see also set of all ordered n-tuples ($\mathbb{R}^n$)
origin, 184
orthocentre, 228
orthogonal basis, 249n, 251–252, 369, 479–480
orthogonal codes, 419–420
orthogonal complement, 370–371, 419, 481
orthogonal diagonalization, 376–382, 486–491
orthogonal hermitian matrix, 401–402
orthogonal lemma, 368, 479
orthogonal matrix, 140, 376–377, 376n, 383
orthogonal projection, 372–374, 482

orthogonal set of vectors, 401, 477–483
orthogonal sets, 249–252, 249n, 368–369
orthogonal vectors, 201, 202, 249, 401–402, 477
orthogonality
   complex matrices, 397–406
   constrained optimization, 431–434
   dot product, 246–249
   eigenvalues, computation of, 393–396
   expansion theorem, 251, 478
   finite fields, 409–411
   Fourier expansion, 251
   Gram-Schmidt orthogonalization algorithm, 369–370, 378, 380, 390, 452, 479
   normalizing the orthogonal set, 250
   orthogonal codes, 419–420
   orthogonal complement, 371–372, 419
   orthogonal diagonalization, 376–382
   orthogonal projection, 372–374
   orthogonal sets, 249–252, 249n, 368–369
   orthogonally similar, 383
   positive definite matrix, 385–389
   principal axis theorem, 377–378
   projection theorem, 273
   Pythagoras' theorem, 250
   QR-algorithm, 395–396
   QR-factorization, 390–392
   quadratic forms, 381, 422–430
   real spectral theorem, 378
   statistical principal component analysis, 434–435
   triangulation theorem, 382
orthogonally diagonalizable, 377
orthogonally similar, 383
orthonormal basis, 480, 488
orthonormal matrix, 376n
orthonormal set, 401, 477
orthonormal vector, 249

P

PageRank, 165–166, 166n
paired samples, 284
parabola, 425
parallel, 191
parallelepiped, 216, 223–224
parallelogram
   area equal to zero, 215–216
   defined, 97, 187n
   determined by geometric vectors, 187
   image, 223
   law, 97, 187–191, 526
   rhombus, 202
parameters, 2, 12
parametric equations of a line, 192
parametric form, 2
parity-check code, 415
parity-check matrixes, 417–419
parity digits, 417
Parseval's formula, 485
particle physics, 433
partitioned into blocks, 64
path of length, 66
Peano, Giuseppe, 288
Pearson, Karl, 284n
Pearson correlation coefficient, 284n
perfect code, 414
period, 329
permutation matrix, 107–110, 384, 453

perpendicular lines, 198–202
physical dynamics, 405n
pigeonhole principle, 538
Pisano, Leonardo, 171
planes, 204–206, 230
Pless, V., 419n
PLU-factorization, 108
point-slope formula, 194
pointwise addition, 291, 292
polar form, 527–529
polynomials
   associated with the linear recurrence, 363
   characteristic polynomial. See characteristic polynomial
   coefficients, 279–280, 290, 545
   companion matrix, 137
   complex roots, 155, 401, 549
   constant, 546
   defined, 145n, 290, 545
   degree of the polynomial, 145n, 290, 546
   distinct degrees, 304
   division algorithm, 548
   equal, 291, 546
   evaluation, 160, 298
   even, 311
   factor theorem, 548
   form, 546
   indeterminate, 546
   interpolating the polynomial, 144–147
   Lagrange polynomials, 323, 479
   leading coefficient, 145n, 290, 546
   least squares approximating polynomials, 277–279
   Legendre polynomials, 480
   as matrix entries, and determinants, 132
   with no root, 548, 549
   nonconstant polynomial with complex coefficients, 269
   odd, 311
   remainder theorem, 547
   root, 155, 298, 393, 523
   root of characteristic polynomial, 152, 325
   Taylor's theorem, 321–322, 321n
   vector space of, 290
   vector spaces, 320–323
   zero polynomial, 290, 546
position vector, 187
positive correlation, 284
positive definite, 385, 431
positive definite matrix, 385–389, 470
positive semidefinite, 435
positive x-shear, 54
power method, 393–394
power sequences, 361
practical problems, 1
preimage, 337
prime, 410, 538
principal argument, 527
principal axes, 378, 381–382, 423
principal axis theorem, 377–378, 401, 404, 432, 490
principal components, 435
principal submatrices, 386
probabilities, 117
probability law, 434
probability theory, 435
product
   complex number, 526
   cross product. See cross product
   determinant of product of matrices, 137–138
   dot product. See dot product
   inner product. See inner product

matrix products, 59–62
matrix-vector products, 44
scalar product. See scalar product
standard inner product, 397–399
theorem, 138, 147
product rule, 358
projection
 linear operator, 493
 linear operators, 220–222
 orthogonal projection, 372–374, 482
projection matrix, 56, 374, 384
projection on U with kernel W, 481
projection theorem, 273, 372–373, 482
projections, 100, 202–204, 370–374
proof
 by contradiction, 538–540
 defined, 536
 direct proof, 536–537
 formal proofs, 537
 reduction to cases, 537–538
proper subspace, 229, 243
pure imaginary numbers, 523
Pythagoras, 194, 537
Pythagoras' theorem, 185, 185n, 194–195, 200, 250, 477, 537

Q
QR-algorithm, 395–396
QR-factorization, 390–392
quadratic equation, 431
quadratic form, 381, 422–430, 471
quadratic formula, 531
quotient, 408

R
r-ball, 413
radian measure, 53n, 97, 527
random variable, 434, 435
range, 338
rank
 linear transformation, 339, 441
 matrix, 14–16, 253–260, 339, 428
 quadratic form, 428
 similarity invariant, 450
 symmetry matrix, 428
 theorem, 255–258, 259
rational numbers, 289n
Raum-Zeit-Materie ("Space-Time-Matter") (Weyl), 288
Rayleigh quotients, 394
real axis, 526
real Jordan canonical form, 520
real numbers, 1, 42, 289n, 291, 397, 401, 409, 523
real parts, 357, 523
real quadratic, 531
real spectral theorem, 378
recurrence, 169
recursive algorithm, 10
recursive sequence, 169
reduced row-echelon form, 9, 10, 12, 14, 88–89
reduced row-echelon matrix, 9, 13
reducible, 462
reduction to cases, 537–538
reflections
 about a line through the origin, 140
 fixed hyperplane, 504
 fixed line, 498
 fixed plane, 501
 isometries, 497–498
 linear operators, 220–222
 linear transformations, 99–100
regular representation, 453

regular stochastic matrix, 122
remainder, 408
remainder theorem, 321, 547
repellor, 165
reproduction rate, 150
restriction, 455
reversed, 6
rhombus, 202
right cancelled invertible matrix, 74
right-hand coordinate systems, 217
right-hand rule, 217
root
 of characteristics polynomial, 152, 325, 326
 of polynomials, 155, 298, 393, 523
 of the quadratic, 531
roots of unity, 530–531
rotations
 about a line through the origin, 451
 about the origin, and orthogonal matrices, 140
 axis, 502
 describing rotations, 97
 fixed axis, 504
 isometries, 497–498, 501–502, 504
 linear operators, 222–223
 linear transformations, 97–98
round-off error, 153n
row-echelon form, 9
row-echelon matrix, 9, 10, 13, 254
row-equivalent matrices, 90
row matrix, 32
row space, 253
rows
 convention, 32
 elementary row operations, 5–7
 (i, j)-entry, 33
 leading 1, 9
 as notations for ordered n-tuples, 239n
 shape of matrix, 32
 Smith normal form, 87–88
 zero rows, 9

S
saddle point, 165
same action, 52, 291, 334
sample
 analysis of, 282
 comparison of two samples, 283–284
 defined, 282
 paired samples, 284
sample correlation coefficient, 284–285
sample mean, 283
sample standard deviation, 283, 283n
sample variance, 283
sample vector, 282
satisfy the relation, 361
scalar, 36, 289, 289n, 409
scalar equation of a plane, 204
scalar matrix, 125
scalar multiple law, 97, 189, 191
scalar multiples, 20, 36, 96
scalar multiplication
 axioms, 289, 292n
 basic properties, 293–294
 closed under, 43
 closed under scalar multiplication, 229
 described, 36–38
 distributive laws, 37
 of functions, 292
 geometric vectors, 189–191
 geometrical description, 96
 transformations, preserving scalar

multiplication, 91
vectors, 289
scalar operator, 332
scalar product
 defined, 198
 elementary row operations, 20
 geometric vectors, 186
scatter diagram, 284
Schmidt, Erhardt, 369n
Schur, Issai, 403, 403n
Schur's theorem, 403–404, 405
Schwarz, Hermann Amandus, 473n
second-order differential equation, 325, 357
Seneta, E., 114n
sequences
 of column vectors, 151
 constant sequences, 361
 equal, 361
 Fibonacci sequence, 171, 171n
 linear recurrences, 168–172
 notation, 360–361
 ordered sequence of real numbers, 42
 power sequences, 361
 recursive sequence, 169
 satisfy the relation, 361
set, 229n
set notation, 230n
set of all ordered n-tuples ($\mathbb{R}^n$)
 closed under addition and scalar multiplication, 43
 complex eigenvalues, 269
 dimension, 240–244
 dot product, 246–249, 397
 expansion theorem, 251–252
 as inner product space, 468
 linear independence, 236–240
 linear operators, 219–224
 n-tuples, 253, 289n
 notation, 42
 orthogonal sets, 249–252
 projection on, 374
 rank of a matrix, 253–260
 rules of matrix arithmetic, 289
 similar matrices, 262–264
 spanning sets, 231–234
 special types of matrices, 289
 standard basis, 93
 subspaces, 231–234, 290
 symmetric matrix, 270–271
Shannon, Claude, 412
shift operator, 364–367
shifting, 396
sign, 127
similar matrices, 262–264
similarity invariant, 450
simple harmonic motions, 328
simplex algorithm, 16, 434
sine, 97, 219, 527
single vector equation, 43
size $m \times n$ matrix, 32
skew-hermitian, 407
skew-symmetric, 42, 459, 492
Smith, Henry John Stephen, 87n
Smith normal form, 87–88
Snell, J., 122n
Solow, D., 536n
solution
 algebraic method, 4, 9
 basic solutions, 22, 23
 best approximation to, 273–277
 consistent system, 2, 15
 general solution, 2, 13
 geometric description, 3
 inconsistent system, 2
 to linear equation, 1–3
 nontrivial solution, 18, 152
 in parametric form, 2

solution to a system, 1–3, 13
 trivial solution, 18
solution to a system, 1
span, 232, 299
spanning sets, 231–234, 299–301
spectral theorem, 40
spectrum, 378
sphere, 433
spring constant, 329
square matrix ($n \times n$ matrix)
 characteristic polynomial, 152, 405
 cofactor matrix, 140
 complex matrix. See complex matrix
 defined, 33
 determinants, 51, 126, 139
 diagonal matrices, 69, 151, 156
 diagonalizable matrix, 156, 448
 diagonalizing matrix, 151
 elementary matrix, 83–89
 hermitian matrix, 400–402
 idempotent, 69
 identity matrix, 46, 50–51
 invariants, 154–156
 inverse. See inverses
 lower triangular matrix, 134
 matrix of an operator, 448
 nilpotent matrix, 167
 orthogonal matrix, 140
 positive definite matrix, 385–389, 470
 regular representation of complex numbers, 453
 scalar matrix, 125
 similarity invariant, 450
 skew-symmetric, 42
 trace, 69, 263
 triangular matrix, 134
 unitary matrix, 402
 upper triangular matrix, 134
staircase form, 9
standard basis, 93, 233, 241, 242, 250, 307, 397
standard deviation, 434
standard generator, 417
standard inner product, 397–399
standard matrix, 445
standard position, 97, 188, 527
state vector, 118, 119–123
statistical principal component analysis, 434–435
steady-state vector, 122
Steinitz Exchange Lemma, 306
Stewart, G.W., 395
stochastic matrices, 113, 119, 122
structure theorem, 499–501
submatrix, 261
subset, 229n
subspace test, 297
subspaces
 basis, 241
 closed under addition, 229
 closed under scalar multiplication, 229
 complex subspace, 406
 defined, 229, 296
 dimension, 240–244
 eigenspace, 267
 fundamental theorem, 240
 image, 337, 338–339
 intersection, 235, 317–318
 invariance theorem, 241
 invariant subspaces, 454–464
 kernel, 338–339
 $m \times n$ matrix, 230–231
 planes and lines through the origin, 230
 projection, 371–372

proper subspace, 229, 243
spanning set, 231–234
subspace test, 297
sum, 235, 317–318
vector spaces, 296–299
zero subspace, 229, 297
subtraction
 complex number, 523–524
 matrix, 35
 vector, 188, 293
sum
 see also addition
 algebraic sum, 28
 complex number, 526
 direct sum, 318, 459–464, 467
 elementary row operations, 20
 geometric vectors, 188
 geometrical description, 96
 matrices of the same size, 34
 matrix addition, 34–36
 of product of matrix entries, 132
 of scalar multiples, 20
 subspaces, 235, 317–318
 subspaces of a vector space, 317–318
 of two vectors, 293
 variances of set of random variables, 435
 of vectors in two subspaces, 459
summation notation, 179, 179n
Sylvester's law of inertia, 428
symmetric bilinear form, 431
symmetric form, 196, 422, 431
symmetric linear operator, 489
symmetric matrix
 absolute value, 270n
 congruence, 427–430
 defined, 40
 index, 428
 orthogonal eigenvectors, 377–381
 positive definite, 385–389
 rank and index, 428
 real eigenvalues, 270–271
syndrome, 419
syndrome decoding, 419
system of first order differential equations. See differential equations
system of linear equations
 algebraic method, 4
 associated homogeneous system, 47
 augmented matrix, 3
 chemical reactions, application to, 29–30
 coefficient matrix, 3
 consistent system, 2, 15
 constant matrix, 3
 defined, 1
 electrical networks, application to, 27–28
 elementary operations, 4–7
 equivalent systems, 4
 gaussian elimination, 13, 14–16
 general solution, 2
 homogeneous equations, 18–23
 inconsistent system, 2
 infinitely many solutions, 3, 3f
 inverses and, 72
 with $m \times n$ coefficient matrix, 44–45
 matrix algebra. See matrix algebra
 matrix multiplication, 61–62
 network flow application, 25–26
 no solution, 2, 3f
 nontrivial solution, 18–19
 normal equations, 273
 positive integers, 30
 rank of a matrix, 14–16

solutions, 1–3, 13
 trivial solution, 18
 unique solution, 3, 3f
systematic generator, 417

T

T-invariant, 454
tail, 186
Taylor's theorem, 321–322, 321n, 447n
tetrahedron, 198
theorems, 540
theory of Hilbert spaces, 369n
third-order differential equation, 325, 357
Thompson, G., 122n
3-dimensional space, 51, 184
time, functions of, 151n
tip, 186
tip-to-tail rule, 187
total variance, 435
trace, 69, 166, 263, 332, 450
trajectory, 164
transformations
 see also linear transformations
 action, 52, 436, 438–439
 composite, 57, 95
 defining, 52
 described, 32, 52
 equal, 52
 identity transformation, 53, 101
 matrix transformation, 53, 426
 zero transformation, 53, 102
transition matrix, 118, 119
transition probabilities, 117, 119
translation, 54, 336, 493
transpose of a matrix, 38–40
transposition, 38–39, 332
triangle
 altitude, 228
 centroid, 198
 hypotenuse, 537
 inequality, 213, 249, 473–474, 525
 median, 198
 orthocentre, 228
triangle inequality, 213, 249, 473–474, 525
triangular matrices, 103–104, 111, 134
triangulation algorithm, 513
triangulation theorem, 382
trigonometric functions, 97
trivial linear combinations, 236, 303
trivial solution, 18

U

uncorrelated, 435
unit ball, 433, 472
unit circle, 97, 472, 527
unit cube, 224
unit square, 224
unit triangular, 111
unit vector, 189, 246, 398–399, 472
unitarily diagonalizable, 403
unitary matrix, 402
upper Hessenberg form, 396
upper triangular matrix, 103, 134, 403

V

Vandermonde determinant, 146
Vandermonde matrix, 133, 364
variance, 282–286, 434
variance formula, 286
variation, 434
vector addition, 289, 526

vector equation of a line, 192
vector equation of a plane, 205
vector geometry
 angle between two vectors, 200
 computer graphics, 225–227
 cross product, 206–209
 defined, 184
 direction vector, 191–192
 geometric vectors. See geometric vectors
 line perpendicular to plane, 198–202
 linear operators, 219–224
 lines in space, 191–194
 planes, 204–206
 projections, 202–204
 symmetric form, 197
 vector equation of a line, 192
vector product, 207
 see also cross product
vector quantities, 186n
vector spaces
 abstract vector space, 288
 axioms, 289, 292, 294
 basic properties, 288–295
 basis, 306
 cancellation, 292
 as category, 351n
 continuous functions, 469
 defined, 289
 differential equations, 325–329
 dimension, 307–308
 direct sum, 459–460
 examples, 288–295
 finite dimensional spaces, 311–318
 infinite dimensional, 312
 introduction of concept, 288
 isomorphic, 347–348
 linear independence, 303–308
 linear recurrences, 360–367
 linear transformations, 331
 polynomials, 290–291, 320–323
 scalar multiplication, basic properties of, 293–294
 set of all ordered n-tuples ($\mathbb{R}^n$). See set of all ordered n-tuples ($\mathbb{R}^n$)
 spanning sets, 299–301
 subspaces, 296–299, 317–318
 theory of vector spaces, 292–293
 3-dimensional space, 184
 zero vector space, 295
vectors
 addition, 289
 arrow representation, 51n
 column vectors, 151
 complex matrices, 397
 coordinate vectors, 189, 207, 224, 233, 437
 defined, 42, 288, 397
 difference of, 293
 direction of, 185
 direction vector, 191–194
 fixed vectors, 505
 geometric vectors. See geometric vectors; vector geometry
 initial state vector, 119
 intrinsic descriptions, 186
 length, 184–185, 246, 398
 matrix recurrence, 151
 matrix-vector multiplication, 43–48
 matrix-vector products, 44
 negative, 43, 289
 nonzero vectors, 191, 231n, 248, 249n
 orthogonal vectors, 201, 202, 249, 401–402, 477
 orthonormal vector, 249, 401

position vector, 201
sample vector, 282
scalar multiplication, 289
set of all ordered n-tuples ($\mathbb{R}^n$). See set of all ordered n-tuples ($\mathbb{R}^n$)
single vector equation, 43
state vector, 118, 119–123
steady-state vector, 122
subtraction, 293
sum of two vectors, 289
unit vector, 189, 246, 398–399
zero n-vector, 43
zero vector, 46, 189n, 229, 236, 289
velocity, 186n
vertices, 66
vibrations, 433
volume
 linear transformations of, 223–224
 of parallelepiped, 216, 223–224

W

Weyl, Hermann, 288
whole number, 536n
Wilf, Herbert S., 166n
Wilkinson, J.M., 395
words, 412
wronskian, 330
Wu, N., 434n

X

x-axis, 184
x-compression, 53
x-expansion, 53
x-shear, 54

Y

y-axis, 184
y-compression, 53
y-expansion, 53

Z

z-axis, 184
zero matrix
 described, 34
 no inverse, 70
 scalar multiplication, 36
zero n-vector, 43
zero polynomial, 290, 546
zero rows, 9
zero subspace, 229, 297
zero transformation, 53, 102, 332
zero vector, 46, 189n, 229, 236, 289
zero vector space, 295

Definitions

Definition 1.1: Elementary Operations	4
Definition 1.2: Elementary Row Operations	5
Definition 1.3: Row-Echelon Form (reduced)	9
Definition 1.4: Rank of a Matrix	14
Definition 1.5: Basic Solutions	22
Definition 2.1: Matrix Addition	34
Definition 2.2: Matrix Scalar Multiplication	36
Definition 2.3: Transpose of a Matrix	38
Definition 2.4: The set $\mathbb{R}^n$ of ordered n-tuples of real numbers	42
Definition 2.5: Matrix-Vector Multiplication	44
Definition 2.6: Dot Product in $\mathbb{R}^n$	48
Definition 2.7: The Identity Matrix	50
Definition 2.8: Matrix Transformation T_A	53
Definition 2.9: Matrix Multiplication	57
Definition 2.10: Block Partition of a Matrix	64
Definition 2.11: Matrix Inverses	70
Definition 2.12: Elementary Matrices	84
Definition 2.13: Linear transformations $\mathbb{R}^n \to \mathbb{R}^m$	91
Definition 2.14: LU-factorization	104
Definition 2.15: Markov Chain	117
Definition 3.1: Cofactors of a Matrix	127
Definition 3.2: Cofactor expansion of a Matrix	128
Definition 3.3: Adjugate of a Matrix	140
Definition 3.4: Eigenvalues and Eigenvectors of a Matrix	151
Definition 3.5: Characteristic Polynomial of a Matrix	152
Definition 3.6: Diagonalizable Matrices	156
Definition 3.7: Multiplicity of an Eigenvalue	158
Definition 3.8: Linear Dynamical System	161
Definition 4.1: Geometric Vectors	186
Definition 4.2: Parallel Vectors in $\mathbb{R}^3$	191
Definition 4.3: Direction Vector of a Line	191
Definition 4.4: Dot Product in $\mathbb{R}^3$	198
Definition 4.5: Orthogonal Vectors in $\mathbb{R}^3$	201
Definition 4.6: Projection in $\mathbb{R}^3$	202
Definition 4.7: Normal Vector in a Plane	204
Definition 4.8: Cross Product	207
Definition 4.9: Linear Operator on $\mathbb{R}^3$	219
Definition 5.1: Subspace of $\mathbb{R}^n$	229
Definition 5.2: Linear Combinations and Span in $\mathbb{R}^n$	232
Definition 5.3: Linear Independence in $\mathbb{R}^n$	236
Definition 5.4: Basis of $\mathbb{R}^n$	241
Definition 5.5: Dimension of a Subspace of $\mathbb{R}^n$	241
Definition 5.6: Length in $\mathbb{R}^n$	246
Definition 5.7: Orthogonal and Orthonormal Sets	249
Definition 5.8: Normalizing an Orthogonal Set	250
Definition 5.9: Normalizing an Orthogonal Set in $\mathbb{R}^n$	250